SYMMETRY 2

International Series in
MODERN APPLIED MATHEMATICS AND COMPUTER SCIENCE
Volume 18

Series Editor: ERVIN Y. RODIN, *Washington University*

(Volumes in the series have also been published as Special Issues of the journal Computers and Mathematics with Applications)

RELATED PERGAMON TITLES
Other Volumes in the Series

JOHNSON
Formal Theories of Politics

MITTNIK
System Theoretic Methods in Economic Modelling 1

HARGITTAI *et al.*
Crystal Symmetries: Shubnikov Centennial Papers

WITTEN
Mathematical Models in Medicine Volume 2

LEE & MACDANIEL
Ocean Acoustic Propagation by Finite Difference Methods

YAVIN & PACHTER
Pursuit-Evasion Differential Games

Other Books of Interest

ENNALS
Artificial Intelligence

JESSHOPE
Parallel Processing

KRISHNA
Recent Advances in X-ray Characterization of Materials II

MALING
Measurements from Maps

RODIN & AVULA
Mathematical Modelling in Science and Technology

Related Journals*

Applied Mathematics Letters
Bulletin of Mathematical Biology
Computers and Graphics
Computers and Mathematics with Applications
Mathematical and Computer Modelling
Pattern Recognition
Progress in Crystal Growth and Characterization
System
Technology in Society

*Sample copies available on request

SYMMETRY 2

Unifying Human Understanding

Edited by

ISTVÁN HARGITTAI

Hungarian Academy of Sciences and Eötvös University, Budapest

PERGAMON PRESS

OXFORD · NEW YORK · BEIJING · FRANKFURT
SÃO PAULO · SYDNEY · TOKYO · TORONTO

U.K.	Pergamon Press plc, Headington Hill Hall, Oxford OX3 0BW, England
U.S.A.	Pergamon Press Inc., Maxwell House, Fairview Park, Elmsford, New York 10523, U.S.A.
PEOPLE'S REPUBLIC OF CHINA	Pergamon Press, Room 4037, Qianmen Hotel, Beijing, People's Republic of China
FEDERAL REPUBLIC OF GERMANY	Pergamon Press GmbH, Hammerweg 6, D-6242 Kronberg, Federal Republic of Germany
BRAZIL	Pergamon Editora Ltda, Rua Eça de Queiros, 346, CEP 04011, Paraiso, São Paulo, Brazil
AUSTRALIA	Pergamon Press Australia Pty Ltd, P.O. Box 544, Potts Point, N.S.W. 2011, Australia
JAPAN	Pergamon Press, 5th Floor, Matsuoka Central Building, 1-7-1 Nishishinjuku, Shinjuku-ku, Tokyo 160, Japan
CANADA	Pergamon Press Canada Ltd, Suite No. 271, 253 College Street, Toronto, Ontario, Canada M5T 1R5

ISBN 0 08 037237 6

Cover illustration designed by Magdolna Hargittai

Published as a special issue of the journal Computers and Mathematics with Applications, Volume 17, Numbers 1–6 and supplied to subscribers as part of their normal subscription.

Library of Congress Cataloguing-in-Publication Data

Symmetry 2: unifying human understanding/edited by István Hargittai.
p. cm.—(International series in modern applied mathematics and computer science; v. 18)
Published as a special issue of the journal Computers and mathematics with applications, v. 17, no. 1–6.
1. Symmetry. I. Hargittai, István.
II. Title: Symmetry two. III. Series.
Q172.5.S95S93 1989 500—dc20 89-16316

British Library Cataloguing in Publication Data

Symmetry: Unifying human understanding.
1. Symmetry
I. Hargittai, Istvan II. Computers and Mathematics with applications III. Series
500

ISBN 0-08-037237-6

Printed in Great Britain by BPCC Wheatons Ltd, Exeter

CONTENTS

PART 2

ERRATA

Page 379, line before the last (equation 7) should be $(3C_2^1, 4C_3^1, 6O_d^-)$.

Page 383, bottom part, the formulae for cyclopentylene, cyclohexylene and dimethine units should be $(>CH-)_5$, $(>CH-)_6$ and $(>CH-HC<)$ respectively.

Page 389, equation (17b), second term should be $(>CH-HC<)q_g$.

Page 390, equation (20a), second term should be $(>CH-HC<)t_g$.

Equation (21a), second term of the first line should be $(>CH-HC<)q_g$ and the second term of the second line should be $(>CH-HC<)t_g$.

Page 391, equation (25a), first term of second line should be $(>CH-HC<)q_g$ and the second term of the third line should be $(>CH-HC<)t_g$.

Page 385, last word on line 4 should read unit instead of unite.

Page 387, Table 1, column 1, the formula for hexadecahedrane on the second and third line should read as follows

$(HC(CH)_3(HCCH)_3$.

$(HC(CH)_3)_3(HCCH)_3)$.

HARGITTAI: Symmetry 2. ISBN 0 08 037237 6

PREFACE

This second volume of Symmetry was initiated in response to the good reception of the first volume, *Symmetry: Unifying Human Understanding* (Ed. I. Hargittai). Pergamon Press (1986). People did not seem to be overwhelmed by the large volume, over-a-thousand pages, but called instead for its continuation. This was the main factor in our decision to embark upon the second volume. We were also pleased and gratified by the good reviews and the recognition that the journal version was named "The Best Single Issue of a Journal" by the Professional and Scholarly Publishing Division of the Association of American Publishers for 1986. Notwithstanding its relatively high price, the book was out of print within two years of its publication.

The purpose of the second volume is the same as that of the first. Our project was envisioned from the very beginning as a contribution to the bridging role of symmetry by bringing together the most diverse fields. An objective was to present the contemporary status of symmetry studies as well as various views on the role and importance of the concept. Having the same purpose for the second volume does not mean just more of the same, but it means also expansion to further fields. Medical sciences and economics are but two of the new areas which were not included in the first volume. In some other fields the extension was in depth rather than in breadth.

The seemingly random ordering of the contributions of the first volume was meant to add emphasis to its interdisciplinarity. This approach was not entirely abandoned in the present volume. We only tried to let the papers flow generally from mathematics and sciences at the beginning towards the arts in later parts. Many of the papers would not lend themselves to any rigorous classification even if we had wished for it, which we certainly did not.

My hopes in sending *Symmetry 2* on its way are two-fold. I would like to see symmetry exercising its ability to help human cognitive processes; and I would like to see symmetry linking various areas of human endeavor.

When completing *Symmetry 1* I was doubtful whether it had accomplished all that had been called for at its conception. At the completion of *Symmetry 2* it is clear to me that it means merely another step in the original project on which we embarked almost five years ago. The fields of relevance and the relevance of various fields seem to expand rather than to close. The concept of symmetry keeps providing impetus for new thoughts and searches.

I am delighted by the growing internationality of the symmetry project. The two volumes have now involved scientists and artists from 23 countries. Since symmetry studies are instrumental in getting to know man's past and culture, the internationality of this endeavor has professional importance beyond good-will and friendship.

I would like to express my appreciation to all contributors to this volume for their dedication and hard work and patience, often involving revisions and other extra work which did not generate the slightest complaint. I am most grateful to all the referees who served unselfishly and with enthusiasm. We have also benefitted from constructive criticism and suggestions from numerous friends and colleagues. This volume is once again the result of wonderful cooperation by many.

I. Hargittai

SYMMETRY 2

Unifying Human Understanding
(Part 1)

Liquid dendritic form. Photograph (1949) courtesy of Gyorgy Kepes.

Computers Math. Applic. Vol. 17, No. 1–3, pp. 1–12, 1989
Printed in Great Britain. All rights reserved

SYMMETRY REVISITED†

M. Senechal

Department of Mathematics, Smith College, Northampton, MA 01063, U.S.A.

Abstract—Rereading Hermann Weyl's now-classic 1952 monograph *Symmetry*, one is struck both by its beauty and by its limitations. Many of the most interesting problems in contemporary symmetry theory concern local configurations, and group theory may not be the only, or the best tool for studying them.

1. INTRODUCTION

In 1901, at a meeting of the British Association for the Advancement of Science, the crystallographers William Barlow and Henry Miers pointed out that the history of the development of the theory of the structure of crystals, which closely parallels the development of the theory of symmetry:

> "...is the history of an attempt to express geometrically the physical properties of crystals, and at each stage of the process an appeal to their known morphological properties has driven the geometrician to widen the scope of his inquiry and to enlarge his definition of homogeneity."

Barlow had only recently completed his derivation of the three-dimensional crystallographic groups. These are the groups of rigid motions, or symmetries, that map a three-dimensional repeating pattern—the abstract model of the structure of an ideal crystal—onto itself. This enumeration, first carried out by the German mathematician Arthur Schoenflies and the Russian crystallographer E. S. Fedorov in 1891, marked the conclusion of a long struggle to define what is meant by crystalline order.

We are so accustomed to thinking of crystals as periodic, modular structures that it comes as a surprise to learn that people once thought otherwise. However, as late as the eighteenth century some respected scientists were still arguing that crystals grow like plants [1]! In their view, the veins and cavities in crystals, which we now regard as structural defects, were channels for the internal distribution of appropriate juices.

Robert Hooke, the seventeenth century English scientist who is famous for his investigations with the microscope, was one of the first to consider the way in which crystal structure might account for crystal form (Fig. 1). By 1822, the French scientist Rene Just Haüy had worked out a highly developed theory of crystal structure in which tiny, identical building blocks were arranged in periodic arrays (Fig. 2); these arrays could be terminated in various ways, accounting for the variations in form of a single crystal species. Debates on the reality of Haüy's building blocks led by the middle of the nineteenth century to the concept of a point lattice (Fig. 3). After the lattices were classified (about 1850) the French mathematician Camille Jordan used the new tool of group theory to enlarge the definition of homogeneity to include repeating patterns with rotational symmetry. Surprisingly, it took many more years for a consensus to emerge that reflections too should be added to the patterns under consideration. In 1891 it was finally established that there are 230 groups of motions in three-dimensional space, and a little later, that there are 17 symmetry groups of repeating patterns in the plane.

It was the famous German mathematician Felix Klein who had suggested to Schoenflies the problem of enumerating the crystallographic groups. Klein was the foremost exponent of the view that the study of geometry should be the study of symmetry groups, or more generally automorphism groups, and he probably realized that the geometry of crystals would be a nice example to illustrate his ideas. The efficacy of group theory for mathematical crystallography was

---

†This paper is based on lectures presented to the New York Academy of Sciences, 6 November 1986, and to the Special Session on "The Mathematical Science of Hermann Weyl: A Centenary Tribute", American Mathematical Society, Amherst, Mass., 26 October 1985.

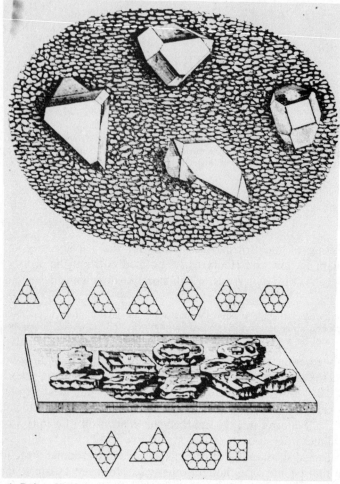

Fig. 1. Robert Hooke's drawings of crystal structure, from *Micrographia* (1665).

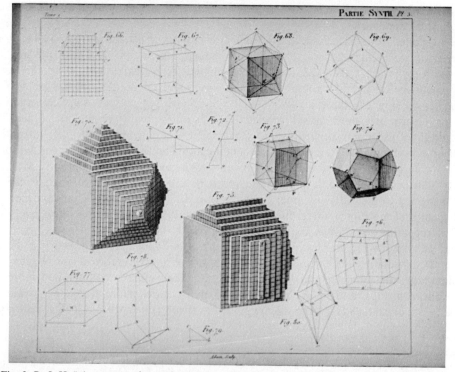

Fig. 2. R. J. Haüy's concept of crystal structure and form, from *Traite de Cristallographie* (1822).

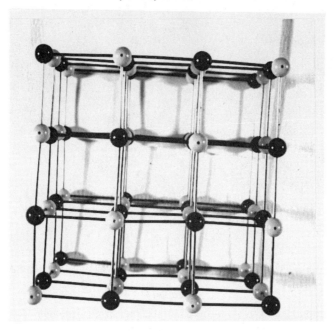

Fig. 3. A point lattice.

strongly supported by the discovery of the diffraction of X-rays by crystals in 1912, one of the great scientific and technical landmarks in our century. Since then, group theory has played a central role in crystallography, because the classification of crystals by symmetry is the first step in the determination of their structure. But even a casual inspection of Vol. 1 of *Symmetry: Unifying Human Understanding* shows that new problems are implicitly challenging group theory's hegemony.

Thus today, 87 years after Barlow and Miers presented their paper, we are again confronted with the need to widen the scope of our inquiry and enlarge our definition of homogeneity, as we are confronted with patterns which, though orderly, do not satisfy the strict requirements of crystalline order. It is not only the famous quasicrystals which suggest this, although they certainly do. The problem is much more general, and it is clear that we must go beyond group theory in order to deal with it. It is less clear what the new tools will turn out to be.

2. CLASSICAL SYMMETRY

A symmetry operation is a rigid motion, a motion that leaves distances between points unchanged. These motions include reflection, rotation, translation and various of their combinations A symmetry operation effects orderly repetition, and orderly repetition can have strong aesthetic appeal. Not surprisingly, symmetry is often found in art, and art museums are wonderful sources of examples of symmetry. For example, let us consider two artistic renderings of pairs of hands.

Figure 4, by Santa Graziani is entitled *"Catch"*. In this pair, one hand is a right hand and the other a left, which is not surprising. But look carefully: these hands are not only related to each other by a mirror reflection, they are also viewed by us as if through a mirror!

In Fig. 5 we see the beautiful sculpture *"Cathedral"* by Rodin. It is at first a surprise to find that here both hands are left hands, related not by reflection but by rotation. I think that the surprise and the rotational symmetry are the sources of the sculpture's compelling beauty.

Mirror symmetry is frequently encountered in art and in science. It is a profound concept, with deep echoes in literature and philosophy. There are still many mysteries of right and left, on the large scale of the structure of the universe, on the small scale of elementary particles, and at all levels in between. Rotational symmetry is found almost everywhere, often combined harmoniously with reflection. We find it in snowflakes, domes, flowers, stained glass windows and pottery. This

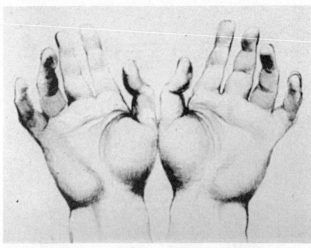

Fig. 4. *"Catch"* from *"Homage to Ingres Series"*, Santa Graziani, Allentown Art Museum.

relation between symmetry and art is more than just pleasing to the eye: it reinforces our tendency
to look for symmetry, and even to find it, where it does not belong and may not exist.

Looking at the plate in Fig. 6, it is easy to see what is meant by a symmetry group. There are
several mirror planes implied by the decorative pattern. If we reflect in two adjacent mirrors
successively, we find that we have in effect rotated the pattern. More generally, every combination
of symmetry operations amounts to another symmetry operation. Thus, the system of symmetries
of an object is "closed"; this is the critical property of a group. Leonardo da Vinci was the first

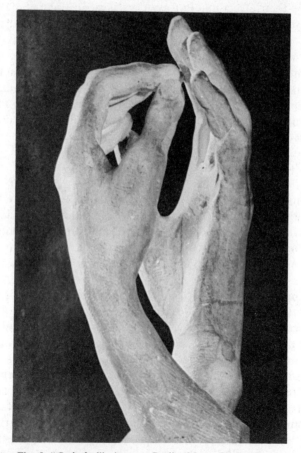

Fig. 5. *"Cathedral"*, Auguste Rodin, Musee Rodin, Paris.

Fig. 6. Symmetrical dish, Allentown Art Museum.

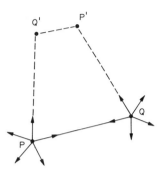

Fig. 7. A plane lattice cannot have five-fold symmetry: the assumption that P and Q are a minimal distance apart is contradicted.

to prove that the symmetry group of this plate is typical for finite two-dimensional objects. The symmetry group of any finite object in the plane consists of reflection in a single mirror, or rotation about a single point, or rotation about a single point combined with reflections in mirrors inclined at equal angles. The symmetry groups of finite three-dimensional polyhedral forms (found in nature as radiolaria, crystals and molecular structure) contain rotations about several axes, reflections in several planes and combinations of rotations and reflections. The possible combinations are severely restricted by the requirement of closure. For example, the groups of rotational symmetries are only those of the n-gonal pyramid, the n-gonal prism, the tetrahedron, the cube and the icosahedron.

The symmetry of infinite repeating patterns includes translations, or shifts through fixed distances. By repeating the shifts over and over we introduce periodicity. The number of symmetry groups for periodic patterns is also restricted; that is why they can be classified into 17 plane groups and 230 space groups. Just as one of the basic ideas of a group is closure, one of the basic features of a periodic pattern is discreteness: two equivalent points cannot be arbitrarily close to one another. Discreteness and closure together are responsible for the restricted symmetries in a pattern.

The plane can be tiled by squares, and also by equilateral triangles and hexagons. The sets of vertices of the first two of these tilings are lattices with, respectively, four-fold, and six-fold rotational symmetry. Regular pentagons do not tile the plane without gaps, and their five-fold symmetry is incompatible with a planar lattice. For, suppose that the points of a lattice had five-fold symmetry. Let P be a lattice point, and Q another lattice point at minimum distance from P. Then P must be surrounded by five equivalent Qs, and Q by five equivalent Ps (Fig. 7). This contradicts the assumption that the distance between P and Q is minimal. Similarly, the order of rotation cannot be greater than 6.

It follows that the only rotational symmetry possible in a two- or three-dimensional repeating pattern is two-, three-, four- and six-fold (and these only in certain restricted combinations). This basic group-theoretic result is known as the crystallographic restriction; Haüy deduced it implicitly in the early 1800s. Recently it has been generalized to the symmetry groups of n-dimensional periodic patterns. The allowable rotations depend upon the dimensions; five-fold symmetry first appears in four-dimensional patterns.

Over the years, crystallographers have adopted periodicity as a defining property of crystal structure. This axiom has been visually reinforced by the example of ornamental patterns, mosaics, tessellations and other periodic designs. One cannot overestimate the impact of this mutual reinforcement of the visual and the conceptual on crystallographic thinking. Even when the details

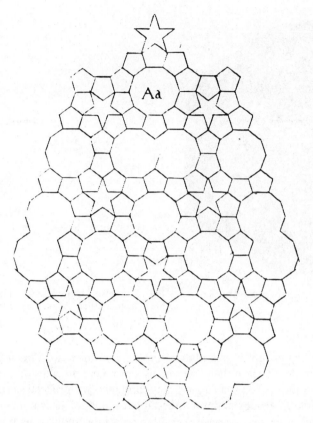

Fig. 8. Kepler's tiling of the plane by pentagons, after his drawing in *Harmonices Mundi* (1611).

of the structure of an actual crystal has seemed unwieldy or untidy, the lattice has always been the framework into which the structure must be fit.

Kepler seems to have been the first to investigate the properties of a nonperiodic pattern. In *Harmonices Mundi*, published in 1619, he discussed an irregular tiling of the plane by regular pentagons (Fig. 8). Kepler concluded that the pattern would never repeat: there would always be "surprises". The study of these surprises seems to have stopped with Kepler; nothing more was done with this pattern until Roger Penrose studied it in the 1970s. In the last few years nonperiodic patterns like this one have been the subject of intense research, raising important questions about the physical basis for the crystallographic restriction and the validity of the crystallographic paradigm.

3. REREADING WEYL

In the years between Kepler and Penrose the mathematical theory of crystalline symmetry was developed to apparent perfection. The classical view of the subject of symmetry was beautifully expounded by the mathematician Hermann Weyl in his famous and beautiful book, *Symmetry* [2]. The book concludes with these words:

> "Symmetry is a vast subject, significant in art and nature. Mathematics lies at its root, and it would be hard to find a better one on which to demonstrate the working of the mathematical intellect."

One might add that it would be hard to find a better book for demonstrating it, either.

The book is based on a series of three richly-illustrated lectures which Weyl delivered at the Institute for Advanced Study on the eve of his retirement in 1952, three years before his death. The lectures were, he said, his swan song. According to *The Reader's Encyclopedia*, the swan song is "the song fabled to be sung by swans at the point of death; hence the last work of a poet, composer, etc". The Encyclopedia goes on to say that "the fable that the swan sings beautifully just before it dies is very ancient, though baseless. Swans do not 'sing' at all, in the ordinary sense of the term". However, the legend is true in this case: Weyl was a swan among mathematicians and *Symmetry* is a song of uncommon beauty.

When first published, *Symmetry* was not just the beautiful exposition of well-known ideas that it seems to us today. In 1952 the concept of symmetry had not yet achieved the preeminence in theoretical physics that it has attained since then, nor had group action clearly emerged as a unifying concept in mathematics. It may be that Weyl's little book, in spite of its nontechnical character or, more likely, because of it, played a role in the direction of science itself.

Rereading it after many years, as I did recently, is still a great pleasure. Again one experiences the joy of finding profound mathematical ideas discussed with literary, philosophical and scientific erudition, as Weyl engages in learned discourse with the great natural philosophers from Leibniz and Newton to Kant and Mach. The book sets an exalted standard for "popular" science writing, at least for the cultivated European intellect of a generation ago.

On the other hand, I also found the book somewhat dated. This is as it should be: one of the hallmarks of good science is that it carries within itself the concepts that make it obsolete.

Weyl argued that symmetry is a tool for the study of complex phenomena; this point of view, first emphasized by Klein, is almost axiomatic in scientific thinking today. Following Klein, Weyl explained how to use the tool in these words:

"Whenever you have structure-endowed entity, try to determine its group of automorphisms."

The problem with this is that while the group of automorphisms can give us information about the structure as a whole—that is, it characterizes its global properties—it does not always help us to understand the reasons why these properties exist. Weyl ignored this problem. However, increasingly, that is what we need to know.

Let us consider some of the problems that suggest a need to go beyond group theory in the study of order. My first example concerns ornamental patterns. Weyl states that:

"The art of ornament contains in implicit form the oldest piece of higher mathematics known to us."

It is certainly true that repeating patterns can be interpreted as nontrivial abstract algebra made concrete. Abstract group-theoretic concepts such as subgroups, cosets, conjugates, normal subgroups, group extensions and permutation representations are hidden in Islamic, Egyptian and other ornamental patterns, like the puzzle pictures that were popular many years ago [3]. But does this mean that the ancients who created these patterns were implicitly doing what we now call group theory? This question leads us directly to the new problems confronting symmetry theory today.

To understand this, let us continue a little further with the enthusiastic Weyl. One of the most-quoted remarks is his assertion that:

"Examples for all 17 groups of symmetry are found among the decorative patterns of antiquity, in particular among the Egyptian ornaments."

This assertion is misleading if it is interpreted to mean, as it frequently has been, that the artists who created the patterns were consciously exploring the 17 possibilities. The anthropologist Dorothy Washburn, who has successfully developed a method of identifying the origins of cultural artifacts by their symmetries, recently noted that:

"the designs in any given culture are organized by just a few symmetries rather than by all classes of the plane pattern symmetries [4]."

This is true even for the civilizations with the greatest wealth of ornamental art. The mathematician Branko Grünbaum went to Spain on a Guggenheim Fellowship a few years ago to study the mosaics of the Alhambra of Granada, Spain. He found only 13 of the 17 there [5]. I have examined two authoritative sources of Egyptian tomb ornamental art, the book *Egyptian Ornament* by Pavla Fortova-Samalova [6] and *The Grammar of Ornament* by Owen Jones [7]; I could find only 11 symmetry types. It is true that all of the 17 groups have appeared somewhere, sometime, in the ornamental art of some civilization, but this is not the point Weyl was trying to make.

This counting may seem to be rather picky, but in fact it gets to the heart of the problem. If the artists were not doing group theory, then what were they doing? Is the symmetry group really the best or only mathematical tool for the study of ornaments (Fig. 9)? Or is it fundamentally misleading?

Fig. 9. An artist coloring tiles.

As Grünbaum has pointed out in another article [8]:

"The main problem is with the very idea of symmetry. There is no basis whatsoever to assume that symmetry— an isometric mapping of the ornament onto itself—was anywhere or at any time motivating artists or craftsmen."

Washburn is a little more equivocal on this point—in another article [9], she notes that some artists seem to have had some regard for the overall pattern. But surely they conceived of the pattern as covering, say, a bowl or a floor, not as being potentially infinite.

The artists appear to have been much more concerned with *local* regularity, the relation of each motif to its neighbors, perhaps over a fairly large area but local nonetheless. Even Escher, who carefully explored and classified the periodic symmetries and tessellations of the plane and fully understood that they were intended to represent infinite patterns, organized his system by means of local adjacencies. (We know this from his notebooks, which Doris Schattschneider has analyzed in detail [10].) This approach is mirrored in the formation of natural structures. Bees building a honeycomb presumably are more concerned with their relation to their nearer neighbors than to farther ones, and we assume that a crystal builds itself to meet the requirements of local atomic forces. Thus, *local geometry* is the driving force in these structures.

For whatever reason, classical symmetry theory, described so eloquently by Weyl, has not been concerned with the problem of the origin of symmetry in any of its manifestations. Most questions of this type are very difficult, and often they are not mathematical. For example, it was physics, not mathematics, that later contradicted Weyl's conclusion that "in all physics nothing has shown up indicating an intrinsic difference of left and right". The reasons for the symmetry of flowers is still beyond our understanding and in any case is a biological problem, the problem of morphogenesis. Nevertheless the question need not be avoided entirely.

Restricting our study of patterns to their classification by symmetry groups may cause us to overlook this important problem. It begs the really interesting question of why patterns generated by local configurations are symmetrical at all. In fact, their global symmetry seems to be quite fortuitous. If we want to understand where their symmetry comes from, why it arises—we must understand why this fortuitous effect occurs.

There is an old problem of natural philosophy here. Seventeenth century scientists wondered whether crystal structures were built according to some "architechtonic principle", or self-assembled by the local interactions of constituent "corpuscles". Robert Boyle opted for the latter in his "Essay on the Origine and Virtues of Gems"; it is curious to find mathematics on the side of architectonics in 1988 [11]!

4. "ENLARGING OUR DEFINITION OF HOMOGENEITY"

Now we are ready to consider some of the geometrical puzzles posed by the local interaction of corpuscles of various kinds.

Euler's famous formula $F + V = E + 2$, which relates the number of faces, vertices and edges of a finite planar graph, can be extended to infinite ones as well. We can use it to determine the infinite tiling-like graphs in the plane in which all the faces are surrounded by their immediate neighbors in exactly the same way (Fig. 10). It turns out that there are precisely eleven of them.

These graphs are known as the Laves nets, after the crystallographer who first enumerated them. As Fig. 10 shows, they can be embedded in the plane very symmetrically. In fact, the faces are symmetrically equivalent—that is, that their symmetry groups include motions which map any chosen face onto any other face. Why should this be so? All that we required in deriving these nets was that all the cells be surrounded by their *immediate* neighbors in the same way. This is a local property, and Euler's formula in and of itself does not have much to say about symmetry.

Grünbaum and Shepherd recently enumerated all the tiling of the plane with symmetrically equivalent tiles. There are 81 types [12]. To carry out this enumeration, they took the Laves nets as their starting point, and used "adjacency symbols" to describe the symmetry of and local configuration about each tile. The symbols were then used to characterize the tilings.

Why are the tiles of these tilings symmetrically equivalent? In this case, local regularity implies global symmetry. This can be proved by using the local criterion for regularity developed by Delone *et al.*, which I discussed in detail in Vol. 1 [13].

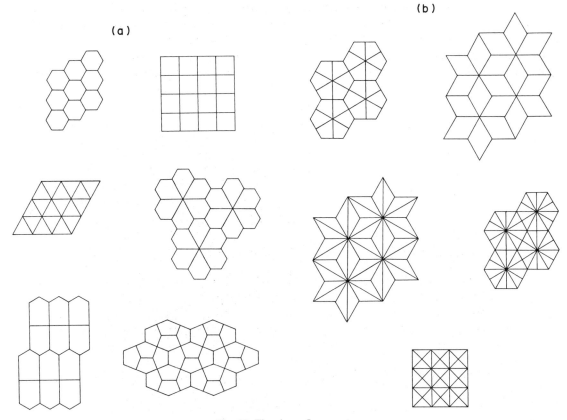

(b)

(a)

Fig. 10. The eleven Laves nets.

The three-dimensional tilings are far from being classified. Adjacency symbols cannot be readily extended to three-dimensional tilings, and even if they could, global symmetry would not necessarily follow. In the first place there is no three-dimensional analogue of the Laves nets, so there is no way to define the adjacency symbol. Moreover, several examples of distinct tilings of 3-space in which nearest neighbor relations are the same have recently been constructed [14]. (This result may help us to understand the variety of structures of some crystals, for example the so-called polytypic layer structures.) Thus the relation between local and global symmetry for three-dimensional tilings is still unresolved.

In fact, as the next two examples show, global symmetry and short-range local order are not necessarily related at all. Let us consider the structure of the so-called spherical viruses. Their polyhedral forms are frequently and happily cited by mathematicians as examples of geometric symmetry in nature. Viruses are built of protein subunits packed around a core of infectious RNA. Some of these structures are helical, others spherical or, more precisely, polyhedral. The study of the latter began with the observation of Watson and Crick in 1956 [15] that X-ray diffraction patterns of crystals of certain viruses indicate that the constituent viruses themselves have a high degree of symmetry. This suggested to Watson and Crick that the symmetry group of such a virus particle was likely to be that of one of the regular solids, that is, the tetrahedral, octahedral or icosahedral group. Further research showed that icosahedral symmetry predominates.

The overall spherical shapes of these viruses serve the crucial function of encapsulating the RNA until the viruses can find suitable hosts to infect. But their specific configurations are the result of local geometric, chemical and kinetic constraints on the constituent subunits. So why should viruses be *symmetrical?* Watson and Crick argued that: "Whenever, on the molecular level, a structure of a definite size and shape has to be built up from smaller units . . . the packing arrangements are likely to be repeated again and again and hence the subunits are likely to be related by symmetry elements."

In other words, Watson and Crick are arguing that viruses do what my young daughter was doing (Fig. 11) and that this repetition of local configurations would lead to global symmetry. But it turns out that this is *not* why the viruses have icosahedral forms!

Another, different challenge to our concept of symmetry is posed by the icosahedral quasicrystals, which have attracted a great deal of attention during the last few years. These crystals, which are alloys of aluminium and manganese, have electron diffraction images with five-fold symmetry (Fig. 12), a symmetry which, as we have seen, is incompatible with translational periodicity in two or three dimensions.

Fig. 11. Diana Senechal playing with tiles.

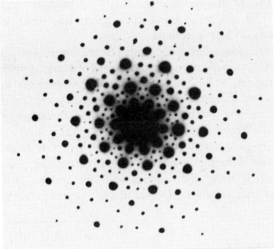

Fig. 12. Diffraction pattern of Schechtmanite. Photograph courtesy of John Cahn.

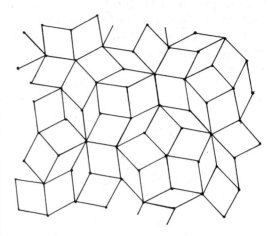

Fig. 13. A Penrose tiling by rhombs.

However, if no periodicity in fact exists, what then is the structure of these crystals? Already in 1981 the British crystallographer Alan Mackay had shown that optical diffraction patterns obtained from a Penrose tiling of the plane exhibit five-fold symmetry [16]. By a great leap of analogical thinking, some physicists suggested that the quasicrystal structure is some sort of three-dimensional analogue of the Penrose tiles.

In one of their several incarnations the planar Penrose tiles are copies of two kinds of rhombs which, when juxtaposed according to certain local "matching rules", tile the plane nonperiodically (Fig. 13). Tilings with these tiles are closely related to Kepler's tiling (Fig. 9); if you look carefully, you will find that throughout the tiling there are rhombs fitted together to form whole or partial star pentagons, decagons and other configurations with five-fold symmetry. The pattern repeats in the sense that these configurations can be found everywhere within it, but the repetition is not periodic.

Several different methods have been used to study the Penrose tilings. One is the original local method of investigating matching rules and what they imply. Group theory plays no role here, since there is no group which prescribes the positions of the tiles. In contrast to this local approach, there are several global methods of studying them. In the most popular one, the Penrose tiles and their three-dimensional analogues are regarded as projections of sections of a regular tiling by unit cubes in a Euclidean space of dimension sufficiently high for five-fold symmetry to be compatible with a translation lattice. It is easy to calculate diffraction patterns in this model since, in effect, conventional symmetry is restored by going to higher dimensions. But this "simplicity" is achieved only at the price of important information about the local structure. We are back to the problem of the craftsmen: like ornaments and crystals, tilings "grow" in two- and three-dimensional space, and it is in this context that we must eventually understand them. First steps toward reconciling the two approaches have recently been made by Katz [17].

Although the quasicrystal structure is still unknown, X-ray and other studies have shown that it is not a "decoration" of a three-dimensional Penrose pattern. But the Penrose tilings suggest one possible direction of research on the mathematical problem of enlarging our concept of order: we might try to include patterns whose images under certain transformations (e.g. diffraction) exhibit recognizable symmetry. It is important to remember, however, that the aesthetic appeal of tilings, even nonperiodic ones, can mislead us; the patterns we are looking for may not turn out to be beautiful.

If we are to understand the formation of orderly patterns, then we can no longer neglect the problem of growth, function, and form. Perhaps we should follow the guidance of D'Arcy W. Thompson, the biologist whose masterpiece *Growth and Form* [18] is, in Weyl's words, "a masterpiece of English literature which combines profound knowledge in geometry, physics and biology with humanistic erudition and scientific insight of unusual originality". Thompson would urge us to look to biology for new inspiration. This is already being done: interesting ideas are percolating into geometry from robotics and from cellular automata.

The virus problem is also very instructive in this regard. After many years research, during which the early, relatively simple, icosahedral models were gradually modified by a series of more complex and sophisticated but still comprehensible structures, it has been discovered that there is no apparent relation between its local configurations and the global icosahedral symmetry: centers of six-fold rotational symmetry function as five-fold centers for the larger structure! In the words of one of the leading investigators in this field from its beginnings, D. L. D. Caspar, "the design has no geometrical rationale but it obviously has biological logic" [19]. That is, the structure is not designed to be beautiful, but rather to fulfill certain dynamic functions, and it is from this point of view that we should seek to understand it. It is likely that viruses will still present some interesting symmetry problems once the biological logic is better understood and symmetry theory has been expanded to include it.

REFERENCES

1. H. Metzger, *La Génèse de la Sciènce des Cristaux*. Blanchard, Paris (1922).
2. H. Weyl, *Symmetry*, Princeton University Press, N.J. (1952).
3. M. Senechal, The algebraic Escher. *Struct. Topology* (in press).
4. D. Washburn, Pattern symmetry and colored repetition in cultural contexts. *Comput. Math. Applic.* **12B**, 767–781 (1986). Reprinted in *Symmetry: Unifying Human Understanding* (Ed. I. Hargittai). Pergamon Press, Oxford (1986).
5. B. Grünbaum, Z. Grünbaum and G. S. Shepherd, Symmetry in Moorish and other ornaments. *Comput. Math. Applic.* **12B**, 641–653 (1986). Reprinted in *Symmetry: Unifying Human Understanding* (Ed. I. Hargittai). Pergamon Press, Oxford (1986).
6. P. Fortova-Samalova, *Egyptian Ornament*. Alan Wingate, London (1963).
7. O. Jones, *The Grammar of Ornament*, Bernard Quaritch, London (1868).
8. B. Grünbaum, The Emperor's new clothes: full regalia, g-string, or nothing? *Math. Intell.* **6**(4), 47–56 (1984).
9. D. Washburn, An anthropological perspective on the "Emperor's Edict". *Math. Intell.* **8**(4), 66–67 (1986).
10. D. Schattschneider, M. C. Escher's classification system for his colored periodic drawings. In *M. C. Escher: Art and Science* (Eds H. S. M. Coxeter, M. Emmer, R. Penrose and M. L. Teuber). North-Holland, Amsterdam (1986).
11. E. F. Keller, The force of the pacemaker concept in theories of aggregation in cellular slime mold. In *Reflections on Gender and Science*. Yale University Press (1985).
12. B. Grünbaum and G. C. Shephard, The eighty-one isohedral tilings of the plane. *Math. Proc. Cambridge Philos. Soc.* **82**, 177–196 (1977).
13. M. Senechal, Geometry and crystal symmetry. *Comput. Math. Applic.* **12B**, 565–578. Reprinted in *Symmetry: Unifying Human Understanding* (Ed. I. Hargittai). Pergamon Press, Oxford (1986).
14. P. Engel, *Comput. Math. Applic.* **12B**. Reprinted in *Symmetry: Unifying Human Understanding* (Ed. I. Hargittai). Pergamon Press, Oxford (1986).
15. J. Watson and F. Crick, The structure of small viruses. *Nature*, **177**, 473–75 (1956).
16. A. Mackay, Crystallography and the Penrose pattern. *Physica* **114A**, 609–613 (1982).
17. A. Katz, Theory of matching rules for the 3 dimensional Penrose tiles. Preprint.
18. D'Arcy W. Thompson, *Growth and Form*. Cambridge University Press (abridged edn), Cambridge (1961).
19. D. L. D. Caspar, Lecture presented to the Shaping Space Conference, Smith College, Northampton, Mass. 7 April 1984. A summary appears in *Shaping Space* (Ed. M. Senechal and G. Fleck). Birkhauser, Boston (1988).

Computers Math. Applic. Vol. 17, No. 1–3, pp. 13–15, 1989
Printed in Great Britain. All rights reserved

0097-4943/89 $3.00 + 0.00

SYMMETRY AT THE FOUNDATIONS OF SCIENCE

J. ROSEN

School of Physics and Astronomy, Raymond and Beverly Sackler Faculty of Exact Sciences,
Tel Aviv University, 69978 Tel Aviv, Israel

Abstract—Reproducibility and predictability, forming the dual foundation of science, are shown to be kinds of symmetry.

1. INTRODUCTION

Science rests firmly on the dual foundation of reproducibility and predictability. Reproducibility means that experiments can be repeated by the same and by other investigators, thus giving data of objective, lasting value. Reproducibility makes science a common human endeavor (rather than, say, a collection of private, incommensurate efforts). Predictability means that order can be found among the phenomena investigated, from which laws can be formulated, predicting the results of new experiments. Then theories can be developed to explain the laws. Predictability makes science our means both to understand and to exploit nature. There seems to be no *a priori* necessity that nature be reproducible or predictable at all, but the fact that we are doing science proves that nature indeed possesses reproducible and predictable aspects. (As for nature's irreproducible or unpredictable aspects, whatever they might be, they lie outside the domain of concern of science.)

We are accustomed to the idea of symmetry in science. Two simple and concrete examples are the spatial symmetry of crystal lattices and the temporal symmetry of periodic processes such as oscillators. What might seem surprising is that science itself is based on symmetry. Indeed, it is the aim of this article to point out that both *reproducibility and predictability are kinds of symmetry*, and thus, as the title indicates, that symmetry lies at the foundations of science.

In the following section we briefly review the meaning of symmetry in rather general terms. Then in Section 3 we consider the meaning of reproducibility in more detail and see that it is indeed a kind of symmetry. Section 4 is devoted to a detailed consideration of the meaning of predictability, from which follows that it too is a kind of symmetry. Mathematical, physical and philosophical technicalities are avoided as far as possible.

2. SYMMETRY

In everyday speech symmetry usually means a balance, a repetition of parts, a regularity of form. More precisely and generally symmetry can be said to be invariance under transformation, i.e. the situation is symmetric if there are one or more changes that can be made that nevertheless leave some aspect of the situation unchanged. Take, for example, a uniform metal equilateral triangle and imagine rotating it by 120° or 240° about its center within its plane. Although a transformation, a change, has been made, the result looks the same and has the same physical properties as the original. Thus, we can say that our piece of metal possesses symmetry under these rotations with respect to external appearance and physical properties. If the triangle were not uniform or had a corner chopped off, it would not possess this symmetry.

Systems that might possess symmetry are not confined to the domain of concrete objects, but may be abstract to the extreme. The transformations involved do not have to be geometric; the imagination is free to roam: space, time, particle–antiparticle, permutation, and on to the abstract. Neither must the invariant aspect be appearance nor physical property, but may be any concrete or abstract aspect of the system under consideration. However, the very least we need for symmetry is the possibility of making a change and some aspect that is immune to this change.

3. REPRODUCIBILITY AS SYMMETRY

Let us express things in terms of experiments and their results. Reproducibility is then commonly defined by the statement that the same experiment always gives the same result. But what is the "same" experiment? Actually each experiment, and we are including here even each run of the same experimental apparatus, is a unique phenomenon. No two experiments are identical. They must differ at least in time (the experiment being repeated in the same lab) or in location (the experiment being duplicated in another lab) and might, and in fact always do, differ in other aspects as well. So when we specify "same" experiment and "same" result, we actually mean "equivalent" in some sense rather than "identical." We cannot even begin to think about reproducibility without permitting ourselves to overlook certain differences, these differences involving time or location as well as various other aspects of experiments.

Consider the difference between two experiments as being expressed by a transformation, the transformation being the change that must be imposed on one experiment in order to make it into the other. Such a transformation might involve temporal displacement, if the experiments are performed at different times. It might (also) involve spatial displacement, if they are (also) performed at different locations. If the experimental setups have different directions in space, the transformation will involve rotation. If they are in different states of motion, a velocity transformation will be involved. We might bend the apparatus. Or we might measure velocity rather than pressure. Etcetera, etcetera.

However, not all possible transformations are what we associate with reproducibility. Let us list those we do. We certainly want temporal displacement, to allow the experiment to be repeated in the same lab, and spatial displacement and rotation, to allow other labs to perform the experiment. The motion of the Earth, its diurnal rotation and annual revolution, require spatial displacement and rotation even for experiments performed in the same lab as well as velocity transformations for those performed at different times or locations. Then, to allow the use of different sets of apparatus, we need replacement by other materials, other atoms, other elementary particles, etc. Due to unavoidably limited experimental precision we must also include small changes in the conditions. And we also need changes in certain aspects of experimental setups, which we will not examine here, over which we have no control practically or in principle. These are the transformations associated with reproducibility that I can think of.

Now that we have collected all transformations we associate with reproducibility (add any I might have left out), we define reproducibility as follows: consider an experiment and its result, consider the experiment obtained by transforming the original one by any transformation belonging to this set and consider the result obtained by transforming the original result by the same transformation. If this transformed result is what is actually obtained by performing the transformed experiment, and if this relation holds for all transformations belonging to the set, we have reproducibility.

As an example, imagine some experiment whose result is a particle appearing at some point in the apparatus some time interval after the switch is turned on. Now imagine repeating the experiment with the same apparatus, in the same direction, in the same state of motion, etc., but 24 h later and at a location 1 km north of the original location. If that particle now appears 24 h later than and 1 km north of its previous appearance, we have evidence that the experiment might be reproducible. (As usual in this business, whereas a single negative result disproves reproducibility, no number of positive results can prove it. A few positive results make us suspect reproducibility; many will convince us; additional positive results will confirm our belief.)

Do you notice symmetry materializing here? Reproducibility is indeed symmetry. We can see that in this way: consider a reproducible experiment and its result. Transform it and its result together by any transformation belonging to the set of transformations we associate with reproducibility. The pair (transformed experiment, transformed result) is of course different from the pair (original experiment, original result), but there is an aspect of the pairs that does not change under the transformation. This aspect is the relation, call it physicality, actuality, reality, or whatever, that *the result is what is actually obtained by performing the experiment.* In other words, the symmetry that is reproducibility is that for any reproducible experiment and its result, the experiment and result obtained from them by any transformation belonging to the above set are also an experiment and its actual result.

4. PREDICTABILITY AS SYMMETRY

Here, too, we express things in terms of experiments and their results. Predictability, then, is that it is possible to predict the results of new experiments. Of course that does not come about through pure inspiration, but is attained by performing experiments, studying their results, finding order and formulating laws.

So imagine we have an experimental setup and run a series of n experiments on it, with experimental inputs $\exp_1, \exp_2, \ldots, \exp_n$, respectively, and corresponding experimental results $\text{res}_1, \text{res}_2, \ldots, \text{res}_n$. We then study these data, apply experience, insight and intuition, perhaps plot them in various ways, and discover order among them. Suppose we find that all the data obey a certain relation, denote it R, such that all the results are related to their respective experiments in the same way. Using function notation, we find that $\text{res}_i = R(\exp_i)$, for $i = 1, \ldots, n$. This relation is a candidate for a law, $\text{res} = R(\exp)$, predicting the result res for *any* experimental input exp. Imagine further that this is indeed the correct law. Then additional experiments will confirm it, and we will find that $\text{res}_i = R(\exp_i)$, also for $i = n + 1, \ldots$, as predicted. Predictability is the existence of such relations for experiments and their results.

For an example of that, consider the experimental setup of a given sphere rolling down a fixed inclined plane, with the experimental procedure of releasing the sphere from rest, letting it roll for any time t, and noting the distance d the sphere rolls in that time. (Here t and d are playing the roles of exp and res, respectively.) Suppose we perform ten experiments, giving the ten data pairs $(t_1, d_1), (t_2, d_2), \ldots, (t_{10}, d_{10})$. We study these data and plot them in various ways. The plot of distance d_i against square of elapsed time t_i^2 looks like all ten points tend to fall on a straight line. That suggests the relation that the distance traveled from rest is proportional to the square of the elapsed time, $d_i = bt_i^2$, for $i = 1, \ldots, 10$. This suggests the law $d = bt^2$ predicting the distance d for *any* time t. As it happens, this law is correct, and all additional experiments confirm it. Thus, the relation of distance to elapsed time is a predictable aspect of the setup.

That predictability is a symmetry can be seen as follows: for a given experimental setup consider all the different experiment–result pairs (exp, res) that have been, will be or could be obtained by performing the experiment. Transform any one of these into any other simply by replacing it. The transformed pair is different from the original one, but the pairs possess an aspect that is not changed by the transformation. This aspect is that *exp and res obey the same relation for all pairs*, namely the relation $\text{res} = R(\exp)$. Put in different words, this symmetry is that for any predictable experiment and its result, the experiment and its result obtained by changing the experimental input obey the same relation as the original experiment and result.

5. CONCLUSION

Following the definition of symmetry as invariance under transformation, we showed that both reproducibility and predictability are kinds of symmetry by showing for each the changes that can be made and the aspect that is immune to these changes. For reproducibility any experiment–result pair can be transformed by any of the set of transformations associated with reproducibility. The invariant aspect is that the result is what is actually obtained by performing the experiment. For predictability any experiment–result pair can be transformed by replacing it with any other pair for the same experimental setup. What is invariant is the relation between experimental input and result.

Since reproducibility and predictability are the two most fundamental foundation stones of science, we see that symmetry not only serves within science but actually lies at its very foundation.

BIBLIOGRAPHY

1. J. Rosen, *Symmetry Discovered*. Cambridge. University Press, Cambridge (1975).
2. J. Rosen, *A Symmetry Primer for Scientists*. Wiley, New York (1983).

Computers Math. Applic. Vol. 17, No. 1–3, pp. 17–32, 1989
Printed in Great Britain. All rights reserved

0097-4943/89 $3.00 + 0.00

SYMMETRY AND CHAOS

A. A. Chernikov, R. Z. Sagdeev, D. A. Usikov and G. M. Zaslavsky

Space Research Institute, Profsoyuznaya 84/32, Moscow 117810, U.S.S.R.

Abstract—The dynamical model of a particle moving in the magnetic field and wave packet field is discussed. Particle trajectories produce tilings with the same quasisymmetry on the phase plane as is, for example, in quasicrystals. This makes it possible to find the dynamical generator of quasisymmetry and to consider the quasisymmetry onset as the result of a weak interaction between a rotational symmetry and a translational symmetry. Such an interaction produces thin channels of chaotic dynamics of the particle in the phase space (stochastic web) which realizes the tiling of the plane with quasisymmetry.

1. INTRODUCTION

Attempts to look into the secrets of the geometry of the world in which we live began far back in time. There are good reasons why the greatest discoveries that drastically changed our concepts of the laws of nature have become milestones along this road. Much work is still being done in the same direction, since it is believed that better understanding of the properties of fields and particles should explain typical features of the micro and macroworld. This is, however, but one side of the medal. Its other side is art, where no smaller efforts were taken to understand the geometric harmony of the world, to unravel the mysteries of patterns, configurations, structures the nature itself creates. These two trends of search have never moved far apart, and from time to time there appeared personalities who managed to make discoveries in both geometry and art. Long is the list from unknown artists of ancient ages to Plato, Archimedes, Leonardo da Vinci, Kepler and Dürer—of those who had the same goal and strove to understand symmetry. Most intricate oriental ornaments are the best examples of works of art to emphasize which certain symmetry laws should be known.

Symmetry of patterns, symmetry of tilings, theory of tilings and dense packings on the plane and in space are only a few related branches in the studies of nature; joined with them are other branches known better (symmetry of crystals, symmetry of structures in fluids) or worse (symmetry of foam, symmetry of clouds). Unity of symmetry laws continuously makes this list longer and longer, adding to it purely applied problems [1, 2] as well.

Simpler laws of symmetry are associated with reflections of figures, with their rotations and translations; for instance, a two-dimensional (plane) ornament invariant to arbitrary translations may have 17 various types (groups) of symmetry. All of them were known by ancient Egyptian craftsmen. Eleven groups are found in the Moorish ornaments of Alhambra (Granada, Spain), 5 groups, additional to these 11, exist in African handicraft articles in Southern Sakhara, and one more, the last group, was found among the Chinese ornaments. All these groups of symmetry are presented in the work of Dutch artist M. C. Escher (1898–1972). Many of these symmetries were independently discovered by him.

It is easy to create tilings which remain symmetric when turned through the angle $\alpha = 2\pi/q$, where q is an integer (q-fold rotation symmetry). However, it is much more difficult to combine both types of symmetry in tilings: translational and rotational. This can only be done for $q = 2$, 3, 4, 6. That is why it was commonly accepted up to 1984 that crystals of the 5-fold symmetry do not exist. In 1984 Shechtman and his group produced—by cooling the alloy of Al and Mn—a new substance with the 5-fold symmetry as the X-ray spectroscopy pattern showed [3]. This alloy, also called Shechtmanite, is somewhat intermediate between crystals and liquids. This was not only a revolutionary step in crystallography, but also gave a new impetus to the studies of symmetry properties in nature.

The plane cannot be densely tiled with regular pentagons only, without gaps, however, the pentagons can be used as the basic elements, and it was Kepler and Dürer who proved it. After Penrose's [4], pentahedric tilings have been thoroughly analyzed by many professionals and amateurs [5, 6]. A world of new possibilities has opened up—the world of nonperiodic tilings with highly ordered-symmetry motifs. What is the origin of new types of symmetry which we call a symmetry of quasicrystal type and could they be associated with real-world physical processes?

SECTION 2

A possible answer to this question was formulated when quite a different problem was being considered at the Space Research Institute, U.S.S.R. Academy of Sciences. We know that the very same symmetry laws may manifest themselves in the phenomena of having nothing in common, at first glance. In 1959 one of the authors (R.Z.S.) described the model of collisionless shock waves and suggested a mechanism by which particles are accelerated before the front of the shock moving perpendicularly to the magnetic field [7] (now this mechanism is known as "surfatron"). If the magnetic field is constant and uniform then, in the plane perpendicular to the magnetic field, particles are moving along circular orbits. Imagine that near a particle there forms a wall parallel to the magnetic field. Then, the particle will be regularly reflected from the wall and drift along it perpendicularly to the magnetic field (Fig. 1), and its orbits will be arcs of a circle. If the wall is moving perpendicularly to the magnetic field the particle gains energy in each collision. The shock front acts as a wall. The magnetic field is a factor ensuring multiple collisions since it is due to the fact that after each collision the particle returns to the shock front. The moving shock wave affects the particle even if it is an untrapped particle, that is, when its energy exceeds the energy potential of the wave. Each time a particle passes above the crest of the wave it loses energy, moving in the same direction as the wave, or gains energy meeting the wave. The balance of these two effects depends on the relationship between the phases of a particle and a wave. If instead of one wave we consider a superposition of many plane waves (wave packet) there forms a very intricate dynamic picture which has been an object of study for many years in the context of various problems in plasma physics and astrophysics. It is one of these cases that is related with the problem of symmetry of plane tilings [8–10]. The wave packet may be arranged so that the particle undergoes very short-lived collisions (kicks) following equal time intervals. If, during the period of particle rotation in the magnetic field, exactly q kicks occur, the resonance of order q is said to take place. The number q, as we shall see below, will be fundamental for describing the dynamics of particles.

SECTION 3

When a particle is moving in complex fields its trajectory could be very intricate, though the "intricacy" may only seem to exist and may result from the combination of several simple motions. Since the trajectory of the particle is uniquely determined by setting the initial coordinates and impulses, it is usually represented in the phase space of all coordinates and impulses of the particle. However, now it is difficult to present the trajectory graphically, that is, the phase space is six-dimensional. Instead, Poincaré mappings are used. An oriented surface or plane is chosen and points of puncture of the plane by the trajectory in a certain direction are marked on it (Fig. 2). Thus, a set of punctured points forms. It is the function with which the position of a point the trajectory punctures on the plane can be estimated from the position of the previous one that specifies the Poincaré mapping:

$$\mathbf{R}_{n+1} = \hat{M}\mathbf{R}_n, \tag{1}$$

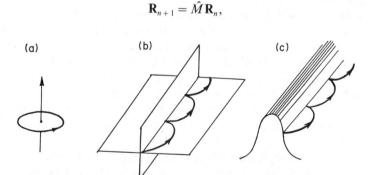

Fig. 1. Particle drift in a magnetic field: (a) free particle rotation; (b) particle drift along the reflecting wall; (c) particle drift along a wave front.

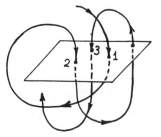

Fig. 2. Poincaré mapping; 1, 2, 3—consequent points of the mapping.

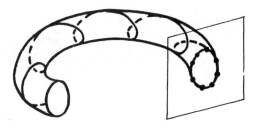

Fig. 3. Invariant curve formation in the plane of torus section.

where $\mathbf{R}_n$ is the vector that determines the position of the nth punctured point on the chosen plane; $\hat{M}$ is the matrix connecting two consecutive punctured points.

A set of punctured points produces an image of the mapping. If all the points of the image, for instance, are located along some closed curve, it means that the trajectory is winding on an invariant torus, whereas the curve itself is also invariant and results from the torus intersection with the plane chosen (Fig. 3). Here the word "invariant" is very important, meaning that the position of the torus and the curve is not time dependent. Different trajectories result for different initial conditions. These trajectories generate different invariant curves on the plane intersecting the tori. All this leads to a certain pattern of invariant curves on the chosen phase plane, which builds a phase portrait of the particle. It is this phase portrait that will help us judge the basic dynamic properties of particles or more complicated systems. For instance, the phase plane of the pendulum is made of closed curves, which represent its oscillations and open curves, which represent its rotations [Fig. 4(a)]. Two different types of motion—oscillations and rotations—are separated by a single curve, a separatrix which belongs to neither.

Extremely essential is the role of the separatrix in the modern theory of dynamic systems. The problem is that the vicinity of the separatrix is unusually sensitive to perturbations, however small they are. Indeed, it is here that a slight "stirring" of the initial condition may replace one type of motion with another. Hence if the pendulum is subjected to some arbitrary perturbation, periodic in time, then for the initial conditions near the separatrix, the pendulum trajectory will appear very unusual. Depending on the difference between the pendulum phase and the external force phase, the pendulum will be either oscillating or rotating. Even the very number of oscillations or rotations—between two changes of the mode—will vary without any order. As the analysis shows, such motion of the pendulum is chaotic as in no way does it differ from a certain random process [11]. Though a regular periodic force acts upon the pendulum, the latter behaves, however, as if this force is random. As a result, the phase portrait of the pendulum subjected to some perturbation has one more type of trajectories—stochastic trajectories.

There is a range of nonzero measures near the separatrix, such that if the initial condition is within this range, the respective trajectory will be stochastic. This region is called a stochastic layer [Fig. 4(b)]. The smaller the perturbation, the thinner the layer. It is a very nontrivial fact that the onset of chaos may occur in a system without any random forces. The studies of several recent decades have demonstrated that there exists no surpassable boundary between the random and the nonrandom. The same physical system may behave regularly (for instance, its coordinates will vary quasiperiodically in time) or randomly (if some of its parameters will be changed a little).

The above described phenomenon of a very complex irregular motion arising in dynamic systems is now usually called chaos. The phenomenon is most widely applied to all branches of modern physics [11]. In the so-called Hamiltonian systems, chaos originates in stochastic layers near the separatrices, destroyed due to perturbations. That is why the analysis of stochastic layers is very instrumental in identifying the properties of the system. It is easy to imagine that in cases more complicated than that of pendulum stochastic, layers form an intricate net of "channels" in the phase space (Fig. 5). This net may be called a stochastic web. It has a finite width. If the initial condition of a particle is such that it is inside one of the web channels, then its further motion is random walkings along web channels. On the contrary, if a particle was within a web cell its trajectory will be regular and its Poincaré mapping will produce an invariant curve inscribed in the web cell. In 1964 Arnold predicted the existence of a stochastic web as a universal property

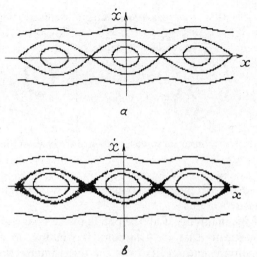

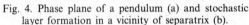

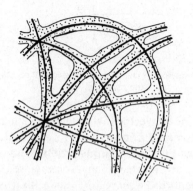

Fig. 4. Phase plane of a pendulum (a) and stochastic layer formation in a vicinity of separatrix (b).

Fig. 5. Separatrix network (bold lines) and stochastic web (the dotted region).

of dynamic systems with more than two degrees of freedom (systems with N degrees of freedom are characterized by N coordinates and N impulses) [12]. It is now evident that a stochastic web, infinite in size, may arise even in the case when a system with only one degree of freedom is perturbed by a periodic force [8, 10]. In particular, it is exactly this case that the already mentioned problem of a charged particle, moving in a constant magnetic field and in the field of an electrostatic wave packet propagating perpendicularly to the magnetic field, can be reduced. In that problem the existence of a qth order resonance is the condition for the web to appear.

The appearance of an infinite stochastic web has several implications. Some of them are obvious. The web may exist at a minimal possible dimensionality of the system as, for $N = 1$, all systems have regular trajectories. That the web is infinite means a possibility of a diffusive acceleration of particles along the web channels up to very high velocities. However, the most unusual properties of the web are associated with its geometry and its symmetry.

SECTION 4

The Poincaré mapping for a particle in the magnetic field and the field of a perpendicular wave packet, for certain conditions, is presented by:

$$\hat{M}_\alpha : \begin{cases} u_{n+1} = (u_n + K \sin v_n)\cos \alpha + v_n \sin \alpha, \\ v_{n+1} = -(u_n + K \sin v_n)\sin \alpha + v_n \cos \alpha, \end{cases} \tag{2}$$

where (v, u) are the coordinates of the vector **R** on the phase plane; K is the parameter proportional to the wave packet amplitude; α is the angle by which a particle turns in the magnetic field between two consecutive kicks caused by the wave packet. The mapping $\hat{M}_\alpha$ relates the coordinate v_n and the particle impulse u_n at the time instant t_n, a kick acts upon the particle, with the coordinate v_{n+1} and the impulse u_{n+1} at the time t_{n+1} of the next kick acting. If all the time intervals between kicks are assumed to be unity, then

$$t_n = n.$$

The Hamiltonian of the particle motion has the form

$$H = \tfrac{1}{2}(\dot{x}^2 + \alpha^2 x^2) - \alpha K \cos x \sum_{n=-\infty}^{+\infty} \delta(t - n) \tag{3}$$

and the equation of motion for the particle is

$$\ddot{x} + \alpha^2 x = \alpha K \sin x \sum_{n=-\infty}^{+\infty} \delta(t - n). \tag{4}$$

It is evident from equations (3) and (4) that they describe the motion of a linear oscillator with the frequency α when subjected to a perturbation in the form of periodical δ-pulses. The same equations describe the motion of a particle in the magnetic field with the Larmor frequency α while the perturbation is a wave packet with an infinite number of harmonics, which are propagating along the x-axis. Comparing the solutions after two successive kicks permits the differential equation of motion, equation (4), to be replaced with a different equation. This gives mapping (2), where $v = x$, $u = \dot{x}/\alpha$.

We deliberately wrote equation (3) for the Hamiltonian H which gives origin to the mapping $\hat{M}_\alpha$. It is obvious from the form of H that its first term (in parentheses) describes only the rotation of the particle. Trajectories of rotation on the phase plane are characterized by a degenerate rotation symmetry. The second term of the Hamiltonian describing the kicks is invariant to the coordinate shift $x \to x + 2\pi m$ (m is the integer). In the absence of the magnetic field particle trajectories in the phase plane would have a translation symmetry like the phase portrait of the pendulum in Fig. 4. Thus, the Hamiltonian H realizes the rotation and translation symmetries interaction. The parameter K characterizes the intensity of this interaction.

For small K the interaction between rotational and translational symmetries is insignificant, and we face the question: what should the phase portrait of a dynamic system look like to implement both symmetries together? Though the parameter K is small, the interaction of symmetries should be most effective if the resonance condition is valid

$$\alpha = 2\pi/q, \tag{5}$$

where q is the integer.†

The above considerations show that we have achieved some success, though no specific results have been obtained yet. By this, we mean that we managed to formulate the problem of symmetries interaction as a problem of determining the phase portrait of a certain dynamic system. In other words, the problem of such tiling of the plane with weakly interacting rotation and translation symmetry is solved by determining invariant sets of mapping (2) over the phase plane (u, v). It only remains to consider these sets.

SECTION 5

If the value q belongs to the set

$$\{q_c\}: q = 2, 3, 4, 6,$$

then and only then the rotation and the translation symmetries may coexist on the plane. The mapping (2) at $q \in \{q_c\}$ generates simple square and hexagonal grids on the plane. Figure 6 shows such grids (for $q = 3$ and $q = 6$ the same structure results which is called a "kagome lattice").

For small values of parameter K in $\hat{M}_\alpha$ and for the resonance values of $\alpha = 2\pi/q$ there are stochastic webs with a crystal structure on the phase plane. The web is not thick but finite. It is generated by a single trajectory. If the initial coordinate of a particle (u_0, v_0) is chosen so that it is lying within the area occupied by the web, then iterations of this point according to mapping (2) will create these pictures illustrated in Fig. 6. If the initial condition (u_0, v_0) is somewhere inside the cells, formed by the web, then on the phase plane there will appear within the cells closed orbits which are cross-sections of invariant tori.

SECTION 6

The mapping $\hat{M}_q \equiv \hat{M}_{\alpha = 2\pi/q}$ has many striking properties. One of them is that the stochastic web exists at any $q \neq 2$ and, obviously, at arbitrarily small K [8, 9]. This means, in particular, that at $q \not\in \{q_c\}$ new kinds of structure should originate. That they appear at resonance values of α [see equation (5)] is the most nontrivial fact in the theory, and even at a first glance (Fig. 7 for $q = 5$

---

†Cases with $\alpha = 2\pi p/q$ (p, q are integers) somewhat change the dynamics of particles, but do not change the symmetry of the phase portrait.

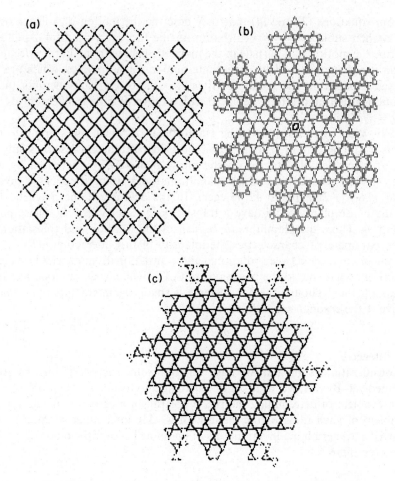

Fig. 6. Square and hexagonal lattices. (a) $q = 4$; (b) $q = 3$; (c) $q = 6$.

and $q = 7$) it seems that we meet with a new type of symmetry.† We will refer to it as a q-fold symmetry. Patterns in Fig. 7 as well as in Fig. 6 are drawn again by a single particle trajectory and this, also amazing, fact implies that, with simple electromagnetic field configurations, we may construct patterns the complexity of which could not have even been imagined until recently. Furthermore, if we now inspect Figs 6(b), 7(a) and 7(c) it becomes obvious that the "snowflakes" imaged there are typical fractals. Hence, the trajectories of a particle, generated by the mapping $\hat{M}_q$, may look like such monster-curves which were earlier assigned to the category of "formal curves".

The webs in Fig. 7 have approximately a rotative symmetry. On the one hand, their approximate character is related with the finite time over which the image is obtained. Owing to this, not all of its details have been obtained during that time. On the other hand, the web thickness is finite. Its width irregularly varies over the plane, and it also slightly perturbs the symmetry. However, the most essential question is which structure, regular or amorphous, originates at $q \notin \{q_c\}$. The answer is obvious after the Fourier spectrum of the web is derived, that is, when its X-ray pattern is obtained. According to Fig. 8, we deal with an ordered structure. The spectral pattern in Fig. 8(c) is very similar to those recorded for real quasicrystals [3]. This is a good reason for us to believe that the symmetry of the web corresponds to a quasicrystal symmetry.

SECTION 7

A more rigorous derivation of this conclusion is possible [9, 13]. We have already mentioned that a stochastic web forms after the separatrix net is destroyed. It may also be said that the separatrix

---

†The "window" in Fig. 7(c) remains open for a long time (for about 10^6 iteration steps).

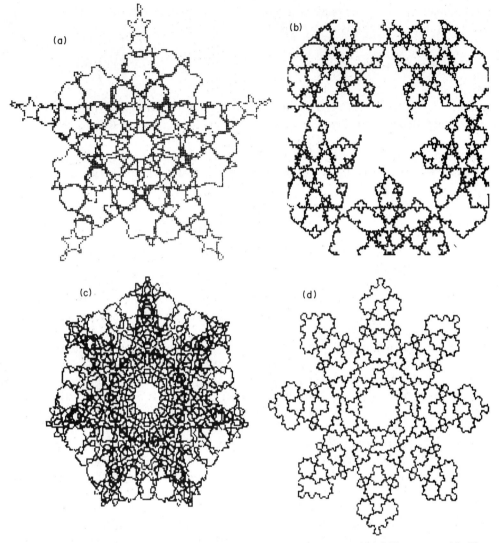

Fig. 7. Stochastic web with 5-fold symmetry (a, b), 7-fold symmetry (c) and 8-fold symmetry (d). The cases (a) and (b) differ in initial conditions.

net constitutes a web skeleton. Thus, we may try to outline this skeleton which should be a certain, infinitely thin frame of the web from which minor features are removed. We may go further still and outline the original dynamic system for which the web skeleton forms a separatrix net. Note that we are consistent when we try to get the skeleton not as a geometric object but rather as an invariant set on the phase plane for a certain appropriate dynamic system.

This program was employed in Refs [9, 13]. Replacing the variables and averaging equation (3) yield a "seed Hamiltonian"

$$H_q = \sum_{j=0}^{q} \cos(\mathbf{Re}_j), \tag{6}$$

where $\mathbf{R} = (v, u)$ and the "hedgehog" $\{\mathbf{e}_j\}$ consists of unit vectors which form a regular q-star

$$\mathbf{e}_j = \left(\cos 2\pi \frac{j}{q}, \sin 2\pi \frac{j}{q} \right). \tag{7}$$

Equation (6) coincides with the one used for the free energy of quasicrystals from phenomenological considerations [14–16]. For instance, for $q = 4$:

$$H_4 = 2(\cos u + \cos v).$$

This Hamiltonian has a square separatrix net. It is often employed in the physics of solids to describe the motion of electrons in crystals. When $q = 3$ or $q = 6$:

$$H_3 = \frac{1}{2} H_6 = \cos u + \cos\left(\frac{1}{2} u + \frac{\sqrt{3}}{2} v\right) + \cos\left(\frac{1}{2} u - \frac{\sqrt{3}}{2} v\right).$$

This Hamiltonian's separatrix net is a kagome lattice. It coincides with the stream function for the thermal Rayleigh–Bénard convection. In this case, hexagonal convective Bénard cells are formed in a fluid.

Thus, we have obtained for $q \in \{q_c\}$ just what we needed: a separatrix net of "seed Hamiltonian" has an ideal 3-, 4- or 6-fold symmetry. However, in the case of quasisymmetry the situation is not that easy.

SECTION 8

One of the main structural elements of H_q are singular points of saddle type. On the phase plane they are joined by specific trajectories, separatrices. Not very often could separatrices form a single net, that is, a web. For $q \in \{q_c\}$ such a net forms, this happens since the same $H_q = E_c$ ($E_c = 0$ for $q = 4$ and $E_c = -1$ for $q = 3$) are assigned to all saddles. In other words, in the case of crystal symmetry all saddles are in the plane of one level $H_q = \text{constant} = E_c$. That is why they may be connected by a single separatrix net.

If $q \not\in \{q_c\}$ the saddles are at different levels of H_q [13]. Owing to this fact, separatrix net common for the entire plane cannot be built in the plane of one particular level of H_q-values. It is quite obvious from Fig. 9, where better resolving power of the picture shows that many intersections within the net on the plane do not exist really. This is very essential for our understanding the origin of the web with quasisymmetry. The Hamiltonian at $q \in \{q_c\}$ has a skeleton with crystal symmetries $\{q_c\}$, but does not have skeletons with quasisymmetries at $q \not\in \{q_c\}$. However, even very small perturbation δH_q blurs small gaps between individual loops in separatrices, and a single unlimited web appears, via which propagation is possible over any arbitrarily long distance.

With this explanation it is relatively easy to obtain a quasiskeleton from H_q in the case of quasisymmetry [13]. To do this, that value E_0 should be determined, to which the peak of the distribution of the number of saddles as a function of $H_q = E$ values corresponds. Next, a set of points should be obtained which belong to the range of levels $E \in (E_0 - \Delta E, E_0 + \Delta E)$. This is equivalent to a weak blurring of separatrices lying at the level E_0. It is the thus derived picture which is a quasiskeleton (Fig. 10).

SECTION 9

The new type of tiling the plane with quasicrystal symmetry, or quasisymmetry, is fully determined by the tiling generator $\hat{M}_q$:

$$\hat{M}_q: \begin{cases} \bar{u} = (u + K \sin v)\cos\dfrac{2\pi}{q} + v \sin\dfrac{2\pi}{q}, \\[2mm] \bar{v} = -(u + K \sin v)\sin\dfrac{2\pi}{q} + v \cos\dfrac{2\pi}{q}. \end{cases} \tag{8}$$

The smaller the K values, the more regular the patterns are.

The symmetry of a tiling may be controlled by varying only one parameter q. This is why we are now able to show what an arbitrary q-sided snowflake can look like, whereas Figs 7(a) and 7(c) given here demonstrate actual possible shapes of a pentagonal snowflake (the problem that was already puzzling Keppler) and of a heptagonal snowflake. The plane cannot be paved with regular q-gons if $q \not\in \{q_c\}$. However, it can be paved so that basic paving elements be almost regular q-gons (or $2q$-gons), q-gonal stars and something else. That "something else" includes other figures; however, a stochastic web is the most important element in the paving process. It is this web that helps to remove various inconsistencies during the paving, though it can be made

Fig. 8. Web skeleton for $q = 5$ (a) and its Fourier spectrum (b).

Fig. 9. Web skeleton for $q = 5$ (a) and its part in the square magnification (b).

arbitrarily thin. Of course, the paving method depends on how thick the web is. The finitely thick web smoothes individual inconsistencies of tilings with quasisymmetries, and the fact that the web is fractal makes us approach quite different possibilities of symmetry analysis. Whatever idea we may now have about a quasicrystal it is unlikely that different symmetries can coexist in this crystal without their interaction. That is why, for instance, that the Penrose tiling of the 5-fold symmetry, which covers the plane only with two different rhombs, may be good approximation to a real structure whose roughness, indeed, leads to fractality. It is for this reason that in such cases the quasisymmetry is preferable to the exact symmetry.

We have seen already that the mapping $\hat{M}_q$ can be less accurate coarse-grained by remvoing some small details. Then there appears in the phase plane a pattern with a rotation symmetry, q-fold for even q and $2q$-fold for odd q (Figs 9 and 10). These patterns are obtained by the same principle as used for mountain relief on maps when regions in the same height range are shown by the same

(a)

(b)

(a)

(b)

Fig. 10. Quasiskeletons for $q = 7$ (a) and $q = 8$ (b).

Fig. 11. Fractal trees for $q = 5$ (a) and $q = 7$ (b), result from the particle diffusion.

color. In the former case the energy of a particle is used instead of height. Thus, derived reliefs have very distinct cluster regions in the form of q- or $2q$-gons. What is more, these clusters are formed by a set of almost straight lines, parallel and turned q times around the center. Fairly essential here is the thickness of lines. If allowed to approach zero, many elements in the structure will disappear. Hence an important aspect of quasisymmetry is its nonideality. Not only does this property spoil the symmetry, it simultaneously leads to quasisymmetry in the mapping. The rotation symmetry in those patterns and the presence of a set of straight lines point to the existence of the long-order though the translational symmetry lacks. On the other hand, aperiodic character of star clusters is similar to that of fluids. The higher the q, the closer the quasicrystal structure is, in some of its properties, to fluids [13].

If the parameter K is made larger in $\hat{M}_q$ the interaction of the two symmetries will become stronger and the stochastic web becomes wider. With its further growth, small cells of the web will be overgrown with the areas of chaotic dynamics and some symmetry properties of the web will be destroyed. Intense Brownian motion of particles beings (in case there are no random forces!)

(a)

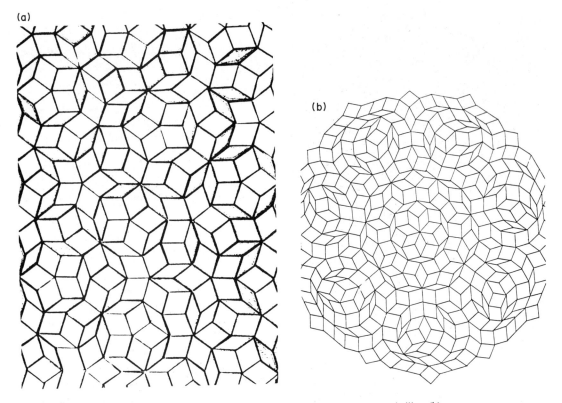

(b)

Fig. 12. Pentagonal tiling (Penrose tiling) (a) and heptagonal tiling (b).

which is imaged in the phase plane (by the trajectory puncturing the plane) as a growing fractal tree. The tree preserves a rotation q-fold symmetry and may be regarded as a result of particle diffusion process (Fig. 11).

Another extreme in simplifying the structures with quasisymmetry is building of parquets of the finite number of rhombs. Examples of pentagonal structure (Penrose tilings according to MacKay [6]) and heptagonal structure [13] are given in Fig. 12. At $q = 5$ the tilings are made with two rhombs, one angle of which is $\pi/5$ or $2\pi/5$. At $q = 7$ three rhombs are used for the tiling whose minor angles are $\pi/7$, $2\pi/7$, $3\pi/7$, respectively. This rule may obviously be extended in the similar manner for arbitrary q.

Tilings in Fig. 12 are quite definitely related to reliefs for $q = 5$ [Fig. 9(a)] and $q = 7$ [Fig. 10(a)]. There exist algorithms of such decoration of reliefs which turn them into parquets (for the case $q = 5$, see Ref. [9]).

SECTION 10

We need one more step which will reveal one of the most fundamental properties of quasisymmetry. The existence of symmetries is associated with a certain procedure for idealizing real processes. To get a quasisymmetry, the level of idealization should not be too high, nor should this idealization be too trivial. Up to now we followed the same sequence: web–relief–parquet. Structural metamorphoses, in the sequence, made the overall picture less detailed, thus, helping to single out very accurately some of its properties.

Let us give one more result of such coarse-graining. It is the determination of X-ray pattern of structural relief of the type illustrated by Fig. 10.

To do this a two-dimensional Fourier transform of a respective relief should be obtained. We now consider, for instance, an area in the central part of the relief at $q = 11$ and its Fourier transform (Fig. 13). Not only does the spectral image show that it is an ordered structure but also that it has the 11-fold rotation symmetry.

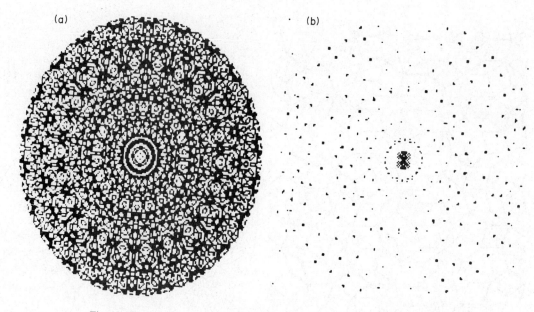

Fig. 13. Central part of the skeleton for $q = 11$ (a) and its Fourier spectrum (b).

We now select an area of relief of the same size and scale in a different part of the plane, far enough from the central part in Fig. 13(a). Figure 14(a) is the example. From what this area looks like we can say nothing about its symmetry, nor about the level of order of the entire relief whose part it is. What is more, it is natural to refer to this pattern as to a texture, rather than a crystal. However, the appropriate Fourier analysis [Fig. 14(b)] gives an amazing result. Shown there is the spectrum of the ordered structure with an 11-fold rotation symmetry which almost completely coincides with the Fourier spectrum of the area in the central part in Fig. 13(b).

This example shows that sufficiently large elements of a quasisymmetric tiling have very similar coarse-grained Fourier spectra. Coarse-graining involves cutting off the harmonics with very small amplitudes and smoothing other harmonics over amplitudes.

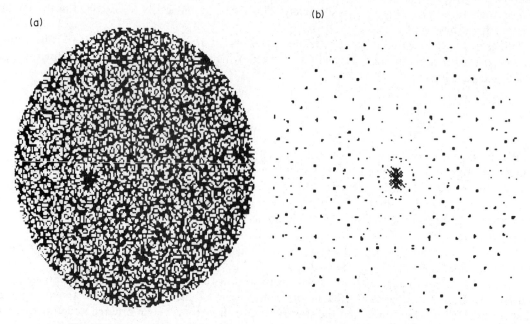

Fig. 14. Noncentral part of the skeleton for $q = 11$ (a) and its Fourier spectrum (b).

SECTION 11

The problem of charged-particle dynamics in the magnetic field and in the field of the plane wave packet is very popular in the theory of charged-particle accelerator, in plasma where oscillations are excited and in some problems of hydrodynamics. This problem seems very specific, nevertheless, a new class of tiling generators with q-fold rotation symmetry developed within it, which is of universal nature. This class is related with the structure of quasicrystals, besides, there is one more important application of tiling generators: these are possible types of pattern in gases and fluids, which are formed at the transition from a laminar (regular) to a chaotic (turbulent) state.

Random walkings of a particle on the plane generate a certain pattern. It grows as snowflakes do or, in a more general case, its growth reminds that of crystals or quasicrystals. Similar problems are also met in the finite automata theory. Hence the tiling generator $\hat{M}_q$ may be an unusually convenient tool for drawing various ornaments. Thus, the problem of the intricate dynamics of a charged particle opens up one more vista, i.e. computer graphics. With a color display all kinds of computer-aided configurations and ornaments can be plotted, if only a specific algorithm can be given which is a pattern generator. Search for most interesting algorithms has become a kind of sport, with its own program-favorites and its own business aspects. The new generator $\hat{M}_q$ may generate a great number of various ornaments with any order of symmetry. This number drastically increases because the patterns that thus appear are fractal. When the scale of the image is changed the typical features of an elementary information-carrying cell on display, that is, of a pixel, also change. Owing to this, the latter may be differently colored. Because of fractality the change in the resolving power of the instrument is accompanied by the change in the structure of the object

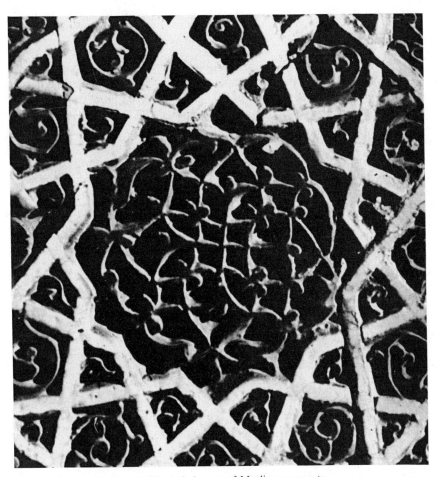

Fig. 15. Typical element of Muslim ornaments.

Fig. 16. "Decoding" of the ornament from Fig. 15 by the 5-fold symmetry skeleton.

it determines. We again come back to the problem of ornaments, though now this variety is described by real physical processes and is implemented in such real objects as quasicrystals, foam, hydrodynamic pattern. In this respect, not only does a new computer graphics produced with mapping $\hat{M}_q$ have a certain aesthetic configuration, but it also reflects the beauty of the real physical world.

Now that we know certain typical features of the tiling generator $\hat{M}_q$, we can approach quite differently the secrets of the art of the Moslem painters of the past who created a tremendous number of various ornaments which make us follow with attention and amazement the extraordinary intricacy and sophistication with which a craftsman drew the fundamental lines in his pattern. If the secrets of symmetric tiling are known it is easy to perceive how a given ornament was drawn. Figure 15 is an example of an element of a fairly typical ornament. It has a local-decagon. The scheme of the ornament is easily obtainable if a relief with the same symmetry and derived from $\hat{M}_q$ ($q = 5$) is applied. The respective decoration procedure is shown in Fig. 16. The painter, however, did not follow the scheme and the 5-fold symmetry was destroyed. The entire ornament (Fig. 17) is like a square lattice. One can find an element of nonperiodic pentagonal ornament in one of the frescoes of the ancient Moorish palace, Alhambra, in Granada.

Examples of the generation of ornaments determined by the symmetry of a "quasicrystal" type or, to be more exact, having certain quasisymmetry, open up new ways for analysis of geometrical properties of nature. One of the striking features of the new forms of symmetry is that, with the help of disorder elements, they produced a long-range order in structures. These disorder elements are the channels of the stochastic web, which smooth over minor disagreements between structure's bricks. Thus, an almost dense packing is achieved with only certain minimal gaps. Dynamic models help to separate the areas of elements packed and the areas of gaps in between according to the type of motion within them. This became possible with the help of two radically opposite types of motion, i.e. regular and random. It is just the essence of a new view on the long-range order organization and it may be due to the presence of chaotic disorder elements. Real physical processes, which include weak interaction of two mutually excluding symmetries, give rise to a weak

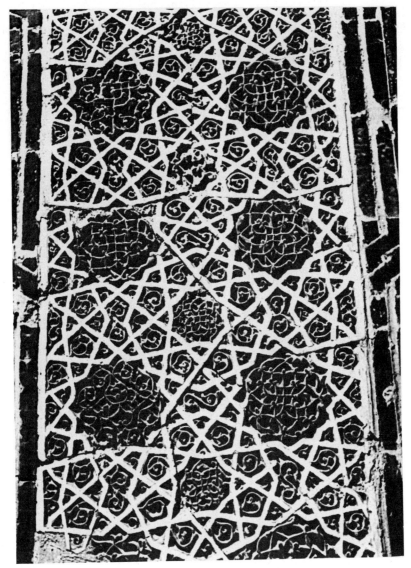

Fig. 17. Periodical tiling which includes the element from Fig. 15.

stochasticity. It is the tribute that should be paid for the maintaining of order in the form of quasisymmetry!

REFERENCES

1. H. Weyl, *Symmetry*. Princeton Univ. Press, N.J. (1952).
2. H. S. M. Coxeter, *Introduction to Geometry*. Wiley, New York (1961).
3. D. Shechtman, I. Blech, D. Gratias and J. W. Cahn, Metallic phase with long-range orientational order and no translational symmetry. *Phys. Rev. Lett.* **53**, 1951–1953 (1984).
4. R. Penrose, The role of aesthetics in pure and applied mathematical research. *Bull. Inst. Math. Its Appl.* **10**, 266–271 (1974). Also *The Mathematical Gardner* (Ed. D. A. Klarner). Prindle, Weber and Schmidt, Boston (1981).
5. M. Gardner, Mathematical games. *Scient. Am.* **236**, 110–121 (1977).
6. A. L. Mackay, De Nive Quinquangula: On the pentagonal snowflake. *Soviet Phys. Crystallogr.* **26**, 517–522 (1981).
7. R. Z. Sagdeev, Collective processes and shockwaves in collisionless plasmas. In *Voprosy Teorii Plasmy* (Ed. M. A. Leontovich), Vol. 4, pp. 20–80. Atomizdat, Moscow (1964).
8. G. M. Zaslavsky, M. Yu. Zakharov, R. Z. Sagdeev, D. A. Usikov and A. A. Chernikov, Stochastic web and diffusion of particles in a magnetic field. *Soviet Phys. JETP* **64**, 294–303 (1986).
9. G. M. Zaslavsky, M. Yu. Zakharov, R. Z. Sagdeev, D. A. Usikov and A. A. Chernikov. Generation of ordered structures with a symmetry axis from a Hamiltonian dynamics. *JETP Lett.* **44**, 451–456 (1986).
10. A. A. Chernikov, R. Z. Sagdeev, D. A. Usikov, M. Yu. Zakharov and G. M. Zaslavsky. Minimal chaos and stochastic webs. *Nature* **326**, 559–563 (1987).

11. G. M. Zaslavsky, *Chaos in Dynamic Systems*. Harwood Academic, Switzerland (1985).
12. V. I. Arnold, On the instability of dynamic systems with many degrees of freedom. *Dokl. Akad. Nauk SSSR* **156,** 9–12 (1964).
13. G. M. Zaslavsky, R. Z. Sagdeev, D. A. Usikov and A. A. Chernikov, The Hamiltonian method for quasicrystal symmetry. Preprint IKI N 1229 (1987). *Phys. Lett.* **125A,** 101–106 (1987).
14. P. A. Kalugin, A. Yu. Kitaev and L. S. Levitov, $Al_{0.86}Mn_{0.14}$-six-dimensional quasicrystal. *Pis'ma Zh. Exp. Teor. Fiz.* **41,** 119–121 (1985).
15. D. Levine and P. J. Steinhardt, Quasicrystals: a new class of ordered structures. *Phys. Rev. Lett.* **53,** 2477–2480 (1984).
16. D. R. Nelson and S. Sachdev, Incommensurate icosahedral density waves in rapidly cooled metals. *Phys. Rev.* **32B,** 689–695 (1985).

Computers Math. Applic. Vol. 17, No. 1–3, pp. 33–48, 1989
Printed in Great Britain. All rights reserved

0097-4943/89 $3.00 + 0.00

THE MAGIC OF THE PENTANGLE:
DYNAMIC SYMMETRY FROM MERLIN TO PENROSE

A. L. LOEB

Department of Visual and Environmental Studies, Carpenter Center for the Visual Arts, Harvard
University, Cambridge, MA 02138, U.S.A.

Abstract—Merlin the Magician instructs his charges Arthur and Kay in the secrets of dynamic symmetry, the logarithmic spiral, the golden fraction, -rectangle, and -triangle, Fibonacci numbers and the pentangle. The pentangle is central to the late fourteenth-century "Sir Gawain and the Green Knight", to medieval sign theory as well as to recent research in quasi-periodic alloy crystals.

1. THE BEETLES

"What does he want with those beetles?" asked Kay. "He is training each of them to track the next one exactly, teaching each always to move at the same constant speed as the other three." replied Arthur. "He keeps mumbling about behaviourist conditioning" added Kay sceptically.

The boys were used to Merlin's preoccupation with the patterns generated by animals [1], but they sensed that this time something special was up. Finally the day arrived when Merlin showed the boys a smooth board, one foot square, which he placed on the ground, with two edges pointing exactly north–south, the other two east–west. "We shall put one beetle at the north–west corner, facing east, the next one at the north–east corner, facing south, the third one at the south–east corner facing west, and finally the fourth one at the remaining corner, facing north" announced Merlin. And then he asked: "How far will each beetle need to travel to reach the nearest beetle he is facing?" "Obviously one foot" declared Kay, who was not usually the first to answer. "Explain!" pressed Merlin. "Well, they started one foot apart, so if they all walk along the edge of the board..." "There won't be anyone there when they get there!" interrupted Arthur. "Of course there will be someone there if they get there!" countered Kay. "What I mean," continued Arthur patiently, "is that the moment one beetle will crawl toward its neighbor, that neighbor will crawl towards its neighbor, and so on, so that each will be tracking a moving target". "Well, then they will never reach each other at all" concluded Kay, irritated at having spoken before he had thought.

Merlin then took a piece of charcoal, and drew a very short arrow at each of the corners of the square, pointing in the direction in which the beetles were originally facing (cf. Fig. 1). He looked gravely at Arthur and Kay, and the boys recognized that twinkle in Merlin's eyes which meant "Maths". The Merlin spoke: "Let us suppose that each beetle had moved a very short distance before realizing that it had to adjust its direction in order to keep tracking its target", and he marked the length of each short arrow "ds", meaning, Arthur supposed, "distance". "Are the beetles any closer to each other than they were initially?" questioned Merlin. The boys took out some sticks to draw in the sand, and after some time Arthur concluded that the new distance between nearest beetles could be found as the hypotenuse of a right triangle whose sides are $(1 - ds)$ and ds:

$$\text{Distance}^2 = (1 - ds)^2 + ds^2. \qquad (1)$$

At this point an argument erupted between the boys, because Kay insisted that the beetles were further apart than had initially been the case because of the additional term ds^2, whereas Arthur argued that the first term in the expression was diminished by more than the effect of the second term. Merlin interfered, however, by suggesting that the boys look at the figure formed by the four beetles in their new positions. "Why, they are still at the corners of a square!" exclaimed Kay, who was always good at visual games. "How does the diagonal of the new square compare with that of the original one?" asked Merlin. "It looks shorter" replied Kay. "Quite so, quite so" mumbled

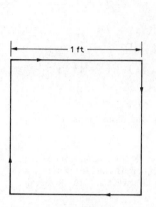

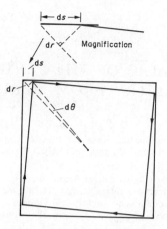

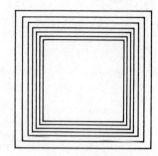

Fig. 1. The beetles' starting positions.

Fig. 2. The beetles occupy corners of a slowly rotating shrinking square.

Fig. 3. The shrinking square as it appeared to the boys in the whirli-jig.

Merlin, pleased, and he dropped a perpendicular from the end of an arrow marked "ds" onto the diagonal of the original square (cf. Fig. 2). And then his eyes assumed that far-away expression which made the boys take note, as he continued:

"Every time the beetles travel a distance ds along an edge of the square whose corners they occupied, they will arrive at the corners of a slightly smaller square which is also slightly rotated with respect to the original one, but whose center is located exactly where the center of the previous square was located. As the beetles keep tracking each other, they will occupy the corners of a shrinking, rotating square. For every distance ds traveled, their distance from the center decreases by an amount dr" he continued, marking "dr" on the wooden square. "Your homework for tomorrow is to find the relation between ds and dr" he concluded, picked up his board and beetles, and disappeared.

The next morning Arthur and Kay turned up triumphantly. Kay had cut out a square piece of parchment. On it, he had drawn two diagonals, and he proudly showed that at any time the angle between the direction of travel of a beetle and the line drawn from that beetle to the center of the square must be forty-five degrees. "Therefore," added Arthur, "the relation between ds and dr is:"

$$ds = \sqrt{2}\, dr. \qquad (2)$$

"Splendid!" exclaimed Merlin, and gave the boys the rest of the day off to watch the jousts in observance of the approaching full moon.

That night the boys were awakened by the kind of soft whisper which carries, and they saw the familiar silhouette of Merlin's cone-shaped hat outlined against the window. In no time they were outside. Merlin had a strange contraption which he called a "whirli-jig", in which he invited them to take a seat. The whirli-jig took to the air, and became suspended over the square board which Merlin had used in the previous day's lesson. At the corners were the four beetles, oriented just as Merlin had said they would be. As the insects began their quest, the whirli-jig turned almost imperceptibly, but at exactly at the same rate as the square whose corners were occupied by the beetles. As a result the square appeared to the boys to shrink, but to keep its edges in the same orientation (cf. Fig. 3). Then the moon disappeared behind the castle, and suddenly all was pitch dark. The beetles, it turned out, were luminescent, and so the boys could follow their progress in the dark. Since apparently the whirli-jig kept turning at the same rate as the square described by the positions of the beetles, the latter appeared to be crawling in a straight line toward the center of the squares. "Now we shall see whether the beetles will reach each other", whispered Kay, but his voice quavered a bit. Something seemed to be happening to the whirli-jig: it appeared to turn faster and faster. "Keep track of the beetles!" shouted Kay, but Arthur was feebly resisting nausea. Faster and faster did the whirli-jig whirl. "Help!" yelled the boys... Then a crash... When they came to, they found themselves in bed.

Fig. 4. *The Beetles' Quest*, collage by Deli Bloembergen. (From the Teaching Collection in the Carpenter Center for the Visual Arts, Harvard University. Reproduced with permission from the artist and the Curator.)

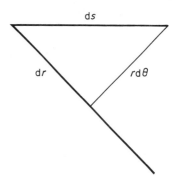

Fig. 5. Detail of Merlin's labeling.

"Well," said Merlin smiling when they met the next morning, "Do the beetles ever catch up with each other?" "Why did we crash?" rebutted Kay peevishly. Merlin explained that the whirli-jig was set to rotate at exactly the same rate as the square whose corners were occupied by the beetles. "But why did the stupid creatures start running around?" complained Kay. "They did not speed up at all: I trained them to keep exactly the same speed throughout" replied Merlin calmly. "Wait a moment!" exclaimed Arthur, "remember that linear speed equals the product of angular velocity and the radius! That means that, in order to maintain constant speed, their angular velocity has to increase as they approach the center of the square!" "Quite so." puffed Merlin. Kay then suggested retraining the beetles to maintain constant *angular* velocity. However, the boys feared that in that case the beetles would need to crawl more and more slowly, so that the experiment might never end. "Actually," found Kay somewhat to his relief, "they will touch shoulders before they reach the center."

This notion touched off a debate between the boys, who feared that at constant speed the beetles would explode as a result of centrifugal force if they were largish, but if small enough to be immune to the centrifugal force, would need to speed up to immense angular velocities which would make their reunion less than comfortable. Merlin put a stop to the argument by asking how far the beetles would have had to travel if they were smaller than any dimension the boys could discern, until finally they did meet, regardless of the inconvenience. This question silenced the boys for the rest of the morning.

It was not until the next afternoon that Arthur reappeared with an answer. He showed that every distance ds along the path of travel equaled $\sqrt{2}$ times amount by which the beetles' distance from the center decreased. Since the beetles had started at the corners of a square whose edgelength is 1 ft, they had started at a distance $\frac{1}{2}\sqrt{2}$ ft from the center. The sum of all little line segments ds would therefore have to be exactly $\sqrt{2}$ times $\frac{1}{2}\sqrt{2}$ ft, or 1 ft! "See, I was right after all!" bragged Kay. "But for the wrong reason!" added Merlin drily.

The lady Delia, who is an artist at weaving fair tapestries, has immortalized the quest of the beetles in the design shown in Fig. 4. It shows the rotating shrinking square, and in outline the spiral path pursued by the beetles. As we know, Merlin remembered the future: he therefore could remember Newton and the calculus. Perhaps archeologists will some day discover the board, 1 ft^2, which Merlin used to instruct Arthur and Kay, and perhaps even, with infrared techniques, show that (cf. Fig. 5) Merlin labeled the perpendicular from the tip of ds to the diagonal $r\,d\theta$, the infinitesimal arc length corresponding to the infinitesimal rotation of the shrinking square through an angle $d\theta$.

As Kay had discovered, the direction of travel was at 45° to the diagonal of the square, so that:

$$r\,\mathrm{d}\theta = -\mathrm{d}r, \quad \frac{\mathrm{d}\theta}{\mathrm{d}r} = -r, \quad \text{and} \quad r = \tfrac{1}{2}\sqrt{2}\,\mathrm{e}^{-\theta}\,\text{ft}.$$

Accordingly, Merlin remembered, the curve generated by the beetles is a logarithmic spiral. It is a tribute to his pedagogic powers that he had Arthur and Kay discover without the benefit of the calculus that the logarithmic spiral, even though it keeps turning indefinitely around the center, is actually finite in length. Kay's visual intuition and Arthur's analytical powers interacted to solve a problem which might have been beyond the power of either to solve by himself.

The next day Merlin gave the boys a Nautilus shell. Arthur proposed putting it in Sir Ector's Zoological Museum (Sir Ector was Kay's father), but Kay wanted to keep it in his room. Merlin, who had anticipated this kind of argument, produced a finely cut diamond, with which he neatly sliced the shell in two halves, giving one half to each of the boys (Fig. 6). The chambers inside the shell were much admired; Merlin explained that every year the Nautilus adds a chamber to his shell. "Does he live in the chamber?" asked Kay. "No", replied Merlin, "this air chamber serves to keep the Nautilus afloat. The volume of each year's chamber has to be proportional to the weight of the Nautilus that year. Accordingly, as the Nautilus grows, the size of the new chamber must be correspondingly larger than that of the previous year". As the boys measured the chambers, they discovered that the volume of each chamber was proportional to the total volume of the shell already there, in other words that the Nautilus's annual rate of growth was proportional to its actual size.

As the Nautilus grows, therefore, so does its rate of growth: the consequences were dizzying. Merlin mumbled something about spiraling inflation, but noticing the boys' bafflement, quickly

Fig. 6. Cross section through a Nautilus shell.

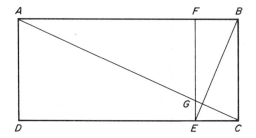

Fig. 7. Rectangle and gnomon.

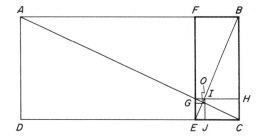

Fig. 8. Continued subdivision of the rectangle into gnomon and a new rectangle geometrically similar to the original one.

recovered himself, and showed them that in old age the Nautilus appeared to be running out of steam, generating faulty small chambers.

The boys now had two spirals, one generated by the bugs, the other by the Nautilus. One spiraled inward, the other outward, but anyone unfamiliar with the process by which the spirals were generated, would not consider this distinction significant. Merlin sent Kay to fetch the parchment which he had used for the homework assignment, and showed that, although different in scale, the geometry drawn there prevailed throughout the beetles' travels: as their distance from the center decreased, their path curved more sharply, and the line element ds decreased, being proportional to the distance r from the center. Merlin proposed calling this type of spiral "proportional growth" spirals, and suggested that the boys look for other examples in nature.

2. DYNAMIC SYMMETRY

Some time passed while the boys were preoccupied with a large hunt, but when they reported for their next lesson, Merlin instructed each to draw a rectangle on the ground, to draw a diagonal joining two vertices of the rectangle, and to drop a perpendicular from a third vertex onto the diagonal, extending it till it met the opposite side of the rectangle. Arthur had chosen to start with a square, but Merlin persuaded him that, although in principle there was nothing wrong with starting with a square, he would not learn much from it, and accordingly ought to postpone consideration of the square till later, when he understood the general example better. Merlin labeled Kay's figure with some letters, so that it looked like Fig. 7. Through the point E he drew a line EF parallel to the short sides of the original rectangle, dividing the latter into two rectangles. The intersection of this line EF with diagonal AC he called G. Of these two rectangles, rectangle $BCEF$ was seen to be geometrically similar to the original rectangle $ABCD$, as their respective sides as well as their diagonals were perpendicular to each other. "When we substract rectangle $BCEF$ from rectangle $ABCD$, to which it is geometrically similar," Merlin lectured, "we are left with rectangle $AFED$, which we call the *gnomon* of rectangle $ABCD$." After a puff on his pipe, which was stuffed with herbs because much to his regret tobacco had not yet been discovered, he addressed Arthur: "Well, Mr Smarty, what would be the gnomon of your square?" Arthur squirmed a while much to Kay's delight, and finally concluded that in that case the line EF would coincide with AD, with the result that the gnomon of the square would have degenerated to a line. Kay, meanwhile had observed that rectangle $BCEF$ already had one diagonal drawn in, as well as a perpendicular onto it, so he continued in analogy to the construction previously performed in rectangle $ABCD$. "Rectangle $CEGH$ is geometrically similar to $ABCD$, and $BHGF$ is the gnomon of $BCEF$" he announced proudly. Drawing IJ parallel to GE, he continued subdividing rectangles into smaller rectangles similar to the original $ABCD$ and their gnomons (Fig. 8).

"You could draw a proportional-growth spiral from A to B to C, then to E, G, I, etc." exclaimed Arthur excitedly. "Eventually you should finish at O," observed Kay. "If you ever get there!" added Arthur with a wink. Then the boys regarded Merlin gravely, for they knew that these momentous discoveries were usually followed by a lecture.

Merlin indeed pontificated: "When we rotate a body, keeping it rigid, i.e. keeping all its dimensions intact, we perform a *symmetry operation*. However, when rotation is accompanied by

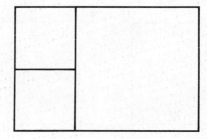

Fig. 9. Kay's assignment.

scaling, that is to say by shrinking or expansion while the object remains geometrically similar to itself, we create a pattern which has *dynamic symmetry* [2]. The Nautilus shell has dynamic symmetry if all its chambers are similar to each other, and are rotated with respect to each other." The boys agreed that the pattern of nested rectangles drawn by Kay also possessed dynamic symmetry.

3. THE SQUARE GNOMON

Next Merlin produced two sheets of parchment ruled with horizontal and vertical lines so that the sheets were divided into small squares. To Arthur he assigned the task of fitting on the little squares a rectangle whose gnomon is a square. "Square Arthur!" mocked Squire Kay, who was, however, quickly diverted by his own assignment: Merlin drew a pair of adjacent small squares on Kay's sheet, producing a 1-by-2 rectangle (cf. Fig. 9).

On the long side of the 1-by-2 rectangle Merlin drew a 2-by-2 square, so that a new rectangle appeared. "2-by-3" commented Kay. "Dynamic symmetry?" questioned Merlin. "No, for 1-by-2 is not geometrically similar to 2-by-3" was the reply. "Your assignment", continued Merlin, "is to add a square to the long side of each new rectangle you generate, and to keep track of the ratio of the lengths of the sides of each new rectangle you construct."

Merlin withdrew to devote himself to some of his mathematical tables, as he knew that the boys would be occupied for a while. Arthur appeared first, all hot and bothered. "It won't fit on the little squares" he complained. He showed Merlin [cf. Fig. 10(a)] that he had called the length of the longer side, *AB*, of the large rectangle *1*: as the boys were not encumbered by the notion of standard measures, they were wont to establish a covenient unit for the problem at hand. For the

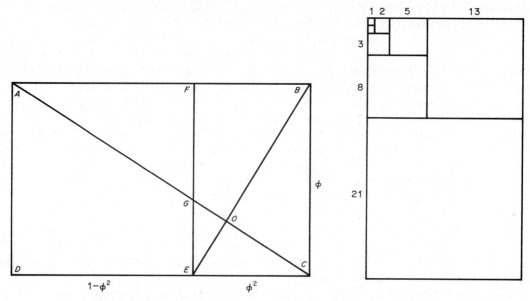

Fig. 10(a). Arthur's solution.   Fig. 10(b). Kay's solution.

short side he used a symbol which we shall approximate by ϕ, this symbol denoting a fraction. Dynamic symmetry tells us that the length of EC relates to that of CB just as that of CB does to BA, the latter being 1. "Therefore the length of EC must be ϕ^2" he added proudly. Since the length DC, like AB, must be 1, that would make the length of DE equal to $1 - \phi^2$. Then, if the gnomon rectangle $AFED$ is to be a square:

$$1 - \phi^2 = \phi$$

or

$$\phi^2 = 1 - \phi. \tag{3}$$

At this point Kay interrupted: his sheet was full. "Patience!" admonished Merlin, and showed Arthur's condition for a square gnomon to Kay. Arthur problem was that he could not construct a rectangle whose edgelengths satisfied this condition. "Well, let us look at Kay's result first", proposed Merlin. Kay had constructed a series of rectangles [Fig. 10(b)], whose sides were in the ratios show in Table 1.

Arthur was moping because he considered Kay's assignment childish, and he felt that his own problem was not getting the attention it deserved. "Do any of these fractions satisfy your condition, Arthur?" asked Merlin, to keep his interest, and told him to call Kay's ratios f, and to tabulate and compare the values of f^2 and $1 - f$ by bringing them over a common denominator. At first Arthur complied sullenly, as the two columns did not appear to bear much relation to each other. Gradually, however, his sulk became transformed to interest. He noted that the difference between the columns was always 1 divided by the square of the length of the longer side of Kay's corresponding rectangle, a quantity which rapidly became very small as one went down the table (cf. Table 2). Accordingly, Kay's rectangles appeared gradually to approach Arthur's condition for a square gnomon! "But how far must we go on with Kay's construction?" asked Arthur rather plaintively.

Table 1. Ratios of sidelengths of Kay's rectangles
1/1
1/2
2/3
3/5
5/8
8/13
13/21

Table 2. Testing Kay's fractions in Arthur's equation

f	f^2	$1 - f$
1/1	1/1	0
1/2	1/4	1/2 = 2/4
2/3	4/9	1/3 = 3/9
3/5	9/25	2/5 = 10/25
5/8	25/64	3/8 = 24/64
8/13	64/169	5/13 = 65/169
13/21	169/441	8/21 = 168/441

4. FIBONACCI'S NUMBERS

To answer Arthur's question, Merlin took the boys to a wooded section of Sir Ector's castle grounds. This was the signal that Merlin was going to "remember": he preferred not to be observed, and the boys knew that they would never be able to share in Merlin's magic if they so much as breathed a word of what they had witnessed. This time Merlin produced some magnificent garments, which he declared to be the height of fashion in Pisa in Italy around the year 1200 A.D. Donning these, he assumed the form of one Leonardo, known as Fibonacci, a mathematician who advocated the use of Arabic numerals rather than the Roman system so revered by the great Uther Pendragon. (As we have observed, Merlin had taught the boys the Arabic system all along!) Leonardo smiled kindly upon a rather frightened Kay, and instructed him to tabulate the lengths of the short sides of his rectangles. Somewhat shakily, Kay produced: 1 1 2 3 5 8 13 21... Arthur, who was a bit jealous of the attention Kay was getting, commented that each number could be calculated as the sum of the previous two. "That follows directly from my construction!" rejoined Kay. Then Leonardo known as Fibonacci slowly faded into the familiar Merlin, who declared that that had been enough excitement for one day.

The next day Arthur turned up early: he had extended Fibonacci's series way out, beyond 200, but noted that he could never quite attain his condition for a square gnomon. "Well," said Merlin, "suppose that your fraction ϕ equalled the ratio p/q, where p and q are both integers. We may assume that p and q have no common factors, because if they did, we would factor them out of the ratio." Merlin then substituted this ratio in Arthur's equation (3), and produced:

$$p(p + q) = q^2. \tag{4}$$

"We know that p and q may not both be even, as they do not share a common factor. Why don't you make a table of all possible combinations of even and odd!" he suggested. Arthur quickly came up with:

p	q
Even	Odd
Odd	Even
Odd	Odd

Kay had joined the discussion. "If p is even, what does that tell us about the left-hand side of equation (4)?" Merlin asked Kay. "Why, it must be even" was the answer. "But then the right-hand side must also be even, and that would make q even as well, which it isn't" said Arthur. "If p is odd, and q is even, then their sum is odd" added Kay. "The left-hand side of the equation will then be the product of two odd numbers, which is odd, but the right-side would be even as q is even, so we once more are led to a contradiction to the equation" Merlin pointed out. "I guess p and q must both be odd" sighed Kay. Arthur, however, observed that in that case their sum would be even, so that the left-hand side would be even and the right-hand side odd, which also was impossible. "We must have overlooked another possibility" thought Kay. "Such as that our assumption that ϕ is expressible as a ratio of two integers might not be valid" suggested Merlin. "That would be the only possibility remaining" concluded Arthur. "That is why I could not fit the rectangle having a square gnomon on the little squares on my parchment." Merlin explained that that rectangle is considered very powerful: therefore it is called the "golden" rectangle, and the irrational fraction ϕ is called the "golden" fraction. The notion that not every fraction is expressible as a *ratio* of integers was somewhat of a shock to the boys. Explaining that such fractions are called irra*tio*nal, but that the golden fraction ϕ can be closely approximated by the ratio of successive Fibonacci numbers, Merlin dismissed them for the day, and withdrew to his table of the powers of the golden fraction.

Using equation (3) he reduced each higher power of the golden fraction to lower powers:

$$\phi = \phi$$
$$\phi^2 = 1 - \phi$$
$$\phi^3 = 2\phi - 1$$
$$\phi^4 = 2 - 3\phi$$
$$\phi^5 = 5\phi - 3$$
$$\phi^6 = 5 - 8\phi$$
etc.

"Since ϕ is positive, the left-hand side of each equation must be positive" he mumbled. But that means that ϕ must be less than 1, greater than 1/2, less than 2/3, more than 3/5, less than 5/8, etc., in other words alternately bounded above and below by ratios of successive Fibonacci numbers." Generalizing, Merlin then found an expression for the nth power of the golden fraction:

$$\phi^n = (-1)^{n+1}(a_n\phi - b_n), \tag{5}$$

where a_n and b_n are constants still to be determined. To determine these two constants, he obtained ϕ^{n+1} from equation (5) in two different manners, and then equated the results. First he rewrote equation (5) substituting $n + 1$ for n:

$$\phi^{n+1} = (-1)^{n+2}(a_{n+1}\phi - b_{n+1}). \tag{6}$$

Secondly, he multiplied both sides of equation (5) by ϕ:

$$\phi^{n+1} = (-1)^{n+1}(a_n\phi^2 - b_n\phi).$$

Since $\phi^2 = 1 - \phi$, this last equation becomes:

$$\phi^{n+1} = (-1)^{n+2}[(a_n + b_n)\phi - a_n]. \tag{7}$$

Equations (6) and (7) are equivalent if $b_{n+1} = a_n$ and $a_{n+1} = a_n + b_n$, so that $a_{n+1} = a_n + a_{n-1}$. This is precisely the recursion relation which Arthur and Kay had found for the Fibonacci numbers; we may conclude that the numbers a_n are in point of fact the Fibonacci numbers, so that

$$\phi^n = (-1)^{n+1}(a_n\phi - a_{n-1}). \tag{8}$$

Since $\phi < 1$, ϕ^n will tend to zero when n becomes very large. As a result the right-hand side of equation (8) will also tend to zero when n becomes very large, so that Merlin assured himself that when n is sufficiently large, ϕ may be closely approximated by the ratio of a_{n-1} to a_n. Moreover, the left-hand side of that equation yields an estimate of the magnitude of the error implied by this approximation. Satisfied that what he had led his young charges to discover was indeed sound, Merlin took a small snifter of his favorite yet-to-be-invented cognac, and drifted off to sleep.

5. THE GOLDEN TRIANGLE

There was some to do with King Lot of Lothian, and its was some time before Arthur and Kay could resume their lessons. But one morning Kay was pounding a peg in the ground, to which he attached a string. A pointed twig tied to the other end of the string enabled Kay to draw fine circles in the sand. Merlin suggested that he draw an isosceles triangle, labeling its apex (where the two sides of equal length meet) C, the base vertices A and B. Next he was instructed to construct a second isosceles triangle having A as its apex, B as one base vertex, and the second base vertex, D on the line segment BC. This Kay could do easily with the aid of his twig-string-and-peg apparatus (Fig. 11).

Merlin noted that, of course, CAB is similar to ABD, and he proposed that ADC be called the gnomon of CAB, in analogy to the rectangle gnomon which the boys had learned about before the King Lot of Lothian to do. "Could the gnomon also be isosceles?" wondered Merlin. Kay called the magnitude of angle ACB "Γ", and, noting that the sum of the angles of a triangle is $180°$ and that angles CAB and CBA are equal, he assigned the latter two angles the values $90° - \frac{1}{2}\Gamma$. The angle ADB necessarily has the same value, as BAD is isosceles, with the result that its supplement, angle CDA equals $90° + \frac{1}{2}\Gamma$. Angle CAD would have to be equal to Γ if CAD is to be isosceles. Since all angles of triangle ACD must add up to $180°$ angle ACB equals $36°$.

Meanwhile Arthur, who had joined them, came up with a different approach. Calling, according to custom, the length of side AC 1, and that of base AB f, he noted that the similarity of triangles ACB and BAD would make the length of line segment BD equal to f^2; since the length of BC would also be 1, line segment CD would have length $1 - f^2$. BAD being isosceles, the length of AD would be the same as that of AB, namely f. Therefore CAD is isosceles only if

$$1 - f^2 = f. \tag{9}$$

"The equation for the golden fraction!" exclaimed Kay, excited. Merlin summarized that, indeed, the isosceles triangle whose apex angle is $36°$ is singularly important because its *base* length equals the product of the golden fraction and its sidelength, and its base angles are just twice the apex angle, namely $72°$. The golden gnomon, on the other hand, has *base* angles of $36°$, whereas the apex angle equals three times that, namely $108°$. The *side* length of the gnomon equals the product of the golden fraction and its baselength, the reverse of the relationship for the golden triangle. "The magic of the golden triangle and the golden gnomon lies in the fact that their angles are all multiples of $36°$, and that their sides are all related to each other by the golden fractions" revealed Merlin. He marked all these relationships on Kay's sand drawing (cf. Fig. 12), and then assigned Arthur for the next day a calculation of the fraction of the area of ACB occupied respectively by its two component triangles, golden triangle BAD and golden gnomon CAD.

To Kay Merlin assigned the task of enumerating all triangles whose angles are integer multiples of $36°$. Kay solved his assignment as follows. He assigned to the desired angles the respective values $i \cdot 36°$, $j \cdot 36°$ and $k \cdot 36°$, where i, j and k are positive integers. Since the sum of these angles must be $180°$, $i + j + k = 5$. Merlin added approvingly: "Without loss in generality we may assume $0 \leqslant i \leqslant j \leqslant k$, for it does not matter which of the three angles is designated by i, which by j, and which by k."

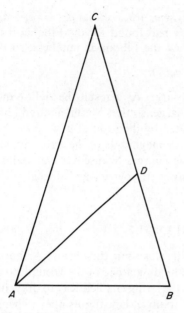

Fig. 11. Two isosceles nested triangles.

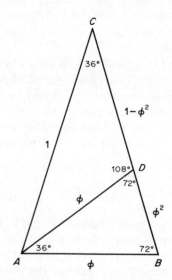

Fig. 12. Golden triangle and golden gnomon.

Kay then found that there are only two solutions to the last equation, namely the combinations (1, 2, 2) and (1, 1, 3), yielding the triangles having respective angular combinations (36°, 72°, 72°) and (36°, 36°, 108°). "The golden triangle and its gnomon!" added Arthur. "Quite so, they are unique." Merlin ruminated.

To follow Arthur's solution to his area assignment, we need to realize that Merlin, not wanting to educate the boys in the prevailing superstitious belief of the supremacy of the square, had introduced areas by the basic relationship that the ratio between the areas of geometrically similar figures equals the square of the ratio between the lengths of corresponding lines in those figures [3]. Accordingly, Arthur began the next morning with the observation that:

$$\text{Area}(ABD)/\text{Area}(CAB) = (\text{Length } BD/\text{Length } AB)^2 = \phi^2.$$

Hence,

$$\text{Area}(CAD)/\text{Area}(CAB) = 1 - \phi^2 = \phi.$$

Thus, the golden fraction represents the fraction of the large golden triangle occupied by the golden gnomon, and the smaller golden triangle has an area equal to the product of the golden fraction and the area of the golden gnomon.

Merlin summarized the boys' sand drawings as follows: when the smaller golden triangle is subtracted from the larger golden triangle, a golden gnomon results. "Now what would happen if we had *added* that gnomon to the larger triangle?" he asked. The boys noted that the sides AD and CD were the same length as the base AB of the larger golden triangle, and therefore came up with the following construction (Fig. 13), noting that the sum was a larger golden gnomon, whose sidelength was 1, and whose baselength $(1 + \phi)$.

Since this golden gnomon had the same sidelength as the original golden triangle, the boys decided to juxtapose them, producing, not too surprisingly, a larger golden triangle, having sidelength $(1 + \phi)$, and baselength 1 (Fig. 14).

Arthur then remembered that for the golden triangle the baselength equaled the product of the golden fraction and the sidelength, so that one would expect:

$$1/(1 + \phi) = \phi. \tag{10}$$

"Is that correct?" he asked with some hesitation. "Why not multiply both sides of that equation by $(1 + \phi)$" suggested Merlin. The result, $1 = \phi + \phi^2$, agreed nicely with the equation first found for the golden fraction [equation (3)]. Suddenly Merlin drew himself up, pulled his cloak around

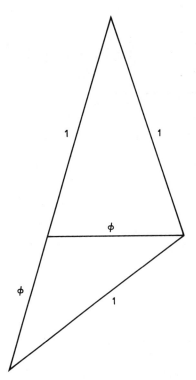

Fig. 13. The sum of a golden triangle and a golden gnomon is a golden gnomon.

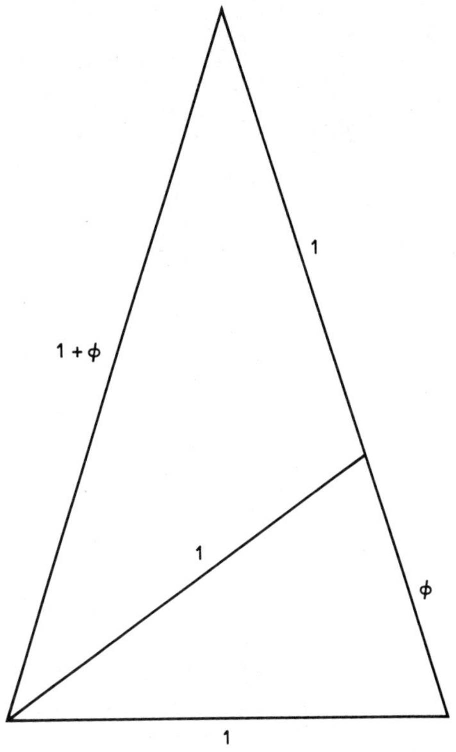

Fig. 14. The sum of a golden triangle and a golden gnomon is a golden triangle.

him, appeared to grow several inches, and intoned: "DIVISION BY THE GOLDEN FRACTION IS EQUIVALENT TO MULTIPLICATION BY THE GOLDEN-FRACTION-PLUS-ONE!" The boys nudged each other and giggled: they were used to Merlin's magical incantations, which were intended for those who were incapable of understanding, and therefore had to be persuaded to accept on faith [4]. They spent the remainder of the day cutting out golden triangles and gnomons whose dimensions were related by the golden fraction, and the next morning they brought in a wealth of designs. One design [Fig. 15(a)] was very irregular: Arthur had carefully marked the lengths of all line segments, checking the baselength to side-length of each triangle. Another design, [Fig. 15(b)] was, on the other hand, very symmetrical. One design in particular pleased Merlin (Fig. 16).

"The pentangle is a very powerful figure." he said. "It is a five-pointed star: it possesses five-fold rotational symmetry. As you can see, it encloses a smaller pentagon, and in turn it is generated by the diagonals of a larger pentagon. The entire structure therefore possesses dynamic symmetry as well, and the diagonals divide each other according to the golden fraction." He then labeled the various distances, using the distance relations which the boys had found for the golden triangles and gnomons. The spaces between the points of the pentangle could be filled by gnomons, with the pentagon resulting. In turn, larger golden triangles can be added to each side of the pentagon, generating a larger pentangle. Since it was shown that larger golden triangles and gnomons can be fashioned out of smaller ones, as long as their dimensions are related by the golden fraction, it would appear that the entire plane can be covered by replicas of a single pair of golden triangle and gnomon. "Except for the central pentagon" reminded Arthur. Kay, however, quickly constructed a pentagon out of a single golden triangle and two gnomons to fill the gap (Fig. 17).

6. SIR GAWAIN

Merlin knew, as the boys did not, that their idyllic existence was not to last much longer. The awesome day arrived when Arthur pulled the sword from the stone and was revealed to be the son of King Uther Pendragon and Queen Ygraine [5]. On the night of that glorious day Arthur

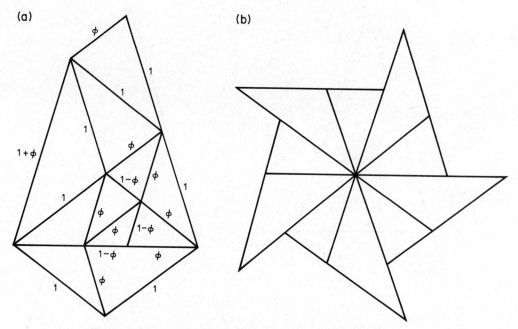

Fig. 15. Designs using golden triangles together with golden gnomons.

was visited by a mysterious woman who did not reveal to him that she was his half sister Morgause; a result of that visit was the birth of Mordred. By the time of Mordred's birth Morgause had married King Lot of Lothian, whose domains included the remote Orkneys, and to whom she bore four more sons, the eldest being Gawain [6].

Merlin hated Morgause for what she had done, knowing that Mordred would eventually destroy Camelot; Morgause never ceased in her attempts to learn Merlin's magic. After King Lot's death she managed through her trickery to injure Merlin and thus to decrease his powers. Arthur became King, married Guinevere and founded the Order of the Round Table at Camelot. When he discovered the intrigues of his half sisters Morgause and Morgan, he sent for them and for Morgause's five young sons. He banished his sisters to a nunnery and educated the boys at his court.

Merlin, meanwhile, had in his dotage become enchanted by a young lady, Nimuwe, to whom he taught his magic. An apt student, she trapped him in his crystal cave, where he remains locked to this day; it was believed that the day will come when Merlin will reappear [7].

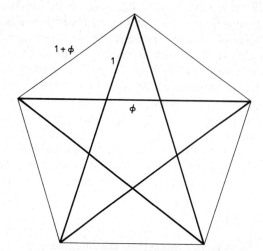

Fig. 16. The pentangle generated from golden triangles and golden gnomons.

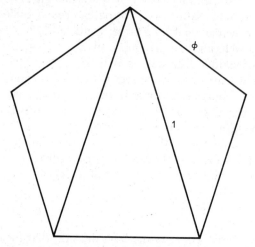

Fig. 17. Pentagon constituted from a golden triangle and two golden gnomons.

Fig. 18. The endless knot on Sir Gawain's shield.

One year, when the Court at Camelot was celebrating the Christmas season, a gigantic green knight appeared [8], challenging the King to combat. Young Gawain begged his uncle Arthur to be allowed to meet the challenge, which was to cut off the knight's head, and to meet him for a return match a year later. Gawain complied, but to everyone's amazement the knight calmly put on his head and departed. Gawain spent most of the year in courtly activities of one kind or other, but by All Saints' Day the time had come for him to take his leave and to meet his challenger at the rendez-vous agreed upon, the Green Chapel. There was great sorrow at Camelot, for Gawain was much beloved, and considered the perfect and pure knight.

When Gawain departed on his perilous journey, Arthur noted that the device on Gawain's shield was an endless knot (Fig. 18). He quickly looked at Sir Kay, who as usual was nearby, and who had also recognized Merlin's pentangle. What Arthur and Kay knew to be a composite of five golden triangles, was considered by others to be a knot with no beginning nor end, hence symbolic of eternity and perfection [9]. They observed that the five straightline segments of the knot divide each other into two portions whose lengths are related to each other by the golden fraction.

We are, unfortunately, not permitted to follow Gawain on his quest after the Green Knight. Just one hundred years ago, in 1888, a Connecticut Yankee named Hank Morgan also traveled back in time to visit Camelot [10]. This opportunity exists just once every century, and we have ourselves taken advantage of it to observe the education of young Arthur and Kay. We are returned to a library in our own century, where we may peruse the story as told by the anonymous poet of the late fourteenth-century. We note that our anonymous poet suspended the action of his story for as much as fifty lines [11] in order to describe Gawain's pentangular device, and commented that what the less sophisticated considered an endless knot, was known by people of learning to be a symmetrical composite [11, pp. 27 and 30]. We note also that this twentieth-century (R. G.) Arthur stresses the centrality of sign theory to medieval thought, and it is significant that the index of his book *Medieval Sign Theory and Sir Gawain and the Green Knight* contains 15 references to the pentangle, two more than to the Green Knight himself!

R. G. Arthur further makes reference to medieval interest in the concept of infinity, and its relation to divinity, citing Nicolas of Cusa [12] and Anselm [13]. He is not the only one [14–19] to emphasize the allegorical aspect of medieval stories, in particular of *Sir Gawain and the Green Knight*. It was expected that these stories would be understood at different levels of meaning: the knowledgeable reader understands symbols such as the pentangle, and is educated or edified by them.

It is in this spirit that I have presented these lessons from Merlin. It is believed that Arthur and Merlin did indeed exist at the time of Saxon invasions of the British Isles, and that they may have descended from Roman aristocracy. They represent an "old religion", perhaps Roman, Mithraic or early Christian. Nennius, a Welsh priest, mentions Arthur in his *Historiae Britonum* around 800 A.D. Arthur, Kay and Mordred appear in Welsh poetry about the same time [20]. Geoffrey of Monmouth brought history and legend together in his *History of the Kings of Britain*, written in Latin around 1135; it is in the later Middle Ages that the Camelot traditions grew and became unified, notably through the writings of the French poet Chrestien de Troyes between *ca* 1170 and 1180, through the anonymous late fourteenth-century *Sir Gawain and the Green Knight*, and through Malory's *Le Morte d'Arthur* in the second half of the fifteenth-century. Notable is Robert de Boron's triology on the Holy Grail, of which the middle book is devoted entirely to Merlin [21]. Our view of Camelot is thus characteristic of the late Middle Ages rather than of the sixth-century when the events are reputed to have taken place: the authors told the story as they felt it ought to have happened. "Sir Gawain and the Green Knight" tells us, beside an interesting story, that in the fourteenth-century there was an acute interest in abstract ideas, in particularly in the symbolic significance of the symmetry of geometric figures as well as in the concept of infinity.

Malory's *Le Morte d'Arthur* received wide circulation as a result of William Caxton's printing at Westminster in 1485, the very year when Henry VII assumed the throne of England. Henry, the first Tudor king, was a Welshman, much interested in symbolism and allegory; it was he who put about the great myth of the Tudor double rose, symbolizing the union of the red Lancastrian and white Yorkist rose [22]. Christopher Morris points out that there are elements of truth in the popular belief that Welshmen are apt to be a little "fey" and yet the possessors of sharp eyes for the main chance, romantics with a curious streak of realism .... Henry Tudor did possess these qualities, and it is not insignificant that he called his eldest son Arthur!

By contrast the twentieth-century Dutch author Jaap ter Haar [23] has attempted to strip the Arthurian legends of what he considered late-medieval romantizations; in doing so, he has deprived the story of precisely the mystical/rational dichotomy which lends the late middle ages a special fascination. Such "magical" and "informed" interpretations of patterns are characteristic as well of our own times: the work of M. C. Escher and R. Buckminster Fuller is being interpreted not only by the "people with learning", as the Gawain poet calls them [24–26] but is also being embraced by groups embracing various belief systems, who appear to be edified and inspired by what they see and read. A significant recognition of this trend is the exhibition shown in the Los Angeles County Museum of Art and the Gemeente Museum in the Hague, entitled *The Spiritual in Art: Abstract Painting 1890–1985* [27].

7. MERLIN UNLEASHED?

Merlin lived and was trapped in a crystal cave. We may therefore conclude that his power was derived from a knowledge of crystals, minerals and metals, an interest not surprising in Cornwall, where much of the legends around Merlin originated, and where the Phoenicians had established a mining and metal industry. The "Morte d'Arthur" commemorates a civilization which had perished; it was followed by what is commonly referred to as "the dark ages". We must not forget, however, that around the Mediterranean the geometry of the Greeks continued to flourish, notably in Alexandria, and that the Arabs made extensive use of mathematics in their architecture and design.

It appears that tessellations of the Euclidean plane (a filling of the plane by means of replicas of only a finite number of forms without any overlaps or interstitial spaces) by golden triangles together with their gnomons was known by Islamic designers [28]. We have seen [cf. Figs 15(b) and 16] that such designs are apt to possess five-fold rotational symmetry. It is known that the existence of a center of five-fold rotational symmetry precludes translational symmetry in that plane [29]. Therefore, if a golden-triangle-gnomon tiling has a center of five-fold rotational symmetry, then such a tiling, even though infinite in extent, cannot be periodic. Such non-periodic tilings appear to have been known to Islamic designers, but they were ignored by crytallographers until a British mathematician of possibly Cornish antecedents, Roger Penrose, published an analysis of them in 1974 [30].

This demonstration of the possibility of non-periodic tilings gave credibility to the discovery by Shectman *et al.* [31] of apparent five-fold rotational symmetry in rapidly cooled aluminum–magnesium alloy crystals. The diffraction patterns of this alloy was sharp; this had been assumed to be a characteristic of periodic crystals. This assumption, however, would be untenable if the alloy crystal did indeed have five-fold rotational symmetry. Intriguing also was the observation of the golden fraction ϕ in the spacing of bright spots in the diffraction pattern [32].

Thus, an understanding of the pentangle and of its components the golden triangle and its gnomon, overcame the superstition that all crystals are periodic [33]. Indeed, it has been shown that three-dimensional analogs of the Penrose tilings will have sharp diffraction patterns, and additional materials have been found having quasi-periodic structure [34, 35].

8. EPILOGUE

It appears as if Merlin does occasionally escape from his crystal cave. The appearance of Arthurian sagas during the waning of the Middle Ages heralded an age of renewed national awareness in England. At a different level, however, we note in the same period the interest in concepts and patterns remarked upon in "Sir Gawain", as well as the rediscovery of linear perspective. Edgerton [36] has noted that this new representation of three-dimensional space has had substantial consequences in the development of an industrial society. R. G. Arthur [37] reminds us that Augustine stressed the distinction between *res* and *signa*, between *thing* and *sign*. The endless knot is *res* as well as *signa*, (R. G.) Arthur reminds us. And so linear perspective, *signa* in painting, was *res* in architecture.

In our own day, the duality of *res* and *signa* emerges in our assignment of different functions to the two different sides of the brain. I tried in my characterization of Arthur and Kay to contrast the predominantly analytical and the primarily intuitive, as I have observed it in my own classroom. The reemergence of the pentangle in the abstract mathematical work of Penrose and in Shectman's research on real materials may be seen as a signa/res duality. Certainly R. G. Arthur's interest in Medieval sign theory, and notably his emphasis on the pentangle remarkably complements the simultaneous work in natural science on quasi-crystalline order.

Acknowledgement—The author gratefully acknowledges the assistance of the Cultureel Centrum De Pauwhof outside The Hague, The Netherlands, where he was able to work on this manuscript and to consult various Dutch sources.

REFERENCES

1. T. H. White, *The Once and Future King*. Collins, London (1958); also *The Book of Merlyn* (Ed. Sylvia Townsend Warner). University of Texas Press, Austin (1977).
2. E. B. Edwards, *Patterns and Design with Dynamic Symmetry*. Dover, New York (1967, reprint of the 1932 edn).
3. A. L. Loeb, Remarks on some elementary volume relations between familiar solids. *Maths Teacher* **58**, 417 (1965). A. L. Loeb, Contribution to R. Buckminster Fuller's *Synergetics*, pp. 832–836. McMillan, New York (1975).
4. R. G. Arthur, *Medieval Sign Theory and Sir Gawain and the Green Knight*. University of Toronto Press, Toronto (1987).
5. Sir Thomas Malory, *Le Morte d'Arthur* (written in 1469–70, Ed. J. Cowen). Penguin Books, Harmondsworth (1969).
6. Mary Stewart, *The Wicked Day*. Morrow, New York (1983).
7. Mary Stewart, *The Crystal Cave* (1970); *The Hollow Hills* (1973) and *The Last Enchantment* (1979). Hodder, London.
8. Anonymous (fourteenth century), *Sir Gawain and the Green Knight* (Eds J. J. R. Tolkien and E. V. Gordon, 2nd edn rev. N. Davis). Clarendon Press, Oxford (1967).
9. *Sir Gawain and the Green Knight* (Eds J. J. R. Tolkien and E. V. Gordon, 2nd edn rev. N. Davis). Lines 657–669: "Now all these five multiples, truly, were fastened together in this knight, and each one was joined into another, so that it had no end, and fixed upon five points that never failed nor came together in any side or separated either, without end at any angle anywhere, ... (T)his knot ... is called the pure pentangle by people with learning, ..."; and lines 626–629: "It is a sign that Solomon once established as ... a figure that has five points, and each line overlaps and locks into another, and everywhere it is endless."
10. Mark Twain (S. L. Clemens), *A Connecticut Yankee at King Arthur's Court*. Webster, New York (1889).
11. R. G. Arthur, *Medieval Sign Theory and Sir Gaiwan and the Green Knight*. University of Toronto Press, Toronto (1987).
12. Nicholas of Cusa, *Of Learned Ignorance* (transl. G. Heron). Routledge, London (1954).
13. Anselm of Canterbury, *Monologion*, cxviii, in *Opera Omnia* (Ed. F. S. Schmitt), Vol. 1, p. 32. Secovii (1938).
14. R. W. Ackerman, Gawain's Shield: Penitential Doctrine in *Gawain and the Green Knight*. *Anglia* **76**, 254–265 (1956).
15. A. Derrickson, The pentangle: guiding star for the Gawain Poet. *Comitatus* **11**, 10–19 (1980).
16. W. O. Evans, Gawain's new pentangle. *Trivium* **3**, 92–94 (1968).
17. V. Y. Haines, *The Fortunate Fall of Sir Gawain*. University Press of America (1982).
18. A. K. Hieatt, Sir Gawain: pentangle, luf-lace, numerical structure. *Pap. Lang. Lit* **4**, 339–359 (1968).
19. G. Morgan, The significance of the pentangle symbolism in *Sir Gawain and the Green Knight*. *Mod. Lang. Rev.* **74**, 769–790 (1979).
20. D. S. Brewer (Ed.), *The Morte d'Arthur*, pp. 4–5. Arnold, London (1968).
21. Robert de Boron: *Merlin, roman du XIIIme siècle* (Ed. Alexandre Micha). Droz, Geneva (1980).
22. C. Morris, *The Tudors*, pp. 56–57. Fontana-Collins, London (1966). First published by Batsford, London (1955).
23. J. ter Haar, *Koning Arthur*. Fibula-van Dischoeck, Bussum, The Netherlands (1962).

24. H. S. M. Coxeter, M. Emmer, R. Penrose and M. L. Teuber (Eds), *M. C. Escher: Art and Science*. North-Holland, Amsterdam (1986).
25. A. C. Edmondson, *A Fuller Explanation, Design Science Collection* (Series Ed. A. L. Loeb). Birkhaeuser, Basel/Boston (1986).
26. C. H. MacGillavry, *Fantasy and Symmetry*, Abrams, New York; originally published as *Symmetry Aspects of M. C. Escher's Periodic Drawings*. International Union of Crystallography and A. Oosthoek, Utrecht, the Netherlands (1965).
27. *The Spiritual in Art: Abstract Painting 1890–1985*. Abbeville Press, New York (1987).
28. W. Chorbachi, Private communication (to be published).
29. A. L. Loeb, *Color and Symmetry*, p. 33. Wiley, New York (1971); also Krieger, New York (1978).
30. R. Penrose, The role of aesthetics in pure and applied mathematics research. *J. Inst. Math. Applic.* **10**, 266–271 (1974). M. Gardner, Mathematical games: extraordinary nonperiodic tiling that enriches the theory of tiles. *Scient. Am.* **236**, 110–121 (1977).
31. D. Shechtman, I. Blech, D. Gratias and J. W. Cahn, A metallic phase with long-ranged orientational order and no translational symmetry. *Phys. Rev. Lett.* **53**, 1951 (1984). D. R. Nelson: Review article. *Scient. Am.* **255**, 42 (1986).
32. J. W. Cahn: Quasiperiodic crystals: a revolution in crystallography. 1985 Von Hippel Lecture. *MRS Bull.* March/April, pp. 9–14 (1986).
33. C. L. Henley, Quasicrystal order, its origins and its consequences: a survey of current models. *Comments Cond. Matter Phys.*, pp. 58–117 (1987).
34. W. Ohashi and F. Spaepen, Stable GaMgZn Quasi-periodic crystals with pentagonal dodecahedral solidification morphology (to be published).
35. C. J. Schneer, Hierarchical symmetry and morphology of snowflakes and related forms (to be published).
36. S. Y. Edgerton, *The Renaissance Rediscovery of Linear Perspective*. Basic Books, New York (1975).
37. R. G. Arthur, *Medieval Sign Theory and Sir Gawain and the Green Knight*, p. 10. University of Toronto Press, Toronto (1987).

Computers Math. Applic. Vol. 17, No. 1–3, pp. 49–57, 1989
Printed in Great Britain. All rights reserved

SYMMETRY, GROUPOIDS AND HIGHER-DIMENSIONAL ANALOGUES†

R. Brown

School of Mathematics, University of Wales, Bangor, Gwynedd LL57 1UT, U.K.

Abstract—We give some history and motivation for the notion of groupoid as a powerful and flexible generalization of the notion of group, stressing the links with symmetry and with paths in dynamical systems. Two-dimensional groupoids are explained, and their possibilities for new notions of symmetry are explored.

The notion of *groupoid* as a generalization of that of group was found by Brandt [1] as a result of his work on the arithmetic of quaternary quadratic forms, generalizing the work of Gauss for the case of binary quadratic forms. This concept of groupoid can now be seen as a significant extension of the range of discourse, allowing for a more flexible and powerful approach to symmetry. The following quotation is from Ref. [2]:

> "The concept of groupoid is one of the means by which the twentieth century reclaims the original domain of application of the group concept. The modern, rigorous definition of a group is far too restrictive for the range of geometrical applications envisaged in the work of Lie. There have thus arisen the concepts of Lie pseudogroup, of differentiable and of Lie groupoid, and of principal bundle—as well as various related concepts such as Lie equation, graded Lie algebra and Lie algebroid—by which mathematics seeks to acquire a precise and rigorous language in which to study the symmetry phenomena associated with geometrical transformations which are only locally defined."

Further, the concept of groupoid has lead to some notions of *many variable groupoid theory* which are uninteresting when restricted to groups alone. These varieties of developments available for groupoids but not for groups open up the possibility of extensive ranges of applications. Our aim in this paper is to give some idea of the background and of sources for these ideas. Information on the algebraic theory of groupoids is given in Ref. [3] (which, unfortunately, is out of print), Ref. [2], which also deals with the topological and differentiable cases, and Ref. [4].

The symmetry of an object is classically described by the group of transformations of the object which leave its geometric structure invariant. For example the symmetry of the square

is described by the dihedral group D_4 whose elements are the eight permutations leaving the square invariant. There is also a description of D_4 by generators and relations. Let x be the rotation of the square through $\pi/4$ anticlockwise, and let y be the reflection in the vertical line through the center of the square. Then D_4 is generated by x and y with the relations $x^4 = y^2 = xyxy = 1$. Such a description by generators and relations is available for all groups, and in many cases gives a convenient way of handling the group.

This use of the group of symmetries fits with Klein's famous Erlanger programme of 1872, which suggested that a geometry should be studied through its group of automorphisms, that is the group of transformations leaving the geometric structure invariant. This programme has had a great influence on the subsequent development of group theory (cf. Ref. [5]).

The notion of *groupoid* arises when one is given not just one structure S but a family $\mathbf{S} = \{S_i\}_{i \in I}$ of structures. It is then appropriate to consider not just the group $G(i)$ of automorphisms of each S_i, but also to consider for all i, j the set $G(i, j)$ of isomorphisms $S_i \to S_j$. An isomorphism $f: S_i \to S_j$

---

†Revised version of a talk entitled "Paths to symmetry" and delivered as the London Mathematical Society Lecture to the *British Association for the Advancement of Science Annual Meeting*, Belfast, August 1987.

may be composed with an isomorphism $g: S_j \rightarrow S_k$ to give an isomorphism $gf: S_i \rightarrow S_k$. There is an identity isomorphism $1_i: S_i \rightarrow S_i$ for all elements i of I, and an isomorphism $f: S_i \rightarrow S_j$ has an inverse $f^{-1}: S_j \rightarrow S_i$ such that $ff^{-1} = 1_j, f^{-1}f = 1_i$. The composition is associative: if hg and gf are defined, then $h(gf)$ and $(hg)f$ are defined and are equal. We write G for the union of all the sets $G(i,j)$ for $i, j \in I$. The important point is that if f, g are elements of G, then the composite or product gf may or may not be defined. Otherwise the rules for the product are exactly similar to those of a group. This groupoid is called the groupoid $G(S)$ of *symmetries* of the family S. For references to the use of this groupoid, cf. Refs [2, 6].

A groupoid may be thought of as a *group with many identities*. Now one of the first propositions of group theory is that a group has only one identity. It is the fact that in a groupoid the product is not always defined which allows for many identities.

The formal definition of a groupoid is now easily given. There is a set G of *arrows* or *elements*, a set I of *vertices* or *objects*; the set G is the union of disjoint sets $G(i,j)$ for all i, j in I, so that if f is an element of G then f has a unique source sf and target tf, and we write $f: sf \rightarrow tf$. Two elements g, f of G are *composable* if and only if $tf = sg$, and then their product gf belongs to $G(sf, tg)$. This is represented by the diagram

$$sf \xrightarrow{\ \ f\ \ } tf = sg$$

with gf and g leading to tg.

The axioms for a groupoid are then a statement in the general situation of the rules we have already written down for our specific example, the groupoid $G(S)$ of symmetries of the family $\mathbf{S} = \{S_i\}_{i \in I}$.

Here are some more examples of groupoids.

Example 1

Let X be any set. Then there is a groupoid also written X called the *fine*, or *null*, groupoid on X. It has only identities 1_x, one for each x in X; these may be composed with themselves so that $1_x 1_x = 1_x$, and there are no other products.

Example 2

Let X be any set. Then there is a groupoid with vertex set X and set of arrows the product set $X \times X$, so that an arrow $x \rightarrow y$ is simply the ordered pair (y, x). The product is then given by

$$(z, y)(y, x) = (z, x).$$

This groupoid looks rather simple, banal and unworthy of consideration! Surprisingly, though, it plays a key role in the theory and applications. One reason is that if G is a subgroupoid of $X \times X$ (the definition of subgroupoid is left to the reader) and G has the same vertex set X as $X \times X$, then G is essentially an equivalence relation on X. That is, for all x in X, $(x, x) \in G$; if $(x, y) \in G$, then $(y, x) \in G$; and if $(z, y), (y, x) \in G$ then $(z, x) \in G$. Now equivalence relations are important in mathematics and science because they formalize the idea of *classification*—two elements have the same classification if and only if they are equivalent. In mathematical terms, we say that equivalence relations formalize the idea of *quotienting*. Thus, it is an important aspect of their applications that groupoids generalize both groups and equivalence relations. This versatility is further shown by the next example.

Example 3

Let H be a group of transformations of a set X. This means that for each h in H and each x in X there is an element $h \cdot x$ in x and the following rules are satisfied:

$$1 \cdot x = x, \quad (h' \cdot h) \cdot x = h' \cdot (h \cdot x),$$

for all x in X and all h, h' in H. In such case we often picture the operation or action of h on x as

$$x \circ \xrightarrow{\quad h \quad} \circ h \cdot x.$$

Since the arrow is here based at x, it would be more accurate to write it as $(h, x): x \to h \cdot x$. This suggests defining a composition

$$(h', h \cdot x)(h, x) = (h'h, x)$$

and it is easily checked that in this way we obtain a groupoid. This groupoid has been called the *actor groupoid*, *covering groupoid* and *affine groupoid* of the action. Another name used is *semidirect product groupoid*, with notation $H \ltimes X$.

A further example of groupoids comes from the consideration of *paths* in a space. It is worthwhile in a book on symmetry to recall this other motivation for both group and groupoid theory, since it is important for the extensions of the notion of groupoid to higher dimensions. The *group of paths*, or *fundamental group*, is due to Poincaré in 1895. It arose from Poincaré's work in celestial mechanics.

Recall that the first major problem of celestial mechanics was to *describe the motion of two heavy bodies moving under the influence only of inverse square gravitational forces*. This is called *the two body problem*. Of course Newton's answer was that the two bodies move so that each centre of gravity describes a conic section with focus the centre of gravity of the two bodies. In the course of verifying this answer, and so giving a theoretical basis for Kepler's laws of planetary motion, Newton was led to develop the basic techniques of the differential and integral calculus.

The next problem was to describe the motion of three bodies under gravitational forces; this is the *three body problem*. It remains unsolved. Poincaré showed that there is no solution which can be given in a finite number of integrations, although of course from any given starting situation, the resulting motion can be computed to an accuracy limited only by computational resources.

As a result of the lack of an analytic solution to the problem, Poincaré sought other kind of interesting questions, and formulated *qualitative* rather than *quantitative* questions. These are of the kind:

  (i) What are the periodic solutions (i.e. those that return to their original positions and velocities)?
  (ii) Is a solution stable (i.e. if you perturb the starting position a bit, does the resulting motion stay similar to the original one)?
  (iii) Is the solution bounded over indefinite time?

For a long time the only periodic solutions to a many body problem was the Lagrange solution, with three equal bodies at the vertices of an equilateral triangle. New periodic solutions were discovered by Davis *et al.* [7] as a result of computer experiments. In any case, these questions are a good illustration of the maxim that in science the interesting question is very often not: "What is the answer?" but instead: "What is the question?". It is unfortunate that in so many aspects of our scientific and mathematical education, the main weight of emphasis is on *knowledge* and *problem solution*, and much less on the practical aspect, which must be first and foremost, *problem formulation*.

Poincaré took the following approach to the three body problem, and similar questions in mechanics. Consider each body as a particle concentrated at the centre of mass. For spherical bodies, this is no loss of generality. The position and velocity of each particle are each given by three coordinates. For three particles, that makes 18 coordinates. However the motion is restricted by the energy equation

$$kinetic\ energy + potential\ energy = constant.$$

So the motion lives not in eighteen-dimensional space, but in a seventeen-dimensional subspace called the *phase space* P_e for this particular energy level e. A particular motion at a given energy e is then a path in the phase space P_e. So we are led to the question: "What different *kinds* of paths are possible in P_e?" The emphasis here is on the words *kinds*. In mathematics, as in any

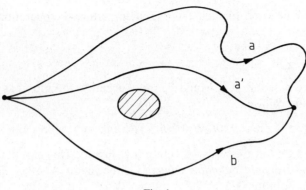

Fig. 1

science, the making of lists, even where possible, is not the only aim. The further aims are *classification* and *understanding*.

A reasonable classification for paths is that of *deformability*. In Fig. 1 it is clear that the path *a* deforms to *a'*, but that *a* does not deform to *b* because of the hole. Poincaré therefore considered equivalence classes of *loops*, that is, paths which start and return to a base point, under the relation of deformability. Two loops may be composed as in Fig. 2 and it is not hard to verify that the classes of paths form under the composition of paths a group, called the *fundamental group* $\pi_1(P_e, x)$ of the space P_e at the base point *x*. The consideration of this group was one of the starting points of the subject of *algebraic topology*, which was first called *analysis situs*. The other starting point was the notion of *homology*, which derived from integration theory.

In many ways, it is unnatural to consider only loops at a point *x*. It is often convenient to consider paths from *x* to another point *y*, say. These also may be composed as in the following diagram:

$$x\bigcirc \overset{a}{\longrightarrow} \underset{y}{\bigcirc} \overset{b}{\longrightarrow} \bigcirc z$$

to give a path *ba*. The classes of paths under deformability then form not a group but a groupoid, called the *fundamental groupoid* $\pi_1 P_e$ of the space P_e. This groupoid contains all the fundamental groups $\pi_1(P_e, x)$ for all base points *x*, as well as information on how they are related. It is this extra information which is particularly valuable in using groupoids.

It was realized by Higgins [3] that there is also for groupoids a convenient theory of generators and relations, similar to that for groups, but with an additional flexibility which enables groupoids to be used for proving purely group theoretic results. That is, even if the main interest is in groups, it is still often convenient to leave the group theoretic world in order to formulate and prove theorems. One reason for this is Example 3, which constructs a semidirect product groupoid from a group action. There are many proofs and applications in group theory which work via a group action, and it is quite often useful to reformulate such methods in terms of groupoids. For examples of this, see Refs [8, 9] as well as Refs [3, 4], and the survey [6].

It now seems likely that in the future group theory will be seen to be a proper subset of groupoid theory, in the sense that a proper account of group theory will need to take account of groupoid methods. At present, it is difficult to find a book on group theory which even mentions the idea. In view of our discussion of paths and celestial mechanics, it is also relevant that groupoids are

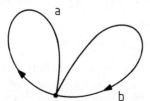 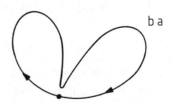

Fig. 2

now being found appropriate for use in the theory of Hamiltonian systems, for example in the study of non-linear commutation relations [10].

For me, one of the chief interests of groupoids has been that they allow for a higher-dimensional theory. The intuitive idea for this came from an attempt to generalize the fundamental groupoid to higher dimensions, with a view to applications in the subject called *homotopy theory*, which is the mathematical study of the theory of deformations of paths and more generally of functions. The universality of this notion of deformation is such that the techniques of homotopy theory are playing an increasing rôle in many applications, for example in particle physics, where again group theory has already proved important.

As explained above, the fundamental groupoid arises from composing classes of paths. The question is, what kind of algebra results from composing squares instead of paths? Such an algebra should give rise to a higher-dimensional group theory. The idea for higher-dimensional algebra is as follows:

Normally we are constrained in our mathematical writing and to some extent in our thinking by the way we write mathematics in formulae along a line. So we write

$$x = yz, \quad x = y + z,$$

or more generally

$$x = x_1 x_2 \cdots x_n.$$

That is, x is the product or composite of n elements $x_1, \ldots, x_n$. The question is, why cannot we write

$$x = \begin{bmatrix} a & b \\ c & d \end{bmatrix},$$

showing that x is some kind of *two-dimensional product*? More generally, we would like to have a product

$$x = \begin{bmatrix} x_{11} & x_{12} \cdots x_{1n} \\ \cdots\cdots\cdots\cdots\cdots \\ x_{m1} & x_{m2} \cdots x_{mn} \end{bmatrix}.$$

You may well ask: "what is new about that?" All you have is a matrix, an array? My point is that x is intended to be *evaluated* as a single element. Further, we are going to allow not only horizontal compositions but also vertical ones, not only

$$[a \quad b] \quad \text{but also} \quad \begin{bmatrix} a \\ c \end{bmatrix}.$$

The reason for trying to do this was the following curious mathematical point. A lot of mathematics is concerned with turning *geometry* into *algebra*. This allows problems to be solved by calculation rather than by special geometric argument. The most famous example is the use in geometry by Descartes in 1637 of the symbolic method used by Viete in the late sixteenth century. In this way Descartes founded *analytic geometry*, and the method of now daily use, that of *Cartesian coordinates*. Throughout the intervening time the algebraicization of geometry has been a spur to the invention of new mathematical techniques. The algebra models the geometry, and the geometry guides the algebra. It is the interaction between these which makes much of our contemporary mathematics.

Consider now a subdivision of the line

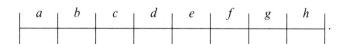

The algebra for this subdivision is an expression of the type

$$x = abcdefgh.$$

Our starting problem is the following. Here is a *geometric* figure, namely a square subdivided:

a	b	c	d	e	f
g	h	i	j	k	l
m	n	p	q	r	s

Fig. 3

Where is the corresponding *algebra*?

It turns out that the definition of what we call a *double groupoid* is not hard [11]. One simply has three classes of elements, *vertices*, *edges* and *squares*. Analogously to the case of groupoids, one makes the obvious geometric relations between these entities, for example that a square has four boundary edges. The edges are assumed to have a partially defined composition operation giving a groupoid. The squares are assumed to have *two* partially defined compositions, described by the diagrams

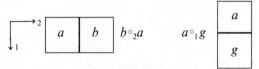

so that the compositions of squares are defined if and only if the appropriate edges coincide, as suggested by the diagrams. It is required that each of the compositions $\circ_1$ and $\circ_2$ are groupoids, and that the boundaries of composites are as suggested by the diagrams. Finally, there is a crucial rule which connects the two compositions of squares, namely that the diagram

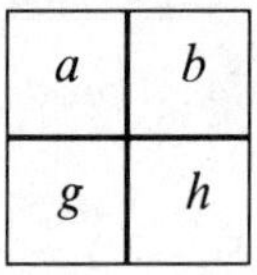

determines only one composition: this amounts to the rule, known as the *interchange law*,

$$(b \circ_1 h) \circ_2 (a \circ_1 g) = (b \circ_2 a) \circ_1 (h \circ_2 g).$$

A consequence is that any matrix composition, for example that shown in Fig. 3, may be evaluated by composing rows first, or columns first, or by first block subdividing and composing each block.

It is worth explaining why this algebra gives little in the group case. Suppose then that all the groupoid structures are in fact group structures, so that all compositions are defined. Then there is an identity e_1 for $\circ_1$ and an identity e_2 for $\circ_2$. Hence

$$e_2 = e_2 \circ_2 e_2 = (e_2 \circ_1 e_1) \circ_2 (e_1 \circ_1 e_2)$$

$$= (e_2 \circ_2 e_1) \circ_1 (e_1 \circ_2 e_2)$$

$$= e_1 \circ_1 e_1 = e_1.$$

That is, the identities for $\circ_1$ and $\circ_2$ coincide, and so may be written simply e. It now follows that if a, b are squares, then

$$a \circ_1 b = (a \circ_2 e) \circ_1 (e \circ_2 b) = (a \circ_1 e) \circ_2 (e \circ_1 b) = a \circ_2 b,$$

so that the two compositions coincide. Hence each may be written simply $\circ$. It now follows that

$$a \circ_b = (e \circ a) \circ (b \circ e) = (e \circ b) \circ (a \circ b) = b \circ a,$$

so that the composition is commutative. This fact was found in the late 1930's and led to the view that a higher-dimensional group theory was not possible. Indeed it was embarrassing that the higher-dimensional theory was less complicated, rather than more complicated, than the one-dimensional theory.

We can now see that while this is true for group theory, the extension to groupoids does lead to a significant theory, and that it was probably a general reluctance to take this extension seriously which prevented these methods from being found earlier.

Even when the idea had been formulated, it took nine years to find reasonable generalizations to higher-dimensions of the fundamental groupoid which would enable the use of higher-dimensional generator and relation types of arguments. Eventually it was found in 1974 by Brown and Higgins [12] that a sensible generalization is obtained not just for a space X, but for a space X with subspaces Y and Z such that Z is contained in Y. Instead of considering paths in X one considers maps of a square

$$\mathbf{I}^2 = \{(s, t): 0 \leqslant s, t \leqslant 1\}$$

such that the edges of $\mathbf{I}^2$ map into Y and the vertices of $\mathbf{I}^2$ map into Z, as shown in the following diagram:

It is the deformation classes of these maps, where under the deformation the edges remain in Y and the vertices *remain fixed*, which form by an obvious gluing process a double groupoid in the sense considered above. There are reasonable generalizations of these ideas to higher-dimensions [13], and even generalizations which allow for more general objects than squares and wider kinds of subdivisions [14]. However, the proofs become quite hard.

In order to give some flavour of the theory, and to show how the additional flexibility of two dimensions does make a difference, I will give an argument which does use the capabilities of the block subdivision and combination.

In dimension 1, an identity can be regarded as a kind of "squashed" element. In terms of paths, an identity is represented by a path which does not move, a stationary journey. However, in dimension 2 there are more kinds of squashing which are available. We like to call such squashed squares *thin* and to use the following notation for some of them [15]:

$$\equiv \quad | \quad | \quad \llcorner \quad \lrcorner \quad \urcorner \quad \ulcorner \quad \square .$$

Here a line denotes a constant or identity edge, and the whole square is to have commuting boundary. For example, some of these squares have edges as shown below:

For example if a is a path, then these maps of squares are defined respectively

$$(s, t) \mapsto a(s), \quad (s, t) \mapsto a(\max\{s, t\}), \quad (s, t) \mapsto a(1 - \max\{s, 1 - t\}),$$

for $0 \leqslant s, t \leqslant 1$. The most important rule on a thin element is that it is completely determined by its (commutative) boundary. This leads to the following curious equations on thin elements in a double groupoid, where we use matrix notation to represent multiple compositions:

The reason for these equations is that both sides are thin and have the same boundaries, hence are equal.

Here is another application. Given a square a we can border it with squashed, or rather, as we have used the word, *thin* elements, to give new elements

We think of ρ, τ as giving rotations through $\pi/2$ in clockwise and anticlockwise directions, respectively. We now give the calculation that $\tau\rho(a) = a$. We have

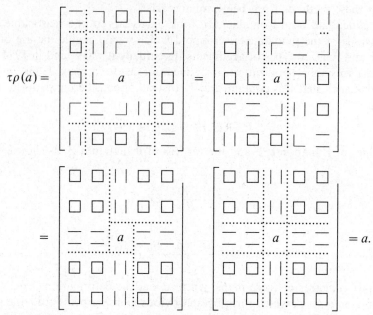

I do not expect you to find that too easy to follow. However, the key aspects of this calculation are resubdivision and recombination in two dimensions, and, as is to be expected, this allows for a greater variety of possibility than in dimension 1. Two-dimensional algebra is, as one would hope and expect, more subtle and complicated than one-dimensional algebra. I leave as a hard exercise for the reader the proof that $\rho^4 = 1$ (cf. Ref. [15]).

Following up these ideas has lead to new results in homotopy theory and has given new constructions in group theory and in other algebraic systems, such as Lie algebras.

One problem that this theory raises is the following. In dimension 1, we have both groups and groupoids arising from notions of symmetry, and from paths. These parallel notions have proved significant in the theory and applications of both groups and groupoids. In dimension 2 we have an algebra of double groupoids and a geometric example arising from using squares as a generalization of paths. *Where then is the two-dimensional notion of symmetry?*

Recent work of Brown and Gilbert [16] has suggested a framework for an answer. From the automorphisms of a structure **S** one obtains a group $G(\mathbf{S})$ of symmetries of **S**. We think of **S** as being zero-dimensional and $G(\mathbf{S})$ as being one-dimensional.

If G is a group, it is standard to consider the group $\text{Aut}(G)$ of automorphisms of the group G. But it is also useful to consider the homorphism $\chi: G \to \text{Aut}(G)$ which sends g in G to the inner automorphism $x \mapsto gxg^{-1}$. Now define a *square* to be a quintuple (cf. Ref. [11])

$$\alpha = \left(g : a\genfrac{}{}{0pt}{}{b}{d}c \right),$$

such that a, b, c, d, are in $\text{Aut}(G)$, g is in G and $\chi(g) = a^{-1}d^{-1}cb$. It is not hard to cook up horizontal and vertical compositions of squares giving the right boundary operations, and so giving a double groupoid [11]. Thus we find that the *automorphisms of a group form in a canonical way part of a double groupoid structure.*

This idea continues certainly to the next dimension [16], and presumably to all cases. That is, we expect an heirarchy of structures of levels given by positive integers, and where the symmetries of a structure of level n form part of a structure of level $n + 1$. There are also expected to be analogues of these ideas for other structures than groups and groupoids; for example, we expect analogous ideas for Lie algebras, associative algebras, commutative algebras, Jordan algebras, etc. Ref. [17]. However, there is much work to be done to realize these ideas.

Note also that associative algebras, Lie algebras and Jordan algebras play an important role in many applications of mathematics to physics. In view of what Wigner has called "the unreasonable success of mathematics in the physical sciences" [18] we can expect that these higher-dimensional structures will also in the end find some key applications in our understanding of natural phenomena.

Remark

The following set of references is brief. The survey article [6] has a bibliography of 160 articles which can serve as entry points to the literature, and which includes references on applications in quantization in physics, and in crystallography.

REFERENCES

1. H. Brandt, Über eine Verallgemeinerung des Gruppenbegriffes. *Math. Ann.* **96**, 360–366 (1926).
2. K. Mackenzie, *Lie Groupoids and Lie Algebroids in Differential Geometry.* London Math. Soc. Lecture Note Series 124. Cambridge University Press, Cambridge (1987).
3. P. J. Higgins, *Categories and Groupoids.* Van Nostrand–Reinhold, New York (1971).
4. R. Brown, *Elements of Modern Topology.* McGraw-Hill, Maidenhead (1968). Republished as *Topology.* Ellis Horwood (1988).
5. H. Wussing, *The Development of the Abstract Group Concept.* M.I.T. Press, Cambridge, Mass. (1984).
6. R. Brown, From groups to groupoids: a brief survey. *Bull. Lond. Math. Soc.* **19**, 113–134 (1987).
7. I. Davis, A. Trueman and D. Williams, Classical periodic solutions of the equal mass $2n$-body problem, $2n$-ion problem, and the n-electron atom problem. *Phys. Lett.* **99A** (1), 15–18 (1983).
8. P. R. Heath and H. Kamps, On exact orbit sequences. *Ill. J. Math.* **26**, 149–154 (1982).
9. P. J. Higgins, Presentations of groupoids with applications to groups. *Proc. Camb. Phil. Soc.* **60**, 7–20 (1964).
10. A. Coste, P. Dazord and A. Weinstein, Groupoids symplectiques. Preprint Universite Claude Bernard-Lyon, 2/A-1987, 62 pp. (1987).
11. R. Brown and C. B. Spencer, Double groupoids and crossed modules. *Cah. Top. Geom. Diff.* **17**, 343–362 (1976).
12. R. Brown and P. J. Higgins, On the connection between the second relative homotopy groups of some related spaces. *Proc. Lond. Math. Soc.* **36** (3), 193–212 (1978).
13. R. Brown and J.-L. Loday, Van Kampen theorems for diagrams of spaces. *Topology* **26**, 311–335 (1987).
14. D. W. Jones, Poly-T-complexes. University of Wales Ph.D. Thesis (1984). A general theory of polyhedral sets and their related T-complexes. *Diss. Math.* **266** (1988).
15. R. Brown, Higher dimensional group theory. In *Low-Dimensional Topology*, London Math. Soc. Lecture Note Series 46, (Eds. R. Brown and T. L. Thickstun), pp. 215-238, Cambridge University Press, Cambridge (1982).
16. R. Brown, and N. D. Gilbert, Algebraic models for 3-types and automorphism structures for crossed modules. *Proc. London Math. Soc.* (in press).
17. G. J. Ellis, Higher dimensional crossed modules of algebras. *J. Pure Appl. Alg.* **52**, 277–282 (1988).
18. E. Wigner, *Symmetries and Reflections: Scientific Essays of Eugene P. Wigner.* Bloomington Indiana Press, Indiana (1987).
19. M. E. Mayer, Groupoids and Lie bigebras in gauge and string theories. In *Differential Geometrical Methods in Theoretical Physics* (Ed. K. Blealer). Reidel, Dordrecht (1988).

Computers Math. Applic. Vol. 17, No. 1–3, pp. 59–71, 1989
Printed in Great Britain. All rights reserved

0097-4943/89 $3.00 + 0.00

TRISECTING AN ORTHOSCHEME

H. S. M. Coxeter

Department of Mathematics, University of Toronto, Toronto M5S 1A1, Canada

Abstract—A tetrahedron having two right angles at each of two vertices was investigated by Lobachevsky (who called it a "pyramid"), Schläfli (who called it an "orthoscheme"), Wythoff (who called it "double-rectangular"), and Schoute (who called its theory "polygonometry"). There is a simple procedure for dissecting such a tetrahedron into three smaller orthoschemes. The two cutting planes meet three of the four faces (which are right-angled triangles) along lines which can easily be described. When the tetrahedron is unfolded so as to put all the faces in one plane, the arrangement of lines suggests an interesting theorem of absolute geometry. When a particular spherical orthoscheme of known volume is dissected into three pieces, and the volumes of these smaller orthoschemes are expressed as definite integrals, the result is a peculiar identity which has not been verified directly. There is a one-parameter family of orthoschemes for which the three smaller orthoschemes are all congruent; the Euclidean member of this family turns out to be related to a very simply frieze pattern of integers.

1. FRIEZE PATTERNS

Let us define a *frieze pattern* to be an arrangement of real numbers in staggered rows so that, if a and d are adjacent in one row, with b between them in the preceding row and c in the following row, then

$$ad - bc = 1,$$

as in a unimodular matrix rotated by $45°$. For instance, we might have an endless row of zeros followed by an endless row of ones and then a periodic row associated with a convex n-gon as follows. Let the n-gon be dissected, by means of $n - 3$ diagonals, into $n - 2$ triangles. Associate the successive vertices with positive integers which count the number of triangles occurring at the vertex. When such a cycle of integers is used for the third row of the frieze pattern (with zeros and ones in the preceding rows), the nth row will be found to consist (like the second) entirely of ones. No other choice of positive integers in the third row will have this remarkable effect [1, p. 181]. For instance, although there are several such frieze patterns when $n > 5$, there is essentially only one when $n = 5$, since the only way to triangulate a pentagon by means of two diagonals is when these diagonals share one vertex. This vertex belongs to all the 3 triàngles; thus the cycle of numbers is 3, 1, 2, 2, 1, and the frieze pattern is:

```
    0     0     0     0     0     0     0     0     0     0
 ···   1     1     1     1     1     1     1     1     1    ···
    3     1     2     2     1     3     1     2     2     1
 ···   2     1     3     1     2     2     1     3     1    ···
    1     1     1     1     1     1     1     1     1     1
 ···   0     0     0     0     0     0     0     0     0    ···
```

2. THE GENERAL ORTHOSCHEME IN ABSOLUTE 3-SPACE

There is a theorem of "absolute" geometry which states the following

Theorem 1

> If two lines AB and CD are such that AB lies in a plane perpendicular to CD, then CD lies in a plane perpendicular to AB.

This is trivial when the two lines intersect. To prove it when they are skew, let C be the point where CD meets the perpendicular plane through AB, and let B be the foot of the perpendicular from C to AB, as in Fig. 1. Extend AB to A′ so that AB = BA′. By considering pairs of congruent triangles [2, pp. 182–183] we deduce, in turn, that CA = CA′, that DA = DA′, and that DB is

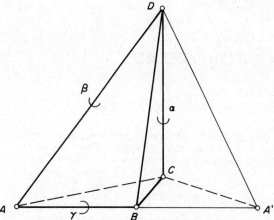

Fig. 1. How two skew lines may be perpendicular.

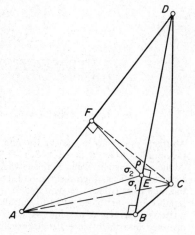

Fig. 2. A trisected orthoscheme.

perpendicular to AA′. Hence AA′ (or AB), being perpendicular to both BC and BD, is perpendicular to the plane BCD (through CD), as desired.

The two skew lines AB and CD, each lying in a plane perpendicular to the other, are naturally said (though not by Euclid, nor by Forder [3, p. 124]) to be *perpendicular*. Then we can define an *orthoscheme* to be a tetrahedron ABCD in which the "successive" edges AB, BC, CD are *mutually* perpendicular.

It follows that all the four faces of an orthoscheme are right-angled triangles, the right angles being ABC, ABD, ACD, BCD. Moreover, the dihedral angles along the edges AC, BC, BD are right angles: say

$$(AC) = (BC) = (BD) = \pi/2.$$

(The symbol (BC) is an abbreviation for Schoute's A(BC)D [4, p. 268], meaning the dihedral angle formed by the planes ABC and BCD.) The remaining dihedral angles

$$\alpha = (CD), \quad \beta = (AD), \quad \gamma = (AB)$$

determine the shape of the orthoscheme (and thus, in the non-Euclidean cases, its shape and size). The symbol

$$(\alpha, \beta, \gamma)$$

is convenient, although it fails to distinguish between the *dextro* and *laevo* varieties of this chiral solid, which are "oppositely congruent". However, (α, β, γ) and (γ, β, α) are "directly congruent", so we can simply regard them as alternative symbols for the same object.

One might well imagine that, if all the four faces of a tetrahedron are right-angled triangles, the tetrahedron must be an orthoscheme. Although this is true in Euclidean and hyperbolic spaces, it is not true in spherical and elliptic spaces [2, p. 142]. The counter-example, noticed by J. B. Wilker, is the "Clifford disphenoid" formed by taking a rectangular Clifford parallelogram and inserting its two diagonals.

3. THE DISSECTION INTO THREE ORTHOSCHEMES

Just as any planar polygon can be dissected into a finite number of right-angled triangles, any polyhedron (Euclidean or non-Euclidean) can be dissected into a finite number of orthoschemes [5, p. 246; 6, p. 241]. In particular, we have the following absolute theorem.

Theorem 2

Any orthoscheme can be dissected into three orthoschemes.

To prove this, we construct the two cutting planes ACE and CEF by drawing CE perpendicular to BD, and then EF perpendicular to AD, as in Fig. 2. Consider the two skew lines AD and CE.

AD lies in the plane ABD, which is perpendicular to CE (since the dihedral angle along BD is a right angle), so CE must lie in a plane perpendicular to AD, namely the plane CEF. Thus, AD is perpendicular to CF. In other words, all the four angles at F (in the planes ABD and ACD) are right angles [7, Section 3.5]. It follows that the two planes ACE and CEF serve to decompose the orthoscheme ABCD into three smaller orthoschemes

$$ABEC, \quad AFEC, \quad DFEC.$$

Since α, β, γ are the angles at C, F, B in the right-angled triangles ACB, CFE, DBC, respectively, these three angles must be acute if the geometry is Euclidean or hyperbolic. However, there is no such restriction in elliptic or spherical geometry. For instance, there is one peculiar kind of orthoscheme, $(\alpha, \beta, \frac{1}{2}\pi)$, in which D is the pole of the plane ABC, and another, $(\alpha, \frac{1}{2}\pi, \gamma)$, in which AB and CD are polar lines. Combining these pecularities, we have $(\alpha, \frac{1}{2}\pi, \frac{1}{2}\pi)$, in which α is the only acute dihedral angle and all the edges except AB have length $\frac{1}{2}\pi$. Finally, $(\frac{1}{2}\pi, \frac{1}{2}\pi, \frac{1}{2}\pi)$ is the *orthant*, all of whose edges and dihedral angles are $\frac{1}{2}\pi$. This, being analogous to the quadrant of a circle, or the octant of a 2-sphere, is one-sixteenth of the whole 3-sphere.

4. TWO LINES CROSSING AT RIGHT ANGLES

The three faces of the orthoscheme ABCD that surround the vertex D form a trihedral angle which can be slit along the edge DC ($=DC'$) and unfolded onto a plane so as to form a pentagon ABCDC' with diagonals DA and DB, as in Fig. 3. The angle EFC is thus straightened into a single line EC' perpendicular to DA, and we can deduce the following theorem (which again is "absolute").

Theorem 3

Let ABCDC' be a pentagon whose sides CD and DC' have the same length while the three angles ABD, BCD, AC'D are right angles. Draw CE perpendicular to the diagonal BD. Then C'E is perpendicular to AD.

This theorem has aroused the interest of several geometers. J. S. Frame uses non-Euclidean trigonometry. J. F. Rigby observes that a proof in the style of the late Friedrich Bachmann [8, p. 93] can be based on Hjelmslev's "semi-rotation" (*Halbdrehung*).

Günter Pickert uses polarity with respect to the circle through C and C' with centre D. This has the "absolute" property that, if P and p are pole and polar, p is perpendicular to PD. Thus the line CE, through C perpendicular to BD, is the polar of B; and the line AB, through B perpendicular to BD, is the polar of E. Since C' and E are the poles of AC' and AB, C'E is the polar of A. Therefore C'E is perpendicular to AD.

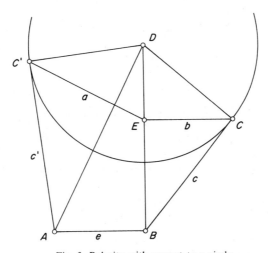

Fig. 3. Polarity with respect to a circle.

5. AN OCTAHEDRON DISSECTED INTO 16 OR 48 ORTHOSCHEMES

In Euclidean or non-Euclidean 3-space, consider a regular octahedron $\{3, 4\}$ with centre C and dihedral angle 2γ (which is $\pi - \sec^{-1} 3 \approx 109°28'16''$ if the space is Euclidean, greater if it is spherical, smaller if it is hyperbolic). Figure 4 shows a square pyramid which is the "top" half of the octahedron, so that ADA′ is one face (an equilateral triangle) and B is the midpoint of the edge AA′. The two "vertical" planes ACD and A′CD dissect the pyramid into four congruent trirectangular tetrahedra, one of which is CAA′D (with its three right angles at C). Another "vertical" plane BCD bisects this tetrahedron, one half being the orthoscheme ABCD, which is thus one-eighth of the pyramid, or one-sixteenth of the octahedron. For comparison with Fig. 1, notice that now the dihedral angles along the edges CD, AD, AB are $\pi/4$, γ, γ, so this orthoscheme is of type $(\pi/4, \gamma, \gamma)$. For comparison with Fig. 2, notice that now E is the centre of the equilateral triangle ADA′ and F is the midpoint of AD, so that the angles at E in the triangle ABD are all equal: $\sigma_1 = \sigma_2 = \rho = \pi/3$.

The octahedron has altogether nine planes of symmetry, three mutually perpendicular (already used) and six "oblique", perpendicularly bisecting pairs of opposite edges. In particular, the perpendicular bisectors of the edges DA′ and DA are the planes ACE and CEF which trisect the orthoscheme ABCD into three *congruent* pieces: the orthoschemes ABEC, AFEC, DFEC, all of type $(\pi/3, \pi/4, \gamma)$. In fact, the plane ACE bisects the right dihedral angle along AC and reflects ABEC into AFEC, while the plane CEF reflects AFEC into DFEC.

Schläfli [5, pp. 263–270; 6, pp. 175, 259] used the symbol $f(\alpha, \beta, \gamma)$ for the volume of the orthoscheme, in terms of the orthant as unit of measurement. In this notation, what we have found is that

$$f(\pi/4, \gamma, \gamma) = 3f(\pi/3, \pi/4, \gamma). \tag{1}$$

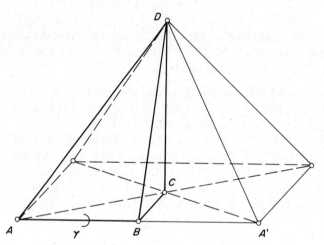

Fig. 4. Dissecting one-half of a regular octahedron.

6. POLYGONOMETRY

Schläfli's investigation of the general orthoscheme was continued by Schoute [4, pp. 267–273], Wythoff [9] and Böhm [10, p. 22]. Schoute named it *Polygonometry*: "the measuring of many angles." My own contribution [11, pp. 124–141] is the observation that these "many angles" can be consistently denoted by 4-digit symbols

$$[s \quad t \quad u \quad v] \quad (-1 \leqslant s < t < u < v \leqslant 4)$$

which are related to the cross ratios of tetrads among six (or more) points on a circle. These cross ratios involve certain 2-digit symbols (u, v) satisfying the symmetrical equation

$$(t, u)(s, v) + (u, s)(t, v) + (s, t)(u, v) = 0, \tag{2}$$

which implies $(u, u) = 0$, $(u, v) + (v, u) = 0$.

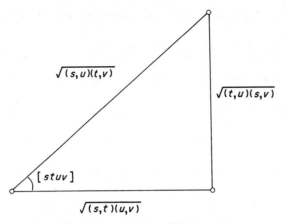

Fig. 5. An application of Pythagoras.

Böhm [10, pp. 22, 44] prefers to let the numbers s, t, u, v run from 0 to 5, instead of -1 to 4.

We shall find it convenient to assume that (u, v) is positive when $u < v$ and to write equation (2), with $s < t < u < v$, in the less symmetrical form

$$(s, t)(u, v) + (t, u)(s, v) = (s, u)(t, v), \qquad (3)$$

so that, by Pythagoras, the square roots of the three products are the sides of a Euclidean right-angled triangle, as in Fig. 5.

It can be verified that all the trigonometrical relations among the 15 "parts" of a *spherical* orthoscheme ABCD (or $A_0A_1A_2A_3$) are satisfied when we define $[s \quad t \quad u \quad v]$ to be the angle opposite to $\sqrt{(t, u)(s, v)}$ in the Euclidean triangle, so that

$$\sin[s \quad t \quad u \quad v] = \sqrt{\frac{(t, u)(s, v)}{(s, u)(t, v)}}, \quad \cos[s \quad t \quad u \quad v] = \sqrt{\frac{(s, t)(u, v)}{(s, u)(t, v)}},$$

$$\tan[s \quad t \quad u \quad v] = \sqrt{\frac{(t, u)(s, v)}{(s, t)(u, v)}}.$$

We name the $\binom{6}{4} = 15$ "parts" as follows:

$$AD = [-1 \quad 0 \quad 3 \quad 4], \qquad BD = [-1 \quad 1 \quad 3 \quad 4], \qquad CD = [-1 \quad 2 \quad 3 \quad 4],$$

$$BC = [-1 \quad 1 \quad 2 \quad 4], \qquad AC = [-1 \quad 0 \quad 2 \quad 4], \qquad AB = [-1 \quad 0 \quad 1 \quad 4],$$

$$\alpha = (CD) = ACB = [-1 \quad 0 \quad 1 \quad 2], \qquad \beta = (AD) = [0 \quad 1 \quad 2 \quad 3],$$

$$\gamma = (AB) = CBD = [1 \quad 2 \quad 3 \quad 4],$$

$$CAD = [0 \quad 2 \quad 3 \quad 4], \qquad BAD = [0 \quad 1 \quad 3 \quad 4], \qquad BAC = [0 \quad 1 \quad 2 \quad 4],$$

$$BDC = [-1 \quad 1 \quad 2 \quad 3], \qquad ADC = [-1 \quad 0 \quad 2 \quad 3], \qquad ADB = [-1 \quad 0 \quad 1 \quad 3].$$

For instance, the relation $\cos ADB \cos BDC = \cos ADC$ (for the trihedron at D) comes from

$$\frac{(-1, 0)(1, 3)}{(-1, 1)(0, 3)} \frac{(-1, 1)(2, 3)}{(-1, 2)(1, 3)} = \frac{(-1, 0)(2, 3)}{(-1, 2)(0, 3)}.$$

The $\binom{6}{2} = 15$ positive 2-digit symbols may conveniently be exhibited as a triangular pattern:

$$(-1, 0) \qquad (0, 1) \qquad (1, 2) \qquad (2, 3) \qquad (3, 4)$$

$$(-1, 1) \qquad (0, 2) \qquad (1, 3) \qquad (2, 4)$$

$$(-1, 2) \qquad (0, 3) \qquad (1, 4)$$

$$(-1, 3) \qquad (0, 4)$$

$$(-1, 4)$$

[12, p. 204; 13, pp. 59–60]. There are enough degrees of freedom to allow the simplifying assumption

$$(u, u + 1) = 1,\tag{4}$$

which makes the first row consist entirely of ones, so that

$$\sec^2\alpha = (-1, 1)(0, 2),\quad \sec^2\beta = (0, 2)(1, 3),\quad \sec^2\gamma = (1, 3)(2, 4),\tag{5}$$

$$\tan^2\alpha = (-1, 2),\quad \tan^2\beta = (0, 3),\quad \tan^2\gamma = (1, 4).\tag{6}$$

Since the shape of the orthoscheme is determined by three suitably chosen edges or angles (such as the dihedral angles α, β, γ), the four entries $(u - 1, u + 1)$ in the second row still admit one degree of freedom. Thus, one of these four numbers may be chosen how we please, and if the dihedral angles are given, the remaining three numbers in the second row can be computed from conditions (5).

By equation (3) with $s = t - 1$ and $v = u + 1$,

$$1 + (t, u)(t - 1, u + 1) = (t - 1, u)(t, u + 1).\tag{7}$$

Hence, any two adjacent entries a and d in one row, with b between them in the preceding row, are followed in the next row by $(ad - 1)/b$. In other words, if the "diamond"

$$\begin{array}{ccc} & b & \\ a & & d \\ & c & \end{array}$$

is part of the pattern, we have $ad - bc = 1$, as in Section 1.

With these simplifications, the orthoscheme (α, β, γ) has

$$\tan^2 AD = (-1, 4)(0, 3),\quad \tan^2 BD = \frac{(-1, 4)(1, 3)}{(-1, 1)},\quad \tan^2 CD = \frac{(-1, 4)}{(-1, 2)},\tag{8}$$

$$\tan^2 BC = \frac{(-1, 4)}{(-1, 1)(2, 4)},\quad \tan^2 AC = \frac{(-1, 4)(0, 2)}{(2, 4)},\quad \tan^2 AB = \frac{(-1, 4)}{(1, 4)},$$

$$\tan^2 CAD = \frac{(0, 4)}{(0, 2)},\quad \tan^2 BAD = (1, 3)(0, 4),\quad \tan^2 BAC = \frac{(0, 4)}{(2, 4)},$$

$$\tan^2 BDC = \frac{(-1, 3)}{(-1, 1)},\quad \tan^2 ADC = (0, 2)(-1, 3),\quad \tan^2 ADB = \frac{(-1, 3)}{(1, 3)}.$$

In particular, the orthoscheme ABCD of Fig. 4, being of type $(\pi/4, \gamma, \gamma)$, satisfies conditions (5) in the form

$$(-1, 1)(0, 2) = 2,\quad (0, 2)(1, 3) = (1, 3)(2, 4) = m,$$

where $m = \sec^2\gamma$, making it natural to choose $(-1, 1) = 2, (0, 2) = (2, 4) = 1, (1, 3) = m$, so that the triangular pattern is

$$\begin{array}{ccccccccc}
1 && 1 && 1 && 1 && 1 \\
& 2 && 1 && m && 1 & \\
&& 1 && m-1 && m-1 && \\
&&& m-2 && m-2 &&& \\
&&&& m-3 &&&&
\end{array}$$

and the "parts" are given by

$$\tan^2 AD = (m-1)(m-3), \quad \tan^2 BD = \frac{m(m-3)}{2}, \quad \tan^2 CD = m-3,$$

$$\tan^2 BC = \frac{m-3}{2}, \qquad \tan^2 AC = m-3, \qquad \tan^2 AB = \frac{m-3}{m-1},$$

$$\tan^2 CAD = m-2, \qquad \tan^2 BAD = m(m-2), \quad \tan^2 BAC = m-2,$$

$$\tan^2 BDC = \frac{m-2}{2}, \qquad \tan^2 ADC = m-2, \qquad \tan^2 ADB = \frac{m-2}{m}.$$

Similarly, the orthoscheme ABEC, which is one-third of ABCD, is of type $(\pi/3, \pi/4, \gamma)$, and we have

$$(-1,1)(0,2) = 4, \quad (0,2)(1,3) = 2, \quad (1,3)(2,4) = m.$$

Now we choose $(-1,1) = (0,2) = 2$, $(1,3) = 1$, $(2,4) = m$, so that the pattern is

$$
\begin{array}{ccccccccc}
1 & & 1 & & 1 & & & 1 & & 1 \\
& 2 & & 2 & & & 1 & & m & \\
& & 3 & & 1 & & & m-1 & & \\
& & & 1 & & & m-2 & & & \\
& & & & m-3 & & & & &
\end{array}
$$

and the parts (with C replaced by E, and D by C) are given by

$$\tan^2 AC = m-3, \qquad \tan^2 BC = \frac{m-3}{2}, \qquad \tan^2 CE = \frac{m-3}{3},$$

$$\tan^2 BE = \frac{m-3}{2m}, \qquad \tan^2 AE = \frac{m-3}{2m}, \qquad \tan^2 AB = \frac{m-3}{m-1},$$

$$\tan^2 CAE = \frac{m-2}{2}, \qquad \tan^2 BAC = m-2, \qquad \tan^2 BAE = \frac{m-2}{m},$$

$$\tan^2 BCE = \frac{1}{2}, \qquad \tan^2 ACE = 2, \qquad \tan^2 ACB = 1.$$

7. LOBACHEVSKY'S "PYRAMID"

So far, we have assumed the geometry to be spherical (or elliptic), so that $\sin \alpha \sin \gamma > \cos \beta$ [12, p. 16; 5, p. 258; 6, p. 248]. Since, by equations (5),

$$\sin^2 \alpha \sin^2 \gamma - \cos^2 \beta$$

$$= \frac{(1,4)}{(1,3)(2,4)} \frac{(-1,2)}{(-1,1)(0,2)} - \frac{1}{(0,2)(1,3)} = \frac{(-1,4)(1,2)}{(1,3)(2,4)(-1,1)(0,2)} = (-1,4)\cos^2 \alpha \cos^2 \gamma, \quad (9)$$

this condition is equivalent to $(-1,4) > 0$. In fact, the same formulae remain valid for *hyperbolic* space if we allow $(-1,4)$ to be negative and make the natural adjustments, that is, divide the pure-imaginary edge-lengths by i and use the conversion

$$\sec ix = \operatorname{sech} x, \quad \tan ix = i \tanh x.$$

Thus, the hyperbolic orthoscheme (α, β, γ) has the same triangular pattern, derived from conditions (5), but now the rule for computing successive rows automatically leads to a negative value for $(-1, 4)$, and the six edges are given by

$$\tanh^2 AD = -(-1, 4)(0, 3), \quad \tanh^2 BD = \frac{-(-1, 4)(1, 3)}{(-1, 1)}, \quad \tanh^2 CD = \frac{-(-1, 4)}{(-1, 2)},$$

$$\tanh^2 BC = \frac{-(-1, 4)}{(-1, 1)(2, 4)}, \quad \tanh^2 AC = \frac{-(-1, 4)(0, 2)}{(2, 4)}, \quad \tanh^2 AB = \frac{-(-1, 4)}{(1, 4)},$$

while the remaining "parts" are the same as in the spherical case.

One more restriction must be applied to the 2-digit symbols to ensure that the orthoscheme ABCD has a finite volume: the dihedral angles at D and A must be the angles of spherical or horospherical triangles, that is,

$$\alpha + \beta \geqslant \pi/2 \quad \text{and} \quad \beta + \gamma \geqslant \pi/2.$$

Thus,

$$\cos^2 \alpha + \cos^2 \beta \leqslant 1 \quad \text{and} \quad \cos^2 \beta + \cos^2 \gamma \leqslant 1.$$

The latter inequality is equivalent to

$$\frac{1}{(0, 2)(1, 3)} + \frac{1}{(1, 3)(2, 4)} \leqslant 1$$

or

$$(0, 2) + (2, 4) \leqslant (0, 2)(1, 3)(2, 4) = (0, 2)\{(1, 2)(3, 4) + (2, 3)(1, 4)\}$$

or

$$(2, 4) \leqslant (0, 2)(1, 4) = (1, 2)(0, 4) + (0, 1)(2, 4)$$

or

$$(0, 4) \geqslant 0. \tag{10}$$

Similarly, the inequality $\cos^2 \alpha + \cos^2 \beta \leqslant 1$ is equivalent to

$$(-1, 3) \geqslant 0. \tag{11}$$

The following expressions for edges all involve $(0, 4)$ or $(-1, 3)$ or both:

$$\operatorname{sech}^2 AD = (-1, 3)(0, 4), \quad \operatorname{sech}^2 BD = (-1, 3)(1, 4), \quad \operatorname{sech}^2 CD = (-1, 3)(2, 4),$$

$$\operatorname{sech}^2 AC = (-1, 2)(0, 4), \quad \operatorname{sech}^2 AB = (-1, 1)(0, 4).$$

Since $\operatorname{sech} \infty = 0$, we conclude that the hyperbolic orthoscheme is *asymptotic* if $(-1, 3)(0, 4) = 0$. More precisely, A *is at infinity of* $(0, 4) = 0$ *and* D *is at infinity if* $(-1, 3) = 0$.

In particular, if $m = 2$, that is, if $\alpha = \pi/4$, the orthoscheme $(\alpha, \alpha, \pi/4)$ is doubly asymptotic (with both A and D at infinity) while $(\alpha, \pi/4, \pi/3)$ is singly asymptotic (with A *or* D at infinity), in agreement with the dissection equation (1).

The hyperbolic orthoscheme was investigated (though not named), long before Schläfli, by Lobachevsky [14, pp. 92–97] who called it simply a "pyramid." For comparison with his treatment, note that we have interchanged his B and C. (He evidently wished the right angle in the "base" ABC to be at C, as in most treatments of trigonometry.) As for the "parts", his

$$a, \quad b, \quad c, \quad h, \quad q, \quad r, \quad \alpha', \quad \beta', \quad \omega, \quad B, \quad \tfrac{1}{2}\pi - \xi, \quad \mu$$

are our

$$BC, \quad AB, \quad AC, \quad CD, \quad BD, \quad AD, \quad BDC, \quad ADB, \quad CAD, \quad \alpha, \quad \beta, \quad \gamma.$$

Moreover, his equation (156) shows that his δ is given by

$$\tan \delta = \cos h' \tan \mathbf{B} = \tanh CD \tan \alpha = \sqrt{\frac{-(-1,4)}{(-1,2)}} \sqrt{(-1,2)}$$

$$= \sqrt{-(-1,4)}. \tag{12}$$

In terms of the "Lobachevsky function"

$$L(x) = \int_0^x \log \sec \theta \, d\theta = x \log 2 + \frac{1}{2} \sum_1^\infty \frac{(-1)^n}{n^2} \sin 2nx \tag{13}$$

[14, pp. 53–54 with a minus sign corrected], he found that the volume P of the orthoscheme (α, β, γ) is given by

$$4P = L\left(\frac{\pi}{2} - \alpha + \delta\right) - L\left(\frac{\pi}{2} - \alpha - \delta\right) - L(\beta + \delta) + L(\beta - \delta)$$

$$+ L\left(\frac{\pi}{2} - \gamma + \delta\right) - L\left(\frac{\pi}{2} - \gamma - \delta\right) - 2L(\delta). \tag{14}$$

It follows from equation (1) that, if $\tan^2 \gamma + \tan^2 \delta = 2$ (arising from $\sec^2 \gamma = m$ and $\tan^2 \delta = 3 - m$), we have

$$2L\left(\frac{\pi}{2} - \gamma + \delta\right) - 2L\left(\frac{\pi}{2} - \gamma - \delta\right) + L(\gamma + \delta) - L(\gamma - \delta)$$

$$= 4L(\delta) + 4L\left(\frac{\pi}{4} + \delta\right) - 4L\left(\frac{\pi}{4} - \delta\right) - 3L\left(\frac{\pi}{6} + \delta\right) + 3L\left(\frac{\pi}{6} - \delta\right).$$

In particular, by setting $\gamma = \delta = \pi/4$ and recalling that $L(\pi/2) = (\pi/2)\log 2$, we see that

$$4L\left(\frac{\pi}{4}\right) - 3L\left(\frac{5\pi}{12}\right) - 3L\left(\frac{\pi}{12}\right) + \frac{\pi}{2}\log 2 = 0$$

[12, p. 15]. By definition (13),

$$L\left(\frac{\pi}{4}\right) = \frac{\pi}{4}\log 2 + \frac{1}{2}\sum_1^\infty \frac{(-1)^n}{n^2} \sin \frac{n\pi}{2}$$

$$= \frac{\pi}{4}\log 2 - \frac{1}{2}\left(1 - \frac{1}{3^2} + \frac{1}{5^2} - \frac{1}{7^2} + \cdots\right),$$

and the volume P of the "symmetrical" orthoscheme $(\pi/4, \pi/4, \pi/4)$ is given by

$$4P = \frac{\pi}{2}\log 2 - 2L\left(\frac{\pi}{4}\right) = 1 - \frac{1}{3^2} + \frac{1}{5^2} - \frac{1}{7^2} + \cdots = 0.915965\ldots$$

[15, pp. 473, 479; 12, p. 20].

8. TRISECTING A EUCLIDEAN ORTHOSCHEME

By regarding Euclidean 3-space as a 3-sphere of infinite radius, we can justify Schläfli's assertion [6, p. 156] that a Euclidean orthoscheme (α, β, γ) satisfies

$$\sin \alpha \sin \gamma = \cos \beta, \tag{15}$$

whence, by equation (9),

$$(-1, 4) = 0. \tag{16}$$

The expressions in equations (8) for the edges are all zero, but the "triangular pattern" yields numbers proportional to the Euclidean lengths if we delete the factor $(-1, 4)$ from each of the expressions for $\tan^2 AD$, etc. Thus, we can write

$$AD = \sqrt{(0, 3)}, \qquad BD = \sqrt{\frac{(1, 3)}{(-1, 1)}}, \quad CD = 1/\sqrt{(-1, 2)},$$

$$BC = 1/\sqrt{(-1, 1)(2, 4)}, \quad AC = \sqrt{\frac{(0, 2)}{(2, 4)}}, \qquad AB = 1/\sqrt{(1, 4)},$$

while the remaining "parts" are the same as in the spherical case. Since the right-angled triangles yield complementary angles, we have $\tan CAD \tan ADC = 1$ and so on, in agreement with $(-1, 3)(0, 4) = 1$, which comes from equation (7) with $t = 0$, $u = 3$.

The "one degree of freedom" can now be used to choose

$$(-1, 3) = (0, 4) = 1 \tag{17}$$

and to exhibit the "triangular pattern" as a fragment of an infinite frieze pattern

$(-1, 0)$	$(0, 1)$	$(1, 2)$	$(2, 3)$	$(3, 4)$	$\cdots$
$(-1, 1)$	$(0, 2)$	$(1, 3)$	$(2, 4)$	$(3, 5)$	
$(-1, 2)$	$(0, 3)$	$(1, 4)$	$(2, 5)$	$(3, 6)$	
$\cdots$ $(-1, 3)$	$(0, 4)$	$(1, 5)$	$(2, 6)$	$(3, 7)$	

in which both the first and last rows consist entirely of ones [13, p. 22]. The basic equation (2) determines a periodicity

$$(s, t) = (t, s + 5) = (s + 5, t + 5), \tag{18}$$

which makes the frieze symmetrical by a glide-reflection: it is of type **1g** in Senechal's notation [16]. In fact [13, p. 56], by equation (3) with $u = s + 1$ and $v = s + 5$, since $(s + 1, s + 5) = 1$ and $(s, s + 5) = 0$, we have

$$(s, t) = (s, s + 1)(t, s + 5) = (t, s + 5).$$

It follows that $(-1, 1) = (1, 4)$, $(-1, 2) = (2, 4)$, and the "parts" of the Euclidean orthoscheme are given by

$$AD^2 = (0, 3), \qquad BD^2 = \frac{(1, 3)}{(1, 4)}, \qquad CD^2 = \frac{1}{(2, 4)},$$

$$BC^2 = \frac{1}{(1, 4)(2, 4)}, \qquad AC^2 = \frac{(0, 2)}{(2, 4)}, \qquad AB^2 = \frac{1}{(1, 4)},$$

$$\tan^2 \alpha = (2, 4), \qquad \tan^2 \beta = (0, 3), \qquad \tan^2 \gamma = (1, 4),$$

$$\tan^2 CAD = \frac{1}{(0, 2)}, \qquad \tan^2 BAD = (1, 3)(0, 4), \quad \tan^2 BAC = \frac{1}{(2, 4)},$$

$$\tan^2 BDC = \frac{1}{(1, 4)}, \qquad \tan^2 ADC = (0, 2), \qquad \tan^2 ADB = \frac{1}{(1, 3)}.$$

In particular, the orthoschemes $(\pi/4, \gamma, \gamma)$ and $(\pi/3, \pi/4, \gamma)$ are Euclidean if $\sec^2 \gamma = m = 3$, so that $\sec 2\gamma = -3$ and $\gamma \approx 54°44'8''$ (see Section 5). It is remarkable that both the triangular patterns

1	1	1	1	1	and	1	1	1	1	1
2	1	3	1			2	2	1	3	
1	2	2				3	1	2		
1	1					1	1			
0						0				

are fragments of the frieze pattern displayed in Section 1. In the case of $(\pi/4, \gamma, \gamma)$, the three perpendicular edges are

$$AB = \sqrt{\frac{1}{(1,4)}} = \sqrt{\frac{1}{2}}, \quad BC = \sqrt{\frac{1}{(1,4)(2,4)}} = \sqrt{\frac{1}{2}}, \quad CD = \sqrt{\frac{1}{(2,4)}} = 1.$$

Trisecting this orthoscheme, we obtain ABEC, of type $(\pi/3, \pi/4, \gamma)$. Using the second triangular pattern, we find the perpendicular edges

$$AB = \sqrt{\frac{1}{(1,4)}} = \sqrt{\frac{1}{2}}, \quad BE = \sqrt{\frac{1}{(1,4)(2,4)}} = \sqrt{\frac{1}{6}}, \quad EC = \sqrt{\frac{1}{(2,4)}} = \sqrt{\frac{1}{3}}.$$

Since the volume of a Euclidean orthoscheme is one-sixth of the product of its three perpendicular edges, we are not surprised to see that

$$BC \times CD = 3BE \times EC.$$

For convenience in constructing a model, we can multiply by $\sqrt{2}$, so that

$$AB = BC = 1, \quad CD = \sqrt{2}, \quad BE = \sqrt{\frac{1}{3}}, \quad EC = \sqrt{\frac{2}{3}}.$$

The only essentially different triangular pattern contained in the frieze of Section 1 is

$$
\begin{array}{ccccccccc}
1 & & 1 & & 1 & & 1 & & 1 \\
& 1 & & 2 & & 2 & & 1 & \\
& & 1 & & 3 & & 1 & & \\
& & & 1 & & 1 & & & \\
& & & & 0 & & & &
\end{array}
$$

This describes the "cubic" orthoscheme $(\pi/4, \pi/3, \pi/4)$ in which $AB = BC = CD$ [13, p. 22]. The planes joining one diameter of a cube to the other three diameters, in turn, dissect the cube into six such orthoschemes: three *dextro* and three *laevo*. Moorhouse [17] noticed that two such orthoschemes, directly congruent (i.e. both *dextro* or both *laevo*), can be so placed that their $4 + 4$ vertices belong to the cube. They then provide a solution for Möbius's problem of constructing a pair of "mutually inscribed" tetrahedra, the four vertices of each lying on the four face-planes of the other [18, p. 258].

9. A CHALLENGING DEFINITE INTEGRAL

Returning to the subject of Section 3, consider the dissection of the spherical orthoscheme $(\pi/3, \pi/3, \pi/3)$ into three pieces (only two of them congruent). Since now

$$\sec^2 \alpha = \sec^2 \beta = \sec^2 \gamma = 4,$$

the appropriate triangular pattern is

$$
\begin{array}{ccccccccc}
1 & & 1 & & 1 & & 1 & & 1 \\
& 2 & & 2 & & 2 & & 2 & \\
& & 3 & & 3 & & 3 & & \\
& & & 4 & & 4 & & & \\
& & & & 5 & & & &
\end{array}
$$

Looking again at Fig. 2, we see that, since now $\beta = \gamma$, the oppositely congruent pieces ABEC and AFEC are of type $(\sigma, \pi/4, \pi/3)$, where $\sigma = \sigma_1 = \sigma_2$, while DFEC is of type $(\rho, \pi/3, \pi/3)$. (We will determine ρ and σ later.)

The triangular patterns for $(\rho, \pi/3, \pi/3)$ and $(\sigma, \pi/4, \pi/3)$ are

$$
\begin{array}{ccccc}
1 & 1 & 1 & 1 & 1 \\
m & & 2 & 2 & 2 \\
2m-1 & & 3 & 3 & \\
3m-2 & & 4 & & \\
4m-3 & & & &
\end{array}
$$

$$
\begin{array}{ccccc}
1 & 1 & 1 & 1 & 1 \\
n & & 1 & 2 & 2 \\
n-1 & & 1 & 3 & \\
n-2 & & 1 & & \\
n-3 & & & &
\end{array}
$$

where $m = \frac{1}{2}\sec^2 \rho$ and $n = \sec^2 \sigma$. Using equations (8) with ABCD changed to ABEC or DFEC, we see that the length a of the common edge EC is given by

$$
\tan^2 a = \frac{(-1,4)}{(-1,2)} = \frac{4m-3}{2m-1} = \frac{2\tan^2 \rho - 1}{\tan^2 \rho} = 2 - \cot^2 \rho
$$

and also

$$
\tan^2 a = \frac{(-1,4)}{(-1,2)} = \frac{n-3}{n-1} = \frac{\tan^2 \sigma - 2}{\tan^2 \sigma} = 1 - 2\cot^2 \sigma.
$$

When $a < \pi/4$, it is convenient to express ρ and σ in terms of

$$
x = \sec 2a = \sec 2\rho - 2 = \sec^2 \sigma - 2, \tag{19}
$$

so that

$$
\rho = \tfrac{1}{2}\sec^{-1}(x+2), \quad \sigma = \sec^{-1}\sqrt{x+2}
$$

and, if we allow the shape of the orthoschemes to change continuously,

$$
2\,d\rho = dx/(x+2)\sqrt{(x+1)(x+3)}, \quad 2\,d\sigma = dx/(x+2)\sqrt{x+1}.
$$

By equation (15), $(\rho, \pi/3, \pi/3)$ is Euclidean when $\sin \rho = \sqrt{\tfrac{1}{3}}$, so that $x = \sec 2\rho - 2 = 1$; and $(\sigma, \pi/4, \pi/3)$ is Euclidean when $\sin \sigma = \sqrt{\tfrac{2}{3}}$, so that $x = \sec^2 \sigma - 2 = 1$ again.

Schläfli [6, pp. 157, 180] showed that, when α varies while β and γ remain constant, the volume of (α, β, γ) has the differential

$$
df(\alpha, \beta, \gamma) = (4/\pi^2)a\,d\alpha,
$$

where a is the length of the edge where the variable dihedral angle α occurs. In the present case, $d\alpha$ is $d\rho$ or $d\sigma$, so, using equations (19), we have

$$
f(\rho, \pi/3, \pi/3) = \frac{1}{\pi^2} \int_1^{\sec 2a} \frac{\sec^{-1} x\,dx}{(x+2)\sqrt{(x+1)(x+3)}}
$$

and

$$
f(\sigma, \pi/4, \pi/3) = \frac{1}{\pi^2} \int_1^{\sec 2a} \frac{\sec^{-1} x\,dx}{(x+2)\sqrt{x+1}}.
$$

When these orthoschemes arise as pieces of $(\pi/3, \pi/3, \pi/3)$, we have not only $\sec 2\rho = \sec^2 \sigma$ (or $\cos 2\rho = \cos^2 \sigma$) but also, as we see in Fig. 2,

$$
\rho + 2\sigma = \pi,
$$

whence $\sec 2\rho = \sec^2 \sigma = 8$ and $\sec 2a = 6$.

On the other hand, the volume of $(\pi/3, \pi/3, \pi/3)$ itself is known to be $2\pi^2/5!$ because this "symmetrical" orthoscheme is the fundamental region for the symmetric group $S_5 \cong [3, 3, 3]$, which is the symmetry group of the regular simplex $\{3, 3, 3\}$ [19, p. 573]. Hence, in terms of the orthant $(\pi/2, \pi/2, \pi/2)$ as unit,

$$f(\pi/3, \pi/3, \pi/3) = 2^4/5! = 2/15,$$

and the decomposition $f(\pi/3, \pi/3, \pi/3) = f(\rho, \pi/3, \pi/3) + 2f(\sigma, \pi/4, \pi/3)$ yields the surprising identity

$$\int_1^6 \frac{\sec^{-1} x}{(x + 2)\sqrt{x + 1}} \left(\frac{1}{\sqrt{x + 3}} + 2 \right) dx = \frac{2}{15} \pi^2$$

[20].

REFERENCES

1. J. H. Conway and H. S. M. Coxeter, Triangulated polygons and frieze patterns. *Mathl Gaz.* **57,** 87–94, 175–186 (1973).
2. H. S. M. Coxeter, *Non-Euclidean Geometry* (5th edn). Univ. of Toronto Press, Toronto (1965).
3. H. G. Forder, *The Foundations of Euclidean Geometry.* Dover, New York (1958).
4. P. H. Schoute, *Mehrdimensionale Geometrie I.* Göschen, Leipzig (1902).
5. L. Schläfli, *Gesammelte Mathematische Abhandlungen I.* Birkhäuser, Basel (1950).
6. L. Schläfli, *Gesammelte Mathematische Abhandlungen II.* Birkhäuser, Basel (1953).
7. H. S. M. Coxeter, Regular and semi-regular polytopes III. *Math. Z.* (in press).
8. F. Bachmann, *Aufbau der Geometrie aus dem Spiegelungsbegriff* (2nd edn). Springer, Berlin (1973).
9. W. A. Wythoff, The rule of Neper in the four-dimensional space. *K. Akad. Wetensch. Amsterdam, Proc. Sect. Sci.* **9,** 529–534 (1906).
10. J. Böhm, Untersuchung des Simplexinhaltes in Räumen konstanter Krümmung beliebiger Dimension. *J. reine angew. Math.* **202,** 16–51 (1959).
11. H. S. M. Coxeter, On Schläfli's generalization of Napier's pentagramma mirificum. *Bull. Calcutta math. Soc.* **28,** 123–144 (1936).
12. H. S. M. Coxeter, *Twelve Geometric Essays.* Southern Illinois Univ. Press, Carbondale (1968).
13. H. S. M. Coxeter, *Regular Complex Polytopes.* Cambridge Univ. Press, Cambridge (1974).
14. N. I. Lobachevsky, *Imaginäre Geometrie und Anwendung der imaginäre Geometrie auf einige Integrale.* Teubner, Leipzig (1904).
15. T. J. I'a. Bromwich, *An Introduction to the Theory of Infinite Series.* Macmillan, London (1908).
16. M. Senechal, Point groups and color symmetry. *Z. Kristallogr.* **142,** 1–23 (1975).
17. G. E. Moorhouse, Problem 917. *Crux Math.* **11,** 99–100 (1985).
18. H. S. M. Coxeter, *Introduction to Geometry* (2nd edn). Wiley, New York (1969).
19. H. S. M. Coxeter, Regular and semi-regular polytopes II. *Math. Z.* **188,** 559–591 (1985).
20. H. S. M. Coxeter, A challenging definite integral. *Am. math. Mon.* **95,** 330 (1988).

Computers Math. Applic. Vol. 17, No. 1–3, pp. 73–88, 1989
Printed in Great Britain

0097-4943/89 $3.00 + 0.00
Pergamon Press plc

DUALITY AND THE DESCARTES DEFICIENCY

P. Hilton

Department of Mathematical Sciences, SUNY Binghamton, Binghamton, NY 13901, U.S.A.

J. Pedersen

Department of Mathematics, Santa Clara University, Santa Clara, CA 95053, U.S.A.

Abstract—Duality is a fundamental symmetry principle in mathematics, appearing in linear algebra, functional analysis, geometry and topology. We describe here one very important aspect of duality which links geometry and topology. Descartes defined the angular deficiency Δ of a convex rectilinear surface homeomorphic to the sphere and showed that $\Delta = 4\pi$. Grünbaum and Shephard recently defined a dual geometric invariant Δ' for such a surface and showed that $\Delta' = 4\pi$. We give a combinatorial interpretation of Δ for any (two-dimensional) polyhedron and deduce, from general duality considerations, an interpretation of Δ'. It then follows that $\Delta = \Delta'$ for any closed rectilinear surface and that $\Delta' = 2\pi\chi$, where χ is the famous Euler characteristic ($\chi = V - E + F$), for any polyhedron. This yields a new insight into the relation $\Delta = 2\pi\chi$, proved by the authors for any closed rectilinear surface (see Ref. [2]).

1. INTRODUCTION

Duality is a concept which features in many parts of mathematics. It is particularly prominent in linear algebra, where—as you may know, *but need not trouble if you do not*—we associate with every vector space V (over the reals $\mathscr{R}$, say) its dual space V'; and with every linear transformation F: $V \to W$ its dual transformation $F': W' \to V'$. Here V' is the set of linear functionals from V to $\mathscr{R}$ and if φ is a linear functional from W to $\mathscr{R}$ then $F'(\varphi)$, as a linear functional from V to $\mathscr{R}$, is given by

$$F'(\varphi)(v) = \varphi F(v), \quad v \in V.$$

Whether or not you are familiar with this prototype of a *duality operator*, (you will not need to be to appreciate what follows!) we would wish to emphasize two points from the above description:

(i) A *reversal* takes place; thus, whereas F goes from V to W, F' goes from W' to V'.

(ii) If V is finite-dimensional, then V' is isomorphic to V and V'' is equal to V.

We may think of duality as a kind of *reflection*. That is, it produces what we may call a mirror-image, in which "arrows" get reversed. Moreover, the image is somehow equivalent to the original, though different; while, if we repeat the reflection process, we get back exactly what we started with.

Duality thus represents a basic aspect of symmetry in mathematics, and so merits attention in this symposium. It appears conspicuously in topology in the study of *combinatorial manifolds*, that is, in that area of topology seen by many to be its prime ingredient. In this article we will only discuss *two-dimensional* manifolds, or *surfaces*; however, we emphasize that dimensionality is an *intrinsic* property of our geometrical configurations, so that our surfaces will usually be part of—or, as we say, *embedded in*—three-dimensional† space. The *combinatorial* nature of our surfaces is expressed by the fact that they are composed of *entities* (or *cells*) of dimensions 0, 1 and 2, described, respectively, as *vertices*, *edges* and *faces*. Thus, a given surface may admit several different combinatorial structures and each structure will be regarded as combinatorially distinct. You should thus be warned that surfaces which we regard as *the same* (i.e. combinatorially equivalent), you may hitherto have regarded as different [see Fig. 1(a)]; and surfaces which we regard as *different* (i.e. combinatorially distinct), you may hitherto have regarded as the same [see Fig. 1(b)].

---

†Plane geometry is only a small part of two-dimensional geometry!

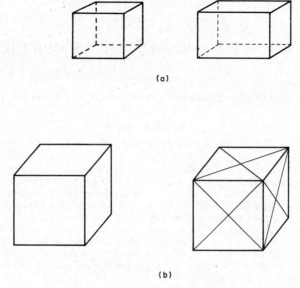

(a)

(b)

Fig. 1. The same or different?

Properties of surfaces are said to be *combinatorially invariant* if combinatorially equivalent surfaces share them in common. Thus, lengths of edges are *not* combinatorial invariants, but the *number* of edges (or of vertices, or of faces) is.

We will be very careful to specify in the subsequent sections of this article just what surfaces we are talking about. Surfaces with a given combinatorial structure are called *polyhedra*, so we will use that term. There is, however, no broad agreement—especially if both geometers and topologists are talking together†—as to just how restrictive the term "polyhedron" should be. We will usually be talking of *closed, connnected, rectilinear surfaces* which we define in the next secton and which we often abbreviate to CCRS. However, we will often permit much more general two-dimensional polyhedra—and we will sometimes refer to the most restrictive definition of all, adopted by many geometers and sanctioned by early history, namely, that a polyhedron is a convex CCRS homeomorphic to the two-dimensional sphere, $\mathbb{S}^2$.

If we confine attention to polyhedra which are CCRS, we may introduce the concept of *dual polyhedra* which we now explain. Basic to the combinatorial structure of even the most general polyhedron is the idea of an *incidence relationship*. A vertex may, or may not, be *incident* with an edge; an edge may, or may not, be *incident* with a face. We write $A \prec \ell \, (\ell \prec \pi)$ if the vertex A is incident with the edge ℓ (if the edge ℓ is incident with the face π). If we are given two polyhedra P and P', and if we may match the vertices, edges and faces of P with the faces, edges and vertices, respectively, of P', in such a way that the incidence relationships are *reversed*, then we say that P and P' are *dual*. This duality thus exhibits the features characteristic of our earlier duality, namely

  (i) A *reversal* takes place; thus entities of P are matched with entities of P' of complementary dimension, and $x \prec y$ in P if and only if $y' \prec x'$ in P' where x is matched with x', and y is matched with y'.

  (ii) If P is a CCRS, then P' is homeomorphic to P and $P'' = P$. (Note that equality here means combinatorial equivalence.)

There is, thus, a built-in symmetry in the study of combinatorial surfaces. In this article we explore only one aspect of that symmetry. Descartes defined the *total angular deficiency* Δ of a convex CCRS P homeomorphic to $\mathbb{S}^2$ and showed that $\Delta(P) = 4\pi$. A scholarly but very readable account of this work of Descartes may be found in Ref. [1]. In Ref. [2] we extended Descartes' definition to an arbitrary CCRS, and showed that $\Delta(P)$, so defined, was a combinatorial invariant. In fact, so restricted, it is a topological invariant since (Theorem 4) we then have $\Delta(P) = 2\pi\chi(P)$,

†PH is a topologist; JP is a geometer.

where χ is the famous *Euler characteristic*, known to be topologically invariant. However, we may observe (see Ref. [3]) that Δ is a combinatorial invariant in a more general domain of discourse than that in which Theorem 4 holds. Thus, there is a ready-made *dual* concept $\Delta'(P)$, and, since $\chi'(P) = \chi(P)$, we must have $\Delta'(P) = \Delta(P)$ for any CCRS P. Since $\Delta(P)$ was originally defined very geometrically, using angular measure, it is natural, indeed irresistible, to seek a geometrical interpretation of $\Delta'(P)$. This was first provided in a very illuminating article by Grünbaum and Shephard [4], only in the case in which P is homeomorphic to $\mathbb{S}^2$. We first extend their interpretation to an arbitrary orientable CCRS P; that is, to any CCRS P embedded in three-dimensional space. Then we give a purely combinatorial interpretation of $\Delta'(P)$, consistent with its geometric interpretation for an orientable CCRS P, consistent with the duality, valid for any two-dimensional polyhedron in the most general sense, and verifying, even in that generality, the relation $\Delta'(P) = 2\pi\chi(P)$.

2. POLYHEDRA

Let us first be precise as to what we will mean by a polyhedron in this article. It is customary to distinguish between a *polygon in the plane* (which is one-dimensional, since it is a closed connected polygonal path, without self-intersections, lying in the two-dimensional plane) and a *polygonal region in the plane* (which is two-dimensional, since it is the union of its one-dimensional boundary and its two-dimensional interior). When we speak of a *polygon* we will mean a one-dimensional figure that lies in a plane and consists of closed, connected, straight sides† hinged together at vertices—and it *does not* include its interior. Similarly, when we speak of a *polyhedron* we will mean a surface (which will usually exist in three-dimensional space) consisting of faces, which are polygonal regions, hinged together at edges. Of course, faces can be hinged together at edges so that they create a surface that partitions three-dimensional space into two pieces, the *inside* and *outside*, so that it then makes sense to talk about points on the *interior* (or *exterior*) of the surface. This, however, is of minor interest to us, since we are only concerned with the *surface* created by the faces and we *do not* include its interior (or its exterior!). Thus, when we speak of a polyhedron we mean a *two-dimensional surface* and not a *solid*. Note, too, that when we speak of a surface, we require that every point have a neighborhood homeomorphic to a disk or—in the case of a boundary point—a semidisk. Thus not all two-dimensional polyhedra in the most general sense are surfaces, for example, the result of joining two cubes at just one vertex would not be a surface.

The most general rectilinear two-dimensional polyhedron considered by topologists—and by us in this article—is a union of polygonal regions (or faces) π, edges ℓ and vertices such that any two of these entities are either disjoint or they intersect in one of these entities; thus, for example, two distinct faces may be disjoint or they may intersect in an edge or a vertex.‡ Figure 2 gives an example of such a polyhedron which certainly would not be regarded as a surface. However, we will adopt this definition as the most general kind of rectilinear two-dimensional polyhedron, so that a statement made about "any polyhedron" will refer to this concept.

If we specialize by insisting that every edge is a side of a face (= polygonal region), we arrive at the idea of a *rectilinear surface*. The Golden Dodecahedron [Fig. 10(b), with each facet triangulated as in Fig 10(d)] is a good—and beautiful—example of a rectilinear surface, but it is not *closed* in the important sense we are about to describe and is therefore excluded from the domain of validity of one of our main theorems (Theorem 4).

The surfaces shown in Figs 3(a), (b), (d) and (e) are closed rectilinear surfaces. Notice that if we think of Fig. 3(d) as having been formed from a cube by adding a square tunnel connecting the front and back faces, as shown in Fig. 3(c), then we see why it was *necessary*, in order to obtain a polyhedron, to add extra edges to the ring-shaped regions that appeared on the front and back

---

†We make an important distinction between the word "side" and the word "edge". If a line joins two adjacent vertices of a polygon we call that line segment a *side* of the polygon. If a line joins two adjacent vertices of a polyhedron (which we will define shortly) we call that line segment an *edge* of the polyhedron.

‡This last condition is unnecessarily restrictive from the combinatorial viewpoint [for example, the braided Golden Dodecahedron of Fig. 10(b) requires extra edges—indicated by dots in Fig. 10(e)—to satisfy the given condition], but it does not restrict the class of surfaces which we want to consider.

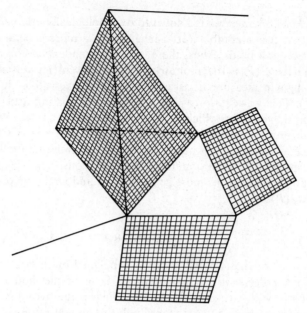

Fig. 2. A two-dimensional rectilinear polyhedron which is definitely not a surface!

of the cube as a result of the tunnel being made. For, otherwise, the ring-shaped regions on the front and back of Fig. 3(c) would not be polygonal regions. Of course, how the new edges are added is not unique—the choice in Fig 3(d) was purely personal and aesthetic—many other possibilities exist. Each of the other two surfaces in Fig. 3 are also clearly defective in some way; Fig. 3(f) is not composed of polygonal faces and the surface in Fig 3(g) has line segments bounding the surface which are not sides of two faces, nor are its "faces"—or *facets*, to distinguish them from true polygonal faces—polygonal regions. We can further subdivide [see Fig. 10(e)] to overcome the second defect, but we will still be left with the first; thus it would then become a rectilinear surface but it would not be *closed*, that is, it would still lack the property that *every edge is a side of exactly two faces*.

It may have surprised you that we allowed the surface shown in Fig 3(e) to be a polyhedron, since it breaks up into two disjoint sets, and might therefore have been thought of as the union of two disjoint polyhedra. If we wish to rule out this possibility, we will require that a polyhedron must be *connected*. This means that the polyhedron is all in one piece; technically, we require that it be possible to get from any vertex to any other vertex on the polyhedron by traversing a sequence of connected edges. So, finally, we arrive at the concept of a CCRS. However, it is worth pointing out that Theorems 2 and 6 do *not* require the connectedness assumption; in fact, we could also dispense with connectedness in the statements of Theorems 4 and 5.

We should also point out that there is one restriction on which we will not insist in this article. It is common among some of the best geometers to require that, in order for a surface to qualify as a polyhedron, it must be possible to create it from a malleable *spherical* ball by "pushing and pulling" it around until its surface consists of flat polygonal faces that correspond with the faces of the surface in question. If this can be done the original surface is said to be *homeomorphic*† *to the two-sphere* $\mathbb{S}^2$. The surfaces shown in Figs 3(a) and (b) satisfy this requirement, and so does the surface in Fig. 3(f)—however (f) is not a *rectilinear* surface because it does not consist of *flat* faces. As you will readily believe, the surface in Fig. 3(c) is *not* homeomorphic to $\mathbb{S}^2$, since it could only have been created by deforming a surface having the shape of an inner tube (or two-dimensional torus $\mathbb{T}^2$), and could not have been created by deforming the surface of a ball. We emphasize again that we do not require that our polyhedra be homeomorphic to $\mathbb{S}^2$. Our definition of a polyhedron includes the surface shown in Fig. 3(d) and, as we will exemplify, it also includes non-convex surfaces as well as non-orientable ones.

---

†More generally, two surfaces are *homeomorphic* if one may be "pushed and pulled" into the shape of the other; any property shared by homeomorphic surfaces is called a *topological invariant* (of surfaces).

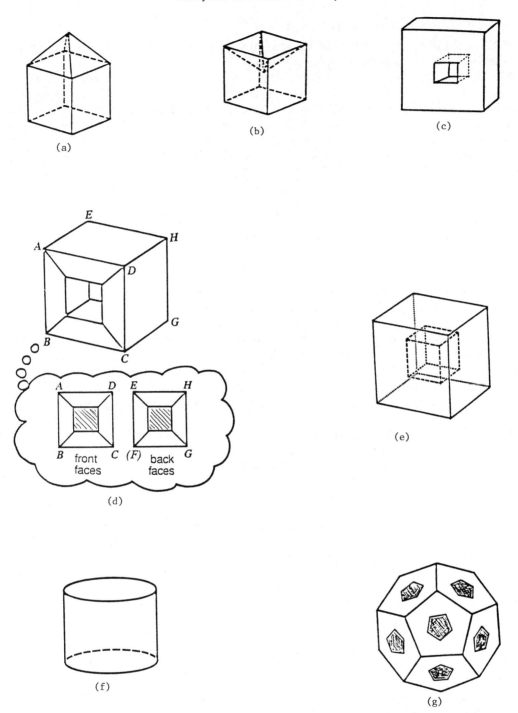

Fig. 3. (a) Box with roof (no attic floor); (b) box with a caved-in roof; (c) cube with tunnel connecting front and back; (d) cube with tunnel as polyhedron; (e) "pregnant cube" (after Pólya); (f) cylinder; (g) regular pentagonal dodecahedron with regular pentagonal holes made in each of its faces.

Notice that our definition of a CCRS includes all of the surfaces (both convex and non-convex) that are homeomorphic to $\mathbb{S}^2$; of course, any polyhedron homeomorphic to $\mathbb{S}^2$ is automatically closed and connected, so it is not necessary to add the connectedness hypothesis if we're content to confine our attention to this restricted class of polyhedra. We are, however, not content to do so! Nevertheless, to begin laying the groundwork for our final results concerning our more general polyhedra, we state a well-known theorem due to Leonhard Euler (1701–1783), concerning these polyhedra.

Theorem 1 (Euler's theorem for polyhedra)

If V, E and F are the number of vertices, edges and faces, respectively for a given convex polyhedron P, which is homeomorphic to $\mathbb{S}^2$, then $V - E + F = 2$.

The reader unfamiliar with this theorem is invited to verify that it is true for the surfaces shown in Figs 3(a) and 6. Observe that the conclusion of Theorem 1 is not claimed (and, in fact, is not true) for the surfaces shown in Figs 3(c)–(e) and 3(g), since those surfaces do not satisfy the hypotheses of the theorem. We do not offer an argument to substantiate Theorem 1 (a heuristic proof may be found in Ref. [5]). A rigorous proof is difficult (one is given in Ref. [6], but only the foolhardy are invited to try to master it). However, we will show, by an elementary argument in Section 5, that Theorem 1 is identical with a very remarkable and even more unintuitive theorem due to Descartes (1596–1650). This latter appears as Theorem 3.

It turns out that the quantity $V - E + F$ is a topological invariant of a two-dimensional polyhedra. It is well-known as the *Euler characteristic* and denoted by χ. Thus,

$$V - E + F = \chi. \tag{1}$$

For example, if the surface is homeomorphic to the torus the value of χ is always 0. You may wish to verify that this is so for the polyhedron shown in Fig. 3(d). It is an unfortunate accident that if the ring-shaped facets on the front and back of Fig. 3(c) are *counted as faces*, then the value obtained for $V - E + F$ is 2 which is, of course, *not* the Euler characteristic of this surface. Indeed, this highly misleading accident leads to many errors in elementary textbooks—surfaces which are given cellular structures which do not satisfy the requirements for a polyhedron are thus erroneously included with those surfaces that are homeomorphic to $\mathbb{S}^2$. Of course, Fig. 3(c) is also homeomorphic to the torus—putting in extra edges doesn't change the topology!

We close this section by introducing the crucial idea of *combinatorial equivalence*. Given two polyhedra P and Q (in the most general sense) we say that P and Q are *combinatorially equivalent*, and write $P \sim Q$, if there is a one–one correspondence ρ between

> the *vertices* of P and the *vertices* of Q
> the *edges* of P and the *edges* of Q
> the *faces* of P and the *faces* of Q

which *preserves the incidence relations*. Here an incidence relation (in P, say) asserts that A is a vertex of the edge ℓ, written $A \prec \ell$, or that ℓ is a side of the face π, written $\ell \prec \pi$. Thus

$$A \prec \ell \Leftrightarrow \rho A \prec \rho \ell$$

$$\ell \prec \pi \Leftrightarrow \rho \ell \prec \rho \pi.$$

Notice, first, that two polygons are combinatorially equivalent if and only if they have the same number of sides, which is as it should be. Notice, next, that combinatorially equivalent polyhedra are certainly homeomorphic, but the converse is utterly false. Thus, if $P \sim Q$ we may "map" P to Q by mapping each entity of P (face, edge or vertex) onto the corresponding entity of Q by a continuous function f which is linear on each entity (see Fig. 4). However, it is plain that the tetrahedron and the cube (see Fig. 6), for example, while being each homeomorphic to $\mathbb{S}^2$ and hence homeomorphic to each other, are certainly not combinatorially equivalent. Indeed, the problem of enumerating all the combinatorial equivalence classes of polyhedra homeomorphic to $\mathbb{S}^2$ remains unsolved to this day.

A property of polyhedra which is shared by all combinatorially equivalent polyhedra is called a *combinatorial invariant*. Thus, c is a combinatorial invariant if $P \sim Q \Rightarrow c(P) = c(Q)$. It is obvious that χ is a combinatorial invariant—for plainly V, E and F are combinatorial invariants. What is, of course, much more interesting and striking is that χ (unlike, V, E and F) is a *topological invariant*, that is, that $\chi(P) = \chi(Q)$ if P and Q are homeomorphic, whether or not $P \sim Q$.

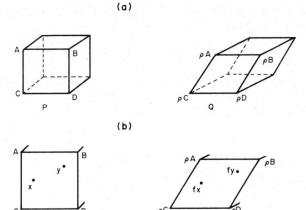

Fig. 4. (a) Combinatorially equivalent polyhedra P and Q, showing the correspondence ρ on one face; (b) extending the correspondance ρ to a continuous function f of the entire polyhedra which is linear on each face.

3. DUAL POLYHEDRA†

What do we mean by the concept of dual polyhedra? We have said that two polyhedra P and P' are *dual* if there exists a one–one correspondence τ between

the *vertices* of P and the *faces* of P'
the *edges* of P and the *edges* of P'

and

the *faces* of P and the *vertices* of P',

which *reverses the incidence relations.* Thus, if A is a vertex, ℓ an edge, and π a face of P, with $A \prec \ell \prec \pi$, then, in P', τA is a face, $\tau \ell$ is an edge, $\tau \pi$ is a vertex, and

$$\tau \pi \prec \tau \ell \prec \tau A.$$

We may call τ the *dualizing symmetry.* Notice that it is completely determined by its values on the *vertices* of P. For if ℓ is the edge of P joining the vertices A and B, then $\tau \ell$ is the common side of the faces τA and τB; and if the face π has vertices A, B, C, ..., then $\tau \pi$ is the (necessarily unique) common vertex of the faces τA, τB, τC, ... Notice, too, that duality is really a relation between (combinatorial) equivalence classes of polyhedra. Thus,

$$P_1 \sim P_2 \Leftrightarrow P_1' \sim P_2';$$

so that we may say that a polyhedron is the dual of its dual, meaning that

$$P \sim P''. \tag{2}$$

This last relation fully justifies us in regarding duality as a particular manifestation of symmetry in mathematics. However, it is especially satisfactory in this connection that we have the following theorem; recall that here a polyhedron is a CCRS.

Theorem 2

Dual polyhedra are homeomorphic.

By introducing the concept of the dual polyhedron, we are able to dualize any combinatorial invariant of polyhedra. Thus, if $c(P)$ is a function defined on the class of polyhedra, with the property that $c(P_1) = c(P_2)$ if $P_1 \sim P_2$, then we may characterize the *dual function* $c'(P)$ by the rule

$$c'(P) = c(P'), \tag{3}$$

provided that P' exists. This will certainly characterize the dual of any clearly combinatorial function c.

—

†In this section a polyhedron is always understood to be a CCRS.

For example, the Euler characteristic $\chi(P) = V - E + F$ is obviously a combinatorial invariant. However, in this case it is plain that $\chi = \chi'$, for

$$\chi'(P) = \chi(P') = V' - E' + F' = F - E + V = \chi(P). \tag{4}$$

Of course, this result is very closely related to Theorem 2, for χ is, in fact, a topological invariant—indeed, it completely characterizes the topological type of orientable polyhedra, that is, those CCRS which can exist in three-dimensional space.

Now we will establish in Section 5 that the Descartes total angular deficiency of a polyhedron, $\Delta(P)$, is a combinatorial invariant [formula (10)]. Thus it has a dual, $\Delta'(P)$. However, it is not entirely satisfactory (to the geometer!) to define a dual invariant by means of equation (3); that is, we would wish to interpret $c'(P)$ directly in terms of the structure of P itself. In Ref. [4], Grünbaum and Shephard propose an elegant direct definition of $\Delta'(P)$, suitable for a polyhedron P homeomorphic to $\mathbb{S}^2$. We will show how to adapt their definition to any orientable polyhedron P; and will then propose a definition [formula (15)] which is far more general in its scope and which clearly exhibits the combinatorial invariance of Δ'.

We close with a very famous example of dual polyhedra (Fig. 5), namely, the hexahedron (cube) H and the octahedron O. Figure 5(a) shows O positioned inside H so that the vertices of O lie at the centers of the faces of H; this defines the dualizing symmetry $\tau: O \rightarrow H$ on the vertices of O and thus completely determines τ. For example, adopting the notation of Fig. 5,

the edge joining vertex $3^\bullet$ and vertex $4^\bullet$ on O must be matched (by the symmetry τ) with the edge joining face $\langle 3 \rangle$ and face $\langle 4 \rangle$ on H.

Likewise,

the vertex surrounded by faces $\langle 1 \rangle$, $\langle 3 \rangle$ and $\langle 4 \rangle$ on H must be matched (by τ^{-1}) with the face surrounded by vertices $1^\bullet$, $3^\bullet$ and $4^\bullet$ on O.

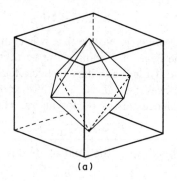

(a)

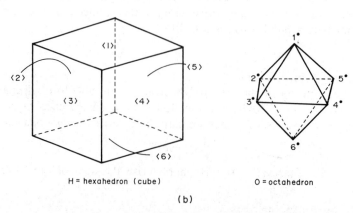

H = hexahedron (cube)   O = octahedron

(b)

Fig. 5. Dual elements are labelled with the same number.

Of course, this example is special. We chose regular models of H and O and adjusted their edge-lengths to fit O into H as described. This enabled us to describe a dualizing symmetry τ very easily and naturally; but it is not necessary that τ be "realizable" in this way. For example, the pair of polyhedra in Figs. 3(a) and 3(b) are equivalent to each other, but they are also dual to each other—they are thus *self-dual*. However, they cannot be embedded in each other or in themselves as drawn (with all edges equal), in the striking manner of H and O.

The regular tetrahedron T is another example of a self-dual polyhedron which, in fact, can be embedded in another regular tetrahedron of suitable edge-length, just as O was embedded in H. Likewise, the regular dodecahedron D and the regular icosahedron I are dual polyhedra related to each other just like H and O above (see Fig 6). Try it for yourselves!

4. DESCARTES ANGULAR DEFICIENCY Δ AND ITS DUAL Δ'

For any polyhedron, in the most general sense, we may take all the faces that come together at a particular vertex and lay them flat so that (a) they leave a gap, (b) they exactly fill up the plane, or (c) they overlap. By way of example, we refer to the regular polyhedra H and O of Fig. 5, and note that for either of those polyhedra, we would get the same arrangement at each vertex, as shown in Figs 7(a) and (b), respectively.

Euclid is credited with the observation that there would always be a gap when the faces surrounding any vertex of a *convex* polyhedron are laid flat in the plane. However, we can observe from the polyhedron in Fig. 1(b) that, although convexity is a sufficient condition for a gap at each vertex, it is not a necessary condition. Descartes (considering only convex polyhedra P) called this

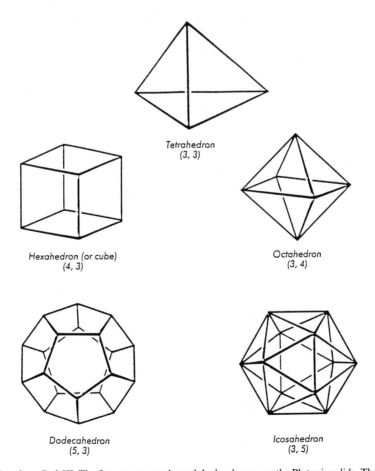

Fig. 6. Taken from Ref. [7]. The five convex regular polyhedra, known as the Platonic solids. The notation (p, q) means that each face is a regular p-gon and q faces come together at each vertex.

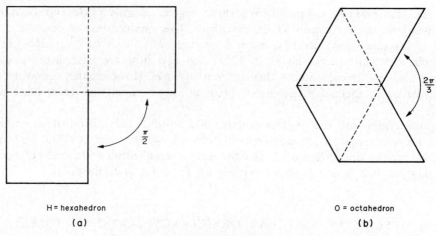

H = hexahedron   O = octahedron

(a)   (b)

Fig. 7. The flattened neighbourhood of a vertex.

gap the *angular deficiency* of P at that particular vertex. If all the angular deficiencies at the individual vertices of a polyhedron P are added together, the resulting quantity $\Delta(P)$ is called the *total angular deficiency of P*. Descartes showed that, if P is homeomorphic to $\mathbb{S}^2$, then $\Delta(P) = 4\pi$. We will shortly show, following the argument in Ref. [5], how to relate this fact to Euler's theorem (Theorem 1). Let us now describe Descartes' result precisely.

We suppose the vertices of the polyhedron P are numbered $1, 2, \ldots, V$; thus we define the angular defect at the ith vertex to be

$$\delta_i = 2\pi - (\text{the sum of the face angles at the } i\text{th vertex}).\tag{5}$$

We then observe that the value of δ_i will be

    (a) positive when there is a gap,
    (b) 0 when the sum of the faces surrounding the ith vertex is exactly 2π,

and

    (c) negative if there is an overlap.

Now defining the total angular deficiency of the polyhedron P to be

$$\Delta(P) = \sum_{i=1}^{V} \delta_i,\tag{6}$$

we may state Descartes' discovery as the following theorem.

Theorem 3 (Descartes' theorem for angular deficiency)

If P is a convex polyhedron homeomorphic to $\mathbb{S}^2$, then

$$\Delta(P) = \sum_{i=1}^{V} \delta_i = 4\pi.$$

This theorem is a fairly deep result (Pólya gives a nice plausibility argument for it in Ref. [5]) and we will not prove it directly. However, as promised, we will show in Section 5, as a special case of a more general result, that Theorem 1 is completely equivalent to Theorem 3. In some sense this equivalence is almost as amazing as the truth of either of the separate results.

Grünbaum and Shephard report (in Ref. [4]) their discovery of a good candidate for the long-hidden dual Δ' of the Descartes deficiency Δ (Theorem 3.1 of Ref. [4]). In their paper they reason that the dual of Δ should tell us something about the "deficiencies of the faces". They then look for a suitable interpretation of that concept. We quote here their definition (which they offer as "one possibility") of the deficiency for a *face* of a convex polyhedron P homeomorphic to $\mathbb{S}^2$:

> "... we shall define $\delta(F, x)$ for a face $F = ABC$... of the convex polyhedron P and for any point x in the interior of P, in the following way. Consider any sphere centered at x, and let $A'B'C'$... be the projection of the face F from x onto this sphere. Let σ be the sum of the angles of F and let σ' be the sum of the angles of the spherical polygon $A'B'C'$.... Then we define the *deficiency* of the face F with respect to the point x as
> $$\delta(F, x) = \sigma' - \sigma. \tag{4}$$
> Then the dual to Descartes' Theorem states that *for all convex 3-dimensional polyhedra P, and all points x in the interior of P,*
> $$\sum \delta(F, x) = 4\pi.$$
> *where the summation is over all the faces of P.*"†

As an example (that requires minimal knowledge of spherical geometry) consider the regular octahedron O (shown in Figs 5 and 6) and locate x at its center. Then a sphere may be constructed about O in such a way that every vertex of O lies on the surface of the sphere. In terms of the earth we may think of the vertices $1^\bullet$ and $6^\bullet$ of Fig. 5(b) as lying on the north and south poles respectively, with the remaining vertices lying on the equator. It is clear that a longitudinal line between the poles crosses the equator at right angles on the globe and, since all vertices of O are surrounded by the same arrangement of identical faces, we see that the projection of the edges of O onto the surface of the sphere produces eight spherical triangles each of which contains *three right spherical angles* (*see* Fig. 8). Thus the defect for one face is

$$3\left(\frac{\pi}{2}\right) - \pi = \frac{\pi}{2},$$

and, summing the defects over all eight faces, we get the expected value of 4π for Δ'.

Grünbaum and Shephard remark that neither Descartes' theorem nor its dual requires that the polyhedron P be convex, provided one is prepared to tolerate negative angular deficiencies at certain vertices of P. A similar remark had appeared in Refs [2, 3].

In Section 5 we will give a simple argument to show that for *any* CCRS $\Delta = \Delta'$. We will then propose a new interpretation of Δ' which agrees with that given by Grünbaum and Shephard for a CCRS homeomorphic to $\mathbb{S}^2$, and satisfies the duality condition (3). In order to obtain our results, and for the sake of completeness, we first give our adaptation of a proof by George Pólya (1887–1985), concerning the equivalence of Theorems 1 and 3.

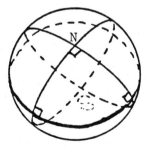

Fig. 8

5. THE RELATIONSHIP BETWEEN χ, Δ AND Δ'

As in Theorem 1, for a given polyhedron P, let V, E, F be the number of vertices, edges and faces, respectively. Likewise, as in Section 4, let the angular deficiency, δ_i, and the total angular deficiency, Δ, be defined by equations (5) and (6), respectively. Furthermore, let

$$S = \text{the total numbers of } sides \text{ for } P,$$

and

$$f_j = \text{the number of } j\text{-gons among } P\text{'s faces.}$$

---

†Recall from the Introduction that we would not regard P as three-dimensional. A polyhedron in Ref. [4] is a convex CCRS homeomorphic to $\mathbb{S}^2$.

Then we know that

$$\sum_{j=3}^{\infty} f_j = F \tag{7}$$

and

$$\sum_{j=3}^{\infty} jf_j = S. \tag{8}$$

We also know that every j-gon contributes $(j-2)\pi$ to the sum of the face angles of P.

Let us now count Ω, the sum of the face angles of P, in two ways:

By faces	By vertices
$\Omega = \sum_{j=3}^{\infty} f_j(j-2)\pi$	$\Omega = \sum_{i=1}^{V} (2\pi - \delta_i) = 2\pi V - \Delta(P),$
	by equation (5).

Thus we have

$$\Delta(P) = 2\pi V - \sum_{j=3}^{\infty} f_j(j-2)\pi.$$

$$= 2\pi V - \pi \sum_{j=3}^{\infty} jf_j + 2\pi \sum_{j=3}^{\infty} f_j. \tag{9}$$

Using equations (7) and (8), we have

$$\Delta(P) = 2\pi V - S\pi + 2\pi F = \pi(2V - S + 2F). \tag{10}$$

So far, this is very general. However, if we now confine attention to those polyhedra P which are CCRS, then we require of P that every edge is the side of exactly two faces. Thus, $S = 2E$, and we obtain the result, for any CCRS P,

$$\Delta(P) = 2\pi(V - E + F) = 2\pi\chi(P).$$

We have now proved that Theorems 1 and 3 are entirely equivalent, but, in fact, we have established a much more general result which may be expressed as follows.

Theorem 4

For any closed connected rectilinear surface P,

$$\Delta(P) = 2\pi\chi(P). \tag{11}$$

As we said, this result, though nothing like as deep as Theorems 1 or 3, is nevertheless very remarkable. For, although Euler's result (Theorem 1) is somewhat surprising, it is quite obvious that χ is a combinatorial invariant.† On the other hand, there is no simple way to see that the sum of all the angular defects of P will remain constant if P is replaced by a combinatorially equivalent polyhedron, for example, if P is pushed and pulled about, since in this circumstance the angular defects at the individual vertices are all liable to change. Thus equation (11) includes the unobvious fact that the *geometric* invariant $\Delta(P)$ is indeed a *combinatorial* invariant.

If we now examine the proof of Theorem 4 (and not merely its conclusion), several significant things are revealed. From equation (9) we see that

    (a) $\Delta(P)$ is completely determined by the number of vertices of P and the number of faces of P of each type,

and

    (b) that $\Delta(P)$ is *always* a multiple of π;

and that these two facts are true for any polyhedron P, not merely a CCRS, and certainly not merely a CCRS homeomorphic to $\mathbb{S}^2$.

---

†In fact, as already stated, it is a topological invariant.

We now turn our attention to Δ'. Grünbaum and Shephard [4] argued that since, summing over faces, $\Sigma \, \sigma' = 2\pi V$, and since $\Sigma \, \sigma$ is simply the sum Ω of the face angles of P, or $2\pi V - \Delta$, it follows that, if P is a CCRS homeomorphic to $\mathbb{S}^2$, then

$$\Delta'(P) = \sum (\sigma' - \sigma)$$
$$= \sum \sigma' - \sum \sigma$$
$$= 2\pi V - [2\pi V - \Delta(P)] = 2\pi V - (2\pi V - 4\pi) = 4\pi. \qquad (12)$$

We point out now that the elegant argument given by Grünbaum and Shephard is applicable to any CCRS embedded in three-dimensional space, provided it is interpreted as simply asserting that $\Delta'(P) = \Delta(P)$. For the only point to be made is that it suffices, in their definition of Δ', to take x to be a general† point not in the surface—after all, a face of a rectilinear surface does not know if the surface is homeomorphic to a sphere! Thus, we claim that the following theorem holds.

Theorem 5

For any closed connected rectilinear surface P embedded in three-dimensional space

$$\Delta(P) = \Delta'(P). \qquad (13)$$

Proof. As above,

$$\Delta'(P) = \sum \sigma' - \sum \sigma = 2\pi V - \Omega;$$

but

$$\Omega = 2\pi V - \Delta(P), \text{ so } \Delta'(P) = \Delta(P).$$

Remark

From our point of view, Theorem 5 is really a vindication of the Grünbaum–Shephard definition rather than a new "fact". For since the dual Δ' *must* satisfy $\Delta'(P) = \Delta(P')$ and since $\chi(P) = \chi(P')$, we have, quite formally,

$$\Delta'(P) = \Delta(P') = 2\pi\chi(P') = 2\pi\chi(P') = 2\pi\chi(P) = \Delta(P).$$

Thus far we have only considered $\Delta'(P)$ where P is a CCRS, so that an edge of P is formed by identifying a side of one face with a side of an adjacent face. If we wish to extend our study of Δ' beyond the case of a CCRS, we must look more closely at the notion dual to that of a side.

What is the dual of the idea of a side? Formally, we may regard a side as a pair (ℓ, π) consisting of an edge ℓ and a face π such that $\ell \prec \pi$. It is plain that the dual notion is a pair (ℓ, A), consisting of an edge ℓ and a vertex A such that $A \prec \ell$. This, however, is the familiar notion of a *ray* emerging from the vertex A. Thus if R is the number of rays of the polyhedron P, then we have the duality relation

$$S(P') = R(P), \qquad (14)$$

asserting that the number of sides of P' is equal to the number of rays of P.

Now equation (10) tells us that $\Delta(P) = \pi(2V - S + 2F)$. Thus

$$\Delta'(P) = \Delta(P') = \pi[2V(P') - S(P') + 2F(P')] = \pi(2F(P) - R(P) + 2V(P)),$$

that is,

$$\Delta'(P) = \pi(2V - R + 2F). \qquad (15)$$

This is our promised new interpretation‡ of Δ'; true, it has the disadvantage that it says nothing about angles, but one must set against this the advantage that it is evidently dual to Δ.

---

† By this we mean that the point is taken in *general position*, so that, when projecting from the point, no degeneration of a face or an edge of P takes place.

‡ Formula (15), regarded as a *definition*, does not require that P' exist!

Moreover, it has another extraordinary advantage, not at all looked for in our search for a dual of Δ. As we have pointed out, the relation $S = 2E$ is special to a CCRS P (actually, connectedness is not required) and it is plain from our proof that this relation is necessary and sufficient for the conclusion of Theorem 4, namely, $\Delta(P) = 2\pi\chi(P)$. Likewise, the relation $R = 2E$ is necessary and sufficient for the dual conclusion, $\Delta'(P) = 2\pi\chi(P)$. However, the relation $R = 2E$ holds for *any* polyhedron P! Thus, while $\Delta(P)$ is a topological invariant only for a CCRS P, we have

Theorem 6

For any two-dimensional polyhedron P,

$$\Delta'(P) = 2\pi\chi(P),$$

and so $\Delta'(P)$ is a topological invariant of P.

6. EXAMPLES

If we define $\Delta'(P)$ by formula (15) to be $\pi(2V - R + 2F)$, then, by Theorem 6, we can certainly extend the validity of the statement $\Delta(P) = \Delta'(P) = 2\pi\chi(P)$ to the broader class of all CCRS, even those which are non-orientable (and so cannot be embedded in ordinary three-dimensional space). As an example we cite the real projective plane as illustrated in Fig. 9. Part (a) of Fig. 9 shows the projective plane drawn in the standard way as a circular disk with diametrically opposite points on the boundary identified. However, to fit our definition of a CCRS, the edges would have to be straightened, as shown in Fig. 9(b). Since the projective plane cannot be regarded as a part of three-dimensional space, it can only be represented on paper with such identifications—they are, of course, not intrinsic to the surface. You may verify that, for the projective plane, $\chi = 1$, and $\Delta = \Delta' = 2\pi$.

We now give an example of a surface that is not closed. Figure 10(a) illustrates one of the famous Platonic solids, the regular pentagonal dodecahedron (see also Fig. 6). The Golden Dodecahedron of Fig. 10(b) may be constructed by braiding together six identical straight strips (complete instructions for how to prepare the strips by simply folding gummed tape appear in Ref. [7]). Although it looks similar to the surface of part (a), it is fundamentally different because it has a pentagonal hole in the center of each of the 12 original faces. We will concern ourselves with the visible surface as shown in Fig. 10(c). If we then introduce new edges on each facet as shown in Fig. 10(d) we obtain a polyhedron 10(e), which we call the Golden Dodecahedron G. This polyhedron is a CRS, that is, it satisfies every condition of being a CCRS except that it is not closed.

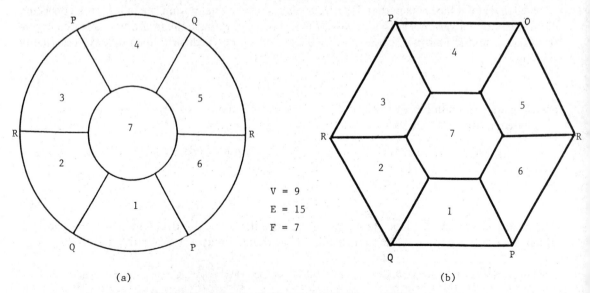

$$V = 9$$
$$E = 15$$
$$F = 7$$

(a)   (b)

Fig. 9. The real projective plane subdivided as a polyhedron.

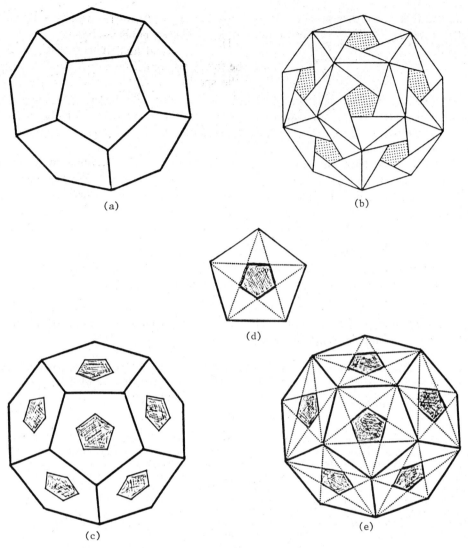

Fig. 10

Let us first compute $\chi(G)$. Notice that if there were no pentagonal holes in the 12 facets of G the value for χ would be 2, by Theorem 1. Thus, since $\chi = V - E + F$ for *any* subdivision of the surface, each time a hole is introduced into a polyhedron χ will be *reduced* by one. Hence $\chi(G) = 2 - 12 = -10$.

We compute $\Delta(G)$ by an appeal to the *meaning* of formula (9) (skeptics may work out the details from the fact that G has 80 vertices and 120 triangular faces). We observe that by removing a pentagon from a facet we *decrease* the face angles which get *subtracted* by 3π; that is, we *increase* Δ by 3π. Hence, since we do this 12 times, we infer that $\Delta(G) = 4\pi + 12(3\pi) = 40\pi$. Notice that we do *not* have the relationship $\Delta(G) = 2\pi\chi(G)$.

The computation of $\Delta'(G)$ comes from formula (15). Since $F = 120$, $R = 420$ and $V = 80$, we have $\Delta'(G) = \pi(240 - 420 + 160) = -20\pi$, which is $2\pi\chi(G)$, as predicted by Theorem 6. That is, we retain $\Delta' = 2\pi\chi$ at the expense of $\Delta' = \Delta$!

Notice that χ is a topological invariant, so that, for example, the value of χ would be the same if we removed triangular holes from each face, instead of pentagonal holes. Indeed, the computation of $\chi(G)$ did not depend on the number of sides of each hole—it only depended on the *number* of holes. The value of χ would be -10, so long as some polygonal hole is introduced into each of the 12 original faces. On the other hand Δ *does* depend on the nature of the holes. For example, if we had removed triangular holes from each of the 12 faces, instead of pentagonal holes, the decrease in the face angles which get subtracted in formula (9) would be π for each hole removed, and this would then give $\Delta = 4\pi + 12\pi = 16\pi$. Thus, Δ is *not* a topological invariant for connected rectilinear surfaces which are not closed—whereas, of course, Δ' is.

7. CONCLUDING REMARKS

We firmly believe that the ideas presented in this article are, and should be, easily accessible to high school students. We do not claim that proofs of Theorems 1–3 should be given to such students; and we are far from believing that we have presented the material exactly as it should be introduced in a teaching situation. However, we do believe that this type of combinatorial geometry is a way—perhaps the best way—to revitalize the teaching of geometry and to place it where it should be in the curriculum.

Explicitly, these ideas have the merits, in addition to their accessibility and intrinsic beauty, of easy verifiability, of surprise, and of natural extension to ideas of fundamental importance in mathematics. To illustrate this last point, we would expect alert students studying these ideas to ask such questions as:

What happens if we drop the condition that all faces be flat and all edges be straight?
What happens if we move beyond two dimensions?

The first question leads us straight (!) into the calculus and differential geometry (there is a beautiful introduction to this development in Ref. [8]); the second question leads us directly into the topology and geometry of higher dimensions.

Acknowledgement—We would like to express our deep appreciation of the very helpful comments of two experts, Branko Grünbaum and Geoffrey Shephard, which have led to substantial improvements in our manuscript.

REFERENCES

1. P. J. Federico, *Descartes on Polyhedra: A Study of the De Solidorum Elementis*. Springer, New York (1982).
2. P. Hilton and J. Pedersen, Descartes, Euler, Poincaré, Pólya and polyhedra. *Enseign. math.* **27**, 327–343 (1981).
3. P. Hilton and J. Pedersen, Discovering, modifying and solving problems: a case study from the contemplation of polyhedra. *Teaching and Learning: A Problem-Solving Focus*, pp. 47–71. NCTM (1987).
4. B. Grünbaum and G. Shephard, A dual for Descartes' theorem on polyhedra. *Math. Gaz.* **71**, 214–216 (1987).
5. G. Pólya, *Mathematical Discovery* (Combined edn) p. 167, Vol. II, pp. 149–156. Wiley, New York (1981).
6. P. Hilton and S. Wylie, *Homology Theory*, p. 167. Cambridge Univ. Press (paperback, 1967).
7. P. Hilton and J. Pedersen, *Build Your Own Polyhedra*. Addison–Wesley, Reading, Mass. (1988).
8. S. S. Chern, From triangles to manifolds. *Am. math. Mon.* **56**(5), 339 (1979).

Computers Math. Applic. Vol. 17, No. 1–3, pp. 89–102, 1989
Printed in Great Britain

0097-4943/89 $3.00 + 0.00
Pergamon Press plc

EXTENDING THE BINOMIAL COEFFICIENTS TO PRESERVE SYMMETRY AND PATTERN

P. Hilton

Department of Mathematical Sciences, SUNY Binghamton, Binghamton, NY 13901, U.S.A.

J. Pedersen

Department of Mathematics, Santa Clara University, Santa Clara, CA 95053, U.S.A.

Abstract—We show how to extend the domain of the binomial coefficients $\binom{n}{r}$ so that n and r may take any integer value. We argue from two directions; on the one hand we wish to preserve symmetry and pattern within Pascal's triangle (thus, creating Pascal's hexagon), and on the other hand we wish the binomial coefficients to preserve their *algebraic* role in terms of the Taylor series and Laurent series expansions of $(1 + x)^n$, valid when $|x| < 1$, $|x| > 1$, respectively.

A geometric configuration within the Pascal triangle, called the Pascal flower, has some extraordinary properties—these properties persist into the hexagon. Moreover, the binomial coefficients may, by the use of the Γ-function, even be extended to all real (or complex) values of n and r, with the conservation of their principal properties.

1. INTRODUCTION

Look at Fig. 1. Observe how the non-zero numbers in the northeast and northwest sectors are related to the non-zero numbers in the southern region. See if you can determine the correct entries for the next ring of numbers round the hexagonal boundary.

The arrangement of the non-zero numbers in the southern region of Fig. 1 is well-known as Pascal's triangle. In this article we will adhere initially to the standard convention of denoting entries in Pascal's triangle by the symbol $\binom{n}{r}$, where n and r are the values of the row and diagonal locations, respectively. Thus, for example, we can read from Fig. 1 that $\binom{5}{2}$ has the value 10. We will, in fact, adopt this notation for the entire array of numbers in Fig. 1 [so that, for example, $\binom{-3}{1} = -3$]. In accordance with the combinatorial interpretation of the symbol $\binom{n}{r}$ when n and r are non-negative integers, we will always read $\binom{n}{r}$ as "n choose r", meaning the number of ways r objects can be selected from a collection of n objects. Of course, when either n or r is a negative integer, this does not have any combinatorial interpretation—but this should not deter us from examining whether or not such symbols can have some sensible interpretation, consistent with the meaning and properties of binomial coefficients in their original, restricted domain of definition. We will attach such a consistent meaning to the entries in Pascal's hexagon (Fig. 2) in Section 2. Meanwhile, we immediately give an important example of what we mean by the preservation of the properties.

A basic property of Pascal's triangle is that all triples of adjacent entries A, B, C positioned in the triangle like this,

$$A* \qquad B*$$

$$C*$$

satisfy what is called the Pascal identity, namely: $A + B = C$. (We express this more explicitly in Section 2.) Notice, then, that this identity also holds elsewhere in the array of numbers in Fig. 1. In fact, there is just one exception—can you find it?

Since the array of numbers in Fig. 1 is an extension of Pascal's triangle we call it the *extended Pascal triangle* or, more simply, the *Pascal hexagon*. Note that the Pascal hexagon can be constructed as follows. We start with the X-shaped array of **1**'s running northwest and northeast in Fig. 1. We then insert new entries (contiguous to existing entries, of course) by simply adhering to the Pascal identity. (This last statement should give you a hint—if you needed it—where to look for the exceptional triple that does not satisfy the Pascal identity.)

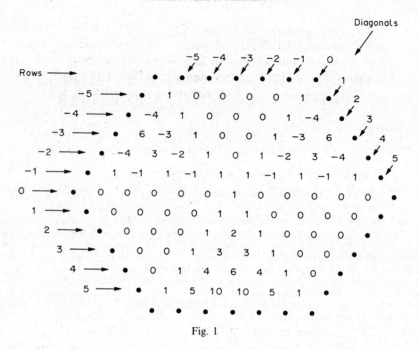

Fig. 1

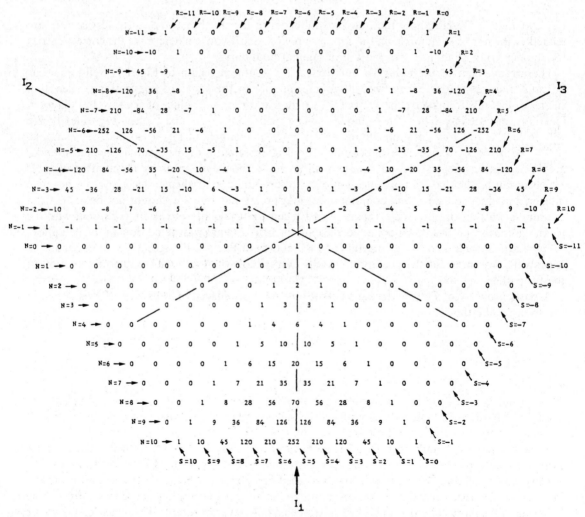

Fig. 2. The Pascal hexagon.

Why should we concern ourselves with the Pascal hexagon? Why should the extra entries be of *any* particular interest? Of course, if you have already had the experience of seeing special relationships between numbers whose locations form patterns in Pascal's triangle (see Refs [1, 2] for examples), then you may be interested to see if the same, or similar, patterns hold in the extended Pascal triangle. In fact, one of the things we do in Sections 3 and 4 is to look at some patterns involving numbers that lie in certain special geometric arrangements within the Pascal hexagon—but if we left our discussion at that point, we would miss an opportunity to develop some lovely mathematics that utilizes both algebraic techniques and symmetry.

In Section 2 we appeal to the algebraic interpretation of the binomial coefficient in order to give important meaning to the entries in the non-zero northeast sector of the Pascal hexagon (which also gives us a way of computing those entries without having to construct any part of the hexagonal array of numbers). In fact, as you will see, it gives much more. Next, our desire to keep the array symmetric about the vertical axis leads us to develop a natural mathematical extension that results in the entries shown in the non-zero northwest sector. Further considerations of symmetry allow us, finally, to complete the hexagon with zero entries. Thus we extend the meaning of $\binom{n}{r}$ over the full range of integers. Moreover, it turns out, not unsurprisingly, that the entries in the northwest region, though determined simply by aesthetic considerations of symmetry and pattern-preservation, have an important significance analogous with the algebraic interpretation of the entries in the northwest region.

Having carried out the extension of $\binom{n}{r}$ to arbitrary integers n, r, we interrupt the discussion which leads to the generalization of the binomial coefficients so that we can examine, in Sections 3 and 4, some remarkable patterns that hold both in the original Pascal triangle and in its extension, the Pascal hexagon.

In Section 5 we return to the mathematical development and show, using the Γ-function $\Gamma(z)$, how the domain of the binomial coefficients may be further extended over the *complex* numbers. We admit that, in general, we have very little opportunity to use an expression like "$\pi + i$ choose e" but we think it is intellectually rewarding to know that $\binom{\pi + i}{e}$ can be *defined*. Moreover, the identities and relationships noticed in the Pascal hexagon continue to be valid in what we may fancifully call the *complex Pascal plane*. We cannot seriously doubt that expressions of the form $\binom{i}{z_2}$ will prove mathematically useful, even if they have not already done so.

2. EXTENDING THE DOMAIN OF THE BINOMIAL COEFFICIENTS OVER THE FULL RANGE OF INTEGERS

In this section we look at two interpretations of $\binom{n}{r}$, where n and r are non-negative integers. We then demonstrate how to define $\binom{n}{r}$, where n is any real number and r is a non-negative integer. After that we restrict n to being an integer and use a symmetry argument to define $\binom{n}{r}$ where r is now an arbitrary integer. Finally, we interpret $\binom{n}{r}$, where r is negative.

At various stages of the arguments that follow the reader may find it helpful to look at a special numerical case. The values of the binomial coefficient may conveniently be read off Fig. 2. Having offered this well-meant advice† let us begin.

We may define the binomial coefficient $\binom{n}{r}$, where n and r are non-negative integers, either *combinatorially* as the number of ways of selecting r objects from a set of n objects, or *algebraically* (and equivalently) as the coefficient of x^r in the binomial expansion of $(1 + x)^n$. So defined, $\binom{n}{r}$ satisfies the following identities.

Symmetry identity:

$$\binom{n}{r} = \binom{n}{n-r}, \quad \text{for } 0 \leqslant r \leqslant n. \tag{1}$$

---

†We have in mind that readers of this volume may well not be mathematical specialists.

Pascal identity:

$$\binom{n}{r} + \binom{n}{r-1} = \binom{n+1}{r}, \quad \text{for } r \geqslant 1. \tag{2}$$

Vanishing identity:

$$\binom{n}{r} = 0, \qquad \text{for } r > n \geqslant 0. \tag{3}$$

In the light of what follows the symmetry identity will also be called the *first reflection identity*.

The algebraic definition above easily extends to allow n to be any real number (while r remains a non-negative integer). For we may consider the formal power series expansion of $(1 + x)^a$, for any real number a and define $\binom{a}{r}$ to be the coefficient of x^r. Notice that identity (1) above does not extend, simply because $\binom{a}{a-r}$ has no meaning unless a is an integer such that $a \geqslant r$. Notice, however, that identity (2) does extend to the following.

Pascal identity:

$$\binom{a}{r} + \binom{a}{r-1} = \binom{a+1}{r}, \quad \text{for } r \geqslant 1. \tag{2'}$$

Notice, finally, that identity (3) is actually *untrue* in the extended domain of definition of the binomial coefficient, that is, if we no longer require $n \geqslant 0$; for example $\binom{-1}{1} = -1$. On the other hand, a new identity makes its appearance, for

$$\binom{-a}{r} = \frac{(-a)(-a-1)\ldots(-a-r+1)}{r!}$$

$$= (-1)^r \frac{(a+r-1)(a+r-2)\ldots a}{r!} = (-1)^r \binom{a+r-1}{r}$$

so that we have the following.

Second reflection identity:

$$\binom{-a}{r} = (-1)^r \binom{a+r-1}{r}, \quad \text{for } r \geqslant 0. \tag{4}$$

We will, until Section 5, confine the domain of the binomial coefficient $\binom{a}{r}$ so that a is an integer. At this point r is a non-negative integer, but we seek to extend the validity of r to the whole set of integers. Thus, we must define $\binom{a}{-r}$, where a is an integer and r is a positive integer.

We are, in fact, guided by the symmetry identity (1) which we will want to be valid over all integer values of the pair n, r. This will oblige us to define

$$\binom{n}{-r} = 0, \qquad \text{for } n \geqslant 0, r \geqslant 1; \tag{5}$$

$$\binom{-n}{-r} = (-1)^{n-r} \binom{r-1}{n-1}, \quad \text{for } r \geqslant n, r \geqslant 1. \tag{6}$$

For

$$\binom{n}{-r} = \binom{n}{n+r}, \quad \text{by identity (1) extended over all integers,}$$

$$= 0, \qquad \text{by identity (3), } n \geqslant 0, r \geqslant 1;$$

so that equation (5) holds; and

$$\binom{-n}{-r} = \binom{-n}{r-n}, \qquad \text{by identity (1) extended over all integers,}$$

$$= (-1)^{n-r}\binom{r-1}{r-n}, \quad \text{by identity (4), for } r \geqslant n,$$

$$= (-1)^{n-r}\binom{r-1}{n-1}, \quad \text{by identity (1),}$$

so that identity (6) holds. In fact, of course, this establishes identity (6) even without the condition $r \geqslant 1$, but $\binom{-n}{-r}$ is already defined if $r \leqslant 0$.

Since

$$\binom{r-1}{n-1} = 0 \quad \text{if } n > r \geqslant 1,$$

we are finally led to extend identity (6) and postulate the identity:

Third reflection identity:

$$\binom{-n}{-r} = (-1)^{n-r}\binom{r-1}{n-1}, \quad \text{for } n \geqslant 1, r \geqslant 1. \tag{7}$$

We have thus extended the domain of the binomial coefficient over the full range of integers. We now proceed to reexamine our identities (1), (4) and (7) so that they, too, apply over the full range of integers.

Let sgn R, where R is an integer, be defined by

$$\text{sgn } R = \begin{cases} 0, & \text{if } R \geqslant 0, \\ 1, & \text{if } R < 0. \end{cases} \tag{8}$$

It is then easy to prove the following symmetry theorem.

Theorem 1 (the symmetry theorem)

The binomial coefficients $\binom{N}{R}$, where N and R are arbitrary integers, satisfy the following identities:

I_1 *(First reflection identity):*

$$\binom{N}{R} = \binom{N}{N-R}.$$

I_2 *(Second reflection identity):*

$$\binom{N}{R} = (-1)^{R + \text{sgn } R}\binom{-N+R-1}{R}.$$

I_3 *(Third reflection identity):*

$$\binom{N}{R} = (-1)^{(N-R) + \text{sgn}(N-R)}\binom{-R-1}{-N-1}.$$

Moreover, these three identities fully determine the value of $\binom{N}{R}$, where N and R are arbitrary integers, given the known values of $\binom{N}{R}$, where N and R are *non-negative* integers.

Figure 2 shows the binomial coefficients $\binom{N}{R}$ for $-11 \leqslant n \leqslant 10$ and $-11 \leqslant r \leqslant 10$. Note that, by I_1, $\binom{N}{R} = \binom{N}{S}$, where $R + S = N$. Notice, too, the geometric interpretation of I_2 and I_3.

We will be content to prove I_2. If $R \geqslant 0$ then I_2 restates equation (4). If $N < R < 0$ or if $N \geqslant 0$, $R < 0$, then both sides of I_2 are zero. If $R \leqslant N < 0$, then

$$\binom{N}{R} = \binom{N}{N-R} = (-1)^{N-R}\binom{-R-1}{N-R},$$

$$\binom{-N+R-1}{R} = \binom{-N+R-1}{-N-1} = (-1)^{N-1}\binom{-R-1}{-N-1} = (-1)^{N-1}\binom{-R-1}{N-R},$$

so that

$$\binom{N}{R} = (-1)^{R+1}\binom{-N+R-1}{R} = (-1)^{R+\operatorname{sgn}R}\binom{-N+R-1}{R}.$$

Let us note explicitly the incidence of zero as a value of $\binom{N}{R}$:

$$\binom{N}{R} = 0 \Leftrightarrow R > N \geqslant 0 \quad \text{or} \quad 0 > R > N \quad \text{or} \quad N \geqslant 0 > R \tag{9}$$

Next we look again at the Pascal identity; we find the following theorem.

Theorem 2

The Pascal identity

$$\binom{N}{R} + \binom{N}{R-1} = \binom{N+1}{R}$$

holds, where N and R are any integers, except that

$$\binom{-1}{0} + \binom{-1}{-1} \neq \binom{0}{0}.$$

Proof. Certainly, by equation (2'), the Pascal identity holds if $R \geqslant 1$. Now if $R = 0$ and $N \neq -1$, it asserts that $1 + 0 = 1$. If $R < 0$, $N \geqslant 0$ or $R < 0$, $N < R - 1$ it asserts that $0 + 0 = 0$. If $R < 0$, $N = R - 1$ it asserts that $0 + 1 = 1$. Thus, we are left with $0 > N > R$; but then, by the second reflection identity,

$$\binom{N}{R} = (-1)^{N-R}\binom{-R-1}{-N-1},$$

$$\binom{N}{R-1} = -(-1)^{N-R}\binom{-R}{-N-1},$$

$$\binom{N+1}{R} = -(-1)^{N-R}\binom{-R-1}{-N-2},$$

so we must show that, if $0 > N > R$, then

$$\binom{-R-1}{-N-1} + \binom{-R-1}{-N-2} = \binom{-R}{-N-1}.$$

However, this is the Pascal identity in an already proven situation if $N \leqslant -2$; and if $N = -1$, $R \leqslant -2$, it asserts that $1 + 0 = 1$.

We now proceed to give the promised algebraic interpretation of $\binom{N}{R}$ if R is negative. Let us write $\binom{N}{-r}$ for $\binom{N}{R}$, so that $r \geqslant 1$. Recall that $\binom{N}{r}$ is the coefficient of x^r in the Taylor series expansion of $(1+x)^N$. Now we may instead construct the Laurent series† expansion of $(1+x)^n$. Thus,

$$(1+x)^N = x^N\left(1+\frac{1}{x}\right)^N = \sum_{s=0}^{\infty}\binom{N}{s}x^{N-s}. \tag{10}$$

---

†Whereas the Taylor series converges if $|x| < 1$, the Laurent series converges if $|x| > 1$; here we assume $N < 0$. The two series are finite and essentially coincide if $N \geqslant 0$.

If we define $\binom{N}{-r}$ to be the coefficient of x^{-r} in this Laurent series expansion, we find

$$\binom{N}{-r} = 0 \quad \text{if } N \geqslant 0, \; r \geqslant 1, \tag{11}$$

since there are no terms involving negative powers of x on the r.h.s. of equation (10) when $N \geqslant 0$; but equation (11) merely restates condition (5). However, if $N < 0$, say $N = -n$, with $n \geqslant 1$, then we have

$$\binom{N}{-r} = 0, \quad r < -N, \qquad \binom{N}{-r} = \binom{N}{N+r}, \quad r \geqslant -N$$

or

$$\binom{-n}{-r} = 0, \quad r < n, \qquad \binom{-n}{-r} = \binom{-n}{-n+r}, \quad r \geqslant n, \tag{12}$$

and this is easily seen to yield precisely the values given by the third reflection identity (7). We are thus entitled to claim that considerations of symmetry have led us to adopt the correct definition of $\binom{N}{R}$ for arbitrary integers N, R.

We close with an important notational remark. From the geometrical viewpoint it is more natural to write $\binom{N}{R}$ as a multinomial coefficient $\binom{N}{RS}$ with $R + S = N$. Then the first reflection identity asserts that

$$I_1: \quad \binom{N}{RS} = \binom{N}{SR}.$$

Actually, we represent the inherent symmetry even better by writing $T = -N - 1$ and replacing $\binom{N}{RS}$ by the three-vector (R, S, T) in the plane $R + S + T + 1 = 0$. Then Theorem 1 reads

$$I_1: \quad (R, S\,T) = (S, R, T)$$

$$I_2: \quad (R, S, T) = (-1)^{R + \operatorname{sgn} R}(R, T, S)$$

$$I_3: \quad (R, S, T) = (-1)^{S + \operatorname{sgn} S}(T, S, R).$$

It is now easy to see that any two of these identities imply the third, and that, if we extend the Pascal triangle to a Pascal hexagon (see Fig. 2), each expresses a reflection principle.

3. SOME PATTERNS IN THE PASCAL HEXAGON

A well-known fact about numbers in Pascal's triangle is that the sum of any two adjacent entries in the $r = 2$ row is a perfect square. Thus, for example, $1 + 3 = 4 = 2^2$ or, using binomial coefficients,

$$\binom{2}{2} + \binom{3}{2} = 2^2.$$

Again, $3 + 6 = 9 = 3^2$ or

$$\binom{3}{2} + \binom{4}{2} = 3^2.$$

In general this property may be stated as

$$\binom{N}{2} + \binom{N+1}{2} = N^2, \quad \text{where } N \geqslant 2. \tag{13}$$

In the light of our extended meaning for the binomial coefficient we might hope that the restriction on N in equation (13) is not required. In fact, you may verify from Fig. 1 that the relationship of equation (13) holds for $N \geqslant -3$. Of course it is also possible to use the formulae of Section 2 to show that equation (13) holds for any integer N—indeed, for any real number N.

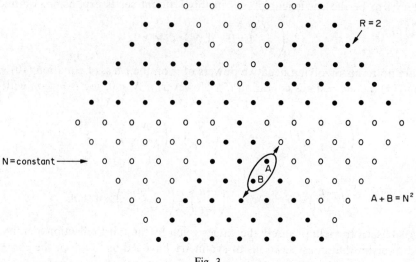

Fig. 3

Figure 3 is an abbreviated version of the Pascal hexagon in which each of the non-zero entries has been replaced by a dot. This cryptic figure illustrates the identity (13). Notice particularly that the arrows in the direction of $R = 2$ tell us that the relationship involving the labeled entries within the ellipse will be valid as long as the ellipse sits on the line $R = 2$. The arrow labeled "$N = $constant" along with the designated value of R gives us the information required to convert the letters within the figure to binomial coefficients. Thus the identity (13) can be *read off* from Fig. 3, except that, because the arrows point in both directions along $R = 2$ we know there is no restriction on N. (A similar convention is adopted in Figs 4 and 5.)

The pattern of Fig. 4 was first shown to us by a student, Allison Fong, then a freshman at Santa Clara University. From Fig. 4 we read off the identity

$$\binom{N+2}{3} - \binom{N}{3} = N^2, \quad \text{for all integers } N. \tag{14}$$

Of course identity (14) can be proved algebraically. However, a proof, for $N \geqslant 0$, which does not rely on algebraic manipulation appears in Ref. [1]. The algebraic proof has the obvious merit, however, of being valid for any real number N.

Figure 5 is an example of a more general pattern. Notice first, that, as this array slides diagonally so that a lies on the line $N - R = 4$, the quantities $(a \times d)/(c \times g)$ and $(a \times e)/(b \times f)$ will *always*

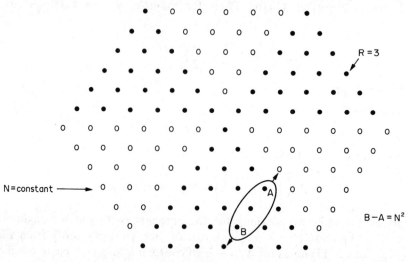

Fig. 4

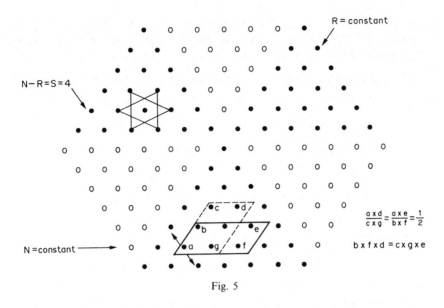

Fig. 5

have the value of 1/2. Second, since the two quantities are always equal we may infer that, in any given instance†

$$b \times f \times d = c \times g \times e.$$

This last relationship is a special case of the *Star of David theorem*: look at the two triangles determined by the vertices b, f, d and c, g, e, respectively to see why this name is appropriate. We have drawn the star in one of its positions.

Let us look at the features of this last example with an eye to seeing how much we can generalize it. It involved two parallelograms which were reflections of each other across a line bisecting the edges of the parallelograms at the vertex nearest a. The sides of the parallelograms were in the directions of constant N and constant R, and the arrangement could slide in the direction of constant S without affecting the ratio of the designated products. The *abef* parallelogram has dimensions 2 by 1 in the N and R directions, respectively, and the *acdg* parallelogram has dimensions 1 by 2 in the N and R directions, respectively. Both parallelograms are *anchored* at a.

Now what are the possibilities for generalizing? We might inquire whether or not the line along which we slide the parallelograms (in this case, $S = 4$) is restricted to the choice in Fig. 5. (The answer is NO, *any* integer value of S will work.) We might also look at parallelograms of other dimensions k by l in the N and R directions, with each anchored at a. Then we might ask:

> Does $(a \times d)/(c \times g)$ always equal $(a \times e)/(b \times f)$, as you slide the arrangement of parallelograms along the line of constant S? And, if so, what is that value?

Of course, if the answer is affirmative (and it is!) we automatically get a generalization of the Star of David theorem.

However, what about other types of parallelograms? The parallelograms in Fig. 5 have sides parallel to the N and R directions and they slide in the S direction. By symmetry there are obviously two other possibilities. *Will the analogous statements hold in those two cases?* The answer is YES! and that is what we will now show, in its complete generality, by introducing a figure we call the *Pascal flower*.‡

---

†Although the values of $b \times f \times d$ and $c \times g \times e$ are always *equal*, regardless of where the arrangement has been moved to (by sliding along the line $S = 4$) they do not remain fixed, under these movements.

‡The participants in a short course we gave, under the auspices of the Mathematical Association of America, at Hope College in Holland, Michigan, were so enamored of this diagram that they named it the *Pascal Tulip*.

However, in this figure and in what follows, we introduce a small notational change in the interests of convenience. We will henceforth designate the directions in the Pascal hexagon by *lower case* letters n, r, s instead of upper case letters N, R, S. Thus, henceforth,

n, r, s, are arbitrary integers, positive or negative, subject to $r + s = n$.

As shown in Fig. 6, there are three types of parallelogram, namely,

P_1 (with sides parallel to the r and s directions),

P_2 (with sides parallel to the s and n directions)

and

P_3 (with sides parallel to the n and r directions).

Each of these parallelograms P_i is shown, with its corresponding reflection P'_i indicated by dotted lines. All of the parallelograms are anchored at $\binom{n}{r}$. By anchoring the parallelograms at $\binom{n}{r}$ we can give a geometric description of the numbers we wish to consider. In each case we define the cross-ratio (or weight), W_i (or W'_i), of the corresponding parallelogram, P_i (or P'_i), to be the product of the binomial coefficient $\binom{n}{r}$ with the binomial coefficient at the vertex opposite $\binom{n}{r}$, divided by the product of the two binomial coefficients at the other vertices of the parallelogram. When n and r are non-negative integers† we can calculate from this definition, and the fact that

$$\binom{n}{r} = \frac{n!}{r!(n-r)!},$$

the following:

$$
\left.
\begin{aligned}
W_1 &= \binom{n}{r}\binom{n-l-k}{r-k} \Big/ \binom{n-k}{r-k}\binom{n-l}{r} = \frac{n!(n-k-l)!}{(n-k)!(n-l)!} \\[2mm]
W_2 &= \binom{n}{r}\binom{n-k}{r-k-l} \Big/ \binom{n}{r-l}\binom{n-k}{r-k} = \frac{(r-k)!(r-l)!}{r!(r-k-l)!} \\[2mm]
W_3 &= \binom{n}{r}\binom{n-l}{r+k} \Big/ \binom{n-l}{r}\binom{n}{r+k} = \frac{(n-r-k)!(n-r-l)!}{(n-r)!(n-r-k-l)!}
\end{aligned}
\right\}. \tag{15}
$$

We now notice several remarkable features. Observe that, for given k and l,

$$
\left.
\begin{aligned}
&W_1 \quad \text{depends only on } n \text{ and is symmetric in } k \text{ and } l, \\
&W_2 \quad \text{depends only on } r \text{ and is symmetric in } k \text{ and } l, \\
&W_3 \quad \text{depends only on } n - r \ (= s) \text{ and is symmetric in } k \text{ and } l.
\end{aligned}
\right\} \tag{16}
$$

The statement about W_3 in observation (16) assures us that the property we noticed about the parallelograms in Fig. 5 does, indeed, generalize to k by l parallelograms sliding along any diagonal line in the S direction. Of course, with each pair P, P' of parallelograms we have a generalized Star of David theorem.

Furthermore, the statements W_1 and W_2, in conditions (16), assure us that there was nothing special about taking the sides of our parallelograms parallel to the r and s directions. The expression for W_1 tells us that we can either reflect, or slide horizontally (in the n direction), a k by l parallelogram, P_1, as in Fig. 6, without changing the value of W_1. Similarly, the expression for W_2 informs us that we can either reflect, or slide diagonally (in the r direction) a k by l parallelogram, P_2, as in Fig. 6, without changing the value of W_2. And, in all of these individual pairs of parallelograms with equal weights (cross-ratios) we can obtain a corresponding Star of David theorem.

---

†If n and r are not both non-negative integers and the entire parallelogram, P_i, lies in one of the non-zero regions (other than the Pascal triangle region), then the reflective properties of Section 2 can be used to show that the dependency and symmetry statements given in condition (16) are still valid. Moreover, using geometrical language, we may transport a parallelogram, in the appropriate direction, from one non-zero zone of the hexagon, across a zero zone, to another non-zero zone and the weight will resume its original value.

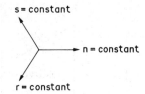

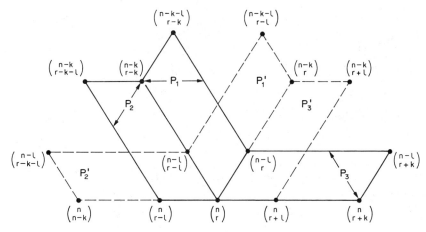

Fig. 6. The Pascal flower.

4. WEIGHTS IN THE PASCAL FLOWER

We note the following identity for extended binomial coefficients:

$$\binom{m}{k}\binom{m-k}{l} = \binom{m}{l}\binom{m-l}{k}. \tag{17}$$

For each side is just the trinomial coefficient

$$\binom{m}{k \quad l \quad m-k-l},$$

which may also be extended to all integers m, k, l, by appeal to symmetry. We rewrite equation (17) as

$$\binom{m-l}{k}\bigg/\binom{m}{k} = \binom{m-k}{l}\bigg/\binom{m}{l}, \tag{18}$$

noting that either side (or both sides) of equation (18) may have the indeterminate form 0/0. We call the common value of equation (18) the *weight function* $\Phi(m, k, l)$, so that

$$\Phi(m; k, l) = \Phi(m; l, k) = \frac{(m-k)!(m-l)!}{m!(m-k-l)!}. \tag{19}$$

We now return to the Pascal flower and recall from equation (15) the *cross-ratios*, W_1, W_2, W_3, for the parallelograms P_1, P_2, P_3, hinged at $\binom{n}{r}$. An immediate consequence of equations (15) is the following theorem.

Theorem 3

$$W_1 = \Phi(n; k, l)^{-1}; \quad W_2 = \Phi(r; k, l); \quad W_3 = \Phi(s; k, l).$$

Thus, W_1 depends only on n and is symmetrical in k and l; W_2 depends only on r and is symmetrical in k and l; W_3 depends only on s and is symmetrical in k and l.

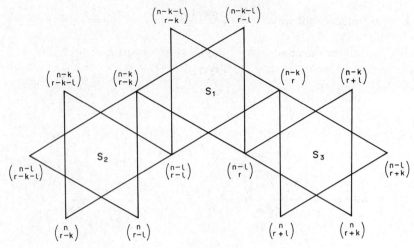

Fig. 7

Of course, since P'_1, P'_2, P'_3 are the reflections of parallelograms P_1, P_2, P_3 (obtained by exchanging k and l), we have

Corollary 3.1

$$W_1 = W'_1$$

$$W_2 = W'_2$$

$$W_3 = W'_3.$$

Each equality in Corollary 3.1 yields a Star of David theorem. Thus, for example, since $W_1 = W'_1$ we infer from equations (15) that

$$\binom{n-l}{r-l}\binom{n-k}{r}\binom{n-k-l}{r-k} = \binom{n-k}{r-k}\binom{n-l}{r}\binom{n-k-l}{r-l}. \tag{20}$$

The geometric significance of equation (20) is seen in Fig. 7. Each side of the equation is the product of the vertices attached to one of the two triangles making up the (slightly distorted) Star of David surrounding S_1. Identities similar to equation (20) can be read off Fig. 7 for the stars surrounding S_2 and S_3, so that we have the following general result.

Theorem 4 (generalized Star of David theorem)

The product of the vertices of one triangle surrounding S_i is equal to the product of the vertices of the other triangle surrounding S_i for $i = 1, 2, 3$ in Fig. 7.

5. EXTENDING THE DOMAIN OF THE BINOMIAL COEFFICIENTS OVER THE COMPLEX NUMBERS

We remark that there is no impediment to extending binomial coefficients $\binom{n}{r}$ far beyond the integers, indeed to the complex numbers. For we may define the factorial function $z!$ by

$$z! = \Gamma(z + 1)$$

here the Γ-function $\Gamma(z)$ is meromorphic in the entire complex plane with simple poles at $z = 0, -1, -2, \ldots$. Thus, the definition

$$\binom{n}{r} = \frac{n!}{r!(n-r)!}$$

yields a function which is regular everywhere for all complex numbers n, r, except where n is a negative integer. As we have seen, it is still possible to define $\binom{n}{r}$ even if n is a negative integer, provided r is also an integer, but we will exclude these singular values of n from the subsequent discussion.

We wish now to establish the analogues of the identities I_1, I_2, I_3 of Section 2 in this extended domain. To this end, we invoke the key identity

$$\Gamma(z)\Gamma(1-z) = \frac{\pi}{\sin \pi z};$$

the form of this identity most useful to us will be

$$z!(-1-z)! = -\frac{\pi}{\sin \pi z}. \tag{21}$$

Reverting to the notation

$$\binom{n}{r} = (r, s, t),$$

where $r + s = n = -t - 1$, we have

$$(r, s, t) = (s, r, t) \tag{22}$$

$$(r, s, t) = \frac{n!}{r!(n-r)!} = \frac{(-t-1)!}{r!s!} = \frac{\sin \pi s}{\sin \pi t} \frac{(-s-1)!}{r!t!} = \frac{\sin \pi s}{\sin \pi t}(r, t, s), \tag{23}$$

by repeated application of equation (21).

Likewise,

$$(r, s, t) = \frac{\sin \pi r}{\sin \pi t}(t, s, r). \tag{24}$$

We note that, if we write $[r, s, t] = (r, s, t)\sin \pi t$, then the three identities (22)–(24) become

$$I_1: \quad [r, s, t] = [s, r, t]$$

$$I_2: \quad [r, s, t] = [r, t, s]$$

$$I_3: \quad [r, s, t] = [t, s, r].$$

It follows that $[r, s, t]$ is invariant under the full symmetric group S_3 acting on the points (r, s, t) of the complex plane $r + s + t + 1 = 0$. We note further that

$$[r, s, t] = (r, s, t)\sin \pi t = \frac{(-t-1)!}{r!s!}\sin \pi t$$

is, indeed, analytic everywhere; for $(-t-1)!$ has simple poles at $t = 0, 1, 2, \ldots$, at which $\sin \pi t$ has zeros. Of course, $(-t-1)! \sin \pi t = -\pi/t!$, by equation (21), so that

$$[r, s, t] = \frac{-\pi}{r!s!t!}, \tag{25}$$

and we may write

$$(r, s, t) = -\frac{\pi}{\sin \pi t}\frac{1}{r!s!t!}.$$

Formula (25) constitutes, of course, an immediate proof of the identities I_1, I_2, I_3.

We may now proceed to generalize Theorem 3. For the identity (17) holds in complete generality, each side being equal to the generalized trinomial coefficient

$$\frac{m!}{k!l!(m-k-l)!}.$$

Thus, the weight function $\Phi(m; k, l)$ is defined for complex values of m, k, l and is symmetrical in k and l; and the formal calculation inherent in the proof of Theorem 3 goes through without change. Likewise, the Pascal identity of Theorem 2 holds for these generalized binomial coefficients, since it only depends on the relation $\Gamma(z) = z\Gamma(z-1)$, which holds so long as $z \neq 0, -1, -2, \ldots$. There is no necessity to formally restate either of these theorems.

Finally, we remark that the analogue of Theorem 3 would hold if we replaced the factorial function by any other function in the definition of the binomial coefficient (see Ref. [1]). Thus, quite generally, we may introduce an arbitrary function $f(z)$ of the complex variable z and consider the "f-binomial coefficient", defined by

$$\binom{n}{r}_f = \frac{f(n)}{f(r)f(n-r)}. \tag{26}$$

We may carry out such an extension of the theory even if the function f is only defined over a restricted domain, for example, over the non-negative integers. We give such an example now.

Of particular interest is the q-analogue of the binomial coefficient, given by

$$f(n) = (q^n - 1)(q^{n-1} - 1)\ldots(q - 1), n \geqslant 1; f(0) = 1. \tag{27}$$

Of course, we would, in this generality, lose the full symmetry inherent in the symbol $[r, s, t]$ above [see equation (25)], since this depended on the special property $\Gamma(z)\Gamma(1-z) = \pi/\sin \pi z$ of the Γ-function. Likewise, we would lose the Pascal identity; in fact, for the q-analogue given by equation (27), we have, instead,

$$q^r \binom{n}{r}_f + \binom{n}{r-1}_f = \binom{n+1}{r}_f.$$

We repeat that, in this final remark, we have not attempted to generalize beyond the usual range in which n, r are integers with $0 \leqslant r \leqslant n$. It may well be worthwhile to attempt to do so.†

The Pascal identity, in fact does generalize to the formula

$$\sum_{i+j=k} \binom{a}{i}\binom{b}{j} = \binom{a+b}{k},$$

which is valid (i) if b and j are positive integers, or (ii) if i and j (and hence k) are positive integers (see Ref. [3]). We do not know the full extent of its validity.

REFERENCES

1. P. Hilton and J. Pedersen, Looking into Pascal's triangle: combinatorics, arithmetic and geometry. *Math. Mag.* **60**, 305–316 (1987).
2. P. Hilton and J. Pedersen, Symmetry in mathematics. *Comput. Math. Applic.* **12B**, 315–328 (1986). Reprinted in *Symmetry: Unifying Human Understanding* (Ed. I. Hargittai). Pergamon Press, Oxford (1986).
3. P. Hilton, J. Pedersen and P. Ross, In mathematics there are no accidents. *Menemui Matematik* **9**, 121–143 (1987).

†See a forthcoming paper by the authors and W. Rosenthal.

Computers Math. Applic. Vol. 17, No. 1–3, pp. 103–115, 1989
Printed in Great Britain. All rights reserved

0097-4943/89 $3.00 + 0.00
Copyright © 1989 Pergamon Press plc

SYMMETRY IN THE SIMPLEST CASE: THE REAL LINE

S. Marcus

Institute of Mathematics, University of Bucureşti, Str. Academiei 14, 70109 Bucureşti, Romania

Abstract—Various types and degrees of symmetry in the set of real numbers are investigated. It is shown that symmetry and its polar opposite, antisymmetry, are submitted to similar restrictions.

POINTS OF (LOCAL) SYMMETRY

In its simplest form, in a one-dimensional Euclidean space, symmetry is defined with respect to one point. Given two points A and B in this space, the symmetric point C of A with respect to B is one such that the distance from C to B is equal to the distance from A to B, whereas the distance from C to A is the sum of these two distances. Given a set X of real numbers, it is symmetric with respect to the point B if for each point A in X its symmetric point with respect to B is also in X. One can weaken this property in the following way: B is a point of local symmetry for the set X if there exists an interval I whose centre is in B, such that if the point A belongs to both I and X, then the symmetric point of A with respect to B also belongs to X. So every point of symmetry for X is a point of local symmetry for X, but the converse is not true.

SYMMETRIC AND LOCALLY SYMMETRIC SETS

The set X is said to be (locally) symmetric if X is (locally) symmetric with respect to every point of X. Every symmetric set is locally symmetric, but, as we shall see, the converse is not true. Examples of symmetric sets: every open interval; every compact interval; the set of integers; the set of rational numbers; the set of algebraic numbers and the set of real numbers. It is not difficult to see that a symmetric finite set cannot have more than one point. For a proof, let us put the elements of a finite set in their increasing order $a_1 < \ldots < a_n$. The symmetric point of a_2 with respect to a_1 is no longer in the set. But any finite set is locally symmetric in a trivial way, because in the interval I with centre in a_i and small enough to contain neither a_{i-1} nor a_{i+1} the condition of symmetry with respect to a_i is trivially fulfilled.

IS ABSENCE OF SYMMETRY RELATED TO DISORDER?

Beyond the interest in itself, symmetry in one dimension is a basic phenomenon which, by its simplicity, is a term of reference and a source of suggestions, models and metaphors for more complex phenomena of symmetry. Common sense tells us that symmetry implies order, whereas its absence implies disorder. This intuitive impression is to a large extent contradicted by a careful examination. It can be proved that antisymmetry, the polar opposite of symmetry, is to the same extent restrictive as symmetry. Moreover, the type of restrictions imposed by symmetry is the same for antisymmetry. In other words, the deep structure of symmetric sets is isomorphic to the deep structure of antisymmetric sets. However, before discussing this fact, we need some conceptual clarifications.

POINTS OF NON-SYMMETRY

Let us see what is the exact meaning of non-symmetry and of antisymmetry. Non-symmetry is the negation of symmetry. So, a set X is non-symmetric if there exists at least one point a of X with respect to which X is not symmetric; i.e. there exists at least one point x in X whose symmetry point with respect to a is not in X. An example of a non-symmetric set is the set of irrational numbers: the symmetric point of $\sqrt{2}$ with respect to $\sqrt{2}/2$ is zero. Another example is any finite set of cardinal number at least equal to two. For instance, the set formed by the points $-1, 0$, and 1 is symmetric with respect to zero, but it is not symmetric with respect to -1 or 1. The point a is not of local symmetry for X if for every interval I centred in a there exists a point x common to I and X, such that its symmetric point with respect to a is not in X. Non-local symmetry implies non-symmetry, but the converse is not true.

Antisymmetry is the polar opposite of symmetry, such as black is opposed to white. Nobody will confuse non-white with black. Non-white may mean red, yellow, black etc. So, a point a is of antisymmetry for the set A if for any point x in X its symmetric point with respect to a does not belong to X. For instance, any rational number is a point of symmetry for the set of irrational numbers, but any irrational number is a point of antisymmetry for the set of rational numbers. The point a is of local antisymmetry for X if there exists an interval I with the centre in a such that for each x common to X and I its symmetric point with respect to a does not belong to X. For instance, if X is the union of the set of negative numbers smaller than -2 and the set of positive numbers, then zero is a point of local antisymmetry for X [we can take as I the interval $(-1, +1)$], but it is not a point of antisymmetry for X.

THREE WAYS IN WHICH SYMMETRY OF A SET
MAY BE TRANSGRESSED

When dealing with a set, the polar opposite of (local) symmetry may be understood in two ways. One way consists of requiring to the set X to have each of its points as point of (local) antisymmetry. We get in this way the concept of (local) antisymmetric set. Another way is to require to X to have no point of (local) symmetry. We get in this way what we call a (local) asymmetric set. It is obvious that (local) antisymmetry implies (local) asymmetry and (local) asymmetry implies (local) non-symmetry. In other words, every (locally) antisymmetric set is (locally) asymmetric and every (locally) asymmetric set is (locally) non-symmetric. The converse is false for each of these implications. (The local variant requires that the considered property P is locally satisfied in each point of the set). Let us take some examples. The set E of prime numbers is not symmetric, because 3 is not a point of symmetry for E [2 belongs to E, but its symmetric number with respect to 3 (which is 4) does not belong to E]. Moreover, E is asymmetric, because no prime number is a point of symmetry for E (let us assume the contrary: there exists a prime number p of symmetry for E; then the symmetric number of 2 with respect to p, which is $2p - 2$, is no longer prime, in contradiction with our hypothesis; so no prime number is of symmetry for E). What about local symmetry? In a trivial sense, any set of integers is locally symmetric, because an interval I small enough and whose centre is an integer contains no other integer; so E is trivially locally symmetric. However, if we consider only intervals containing at least one prime number other than the centre of symmetry, then it is clear that there exists in E points of local symmetry; for instance, 5 is such a point, because 3 and 7 are prime numbers symmetric with respect to 5. It follows that E is not locally asymmetric. An example of a perfect set which is locally asymmetric is the Cantor triadic set [containing those numbers in the interval $(0,1)$ which have an infinite triadic development with no occurrence of the digit 1]. We have proved this result in Ref. [1].

ANTISYMMETRIC SETS

In order to give an example of antisymmetric set we shall use what is called a Hamel basis for the set R of real numbers. Such a basis, let us denote it by H, is a set of real numbers having the following properties: $1°$ any finite subset $H_1 = \{x_1, \ldots, x_n\}$ of H is rationally independent, i.e. if $r_1 x_1 + r_2 x_2 + \cdots + r_n x_n = 0$, where $r_1, r_2, \ldots, r_n$ are rational numbers, then $r_1 = r_2 = \cdots = r_n = 0$; $2°$ for any real number x which is not in H there exist a finite subset $H_2 = \{y_1, y_2, \ldots, y_m\}$ of H and m rational numbers $r_1, r_2, \ldots, r_m$ such that $x = r_1 y_1 + r_2 y_2 + \cdots + r_m y_m$. The existence of a set with properties $1°$ and $2°$ has been proved by Hamel. It is obvious that H is antisymmetric. Indeed, let $x \in H, y \in H$. The symmetric point of y with respect to x is $z = 2y - x$, which cannot belong to H, because the set $\{x, y, z\}$, as a subset of H, is rationally independent.

It follows that H is also locally antisymmetric.

Another example of a (locally) antisymmetric set is the algebraically independent set defined by John von Neumann. The set A of real numbers is algebraically independent if for any finite subset $\{a_1, a_2, \ldots, a_n\}$ of A the relation $P(a_1, a_2, \ldots, a_n) = 0$ [where $P(x_1, x_2, \ldots, x_n)$ is a polynome whose coefficients are integers] is possible only when all coefficients are equal to zero. Von Neumann has given an example of an algebraically independent set having the power of the continuum (see Ref. [2]). It is easy to see that such a set is (locally) antisymmetric.

A third example of a (locally) antisymmetric set is any anticonvex set in the sense of W.A.J. Luxemburg (the concept is discussed in Ref. [3]). A set S of real numbers is called anticonvex whenever for x and y in S, x different from y, the number $(x + y)/2$ does not belong to S. It is easy to see that any Hamel basis and any algebraically independent set are anticonvex sets. In a different context and in a different terminology, the notion of an anticonvex set was considered by H. Steinhaus, when he raised the following problem. What is known about a linear set A such that no point at the same distance with respect to two points in A belongs to A?

It is easy to see that a real set is antisymmetric if and only if it is anticonvex, so these two properties are equivalent. Another possible formulation of the same property is the following: the set contains no three points in arithmetic progression.

SYMMETRY OF A SET A WITH RESPECT TO POINTS WHICH DO NOT BELONG TO A

So far, symmetry of a set A was considered only with respect to a point which belongs to A. However, very often symmetry takes place with respect to points which do not belong to the set. An example in this respect is the set of irrational numbers, which is locally asymmetric (but has no point of local antisymmetry); this set is symmetric with respect to every rational point. This is possible because irrationality is not invariant by addition, whereas the sum of two numbers, one rational, the other irrational, is always irrational. An example of another type is the set Z of integers, which is symmetric; but there are numbers x which do not belong to Z, such that Z is symmetric with respect to x. Such a number x is the arithmetic mean of two consecutive integers. As we shall see later, symmetry of a set with respect to all types of points leads to a more restrictive property of symmetry; it will be called strong symmetry.

MORE SOPHISTICATED EXAMPLES OF SYMMETRIC SETS

An interesting example of symmetric set is furnished by the so-called homogeneous sets introduced by Borel [4]. The real set E is homogeneous if for any x, y in E ($x \neq y$) the translation xy transforms E in itself. Obviously, the set Q of rational numbers is homogeneous, but it is also countable. Borel gives an example of an homogeneous non-countable set of Lebesgue measure zero (a real set E is of Lebesgue measure zero if for any strictly positive number a there exists a sequence $I_1, I_2, \ldots, I_n, \ldots,$ of open intervals having the following two properties: $1°$ the set E is contained in the union of the intervals $I_1, I_2, \ldots, I_n, \ldots$. $2°$ the sum of the series formed with the lengths of the intervals $I_1, I_2, \ldots, I_n, \ldots,$ is smaller than a; any countable set is of Lebesgue measure zero, but the converse is not true). Borel raised the problem whether there exists a decomposition of R into a finite number $n > 1$ or into an infinite denumerable number of homogeneous mutually disjoint sets. In Ref. [5], Erdös and Marcus have shown that the answer is negative with respect to finite decompositions and affirmative with respect to countable decompositions.

Another type of symmetric sets are the Jensen convex sets. A real set X is convex in the sense of Jensen or J-convex if for any x, y in X the number $(x + y)/2$ is also in X. We have investigated the structure of these sets in Refs [6, 7], where it is proved that every J-convex set containing no interval is of interior Lebesgue measure (this concept will be defined later) equal to zero. A real set X has the Baire property if it is of the form $(A - B) \cup C$, where A is open, whereas B and C are meagre sets; a set is meagre if it is a countable union of non-dense sets; a non-dense set is one whose closure has its complementary everywhere dense, i.e. intersects any open interval. It is proved that every set with the Baire property, which is contained in a J-convex set whose complementary set is everywhere dense, is a meagre set.

STRUCTURAL IDENTITY OF SYMMETRY AND ANTISYMMETRY

In Ref. [8] we have shown that: 1° any Lebesgue measurable real set X for which every point of X is of local antisymmetry is of Lebesgue measure zero. The concept of Lebesgue measurability may be obtained as follows: the measure of an open set G is the sum of the series formed with the lengths of the intervals forming G. The exterior measure of a set X is the greatest lower bound of the measures of open sets containing X. The measure of a closed set F contained in a bounded interval I is the difference between the length of I and the measure of the open set $I - F$. The interior measure of a bounded set X is the least upper bound of the measures of closed sets contained in X. The interior measure of a set cannot be greater than its exterior measure; if they are equal, then the set X is said to be Lebesgue measurable). It is important to precise that Lebesgue measurability is a very general property. The existence of a real set which is not Lebesgue measurable is equivalent to the axiom of choice due to Ernst Zermelo, which asserts that given a collection of non-empty pairwise disjoint sets, there exists as set containing exactly one element from every set of the collection.

What about the measure-theoretic structure of a locally symmetric set? Generally speaking, there is no restriction concerning the measure of such a set, because the elementary example of the set R of real numbers shows that its measure may be infinite. However, the non-trivial case is that of a set having a void interior (the interior of a set is the largest open set contained in the set). We have proved [6] that any Lebesgue measurable set X whose interior is void and for which every point in X is of local symmetry is of Lebesgue measure zero. The apparent difference with respect to the measure-theoretic structure of an antisymmetric set is the requirement of a void interior; but it is easy to see that this requirement is obligatory fulfilled by every antisymmetric set (an open real set is a union of intervals, whereas an interval is never an antisymmetric set). So, from a measure-theoretic view point, symmetry and antisymmetry are submitted to the same restriction.

There is a topological analogy of the above results. In Ref. [8] we have proved that any real set X having the Baire property and for which every point of X is of local antisymmetry is a meagre set. In Ref. [6] we have shown that every set X with the Baire property, with a void interior and which is locally symmetric is a meagre set. So, from a descriptive-topological view point, symmetry and antisymmetry are submitted to the same restriction. The structural identity of symmetry and antisymmetry is in this way completely proved.

Some of the results in this sections are consistent with some theorems about J-convex sets in the preceding section.

HOW RESTRICTIVE ARE ASYMMETRIC SETS?

A natural question: Is asymmetry subjected to a restriction similar to that operating on antisymmetry? The example of the set of irrational numbers shows that the answer is negative. Indeed this set is asymmetric, has a void interior, but its Lebesgue measure, in any interval, is strictly positive; it has the Baire property, but it is not a meagre set. The only similarity with antisymmetry remains the following: any local asymmetric set has a void interior. To see this, it is enough to observe that if a real set X contains on open interval I, then any point in I is of local symmetry for X.

DISORDER MEANS COMBINATION OF SYMMETRY AND NON-SYMMETRY

If, as we have seen, symmetry and antisymmetry are equally submitted to constraints and the corresponding restrictions are of the same type, then they both define a certain order whose negation should be looked for somewhere between the polar opposites. So, we need a typology of intermediate situations, one of which being, for instance, the phenomenon of asymmetry, described in the preceding paragraph. However, it is difficult to say exactly where is the "middle of the way" between symmetry and antisymmetry, what should mean that a set is not "too symmetric" and not "too antisymmetric" at the same time. One possibility is to require a set A

to contain no point of local symmetry and no point of local antisymmetry. This is just the case of the set of irrational numbers. Another possibility is to require A to have in each of its portions both a point of local symmetry and a point of local antisymmetry. If we denote by B the set of points of A which are of local symmetry for A and by C the set of points of A which are of local antisymmetry for A, we may expect that many logical structural possibilities for B and C are in fact impossible. A general problem would be here to characterize those pairs $\langle B, C \rangle$ of disjoint real sets, for which B is the set of points of local symmetry, whereas C is the set of points of local antisymmetry for the union of B and C. A more general problem would be the following: Given two disjoint real sets B and C and a set A containing the union of B and C, find necessary and sufficient conditions for A to be locally symmetric exactly at the points of B and locally antisymmetric exactly at the points of C. Another important problem is that obtained when we replace in the above formulation "antisymmetric" by "asymmetric" or "non-symmetric".

SYMMETRY MODULO A NEGLIGIBLE SET

Another way to approach the problem of symmetry of A is to consider the possibility that only a subset A_1 of A be (locally) symmetric or (locally) asymmetric with respect to a given point of A. This situation is interesting mainly when the set $A - A_1$ is, in some sense, negligible. There are three main types of negligible sets: the finite sets and the countable sets are considered negligible from the viewpoint of cardinality (richness); the sets of (Lebesgue) measure zero are negligible from the metric point of view; the meagre sets are negligible from a descriptive-topological viewpoint. Any countable set is both meagre and of measure zero, but a meagre set (and even a non-dense set, such as the Cantor triadic set) and a set of measure zero may have the power of the continuum (i.e., they may have the richness of the set R of real numbers). The common features of these three classes of negligible sets are: $1°$ they are hereditary classes, i.e. each subset of a set in the class belongs to the class; $2°$ they are countably additive classes, i.e. any countable union of sets in the same class is also a set in the class.

Now let us denote by $\mathcal{P}$ a hereditary and countably additive class of real sets. The set A is $\mathcal{P}$-locally symmetric at x if there exists a set P in $\mathcal{P}$ such that x is a point of local symmetry for $A - P$. A similar definition for $\mathcal{P}$-local antisymmetry.

Another way to involve negligible sets in the study of symmetry is the following. The set A is said to be $\mathcal{P}$-locally symmetric if there exists a set P in $\mathcal{P}$ such that every point in $A - P$ is of local symmetry for A. A similar definition for $\mathcal{P}$-local antisymmetry. Such concepts have not yet been investigated.

STRONG SYMMETRY

Symmetry of a set may be described in terms of the symmetry of a function. A function $f:R \to R$ is said to be symmetric at a if for every x in R we have $f(x) = f(2a - x)$; f is said to be locally symmetric at a if there exists an interval I centred in a such that $f(x) = f(2a - x)$ for every x in I [9–11]. In Ref. [11] Rusza proves that for every function $f:R \to A$ locally symmetric in every point there exists a real number q in A such that the set of points x where $f(x)$ is different from q is countable. This means that a locally symmetric function on R is very near to a constant.

From (locally) symmetric functions we move to what we call *strong (locally) symmetric sets*. If X is a real set, let us define the characteristic function f_x of X by $f(x) = 1$ if x is in X and $f(x) = 0$ if x is not in X. We will say that X is strongly (locally) symmetric if f_x is (locally) symmetric. Every strongly (locally) symmetric set is (locally) symmetric, but the converse is not true, as is shown by the set of rational numbers. Answering a query by Evans and Weil [12], Rusza proves that for every strongly locally symmetric subset S of R either the closure of S or the closure of $R - S$ is countable. Clearly, this is not true for any locally symmetric set; for instance, for the set Q of rational numbers both its closure and the closure of its complementary set are R, thus they are not countable.

A function $f:I \to R$ is said to be symmetrically continuous in x if $\lim[f(x + h) - f(x - h)] = 0$ for $h \to 0$. This concept was introduced by Hausdorff [13], who asked whether the set D_f of points of discontinuities of a symmetrically continuous function on R may be uncountable. An affirmative answer to this question was obtained by Preiss [14]. Another question raised by Hausdorff was whether for any real set A which is a countable union of closed sets there exists a symmetrically continuous function f for which $D_f = A$. A negative answer to this question was given by Fried [15], who proved that A must be a meagre set. This implies that the set C_f of points of continuity is dense. Now let us take as f the characteristic function of a strongly locally symmetric set X. As we have seen, such a function f is locally symmetric in each point. However, local symmetry of f implies symmetric continuity of f [because $f(x + h) - f(x - h) = 0$], so f has in every interval a point of continuity. For a characteristic function, this means the existence of a dense set of open intervals where f is constant. As it was shown by Davies [9], this yields that f is constant on an open set whose complementary with respect to R is countable; but this implies the Lebesgue measurability of the set X.

Now let us recall that among the non-trivial examples of symmetric sets were the homogeneous sets in the sense of Borel and the Jensen convex sets. Homogeneous sets are involved in some decomposition theorems of Borel [4] which proposed in this respect two problems: "La question reste ouverte de savoir si le continu peut-etre décomposé en un nombre fini ou une infinité dénombrable d'ensembles homogènes égaux, qui ne pourraient être de mesure nulle si leur nombre est fini et ne seraient pas mesurables si leur infinité est dénombrable." Erdös and Marcus [5] have proved that the answer is negative in the finite case (> 1) and affirmative in the infinite case. We may ask whether in these two theorems we can choose the corresponding homogeneous sets to be strongly locally symmetric. The answer is affirmative for the finite case, but negative for the infinite case. Indeed, since strong local symmetry of a set implies its measurability and since Erdös and Marcus [11] have shown that every homogeneous real set other than R is of interior measure equal to zero, it follows that any homogeneous strongly locally symmetric set other than R is of Lebesgue measure zero. However, no finite union of sets of measure zero can be equal to R. On the other hand, a denumerable union of sets of measure zero cannot be equal to R, which is not of measure zero.

Measurability of strongly locally symmetric sets is an important aspect of their regularity. Another regularity of these sets is related to their topological aspect. Since, following the Davies argument, mentioned above, any locally symmetric function is constant on an open set whose complementary is countable, it follows that strongly locally symmetric sets have the Baire property (i.e. they are of the form $G - A \cup B$, where G is open whereas A and B are meagre sets).

STRONG ASYMMETRY AND STRONG ANTISYMMETRY

Local asymmetry of a set X was defined by absence in X of any point of local symmetry. Strong local asymmetry on the interval I should be defined by the property: no point in I is of local symmetry for X. Local antisymmetry of X was defined as local antisymmetry of X with respect to each of its points. Strong local antisymmetry of X on an interval I should be defined as local antisymmetry of X with respect to every point in I (belonging or not to X).

Is strong antisymmetry really possible? The answer is obviously negative, because given two elements x and y of a set, they are symmetric with respect to $(x + y)/2$. As a special remark, no hope to have a Hamel basis which is strongly antisymmetric or an anticonvex set with this property (let us recall here that anticonvex sets were introduced by Luxemburg and discussed in Ref. [3]). What about strong local antisymmetry? Let us assume the existence of a set A which is strongly locally antisymmetric on the interval I. Such a set is locally antisymmetric on I, so, in view of a preceding theorem, the interior of A is empty in I, so $I - A$ is dense in I. On the other hand, if there exists a subinterval J of I whose intersection with A is empty, then A is no longer locally antisymmetric in J. But this contradicts the hypothesis. It follows that A is dense in I. It is thus proved that both A and $I - A$ are dense in $I - A$. Another consequence of strong local antisymmetry is that the cardinal of A is never greater than the cardinal of $I - A$; a similar situation for their measures (if they exist). The characteristic function of A has no point of local symmetry; moreover, it has no point of symmetric continuity (since the difference $f(x + h) - f(x - h)$ alternates the values 1 and 0 when h is approaching zero).

The reader is challenged to check whether such a set A really exists.

SYMMETRIC CONTINUITY

This concept, already used in the preceding sections, deserves more attention. Two main problems were considered; one concerns the structure of the set D_f of discontinuities of a symmetrically continuous function on an open interval I, the other is related to the structure of the set S_g of points of symmetric continuity of an arbitrary real function g of a real variable. Let us first remark the elementary fact that continuity implies symmetric continuity. Continuity of f in x means $f(x + h) - f(x)$ is approaching zero when h is approaching zero. However, $f(x + h) - f(x - h) = f(x + h) - f(x) + f(x) - f(x - h)$, so if $f(x + h) - f(x)$ is approaching zero, the same happens with $f(x + h) - f(x - h)$. The converse is not true. Dirichlet function, equal to zero for x irrational and to one for x rational, is discontinuous at every point, but it is symmetrically continuous at each rational point. Against this discrepancy, we feel that symmetric continuity could have a global influence; such a function could not be "too discontinuous". This feeling is confirmed by a result of Preiss [14], stating that symmetric continuity in the interval I implies continuity almost everywhere in I (i.e. the set D_f is of Lebesgue measure zero). It is also known that D_f is a meagre set [15]. Thus, for any symmetrically continuous function in I the set D_f is both meagre and of measure zero. Since D_f is anyway a countable union of closed sets, the following problem appears naturally: given a meagre set A of measure zero which is a countable union of closed sets, does there exist an everywhere symmetrically continuous function f whose set D_f of points of discontinuity is equal to A? The answer is negative, because we have shown [1] that there exists no symmetrically continuous function f for which D_f is the Cantor ternary set. We recall that this set is formed by those numbers x between 0 and 1 whose development in base 3 contains no occurrence of the digit 1. Cantor's ternary set is non-dense, perfect(= closed and dense in itself) and of measure zero.

Since D_f is both meagre and of measure zero, it is interesting to direct attention towards a new class of negligible sets called porous sets, introduced by Dolženko [16] (see also Zajicek [17] and defined as follows. Let E be a real set. Denote by $l(E, x, d)$ the length of the longest interval in $(x - d, x + d)$, which is disjoint with E. The porosity of E at the point x is defined by the superior limit of $l(E, x, d)/d$ when $d \to 0$. The set E is said to be porous at x if its porosity at x is strictly positive. The set E itself is porous if it is porous at each of its points. A σ-porous set is a countable union of porous sets. Any σ-porous set is meagre and of measure zero, but the converse is not true. Pu [18] asked whether given a real porous set E which is a countable union of closed sets, there always exists a symmetrically continuous function f such that $D_f = E$. Since Cantor's ternary set is porous, the answer to Pu's question is negative.

The following problem remains open: characterize the set D_f of points of discontinuity of a symmetrically continuous function $f : R \to R$.

THE SET OF POINTS OF SYMMETRIC CONTINUITY

Let us refer now to the other problem quoted in the introduction of the preceding section. Given a function $f : R \to R$, what can be said about the structure of the set S_f of its points of symmetric continuity? In Ref. [19] we have proved that given an arbitrary set A which is a countable intersection of open sets, there exists a function f whose set S_f is equal to A. This result is similar to what happens with the usual property of continuity: given an arbitrary set A which is a countable intersection of open sets, there exists a function f whose set C_f of points of continuity is equal to A. However, here the converse is also true: for any function f, the set C_f is a countable intersection of open sets. What about the corresponding situation for symmetric continuity? Is, for any function f, the set S_f a countable intersection of open sets? The answer is negative. An example in this respect is the Dirichlet function, the characteristic function of the set of rational numbers. For this function, the set S_f is equal to the set of rational numbers, which is not a countable intersection of open sets. So, the following problem remains open: find a necessary and sufficient condition for a set A to be the set S_f of a suitable function.

OTHER ASPECTS OF SYMMETRIC CONTINUITY

One of the strongest example of discrepancy between continuity and symmetry continuity is due to Erdös [20], who has shown that if the continuum hypothesis (which asserts that there is no cardinal number greater than that of countable sets and smaller than that of the set of real numbers) is assumed, then there exists an additive subgroup G of R of measure zero, such that $G \cap I$ is a meagre set for no interval I. The characteristic function of G is symmetrically continuous at each point of G, but continuous nowhere.

In Ref. [21] it is shown that if a sequence of symmetrically continuous functions on R converges uniformly on R to the function $f: R \to R$, then f is symmetrically continuous. Since the limit of a pointwise convergent sequence of symmetrically continuous functions is not always symmetrically continuous, it would be interesting to investigate a possible "symmetric classification" of functions; i.e. a classification where the functions of class zero are the symmetrically continuous functions, whereas the pointwise limit of a sequence of functions of class n is a function of class $n + 1$. How effective is such a classification? For what values of n do there exist functions whose class is exactly n?.

In the same order of ideas we may look for a characterization of that type of convergence which could be called symmetric convergence. How to define it in order to have the following theorem: the limit of a convergent sequence of symmetrically continuous functions on R is a symmetrically continuous function on R if and only if the convergence is symmetric? This question was answered by Popescu [22].

LOCALLY SYMMETRIC FUNCTIONS

A function $f: R \to R$ is said to be locally symmetric at the point x in R if there exists a positive number δ_x such that for every h strictly between 0 and δ_x the equality $f(x - h) = f(x + h)$ holds. In Ref. [21] it is proved that every locally symmetric function on R is the limit of a sequence of continuous functions on R. Thus, a classification starting with the locally symmetric functions as the class zero will remain within the framework of the classical classification starting with continuous functions as functions of class zero. How effective is such a locally symmetric classification? Do there exist, for every n, functions of the exact class n? Since the limit function of a pointwise convergent sequence of locally symmetric functions is constant on the complement of a countable set [21], it follows that higher locally symmetric classes contain only functions which are constant, excepting a countable set.

A few words about the temptation to define a property of uniform local symmetry as follows: there exists a positive number δ such that $f(x - h) = f(x + h)$ for any x and any h with $0 < h < \delta$. It is easy to see that the only functions satisfying this property are the constant ones. Thus, *non-trivial local symmetry is never uniform*. A natural problem arises: for a locally symmetric function f what properties of a real set A will guarantee that from uniform local symmetry on $R - A$ follows the identification of f to a constant on R?

NON-TRIVIAL SYMMETRIC CONTINUITY IS NEVER UNIFORM

Trying to impose to symmetric continuity a condition of uniformity, we get the following definition: a function $f: I \to R$ (where I is an arbitrary real interval) is said to be uniformly symmetric continuous if for every number $\varepsilon > 0$ there exists a number $\eta > 0$ such that if $|h| < \eta$ and both $x - h$ and $x + h$ are in I, then $|f(x + h) - f(x - h)| < \varepsilon$ for any x in I. In contrast with symmetric continuity, uniform symmetric continuity is meaningful not only on open intervals, but on an arbitrary interval. This fact is a consequence of a more general one: *uniform symmetric continuity is equivalent to uniform continuity*. Indeed, we can reformulate the new definition as follows: $f: I \to R$ is uniformly symmetric continuous if for every $\varepsilon > 0$ there exists $\eta > 0$ such that for any pair x, y in I satisfying $|x - y| < 2\eta$ we have $|f(y) - f(x)| < \varepsilon$. In other words: *non-trivial symmetric continuity is never uniform*.

Let $f: I \to R$ (I open) be a symmetrically continuous function. Let A be a subset of I on which the symmetric continuity of F is uniform; i.e. for every $\varepsilon > 0$ there exists $\eta > 0$ such that if $|h| < \eta$ then $|f(x + h) - f(x - h)| < \varepsilon$ for every x in A. Obviously, if $I - A$ is finite, then f is uniformly symmetric continuous on I, i.e. it is uniformly continuous on I. It seems interesting to determine cases when, under the conditions given above, with $I - A$ infinite, uniform symmetric continuity on A implies uniform continuity on I. It would also be interesting to find another type of uniformity, which transgresses the trivial case.

SYMMETRY STRUCTURE OF A SET AND OF A FUNCTION: AN OPEN PROBLEM

Given a set A, under what necessary and sufficient conditions there exists another set B containing A and such that B is locally symmetric with respect to every point in A and only with respect to these points? (In order to avoid trivial situations, we exclude from consideration isolated points.)

We will not answer this question in the general case, but we will give an example showing the surprising nature of this problem. Namely, we will prove that given a compact interval $[a, b]$ there exists no set B for which the above set A is $[a, b]$. Indeed, if such a set B would exist, then b would be a point of local symmetry for B. This implies the existence of a point c between a and b, such that the symmetric point with respect to b, of any point in $B \cap (c, 2b - c)$ belongs to B. However, $[a, b]$ is contained in B and contains the interval (c, b), thus the symmetric point with respect to b of any point in (c, b) belongs to B. It follows that the interval $(b, 2b - c)$ is contained in B, thus b is interior to B. It results that any point between b and $2b - c$ is a point of local symmetry for B; but this contradicts the hypothesis that A contains *all* points of local symmetry for B. Thus, no set B exists having A as its set of points of local symmetry. The result remains valid in the more general case when A is a finite union of compact intervals.

When dealing with symmetric continuity, we are forced to work with functions defined on an open interval. However, we may consider a compact interval $[a, b]$ contained in the open interval (c, d) and ask whether there exists a function defined on (c, d) and symmetric continuous at x if and only if x belongs to $[a, d]$. The answer is affirmative. Indeed, starting from a function which is symmetrically continuous on (a, b) and for which $f(b - h)$ tends to zero when h tends to zero, we may define a function g on (c, d) by $g(x) = f(x)$ if x belongs to $[a, b]$ and $g(x) = f(b - h) + \varphi(x), h > 0$, where $x = b + h$ [we assume that (c, d) is smaller than (a, b)] whereas φ is a function which is symmetrically continuous at no point in (b, c), tends to zero when $x \to b$.

SYMMETRIC DERIVATIVES: THEIR HISTORICAL ORIGIN

The concept of symmetric derivative is historically the first in the study of symmetry in mathematical analysis. The initial motivation is related to the study of trigonometric series in the last two centuries (Riemann, Lebesgue, Fatou, Denjoy etc.). A trigonometric series has as its general term the expression $a_n \cos nx + b_n \sin nx$ and is the main tool in the investigation of periodic phenomena in physics. Under adequate circumstances (for instance, uniform convergence) a trigonometric series is the Fourier series of its sum, i.e. the coefficients a_n and b_n are expressed by means of the integral of the sum $f(x)$ multiplied to $\cos nx$ and $\sin nx$, respectively (with a constant factor). The dream of mathematicians interested in this problem was to transform every convergent trigonometric series in a Fourier series of its sum. However, the concept of a Fourier series is dependent on the considered concept of integral. With the classical Riemann integral, many convergent trigonometric series are not Fourier series. At the beginning of our century, Lebesgue introduced a new concept of the integral, more general than the Riemann integral. With the Lebesgue integral the class of Fourier series is considerably enlarged, but still remains a convergent trigonometric series which is not a Fourier series. A decisive attempt to transgress this gap was made by Arnaud Denjoy, who introduced in 1912 a concept of integral (more general than that of Lebesgue) allowing to integrate any finite derivative, by means of a Newton–Leibniz formula type. Unfortunately, although the class of Fourier–Denjoy series is considerably larger than that of Fourier–Lebesgue series, there still remains some convergent trigonometric series which are not Denjoy–Fourier series. An example in this respect is obtained by a series of general term $a_n \sin nx$, where (a_n) is a decreasing sequence with limit equal to zero and such that the series of general term a_n/n is divergent. The sum of this series is not integrable in the sense of Denjoy.

Due to this failure, the problem to express the coefficients of a convergent trigonometric series be means of its sum took another orientation. The classical problem of finding a primitive of a continuous function was replaced by another one, where instead the ordinary derivative some other derivatives are considered. This is the moment when the idea of symmetry penetrates into the study of the trigonometric series. In fact, this idea appears very early, with Riemann, but it only takes a systematic development in our century. Two main types of symmetric derivatives are considered; they will be presented in the next sections. By means of these, Denjoy was able to completely solve the problem of coefficients of a convergent trigonometric series. Such a solution by means of a "primitive of the second order" was announced by Denjoy in 1921, but the details came very late, in a series of volumes published by the same author during the 1950s. Some other solutions of the coefficients problem came concomitantly.

SYMMETRIC DERIVATIVE

The symmetric derivative of the function $f:I \to R$ at the point x in I is the limit of $[f(x + h) - f(x - h)]/2h$, where $h \to 0$ (if this limit exists; if not, we consider the upper and the lower limit of the same expression and we get the upper and the lower symmetric derivative of f at x). We denote the symmetric derivative at x (the upper and the lower symmetric derivatives) by $f^s(x)$ (by $\bar{f}^s(x)$ and $\underline{f}^s(x)$, respectively). In 1928, in an article published in *Fundamenta Mathematica*, Sierpinski posed the question whether there is a non-measurable function whose symmetric derivative is equal to zero at every x. In 1971, Preiss [14] gave a negative answer, by showing that a real function having a finite symmetric derivative at every x is continuous almost everywhere, thus measurable. Larson posed then a new problem: is f measurable if $f^s(x)$ exists (finite or not) at every x? An affirmative answer to this question is obtained by Uher [23]; a second proof of the same result is given by Pu [24].

From the regularity (= measurability) of a function with symmetric derivative we move to the regularity of the symmetric derivative. Improving a previous result by F. M. Filipczak, Larson [25] shows that an arbitrary symmetric derivative is in the first class of Baire (i.e., it is the limit of a sequence of continuous functions).

SYMMETRIC DERIVATIVE VERSUS ORDINARY DERIVATIVE

An important problem is to establish minimal conditions under which symmetric derivative is ordinary derivative. The existence of the latter implies the existence of the former and their equality [because $f(x + h) - f(x - h) = f(x + h) - f(x) + f(x) - f(x - h)$], but the converse is not true. However, globally the converse may be partially true. For instance, a classical result asserts that if f is a continuous real valued function whose symmetrical derivative $f_s'(x)$ exists everywhere, then $f'(x)$ exists exception on a set which is both meagre and of measure zero. An important class of functions having symmetric derivative everywhere is formed by those functions f which have at every x finite right and left derivatives (the difference of two concave upward functions is an example in this respect); for any function of this type, the exceptional set where the ordinary derivative $f'(x)$ does not exist is at most countable. In the general case of existence of symmetric derivative everywhere, situation is different. Foran has given an example of a continuous function f such that $f_s'(x)$ exists everywhere but $f'(x)$ fails to exists on an uncountable set [10]. However, as Foran shows, the exceptional set of his example is "small", i.e. of Hausdorff dimension zero [it is an enumerable set along with a perfect set which can be covered with 2^{n-1} intervals of size $2^n(n!)^{-2}$]. Foran raises the question whether the dimension of the exceptional set can be increased or whether any perfect set of measure zero can be the exceptional set.

SYMMETRIC DERIVATIVE AND MONOTONICITY

It is known from elementary mathematical analysis that for any differentiable function on an interval I monotonicity is equivalent to the fact that the derivative does not change its sign on I. Further investigations have shown that monotonicity is a consequence of some weaker conditions than those mentioned above. Improvements concern the possibility to ignore the behaviour of the derivative on some negligible sets and/or to replace the derivative by upper or lower derivative or by one of the derivates of the function i.e. the upper and lower limits at right and at left of $[f(y) - f(x)/(y - x)]$ when y is approaching x.

In the light of the above facts, a natural question arises. Can we obtain sufficient conditions of monotonicity expressed in terms of symmetric derivative? Several results have been obtained in this respect, but we shall quote only one of them, which seems to be one of the most significant. Evans [26] has proved that for any function $f:R \to R$ with the Darboux property [for any two points u and v and for any value λ between $f(u)$ and $f(v)$ there exists between u and v a point w such that $f(w) = \lambda$] and of the first Baire class, for which the lower symmetric derivative $\underline{f^s}(x)$ is negative for no x in R the function f is non-decreasing if and only if f is measurable and $\liminf f(t) \leqslant f(x) \leqslant \limsup f(t)$ when $t \to x$. In order to understand the generality and power of this result, let us observe that any continuous function is an ordinary derivative, any finite derivative is of first Baire class and has the Darboux property and any function in the classification of Baire is measurable.

Another significant result in the area of symmetry and monotonicity belongs to Belna *et al.*[27]. A real valued function defined on the real line R is said to be *non-decreasing* at x if there exists a positive number δ_x such that $f(x - h) \leqslant f(x) \leqslant f(x + h)$, for all $0 < h < \delta_x$; f is said to be *symmetrically non-decreasing* at x if there exists a positive number δ_x such that $f(x - h) \leqslant f(x + h)$ for all $0 < h < \delta_x$. Let us put $M = \{x; f \text{ is non-decreasing at } x\}$ and $S = \{x; f \text{ is symmetrically non-decreasing at } x\}$. M is contained in S. Belna *et al.* show that the discrepancy between M and S is not important. The set $S - M$ is of measure zero for any measurable function f; $S - M$ is a meagre set for any function f fulfilling the Denjoy property: if the set $A = \{x; \alpha < f(x) < \beta, a < x < b\}$ is not empty, then A is of positive measure.

SCHWARTZ SYMMETRIC FUNCTIONS

Another type of derivative imposed by the study of trigonometric series is the direct second order derivative or Schwartz derivative defined by

$$D^2f(x) = \lim_{h \to 0} \frac{f(x + h) + f(x - h) - 2f(x)}{4h^2}.$$

A concrete problem related to this derivative is to find a continuous function f when we know its direct second order derivative D^2f.

It was proved that if f is a function of the first Baire class, if it has the Darboux property and if $D^2f(x)$ is negative at no point in R, then f is convex (see, for instance, Larson [25]). This result stresses the way in which Schwartz symmetric derivative (called sometimes Riemann derivative or second order symmetric derivative) generalizes the ordinary second order derivative.

How discontinuous can be a function f having at each point a finite Schwartz symmetric derivative? Tran [28] has recently shown that such a function may be measurable, but at the same time discontinuous on an uncountable set. Pu [29] has generalized a result following which for any measurable function f having everywhere a finite Schwartz symmetric derivative the set of points where the upper left derivate is different from the upper right derivate or the lower left derivate is different from the lower right derivate is a meagre set.

OTHER RESULTS, PROBLEMS AND SUGGESTIONS

Since important ideas of symmetry were inspired by the theory of trigonometric series and since these series are mainly related to periodic functions, it seems interesting to check possible relations between local symmetric sets and functions on the one hand and periodic sets and periodic and almost periodic functions, on the other hand. We recall that a real valued function f of a real variable is almost periodic (in the sense of Harald Bohr) if for any $a > 0$ there exists $b > 0$ such that every interval of length b contains at least one point y for which we have $|f(x + y) - f(x)| < a$ for any x in R. Bohr defined this property for continuous functions, but the condition of continuity can be sometimes dropped or weaken.

The *symmetric oscillation* $\omega_s(f, x, y)$ of f on the interval $[x - y, x + y]$ is defined as the supremum of the set $\{|f(x + h) - f(x - h)|; |h| \leqslant y\}$. The symmetric oscillation $\omega_s(f, x)$ of f at x is defined as the infimum of the set $\{\omega_s(f, x, y); y \text{ in } R_+\}$. It is easy to see that $\omega_s(f, x, y)$ is non-decreasing with respect to y. So, we have $\omega_s(f, x) = \lim \omega_s(f, x, y)$ when $y \to 0$. What is the behaviour of the symmetric oscillation $\omega_s(f, x)$? It is equal to zero if and only if f is symmetrically continuous at x. For the usual oscillation $\omega(f, x)$, defined as the infimum of the oscillations of f on various intervals centred at x, it is known that $\omega^3(f, x) = \omega^2(f, x)$ for any $f : R \to R$, and for any x in R ($\omega^2(f, x) = \omega[\omega(f, x), x]$, $\omega^3(f, x) = \omega[\omega^2(f, x), x]$). For other types of oscillation, obtained by neglecting various types of exceptional sets (countable, meagre, of measure zero, finite etc.). Blumberg has shown that in most cases oscillations of order higher than two are not identical [30]. When only the value $f(x)$ is ignored, we get the so-called reduced oscillation $\omega_0(f, x)$, investigated by Froda [31], who showed that for any positive integer n we may have $\omega_0^n(f, x)$ different from $\omega_0^{n+1}(f, x)$. In the light of these situations and taking them as a term of reference, it would be interesting to investigate the iterated symmetric oscillations of an arbitrary function and of some special classes of functions.

Another interesting topic could be the investigation of the possible relations between locally symmetric functions and locally recurrent functions [32, 33] defined by the property that every deleted neighbourhood of x contains a point y such that $f(x) = f(y)$.

Erdös [34] has proved that given an infinite set S in the k-dimensional Euclidean space, there exists a subset S_1 of S having the same cardinal as S, such that all the distances between points of S_1 are distinct. When $k = 1$, the property of S_1 is stronger than that of local antisymmetry. Another result (by Erdös and Kakutani) asserts that the equality between the cardinal of the continuum and the cardinal aleph one is equivalent to the statement that the real line is the union of countably many Hamel bases [34]. Let us recall that any Hamel basis is an antisymmetric set, so we find out the possibility to express the real line as a countable union of countably many antisymmetric sets. On the other hand, as we have seen, the real line is also the union of countably many symmetric sets. We get in this way a new structural analogy between symmetry and antisymmetry.

Recently, we have received from Janusz Jaskula and Bozena Szkopinska a letter announcing some new results concerning the set of points of symmetric continuity of an arbitrary function. These authors obtain some necessary conditions, on the one hand, and some sufficient conditions, on the other hand, that a given set be the set of points of symmetric continuity for some function $f : R \to R$. The concept of symmetric oscillation introduced in Ref. [35] is used in this respect. It is shown the existence of a set which is not the set of all points of symmetric continuity for any function $f : R \to R$. The considered example is, at the same time, an example of a function $f : R \to R$ for which the set of points of symmetric continuity is not measurable (this is done without using the continuum hypothesis, as in Erdös example quoted above). Let us also recall that Belna [36] has shown that for any function $f : R \to R$ the set of points where f is both discontinuous and symmetrically continuous is of interior measure zero.

Larson has published a survey article on symmetrical real analysis, mainly towards the problems of symmetric differentiability. Among the open problems he mentions, let us quote the following: "It is unknown whether symmetrically continuous functions are in any Baire class." [25]

SOME CONCEPTUAL AND PHILOSOPHICAL DIFFICULTIES

Symmetry is a kind of constraint, of restriction, of regularity. At the same time, symmetry is a kind of repetition. Since any regularity is a kind of repetition, there is a tendency to extend the status of symmetry to any regularity. But regularity means presence of some rules. So, we arrive at the paradoxical moment of our concern: rules are equally unavoidable when dealing with various types of non-symmetry. This happens because imposing the absence of a definite rule is still a rule. The culminant point of this situation is represented by antisymmetry, where the basic restrictions are identical to those operating on symmetry (as it was shown above).

How can we transgress this difficulty? We are afraid no answer in this respect exists. A similar experience is significant for our problem too: various attempts to define, to characterize randomness. Randomness may mean very different things: absence of rules; generality (arbitrariness); imprevisibility; high complexity; great entropy; chaotic behaviour etc. Although these aspects interfere, they are not mutually equivalent. When dealing with randomness, we have to precise everytime in what sense is it considered. Many confusions follow from imprecision in this respect. There is no example of a intuitively satisfactory random sequence because in order to make explicit such an example we need a rule. The same happens in the field of symmetry, where no total absence of symmetry can be made explicit. Beyond its proper, precise mathematical meaning, symmetry may mean order, regularity, law, repetition etc. These meanings are not mutually equivalent and, in fact, their mathematical models are different.

REFERENCES

1. S. Marcus, Sur un problème de F. Hausdorff concernant les fonctions symmétriques continues. *Bull. Acad. pol. Sci.* (classe III) **4**(4), 201–205 (1956).
2. S. Marcus, Sur un théorème de F. B. Jones. Sur un théorème de S. Kurepa. *Bull. math. Soc. Sci. math. Phys Repub. pop. roum.*, **1**(4). 433–434. (1957).
3. S. Marcus, On anticonvex sets. *Revue Roum. math. Pure Applic.* **13**(9), 1399–1401 (1968).
4. E. Borel, Les ensembles homogènes. *C. r. hebd. Séanc. Acad. Sci. Paris* **222**, 617–618 (1946).
5. P. Erdös and S. Marcus, Sur la décomposition de l'espace euclidien en ensembles homogènes. *Acta math. hung.* **8** (3–4), 443–452 (1957).
6. S. Marcus, Fonctions convexes et fonctions internes. *Bull. Sci. math.* **81**(2), 1–5 (1957).
7. S. Marcus, Sur une classe de fonctions définies par des inégalités introduites par M. A. Császar. *Acta Sci. math.* **19**(34), 192–218 (1958).
8. S. Marcus, Sur un problème de la théorie de la mesure de H. Steinhaus et S. Ruziewicz. *Bull. Acad. Pol. Sci.* (classe III) **4**(4), 197–199 (1956).
9. R. O. Davies, Symmetric sets are measurable. *Real Analysis Exch.* **4**(1), 87–89 (1978/79).
10. J. Foran, The symmetric and ordinary derivative. *Real Analysis Exch.* **2**(2), 105–108 (1977).
11. T. Z. Ruzsa, Locally symmetric functions. *Real Analysis Exch.* **4**(1), 84–86 (1978/79).
12. M. J. Evans and C. E. Weil, Query 37. *Real Analysis Exch.* **3**, 107 (1977/78).
13. F. Hausdorff, Problème No. 62. *Fundam. Math.* **25**, 578 (1935).
14. D. Preiss, A note on symmetrically continuous functions. *Čas. Pěst. Mat.* **96**, 262–264 (1971).
15. H. Fried, Über die symmetrische Stetigkeit von Funktionen. *Fundam. Math.* **29**, 134–137 (1937).
16. E. P. Dolženko, Boundary properties of arbitrary functions (in Russian). *Izv. Akad. Nauk SSSR* **31**(1), 1–12 (1967).
17. L. Zajicek, Sets of σ-porosity and sets of σ-porosity (q). *Čas. Pěst. Math.* **101**, 350–359 (1976).
18. H. W. Pu, Personal letter. 4 November (1983).
19. S. Marcus, Mulţimile F$_\sigma$ şi continuitatea simetrică. *Bul. Şti. Acad. Rep. Pop. rom. seria Matem. Fiz.* **7**(4), 871–886 (1955).
20. P. Erdös, Some remarks on subgroups of real numbers. *Colloquium math.* **42**, 119–120 (1979).
21. P. Kostyrko, T. Neubrunn, J. Smital and T. Salát, On locally symmetric and symmetrically continuous functions. *Real Analysis Exch.* **6**, 67–76 (1980/81).
22. S. Popescu, On symmetric continuity. *Noesis* **12** (1986).
23. I. Uher, Symmetrically differentiable functions are differentiable almost everywhere. *Real Analysis Exch.* **8** (1), 252–260 (1982/83).
24. H. W. Pu, Measurability of real functions having symmetric derivatives everywhere. *Real Analysis Exch.* **10**(1), 214–219 (1984/85).
25. L. Larson, Symmetric real analysis: a survey. *Real Analysis Exch.* **9**(1), 154–178 (1983/84).
26. M. S. Evans, A symmetric condition for monotonicity. *Real Analysis Exch.* **3**(12), 98–99 (1977/78).
27. C. L. Belna, M. J. Evans and P. D. Humke, Symmetric monotonicity. *Real Analysis Exch.* **2**(2), 120–124 (1977).
28. T. Tran, Symmetric functions whose sets of points of discontinuity is uncountable. *Real Analysis Exch.* **12**(2), 496–509 (1986/87).
29. H. W. Pu, Derivates for symmetric functions. *Real Analysis Exch.* **4**(2), 164–166 (1978/79).
30. H. Blumberg, Certain general properties of functions. *Ann math.* **18**, 147–160 (1917).
31. A. Froda, Aspura distribuţiei discontinuităţilor funcţiilor reale. *Comunle Acad. Rep. pop. rom.* **5**(1), 31–37 (1955).
32. K. A. Bush, Locally recurrent functions. *Am. math. Mon.* **69**, 199–206 (1962).
33. S. Marcus, On locally recurrent functions. *Am. math. Mon.* **70**(8), 822–826 (1963).
34. P. Erdös, Set theoretic, measure theoretic, combinatorial, and number theoretic problems concerning point sets in Euclidean space. *Real Analysis Exch.* **4**(2), 113–138 (1978/79).
35. S. Marcus, Symmetry in one dimension and symmetric continuity. *Noesis* **12**, 98–103 (1986).
36. C. L. Belna, Symmetric continuity of real-functions. *Proc. Am. math. Soc* **87**(1), 99–102 (1983).

Computers Math. Applic. Vol. 17, No. 1–3, pp. 117–124, 1989
Printed in Great Britain. All rights reserved

0097-4943/89 $3.00 + 0.00

DELICATE SYMMETRY

J. Pach

Mathematical Institute, Hungarian Academy of Sciences, Budapest, P.O.Box 127, H-1364, Hungary and
Courant Institute, New York University, 251 Mercer St, New York, NY 10012, U.S.A.

Abstract—It is illustrated by a few mathematical results (mainly from combinatorics and discrete geometry) that

    (i) even the most chaotic structures necessarily contain relatively large symmetric substructures;
    (ii) minor perturbations may lead to radical transformations of a symmetrical system;
    (iii) some properties of symmetrical structures are stable.

We give a new elementary proof of (a generalization of) the following theorem conjectured by M. Freedman. Let S denote the set of all lattice points in the plane with integer coordinates, and consider a mapping from S into the plane satisfying some Lipschitz condition. Then any sufficiently large disc contains at least one point which is the image of a lattice point.

"But what about the soul? What is better: if chaos prevails in it, or if order and harmony?" (Socrates/Plato)

1. INTRODUCTION

There has always been something mysterious about symmetry. Artifacts and relics of various cultures from all ages abound with wonderful symmetric patterns and figures. They are pleasing to the eye and the soul, in fact, they seem to embody our eternal human aspiration for perfection and harmony. The Pythagoreans attributed special power to the circle and the sphere, because these are the only shapes with full rotational symmetry. For similar reasons, Aristotle assumed the Earth to be round and Kepler suspected some sort of mysterious connection between the five regular platonic bodies and the orbits of the planets known at his time. These thoughts, however superstitious they may appear, truly reflect the belief shared by most scientists that the basic laws of Nature must be simple and harmonic. The discovery of a number of transformations that leave the physical laws and phenomena unchanged, i.e. the basic principles of symmetry, certainly rank among the greatest achievements of modern science. Mathematicians cherish a belief that they possess a uniform language capable of describing a very wide range of concepts and phenomena related to symmetry. However, is it conceivable that the key to the mystery lies in a language?

In this paper we make an attempt to illustrate three different aspects of the stability of symmetrical structures by a few examples taken from combinatorics and discrete geometry. The last section is slightly more technical than the first two, for it contains a new elementary proof of a result conjectured by Michael Freedman.

2. TOTAL CHAOS IS IMPOSSIBLE

Investigating how friendships develop in school communities, some 30-years ago the wellknown Hungarian sociologist Alexander Szalay discovered that in the classes he tested there were surprisingly large groups of students who either mutually liked or mutually disliked each other. Instead of looking for psychological reasons which might explain this phenomenon, he started drawing graphs whose nodes represent students, two nodes being joined by an edge, if the corresponding students like each other. Having an excellent mind, he soon came up with an accurate proof of the fact that, for any natural number n, there exists another number $R(n)$ with the property that every graph with at least $R(n)$ nodes contains either a complete or an empty subgraph with n nodes. However, this theorem had been established before by F. Ramsey, and it turned out to be one of the central results in modern combinatorics. (For a detailed account on this subject consult Ref. [1].)

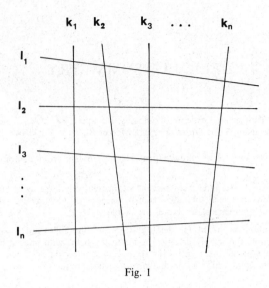

Fig. 1

Another classical theorem of similar spirit is due to van der Waerden [2]. It states that, no matter how we divide the set of all natural numbers into two (or, in fact, into any finite number of) classes, we can always find arbitrarily long arithmetic progressions, all of whose elements belong to the same class.

Our third example is taken from geometry. Two bundles of straight lines $\{l_1, l_2, \ldots, l_n\}$ and $\{k_1, k_2, \ldots, k_n\}$ are said to form a gridlike configuration, if they meet in the manner shown in Fig. 1. (The l_is need not be parallel to each other, but they have to cross the k_js in the order prescribed by their numbering.) In Ref. [13] it is proved that for every natural number n there exists a number $G(n)$ such that, in any arrangement of at least $G(n)$ lines in the plane such that no two of them are parallel, one can find two n-element bundles forming a gridlike configuration.

With a little profane wording, we can say that the common moral of these results is that *all structures* (even the most chaotic ones) *contain symmetric* (i.e. highly regular) *substructures*. Since many important results from various branches of mathematics can be paraphrased in a similar form, this statement can rightly be called a *metatheorem*.

It might be interesting to point out that there is an intimate relationship between *chaotic* and *random* structures. Let us think of the chaos manifested in dadaist poems (composed by putting randomly chosen words one after the other) or the art of the so-called action-painters. More seriously, let us recall the remarkable fact that the investigation of the chaotic motion of gas molecules led to the birth of mathematical statistics and of the theory of random processes. Thermodynamics and agitation cannot be understood without the laws of chance. As far as mathematics is concerned, it was realized towards the middle of this century that many different highly irregular mathematical objects can only be constructed with the help of random methods [4]. In particular, the best lower bounds for the numbers $R(n)$ in Ramsey's above-mentioned theorem were also obtained in this way.

It is a well-known result in elementary probability theory that there is a number $F(n)$ with the property that, if we flip a coin at least $F(n)$ times, then we get n consecutive heads (or n consecutive tails) with probability greater than 0.99. Note that the inevitable occurrence of regular substructures in random constructions is in perfect agreement with our metatheorem (and may well be responsible for the genesis of life on earth).

An important common feature of the above examples is that the size of the regular substructures, whose existence is guaranteed by the theorems, is usually much smaller than the size of the original structures. In some extreme cases it can even happen that the regularity is confined to a single point. A typical instance of such a result is the following "fixed point theorem": no matter how we comb a hedgehog, we can always find a spine which sticks out from its body perpendicularly. (The hedgehog has a convex smooth body completely covered by spines, i.e. by straightline segments. We also assume that, if two spines are rooted close to each other, then they are nearly parallel, i.e. the direction of the spines is a continuous function defined on the surface/skin of the hedgehog.) The proof of this result is based on the following useful topological lemma of Sperner (see e.g. Ref. [5]).

Let C be a simple closed Jordan curve (say, a circle) and let r, b, g denote three distinct points of C. Assume that C and its interior are coloured by three colours (red, blue, green), i.e. it is covered by three (not necessarily disjoint) compact sets R, B, G so that

(i) $r \in R$, $b \in B$, $g \in G$;
(ii) $\overline{rb} \in R \cup B$, $\overline{bg} \in B \cup G$, $\overline{gr} \in G \cup R$,

where $\overline{rb}$ denotes the portion of C between r and b, that does not contain g.

Then there is a point getting all three colours, i.e. $R \cap B \cap G$ is nonempty.

3. FRAGILE SYMMETRY

If there are no strong forces acting in a physical system, then the particles tend to move randomly, chaotically. The most probable state of the system is the one, in which the disorder is maximal. This universal tendency towards chaos is at the bottom of Maxwell's laws of thermodynamics.

However, if we put the matter under great pressure or we cool it down, then the forces between the particles overpower the agitation and a new phenomenon appears: the particles start forming regular configurations, symmetric molecules, crystals. Only those formations survive, which are stable. (In physics this usually means that they locally minimize the energy level of the system.) The crystals of many different metals (aluminium, copper, iron, magnesium etc.) consist of regular hexagonal layers of atoms. (One such layer is shown in Fig. 2.) The centres of the atoms are said to form a *hexagonal lattice*.

In almost all domains of life we frequently come across beautiful, symmetric, perfectly balanced situations and arrangements which helplessly collapse, due to their *instability*. (Just try to balance a match on your finger, or build a house of cards!) Catastrophe theory and the theory of differential equations provide a large number of mathematical examples showing that minor changes in the circumstances may provoke radical transformations of the whole system. We would like to present here two similar examples from the field of discrete geometry. In both cases the symmetric configuration, we start with, is the hexagonal lattice.

A system of pairwise disjoint open discs in the plane is called a circle packing. Almost a century ago A. Thue proved that the densest packing of *equal* circles in the plane is the so-called *regular packing* shown in Fig. 2. (See e.g. Ref. [6].) In this arrangement every circle has exactly six neighbours. What happens, if we slightly perturb this system, while making sure that the incidence structure remains unchanged? Is it possible that we obtain a new arrangement, in which the circles are not equal any more, but their radii vary (say from 10^{-10} to 10^{10})? László Fejes Tóth conjectured that the answer to this question is in the negative.

The following (stronger) theorem was proved in Ref. [7]: if all elements of a circle packing have at least six neighbours, then it is either the regular packing, or it contains arbitrarily small circles. In other words, if we slightly change the size of just one element of the system, then that necessarily leads to dramatic changes in the size of infinitely many others. (Recently, a related result has been established by B. Rodin and D. Sullivan.)

Note that a similar sort of instability is expressed in the following well-known fact of elementary calculus: a convex function is either constant or unbounded. A discrete version of this statement plays an interesting role in the theory of Markov processes. Let us assign a real number $f(x)$ to each vertex x of the hexagonal lattice, so that $f(x)$ is at most as large as the average of the values assigned to its neighbours. That is, for every x,

$$f(x) \leqslant \frac{1}{6} \sum_{i=1}^{6} f(x_i),$$

where $x_1, \ldots, x_6$ denote the neighbours of x. Then $f(x)$ is either a constant function or it is not bounded from above. (See e.g. Ref. [8].)

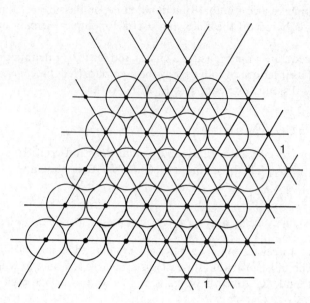

Fig. 2

Next we describe another way of creating chaos from symmetry. Let us fix two points of the plane, A and B at distance 1 from each other, and draw two unit circles around them. They will meet in two points. If we draw two unit circles around these points, too, then we get six new intersection points. In general, at every step of our algorithm we add to our picture all unit circles whose centres are intersection points of previously defined circles, and which have not been drawn in before.

Taking a few steps, we obtain a picture similar to Fig. 3. If we follow this procedure beyond any limit, then the centres of our circles will form a hexagonal lattice.

What happens if the distance of A and B, the centres of the two initial unit circles, slightly differs from 1? Let $S(A, B)$ denote the set of centres of all unit circles produced by our (infinite) drawing algorithm starting with A and B. An (infinite) point set S is said to be *everywhere dense* in the plane if every circular disc contains at least one element of S. The following conjecture of Fejes Tóth was proved in Ref. [9]. Let A and B be two points of the plane having mutual distance $d \leqslant 2$. If $d = 1, 2$ or $\sqrt{3}$, then $S(A, B)$ is the regular hexagonal lattice. Otherwise, $S(A, B)$ is everywhere dense in the plane.

This can again be regarded as an instability result: if we slightly move away from the value $d = 1$, then the simple, symmetrical structure of $S(A, B)$ fatally collapses.

Fig. 3

4. FLEXIBLE SYMMETRY

If we melt a piece of crystallized pure metal, then its regular structure disintegrates and conspicuous changes take place. On the other hand, many important characteristics of the material remain almost unchanged. In particular, at the beginning most atoms do not get very far from their original positions in the lattice, no large holes arise in the system, and its total volume does not essentially increase either.

In this section we would like to give some mathematical examples showing that a slight deviation from symmetry may leave some important structural properties of the system invariant. Our "guinea pig", the symmetrical system we are marring, is again the hexagonal lattice on Fig. 2.

It is a well-known fact, which follows by elementary number theoretic arguments, that there is a line in the plane such that the orthogonal projection into this line takes the vertex set of the hexagonal lattice into a point set everywhere dense in the line. In other words, there is a direction such that any parallel strip of positive width, pointing to this direction, contains at least one lattice point. That is, the hexagonal lattice is not transparent in this direction. (In fact, any direction not determined by two lattice points will do.) The following statement (see Ref. [10]) shows that this interesting feature of the lattice will be preserved, if we slightly disturb its structure without creating large empty holes. Given any positive number r and any (infinite) point set S in the plane with the property that every disc of radius r contains at least one element of S, one can always find a line such that the orthogonal projection of S into this line is everywhere dense.

To describe another example of a result showing the stability of some property of the hexagonal lattice, we need a little preparation. For any two points x and y in the plane, let $|x - y|$ denote their Euclidean distance. Let S be an infinite subset of $\mathbb{R}^2$ (the plane), and let $\lambda \leqslant \Lambda$ be two positive numbers. A mapping f from S into $\mathbb{R}^2$ is said to satisfy the *Lipschitz condition* with coefficients λ and Λ, if

$$\lambda \leqslant \frac{|f(x) - f(y)|}{|x - y|} \leqslant \Lambda,$$

for any pair of distinct elements (x, y) of S. Roughly speaking, this means that f does not radically change the distances between the elements of S.

Michael Freedman [11] has recently conjectured that, if S is the hexagonal lattice and f is a mapping from S into $\mathbb{R}^2$ satisfying some Lipschitz condition, then any sufficiently large disc in the plane contains at least one element of $f(S)$. That is, there are no large empty holes in the image of S. Since the hexagonal lattice does not contain large holes either, this also can be regarded as some kind of stability property.

In the sequel we are going to sketch a new elementary proof of (a generalization of) this conjecture. At this point we say goodbye to the reader not interested in the mathematical details.

Theorem

Let d, λ, Λ be positive numbers, let S be a point set in the plane with the property that every open disc of diameter d contains at least one element of S, and let $f : s \to \mathbb{R}^2$ be a mapping satisfying the Lipschitz condition with coefficients λ and Λ.

Then every open disc of diameter $20(\Lambda^2/\lambda)d$ contains at least one point belonging to $f(S)$.

We may obviously assume without loss of generality that $d = \lambda = 1$, otherwise we can rescale our picture. We also suppose for simplicity that Λ is an integer. For any natural number $n \geqslant 3$, take two concentric circles around the origin with radii n and $n + 1$, respectively. Let us divide the ring R^n between these two circles into $6n$ congruent pieces by $6n$ rays starting from the origin, as shown in Fig. 4. These pieces will be denoted by $R_1^n, R_2^n, \ldots, R_{6n}^n$. In view of the fact that each R_i^n contains a disc of unit diameter, our assumptions guarantee that we can pick a point $s_i^n \in S$ lying in the interior of R_i^n. Note that $|s_i^n - s_{i+1}^n| \leqslant 3$, hence $|f(s_i^n) - f(s_{i+1}^n)| \leqslant 3\Lambda$, for every i.

Let $(t_1^n, t_2^n, \ldots, t_{k(n)}^n)$ be a minimal circular subsequence of $(s_1^n, s_2^n, \ldots, s_{6n}^n)$, with the property that $|f(t_i^n) - f(t_{i+1}^n)| \leqslant 3\Lambda$, for all $1 \leqslant i \leqslant k(n)$.

Claim 1

The points $f(t_1^n), f(t_2^n), \ldots, f(t_{k(n)}^n)$ in this circular order determine a simple (i.e. nonselfintersecting) closed polygon, which will be denoted by P^n.

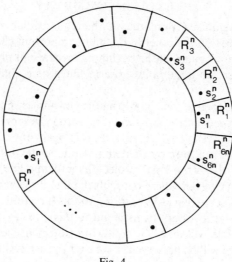

Fig. 4

Proof. Assume, in order to obtain a contradiction, that the segments $f(t_i^n)f(t_{i+1}^n)$ and $f(t_j^n)f(t_{j+1}^n)$ intersect each other for some $i \neq j$. Then either $|f(t_i^n) - f(t_j^n)| \leqslant 3\Lambda$ or $|f(t_{i+1}^n) - f(t_{j+1}^n)| \leqslant 3\Lambda$. Hence either all points between $f(t_i^n)$ and $f(t_j^n)$, or all points between $f(t_{i+1}^n)$ and $f(t_{j+1}^n)$ can be deleted from the sequence without violating the conditions, which contradicts the minimality of P^n.   ☐

Claim 2

If $n \geqslant 3\Lambda$, then P^n contains a disc of radius $n/3$ in its interior.

Proof. Let ρ be the radius of the largest disc inscribed in P^n. Since no edge of P^n is longer than 3Λ, for any x inside P^n we can find at least one vertex $f(t_i^n)$ with $|x - f(t_i^n)| \leqslant \rho + \frac{3}{2}\Lambda$. If such a vertex exists with $t_i^n \in \bigcup_{m=1}^{2n} R_m^n$, then let us colour x red. If $t_i^n \in \bigcup_{m=2n+1}^{4n} R_m^n$ or $t_i^n \in \bigcup_{m=4n+1}^{6n} R_m^n$, then we colour x blue or green, respectively. Note that some points may get more than one colour. It is easy to see that this colouring satisfies the conditions of Sperner's lemma (stated at the end of Section 2), thus we can conclude that there is a point x_0 in P^n which gets all three colours. This means that

$$|x_0 - f(t_i^n)| \leqslant \rho + \tfrac{3}{2}\Lambda, \quad \text{for some} \quad t_i^n \in \bigcup_{m=1}^{2n} R_m^n,$$

$$|x_0 - f(t_j^n)| \leqslant \rho + \tfrac{3}{2}\Lambda, \quad \text{for some} \quad t_j^n \in \bigcup_{m=2n+1}^{4n} R_m^n,$$

$$|x_0 - f(t_k^n)| \leqslant \rho + \tfrac{3}{2}\Lambda, \quad \text{for some} \quad t_k^n \in \bigcup_{m=4n+1}^{6n} R_m^n.$$

Hence we obtain

$$|f(t_i^n) - f(t_j^n)|, |f(t_j^n) - f(t_k^n)|, |f(t_k^n) - f(t_i^n)| \leqslant 2\rho + 3\Lambda.$$

Observe that at least one of the three distances determined by t_i^n, t_j^n, t_k^n is at least $\sqrt{3}n$, say, $|t_i^n - t_j^n| \geqslant \sqrt{3}n$. Thus

$$\sqrt{3}n \leqslant |t_i^n - t_j^n| \leqslant |f(t_i^n) - f(t_j^n)| \leqslant 2\rho + 3\Lambda,$$

which implies $\rho \geqslant n/3$.   ☐

Claim 3

Let $n \geqslant 20\,\Lambda^2$ be fixed, $m = n + 4\Lambda$. Then

- (i) P^m contains P^n in its interior;
- (ii) the minimal distance between P^n and P^m is at least Λ;
- (iii) every disc of diameter $20\Lambda^2$, whose centre is in the ring bounded by P^n and P^m, contains at least one vertex of P^n.

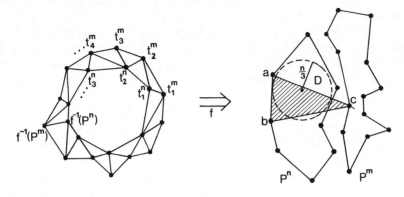

Fig. 5

Proof. By our assumptions, for all i and j

$$|f(t_i^n) - f(t_j^m)| \geqslant |t_i^n - t_j^m| \geqslant m - n - 1 \geqslant 4\Lambda - 1.$$

In view of the fact that the sidelengths of P^n and P^m are at most 3Λ, this implies condition (ii). In particular, P^n and P^m do not intersect each other.

To show condition (i), it is sufficient to exclude the following two possibilities:

(1) P^n contains P^m in its interior;
(2) the interiors of P^n and P^m are disjoint.

We can obviously choose an infinite sequence of distinct elements $t_1^m = s_1, s_2, s_3, \ldots \in S$, whose distances from the origin are larger than m, and $|s_k - s_{k+1}| \leqslant 3$, hence $|f(s_k) - f(s_{k+1})| \leqslant 3\Lambda$, for every k. Further, we can ensure that $|s_k - s_j| > 1$, hence $|f(s_k) - f(s_j)| > 1$, for all k and j. Assume possibility (1). Then $f(s_1) = f(t_1^m) \in P^m \subseteq \operatorname{int} P^n$. Suppose that $f(s_k) \in \operatorname{int} P^n$ but $f(s_{k+1}) \notin \operatorname{int} P^n$ for some k. Then there exists a vertex $f(t_i^n)$ of P^n such that either $|f(s_k) - f(t_i^n)| < 3\Lambda$ or $|f(s_{k+1}) - f(t_i^n)| < 3\Lambda$. On the other hand,

$$|f(s_k) - f(t_i^n)| \geqslant |s_k - t_i^n| \geqslant m - n - 1 \geqslant 4\Lambda - 1,$$

$$|f(s_{k+1}) - f(t_i^n)| \geqslant |s_{k+1} - t_i^n| \geqslant m - n - 1 \geqslant 4\Lambda - 1.$$

This contradiction implies that $f(s_k) \in \operatorname{int} P^n$, for every k. Since $\operatorname{int} P^n$ is a bounded set, there are j and k such that

$$|f(s_k) - f(s_j)| < 1,$$

which is impossible. Hence possibility (1) cannot hold.

Assume next possibility (2), and consider the two closed polygons $f^{-1}(P^n) = t_1^n t_2^n \ldots t_{k(n)}^n$ and $f^{-1}(P^m) = t_1^m t_2^m \ldots t_{k(m)}^m$. It is clear from the definitions that the ringlike region bounded by $f^{-1}(P^n)$ and $f^{-1}(P^m)$ can be triangulated (without introducing new vertices) so that the diameter of every triangle is at most 10Λ. (See Fig. 5.) Let us notice that f carries this triangulation into a collection of triangles whose union covers the interior of P^n. In particular, we can find a triangle abc having diameter at most $10\Lambda^2$ (by the Lipschitz condition), which is covering the centre of the disc D of radius $n/3$ contained in P^n (cf. Claim 2). Since a,b,c are not contained in D, we obtain $10\Lambda^2 \geqslant \sqrt{3}\frac{n}{3}$, contradicting our assumption on n. This proves condition (i).

The proof of condition (iii) is very similar to that of condition (i), and it is left to the reader. $\square$

Let us consider now the partition of the plane into ringlike regions, determined by the polygons $P^{20\Lambda^2 + k\Lambda}$ $(k = 0, 1, 2, \ldots)$. Part (ii) of Claim 3 implies that every point of the plane is either in $P^{20\Lambda^2}$ or it is contained in the ring bounded by $P^{20\Lambda^2 + k\Lambda}$ and $P^{20\Lambda^2 + (k+1)\Lambda}$, for some k. On the other hand, by condition (iii), every disc of diameter $20\Lambda^2$, whose centre is not in $P^{20\Lambda^2}$ contains at least one element of $f(S)$. If we repeat the whole argument starting with a new origin which is far away from the old one, then we will find that the discs centred in $P^{20\Lambda^2}$ are not exceptional. This completes the proof of our theorem.

Note that with a little more care the value $20(\Lambda^2/\lambda)d$ in the theorem can be improved to $20(\Lambda/\lambda)d$. Somewhat weaker results relying on advanced analytical techniques have been recently found by Freedman and some of his students. According to the first result mentioned in this section, our theorem has the following corollary.

Corollary

Let d, λ, Λ be positive numbers, let S be a point set in the plane with the property that every open disc of diameter d contains at least one element of S, and let $f: S \to \mathbb{R}^2$ be a mapping satisfying the Lipschitz condition with coefficients λ and Λ.

Then one can find a line such that the orthogonal projection of $f(S)$ into this line is everywhere dense.

All results of this section can be generalized to higher dimensions.

REFERENCES

1. R. Graham, B. Rothschild and J. Spencer, *Ramsey Theory*. Wiley-Interscience, New York (1980).
2. B. L. van der Waerden, How the proof of Baudet's conjecture was found. In *Studies in Pure Math*. (Ed. L. Mirsky) pp. 251–260. Academic Press, New York (1971).
3. J. Pach and M. Sharir, The upper envelope of piecewise linear functions and the boundary of a region enclosed by convex plates: Combinatorial analysis. *Discr. Comput. Geom.* **3**, (1988).
4. P. Erdös and J. Spencer, *Probabilistic Methods in Combinatorics*. Academic Press, New York (1974).
5. E. M. Patterson, *Topology*. Oliver and Boyd, London (1956).
6. L. Fejes Tóth, *Lagerungen in der Ebene auf der Kugel und im Raum* (2. Aufl.). Springer, New York (1972).
7. I. Bárány, Z. Füredi and J. Pach, Discrete convex functions and proof of the six circle conjecture of Fejes Tóth. *Can. J. Math.* **36**, 569–576 (1984).
8. E. B. Dynkin and A. A. Juschkewitsch, *Sätze und Aufgaben über Markoffsche Prozesse*. Springer, Berlin (1969).
9. K. Bezdek and J. Pach, A point set everywhere dense in the plane. *Elem. Math.* **40**, 81–84 (1985).
10. J. Beck, F. Galvin and J. Pach, Advanced Problem No. 6421. *Am. math. Mon.* **90**, 134 (1983).
11. M. Freedman, Personal communication.

Computers Math. Applic. Vol. 17, No. 1–3, pp. 125–132, 1989
Printed in Great Britain. All rights reserved

HALLEY MAPS FOR A TRIGONOMETRIC AND RATIONAL FUNCTION

C. A. Pickover

IBM Thomas J. Watson Research Center, Yorktown Heights, NY 10598, U.S.A.

Abstract—Among the methods available for the characterization of complicated mathematical and physical phenomena, computers with graphics are emerging as an important tool. In this article, I present computational and graphical results on Halley's method for one-parameter functions of the form $\zeta(\zeta^\alpha - 1) = 0$ and $\sin(\zeta) = 0$ in order to gain insight as to where the method can be relied upon and where it behaves strangely. The resulting plots reveal a visually striking and intricate class of patterns indicating behavior ranging from stable attractive and repulsive points to chaos. Iterative approximation methods such a Halley's method occur frequently in science and engineering.

The use of computational techniques based on the use of recurrence relationships can be traced back to the dawn of mathematics. The Babylonians used such techniques to compute the square root of a positive number, and the Greeks to approximate π [1]. Today virtually every important special function of mathematical physics may be computed by recurrence formulas. The *goal* of this article is to give a flavor of the subject of recurrence relations and chaos, and the computer graphics reveal not only the beauty that can be found in such relationships but also provide insight into their behavior. Research over the past decade has made it clear that many systems of physical, biological, electrical and chemical interest exhibit highly unstable or chaotic behavior [1, 2]. Chaos theory often involves the study of how complicated behavior can arise in systems which are based on simple rules, and how minute changes in the input of a system can lead to large differences in the output.

Numerical methods used numbers to simulate mathematical processes, which in turn usually simulate real-world situations [3]. The choice of a particular algorithm influences not only the process of computing but also how we are to understand the results when they are obtained. In this article, I address the process of solving equations of the form $f(x) = 0$. The problem of finding the zero's of a continuous function by iterative methods occurs frequently in science and engineering [3–6]. These approximation techniques start with a guess and successively improve upon it with a repetition of similar steps. The graphs in this paper give an indication of how well one of these iterative methods, Halley's method, works in order to gain insight as to where Halley's method can be relied upon and where it behaves strangely. Halley's method is of interest theoretically because it converges rapidly relative to many other methods. Interesting past works includes a study of the iterates of a related method, Newton's method [6], for cubic polynomials. Other work suggests that computer graphics can play a role in helping mathematicians form the intuitions needed to prove theorems about convergence of points in the complex plane [7]. Some new features of my paper include the focus on high-resolution graphics characterizing chaotic aspects of the behavior of Halley's method applied to functions with a root at the origin, unusual convergence tests, and the application of image processing techniques and other graphical methods in order to reveal the subtle structures of the maps.

Let $F(\zeta)$ be a complex-valued function of the complex variable ζ. The *Halley map* is the function

$$H(\zeta)\colon \zeta_{n+1} = \zeta_n - \left[\frac{F(\zeta_n)}{F'(\zeta_n) - \left(\dfrac{F''(\zeta_n)F(\zeta_n)}{2F'(\zeta_n)} \right)} \right]. \tag{1}$$

This iteration is used to find the zeros of F and is clearly derived in Ref. [5]. In brief, we may develop the Halley method by truncating the Taylor series expansion of $F(\zeta)$ about a point ζ_n after the second derivative:

$$F(\zeta) = F(\zeta_n) + F'(\zeta_n)(\zeta - \zeta_n) + \left(\frac{F''(\zeta_n)(\zeta - \zeta_n)^2}{2} \right). \tag{2}$$

If we substitute $\zeta = \zeta_{n+1}$ and assume ζ_{n+1} is a good approximation to the root so that $F(\zeta_{n+1}) = 0$ we get

$$0 = F(\zeta_n) + F'(\zeta_n)(\zeta_{n+1} - \zeta_n) + \left(\frac{F''(\zeta_n)(\zeta_{n+1} - \zeta_n)^2}{2} \right). \tag{3}$$

To finish the derivation, solve for ζ_{n+1}.

I consider functions, F, that are analytic in the complex plane $\mathbb{C}$. "ζ_z" is a zero of F and a fixed point of $H : H(\zeta_z) = \zeta_z$. The *basin of attraction* of ζ_z is the set of all points whose forward orbits by H converge to ζ_z.

To simplify the discussion, I first consider, as an example, the one-parameter polynomial

$$\zeta(\zeta^6 - 1) = 0. \tag{4}$$

Polynomial problems occur frequently in practice, and polynomials are useful for theoretical study since a polynomial of degree M has exactly M zeros, and we therefore know when we have found all the zeros. This polynomial has seven roots; one is at the origin, and the others are at $\zeta = e^{2\pi i m/6}$. I also studied the simple trigonometric function $\sin(\zeta) = 0$ which has periodic roots on the real axis. In this article, the graphical behavior of Halley's method results from forward iteration. An initial point on the complex plane is selected and iterated N times. Traditionally a point is considered to have converged if

$$|\zeta_{n+1} - \zeta_n|^2 < \epsilon, \tag{5}$$

where ϵ is a small value. However in this article the following test was used (discussed later):

$$\| (\zeta_n + 1)|^2 - |\zeta_n|^2 | < \epsilon. \tag{6}$$

To verify that this criterion for ending the iteration has actually allowed the system to converge to a root, $|F(\zeta)| < \epsilon$ was used in conjunction with equation (6) producing visually identical plots. The value of ϵ was 0.0001. The iteration in equation (1) was performed on four million initial parameter values in a 2000-by-2000 point square grid.

Three types of plots are presented. One is a bi-level plot, created by plotting a black dot if $n = 0 \bmod(2)$. This operation creates contour lines and helps to visually emphasize different regions of behavior of the function. The second type of graph maps the value of n to darkness on the graph, thereby showing relative rates of convergence within each basin of attraction. *Histogram equalization* (a digital picture-processing technique) was performed in order to visually bring out features in the map. Histogram equalization takes a raster of intensities, plots the number of times each intensity occurs, and then creates a mapping from the original intensities to a new set so that each intensity level occurs with approximately equal frequency. Finally the map is halftoned using *damped error diffusion* [8]. The third plot maps iteration to height on a 3-D map. The x–y plane for these maps represents the complex plane.

Figure 1 shows a graph of Halley's map for $\zeta(\zeta^6 - 1) = 0$. The basins of attraction for the roots of the equation are displayed for various initial values of (ζ_0) in the complex plane (between -2.5 and 2.5 in the real and imaginary directions). The six central white regions, and the region at the origin, contain the roots and correspond to starting points where convergence is achieved rapidly (within three iterations). Initial "guesses" in the tearshaped basins fanning out from the roots are "safe"; that is, any starting points selected from these regions come close to a root within a small number of iterations. Black regions converge much more slowly (about 50 iterations), and behavior on the black radial boundary region is considerably more complicated. These borders consist of elaborate swirls that can pull Halley's method into any one of the seven roots. In this vicinity, a tiny shift in starting point can lead to widely divergent results.

Figure 2 is a contour plot of the same region of the complex plane as a Fig. 1. Note the complicated behavior along the boundary regions and the various "nodules" along the high-iteration radial branches. The use of equation (6) produces the whisker-like projections around each contour, and these whiskers generally point to the root (or to regions of fast convergence). Therefore, directionality now can be easily understood by observing the contour plots. Some contours do not contain whiskers, and these are regions which converge to the root at the origin. Figure 3 is a magnification of one of the nodules near the origin in Fig. 2 and gives a high-resolution

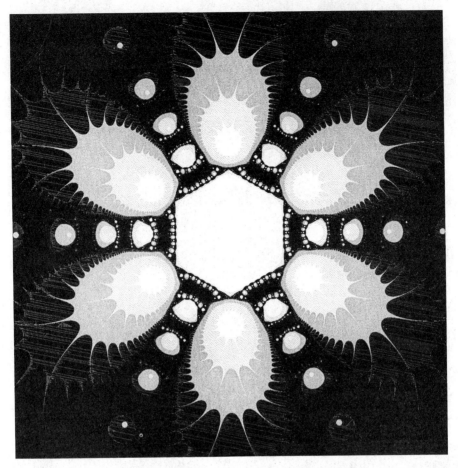

Fig. 1. Halley map portrayed for $\zeta(\zeta^\alpha - 1) = 0$, where $\alpha = 6$. The basins of attraction for the roots of the equation are displayed graphically for various initial values of ζ_0 in the complex plane. Light regions correspond to starting points where convergence is rapidly achieved (see text).

visual indication of the complexity of the behavior of the Halley map when applied to a simple function. The several large "bull's-eye" region converge rapidly to a solution, and by testing the value of ζ_n after N iterations, one can determine to which root these areas converge. Using interactive computer graphics routines, one can simply point at the picture and extract the root. The results indicate that nearby points have different fates upon iteration. For example, the following diagram indicates the final fates for points in the bull's-eye regions.

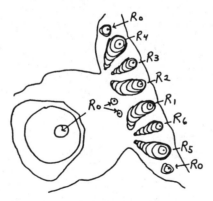

The roots are: $R_0 = (0, 0i)$, $R_1 = (1/2 + i\sqrt{3}/2)$, $R_2 = (1 + 0i)$, $R_3 = (1/2 - i\sqrt{3}/2)$, $R_4 = (-1/2 - i\sqrt{3}/2)$, $R_5 = (-1 + 0i)$, $R_6 = (-1/2 + i\sqrt{3}/2)$.

Figure 3 also reveals minature copies of the nodules and copies of the pattern in Fig. 2. I have found that this "self-similarity" on all scales is characteristic of Halley's plot for polynomial

Fig. 3. Magnification of one of the nodules near the origin in Fig. 2. Successive close-ups reveal self-similarity in the pictures: the pattern of nodules seems to repeat on all sizes scales. This is a characteristic property of "Julia sets" which are described in Ref. [2].

Fig. 2. Contour plot for the same region of the complex plane as in Fig. 1. Note the complicated behavior along the boundary regions between roots and the various "nodules" along these high-iteration radial branches.

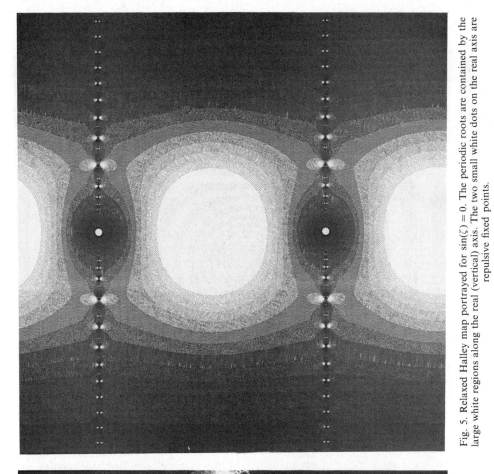

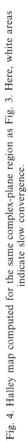

Fig. 5. Relaxed Halley map portrayed for $\sin(\zeta) = 0$. The periodic roots are contained by the large white regions along the real (vertical) axis. The two small white dots on the real axis are repulsive fixed points.

Fig. 4. Halley map computed for the same complex-plane region as Fig. 3. Here, white areas indicate slow convergence.

Fig. 6. Contour plot for $\sin(\zeta) = 0$. Fixed points are at the enter of the concentric rings.

equations. Figure 4 is computed for the same region as Fig. 3 and, like Fig. 1, indicates the behavior of the function in gradations of intensity which make visually obvious the relative speed of convergence of different starting points. Here dark regions indicate areas of rapid convergence. The complexity and richness of resultant forms contrasts with the simplicity of the formula being solved.

Figures 5–9 are plots for $\sin(\zeta) = 0$. To solve for the roots of this function, I used

$$H(\zeta): \quad \zeta_{n+1} = \zeta_n - \lambda \left[\frac{F(\zeta_n)}{F'(\zeta_n) - \left(\dfrac{F''(\zeta_n) F(\zeta_n)}{2F'(\zeta_n)} \right)} \right]. \tag{7}$$

The coefficient λ in the modified Halley's method is known as a *relaxation coefficient*, and is used to control stability of convergence where the method may be susceptible to overshoot. $\lambda = 0.1$ was used. Decreasing λ from 1 damps the Halley step and enlarges the domain of monotonic

Fig. 7. Close-up contour plot near the thin chaotic boundary regions of Fig. 6.

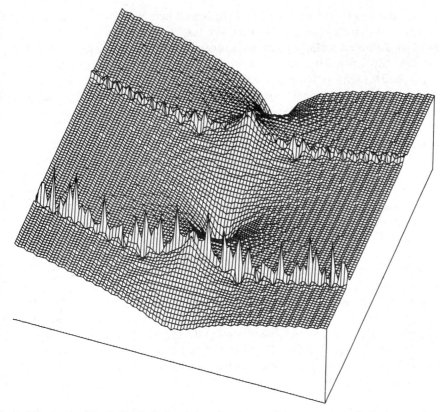

Fig. 8. 3-D plot for Fig. 5. Height indicates iteration (areas of slow convergence). Both attractive and repulsive fixed points are indicated by the wells and broad peaks, respectively. The thin ridges are in the region of chaotic fragmentation between the roots.

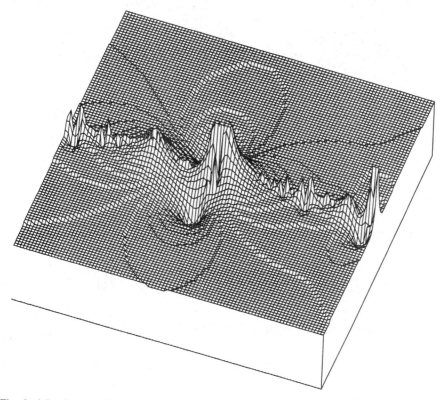

Fig. 9. 3-D close-up of a piece of a chaotic boundary ridge for $\sin(\zeta) = 0.$, Wells indicate rapid convergence.

convergence (also tending to reduce the size of the area where self-similar chaotic fragmentation occurs between roots). I have used the convergence test in equation (5) for these figures. Roots are ecompassed by the broad white regions and are separated by periodic thin chaotic regions. The 3-D plots show the attractive fixed points for the roots as holes in a surface since these represent low iteration points. In addition to attractive fixed points at the roots, the Halley map has repulsive fixed points where $F'(\zeta) = 0$. For $\sin(\zeta)$ we can see these repulsive fixed points most clearly in the 3-D plots (as broad peaks) and in the half-tone plot as dark regions surrounding a small white fixed point.

To help characterize chaotic physical and mathematical phenomena, computer graphics can be used to produce visual representations with a spectrum of perspectives (for several papers by the author, see Refs [9, 10]). In this paper, Halley maps of the equations $\zeta(\zeta^\alpha - 1) = 0$ and $\sin(\zeta) = 0$ are presented, and their behavior ranges from stable attractive and repulsive points to chaotic fluctuations. The system becomes irregular in well-defined regions. The chaotic portions of the maps, while exhibiting complicated behavior, is composed of various underlying self-similar structures. The beauty and complexity of these drawings correspond to behavior which no one could fully have appreciated or suspected before the age of the computer. This complexity makes it difficult to objectively characterize structures such as these, and, therefore, it is useful to develop graphics systems which allow the maps to be followed in a qualitative and quantitative way. Provocative avenues of future research include extension to nonpolynomial equations and to related root-finding numerical methods such as Muller's method, Aitken's method, and the secant method. A report such as this can only be viewed as introductory; however, it is hoped that the described techniques will provide a useful tool and stimulate future studies in the characterization of complicated behavior of numerical methods—which are being used in many branches of modern science with increasing frequency.

REFERENCES

1. J. Wimp, *Computation with Recurrence Relations*. Pitman Publishing, Boston (1984); D. Campbell, J. Crutchfield, D. Farmer and E. Jen, Experimental mathematics: the role of computation in nonlinear science. *Commun. ACM.* **28**, 374–389 (1985); A. Fisher, Chaos: the ultimate asymmetry. *Mosaic* **16**, 24–30 (1985); I. Peterson, Portraits of equations. *Sci. News* **132**(12), 184–186 (1987); B. Mandlebrot, *The Fractal Geometry of Nature*. Freeman, New York (1983).
2. R. Devaney, Chaotic bursts in nonlinear dynamical systems. *Science* **235**, 342–345 (1986).
3. R. Hamming, *Numerical Methods for Scientists and Engineers*. Dover Publications, New York (1973).
4. I. Peterson, Zeroing in on chaos. *Sci. News* **131**, 137–139 (1987); R. May, Simple mathematical models with very complicated dynamics. *Nature* **261**, 459–467 (1976).
5. W. Grove, *Brief Numerical Methods*. Prentice-Hall, Englewood Cliffs, N.J. (1966).
6. H. Benzinger, S. Burns and J. Palmore, Chaotic complex dynamics and Newton's method. *Phys. Lett. A* **119**, 441–445 (1987).
7. J. Crutchfield, J. Farmer and N. Packard, Chaos. *Sci. Am.* **255**, 46–57 (1986) C. Pickover and E. Khorasani, Computer graphics generated from the iteration of algebraic transformations in the complex plane. *Comput. Graph.* **9**, 147–151 (1985).
8. W. Newman and R. Sproull, *Principles of Interactive Computer Graphics*. McGraw-Hill New York (1979).
9. For papers in the author's ten-part "Mathematics and Beauty" series, see, for example: C. Pickover Computers, pattern, chaos, and beauty. IBM Research Report RC 12281 (1986) (Order from: IBM Watson-Distribution, Yorktown Hts, NY 10598). C. Pickover, Mathematics and beauty: time-discrete phase planes associated with the cyclic system, $\{\dot{x}(t) = -f(y(t)), \dot{y}(t) = f(x(t))\}$. *Comput. Graph.* **11**(2), 217–226 (1987); C. Pickover, Biomorphs: computer displays of biological forms generated from mathematical feed back loops. *Comput. Graph. Forum* **5**(4), 313–316 (1987).
10. For other papers by the author on the use of graphic representations for making complicated data easier to understand, see, for example: C. Pickover, Spectrographic representations of globular protein breathing motions. *Science* **233**, 181 (1984); C. Pickover, The use of symmetrized-dot patterns characterizing speech waveforms. *J. Acoust Soc. Am.* **80**, 955-960 (1984). C. Pickover, On the educational uses of computer-generated cartoon faces. *J. Educational Tech. Syst.* **13**, 185–198 (1985); C. Pickover, Frequency representations of D.N.A. sequences: application to a bladder cancer gene. *J. Molec. Graph.* **2**, 50 (1984); C. Pickover, Representation of melody patterns using topographic spectral distribution maps. *Computer Music J.* **10**, 72–78 (1986); C. Pickover, Computer-drawn faces characterizing nucleic acid sequences. *J. Molec. Graph.* **2**, 107–110 (1985); C. Pickover, D.N.A. Vectograms: representation of cancer gene sequences as movements along a 2-D cellular lattice. *IBM J. Res. Dev.* **31**, 11–119 (1987); C. Pickover, The use of random-dot displays in the study of biomolecular conformation. *J. Molec. Graph.* **2**, 34 (1984).

Computers Math. Applic. Vol. 17, No. 1–3, pp. 133–136, 1989
Printed in Great Britain. All rights reserved

0097-4943/89 $3.00 + 0.00
Copyright © 1989 Pergamon Press plc

A CLASS OF SYMMETRIC POLYTOPES

A. Hill and D. G. Larman

Department of Mathematics, University College London, Gower Street, London WC1E 6BT, England

Abstract—In E^3 a polytope which possesses two facets which can be interchanged always possesses a second pair. However, this is not so in E^n, $n \geq 4$.

INTRODUCTION

This paper is the outcome of an empirical observation of one of us (A.H.) that whenever a 3-polytope possesses a pair of facets which could be interchanged then it always seemed to possess a second pair. Our first theorem proves that this must always be the case. However, we shall also show that in E^d, $d \geq 4$, there exists a d-polytope which has exactly two facets which can be interchanged. It would be interesting to characterize all such polytopes which seems to belong, as the proof of Theorem 1 shows, to a somewhat limited class.

If we dualize our observations in E^3 we are led to consider 3-polytopes with pairs of vertices which can be interchanged. Using Steinitz's theorem, Theorem 1 also holds for 3-connected planar graphs. However, we shall show that the result does not hold for all 3-connected graphs with at least 7 vertices.

Theorem 1

Let P be a polytope in E^3 which possesses a pair of facets which can be interchanged by an isomorphism of the face lattice. Then P possesses at least two such pairs.

Example 1

There exists in E^d, $d \geq 4$, a d-polytope P which possesses only one pair of facets which can be interchanged by an isomorphism of the face lattice of P.

The first part of Theorem 2 is the dual version of Theorem 1 (using Steinitz's theorem).

Theorem 2

Let G be a 3-connected planar graph which possesses a pair of vertices which can be interchanged by an isomorphism of the graph. Then there are at least two such pairs. Further, the result holds for all 3-connected graphs with at most 6 vertices, but there exists a 3-connected graph with 7 vertices with only one pair of interchangeable vertices.

Proof of Theorem 1. Let P^* be a 3-polytope and let ϕ^* be an isomorphism which interchanges two facets A, B of P^*. Let P be the dual of P^* and let ϕ be the isomorphism of the face lattice which interchanges the corresponding two vertices $\mathbf{a}$, $\mathbf{b}$ of P. We shall suppose, for the moment, that P is a d-polytope, $d \geq 3$, and specialize to $d = 3$ only when necessary.

We claim that we can assume that any vertex $\mathbf{c}$ of P which is joined to $\mathbf{a}$ by an edge is also joined to $\mathbf{b}$ by an edge. If not, then $\phi(\mathbf{c})$ is joined to $\mathbf{b}$ by an edge but not to $\mathbf{a}$. In particular $\phi(\mathbf{c}) \neq \mathbf{c}$. Indeed, repeating this process, we have that $\phi^{2m}(\mathbf{c})$ is joined to $\mathbf{a}$ by an edge but not to $\mathbf{b}$, and that $\phi^{2m+1}(\mathbf{c})$ is joined to $\mathbf{b}$ by an edge but not to $\mathbf{a}$. Eventually there exists m, n; $n > m$ such that $\phi^{2n}(\mathbf{c}) = \phi^{2m}(\mathbf{c})$. If we suppose that $n - m$ is minimal then $\phi^{n-m}(\mathbf{c}) \neq \mathbf{c}$. For, if $\phi^{n-m}(\mathbf{c}) = \mathbf{c}$ then $n - m$ must be odd (by the minimality of $n - m$) but then $\phi^{n-m}(\mathbf{c})$ is joined to $\mathbf{b}$ by an edge but $\mathbf{c}$ is not, i.e. $\phi^{n-m}(\mathbf{c}) \neq \mathbf{c}$. So ϕ^{n-m} is an isomorphism of P which interchanges $\mathbf{c}$ and $\phi^{n-m}(\mathbf{c})$ as required. Hence we may assume that any vertex $\mathbf{c}$ of P which is joined to $\mathbf{a}$ by an edge is also joined to $\mathbf{b}$ by an edge.

Let $\mathbf{c}_1, \ldots, \mathbf{c}_k$, $k \geq 2$ be a list of the vertices joined to both $\mathbf{a}$ and $\mathbf{b}$ by an edge. Let F be a facet of P containing $\mathbf{a}$ and let $[\mathbf{a}, \mathbf{c}_1], \ldots, [\mathbf{a}, \mathbf{c}_l]$ (possibly together with $[\mathbf{a}, \mathbf{b}]$) be a list of the edges containing **a** and contained in F. We claim that we may suppose that $[\mathbf{b}, \mathbf{c}_1], \ldots, [\mathbf{b}, \mathbf{c}_t]$ (possibly together with $[\mathbf{a}, \mathbf{b}]$) form the list of edges containing **b** and contained in some facet of G of P.

If not, then F cannot contain **b** for, if it did, then $[\mathbf{b}, \mathbf{c}_1], \ldots, [\mathbf{b}, \mathbf{c}_t]$ are the edges of F emanating from **b**. Hence $\phi(F) \neq F$. Eventually, there exists m, n, $n > m$, $n - m$ minimal such that $\phi^{2(n-m)}F = F$. So

$$\{\mathbf{c}_1, \ldots, \mathbf{c}_t\} = \{\phi^{2(n-m)}(\mathbf{c}_1), \ldots, \phi^{(n-m)}(\mathbf{c}_t)\}.$$

So for each j, $j = 1, \ldots, t$ there will be a least positive integer k_j such that $\phi^{2k_j}(\mathbf{c}_j) = \mathbf{c}_j$. If k_j were even, then $\phi^{k_j}(\mathbf{c}_j) \neq \mathbf{c}_j$ and so ϕ^{k_j} would be the required isomorphism. So each k_j can be supposed odd.

If there exists j, $1 \leqslant j \leqslant t$, with $\phi^{k_j}(\mathbf{c}_j) \neq \mathbf{c}_j$ then ϕ^{k_j} is the required isomorphism interchanging $\mathbf{c}_j$ and $\phi^{k_j}(\mathbf{c}_j)$. So we may suppose that $\phi^{k_j}(\mathbf{c}_j) = \mathbf{c}_j$, $j = 1, \ldots, t$. So $\phi^S(F)$ is a facet of P containing **b**, as required. Hence, we may suppose that $[\mathbf{b}, \mathbf{c}_1], \ldots, [\mathbf{b}, \mathbf{c}_t]$ (possibly together with $[\mathbf{a}, \mathbf{b}]$) form the list of edges containing **b** and containing in some facet G of P. If F and G are distinct then F is the convex hull of $\mathbf{a}, \mathbf{c}_1, \ldots, \mathbf{c}_t$ and G is the convex hull of $\mathbf{a}, \mathbf{b}, \mathbf{c}_1, \ldots, \mathbf{c}_t$. If $F = G$ then F is the convex hull of $\mathbf{a}, \mathbf{b}, \mathbf{c}_1, \ldots, \mathbf{c}_t$. From above, it also follows that P is the convex hull of $\mathbf{a}, \mathbf{b}, \mathbf{c}_1, \ldots, \mathbf{c}_t$ and the facets of P take one of the three forms $\mathrm{conv}\{\mathbf{a}, \mathbf{c}_1, \ldots, \mathbf{c}_t\}$, $\mathrm{conv}\{\mathbf{b}, \mathbf{c}_1, \ldots, \mathbf{c}_t\}$ and $\mathrm{conv}\{\mathbf{a}, \mathbf{c}_1, \ldots, \mathbf{c}_t\}$.

If we now dualize the situation there are two possible cases arising.

(i) [a b] is not an edge of P

In this case P^* is a cylinder with bottom facet A and top facet B, where A and B are the duals of **a** and **b**. For P^* in E^3 any two consecutive side facets are interchangeable under an isomorphism of the face lattice of P^*. An isomorphism will be the (combinatorial) reflection in their common edge.

(ii) [a b] is an edge of P

In this case P^* is the convex hull of A and B (where A and B are similar facets) with $A \cap B$ a $d - 2$ face of P^*, i.e. P^* is almost a cylinder with bottom facet A and top facet B except that A meets B in a $d - 2$ face $A \cap B$.

For P^* in E^3 the two side facets adjacent to $A \cap B$ are interchangeable under an isomorphism of the face lattice of P^*. An isomorphism will be the (combinatorial) reflection in $A \cap B$.

This completes the proof of Theorem 1.

Construction of Example 1

In order to construct the example let us continue the analysis of Theorem 1. An obvious candidate would be a cylinder P^* in E^d whose bottom facet A (and hence top face B) have no isomorphisms of their face lattice which interchanges $d - 2$ faces. Then we could interchange A and B but there remains the possibility that there is some other isomorphism λ which interchanges some other pair of facets (and necessarily $\lambda A \neq A$ or B, or $\lambda B \neq B$ or A). To prevent this occurring we ensure that the number of $d - 2$ faces of A (and B) exceed those of the side facets.

So dualizing this argument we need to construct in E^{d-1}, $d \geqslant 4$, a $d - 1$ polytope Q^{d-1} with no interchangeable vertices and whose total number of vertices exceed the maximal valence of any one vertex by at least 3. This we do by induction. For E^3, Q^3 is as in Fig. 1.

We show that no pair of vertices can be interchanged (by exhaustion).

b is fixed. **b** is the only six valent vertex.

a, d are fixed. The only two five valent vertices are **a** and **d**. However, **d** lies on a facet with 7 vertices and **a** does not. Hence **a** and **d** cannot be interchanged.

g is fixed. **g** is the only vertex with edges to **a**, **d** and **b**.

c is fixed. **g** is the only other vertex with edges to **a** and **d**.

j is fixed. **d** is the only other vertex with edges to **a** and **c**.

i is fixed. **i** is the only vertex with edges to **a**, **b**, **j**.

h is fixed. **h** is the only vertex amongst **h**, **e**, **f** with an edge to **i**.

f is fixed. **f** is the only vertex amongst **e**, **f** with an edge to **h**.

e is fixed. All the other vertices of Q^3 have now been shown to be fixed.

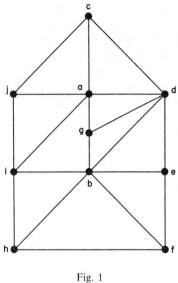

Fig. 1

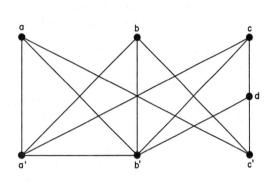

Fig. 2

So no pair of vertices of Q^3, which is a 3-polytope with 10 vertices and the maximum valence of its vertices is 6, can be interchanged by an isomorphism of the face lattice of Q^3.

Suppose now that a $d-1$ polytope Q^{d-1} has been constructed, $d \geqslant 4$, with no interchangeable vertices. Suppose further that Q^{d-1} has X_{d-1} vertices with maximum valence Y_{d-1}, where $Y_{d-1} + 3 \leqslant X_{d-1}$. As $X_3 = 10$ and $Y_3 = 6$, this is true for $d = 4$.

For the construction of Q^d we suppose that Q^{d-1} has centroid $\mathbf{O}$ and lies in the coordinate hyperplane $x_d = 0$ of E^d. Now, if $\mathbf{e}_d$ is the dth unit vector let

$$Q^d = \operatorname{conv}\left\{\mathbf{O}, \frac{1}{2}\mathbf{e}_d + \frac{1}{\sqrt{2}}Q^{d-1}, \mathbf{e}_d + Q^{d-1}\right\}.$$

The vertices of Q^d are $\mathbf{O}$ and the vertices of the two copies of Q^{d-1}, i.e. Q^d has $2X_{d-1} + 1$ vertices. The valence of $\mathbf{O}$ is X_{d-1} and the valences of any other vertex is at most two more than its valence in (the copy of) Q^{d-1}. So $Y_d = X_{d-1}$, $d \geqslant 4$ and hence $Y_d + 3 \leqslant X_d$. Thus, Q^d has been constructed inductively. We claim that there is no isomorphism of the face lattice which interchanges two vertices. Firstly the vertex $\mathbf{O}$ has uniquely the maximum valence and hence is fixed. There are no edges joining $\mathbf{O}$ to any vertex within $\mathbf{e}_d + Q^{d-1}$. Consequently any such isomorphisms would have to permute the vertices of

$$\frac{1}{2}\mathbf{e}_d + \frac{1}{\sqrt{2}}Q^{d-1}$$

and $\mathbf{e}_d + Q^{d-1}$ separately. By the inductive assumption on Q^{d-1} this is only possible if all the vertices remain fixed.

Finally taking the dual of Q^{d-1}, say $(Q^{d-1})^*$ and taking the cylinder in E^d over $(Q^{d-1})^*$ we obtain an example of a d-polytope, $d \geqslant 4$ in which exactly two facets can be interchanged by an isomorphism of the face lattice.

Proof of Theorem 2. The first part of Theorem 2 follows from Theorem 1 using Steinitz's theorem. So if a 3-connected graph G has just one interchangeable pair of vertices then it must be non-planar. Consequently G must contain a refinement of at least one of the two Kuratowski graphs.

If G has 5 vertices it is the complete 5-graph which has all pairs of vertices interchangeable. If G has 6 vertices then it is either the complete (3,3) bipartite graph, in which case all pairs of vertices are interchangeable, or G contains a refinement of the completed 5-graph C_5. In this latter case we consider G as a vertex $\mathbf{v}$ being added (with edges) to C_5. There are two possibilities.

(i) $\mathbf{v}$ does not lie on the edges of C_5 and hence $\mathbf{v}$ is joined by edges to at least 3 of the 5 vertices of C_5. Then all the vertices of C_5 joined by an edge to $\mathbf{v}$ are interchangeable as are all the vertices of C_5 which are not joined by an edge to $\mathbf{v}$. This yields at least two interchangeable pairs.

(ii) **v** lies on one of the edges, [**a**, **b**] say and so **v** is joined to at least one other vertex of C_5. In this case **a** and **b** are interchangeable. Also amongst the other three vertices **c**, **d**, **e** there will be a pair which are either both joined to **v** or both are not joined to **v**. Such a pair is also interchangeable. So again there are two interchangeable pairs.

To complete the proof of Theorem 2, we construct a 3-connected graph G with 7 vertices which possesses exactly one pair of interchangeable vertices. This graph is illustrated in Fig. 2.

Clearly **a**, **b** are interchangeable and we claim that there are no other pairs of interchangeable vertices.

Firstly any isomorphism ϕ of the graph must fix **a′** and **b′** since they are respectively the only 4 and 5 valent vertices.

Since of the five remaining vertices **a**, **b**, **c**, **d**, **c′** only **c′** is not joined to **b′**, ϕ must also fix **c′**.

We next show that ϕ fixes **d**. If not then ϕ**d** must be one of **a**, **b**, **c**. If ϕ**d** = **a** (or **b**) then ϕ**c**, which also has to be one of **a**, **b**, **c**, **d** is not joined by an edge to ϕ**d**, whereas [**c**, **d**] is an edge; which is impossible. If ϕ**d** = **c** then again since [**c**, **d**] is an edge, ϕ**c** = **d**. However, [**c**, **a′**] is an edge but [ϕ**c**, ϕ**a′**] = [**d**, **a′**] is not. So ϕ fixes **d**.

It remains to show that ϕ**c** = **c**. If ϕ**c** = **a** then [**c**, **d**] is an edge but [ϕ**c**, ϕ**d**] = [**a**, **d**] is not, which is impossible. So ϕ**c** = **c** which completes the proof of Theorem 2.

REFERENCES

1. B. Grünbaum, *Convex Polytopes*. Interscience, London (1967).
2. A. F. Hawkins, A. C. Hill, J. E. Reeve and J. A. Tyrrell, On certain polyhedra. *Math. Gaz.* L (372), 140–144 (1966).

Computers Math. Applic. Vol. 17, No. 1–3, pp. 137–145, 1989
Printed in Great Britain. All rights reserved

0097-4943/89 $3.00 + 0.00

ARRANGEMENTS OF MINIMAL VARIANCE— MULTIDIMENSIONAL SCALING IN THE SYMMETRICAL CASE

L. Telegdi

Computer and Automation Institute, Hungarian Academy of Sciences, Budapest,
P.O. Box 63, H-1502, Hungary

Abstract—We deal with the problem of how to arrange n points in the plane with a given mean and minimal variance. Ordinary and multiple multidimensional scaling are outlined, investigated in the symmetrical case and applied to congenital abnormalities.

1. ARRANGEMENT OF POINTS IN THE PLANE WITH DISTANCE OF MINIMAL VARIANCE

Certain extremum requirements imply regularity. This was illustrated by some examples in Ref. [1]: the area of an n-gon of given perimeter is maximized by the regular n-gon; among all convex polyhedra containing a ball, the circumscribed cube has the least total edge-length etc. In what follows we deal with the problem how to arrange n points in the plane with a given mean and minimal variance. (The solution conjectured and given below is not regular for $n > 6$, but it is symmetrical and consists of regular parts.) This problem is equivalent with the following: determine the points $x_1, x_2, \ldots, x_n$ in the plane so as to minimize the sum of squares

$$V = \sum_{i=1}^{n-1} \sum_{j=i+1}^{n} (K - \|x_i - x_j\|)^2$$

of the deviations of the distances between pairs of them from a given constant K. Moreover, we would like to determine the value $\tilde{V}$ of V for this optimal configuration. The latter formulation of the problem suggests its connection with mechanics: V can be regarded as the potential of C_n^2 springs with length K and spring constant 2 between each pair x_i, x_j of points.

As V depends on the x_is only through their distances, the configuration obtained for the solution of the problem is obviously indeterminate with respect to translation, rotation and reflection. Almost the same can be true for uniform expansion or contraction: they result in similar configurations, and obviously $\tilde{V}(K) = K^2 \tilde{V}(1)$. Therefore, we assume that $K = 1$.

The general solution of the problem is not known. As V is the function of the x_is, i.e.— denoting by x_{ik} the kth coordinate of the ith point—that of the unknown variables $x_{11}, x_{12}, x_{21}, x_{22}, \ldots, x_{n1}, x_{n2}$, a solution can be obtained by the minimization of a $2n$-variate function without constraint. Such minimizing procedures are usually gradient- or Fletcher-type algorithms, which are not able to distinguish between a local and a global minimum, therefore they need a good initial configuration.

Let $z_1, z_2, \ldots, z_n$ be the vertices of the $(n-1)$-dimensional regular unit n-hedron, then obviously V is 0, i.e. minimal for these z_is. Let

$$\mathbf{I} = \begin{bmatrix} 1 & 0 & \cdots & 0 \\ 0 & 1 & \cdots & 0 \\ \cdot & \cdot & & \cdot \\ \cdot & \cdot & & \cdot \\ \cdot & \cdot & & \cdot \\ 0 & 0 & \cdots & 1 \end{bmatrix}$$

be the identity,

$$
\mathbf{J} = \begin{bmatrix}
1 & 1 & \cdots & 1 \\
1 & 1 & \cdots & 1 \\
\cdot & \cdot & & \cdot \\
\cdot & \cdot & & \cdot \\
\cdot & \cdot & & \cdot \\
1 & 1 & \cdots & 1
\end{bmatrix}
$$

the unit matrix, and

$$
\mathbf{B} = \frac{1}{2}\left(\mathbf{I} - \frac{1}{n}\mathbf{J}\right).
$$

Let $\lambda_1 \geqslant \lambda_2 > 0$ be the two largest eigenvalues of $\mathbf{B}$.

Definition

Let $\mathbf{v}_{(i)}$ be the ith eigenvector of $\mathbf{B}$,

$$
\mathbf{B}\mathbf{v}_{(i)} = \lambda_i v_{(i)},
$$

normalized according to $\mathbf{v}_{(i)}^{\mathrm{T}}\mathbf{v}_{(i)} = \lambda_i$, $i = 1, 2$. The rows of the matrix $[\mathbf{v}_{(1)}, \mathbf{v}_{(2)}]$ are called the *principal coordinates* of the configuration $\mathbf{Z} = (\mathbf{z}_1, \mathbf{z}_2, \ldots, \mathbf{z}_n)$ in two dimensions.

Remark

$\lambda_1 = \lambda_2 = \frac{1}{2}$; $\mathbf{v}_{(1)}$ and $\mathbf{v}_{(2)}$ are two arbitrary orthogonal vectors of the $(n-1)$-dimensional subspace orthogonal to the vector $(1, 1, \ldots, 1)^{\mathrm{T}}$.

Theorem

Amongst all projections of $\mathbf{Z}$ onto planes, the quantity

$$
\sum_{i=1}^{n-1} \sum_{j=i+1}^{n} \left[1 - \|\mathbf{x}_i - \mathbf{x}_j\|^2 \right]
$$

(which is positive, because projecting a configuration reduces the interpoint distances) is minimized when $\mathbf{Z}$ is projected onto its principal coordinates in two dimensions [2].

This theorem suggests a possible choice of initial configuration: choose the configuration in the plane whose coordinates are determined by the first two eigenvectors of $\mathbf{B}$. V can be minimized iteratively, starting with this configuration. The iteration can be the repeated application of the following procedure: the negative gradient vector of the $2n$-variate function $V = V(\mathbf{x}_1, \mathbf{x}_2, \ldots, \mathbf{x}_n)$ is determined (explicitly calculated), and by performing a line search in the direction of this vector the new coordinates of the points $\mathbf{x}_i$ are calculated ($i = 1, 2, \ldots, n$). At computer realization it is worthwhile taking the gradients of the functions

$$
\sum_{\substack{j=1 \\ j \neq i}}^{n} \left(1 - \|\mathbf{x}_i - \mathbf{x}_j\| \right)^2, \quad i = 1, 2, \ldots, n,
$$

instead of the gradient of V (moving only one point at the same time instead of n points. By this both computer time and memory demand decrease).

Even in case of the above initialization it can occur that only a local minimum of V is produced. The danger of its occurrence can be decreased in such a way that sum of qth powers is written instead of sum of squares in V, and the value of q is changed in the course of the algorithm. According to computational experiences, the algorithm is the most effective if after the initialization q is chosen for 3, then as soon as V already decreases only to a small extent (the points $\mathbf{x}_i$ change scarcely), its value is taken first for 2, then for 1, and finally again for 2.

Proceeding in the above fashion, the algorithm gives arrangements in which the points are situated—for $n = 3, 4, \ldots, 65$—on vertices of concentric regular polygons (regular polygons inscribed into concentric circles). For example, for $n = 9$, 16 and 23 we have, respectively, (i) a regular octagon and its centre; (ii) a regular 12-gon and within this a concentric square; (iii) a

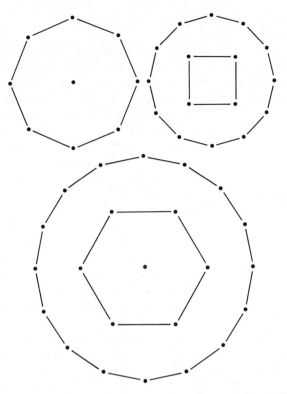

Fig. 1. Arrangements of minimal variance for $n = 9$, 16 and 23.

regular 16-gon, within this a concentric regular hexagon, and their common centre (Fig. 1). The numbers m_v of vertices of the various polygons for some other ns are exhibited in Table 1 (from outside inwards). For the final value $\tilde{V}$ of the term V the equation

$$\tilde{V} = B_n C^2_{n-2},\qquad(1)$$

and here for B_n under $4 \leqslant n \leqslant 65$ the inequality

$$0.1716 \leqslant B_n \leqslant 0.1807,$$

was obtained

$$(0.1716 = 3 - 2\sqrt{2} = B_4;\ 0.1807 = \frac{75 - 36\sqrt{3}}{70} = B_7;$$

as the value of n approaches 65, the value of B_n approaches 0.176).

Table 1. Numbers of vertices of polygons

n	m_1	m_2	m_3	m_4
3	3	—	—	—
4	4	—	—	—
5	5	—	—	—
6	6	—	—	—
7	6	1	—	—
10	9	1	—	—
15	12	3	—	—
20	15	5	—	—
25	17	7	1	—
30	19	9	2	—
35	21	10	4	—
40	24	11	5	—
45	25	12	7	1
50	28	13	8	1
55	29	15	8	3
60	30	16	10	4
65	32	17	12	4

2. MULTIDIMENSIONAL SCALING

Let us assume that we are investigating M objects and n arbitrary variables characterizing them. Most of multivariate statistical methods work with data of the variables observed on the objects. In contrast with these methods, in case of *multidimensional scaling* (MDS) we can't—or don't want to—observe directly the data points as M points in n-dimensional Euclidean space, but we have only indirect information about them. This information may concern the *distance* (*dissimilarity*) or—on the contrary—*proximity* (*similarity*) of the objects or/and variables. MDS deals with the following problem: how can the objects or/and variables be drawn in the space on the basis of an $(M \times M)$, $(n \times n)$ or $(M \times n)$ distance of similarity matrix (data matrix of MDS), in other words how can an M-tuple or/and an n-tuple of points be constructed in low-dimensional Euclidean space with the property that Euclidean distance of the points should reflect distance (dissimilarity) of the objects or/and variables as well as possible?

Tasks involving MDS can be classified according to a few organizing concepts. A major one is whether the data of MDS (i.e. the distances or similarities) represent one or two sets of things. If they represent one set of things (either the objects, or the variables), they are called *one-mode*, if two sets of things (the objects *and* the variables), they are called *two-mode*.

In case of MDS with one-mode data the set of things represented by the rows of the data matrix of MDS is the same as the set of things represented by the columns. In such a case the data matrix is square and symmetric. Depending on whether the rows and columns represent the objects or the variables, we speak of MDS of the objects or that of the variables.

In case of MDS with two-mode data the set of things represented by the rows differs from the set of things represented by the columns. In such a case the data matrix is rectangular. MDS with two-mode data is called also *multidimensional unfolding* (MDU). In what follows we deal with the MDS of the variables (i.e. MDS with one-mode data, representing the variables). For more details on MDS [3, 4], Chap. 14 of Ref. [2], Chap. 5 of Ref. [5], furthermore Refs [6–8] are recommended to the reader.

Let $\mathbf{D} = [d_{ij}]$ be an arbitrary *distance matrix* (it means the following: $\mathbf{D}$ is symmetric, and

$$d_{ii} = 0, \quad d_{ij} \geqslant 0, \quad i \neq j).$$

k-dimensional points $\mathbf{x}_1, \mathbf{x}_2, \ldots, \mathbf{x}_n$ are to be determined in such a way that denoting by $\hat{d}_{ij}$ the Euclidean distance of $\mathbf{x}_i$ and $\mathbf{x}_j$, the matrix $[\hat{d}_{ij}]$ should be "similar" to $\mathbf{D}$ in some sense. Usually not only the points $\mathbf{x}_i$, but also the dimension k is unknown. In practice, this latter one is mostly chosen for 1, 2 or 3, because then the variables are in fact drawn by the points. In many cases there is a point configuration in some p-dimensional Euclidean space R^p the interpoint distance matrix of which is just $\mathbf{D}$ (i.e. $\mathbf{D}$ is *Euclidean*). This configuration can be accepted as solution to the MDS problem only if p can be chosen for k. However, in practice p is usually too large for this.

The deviation between an arbitrary distance matrix $[d_{ij}]$ and the Euclidean distance matrix $[\hat{d}_{ij}]$ of some point configuration $\mathbf{X}$ in the course of MDS is measured mostly by one of the following terms:

$$\sum_{i=1}^{n-1} \sum_{j=i+1}^{n} (d_{ij} - \hat{d}_{ij})^2, \tag{2}$$

$$\sum_{i=1}^{n-1} \sum_{j=i+1}^{n} c_{ij}(d_{ij} - \hat{d}_{ij})^2, \tag{3}$$

$$\sum_{i=1}^{n-1} \sum_{j=i+1}^{n} (d_{ij} - \hat{d}_{ij})^2 \bigg/ \sum_{i=1}^{n-1} \sum_{j=i+1}^{n} \hat{d}_{ij}^2$$

[the c_{ij}s are given weighting factors; expression (2) is the special case of expression (3) belonging to $c_{ij} \equiv 1$]. Each of the above terms is the function of $\mathbf{X}$, i.e. an $(n \times k)$-variate function. Therefore the MDS problem is solved similarly to the algorithm described in the previous section. The problem of the arrangement of minimal variance is equivalent to MDS under $k = 2$, $d_{ij} \equiv 1$ and deviation measure (2).

3. MULTIPLE MDS

Let $y_1, y_2, \ldots, y_M$ be arbitrary objects characterized by the *dichotomous* variables $W_1, W_2, \ldots, W_n$ indicating the presence or absence of the characters $A_1, A_2, \ldots, A_n$. Let us assume that our task is MDS of the variables. In accordance with the previous section it means the following: distances are constructed between the variables, and the variables are to be put in low-dimensional Euclidean space in such a way that Euclidean distances of the points corresponding to the variables should differ from distances of the variables to as small extent as possible. In order to be well scalable, the variables must be consistent in the following sense: if two variables are near to a third one, they must be near to one another too. Let us assume for instance (Example 1) that $n = 3$, $M = 44$, the first 19 objects have the characters A_1 and A_2, the following 14 ones A_1 and A_3, and the last 11 ones A_2 and A_3. To these objects the distance matrix

$$\begin{bmatrix} 0 & 3 & 4 \\ 3 & 0 & 5 \\ 4 & 5 & 0 \end{bmatrix}$$

of the variables can be assigned. On its basis the variables can be well scaled (see Fig. 2). Let us assume now (Example 2) that $M = 33$, the first 19 objects have the characters A_1 and A_2, and the other 14 ones A_1 and A_3. To these objects the distance matrix

$$\begin{bmatrix} 0 & 3 & 4 \\ 3 & 0 & 60 \\ 4 & 60 & 0 \end{bmatrix}$$

of the variables can be assigned. On its basis the variables can be scaled only badly. Let us try therefore to scale them on the basis of the first 19 and the other 14 objects separately! To these sets of objects (which can be regarded as clusters) the distance matrices

$$\begin{bmatrix} 0 & 3 & 60 \\ 3 & 0 & 60 \\ 60 & 60 & 0 \end{bmatrix} \quad \text{and} \quad \begin{bmatrix} 0 & 60 & 4 \\ 60 & 0 & 60 \\ 4 & 60 & 0 \end{bmatrix}$$

of the variables can be assigned. On their basis the variables—on two planes!—can already be well scaled (see Fig. 3).

Multiple multidimensional scaling (MMDS), see Ref. [9], deals with cases similar to Example 2, with the problem arising if the consistency is not fulfilled: the objects are to be divided into disjoint clusters as homogeneously as possible, where the homogeneity of a cluster is measured by the goodness of (ordinary) MDS of the variables under it. We would like to determine the positive integer p, the disjoint clusters

$$Y_1, Y_2, \ldots, Y_p \subset \{y_1, y_2, \ldots, y_M\} = Y$$

of the objects with the property that

$$\bigcup_{m=1}^{p} Y_m = Y,$$

furthermore, the points $\mathbf{x}_i^{(m)}$ of R^k ($i = 1, 2, \ldots, n; m = 1, 2, \ldots, p$) which represent the variables in the sense that the closeness of the points $\mathbf{x}_i^{(m)}$ and $\mathbf{x}_j^{(m)}$ corresponds to the proximity of the variables W_i and W_j under Y_m.

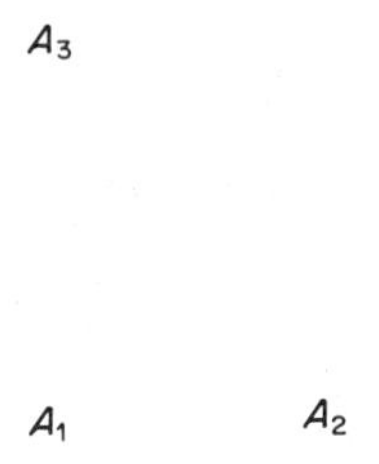

A_3

$A_1 \qquad A_2$

Fig. 2. MDS in the consistent case (Example 1).

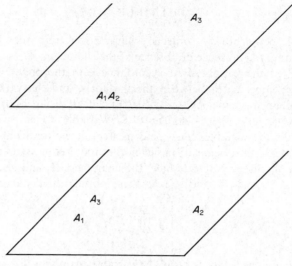

Fig. 3. MDS on two planes in the inconsistent case (Example 2).

Let e_{gi} be the value—1 or 0—of the variable W_i observed on the object y_g ($g = 1, 2, \ldots, M$; $i = 1, 2, \ldots, n$), and let

$$\mathbf{e}_g = [e_{g1}, e_{g2}, \ldots, e_{gn}]^{\mathrm{T}}$$

($g = 1, 2, \ldots, M$). As objects with the same $\mathbf{e}_g$ are indistinguishable, let us assume that

$$\mathbf{e}_{g_1} = \mathbf{e}_{g_2} \Rightarrow g_1 = g_2,$$

and the objects have multiplicities. As the allocation of the various objects must obviously depend on the variables having value 1 on them, MMDS can't do anything with the objects on which at most one variable has value 1. Therefore, let us assume that on each object at least two variables have value 1. For arbitrary combination G_g of characters let the number of characters belonging to G_g be called the *side* of G_g, furthermore, let us denote by $O(G_g)$ the number of objects which have the characters belonging to G_g, but others not, and by N the number of objects different with respect to the characters and having at least two of them. Then each object y_g corresponds to a character combination G_g with side not less than 2, and has multiplicity $O(G_g)$ ($g = 1, 2, \ldots, N$). Let

$$n_{ij}^{(m)} = \sum_{\substack{g: y_g \in Y_m \\ e_{gi}, e_{gj} = 1}} O(G_g) \tag{4}$$

(the number of those objects of the mth cluster in which A_i and A_j are present), and

$$E^{(m)} = \sum_{i=1}^{n-1} \sum_{j=i+1}^{n} [n_{ij}^{(m)} \|\mathbf{x}_i^{(m)} - \mathbf{x}_j^{(m)}\|^2 + (K - \|\mathbf{x}_i^{(m)} - \mathbf{x}_j^{(m)}\|)^2],$$

then

$$E = \sum_{n=1}^{p} E^{(m)},$$

which is the MMDS function (K is an appropriate constant), is minimized by the alternate application of the following two procedures: (1) optimal classification of the objects among the actual clusters; (2) optimal allocation of the variables for the actual classification of the objects.

In the second procedure for $m = 1, 2, \ldots, p$ $E^{(m)}$ is (in the course of the various steps not minimized, but only) decreased similarly to the algorithm described in Section 1. Let the distances

$$d_{ij}^{(m)} = \frac{K}{n_{ij}^{(m)} + 1}$$

between the variables and the corresponding weights $c_{ij}^{(m)} = n_{ij}^{(m)} + 1$ be introduced, and let

$$V^{(m)} = \sum_{i=1}^{n-1} \sum_{j=i+1}^{n} c_{ij}^{(m)} [d_{ij}^{(m)} - \|\mathbf{x}_i^{(m)} - \mathbf{x}_j^{(m)}\|]^2,$$

then $[E^{(m)} - V^{(m)}]$ does not change in these, gradient steps (see Ref. [9]), thus MMDS is the generalization of MDS.

In the first procedure for $g = 1, 2, \ldots, N$

$$U_g^{(m)} = \sum_{\substack{i=1 \\ e_{gi}=1}}^{n-1} \sum_{\substack{j=i+1 \\ e_{gj}=1}}^{n} \|\mathbf{x}_i^{(m)} - \mathbf{x}_j^{(m)}\|^2$$

is minimized in m. Ref. [9] proves that E is monotone decreasing in this step (in the gradient steps obviously).

Let us denote by T the algorithm which minimizes $U_g^{(m)}$ in m in the first procedure and decreases $E^{(m)}$ in the second procedure. In the course of T some clusters can become almost or entirely empty. It is reasonable to cease such clusters (by which p naturally decreases) in such a way that some object y_g having belonged to them is put into that cluster of the remaining ones under which $U_g^{(m)}$ is minimal. For—among others—this reason it is worthwhile choosing the initial value of p large. Initial clusters can be obtained, e.g. by applying the k-means (see Ref. [10]), more precisely in our case p-means method to the objects y_g on the basis of the vectors $\mathbf{e}_g$. Initial point configurations under the various initial clusters can be chosen similarly to Section 1.

It would be good to evaluate MMDS of some data field from as many as two points of view. The mathematical evaluation of the goodness of MMDS compared with other clustering methods is problematic theoretically. Namely the various clustering methods differ from one another decisively just in the criteria they give to the goodness of clustering. According to its own criterion each method is the best, however the criteria can't be compared objectively. (For more thorough investigations of the comparison of clusterings see, e.g. Refs [11, 12].) One can evaluate how well the objects of the concrete data field can be clustered with respect to MMDS theoretically in the following way. Let us assume that at the end of MMDS of the data field under some K the number of clusters is p and the value of E is

$$E^* = E^*(n, \mu, p, K),$$

where

$$\mu = \frac{\sum_{m=1}^{p} \sum_{i=1}^{n-1} \sum_{j=i+1}^{n} n_{ij}^{(m)}}{C_n^2} = \frac{\sum_{i=1}^{n-1} \sum_{j=i+1}^{n} O_T(i, j)}{C_n^2},$$

$O_T(i, j)$ denotes the number of objects which have the characters A_i, A_j, and possibly others too. Let us consider the respective data fields which belong to the pair (n, μ) of values, and are optimal and pessimal with respect to MMDS. Let us denote by E_{opt} and E_{pess} the values of the term E at the end of MMDS of these data fields under the above values of p and K. One can characterize how good the concrete data field is with respect to MMDS in a natural way by the term

$$\frac{E^* - E_{\text{opt}}}{E_{\text{pess}} - E_{\text{opt}}}, \tag{5}$$

which can't be less than 0 and greater than 1. Its determination would need the knowledge of the values E_{opt} and E_{pess}. However they are unknown, because the optimal and pessimal data fields are unknown too. In the next section we will consider a data field which is though not pessimal, but from the point of view of clustering bad. Let us denote by E_{bad} the value of E at the end of MMDS of this data field under the above p and K. In a fortunate case $(E_{\text{bad}} - E_{\text{pess}})$ and E_{opt} are small enough, thus expression (5) can be substituted for (E^*/E_{bad}).

4. MMDS IN THE SYMMETRICAL CASE

In the operation of MMDS the function E characterizing MMDS has a decisive role. It was specified in such a way that it depends only on the joint occurrences of the various pairs of characters. It implies that if the probabilities of the occurrences of the various characters are different, the procedure doesn't become aware of independence: it brings two characters close to one another even if they occur often together only because both of them are frequent. (However, this is on purpose: in many cases—e.g. in the statistical investigation of congenital abnormalities, out of which MMDS has grown—it is an important requirement that typical character combinations should not remain unobserved.) One can easily accomplish that the procedure should become aware of independence: in definition (4) of $n_{ij}^{(m)}$ the sum is to be divided by the term

$$\sum_{\substack{g:y_g \in Y_m \\ e_{gi}=1}} O(G_g) \times \sum_{\substack{g:y_g \in Y_m \\ e_{gj}=1}} O(G_g) \Bigg/ \sum_{g:y_g \in Y_m} O(G_g).$$

However, if the probabilities of the occurrences of the various characters are equal to each other, it is not necessary. Therefore, independence was investigated in this, symmetrical case. Particularly, a data field with the property

$$O_T(i, j) \equiv \mu$$

$(1 \leqslant i < j \leqslant n)$ was considered. Then taking the number p of the clusters of objects constant, MMDS gave clusters of objects under which no cluster of characters became distinct and the points corresponding to the characters were situated symmetrically, according to Section 1. For the value $\tilde{E}(n, \mu, p, K)$ of the term E at the end of MMDS of the data field under given values of n, μ, p and K, the equation

$$\tilde{E}(n, \mu, p, K) = \frac{pK^2}{\mu + p} (p\tilde{V} + C_n^2 \mu)$$

was obtained [for $\tilde{V}$ see expression (1)].

5. MMDS OF CONGENITAL ABNORMALITIES

From the conception until the birth structural defects may develop in the embryo and fetus. Such a defect is called *congenital abnormality* (CA). Within the CAs the *multiple congenital abnormalities* (MCAs), which are the concurrences of two or more different CAs in the same person, have a special importance. A major purpose of their statistical analysis, reported in details in Ref. [13], was to explain the possible cause(s) of combination of CAs within MCAs. The analysis was based on children born in Hungary 1970–1976 (Data field 1) and 1977–1982 (Data field 2). Under Data field 1 $n = 40$, $M = 1,186,776$ and $N = 881$, under Data field 2 $n = 45$, $M = 937,320$ and $N = 867$.

The statistical investigation of MCAs can be based on different alternative hypotheses. There are two models which are reasonable, general and effective. The *Gaussian threshold model* assumes that for any member of the population there is a measure of any CA which can be expressed in a real number. In other words, a background variable L_i, the so-called liability is assigned to A_i. According to the model, the joint distribution of the L_is are multidimensional Gaussian. The fact that somebody has A_i mean that his (or her) liability exceeds a threshold T_i characteristic for the population. The *mixture model* assumes that the probability distribution of MCAs is the mixture of distributions having the property that in each of them the CAs are independent.

One of the main aims of the statistical investigation of MCAs was the classification of children with CAs on the basis of the occurrences of the CAs, and by this the detection of characteristic CA combinations. MMDS was elaborated for solving this task. We can ascertain its adequacy in such a way that to clusters of CAs generated randomly or produced systematically we generate a random data field, and examine whether the classification of the generated children by MMDS has the property that under the various clusters of children the clusters of CAs become distinct. For generating the random children we must first, of course, specify the hypothesis and model (describing MCAs) on the basis of which the generation will be performed. The problem of the inconsistency of the variables W_i requiring MMDS doesn't arise under the Gaussian threshold model, therefore we generated data fields on the basis of the mixture model. Their MMDS gave favourable results, ascertaining the adequacy of MMDS of the "real" data fields [13].

REFERENCES

1. L. Fejes Tóth, Symmetry induced by economy. *Comput. Math. Applic.* **12B,** 83–91 (1986). Reprinted in *Symmetry: Unifying Human Understanding* (Ed. I. Hargittai). Pergamon Press, Oxford (1986).
2. K. V. Mardia, J. T. Kent and J. M. Bibby, *Multivariate Analysis.* Academic Press, New York (1979).
3. J. B. Kruskal and M. Wish, *Multidimensional Scaling.* Sage Publications, Beverly Hills (1978).
4. S. S. Schiffman, M. L. Reynolds and F. W. Young, *Introduction to Multidimensional Scaling.* Academic Press, New York (1981).
5. A. D. Gordon, *Classification.* Chapman and Hall, London (1981).
6. J. De Leeuw and W. Heiser, Theory of multidimensional scaling, In *Handbook of Statistics, Vol. 2: Classification, Pattern Recognition and Reduction of Dimensionality* (Eds P. R. Krishnaiah and L. N. Kanal), pp. 285–316. North-Holland, Amsterdam (1982).
7. M. Wish and J. D. Carroll, Multidimensional scaling and its applications, In *Handbook of Statistics, Vol. 2: Classification, Pattern Recognition and Reduction of Dimensionality* (Eds P. R. Krishnaiah and L. N. Kanal), pp. 317–345. North-Holland, Amsterdam (1982).
8. J. Gower, Multivariate analysis: ordination, multidimensional scaling and allied topics, In *Handbook of Applicable Mathematics, Vol. VI: Statistics,* Part B (Ed. E. Lloyd), pp. 727–781. Wiley-Interscience, Chicester (1984).
9. L. Telegdi, Multiple multidimensional scaling: a new approach to the analysis of multidimensional contingency tables with application to congenital abnormalities. *Metron* **40,** 277–288 (1982).
10. J. A. Hartigan, *Clustering Algorithms.* Wiley, New York (1975).
11. L. J. Hubert and P. Arabie, Comparing partitions. *J. Classification* **2,** 193–218 (1985).
12. J. P. Barthelemy, B. Leclerc and B. Monjardet, On the use of ordered sets in problems of comparison and consensus of classifications. *J. Classification* **3,** 187–224 (1986).
13. A. Czeizel, L. Telegdi and G. Tusnády, *Multiple Congenital Abnormalities.* Publishing House of the Hungarian Academy of Sciences, Budapest (1988).

Computers Math. Applic. Vol. 17, No. 1–3, pp. 147–165, 1989
Printed in Great Britain. All rights reserved

0097-4943/89 $3.00 + 0.00

TILINGS BY REGULAR POLYGONS—II

A CATALOG OF TILINGS

D. Chavey†

Computer Sciences Department, University of Wisconsin-Madison, 1210 West Dayton Street, Madison,
WI 53706, U.S.A.

Abstract—Several classification theorems involving highly symmetric tilings by regular polygons have been established recently. This paper surveys that work and gives drawings of these tilings—many of which were not shown in the original papers. Included are all tilings with at most three symmetry classes (orbits) of tiles, vertices or edges and those tilings which satisfy certain homogeneity criteria; i.e. tilings where locally congruent portions of the tiling are always equivalent under a global symmetry of the tiling.

INTRODUCTION

Tilings of the plane which use only regular polygons include many of the most symmetric tilings, and many of the most beautiful tilings. Such tilings have been known since antiquity, but most of the classification results involving them have appeared only in the last two decades. These classification results are scattered across several papers, and some of the results, unfortunately, appear without the pictures of the tilings which provide their greatest appeal. The purpose of this work is to combine these results into a single catalog of highly regular tilings; to describe the various natures of their regularities; and to provide pictures of these tilings. It should be noted that many of these tilings also appear in Grünbaum and Shephard [1], who go into much greater depth discussing regularity of more general types of tilings. The reader is also referred to this work for other types of tilings and for precise definitions of the terminology associated with tilings.

In addition to using only regular polygons, all tilings in this paper are assumed to be "edge-to-edge", i.e. whenever two polygons intersect at more than one point, they share an edge. This means that tiles cannot meet at edges that are offset with respect to each other (such as the triangles in Fig. 1), and means that all of the polygons must have the same edge length (unlike the polygons in Fig. 1). Although this restriction eliminates many fascinating tilings, it makes the classification theorems much more tractable. Very few classification theorems have been established for non edge-to-edge tilings by regular polygons. The reader is referred to Section 2.4 of Grünbaum and Shephard [1] for a survey of most of the available results and pictures of many of the relevant tilings. The reader is warned that some of the statements made later in the present paper are not correct for tilings more general than the edge-to-edge tilings by regular polygons.

REGULARITY PROPERTIES

A tiling has associated with it in a natural way not only the constituent *tiles*, but also the *edges* and *vertices* where these tiles meet. Collectively the tiles, vertices and edges of a tiling are called the *elements* of that tiling. One measure of the regularity of a tiling is how many equivalence classes of these elements the tiling has under the symmetry group of the tiling. In the language of group theory, these equivalence classes are the *orbits* of the elements under the symmetry group. We say that two elements of the tiling are *symmetric with* each other if they are in the same orbit; i.e. one can be carried into the other by a symmetry of the tiling. The number of orbits of the tiling elements is finite if and only if the tiling is *periodic*, i.e. it has translational symmetries in at least two different

---

†Current address: Department of Mathematics and Computer Science, Beloit College, Beloit, WI 53511, U.S.A.

148   D. CHAVEY

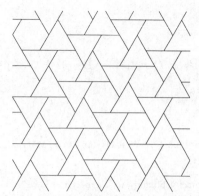

Fig. 1. A non edge-to-edge tiling by regular polygons. This tiling is homogeneous with respect to tiles, vertices and edges; and is strongly edge-homogeneous.

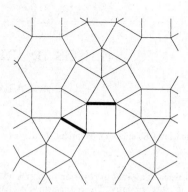

Fig. 2. The two bold edges have the same simple edge type, but different edge types.

directions. Those tilings with only a few different orbits of one or more of the tiling elements are the most regular or most symmetric tilings. The tilings (edge-to-edge tilings by regular polygons) which have at most three orbits of one of the tiling elements have all been classified, and the number of these tilings is listed in Table 1, along with references to the original results. The author [2] has shown that the number of orbits of vertices is always bounded by the number of orbits of edges; consequently the results of the last row of Table 1 can be established from examining the tilings corresponding to the results of the middle row of this table.

The discovery of the tilings with a single orbit of tiles, listed in Table 1 as "ancient" is lost in antiquity, but they were almost certainly known to the early Greek geometers. The Pythagoreans were least aware that the only ways to combine regular polygons of one type at a vertex were three hexagons, four squares, or six triangles (see Ref. [3]), and hence must have known of these tilings. Pappus of Alexandria in the third century A.D. writes of the classification of these tilings in the preface of his *Collection* (quoted in full in Ref. [4]) as if it was an old, well-known fact. In addition, Pappus is aware of the need for the assumption that the tilings are edge-to-edge, and is careful to state this.

We should note that we give credit to Kepler for the classification of the tilings with only one orbit of vertices. Although he found all of the tilings, his proof that there are no others is incorrect. In classifying the possible vertex types which can be constructed from polygons (regardless of whether they can be extended to tilings), he misses the vertex types 3.7.42, 3.8.24 and 3.9.18 (using the notation to be defined later). This is a fairly minor point, since his Theorem 17 can be used to show that these vertex types cannot occur in a tiling. Nevertheless, the first fully correct proof appears to be due to Sommerville [5].

Another way of viewing a tiling as "regular" is what we describe as *homogeneous*: any two tiling elements of some type which "might" be symmetric with each other actually are. For example, a triangle obviously cannot be symmetric with a square, but two different triangles could, conceivably, be symmetric with each other. Thus, a tiling would be described as *tile-homogeneous* if any two congruent polygons were in the same orbit; i.e. were symmetric with each other. Figure 1, although not an edge-to-edge tiling, is an example of a tile-homogeneous tiling; all triangles are equivalent, and all hexagons are equivalent. Notice that homogeneity can be thought

Table 1. Classification results

	One orbit	Two orbits	Three orbits
Tiles	3 [Ancient]	13 [6]	25 [7]
Vertices	11 [8]	20 [9]	61 [10, 11]
Edges	4 [12]	4 [10, 13]	10 [10, 13]

of in terms of local/global properties: whenever two areas of the tiling (e.g. tiles) locally look the same, there is a global symmetry which also recognizes that "sameness".

To define vertex-homogeneity, we first define the *vertex figure* of a vertex in a tiling to be the union of all edges incident to that vertex. This is the obvious "local region" for a vertex. A tiling is then *vertex-homogeneous* if any two vertices with congruent vertex figures are symmetric with each other.†

The correct definition of edge-homogeneous is not quite as obvious. If one views an edge as connecting the two tiles which contain it, then an "edge figure" consists of an edge and its two incident tiles. With this viewpoint one would think of the two bold edges of Fig. 2 as being "locally" the same. On the other hand, if we think of an edge as connecting the two vertices at its ends, then an "edge figure" consists of an edge together with all edges which meet it. Under this viewpoint, the two bold edges of Fig. 2 are "locally" different. For this survey we refer to an edge and its two incident tiles as a "simple edge figure", and call an edge with its two incident vertex figures an "edge figure". A tiling in which any two congruent edge figures are symmetric is called *edge-homogeneous* while a tiling in which any two congruent simple edge figures are symmetric is called *strongly edge-homogeneous*. This terminology reflects the author's belief in what the "correct" definition of edge-homogeneous should be. It is easily verified that a tiling which is strongly edge-homogeneous is also edge-homogeneous (the converse is false).

Most of the homogeneous tilings have been classified. The 135 vertex-homogeneous tilings were classified by Krötenheerdt [9, 14, 15]. The 22 tile-homogeneous tilings were classified by DeBroey and Landuyt [6]. The 22 strongly edge-homogeneous tilings were classified by the author [10]. The edge-homogeneous tilings have not been classified, and it is expected that there would be a very large number of such tilings. It is worth noting that there are some relationships between the various types of homogeneity. All of the strongly edge-homogeneous tilings are also vertex-homogeneous (a direct proof of this fact is the basis of the classification of such tilings). All but one of the tile-homogeneous tilings are also vertex-homogeneous. (This exceptional tiling is also not edge-homogeneous; it is the tiling of Fig. 9.)

It is the author's feeling that part of the success of a classification theorem in the theory of tilings can be measured by the beauty of the new tilings discovered. From this viewpoint it would seem that the two most successful classification theorems for edge-to-edge tilings by regular polygons are the tilings with a single vertex orbit [8] and the vertex-homogeneous tilings [9, 14, 15]. This latter work included the classification of all tilings with two vertex orbits. The author suspects that several other beautiful tilings could be found among the edge-homogeneous tilings.

Another measure of the success of a classification theorem might be its completeness. It is always possible to imagine further classification theorems which would extend Table 1 a little further. The classification of the homogeneous tilings is more of a "final" classification, it is difficult (and rather artificial) to try to extend these definitions to broader classes of tilings.

THE TILINGS

On the following pages are drawings of all the tilings which arise from the classification theorems mentioned in the previous section. In general, the number of vertex orbits seems to be a good measure of the complexity of the tiling. Consequently, the tilings are arranged according to the number of vertex orbits in the tiling, labelled according to the vertex types which appear, and a representative of each vertex orbit is marked by a bold circle. The *vertex type* of a vertex, as used to label the tilings, is a listing, in either clockwise or counter-clockwise order, of all the polygons which meet the vertex. For example, a vertex of type 3.4.6.4 meets, in order, a 3-gon (triangle), a 4-gon (square), a 6-gon (hexagon), and another 4-gon. Of all possible such labellings of a vertex, the vertex type is that one which precedes all others lexicographically; e.g. we use 3.4.6.4 in preference to 4.6.4.3, even though they describe the same vertex. For conciseness, we list a vertex of type 3.3.3.3.6 as $3^4.6$, and similarly for other types. In addition to the labelling of tilings by the

---

†Grünbaum and Shephard [1] use the term "homogeneous" for vertex-homogeneous and "equitransitive" for tile-homogeneous. They do not define a notion of edge-homogeneous.

vertex types of each orbit, we also use subscripts when necessary to distinguish two different tilings with otherwise similar labellings; e.g. $(3^6; 3^3.4^2)_1$ vs $(3^6; 3^3.4^2)_2$. Within a figure, the tilings are arranged lexicographically according to this labelling.

The number of orbits of each tile element is listed with each tiling; e.g. a tiling with two vertex orbits, three tile orbits, and four edge orbits would have this fact noted by the legend "$v = 2$; $t = 3$; $e = 4$" In the terminology of Ref. [1], such a tiling would be called "2-isogonal, 3-isohedral, and 4-isotoxal". For readers familiar with this terminology, the captions of the figures also use these words. Although representatives of the tile orbits and edge orbits are not noted (as with the vertices), the former is usually easy to identify, and the latter can usually be found with only a little work once you note that representatives of all edge orbits must be incident to one (or more) of the marked vertices.

Figure 3 shows the so-called "Platonic" tilings, those whose symmetry group is transitive on the tiles (i.e. those with $t = 1$). Figure 4 shows the other "Archimedean" tilings, those whose symmetry group is transitive on the vertices ($v = 1$, but $t > 1$). Normally, "Archimedean" only means that all of the vertex figures are congruent. For the tilings considered here, this can be shown to imply that the vertex figures are all symmetric with each other; i.e. tilings with only one vertex figure are necessarily vertex-homogeneous. Figure 5 shows the tilings with $v = 2$, all of which are vertex-homogeneous. The tilings with $v = 3$ are split into two groups: those in Fig. 6 are vertex-homogeneous; those in Fig. 7 are not (hence those in Fig. 7 have two types of vertex figures, but three vertex orbits). Figures 3–7 also include all tilings with $e \leqslant 3$, since (see Ref. [2]) $e \geqslant v$. Figure 8 shows the vertex-homogeneous tilings which have four or more vertex orbits. Figure 9 shows the unique tiling which is tile-homogeneous but not vertex-homogeneous. This is also the only tiling with two orbits of tiles which is not pictured previously. Finally Fig. 10 shows the remaining nine tilings which have three orbits of tiles.

If a tiling satisfies any of the homogeneity criteria, this is also noted, except where all (or nearly all) of the tilings in some figure satisfy one of the homogeneity criteria. In this case, the homogeneity properties may be just noted in the caption to the figure (and not individually). Further, since strongly edge-homogeneous tilings are necessarily edge-homogeneous, the latter fact is not mentioned when both are true. For conciseness, "strongly edge-homogeneous" is abbreviated as "S edge-homogeneous".

Finally, the symmetry group of each tiling is listed, using the standard crystallographic notation. Further details on the groups corresponding to this notation can be found in several places; e.g. Ref. [1, Section 1.4]. Of the 17 two-dimensional symmetry groups, all but four of them occur as the symmetry group of a tiling included here. Each of the other four groups also can arise as the symmetry group of an edge-to-edge tiling by regular polygons, but we cannot say for sure what the "simplest" of such tilings are. It is worth noting that some of the tilings in the following pictures occur in "enantiomorphic" forms; i.e. the mirror image of the tiling cannot be superimposed on the original. One example is the first tiling in Fig. 4. For the purposes of the drawings in this paper and for the numbers listed in Table 1, we do not view two enantiomorphic forms of a tiling as different tilings. Only nine of the 165 tilings included here have different enantiomorphic forms. These can easily be found by noting that a tiling has two such forms if and only if its symmetry group contains no reflections or glide-reflections; i.e. if and only if the crystallographic name of the symmetry group contains neither an "m" nor a "g". The number of "enantiomorphically different" tilings with at most three orbits of some tile element is given in Table 2.

Table 2. Classification results for enantiomorphically different tilings

	One orbit	Two orbits	Three orbits
Tiles	3	14	26
Vertices	12	21	65
Edges	4	4	11

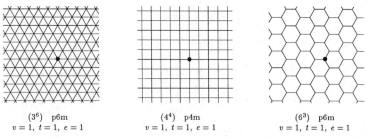

(3⁶) p6m
$v = 1$, $t = 1$, $e = 1$

(4⁴) p4m
$v = 1$, $t = 1$, $e = 1$

(6³) p6m
$v = 1$, $t = 1$, $e = 1$

Fig. 3. The three "Platonic" tilings, i.e. those with $t = 1$ (also called "isohedral" tilings). These tilings are all homogeneous with respect to vertices, tiles and edges, and are strongly edge-homogeneous.

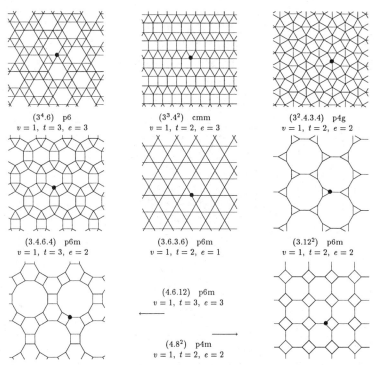

(3⁴.6) p6
$v = 1$, $t = 3$, $e = 3$

(3³.4²) cmm
$v = 1$, $t = 2$, $e = 3$

(3².4.3.4) p4g
$v = 1$, $t = 2$, $e = 2$

(3.4.6.4) p6m
$v = 1$, $t = 3$, $e = 2$

(3.6.3.6) p6m
$v = 1$, $t = 2$, $e = 1$

(3.12²) p6m
$v = 1$, $t = 2$, $e = 2$

(4.6.12) p6m
$v = 1$, $t = 3$, $e = 3$

(4.8²) p4m
$v = 1$, $t = 2$, $e = 2$

Fig. 4. The eight "Archimedean" tilings which are not Platonic; i.e. those with $v = 1$ (also called "isogonal" tilings) but where $t > 1$. These tilings are all homogeneous with respect to both vertices and edges. All but the first tiling are tile-homogeneous and strongly edge-homogeneous.

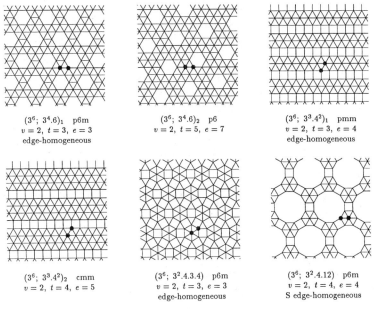

(3⁶; 3⁴.6)₁ p6m
$v = 2$, $t = 3$, $e = 3$
edge-homogeneous

(3⁶; 3⁴.6)₂ p6
$v = 2$, $t = 5$, $e = 7$

(3⁶; 3³.4²)₁ pmm
$v = 2$, $t = 3$, $e = 4$
edge-homogeneous

(3⁶; 3³.4²)₂ cmm
$v = 2$, $t = 4$, $e = 5$

(3⁶; 3².4.3.4) p6m
$v = 2$, $t = 3$, $e = 3$
edge-homogeneous

(3⁶; 3².4.12) p6m
$v = 2$, $t = 4$, $e = 4$
S edge-homogeneous

Fig. 5—*continued overleaf.*

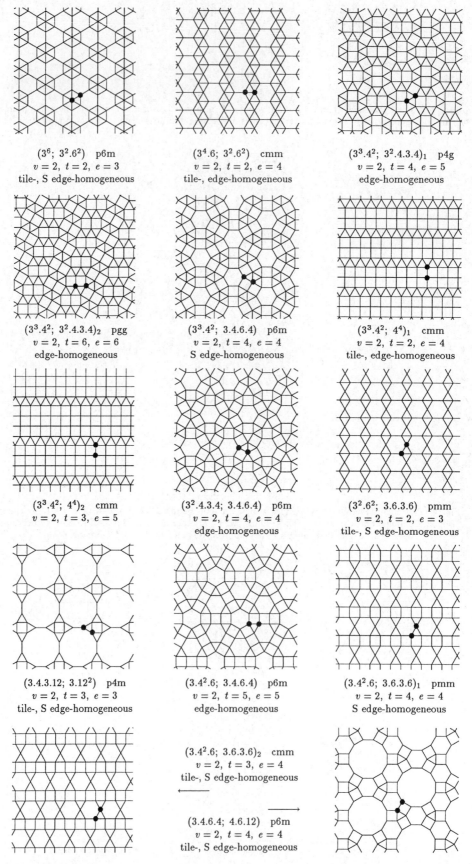

$(3^6; 3^2.6^2)$ p6m
$v = 2, t = 2, e = 3$
tile-, S edge-homogeneous

$(3^4.6; 3^2.6^2)$ cmm
$v = 2, t = 2, e = 4$
tile-, edge-homogeneous

$(3^3.4^2; 3^2.4.3.4)_1$ p4g
$v = 2, t = 4, e = 5$
edge-homogeneous

$(3^3.4^2; 3^2.4.3.4)_2$ pgg
$v = 2, t = 6, e = 6$
edge-homogeneous

$(3^3.4^2; 3.4.6.4)$ p6m
$v = 2, t = 4, e = 4$
S edge-homogeneous

$(3^3.4^2; 4^4)_1$ cmm
$v = 2, t = 2, e = 4$
tile-, edge-homogeneous

$(3^3.4^2; 4^4)_2$ cmm
$v = 2, t = 3, e = 5$

$(3^2.4.3.4; 3.4.6.4)$ p6m
$v = 2, t = 4, e = 4$
edge-homogeneous

$(3^2.6^2; 3.6.3.6)$ pmm
$v = 2, t = 2, e = 3$
tile-, S edge-homogeneous

$(3.4.3.12; 3.12^2)$ p4m
$v = 2, t = 3, e = 3$
tile-, S edge-homogeneous

$(3.4^2.6; 3.4.6.4)$ p6m
$v = 2, t = 5, e = 5$
edge-homogeneous

$(3.4^2.6; 3.6.3.6)_1$ pmm
$v = 2, t = 4, e = 4$
S edge-homogeneous

$(3.4^2.6; 3.6.3.6)_2$ cmm
$v = 2, t = 3, e = 4$
tile-, S edge-homogeneous

←

——→

$(3.4.6.4; 4.6.12)$ p6m
$v = 2, t = 4, e = 4$
tile-, S edge-homogeneous

Fig. 5. The 20 2-isogonal tilings; i.e. those with $v = 2$. All of these tilings are vertex-homogeneous.

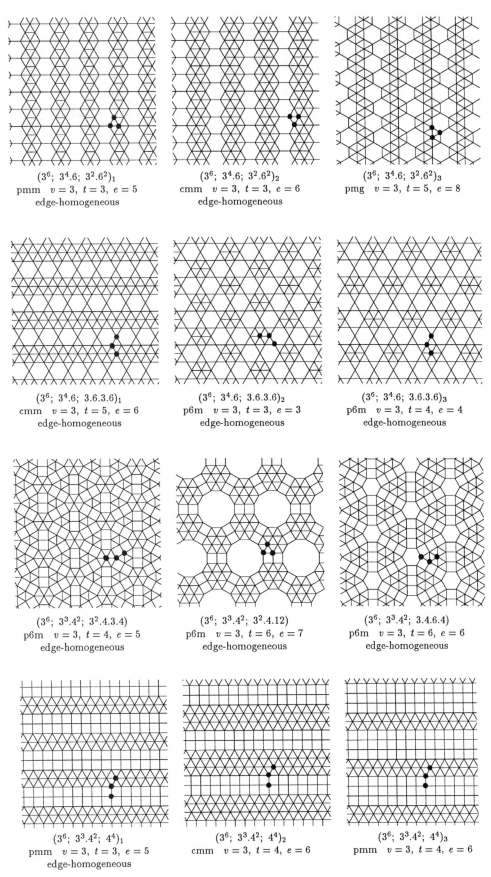

$(3^6; 3^4.6; 3^2.6^2)_1$
pmm $v = 3$, $t = 3$, $e = 5$
edge-homogeneous

$(3^6; 3^4.6; 3^2.6^2)_2$
cmm $v = 3$, $t = 3$, $e = 6$
edge-homogeneous

$(3^6; 3^4.6; 3^2.6^2)_3$
pmg $v = 3$, $t = 5$, $e = 8$

$(3^6; 3^4.6; 3.6.3.6)_1$
cmm $v = 3$, $t = 5$, $e = 6$
edge-homogeneous

$(3^6; 3^4.6; 3.6.3.6)_2$
p6m $v = 3$, $t = 3$, $e = 3$
edge-homogeneous

$(3^6; 3^4.6; 3.6.3.6)_3$
p6m $v = 3$, $t = 4$, $e = 4$
edge-homogeneous

$(3^6; 3^3.4^2; 3^2.4.3.4)$
p6m $v = 3$, $t = 4$, $e = 5$
edge-homogeneous

$(3^6; 3^3.4^2; 3^2.4.12)$
p6m $v = 3$, $t = 6$, $e = 7$
edge-homogeneous

$(3^6; 3^3.4^2; 3.4.6.4)$
p6m $v = 3$, $t = 6$, $e = 6$
edge-homogeneous

$(3^6; 3^3.4^2; 4^4)_1$
pmm $v = 3$, $t = 3$, $e = 5$
edge-homogeneous

$(3^6; 3^3.4^2; 4^4)_2$
cmm $v = 3$, $t = 4$, $e = 6$

$(3^6; 3^3.4^2; 4^4)_3$
pmm $v = 3$, $t = 4$, $e = 6$

Fig. 6—*continued overleaf.*

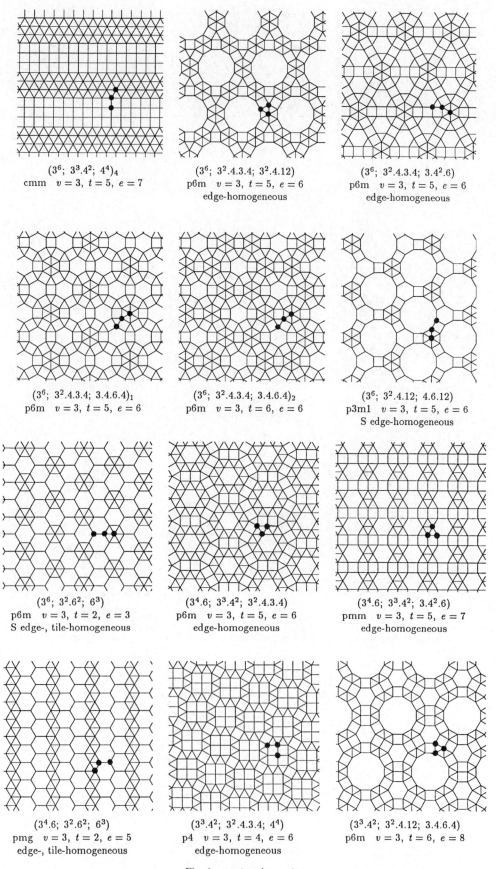

$(3^6; 3^3.4^2; 4^4)_4$
cmm $v = 3, t = 5, e = 7$

$(3^6; 3^2.4.3.4; 3^2.4.12)$
p6m $v = 3, t = 5, e = 6$
edge-homogeneous

$(3^6; 3^2.4.3.4; 3.4^2.6)$
p6m $v = 3, t = 5, e = 6$
edge-homogeneous

$(3^6; 3^2.4.3.4; 3.4.6.4)_1$
p6m $v = 3, t = 5, e = 6$

$(3^6; 3^2.4.3.4; 3.4.6.4)_2$
p6m $v = 3, t = 6, e = 6$

$(3^6; 3^2.4.12; 4.6.12)$
p3m1 $v = 3, t = 5, e = 6$
S edge-homogeneous

$(3^6; 3^2.6^2; 6^3)$
p6m $v = 3, t = 2, e = 3$
S edge-, tile-homogeneous

$(3^4.6; 3^3.4^2; 3^2.4.3.4)$
p6m $v = 3, t = 5, e = 6$
edge-homogeneous

$(3^4.6; 3^3.4^2; 3.4^2.6)$
pmm $v = 3, t = 5, e = 7$
edge-homogeneous

$(3^4.6; 3^2.6^2; 6^3)$
pmg $v = 3, t = 2, e = 5$
edge-, tile-homogeneous

$(3^3.4^2; 3^2.4.3.4; 4^4)$
p4 $v = 3, t = 4, e = 6$
edge-homogeneous

$(3^3.4^2; 3^2.4.12; 3.4.6.4)$
p6m $v = 3, t = 6, e = 8$

Fig. 6—*continued opposite*.

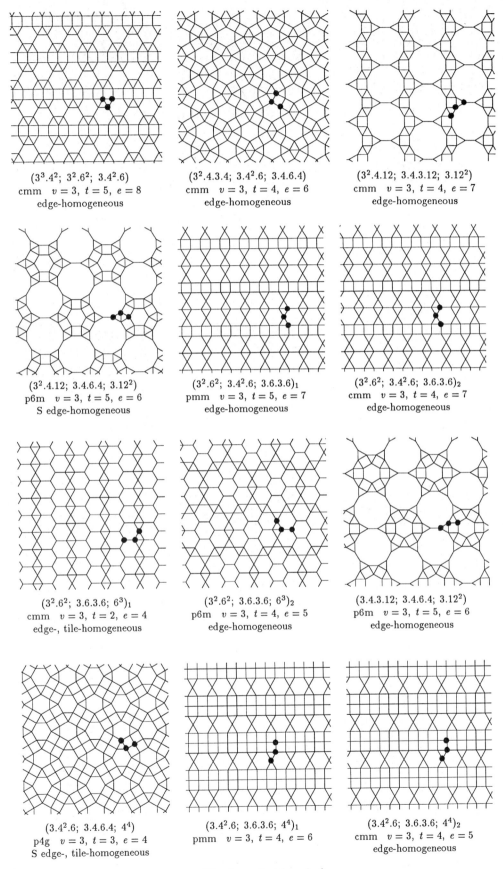

$(3^3.4^2;\ 3^2.6^2;\ 3.4^2.6)$
cmm $v = 3,\ t = 5,\ e = 8$
edge-homogeneous

$(3^2.4.3.4;\ 3.4^2.6;\ 3.4.6.4)$
cmm $v = 3,\ t = 4,\ e = 6$
edge-homogeneous

$(3^2.4.12;\ 3.4.3.12;\ 3.12^2)$
cmm $v = 3,\ t = 4,\ e = 7$
edge-homogeneous

$(3^2.4.12;\ 3.4.6.4;\ 3.12^2)$
p6m $v = 3,\ t = 5,\ e = 6$
S edge-homogeneous

$(3^2.6^2;\ 3.4^2.6;\ 3.6.3.6)_1$
pmm $v = 3,\ t = 5,\ e = 7$
edge-homogeneous

$(3^2.6^2;\ 3.4^2.6;\ 3.6.3.6)_2$
cmm $v = 3,\ t = 4,\ e = 7$
edge-homogeneous

$(3^2.6^2;\ 3.6.3.6;\ 6^3)_1$
cmm $v = 3,\ t = 2,\ e = 4$
edge-, tile-homogeneous

$(3^2.6^2;\ 3.6.3.6;\ 6^3)_2$
p6m $v = 3,\ t = 4,\ e = 5$
edge-homogeneous

$(3.4.3.12;\ 3.4.6.4;\ 3.12^2)$
p6m $v = 3,\ t = 5,\ e = 6$
edge-homogeneous

$(3.4^2.6;\ 3.4.6.4;\ 4^4)$
p4g $v = 3,\ t = 3,\ e = 4$
S edge-, tile-homogeneous

$(3.4^2.6;\ 3.6.3.6;\ 4^4)_1$
pmm $v = 3,\ t = 4,\ e = 6$

$(3.4^2.6;\ 3.6.3.6;\ 4^4)_2$
cmm $v = 3,\ t = 4,\ e = 5$
edge-homogeneous

Fig. 6—*continued overleaf.*

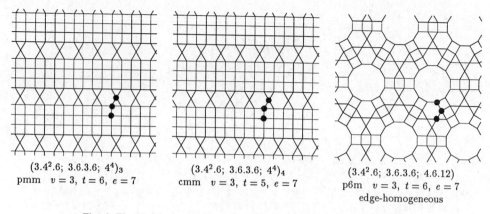

$(3.4^2.6; 3.6.3.6; 4^4)_3$
pmm $v = 3$, $t = 6$, $e = 7$

$(3.4^2.6; 3.6.3.6; 4^4)_4$
cmm $v = 3$, $t = 5$, $e = 7$

$(3.4^2.6; 3.6.3.6; 4.6.12)$
p6m $v = 3$, $t = 6$, $e = 7$
edge-homogeneous

Fig. 6. The 39 3-isogonal tilings ($v = 3$) which are vertex-homogeneous.

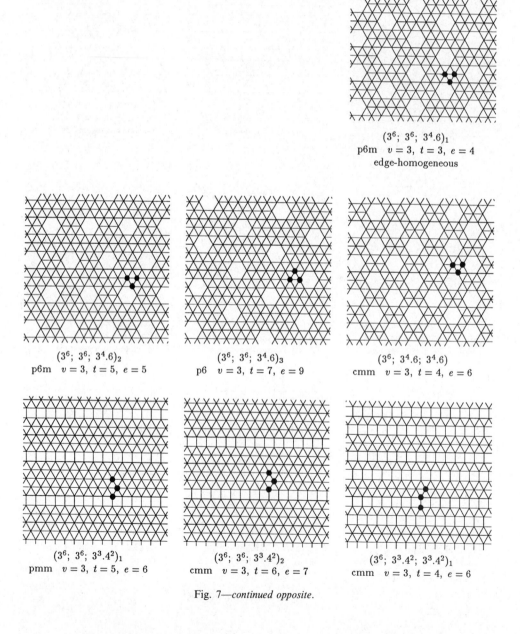

$(3^6; 3^6; 3^4.6)_1$
p6m $v = 3$, $t = 3$, $e = 4$
edge-homogeneous

$(3^6; 3^6; 3^4.6)_2$
p6m $v = 3$, $t = 5$, $e = 5$

$(3^6; 3^6; 3^4.6)_3$
p6 $v = 3$, $t = 7$, $e = 9$

$(3^6; 3^4.6; 3^4.6)$
cmm $v = 3$, $t = 4$, $e = 6$

$(3^6; 3^6; 3^3.4^2)_1$
pmm $v = 3$, $t = 5$, $e = 6$

$(3^6; 3^6; 3^3.4^2)_2$
cmm $v = 3$, $t = 6$, $e = 7$

$(3^6; 3^3.4^2; 3^3.4^2)_1$
cmm $v = 3$, $t = 4$, $e = 6$

Fig. 7—continued opposite.

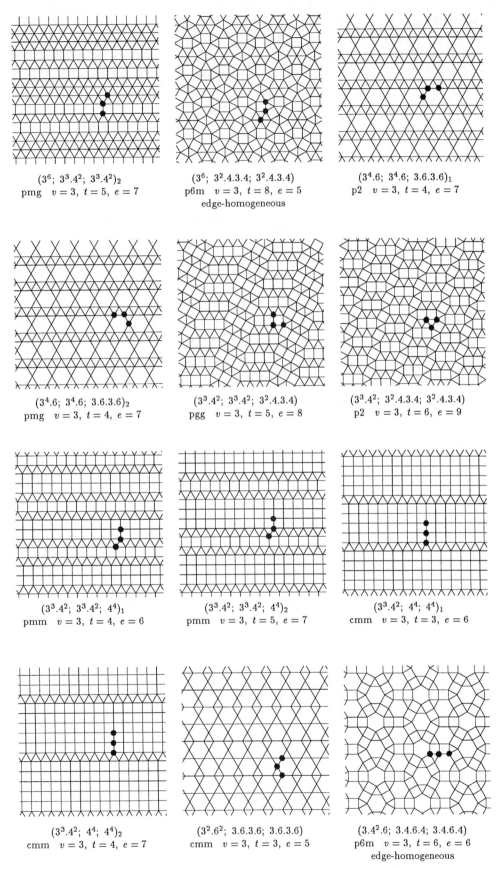

$(3^6; 3^3.4^2; 3^3.4^2)_2$
pmg $v = 3, t = 5, e = 7$

$(3^6; 3^2.4.3.4; 3^2.4.3.4)$
p6m $v = 3, t = 8, e = 5$
edge-homogeneous

$(3^4.6; 3^4.6; 3.6.3.6)_1$
p2 $v = 3, t = 4, e = 7$

$(3^4.6; 3^4.6; 3.6.3.6)_2$
pmg $v = 3, t = 4, e = 7$

$(3^3.4^2; 3^3.4^2; 3^2.4.3.4)$
pgg $v = 3, t = 5, e = 8$

$(3^3.4^2; 3^2.4.3.4; 3^2.4.3.4)$
p2 $v = 3, t = 6, e = 9$

$(3^3.4^2; 3^3.4^2; 4^4)_1$
pmm $v = 3, t = 4, e = 6$

$(3^3.4^2; 3^3.4^2; 4^4)_2$
pmm $v = 3, t = 5, e = 7$

$(3^3.4^2; 4^4; 4^4)_1$
cmm $v = 3, t = 3, e = 6$

$(3^3.4^2; 4^4; 4^4)_2$
cmm $v = 3, t = 4, e = 7$

$(3^2.6^2; 3.6.3.6; 3.6.3.6)$
cmm $v = 3, t = 3, e = 5$

$(3.4^2.6; 3.4.6.4; 3.4.6.4)$
p6m $v = 3, t = 6, e = 6$
edge-homogeneous

Fig. 7—*continued overleaf.*

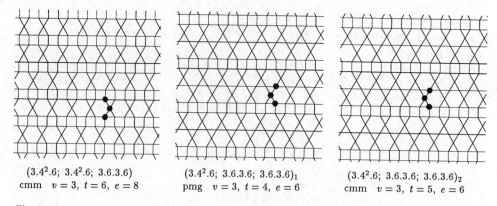

$(3.4^2.6; \; 3.4^2.6; \; 3.6.3.6)$
cmm $v = 3, \; t = 6, \; e = 8$

$(3.4^2.6; \; 3.6.3.6; \; 3.6.3.6)_1$
pmg $v = 3, \; t = 4, \; e = 6$

$(3.4^2.6; \; 3.6.3.6; \; 3.6.3.6)_2$
cmm $v = 3, \; t = 5, \; e = 6$

Fig. 7. The 22 3-isogonal tilings ($v = 3$) which are not vertex-homogeneous. These tilings all have two distinct vertex figures, but three orbits of vertices.

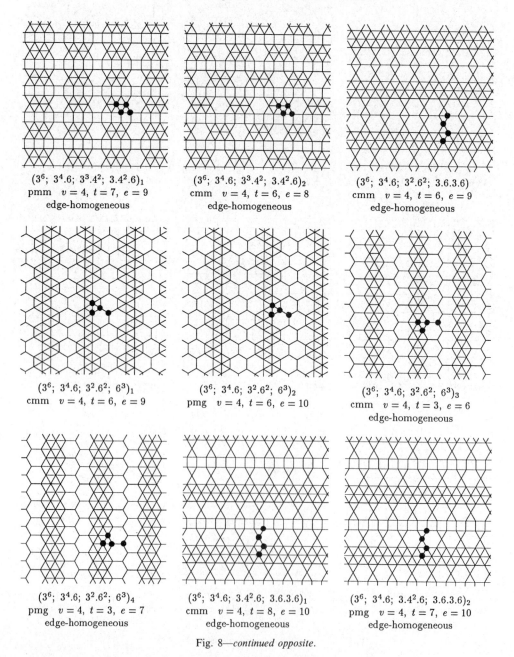

$(3^6; \; 3^4.6; \; 3^3.4^2; \; 3.4^2.6)_1$
pmm $v = 4, \; t = 7, \; e = 9$
edge-homogeneous

$(3^6; \; 3^4.6; \; 3^3.4^2; \; 3.4^2.6)_2$
cmm $v = 4, \; t = 6, \; e = 8$
edge-homogeneous

$(3^6; \; 3^4.6; \; 3^2.6^2; \; 3.6.3.6)$
cmm $v = 4, \; t = 6, \; e = 9$
edge-homogeneous

$(3^6; \; 3^4.6; \; 3^2.6^2; \; 6^3)_1$
cmm $v = 4, \; t = 6, \; e = 9$

$(3^6; \; 3^4.6; \; 3^2.6^2; \; 6^3)_2$
pmg $v = 4, \; t = 6, \; e = 10$

$(3^6; \; 3^4.6; \; 3^2.6^2; \; 6^3)_3$
cmm $v = 4, \; t = 3, \; e = 6$
edge-homogeneous

$(3^6; \; 3^4.6; \; 3^2.6^2; \; 6^3)_4$
pmg $v = 4, \; t = 3, \; e = 7$
edge-homogeneous

$(3^6; \; 3^4.6; \; 3.4^2.6; \; 3.6.3.6)_1$
cmm $v = 4, \; t = 8, \; e = 10$
edge-homogeneous

$(3^6; \; 3^4.6; \; 3.4^2.6; \; 3.6.3.6)_2$
pmg $v = 4, \; t = 7, \; e = 10$
edge-homogeneous

Fig. 8—continued opposite.

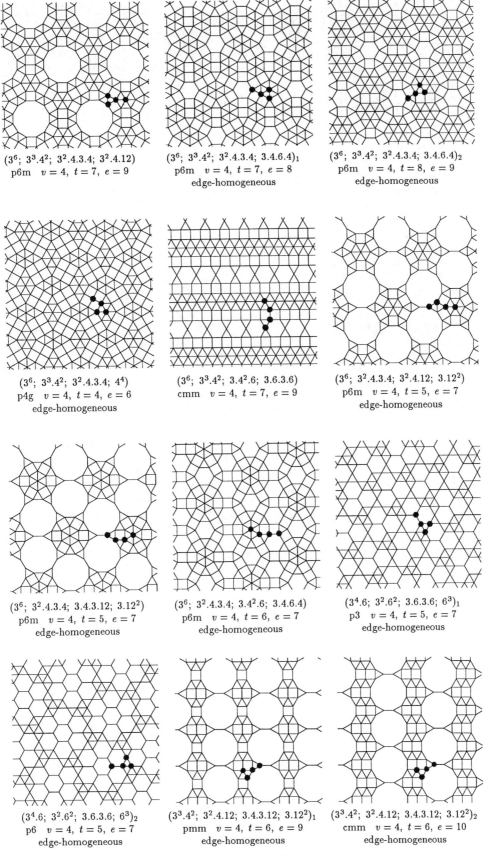

$(3^6; 3^3.4^2; 3^2.4.3.4; 3^2.4.12)$
p6m $v = 4$, $t = 7$, $e = 9$

$(3^6; 3^3.4^2; 3^2.4.3.4; 3.4.6.4)_1$
p6m $v = 4$, $t = 7$, $e = 8$
edge-homogeneous

$(3^6; 3^3.4^2; 3^2.4.3.4; 3.4.6.4)_2$
p6m $v = 4$, $t = 8$, $e = 9$
edge-homogeneous

$(3^6; 3^3.4^2; 3^2.4.3.4; 4^4)$
p4g $v = 4$, $t = 4$, $e = 6$
edge-homogeneous

$(3^6; 3^3.4^2; 3.4^2.6; 3.6.3.6)$
cmm $v = 4$, $t = 7$, $e = 9$

$(3^6; 3^2.4.3.4; 3^2.4.12; 3.12^2)$
p6m $v = 4$, $t = 5$, $e = 7$
edge-homogeneous

$(3^6; 3^2.4.3.4; 3.4.3.12; 3.12^2)$
p6m $v = 4$, $t = 5$, $e = 7$
edge-homogeneous

$(3^6; 3^2.4.3.4; 3.4^2.6; 3.4.6.4)$
p6m $v = 4$, $t = 6$, $e = 7$
edge-homogeneous

$(3^4.6; 3^2.6^2; 3.6.3.6; 6^3)_1$
p3 $v = 4$, $t = 5$, $e = 7$
edge-homogeneous

$(3^4.6; 3^2.6^2; 3.6.3.6; 6^3)_2$
p6 $v = 4$, $t = 5$, $e = 7$
edge-homogeneous

$(3^3.4^2; 3^2.4.12; 3.4.3.12; 3.12^2)_1$
pmm $v = 4$, $t = 6$, $e = 9$
edge-homogeneous

$(3^3.4^2; 3^2.4.12; 3.4.3.12; 3.12^2)_2$
cmm $v = 4$, $t = 6$, $e = 10$
edge-homogeneous

Fig. 8—*continued overleaf.*

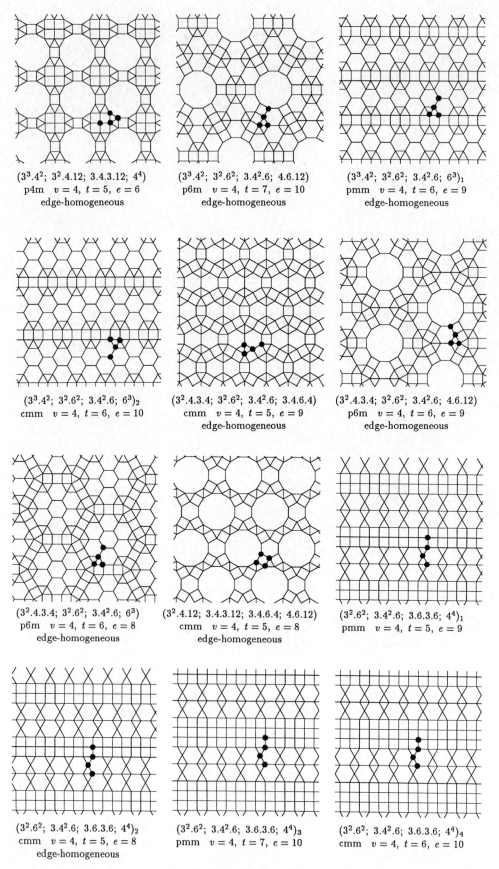

$(3^3.4^2; 3^2.4.12; 3.4.3.12; 4^4)$
p4m $v = 4$, $t = 5$, $e = 6$
edge-homogeneous

$(3^3.4^2; 3^2.6^2; 3.4^2.6; 4.6.12)$
p6m $v = 4$, $t = 7$, $e = 10$
edge-homogeneous

$(3^3.4^2; 3^2.6^2; 3.4^2.6; 6^3)_1$
pmm $v = 4$, $t = 6$, $e = 9$
edge-homogeneous

$(3^3.4^2; 3^2.6^2; 3.4^2.6; 6^3)_2$
cmm $v = 4$, $t = 6$, $e = 10$

$(3^2.4.3.4; 3^2.6^2; 3.4^2.6; 3.4.6.4)$
cmm $v = 4$, $t = 5$, $e = 9$
edge-homogeneous

$(3^2.4.3.4; 3^2.6^2; 3.4^2.6; 4.6.12)$
p6m $v = 4$, $t = 6$, $e = 9$
edge-homogeneous

$(3^2.4.3.4; 3^2.6^2; 3.4^2.6; 6^3)$
p6m $v = 4$, $t = 6$, $e = 8$
edge-homogeneous

$(3^2.4.12; 3.4.3.12; 3.4.6.4; 4.6.12)$
cmm $v = 4$, $t = 5$, $e = 8$
edge-homogeneous

$(3^2.6^2; 3.4^2.6; 3.6.3.6; 4^4)_1$
pmm $v = 4$, $t = 5$, $e = 9$

$(3^2.6^2; 3.4^2.6; 3.6.3.6; 4^4)_2$
cmm $v = 4$, $t = 5$, $e = 8$
edge-homogeneous

$(3^2.6^2; 3.4^2.6; 3.6.3.6; 4^4)_3$
pmm $v = 4$, $t = 7$, $e = 10$

$(3^2.6^2; 3.4^2.6; 3.6.3.6; 4^4)_4$
cmm $v = 4$, $t = 6$, $e = 10$

Fig. 8—*continued opposite.*

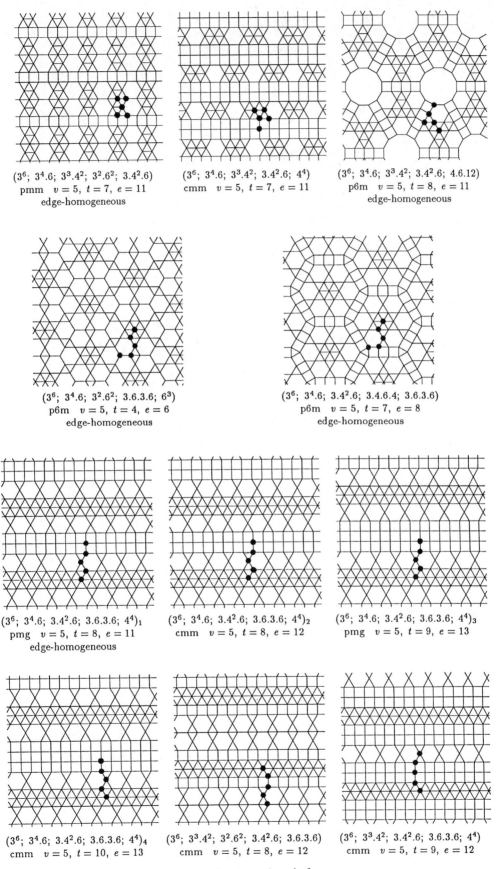

$(3^6; 3^4.6; 3^3.4^2; 3^2.6^2; 3.4^2.6)$
pmm $v = 5, t = 7, e = 11$
edge-homogeneous

$(3^6; 3^4.6; 3^3.4^2; 3.4^2.6; 4^4)$
cmm $v = 5, t = 7, e = 11$

$(3^6; 3^4.6; 3^3.4^2; 3.4^2.6; 4.6.12)$
p6m $v = 5, t = 8, e = 11$
edge-homogeneous

$(3^6; 3^4.6; 3^2.6^2; 3.6.3.6; 6^3)$
p6m $v = 5, t = 4, e = 6$
edge-homogeneous

$(3^6; 3^4.6; 3.4^2.6; 3.4.6.4; 3.6.3.6)$
p6m $v = 5, t = 7, e = 8$
edge-homogeneous

$(3^6; 3^4.6; 3.4^2.6; 3.6.3.6; 4^4)_1$
pmg $v = 5, t = 8, e = 11$
edge-homogeneous

$(3^6; 3^4.6; 3.4^2.6; 3.6.3.6; 4^4)_2$
cmm $v = 5, t = 8, e = 12$

$(3^6; 3^4.6; 3.4^2.6; 3.6.3.6; 4^4)_3$
pmg $v = 5, t = 9, e = 13$

$(3^6; 3^4.6; 3.4^2.6; 3.6.3.6; 4^4)_4$
cmm $v = 5, t = 10, e = 13$

$(3^6; 3^3.4^2; 3^2.6^2; 3.4^2.6; 3.6.3.6)$
cmm $v = 5, t = 8, e = 12$

$(3^6; 3^3.4^2; 3.4^2.6; 3.6.3.6; 4^4)$
cmm $v = 5, t = 9, e = 12$

Fig 8—*continued overleaf.*

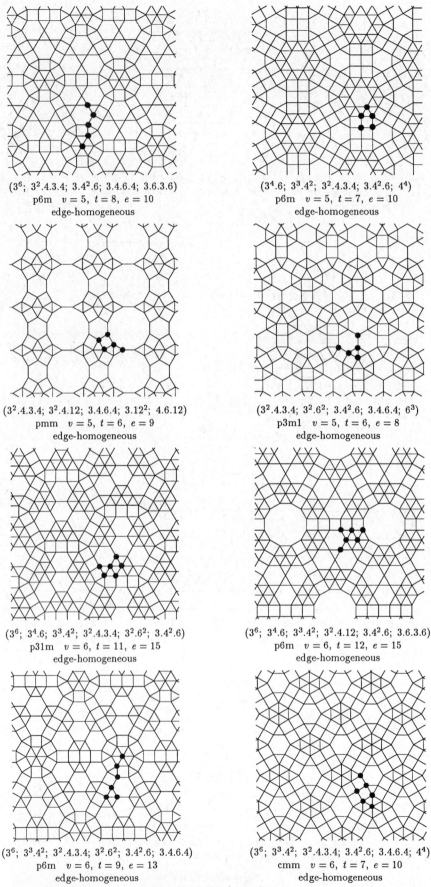

$(3^6;\ 3^2.4.3.4;\ 3.4^2.6;\ 3.4.6.4;\ 3.6.3.6)$
p6m $v = 5,\ t = 8,\ e = 10$
edge-homogeneous

$(3^4.6;\ 3^3.4^2;\ 3^2.4.3.4;\ 3.4^2.6;\ 4^4)$
p6m $v = 5,\ t = 7,\ e = 10$
edge-homogeneous

$(3^2.4.3.4;\ 3^2.4.12;\ 3.4.6.4;\ 3.12^2;\ 4.6.12)$
pmm $v = 5,\ t = 6,\ e = 9$
edge-homogeneous

$(3^2.4.3.4;\ 3^2.6^2;\ 3.4^2.6;\ 3.4.6.4;\ 6^3)$
p3m1 $v = 5,\ t = 6,\ e = 8$
edge-homogeneous

$(3^6;\ 3^4.6;\ 3^3.4^2;\ 3^2.4.3.4;\ 3^2.6^2;\ 3.4^2.6)$
p31m $v = 6,\ t = 11,\ e = 15$
edge-homogeneous

$(3^6;\ 3^4.6;\ 3^3.4^2;\ 3^2.4.12;\ 3.4^2.6;\ 3.6.3.6)$
p6m $v = 6,\ t = 12,\ e = 15$
edge-homogeneous

$(3^6;\ 3^3.4^2;\ 3^2.4.3.4;\ 3^2.6^2;\ 3.4^2.6;\ 3.4.6.4)$
p6m $v = 6,\ t = 9,\ e = 13$
edge-homogeneous

$(3^6;\ 3^3.4^2;\ 3^2.4.3.4;\ 3.4^2.6;\ 3.4.6.4;\ 4^4)$
cmm $v = 6,\ t = 7,\ e = 10$
edge-homogeneous

Fig. 8—*continued opposite.*

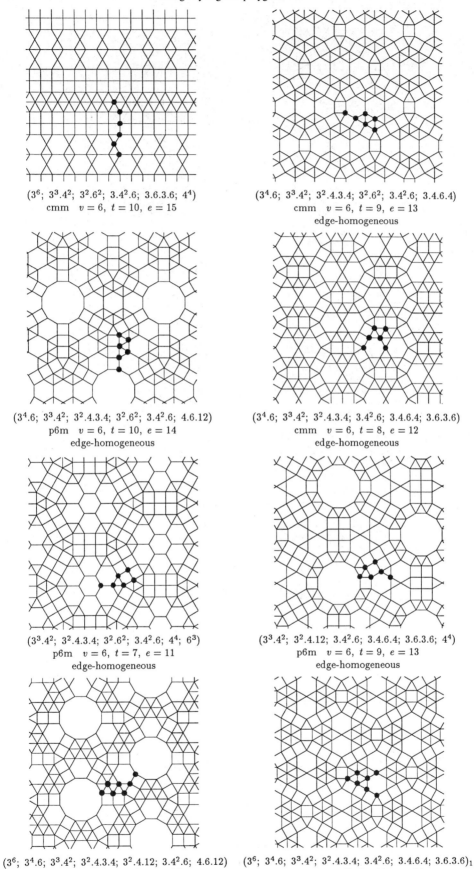

$(3^6; 3^3.4^2; 3^2.6^2; 3.4^2.6; 3.6.3.6; 4^4)$
cmm $v = 6$, $t = 10$, $e = 15$

$(3^4.6; 3^3.4^2; 3^2.4.3.4; 3^2.6^2; 3.4^2.6; 3.4.6.4)$
cmm $v = 6$, $t = 9$, $e = 13$
edge-homogeneous

$(3^4.6; 3^3.4^2; 3^2.4.3.4; 3^2.6^2; 3.4^2.6; 4.6.12)$
p6m $v = 6$, $t = 10$, $e = 14$
edge-homogeneous

$(3^4.6; 3^3.4^2; 3^2.4.3.4; 3.4^2.6; 3.4.6.4; 3.6.3.6)$
cmm $v = 6$, $t = 8$, $e = 12$
edge-homogeneous

$(3^3.4^2; 3^2.4.3.4; 3^2.6^2; 3.4^2.6; 4^4; 6^3)$
p6m $v = 6$, $t = 7$, $e = 11$
edge-homogeneous

$(3^3.4^2; 3^2.4.12; 3.4^2.6; 3.4.6.4; 3.6.3.6; 4^4)$
p6m $v = 6$, $t = 9$, $e = 13$
edge-homogeneous

$(3^6; 3^4.6; 3^3.4^2; 3^2.4.3.4; 3^2.4.12; 3.4^2.6; 4.6.12)$
p31m $v = 7$, $t = 11$, $e = 16$
edge-homogeneous

$(3^6; 3^4.6; 3^3.4^2; 3^2.4.3.4; 3.4^2.6; 3.4.6.4; 3.6.3.6)_1$
cmm $v = 7$, $t = 11$, $e = 14$
edge-homogeneous

Fig. 8—*continued overleaf.*

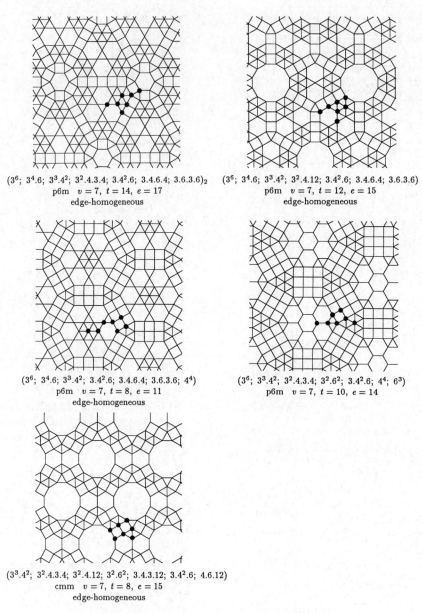

$(3^6; 3^4.6; 3^3.4^2; 3^2.4.3.4; 3.4^2.6; 3.4.6.4; 3.6.3.6)_2$
p6m $v = 7$, $t = 14$, $e = 17$
edge-homogeneous

$(3^6; 3^4.6; 3^3.4^2; 3^2.4.12; 3.4^2.6; 3.4.6.4; 3.6.3.6)$
p6m $v = 7$, $t = 12$, $e = 15$
edge-homogeneous

$(3^6; 3^4.6; 3^3.4^2; 3.4^2.6; 3.4.6.4; 3.6.3.6; 4^4)$
p6m $v = 7$, $t = 8$, $e = 11$
edge-homogeneous

$(3^6; 3^3.4^2; 3^2.4.3.4; 3^2.6^2; 3.4^2.6; 4^4; 6^3)$
p6m $v = 7$, $t = 10$, $e = 14$

$(3^3.4^2; 3^2.4.3.4; 3^2.4.12; 3^2.6^2; 3.4.3.12; 3.4^2.6; 4.6.12)$
cmm $v = 7$, $t = 8$, $e = 15$
edge-homogeneous

Fig. 8. The vertex-homogeneous tilings with $v \geqslant 4$.

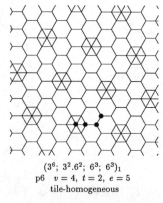

$(3^6; 3^2.6^2; 6^3; 6^3)_1$
p6 $v = 4$, $t = 2$, $e = 5$
tile-homogeneous

Fig. 9. The unique tile-homogeneous tiling which is not vertex-homogeneous. This is also the unique tiling with $t = 2$ which is not included in Figs 3–5.

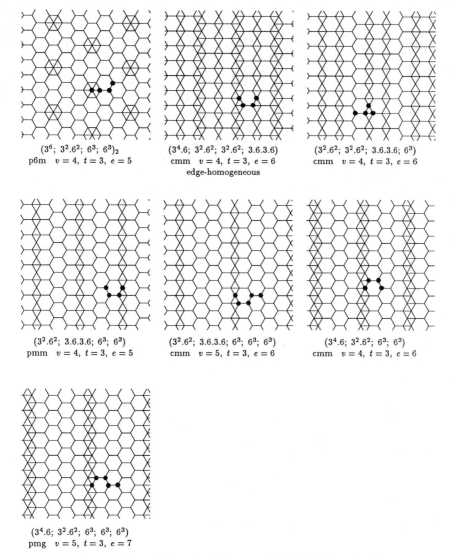

$(3^6; 3^2.6^2; 6^3; 6^3)_2$
p6m $v = 4, t = 3, e = 5$

$(3^4.6; 3^2.6^2; 3^2.6^2; 3.6.3.6)$
cmm $v = 4, t = 3, e = 6$
edge-homogeneous

$(3^2.6^2; 3^2.6^2; 3.6.3.6; 6^3)$
cmm $v = 4, t = 3, e = 6$

$(3^2.6^2; 3.6.3.6; 6^3; 6^3)$
pmm $v = 4, t = 3, e = 5$

$(3^2.6^2; 3.6.3.6; 6^3; 6^3; 6^3)$
cmm $v = 5, t = 3, e = 6$

$(3^4.6; 3^2.6^2; 6^3; 6^3)$
cmm $v = 4, t = 3, e = 6$

$(3^4.6; 3^2.6^2; 6^3; 6^3; 6^3)$
pmg $v = 5, t = 3, e = 7$

Fig. 10. The seven tilings with $t = 3$ which are not included in the earlier figures.

REFERENCES

1. B. Grünbaum and G. C. Shephard, *Tilings and Patterns*. Freeman, San Francisco, Calif. (1987).
2. D. Chavey, Periodic tilings and tilings by regular polygons I: Bounds on the number of orbits of vertices, edges and tiles. *Mitt. math. Semin. Giessen* **164**(2), 37–50 (1984).
3. T. Heath, *Euclid. Elements*, Vol. II (1947).
4. T. Heath, *A History of Greek Mathematics*, Vol. II, commentary on Book IV, Prop. 10. Clarendon Press, Oxford (1921).
5. D. M. Y. Sommerville, Semi-regular networks of the plane in absolute geometry. *Trans. R. Soc. Edinb.* **41**, 725–747 + 12 plates (1905).
6. I. DeBroey and F. Landuyt, Equitransitive edge-to-edge tilings by regular convex polygons. *Geom. Dedicata* pp. 47–60 (1981).
7. D. Chavey, Tilings by regular polygons VII: Tile regularity (in press).
8. J. Kepler, *Harmonice Mundi, Lincii* (1619). German translation: M. Caspar (1939). Also Johannes Kepler Gesammelte Werke. (Ed. M. Caspar), Band VI. Beck, Munich, (1940).
9. O. Krötenheerdt, Die homogenen Mosaike n-ter Ordnung in der euklidischen Ebene. I, Wiss. Z. Martin-Luther-Univ. Halle-Wittenberg. *Math.-natur. Reihe* **18**, 273–290 (1969).
10. D. Chavey, Periodic tilings and tilings by regular polygons. Ph.D. Thesis, Univ. of Wisconsin-Madison (1984).
11. D. Chavey, Tilings by regular polygons V: Vertex regularity (in press).
12. B. Grünbaum and G. C. Shephard, Isotoxal tilings. *Pacif. J. Math.* **76**, 407–430 (1978).
13. D. Chavey, Tilings by regular polygons VI: Edge regularity (in press).
14. O. Krötenheerdt, Die homogenen Mosaike n-ter Ordnung in der euklidischen Ebene. II, Wiss. Z. Martin-Luther-Univ. Halle-Wittenberg. *Math.-natur. Reihe* **19**, 19–38 (1970).
15. O. Krötenheerdt, Die homogenen Mosaike n-ter Ordnung in der euklidischen Ebene. II, Wiss. Z. Martin-Luther-Univ. Halle-Wittenberg. *Math.-natur. Reihe* **19**, 97-122 (1970).

Computers Math. Applic. Vol. 17, No. 1–3, pp. 167–175, 1989
Printed in Great Britain. All rights reserved

SYMMETRY AND POLYHEDRAL STELLATION—Ia†

G. M. Fleurent

Abdij der Norbertijnen, Abdijstraat 1, 3281 Averbode, Belgium

Abstract—In this article polyhedral symmetry provides the basis for an investigation of the stellation pattern of the icositetrahedron, the dual of the snub cube. This is an asymmetrical pattern. Nevertheless, it reveals itself as having a richness of symmetrical properties which at first sight is far from being obvious.

We begin our investigation with a classification of points and then proceed to the enumeration and classification of lines in the stellation pattern. These are then used to show how the points and lines of the stellation pattern are related to the cell structure generated by the intersecting planes in space. Finally we show how stellated forms can thus be derived.

INTRODUCTION

In *The Fifty-nine Icosahedra* Coxeter *et al.* [1] study the structure of the stellation pattern of the icosahedron. The pattern of this platonic solid is quite simple and has a trigonal symmetry. In that work the authors elaborate two methods to analyze the stellation pattern and to derive the stellated forms.

In this article we direct our attention to the stellation pattern of the pentagonal icositetrahedron, the dual of the snub cube. We ask: What is the symmetrical structure of this pattern and how are stellated forms derived from it?

Generally speaking there are symmetrical stellation patterns with axial reflection and there are asymmetrical patterns. That of the dual of the snub cube is a very simple asymmetrical pattern. All the stellation patterns of the uniform polyhedra used by Wenninger in *Dual Models* [2] may be studied by the method elaborated in this article. Such an extensive study, however, must remain as a topic for future research. Thus it is beyond the scope of the present article.

We believe that what we offer here is a new method for the construction of stellated models by means of the stellation pattern. We exhibit only a few stellated forms chosen in such a way that we have sufficient insight into the structure of the whole stellation pattern.

THE STRUCTURE OF THE ASYMMETRICAL STELLATION PATTERN

(a) Classification of points

Some special points are marked in the stellatin pattern of the pentagonal icositetrahedron. The following symbols are used:

⊢—This symbol marks a digonal point. In Fig. 1 P on line 9 is such a point. It is the point of reflection for all the other points on line 9; namely, equal line segments lie on either side of this point. We call P a midpoint, and we call line 9 an isolated line.

●—This symbol marks a trigonal point. In Fig. 1 F on lines 10 and 11 is such a point. It is the point of intersection of lines built up symmetrically of correspondingly equal line segments. It is also an intersecting point of two lines which we call a pair of lines. In Fig. 1 lines 10 and 11 form such a pair. We refer to such a pair as 10–11.

✦—This symbol marks a tetragonal point. This is an intersection point of three lines. Two lines form a pair and are built up symmetrically with respect to the point of intersection. The third line is an isolated line which is built up symmetrically with respect to the intersecting point as a midpoint. In Fig. 1 J on lines 7, 8 and 19 marks such a tetragonal point. We refer to such a triple of lines as 7–8, 19.

---

†Professor M. J. Wenninger (St John's Abbey, Collegeville, MN 56321, U.S.A.) has revised and edited the text of this paper. His kind and valuable contribution is gratefully acknowledged.

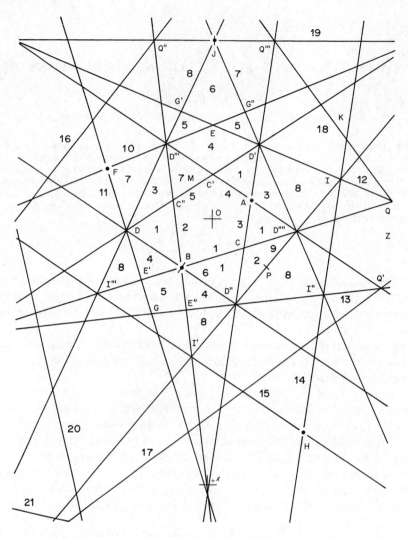

Fig. 1. The central portion of the stellation pattern for the pentagonal icositetrahedron, dual of the snub cube.

At first glance several parallel lines in the stellation pattern attract attention, for example the lines 4, 6 and 15 in Fig. 1. These form open cells, eventually, belonging to stellated forms extending to infinity.

It is more important to observe that the stellation pattern is built up by pairs of lines and/or isolated lines. Because of the symmetry with respect to the point of intersection or the midpoint the lines of a pair or of the isolated line are built up by segments of lines which are equal in pairs. The midpoint of an isolated line may be regarded as the foot of the normal from the origin. In Fig. 1 0 is the origin of a two-dimensional coordinate system.

We place a model of the dual polyhedron on the stellation pattern. Then we observe that the vertices or digonal, trigonal and tetragonal points of the polyhedron can be projected by gnomonic or central projection onto the digonal, trigonal and tetragonal points of the stellation pattern. In Fig. 3 for example Q is the gnomonic or central projection of P with 0_3 the center of the polyhedron as the center of projection.

The number of each of these special points corresponds to the number of axes of rotation of the polyhedron. So for the dual of the snub cube there are six digonal points, four trigonal points and three tetragonal points. When the distance of a tetragonal point from the center of the stellation pattern is greater, this implies that the vertex of a tetragonal pyramid is higher in the stellated form.

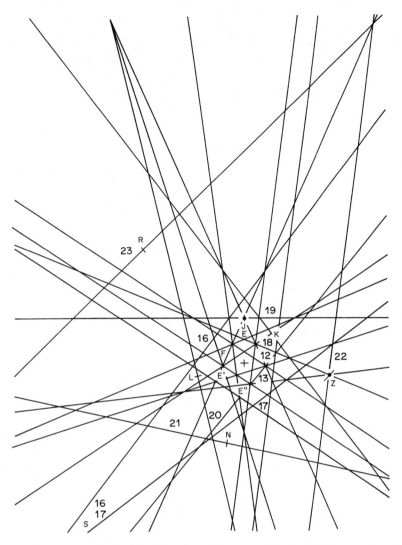

Fig. 2. The same pattern as that in Fig. 1, but shown here on a reduced scale, thus revealing all 23 lines of intersection.

It should be known in general there are three kinds of symmetry axes for symmetrical polyhedra; namely, digonal axes in the tetrahedral, octahedral and icosahedral symmetry group, trigonal axes in each of these groups and finally trigonal, tetragonal and pentagonal axes distributed according to the three symmetry groups. Table 1 gives the number of axes of symmetry in each of the symmetry groups.

(b) Enumeration of lines

The adapted Schläfli symbol used by Coxeter [3] for snub cube is $s(3/4)$. In order to indicate the dual solid we add an asterisk: $s(3/4)^*$. Classifying the lines acording to their distances from the origin of the stellation pattern we obtain an obvious numbering of these lines. See Fig. 1 where the numbering of the lines is in bold type.

The central pentagon, forming one face of the icositetrahedron, is made up of lines 1, 2, 3, 4, 5. Next there are three lines passing through a tetragonal point, namely one pair of lines and one isolated line. There are three such tetragonal points, namely 1–2, 6; 7–8, 19; 12–13, 22. Two lines pass through a trigonal point. These belong to one pair. There are four such trigonal points, namely 3–4; 10–11; 14–15; 16–17. One line passes through a digonal point. There are six such isolated lines which are not connected with tetragonal points, namely 5, 9, 18, 20, 21, 23.

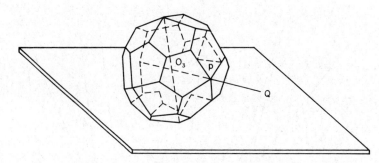

Fig. 3. An example of gnomonic or central projection, showing how a line is drawn from the center of the polyhedron through one of its vertices to intersect the plane of the stellation pattern.

Summary.

7 pairs of lines giving	14 lines
9 isolated lines giving	9 lines
Total:	23 lines

The face of the dual, coinciding with the stellation pattern, does not form a line and therefore it does not enter into the enumeration of lines.

(c) Stellated forms related to the stellation pattern

We will now explain the correspondence between faces of the original polyhedron and lines of the stellation pattern. If we place a model of $s(3/4)^*$ centrally in the stellation pattern so that the face of the base coincides with the central pentagon, each face of the solid when produced intersects the face of the base when produced. The lines of intersection belong to the stellation pattern. Each edge of the polyhedron is the intersection of two adjacent faces of the polyhedron. The edge when produced intersects the face of the base when produced at a point. This is a point of intersection of two lines of the stellation pattern corresponding to the two adjacent faces. We number each face of the polyhedron according to the number of the corresponding lines of the stellation pattern.

First we analyze a tetragonal point, for example the intersection of lines 7, 8 and 19. The face of the base and the faces 7, 8 and 19 constitute a tetragonal pyramid, the axis of which passes through a tetragonal vertex of $s(3/4)^*$. The vertex of the pyramid coincides with a tetragonal point of the stellation pattern. The four faces mentioned above have with respect to the tetragonal vertex of $s(3/4)^*$ a rotational symmetry of 90°. This holds also for the lines 1–2, 6 and 12–13, 22.

Now we analyze a trigonal point, namely the intersection of lines 10–11. The face of the base and the faces 10 and 11 constitute a trigonal pyramid, the axis of which passes through a trigonal vertex of $s(3/4)^*$. The vertex of the pyramid coincides with a trigonal point in the stellation pattern. The three faces mentioned above have with respect to the trigonal vertex of $s(3/4)^*$ a rotational symmetry of 120°. This holds also for the lines 3–4, 14–15 and 16–17.

Finally we analyze a digonal point such as P on line 9. The face of the base and face 9 constitute a dihedral angle, the line of intersection of which coincides with line 9 of the stellation pattern. The normal from the center of $s(3/4)^*$ to this line of intersection passes through a digonal point of the stellation pattern. The two faces mentioned above have with respect to the digonal point of $s(3/4)^*$ a rotational symmetry of 180°. This holds also for the lines 5, 18, 20, 21 and 23.

Table 1

Symmetry group	Axes of symmetry		
Tetrahedral	3 Digonal	2 Trigonal	2 Trigonal
Octahedral	6 Digonal	4 Trigonal	3 Tetragonal
Icosahedral	15 Digonal	10 Trigonal	6 Pentagonal

ANALYSIS OF THE STELLATION PATTERN

(a) Points and lines

One vertex or one point of intersection of a stellated form is the intersection of n planes. The point belongs to n planes and in every one it is situated in a different way. If it is a point of symmetry then it takes the same position in every plane. In the former case there are n points in n different places in the stellation pattern but equally distant from the center. This means the n points are concentric, having 0 as the center. For example in Fig. 1 D, D′, D″, D‴ are concentric. They belong to one point or one vertex of the first stellation whose facial plane is shown in Fig. 5.

Two planes of a stellated form intersect in one straight line. This line of intersection belongs to two planes and this line is situated in two different ways on the stellation pattern equally distant from the center. On this line of intersection there is a line segment which is a part of the perimeter of a facial plane of the stellated form. We call it an edge; for example the line segment AB in Fig. 4. If this line segment is situated on the interior part of a facial plane, then we call it an interior line of intersection; for example AE in Fig. 4. We use the term "facet" for a visible region of a facial plane. For example in Fig. 4 a facet is shown with the numbers 4, 1a and one part marked 1b. A line of intersection of the stellated form contains several points of intersection. The two lines of the stellation pattern are built up of line segments in correspondingly equal pairs and come in

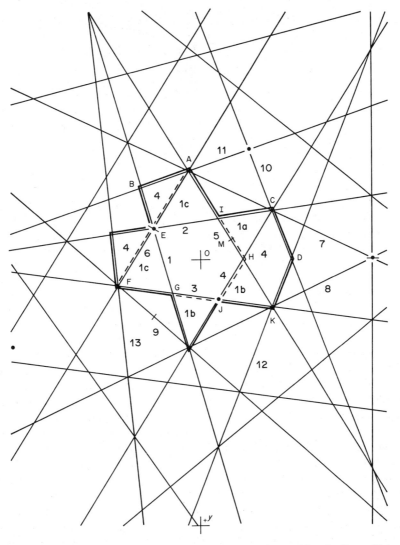

Fig. 4. The facial plane of a stellated form, showing how facets, edges and interior lines of intersection appear in the stellation pattern.

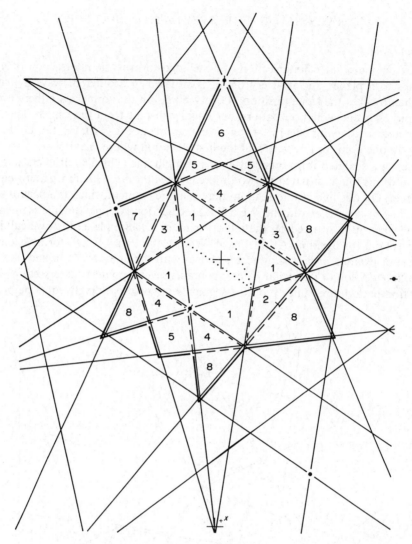

Fig. 5. The stellation pattern, showing the positions of the face of the original polyhedron and its 1st (---),
2nd (——) and 3rd (——) stellations facial planes. Faces of $s\{3/4\}^*$ (···).

the same succession. Also corresponding points will have the same number of intersecting lines with
reference to the number of planes passing through a point of intersection of the stellated form.
If the point is a point of symmetry, then the two lines of the stellation pattern will intersect one
another in a corresponding point of symmetry.

The line of intersection of two planes of the stellated form may be a line having one point
of two-fold rotational symmetry. Corresponding to this line we find in the stellation pattern an
isolated line, the foot of the normal from the origin being a point of symmetry of the line. As
already mentioned we call this a midpoint. With respect to the point of symmetry both the line
of intersection of the stellated form and the isolated line are built up of line segments in equal pairs.
The central line segment appears only once. On corresponding points of the isolated line we find
that the same number of intersecting lines will meet.

One part of the stellation pattern which may be composed of one or more plane cells corresponds
to a visible region of a facial plane. From a given stellated form a new one may be deduced by
adding one or more cells in space. In order to add a three-dimensional cell, we have to know which
are the faces of the cell to be joined and how they are situated as plane cells in the stellation pattern.
Those points of the stellation pattern which constitute a point of intersection of a stellated form
will be called connected points. We need to elaborate a method for finding the connected points
of a stellation pattern. This now follows.

(b) Connected points and cells

First we calculate the equations of the normal to the lines of the stellation pattern and classify those lines according to their increasing distances from the center. If we do not wish to calculate these distances, we may use a compass to determine which are at the same distance from the center. Two lines equally distant from the center form a pair, their intersection being a point of symmetry. If there are three lines equally distant from the center, then one is an isolated line and the other two form a pair; for example in Fig. 1 the lines 7–8, and 9 and also the lines 16–17, 18. If there are four lines equally distant from the center, then they form two pairs; for example in Fig. 1 the lines 1–2, 3–4 and the lines 10–11, 12–13.

It is possible to calculate the polar coordinates of the points of intersection of all the lines of the stellation pattern and to classify those points according to their increasing distances from the center. As a result we get the connected points. These may also be found by analyzing the stellation pattern. The arrangement shown below illustrates how the points E, E′, E″ are connected.

$$E\ 10;\quad E'\ 11\quad 1$$
$$E\ 12;\quad E''\ 13\quad 2.$$

Here E is the starting point. It is the intersection of lines 10 and 12. E′ lies on line 11 just as E lies on line 10, with F the point of symmetry. E″ lies on line 13 just as E lies on line 12, with Z the point of symmetry. (Lines 12 and 13 meet at Z outside Fig. 1. It is shown in Fig. 2.) E′ also lies on line 1. E″ also lies on line 2. The lines 10–11 and 12–13 form pairs. Thus, the three points are connected if the two lines 1 and 2 form a pair. This condition is satisfied. Thus the points E, E′, E″ are also concentric; namely, they lie on a circle whose center is 0, the origin.

Corresponding line segments of an isolated line may also be used; for example, the isolated line 5 for the connected points C, C′ and C″.

In the following example two isolated lines are involved and three lines pass through each point of intersection. The arrangement shown below is a bit more complicated.

$$Q\ 1;\quad Q''\ 2\quad 16\ \text{and}\ 19$$
$$Q\ 12;\quad Q'\ 13\quad 4\ \text{and}\ 17$$
$$Q\ 18;\quad Q'''\ 18\quad 3\ \text{and}\ 19$$

Here Q is the starting point and it is the intersection of lines 1, 12 and 18:

Q″   lies on line 2 just as Q lies on line 1;
Q′   lies on line 13 just as Q lies on line 12;
Q‴   lies on line 18 just as Q lies on line 18;
Q″   also lies on lines 16 and 19;
Q′   also lies on lines 4 and 17;
Q‴   also lies on lines 3 and 19;

The lines 3–4, also 16–17 form pairs. Line 19 is an isolated line. Thus, the points Q, Q′, Q″ and Q‴ are connected and concentric. A table containing the three lines of each point of intersection will demonstrate sufficiently that connected points are involved.

The following lines occur once in Table 2: 1–2, 3–4, 12–13, 16–17.

The isolated lines 18 and 19 each occur twice.

(c) Lines and cells

By means of the connected points a new three-dimensional cell can be added to a given stellated form. All the plane cells with the same number in Fig. 1 constitute the faces of one new cell in space. The base of that cell in space belongs to a previous stellated form.

Table 2

Q	1	12	18
Q′	4	13	17
Q″	2	16	19
Q‴	3	18	19

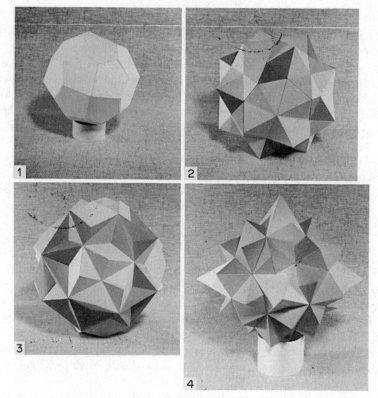

Plates 1–4.

For example in order to join to a stellated form the three-dimensional cell whose number is 8 we start with the point I and then deduce the connected points I′, I″ and I‴. By this the necessary faces of the cell are found in the stellation pattern.

Two line segments correspond to every new edge. These are found in the set of the four plane cells. For example in Fig. 1 the segment ID′ on line 12 and I″D″ on line 13 are two line segments in two of the four cells with the number 8. The central part of an isolated line will appear once. For example D″D‴ on line 9 is the base of one of the four cells with the number 8. Every interior line of intersection by which a new three-dimensional cell is joined to a previous cell is found as a line segment that will appear once in one of the four plane cells. For example the interior line of intersection D′D‴ on line 7 separates two cells 3 and 8. The corresponding line segment DD‴ separates two cells 3 and 7. Thus, the interior line of intersection D′D‴ is the base of one of the four cells 8. The interior lines of intersection are closer to the center than the edges.

Two corresponding line segments of the stellation pattern belonging to a pair of lines or an isolated line separate four cells, two of them having the same number and two of them a different number. With reference to the center they are situated on the same side of the lines. In the same example there are two cells 3 and one cell 7 and one cell 8. The two cells 3 are both nearer to the center.

The central segment of an isolated line separates two cells of different number. For example the line segment D″D‴ on line 9 separates the cells 2 and 8. The cells situated around the vertices of the pyramids which coincide with the points of symmetry of the stellation pattern are arranged differently, for example at the point J in Fig. 1.

(d) Construction of cells

We are now ready to construct the first eight three-dimensional cells.

The central polyhedron. There are in the central pentagon of the stellation pattern no points connected with A which is a trigonal point nor with B which is a tetragonal point. C, C′ and C″ are connected points.

Cell 1. Starting with D we find D′, D″, D‴ and D⁗ to be connected. The three-dimensional cell has the central pentagon for its base and five cells for its faces. Two of the five cells are situated at a tetragonal point, two cells are at a trigonal point and one cell at a digonal point. See Fig. 4 where the cell 1a lies at a digonal point, the two cells 1b lie at a trigonal point, and the two cells 1c lie at a tetragonal point.

Cell 2. The line segment D″D⁗ is the central part of the isolated line 9. No other line segment of the stellation pattern corresponds to it. Because this unique line segment contains a digonal point it must be the boundary of a cell in space having cell 2 twice for new faces. In the stellation pattern the edge D″D⁗ of two-fold symmetry is one level higher than C′C″ of the central polyhedron.

Cell 3. The line segments D′D⁗ and DD‴ are equal, belonging to the lines 7–8. Those line segments are the edges of two plane cells with the number 3 needed to construct the three-dimensional cell 3.

Cell 4. The arrangement E, E′ and E″ was treated above. These three connected points indicate three plane cells in the stellation pattern. Equal line segments appear on the lines 10–11 and 12–13. The two remaining equal segments are on the lines 1–2.

Cell 5. G is the point of intersection of lines 11 and 13. We locate the connected points G′ and G″ and in this way we find the necessary line segments on lines 7–8 and on lines 10–11 and on lines 12–13. Thus we find three plane cells having the number 5.

Cell 6. Two equal line segments on the lines 7–8 form one plane cell at the point J which is one face of a tetragonal pyramid one level higher than the one at point B.

Cell 7. The same is repeated at the trigonal point F by two equal line segments on the lines 10–11.

Cell 8. The four connected points I, I′, I″ and I‴ indicate four cells in the stellation pattern which form the faces of one new cell in space. The equal line segments are on the lines 1–2, 7–8, 12–13 and the isolated line 9.

By combination of these eight cells many stellated forms can be built. The face of the icositetrahedron along with the facial planes of the first three stellated forms is shown in Fig. 5. Models are shown Plates 1–4.

REFERENCES

1. H. S. M. Coxeter, P. Du Val, H. T. Flather and J. F. Petrie. *The Fifty-nine Icosahedra*. Springer, New York (1982).
2. M. J. Wenninger, *Dual Models*. Cambridge University Press, New York (1983).
3. H. S. M. Coxeter, Regular and semi-regular polytopes. *Math. Z.* **46,** 380–407 (1940).

Computers Math. Applic. Vol. 17, No. 1–3, pp. 177–193, 1989
Printed in Great Britain. All rights reserved

SYMMETRY AND POLYHEDRAL STELLATION—Ib†

G. M. Fleurent

Abdij der Norbertijnen, Abdijstraat 1, 3281 Averbode, Belgium

Abstract—Dual stellated forms and, among its applications, asymmetrical stellation patterns are discussed in detail in this contribution which is a sequel to the preceding paper.

MODEL MAKING BY MEANS OF THE STELLATION PATTERN

(a) Classification of stellated forms

The levels of the tetragonal and trigonal pyramids and digonal points may, in the case of $s(3/4)*$, be combined in $3*4*6 = 72$ possibilities. If our exploration is exhaustive then only 49 stellated forms may be built. Corresponding to each combination there is a set of stellated forms. Each set can be denoted by three capital letters. On the axis of four-fold rotation, there are three levels, to which correspond the capital letters A, B and C, A being the lowest level. On the axis of three-fold rotation, there are four levels with the letters A–D and on the axis of two-fold rotation, there are six levels with the letters A–F.

In the symbol, belonging to a set of stellated forms, the first letter refers to the level on the axis of four-fold rotation, the second letter refers to the level on the axis of three-fold rotation and the third letter refers to the level on the axis of two-fold rotation. If different stellated forms of the same set, denoted therefore by the same symbol, have to be distinguished, a number may be added to the reference symbol.

At the top of Table 1 the three levels of the axis of four-fold rotation are indicated. On the left-hand side there are the four levels of the axis of three-fold rotation. Within the table we obtain for each combination of the axes of four- and three-fold rotation, the kind of levels, also their number, of the axis of two-fold rotation. In the following models one of each set is presented in such a way that the lowest level on the axis of two-fold rotation, is chosen. By this criterion we get 12 models [see Table 2, Plates 6–17 and Figs 6–17‡ (stellation patterns)]. After construction, the 12 models may be displayed according to their place in the matrix, illustrating in this way the evolution of the levels of the pyramids and the digonal dihedral angles.

Besides these 12 models there are others of interest: the first stellation, the last stellation and a series of three stellated forms, selected in such a way that they illustrate the notion of stellation. Plates 1–4, also the stellation pattern of Fig. 5, represent: $s(3/4)*$, the first stellation, a second stellation and a third stellation. Each of the stellated forms adds a layer of three-dimensional cells to the previous stellated form.

(b) Line segments and stellated forms

By means of the model of Fig. 4 (see Plate 5), on the surface of which lines of the stellation pattern are drawn, we will illustrate how lines of the stellation pattern function in a model in space. Compare the plate and the stellation pattern. The stellated form is built up of the three-dimensional cells 1 and 4. The surface is in the stellation pattern built up by three facets, of which one contains three plane cells, namely 1a, 1b and 4.

A facet is bounded by the edges and the interior lines of intersection. In the stellation pattern edges are drawn by using full lines, for example AB in Fig. 4. Interior lines of intersection are drawn by using dotted lines, for example AE in Fig. 4. Even the perimeter of a face, not belonging to a facet, is drawn using heavy lines, for example AI in Fig. 4.

Line segments have a four-fold function.

1. Two edges in the stellation pattern correspond to one edge of the stellated form. The edges

---

†Professor M. J. Wenninger (St John's Abbey, Collegeville, MN 56321, U.S.A.) has revised and edited the text of this paper. His kind and valuable contribution is gratefully acknowledged.

‡The illustrations of this paper are numbered consecutively following those of the preceding paper [1].

Table 1. Constructible stellated forms of
$s(3/4)^*$ according to the matrix $(3*4*6)$

	A	B	C
A	A to C 3	A to C 3	B to C 2
B	A to F 6	A to F 6	B to F 5
C	B to F 5	B to F 5	B to F 5
D	D to F 3	D to F 3	D to F 3

Horizontally—axis of four-fold rotation with 3 levels: A, B, C; vertically—axis of three-fold rotation with 4 levels: A, B, C, D; tabulated—axis of two-fold rotation with 6 levels: A, B, C, D, E, F.

Table 2. Constructed stellated forms of the matrix $(3*4*6)$

	A	B	C
A	AAA	BAA	CAB
B	ABA	BBA	CBB
C	ACB	BCB	CCB
D	ADD	BDD	CDD

of the pattern lie in a corresponding way on a pair of lines or symmetrically on an isolated line. In the example: the heavy line segments CD and AB on lines 10–11.

2. Two interior lines of intersection in the stellation pattern, which in a corresponding way belong to a pair of lines or lie symmetrically on an isolated line, correspond to one line of intersection of the stellated form, lying on the innerside of a visible facet of the facial plane. In the example: the dotted lines JG and JH on lines 3–4. The central part of an isolated line appears only once but may be produced symmetrically by different line segments. In the example: the dotted line HI on line 5.

3. On the surface of the model of Plate 5 lines are drawn. Such a line belongs in a different way to two planes, because it is a line of intersection of a visible facet of the surface with an invisible, interior portion of a plane. Two equal line segments in the stellation pattern, lying in an corresponding way on a pair of lines or symmetrically on an isolated line, correspond to that line of the stellated form. One line segment lies within a facet. For example, in Fig. 4 CH. The other one is an invisible portion of the perimeter. For example, in Fig. 4 FG. The line segments FG and CH lie on lines 3–4. An other example, in Fig. 4 the non-dotted line segments AI and HK on line 5.

4. Interior to the stellated form there are lines of intersection of two planes. These lines are cut by the surface. Two equal line segments in the central part of the stellation pattern correspond to such a line of intersection in the stellated form. These belong in a corresponding way to a pair of lines or they are symmetrically on an isolated line. For example, in Fig. 4 the fine lines EG and EI on lines 1–2 within the face.

By looking for the facets of a stellated form these four rules are a good tool and each line of the face has to be tested with them.

(c) Model making

When we want to construct a stellated form of which a certain number of three-dimensional cells are given, so that the plane cells are situated in the stellation pattern, we have to look for the interior

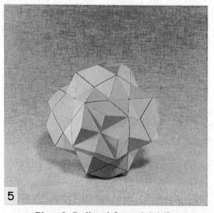

Plate 5. Stellated form AAA/2.

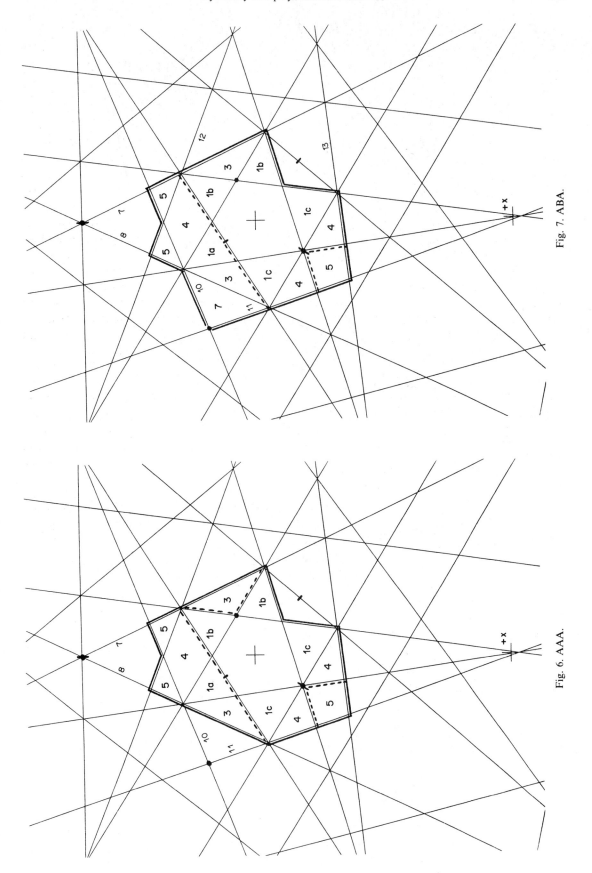

Fig. 7. ABA.

Fig. 6. AAA.

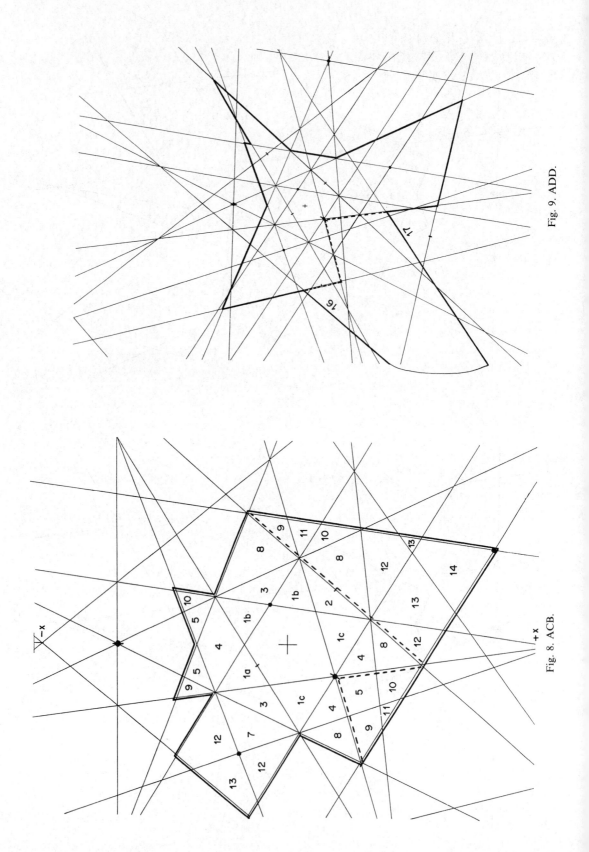

Fig. 9. ADD.

Fig. 8. ACB.

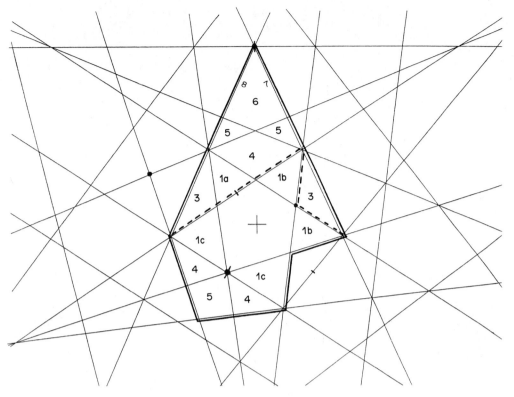

Fig. 10. BAA.

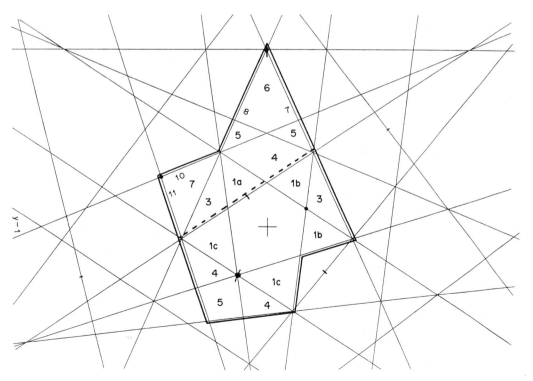

Fig. 11. BBA.

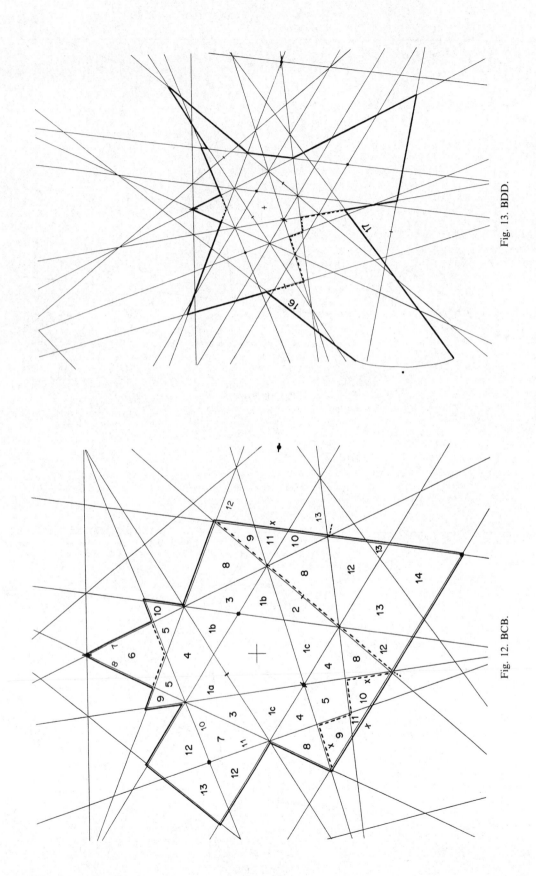

Fig. 13. BDD.

Fig. 12. BCB.

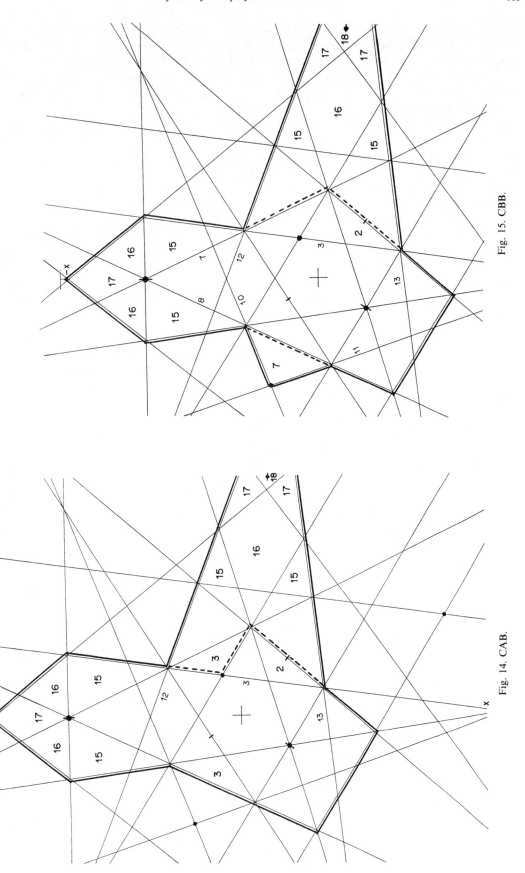

Fig. 15. CBB.

Fig. 14. CAB.

lines of intersection, drawn with dotted lines, within the perimeter of the face. Some hints are given here. Experience will supply the rest. We refer to the stellation patterns and the plates of the 12 models which have been selected. The perimeter of the desired stellated form is given with the plane cells, put together by means of the connected points. Now we point out for every axis of symmetry the highest level, lying within or on the perimeter. When the pyramids are obvious and uninterrupted the solution is easy to find.

In the case of CAB and CBB the lines 12–13 form on the axis of four-fold rotation a pyramid of level C. These two lines have at the foot of the pyramid a breakpoint, because of the intersection with line 3. These points are to be the starting points of interior lines of intersection (see dotted lines). Corresponding to such a breakpoint in the stellation pattern we get in the stellated form a point of intersection of at least three planes, which intersect one another by interior lines of intersection. The solution is obvious in the case of CAB, because of the place of the trigonal and digonal points. Regarding CBB line 3 is excluded, because it contains a trigonal point of lower level. Therefore we look at the feet of the trigonal pyramid on the lines 10–11. These indicate the only possible solution for the interior lines of intersection, which have to be corresponding, equal line segments on the lines 7–8.

The analysis is as easy for ADD, BDD and CDD. The trigonal pyramid of level D is obvious and formed of the lines 16–17. It is amazing to verify how the tetragonal pyramids B and C are structured.

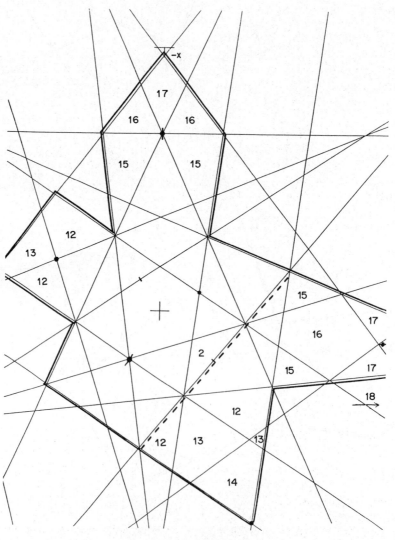

Fig. 16. CCB.

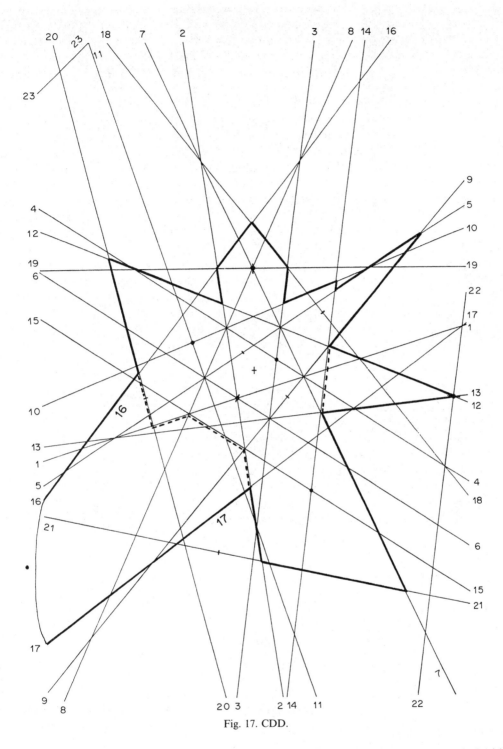

Fig. 17. CDD.

At **BAA** we consider the corresponding breakpoints of lines 7–8. At **AAA** there is besides an unfinished tetragonal pyramid. **ACB** resembles **BAA**. At **BCB** the structure of the tetragonal pyramid, see lines 7–8, is interesting. The feet of this pyramid indicate how by corresponding, equal segments on the lines 10–11 and 12–13 the structure of the tetragonal pyramid becomes possible. Also the three cells 5 prove that.

BBA and **CCB** illustrate another case. There is at **BBA** on the lines 8 and 10 a common breakpoint; in fact this is a common foot of the trigonal and tetragonal pyramids. In the stellated form such a breakpoint is the intersection of three planes, but the line of intersection is common to two

Plate 6. AAA.

Plate 10. BAA.

Plate 14. CAB.

Plate 7. ABA.

Plate 11. BBA.

Plate 15. CBB.

Plate 8. ACB.

Plate 12. BCB.

Plate 16. CCB.

Plate 9. ADD (left and right stellated form).

Plate 13. BDD (with truncated trigonal pyramids).

Plate 17. CDD (with truncated trigonal pyramids).

planes which do not coincide with the plane of the stellation pattern. At this breakpoint we will not find an interior line of intersection (a dotted line). We have to look at the corresponding points on the lines 7 and 11, which form pairs with 8 and 10. Now the solution is obvious. We may analyse CCB analogously. AT ABA the basis of the trigonal pyramid indicates the solution. The point of intersection of the lines 8 and 10 refer to the corresponding points on the lines 7 and 11, by which an interior line of intersection has to be constructed. The structure of the tetragonal pyramid is possible only in one way, if we test the pairs 10–11 and 12–13 with rule 3, already explained.

So far we have the explanation of the 12 models of the matrix, having obvious pyramids, without superfluous, extra cells in space, as on Plate 4.

Using these principles we can build up the duals of the uniform polyhedra, if an asymmetrical stellation pattern is required [2].

By applying the Dorman Luke construction on the uniform polyhedron we get the theoretical shape of a face of the dual. In fact this shape does not delineate the complete intersection of a plane with all the other planes of the model.

We consider the points of intersection of all the lines with the perimeter, as given by the Dorman Luke construction. We complete all pairs of lines and isolated lines by looking for the points connected with these points of intersection. We obtain the same by searching for all corresponding and equal line segments. As a result we get the complete, real intersection of a plane. We call it the extended face. Now it is possible to apply the four rules and the analyses of the pyramids, given above, in order to deduce the interior lines of intersection of the facets. Figure 18 represents the adapted and correct stellation pattern for W.116-dual. Compare with Wenninger [2, p. 126]. By these duals there is on every axis of symmetry one visible pyramid, but there are also visible pyramids not lying on axes of symmetry. See W.116-dual in Ref. [2].

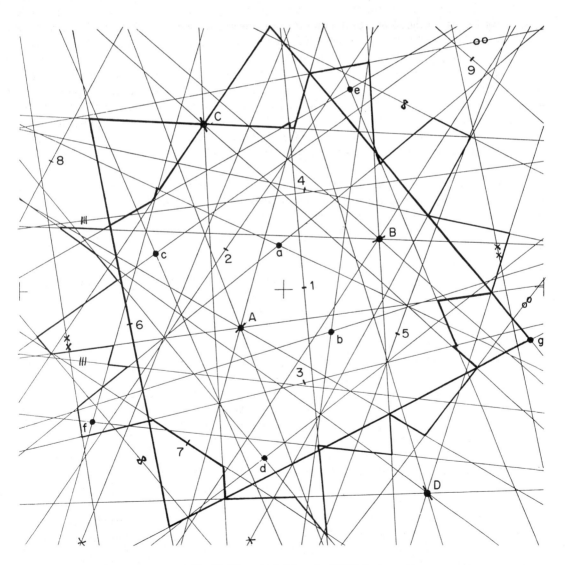

Fig. 18. The extended face for W.116-dual.

SYMMETRICAL COMPOUNDS

(a) The extended stellation pattern

With each asymmetrical polyhedron it is always possible to construct a left and right model. Such a pair of models are called enantiomorphous. The right model can be deduced from the left one by reflexion in a single plane of symmetry or in all planes of symmetry [3, pp. 24–25]. We can easily think of a model of both solids, the left one and the right one, constructed around a common innersphere. The result is called a compound. This term refers to all polyhedra, constructed one intersecting another.

In order to construct the left or right model, we need for the octahedral group the stellation pattern with 23 lines. The compound is derived from the stellation pattern with 46 lines. Analogous to group theory [3, p. 25] we propose to call the stellation pattern with 23 lines the original stellation pattern and the one with 46 lines the extended stellation pattern. The extended stellation pattern contains the 23 lines of the original stellation pattern, completed with 23 new lines, which we call related lines. In the same way we can speak of a related stellation pattern and a related polyhedron. See Fig. 19, the extended stellation pattern of $s(3/4)^*$; the lines with number are the related lines.

The extended stellation pattern may be considered as being deduced from the central polyhedron, which is the intersection of the left and right one. In that solid of intersection planes are parallel. Two parallel lines in the stellation pattern correspond to a pair of parallel planes that belongs to the solid of intersection. The product of the distances to the centre of a pair of parallel lines is constant and equal to the square of the radius of the inscribed sphere of the dual polyhedron, from which the stellation pattern is deduced. Similarly, a related line far from the centre corresponds to a line of the original stellation pattern, lying near to the centre. Inverse, if a line of the original stellation pattern is far from the centre, then the related line is near to the centre. By this property, the product of the distances is equal to the square of the radius of the inscribed sphere, we can deduce the related lines from the lines of the original pattern, when we are calculating the extended stellation pattern. There are two kinds of related lines: 9 isolated lines, corresponding to the 9 isolated lines of the original stellation pattern and related to the 9 planes of symmetry of the dual polyhedron, which we call lines of symmetry; 14 lines, forming 7 pairs, which correspond to the 7 pairs of the original stellation pattern.

Previously we described how the symmetric points of a polyhedron are situated in the original stellation pattern by gnomical projection. Points of symmetry of a dual polyhedron lie in planes of symmetry and these planes cut an inscribed sphere in great circles. Great circles are transformed into lines by gnomical projection. We ask: why do those lines belong to the related stellation pattern?

With respect to a plane of symmetry a new plane is taken, symmetrical to the face of the base, which is the central polygon of the original stellation pattern. The new plane contains by definition a face of the related polyhedron, and when this plane cuts the plane of the stellation pattern we get a line of the related stellation pattern. Now the three planes (the plane of the stellation pattern, the plane of symmetry, the new related plane) cut each other in one line of intersection, and because of the gnomical projection applied to the plane of symmetry, that line of intersection contains several points of symmetry. We can repeat this construction with respect to each plane of symmetry. So we get in the related stellation pattern 9 lines of symmetry related to the 9 planes of symmetry. On each line of symmetry are four points of symmetry. Superimposed onto the original stellation pattern, those 9 related lines connect the digonal, trigonal and tetragonal points. This yields a plane tesselation, being a distorted, gnomical projection of half of the spherical Möbius-tesselation. For the periodic sequence of points of symmetry, see Ref. [4, p. 65].

Till now we discussed why nine lines, which are related to the 9 planes of symmetry, belong to the extended stellation pattern. This is the main point. We omit the discussion why they are related to the 9 isolated lines of the original pattern, because it is accidental and there are two kinds of isolated lines.

The lines of the original stellation pattern, also the 7 pairs of related lines, are divided according to corresponding and equal segments. None of the 9 symmetrical lines has in the extended stellation pattern a symmetrical partition.

The 14 related lines form 7 pairs, which are not divided in equal segments in the usual way, that is according to symmetrical points, also not according to their point of intersection. These 14 lines do not pass through points of symmetry. We must construct the two normals of a pair of related lines. The segments on the first line, taken from the foot of the normal in the direction of their point of intersection, correspond and are equal to the segments on the other line, taken from the foot of the normal in the other direction.

(b) The derivation of compounds

For a given stellated form we assume that it is known how the face is situated in the original stellation pattern and we begin with drawing the facets. Now we keep in mind that in every vertex of the compound the edges are doubled.

At the beginning of this section it was clearly stated that the compound is generated by reflexion of a left or right stellated form with one or all the planes of symmetry. Now it is obvious that all the lines of symmetry will split the facets in order to build up the related polyhedron by reflexion. On those lines of symmetry line segments are cut by the perimeter of the facets. These segments, lying within the facets, appear only once in the stellation pattern. By building the model of the compound the remaining part of the facet is constructed as a left and right element one against the other, separated by a segment of a line of symmetry.

We can look at three compounds. See Plates 18–20 and stellation patterns 20–21. The compounds ADD and CCB have redoubling of respectively one or two pyramids. The edges of the pyramids of both compounds are lines of the original stellation pattern. The lines of intersection of the faces of the pyramids are lines of symmetry. In the case of ADD the base of the pyramid too is cut by lines of symmetry. In the case of CCB the bases of the pyramids are cut by shortened lines segments of the original stellation pattern.

AAA/2 is the compound of the stellated form of Fig. 4 or Plate 5. It is amazing to find where equal segments of interior lines of intersection, taken on a pair of related lines, are situated in the extended stellation pattern. See Fig. 21. The line segments a are on the related pair 16–17, the segments b on the related pair 14–15. We can verify how those segments lie according to the point

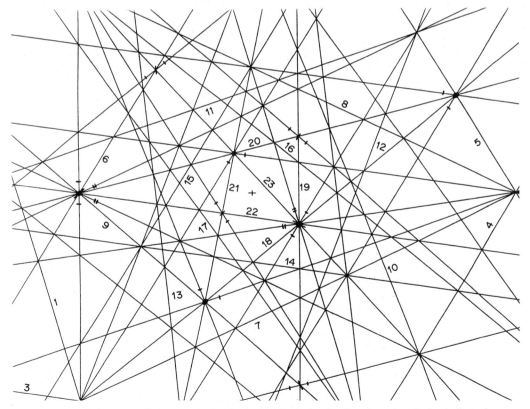

Fig. 19. The extended stellation pattern for the dual of the snub cube. Supplementary dashes indicate the pairs of lines or the isolated lines of the original stellation pattern.

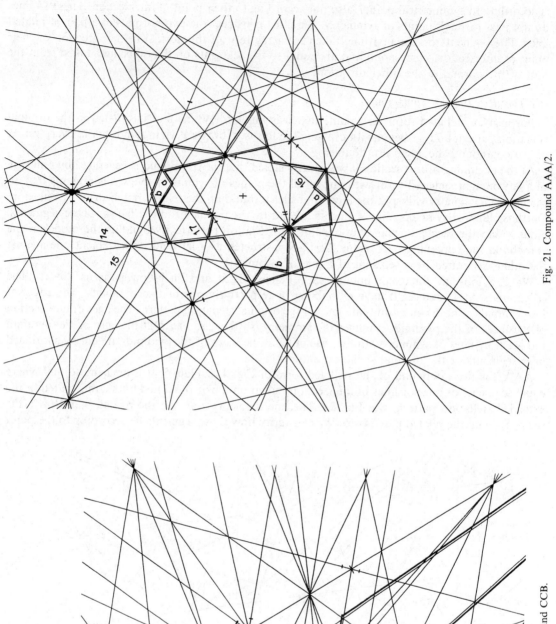

Fig. 21. Compound AAA/2.

Fig. 20. Compounds ADD and CCB.

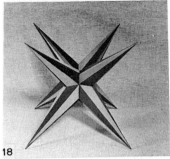
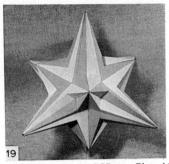
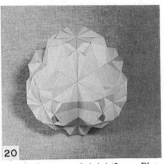

Plate 18. Compound ADD; see Plate 9.   Plate 19. Compound CCB; see Plate 16.   Plate 20. Compound AAA/2; see Plate 5.

of intersection of the pairs and according to the feet of the normals. By building up the model of the compound two different facets are joined by the segment a. One facet lies at the segment a on the line 16, an other, but a different one lies at the segment a on the line 17. One facet belongs to the left polyhedron, the different one belongs to the right polyhedron. However, this alternates. The same holds for the two different facets, lying at the segment b of the lines 14–15.

APPLICATIONS

(a) The asymmetrical stellation patterns

The asymmetrical stellation patterns, to which this study is relevant, are of two categories:

—duals of snubs and variants;
—duals of truncated quasi-regular polyhedra and variants.

The dual snubs $s(3/4)^*$ and $s(3/5)^*$ have, respectively, 23 and 59 lines in the stellation pattern. They can be understood through this study. This holds also for the following variants of $s(3/5)^*$, having each 59 lines: W112; W111 and W114; W113, 116, 117, see Ref. [2]. Of all these polyhedra compounds of enantiomorphous stellated forms may be built with stellation patterns of 46 lines (octahedral group) or 118 lines (icosahedral group), which are analysed by following the explanation of the previous section.

The stellation pattern for the hexakis octahedron or $t(3/4)^*$ has 48 lines. Variants: W79 and W93. The stellation pattern for the hexakis icosahedron or $t(3/5)^*$ has 118 lines. Variants: W84, W98 and W108. These stellation patterns are analysed in the same way as the extended stellation patterns of a snub. In the centre of the pattern there is a irregular triangle, surrounded by an irregualr pentagon. Indeed, the pattern contains the stellation pattern for a variant of a dual snub. Referring to a variant of $s(m/n)^*$ we may speak of the original stellation pattern. Referring to a variant of $t(m/n)^*$ we may speak of the extended stellation pattern.

(b) The symmetrical stellation patterns

The symmetrical stellation patterns have one or more axial reflexions. For instance: the pattern for the rhombic dodecahedron or $(3/4)^*$ has two axes of reflexion. Mostly there is only one axis of reflexion. For a complete enumeration of symmetrical patterns for duals, the reader is referred to Ref. [2].

We choose the dual snub W115 as an instance of a symmetrical stellation pattern with one axis of reflexion. We call this the central axis. In Fig. 22 we see the central part of the stellation pattern for W115-dual. This polyhedron has 60 faces in parallel pairs. Neither the face, parallel with the base, nor the base itself have any line in the pattern. Hence, there are 58 lines.

One axis of two-fold rotation is parallel to the base and does not cut the stellation pattern in a digonal point. Therefore, there are only 14 digonal points and 14 isolated lines.

Now we count the pairs of lines and the isolated lines:

Through 6 pentagonal points pass $6*4 = 24$ lines.
Through 10 trigonal points pass $10*2 = 20$ lines.
Through 14 digonal points pass $14*1 = 14$ lines.

So we get a total of 58 lines.

Fig. 22. The central portion of the symmetrical stellation pattern for W.115-dual.

A polyhedron of the icosahedral group has 15 planes of symmetry. One plane of symmetry of W.115-dual is perpendicular to the stellation pattern. The line of intersection coincides with the central axis, which is not a line of the stellation pattern. Therefore, there are only 14 lines of symmetry between the 58 lines. Each of these 14 lines passes through several points of symmetry. However, these lines form pairs of lines or isolated lines in those points of symmetry which lie on the central axis.

The 14 lines of symmetry are distributed on the central axis as follows:

  Through 2 pentagonal points pass 4 pairs of lines. So we get 8 lines.
  Through 2 trigonal points pass 2 pairs of lines. So we get 4 lines.
  Through 2 digonal points pass 2 isolated lines. So we get 2 lines.

Each of the remaining 44 lines of symmetry passes through one point of symmetry. This is a point of symmetry for a pair of lines or an isolated line.

An edge or an interior line of intersection in a facet may be a line segment on a line of symmetry. That segment lies in a plane of symmetry of the stellated form. This form is symmetrical with respect to that segment.

We have to adapt the matrix related to the levels of the vertices of the pyramids. Indeed: several points of symmetry are symmetrical with the central axis. Such a pair of points corresponds to one level. A point of symmetry on the central axis corresponds also to one level. Therefore the total number of combinations of levels will be less than that of a asymmetrical pattern. So we get for the 6 pentagonal points $2 + 2 = 4$ levels, for the 10 trigonal points $4 + 2 = 6$ levels, for the 14 digonal points $6 + 2 = 8$ levels. In order to know all the combinations of levels of vertices we construct a matrix of order ($4*6*8$), which contains 192 elements. Not all of these elements correspond to a set of stellated forms.

It is not because the stellation pattern has a symmetrical shape, that the faces of polyhedra or stellated forms lie symmetrically in the stellation pattern. The face of W115-dual has not a symmetrical shape and is not symmetrically situated according to the axis of symmetry. W119-dual is deduced from the same stellation pattern as W115-dual. The face however has a symmetrical shape and is symmetrically situated in the stellation pattern.

This kind of duals has no extended stellation patterns. Enantiomorphous stellated forms and their compounds however may be derived from a symmetrical stellation pattern. But these are not real compounds, because the faces of two enantiomorphous stellated forms coincide and they form the face of one new solid. See Ref. [5].

Acknowledgements—The author is indebted to Jan Cuypers for computer drawings of the stellation patterns used in Figs 9, 13, 17, 19–21.

REFERENCES

1. G. M. Fleurent, Symmetry and polyhedral stellation—Ia. *Comput. Math. Applic.* **17**(1–3), 167–175 (1989).
2. M. Wenninger, *Dual Models.* Cambridge Univ. Press, New York (1983).
3. F. Klein, Lectures on the icosahedron.
4. H. S. M. Coxeter, Regular and semi-regular polytopes. *Math. Z.* **46**, 380–407 (1940).
5. H. S. M. Coxeter, The fifty-nine Icosahedron. Springer, New York (1982).

Computers Math. Applic. Vol. 17, No. 1–3, pp. 195–201, 1989
Printed in Great Britain. All rights reserved

0097-4943/89 $3.00 + 0.00

SYMMETRY AND POLYHEDRAL STELLATION—II

P. W. Messer

3344 West Grace Avenue, Mequon, WI 53092, U.S.A.

M. J. Wenninger

St John's Abbey, Collegeville, MN 56321, U.S.A.

Abstract—In this article polyhedral symmetry provides the basis for an investigation of the stellation pattern of the deltoidal hexecontahedron, also known as the trapezoidal hexecontahedron, the dual of the rhombicosidodecahedron. This pattern reveals a bilateral symmetry which is indeed rich in the variety of its complexity. It belongs to the icosahedral symmetry group, and as such we expect to find in it a wealth of beautiful shapes.

We begin our investigation with a notation identifying the lines of the pattern and the related line segments. The plane cells, also called elementary regions, are then identified. With this information we turn to an analysis of the connectivity of both plane cells and cells in three dimensions. Finally we show how stellated forms can thus be derived.

INTRODUCTION

In "Symmetry and polyhedral stellation—Ia" [1] an investigation was presented for a very simple asymmetrical stellation pattern, that of the dual of the snub cube, belonging to the octahedral symmetry group. In this article we turn our attention to a rather more complex pattern, one having bilateral symmetry and belonging to the icosahedral symmetry group. Since Fleurent in the first article and Messer in this article carried out their investigations independently, we find that they each adopt a unique notation for identifying lines and related parts. Yet there is an interesting similarity in the final results.

The purpose of this article is to offer a method of approach for investigating the stellation pattern of the deltoidal hexecontahedron, the dual of the rhombicosidodecahedron. In *Dual Models* [2, pp. 70–81] Wenninger showed how rich this pattern is for deriving duals of many non-convex uniform polyhedra. It is this richness that prompted him to suggest a deeper investigation into this pattern. The non-convex duals are not necessarily the most beautiful or aesthetically pleasing shapes. This pattern undoubtedly has hidden within it an abundance of stellated forms waiting to be discovered.

We believe the method used in this article can be generally extended to an investigation of other convex duals. Such an extensive study, however, must remain as a topic for future research, both for this example as well as for the one presented in the previous article.

THE STRUCTURE OF A SYMMETRICAL STELLATION PATTERN

(a) Notations and methods

This work on the stellation pattern of the deltoidal hexecontahedron introduces various notations and methods useful for the model builder. For a stellation pattern not having bilateral symmetry some further modifications will be necessary.

First we label the lines of the stellation pattern while observing symmetry about the y-axis. See Fig. 1. Upper case letters (A, B, C, ...) are used for lines whose y-intercept is above the x-axis. Lower case letters (a, b, c, ...) are used for lines whose y-intercept lies below the x-axis. The alphabetical order of labeling is shown in Fig. 1 for a part of the stellation pattern.

If more than one letter exists at a y-intercept the labeling begins with the greatest magnitude of slope. The y-intercept separates the positive part from the negative part of any labeled line. The $(+)$ part is directed toward the positive y-axis and the $(-)$ part is directed toward the negative y-axis. Lines perpendicular to the y-axis have both parts positive. For example see how lines B, b, c are signed. When considering enantiomorphs it may be advantageous to distinguish left (L) from right (R). For example: $+B^R$, $+B^L$, $-B^R$, $-B^L$, $+C^R$, $-C^L$.

195

Fig. 1

It is often very useful to know the "interior" dihedral angle associated with each labeled line. A necessary but not a sufficient condition for the connection of two lines in three dimensions is that they both have the same interior dihedral angle. Clearly, a line having a unique dihedral angle can only connect with itself. Lines are associated with the same dihedral angle if and only if they are equidistant from the center of the stellation pattern.

Elementary line segments are consecutively numbered along labeled lines in a (+) and (−) direction, each beginning at the y-intercept. For example ... − A3, − A2, − A1, + A1, + A2, + A3, ... are labels for line segments along the line A. A stellation pattern may show lines that are parallel to the y-axis. Such lines may be labeled using Greek letters. Analogously, the (+) part is directed toward the positive y-axis and the (−) part is directed toward the negative y-axis. A segment may be numbered ±1 if it is perpendicularly bisected by any coordinate axis.

Having information on dihedral angles and exact lengths of the segments in correct sequence along a line allows us to decide which lines or parts of lines connect in three dimensions. Thus we can represent matchings of segments using simple segmented line graphs. For example for the line f, where the obtuse interior dihedral angle is 143.0767703°, we have:

$$
\begin{array}{ccccccccccccc}
- & - & - & - & - & - & - & + & + & + & + & + & + \\
\text{f } 8 & 7 & 6 & 5 & 4 & 3 & 2 & 1 & 1 & 2 & 3 & 4 & 5 & 6 \text{ f} \\
\text{f } 20 & 19 & 18 & 17 & 16 & 15 & 14 & 13 & 12 & 11 & 10 & 9 & 8 & 7 \text{ f} \\
+ & + & + & + & + & + & + & + & + & + & + & + & + &
\end{array}
$$

For lines like A the matching can only be with another line of the same kind, so for example $-A3^R$ matches only $-A3^L$.

Elementary regions are designated by a layer number followed by a letter. The single core region is labeled zero (0). Letters are in the order a, b, c, d, ... z, A, B, ... and indicate which surface member of the given layer is used. Elementary regions for each layer embedded in the stellation pattern are labeled in an alphabetical, clockwise order on the right-hand side of the y-axis. The order is repeated counter-clockwise on the left-hand side. For example: 3a, 3b, 3c, ... in clockwise and counter-clockwise direction. In general, the order is repeated between lines of bilateral symmetry when more than one exists.

When constructing a main-line stellation (complete layer) it is useful to distinguish elementary segments that form convex edges from those segments that form the concave edges. When examining the stellation pattern for the set of regions present on the outside of a given layer, only those segments belonging to the outermost outline will become convex edges.

We are now able to select correct regions and segment matchings for any main-line stellation. Simple extensions of these methods will provide ways to discover non-main-line or "composite" stellations. A simple example is given here for finding properties of the cells of the 2nd layer of the deltoidal hexecontahedron. Our primary concern is to match convex segments which make up the outermost outline of the 2nd layer embedded in the stellation pattern.

In the 2nd layer we have four region types: 2a–2d. Begin with any region, say 2d and graph all convex segment connections until there are no "loose" ends. Keep checking for correct segment correspondence.

Examining the stellation pattern we note that −F8 belongs to a convex segment of region 2b. Therefore we insert 2b and all of its convex segments into the graph: since +a2 only matches +a2 we have no more loose ends. See Fig. 2.

Regions 2a and 2c are unused, so begin with, say, 2c. In the same way the graph finally emerges as Fig. 3.

In conclusion the 2nd layer consists of two different cell types. Type #1 contains outer regions (a, c) and type #2 contains outer regions (b, d).

These "graphs of connectivity" reveal qualitative information about the kinds of surface regions for each cell. However, quantitative information may be missing. For example, beginning with region 3c of the 3rd layer we graph the following which is theoretically correct: rather than two regions we actually need six (enantiomorphous) regions of 3c in order to complete the exterior of this cell. We could have easily deduced this fact had we known that an axis of three-fold symmetry passes through the convex, outermost vertex of region 3c. This speculation leads to the next step.

Within the stellation pattern of the deltoidal hexecontahedron we can apply some general rules for locating intersection points corresponding to three-fold and five-fold symmetry. Such points shall be called "axis points." The location of two-fold axis points then becomes self-evident. Beginning with the core or central region we quickly see which of its points correspond to three- and five-fold symmetry. Notice how some intersection points located more peripherally in the pattern will radiate many regions of consecutively higher layer numbers. Suspect such points as

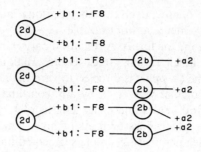

Fig. 2

Fig. 3   Fig. 4

being axis points. The rule is: layers are numbered consecutively without interruption at the next, more peripheral site of n-fold symmetry. A partial tabulation is given in Table 1.

We thus discover the relationship that the number of layers that radiate from a particular n-fold axis point is a multiple of n. A corollary is that the number of different kinds of cells that radiate three dimensionally from a particular n-fold axis is the same multiple of n. However some inconsistencies arise when searching the pattern for two-fold axis points in the case of the deltoidal hexecontahedron. For example, a two-fold axis point may exist at the midpoint of an elementary segment rather than at its endpoint. See Table 2 for a partial tabulation of two-fold axis points.

Model makers may wish to learn the under surfaces of cells which cannot be seen for cases limited to non-reentrant stellations. From such information we can chart graphs of cell connectivity which are useful for cataloguing many "composite" stellations, i.e. non-main-line forms. Further clarification of terminology is needed first.

There are many synonyms for the two kinds of cell surfaces:

outer = exterior = top = exposed; inner = interior = bottom = hidden.

Considering the stellation pattern, a convex segment of any elementary region borders a region belonging to the next highest layers. Remember that convex segments make up the outermost

Table 1

n-fold	Axis point	Location section of lines	Radiating layers	Total number of layers at axis point
Three-fold	1st	a and a	0–2nd	3
	2nd	−B and +a	3rd–8th	6 (multiple of 3)
	3rd	f and f; +f and +A	9th–17th	9
Five-fold	1st	A and A	0–4th	5
	2nd	−a and −A	5th–14th	10 (multiple of 5)

Table 2

n-fold	Axis point	Located at intersection of lines:	Radiating layers	Total number of layers at axis point
Two-fold	1st	+a and −A	0–3rd	4
	2nd	c and y-axis	4th–5th	2
	3rd	+B and −N	6th–9th	4 (multiple of 2)
	4th	F and F; +b and −B	10th–15th	6

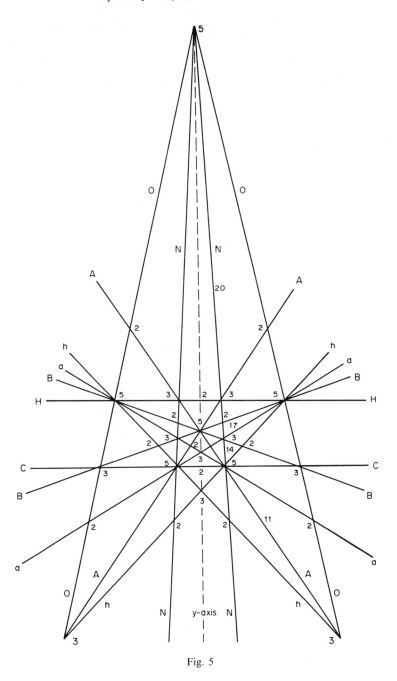

Fig. 5

outline of a given layer embedded in the stellation pattern. Similarly a concave segment of a given region borders a region belonging to the next lowest layer. The consistent method of matching segments whether convex or concave was discussed previously.

We are now ready to state a rule for solving inner cell surfaces: if the concave segment of one region matches the convex segment of another region belonging to next lower layer, then the first region is exterior and the second region is interior for the same cell.

Let us try a simple example: we are given cell designations for the 3rd layer (outer regions): (3ae), (3bdf) and (3c). We want the inner regions for each. Here is one approach.

The inner regions must come from the set of regions (2a–d). We need only to pick one convex segment from each of these regions. Begin alphabetically with region (2a). It has a convex segment +B1 which must match another +B1. Segment +B1 is also a concave segment for region (3a).

Therefore, place the inner (2a) below the outer set of regions containing (3a), namely (3ac); like so:

$$\begin{array}{l} 3ac \\ 2a \ . \end{array}$$

Next pick convex $-F8$ from region (2b) which must match concave $+b1$. Since concave $+b1$ belongs to region (3f) we place (2b) below (3bdf). Of course it would have been faster to pick convex $+a2$ from region (2b). A matching concave $+a2$ in that case would quickly point to the ajacent region (3d), again part of (3bdf). Similarly we find:

$$\text{convex } +b2 \text{ of region (2c)} = \text{concave } -F7 \text{ of region (3b), thus } \frac{3bdf}{2bc} \text{ so far.}$$

$$\text{convex } +b1 \text{ of region (2d)} = \text{concave } -F8 \text{ of region (3c), thus } \frac{3c}{2d}.$$

In summary we found the kinds of regions that make the inner surfaces.

The cells are:

$$\frac{3ae}{2a} \ , \ \frac{3bdf}{2bc} \ , \ \frac{3c}{2d} \ .$$

It lies beyond the scope of this article to list all the inner and outer regions of the complete list of 225 cells of the stellated deltoidal hexecontahedron.

By way of summary and conclusion we offer here some commentary on Fig. 5. This shows the primary regions and axis points for the stellation pattern of the deltoidal hexecontahedron. The figure shows labeled primary lines and the symmetry line (the y-axis) embedded in the complete stellation pattern (for simplicity omitted here). The intersections are n-fold axis points indicated by the number n. Note that consistently n lines meet at an n-fold axis point. Also note the remarkable dissection of the plane into "symmetry regions"—("2–3–5" type triangles). Another property may be observed. Moving along any given line, the n numbers follows a recurrent sequence which can start at any location and direction of a cyclical representation:

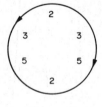

Scheme 1

Primary regions are defined as those bound areas that remain after all symmetry lines are omitted. We see that most symmetry regions are exactly the same as primary regions. Because a primary region has a complete set of n-axis points it can exist as the only kind of region on the exterior of a fully symmetric, non-reentrant stellation. Such stellations are conveniently termed "primary stellations". However, there is some overlap in terminology. Some primaries are also main-liners, if the selected primary region is composed of only one kind of elementary region. Other primaries are composite, i.e. non-main-line forms.

A systematic way of counting and labeling primaries is to consecutively number the primary regions radiating from axis points of the highest n value. In Fig. 5 there are six distinct five-fold axis points. Radiating from them we can label 21 different primary regions. Almost immediately we arrive at the necessary nets for the complete set of 21 primary stellations of the deltoidal hexecontahedron.

Plates 1–5 show some models of these primary stellations.

Plate 1. Model #11 (derived from region #11 of Fig. 5).

Plate 2. Model #14 (derived from region #14 of Fig. 5).

Plate 3. Model #17 (derived from region #17 of Fig. 5).

Plate 4. Model #20 (derived from region #20 of Fig. 5).

Plate 5. Model #11/20 (derived from regions #11 and #20 of Fig. 5).

REFERENCES

1. G. M. Fleurent, Symmetry and polyhedral stellation—Ia. *Comput. Math. Applic.* **17,** (1–3), 167–175 (1989).
2. M. J. Wenninger, *Dual Models.* Cambridge Univ. Press, New York (1983).

Computers Math. Applic. Vol. 17, No. 1–3, pp. 203–250, 1989
Printed in Great Britain. All rights reserved

THE COMPLETE SET OF JITTERBUG TRANSFORMERS AND THE ANALYSIS OF THEIR MOTION

H. F. Verheyen

De Reep, 19, 2230 Schilde, Belgium

Abstract—During many decades and in different locations people have been fascinated with the remarkable variety of ideas and inventions of Richard Buckminster Fuller. Some admire him highly for his versatility while others despise him for not having been a well-outlined architect, engineer or mathematician. And yet in any field his genius revealed itself in original concepts so many in number, that hardly any of these could have been fully worked out during his lifetime. Among these is the Jitterbug, by many considered merely as a geometrical gadget with no further use than performing an attractive transformation between some polyhedra. However, the Jitterbug inspired others to establish similar transformations between some more polyhedra, and there have been publications on these. In these transformations, one could observe Jitterbug-like structures, although they were not studied as sets on their own, but merely appearing while one polyhedron transforms into another one. Clearly there must exist a number of Jitterbug-like transformers and this number must be found when an appropriate definition is applied within the groups of symmetry.

This opens a whole field of investigation one can compare with the study of uniform polyhedra during its history. Since the previous century many new discoveries had been made, but not until 1954 was this matter mathematically dealt with, and in a way that the entire number of uniform polyhedra has been established.

A similar approach is handled in this article, in which the Jitterbug-like set is first defined with respect to its group of symmetries. Then, an enumeration is carried out, resulting in the existence of two infinite classes in the dihedral groups, and 20 types in the tetra-, octa- and icosahedral groups of symmetry. Consequently, the geometrical properties are outlined, and peculiarities explained.

The article ends with an array of applications in architecture, engineering, art and mathematics, such as Fuller would have wanted it. Finally, since the Jitterbug-like transformer needs a new definition to complete the knowledge of its number and its full motion, a new name has been chosen here: dipolygonid.

1. INTRODUCTION

In various publications and lectures, Richard Buckminster Fuller introduced a geometrical structure which he called the "Jitterbug", a set of eight identical regular triangles connected to one another by the vertices [Fig. 1(a)].

The structure is able to perform a symmetrical ex- and impansion motion, illustrating a transformation between the octahedron and cuboctahedron. As Clinton observed in his paper on expanding rigid structures [1], each triangle is subject to a translation-rotation along its symmetry axis. When starting from the position in the octahedron, these axes are the four triangular symmetry axes of the octahedron. When describing cylinders about the triangles along the axes, each vertex common to two triangles moves along the intersecting curve of the two cylinders.

Both Clinton and Stuart [2] extend the Jitterbug transformation by starting this process at other Platonic solids and some Archimedean solids too, thus establishing a number of Jitterbug-like transformers. The question arises now: how can these transformers be geometrically defined resulting in a complete classification, and what is their number?

2. MATHEMATICAL APPROACH

In attempting to create a geometrical definition suitable for the Fuller–Stuart–Clinton transformations some preliminary observations appear:

1. Each transformation starts from a Platonic solid† or an Archimedean solid,‡ while certain rotational symmetries of that solid remain.§
2. The transformers are not symmetrical with respect to reflections which leave the solid invariant.
3. The structures are composed of one or two types of polygons, all having equal edge length.¶
4. The motion of the vertices is along the intersecting curve of two circumscribed cylinders.††

The previous observations make clear that the transformers have to be studied in connection with groups of rotations. Structures having a group of symmetries are usually best understood by extracting a fundamental part in accordance to a generating subset whose number of elements is minimal.‡‡

3. POLYGON–DIPOLYGON–DIPOLYGONID

3.1. Polygon

A polygon can also be defined by a rotation and a point, in an equivalent way to the definitions of Grünbaum [4] and Coxeter [5]. The advantage here is the consistency in definitions when generating higher groups of symmetries. Let A be a rotation and P a point. The image of the line segment $a = [P, A(P)]$ over gen$\{A\}$ (the cyclic group of rotations generated by A) is the regular polygon *produced* by A in P.

If $\#\,\text{gen}\{A\}$ is finite, say s, the polygon is composed of s vertices and s edges, all equal in length. If gen$\{A\}$ is discrete, the polygon is an infinite regular polygon.

For all points P outside of r_A, axis of A, the plane through P and perpendicular to r_A intersects r_A in M, whereas in the triangle $\triangle P, M, A(P)$ the angle $\angle P, M, A(P)$ is invariant and denoted by ψ, and will be referred to as the *central angle* of the rotation A.

Obviously $0 \leqslant \psi \leqslant \pi$.

Let A be distinct from the identity. Among the rotations of gen$\{A\}$, R and R^{-1} are those two whose central angle is the smallest besides the central angle 0 of the identity in the group. If gen$\{A\}$ is finite, the central angle of R and R^{-1} is $2\pi/s$. Then, clearly R an R^{-1} are generators of gen$\{A\}$. Let

$$A = R^d (1 \leqslant d < s) \rightarrow A = (R^{-1})^{s-d}.$$

Let the notations of R and R^{-1} be chosen such that

$$d \leqslant s - d \rightarrow d \leqslant \frac{s}{2}.$$

d is called the *density* of the polygon, and $m = s/d$ the *polygonal value* of the rotation A. The notation for the polygon is $\{m\}$, and $m \geqslant 2$, for $d \leqslant s/2 \rightarrow 2 \leqslant s/d$.§§

---

†Clinton started a face-transformation from each of the five Platonic solids, although the octahedron provided the only coherent transformer: the Jitterbug. To be coherent, the remaining transformers had to be composed of coplanar pairs of polygons, as will be illustrated further. Even then, the tetrahedral transformer, composed of four coplanar pairs of triangles, would only be another appearance of the Jitterbug in one of its positions below the non-convex phase.

‡Stuart started a transformation from a semi-regular solid: the cuboctahedron. The transformation from cuboctahedron to rhombicuboctahedron is nicely shown on a flipmovie by the pages. However, the motion is restricted to the convex phase.

§Namely the rotations determined by those axes, along which the polygons transform. However, the transformers have all a symmetry group of rotations, e.g. Stuart's transformer of the cuboctahedron has the octahedral group of rotations (S_4) as its symmetry group, although the Jitterbug, which starts from the octahedron, has the tetrahedral group of rotations as its symmetry group (A_4), and not the octahedral. This can be understood by considering the octahedron as the semi-regular polyhedron of the tetrahedral group of isometries $(S_4 A_4)$, just like the cuboctahedron is in the octahedral group of symmetries $(S_4 \times I)$.

¶Even in Clinton's edge transformations of the five Platonic solids the edges are merely digons.

††Or at least part of it, since the Fuller–Stuart–Clinton transformers are only determined as being convex, i.e. the motion ceases when the faces start intersecting each other.

‡‡To be compared to the Coxeter–Wythoff construction of the polyhedral kaleidoscope, which represents a set of three reflections as generating the group of isometries of a polyhedron. The kaleidoscope is associated with a fundamental region wherein the fundamental part of the polyhedron is conceived [3].

§§$\{m\}$ is the Coxeter notation given to a regular polygon [5], while the polygonal value [6] implies a definition for density of a polygon, conform to Coxeter's approach [5].

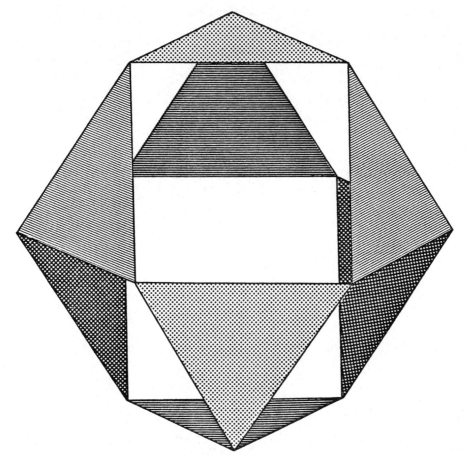

Fig. 1(a). The Jitterbug.

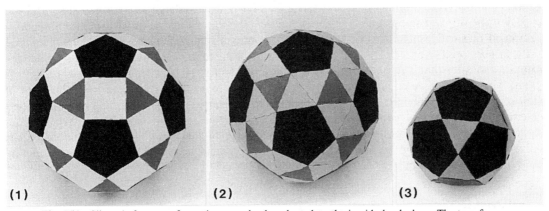

Fig. 1(b). Clinton's face-transformation can also be adapted to the icosidodecahedron. The transformation is illustrated in the (1) rhombicosidodecahedron, (2) snub dodecahedron and (3) icosidodecahedron.

Hence, if m is a natural number > 2, $\{m\}$ is the regular convex polygon composed of m edges and m vertices. If m is rational, $\{m\}$ is a regular star polygon of order s and density s/m.

Since $A = R^d$, the central angle of A can be calculated:

$$\psi_A = \frac{2\pi}{s} \cdot d = \frac{2\pi}{\dfrac{s}{d}} = \frac{2\pi}{m}.$$

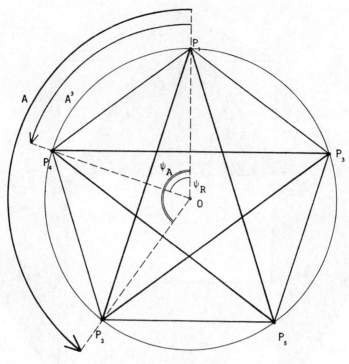

Fig. 2

Figure 2 illustrates an example where A has a central angle of $4\pi/5$, gen$\{A\} = \{E, A, A^2, A^3, A^4\}$ and $s = 5, d = 2$ and R is the rotation of a central angle $2\pi/5$.

If $s = 1$, the polygon $\{1\}$ is a monogon, i.e. composed of 1 vertex and an edge of length 0.†

If $s = 2$, the polygon $\{2\}$ is a digon, being composed of two collapsing edges and having two vertices P and $A(P)$. Although degenerate, the digon will be considered as a real polygon, and will appear to be of great importance further on. When $P \in r_A$, axis of A, the degenerate polygon is composed of s collapsing edges of length 0, and contains s coinciding vertices P.

From here on, the order of A will be supposed to be finite and > 1. The polygonal value of a rotation indicates that any polygon produced in any point is of type $\{m\}$.

The whole process described here means no more than looking at a polygon from the point of view of rotational symmetry. Clearly A^{-1} produces an identical polygon $\{m\}$ in P. The definition determines a regular polygon as being a closed, broken line segment in a plane, perpendicular to r_A. The polygon indicates a maximal bounded subset of the plane, which will be called the polygonal face, or simply the face of the polygon. Polygonal faces will be used for constructing polyhedra [7].

3.2. Dipolygon

Each group of rotations has a point of invariancy O, being the one point of intersection of all axes of the rotations. Hence, the next step in the construction of a solid definition is to consider two rotations A and B whose axes intersect in one point O.

A produces $\{m\}$ in P, and B produces $\{n\}$ in P. This pair of polygons will be called a *dipolygon*, produced by the base $\{A, B\}$ in P (Fig. 3).

Consequently three more bases produce the same dipolygon in P: $\{A, B^{-1}\}, \{A^{-1}, B\}$ and $\{A^{-1}, B^{-1}\}$ (Fig. 4).

Dipolygonal specifications. A has order s, polygonal value m; B has order t, polygonal value n.

The dipolygonal angle θ: r_A and r_B form in O the angles θ and $\pi - \theta$, where the choice for this

---

†When A is the identity, E.

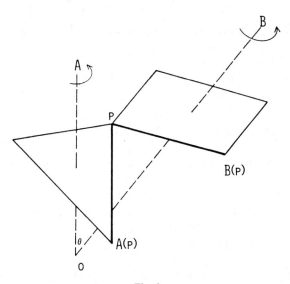

Fig. 3

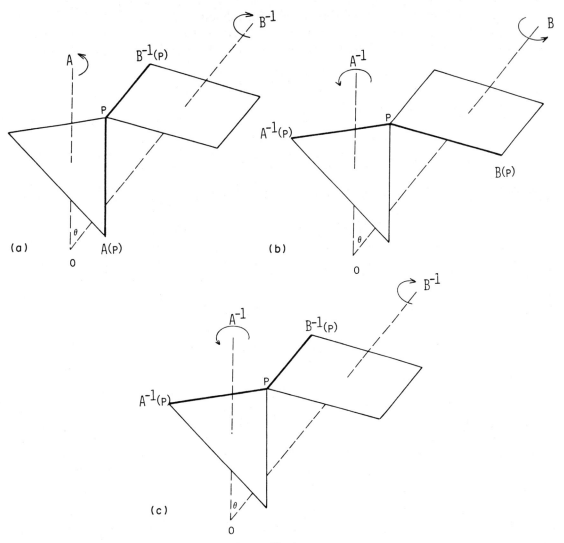

Fig. 4

notation is such that:

$$\theta \leqslant \pi - \theta \to 2\theta \leqslant \pi \to \theta \leqslant \frac{\pi}{2} \to 0 < \theta \leqslant \frac{\pi}{2} \leqslant \pi - \theta < \pi$$

$\{m\}$ has center M
$\{n\}$ has center N.

Also, the following properties exist: (a) the planes determined by the polygons $\{m\}$ and $\{n\}$, are perpendicular to r_A and r_B resp., in the points M and N resp.; (b) the dihedral angle of these planes is θ (or the suppl. $\pi - \theta$).

Now consider the circumscribed cylinders of $\{m\}$ and $\{n\}$. These are simply defined by r_A and P, and r_B and P. When the radii of the cylinders have different length, the intersection will be composed of two distinct curves, which have central inverse symmetry in O [Fig. 5(a)]. P belongs to one of these curves. When the cylinders are equiradial ($R_A = R_B$), the two curves have two common points, altogether forming two intersecting ellipses that have a common smaller axis [Fig. 5(b)]. Still here, the intersection can be considered as being composed of two centrally inverse curves [Fig. 5(c)].

When P moves along the curve to which it belongs, even in the case of equiradial cylinders, in each position a dipolygon can be produced by the base $\{A, B\}$. Such a dipolygon is a transformed image of the first dipolygon, by a translation rotation of $\{m\}$ along r_A, and of $\{n\}$ along r_B. This transformation will be called the *uniform motion* of the dipolygon.

An equation for the intersection of the cylinders is calculated when a left-oriented base (X, Y, Z) is chosen as in Fig. 6. A and B are chosen such that for the radii R_A and R_B of the cylinders:

$$R_A \gtrless R_B; \quad \theta = \measuredangle|\overline{OM}, \overline{ON}|; \quad \mu = \measuredangle|\overline{X}^+, \overline{OP_1}|.$$

The equations of the cylinders are:

$$C_A \to x^2 + y^2 = R_A^2$$
$$C_B \to \cos^2\theta \cdot x^2 + y^2 + \sin^2\theta \cdot z^2 - 2\sin\theta \cdot \cos\theta \cdot xz = R_B^2.$$

The parameter equation of $C_A \cap C_B$ is:

$$x = R_A \cos\mu$$
$$y = R_A \sin\mu$$
$$z = \frac{R_A \cos\theta \cdot \cos\mu \pm \sqrt{R_B^2 - R_A^2 \cdot \sin^2\mu}}{\sin\theta},$$

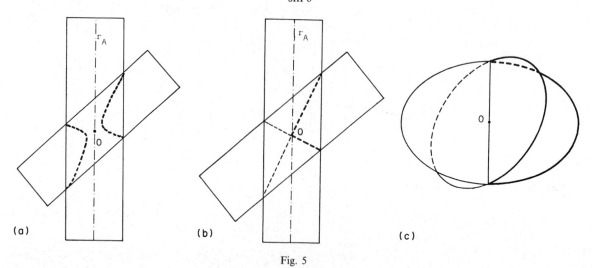

(a)   (b)   (c)

Fig. 5

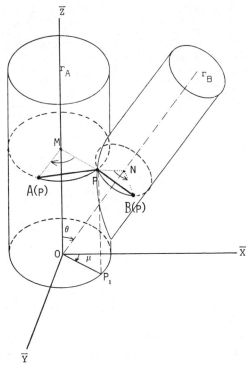

Fig. 6

where $|\sin \mu| \leqslant R_B/R_A$. Hence, the curve along which P moves is indicated by:

$$0 \leqslant |\mu| \leqslant \sin^{-1} \frac{R_B}{R_A} = \mu_c,$$

while the second curve of the intersection is given by

$$\mu + \pi.$$

The curve along which P moves will be referred to as the *path* of the uniform motion.

The sign $(\pm)$ before the square root provides two values for z with each value for μ, except when $|\mu| = \mu_c$. The part of the curve determined by $(+)$ is called the upper half of the path, and the $(-)$ part the lower half.

The axes r_A and r_B determine a plane ω. If S denotes the reflection in ω, S leaves both cylinders and the path invariant. Some special positions with respect to ω will be observed.

(a) *Chiral positions.* The dipolygon produced by $\{A, B\}$ in $S(P)$ is the reflected image of the one in P. These dipolygons are enantiomorphous positions of each other (dextro and laevo). Since A and A^{-1}, B and B^{-1} are transformed operations by $S(A^{-1} = SAS, B^{-1} = SBS)$, all the properties of a dipolygon with respect to its base hold for the laevo position, provided the base is inversed, i.e. the rotations are inversed.

(b) *Extreme positions.* if $P \in \omega$, $S(P) = P$. Such a position is self-enantiomorphous and ω is a symmetry plane of this position of the dipolygon. There are two such positions on the path, called the extreme positions. The plane ω is perpendicular to $\{m\}$ and $\{n\}$, while $M, N, P, O \in \omega$.

In an extreme position $\mu = 0$, and the coordinates of P become:

$$(R_A, 0, R_A \cot \theta \pm R_B \operatorname{cosec} \theta).$$

The extreme position in the upper half is called the maximum position. Here

$$\measuredangle M, P_{\max}, N = \pi - \theta \text{ (right or obtuse).}$$

The extreme position in the lower half is called the minimum position. Here

$$\measuredangle M, P_{\min}, N = \theta \text{ (right or sharp).}$$

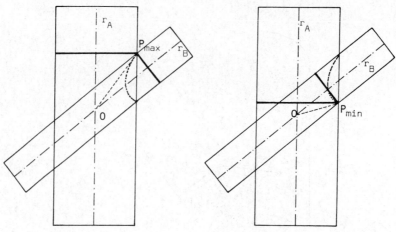

Fig. 7

However, if $\theta = \pi/2$, the plane γ through r_B and perpendicular to ω is a symmetry plane for the two cylinders and hence for the path. Then, each of the extreme positions may represent the maximum or minimum position.

From the equations follows: when the dipolygon moves along the path between two extreme positions, $\{n\}$ describes a rotation over π. Halfway, when $\{n\}$ has rotated over $\pi/2$, the rotation of $\{m\}$ has reached its maximal angle

$$|\mu| = \mu_c.$$

The chiral positions in μ_c and $-\mu_c$ are called the central positions of the dipolygon.

From the equations also follows:

(a) The translation of $\{m\}$ happens between the extreme positions, where the sense reverses; the translation of $\{n\}$ happens between the extreme positions and the central positions. In each of these positions, the sense reverses.

(b) The rotation of $\{m\}$ happens between central positions, where the sense reverses; the rotation of $\{n\}$ is in one sense all along the path.

Hence, while passing a central position $\{m\}$ keeps its sense of translation, but reverses its sense of rotation, while $\{n\}$ reverses its sense of translation, but keeps its sense of rotation.

When $R_A = R_B, \mu_c = \pi/2$. Then, in a central position M coincides with O and since PN and r_B are perpendicular, also N coincides with M and O. The points of the path, in which the dipolygon is in a central position are the intersecting points of the upper and lower-half ellipses on their common smaller axis. If also the dipolygon is regular, i.e. being composed of two congruent polygons, two more planes of special interest can be observed: α and β, bisector planes of r_A and r_B, chosen such that:

$$P_{\max} \in \alpha, \quad P_{\min} \in \beta.$$

Thus, α is a symmetry plane for the dipolygon in the upper half-ellipse, while β is for the lower half-ellipse.

3.3. Dipolygonid

If G denotes the group of rotations generated by A and B, all the axes of the rotations of G intersect in O, point of invariancy of G. G may be discrete or finite.

The image of the dipolygon over G is called the *dipolygonid, produced* by the base $\{A, B\}$ in P (Fig. 8).

By this construction the dipolygonid is invariant over G. Clearly the dipolygonid is also produced by the other three bases:

$$\{A^{-1}, B\}, \{A, B^{-1}\}, \{A^{-1}, B^{-1}\}.$$

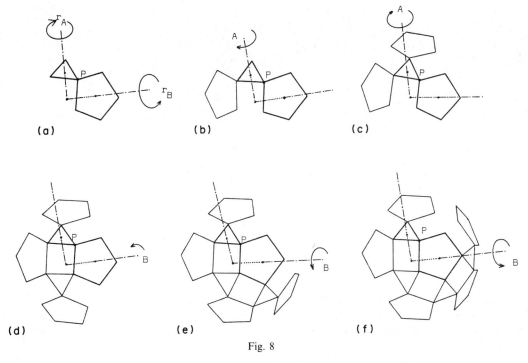

Fig. 8

Moreover, it can be produced in any vertex $R(P)$, where $R \in G$. The base $\{A, B\}$ is then transformed by R into a set $\{RAR^{-1}, RBR^{-1}\}$ which act like a base for the transformed dipolygon in $R(P)$.

Since $\{RAR^{-1}, RBR^{-1}\}$ clearly is a generating set of G too, the dipolygonid is also produced in $R(P)$ by the base $\{RAR^{-1}, RBR^{-1}\}$.

The smallest subset of the dipolygon in P whose image over G is the dipolygonid is the couple of edges a and b (Fig. 10). This two-edge is the fundamental part of the dipolygonid with respect to G.

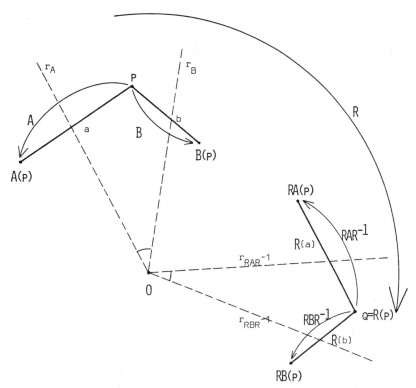

Fig. 9

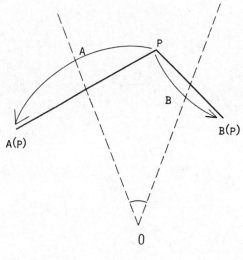

Fig. 10

Each position of the dipolygon produced by $\{A, B\}$ in P has its image over G, thus creating a dipolygonid. When the dipolygon describes its uniform motion along the path, all the transformed dipolygons in $R(P)$ do simultaneously. This motion is called the uniform motion of the dipolygonid (Fig. 11).

4. JITTERBUG AND DIPOLYGONIDS

It can easily be seen that this definition for a dipolygonid is appropriate to the Fuller–Stuart–Clinton transformers, holding the following generalizations:
 —Since G may be discrete, so will the dipolygonid be composed of an infinite number of polygons, yet able to perform the transformations.†
 —The dipolygonid has not necessarily one edge length. Depending on the position of P, the dipolygonid may have two different edge lengths.
 —Along the path of uniform motion the dipolygonid becomes non-convex. In fact the uniform motion rather resembles a pulsating motion from maximum position over central-1, minimum, central-2 to maximum position.

5. FINITE DIPOLYGONIDS

Our next concern is the search for dipolygonids which are composed of a finite number of polygons and vertices. This will be the case if G is finite. Then, the question is brought back to the investigation of the finite groups of rotations, which are well established [5].

If G is finite, the order will be denoted by g. Then, the dipolygonid is composed of g vertices theoretically, and hence, of g dipolygons.

G will be finite, only if A and B belong to one of the finite groups of rotations and generate a subgroup at least. This will solely depend of the orders of A and B, and the angle θ formed by their axes in O. Since r_A and r_B are distinct, however, the cyclic groups can be excluded.

Table 1 classifies the remaining finite groups of rotations by their number of elements and the conjugate maximal cyclic subgroups. The latter numbers indicate the types of axes in Table 2.

In Table 3, a summary is given of all the possible angles between axes of a group. An axis is represented by the order of the maximal cyclic subgroup of rotations, having this axis.

---

†The major part of Clinton's paper deals with transformations (face, edge and vertex) of flat tesselations. Those are composed of an infinite number of polygons. Such a transformer can be considered as a dipolygonid whose base $\{A, B\}$ has its axes intersecting at infinity. Then, $R_A /\!/ R_B$ and the dipolygonid will be planar, and G discrete.

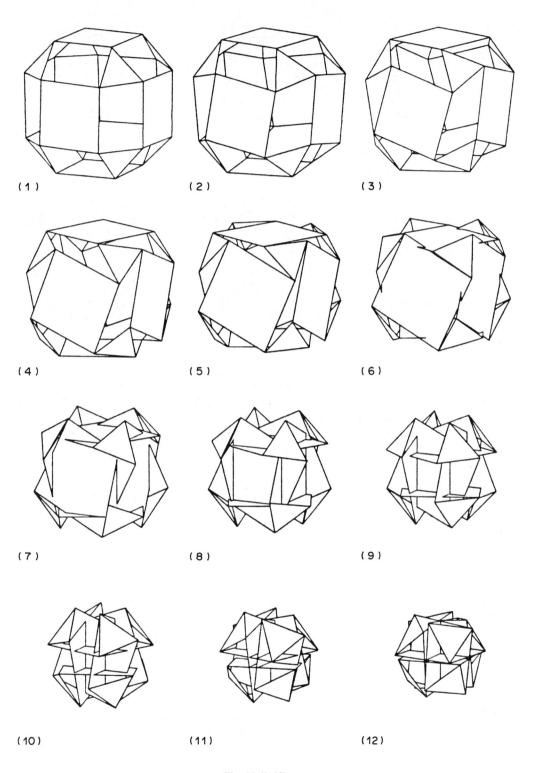

(1)　　　　　　(2)　　　　　　(3)

(4)　　　　　　(5)　　　　　　(6)

(7)　　　　　　(8)　　　　　　(9)

(10)　　　　　　(11)　　　　　　(12)

Fig. 11 (1–12)

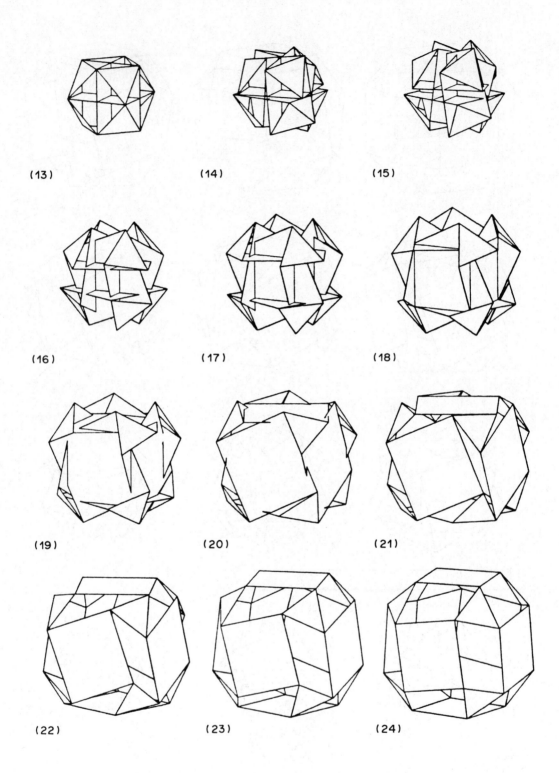

Fig. 11 (13–24)

Table 1

GROUP	ORDER	CONJUGATE CYCLIC SUBGROUPS	
		number	order
Dihedral	2n	n	2
		1	n
Tetrahedral	12	3	2
		4	3
Octahedral	24	6	2
		4	3
		3	4
Icosahedral	60	15	2
		10	3
		6	5

Table 2

GROUP	NUMBER OF AXES
Dihedral (D_n)	n + 1
Tetrahedral	7
Octahedral	13
Icosahedral	31

Out of Table 3 a summary can be realized of all possible pairs of rotations in the dihedral, tetra-, octa- and icosahedral groups with respect to the order of these rotations and the angle of their axes. Table 4 shows this summary, together with the groups that are generated by these pairs.

In C_5, the cyclic group of rotations of order 5, four elements occur of order 5, namely the rotations of polygonal value 5 and 5/2.†

All other pairs of rotations besides those mentioned in Table 4 result as generators of discrete groups of rotations. Apart of the two classes of pairs generating the dihedral groups, there are two types of pairs in the tetrahedral, four in the octahedral, and 14 in the icosahedral group of rotations.

Hence, there is a corresponding amount of dipolygonid types, when the pairs of generators represent a base. If g represents the order of G, s of A, and t of B, the number of the dipolygonid's elements is: vertices—g; polygons $\{m\}$ — g/s; polygons $\{n\}$ — g/t; edges—$2g$.

The following notation will be used for a finite dipolygonid:

$$\frac{g}{s}\{m\} + \frac{g}{t}\{n\}|\theta\ddagger$$

6. CLASSIFICATION OF THE FINITE DIPOLYGONIDS

6.1. Dihedral (D_n)

Two infinite classes are associated to each of the infinite classes of bases:

(A) $$n\{2\} + n\{2\}|\frac{k}{n}\cdot 180°$$

where $n = 2$ and $k = 1$, or $n > 2$ and $0 < k < n/2$ (and k and n are coprime). These dipolygonids have the shape of non-planar polygonal zigzag lines. If both sets of digons have equal length, they represent "Petrie-polygons" [3, 5].

(B) $$n\{2\} + 2\left\{\frac{n}{k}\right\}|90°.$$

These dipolygonids have the shape of rectangular $\{n/k\}$-prisms in extreme positions.

---

†Namely the rotations over $\pm 2\pi/5$ (polygonal value 5) and over $\pm 4\pi/5$ (polygonal value $\frac{5}{2}$).
‡In accordance to the notation of a uniform polyhedron by indicating the numbers of different types of polygons as a summation [8].

Table 3

GROUP	SYMBOL	AXES	ANGLE	(SUPPLEMENT)
DIHEDRAL (n=2,3,...)	D_n	2 ∿ 2	$\frac{k}{n} \cdot 180°$, $k \in \mathbf{N}$, $0 < k \leq \frac{n}{2}$	
		2 ∿ n	90°	(90°)
TETRAHEDRAL	A_4	2 ∿ 2	90°	(90°)
		2 ∿ 3	54°44'08"	(125°15'52")
		3 ∿ 3	70°31'44"	(109°28'16")
OCTAHEDRAL	S_4	2 ∿ 2	60° 90°	(120°) (90°)
		2 ∿ 3	35°15'52" 90°	(144°44'08") (90°)
		2 ∿ 4	45° 90°	(135°) (90°)
		3 ∿ 3	70°31'44"	(109°28'16")
		3 ∿ 4	54°44'08"	(125°15'52")
		4 ∿ 4	90°	(90°)
ICOSAHEDRAL	A_5	2 ∿ 2	36° 60° 72° 90°	(144°) (120°) (108°) (90°)
		2 ∿ 3	20°54'19" 54°44'08" 69°05'42" 90°	(159°05'41") (125°15'52") (110°54'18") (90°)
		2 ∿ 5	31°43'03" 58°16'57" 90°	(148°16'57") (121°43'03") (90°)
		3 ∿ 3	41°48'37" 70°31'44"	(138°11'23") (109°28'16")
		3 ∿ 5	37°22'39" 79°11'16"	(142°37'21") (100°48'44")
		5 ∿ 5	63°26'06"	(116°33'54")

Table 4

A ∿ B			θ		gen{A,B}
2	∿	2	$\frac{k}{n} \cdot 180°$ †		D_n
2	∿	3	20°54'19"		A_5
			35°15'52"		S_4
			54°44'08"		A_4
			69°05'42"		A_5
2	∿	4	45°		S_4
2	∿	5	31°43'03"		A_5
			58°16'57"		A_5
2	∿	n	90°		D_n
3	∿	3	41°48'37"		A_5
			70°31'44"		A_4
3	∿	4	54°44'08"		S_4
3	∿	5	37°22'39"		A_5
			79°11'16"		A_5
4	∿	4	90°		S_4
5	∿	5	63°26'06"		A_5

†$n \geqslant 2$: if $n = 2, k = 1$; if $n \geqslant 3, 1 \leqslant k < n/2, k$ and n coprime.

The example $3\{2\} + 2\{3\}|90°$ in D_3 is illustrated in Fig. 12.

6.2. Tetrahedral

See Fig. 13; $T_1–T_2$.

6.3. Octahedral

See Fig. 14; $O_1–O_4$.

6.4. Icosahedral

See Fig. 15; $I_1–I_{14}$.

Since the number of polygons $\{m\}$ is g/s, as is also the index of the cyclic subgroup of order s in G (the number of s-fold axes in G), clearly the number of polygons along one axis is known.

For example, in I_3, 12 pentagons occur along 6 five-fold axes. Hence, two pentagons have a common axis. The question rises: do these polygons occur on the same side or different sides of O?

The answer is easily found by observing the two-fold rotation whose axis is perpendicular to ω. If this rotation is in G, it maps the dipolygon within the dipolygonid on the opposite side of both axes of the dipolygon's base. (See Diagram 1.)

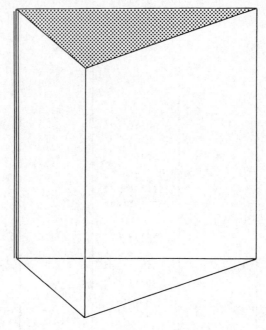

Fig. 12

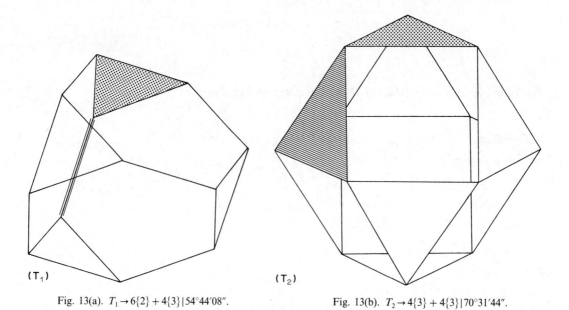

(T₁)

(T₂)

Fig. 13(a). $T_1 \rightarrow 6\{2\} + 4\{3\} | 54°44'08''$.   Fig. 13(b). $T_2 \rightarrow 4\{3\} + 4\{3\} | 70°31'44''$.

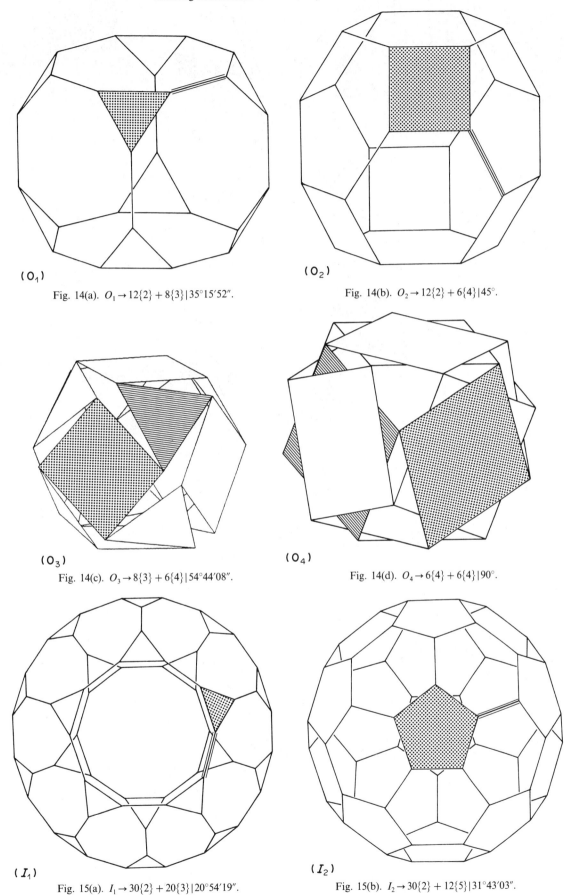

(O_1)

Fig. 14(a). $O_1 \rightarrow 12\{2\} + 8\{3\} \mid 35°15'52''$.

(O_2)

Fig. 14(b). $O_2 \rightarrow 12\{2\} + 6\{4\} \mid 45°$.

(O_3)

Fig. 14(c). $O_3 \rightarrow 8\{3\} + 6\{4\} \mid 54°44'08''$.

(O_4)

Fig. 14(d). $O_4 \rightarrow 6\{4\} + 6\{4\} \mid 90°$.

(I_1)

Fig. 15(a). $I_1 \rightarrow 30\{2\} + 20\{3\} \mid 20°54'19''$.

(I_2)

Fig. 15(b). $I_2 \rightarrow 30\{2\} + 12\{5\} \mid 31°43'03''$.

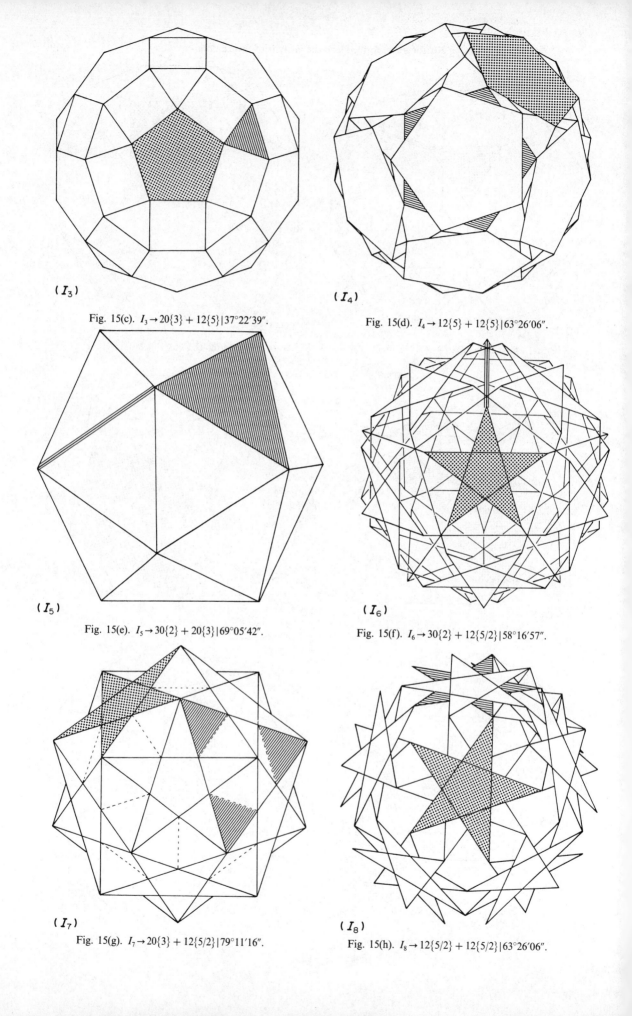

(I_3)

Fig. 15(c). $I_3 \rightarrow 20\{3\} + 12\{5\}\,|\,37°22'39''$.

(I_4)

Fig. 15(d). $I_4 \rightarrow 12\{5\} + 12\{5\}\,|\,63°26'06''$.

(I_5)

Fig. 15(e). $I_5 \rightarrow 30\{2\} + 20\{3\}\,|\,69°05'42''$.

(I_6)

Fig. 15(f). $I_6 \rightarrow 30\{2\} + 12\{5/2\}\,|\,58°16'57''$.

(I_7)

Fig. 15(g). $I_7 \rightarrow 20\{3\} + 12\{5/2\}\,|\,79°11'16''$.

(I_8)

Fig. 15(h). $I_8 \rightarrow 12\{5/2\} + 12\{5/2\}\,|\,63°26'06''$.

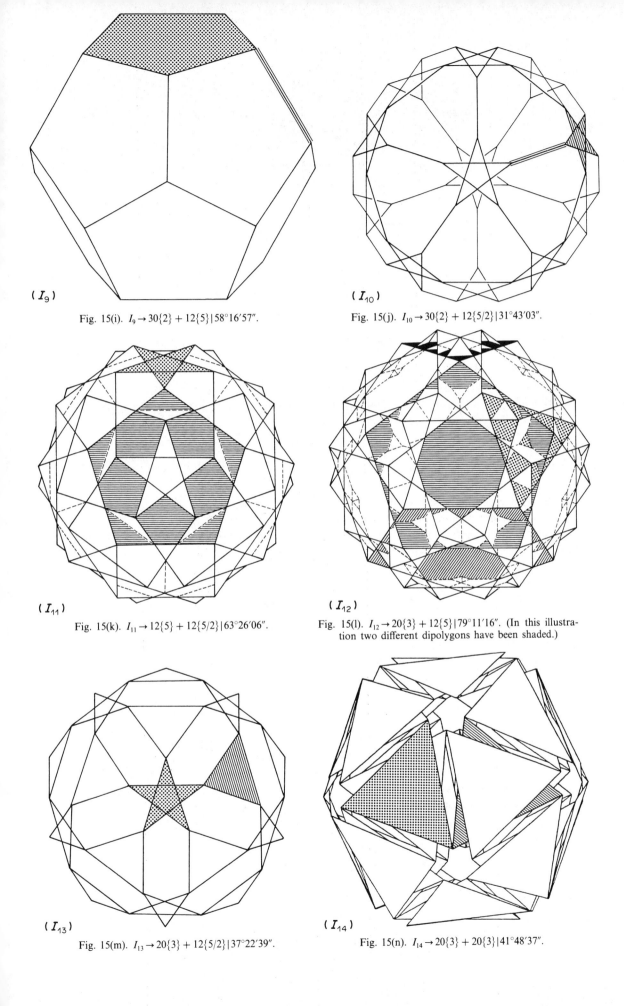

(I_9)

Fig. 15(i). $I_9 \rightarrow 30\{2\} + 12\{5\}\,|\,58°16'57''$.

(I_{10})

Fig. 15(j). $I_{10} \rightarrow 30\{2\} + 12\{5/2\}\,|\,31°43'03''$.

(I_{11})

Fig. 15(k). $I_{11} \rightarrow 12\{5\} + 12\{5/2\}\,|\,63°26'06''$.

(I_{12})

Fig. 15(l). $I_{12} \rightarrow 20\{3\} + 12\{5\}\,|\,79°11'16''$. (In this illustration two different dipolygons have been shaded.)

(I_{13})

Fig. 15(m). $I_{13} \rightarrow 20\{3\} + 12\{5/2\}\,|\,37°22'39''$.

(I_{14})

Fig. 15(n). $I_{14} \rightarrow 20\{3\} + 20\{3\}\,|\,41°48'37''$.

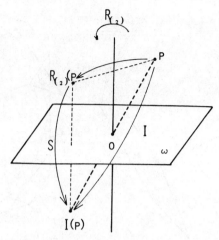

Diagram 1. The half-turn R is about an axis perpendicular to ω. Then $I = SR = RS$.

From Table 1 is found that the only exceptions for the two-fold rotation being in G are for the bases of type:

1. In D_n: $2 \sim 2 \mid \dfrac{k}{n} \cdot 180° \left(0 < k < \dfrac{n}{2}, n \text{ odd, } k \text{ and } n \text{ coprime} \right)$

2. In A_4: $3 \sim 3 \mid 70°31'44''$

which correspond to the dipolygonids of class A in the odd dihedral groups D_n, and the dipolygonid T_2.

In the first case, the polygons of type $\{m\}$ and $\{n\}$, sharing a common axis of order 2 are two digons of different edge length (although they may be equal too).

In the second case, two triangles of different edge length (or equal) share a common three-fold axis.

7. ASPECTS OF UNIFORM MOTION

Together with the dipolygon in P, and all the dipolygons in the points $R(P)$, where $R \in \text{gen}\{A, B\}$, the dipolygonid describes its uniform motion. Special positions of the dipolygon have their equivalent special positions of the dipolygonid.

7.1. Chiral positions

Each step in the production of a dipolygonid in $S(P)$ by the laevo base $\{A^{-1}, B^{-1}\}$ is the reflected operation under S of a step in the production of the dipolygonid in P by $\{A, B\}$. However, the dipolygonid in $S(P)$ is also produced by $\{A, B\}$. The set of both dipolygonids is also the image over G of the chiral dipolygons in P and $S(P)$, which are reflected images by S. Then, the set of dipolygonids is invariant over $G_S = \text{gen}\{A, B, S\}$, an extended group of G [9].

The dipolygonids are reflected images of each other under any reflection of G_S, and therefore can be called enantiomorphous (dextro and laevo) too.

7.2. Extreme positions

Maximum and minimum positions of the dipolygonid correspond with those of the dipolygon.

7.3. Central positions

A central position of the dipolygonid is produced when the dipolygon is in a central position. Figure 16 illustrates O_4 from the maximum to central position in steps.

Besides the two exceptions mentioned in Section 6, all the finite dipolygonids are composed of opposite polygons $\{m\}$ and $\{m\}'$ along the s-fold axes, and $\{n\}$ and $\{n\}'$ along the t-fold axes of $G = \text{gen}\{A, B\}$.

Since the centers of the polygons, M and M', and N and N' each remain on equal distances of O during the uniform motion of the dipolygonid, the property holds for each position. During the uniform motion M or N may coincide with O.

Since $\{n\}$ reverses its sense of translation in a central position, N would only coincide with O when $R_A = R_B$ in both central positions. However, the sense of the translation of $\{m\}$ reverses in the extreme positions and hence, M may coincide with O, depending on R_A and R_B, or analogue, depending on the edge lengths of $\{m\}$ and $\{n\}$.

When $R_A = R_B$, M and N of the dipolygon coincide with O, and hence, so will all of the polygons have their center coinciding with O.

Before going into more details of equiradial dipolygonids, some more understanding of the symmetries in extended groups of isometries is needed.

8. EXTENDED GROUPS OF ISOMETRIES

Besides the finite groups of rotations, the remaining groups of isometries are distinguished by their containing of the central inversion in O (denoted by I) or not [5].

(a) *Groups containing* I. These groups all are the direct product of a finite group of rotations and the group $\{E, I\}$, which in Table 5 is denoted by the abbreviated symbol I.

(b) *Groups that do not contain* I. These groups are called "mixed groups". If G' is a group of rotations of order $2n$, containing a subgroup G of order n, the mixed group is $G \cup (G' - G)I$, which in Table 5 is denoted by $G'G$.

Each of these groups has an even order $2g$, and contains a subgroup of rotations of order g. The number of rotatory inversions (products of a rotation of G and I) is also g. A rotatory inversion is either a rotatory reflection (a product of a rotation and a reflection) or a reflection. Such a reflection is then the product of a half-turn (a rotation of order 2) and I, hence, the number of reflections in the extended group is the number of half-turns in G (case a) or the number of half-turns in the complement $G' - G$ (case b).

The remaining number is that of the rotatory reflections (see Table 6).

Table 5

SYMBOL	ORDER		
$C_n \times I$	$2n$	$(n=1,2,..)$	
$D_n \times I$	$4n$	$(n=2,3,..)$	†
$A_4 \times I$	24		
$S_4 \times I$	48		
$A_5 \times I$	120		
$C_{2n}C_n$	$2n$	$(n=1,2,..)$	
D_nC_n	$2n$	$(n=1,2,..)$	‡
$D_{2n}D_n$	$4n$	$(n=1,2,..)$	§
S_4A_4	24		

†$D_1 \cong C_2 \to D_1 \times I \cong C_2 \times I$
‡D_1C_1 is generated by one reflection.
§D_2D_1 has order 4, and is generated by two orthogonal reflections.

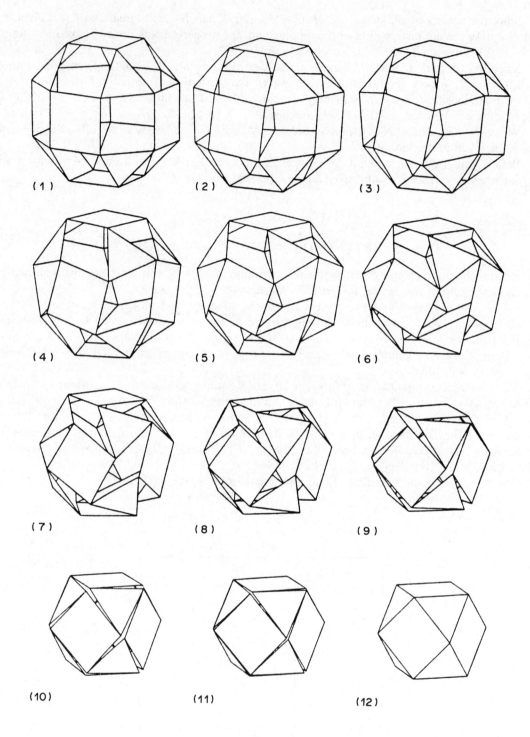

Fig. 16 (1–12)

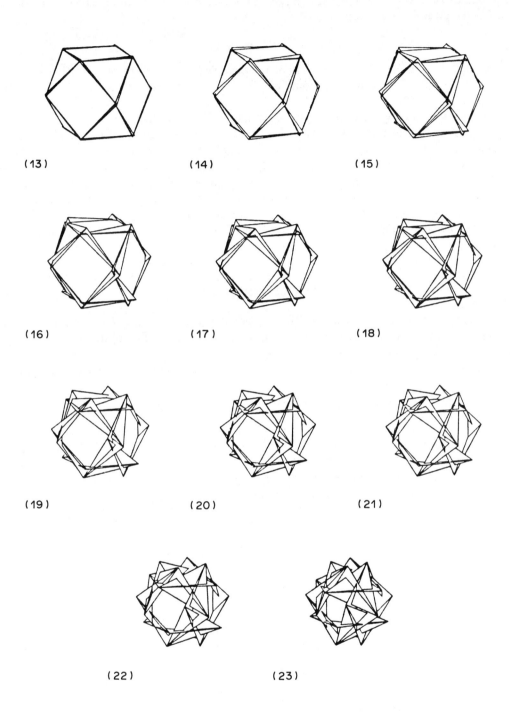

Fig. 16 (13–23)

8.1. Dipolygonal generators

The sets of isometries: $\{A, B\}$; $\{A, B, S\}$; $\{A, B, I\}$, each are called dipolygonal generator sets of symmetry groups. The group

$$F = \text{gen}\{A, B, C\},$$

where $C \in \{S, I\}$, is either discrete, or one of the classified finite groups of isometries containing $G = \text{gen}\{A, B\}$. If F is finite, it contains G as a subgroup of index 2. Then clearly $F = G \cup GC = G \cup CG$.

(1). $C = I$. F is clearly a group of type a, namely a direct product $G \times I$.

(2). $C = S$. The groups in Table 5 have to be checked for their containing of S, with respect to the choice of $\{A, B\}$. This can be obtained from the information given in Table 3, by checking if the two-fold rotation whose axis is perpendicular to ω belongs to G, $G' - G$ or to none of both. It is then also found out whether F is a direct product or a mixed group. From Table 3 also can be seen which axes are perpendicular to the axis of the two-fold rotation, and hence, which are lying in ω. From Table 7 is seen how $D_{2n}D_n$ (n is even) has no dipolygonal generators. It has no reflection plane containing more than one axis, and does not contain I. The results are shown in Table 8.

9. REGULAR DIPOLYGONIDS

In a regular dipolygon (Section 3.2), the polygonal values of A and B, m and n, are equal. The upper half-ellipse of the path lies in α, and v, the intersecting line of α and ω, is along the greater axis of the ellipse.

The reflection V in α is a symmetry operation for the dipolygon in the upper half, since for any P:

$$VB(P) = A(P) \quad \text{or} \quad VB(P) = A^{-1}(P).$$

Table 6

GROUP	ROTATIONS	REFLECTIONS	ROTATORY REFLECTIONS
$C_n \times I$			
$\quad$ n even	n	1	n−1
$\quad$ n odd	n	0	n
$C_{2n}C_n$			
$\quad$ n even	n	0	n
$\quad$ n odd	n	1	n−1
D_nC_n	n	n	0
$D_n \times I$			
$\quad$ n even	2n	n+1	n−1
$\quad$ n odd	2n	n	n
$D_{2n}D_n$			
$\quad$ n even	2n	n	n
$\quad$ n odd	2n	n+1	n−1
$A_4 \times I$	12	3	9
$S_4 A_4$	12	6	6
$S_4 \times I$	24	9	15
$A_5 \times I$	60	15	45

Table 7

GROUP	SYMMETRY PLANES	COPLANAR AXES (NUMBER)
$C_n \times I$	0/1	(0)
$C_{2n}C_n$	0/1	(0)
D_nC_n	n	n (1)
$D_n \times I$		
n even	1	2 (n)
	n	2 (1), n (1)
n odd	n	n (1)
$D_{2n}D_n$		
n even	n	n (1)
n odd	1	2 (n)
	n	2 (1), n (1)
$A_4 \times I$	3	2 (2)
S_4A_4	6	2 (1), 3 (2)
$S_4 \times I$	3	2 (2), 4 (2)
	6	2 (1), 3 (2), 4 (1)
$A_5 \times I$	15	2 (2), 3 (2), 5 (2)

The choice of A and A^{-1} can be determined such that

$$VB(P) = A(P)$$

Then clearly, V is also a symmetry operation for the dipolygonid being the image of the dipolygon over G. Hence, the dipolygonid is invariant over

$$\text{gen}\{A, B, V\}.$$

Table 8

GROUP	ADDED GENERATOR	EXTENDED GROUP
D_n	S	$D_{2n}D_n$ (n odd)
		$D_n \times I$ (n even)
	I	$D_n \times I$
A_4	S	$S_4 A_4$
	I	$A_4 \times I$
S_4	S	$S_4 \times I$
	I	$S_4 \times I$
A_5	S	$A_5 \times I$
	I	$A_5 \times I$

Analogue, the reflection W in β is a symmetry operation for the dipolygonid in the lower half, whose symmetry group is then:

$$\text{gen}\{A, B, W\}.$$

To determine these groups of isometries when G is finite, one has to observe if these generated groups contain S and/or I, or none of both.

In the upper half:

$$S \in \text{gen}\{A, B, V\} \Leftrightarrow SV \in \text{gen}\{A, B, V\},$$

SV is the two-fold rotation along v, hence, $SV \in G$. From Table 3 is found if this two-fold rotation is in G. If so, S is a symmetry operation for the dipolygonid, which means it is self-enantiomorphous, and composed of coplanar pairs of polygons $\{m\}$, each along one side at least, of an axis of order s.

The dipolygonid's symmetry group in this half is $\text{gen}\{A, B, S\}$. Analogue, this property occurs in the lower half if, and only if $SW \in G$. Then,

$$SW \in G \Leftrightarrow SVW = I \in \text{gen}\{A, B, S\}.$$

9.1. Conclusions

9.1.1. $\{SV, SW\} \in G$. The symmetry group of the dipolygonid contains S and I. Therefore, it has central inverse symmetry all over the uniform motion, where it is composed of coplanar pairs of polygons $\{m\}$, distributed along both sides of all the s-fold axes of G.

The uniform motion can be analysed when two coplanar pairs forming chiral dipolygons in the upper half are denoted by:

$$\{m\}_1 \text{ and } \{n\}_2/\{m\}_2 \text{ and } \{n\}_1$$

and on the opposite side:

$$\{m\}_1' \text{ and } \{n\}_2'/\{m\}_2' \text{ and } \{n\}_1'.$$

When the dipolygonid passes the central position, the polygons $\{m\}$ and $\{m\}'$ reverse their sense of rotation, and keep their sense of translation, while the polygons $\{n\}$ and $\{n\}'$ keep their sense of rotation, but reverse their sense of translation.

In a central position, the following sets of polygons are coplanar:

$$\{\{m\}_1, \{m\}_1', \{m\}_2, \{m\}_2'\}$$
$$\{\{n\}_1, \{n\}_1', \{n\}_2, \{n\}_2'\}.$$

Also, continuing over the lower half, the new coplanar pairs become:

$$\{m\}_1 \text{ and } \{n\}_2'/\{m\}_2' \text{ and } \{n\}_1$$

and opposite:

$$\{m\}_1' \text{ and } \{n\}_2/\{m\}_2 \text{ and } \{n\}_1',$$

which actually means the couples "have changed partners".

9.1.2. $SV \in G, SW \notin G$. The dipolygonid's symmetry group in the upper half is $\text{gen}\{A, B, S\}$, which does not contain I, hence, it is a mixed group. The dipolygonid is composed of coplanar pairs of polygons $\{m\}$, distributed along one of both sides of the s-fold axes of G.

The symmetry group in the lower half, $\text{gen}\{A, B, W\}$, may or may not contain I. In any case, the dipolygonid is composed of single polygons $\{m\}$, distributed along both sides of the s-fold axes of G. If I is not within the greater symmetry group, obviously the dipolygonid has no central inverse symmetry. If I is, the dipolygonid's symmetry group is $\text{gen}\{A, B, I\}$.

9.1.3. $SV \notin G, SW \in G$. This is the analogue situation of Section 9.1.2., when upper and lower halves are interchanged.

9.1.4. $SV \notin G, SW \notin G$. The dipolygonid's symmetry groups in upper and lower halves do not contain S and hence, it is composed of single polygons.

When a regular dipolygonid has central inverse symmetry in one of the halves of the path of uniform motion, two opposite polygons rotate in the same sense. If not, provided there are opposite polygons, they rotate in opposite senses.

The uniform motion of the regular dipolygonids in the finite groups of rotations will now be separately analysed. The position of R_A and R_B in ω, together with the axes v and w, provided they belong to G, are shown in Figs 17, 19, 21, 22. These figures are obtained from Tables 3 and 7, and indicate one of the four cases higher described.

9.2. Dihedral

The regular dipolygonids are of the type $n\{2\} + n\{2\} | k/n \cdot 180°$ in D_n, where $n \geqslant 2$, and $k = 1$ (when $n = 2$), or $1 \leqslant k < n/2$ where k and n are coprime. *These are the Petrie-polygons (non-planar zigzag lines).* The bisector line of R_A and R_B in the upper half is v, where the angle between v and R_A is given by:

$$\frac{k}{2n} \cdot 180°$$

According to Table 3, v represents a half-turn of D_n only if $k/2$ is a natural number, specifying to the conditions given there. This will be, when k is even.

The bisector line w in the lower half forms, together with R_A, an angle

$$\frac{n-k}{2n} \cdot 180°$$

and will represent the axis of a two-fold rotation of D_n only when $n - k$ is even.

9.2.1. n is odd, k is odd. Then, $n - k$ is even, referring to case 9.1.3.

Lower half—the symmetry group is gen$\{A, B, S\} = D_{2n}D_n$ (Table 8). There are n coplanar pairs of digons, each along one side of the n two-fold axes of D_n [Diagram 2(a)].

Upper half—the extended symmetry group not containing S is $D_n \times I$. There are n pairs of centrally inverse digons along each side of the n two-fold axes [Diagram 2(b)].

9.2.2. n is odd, k is even. This refers to case 9.1.2.

The previous situation is found, provided upper and lower halves are interchanged.

9.2.3. n is even. Since k and n are coprime, both k and $n - k$ are odd, which refers to case 9.1.4.

Since the symmetry group must not contain S, and $D_n \times I$ contains both S and I (Table 8), it must be $D_{2n}D_n$ in both halves.

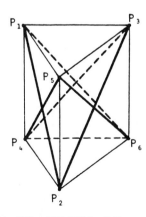

Diagram 2(a). $3\{2\} + 3\{2\} | 60°$ in $D_6 D_3$: a position in the lower half, described within a triangular prism.

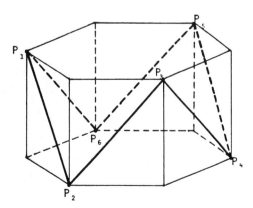

Diagram 2(b). $3\{2\} + 3\{2\} | 60°$ in $D_3 \times I$: a position in the upper half, described within a hexagonal prism.

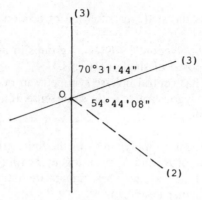

Fig. 17

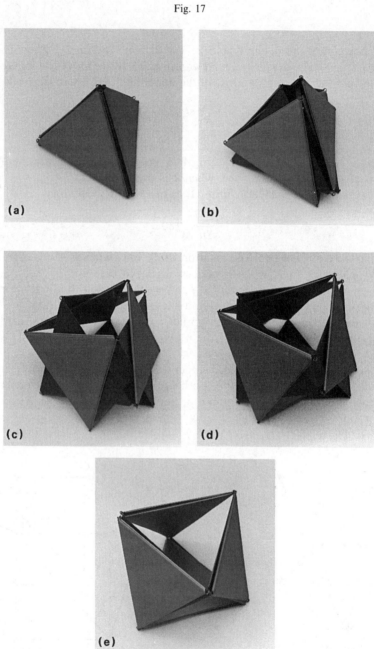

Fig. 18. The convex part of the motion of $4\{3\} + 4\{3\}|70°31'44''$ in S_4A_4 (the lower half) illustrated by a vinyl model in five steps. The minimum position is in (e).

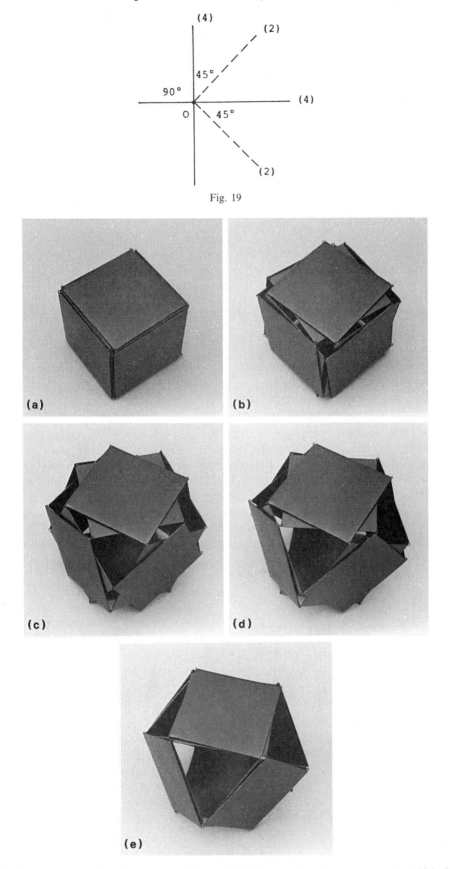

Fig. 19

Fig. 20. The convex part of the motion of $6\{4\} + 6\{4\}\,|\,90°$ in $S_4 \times I$, in either lower or upper half, in five steps. The extreme position, illustrated by this cardboard model is in (e).

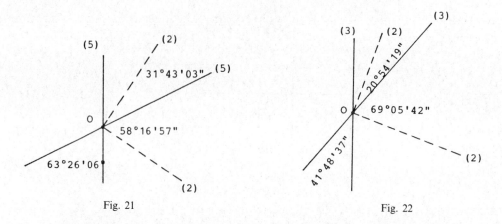

Fig. 21   Fig. 22

n pairs of opposite digons appear along both side of the n two-fold axes, each pair being invariant over the half-turn within the cyclic subgroup of order n in D_n.

One peculiar appearance of the regular dihedral dipolygonids occurs in the central positions: since all centers of the $2n$ digons coincide with O, all the digons collapse along the n-fold axis of D_n, as this axis is the intersection of all the planes, perpendicular to the two-fold axes.

9.3. Tetrahedral

The regular dipolygonid in A_4 is of the type

$$4\{3\} + 4\{3\}\,|\,70°31'44''.$$

The position of the axes refers to case 9.1.3 (Fig. 17).

Lower half—the symmetry group gen$\{A, B, S\} = S_4A_4$ (Table 8). There are four coplanar pairs of triangles, each lying along one side of the four three-fold axes of A_4. (Fig. 18).

Upper half—the extended group not containing S is $A_4 \times I$, which is the dipolygonid's symmetry group. There are four pairs of centrally inverse triangles, each along both sides of the four three-fold axes. The convex part of the uniform motion is Fuller's Jitterbug [Fig. 1(a)].

In the central positions, the four coplanar pairs of triangles have their centers coinciding with O. Such a pair of centrally inverse triangles has its vertices sharing with a regular hexagon. One can picture such a position within the four hexagons found within the cuboctahedron.

The symmetry group in a central position is the direct product of the groups S_4A_4 and $A_4 \times I$, which is $S_4 \times I$, the only group found in Table 5 containing both groups.

9.4. Octahedral

The regular dipolygonid in S_4 is of the type

$$6\{4\} + 6\{4\}\,|\,90°.$$

The position of the axes refers to case 9.1.1 (Fig. 19).

The symmetry group of the dipolygonid containing both S an I (Table 8) is $S_4 \times I$.

In each position there are six coplanar pairs of squares distributed along both sides of the three four-fold axes. The set of squares can also be considered as containing six centrally inverse (hence: parallel) pairs. In a central position, $PO \perp \omega$ and opposite pairs collapse all four with one square. Like explained under case 1, here the "change partners" effect takes place.

The central position can easily be pictured in the location of the three mutually perpendicular squares within the octahedron.

Moreover, since $\theta = 90°$, the positions in upper and lower halves are equivalent (Fig. 20).

9.5. Icosahedral

There are three types of regular dipolygonids in A_5.

9.5.1. $12\{5\} + 12\{5\}\,|\,63°26'06''$. The position of the axes refers to case 9.1.1 (Fig. 21). The symmetry group (Table 8) is $A_5 \times I$.

In each position there are 12 coplanar pairs of pentagons distributed along both sides of the six five-fold axes of A_5.

Since the half-turn, whose axis is perpendicular to ω, is in A_5, opposite pairs of pentagons are invariant over this half-turn, as well as over the central inversion. Hence, they are parallel pairs.

In central positions, opposite parallel and centrally inverse pairs coincide with the two pentagons, sharing the 10 vertices of a regular decagon. Here, the "change partners" effect takes place.

A central position can easily be pictured in the location of the six central decagons within an icosidodecahedron.

9.5.2. $12\{5/2\} + 12\{5/2\}|63°26'06''$. The dipolygonid shares its axes with the previous one. Hence, all the results hold when the pentagons are replaced by pentagrams.

9.5.3. $20\{3\} + 20\{3\}|41°48'37''$. The "Vampire".† The position of the axes refers to case 9.1.1 (Fig. 22). The symmetry group is equally $A_5 \times I$.

In each position there are 20 coplanar pairs of triangles distributed along both sides of the 10 three-fold axes of A_5. For the same reason as in (1) and (2), opposite pairs of triangles are parallel, and centrally inverse. In central positions opposite pairs coincide with two triangles sharing the vertices of a regular hexagon. Also here, the "change partners" effect takes place, when passing that position.

The situation of the triangles can be less easily, but nevertheless correctly, pictured in the location of the 10 central hexagons within the small dodecahemicosahedron [8].

10. EQUIRADIAL DIPOLYGONIDS

Besides the regular dipolygonids of the previous paragraph, for which $R_A = R_B$, there is one more finite equiradial dipolygonid having a particular appearance:

$$12\{5\} + 12\left\{\frac{5}{2}\right\}|63°26'06''.$$

The general dipolygonid of this type, where $R_A \neq R_B$ is composed of opposite pairs of pentagons and pentagrams sharing one five-fold axis. The pentagons and pentagrams are distributed at different distances of O.

However, when $R_A = R_B$, the distances are equal: since the half-turns SV and SW are elements of A_5, the dipolygonid is composed of 12 coplanar pairs of pentagons–pentagrams. The uniform motion is analogue with the first regular dipolygonid in A_5 (Section 9.5.1), when one pentagon in a pair is replaced by a pentagram.

11. PAIRS OF CHIRAL DIPOLYGONIDS

Pairs of chiral dipolygonids have an extended symmetry group, namely $\text{gen}\{A, B, S\}$. In Section 9 was established where the regular dipolygonids are self-enantiomorphous. The regular dipolygonid of A_4 is so in the lower half, but not in the upper half. A pair of chiral such dipolygonids in the upper half is composed of eight pairs of coplanar triangles distributed along the opposite sides of each of the four three-fold axes, and its symmetry group is the direct product of $A_4 \times I$ and $\{S, E\}$, which, according to Table 5 is $S_4 \times I$. Each chiral pair of dipolygonids is composed of coplanar polygons $\{m\} - \{m\}'$ and $\{n\} - \{n\}'$.

---

†I built the first model of this dipolygonid in 1979, when I was visiting Magnus J. Wenninger in the Benedictine Monastery in the Bahamas, where I was taught techniques of model making by him. When Wenninger saw the model, he spontaneously baptized it "Verheyen's Vampire" following the tradition among polyhedronists to give horror-names to their weirdest creations. (cf. "Miller's Monster", a complicated non-convex snub polyhedron [3], and "Skilling's Spectre", the Great Disnub Dirhombidodecahedron [10]).

Fig. 23. The convex part of the motion of $20\{3\} + 20\{3\}|41°48'37''$ in $A_5 \times I$ (upper half), illustrated by a metal model in six steps, where (g) had to show the 7th step, namely the maximum position. However, since the proper type of connector (Fig. 24) is replaced here by an ordinary ring, the rigid model looses its rigidity nearing the maximum position, and becomes completely floppy (g).

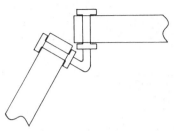

Fig. 24. This type of connector assures the dipolygonal angle θ.

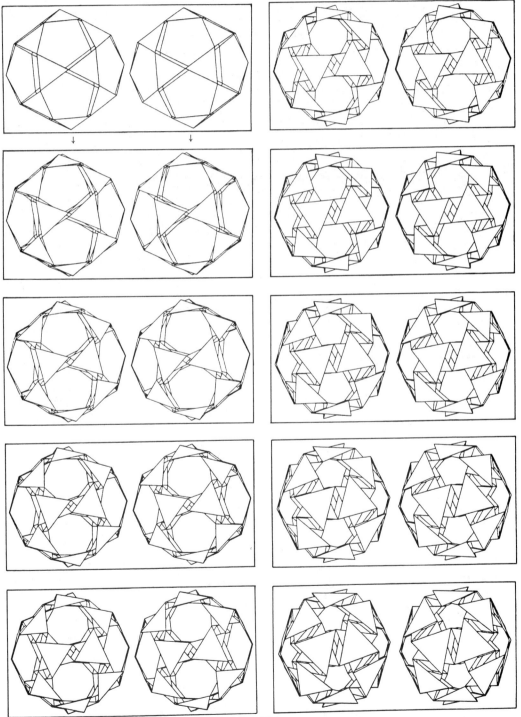

Fig. 25.1.

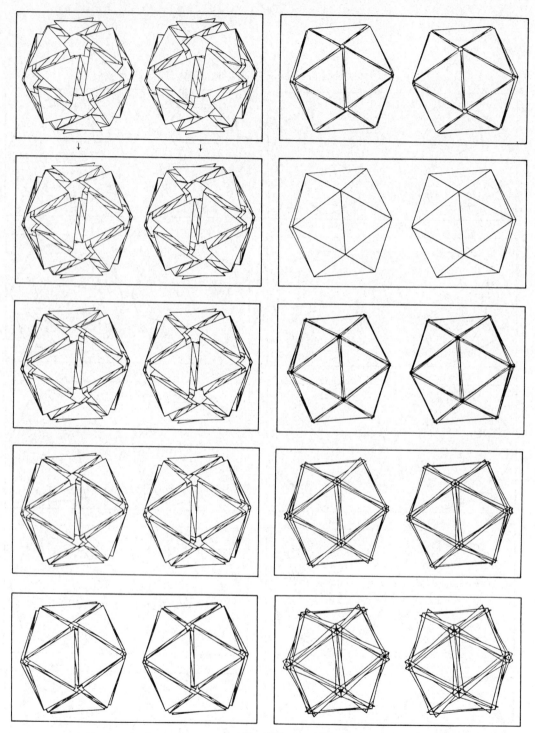

Fig. 25.2.

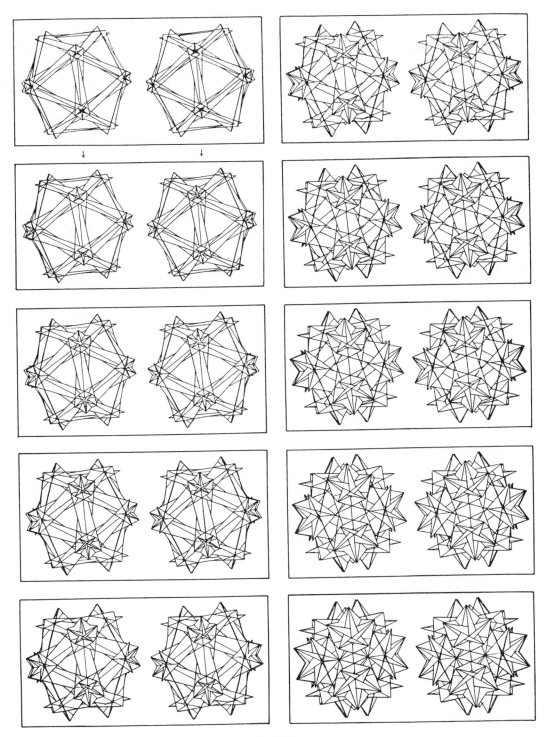

Fig. 25.3.

Fig. 25.4.

Fig. 25.5.

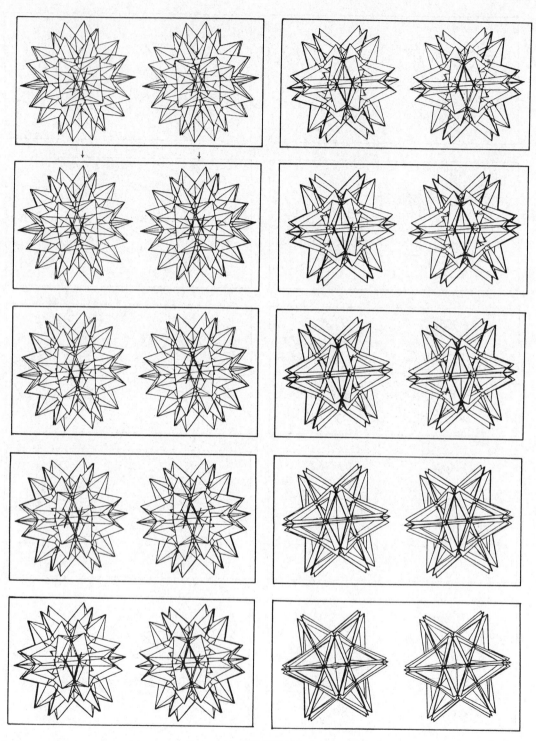

Fig. 25.6.

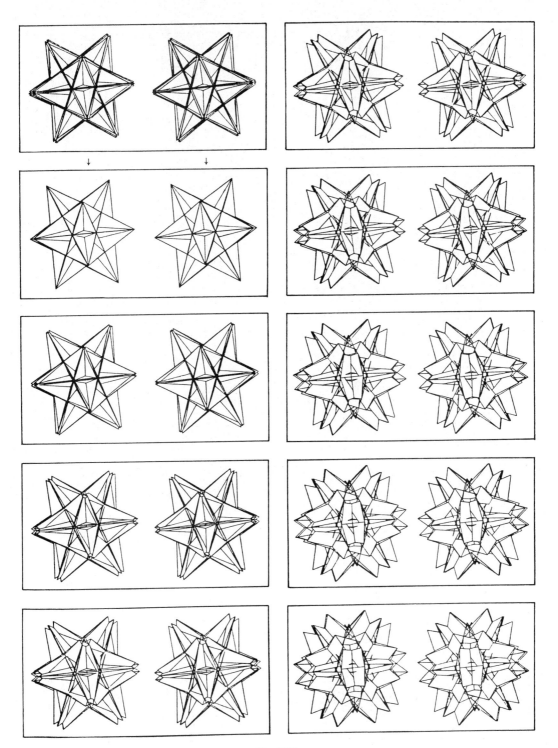

Fig. 25.7.

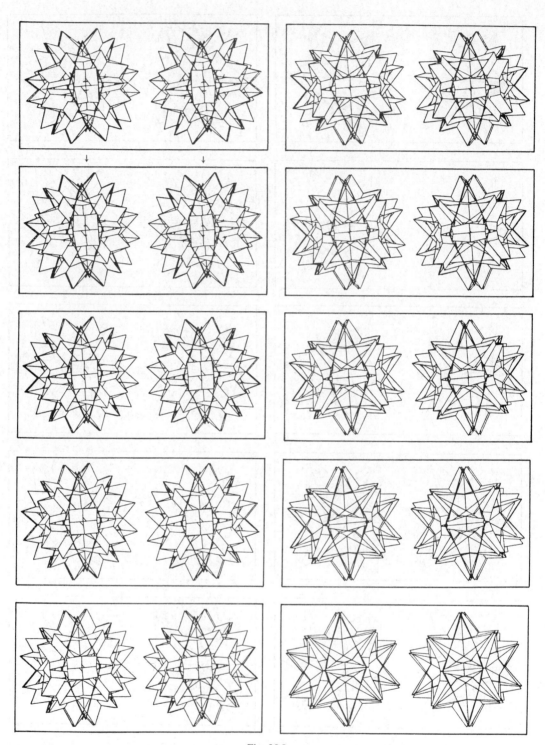

Fig. 25.8.

Fig. 25.9.

Fig. 25. Set of stereographic, computer-generated drawings, illustrating the uniform motion of the Vampire from maximum to minimum position. ©: central position.

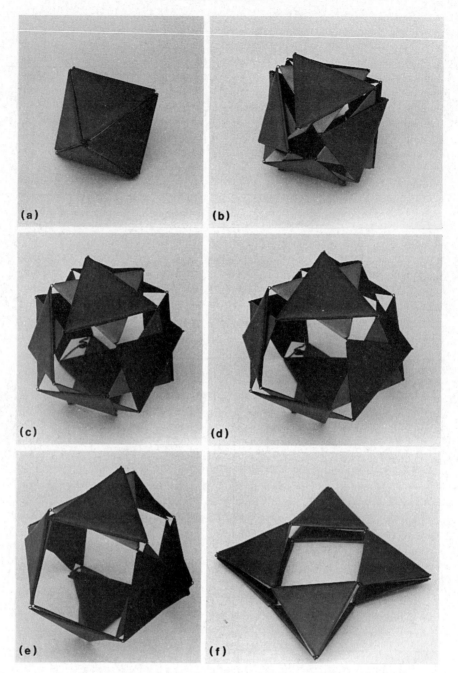

Fig. 26. Motion of a set of two chiral dipolygonids $4\{3\} + 4\{3\} | 70°31'44''$ ("Jitterbugs") in $S_4 \times I$, restricted to the convex part, as illustrated here by a vinyl model. As in Fig. 23, the maximum position in (f) is entirely floppy.

As an example, the pair of chiral $8\{3\} + 6\{4\} | 54°44'08''$ is $S_4 \times I$ is composed of eight coplanar pairs of triangles and six coplanar pairs of squares. When $R_A = R_B$, clearly there is no "change partners" effect in a central position.

12. THE USE OF DIPOLYGONIDS

The main interest of the dipolygonids is found in an easy to understand visual approach to construct the complete set of uniform polyhedra, in total 75 [3] + 1 [10]. The details are not within

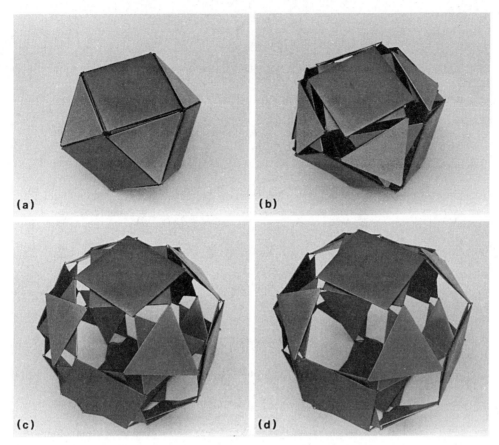

Fig. 27. Model of the two chiral dipolygonids $8\{3\} + 6\{4\}\,|\,54°44'08''$ of equal edge length in cardboard.
It illustrates the convex part of the uniform motion in the upper half, until it loses the rigidity in the
maximum position.

the realm of this paper, however, it can be stated that the convex and non-convex snub polyhedra
come out in a natural way during the transformation from one position into another, caused by
the uniform motion of triads of dipolygonids (Fig. 28).

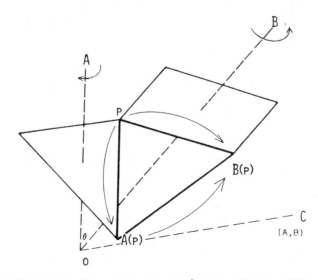

Fig. 28. Triad of dipolygonids. The rotation $C = BA^{-1}$ maps $A(P)$ into $B(P)$ and produces a
polygon $\{k\}$.

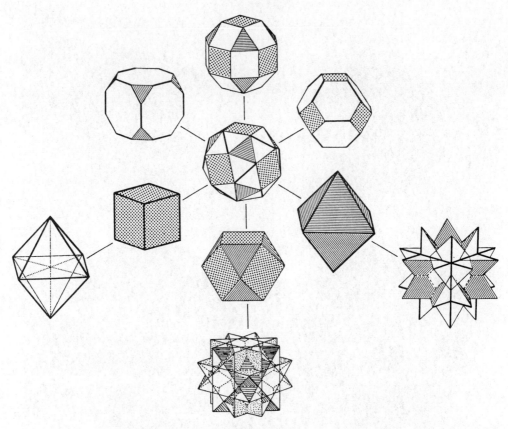

Fig. 29. In the snub cube, three dipolygonids of a basic triad meet.

There can be found 14 basic triads of dipolygonids in A_4, S_4 and A_5 [6]. In a snub polyhedron, three dipolygonids of a triad meet.

Other applications of dipolygonids can be found in the construction of transformable space frames that are basically space filling lightweight frames, like the impandable edge cube (ICU) and the impandable rhombic edge dodecahedron (IRODO) [11] (Figs 30 and 31).

An example of a spherical variation of an impandable rhombic edge triacontahedron is found at the Space Research Center in Sydney, Australia, where it stands as a reminder of the conference cited in Ref. [12] (Fig. 33).

Dipolygonid models can also be used as space fillers which transform into other space fillers in each position.

Such examples are:

(a) The regular $6\{4\} + 6\{4\}|90°$ model in $S_4 \times I$, as illustrated in Fig. 20. Twenty-seven of these are used to construct the space filling shown in Fig. 33.

(b) The regular $4\{3\} + 4\{3\}|70°31'44''$ of A_4, used in:

(1) the pair of chiral dipolygonids in the upper half, like the model in Fig. 26;
(2) the dipolygonid in the lower half, like the model in Fig. 18.

When the model (1) shares a pair of triangles of a model (2), the expandable pyramid can be constructed as shown in Fig. 34.

Figure 35 is a variation of the ICU (Fig. 30) when triangular pyramids are replaced by sphere packings. And finally, $6\{4\} + 6\{4\}|90°$ in $S_4 \times I$ stood model for a piece of furniture, a salon table that is able to transform into a glass and bottle closet, by a push on a button [13], (see Fig. 36).

Fig. 30. Model of ICU with wide middle part in wood.

Fig. 31. Impansion of IRODO in four steps. This wooden structure is extremely rigid in all positions.

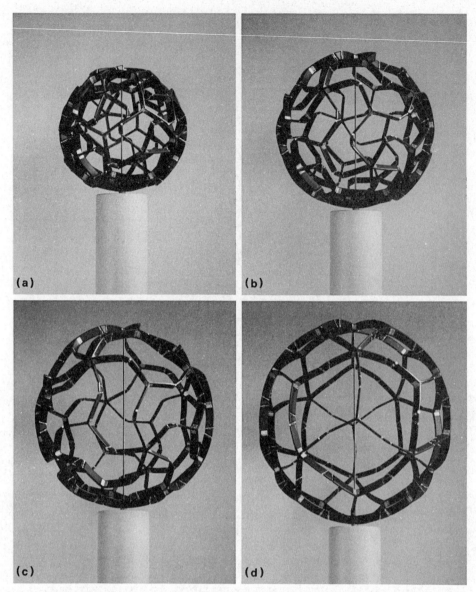

Fig. 32. SPHEROTRAQ in Sydney (University of NSW): motorized model in wood and metal, illustrating a pulsating spherical construction of 120 hinges, rigid in each position.

Fig. 33

Fig. 34. Expansion of the pyramid in Figs (a) and (b). (c): rotated view of (b).

Fig. 35

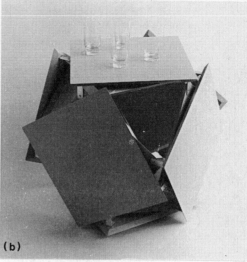

Fig. 36

Acknowledgements—The author wishes to express his gratitude to Dr John Skilling, Department of Applied Mathematics and Theoretical Physics at the University of Cambridge, who developed a special computer program for the uniform motion of dipolygonids to provide the graphics for Figs 11, 16 and 25.

REFERENCES

1. J. D. Clinton, Advanced structural geometry studies, Part 2. A geometric transformation concept for expanding rigid structures. Southern Ill. Univ., NASA Report CR-1735, Washington D.C. (1971).
2. R. D. Stuart, *Polyhedral and Mosaic Transformations*. Student Publications of the School of Design, Univ. of North Carolina, Vol 12, No. 1 (1963).
3. H. S. M. Coxeter, M. S. Longuet-Higgins and J. C. P. Miller, Uniform polyhedra. *Phil. Trans. R. Soc.* **246A**, 401–450 (1954).
4. B. Grünbaum, Regular polyhedra—old and new. *Equationes Math.* **16**, 1–20 (1977).
5. H. S. M. Coxeter, *Introduction to Geometry*. Wiley, New York (1961).
6. H. F. Verheyen, *Dipolygonids*. UIA, Univ. of Antwerp, Report 79–40 (1979).
7. H. F. Verheyen, Expandable structures based on dipolygonids. *Proc. 3rd Int. Conf. Space Structures*. Elsevier, London (1984).
8. M. J. Wenninger, *Polyhedron Models*. Cambridge Univ. Press, New York (1971).
9. F. Klein, *The Icosahedron and the Solution of Equations of the Fifth Degree* (2nd edn). Dover, New York.
10. J. Skilling, The complete set of uniform polyhedra. *Phil. Trans. R. Soc.* **278A**, 111–135 (1975).
11. H. F. Verheyen, Space frames of variable geometry. *Space Structs* **1**, 199–208 (1985).
12. H. F. Verheyen, A pulsating globe. *Proc. 1st Int. Conf. Lightweight Structures in Architecture*. Vol. 1, pp. 476–478. Unisearch, Sydney (1986).
13. H. F. Verheyen, Applications de Géométrie Variable. *Courr. Bois* **78**, 41–44, Brussels (1987).

Computers Math. Applic. Vol. 17, No. 1–3, pp. 251–254, 1989
Printed in Great Britain. All rights reserved

0097-4943/89 $3.00 + 0.00

A GEOMETRICAL ANALOGUE OF THE PHASE TRANSFORMATION OF CRYSTALS

G. Fejes Tóth and L. Fejes Tóth

Mathematical Institute, Hungarian Academy of Sciences, Budapest, P.O. Box 127, H-1364, Hungary

Abstract—A geometrical problem is discussed whose solution shows a close analogy to the transformation of one allotropic crystalline form to another.

From the viewpoint of classical crystallography the intrinsic structure of crystals may be considered as one of the most perfect physical embodiments of the concept of symmetry. However, since the discovery of the 230 crystallographic spacegroups, interest gradually shifted from static, ideal crystals to the changes and to various kinds of imperfections in crystals which often show a peculiar interplay of different symmetry elements.

Grünbaum and Shephard [1, p. 52] pointed to "the mathematical analogue of the well-known physical effect of introducing a foreign atom into a crystal". In what follows we shall discuss a geometrical analogue of another phenomenon: the transformation of a crystalline form to another.

The process of transformation of ice into water belongs to highschool curriculum. It is less generally known that there are also phase transformations from solid to solid state. As an example we consider one of the most common metals: iron.

Although in its appearance common iron seems to be amorphous, it consists of a conglomerate of microscopic crystals. Bigger iron crystals are rare. They come into being only under special conditions. Below a certain temperature t_0, which is approx. 906°C, the iron atoms are arranged in a body-centred cubic lattice [2]. What happens if we start slowly heating a piece of iron of temperature $t < t_0$? Reaching the temperature t_0 the temperature stops to rise, but parts of the atoms start to rearrange into a face-centred cubic lattice. Along with the energy transmitted in form of heat the number of the atoms in the new phase increases until all the atoms have rearranged, and the temperature continues to rise.

The lattice structure of crystals is supposed to originate from the tendency of minimizing the energy due to the forces which interact between the atoms. However, we are immensely far from being able to solve such a minimum problem mathematically. Therefore, it seems to be of some importance to study simpler extremum problems concerning arrangements of a great number of points or sets of points which lead to lattices. Particular interest is due to the problem which we are going to discuss because of the close analogy between the solution and the state of transformation from one phase of a crystal into another: (i) the solution is composed of two lattices and (ii) depending on a parameter the proportion of the number of points contained in the two lattices changes, but the structure of the lattices remains unchanged.

We stress that at this point the analogy breaks: the geometric problem has no sensible physical interpretation.

We start our geometric consideration with the problem of the densest packing of equal circles. Roughly speaking, the problem is to accommodate in a "big" square, or in any other big "container" as many non-overlapping unit circles as possible. Since the exact solution depends on the shape and size of the container in an unmanageable way, we are interested only in the asympototic behaviour of the best arrangement in the limiting case as the container goes over into the whole plane. The notion of the *density* of an infinite family of discs scattered in some way in the plane enables us to phrase the problem precisely. The density is defined by a limiting value which may be interpreted as the quotient of the total area of the discs and the area of the whole plane. (For the exact definition see Refs [3–5].) If no two discs overlap then the discs are said to form a *packing*.

Now the problem is to find a packing of equal circular discs for which the density attains its maximum. A solution to this problem is given by the packing in which each circle is touched

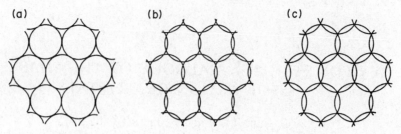

Fig. 1. Arrangements of circles with prescribed density covering the greatest possible part of the plane.

by six others [3, 6] [Fig. 1(a)]. The density of this packing is equal to the area of a circle divided by the area of the circumscribed regular hexagon, i.e. $\pi/\sqrt{12} \approx 0.907$. We shall call this packing "the" solution to our problem, although there are infinitely many circle-packings with the same density containing irregularities analogous to those which occur in real crystals.

A dual counterpart of the problem of the densest packing of circles is the problem of the thinnest covering of the plane with equal circles. A set of discs is said to form a covering if each point of the plane belongs to the interior or to the boundary of at least one disc. The problem is to find among all coverings with equal circles that one of minimal density. The covering shown by Fig. 1(c) in which each circle is intersected by exactly six others at the vertices of a regular hexagon yields "the" solution to this problem [7]. The density of this covering equals the area of a circle divided by the area of the inscribed regular hexagon, i.e. $2\pi/\sqrt{27} \approx 1.209$.

The problem of the densest circle-packing and the problem of the thinnest circle-covering can be united in one problem: distribute equal circles in the plane with a given density d so as to cover the greatest possible part of the plane. In order to formulate the problem exactly we first define the density of a point-set S similarly to the density of a set of discs. Intuitively speaking the density of S is the quotient of the area of S and the area of the plane. Now we define the *covering measure* δ of a set of discs as the density of the point-set union of the discs. The problem is to find the maximum $\Delta(d)$ of δ for all sets of equal circles of density d.

For $d \leqslant \pi/\sqrt{12}$ we can arrange the circles without overlapping, and for $d \geqslant 2\pi/\sqrt{27}$ so that they cover the plane completely. Therefore we have $\Delta(d) = d$ for $d \leqslant \pi/\sqrt{12}$, and $\Delta(d) = 1$, for $d \geqslant 2\pi/\sqrt{27}$. For $\pi/\sqrt{12} < d < 2\pi/\sqrt{27}$ the extremal arrangement arises by blowing up the circles in the densest packing concentrically [3] [Fig. 1(b)]. The graph of $\Delta(d)$ is exhibited in Fig. 2.

Now we consider the generalization of the last problem which arises by letting the part of the circles be played by translates of a centro-symmetric convex disc c.

According to a remarkable theorem of Dowker [8], among the hexagons of least area containing c, there is one, say H, which is centro-symmetric. Similarly, among the hexagons of maximal area contained in c there is one, say h, which is centro-symmetric. On the other hand, it is known [3] that the densest packing of translates of c arises by tiling the plane with translates of H and

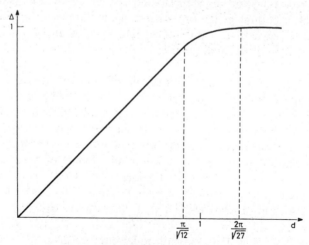

Fig. 2. Graph of the function $\Delta(d)$.

inscribing into each hexagon a translate of c. Similarly, the thinnest covering with translates of c arises by tiling the plane with translates of h and circumscribing about them translates of c. Therefore both in the densest packing and in the thinnest covering by translates of c the discs form a lattice. At first glance we would expect the same for the arrangements of translates of c with maximal covering measure for intermediate densities. A closer investigation [9] has revealed that this is not always so.

We keep the notation $\Delta(d)$ used in the case of circles for the maximum of the covering measure extended over all sets of translates of c of density d. We shall give an explicit formula for $\Delta(d)$ in terms of a function associated with c.

Let c be of unit area. Let $\chi = \chi(x)$ be a centro-symmetric hexagon of area x which covers the greatest possible part of c. Let $a(x)$ be the area of the intersection $\chi \cup c$.

If c is a circle then χ is a regular hexagon concentric with c, and the function $a(x)$ is concave. However, there are centrally symmetrical convex discs for which $a(x)$ is not concave [10]. Let $\bar{a}(x)$ be the least concave function not less than $a(x)$. The solution to our problem is substantially given by the theorem which claims that

$$\Delta(d) = d\bar{a}(1/d). \tag{1}$$

This theorem states, on the one hand, that the covering measure of any set of translates of c of density d is at most $d\bar{a}(1/d)$, on the other hand, that there is a set of translates of c with density d and covering measure $d\bar{a}(1/d)$. The first statement follows from Theorem 1 in Ref. [9] and from the fact that for those values x for which $a(x) = \bar{a}(x)$ there is a centro-symmetric hexagon χ of area x concentric with c such that $\chi \cup c$ has area $a(x)$. This last proposition can be proved similarly as Lemma 1 in Ref. [9].

In order to construct a set of translates of c of density d and covering measure $\Delta(d)$ observe that tiling with translates of χ generates a lattice Λ of translates of c with density $d = 1/x$ and covering measure $\delta = a(x)/x = da(1/d)$.

Let us first consider the case when $\bar{a}(1/d) = a(1/d)$. Now we have, according to equation (1), $\delta = \Delta(d)$, which means that the lattice Λ represents the extremal arrangement.

Let us now suppose that $a(1/d) < \bar{a}(1/d)$. Then there is an interval (x_1, x_2) containing $1/d$ in which $\bar{a}(x)$ is linear and $a(x_i) = \bar{a}(x_i)$, $i = 1, 2$. Now we can construct two lattices Λ_i with densities $d_i = 1/x_i$ and covering measures $\delta_i = d_i a(1/d_i)$. The optimal arrangement will be obtained by a suitable combination of Λ_1 and Λ_2.

Let μ_1 and μ_2 be positive numbers such that $\mu_1 + \mu_2 = 1$ and $1/d = \mu_1/d_1 + \mu_2/d_2$. Then $\bar{a}(1/d) = \mu_1 a(1/d_1) + \mu_2 a(1/d_2)$. We divide the plane into two parts E_1 and E_2 satisfying the following conditions: (i) The density of E_i is equal to $d\mu_i/d_i$. (ii) The density of the parallel set of the common boundary of E_1 and E_2 at distance 1 is equal to zero. [The parallel set of a pointset S at distance r is the union of all points whose distance from a point of S is at most r. It is not difficult to show that if the density $d(S, r)$ of a parallel set of a set S at some distance r is zero then so is $d(S, x)$ for any positive x.]

The above conditions are obviously fulfilled if, e.g. E_1 and E_2 are complementary angular regions of angles $2\pi d\mu_1/d_1$ and $2\pi d\mu_2/d_2$.

Condition (ii) implies that the density of the set of those discs of Λ_i which intersect the common boundary of E_1 and E_2 is equal to zero. It follows that the set S_i of those discs of Λ_i which lie in E_i has density $d_i \cdot d\mu_i/d_i$ and covering measure $d_i a(1/d_i) \cdot d\mu_i/d_i$. Therefore, the set $S = S_1 \cup S_2$ which consists of the discs of S_1 and the discs of S_2 has density $d_1 \cdot d\mu_1/d_1 + d_2 \cdot d\mu_2/d_2 = d$ and covering measure $d_1 a(1/d_1) \cdot d\mu_1/d_1 + d_2 a(1/d_2) \cdot d\mu_2/d_2 = d\bar{a}(1/d)$. Referring again to equation (1) we see that the maximal covering measure is attained by S.

We recapitulate the case when the centro-symmetric convex disc c is such that the function $a(x)$ is not concave. Then for a value d such that $a(1/d) < \bar{a}(1/d)$ the arrangement of translates of c of density d with maximal covering measure can be described as follows. We superimpose two different lattices Λ_1 and Λ_2 of translates of c, divide the plane under certain conditions into two parts E_1 and E_2, and take the union of those discs of Λ_1 which lie in E_1 and those discs of Λ_2 which lie in E_2.

Along with d the frontier between E_1 and E_2 varies. Condition (ii) allows the frontier to imitate various kinds of "interfaces" between two crystalline phases and their motions.

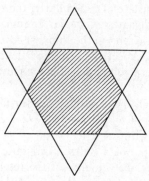

Fig. 3. Triangles crossing each other.

Finally we note that the theorem expressed by equation (1) holds under much more general conditions. We say that two convex discs *cross* if removing their intersection causes both discs to fall into disjoint components (Fig. 3). Instead of translates of c we can consider congruent copies of c no two of which cross each other. Then the parallel orientation of the discs automatically ensues by the requirement of maximizing the covering measure [9]. (Probably the condition that the discs are not allowed to cross is also superfluous, but it plays a central part in the proof, and so far we didn't succeed to get rid of it.)

Under the above more general condition, the discs in the extremal arrangement needn't be all translates of each other. In each connected component of E_1 and E_2 the discs belong to a lattice congruent to, but not necessarily identical with Λ_1 and Λ_2, respectively. This makes the analogy between our problem and the phase transformation of crystals even closer.

Acknowledgement—The authors are grateful to Dr T. Ungár for several helpful discussions.

REFERENCES

1. B. Grünbaum and G. C. Shephard, *Tilings and Patterns.* Freeman, New York (1987).
2. E. Hornbogen, Physical metallurgy of steels. In *Physical Metallurgy* (Eds R. W. Cahn and P. Haaren), 3rd edn, Chap. 3. North-Holland, Amsterdam (1983).
3. L. Fejes Tóth, *Lagerungen in der Ebene, auf der Kugel und im Raum* (2nd edn). Springer, New York (1972).
4. H. Groemer, Existenzsätze für Lagerungen im Euklidischen Raum. *Math. Z.* **81,** 260–278 (1963).
5. C. A. Rogers, *Packing and Covering.* Cambridge Univ. Press, Cambridge (1964).
6. A. Thue, Über die dichteste Zusammenstellung von kongruenten Kreisen in der Ebene. *Christiania Vid. Selsk. Skr.* **1,** 3–9 (1910).
7. R. Kershner, The number of circles covering a set. *Am. J. Math.* **61,** 665–671 (1939).
8. C. H. Dowker, On minimum circumscribed polygons. *Bull. Am. math. Soc.* **50,** 120–122 (1944).
9. G. Fejes Tóth, Covering the plane by convex discs. *Acta math. hung.* **23,** 263–274 (1972).
10. G. Fejes Tóth, On the intersectin of a convex disc and a polygon. *Acta math. hung.* **29,** 149–153 (1977).

Computers Math. Applic. Vol. 17, No. 1–3, pp. 255–278, 1989
Printed in Great Britain. All rights reserved

0097-4943/89 $3.00 + 0.00

COMPUTERS AND GROUP-THEORETICAL METHODS FOR STUDYING STRUCTURAL PHASE TRANSITIONS

G. M. Chechin

Department of Physics, Rostov State University, 344006 Rostov-on-Don, U.S.S.R.

Abstract—The present work is a review of a complex of programs developed at Rostov University for a group-theoretical analysis of phase transitions in crystals. A comparative analysis of various approaches and results from different authors is also presented.

1. INTRODUCTION

The group-theoretical methods are widely used in the investigation of phase transitions in crystals ever since the classical works of Landau and Lifshitz [1–3]. During the last three decades considerable progress has been made in the phenomenological theory of phase transitions. This progress has been facilitated by the fact that the group-theoretical analysis was singled out as an independent stage preceding the thermodynamic study itself. The procedure of minimization of the nonequilibrium thermodynamic potential ϕ employed in the Landau theory proves, in the case of the multicomponent order parameters, in fact, unfeasible without prior determination of the general form of its extreme points. On the other hand, the group-theoretical methods allow one to find the explicit form of the order parameters corresponding extremums of ϕ [4–6], which drastically simplifies the subsequent thermodynamic investigation of the lower-symmetry phases associated with these parameters. The group-theoretical analysis is, of course, a cumbersome enough procedure requiring, in the case of multicomponent order parameters, the use of a computer. In the last 15 years, a number of algorithms as well as computer programs corresponding to these have been devised which have introduced automatism into the group-theoretical studies of phase transitions in crystals. In 1973, such work was started at Rostov University under the guidance of the present author. A program complex worked out by our group is actually based on the stationary vectors method originating from the work by Gufan and Sakhnenko [4, 7, 8] (a similar approach was developed also in Refs [9, 10]). The results of the calculations were published in Refs [11–19]. Another program complex for group-theoretical analysis of phase transitions in crystal was developed by Hatch and Stokes [20, 21], their programs, as far as we know, are essentially based on the Birman criterion [22].

The present work is a review aiming to acquaint the reader with the aforementioned complex of programs worked out at Rostov University. We should like to note that undergraduates of our Department of Physics have taken an active part in its development. We also present herein a comparative analysis of informativeness of the approaches based, respectively, on the notion of stationary vectors and on the Birman criterion as well as a comparison of the results obtained by means of our programs with the results of other authors.

2. GROUP-THEORETICAL METHODS FOR ANALYSIS OF PHASE TRANSITIONS

In the Landau theory of phase transitions, a crystal is described by a certain density function $\rho(\mathbf{r})$ of the point $\mathbf{r}$ of the three-dimensional Euclidean space. Its physical sense can be different (e.g. $\rho(\mathbf{r})$ may be considered as a time-averaged density of distribution of the electric charge in the crystal) because for the Landau theory only symmetry properties of $\rho(\mathbf{r})$ are of importance. In second order phase transitions, in spite of the continuous changes of the $\rho(\mathbf{r})$ function, the symmetry of the crystal changes jumpwise, and, there exists a subgroup relationship $G_D \subset G$ between the space group G of the crystal before transition and the space grup G_D after it. The purely geometrical considerations seem to warrant the conclusion that in consequence of a phase

255

transition, a low-symmetry phase may arise having any conceivable subgroup G_D of the group G, for example, a phase characterized by arbitrary increase in the volume V' of the primitive cell of the G_D-phase with respect to the volume V of the primitive cell of the original G-phase. However, the Landau theory is a thermodynamic theory, which imposes quite rigid restrictions upon G_D groups of the low-symmetry phases.

Following Landau, expand the density function $\rho_D(\mathbf{r})$ corresponding to a low-symmetry phase in the complete set of basis functions of all irreducible representations (IR) of the G-group (the possibility of such an expansion is purely group-theoretical)

$$\rho_D(\mathbf{r}) = \sum_j \sum_{i=1}^{n_j} C_i^j \varphi_i^j(\mathbf{r}). \tag{1}$$

Here the index j numbers the full IR's Γ_j of dimensionality n_j of the space group G and the index i numbers their basis functions $\varphi_i(\mathbf{r})$; the coefficients C_i^j are the components of order parameters, they depend on thermodynamic conditions, in particular, on such variables as the pressure p and the temperature T: $C_i^j = C_i^j(p, T)$. In the neighborhood of the continuous phase transition, all these coefficients, except that one which corresponds to the identical IR,† are small quantities of various orders of smallness vanishing at the very point of transition (p_c, T_c).

Let us introduce for each IR Γ_j the corresponding vector of the coefficients C_i^j from formula (1) $\mathbf{C}^j = (C_1^j, \ldots, C_{n_j}^j)$ as well as the vector of the basis $\boldsymbol{\phi} = (\varphi_i^j(\mathbf{r}), \ldots, \varphi_{n_j}^j(\mathbf{r}))$ whose components are the basis functions of this representation. Then formula (1) may be written as a formal scalar product

$$\rho_D(\mathbf{r}) = \sum_j (\mathbf{C}^j, \boldsymbol{\phi}^j(\mathbf{r})) = \sum_j \rho_D^j(\mathbf{r}). \tag{2}$$

Thus, the function $\rho_D(\mathbf{r})$ is now represented as a sum of contributions by each of the IR's of the group G. A space group has, however, an infinite number of IR's. Indeed, the full IR's are characterized by the wave vector $\mathbf{k}$ taking on a continuous set of values in the Brillouin zone and by the index of the so-called "small" IR, i.e. a representation of the wave vector group.

One of the most important corollaries of the Landau theory is the conception of one irreducible representation. Precisely the thermodynamic principles have served as a basis for the demonstration that for a line of the second-order phase transitions to exist in the plane of two thermodynamic variables (such as p and T), it is required that the crystal lattice lose its stability with respect to one, hereafter referred to as critical, IR Γ_0 of the group G. This IR corresponds to the critical (primary) order parameter $\mathbf{C}^0$. The remaining, noncritical (secondary) order parameters $\mathbf{C}^j$ $(j \neq 0)$ from formula (2) corresponding to the noncritical IR's Γ_j are, in the vicinity of a transition point, the quantities of a higher order of smallness in comparison with the critical order parameter [19], they will be considered below. Thus, the conception of one IR helps drastically reduce the number of degrees of freedom responsible for the loss of stability of the cristal lattice in a phase transition, and, out of the infinite number of addends in formula (2), one may, near the transition point, take into account solely the contribution by one critical IR.

The next important principle of the phenomenological theory of phase transitions is the criterion, established by Lifshitz, of stability of the low-symmetry phase to long-wave fluctuations of the density function. It is a criterion of the transition into the G_D phase commensurate with the G-phase—at the transition point, the primitive cell is increased an integer number of times as contrasted with the original phase. The Lifshitz criterion, hereafter called hard, states that the antisymmetrized square of the critical IR and the representation according to which the usual three-dimensional vectors are transformed (this representation may be reducible) must not have common IR's of the group G. More convenient, from a practical viewpoint, is a less categorical statement following from the above criterion, which is often called the "soft" Lifshitz condition [10]. It amounts to the assertion that the wave vector $\mathbf{k}_0$ associated with the critical IR must

---

†Note that in the Landau theory one usually works with the function $\delta\rho(\mathbf{r})$, which equals the difference between $\rho_0(\mathbf{r})$, corresponding to the high-symmetry phase, and the function $\rho_D(\mathbf{r})$, thereby excluding the contribution by the identical IR. Since in the description of the structure of a lower-symmetry phase (see Section 5) this contribution must anyway be taking into account, we find it expedient to write the formula (1) directly for the function $\rho_D(\mathbf{r})$.

correspond to the point of the high-symmetry in the Brillouin zone (in the infinitesimally small vicinity of this point, there are no vectors **k** that would correspond to the same symmetry as that of the vector $\mathbf{k}_0$). The number of points of the high-symmetry is relatively small for any space group, and it is of crucial importance that these points correspond to the vectors **k** whose components have, in the expansion with respect to the vectors of the reciprocal lattice, the form of the proper fractions m/n with comparatively small denominators ($n = 2, 3, 4$). This means that for any space group the volume V' of the primitive cell of the low-symmetry phase must not exceed by more than 32 times the volume V of the primitive cell of the parent phase. For example, the group $O_h^7 = Fd3m$ corresponds to a cubic face-centred lattice for which there exist in the Brillouin zone only four points of the high-symmetry, namely Γ, X, L and W. A certain number of full IR's of the group O_h^7 correspond to each of these (22 representations altogether with dimensionalities 1, 2, 3, 4, 6, 8, 12). Two twelve-dimensional IR's correspond to the point W, and they both, while satisfying the soft, violate the hard Lifshitz condition. Thus, only from the Lifshitz criterion may we infer that for the crystals with the space symmetry O_h^7 merely the following variations of the volume of the primitive cell are possible in consequence of the second-order phase transitions: $V'/V = 2, 4, 8$.

In connection with the conception of one IR the problem arises regarding the classification of all continuous phase transitions by those IR's of the space group of a crystal which induce these transitions (critical IR's). Since the number of the IR's satisfying the Lifshitz condition is relatively small, many authors (see, for example, Refs [8, 18–29]) tabulate for each IR all the lower-symmetry G_D-phases induced by it. It is to be noted that in this case also those phases get into the list of the G_D-phases which emerge as a result of the quasicontinuous transitions.† This may occur even then when the critical representation satisfies not only the Lifshitz criterion, but also the well-known Landau criterion of the second-order transition, which requires that there should be no third-degree invariants in the expansion of a thermodynamic potential. This fact can be explained by that the conditions of Lifshitz and Landau are not sufficient for a transition to be second-order (see, for example Refs [5, 6]). One may add that the experimental research has shown the majority of the continuous phase transitions to be really induced by one IR. In those relatively rare cases when this is not so, there often are some additional reasons of symmetry character responsible for reducibility of the critical representation [30]. There are, however, exceptions to this rule, too [31].

Of the methods employed for the classification of the phase transitions by the IR's of the group G, the most effective are those based on the idea of stationary vectors [4, 5, 32, 33] and on the Birman criterion [22, 34], hereafter called the SV-method and the BC-method, respectively. Let us now consider in more detail the idea behind the SV-methods following the treatment of this problem in Ref. [19]. Since the function $\rho_D(\mathbf{r})$ fully determines the symmetry of the G_D-phase, it must be invariant under all elements $g \in G_D$. In more precise terms, this may be expressed as follows. The action of the element $g \in G$ on the argument of an arbitrary function $f(\mathbf{r})$ entails its change which may be represented as a result of acting upon it with a certain operator $\hat{g}$ which corresponds to the element g, according to the well-known rule $\hat{g}f(\mathbf{r}) = f(g^{-1}\mathbf{r})$. From the condition of invariance of $\rho_D(\mathbf{r})$ we have

$$\hat{g}\rho_D(\mathbf{r}) = \rho_D(\mathbf{r}), \quad g \in G_D. \tag{3}$$

Hence, taking into account the relationship (2), we arrive at

$$\hat{g}\rho_D(\mathbf{r}) = \sum_j (\mathbf{C}^j, \hat{g}\boldsymbol{\phi}^j(\mathbf{r})) = \sum_j (\mathbf{C}^j, \tilde{\mathcal{U}}^j(g)\boldsymbol{\phi}^j(\mathbf{r})) = \sum_j (\mathcal{U}^j(g)\mathbf{C}^j, \boldsymbol{\phi}^j(\mathbf{r})). \tag{4}$$

Here we have made use of the relationship $\hat{g}\boldsymbol{\phi}^j = \tilde{\mathcal{U}}^j(g)\boldsymbol{\phi}^j$ which means that under the action of the operator $\hat{g}$ on any basis function of the IR it is transformed into a certain linear combination of the basis functions of the same IR. The coefficients of this combination are the elements of the corresponding column of the matrix $\mathcal{U}^j(g)$, which, by the definition of the representation, is associated with the element $g \in G$. In the derivation of relationship (4) we also took account of

---

†In quasicontinuous phase transitions, the order parameter changes at the transition point jumpwise, though this jump may, in a certain sense, be assumed to be small.

the known property of the scalar product of the vectors which permits transfer of the matrix action from one vector of the product onto another.

A comparison of expressions (3) and (4) yields the following basic relationship of stationarity of the order parameter corresponding to the IR Γ_j:

$$\mathscr{U}^j(g)\,\mathbf{C}^j = \mathbf{C}^j, \quad g \in G_D \tag{5}$$

Thus, the vector $\mathbf{C}^j$ is the common stationary vector (i.e.) an eigenvector with the eigenvalue of unity) of those matrices $\mathscr{U}^j(g)$ of the IR Γ_j which correspond to all $g \in G_D$. This set of matrices is called the restriction of the IR Γ_j of the group G to its subgroup G_D, it is designated as $\Gamma_j \!\downarrow\! G_D$. Being the general solution to a system of homogeneous linear equations, the stationary vector depends on a certain number of arbitrary parameters which we denote by $a, b, c, \ldots$, hence it singles out a certain subspace in the space of the representation, whose dimensionality equals the number of these arbitrary parameters. The knowledge of the stationary vector $\mathbf{C}^j$ and of the basis functions of the IR Γ_j fully determines the contribution by this representation to the function $\rho_D(\mathbf{r})$. For example, if for a certain six-dimensional IR Γ_j the stationary vector has the form $\mathbf{C}^j = (a, b, -a, b, c, -c)$, then the contribution by this IR to $\rho_D(\mathbf{r})$ will be

$$\rho_D(\mathbf{r}) = a\,\varphi_1^j + b\varphi_2^j - a\varphi_3^j + b\varphi_4^j + c\varphi_5^j - c\varphi_6^j. \tag{6}$$

The three arbitrary parameters (a, b, c), on which this vector depends, cannot be determined on the basis of symmetry considerations. They may, e.g. be found by means of the minimization of the non-equilibrium thermodynamic potential ϕ, provided that one knows concrete coefficients of its expansion in the components of the order parameter. Thus, a stationary vector furnishes the most general explicit form of the order parameter which can be determined proceeding from purely symmetry-related considerations.

Let us take the critical IR Γ_0 inducing the phase transitions $G \to G_D$. Omitting for brevity the index $j = 0$, we may rewrite for this IR equation (5) as

$$\mathscr{U}(g)\mathbf{C} = \mathbf{C}, \quad g \in G_D. \tag{7}$$

Between the stationary vectors $\mathbf{C}$ of the critical IR and the symmetry groups G_D a one-to-one relationship exists. Indeed, by selecting all the matrices of the IR Γ_0 which leave invariant a certain vector $\mathbf{C}$ in the representation space and collecting all elements g of the group G corresponding to these matrices, one finds the complete set of the elements of the subgroup $G_D = G_D[\mathbf{C}]$. On the other hand, if the subgroup G_D is known, then one may find for it from equation (7) a quite definite stationary vector $\mathbf{C} = \mathbf{C}[G_D]$. The algorithm of the stationary vectors method is based on just this relationship between the stationary vectors and the subgroups G_D of the group G; it consists of the following three steps:

(1) Construct matrices of a given critical IR Γ_0;
(2) Find by a specific procedure (see Section 4) all stationary vectors possible for the given IR (they single out certain sets of matrices of this IR);
(3) Find for each such stationary vector all elements $g \in G$ whose matrices leave it invariant and identify the obtained set of symmetry elements as one of the 230 space groups G_D.

Note in this connection that if there is no need to study the domains of one and the same phase, then out of the set of all possible stationary vectors one should select only the vectors nonequivalent with respect to one another. We call equivalent those two vectors C and C' which are transformed into each other under the action of at least one matrix $\mathscr{U}(g_0)$ of the representation Γ_0: $\mathbf{C}' = \mathscr{U}(g_0)\mathbf{C}$. It is not difficult to verify that the sets of the matrixes $H[\mathbf{C}]$ and $H[\mathbf{C}']$ that leaves these vectors invariant are related by the equation $H[\mathbf{C}'] = \mathscr{U}(g_0) \cdot H[\mathbf{C}] \cdot \mathscr{U}^{-1}(g_0)$. Then, for the subgroups of G corresponding to these vectors we have

$$G_D[\mathbf{C}'] = g_0 \cdot G_D[\mathbf{C}] \cdot g_0^{-1} \tag{8}$$

Thus, the subgroups $G_D[\mathbf{C}]$ and $G_D[\mathbf{C}']$ are equivalent in the conventional crystallographic sense, i.e. they are the domains of one and the same low-symmetry phase. Such definition of equivalence was accepted in all our publications. In the first works of Hatch and Stokes a somewhat different definition of equivalence was used, but later they, too, accepted definition (8) [35]. It is to be noted that if this definition is used, the enantiomorphous phases may prove to be different domains of the same G_D-phase [12, 35]. The three-above described steps of the SV-method have been covered by the following programs of our complex: REPS, VECTOR and GROUP, respectively.

Let us now compare the SV-method with the methods based on the Birman criterion. The former rests on the notion of a stationary vector while the latter on that of a "subduction number", i.e. the number showing how many times an identity IR of the group G_D enters into the restriction $\Gamma_j \downarrow G_D$. However, what is the relationship between the two notions?

Let among n components of the stationary vector $\mathbf{C}^j = (C_1, C_2, \ldots, C_n)$ of the IR Γ_j, which corresponds to the G_D-phase, be only m independent ones. By setting successively all these arbitrary constants except one at zero, we obtain m linearly independent vectors $\mathbf{V}_1, \ldots, \mathbf{V}_m$ in the space of the order parameter $\mathbf{C}^j$. By virtue of the formula

$$\rho_D^j(\mathbf{r}) = \sum_{i=1}^{n_j} C_i^j \varphi_i^j(\mathbf{r}) \tag{9}$$

the vectors $\mathbf{V}_1, \ldots, \mathbf{V}_m$ define the independent linear combinations of the basis functions $\varphi_i^j(\mathbf{r})$ invariant with regard to all elements of the group G_D. In other words, each of the vectors $\mathbf{V}_1, \ldots, \mathbf{V}_m$ defines that linear combination of the basis functions of the IR Γ_j which is a basis function of the identity representation of the subgroup G_D. Thus, the subduction number in the Birman method equals the parametricity of the stationary vector, i.e. the dimensionality of the subspace which it determines (it simply equals the number of the different letters a, b, c, $\ldots$, used for writing down the stationary vector). Consequently, the methods based on the Birman criterion give only partial information on the stationary vectors, viz. the number of their independent parameters. The SV-method, on the other hand, yields the explicit form of the stationary vectors, which is needed for solving most problems of thermodynamics (since these vectors correspond to extremum points of the thermodynamic potential ϕ) and of structural analysis [from the formulas of type (1), one may determine by means of the stationary vectors, when the basis funtions $\varphi_i(\mathbf{r})$ are known, the general form of the density function $\rho_D(\mathbf{r})$, i.e. the structure of the low-symmetry phase]. Seeing that the SV-method allows one to solve these problems, we think that it is not only more informative, but also methodologically more consistent, albeit more complex, than the methods based on the Birman criterion. We should like to note that the simplicity of the latter methods is, to a considerable degree, connected with the fact that in order to calculate the subduction number, one has to know only the characters of the IR's, but not the explicit form of their matrices, which is required for the SV-method.

So far we have been considering the first problem of the group-theoretical analysis of phase transitions in crystals, namely, the determination of possible groups of symmetry G_D of the low-symmetry phases induced by the given (critical) IR. Most works are concerned with the solution to just this problem. However, there exists also the second problem of the group-theoretical analysis of phase transitions which consists in the determination of the structure of the G_D-phases. This problem was examined in Ref. [19] and a general method for its solution was suggested there. This method (see Section 6) allows one to find for each stationary vector $\mathbf{C}$ of the critical IR all the stationary vectors $\mathbf{C}^j$, corresponding to it, of the noncritical IR's which make a contribution to the function $\rho_D(\mathbf{r})$ describing the structure of the low-symmetry phase. The complete set of the critical (primary) and the noncritical (secondary) order parameters forms a "condensate" of the stationary vectors (order parameters). It is possible to evaluate the relative smallness of the critical and various noncritical order parameters near a phase transition point (p_c, T_c) in terms of the Landau theory [19]. It has turned out that the temperature dependence of the components of the critical order parameter has the form $|T - T_c|^{1/2}$, while that of the noncritical ones: $|T - T_c|^{m/2}$, where m is the minimal direct symmetrized power of the critical IR Γ_0 which contains the given noncritical IR Γ_j. Thus, in order to find the structure of the low-symmetry phase, i.e. the function $\rho_D(\mathbf{r})$ from the formula (1), one has to know the complete condensate of the stationary vectors for this phase as well as the basis functions of the critical IR and of the noncritical IR's corresponding to it. For the calculation of these quantities the respective program CONDENSATE and BASIS were written.

The function $\rho_D(\mathbf{r})$ constructed with the aid of the complete condensate provides the most general form of the structure of the low-symmetry G_D-phase allowed by its symmetry. The same information may also be obtained by means of a simpler crystallographic method [36]. Although the latter method, in contrast to the former, does not permit the evaluation of relative magnitudes of the critical and the noncritical order parameters near a phase transition point, it still may be used both to control the results derived with the aid of the complete condensate of stationary vectors and to obtain information on the lowering of the local symmetry at each point of the orbits of the group G of the original phase which occurs as a result of the phase transition $G \to G_D$. The relevant program ORBIT will be described in Section 8.

After this brief survey of the group-theoretical methods used in the analysis of the structural phase transitions, we turn to the consideration of the algorithms implemented in six programs mentioned above. As it is not possible to describe in one article all these algorithms with sufficient thoroughness, we shall examine in detail only those, which are the most essential for the SV-method (as an example, we consider obtaining all low-symmetry phases for the four-dimensional IR of the two-dimensional group $p4bm$ [18]), touching briefly on only some distinctive features of other programs. While describing the algorithms, we make references to the works where the results were published, obtained with the aid of the programs based on these algorithms, and in the closing Section 9 we compare the results with those of other authors.

3. OBTAINING FULL IR'S OF THE SPACE GROUPS

The group-theoretical approach to the investigation of phase transitions described in the previous Section is based on the knowledge of the full IR's [37, 38] of the space group G. They are numbered by the wave vector $\mathbf{k}$ and by the number of the small IR, i.e. the representation of the wave vector group $G_\mathbf{k}$. Unlike the small IR's, the full ones are characterized by the whole star of the wave vector, they contain matrices corresponding to all the elements of G. Incidentally, a large number of problems of solid state physics, such as the construction of a symmetrized basis for the study of the electron energy spectrum in a crystal requires the use of only the small IR's. This has resulted in that publications entitled *Tables of Irreducible Representations of the Space Groups* (see, for example, the Refs [39–43]) customarily contain only the small IR's, occasionally, merely the characters of these. Some programs for the construction of small IR's have been developed by means of a computer [44, 45]. Since the symmetry classification of phase transitions in crystals requires knowledge of the explicit form of matrices for all full IR's of 230 space groups, we, together with V. N. Raspopov and V. P. Popov, worked out in 1974 a specific program REPS). In 1981 it was modified in cooperation with M. G. Glumov.

This program uses a standard algorithm for inducing the full IR's from the small ones (see, for example Refs [38, 46]). The latter can be found with the aid of the ray IR's of the wave vector group $G_\mathbf{k}$. Their construction is the most cumbersome part of the program, it is based, in its last version, on the work [42]. When constructing the IR's of the space groups, it is necessary to perfom, along with the usual calculational procedures, various algebraic manipulations, e.g. work with the elements of the space groups, find the stars of wave vectors etc. These transformations are performed in the program REPS by means of the special coding of the elements of the point groups of symmetry as well as by using certain concrete realizations of the Bravais lattices and of the reciprocal lattices corresponding to them. These expedients allowed one to reduce the algebraic operations to purely arithmetic ones and to evolve concise procedures, suitable for all lattices, for constructing the stars, obtaining all the elements of the space groups with the aid of their generators, for building the block structure of full representations etc. The program REPS is provided with a special block for controlling the correctness of the construction of the little IR matrices, which checks on the correspondence between the multiplication table of the matrices of the representation and that of the elements of the group $G_\mathbf{k}$. Some details of the first version of the program are given in Ref. [47]. Most procedures of REPS were subsequently utilized in other parts of the program complex for the investigation of phase transitions in crystals.

The program REPS uses the following permanent information (a) the table of generators for all 230 space groups; (b) the basis vectors of 14 Bravais lattices and of reciprocal lattices corresponding to them and (c) the matrices of the ray IR's for the standard factor systems [42] of 18 point groups, nonisomorphous with respect to one another and associated with 230 space groups in a special "genesis" form [48, 49]. As input information, it is sufficient to use only the number of the space group and the wave vector, for which the full IR's are to be constructed.

The REPS program prints out: (a) the elements of the space group (rather, the representatives in its decomposition into cosets with respect to the subgroup of translations T); (b) the star of the wave vector $\mathbf{k}$ with the indication of both the elements of the point group of the crystal, which generate its arms, and the elements, which transform $\mathbf{k}$ to $\mathbf{k} + \mathbf{b}$, where $\mathbf{b}$ is a certain vector of the reciprocal lattice (i.e. the elements of $G_{\mathbf{k}}$); (c) the matrix of the ray representations of the point group corresponding to $G_{\mathbf{k}}$ for the factor system determined by the vector $\mathbf{k}$ and by the set of improper translations of the space group under consideration and (d) the matrices of the full IR either for the generators of the space group or for all representatives of its cosets with respect to the subgroup T.

Using REPS we composed the tables of the full IR's of 230 space groups for the points of the high symmetry in the Brillouin zone (as has already been mentioned, only such IR's can induce commensurate phase transitions). The results of this computation have been used in all our further work devoted to the group-theoretical analysis of phase transitions in crystals. In particular, these results helped us select the images (L-groups, in the terminology of Ref. [4, 5]) of the four-dimensional [17], six-dimensional [11] and other IR's of all space groups. The classification of the IR's by the images plays an important part in the theory of phase transitions (see, for example, Refs [5, 6, 10]), since identical sets of stationary vectors and identical bases of invariants correspond to all representations with the same image. Tables of the full IR's, albeit incomplete, were published—for all groups of the hexagonal syngony this was done in Ref. [50], and of the cubic syngony in Refs [48, 49]. In the two last works, the so-called genesis form of the tables was proposed, which makes it possible to trace the genesis of the full IR's of the groups from the full IR's of some of their subgroups. At present, we are aware of only one more variant of the tables of full IR's, also for the points of the high-symmetry in the Brillouin zone, done by hand [51].

4. OBTAINING A SET OF NONEQUIVALENT STATIONARY VECTORS

4.1. General description of the algorithm

The most laborious step in the finding of all possible symmetry groups of low-symmetry phases induced by the multidimensional critical IR Γ_0 is the obtaining of the set of stationary vectors. This set is, apparently, determined solely by the set of different matrices of the IR Γ_0, i.e. by its image $I(\Gamma_0)$. The principal of algorithm, used in our programs, for obtaining the set of the stationary vectors consists of a number of steps.

1. Find for each matrix $\mathcal{U}(g) \in I(\Gamma_0)$ its stationary vector from the equation

$$\mathcal{U}(g)\, \mathbf{C} = \mathbf{C}. \tag{10}$$

Some of the stationary vectors may prove to be zero vectors, some will be coinciding. Denote by V_1 the set of all nonzero vectors, different from one another, obtained at this stage. Each of the vectors of the set V_1 is, generally speaking, left invariant not only by the matrix for which it was obtained from equation (10), but also by some other matrices (their own stationary vectors contain this stationary vector as a particular case), whose totality forms, apparently, a matrix subgroup of the group $I(\Gamma_0)$. However, by far not all its matrix subgroups can be singled out in this way. Indeed, let us consider two stationary vectors $\mathbf{C}_i$ and $\mathbf{C}_j$, associated with the subgroups $H[\mathbf{C}_i]$ and $H[\mathbf{C}_j]$ of $I(\Gamma_0)$. Each of these vectors depends on a certain number of the arbitrary constants and, thus, determines a certain subspace in the representation space invariant with respect to $H[\mathbf{C}_i]$ and $H[\mathbf{C}_j]$, respectively. The intersection of these subspaces $\mathbf{C}_i \cap \mathbf{C}_j$ is, in fact, a subspace of lower-dimensionality, which is invariant with respect to both all the matrices from $H[\mathbf{C}_i]$ and $H[\mathbf{C}_j]$ and all their products. Hence, the intersection of the vectors $\mathbf{C}_i$ and $\mathbf{C}_j$ gives rise to the stationary vector $\mathbf{C}_k = \mathbf{C}_i \cap \mathbf{C}_j$, which is, in the general case, associated with the new subgroup $H[\mathbf{C}_k]$. The foregoing has led us to the second step of the SV-method.

2. Find all possible intersections of the vectors from the set of the stationary vectors V_1 and augment this set with these.

All these intersections may conveniently be found by performing successive pairwise intersections. After carrying out all pairwise intersections of the stationary vectors from the set V_1 we obtain the set V_2 of new stationary vectors, different from those of V_1. Intersect each of the vectors of the set V_2 with each of the vectors of the set V_1 (obtaining in this way triple intersections of the stationary vectors) and then all the vectors from V_2 with one another (which gives quadruple intersections). Denote the set of these newly found stationary vectors by V_3 and find thier pairwise intersections with one another and with the already available sets of the vectors from V_1 and V_2. Carry on this procedure until new stationary vectors cease to appear. This ends the second step of the algorithm.

Each of the stationary vectors $\mathbf{C}_i$ obtained determines its, automatically maximal, matrix subgroup $H[\mathbf{C}_i]$ of the group $I[\Gamma_0]$. This means that in the total set of the stationary vectors there will not be such that would be particular cases of one another generating at the same time the equal matrix subgroup of $I(\Gamma_0)$.

So as to avoid obtaining nonmaximal subgroups $G_D \in G$, a special chain criterion was suggested in the BC-methods [34].

The complete set of the stationary vectors obtained corresponds to the set of all possible domains of all low-symmetry G_D-phases induced by a given IR Γ_0. We have to know precisely this set when dealing with certain physical problems. When, however, we restrict ourselves, as is usually done, to finding only possible groups G_D of the lower-symmetry phases, then one more step becomes necessary in the SV-method.

3. Let us select out of the complete set of stationary vectors a set of the vectors nonequivalent with respect to one another. Take with this aim a certain stationary vector $\mathbf{C}_i$ and act successively upon it with all matrices from $I(\Gamma_0)$. The stationary vectors obtained as a result correspond to the domains $G_D[\mathbf{C}_i]$, they must, therefore, be crossed off the set of the stationary vectors. After this selection procedure performed for all vectors in the complete set, we obtain a set of the stationary vectors of the critical IR Γ_0 under consideration, which are nonequivalent with respect to one another.

4.2. Some details of the algorithm

When carrying out the three steps of the algorithm, one should bear in mind that the matrices of the IR's of the space groups possess a certain structure. Large majority of these IR's (over 98%) are constructed of matrices with a structure which we call the structure of type S1—there is in each column and each row of this matrix only one nonzero element (whose modulus equals unity in virtue of its unitarity). The rest of IR's of the space groups have the S2-structure (see below). The task of finding the stationary vectors of the individual matrices and of their intersections requires the solution of certain systems of linear equations. It would be not reasonable to try to solve them by the general methods of linear algebra in view of the special structure characterizing the matrices of IR's of the space groups. Let us consider this point in more detail.

Represent the matrix A of the S1-structure as a unity of its "structural" (P) and argumentary (μ) parts—$A = \{^P_\mu\}$. In the structural part, indicate successively for each row of the matrix A the number of its column in which there is a nonzero element, which is equivalent to the determination of a certain permutation P of n numbers $P = (j_1, j_2, \ldots, j_n)$. In the argumentary part, indicate the values of nonzero elements μ_i of the matrix A, which correspond to its successive rows—$\mu = (\mu_1, \mu_2, \ldots, \mu_n)$, i.e. μ is a number (complex in the general case) from the ith row of the matrix. For example

$$A = \left\{ \begin{array}{c|c|c} & & \mu_1 \\ \hline \mu_2 & & \\ \hline & \mu_2 & \end{array} \right\} \longrightarrow \left\{ \begin{array}{l} P = (3, 1, 2) \\ \mu = (\mu_1, \mu_2, \mu_3) \end{array} \right\}.$$

Let us first of all consider the question of finding the stationary vector $\mathbf{X}$ of a matrix having the S1-structure. By definition the vector $\mathbf{X} = (X_1, \ldots, X_n)$ is the general solution to the equation $A\mathbf{X} = \mathbf{X}$ which is equivalent to the system of the linear scalar equations

$$\mu_i x_j = x_i. \tag{11}$$

Here $i = 1, 2, \ldots, n$ and $j = j(i)$ is determined by the permutation P, i.e. j is the number of the column of the matrix A in which the nonzero element μ_i stands in its ith row. Generally, the permutation P may be decomposed into a certain number of the cycles, each of which generates a chain of linked equations. The result is that system (11) breaks up into a certain number of independent subsystems. Applying in regard to each such chain (subsystem) the method of successive elimination of unknowns, we arrive at one equation of the form

$$\Omega_k X_k = X_k, \tag{12}$$

where Ω_k is the product of the factors μ_i corresponding to all elements of a given chain and k is the number of its last equation. If $\Omega_k = 1$, equation (12) is satisfied identically and $X_k = a$, where a is an arbitrary parameter. In this case, all components of the vector $\mathbf{X}$, contained in this chain, also depend on only this parameter. However, if $\Omega_k \neq 1$, $X_k = 0$, hence all other X_i of this chain equal zero. Thus, the stationary vector depends, in the general case, on several arbitrary parameters whose number (P_x) is determined by the number and the type of the chains of system (11)

$$\mathbf{X} = \sum_{k=1}^{P_x} a_k \mathbf{W}_k. \tag{13}$$

(The number P_x equals, as was noted above, the subduction number in the BC-methods.) Here $\mathbf{W}_k$ are the vectors determining the fundamental solution to system (11). For example, the stationary vector $\mathbf{X}$ of the matrix

$$A = \begin{Bmatrix} 3, & 4, & 1, & 2, & 6, & 5 \\ -1, & 1, & -1, & -1, & 1, & 1 \end{Bmatrix}$$

has the form

$$\mathbf{X} = \begin{pmatrix} a \\ 0 \\ -a \\ 0 \\ b \\ b \end{pmatrix} = a \begin{pmatrix} 1 \\ 0 \\ -1 \\ 0 \\ 0 \\ 0 \end{pmatrix} + b \begin{pmatrix} 0 \\ 0 \\ 0 \\ 0 \\ 1 \\ 1 \end{pmatrix}.$$

The vectors $\mathbf{W}_k$ are the orthogonal eigenvectors of the matrix A, which correspond to unity eigenvalue.

In virtue of the peculiar structure of stationary vectors of the S1-structure matrices, a simple method for finding their intersections can be suggested. Its idea may be explained by the following example. Let $\mathbf{X} = (a, b, b, -a, c, -c)$ and $\mathbf{y} = (-p, q, q, p, -q, p)$ where a, b, c, p, q are arbitrary parameters. The finding of intersections of these vectors amounts to the solution to the following system of linear equations:

$$a \begin{pmatrix} 1 \\ 0 \\ 0 \\ -1 \\ 0 \\ 0 \end{pmatrix} + b \begin{pmatrix} 0 \\ 1 \\ 1 \\ 0 \\ 0 \\ 0 \end{pmatrix} + c \begin{pmatrix} 0 \\ 0 \\ 0 \\ 0 \\ 1 \\ -1 \end{pmatrix} = p \begin{pmatrix} -1 \\ 0 \\ 0 \\ 1 \\ 0 \\ 1 \end{pmatrix} + q \begin{pmatrix} 0 \\ 1 \\ 1 \\ 0 \\ -1 \\ 0 \end{pmatrix}. \tag{14}$$

Unlike system (11), now each unknown is contained in several equations, however, the essential point is that each equation contains only two such unknowns—one from the vector $\mathbf{X}$, another from $\mathbf{Y}$. Taking account of this special feature of system (14) and applying the method of successive elimination of unknowns, one may obtain an effective "chain" algorithm of its solution similar to the above-considered algorithm for finding stationary vectors of indivudal matrices [52]. The solution of system (14) provides the vector (written for convenience in a row form)

$$\mathbf{Z} = (a, -a, -a, -a, a, -a). \tag{15}$$

The number of its unknowns P_z, apparently, satisfies the relationship $P_z \leqslant \min(P_x, P_y)$, where P_x and P_y are the numbers of the arbitrary parameters of the vectors $\mathbf{X}$ and $\mathbf{Y}$ being intersected. Note that in the case of one-parameter stationary vectors there is no need to find their intersections with other stationary vectors since this operation results in the obtaining of either the zero or the original one-parameter stationary vector.

Furthermore, the peculiar structures characterizing matrices of the IR's of the space groups and the stationary vectors corresponding to these allows the third step of the SV-method to be substantially simplified. Let the two vectors $\mathbf{X}$ and $\mathbf{Y}$ be checked on equivalence. Let us act on $\mathbf{X}$ with a certain matrix $\mathcal{U}_0$ from the set $\mathcal{U}nd$ of nondiagonal matrices of the representation Γ_0. If the structural part of the vector $\mathbf{X}'$ so obtained does not coincide with the structural part of $\mathbf{Y}$, clearly, the vector $\mathbf{X}$ cannot be transformed into $\mathbf{Y}$ by any matrix of the coset, whose representative is the matrix $\mathcal{U}_0$, in the decomposition of $I(\Gamma_0)$ with respect to the subgroup of diagonal matrices $\mathcal{U}d$ of this image. Indeed, all elements of the given coset differ from one another only in the multiplication by different diagonal matrices, when, however, these matrices act on the vector, only its argumentary but not the structural part is changed. Thus, one has to act on the stationary vector by diagonal matrices from the set $\mathcal{U}d$ only then when the structural part of $\mathbf{X}$ coincides with the structural part of $\mathbf{Y}$.

As noted above, only a small percentage of the IR's of the space group contain matrices with the S2-structure (this is immediately evident from the inspection of the tables of representations of the wave vector groups [39, 51]). The block structure of these matrices is of the S1-type, but there, instead of nonzero elements (i.e. one-dimensional matrices μ_i from $\mu = (\mu_1, \mu_2, \ldots, \mu_n)$) stand two-dimensional unitary matrices ω_i. The matrices of S2-structure may be written in a form completely analogous to that for the matrices of the S1-structure: $A = \{^P_\omega\}$. Here, $\omega = (\omega_1, \omega_2, \ldots, \omega_n)$ is the set of the two-dimensional blocks corresponding to successive "macrorows" of the matrix A. This form of writing clearly shows that the method for obtaining the stationary vectors for the type S1 matrices can readily be generalized to cover the case of the matrices of the type S2. More fully this question was dealt with in Ref. [52].

Not infrequently, it may be desirable or necessary to change from the complex form of the IR to the real one (in particular, this would be necessary when passing to the "physically irreducible" representation). In such a transformation, the original S1-type IR may turn into the representation of the S2-type. In this case, the rational approach would be to find the set of the nonequivalent stationary vectors for the original IR of the S1-type and only then to perform transformation of all vectors $\mathbf{C}_i$ of this set in accordance with the unitary transformation U of the original representation such that $\mathbf{C}'_i = U\mathbf{C}_i$. Here the vectors $\mathbf{C}_i$ will also be real.

Based on the above-described alogrithm for constructing the set of nonequivalent stationary vectors for the IR's of the S1-structure, we wrote back in 1974 together with Petrenko the program VECTOR (its text was later published in Ref. [52]). Group-theoretical analysis has been carried out with the aid of this program in all our subsequent works.

4.3. An example

As an example of the application of the algorithm described, let us find the set of nonequivalent vectors for the four-dimensional IR of the two-dimensional group $P4bm = P4g$ (it corresponds to the point X of the planar Brillouin zone for the square lattice). This group is determined by four generators $\mathbf{a}_1$, $\mathbf{a}_2$, $(h_{14}|0, 0)$, $(h_{26}|1/2, 1/2)$. The first two of them determine the planar Bravais lattice, the third is a rotation by 90° about the Z axis and the fourth represents the glide plane normal to the X axis. Following the number of the point group element taken from Ref. [39], there is given improper translation, corresponding to it, which is defined by the components of the expansion in the basis vectors $\mathbf{a}_1$, $\mathbf{a}_2$ of the Bravais lattice. Next we take from Ref. [18] the matrices of the full IR's of the group under consideration.

By using the matrices of the translations $\mathbf{a}_1$, $\mathbf{a}_2$ given in Table 1, find all different matrices of translational elements of the group $P4g$ and denote this set by $\mathcal{U}d$ (see Table 2). Find, also with the aid of Table 1, the representatives of all cosets in the composition of $P4g$ with respect to its subgroup of proper translations as well as the set of $\mathcal{U}nd$ of the matrices corresponding to these representatives (see Table 3).

For the reader's convenience, we present here the definition (taken from Ref. [39]) of every point element of the symmetry given in this table by indicating the result of its action on an arbitrary point of the plane having the coordinates (x, y): $h_1 = (x, y)$, $h_{14} = (\bar{y}, x)$, $h_4 = (\bar{x}, \bar{y})$, $h_{15} = (y, \bar{x})$, $h_{26} = (\bar{x}, y)$, $h_{27} = (x, \bar{y})$, $h_{37} = (y, x)$, $h_{40} = (\bar{y}, \bar{x})$. The first four elements correspond to rotations about the four-fold axis Z, the rest correspond to reflections in the coordinate (h_{26}, h_{27}) and diagonal (h_{37}, h_{40}) planes containing this axis. Every matrix of the IR may be represented as product of a certain matrix from Table 2 and a certain matrix from Table 3 whose numbers will be written with a hyphen market to indicate the full number of the matrix of the IR.

Let us now illustrate the process of obtaining the stationary vectors of the matrices of the representation under consideration. For example, equation (10) gives for the matrix 2–5

$$
\begin{bmatrix}
 & & -1 & \\
 & & & -1 \\
1 & & & \\
 & -1 & &
\end{bmatrix}
\begin{bmatrix}
a \\ b \\ c \\ d
\end{bmatrix}
=
\begin{bmatrix}
a \\ b \\ c \\ d
\end{bmatrix}.
$$

From this we have $-c = a$, $-d = b$, $a = c$, $-b = d$. Consequently, $a = c = 0$, $d = -b$. Thus, the stationary vector of the matrix 2–5 is of one parameter type and, written in a row, it has the form $(0, b, 0, -b)$, where b is an arbitrary constant. Analogously, for the matrix 1–8 the following equations of type (11) will be derived $d = a$, $-c = b$, $-b = c$, $a = d$. Hence, this matrix is associated with a two-parameter stationary vector of the form $(a, b, -b, a)$. Find now successively for each of the 32 matrices of the image $I(X_1)$ its own stationary vector. Some of them prove to be zero, some are identical. All 19 different stationary vectors of the matrices of the IR X_1 are given in Table 4.

Let us now consider the next step of the SV-method and find the intersections of the stationary vectors. Since, as was already noted, intersection of a one-parameter stationary vector with any other stationary vector produces trivial result, it would, in this case, suffice to try out only the intersections of ten two-parameter vectors among one another (intersection of the four-parameter vector (a, b, c, d) of the general form with any other vector gives simply the last vector). When intersection of two stationary vectors is being sought, one has to bear in mind that the letters a, b, $c, \ldots$ on which their form depends denote some arbitrary constants and must, therefore, be assumed different for the different stationary vectors. Then the finding of an intersection will amount to the equalizing of these vectors. For example, for the vectors Nos. 2 and 8 from Table 4, we obtain the equation $(a, 0, b, 0) = (a', b', 0, 0)$, which yields $a' = a$, $b' = 0$ and $b = 0$. Thus, intersection of these vectors leads to the new stationary vector $(a, 0, 0, 0)$.

Trying out of all pairwise intersections of the vectors from Table 4 produces 8 new stationary vectors, listed in Table 5, which supplement the set of the stationary vectors of Table 4. As all of them are of one-parameter type, their further intersections with one another and with the vectors of Table 4 will not produce any new stationary vectors, hence the procedure of finding all possible intersections ends at that. We have thus obtained for the IR X_1 of the group $P4g$ a set of 28 different stationary vectors.

Let now the next step of the SV-method be considered. Elimination of the equivalent stationary vectors can be subsequently simplified by taking account of the fact that only those vectors can be equivalent which have equal number of arbitrary parameters and equal number of zeros (the latter property follows from the S1-structure of the matrices of the IR's). Here again, one has to take into account the arbitrariness of the constants a, b, c. For example, the action on the stationary vector No. 2 from Table 4 by the matrix 1–2 gives the vector $(0, a, 0, -b)$ which will be equivalent to the third vector from Table 4 $(0, a, 0, b)$, since the sign before the arbitrary parameter (b) is, in our case, of no significance. By crossing all equivalent vectors off the set of altogether 27 vectors (Tables 4 and 5) we arrive at the final set of nine nonequivalent vectors presented in Table 6. Thus, the IR X_1 of the two-dimensional group $P4g$ induces nine lower-symmetry phases.

4.4. An alternative algorithm

The algorithm described in Section 4.1 allows one to obtain the complete set of nonequivalent stationary vectors of a given IR Γ_0 and, consequently, to single out all nonequivalent matrix subgroups of the image $I(\Gamma_0)$ of this representation. Notwithstanding the obvious advantages of

Table 1. Generators of the four-dimensional IR X_1 of the group $P4g$

	$\mathbf{a}_1$	$\mathbf{a}_2$	$(h_{14}|0,0)$	$(h_{26}|1/2,1/2)$

Table 2. The matrices of the image of the IR X_1 corresponding to the proper translations

1	2	3	4
0	$\mathbf{a}_1$	$\mathbf{a}_2$	$\mathbf{a}_1 + \mathbf{a}_2$

Table 3. The matrices of the IR X_1 corresponding to the representatives in the decomposition of $P4g$ with respect to its translational subgroup

1	2	3	4	5	6	7	8
$(h_1|0,0)$	$(h_{14}|0,0)$	$(h_4|0,0)$	$(h_{15}|0,0)$	$(h_{26}|\tfrac{1}{2},\tfrac{1}{2})$	$(h_{40}|-\tfrac{1}{2},\tfrac{1}{2})$	$(h_{27}|-\tfrac{1}{2},-\tfrac{1}{2})$	$(h_{37}|\tfrac{1}{2},-\tfrac{1}{2})$

Table 4. The set of different stationary vectors of the individual matrices of the IR X_1

							Numbers of stationary vectors											
1	2	3	4	5	6	7	8	9	10	11	12	13	14	15	16	17	18	19
a	a	0	a	0	0	a	a	a	a	0	0	0	a	a	a	a	a	a
b	0	a	0	a	$-a$	$-a$	$-b$	0	0	b	b	b	a	b	0	0	$-b$	b
c	0	0	a	a	a	0	0	0	b	b	0	0	b	a	0	$-a$	a	c
d	0	b	0	a	$-a$	0	0	b	0	b	b	$-b$	a	0	$-a$	0	a	$-a$

Table 5. Pairwise intersections of the stationary vectors from Table 4

			Numbers of stationary vectors				
20	21	22	23	24	25	26	27
a	0	0	a	0	a	a	0
0	0	a	0	a	0	0	a
0	a	0	0	0	0	a	$-a$
0	b	0	b	a	$-a$	0	0

Table 6. The set of nonequivalent stationary vectors for the IR X_1 of the group $P4g$

1	2	3	4	5	6	7	8	9
a	a	a	a	a	a	a	a	a
a	0	0	a	a	a	0	b	b
0	a	0	0	a	0	b	0	c
a	0	b	0	0	b	0	0	d

this algorithm, one has to look for alternative algorithms for obtaining a set of nonequivalent stationary vectors when, having limited computer resources, we are confronted with a large dimensionality of the representation and a great number of different matrices contained in it. Following Ref. [52], we describe here one of these alternatives.

Until now, the step of finding the stationary vectors was completely separated from the step of the search for G_D. Partial coincidence of these stages may, in certain cases, yield some calculational advantages. Let G and G_D be the space groups of the high-symmetry and the lower-symmetry phases, respectively, and G^0 and G_D^0—the crystal classes corresponding to these groups. The following problem is posed: find the space groups G_D induced by a given IR of the group G that would correspond to a fixed crystal class G_D^0. This problem, called hereafter B, may be of interest in its own right. Once, however, we have solved a problem A on the listing of all possible G_D for the given IR of the group G, it would suffice to solve the problem B for all $G_D^0 \in G^0$. In this procedure, one should take into account different ways of positioning G_D^0 in G^0, e.g. if $G^0 = O_h$, then the subgroup with $G_D^0 = D_2$ may be realized using either the set of the elements h_1, h_2, h_3, h_4 or the set h_1, h_4, h_{13}, h_{16}. In the former case, all second-order axes are coordinate, in the latter, two of them are diagonal. All possible crystallographically nonequivalent ways of positioning of the point groups of the symmetry G_D^0 in G^0 can be listed beforehand, they have to be successively tried out in search of all space groups G_D of the lower-symmetry phases induced by the given IR.

Let us make use of the decomposition of the group G into cosets with respect to its translational subgroup T

$$G = \sum_i {}^\oplus g_i T.$$

Here, the elements $g_i = (h_i | \alpha_i)$ are the representatives of the cosets. Let a computer combine the representatives g_i of the cosets corresponding to $h_i \in G_D^0$ with the proper translations associated with all different matrices of the IR. Let us find for each such combination of the generators the matrices corresponding to them and construct their common stationary vector. If this stationary vector proves to be nonzero, it will, in effect, mean that we have obtained the admissible variant of G_D as well as the explicit form of the order parameter generating it. By this procedure, one may find all possible G_D for fixed positionings of G_D^0 in G^0. Trying out all such settings and then the different G_D^0 groups will lead to the solution of the problems A and B.

It is to be noted that the algorithms described above may be used not only for the irreducible, but also for the reducible representations (a relevant problem arises when the phase transitions are studied that are described by several order parameters).

5. IDENTIFICATION OF SUBGROUPS OF A SPACE GROUP

5.1. General description of the algorithm

Consider a set of the symmetry elements of the original space group G corresponding to those matrices of the critical IR which leave a certain stationary vector C invariant. This set $G_D[C]$ is a subgroup of G, hence it is one of 230 space groups. Now the problem of identification of G_D arises. We aim to identify it by means of a computer, which involves some specific difficulties, e.g. the diagrams of space groups often used by crystallographers cannot be employed, uniformity is required in the description of each group etc. We shall first discuss some relevant points, which are to be taken into account in the identification of subgroups of the space groups, and then briefly describe the algorithm of the program GROUP which provides for the solution to the identification problem.

1. The loss, in the transition from G to G_D, of some translations and elements of the point group of symmetry is usually accompanied by a change of the Bravais lattice. Let us denote by T, $\mathbf{a}_i$ and $\mathbf{A}_i$ ($i = 1, 2, 3$) the Bravais lattice and the edges of the primitive and the elementary cell of the high-symmetry phase, respectively, and by $T_D, \mathbf{a}_i'$ and $\mathbf{A}_i'$—the corresponding characteristics of the lower-symmetry phase. The type of the Bravais lattice may change due to disappearance in the phase transition of a number of sites of the set T. This, however, may

also occur when all old translations are preserved (a transition without the change in the volume of the primitive cell), but some elements of the point group of crystal symmetry vanish. Let, for example, the point groups of the space groups G and G_D be C_{4v} and C_{2v}, respectively. In this transition, owing to the disappearance of the four-fold axis, the symmetry element vanishes that "holds together" the tetragonal lattice. This results in a deformation of the lattice near the phase transition point which transforms it into one of the orthorhombic syngony lattices (at the transition point $a_1' = a_2'$, while away from it these vectors become different). It should be noted that the lattices listed in Refs [2, 53] simply describe a set of the sites "surviving" a phase transition, while the true type of the Bravais lattice can be found only through an analysis of the remaining elements of the point symmetry and it often is quite different from that given in the references mentioned.

2. In the analytical description of an infinite space group, usually a finite set of the representatives of all cosets $g_i = (h_i | \alpha_i)$ is taken in the decomposition of G_D with respect to its translational subgroup T_D. The set of the elements h_i is unambiguously fixed by the corresponding point group of the crystal symmetry G_D^0, while the choice of α_i is not rigidly fixed. In the first place, α_i depend on the setting of the center of the space group, which, after the set of the elements belonging to G_D has been singled out of G, may prove to be different from the commonly accepted one. As is known, the shift of the group center by the vector $\mathbf{R}$ leads to the following transformation of the space group elements:

$$g_i' = (h_i' | \alpha_i') = (h_i | \alpha_i + h_i \mathbf{R} - \mathbf{R}) \tag{16}$$

(if, in addition to this shift, rotation of the coordinate axes takes place, then $h_i' \neq h_i$). This transformation is made use of, when the symbol G_D should be brought to the standard form.

3. The improper translations α_i also depend on the choice of a representative of the coset in the decomposition of G_D with respect to the subgroup T_D. It is important to take this choice into account when there are present proper translations lying obliquely to the characteristic direction of the element h_i. In crystallographic terms this gives rise to an "alternation" of the elements of various types in the group (e.g. screw axes alternate with the simple axes, reflection planes with the glide planes of a certain type etc.). From the viewpoint of the group-theoretical description, this means that in one and the same coset there are, owing to the addition to its representative of various proper translations from T_D, space group elements of various types.

4. The international symbol of the space group, by which the identification of G_D is made, may change not only when the alternation of the type of symmetry elements is taken into account, but also when the coordinate axes are transformed. The identification of G_D is also complicated by the fact that its crystal class may have different positions with regard to the crystal class of the original symmetry group.

We now briefly describe the main steps of the identification of the space group G_D used in the program GROUP. (Its first version was written in 1977 in collaboration with S. V. Tsybulya, the second—in 1987—together with V. V. Krivtsova.)

Step 1. Determination of syngony. Let us denote by P_k the order of the crystal class G_D^0 of the group G_D. The identification of syngony with the aid of a computer is performed through carrying out a number of successive checks. The triclinic and monoclinic syngonies are singled out by the value of P_k and by the presence or absence of inversion in G_D^0, the tetragonal, trigonal and hexagonal syngonies—by the presence of only one axis of the highest order as well as the value of this order, and the cubic syngony—by the presence of several three-fold axes. In case the syngony is not determined as a result of these checks, it is orthorhombic.

Step 2. The determination of a Bravais cell. The arms $\mathbf{K}_i$ of the Lifshitz star may be written as

$$\mathbf{K}_i = \sum_{j=1}^{3} \mu_{ij} \mathbf{b}_j,$$

where $\mathbf{b}_j$ are the vectors of the reciprocal lattice and the coefficients

$$\mu_{ij} = \frac{m_{ij}}{n_{ij}}$$

are proper fractions with n_{ij} taking only the values 2, 3, 4. Let n_j be the least common multiple of all n_{ij} corresponding to the fixed vector $\mathbf{b}_j$. Determine now a certain enlarged elementary cell (EEC) of the original phase making use of three basis vectors $\mathbf{A}_j = n_j \mathbf{a}_j$. From the form of the matrices of proper translations it is clear that $M(\mathbf{A}_j)$ is a unity matrix, which means that the EEC will certainly be a cell repeatability not only in the original, but also in the lower-symmetry phase (though, generally, not of minimal volume).

Let us assume that within an EEC there have been N sites of the lattice T and N_D sites of the lattice T_D. Clearly, the density of the Bravais lattice sites has, as a result of the phase transition $G \to G_D$, been reduced N/N_D times, which means that the volume of the primitive cell has been increased N/N_D times, i.e. $V'/V = N/N_D$.

Let us now, by repeating the EEC, build using a computer a fragment of the sites of the lattice T_D about the origin of the coordinates (we shall call the set of these sites a σ-set), sufficient for the choice of the basis vectors $\mathbf{A}_1'$, $\mathbf{A}_2'$, $\mathbf{A}_3'$ of the Bravais cell for this lattice. The vectors $\mathbf{A}_1'$, $\mathbf{A}_2'$, $\mathbf{A}_3'$ are determined by the three sites, closest to the origin of the coordinates, along the axes X, Y and Z, which are selected in accordance with the usual crystallographic rules. For example, the Z axis must coincide with the higher-order axis, if there is only one such axis for the given syngony. The X axis is, in this case, directed toward the nearest site lying in the plane normal to the Z axis and the axis Y lies at the angle of $90°$ to the X axis for the tetragonal and of $120°$ for the hexagonal and rhombohedral syngonies.

Step 3. Determination of the centering type of the Bravais cell. Following the determination of the Bravais cell edges $\mathbf{A}_1'$, $\mathbf{A}_2'$, $\mathbf{A}_3'$ of the lower-symmetry phase, we have to find the centering type of this cell. The σ-set of the sites is inspected in order to determine whether there is among these at least one site possessing noninteger, in fractions of $|\mathbf{A}_1'|$, $|\mathbf{A}_2'|$ and $|\mathbf{A}_3'|$, projections onto the axes x, y and z, respectively. There are formed the logical variables xy, xz and yz, which take on truth values in the case of the centering of the corresponding face of the Bravais cell, as well as the variables xyz having the truth value for the centering of its volume. Depending on the truth or falsehood of the above-mentioned variables, the type of the Bravais lattice is determined (P, I, F, C, A, B, H, R in international notation). The trigonal lattice R is considered in the hexagonal system of coordinates. The Bravais cell corresponding to it is, in contrast to the cell of the hexagonal lattice (H), "twice centered."

Step 4. Determination of the international symbol. The final identification of the space group G_D in the program **GROUP** is made with the use of its international symbol. The rules of writing it for each of the syngonies are well-known [10, 54]. In order to construct the symbol of G_D, it is necessary to determine the types of the axes and the symmetry planes entering into this symbol as generators of the group.

For the determination of the type of the axis, project the improper translation of the corresponding rotational element onto this axis and find which part it is of the modulus of the minimal proper translation along the axis under consideration. To eliminate ambiguity in the symbol introduced by alternation of the axis types, one should successively add to the improper translation the proper translations, corresponding to all sites of the σ-set, after which the identification of the axis type is repeated. In such a way, we try to obtain the least projection of the improper translation onto the given axis. An analogous procedure is made use of, when determining the type of the plane, which, too, may change because of alternation. The improper translation corresponding to the plane is projected onto it, and the components of this projection onto basis vectors of the Bravais cell are determined. An analysis is then made to ascertain along which axes there occurs improper gliding and what is its value (1/2 or 1/4 with respect to the proper translation along this axis). After this, all proper translations of the σ-set are successively added to the original improper translation and the procedure of identification of the plane type is repeated. By this procedure a type of the plane with higher priority is selected (it is assumed that the priority is diminished from the left to the right of the standard list of the possible plane types m, n, d, a, b, c).

There is one exception regarding the simplification of the space group symbol by taking account of the alternation of the types of the symmetry elements. The symbol of the group $D_2^9 = I2_1 2_1 2_1$ coincides, after above considered simplification, with that of the group $D_2^8 = I222$, still, these groups are different, since in the latter case the second-order axes do intersect, while in the former they do not. To differentiate these cases, the computer finds the product of two axes of the second-order and determines whether this product is a simple (2) or a screw (2_1) axis; in the former case $G_D = I222$ and in the latter $G_D = I2_1 2_1 2_1$. An analogous check can be carried out also with the groups $T^3 = I23$ and $T^5 = I2_1 3$.

From the foregoing it is evident that the program GROUP is based on a purely crystallographic method of identification of the subgroups of a space group. In Ref. [55] another algorithm of identification is described based on analytical transformations.

5.2. Examples

Let us now return to the case of the four-dimensional IR X_1 of the group $P4g$ and identify the low-symmetry phases induced by it. Making use of Table 1, Tables 2 and 3 may be obtained, where over each matrix of the representation X_1 an element of the space group is indicated corresponding to this matrix. For example, the matrix No. 6 from Table 3 is obtained via multiplication of the matrix No. 2 form the right by the matrix No. 5. Multiplication in the same order of the corresponding symmetry elements gives $(h_{14}|0, 0)$ $(h_{26}|\frac{1}{2}, \frac{1}{2}) = (h_{40}|-\frac{1}{2}, \frac{1}{2})$. Now we make use of Tables 2 and 3 to find the G_D-groups of the low-symmetry phases corresponding to each of the nine stationary vectors of the critical IR X_1 under consideration obtained in the previous section.

1. $\mathbf{C}_1 = (a, 0, 0, a)$. None of the matrices corresponding to proper translations from Table 2 leaves this vector invariant. However, from the appearance of the two-arm star corresponding to the point X of the Brillouin zone $[\mathbf{k}_1 = (0, \frac{1}{2}); \mathbf{k}_2 = (\frac{1}{2}, 0)]$ it is clear that the double proper translations $2\mathbf{a}_1, 2\mathbf{a}_2, 2\mathbf{a}_1 + 2\mathbf{a}_2$ etc. "survive" the phase transition. In Fig. 1 the sites of the Bravais lattice T_D of the lower symmetry phase are indicated by thick dots. Out of four proper translations of the original G-phase, the EEC with the periods $2\mathbf{a}_1, 2\mathbf{a}_2$ (see Table 2) has only one site surviving in the G_D-phase, namely the zeroth. Thus, the site density in the plane is lowered four times and, accordingly, the ratio between the volumes of the primitive cells of the low-symmetry and the parent phase (V' and V, respectively) $V'/V = 4$.

Find now the group G elements surviving in the G_D-phase, which are different from the purely translational elements (in more precise terms, we search for the representatives of the cosets in the decomposition of G_D with respect to its translational subgroup T_D). It can easily be found that the vector $\mathbf{C}_1$ is left invariant by the following four matrices of the critical IR: 1–1, 2–3, 2–6 and 1–8 (may we recall that the number before the dash indicates the matrix from Table 2 and that after it–from Table 3). The following elements of the group $P4g$ correspond to these matrices:

$$(h_1|0, 0); \quad (h_4|1, 0); \quad (h_{40}|\tfrac{1}{2}, \tfrac{1}{2}); \quad (h_{37}|\tfrac{1}{2}, -\tfrac{1}{2}). \tag{17}$$

For example, the product of the elements $(h_1|\mathbf{a}_1) \equiv (h_1|1, 0)$ and $(h_{40}|-\frac{1}{2}, \frac{1}{2})$, which is the element $(h_{40}|\frac{1}{2}, \frac{1}{2})$, corresponds to the matrix 2–6. The set of the elements h_i from elements (17) determines the class $mm2$ which belongs to orthorhombic syngony. For this syngony, the basic periods of the G_D-phase (the edges of the Bravais cell) must be perpendicular to the planes h_{40} and h_{37}. They are shown in Fig. 1 and have the form: $\mathbf{A}'_1 = 2\mathbf{a}_1 + 2\mathbf{a}_2$; $\mathbf{A}'_2 = -2\mathbf{a}_1 + 2\mathbf{a}_2$. It is evident from the figure that the low-symmetry phase has a centered cell (the Bravais lattice of type C).

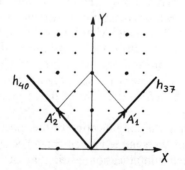

Fig. 1. Orthorhombic lattice of type C for G_D with $V'/V = 4$.

The international symbol of a space group is, as known, constructed from the symbol of the crystal class ($mm2$ in the present case) by adding to it the type of the Bravais lattice and by indicating the type of the elements entering into this symbol, which is determined by the improper translations corresponding to them. The two-fold axis is in this case simple since the improper translation $\boldsymbol{\alpha}_4 = (1, 0)$, corresponding to the element h_4, does not give a projection onto the direction of this axis (the axis h_4 is normal to the plane of the two-dimensional lattice). It is easy to see that the improper translation $\boldsymbol{\alpha}_{40} = (\frac{1}{2}, \frac{1}{2})$, corresponding to the plane h_{40}, is normal to it and, consequently, it is a mirror plane of the type m. Thus, we have obtained the international symbol of the G_D under consideration: $Cmm2$.

For X-ray, neutron-diffraction and other structural studies, useful may prove not only the knowledge of the symbol of the G_D but also of the explicit form of its setting in the group $P4g$ of the parent phase. For this purpose, it would be sufficient to know the transformation of the coordinates' origin, which brings the set of the G_D-group elements to the form consistent with the international symbol obtained above. Having noted that in this symbol ($Cmm2$) the two-fold axis is not accompanied by any improper translation, let us demand that, upon the shift of the system of coordinates according to condition (16), the improper translation $\boldsymbol{\alpha}_4 = (1, 0)$ of the element h_4 disappear: $\boldsymbol{\alpha}_4' = (0, 0)$. Taking account of that $\mathbf{R} = (x, y)$ in condition (16) is an unknown vector, we obtain the equations: $-2x + 1 = 0$; $-2y = 0$, wherefrom $\mathbf{R} = (\frac{1}{2}, 0)$. From equation (16) written for h_{37} and h_{40}, it is easy to see that the improper translations for these elements also disappear due to the obtained shift in the origin of coordinates and thus we come again to the symbol $C2mm$, this time by means of an analytical method.

2. $\mathbf{C}_2 = (a, 0, a, 0)$. Out of four matrices of proper translations (see Table 2)), two—the first and the second—leave this vector invariant. Hence, the density of the sites of the Bravais lattice is decreased twice ($V'/V = 2$). The relevant lattice is shown in Fig. 2: $\mathbf{A}_1' = \mathbf{a}_1$; $\mathbf{A}_2' = 2\mathbf{a}_2$. Note that when searching for the matrices of the "rotational" elements which leave the stationary vector invariant, we act successively upon it first with the matrices from Table 3. If they change the structure of this vector [as doses, e.g. the matrix No. 2, which transforms $\mathbf{C}_2$ into the vector $(0, a, 0, -a)$], then the variants of their combination with the matrices of the proper translations from Table 2 need not be considered. For the vector $\mathbf{C}_2$, there is only one matrix, which leaves it invariant, namely, 3–7 corresponding to the element $(h_{27} | -\frac{1}{2}, \frac{1}{2})$. This element is, actually, a representative of the second coset in the decomposition of G_D with respect to its translational subgroup T_D, already obtained, which is determined by the vectors $\mathbf{A}_1' = \mathbf{a}_1$, $\mathbf{A}_2' = 2\mathbf{a}_2$

$$G_D = T_D + (h_{27} | -\tfrac{1}{2}, \tfrac{1}{2}) \, T_D. \tag{18}$$

The element h_{27} determines the plane normal to the axis Y so that the component of the improper translation $\boldsymbol{\alpha}_{27}$ along this axis does not affect the character of this plane. On the other hand, the component $\boldsymbol{\alpha}_{27}$ along the axis X turns the plane h_{27} into a glide plane of the type a. Thus, the symbol of the G_D-group for the stationary vector $\mathbf{C}_2$ has the form $P1a1 = Pg$.

The shift of the origin of coordinates that we are seeking must provide for the standard analytical description of the group, i.e. it must set at zero the projection of the translation

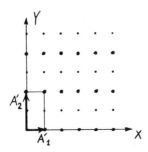

Fig. 2. Orthorhombic lattice of type P for G_D with $V'/V = 2$.

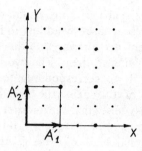

Fig. 3. Square lattice for G_D with $V'/V = 4$.

of h_{27} along the Y axis. From the equation $\alpha'_{27} = \alpha_{27} + h_{27}\mathbf{R} - \mathbf{R} = 0$ we derive $\mathbf{R} = (0, \frac{1}{4})$, which yields $(h_{27}|-\frac{1}{2}, 0)$. However, in virtue of equation (18), one may choose any other element of the second coset as its representative. By adding to it the new proper translation $\mathbf{A}'_1$, we get $(h_{27}|\frac{1}{2}\mathbf{A}'_1)$, which corresponds to the glide plane of type a in standard notation.

3. $\mathbf{C}_3 = (a, a, 0, 0)$. The Bravais lattice is determined by the vectors $\mathbf{A}'_1 = 2\mathbf{a}_1$, $\mathbf{A}'_2 = 2\mathbf{a}_2$ (Fig. 3), $V'/V = 4$, $G_D = P4$.

4. $\mathbf{C}_4 = (a, 0, 0, 0)$. The Bravais lattice is shown in Fig. 2. $\mathbf{A}'_1 = \mathbf{a}_1$; $\mathbf{A}'_2 = 2\mathbf{a}_2$, $V'/V = 2$, $G_D = P112$.

5. $\mathbf{C}_5 = (a, 0, 0, b)$. Clearly, the vector $\mathbf{C}_1 = (a, 0, 0, a)$ considered earlier is a particular case of the vector $\mathbf{C}_5$: we may obtain $\mathbf{C}_1$ from it by assuming $b = a$. Therefore, only the matrices of some elements of the set (17) can leave $\mathbf{C}_5$ invariant. The only such matrix is 2–3 which corresponds to the element $(h_4|1, 0)$. The sites surviving in G_D are shown in Figs 1 or 3. Since this case corresponds to monoclinic space syngony (the second-order axis lies normal to the planar lattice), $\mathbf{A}'_1$ is determined by the vector from the origin of coordinates to the nearest site in the plane perpendicular to h_4, while $\mathbf{A}'_2$ is determined by the another nearest site whose vector is noncollinear with $\mathbf{A}'_1$. We may thus assert that the lattice T_D is, for this case, shown in Fig. 3. So we have $G_D = P112$, $\mathbf{R} = (\frac{1}{2}, 0)$.

6. $\mathbf{C}_6 = (a, 0, b, 0)$. A particular case of this vector $\mathbf{C}_2 = (a, 0, a, 0)$ has already been considered. The corresponding T_D is given in Fig. 2. However, the matrix of the element $(h_{27}|-\frac{1}{2}, \frac{1}{2})$, which left $\mathbf{C}_2$ invariant, does not do so in regard to the vector $\mathbf{C}_6$. The latter corresponds therefore to $G_D = P1$, $V'/V = 2$, $\mathbf{R} = (0, 0)$.

7. $\mathbf{C}_7 = (a, b, 0, 0)$. T_D is shown in Fig. 3, $V'/V = 4$, $G_D = P112$, $\mathbf{R} = (0, 0)$.

8. $\mathbf{C}_8 = (a, b, b, a)$. T_D is shown in Fig. 1, $V'/V = 4$, $\mathbf{R}$ has, in the general case, the form $\mathbf{R} = (\frac{1}{2} - y, y)$. As it is sufficient to choose any shift of the origin of coordinates bringing G_D to the standard description, it would suffice to set $y = 0$ so that $\mathbf{R} = (\frac{1}{2}, 0)$. $G_D = C1m1$.

9. $\mathbf{C}_9 = (a, b, c, d)$. This is a trivial case. $G_D = P1$, $V'/V = 4$.

6. OBTAINING THE COMPLETE CONDENSATE OF STATIONARY VECTORS

The symmetry group $G_D[\mathbf{C}^0]$ of a low-symmetry phase is, as was noted in Section 2 of the present work, fully determined by the corresponding stationary vector $\mathbf{C}^0$ of the critical IR Γ_0. However, the structure of this low-symmetry phase defined by formula (1) depends not only on the critical IR, rather, it receives contributions from some other, noncritical IR's Γ_j of the group G. In order to find the explicit form of these contributions to the structure of a fixed G_D-phase, it is necessary to obtain the complete condensate of the stationary vectors (order parameters) corresponding to this phase. In calculational terms, the problem may be set as follows. Since the subgroup G_D (with the concrete manner of its positioning in the group G taken into account) is now assumed to be known, it is required to single out all those IR's Γ_j which correspond to the nonzero stationary vectors common to all matrices of their restrictions $\Gamma_j{\downarrow}G_D$ to the subgroup G_D. With the aim of solving this problem we wrote, in collaboration with V. N. Kesoretskikh, in 1982 the program

CONDENSATE which is, in fact, a somewhat modified program VECTOR. Its input data comprise the set of the IR's under consideration and the number of the critical IR. The program finds all stationary vectors of this critical IR as was described in Section 4 (it is also possible to input a previously calculated set of these vectors) and then for each of these it selects the numbers of the matrices which enter into the restriction $\Gamma_0 \downarrow G_D$. After this computer considers the matrices with the same numbers but belonging to each of the other given IR's (i.e. the matrices of the restriction $\Gamma_j \downarrow G_D$) and finds the stationary vectors common to all matrices of $\Gamma_j \downarrow G_D$. The IR Γ_j, which produces the nonzero stationary vector $\mathbf{C}^j$ of the restriction $\Gamma_j \downarrow G_D$, makes a contribution to the complete condensate. Thereby we also obtain the explicit form of each of the noncritical stationary vectors $\mathbf{C}^j$ that correspond to the critical stationary vector $\mathbf{C}^0$.

The program CONDENSATE has a special block which calculates multiplicities of the entering of each of the noncritical IR's into the symmetrized direct product of the critical IR Γ_0. As was noted in Section 2, this information may be used for an evaluation of smallness of the components belonging to the critical and various noncritical order parameters near a phase transition point.

The described procedure of the obtaining and analysis of the complete condensate was proposed in Ref. [19]. It was used there for finding the condensate of order parameters of the low-symmetry phases induced by the IR's of the group O_h^7. Analogous results for the phase transitions in crystals with the space group O_h^1 were given in Ref. [56]. A concrete example of obtaining the complete condensate for the point group O_h is considered in detail in Ref. [57], where there is also discussed a symmetry-based approach to the determination of a critical representation corresponding to the given symmetry lowering occurring as a result of a phase transition. In Refs [36], [57], a link was established between the notion of the complete condensate of stationary vectors and the theory of color symmetry. Based on this relationship, a number of theorems on complete condensate was proved, in particular, a criterion of its completeness was suggested which substantially simplifies the obtaining of the condensate and may, to a certain extent, verify correctness of its construction by means of a computer.

7. CONSTRUCTION OF BASIS FUNCTIONS OF THE IR's OF SPACE GROUPS

In order to find, by means of formula (1), the structure of the low-symmetry G_D-phase, it is necessary to know, apart from the complete condensate of the stationary vectors, the basis functions of the critical and all noncritical IR's of the group G. The basis functions cannot be determined by the IR alone, they also depend on the type of the phase transition (displacement, order–disorder, orientational, magnetic etc. transitions) as well as on crystal structure in the original G-phase, i.e. on the distribution of its atoms over the orbits of the space group G. Since these orbits are symmetry-independent, the problem arises of constructing the basis functions for the most frequently occurring transition types for all the orbits and all the IR's of each of the 230 space groups. Our program FONON, which was written in collaboration with S. I. Ulyanova and G. A. Lisochenko, is capable of constructing scalar, vectorial and pseudovectorial basis of the IR's of space groups. It implements the method of projection operators described in Refs [6, 30].

The program requires the following input data (a) the code of the transition type; (b) the starting point of the orbit; (c) generators of the space group; (d) generators of the wave vector group and matrices of the ray IR's corresponding to them (e.g. from the Ref. [39]). Its output yields all variants of the bases of the given IR Γ_j (e.g. for the displacement type transitions the number of such bases equals the number of times which Γ_j enters into the mechanical representation (corresponding to the given orbit of the space group).

The program FONON was used for obtaining the basis functions of a number of space gorups. In Ref. [58], the basis functions were found for transitions of the displacement and ordering types for most orbits of the O_h^7-group and for the IR's corresponding to high-symmetry points in the Brillouin zone. In the same publication, an alternative method is described for constructing the basis functions of IR's of the space groups, which is, in effect, based on the definition of a group representation. This method is more straightforward than the method of projection operators, we have used it for the verification of the computer results.

8. SPLITTING OF THE ORBITS OF A SPACE GROUP
INDUCED BY A PHASE TRANSITION

So as to acquire complete symmetry-related information on the local properties of atomic positions in the low-symmetry G_D-phase, we together with V. V. Krivtsova, developed the program ORBIT. The results that may be obtained with the aid of this program are valuable in themselves, on the other hand, they may be used to verify the correctness of the construction of the complete condensate of normal modes according to formula (1).

The atoms of a crystal occupy in the original phase a certain set of orbits of the space group G. All sites of each orbit in the group G are equivalent being linked by certain symmetry elements $g_i \in G$. However, due to the symmetry lowering occurring in consequence of the phase transition $G \rightarrow G_D$, a given orbit R of the group G may split into a certain set of independent orbits R_D^j of its subgroup G_D. Every R_D^j is characterized by its stabilizer S_D^j (i.e. a group of the local symmetry of its sites), which is a subgroup of the stabilizer S of the original orbit R. We should like to point out the character of the information which can be gained when the type of the orbit splitting is known. For this purpose we use as an example the transition $O_h^7 \rightarrow C_{2h}^3$ with quadrupling of the primitive cell volume (the low-symmetry phase C_{2h}^3 is induced by the six-dimensional IR X_3 of the group O_h^7 [19]). As a result of this transition, the orbit $R = 16$ (d) with the stabilizer $S = D_{3d}$ is split within an EEC, constructed by doubling all periods of the primitive cell of the original phase as follows:

$$R = 2(2)C_{2h} + 1(4)C_i + 1(4)C_s + 1(4)C_2 + 2(8)C_1. \qquad (19)$$

Addition on the r.h.s. of this formula signifies union of the sets R_D^j. The number before the round bracket indicates the number of the orbits R_D^j of this type. The brackets enclose the number of the points of the orbit R_D^j, which lie within the EEC. After the brackets, the stabilizer of the orbit is given.

Thus, the atoms occupying the positions $16(d)$ of the group O_h^7 become, owing to the phase transition, physically unequivalent—they divide into seven different types, which may, in principle, be verified by such methods as EPR, NMR and others [59, 60]. Formula (19) permits a quite definite judgement as to the structure of the low-symmetry phase having the symmetry C_{2h}^3. For example, in displacement type transitions the atoms occupying the first three orbits cannot be displaced because their stabilizers (C_{2h} and C_i) contain inversion; the atoms belonging to the orbit $1(4)C_s$ can be displaced in the plane in any direction, while those of $1(4)C_2$—only along the two-fold axis. The last R_D^j are general orbits of the group C_{2h}^3 (their stabilizers contain the identity element only), and no additional symmetry restrictions arise in regard to possible displacement directions of the atoms corresponding to these orbits.

Practically, the splitting of the orbit R may be performed in the following manner. Take an arbitrary point $\mathbf{r}_1 \in R$ and act on it with all elements of the group G_D. The set of the different points obtained forms the first orbit R_D^1 and those elements of the symmetry, under whose action the point $\mathbf{r}_1$ does not change its position, enter into the stabilizer S_D^1. Cross all the points of the set R_D^1 off the points of the orbit R and act upon the first of the remaining points with all elements of the group G_D, thereby singling out the second orbit R_D^2 and its stabilizer S_D^2 and so forth. Precisely this principle underlies the program ORBIT. Intermediate information from the program GROUP serves as its intput and its output provides a scheme of type (19) splitting of every orbit of the group G and indicates the starting point for every orbit of the group G_D.

A relationship between the splitting of the orbit of a space group occurring in a phase transition and the theory of color symmetry was considered in Refs [36, 57]. The results of the orbit splitting for some concrete groups as well as the use of this information in the theory of structural phase transitions will be dealt with in our subsequent publications.

9. A COMPARISON BETWEEN VARIOUS APPROACHES TO
THE GROUP-THEORETICAL ANALYSIS OF PHASE
TRANSITIONS IN CRYSTALS

In conclusion, we should like to compare briefly the approach and the results considered in the present article with the methods and the results reported by other authors. Most of the works on the group-theoretical studies of phase transitions in crystal are devoted to the solution of the first problem considered herein, viz. listing and classification of possible symmetry groups G_D of the low-symmetry phases arising in consequence of continuous phase transitions. All these works may be divided into the following groups.

1. The works which aim to obtain only partial information on the G_D-phases. For the second-order phase transitions in three-dimensional structures, possible translational changes of symmetry are listed in Refs [2, 53], and the changes in the point groups of symmetry—in Ref. [32]. Analogous problems in regard to diperiodic systems have been solved in Refs [61–64]. The next step has been taken in Refs [65, 66] where possible G_D-groups are classified by the so-called phase transition channels (each such channel is characterizerd by a set of arms of the wave vector star, to which nonzero components of the stationary vectors correspond). In all these works, no use is made of irreducible representations of the space groups, which is why they can provide only partial information on symmetry changes in a phase transition.

2. The works, which do make use of the technique of irreducible representations of the space groups, where a procedure is employed originating from the classical version of the Landau theory, namely, the minimization of the nonequilibrium thermodynamic potential ϕ written in the form of a certain truncated series in the components of the order parameter. On account of the restriction of the power series to only its lower terms, many symmetry-allowed G_D-phases are lost. This group may be represented by such works as Ref. [24] for the case of the space groups or Ref. [67] for two-dimensional systems.

3. The works based on the Birman criterion [22, 23, 25, 26]. There the results are obtained by using not the explicit form of the IR's, but rather the characters of these representations. Despite the basic simplicity of these methods, their manual implementation has often resulted in errors and omissions of various G_D. As an example, we may refer to a discussion in Ref. [24, 25] concerned with completeness of the list of low-symmetry phases for the group O_h^3 or to numerous errors made in Ref. [26] when obtaining the G_D for the two-dimensional structures. These errors came to light after a comparison with the results of a computer calculation carried out in Refs [18, 21] (see below).

4. The works [8, 27–29], based on the notion of the stationary vectors, done manually. Only the work [8] may be regarded as sufficiently correct (a few errors occurring there were corrected in Refs [20, 56] after a comparison with the computer calculations). The work [28] is limited to the listing of the phase transitions associated with one arm of the wave vector star. Other publications contain quite a few errors and omissions, for example, the authors of Ref. [20] found in Ref. [29] more than one hundred errors.

5. The works, carried out with the aid of a computer, based either on the Birman criterion or on the stationary vectors method. The group of Hatch and Stokes employs the former method, our group—the latter. A comparison of the lists of possible low-symmetry phases obtained by these two methods has shown that they coincide almost completely (except for some insignificant slips). Such a comparison was made possible by, for example, the publication of the lists of the G_D-phases for all 80 diperiodic space groups of symmetry [18, 21] as well as for some space groups. Of course, a calculation by means of a computer does not in itself guarantee correctness of the results. There may be deep-hidden mistakes in the program, errors in the input data etc. As an example of such an error, we may point to the list, pubished by us in Ref. [12], of low-symmetry phases for the eight-dimensional IR of the group O_h^7 corresponding to the point L in the Brillouin zone. The mistakes in this listing occurred because of the incorrect matrices of the corresponding IR that were input into the computer—this was later pointed out in Ref. [69]. Having checked the computation, we found, in addition to the list manually obtained in Ref. [69], some more G_D-phases. On this occasion, we should like to present a corrected list of the low-symmetry phases for this case (Table 7). This Table, moreover, illustrates the typical form of the results obtained by the SV-method. Errors, which do occur in various publications, point convincingly to the necessity to conduct group-theoretical calculations for multidimensional IR's of the space groups with the aid of computers. The use of different methods and participation of different authors are, certainly, desirable for such computer calculations.

Table 7. Complete set of the G_D-phases induced by eight-dimensional IR's of the group O_h^7 of the wave vector $\mathbf{k}_9 = \frac{1}{2}(\mathbf{b}_1 + \mathbf{b}_2 + \mathbf{b}_3)$

C	G_D 9/5	G_D 9/6	V'/V	nh	n
$a\ 0\ 0\ 0\ 0\ 0\ a\ 0$	$Cmcm$	$Cmca$	4	6	24
$0\ a\ 0\ 0\ 0\ 0\ 0\ a$	$Cmca$	$Cmcm$	4	6	24
$a\ 0\ a\ 0\ a\ 0\ a\ 0$	$I\bar{4}m2$	$I\bar{4}c2$	8	6	48
$0\ a\ 0\ a\ 0\ a\ 0\ a$	$I\bar{4}c2$	$I\bar{4}m2$	8	6	48
$0\ 0\ 0\ 0\ 0\ 0\ a\ 0$	$B2/m$	$B2/b$	2	12	24
$0\ 0\ 0\ 0\ 0\ 0\ 0\ a$	$B2/b$	$B2/m$	2	12	24
$a\ b\ a\ b\ a\ b\ a\ b$	$F222$	$F222$	8	12	96
$a\ 0\ b\ 0\ b\ 0\ a\ 0$	$Imm2$	$Iba2$	8	12	96
$a\ b\ 0\ 0\ 0\ 0\ a\,{-}b$	$B2/m$	$B2/m$	4	12	48
$a\ 0\ 0\ b\ 0\ b\ a\ 0$	$Ima2$	$Ima2$	8	12	96
$0\ a\ 0\ b\ 0\ b\ 0\ a$	$Iba2$	$Imm2$	8	12	96
$a\ 0\ 0\ 0\ 0\ 0\ b\ 0$	$P2_1/m$	$P2_1/b$	4	12	48
$0\ a\ 0\ 0\ 0\ 0\ 0\ b$	$P2_1/b$	$P2_1/m$	4	12	48
$0\ 0\ 0\ 0\ a\ b\ a\ b$	$B2/b$	$B2/b$	4	12	48
$0\ 0\ 0\ 0\ 0\ 0\ a\ b$	$P\bar{1}$	$P\bar{1}$	2	24	48
$a\ b\ a\,{-}b\ a\,{-}b\ a\ b$	$I\bar{4}$	$I\bar{4}$	8	12	96
$0\ a\ 0\ 0\ 0\ 0\ b\ 0$	$P2/b$	$P2/b$	4	12	48
$a\ 0\ b\ c\,{-}b\ c\ 0\ 0$	$B2/m$	$B2/b$	8	12	96
$0\ a\ b\ c\,{-}b\ c\ 0\ 0$	$B2/b$	$B2/m$	8	12	96
$0\ 0\ 0\ 0\ a\ b\ c\ d$	$P\bar{1}$	$P\bar{1}$	4	24	96
$a\ b\ c\ d\ c\ d\ a\ b$	$B2$	$B2$	8	24	192
$a\ 0\ b\ c\ b\,{-}c\ d\ 0$	Bm	Bb	8	24	192
$0\ a\ b\ c\ b\,{-}c\ 0\ d$	Bb	Bm	8	24	192
$a\ 0\ b\ c\,{-}b\ c\ 0\ d$	$B2$	$B2$	8	24	192
$a\ b\ c\ d\ e\ f\ 0\ 0$	$P\bar{1}$	$P\bar{1}$	8	24	192
$a\ b\ c\ d\ e\ f\ g\ h$	$P1$	$P1$	8	48	384

$(0\ 0,\ w\mathbf{x},\ \tilde{w}\mathbf{x})$:					
$\mathbf{x} = (a, b)$	$R\bar{3}$	$R\bar{3}$	8	8	64
$\mathbf{x} = (a, -\sqrt{3}a)$	$R\bar{3}m$	$R\bar{3}c$	8	4	32
$\mathbf{x} = (\sqrt{3}a, a)$	$R\bar{3}c$	$R\bar{3}m$	8	4	32

Note. This table contains 29 stationary vectors **C** and the corresponding symmetry groups G_D for the IR's 9/5 and 9/6 whose matrices are given in Ref. [19]. The symbols $w\mathbf{x}$ and $\tilde{w}\mathbf{x}$ denote the products of the two-dimensional matrix

$$w = \begin{pmatrix} -1/2 & \sqrt{3}/2 \\ -\sqrt{3}/2 & -1/2 \end{pmatrix}$$

as well as the matrix $\tilde{w}$, transposed with respect to it, and the vectors **x** given in three last rows of this table. V'/V is the ratio between the volumes of the primitive cells before and after phase transition (it equals the number of possible "translational" domains); nh is the number of "rotational" domains and n is the total number of the possible domain types.

On the second problem of the group-theoretical analysis, i.e. the determination of the structure of the low-symmetry phases, only few works are known to us. One may point, for instance, to Ref. [70] where for the case $\mathbf{k} = 0$ a method is proposed, based on the Birman criterion, for obtaining a list of the IR's Γ_j, which make a contribution to the complete condensate, or to Ref. [71] where the case of the magnetic phase transitions is treated and an evaluation is made of the temperature dependence of noncritical order parameters. Such an evaluation may, however, lead to erroneous results, for the reason given in Ref. [19].

In conclusion, let it be noted that, when writing the programs mentioned in the present article, we used the following algorithmic languages: ALGOL-60, PL-1 and PASCAL.

REFERENCES

1. L. D. Landau, On the theory of phase transitions. *Zh. éksp. teor. Fiz.* **7**, 19–40 (1937).
2. E. M. Lifshitz, On the theory of phase transitions of the second order. *J. Phys.* **6**, 61–74 (1942).
3. L. D. Landau and E. M. Lifshitz, *Statistical Physics*, Part 1. Pergamon, Oxford (1980).
4. Yu. M. Gufan, On the theory of phase transitions with the multi-component order parameters (in Russian). *Fizika tverd. Tela* **13**, 225–230 (1971).
5. Yu. M. Gufan, *Theory of Structural Phase Transitions* (in Russian). Nauka, Moscow (1982).
6. Yu. A. Izyumov and V. N. Syromyatnikov, *Phase Transitions and Crystal Symmetry* (in Russian). Nauka, Moscow (1984).
7. Yu. M. Gufan and V. P. Sakhnenko, On phase transitions with two- and three-components order parameters (in Russian). *Zh. éksp. teor. Fiz.* **63**, 1909–1918 (1972).
8. E. B. Vinberg, Yu. M. Gufan, V. P. Sakhnenko and Yu. I. Sirotin, On symmetry changes due to the phase transitions in the crystals with O_h^1 space group (in Russian). *Kristallografiya* **19**, 21–26 (1974).

9. L. Michel, Symmetry defects and broken symmetry. Configurations. Hidden symmetry. *Rev. mod. Phys.* **52**, 617–651 (1980).

10. Yu. I. Sirotin and M. P. Shaskolskaya, *Principles of Physics of Crystals* (in Russian). Nauka, Moscow (1975).

11. Yu. M. Gufan and G. M. Chechin, On geometrical restrictions concerning the choice of the parent phase in the case of a six-component order parameter (in Russian). *Kristallografiya* **25**, 453–459 (1980).

12. V. P. Sakhnenko, V. M. Talanov and G. M. Chechin, Phase transitions allowed by symmetry in the crystals of O_h^7 space group (in Russian). Manuscript deposited in VINITI, No. 638–82, pp. 1–25 (1981).

13. Yu. M. Gufan, V. P. Dmitriev, V. P. Popov and G. M. Chechin, Structures of ordered alloys with hexagonal close packing (in Russian). *Fizika Metall.* **46**, 1133–1142 (1978).

14. Yu. M. Gufan, V. P. Dmitriev, V. P. Popov and G. M. Chechin. On order-disorder type transitions in alloys with cubic face-centered closely packed structure (in Russian). *Fizika tverd. Tela* **21**, 554–561 (1979).

15. E. V. Gorbunov, Yu. M. Gufan, N. A. Petrenko and G. M. Chechin. On the theory of phase transitions with a multicomponent order parameter and some questions of the theory of transitions in boracites (in Russian). *Kristallografiya* **26**, 8–11 (1981).

16. V. P. Dmitriev, Yu. M. Gufan, V. P. Popov and G. M. Chechin. Structure of ordered alloys with hexagonal close packing. *Phys. Status Solidi (a)* **57**, 59–66 (1980).

17. Yu. M. Gufan and V. P. Popov, On the theory of the phase transitions described by a four-component order parameter (in Russian). *Kristallografiya* **25**, 921–929 (1980).

18. V. P. Sakhnenko, G. M. Chechin, M. G. Glumov and N. B. Martynenko, Phase transitions in objects described by diperiodic symmetry groups (in Russian). Manuscript deposited in VINITI, No. 222–83 (1980).

19. V. P. Sakhnenko, V. M. Talanov and G. M. Chechin, Group theory analysis of the complete condensate, arising upon structural phase transitions (in Russian). *Fizika Metall.* **62**, 847–856 (1986).

20. H. T. Stokes and D. M. Hatch, Group–subgroup structural phase transitions: a comparison with existing tables. *Phys. Rev.* **B30**, 4962–4967 (1984).

21. D. M. Hatch and H. T. Stokes, Symmetry-restricted phase transitions in two-dimensional solids. *Phys. Rev.* **B30**, 5156–5166 (1984).

22. J. L. Birman, Simplified theory of symmetry change in second order phase transitions: application to V_3Si. *Phys. Rev. Lett.* **17**, 1216–1219 (1966).

23. F. E. Goldrich and J. L. Birman, Theory of symmetry change in second-order phase transitions in Perovskite structure. *Phys. Rev.* **167**, 528–532 (1968).

24. W. Bociek and J. Lorenc. Landau's approach to symmetry changes in $A15(O_h^3$-$Pm3n)$ structure. *Phys. Rev.* **B25**, 2012–2014 (1982).

25. M. V. Jarić and J. L. Birman, Group theory of phase transitions in A-15 O_h^3-$Pm3n$ structure. *Phys. Rev.* **B16**, 2564–2568 (1977).

26. S. Deonarine and J. L. Birman, Symmetry change in continuous phase transitions in two-dimensional systems. *Phys. Rev.* **B27**, 2855–2867 (1983).

27. V. I. Zinenko and S. V. Misiul, Possible phase transitions in the crystals with O_h^5 space group (in Russian). Manuscript deposited in VINITI, No. 313-78 (1978).

28. M. Sutton and R. L. Armstrong, Symmetry restrictions of phase transitions imposed by group-subgroup structure. *Phys. Rev.* **B25**, 1813–1821 (1982).

29. M. H. Ben Ghozlen and Y. Mlik, Structural phase transitions in crystals with $Fm3m$ symmetry. *J. Phys.* **C16**, 4365–4381 (1983).

30. Yu. A. Iziumov, V. E. Naish and R. P. Ozerov, *Neitronografic Research of Magnetic Materials* (in Russian), Vol. 2. Atomizdat, Moscow (1981).

31. W. Cochran and A. Zia, Structure and dynamics of Perovskite-type Crystals. *Phys. Status Solidi* **25**, 273–283 (1968).

32. V. L. Indenbom, Phase transitions without changing of the number of atoms in primitive cell (in Russian). *Kristallografiya* **5**, 115–125 (1960).

33. I. S. Zheludev and L. A. Shuvalov, Ferroelectric phase transitions and crystal symmetry (in Russian). *Kristallografiya* **1**, 681–688 (1956).

34. M. V. Jarić, Spontaneous symmetry breaking and chain criterion. *Phys. Rev.* **B23**, 3460–3463 (1981).

35. H. T. Stokes and D. M. Hatch, Equivalence among isotropy subgroups of space groups. *Phys. Rev.* **B31**, 7462–7464 (1985).

36. V. A. Koptsik and G. M. Chechin, Color symmetry and space group representations in the theory of phase transitions. In *Group Theoretical Methods in Physics.* (*Proc. third Semin.*) (in Russian), Vol. 1, pp. 695–710. Nauka, Moscow (1986).

37. J. L. Birman, *Theory of Crystal Space Groups and Infra-red and Raman Lattice Processes of Insulating Crystals.* Springer, New York (1974).

38. G. L. Bir and G. E. Pikus, *Symmetry and Deformation Effects in Semiconductors* (in Russian). Nauka, Moscow (1972).

39. O. V. Kovalev, *Irreducible Representations of Space Groups* (in Russian). AN USSR, Kiev (1961).

40. D. K. Faddeyev, *Tables of Principal Unitary Representations of Fedorov Groups* (in Russian). AN SSSR, Moscow (1961).

41. S. C. Miller and W. F. Love, *Tables of Irreducible Representations of Space Groups and Co-representation of Magnetic Space Groups.* Pruett, Boulder, Colorado (1967).

42. B. I. Rezer, R. F. Yegorov and V. K. Zvezdin, Space group representations, 2 Tables (in Russian). Manuscript deposited in VINITI, No. 568–79, 1–65 (1978).

43. J. Zak, A. Cacher, M. Gluck and Y. Gur, *The Irreducible Representations of Space Groups.* Benjamin, New York (1969).

44. N. Neto, Numerical calculations of the irreducible representations of space groups. *Comput. Phys. Communs* **9**, 231–246 (1975).

45. T. G. Worlton, Irreducible multiplier representations. *Comput. Phys. Communs* **6**, 149–155 (1973).

46. G. Ya. Lyubarskii, *The Application of the Group Theory in Physics.* Pergamon, New York (1960).

47. G. M. Chechin, V. N. Raspopov and V. P. Popov, On obtaining irreducible representations of space groups with the aid of a computer (in Russian). *Bull. N.-Caucasian scient. Cent.* **3**, 29–32 (1979).

48. G. M. Chechin and V. P. Popov, The tables of full irreducible representations of space groups. 1. Cubic syngony (in Russian). Manuscript deposited in VINITI, No. 3556-80 (1980).

49. G. M. Chechin and V. P. Popov, Concise tables of full irreducible representations of space groups, In *Group-theoretical Methods in Physics* (*Proc. Second Semin.*) (in Russian), Vol. 1, pp. 105–113. Nauka, Moscow (1980).

50. G. M. Chechin, V. P. Popov and V. N. Raspopov, Irreducible representations of hexagonal space groups (in Russian). *Kristallografiya* **25**, 661–674 (1980).

51. O. V. Kovalev, *Irreducible and Induced Representations and Corepresentations of Fedorov Groups* (in Russian). Nauka, Moscow, (1986).

52. G. M. Chechin and N. A. Petrenko, The methods of obtaining the stationary vectors of space group representations in group-theoretical analysis of phase transitions in crystals (in Russian). Manuscript deposited in VINITI, No. 2655-83 (1983).

53. V. E. Naish and V. N. Syromyatnikov, Changes in translational symmetry upon structural phase transitions in crystals (in Russian). *Kristallografiya* **21**, 1085–1092 (1976).

54. *International Tables for Crystallography* (Ed. T. Hahn). Reidel, Dordrecht, Holland (1983).

55. D. M. Hatch and H. T. Stokes, Practical algorithm for identifying subgroups of space groups. *Phys. Rev.* **B31**, 2908–2912 (1985).

56. T. I. Ivanova, V. N. Kesoretskikh, V. P. Sakhnenko and G. M. Chechin. Group-theoretical analysis of the structure of low-symmetry phases arising upon phase transitions in crystals with the space group O_h^1 (in Russian). Manuscript deposited in VINITI, No. 5263-B86 (1986).

57. G. M. Chechin and V. A. Koptsik, Relation between multidimensional representations of the Fedorov groups and the groups of color symmetry. *Comput. Math. Applic.* **16**(5–8), 521–536 (1988).

58. V. P. Sakhnenko, V. M. Talanov, G. M. Chechin and S. I. Ulyanova, Possible phase transitions and atomic displacements in crystals of the O_h^7 group. 2. An analysis of mechanical and permutational representations (in Russian). Manuscript deposited in VINITI, No. 6379-83, 1–61 (1983).

59. N. M. Nizamutdinov, G. R. Bulka, N. M. Gainullina and V. M. Vinokurov, Symmetry of defects distribution in crystal and properties of electron paramagnetic resonance spectrum. In *Physical Properties of Minerals and Mountainous Rocks.* (in Russian) pp. 3–48. Kazan Univ., Kazan (1976).

60. G. R. Bulka, V. M. Vinokurov, N. M. Nizamutdinov and N. M. Hazanova, Dissymmetrization of crystals: theory and experiment. *Phys. Chem. Miner.* **6**, 283–293 (1980).

61. I. P. Ipatova, Yu. E. Kitayev and A. V. Subashiev, Changes of symmetry in surface second-order phase transitions (in Russian). *Zh. éksp. teor. Fiz.* **32**, 587–590 (1980).

62. I. P. Ipatova, Yu. E. Kitayev and A. V. Subashiev, Translational symmetry changes in second-order phase transitions on clean crystal surface. *Surf. Sci.* **110**, 543–554 (1981).

63. I. P. Ipatova, Yu. E. Kitayev and A. V. Subashiev, Surface second-order phase transitions conserving the number of atoms in a unit cell (in Russian). *Fizika tverd. Tela* **24**, 3311–3317 (1982).

64. I. P. Ipatova and Yu. E. Kitayev, Landau theory of second-order phase transitions on solid surfaces. *Prog. Surf. Sci.* **18**, 189–246 (1985).

65. V. E. Naish and V. N. Syromyatnikov, On possible changes in the crystal symmetry in structural phase transitions (in Russian). *Kristallografiya* **22**, 7–13 (1977).

66. V. E. Naish, S. B. Petrov and V. N. Syromyatnikov, Subgroups of the space groups. 2. The subgroups with an increase in cell volume (in Russian). Manuscript deposited in VINITI, No. 486–477 (1977).

67. L. A. Maksimov, I. Ya. Polishchuk and V. A. Somenkov. Complete classification of second-order phase transitions in two—dimensional systems. *Solid. St. Communs.* **44**, 163–165 (1982).

68. D. M. Hatch and H. T. Stokes, Phase transitions in solids of diperiodic symmetry. *Phys. Rev.* **B31**, 4350–4354 (1985).

69. O. V. Kovalev. On the possibility in principle of obtaining all subgroups of the Fedorov groups (in Russian). *Kristallografiya* **29**, 421–425 (1984).

70. Yu A. Izyumov, V. E. Naish and V. N. Syromiatnikov, Symmetry analysis of structural transitions in La_3S_4 and La_3Se_4 (in Russian). *Kristallografiya* **24**, 1115–1121 (1979).

71. J. Dimmock, The theory of second-order phase transitions. *Phys. Rev.* **130**, 1337–1344 (1963).

Computers Math. Applic. Vol. 17, No. 1–3, pp. 279–299, 1989
Printed in Great Britain. All rights reserved

0097-4943/89 $3.00 + 0.00
Copyright © 1989 Pergamon Press plc

STRUCTURAL THEORY OF SPACE–TIME AND INTRAPOINT SYMMETRY

A. V. Malikov

Institute of Mineralogy, Geochemistry and Crystallochemistry of Rare Elements,
Sadovnicheskaya nab. 71, Moscow 113127, U.S.S.R.

Abstract—The paper presents the space–time theory, based on identifying non-geometrical intrapoint structure and symmetry. It is shown that the structure defines topological (three-dimensionality, continuity, cohesion) and symmetrical (homogeneity, isotropy) characteristics of space and their relationship with the multilevel structure of the universe. Using the *CPT*-theorem the author gives the dynamic interpretation of the intrapoint symmetry and the electromagnetic origin of topological properties of space–time and of the structural hierarchy. Taking oceanic ferro-manganese concretions as an example the author demonstrates the euristic significance of methods of the theory for deciphering the process of natural bodies formation with particularly complicated structure.

INTRODUCTION

The progress of modern science signifies its penetration into the mysteries of macro- and micro-world. The discovery of the cell structure of the universe on the bigger scale, and the quark structure of the matter on the smaller scale reminds us of the infinity of cosmos and the diversity of its structural manifestations. Along with the expanding bounds of the cognizable cosmos, new types of the structures are being discovered in the mega-world. What used to be a sample of homogeneity yesterday, perfect crystals, chemically pure liquids, now becomes, exposed by new experimental techniques, a true kaleidoscope of alternating supermolecular structures. There is no doubt as for the complexity of the structural pattern of the physical vacuum. The atomic structure of the matter is no longer a scientific fetish but rather a natural count-down for the structural levels of the matter.

The concept of structural levels is virtually ubiquitous in the modern biological science, leading to global interdisciplinary conclusions [1–2]. The generalized crystallography closely related with the theory of symmetry uses with success concepts of packing, order and structural hierarchy to discuss complex natural structures [3–5]. The ever growing attention is attributed to a special type of the hierarchical ordering, the fractal structures [6].

In addition, topological and symmetrical properties of the space and time act in dissonance with the structural heterogeneity and hierarchy (multiple levels) of the surrounding world. Physical space is three-dimensional, continuous, homogeneous and isotropic; time is asymmetrical, and also homogeneous and continuous [7, 8]. Related with the homogeneity of time is the law of conservation of energy; with the homogeneity of space, the impulse law, and with the isotropy of space, the law of conservation of momentum [9]. Certain relativity of space and time is true only of their metrical properties. As for their topological properties, space and time retain their absoluteness. Besides, neither relativity theory nor any other theory has led to any deep penetration into the nature of topological or symmetrical properties of space and time, the more so into the cause-effect relation between these properties and the structural multilevelness and heterogeneity of physical world (the universe).

It is this global problem in general that the non-metrical structural theory of space and time (space-time) founded in Refs [10–11] is devoted to. The major concept of the theory is the intrapoint structure, i.e. the non-geometrical structure of the world point as a unit physical event [8]. Mathematically, this structure is described with a separate group of discrete symmetry named the intrapoint symmetry and logically expanding the concepts of the point and spatial symmetry. Ideas of the intrapoint structure and symmetry have been used to develop a new approach to problems of three-dimensional space, its homogeneity and isotropy, as well as the enantiomorphism of the material objects, i.e. their reflection symmetry antipodes [4, 5]. Dynamic interpretation of intrapoint symmetry, in its turn, allows one to visualize the electromagnetic origin of the topological properties of space–time.

The theory includes three stages of a mathematized description of the structural organization of the universe substratum (medium). In the order of the growing adequacy and completeness, the descriptions may be named, by convention: (1) the universe as a multilevel system of contacting natural regions (the multilevel topological structure of this system is described); (2) the universe as a multilevel material space (the superstructure is characterized, realized on the above topological structure; the superstructural unit—that is, the intrapoint structure); (3) the universe as time and light (a possibility is discussed of an electrodynamic interpretation of this superstructure as energetic space–time).

The theory may help to specify the classic spatial-temporal varieties attributing on the physical basis non-geometrical structures to their points. Thus, the number of the outer axiomatic conditions imposed on the spatial-temporal varieties may be drastically cut down. Such specialization would fully agree with the statement by Penrose that the description of space and time requires not so much the general but rather a specific formalism [7]. Indeed, the universe is given to us in a specific and unique form.

THE UNIVERSE AS A MULTILEVEL SYSTEM OF CONTACTING NATURAL REGIONS

> Thus do these organs of the world proceed,
> As thou perceivest now, from grade to grade;
> Since from above they take, and act beneath.†
> DANTE

A. Conceptualization

Primary concepts and principles of the theory

We shall call the natural region each of the regions of the matter or a field bounded, on conditions of various parameters, by the physical surface of division. Some of the examples of such regions are effective atoms, crystals, polycrystalline bodies, the earth's shells, sectors of inter-planetary space identified by their magnetic field patterns.

The contact boundary of the natural regions, the natural boundary, always has a specific inner structure, the electronic, atomic one, etc. As a rule, it differs topologically from the respective structures of the neighbouring regions. The example is the graph model (Fig. 1) of the atomic structure of an intergranular boundary in the germanium polycrystal derived from the data of Ref. [12].

If we abstract from the concrete material substratum (matter, field) of the natural regions, we shall consider them at a randomly selected moment in the existence of the universe. We assume fulfilled the following initial principles of the theory:

  (i) any natural region or boundary is contained in some other natural region, and each natural region or boundary contains another natural region;
  (ii) for any other natural regions there is always a third natural region containing those two.

It is important that these principles result from the generalization and extrapolation of basic experimental data on the topology of space structure within the observable limits.

The framework of the universe concept

Initial principles of the theory suggest that it is necessary to regard the observable universe as a multilevel (infinite level, in the limit) system of contacting natural regions. Or, in other words, as a multilevel bound packing of natural regions physically realized by the cosmos itself; it is to be referred to further on as the framework of the universe. Thus, we can, at this higher stage, reiterate the statement by Aristotle that "the body of the universe cannot be uniform in any other

---

†Cited here and below: *"The Divine Comedy"* by Dante Aligheri (translated by Henry W. Longfellow).

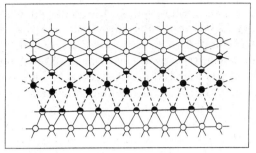

Fig. 1. A topological model (graph) of the atomic structure of germanium crystal intergrain boundary. Circles correspond to atoms (the black ones, to the atom inside the boundary, the white ones, to the atom inside the crystal), lines, to atom connections.

way but through the contacts of its parts".† We shall also go back to the basic notions of Lobachevsky's geometry stating: "Contact joins two bodies in one. Thus all the bodies are visualized as part of a single whole, the Space".‡

Spatial structures in the form of one- or multilevel packings are seen now as physically isolated fragments of the framework of the universe. So, the principle of the packing whose fundamental significance has been shown by the general crystallography and the theory of symmetry [3–5], acquires the truly cosmic importance.

Consequences of the framework of the universe concept

First, this concept suggests a specific trichotomic-hierarchical structure of the universe, or, more precisely, its framework. It makes possible a multiple and simultaneous hierarchical segmentation of: (1) natural regions of the universe; (2) their contact boundaries, i.e. elements of the "contact with" relation; (3) the neighbourhood of these regions, i.e. elements of the "contained in" relation which are responsible for the given neighbourhood in the sense as below. Thus, spatial hierarchies studied by astronomy, geology, biology and other natural sciences act as local manifestations of a more uniform trichotomic-hierarchical structure "penetrating" the cosmos. This structure is a supposed factor for the structural stability of the universe as a whole, and statistic instability of its specific material realizations, or otherwise the source of self-development of the universe.

Second, the framework of the universe concept specifies the nature of the "contact with" and "contained in" relations. In a sense they prove to be mutually opposing and dual. Mathematically, the "contact with" relation is symmetrical though, on the whole, non-transitive. In contrast, the "contained in" relation is transitive but not symmetrical. The statement: "natural region p_1 is contained in natural region p_2" is equivalent to the statement: "p_1 contacts with natural region p_3 which is, by definition, the neighbourhood of region p_1 in p_2 and contains all those and only those regions that are contained in p_2 but not in p_1". The equivalence established is a simple logical consequence of the structural organization of the framework of the universe.

Third, the developed conceptual model illustrates such notion of the universe where natural regions and boundaries are real physical entities irrespective of the method of their geometrical, or rather, pointwise-metrical description. The "contain in" and "contact with" relations act here as primary notions directly reflecting the results of the experiment. They have been used to introduce the notions of the "world point", "three-dimensional space", "enantiomorphism", etc. which is a distinctive feature of the proposed theory [10].

B. Mathematical Models

A graph of the framework of the universe

We shall use the terminology of the graph both for the description of binary relations and for their geometrical interpretation [13].

We specify graph R for the set P of all natural regions of the universe, i.e. we correlate one to one the set of all vertexes in $R–P$. Two vertexes in R are joined by an edge when and only when

---

†Aristotle, "Physics".
‡N. Lobachevsky, "On the basic notions of geometry".

the corresponding natural regions contact with each other (such edge is referred to as "contact relation") or when one is contained inside the other (the "inclusion relation").

Algebraically, $R = R_+ \cup R_-$ and $R_+ \cap R_- = \emptyset$, where R_+ and R_- are binary relations, "contained with" and "contained in" respectively, given on P. More precisely, R_+ is the set of all those and only those ordered pairs (edges) of the type $(p, p') \in P^2$ where region p contacts with region p'. The sequence of regions p, p' is apparently unimportant, i.e. the contact relation (p, p') is an unoriented edge of graph R. Similarly, R_- is the set of all those and only those ordered pairs of the type $(p, 'p) \in P^2$ where region p contains region $'p$ or is contained in $'p$. For definiteness, we assume here and henceforth that p is contained in $'p$, i.e. the inclusion relation $(p, 'p)$ is the oriented edge of graph R. Loop edges in R are excluded.

By definition, graph R is a non-metrical model of the universe's framework describing its multilevel topological structure. Such definition is useful if only because it leads to a unified modelling technique applicable to complex natural packings (framework fragments) in the form of appropriately selected parts of graph R.

Models of the fragments of the framework of the universe

Of primary importance are A-graphs, which are subgraphs of graph R describing the topology of bound packings of the same level $Pl_{(1)}$ of natural regions, i.e. the systems of the contacting regions whose inner structure is not taken into account. Vertexes of A-graph correspond to natural regions while the edges, to the contact boundaries between these regions in packing $Pl_{(1)}$. Figure 1 shows the A-graph of effective atoms packing in the polycrystal. It is an example of a crystallographic packing of the $Pl_{(1)}$ type whose topology is closely related with the laws of crystals symmetry [14].

There are other important realizations of A-graphs. These are A-graphs of molecules as a packing of effective atoms [15, 16], molecular associates as a packing of molecules [10], crystalline aggregates as a packing of crystals [17, 18], biological bodies as a packing of cells [19], etc. Using finite A-graphs we have also introduced B-graphs for a possible existence in $Pl_{(1)}$ of a multicentre (q-centre, $q \geqslant 3$) contact, i.e. a common contact boundary of all q natural regions and not more [10].

The amount of vertexes of B-graph equals the number of all edges of A-graphs, and two vertexes are joined by the edge if and only if the corresponding boundaries in packing $Pl_{(1)}$ intersect. Figures 2 (b) and (c) shows diagrams of A-, B-graphs of the packing of four contacting (fused) fine- and thick-platy crystals of $BaSO_4$. Figure 2 suggests that B-graphs show well q-centre contacts, while in the $Pl_{(1)}$ type of the packing such contact may exist when and only when the respective B-graph contains a complete subgraph of the dimensions $1/2q(q-1)$. Thus, a "compact" packing [Fig. 2 (c)] has two 3-centre contacts, while a "loose" packing [Fig. 2 (b)] lacks them. In relation to molecules, B-graph plotting corresponds to an illustrative analysis of the topology of molecular systems with q-centre contacts of effective atoms describing multicentre chemical bonds. The euristic value of A-, B-graphs is evident from Fig. 1 which lists all the theoretically possible combinations of A-, B-graphs for the system of four contacting regions [17].

Plotting of A-graphs for various structural levels of a single material formation topologically characterizes it as a multilevel (N-level, $N \geqslant 2$) packing $Pl_{(N)}$ of natural regions. Such characteristic suggests an outline of the structure for especially complex natural macro-bodies, their true nature and origin.

A good example of intricately organized macro-bodies is provided by sea ferromanganese concretions. They form vast industrial deposits on the world ocean floor. The former belief that the concretions resulted from the fusion of colloidal particles of Fe- and Mn-oxides dispersed in the ocean has proved wrong. Application of topological methods has shown that concretions form in two stages [20]. First a protoconcretion forms of possibly metastable ferrugenous rhodochrosite. This is followed by a full replacement by finely dispersed oxidized Fe-, Mn-minerals in the protoconcretion body, with its macro-structure inherited. It means that the sea ferromanganese concretion is a multiphase pseudomorphose after a macro-crystalline aggregate [20].

Figures 3 and 4 show a topological model of the oceanic concretion with the typical globular structure in the form of 2-level $Pl_{(2)}$ packing. A-graphs describing this packing at the macro- and micro-levels shall be called A^m- and A_m-graphs, respectively.

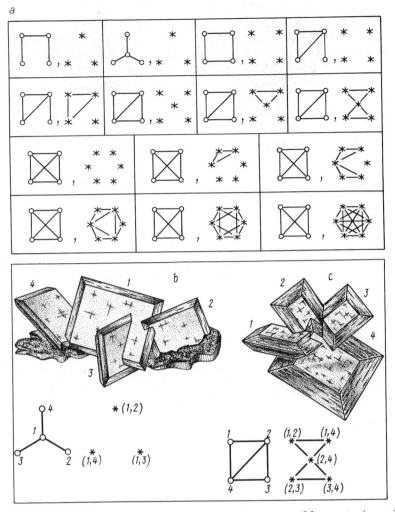

Fig. 2. Theoretically possible combinations of A-, B-graphs for a system of four contacting regions (a); A-, B-graphs and described by them accretions of fine (b) and thick-tabular (c) crystals of $BaSO_4$. Circles show vertexes of the A-graph; asterisks show vertexes of the B-graph.

Figure 4 (b) shows the A^m-graph of the protoconcretion as an idealized packing of globules (with a nucleus at the centre of the packing, e.g. a shark tooth). A typical aggregate of three such globules is shown in Fig. 3 (a). It also shows the A^m-graph of this aggregate; the dashed line shows the A^m-graph of a hypothetical growth which differs from the initial one only in that it lacks any visible distortion in the appearance of two globules in the region where they are the closest to each other. Transition from the triangular to the A-graph of the depicted fusion illustrates well the indicated effect of mutual distortion and non-fusion of globules as a feature of a truly crystalline origin of the concretional macro-structures [18, 20].

Similarly, the type of transition from A^m-graph in Fig. 4 (a) to A^m-graph in Fig. 4 (b), describes this effect for the concretion as a whole and explains the origin of the bigger part of the pore space in ferromanganese concretions. Pseudomorphization of the protoconcretion (oxidation of ferrugenous rhodochrosite) is illustrated by the change in the colour of vertexes from the white in the A^m-graph in Fig. 4 (b) to the corresponding black in the isomorphous A^m-graph in Fig. 4 (c). It means that the black vertex of the isomorphous A^m-graph in Figs 4 (c) and 3 (b) corresponds to a globule altered from inside, its generalized form shown in Fig. 3 (b). Using A_m-graphs, it is possible to understand the restructuring in supermolecular structures under pseudomorphization. In Fig. 4 (b) and (a), a randomly selected vertex of A^m-graph is correlated with a complete A_m-graph describing the globule as an idealized packing of block-monocrystalline beams. The beams have a single centre at which they all contact with each other [Fig. 3 (a)]. The arrows show that each of the A_m-graph vertexes joins a corresponding one fo the A^m-graph with an inclusion relation.

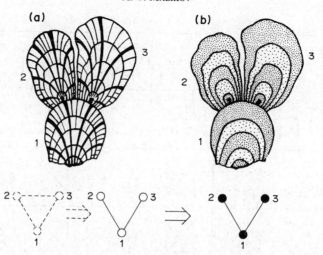

Fig. 3. Typical accretions of three globules of oceanic concretions, and the A-graph describing them. Vertexes of A-graphs are marked by circles. The white circle corresponds to the globule of a protoconcretion (a), the black one, to the globule of the globule of a ferromanganese concretion (b).

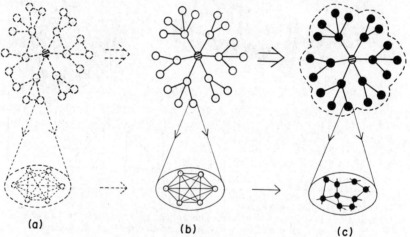

Fig. 4. A two-level topological model describing the formation of an oceanic ferromanganese concretion. The vertexes of the A^m-graph are shown with large circles, those of the A_m-graph by small circles. The shaded circle corresponds to the nucleus of the concretion.

In Fig. 4 (c), the A_m-graph correlated with the A^m-graph vertex describes a changed globule as a loose packing of the composing micro-particles. Such description according to Fig. 3 shows that pseudomorphisation at the micro-level results in the destruction of anisotropic block-monocrystalline beams and in the formation of the isotropized loose aggregate. Dashed lines in Fig. 4 (c) traces the outline of the central section of the concretion.

This example shows that multilevel topological modelling facilitates interpretation and graphic presentation of the process generating complex natural objects. Bound parts of the graph R act here as adequate models for the fragments of the universe's framework. They have the necessary precision to outline what there is essential in the structural independence of these fragments.

Algebraic model of the framework of the universe

The new model is based on the graph one, but, in contrast, it describes more unambiguously the topological and trichotomic-hierarchical structures of the framework of the universe.

Determination of the graph R is apparently equivalent to assuming the ensemble (family) of the type:

$$SR = \{\{p, r_+, r_-\} : p \in P \text{ and } r_+, r_- \in R\}, \tag{1}$$

where $r_+ = (p, p')$ and $r_- = (p, {}'p)$ are the contact relation and the inclusion relation, respectively, whose components are represented by the region p. In the graph R, each of the elements of $s = \{p, r_+, r_-\}$ from SR is equivalent to a bound part described by the diagram:

(2)

where the vertexes correspond to the regions ${}'p, p, p' \in P$. The edges in the diagram show that region ${}'p$ (the larger white circle) contains region p (the smaller white circle) contacting region p' (the black circle).

As is evident, each element $s \in SR$ carries information on the actual existence of region $p \in s$ together with a boundary (modelled by the contact relation $r_+ \in s$) and a neighbourhood (assumed by the inclusion relation $r_- \in s$). In this connection, the set of the type $s = \{p, r_+, r_-\}$ will be called further on the unit subsystem of the universe's framework. We characterize the SR ensemble as an algebraic model of this framework adequately describing its trichotomic-hierarchical and topological structures.

Let $\dot{s} = \{\dot{p}, \dot{r}_+, \dot{r}_-\}$ be an arbitrarily fixed subsystem from SR, where $\dot{r}_+ = (\dot{p}, \dot{p}')$ and $\dot{r}_- = (\dot{p}, {}'\dot{p})$. It will be seen then that in SR there are three subsets, of no less than continual cardinality, of the type:

$$S\dot{p} = \{\{p, r_+, r_-\}: p \in P_{\dot{p}} \text{ and } r_+, r_- \in R\};$$ (3)

$$S\dot{r}_+ = \{\{p, r_+, r_-\}: p \in P_{\dot{r}_+} \text{ and } r_+, r_- \in R\};$$ (4)

$$S\dot{r}_- = \{\{p, r_+, r_-\}: p \in P_{\dot{r}_-} \text{ and } r_+, r_- \in R\},$$ (5)

where $P_{\dot{p}}$, $P_{\dot{r}_+}$, $P_{\dot{r}_-}$ are the sets of all regions from P which are contained in (1) the region $\dot{p} \in \dot{s}$; (2) the natural contact boundary between regions $\dot{p}$ and $\dot{p}'$ and (3) the region ${}'\dot{p}$, respectively, and contain the region $\dot{p}$.

The existence of sets (3)–(5) guarantees the initial principles of the theory. Transition from $\dot{p}$ to $S\dot{p}$ describes the hierarchical segmentation of region $\dot{p}$ as transformation (development) of vertex $p \equiv \dot{p}$ of diagram (2) in the subgraph of graph R that characterizes the inner structure of the region $\dot{p}$. Similarly, transitions from r_+ to $S\dot{r}_+$ and from r_- to $S\dot{r}_-$ describe the respective development of contact relation $r_+ \equiv \dot{r}_+$ and inclusion relation $r_- \equiv \dot{r}_-$ of this diagram. From the existence of sets (3)–(5) follows the major characteristics of the trichotomic-hierarchical structure.

Ensemble SR equals the union of all components of a random term of the recurrent sequence $\{B_n(SR)\}$, in which

$$B_n(SR) = \begin{cases} \{SR\}, & \text{if } n = 0 \\ \{E_{i_n}(S_{n-1}): S_{n-1} \in B_{n-1}(SR) \text{ and } i_n \in \{1, 2, 3\}\}, \\ \text{if } n \in \{1, 2, \ldots\}, \end{cases}$$ (6)

where $E_{i_n}(S_{n-1})$ is the range of the values of reduction of such multivalued representation on the set S_{n-1} that

$$E_1(\dot{s}) = S\dot{p}, \quad E_2(\dot{s}) = S\dot{r}_+, \quad E_3(\dot{s}) = S\dot{r}_-,$$

for every fixed $\dot{s}$ from SR. The cardinality of the set $B_n(SR)$ is 3^n.

Further, for all $s \in SR$, and only for those, we assume

$$t_1(\dot{s}) = \{s, \dot{s}: s \cap \dot{s} = \{\dot{p}\}\}, \quad t_2(\dot{s}) = \{s, \dot{s}: s \cap \dot{s} = \{\dot{r}_+, \dot{r}_-\}\},$$

$$t_3(\dot{s}) = \{s, \dot{s}: s \cap \dot{s} = \{\dot{r}_+\}\}, \quad t_4(\dot{s}) = \{s, \dot{s}: s \cap \dot{s} = \{\dot{p}, \dot{r}_-\}\},$$ (7)

$$t_5(\dot{s}) = \{s, \dot{s}: s \cap \dot{s} = \{\dot{r}_-\}\}, \quad t_6(\dot{s}) = \{s, \dot{s}: s \cap \dot{s} = \{\dot{p}. \dot{r}_+\}\},$$

$$t_7(\dot{s}) = \{s, \dot{s}: s = \dot{s}\}, \quad t_8(\dot{s}) = \{s: s \cap \dot{s} = \emptyset\},$$

and it is apparent that $\{t_j(\dot{s}): j \in \{1, 2, \ldots, 8\}\}$ is the division of the ensemble SR.

We shall consider for each fixed $j \in \{1, 2, \ldots, 8\}$, a new family $T_j S$ composed of all those and only those subsets of ensemble SR which, together with each of their points s contain the respective set $t_j(s)$. Using the notions of the general topology [21], we verify explicitly the statement:

For each fixed $j \in \{1, 2, \ldots, 7\}$ family $T_j S$ is the topology in ensemble SR, and with $j = 7$ the set $\{t_j(s): s \in SR\}$ is its base. The topological space $(SR, T_j S)$ is always Hausdorffian also only with $j = 7$.

Thus valid on the ensemble SR are the topologies naturally determined by the structure of unit subsystems of the framework of the universe. Equations relating conditions (6) and (7), i.e. the trichotomic-hierarchical and topological structures, are presented in Ref. [11].

THE UNIVERSE AS MULTILEVEL MATERIAL SPACE

"A Point beheld I, that was raying out
Light so acute, the sight which it enkindles
Must close perforce before such great acuteness."
....
"... said: 'From that Point
Dependent is the heaven and nature all...'
DANTE

A. Conceptualization

World point and the framework of the universe

The proposed concept is based on the statement that there is a unique non-geometrical structure of world point that determines the topological and symmetrical properties of space. This thesis, on the face of it, strongly contradicts both the Euclidean understanding of the point and Gilbert's concept that the point is an element of space, unidentifiable but subordinate to a certain system of axioms [10, 11]. The following hypothetic experiment though may be basic for the substantiation of the author's thesis, as well as showing that the second statement is erroneous.

We shall place one hypothetical observer in each unit subsystem $s \in SR$ of the universe's framework. The observer will be able to determine whether he is in the region $p \in s$, at its limits (or, otherwise, contact relation $r_+ \in s$) or in the neighbourhood (inclusion relation $r_- \in s$) of this region. Let us imagine that all the observers have left their subsystems and are moving away as a result of the alternation of their posts.

As they become more remote, each of the observers ceases to notice the difference between elements p, r_+, r_- of that subsystem s, where he was initially. The indicated elements are as if merged in a single new object. Taking into consideration the non-discernibility of elements p, r_+, r_- in this object, on the one hand, and the fact that, on the other, the three elements exist in it all at the same time, the observer comes naturally to regarding the observed object as a single physical event $g(s)$.

Following the transition from ensemble SR to set GR of all events of the type $g(s)$, the observers notice an important circumstance. All the features of the universe modelled by the ensemble SR are retained in GR. At the same time, each event of the $g(s)$ type, i.e. the element of the set GR, signifies the existence of a new physical reality connected with the discussed situation of non-discernibility. The presence of this reality was not considered while constructing the ensemble SR. Their sensual perception would lead the observers to distinct correlation of the indicated reality to the point of physical space, or, rather, to the world point.

The exact definition of world point as an event of the $g(s)$ type will be only mathematical and considering that: (1) world point is a unit physical event manifested in its relation to the observer, instrument or other effect; (2) this event is realized in the simultaneous existence and non-discernibility, corresponding to the point of region p, its boundary (contact relation r_+) and neighbourhood (inclusion relation r_-); (3) objects p, r_+, r_- are representable by three structurally heterogeneous parts belonging to the system of events of the same kind.

The concept of world point therefore implicitly takes into account the trichotomic-hierarchical structure of the universe and corresponds to the fusion of three notions, those of the natural region, its boundary (contact relation) and neighbourhood (inclusion relation). In other words, the world point is a mirror of the universe associated with a single cosmologic principle of the trichotomic-hierarchical "development" of points into worlds and "convolution" of worlds into points.

If is this analogy that allows one to see in Leibnitz's monads a germ of the sturctural approach to world point. Since such a point does not have a geometrical structure, though showing topologically the structural position of some region in the universe, the Euclidean definition of the point as having no parts but having location, becomes more clear. As any natural region can be selected as a region for world point, there is also no significant discrepancy with the Gilbert's concept of the point.

Physical treatment of world point as a unit event is enhanced here radically owing to the indication of the uniform type of the structure of these events [8, 22]. The dependence of the world point on the surrounding points, their non-geometrical nature and hierarchical subordination extend the new understanding of the point on the basic notions of the crystallographic point [4, 5], on the points of layered and multidimensional spatial-temporal varieties [23, 24].

To forestall the mathematic definition of world point, we shall identify the range of its major consequences.

Intrapoint structure and symmetry

By nature, world point has no geometrical structure but is a carrier of a unique non-geometrical, intrapoint structure. The intrapoint structure is: (1) unobservable (obscure) but physically real since it implements the structure of the materially registered event; (2) ternary, since there are three and only three different objects, p, r_+ and r_- existing in this event; (3) superstructurally unique as it is a structure of a higher order on the unit structure (subsystem) $s = \{p, r_+, r_-\}$ with structural elements p, r_+, r_-; (4) scale-invariant, since each of the elements p, r_+ and r_- can be represented by a set of independent realizations of intrapoint structure (the possibility is thus not excluded of representing this structure by a definite Mandelbrot's set); (5) symmetrical, as it has a finite series of its states, the transitions between which do not alter the structure itself.

The latter of the above features defines the notion of the intrapoint symmetry related with the invariance of world point with respect to formal mutual transformations of the objects p, r_+, r_- corresponding to the point. Any two world points are structurally isomorphous (undistinguishable) in the sense that the describing groups of intrapoint symmetry are isomorphous. Material polymorphism of world points opposes their structural isomorphism. The polymorphism establishes the difference between the points with respect to the supposed difference of the corresponding regions and/or boundaries, and/or neighbourhoods.

Thus, we have come to the understanding of the deep connection between the structural heterogeneity, multilevelness of the universe and topological (three-dimensionality, cohesion, continuity), symmetrical (homogeneity, isotropy) properties of physical space.

A model of structural organization of space-matter

The set of world points (described as GR) composes the multilevel material space of the cosmos, i.e. the single space-matter. Physical space regarded irrespective of the matter is represented as a specific superstructure realized on the framework of the universe's structure. The superstructural unit is the intrapoint structure. The definition of superstructural units includes presentation of an undisrupted bond between them that would correspond to the topological cohesion and continuity of space. Ternarity, symmetry and scale invariance of each of such units are extended by structural isomorphism onto all the units of the superstructure. The distribution of ternarity may be easily correlated with the three-dimensionality of space; that of symmetry, with the homogeneity and isotropy, while that of scale invariance, with the retention of topological and symmetrical features of space with respect to changes in the scale at which it is examined.

At the same time, material polymorphism of superstructural units, or rather, the world points proper, control the trichotomic-hierarchical structure of the universe's framework which reflects cosmos's structural heterogeneity and multilevelness. It might appear paradoxical, but this multilevelness and heterogeneity, on the one hand, and topological and symmetrical features of space, on the other, prove to be different aspects of the intrapoint structure manifestation. A point, like a gene, contains the bases of both the homogeneity and heterogeneity of the world, and their interrelation law (Fig. 5).

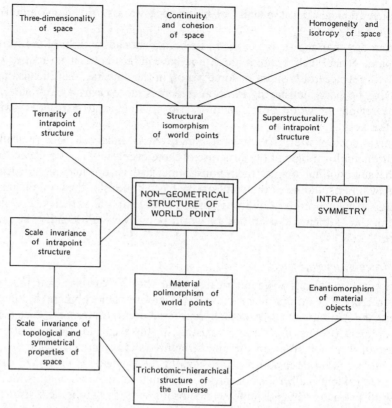

Fig. 5. A conceptual model of the structural organization of multilevel material space: interrelation of basic notions.

A few notes to the diagram in Fig. 5

1. The three-dimensionality of physical space is the major and the least understood of its topological properties [26]. The natural substantiation of this feature though follows from the discussion of the ternarity of the set of all world points of the space-matter, i.e. the set *GR* which is mathematically the ternary relation. The explicit understanding of the dimensionality of physical space as 3-arity of this type of relation links the three-dimensional phenomenon with the system of empirically simpler notions: those of the ternarity of intrapoint structure, structural iso-morphism, trichotomic-hierarchical structure of the universe, etc.

This substantiation leads in theory to the experimentally observed picture of invariance of physical space three-dimensionality relative to the examination scale in macro- and mega-world. It gives an answer to the question by Einstein as to how to retain major features of four-dimensionality if the continuity of space–time is rejected [26], providing a broader sense to spatial local coordinates of the world point [11]. The new understanding of the three-dimensionality agrees also with the following basic result of theoretic studies going back to Ehrenfest and Whitrow on the dimensionality of space: atomic, planetory and other material systems lose the observed structural stability if we assume that with the type of physical laws persisting, the systems are realized in the space of $n \neq 3$ dimensions [25].

2. Orientability of physical space, or more precisely, the spatial variety, is the possibility of continuous division of spatial axes' bases into the right- and left-hand ones [11]. Orientability of space is the necessary condition of enantiomorphism (chirality) of material objects. Enantio-morphism of physical media (crystals, liquids and gases in whose atoms the parity of electro-magnetic effects is disturbed) is reflected in their optical activity. Enantiomorphism of the continuum is introduced either by ascribing their points asymmetrical surrounding [5], or by indicating *a priori* the right- or left-handedness as a non-geometrical feature of the point [27].

The proposed concept takes into consideration both approaches regarding enantiomorphous media as space-matter regions where the respective world points are materially equivalent (not polymorphous) but their intrapoint structures belong, in a certain non-geometric sense, to the right-hand and left-hand states.† In other words, the very fact of the physical media enantio-morphism may be experimentally proving the validity of the model of intrapoint structure.

3. Analysis of the diagram in Fig. 5 suggests another paradoxical conclusion: that major notions of classic mechanics (material point, the degrees of freedom number, integrals of the movement, etc.) implicitly take into consideration the structural multilevel organaization of cosmos. Actually, even the system-structural approach discussed irrespectively of the notions of space and time borrowed from the outside does not include such a high rank of understanding of cosmos cohesion that is contained implicitly in mechanics. This approach did not have an equivalent to the notion of world point existing at one time both as an element and as a relation. However, it was the development of the systematic-structural approach that demanded the search for isomorphism and polymorphism of phenomena [28] that led to the explicit determination of the intrapoint structure previously unknown to physics. Such is the dialectics of cognition which gives the priority to studies of structural isomorphism and material polymorphism of world points as a possible basis for all physically real spatial structures and symmetries.

B. Mathematical Models

Mathematical definition of the world point

It would seem that according to our mental experiment, the world point, i.e. a physical event of the type $g(s)$, where $s = \{p, r_+, r_-\}$ can be defined as: $g(s)$ with equal probability is region p, or contact relation r_+, or inclusion relation r_-. Assuming a pariwise incompatibility or respective unit events would mean that $g(s)$ with probability $1/3$ is p, or r_+, or r_-. Such definition would take into account the fact that elements p, r_+, r_- are unidentifiable at point $g(s)$, but would contradict their existence at the point with probability 1, as prescribes the experiment.

This contradiction when surmounted within the logic law of "*tertium non datur*" leads to the following adequate definition of world point $g(s)$. Let $s = \{p, r_+, r_-\}$ be an arbitrarily fixed element of the set SR, while σ, β, δ are any three, pairwise-different numbers from the set $\theta = \{1, 2, 3\}$. Then, we shall assume with respect to s and arbitrarily fixed α, β, $\delta \in \theta$, an ordered triplet (vector) $g(s) = (g_1(s), g_2(s), g_3(s))$, whose component $g(s)$ will be defined as the following system of pairwise-incompatible events:

$g_\alpha(s)$ with probability (prob.) $1/3$ is region p, if
$$\begin{cases} g_\beta(s) \text{ with prob.}1/3 \text{ is } r_+ \\ \text{and } g_\delta(s) \text{ with prob.}1/3 \text{ is } r_- \\ \qquad\qquad \text{or} \\ g_\beta(s) \text{ with prob.}1/3 \text{ is } r_- \\ \text{and } g_\delta(s) \text{ with prob.}1/3 \text{ is } r_+, \end{cases}$$

or

$g_\alpha(s)$ with prob.$1/3$ is contact relation r_+, if
$$\begin{cases} g_\beta(s) \text{ with prob.}1/3 \text{ is } p \\ \text{and } g_\delta(s) \text{ with prob.}1/3 \text{ is } r_- \\ \qquad\qquad \text{or} \\ g_\beta(s) \text{ with prob.}1/3 \text{ is } r_- \\ \text{and } g_\delta(s) \text{ with prob.}1/3 \text{ is } p, \end{cases}$$

or

$g_\alpha(s)$ with prob.$1/3$ is inclusion relation r_-, if
$$\begin{cases} g_\beta(s) \text{ with prob.}1/3 \text{ is } p \\ \text{and } g_\delta(s) \text{ with prob.}1/3 \text{ is } r_+ \\ \qquad\qquad \text{or} \\ g_\beta(s) \text{ with prob.}1/3 \text{ is } r_+ \\ \text{and } g_\delta(s) \text{ with prob.}1/3 \text{ is } p. \end{cases}$$

---

†The validity of such an approach might have been foreseen by P. Curie. Anyway, V. Vernadsky (citing M. Curie) noted in connection with the problem of dissymmetry of biological systems that P. Curie shortly before his death called dissymmetry the state of space; Vernadsky "Meditations of a naturalist".

Note that this definition of world point $g(s)$ is valid with every fixed set of (α, β, δ), e.g. when $(\alpha, \beta, \delta) = (1, 2, 3)$.

Graphically, vector $g(s)$ can be presented as a single event characterized by six independent situations ("resonance structures") realized simultaneously with the probability of 1/6. These situations are shown schematically by the alternations in the diagram:

$$
\begin{array}{ccc}
[p, r_+, r_-] & & [p, r_-, r_+] \\
\nwarrow \tfrac{1}{6} & \nearrow \tfrac{1}{6} & \\
\end{array}
$$

$$[r_+, p, r_-] \leftarrow \tfrac{1}{6} \rightarrow (g_1(s), g_2(s), g_3(s)) \leftarrow \tfrac{1}{6} \rightarrow [r_-, p, r_+]$$

$$
\begin{array}{ccc}
\swarrow \tfrac{1}{6} & \nwarrow \tfrac{1}{6} & \\
[r_+, r_-, p] & & [r_-, r_+, p]
\end{array}
\tag{8}
$$

Intrapoint structure is the vector of the type $g(s)$, whose elements of the set $s = \{p, r_+, r_-\}$ are assumed indefinite in any of the senses, physical and (or) mathematical.

Note that the presented definition of the world point if necessary can be modified by various means (e.g. by replacing in it the contact relation $r_+ \in s$ by the set $\{r_+\}$ of all contact relations incidental to vertex $p \in s$ of graph R), retaining the basic framework construction.

Structural model of physical space

We shall consider the set $GR = \{g(s): s \in SR\}$, i.e. the range of values of the function $g: SR \rightarrow GR$, correlating each of s elements from SR one-to-one with the corresponding vector $g(s) = (g_1(s), g_2(s), g_3(s))$. It is apparent that GR is a ternary relation, i.e.

$$GR \subset G_R^{(1)} \times G_R^{(2)} \times G_R^{(3)} \subset \bar{G}_R^3, \quad \bar{G}_R = \bigcup_{\alpha \in \theta} G_R^{(\alpha)}, \tag{9}$$

where $G_R^{(\alpha)}$ is the set of α-components of all vectors from GR. Each element from $\bar{G}_R$ has a limited physical sense, as it is an event existing in reality only in relation GR. In contrary, the GR relation is a structural non-metrical model of physical (material) space including, together with new data, all the information on the universe containing in SR. Function g retains the most important features of the ensemble SR in GR.

In particular, function g for each fixed $j \in \{1, 2, \ldots, 7\}$ is the homeomorphism of topological space $(SR, T_j S)$ on representative space $(GR, T_j G)$ with topology $T_j G = \{g(t_j): t_j \in T_j S\}$, where $g(t_j)$ is the range of values of reduction g on the open set t_j. It is apparently possible to find for any intersecting $s, s' \in SR$ such index $j \in \{1, 2, \ldots, 7\}$ and the set $t_j \in T_j S$ that $s, s' \in t_j$ and $g(s), g(s) \in g(t_j)$. Thus, intersection $s \cap s' \neq \emptyset$ is placed in correspondence with the definite type of topological cohesion between points $g(s), g(s')$. The events $g(s), g(s')$ may be said to penetrate each other; it is possible to give a probability-topological assessment of the mutual interpenetration degree responsible for the continuity and cohesion of space.

Similar to ensemble SR, relation GR is equal to the union of all the elements for an arbitrary term of the sequence $\{B_n(GR)\}$, in which

$$
B_n(GR) = \begin{cases}
\{GR\}, & \text{if } n = 0 \\
\{g[E_{i_n}(S_{n-1})]: S_{n-1} \in B_{n-1}(SR) \text{ and } i_n \in \{1, 2, 3\}\}, \\
\quad \text{if } n \in \{1, 2, \ldots\},
\end{cases}
\tag{10}
$$

where $g[E_{i_n}(S_{n-1})]$ is the range of values of reduction of function g on the respective family $E_{i_n}(S_{n-1}) \in B_n(SR)$. However, in contrast to condition (6), expression (10) describes the trychotomic-hierarchical structure of the universe represented now as multilevel material space.

Comparison of these and other properties of formal structures (1) and (9) shows that relation GR is a unique object for a mathematical investigation.

In this connection, we shall consider also an infinite union of the type

$$\hat{P} = P \cup 2^P \cup 2^{2^P} \cup 2^{2^{2^P}} \cup \ldots,$$

where 2^P is the set of all subsets of the set P, 2^{2^P} is the set of all subsets of the set 2^P, etc. All ordinary mathematical structures (functions, relations, etc.) connected with a certain set of individuals (in our case—with P) are known to belong to the corresponding infinite union [29]. Indeed, $R, SR \in \hat{P}$, as $R \subset 2^{2^P}$, and $SR \subset 2^{2^{2^P}}$. Construction of the relation GR was connected ultimately only with individuals P, but $GR \notin \hat{P}$.

Thus we can see that GR is a unique mathematical superstructure given on the structure SR. A specific feature of this superstructure is that it is realized and at the same time realizes a multilevel topological structure of the universe's framework.

Coordination of the structural and classic models of physical space

Important properties of the model GR are suggested by its coordination with three-dimensional spatial variety M, the major mathematical model of physical space [8]. The variety is a topological space for each point of which there is a neighbourhood, homeomorphic (topologically equivalent) to n-dimensional Euclidean space E^n. The number n is called the dimensionality of the diversity and in the case of spatial variety M is postulated as $n = 3$.

Consideration of the points of the type $g(s) \in GR$ as elements of variety M corresponds to the global reduction on M which can be realized with the consideration of a relative indivisibility of the world point [10, 11]. This reduction makes it possible to establish an adequate (in the structural sense) correspondence between the arbitrarily fixed point $g(\dot{s}) = (g_1(\dot{s}), g_2(\dot{s}), g(\dot{s}))$ from the variety relation $M \subset GR$ and its spatial local coordinates $(x_1, x_2, x_3) \in E^3$ according to the law:

$$(x_1, x_2, x_3) = (\omega(g_1(\dot{s})), \omega(g_2(\dot{s})), \omega(g_3(\dot{s})) = \omega(g_1(\dot{s}), g_2(\dot{s}), g_3(\dot{s})) = \omega(g(\dot{s})), \tag{11}$$

where ω is the functional prescribing homeo- and isomorphism simultaneously of a certain neighbourhood of the point $g(s)$ to Euclidean space E^3.

Thus, the three-dimensionality of the variety M with the introduced global reduction $M \subset GR$ corresponds to the 3-arity of M relation. The correctness of such correspondence is emphasized by the known theorem by Brouwer:

unempty open set in E^n cannot be homeomorphous to an open set in E^m with $m \neq n$.

In $M \subset GR$ lines and surfaces can be analytically (parametrically) assumed with the help of spatial local coordinates. The arity of the system of events from M describing these geometrical objects will be certainly equal to 3, which does not exclude though a possibility of determining their differences with respect to topological dimensionality.

Intrapoint symmetry, enantiomorphism and anti-equality of world points

Let us consider symmetrical groups S_3^s and S_3^θ of the third order working for the sets $s = \{p, r_+, r_-\}$ and $\theta = \{1, 2, 3\}$ respectively. The point of the type $g(s)$ is invariant relative to any transformations from groups $\Gamma_{g(s)}^D = S_3^s \otimes S_3^\theta$, $\Gamma_{g(s)}^L = S_3^\theta \otimes S_3^s$ ($\otimes$ is the sign of direct product of groups) describing its intrapoint structure. Groups of the type $\Gamma_{g(s)}^D$, $\Gamma_{g(s)}^L$ are assumed by definition, the groups of intrapoint symmetry.

The correctness of this definition is connected with the following simple property: for any two points $g(s_1)$, $g(s_2) \in GR$, groups $\Gamma_{g(s_1)}^D$, $\Gamma_{g(s_1)}^L$, $\Gamma_{g(s_2)}^D$, $\Gamma_{g(s_2)}^L$ are pairwise isomorphic. Hence, world points of the type $g(s)$ are really structurally isomorphous with each other in the sense of the given definition of intrapoint symmetry groups. Two cases of structural isomorphism are possible: (1) $s_1 \neq s_2$, i.e. points $g(s_1)$, $g(s_2)$ are materially polymorphous; (2) $s_1 = s_2$, i.e. points $g(s_1)$, $g(s_2)$ are materially equivalent.

Feasibility of joint, groups $\Gamma_{g(s)}^D$ and $\Gamma_{g(s)}^L$, consideration follows from the graph model of world point. In agreement with diagram (8), each of the diagrams in Fig. 6 is exactly the graph model of point $g(s)$ in one of its states. The vertex of the diagram corresponds to the object from the set $s = \{p, r_+, r_-\}$ and two vertexes are joined by the edge when and only when the corresponding objects belong to one and the same rearrangement of elements $p, r_+, r_- \in s$. The number of the vertex is the place of the object in the rearrangement; for definiteness, second-degree vertexes

Fig. 6. Theoretically possible types of world points enantiomorphism (unit physical events). The points
are modelled by diagrams, their vertexes are shown by circles: the black circle signifies a natural region,
the shaded one shows the contact relation, and the white one, the inclusion relation.

are given the same number. It can be shown that there exists $|\Gamma^D_{g(s)}| = |\Gamma^L_{g(s)}| = 36$ ($||$ is the
exponent sign of the group) of the diagrams of the indicated form: 12 of them are shown in Fig. 6,
the rest are obtained as a result of renumeration of the above diagrams' vertexes. Diagrams
of the right (left) vertical column in Fig. 6 can be transformed by any operation, say, from
group $\Gamma^D_{g(s)}$ ($\Gamma^L_{g(s)}$, respectively) in diagrams of the same form, a column in particular, whereas
operations from a certain group of isomorphisms of $\Gamma^D_{g(s)}$ group to $\Gamma^L_{g(s)}$ one related diagrams of
different columns.

Now we shall consider two materially equivalent points $g(s)$, $g(s') \in GR$, i.e. $g(s) = g(s')$,
$s = s' = \{p, r_+, r_-\}$, mutually transformable by any operation from $\Gamma^D_{g(s)}$ or $\Gamma^L_{g(s)}$. If the operation
chosen aligns the mirror-symmetric diagrams in Fig. 6, then points $g(s)$, $g(s')$ are enantiomorphic
with respect to it, i.e. are in the enantiomorphic state relative to each other. Due to structural
isomorphism of world points, enantiomorphism of intrapoint structures can also be assumed on
the analogy, including materially polymorphous point from GR.

If follows from the above: (1) orientability of the spatial variety with points of the $g(s)$ type;
(2) possibility for the physical realization of world points enantiomorphism with the consideration of
its types listed in Fig. 6. This realization can be visualized as a form of a complex event determined
by the specific features of intrapoint structure manifestations. We shall illustrate this below.

Let us assume that physically enantiomorphous points $g(s)$, $g(s')$ are related by the operation

$$\gamma^L = \left(\begin{pmatrix} 123 \\ 213 \end{pmatrix}, \begin{pmatrix} p\, r_+\, r_- \\ p\, r_+\, r_- \end{pmatrix} \right)$$

from $\Gamma^L_{g(s)}$. Then we may come to treating γ^L as an event where the first (second) component of
the $g(s)$ point always acts with respect to something (observer, instrument, effect) as the second
(first, respectively) component acts in point $g(s)$ with respect to the same thing. The third
component in $g(s)$ acts always as in point $g(s')$.

We shall now show that non-geometrical operation γ^L can be expressed analytically-naturally
as a classic operation of mirror reflection. For that, we shall consider $g(s)$, $g(s')$ points of the
spatial variety $M \subset GR$. As $g(s) = g(s')$, then $(g_1(s), g_2(s), g_3(s)) = (g_1(s'), g_2(s'), g_3(s'))$ and, therefore $g_1(s) = g_1(s')$, $g_2(s) = g_2(s')$, $g_3(s) = g_3(s')$. Operations

$$\gamma^L = \left(\begin{pmatrix} 123 \\ 213 \end{pmatrix}, \begin{pmatrix} p\, r_+\, r_- \\ p\, r_+\, r_- \end{pmatrix} \right),$$

in agreement with equations (11), the transition from the "right-hand" local system of coordinates in M to the "left-hand" one will correspond by the law:

$$g_1(s) \longleftrightarrow g_2(s'), \quad g_2(s) \longleftrightarrow g_1(s'), \quad g_3(s) \longleftrightarrow g_3(s')$$

$$\downarrow{\omega_D} \qquad \downarrow{\omega_L} \qquad \downarrow{\omega_D} \qquad \downarrow{\omega_L} \qquad \downarrow{\omega_D} \qquad \downarrow{\omega_L} \tag{12}$$

$$x_1 \longleftrightarrow x_2' \quad\quad x_2 \longleftrightarrow x_1' \quad\quad x_3 \longleftrightarrow x_3'$$

Vector (x_1, x_2, x_3) is spatial coordinates of the point $g(s)$ in the right local system of coordinates prescribed by some neighbourhood of the point $g(s)$ and by the homeomorphism ω_D of this neighbourhood on E^3. The set x_1', x_2', x_3' includes the point $g(s')$ coordinates in the corresponding left system of coordinates; with the necessary condition that $x_1 = x_1'$, $x_2 = x_2'$, $x_3 = x_3'$, i.e. scheme (12) describes a mirror reflection of local coordinate systems in M.

We shall assume, finally, that the points $g(s)$, and $g(s')$ are connected by operation

$$\gamma^D = \left(\begin{pmatrix} p\, r_+\, r_- \\ r_+\, p\, r_- \end{pmatrix}, \begin{pmatrix} 123 \\ 123 \end{pmatrix} \right)$$

from $\Gamma^D_{g(s)}$. Then we may treat γ^D as an event where region p (contact relation r_+) acts in point $g(s)$ with regard to something always as the contact relation (region p, respectively) acts with respect to the same thing in the point $g(s')$. Inclusion relation r_- in $g(s)$ manifests itself always as in $g(s')$.

Due to the isomorphism of groups S_3^θ and S_3^s, on the one hand, and of groups $\Gamma^D_{g(s)}$ and $\Gamma^L_{g(s)}$, on the other, it may be shown that operation γ^D is also analytically representable as a mirror reflection of local coordinate systems in M. However, if we establish [by rule (12)] the correspondence $\gamma^L \longleftrightarrow m$, where m is the mirror reflection operation (P-transformation), then under this correspondence $\gamma^D \longleftrightarrow\!\!\!\!/\ m$. On the contrary, if we assume that $\gamma^D \longleftrightarrow m$, then $\gamma^L \longleftrightarrow\!\!\!\!/\ m$. We would seem to arrive at a controversial result that there exist two mutually exclusive types of mirror equality.

Using statement $\underline{1}$ of anti-identification (an example is the operator of charge conjugation, C-transformation) of Heesch–Shubnikov black–white symmetry, it is evident that actually we get a more interesting result:

$$\begin{array}{cc} (m \longleftrightarrow \gamma^D) & (\gamma^D \longleftrightarrow \underline{1}) \\ & \times \\ (m \longleftrightarrow \gamma^L) & (\gamma^L \longleftrightarrow \underline{1}). \end{array} \tag{13}$$

Following from result (13) are: relativity of world points mirror equality and consistent anti-equality, emphasized in a different form in the complete symmetry theory [27].

The relativity of these operations makes difficult to express them analytically. The given interpretation of the transformations γ^L and γ^D as peculiar physical events rules out the dependence of the type of the mentioned events upon the way of introduction of the system of coordinates. But the very fact of the alternative result (13) suggests that if the event γ^L marks optically active particles, then the event γ^D explains physically similar but still a polar phenomenon—the existence of matter antiparticles (C-symmetry). The application of the CPT-theorem enables us to provide further arguments in favour of this conclusion.

THE UNIVERSE AS TIME AND LIGHT†

"Within the deep and luminous subsistence
Of the High Light appeared to me three circles
Of threefold colour and of one dimension,

And by the second seemed the first reflected
As Iris is by Iris, and the third
Seemed fire that equally from both is breathed"
....

"O Light Eterne, sole in thyself that dwellest,
Sole knowest thyself, and, known unto thyself.
And knowing, lovest and smilest on thyself."

DANTE

†The models below are but tentative and subject to further specification.

World point and light signal

The idea that "all that exists in the world is of electromagnetic origin" is not news, the words belonged to Poincaré.† Can this idea be carried into effect by the developed structural concept of the world point as a unit physical events?

We shall get a positive answer to this question if we realize that such an event by its nature is identifiable with a singular light signal, or more precisely, with electromagnetic signal to be called further on the light signal. This signal is certainly to be presented as a quantum of light, but we shall use the notion of signal as the primary one as it is done in the relativity theory or in the information theories.

Intrapoint symmetry is so none other but the statically considered non-geometrical structure of the quantum of light generated by three other types of light signals: p-, r_+- and r_-- signals. These signals are emitted by region p, boundary r_+ and province r_- corresponding to the point (quantum) and are in their turn, identifiable with three structurally heterogenous systems of the similar quanta. Now we see that the previously described structural model of physical space is as a matter of fact a static model of the non-metrical electromagnetic field structure. Topological and symmterical properties of space, just as the trichotomic hierarchical organization of the universe appear but the major characteristics of the structure taken apart from their electromagnetic origin.

However, the quantum of light can't appear in the quiescent state, the static model of intrapoint structure must be compared with the dynamic (electrodynamic) model. The dynamics of the intrapoint structure is a major factor for the union of space, time and field. It is the main aspect of this union that characterizes the pseudo-Euclidean space–time (Minkowski's world). In terms of topology the dynamics allows us to realize from a new point of view the famous programme by Einstein concerning the union of matter and field on structure basis: "We could regard matter as the regions in space where the field is extremely strong. In this way a new philosophical background could be created. Its final aim would be the explanation of all events in nature by structure laws valid always and everywhere."‡

Dynamical interpretation of intrapoint symmetry and CPT-theorem

How does alignment of non-discernibility take place and the retaining of indivduality of p-, r_+-, r_-- signals in the point, i.e. of what makes it the point proper, an independent light signal. The answer is suggested by the quantum theory of field: it is by transmutation of p-, r_+-, r_-- signals by means of light quanta exchange. The symmetry of intrapoint structure requires the parallel quanta exchange, quanta are of the type p-, r_+-, r_-- signals between the initial exchange quanta signals.

Hence the intrapoint symmetry acquires the important dynamic content. Its every non-identical operation (intrapoint transformation) suggests a definite mechanism of the quantum of light generation. Due to its group nature this process is strictly periodical, which is comparable to the wave properties of light.

We can present an algebraic model of the above-described process. Unlike the static probability model of intrapoint structure, this is a deterministic multistage dynamic model.

The dynamic state of the intrapoint structure are evidently controlled by those intrapoint transformations which take place within it. That is why singular p-, r_+-, r_-- signals (quanta) are identifiable with the corresponding transformations $\gamma_p, \gamma_{r_+}, \gamma_{r_-} \in \Gamma$ [Γ is the group of the type $\Gamma_{g(s)} \in \{\Gamma^D_{g(s)}, \Gamma^L_{g(s)}\}$ relative to which the point $g(s)$ and symbols D, L are assumed indefinite). In case of such identification the process of transmutation of p-, r_+-, r_-- signals, i.e. light quantum generation can be easily described by means of the conjugate diagrams in Fig. 7.

The process described in Fig.7($+$) consists of the transformation of quantum γ_p into quantum γ_{r_+} (shown by the straight arrow) produced by the emission (the wavy arrow) of quantum γ_p of quantum γ_{pr_+}. In its turn quantum γ_{r_+}, emitting quantum $\gamma_{r_+r_-}$ is transformed into γ_{r_-}. Finally quantum γ_{r_-} emitting quantum γ_{r_-p} is transformed in quantum γ_p. Simultaneously quantum γ_{r_-p} emitting quantum γ_p is transformed in γ_{pr_+}, etc. A similar description is possible in terms of quanta

---

†Poincaré, "On the dynamics of electron" (1906).
‡A. Einstein, "The evolution of physics: The growth of ideas from early concepts to relativity and quanta", with L. Infeld.

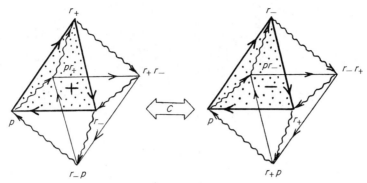

Fig. 7. A model of the generation of the quantum of light as the structure of p-, r_+-, r_-- signals transmutation. Letters show lower indices of the corresponding transformation.

absorption. The non-Abelian type of the group Γ leads to the existence of two conjugate processes described on the whole by the following equations:

$$\gamma_p = \gamma_{r_-p} * \gamma_{r_-}, \quad \gamma_p = \gamma_{r_+p} * \gamma_{r_+}, \quad \gamma_{r_-p} = \gamma_{pr_-}^{-1},$$

$$\gamma_{r_+} = \gamma_{pr_+} * \gamma_p, \quad \gamma_{r_+} = \gamma_{r_-r_+} * \gamma_{r_-}, \quad \gamma_{pr_+} = \gamma_{r_+p}^{-1}, \tag{14}$$

$$\gamma_{r_-} = \gamma_{r_+r_-} * \gamma_{r_+}, \quad \gamma_{r_-} = \gamma_{pr_-} * \gamma_p, \quad \gamma_{r_+r_-} = \gamma_{r_-r_+}^{-1}.$$

where $*$ is the sign of group multiplication in Γ.

Note that according to CPT-theorem, simultaneous charge conjugation (C-transformation), mirror reflection (P-transformation) and time reversal do not change the initial state of the physical system. Each diagram in Fig. 7 is invariant with respect to the simultaneous time reversal in the process given in the diagram, i.e. the reversal direction of all the arrows in the diagram (T-transformation), the mirror reflection of the process (P-transformation), and its conjugation as in Fig. 7. According to the CPT-theorem this is the conjugation which corresponds to the C-transformation.

Thus provided that $\gamma_p \longleftrightarrow p$, $\gamma_{r_+} \longleftrightarrow r_+$, $\gamma_{r_-} \longleftrightarrow r_-$ then, indeed, the operations

$$\left(\begin{pmatrix} p & r_+ & r_- \\ r_+ & p & r_- \end{pmatrix}, \begin{pmatrix} 123 \\ 123 \end{pmatrix} \right), \quad \left(\begin{pmatrix} p & r_+ & r_- \\ p & r_- & r_+ \end{pmatrix}, \begin{pmatrix} 123 \\ 123 \end{pmatrix} \right), \quad \left(\begin{pmatrix} p & r_+ & r_- \\ r_- & r_+ & p \end{pmatrix}, \begin{pmatrix} 123 \\ 123 \end{pmatrix} \right)$$

might be consistent with the C-transformation.

Structural background of the relativist picture of space–time

We shall consider the value $v(\gamma)$ of intrapoint transformations (relative frequency) occurred in a certain point $g(s)$ under the condition that in the "standard" point one intrapoint transformation took place.

It is clear that $v(\gamma) = v(\tilde{e}) + v(e)$, where $v(e)$ and $v(\tilde{e})$ are the number of intraplate transformations respectively accompanied and not accompanied by the recurrence of the mechanism of these transformations. So in case $v(\gamma) > 0$, there may be only three possibilities: (1) $v(\gamma) = v(e)$, i.e. $v(\tilde{e}) = 0$; (2) $v(\tilde{e}) \neq 0$, $v(e) \neq 0$; (3) $v(\gamma) = v(\tilde{e})$, i.e. $v(e) = 0$. We shall characterize these conditions with multigraphs of H_Γ. The oerlex at H_Γ will correspond by definition, to a certain operation from $\Gamma \setminus \{e\}$; the edge will correspond to the transformation of one such operation to another, i.e. again the operation from $\Gamma \setminus \{e\}$; the loop corresponds to the transformation of the operation (vertex) in itself, i.e. e unit of Γ group.

Figure 8(a) shows the multigraph H_Γ defined by the condition $v(\gamma) = v(e)$. The number of loops in this multigraph equals to $v(e) - 1$ [it is accepted conventionally that $v(e) = 6$], i.e. the number of identical intrapoint transformations generated by $v(e)$—multiple recurrence (recycling), stability of certain mechanism of non-identical transformations (intrapoint) in $g(s)$. The loops are not oriented as temporal relations in this situation and are totally indefinite. We can only state that point $g(s)$ selfabsorbs ficticious e-quanta, each of them being described by the unit e of the group Γ. The condition $v(\gamma) = v(e)$ hence characterizes the concentration (localization) of energy in the point $g(s)$, i.e. intrinsically defines $g(s)$ as a point particle (material point).

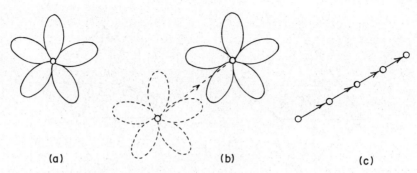

Fig. 8. The multigraph H_Γ, describing: (a) the point particle; (b) the particle translation and (c) field extension.

Figure 8(c) shows the graph H_Γ defined by the condition $v(\gamma) = v(\tilde{e})$. Under this condition constant delocalization of energy takes place in the point $g(s)$: it is found at the lowest energetic state. The given condition hence defines intrinsically the point $g(s)$ as a virtual particle of vacuum matter. Directional alteration of the mechanism of intrapoint transformations in the point of this type can be regarded as the propagation of electromagnetic field in vacuum. Graph H_Γ edges are oriented since temporal relations in this case are fully determined by the relativistic causality.

Finally, Figure 8(b) shows an example of the multigraph H_Γ characterized by the condition $v(e) \neq 0$, $v(\tilde{e}) \neq 0$. Using situations in Figs. 8 (a) and (c) this multigraph can describe uniform and rectilinear motion of a material point in vacuum as a spontaneous process disturbing the state of stability of the given mechanism of the intrapoint transformations in one point and the consecutive exitation of the same state in another point.

The major consequences of this description is (as may be easily shown) the relativistic principle of relativity.

So we see that the topological consideration (in the sense of the H_Γ model) of the intrapoint dynamics features may serve as a basis for the structural, non-metrical background of the relativistic space–time picture. In particular relativistic-causal argumentation the time arrow going back to Bohr and Rosenfeld can be considerably enhanced [30].

Indeed the properties of such intrapoint rhythm defined by the condition $v(\gamma) = v(\tilde{e})$ best agree with the nature of time. Time is homogenous since each intrapoint transformation corresponds to the process of the same type—generation of the quantum of light as the structure of mutual transportation of p-, r_+-, r_-- signals. Time is continuous, as each p-, r_+-, r_-- signal can be represented as the continuum of processes of the same type. Time is asymmetrical as the process of generation of a quantum of light as a dynamic structure is irreversible in its essense. Indeed the hypothetic reversibility of this process in point $g(s)$ means none but recycling of the mechanism of intrapoint transformations, i.e. localization of energy in $g(s)$. This cycling itself may be represented as an irreversible but limited process of generation of real quanta in the system

$$\bigcup_i g\left(E_i(s)\right) = g\left(\bigcup_i E_i(s)\right),$$

where $i = 1, 2, 3$, preserving the structure of this particle.

In other words the electrodynamic nature of the intrapoint structure internally defines the only universal scale of time $t = \lambda/v$ and distances $l = \lambda = c/v$, where v = frequency of electromagnetic fluctuations, which is consistent with the value $v(\gamma)$, and c = the speed of light in vacuum.

Structural, genetic and energy concepts of space–time

The structural representation of space–time is defined by the following relation F_{str}:

$$F_{str} \subset GR \times \Gamma,$$

where $\times$ is the Cartesian product. Arranged in order $3 + 1$-multiple (i.e. a pair, the first component of which is the ordered triple) of the type $(g_1(s), g_2(s), g_3(s), \gamma) \in F_{str}$ describes the world point $g(s) = (g_1(s), g_2(s), g_3(s))$, being in the state of intrapoint transformation $\gamma \in \Gamma$. If needed we may naturally assume that γ in $(g(s), \gamma)$ ranges over the sequence of values from Γ corresponding to the world line, for example. F_{str} viewed as a carrier of spatial-temporal variety excludes the necessity of postulating the major topological properties (3 + 1-dimensions, orientation etc.) of this variety as they can be deduced from F_{str}.

The genetic representation of space–time is given by the multivalued representation F_{gen}:

$$F_{gen}: \Gamma^3 \rightarrow \Gamma \ (\text{or} \ F_{gen} \subset \Gamma^3 \times \Gamma).$$

Each 3 + 1-multiple from F_{gen} has by definition, the form $((\gamma_p, \gamma_{r_+}, \gamma_{r_-}), \gamma)$, i.e. the representation of F_{gen} correlates the triplet $(\gamma_p, \gamma_{r_+}, \gamma_{r_-}) \in \Gamma^3$ as the p-, r_+-, r_-- signals the transformation $\gamma \in \Gamma$ as the quantum of light generated by the mutual transformations of these signals. F_{gen} includes all the structural information contained in F_{str}, as there is correspondence between vectors $(g_1(s), g_2(s), g_3(s))$ and $(\gamma_p, \gamma_{r_+}, \gamma_{r_-})$:

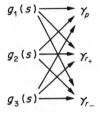

to be understood dynamically [equations (14)]. At the same time, the model F_{gen} characterizes space–time as a unique quasi-growth structure. Space is to be viewed as a structure of transmutation of the quanta of light, while time as a process of generation of the light quanta of this structure.

The energetic representation of space–time follows from the genetic one and is given by the multivalued representation of F_{energ}:

$$F_{energ}: \Lambda^3 \rightarrow \Lambda (\text{or} \ F_{energ} \subset \Lambda^3 \times \Lambda),$$

where Λ is the set of all possible values of the electromagnetic spectrum. Each 3 + 1-multiple from F_{energ} has, by definition, the form:

$$\left(\left(\frac{c}{v_p}, \frac{c}{v_{r_+}}, \frac{c}{v_{r_-}} \right), \frac{1}{v_\gamma} \right) = \left((\lambda_p, \lambda_{r_+}, \lambda_{r_-}), \frac{1}{v_\gamma} \right) = ((x_1, x_2, x_3), t), \tag{15}$$

where $0 < \lambda_p, \lambda_{r_+}, \lambda_{r_-}, \lambda_\gamma < \infty$, i.e. the representation of F_{energ} correlates p-, r_+-, r_-- signals with the wave-lengths of $\lambda_p, \lambda_{r_+}, \lambda_{r_-}$ to the quantum of light γ, controlled by the frequency v_γ and generated by the transmutation of these signals.

It should be noted that the model F_{energ} has the intrinsic pseudo-Euclidean metrics (interval). Indeed, in view of equations (15) for the quadric surface of $\lambda^2 = c^2 t^2 - x_1^2 - x_2^2 - x_3^2$ in relation to F_{energ} we have:

$$\lambda^2 = \lambda_\gamma^2 - \lambda_p^2 - \lambda_{r_+}^2 - \lambda_{r_-}^2. \tag{16}$$

Admitting that $E = hv$, where h is Plank's constant, E is the energy of the quanta from equation (16), we get:

$$E^{-2} = E_\gamma^{-2} - E_p^{-2} - E_{r_+}^{-2} - E_{r_-}^{-2}.$$

For the light signal $\lambda = 0$, i.e.

$$\lambda_\gamma^2 = \lambda_p^2 + \lambda_{r_+}^2 + \lambda_{r_-}^2.$$

In view of the above it is clear that the light signal with the wave-length of γ is generated by the transmutation of signals with the wave-lengths of $\gamma_p, \gamma_{r_+}, \gamma_{r_-}$ and only in case this condition is satisfied. In the case when the events are connected by the space-like interval, $\lambda^2 < 0$, i.e.

$$\lambda_\gamma^2 < \lambda_p^2 + \lambda_{r_+}^2 + \lambda_{r_-}^2.$$

Consequently the structural contents of the relativity of synchronism means that under the mentioned condition the formation of the transmutation structure of p-, r_+-, r_--signals is impossible. Mathematically it implies a limitation on the representation F_{energ}, graph describing F_{energ} as a binary relation is not complete.

Finally we shall discuss the events related by the time-like interval $\lambda^2 > 0$, i.e.

$$\lambda_\gamma^2 > \lambda_p^2 + \lambda_{r_+}^2 + \lambda_{r_-}^2$$

$$E_\gamma^{-2} > E_p^{-2} + E_{r_+}^{-2} + E_{r_-}^{-2}.$$

These conditions define the particle as a wave with the length $l = \lambda$ turned to the point. Mathematically the values l and $E = (E_\gamma^{-2} - E_p^{-2} - E_{r_+}^{-2} - E_{r_-}^{-2})^{-1/2}$ are sums of loop weights of the multigraph [Fig. 8(a)], invariant with reference to Lorentz transformations.

Unlike the Kaluza–Klein theories the new image of the particle does not need introducing additional hidden measurements of space–time. The particle in F_{energ} gives two polar states for the cone of light: one state corresponds to movement of the particle in space (its coordinates $\lambda_p, \lambda_{r_+}, \lambda_{r_-}$), the other one to movement inside the particle (i.e. to transition to space coordinates of the type $\lambda_{p'}, \lambda_{r'_+}, \lambda_{r'_-}$, where $\lambda_{p'}^2 + \lambda_{r'_+}^2 + \lambda_{r'_-}^2 = \lambda^2$). It should be noted that within F_{energ} both cones merged to form the single one. However there might be that the interaction of these cones of light will lead to the gravitation as a secondary structure on the electromagnetic field.

Though the given models are regarded as provisional we can conclude that the geometrical, physical and structural-systematic description of reality has a unique basis—the structure of the quantum of light. It is from the unification of these descriptions that we should considerably broaden our knowledge about the light. We have tried to show that the quantum of light is possible to consider as the structure-process of the light p-, r_+-, r_-- signals transmutation, with a peculiar symmetry between heterogeneity and homogeneity of the world and the asymmetry of its development. There is no doubt that further elaboration of these conceptions will require reconsideration of the medieval ideas regarding the Trinity, as well as regarding Light and Hierarchy ("*Corpus'a Areopagiticum*", Dante Alighieri, Nicolaus Cusanus and others), and in particular the relationship between science, religion and art.

REFERENCES

1. L. von Bertalanffy, *General Systems Theory*. Braziller, New York (1968).
2. M. D. Mesarovic, D. Macko and Y. Takahara. *Theory of Hierarchical Multilevel Systems*. Academic Press, New York (1970).
3. A. L. Mackay, Generalized crystallography. *Comput. Math. Applic.* **12B**, 21–37 (1986). Reprinted in *Symmetry: Unifying Human Understanding* (Ed. I. Hargittai). Pergamon Press, Oxford (1986).
4. A. V. Shubnikov and V. A. Koptsik, *Symmetry in Science and Art*. Plenum Press, New York (1974).
5. B. K. Vainshtein, *Modern Crystallography* Vol. I. *Springer Series in Solid-State Sciences*, Vol. 15. Springer, New York (1981).
6. B. B. Mandelbrot, *The Fractal Geometry of Nature*. Freeman, New York (1983).
7. R. Penrose and W. Rindler, *Spinors and Space–Time*, Vol. 1. Cambridge Univ. Press, Cambridge (1984).
8. D. I. Blokhintsev, *Space and Time in the Microworld* (in Russian). Nauka, Moscow (1982).
9. L. D. Landau and E. M. Lifshitz, *Course of Theoretical Physics*, Vol. 1. Pergamon Press, Oxford (1976).
10. A. V. Malikov, Structural theory of physical space and enantiomorphism of material objects, In *Principles of Symmetry and Systemology in Chemistry* (in Russian), pp. 107–121. Moscow Univ. Publ., Moscow (1987).
11. A. V. Malikov, Theory of hierarchical spaces. In *System. Symmetry. Harmony* (in Russian), pp. 171–190. Mysl', Moscow (1988).
12. O. L. Krivanek, S. Isoda and K. Kobayashi, Lattice imaging of a grain boundary in crystalline germanium. *Phil. Mag.* **36**, 931–940 (1977).
13. O. Ore, *Theory of Graphs*. Am. math. Soc., Rhode Island (1962).
14. R. V. Galiulin, *Crystallographic Geometry* (in Russian). Nauka, Moscow (1984).
15. R. B. King (Ed.) *Chemical Applications of Topology and Graph Theory*. Elsevier, Amsterdam (1983).
16. N. S. Zefirov, S. S. Trach and O. C. Chizhov, Framework and polycyclic compounds. A molecular design based on the principle of isomorphic substitution (in Russian). VINITI, Moscow, (1979).
17. A. V. Malikov, On the modelling of regularities of the grain contact in mineral intergrowths. *Dokl. Akad. Nauk SSSR* **280**, 878–880 (1985).
18. A. V. Malikov, Effect of decrease of combinatory-topological symmetry in aggregates of crystals. *Dokl. Akad. Nauk SSSR* **293**, 868–871 (1987).
19. E. V. Presnov and V. V. Isaeva, *Topology Reconstruction During Morphogenesis* (in Russian). Nauka, Moscow (1985).
20. N. F. Chelishchev and A. V. Malikov, On pseudomorphic nature of underwater ferro-manganese concretions and crusts. *Dokl. Akad. Nauk SSSR* **298**, 698–701 (1988).

21. J. L. Kelley, *General Topology*. Van Nostrand, New York (1957).
22. R. P. Geroch, E. H. Kronheimer and R. Penrose, Ideal points in space–time. *Proc. R. Soc.* **A 327,** 545–567 (1972).
23. M. Daniel and C. M. Viallet, The geometrical setting of gauge theories of the Yang–Mills type. *Rev. mod. Phys.* **52,** 175–197 (1980).
24. A. Chodos, Kaluza-Klein, theories: overview. *Communs. Nucl. Part. Phys.* **13,** 171–181 (1984).
25. F. R. Tangherlini, Schwarzschild field in *n* dimensions and the dimensionality of space problem. *Nuovo Cim.* **27,** 636–651 (1963).
26. G. E. Gorelik, *Space Dimensionality* (in Russian). Moscow Univ. Publ., Moscow (1983).
27. I. S. Zheludev, Space and time inversion in physical crystallography. *Acta Crystallogr.* **A 42,** 122–127 (1986).
28. Yu. A. Urmantsev, Symmetry of system and system of symmetry. *Comput. Math. Applic.* **12B,** 279–405 (1986). Reprinted in *Symmetry: Unifying Human Understanding* (Ed. I. Hargittai). Pergamon Press, Oxford (1986).
29. A. Robinson and E. Zakon, A set-theoretical characterization of enlargements in applications of model theory to algebra, analysis and probability theory. *Proc. Int. Symp. Nonstandard Analysis.* Holt-Rinehart and Winston, Eastbourne, pp. 109–122 (1969).
30. L. Brillouin and L. Rosenfeld, Note Complémentaire. *Revue Métaphys. Morale* **2,** 247 (1962).

Computers Math. Applic. Vol. 17, No. 1–3, pp. 301–320, 1989
Printed in Great Britain

0097-4943/89 $3.00 + 0.00
Pergamon Press plc

VISUAL AND HIDDEN SYMMETRY IN GEOMETRY†‡

A. T. FOMENKO

Department of Higher Geometry and Topology, Faculty of Mechanics and Mathematics, Moscow State
University, Moscow 119899, U.S.S.R.

Abstract—A symmetry appears in modern geometry and its numerous applications both in explicit form
(visually trivial) and sometimes, through sufficiently complex mathematical transformations, in veiled,
hidden form. This current paper is made up of commentaries to a series of mathematical, graphical works
of the author. In these works an attempt was made to show some occurrences of symmetry in geometry,
topology and mechanics.

1. CRITICAL POINTS OF SMOOTH FUNCTIONS ON THREE-DIMENSIONAL MANIFOLDS

So called critical points of potentials (minima, maxima and saddle points) corresponding to
equilibrium positions of the system (stable or unstable) play an important role in physics and
mechanics. With any critical point of a function, f (for example, f can be taken as potential energy)
there are connected separatrices of the gradient vector field, grad f (see Fig. 1), i.e. integral
trajectories of the field, entering or emanating from the point. If f is invariant with respect to some
group (of symmetries) then critical points are "reproduced" by actions of this group (the image
at a critical point is again a critical point). If the group is discrete (as in Fig. 1), then an orbit of
each critical point consists of a discrete set of critical points. If the point is a saddle point, then
in the three-dimensional case the set of its separatrices is divided into two types. Separatrices of
the first kind form a two-dimensional disc (the "bell" in Fig. 1) while separatrices of the second
type form a one-dimensional smooth curve (the vertical axis of the bell on the figure). For example,
along the bell the function is decreasing, while along its vertical axis the function is increasing.
Readers who wish to learn more about this problem are directed to Ref. [1].

2. TWO-ADIC SOLENOID

This geometric object is well-known in topology and algebraic geometry (see for example Refs
[2, 3]). One should look at the torus (bagel), then wind the second torus twice around the axis of
the first torus. Then in a similar way, embed the third torus into the second torus by winding it
twice around its axis (i.e. $2^2 = 4$ times wound around the axis of the first torus) and so on *ad
infinitum*. As a result we obtain an infinite series of tori embedded in each other (they are becoming
thinner and thinner), each of which is twice wound around the axis of the previous torus. Consider
further "the limit" of all these tori. As a result, we will get an interesting topological space
possessing many properties important for geometry. In Fig. 2, only nine tori are drawn representing
the beginning of this process. These tori are marked by Roman numbers. For the convenience of
the audience, in each torus holes are made (a part of the torus is cut) through which successive
tori are seen. This example demonstrates to us the hidden symmetries.

3. ALGEBRAIC SURFACE IN TWO-DIMENSIONAL COMPLEX SPACE

So called Riemannian surfaces of algebraic functions (i.e. solutions of polynomial equations) are
well-known in geometry. With each polynomial equation of the form

$$f(w, z) = \sum_{K=1}^{n} a_K(w) z^k = 0$$

---

†Translated from Russian by Drs V. V. Goldberg and J. Kappraff, both Professors of Mathematics at the Department
of Mathematics, New Jersey Institute of Technology, Newark, NJ 07102, U.S.A.
‡To avoid extensive delays, this paper has been published without the author's corrections.

[where (w, z) is a point of the complex space $\mathbb{C}^2$]. Naturally connected to this polynomial is the well-known Galois group which can be considered as the group of hidden symmetries of the given equation (see for example Ref. [4]). This group can be described in terms of the Riemannian surface associated with the given equation. If we change the coefficients of the equation, the Riemannian surface will be somehow deformed in the space. In Fig. 2, such a deformation is shown. Here the surface corresponds to the equation

$$w^2 - (z - a)(z - b)(z - c)(z - d) = 0.$$

For the deformation presented in Fig. 3, either two roots a and b merge (become multiple) or else two other roots c and d merge. In this case the Riemannian surface is a torus (in Fig. 3 two spheres are shown connected by two pipes to generate the torus). When two of the roots merge, one of the two spheres blows up while the second one becomes smaller. When the other two roots merge, the spheres exchange their positions.

4. TWO-DIMENSIONAL SPHERE, POISSON SPHERE, NUTATION AND PRECESSION

The integrability of the equations of motion of a solid body with mass have been studied in classical mechanics over many years (starting from Euler, Lagrange, Kovalevskaya, Clebtsch, Goryachev, Chaplygin and others). For the purpose of a visual geometrical description of motions of the body, special trajectories are used that are located on a two-dimensional sphere called the Poisson sphere. In order to get a visual picture the reader can imagine one of the important motions of the solid body as a gyroscope motion or as a motion of a top with mass whose rotational axis is anchored at some fixed point. A symmetric top (so called Lagrange top) is a particular case of integrability. It turns out that the equations of motion indicated above can be integrated only in the case when the solid body under investigation possesses some symmetries (for example, when this body is a surface of revolution). The search for such symmetries is a difficult problem of mechanics and Hamiltonian geometry. For some modern aspects of the theory of such symmetries see for example Ref. [5].

5. HEEGARD DIAGRAMS OF THREE-DIMENSIONAL MANIFOLDS

Rolling without slipping

One of the interesting occurrences of symmetry in mechanics is the rolling of a solid body without slipping along a rough two-dimensional surface (possibly of complex profile). The results of this process are studied by the mechanics of non-holonomic systems (i.e. the systems with non-holonomic constraints). Recently, deep and interesting connections of this subject with Lie groups were discovered (that is to say, with some definite groups with symmetry). At the same time Fig. 5 helps to see the process of formation of three-dimensional manifolds from "elementary bricks". In topology, the scheme of gluing such bricks is called the Heegard diagram.

6. ALGEBRAIC FUNCTIONS AND FUNDAMENTAL DOMAINS OF ACTION OF DISCRETE GROUPS

In Fig. 6, there is drawn a level surface of a polynomial function (of higher order) given in three-dimensional space. This concrete function has a discrete symmetry group and therefore its level surface is invariant with respect to actions of this group. On a level surface, it is possible to define the fundamental domain of actions of the group, i.e. the minimal set with the property that all possible shifts by different elements of the group cover the whole surface. For visualizing purposes this fundamental domain is represented in the form of a rectangle. Applying all possible elements of the group to this rectangle "reproduce" it, we shift it and tile the surface by congruent domains. At the same time this mathematical fantasy is considered as a possible illustration of the well-known chess playing scene between Voland and Behemoth in the famous novel *Master and Margarita* by M. A. Bulgakov. (At the end of the game, Behemoth loses to Voland.)

Fig. 1. Critical points of smooth functions on three-dimensional manifolds.

Fig. 2. Two-adic solenoid.

Fig. 3. Algebraic surface in two-dimensional complex space.

Fig. 4. Two-dimensional sphere, Poisson sphere, nutation and precession.

Fig. 5. Heegard diagrams of three-dimensional manifolds.

Fig. 6. Algebraic functions and fundamental domains of action of discrete groups.

308

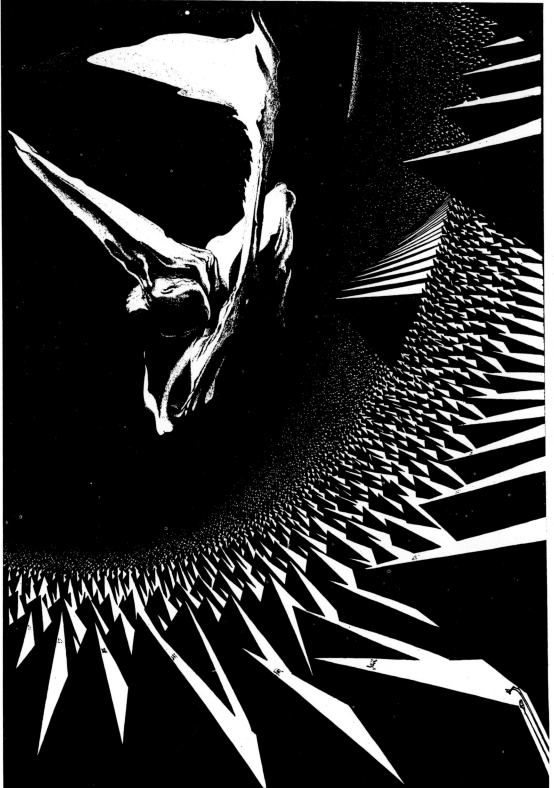

Fig. 7. Crystal structures. The partitioning of the space into simplexes.

Fig. 8. Crystal structures. Cubic partitionings of polyhedra.

7. CRYSTAL STRUCTURES. THE PARTITIONING OF THE SPACE INTO SIMPLEXES

The study of crystal structures (the mechanism of growing crystals, its geometry, etc.) is one of the important parts of modern physics and chemistry. One of such infinite crystal structures is shown in Fig. 7. In topology, the partitioning of a space into "elementary objects" each of which is a simplex is often used. A one-dimensional simplex is a segment, a two-dimensional simplex is a triangle, a three-dimensional simplex is a tetrahedron (three sided pyramid), etc. Three-dimensional simplexes can be added (by identifying their faces) together to form more complex objects, polyhedra. The language of simplexes is very useful in topology. It allows topological invariants (the homology group) of polyhedra to be computed rapidly. The group of symmetries of regular tetrahedra (regular simplex) is one of the interesting finite groups closely connected with crystallography.

8. CRYSTAL STRUCTURES. CUBIC PARTITIONINGS OF POLYHEDRA

In Fig. 8 another method of partitioning of space into elementary parts is presented. Parallelopipeds and their different modifications can be taken as such "bricks". Such partitionings (and their generalizations) appear in topology when the group of cubic homology and co-homologies is defined. These groups [as well as the groups of simplicial (co) homology] are topological invariants of the space (polyhedra). Namely, if two polyhedra are homotopically equivalent then their groups of (co) homologies coincide. The appearance of such groups is one of the illustrations of hidden symmetries in topology. For details see Ref. [6].

9. SYMMETRIC SPACES

Transformation groups

One of the most important classes of manifolds is the symmetric spaces, i.e. the spaces possessing a large group of symmetries (a large group of motions). A space is said to be symmetric if it is transferred into itself by each symmetry keeping an arbitrary point of the space fixed (and rotating by a half-turn geodesic lines passing through the fixed point). The theory of such spaces is a particular case of the general theory of Lie groups and homogeneous spaces since each symmetric space uniquely determines its group of symmetries and some subgroup in it (the isotropy group). The symmetric spaces have an invariant symmetric Riemannian metric. If this metric is altered (for example, perturbed in a neighborhood of the point as it is shown on Fig. 9) then the space ceases to be symmetric. In Fig. 9, an infinite series of standard spheres is shown which are the simplest symmetric spaces. The symmetry here appears visually explicitly as the "symmetry of a geometric object". (See for example Ref. [7].)

10. SYMMETRY OF PLANE WAVES

Scandinavian myths

The picture of the propagation of symmetric waves from a point source of variable density is presented (the amplitude of the wave disturbance changes with time). This implies that the wave-length changes. Similar pictures appear on diffraction cells during optical experiments. The picture presented in Fig. 10 can be also treated as the graph of a function with axial symmetry. Because of this, critical points of the function are not isolated (except the central one) and fill the critical circles. The visual image in Fig. 10 can also be associated with the well-known legend about the goddess of earth, Herde and the god Wotan who came to Herde with questions about the future.

11. CUBIC LATTICES

Probability processes, chaotic walk and crystal destruction

A cubic lattice composed of cubes in three-dimensional space is drawn in Fig. 11. Far from the viewer, the lattice still preserves its original regular form while in the foreground its destruction and loss of symmetry have already begun. Such lattices play an essential role in the theory of crystallographic groups. Each lattice (with a definite type of symmetry) defines a group whose properties reflect the properties of the lattice and the corresponding crystals sufficiently complete. On the other hand the graphical work, Fig. 11, allows one to get a visual image of one of the important ideas of probability theory. Consider a die (with numbers arranged randomly on its faces). Locate on each face a human being who has another die in his hand each face of which contains a human being (of smaller size), who also has a die in his hand, and so on to infinity. Human beings on each consecutive level must decrease their sizes. After this, each of these humans (there are infinitely many) starts, independently of the others, tossing his die randomly. As a result, all dice will undergo chaotic motion which will be complicated more and more when the movement is from one set of dice on one level to dice on the successive level. At the same time, it turns out that by consideration of such chaotic motions it appears as if the symmetries disappear; it is possible to find in such systems, hidden symmetries appearing on higher levels. These hidden symmetries can govern the solutions of stochastic differential equations sometimes possessing groups of symmetries. See for example Ref. [8].

12. CHAOS

Chaos is the opposite of symmetry; it is the complete destruction of any possible order in the object's structure. At the same time (see example 11 above), while studying complex stochastic processes, it is sometimes possible to find in them symmetries of higher orders. It is difficult to watch them visually. Such symmetries appear in the form of the symmetry group of the corresponding differential equations describing the process. For example the chaotic motion of particles of liquid can obey a general globally ordered motion described by sufficiently simple equations. The chaos manifests itself in so-called "strange attractors", limiting sets of dynamical systems. The interesting game of chaos and symmetry is observed in fractal theory (see for example Ref. [9]).

13. HIDDEN SYMMETRICS

Decimal expansions of the numbers π and e

Let's consider the decimal expansions of the famous numbers π and e: $\pi = 3.1415\ldots$, and $e = 2.7182\ldots$. In Fig. 13, each digit of this expansion is placed in its own separate square, inside of which there are located exactly as many black circles as the corresponding digit. When the horizontal row in Fig. 13 is completed, the decimal expansion jumps to the next row *ad infinitum*. The number π is presented on the front face of the prism while the number e is on the lateral face. The series of black spots appears as random to the eye, completely chaotic, and this impression corresponds to the mathematical reality. In a more exact sense, the series of digits in the expansions of π and e represent random series (the exact mathematical proof of this fact is very non-trivial; see for example works of A. N. Kolmogorov). At the same time appears as one of the consequences of the symmetry of the standard planar circumference. The length of circumference of radius one is equal to 2π. If we alter the metric on the circumference, we destroy the symmetry and, in general, obtain another length of circumference (no longer connected with the number).

FIGURE 14

Figure 14 is an attempt at a modern geometric refraction of ideas stimulated by middle age artists (in particular Breugel, Durer, etc.) They were interested in the theory of perspective and they developed principals of the correct rendering of objects, and also groped to find mathematical mechanisms governing optical illusions. An example is a famous engraving *Alchemist* by Breugel. Figure 14 can be considered as a mathematical variation on this theme the main content of which is a visual modeling of the idea of mathematical infinity. Of course, many scientific ideas of the authors of the middle ages lost their actuality, nevertheless many contemporary mathematical ideas have organically grown from the soil that was fertilized by them.

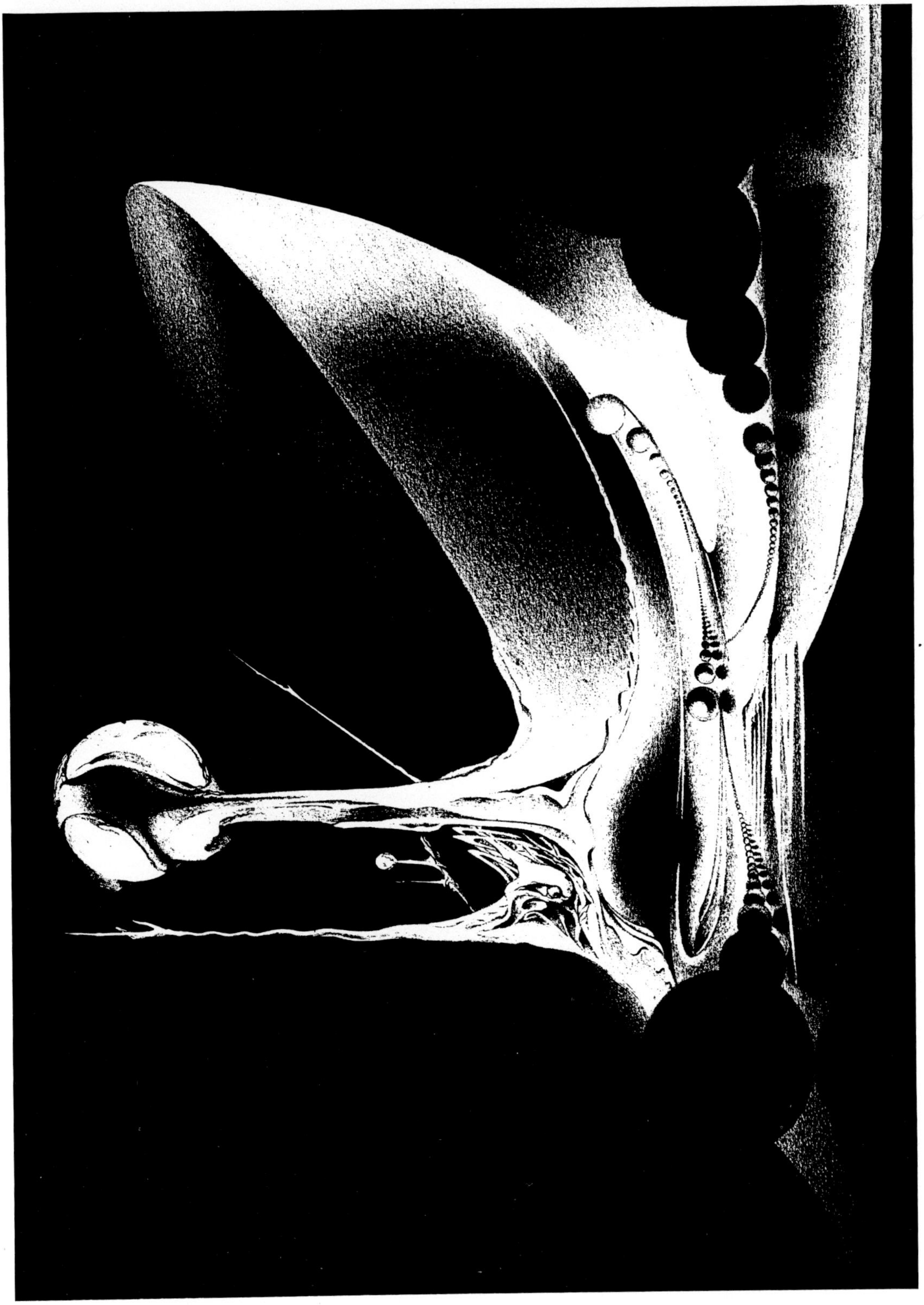

Fig. 9. Symmetric spaces.

313

Fig. 10. Symmetry of plane waves.

Fig. 11. Cubic lattices.

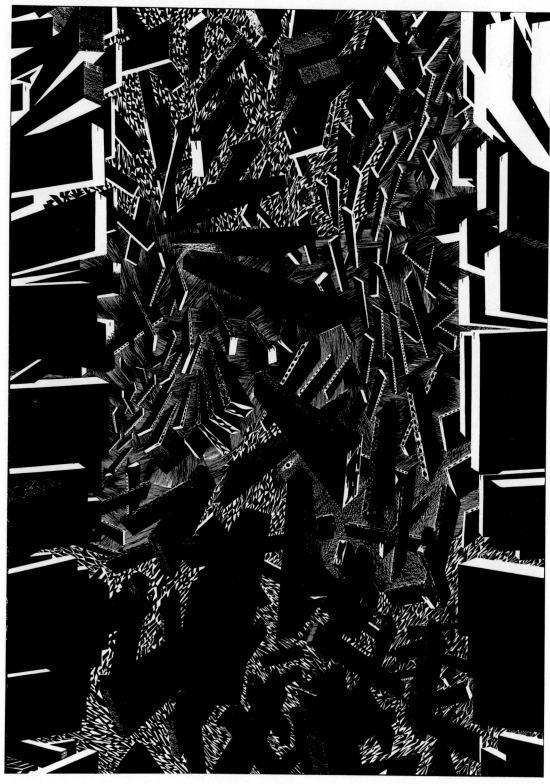

Fig. 12. Chaos.

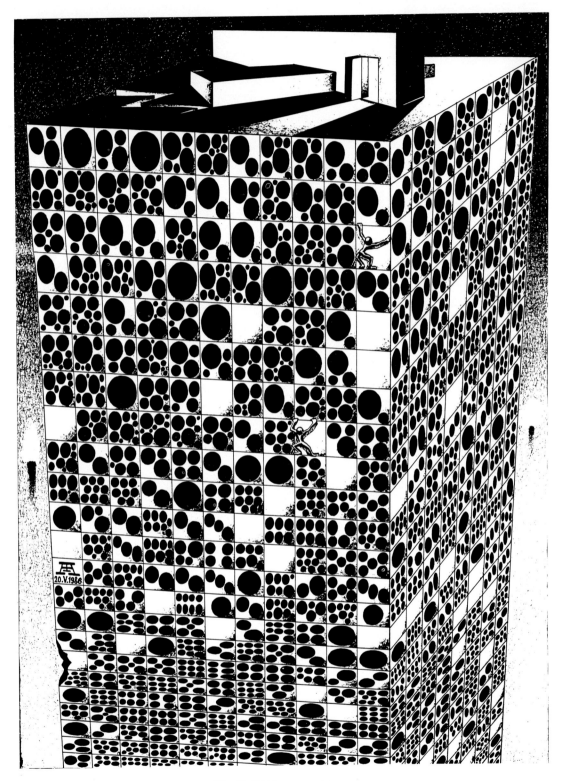

Fig. 13. Hidden symmetrics.

Fig. 14

Fig. 15. Mathematical fantasy.

15. MATHEMATICAL FANTASY

Today we examine our surroundings trying to understand hidden laws governing events. A changing infinite series of geometric objects often hides one symmetry or another which is the manifestation of some physical laws.

REFERENCES

1. B. A. Dubrovin, A. T. Fomenko and S. P. Novikov, *Modern Geometry—Methods and Applications*, Vol. 93 (Part 1) (1984); Vol. 104 (Part 2) (1985). Springer, New York.
2. N. Koblitz, *p-Adic Analysis: A Short Course on Recent Work. London Math. Soc. Lect. Notes. Series*, 46. Cambridge Univ. Press, (1980).
3. N. Koblitz, *p-Adic Numbers, p-Adic Analysis, and Zeta-Functions*. Vol. 85 Springer, New York, (1977).
4. N. Koblitz, *Introduction to Elliptic Curves and Modular Forms* Vol. 97. Springer, New York (1984).
5. A. T. Fomenko and V. V. Trofimov, *Integrable Systems on Lie Algebras and Symmetric Spaces. Advanced Studies in Contemporary Mathematics*, Vol. 2. Gordon and Breach (1988).
6. A. T. Fomenko, L. B. Fuchs and V. L. Gutenmacher, *Homotopic Topology*. Akadémiai Kiadó, Budapest (1986).
7. O. Loos, *Symmetric Spaces*, Vols 1 and 2. Benjamen, New York (1969).
8. A. N. Shiryayev, *Probability*, Vol. 95. Springer, New York (1984).
9. B. B. Mandelbrot, *The Fractal Geometry of Nature*. Freeman, San Francisco, Calif. (1977).

Computers Math. Applic. Vol. 17, No. 1–3, pp. 321–336, 1989
Printed in Great Britain. All rights reserved

0097-4943/89 $3.00 + 0.00

MATHEMATICS AND BEAUTY—VIII

TESSELATION AUTOMATA DERIVED FROM A SINGLE DEFECT

C. A. PICKOVER

IBM Thomas J. Watson Research Center, Yorktown Heights, NY 10598, U.S.A.

Abstract—To help characterize complicated physical and mathematical structures and phenomena, computers with graphics can be used to produce visual representations with a spectrum of perspectives. In this paper, unusual "tesselation automata" (TA) are presented which grow according to certain symmetrical recursive rules. TA are a class of simple mathematical systems which exhibit complex behavior and which are becoming important as models for a variety of physical processes. This paper differs from others in that its focuses on symmetrical TA derived from a single defect, and reader involvement is encouraged by giving "recipes" for the various chaotic forms which represent a visually striking and intricate class of shapes.

INTRODUCTION

"Some people can read a musical score and in their minds hear the music . . . . Others can see, in their mind's eye, great beauty and structure in certain mathematical functions . . . . Lesser folk, like me, need to hear music played and see numbers rendered to appreciate their structures."

P. B. SCHROEDER

Today, there are several scientific fields devoted to the study of how complicated behavior can arise in systems from simple rules and how minute changes in the input of nonlinear systems can lead to large differences in the output; such fields include chaos and tesselation automata (TA) theory. In this paper, I briefly discuss some empirical results obtained by experimentation with a particular class of symmetrical TA. Some of the resulting patterns are reminiscent of the planar ornaments of a variety of cultures (ornaments with a repeating motif in at least two nonparallel directions).

"Tesselation automata" are a class of simple mathematical systems which are becoming important as models for a variety of physical processes. Referred to variously as "cellular automata", "homogeneous structures", "cellular structures" and "iterative arrays", they have been applied to and reintroduced for a wide variety of purposes [1–4]. The term "tesselation" is used in this paper for the following reasons: when a floor is covered with tiles, a symmetrical and repetitive pattern is often formed—straight edges being more common then curved ones. Such a division of a plane into polygons, regular or irregular, is called a "tesselation"—and I have chosen "tesselation" here to emphasize these geometric aspects often found in the figures in this paper.

Usually TA consist of a grid of cells which can exist in two states, occupied or unoccupied. The occupancy of one cell is determined from a simple mathematical analysis of the occupancy of neighbor cells. One popular set of rules is set forth in what has become known as the game of "LIFE" [2]. Though the rules governing the creation of TA are simple, the patterns they produce are very complicated and sometimes seem almost random, like a turbulent fluid flow or the output of a cryptographic system.

The term "chaos" is often used to describe the complicated behavior of nonlinear systems, and TA are useful in describing certain aspects of dynamical systems exhibiting irregular ("chaotic") behavior [5, 6]. Other simple algorithms studied by the author which produce interesting and complicated behavior are described in Ref. [7]. Apart from their curious mathematical properties, many nonlinear maps now have an immense attraction to physicists, because of the role they play in understanding certain phase transitions and other chaotic natural phenomenon [5].

The present paper is number eight in a "Mathematics and Beauty" series [7] which presents aesthetically appealing and mathematically interesting patterns derived from simple functions. The resulting pictures should be of interest to a range of scientists as well as home-computer artists.

MOTIVATION

One goal of this paper is to demonstrate and emphasize the role of recursive algorithms in generating complex forms and to show the reader how to create such shapes using a computer. Another goal is to demonstrate how research in simple mathematical formulas can reveal an inexhaustible new reservoir of magnificent shapes and images. Indeed, structures produced by these equations include shapes of startling intricacy. The graphics experiments presented, with the variety of accompanying parameters, are good ways to show the complexity of the behavior. This paper differs from others in that its focuses on TA derived from a single defect (explained below) using symmetrical rules, and that reader involvement is encouraged by giving "recipes" for the various chaotic forms which represent a visually striking and intricate class of shapes.

METHOD AND OBSERVATIONS

TA are mathematical idealizations of physical systems in which space and time are discrete [1]. Here I present unusual patterns exhibited by figures "growing" according to certain recursive rules. The growth occurs in a plane subdivided into regular square tiles. Note, in particular, that with the rules of growth in this paper, the figures will continue increasing in size indefinitely as time progresses. In each of my cases, the starting configuration is only 1 occupied square, which can be thought of a single defect (or perturbation) in a lattice of all 0s, represented by:

$$
\begin{bmatrix}
0 & 0 & 0 & 0 & 0 \\
0 & 0 & 0 & 0 & 0 \\
0 & 0 & 1 & 0 & 0 \\
0 & 0 & 0 & 0 & 0 \\
0 & 0 & 0 & 0 & 0
\end{bmatrix}
\tag{1}
$$

TA Type 1

This is the simplest system to set up, yet the behavior is still interesting. Given the nth generation, I define the $(n + 1)$th as follows. A square of the next generation is formed if it is orthogonally contiguous to one and only one square of the current generation. Starting with the pattern in equation (1) for $n = 1$ pattern for $n = 2$ would be:

$$
\begin{bmatrix}
0 & 0 & 0 & 0 & 0 \\
0 & 0 & 1 & 0 & 0 \\
0 & 1 & 1 & 1 & 0 \\
0 & 0 & 1 & 0 & 0 \\
0 & 0 & 0 & 0 & 0
\end{bmatrix} .
\tag{2}
$$

Figure 1 indicates the results at $n = 200$. This TA is similar to that described in Refs [2, 4]. Note that no "death's" of squares occur (i.e. no $1 \rightarrow 0$ transitions can occur; deaths are employed in many CA experiments [2]). Note also that on the four perpendicular axes [which go through $(0, 0)$], all the squares will be present. These are the stems from which branching occurs.

TA Type 2. Time dependence of rules

A. "Mod 2" TA. Given the nth generation, I define the $(n + 1)$th as follows. A square of the next generation is formed if:

1. It is orthogonally contiguous to one and only one square of the current generation for even n (i.e. $n \bmod 2 = 0$).

2. It is contiguous to one and only one square of the current generation, where the local neighborhood is both orthogonal and diagonal, for odd n ($n \bmod 2 = 1$).

In other words, for (n mod $2 = 0$)

$$\text{if} \sum C_{\text{orth}} = 1 \to C_{ij} = 1,\tag{3}$$

where

$$C_{\text{orth}} = [C_{i,j+1}, C_{ij-1}, C_{i+1,j}, C_{i-1,j}].\tag{4}$$

For (n mod $2 = 1$)

$$\text{if} \sum C_{\text{orth-diag}} = 1 \to C_{ij} = 1,\tag{5}$$

where

$$C_{\text{orth-diag}} = [C_{i,j+1}, C_{ij-1}, C_{i+1,j}, C_{i-1,j} C_{I+1,j+1}, C_{i-1,j-1}, C_{i-1,j+1}, C_{i+1,j-1}].\tag{6}$$

Notice the discrete symmetrical planes running through these TA. For example, see the planes in Figs 2 and 3.

We can use this observation to get a visual idea of resultant patterns, for large n, in a multi-defect system (see TA Type 3).

B. *"Mod 6" TA*. Given the nth generation, I define the ($n + 1$)th in a same manner as for Type 2A, except that n mod $6 = 0$ vs n mod $6 \neq 0$ determines the temporal evolution of the pattern (Fig. 4).

TA Type 3. Contests between defects

More than one initial defect can be placed on a large infinite lattice. We can let them each grow and finally merge (and compete) according to a set of rules. Figure 5 is a TA of Type 2A, and it shows three defects after just a few generations (this figure is magnified relative to others). Figure 6 shows the growth for large n.

To help see the numerous symmetry planes and to get an idea about the shape of the figure as it evolves, the reader can draw the primary radiating symmetrical discrete planes [see Type 2A] for example, see Fig. 6. For a recent fascinating article on competition of *TA rules*, see Ref. [8] which models biological phenomena of competition and selection by TA "subrule competition".

Type 4. Defects in a centered rectangular lattice

In this type of TA, a single defect is placed in a lattice of the form:

$$\begin{bmatrix} 0 & 1 & 0 & 1 & 0 \\ 1 & 0 & 1 & 0 & 1 \\ 0 & 1 & 0 & 1 & 0 \\ 1 & 0 & 1 & 0 & 1 \\ 0 & 1 & 0 & 1 & 0 \end{bmatrix}.\tag{7}$$

This is known as a "centered rectangular lattice" [9]. In some experiments, two different background lattices with adjacent boundaries are used, and the defect propagates from its beginning point in the centered rectangular lattice through the interface into the second lattice defined by:

$$\begin{bmatrix} 0 & 1 & 0 & 1 & 0 \\ 0 & 1 & 0 & 1 & 0 \\ 0 & 1 & 0 & 1 & 0 \\ 0 & 1 & 0 & 1 & 0 \\ 0 & 1 & 0 & 1 & 0 \end{bmatrix}\tag{8}$$

(known simply as a "rectangular lattice"). Adding a defect to these two-phase systems bears some similarity to seeding supersaturated solutions and watching the crystallization process grow and "hit" the boundary of a solution with a different composition. In the examples in this paper, the two phases are also reminiscent of metal–metal interfaces—such a silicon 100 (centered rectangular) and chromium (rectangular). Note that with no defect present, the rules described have no effect on either lattice! Only when the defect is placed in the lattice does any growth occur.

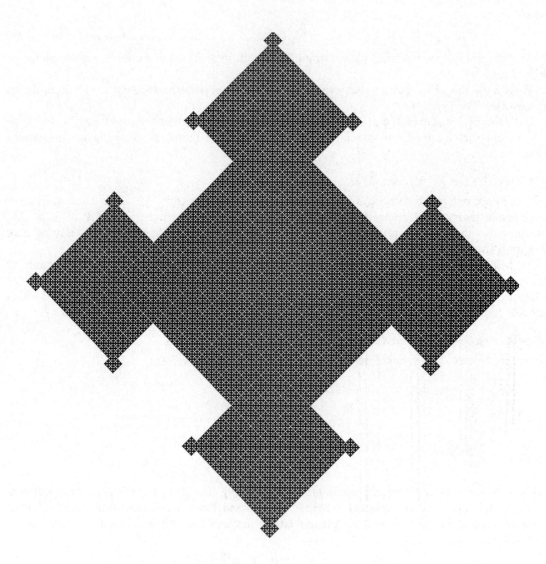

Fig. 1. TA Type 1 "growing" for 200 generations, starting with a single seen in the center of this figure.

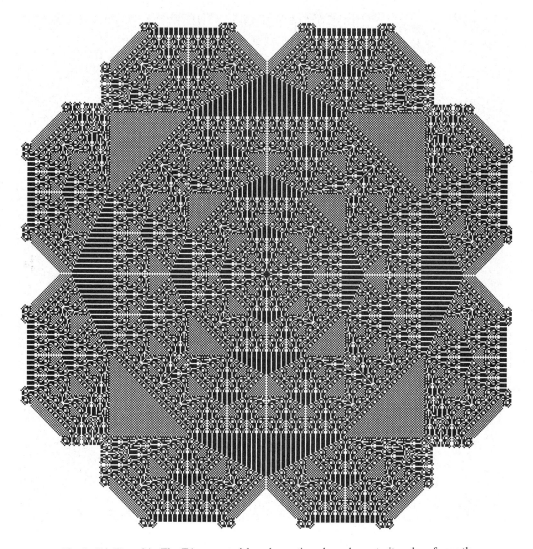

Fig. 2. TA Type 2A. The TA presented here has a time dependency to its rules of growth.

Fig. 3. Same as Fig. 2, but plotted as its negative.

Fig. 4. TA Type 2B, with time dependent growth.

Fig. 5. Multi-defect system composed of three initial seeds of TA Type 2 (figure is magnified relative to Fig. 6).

Fig. 6. Same as Fig. 5, except computed for longer time.

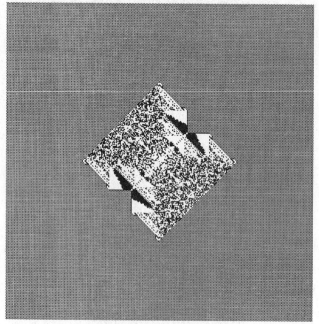

Fig. 7. TA Type 4A defect which has been growing from a center position in a centered rectangular lattice (seen as a diffuse grey background at this resolution). Without the presence of the defect, the rules have no effect on the lattice.

The rules for growth of the defects are as follows (note that deaths of cells can occur in these systems):

A. TA Type 4A.

$$\text{if } \sum C = 3 \wedge \text{ if } C_{i,j} = 0 \to C_{i,j} = 1, \tag{9}$$

$$\text{if } \sum C = 3 \wedge \text{ if } C_{i,j} = 1 \to C_{i,j} = 0, \tag{10}$$

where

$$C = [C_{i,j+1}, C_{i,j-1}, C_{i+1j}, C_{i-1j}, C_{i+1j+1}, C_{i-1j-1}]. \tag{11}$$

The symbol $\wedge$ denotes a logical "and". Figure 7 shows a defect which has been growing from a central position in a centered rectangular lattice (which is seen as a diffuse grey background at this resolution). Figures 8(a)–(d) show the propagation of the defect through a two-phase boundary. Note that the propagation behavior is visually different once in the second layer. For example, notice that the growth in the bottom layer appears to be constrained to planes $0°$ and $60°$ with respect to the lattice.

The introduction of "germ" cells appears to be useful in simulating real nucleation processes. An interesting paper in the literature describes solid–solid phase transformations of shape memory alloys, such as Cu–Zn–Al, using a 1-D cellular automata approach [10]. In this investigation, each cell represents several hundred atoms.

The search for multiphase systems, such as the ones in this paper, which are unaffected by a rule system *until* a defect is added, remains a provocative avenue of future research.

B. TA Type 4B. This case (see Fig. 9) is the same as the subset 4A, except that

$$\text{if } \sum C = 3 \wedge C_{i,j} = 1 \to C_{i,j} = 0 \tag{12}$$

$$\text{if } \sum C = 3 \wedge C_{i,j} = 0 \to C_{i,j} = 1 \tag{13}$$

$$\text{if } \sum C \neq 3 \wedge C_{i,j} = 1 \to C_{i,j} = 1 \tag{14}$$

$$\text{if } \sum C \neq 3 \wedge C_{i,j} = 0 \to C_{i,j} = 0. \tag{15}$$

Type 5. Larger local neighborhood

In TA Types 1–4, the neighborhood was defined as being within one cell of the center cell under consideration. In this system, the local neighborhood is larger. The rule is as follows:

$$\text{if } \sum C = 0 (\text{mod } 2) \to C_{i,j} = 0, \tag{16}$$

$$\text{if } \sum C \neq 0 (\text{mod } 2) \to C_{i,j} = 1, \tag{17}$$

where

$$C = [C_{i-2,j+2}, C_{i+2,j+2}, C_{i,j+1}, C_{i-1,j}, C_{i+1,j}. C_{i,j-1}, C_{i-2,j-2}, C_{i+2,j-2}]. \tag{18}$$

Figures 10(a)–(e) show the evolution of a two-state background defined by the lattices in equations (7) and (8) for several different snapshots in time. Unlike Type 4, the background without a defect is disturbed by this rule-set. Notice the visually unusual behavior of this system with both symmetry and stochasticity present. Also note the interesting growth of the two defects which have been placed next to each other in the top layer.

SUMMARY AND CONCLUSIONS

"Blindness to the aesthetic element in mathematics is widespread and can account for a feeling that mathematics is dry as dust, as exciting as a telephone took. . . . On the contrary, appreciation of this element makes the subject live in a wonderful manner and burn as no other creation of the human mind seems to do."

P. J. Davis and R. Hersch

Among the methods available for the characterization of complicated artistic, mathematical and natural phenomena, computers with graphics are emerging as an important tool (for several papers by this author, see Ref. [11]). In natural phenomena, there are examples of complicated and ordered

Fig. 8(a). Magnified picture of the beginning of propagation of a Type 4A defect through a two-phase system. The top phase is a centered rectangular lattice, while the bottom phase is a rectangular lattice.

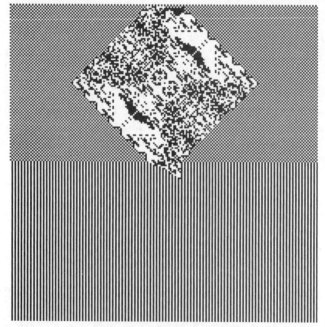

Fig. 8(b). Same as Fig. 8(a) except less magnified and computed for 60 generations. The defect has just "broken through" the boundary.

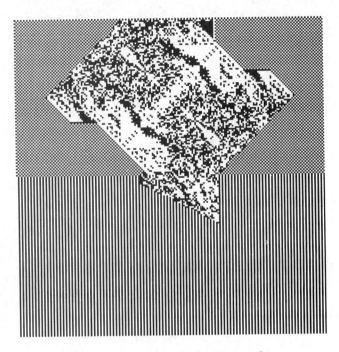

Fig. 8(c). Same as Fig. 8(b), for 80 generations.

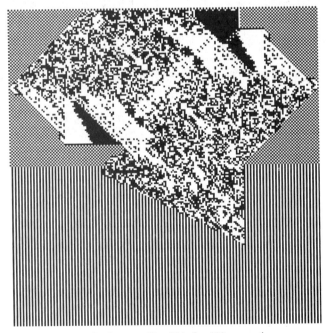

Fig. 8(d). Same as Fig. 8(c), for 100 generations.

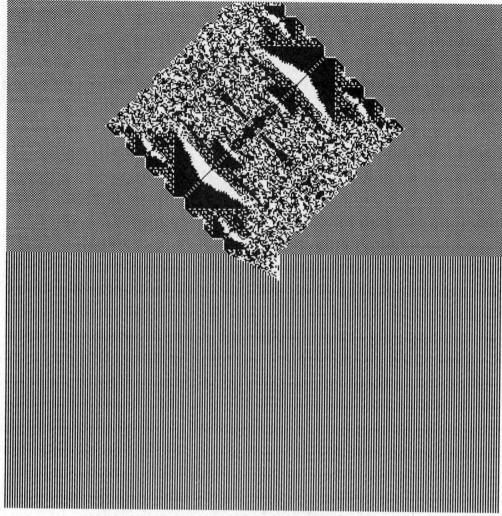

Fig. 9(a). Propagation for TA. Type 4B in a two-phase system ($n = 120$).

structures arising spontaneously from "disordered" states and examples include: snowflakes, patterns of flow in turbulent fluids, and biological systems. As Wolfram points out [1], TA are sufficiently simple to allow detailed mathematical analysis, yet sufficiently complex to exhibit a wide variety of complicated phenomena, and they can perhaps serve as models for some real processes in nature.

In contrast to previous systems where mathematical and aesthetic beauty relies on the use of imaginary numbers [12], there calculations use integers—which also facilitates their study with programming languages having no complex data types on small personal computers. The forms in this paper contain both symmetry and stochasticity, and the richness of resultant forms contrasts with the simplicity of the generating formula. Running TA at high speeds on a computer lets observers actually see the process of growth.

TA portraits contain a beauty and complexity which corresponds to behavior which mathematicians were not able to fully appreciate before the age of computer graphics. This complexity makes it difficult to objectively characterize structures such as these, and, therefore, it is useful to develop graphics systems which allow the maps to be followed in a qualitative and quantitative way. The TA graphics program allows the researcher to display patterns for a specified length of time and for different rule systems.

Some of these figures contain what is known as nonstandard scaling symmetry, also called dilation symmetry, i.e. invariance under changes of scale (for a classification of the various forms of self-similarity symmetries, see the second reference in Ref [12]). For example, if we look at any

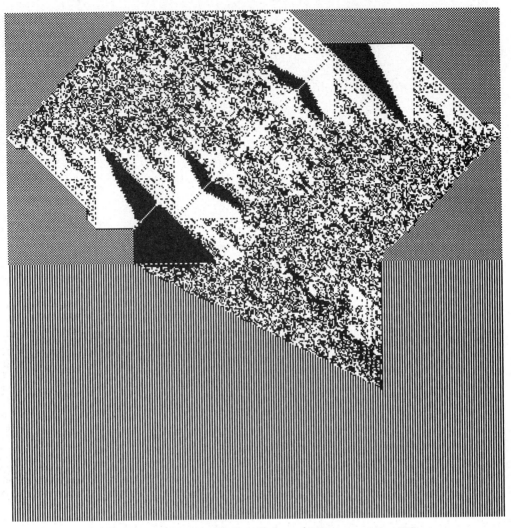

Fig. 9(b). Propagation for TA Type 4B in a two-phase system ($n = 200$).

one of the geometric motifs we notice that the same basic shape is found at another place in another size. Dilation symmetry is sometimes expressed by the formula ($\mathbf{r} \rightarrow a\mathbf{r}$). Thus, an expanded piece of some TA can be moved in such a way as to make it coincide with the entire TA, and this operation can be performed in an infinite number of ways. Other more trivial symmetries in the figures include the bilateral symmetries and the various rotation axes and other mirror planes in the TA. Note the dilation symmetry has been discovered and applied in different kinds of phenomenon in condensed matter physics, diffusion, polymer growth and percolation clusters. One example given by Kadanoff [13] is petroleum-bearing rock layers. These typically contain fluid-filled pores of many sizes, which, as Kadanoff points out, might be effectively understood as 2-D fractal networks known as gaskets [13], and I would add that TA may also serve as visual and physical models for these types of structures.

These figures may also have a practical importance in that they can provide models for materials scientists to build entirely new structures with entirely new properties [14]. For example, Gordon *et al.* [14] have created wire networks on the micron size scale similar to some of these figures with repeating triangles. The area of their smallest triangle was $1.38 \pm 0.01 \, \mu\text{m}^2$, and they have investigated many unusual properties of their superconducting network in a magnetic field (see their paper for details).

From an artistic standpoint, TA provide a vast and deep reservoir from which artists can draw. The computer is a machine which, when guided by an artist, can render images of captivating power and beauty. New "recipes", such as those outlined here, interact with such traditional elements as

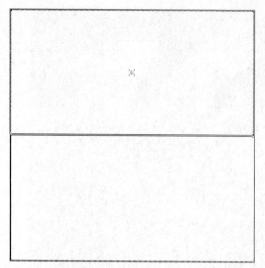

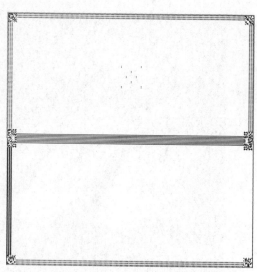

Fig. 10(a). Evolution of a two-state background defined by the lattices in equations (7) and (8). Two adjacent defects have been placed in the top layer, and the result for two generations ($n = 2$) is shown.

Fig. 10(b). Figure after 8 generations ($n = 8$).

Fig. 10(c). $n = 20$.

Fig. 10(d). $n = 40$.

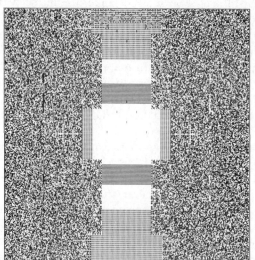

Fig. 10(e). $n = 80$. Note that the patterns, previously well ordered, appears to be on the route to "chaos".

form, shading and color to produce futuristic images and effects. The recipes function as the artist's helper, quickly taking care of much of the repetitive and sometimes tedious detail. By creating an environment of advanced computer graphics, artists with access to computers will gradually change our perception of art.

Also from a purely artistic standpoint, some of the figures in this paper are reminiscent of Persian carpet designs [15], ceramic tile mosaics [15], Peruvian striped fabrics [15], brick patterns from certain Mosques [16], and the symmetry in Moorish ornamental patterns [17]:

Scheme 1

The idea of investigating the ornaments and decorations of various cultures by consideration of their symmetry groups appears to have originated with Polya [15, 18]. This artistic resemblance is due to the complicated symmetries produced by the algorithm, and it is suggested that the reader explore the various parameters to achieve artistic control of the visual effect most desired.

In summary, all the TA shown here have an infinite variety of shapes, and although the equations seem to display what might be called "bizarre" behavior, there nevertheless seems to be a limited repertory of recurrent patterns. A report such as this can only be viewed as introductory. However, it is hoped that the techniques, equations, and systems will provide a useful tool and stimulate future studies in the graphic characterization of the morphologically rich structures produced by relatively simple generating formula.

Acknowledgement—I owe a special debt of gratitude to Charles Bennett for introducing me to TA Type 4B.

REFERENCES

1. S. Wolfram, Statistical mechanics of cellular automata. *Rev. mod. Phy.* **55**, 601–644 (1983).
2. W. Poundstone, *The Recursive Universe*. Morrow, New York (1985).
3. S. Levy, The portable universe: getting to the heart of the matter with cellular automata. *Whole Earth Rev.* **49**, 42–48 (1985).
4. R. Schrandt and S. Ulam, On recursively defined geometrical objects and patterns of growth, In *Essays on Cellular Automata* (Ed. A. Burks). Univ. of Illinois Press, Chicago (1970).
5. A. Fisher, Chaos: the ultimate asymmetry. *Mosaic* **16**, 24–30 (1985).
6. D. Campbell, J. Crutchfield, D. Farmer and E. Jen, Experimental mathematics: the role of computation in nonlinear science. *Commun. ACM* **28**, 374–389 (1985).
7. C. Pickover, Mathematics and beauty: time-discrete phase planes associated with the cyclic system, $\{\dot{x}(t) = -f(y(t))$, $\dot{y}(t) = f(x(t))\}$. *Comput. Graphics* **11**(2), 217–226 (1987); Biomorphs: computer displays of biological forms generated from mathematical feed back loops. *Comput. Graphics Forum* **5**(4), 313–316 (1987); Blooming integers. *Comput. Graphics W.* **10**(3), 54–57 (1987); Graphics, bifurcation, order and chaos. *Comput. Graphics Forum* **6**, 26–33 (1987); Computers, pattern, chaos, and beauty. IBM Research Report RC 12281 (Order from: IBM Watson-Distribution, Yorktown Hts, NY 10598) (1986).
8. D. Brown, Competition of cellular automata rules. *Complex Systems* **1**, 169–180 (1987).
9. E. Lockwood and R. Macmillan, *Geometric Symmetry*. Cambridge Univ. Press, New York (1978).
10. D. Maeder, The free energy concept in cellular automaton models of solid-solid phase transitions. *Complex Systems* **1**, 131–144 (1987).
11. C. Pickover, The use of computer-drawn faces as an educational aid in the presentation of statistical concepts. *Comput. Graphics* **8**, 163–166 (1984); C. The use of symmetrized-dot patterns characterizing speech waveforms, *J. acoust Soc. Am.* **80**, 955–960 (1984); On the educational uses of computer-generated cartoon faces. *J. educ. Tech. Syst.* **13**, 185–198 (1985); Frequency representations of DNA sequences: application to a bladder cancer gene. *J. Molec. Graphics* **2**, 50 (1984): Representation of melody patterns using topographic spectral distribution map. *Comput. Music J.* **10**, 72–78

(1986); Computer-drawn faces characterizing nucleic acid sequences. *J. Molec. Graphics* **2,** 107–110 (1985); DNA Vectorgrams: representation of cancer gene sequences as movements along a 2-D cellular lattice. *IBM J. Res. Dev.* **31,** 111–119 (1987); The use of random-dot displays in the study of biomolecular conformation. *J. Molec. Graphics* **2,** 34 (1984).

12. C. Pickover and E. Khorasani, Computer graphics generated from the iteration of algebraic transformations in the complex plane. *Comput. Graphics* **9,** 147–151 (1985); A. Fractal structure of speech waveform graphs. *Comput. Graphics* **10,** 51–61 (1986); C. Pickover, A Monte Carlo approach for ε placement in waveform fractal-dimension calculation. *Comput. Graphics Forum* **5**(3), 203–209 (1986); C. Pickover, What is chaos? *J. Chaos Graphics* **1,** 1–2 (1987). (Note: *J. Chaos Graphics* is an informal journal devoted to the beautiful aspects of chaos, and copies are available from the author); C. Pickover, *Computers, Pattern, Chaos and Beauty.* Springer, Berlin (1989).

13. L. Kadanoff, Fractals: where's the physics? *Physics Today* **Feb.,** 6–7 (1986).

14. J. Gordon, A. Goldman and J. Maps, Superconducting-normal phase boundary of a fractal network in a magnetic field. *Phys. Rev. Lett.* **56,** 2280–2283 (1986).

15. B. Grünbaum, Z. Grünbaum and G. Shephard, Symmetry in Moorish and other ornaments. *Comput. Math Applic.* **12B,** 641–653 (1986). Reprinted in *Symmetry: Unifying Human Understanding* (Ed. I. Hargittai). Pergamon Press, Oxford (1986). A. Dowlatshahi, *Persian Designs and Motifs.* Dover, New York (1979).

16. E. Makovicky, Symmetrology of art: coloured and generalized symmetries. *Comput. Math. Applic.* **12B,** 949–980 (1986). Reprinted in *Symmetry: Unifying Human Understanding* (Ed. I. Hargittai). Pergamon Press, Oxford (1986).

17. E. Rozsa, Symmetry in Muslim arts. *Comput. Math. Applic.* **12B,** 725–750 (1986). Reprinted in *Symmetry: Unifying Human Understanding* (Ed. I. Hargittai). Pergamon Press. Oxford (1986).

18. G. Polya, Uber die analogie der kristallsymmetrie in der ebene. *Z. Kristallogr.* **60,** 278–282 (1924).

APPENDIX

Recipe for Picture Computation

In order to encourage reader involvement, the following pseudocode is given. Typical parameter constants are given within the code. Readers are encouraged to modify the equations to create a variety of patterns of their own design. Initially, the C array is 0 for all its elements, except for a value of 1 placed in its center. For the program below, a temporary array, Ctemp, is used to save the new results of each generation. The routine below would be called $n = 200$ times in a typical simulation.

```
ALGORITHM:  TA GENERATION (TYPE 2A )
INPUT:      1 DEFECT, Centered in the C array
OUTPUT:     TA PATTERN
TYPICAL PARAMETER VALUES:
            Size = 400
            N is the generation counter - goes from 1 to 200


do i = 2 to size-1;                      (* X - direction            *)
   do j = 2 to size-1;                   (* Y - direction            *)
    if C(i,j) = 0 then do;               (* Test for vacancy         *)
     if mod(n,2) = 0 then                (* Test for even number *)
       sum = C(i,j+1)+C(i,j-1)+C(i+1,j)+C(i-1,j);
     else
       sum = C(i,j+1)+C(i,j-1)+C(i+1,j)+C(i-1,j) +
       C(I+1,j+1) + C(i-1,j-1)+ C(I-1,j+1)+C(i+1,j-1);
     if sum = 1 then Ctemp(i,j) = 1;
   end;                                  (* End j loop          *)
end;                                     (* End i loop          *)
```

Program 1

Computers Math. Applic. Vol. 17, No. 1–3, pp. 337–339, 1989
Printed in Great Britain. All rights reserved

0097-4943/89 $3.00 + 0.00
Copyright © 1989 Pergamon Press plc

INTERPRETATION OF SO-CALLED ICOSAHEDRAL AND DECAGONAL QUASICRYSTALS OF ALLOYS SHOWING APPARENT ICOSAHEDRAL SYMMETRY ELEMENTS AS TWINS OF AN 820-ATOM CUBIC CRYSTAL

L. Pauling

Linus Pauling Institute of Science and Medicine, 440 Page Mill Road, Palo Alto, CA 94306, U.S.A.

Abstract—A summary is presented of the arguments indicating that the so-called icosahedral and decagonal quasicrystals of $MnAl_6$ and other alloys are twins of a cubic crystal with 820 atoms in the unit cube such that the complex of small individual crystallites has icosahedral or decagonal symmetry. The proposed 820-atom cubic crystal has a structure that is similar to that of other intermetallic compounds with complex structures involving clusters of metal atoms with approximate icosahedral symmetry.

In the fall of 1984 a paper was published in *Physical Review Letters* by Shechtman *et al.* in which they reported that by rapidly chilling a molten alloy with composition approx. $MnAl_6$ a solid alloy was obtained that gave electron diffraction photographs with five-, three- and two-fold axes and planes of symmetry corresponding to the holohedral icosahedral point group [1]. This report excited great interest, partially because crystallographers have believed for a long time that it is not possible for crystals to have a five-fold axis of symmetry, and partially because the alloys seemed to be of a new kind, which might well have some interesting and commercially valuable physical and electronic properties. I estimated that by the end of 1986 over 500 papers about the so-called icosahedral and decagonal quasicrystals had been published.

A number of different sets of authors had reported that they had succeeded in showing how aggregates of atoms could exist that would give diffraction patterns with sharp diffraction maxima and icosahedral or decahedral symmetry. Some of the theories involved an icosahedral unit, such as an icosahedron of 12 atoms with or without another atom in its center of a cluster with icosahedral symmetry involving a larger number of atoms, repeated in space, perhaps by sharing faces or edges or corners, in such a way that all of the clusters of atoms are in parallel orientation, but with some randomness of such a nature that the clusters are not related to one another by the three translational symmetry operations characteristic of any of the Bravais lattices. Another way of describing a possible structure involves Penrose tiling. It was knonw that a two-dimensional arrangement of points could be obtained by use of an oblate and a prolate rhomb, with equal edges. These rhombs are arranged so as to fill two-dimensional space. There are some limitations on how the rhombs are associated with one another, such that no translational operations appear, and no true five-fold axis, but only a statistical five-fold symmetry. The same kind of structure can be constructed in three-dimensional space, with use of an oblate and a prolate rhombohedron. A third way involves the consideration of possible atomic arrangement in hyper-dimensional space. Crystals with icosahedral symmetry can be constructed in four- or five- or six-dimensional space, and the authors using this treatment described a three-dimensional quasicrystal as a cross section of one of the hyper-dimensional crystals. All of these theories have in common the fact that in three-dimensional space the vectors associated with pairs of atoms constitute a finite set, with, however, the same vector not being repeated by a translational symmetry operation. These theories account in a general way for the observed electron diffraction photographs. In particular, the characteristic spacings corresponding to the diffraction maxima are often in the ratio of a power of τ, the golden mean, identified more than 2000 years ago as $(5^{1/2} + 1)/2 = 1.6180\ldots$ It is the ratio of the diameter to the side of a regular pentagon, and appears in the dimensional relationships of the icosahedron and other polyhedra based on the icosahedron and pentagonal dodecahedron.

My own interest in the question of the nature of these quasicrystals developed rather slowly, but after some time I decided to follow up the suggestion made and rejected by the original investigators that the quasicrystals might consist of repeatedly twinned cubic crystals, twinned for example by having an icosahedral seed that determined their relative orientation. Making use of my knowledge of values of the metallic radii and of the known tendency of intermetallic compounds involving metals with radii differing by 10 or 20%, to form icosahedral clusters, I formulated a possible structure involving about 1180 atoms in a face-centered cubic arrangement [2]. I reported that this cubic crystal, with edge 25.73 Å for $MnAl_6$, accounted reasonalbly well for the X-ray powder pattern.

It was pointed out to me by other investigators that, although the unit accounted in a moderately satisfactory way for the X-ray powder diffraction pattern, it did not seem to account for the electron diffraction patterns and the high-resolution electron micrographs. I continued to work on the problem for about a year. Finally I realized that the principal diffraction maxima on the five-fold electron diffraction pattern of $MnAl_6$ could be accounted for as orders of $h \cdot h \cdot 0$, the strongest spots being $3 \cdot 3 \cdot 0$, $5 \cdot 5 \cdot 0$, $8 \cdot 8 \cdot 0$, $13 \cdot 13 \cdot 0$, $16 \cdot 16 \cdot 0$ and $21 \cdot 21 \cdot 0$. Except for $16 \cdot 16 \cdot 0$, the values of h, 3, 5, 8, 13, 21, are Fibonacci numbers, and ratios of successive pairs are close to the golden number 1.618.... The value of the edge of unit cube required for this identification is 23.36 Å for $MnAl_6$. I found that all of the diffraction maxima on the electron diffraction photographs could be accounted for by a unit this size, containing 820 atoms, some of them by double diffraction. Moreover, the X-ray powder diffraction pattern was also in as good agreement with this cubic unit as with the larger one [3].

The primitive cubic unit containing 820 atoms consists of icosahedral clusters, each of 104 atoms, at the eight positions of the β–W structure. In this structure there are two kinds of clusters. Those of the first kind, at $0\,0\,0$ and $\frac{1}{2}\frac{1}{2}\frac{1}{2}$, are surrounded by 12 of the second kind, at $\frac{1}{2}\frac{1}{4}0$, $0\frac{1}{2}\frac{1}{4}$, $\frac{1}{4}0\frac{1}{2}$, $\frac{1}{2}\frac{3}{4}0$, $0\frac{1}{2}\frac{3}{4}$, $\frac{3}{4}0\frac{1}{2}$. The clusters of the second kind have ligancy 14, with two shorter interatomic distances, four (two clusters of the first kind) of intermediate length, and eight longer ones, the values being 11.68, 13.06 and 14.01 Å. The proposed structure of the 104-atom cluster is that involving 20 condensed Friauf polyhedra. The principal ligancies of the atoms are 12 and 16. This cluster is found in the 162-atom body-centered crystal $Mg_{32}(Al, Zn)_{49}$, the structure of which was discovered in 1952 [4]. All eight of the clusters in the unit have the same orientation relative to the cubic axes.

The process of twinning that has been proposed is that one of the icosahedral clusters in the microcrystal with the 820-atom structure can serve as the seed for another crystal, in which a three-fold axis of the icosahedral cluster is coincident with the three-fold axis of the new cubic crystal that rose from the seed, with each of the three-fold axes lying in the plane of symmetry. All of the clusters in the twinned crystal have the same orientation as those in the original crystal. Repeated twinning in this way can lead to a grain, containing perhaps 10^9 microcrystals, with 20 different orientations, corresponding to the icosahedral group, and with the clusters in all of the grains essentially in parallel orientation. This sort of twinning provides a mechanism for the formation of a larger structure with apparent icosahedral symmetry.

One interesting observation, made on crystallographically oriented icosahedral-phase material by implanting Mn ions directly into single crystal Al substrates by measurements of the positions of high-resolution X-ray diffraction peaks, is that the interplanar distances are not in the ratios of the Fibonacci numbers, as would be required by a perfect cubic crystal, and also are not in the ratios of the golden number, as required by icosahedral symmetry, but are in between, somewhat closer to the golden number than to the Fibonacci number ratios. Many calculations have been made of the diffraction patterns to be expected from structures involving icosahedral complexes in parallel orientations but without three-dimensional translational operations.

I have pointed out that the diffraction maxima from an array of cubic crystals of small size, perhaps 500 Å on edge, would be broad, and that the intensity distribution across the broad peak would be modulated in two ways by icosahedral interactions. First, the structure factor of the icosahedral 104-atom complex in general has its maximum values in positions different from those vectors in the cubic reciprocal lattice. Second, the diffracted beams from the different microcrystals would interact to produce maxima in positions somewhat displaced from those for the perfect cubic crystal. So far no detailed calculations have been made along these lines, but it seems to me likely that all of the diffraction observations can be explained on the basis of the twinned cubic-crystal model.

These so-called icosahedral quasicrystals are usually made by rapid quenching of the molten alloy. I think that the 104-atom complexes are already present in the molten alloy, and that if it were to be quenched extremely rapidly a supercooled liquid (a metallic glass) would be formed. Somewhat slower quenching might lead to an alloy in which the clusters interact with one another in such a way that they are kept parallel, but that otherwise the arrangement is similar to that in the supercooled liquid. With a still smaller rate of quenching, very small microcrystals with the β–W structure could begin to form, perhaps so small as not to give well-defined X-ray powder lines. With still slower quenching the 500-Å microcrystal, with icosahedral twinning, would result. Various properties of the metallic glass, the icosahedral quasicrystals and crystalline materials with the same composition have been reported, and usually the icosahedral quasicrystals are found to be intermediate between the metallic glass and the crystalline material, as suggested by the foregoing argument.

As a crystallographer, with 65 years of experience in X-ray crystallography, I am pleased that the problem of the so-called icosahedral quasicrystals has been resolved in this way. Crystallographers have believed for many years that crystals cannot have five-fold axes of symmetry. In my model the grains with apparent icosahedral symmetry consist of cubic crystals that have a conventional structure, but that have, by repeated twinning determined by the approximate icosahedral structure of the 104-atom clusters, arranged themselves into an aggregate of microcrystals that shows icosahedral symmetry.

REFERENCES

1. D. Shechtman, I. Blech, D. Gratias and J. W. Cahn, Metallic phase with long-range orientational order and no translational symmetry. *Phys. Rev. Lett.* **53**, 1951–1953 (1984).
2. L. Pauling, Apparent icosahedral symmetry is due to directed multiple twinning of cubic crystals. *Nature* **317**, 512–514 (1985).
3. L. Pauling, So-called icosahedral and decagonal quasicrystals are twins of an 820-atom cubic crystal. *Phys. Rev. Lett.* **58**, 365–368 (1987).
4. F. Bergman, J. L. T. Waugh and L. Pauling, Crystal structure of the intermetallic compound $Mg_{32}(Al, Zn)_{49}$ and related phases. *Nature* **169**, 1057–1058 (1952).

Computers Math. Applic. Vol. 17, No. 1–3, pp. 341–354, 1989
Printed in Great Britain. All rights reserved

LOCAL PSEUDOSYMMETRY IN SIMPLE LIQUIDS†

J. L. FINNEY

Crystallography Department, Birkbeck College, Malet Street, London WC1E 7HX, England

Abstract—At first sight, liquids appear to be entirely devoid of symmetry: unlike crystals, there is no unit cell, and hence the space group operations that "build up" the unit cell contents from the asymmetric unit, and the subsequent unit cell translations that generate the whole crystal, are both irrelevant to liquids. It is, however, both possible and illuminating to consider the atomic arrangements in both liquids and crystals in a parallel manner. Starting from the basic elementary unit—a soft sphere for simple crystals and liquids—both kinds of structure can be "built up" through local "sub-unit" aggregates (tetrahedra and octahedra), which are then arranged in face-sharing mode in larger "super-units"; these are then embedded in the rest of the assembly to fill space. In the crystal case, the build-up operations are clearly defined local symmetry operations, which are applied in a regular sequence, and the embedding process is automatically satisfied by the unit cell translation operations. For the liquid, the same basic tetrahedral and octahedral sub-units can be used; provided the metrical identity requirements are relaxed, these can be arranged to form a finite number of topologically distinct "super-units". The subsequent embedding of these super-units—which appears only to be possible with the help of other "defect" polyhedra—follows local symmetry operation rules which are as yet unknown. Understanding these local build-up symmetry operations, and their *non-regular* rules of application, would constitute a large part of the statistical geometrical structural theory of liquids called for by Bernal thirty years ago.

1. INTRODUCTION

Symmetry is not a concept we would immediately think of as applicable to liquids. Rather, a crystallographer would use concepts of symmetry in dealing with crystals, in which the atoms or molecules are arranged in regular repeating units. In crystals, symmetry can be considered at two levels. First, there is the long range order, which relates to the repetition in three directions in space of a basic unit, the unit cell. This repetition of a unit cell—by operations of translational symmetry—is perhaps the hallmark of crystals. At the level of the elementary unit cell, we may find one or more atoms, molecules, or groups of molecules, which we term the asymmetric unit. If there is more than one asymmetric unit in the unit cell, then they are related to one another by further symmetry operations of rotation, translation, or "mixed" rotations and translations embodied in screw axes. Using these symmetry operations, therefore, we can build up, or generate, the whole crystal from the elementary asymmetric unit by (a) generating the contents of the unit cell by applying to the asymmetric unit one of the 200 allowable combinations of symmetry operations that are termed space groups, followed by (b) translation of the resulting unit cell by translational symmetry operations, in the three directions, and with the respective magnitudes, indicated by the three basis vectors that define the unit cell.

The use of symmetry operations to describe crystals gives us an extremely powerful and extensively used device to assist in theoretical calculations on crystals. Using it, we can calculate various properties—e.g. thermodynamic properties—of ideal crystals, and of imperfect crystals by considering the effects of departures from perfection. For electronic properties, the Bloch theorem is a device which relies on the underlying symmetry of the crystal lattice. Use of symmetry is in fact so powerful a device that, as far as crystalline solids were concerned, there was little incentive to develop alternative routes to relating structure and properties. Consequently, our ability to relate properties to structure in *non-crystalline* condensed matter is rudimentary.

The attractiveness of using the underlying symmetry was so great that it was used in developing early liquid theories, which considered the liquid as a disordered solid, for example as in cell models and their derivatives. This approach is now recognized to have been erroneous; it underestimated the disorder (and hence the entropy) of the liquid, and it was unable to account for the liquid's fluidity and the existence of supercooling. The diffraction pattern of a liquid is also qualitatively different from that of even a highly disordered crystal, consisting of a small number of diffuse rings

---

†In celebration of the election of Alan Mackay to the Fellowship of the Royal Society.

rather than the much larger number of sharper rings found in diffraction patterns from powders. Even considering the increase in line width as average crystallite size is reduced fails to allow us to describe adequately a liquid in terms of a microcrystalline solid—to account for the very diffuse rings by a microcrystalline model, the effective crystallites would have to be so small (of the order of only a few unit cells) that conceptual problems and inconsistencies would arise.

We are thus led to abandon crystal-derived models, and move to our present conceptual view of simple liquids of spherical atoms. According to Bernal, such liquids are "homogeneous, coherent and essentially irregular assemblages of molecules containing no crystalline regions" [1]. This concept is perhaps most easily appreciated by referring to Fig. 1, which shows a physical realization of this ideal model of a liquid—the irregular "heap" of spherical molecules at the top of the figure. This contrasts with the lower, regular "pile" of spheres, arranged in an orderly, regularly repeating array, that relates to the corresponding highly symmetric crystal. Models such as that shown in the figure can be examined in detail, and a quantitative description developed [1, 2]. We learn that such "random packings" can be made with densities about 15% less than those of the corresponding close-packed crystal (a density change that is consistent with experimental results on melting and freezing of simple inert gas liquids), that the constituent atoms have a high coordination number (say 8–10 depending upon the exact definition of coordination number), that the assembly is one of high entropy (though this is still not easy to calculate from the models), *and there is no long-range ordered structures*. We thus have an appealing picture not only of the essential disorder of a simple liquid, but also a conceptual explanation of fluidity, supercooling, and the loss of order on melting. It leads to a quantitative explanation of the volume change on melting, and the characteristic liquid diffraction pattern. The model is also, as stated by Ziman [3], "seen to be the key to any qualitative or quantitative understanding of the physics of liquids".

Although this model is extremely useful conceptually, we have serious problems when we try to relate properties to structure: the model has no long range order, and thus we are unable to use all the standard armoury of condensed matter theory that is based upon the existence of the translational symmetry of a lattice. Although we can build both relatively crude but illustrative models in the laboratory, or increasingly sophisticated ones in the computer, from which we can extract numerically quantitative information for comparing with experiment, we are far from having a theoretical framework to facilitate analytical calculations. We appear to have rejected the application of symmetry in our model; consequently therefore, we are forced to reject also the associated powerful mathematical tools. We are in need of an adequate theory of liquids which is qualitatively different from lattice-based theories of solids, and it is perhaps only now that such is becoming possible through maximum entropy developments of information theory [4].

Fig. 1. Bernal's random close-packing model of a simple liquid (upper part), contrasting with (lower part) the regular "pile" of hard spheres relating to the ideal crystal.

In simple liquids, therefore, we see that both levels of symmetry discussed with respect to crystals at the start of this section are absent. First, translational elements of symmetry are absent from the liquid—necessarily so as long range order is inconsistent with our understanding of the essential liquid-likeness of liquids. Secondly, the space group operations that fill the crystal unit cell with one or more asymmetric units would also seem to be irrelevant to the liquid—there just is no unit cell. This initial discussion therefore implies that concepts of symmetry are in principle inapplicable to liquids at the molecular level. The only obvious—and essentially trivial—thing we appear to be able to say about symmetry in liquids, is that the diffraction pattern—and hence the time-averaged structure—is spherically symmetric. This is, however, little more than a manifestation of the macroscopic isotropy of simple liquids, and is hence really a trivial, and not very helpful, statement.

Looked at in a different way, however, regularities *do* appear in the structures of simple liquids. We discuss in this article the nature of these regularities, and how they might be considered in terms which also relate to symmetry operations in crystals.

2. BUILDING UP A LIQUID FROM ITS ELEMENTARY UNIT

An adequate structural theory of liquids must be able to connect the arrangements of molecules with the macroscopic thermodynamic properties of the liquid. As we have already discussed, we can construct the complete crystal by a "build-up" process, which consists of a sequence of symmetry operations operating on an asymmetric unit (atom, molecule, or group of molecules). Initially, the space group operations relate the asymmetric unit to the unit cell, and then lattice translations take over to relate the unit cell to the macroscopic crystal. This two-step build-up process thus relates the elementary asymmetric unit to its environment (which is of course made up of a regular repeating arrangement of units identical to itself); put in a slightly different way, the symmetry operations tell us how the elementary unit fits into the constraints exerted by its highly ordered surroundings.

That these surroundings are regular, and ordered, is a characteristic of the crystal. However, the build-up *process* itself does not depend upon using strictly defined symmetry operations. Thus, although our model of a simple liquid has removed the ordering requirement that was imposed on the crystal, we can still use a parallel build-up process to construct the instantaneous arrangement of the liquid atoms starting from its elementary unit—a soft sphere. As the *constraint* of symmetry has been removed, however, the rules governing this build-up procedure will be different from those symmetry operations in the crystal case; *there will, however, still be rules which will describe the spatial relationship between a particular elementary unit and its (non-crystalline) environment.* It is uncovering these rules which is perhaps the central requirement of an adequate theory of liquid structure.

How might we set about understanding these rules, and at what level, if at all, are symmetry considerations relevant?

3. SHORT-RANGE ORDER IN SIMPLE LIQUIDS

In trying to understand how a spherical atom in a liquid relates to its environment—i.e. the operations relevant to the build-up process for the liquid—we might consider first the interactions between neighbouring atoms at short range. In the general case, these interactions can of course be of various kinds, including a variety of attractive forces—e.g. ionic, van der Waals, and hydrogen bonding—but also repulsive forces. These latter can often be considered as dominating the structure of condensed phases in general. Put slightly differently, volume exclusion may control the essential structure of a condensed phase, in which the atoms or molecules are in close contact with their neighbours.

This is a well-worn principle in rationalizing crystal structures: in constructing a crystal from, for example, spherical atoms, the symmetry operations used to build up first the unit cell from the asymmetric unit, and secondly the complete crystal from the unit cell must be such as not to violate the sizes of the spherical atoms involved. There must be no atomic overlap created by the build-up symmetry operations at either level. We are thus led to the classic close-packed structures of spherical atoms that are found in inert gas solids, many metals, and also a whole host of silicate and other oxide structures. In three dimensions, these structures are constructed by arranging spheres such that the space group operations which build up the unit cell do not lead to overlap of neighbouring spheres. This unit cell must then be capable of being "embedded" into a surrounding environment of identical unit cells, again in such a way that overlap repulsions—volume exclusion constraints—are not violated. If all these constraints are fulfilled—i.e. we use a particular set of symmetry operations for a given case—we can build up an essentially infinite crystal.

Now consider in a similar way an attempt to build up the instantaneous structure of a simple liquid. We note first that the local atomic level constraints of volume exclusion are unaltered from the crystal case. We still need to arrange the atoms locally in some way such that (i) the atoms do not violate volume exclusion, and (ii) the particular set of operations we use in our build-up process results in local structures (clusters of atoms) that can be "embedded" in the surrounding neighbourhood without violating volume exclusion. In the crystal, this embedding operation is assured by a unit cell of suitable symmetry; in the liquid, no similar unit cell concept applies. The surroundings in which our local arrangement of non-overlapping spheres must be embedded is thus no longer one of identical units related by translational symmetry operations, but severe overlap repulsion constraints remain which our build-up process must somehow satisfy.

In principle, the removal of the lattice constraint—the requirement of translational symmetry—gives much more structural freedom in the local atomic arrangements in a liquid than are allowable in the crystal. In other words, our first step in the build-up process can—and in fact, to satisfy the structural irregularity in, and hence higher entropy of, the liquid, *must*—result in *several different* kinds of local atomic clusters, in contrast to the crystal, where symmetry restricts us to a limited and very small number of cluster types.

The second step in our liquid build-up process is to arrange these different kinds of atomic clusters produced by the first step in such a way as to fill space without violating volume exclusion. As we have already stated, the embedding problem for crystals is solved by making use of symmetry; for liquids, we cannot follow this avenue, and we are left with a much more difficult problem to solve in the absence of appeals to symmetry. We need to understand the rules which govern the ways in which we can pack together these various local units of spherical atoms without creating *either* a repeating assembly *or* atomic overlaps, and yet result in a structure of sufficiently high density. The operations which we apply in this second part of the build-up process must result in an embedding of these local clusters into an assembly which shows some kind of statistical identity between different sample volumes over the medium and long range. As we have said before, understanding these operations and their rules of application would in effect amount to a structural theory of the liquid state.

4. LOCAL SUB-UNITS IN CLOSE-PACKED STRUCTURES

We consider now the kinds of local clusters of spherical atoms—which we will call building "sub-units"—that are likely to result from the very first step in our build-up process from the spherical atom to the extended liquid. Consider first the possible ways in which spheres can be arranged in small, locally dense, sub-unit clusters. Clearly, a tetrahedral arrangement of four spheres satisfies the conditions of volume exclusion (the spheres do not overlap), maximizes the attractive interactions between four atoms, and gives a very dense structure (equivalent packing density about about 0.77). If we now continue to the second step of the build-up process, and try to build-up a *crystal* from tetrahedral units alone, we find we cannot do it—this local tetrahedral sub-unit itself is not consistent with the translational symmetry operations necessary to build up a crystal without violating overlap repulsions. Similarly, if we try the same for a *liquid*, we fail also to build up a non-crystalline densely packed assembly of tetrahedra. Even though we do not use standard symmetry operations in the attempt to pack tetrahedra, we end up either violating the volume exclusion, or have to leave sufficient empty volume that a high density cannot be achieved.

For guidance on how to proceed for the crystal case, we naturally look to close-packed spherical structures such as the face centred cubic arrangement found in copper crystals, or the hexagonal close packing found in some metals. In both these cases, we indeed do find a dominance of tetrahedral sub-units. In packing them together, however, to form a crystal, we need the help of a second—less dense—local arrangement of spheres, namely the octahedron. In close-packed crystals, these two polyhedra pack together in a regular manner, in the ratio of two tetrahedra to one octahedron. This reduces the density from what we would have if we *could* pack tetrahedra on their own, but the resulting packing density (about 0.74) is still high, and probably the maximum we can achieve.

We can understand why this packing of tetrahedral and octahedral sub-units is a good candidate for crystal structure construction by considering the dihedral angles of both these and other polyhedra that might be candidates for packing together. In order to fill space locally by packing together small numbers of the same or different polyhedra (to form what we call "super-units"), the sum of the dihedral angles around every edge in the packing of polyhedra must add up to 2π rad. A larger sum of angles would imply violation of volume exclusion, a sum lower than 2π would leave empty space. Dihedral angles for the tetrahedron and octahedron are 70.53° and 109.47°, respectively. These numbers tell us that we cannot fill space with either tetrahedra or octahedra on their own (neither dihedral angle divides exactly into 2π). That the sum of the dihedral angles for the two different polyhedra is 180° tells us immediately that, provided we pack the tetrahedra and octahedra in a particular way, we will have a good chance of filling space. We succeed in so doing in the regular combination of tetrahedra and octahedra that make up close-packed crystal structures. Our build-up process fits together *first* atoms to form tetrahedral and octahedral sub-units, *then* plugs these together in a very specific way to form slightly larger "super-units". *Finally*, these super-units themselves can be fitted together by a three-dimensional jigsaw-puzzle assembly process (using translational symmetry operations) to fill space with the resulting crystal.

These three steps make up the operations of our build-up process from the atom to the infinite crystal. The process involves symmetry operations applied in a known, regular way.

5. A TWO-DIMENSIONAL ANALOGUE—KAWAMURA'S MODEL

If tetrahedra and octahedra are natural local sub-units for crystals of spherical atoms, and their geometries allow them to pack together to fill space, we might ask the question: could not the instantaneous structure of a simple liquid be considered also in terms of the same two basic units? These might be fitted together locally in different ways to form a variety of larger "super-units" which would pack together in three-dimensional jigsaw-puzzle style, but in a way which would not apply translational symmetry operations to the super-units to generate the space-filling structure. If we go back to Bernal's original papers, we find that, when considered in terms of local arrangements of spheres, there are indeed a large number of polyhedra which can be considered as more or less tetrahedra and octahedra. Can we therefore take this suggestion further?

Before considering further this possibility with respect to three-dimensional simple liquids, it is instructive to consider a two-dimensional analogue developed by Kawamura [5] with respect to two-dimensional crystals and liquids. Just as in three dimensions, where we considered those polyhedral arrangements of packed spheres which could themselves pack together to fill space, we may be able to approach by similar arguments crystal and liquid-like systems in the plane. In addressing this problem in two dimensions, Kawamura proposed using two basic idealized sub-units, namely the equilateral triangle and the square. The internal vertex angles of these figures being $\pi/3$ and $\pi/2$, respectively, the two kinds of figure can themselves separately form crystalline—and only crystalline—arrangements, with six triangles meeting at a vertex to give a plane-filling total angle of $6 \times \pi/3 = 2\pi$, or four squares again giving a total angle around each vertex of 2π.

If we now mix the two units, we find that we can also obtain plane-filling arrangements, but only by constructing super-units under certain very definite rules, namely that the angle at each vertex is always 2π. With vertex angles of only $\pi/3$ and $\pi/2$ available, each vertex must, for plane filling, be formed from triangles (T) and squares (S) in the proportions of $2S$ and $3T$. If we also allow local regions in which all square and all triangular vertices can exist locally ($4S$ and $6T$ super-units), we can build up a disordered structure as a first order model of a two-dimensional fluid (see for example Fig. 2). In fact, we have *four* possible vertex configurations that we can use in our

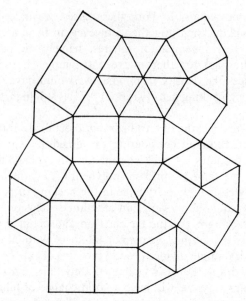

Fig. 2. A two-dimensional liquid after Kawamura, showing how equilateral triangles and squares can be packed together to fill the plane. Note the limited number of local arrangements possible, and the two different $2S3T$ super-units $SSTTT$ and $STSTT$.

two-dimensional liquid build-up process, as the mixed $2S3T$ configuration can be ordered cyclically around the vertex in two different ways, namely $SSTTT$ and $STSTT$ (both of which can be found illustrated in Fig. 2).

With these basic sub-units, we can build up crystals, which obey the long-range constraints imposed by symmetry operations, and also "liquids", which are built up of the same local sub-units, but through a *mixture* of *different* super-units. In the "liquid" case, we still need to continue the build-up operation to generate the filling of the plane, but this involves operations other than the translational symmetry operations which handle the embedding of the local super- (or sub-)units in the crystalline cases. The build-up operations in the "liquid" case must also be well defined, but not in terms of symmetry operations applied in a regular manner. By looking at the constraints imposed on the liquid build-up operation from the basic sub- and super-units, we can estimate configurational entropies of the system, and begin to examine the nature of the "embedding" problem—how we can satisfy the *medium range* constraints which must be obeyed if we are to fill space effectively to infinity.

6. PACKING TETRAHEDRA AND OCTAHEDRA IN THREE DIMENSIONS

Returning now to the three-dimensional case, can we extend Kawamura's ideas, and treat the packing in simple liquid structures in terms of (i) a few basic sub-units which can (ii) be combined locally to form a *variety* of super-units which can then (iii) be packed together to fill space? As candidates for suitable subunits for our build-up process, we naturally look initially at those polyhedra that give high packing density at the local level, and this leads us again to tetrahedra and octahedra. When we look at the dihedral angles, however, we find there is a problem as we indicated at the end of Section 4; the only combinations of tetrahedra (T) and octahedra (O) that will add to a total dihedral angle of 2π is two of each ($2T2O$). We cannot even, as we discussed earlier, pack tetrahedra or octahedra on their own to fill space, although in the case of the tetrahedra, we can place five togther along a common vertex, resulting in only a small dihedral angle deficit of about 7°. In the allowable crystal super-unit of $2T2O$, we do have, as in the two-dimensional analogue, an additional "degree of freedom" from the order in which we place the tetrahedra and octahedra around the common edge—either $TTOO$ or $TOTO$. Is this enough variability to give us the possibility of building up a non-crystalline arrangement of these two polyhedra?

The answer, it appears, is no—we have not been able to find ways of combining these two units both locally and in extended structures so as to fill space, observe the severe restrictions consequent upon volume exclusion, and retain the essential irregularity of the liquid structure. The only packings we can construct are automatically consistent with long-range symmetry operations, and we end up with crystals. There is enough variability in super-units to produce two different crystal structures (face centred cubic and hexagonal close packed) but that appears to be the limit to the structural variation possible.

7. PACKINGS OF SOFT SPHERES

To use the build-up procedure to obtain a space-filling liquid-like structure from our preferred tetrahedral and octahedral sub-units, it appears that we must compromise somewhere. Otherwise, we would have to abandon altogether the attempt to understand the non-crystalline packing in these terms. We could perhaps abandon the tetrahedron and octahedron as basic building sub-units, but this would be inadvisable for two main reasons. First, we would be abandoning units that are consistent with volume exclusions, and yet are dense enough to give the high densities we know simple liquids have. Secondly, Bernal's analysis of local polyhedral units showed a strong domination of tetrahedra (73% in number occuping 48% of the volume) and octahedra (actually counted as half octahedra) (present as 10% by number and occupying 27% of the volume) [1]. With such empirical evidence, we should perhaps look in a slightly different direction to resolve the apparent problem.

The resolution comes, it seems, from moving away from perfection in the polyhedra—in effect, we reduce the symmetries of the local sub-units. We have already noted that five tetrahedra with a common edge *almost* fit together, leaving a small deficit of about 7°. Hence, if we were to (i) allow small metric differences between different polyhedra of the same type, and (ii) within the same polyhedron allow a small inequality of the side lengths (and hence of the various angles including the dihedral angles), then we could, for example, fit together into a super-unit five tetrahedra sharing a common edge to fill space locally. We could in fact do more, and arrange 20 slightly distorted tetrahedra around a common vertex, and obtain an icosahedral structure, although, as we have discussed elsewhere, and will consider briefly below, appealing as this idea initially is as a model of non-crystalline packings, the evidence for it in simple liquid models is extremely weak [6].

Allowing a certain degree of distortion in these two basic building blocks, we make available to the build-up process several different super-units, and hence raise the possibility of building up non-crystalline arrangements. Kawamura's $6T$, $4S$, and two distinct kinds of $3T2S$ vertices gave sufficient freedom to construct a non-crystalline two-dimensional packing; the hypothesis we now propose is that a similar pathway may be available in the three-dimensional case if we allow small deviations from metric identity between similar polyhedral sub-units. For example, allowing distortions that result in dihedral angle deviations from the ideal of $\pm 7°$, we can, in addition to the two $2T2O$ configurations allowed for super-units for ideal regular tetrahedra and octahedra, have $5T$ and $4T1O$ super-unit configurations that could be arranged around a common edge. This additional flexibility of itself raises the possibility of building up non-crystalline structures that *may* possibly fill space. Allowing an additional 3.5° variation in the dihedral angles of both units, we increase further the available common edge cluster super-units by the addition of $6T$, $3T1O$, $1T3O$, and $3O$ (the last two and the $5T$ and $6T$ configurations are not mutually exclusive—they are made up of polyhedra with different degrees of distortion). The possibilities of building up non-crystalline arrangements in the latter case are thus even greater than in the first, and both appear to have more flexibility than the two dimensional analogue.

The discussion so far has considered only the possible local super-units, and hence only the initial part of the build-up process. No consideration has as yet been given to the next part of the process, the "embedding" problem. Can we arrange in three-dimensional jigsaw-puzzle style, these various super-units to fill space, or, in attempting this, will we still be too restricted by volume exclusion constraints to be able to fill space completely without leaving any voids?

8. EVIDENCE FROM MODEL STUDIES

The easiest way of examining this problem is, in the first instance, to look at models that have been constructed, and elucidate the local polyhedral structures (sub-units and super-units) that are found there, and which are therefore consistent with solutions to our embedding problem. We have already mentioned Bernal's pioneering study [1] in which 75% of the volume of a non-crystalline sphere packing (the so-called random close-packed structure) was made up of sub-units identified as tetrahedra or octahedra. Even though his criteria for identification of tetrahedra and octahedra in a hard-sphere—and hence very spatially-restricted—packing were perhaps rather flexible, he still found that 25% of the volume of the model could not be accounted for by distorted tetrahedra or octahedra. This "missing" volume was made up of larger, more volume-expensive polyhedra which were identified as trigonal prisms, Archimedian antiprisms, and tetragonal dodecahedra.

A slightly different way of looking at the same problem is to construct a packing of *soft* spheres, and examine the geometry of the resulting model in terms of tetrahedral and octahedral units of a known degree of distortion. This is perhaps done most simply by relaxing a hard sphere packing in the computer, after first assigning to each sphere a soft repulsive core. By this means, the softness of the repulsions—and hence the deviation from perfection of the polyhedra—can be varied, and the results of such variation assessed.

In one of our earlier studies, using the radius of the largest sphere that could be placed in the interstices of the relaxed models as a criterion for identifying tetrahedral and octahedral holes, a clear simplification of the description of the system could be found as the softness of the sphere was increased. As shown in Fig. 3(a), if we plot the frequency of interstice radius for a hard sphere system, we obtain a single-peaked distribution which tells us that polyhedra of a wide variety of shapes and sizes are found from the tetrahedral (ideal interstice radius ratio 0.225), through the octahedral (ideal radius ratio 0.414) to even larger holes. If we perform the same analysis on a packing relaxed under a soft Morse potential, we obtain the dramatically different result of Fig. 3(b), where, in place of the single-peaked function of the hard sphere case, we have a clear bimodality, with the two peaks centred close to the ideal tetrahedral and octahedral radius ratios. Moreover, the numbers of distorted tetrahedra are, at about 75%, close to the results of Bernal's analysis of the hard sphere case. The number of octahedra are, however, approximately doubled.

Thus, we might conclude that, with the necessary modification of allowing *distorted* tetrahedral and octahedral sub-units, the ideal liquid structure may perhaps be realistically described in terms of a non-crystalline, and hence not translational symmetry controlled, packing of tetrahedra and

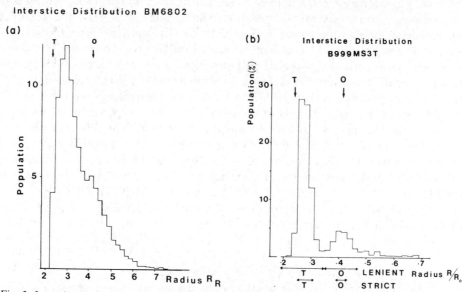

Fig. 3. Interstice distributions for random packed structures of (a) hard spheres and (b) soft spheres (truncated Morse 3 potential). Ideal tetrahedron and octahedron radius ratios are indicated, as are the radius ratio regions for distorted tetrahedra and octahedra according to two different criteria (see Ref. [6] for full details).

octahedra. In allowing a degree of softness, we remove the horrendously strict constraints on plugging together perfect tetrahedral and octahedral sub-units to fill space without overlaps. This gives us more flexibility in local permitted super-unit configurations, allowing the system to gain entropy not only through variations in near neighbour distances, but also through variations in local topological arrangements as super-units of the sub-unit building blocks. Comparing with a real crystal, the neighbour distance variations in our distorted sub-units *are* in fact also present instantaneously through thermal vibrations, and hence give an entropic contribution. For the liquid case, the disorder from the different possible ways of arranging these local distorted sub-units gives rise to an entropic contribution with configurational origins.

Inspecting again the hard sphere interstice distributions (Fig. 3), we can imply, from the change to a bimodal distribution as we soften the spheres, that the polyhedral sub-units approach more closely topological identity as (distorted) tetrahedra and octahedra. From the approximate doubling of the octahedral component over Bernal's hard sphere statistics, we might conclude that his larger, less regular polyhedra are being "squeezed out", and transforming largely to octahedra. However, not all the larger polyhedra are squeezed out in the soft sphere model, as is indicated by the existence of a high radius ratio tail to the population distribution of Fig. 3b. Thus, we are forced to conclude that the intermediate range packing constraints—the restrictions imposed on the arrangement of the various super-units by the embedding problem—are still sufficiently severe to prevent us constructing a model of a non-crystalline assembly from distorted tetrahedra and octahedra alone. Small quantities of large polyhedra are required, holes which we might prefer to refer to as necessary "defects" in a packing which otherwise consists of distorted tetrahedral and octahedral sub-units, arranged into locally dense super-units in a limited number of different ways.

Although these larger holes are only a small fraction of the total number of polyhedra in both the hard and soft sphere cases, their larger size means they will account for a significantly larger fraction of the total volume. For the hard sphere packing, we estimate about 25% of the total volume is made up of these larger polyhedral holes, while the figure for the soft sphere case is lower at about 20%. In both hard and soft sphere systems, the total tetrahedral volume is about the same, volume in the soft sphere case being transferred from the large hole to octahedral hole classification. Thus, our comment above concerning the increased freedom from the softness of the potential leading to an increase in the octahedral holes at the expense of the larger polyhedral clusters is further borne out.

From these empirical model studies, therefore, we can tentatively conclude that in the non-crystalline simple liquid model, although tetrahedral and octahedral sub-units dominate, as they do also in close-packed crystals, we still appear to be unable to fill three-dimensional space with these two basic building blocks alone, even taking into account significant distortions of the tetrahedra and octahedra. Although there are several different ways in which the two units can *locally* fill space by clustering around a common edge to give topologically distinct super-units, and hence yield structures that are consistent with the necessary disorder expected in a liquid structure, these locally dense super-units cannot, it seems, themselves be matched perfectly to each other, even with the significant distortions—and hence flexibility—built in by the soft Morse potential. Put another way, the distorted tetrahedron/octahedron system does not seem to solve the embedding problem—these local, different T/O super-units just cannot pack together without leaving significant volume to be accounted for by larger, less regular polyhedra. It seems possibly useful to consider these larger polyhedra ultimately as *defects* in the packing of distorted tetrahedral and octahedral sub-units. We do not seem to be able, however, to rearrange the structure to transform these "defect" polyhedra into polyhedra we could identify as even fairly highly distorted tetrahedra or octahedra. We can improve significantly on the hard sphere case by softening the potential acting between a pair of neighbouring spheres, an operation that does transform some of the larger polyhedra into octahedra, and in doing so reduces the total "defect volume" from 25 to 20%. Bearing in mind the softness of the Morse potential used, it seems unlikely that we would be able to improve very much on the Morse case while we continue to use a potential that has some reasonable physical justification. It would be an interesting exercise to relax a hard sphere model under other, perhaps rather artificial, even softer potentials, to see if further defect volume could be squashed out, but it would seem unlikely that such systems would be particularly applicable to real liquids.

9. INTERMEDIATE RANGE ORDER

Local arrangements of tetrahedra and octahedra

In order to begin to understand the nature of, and perhaps the restrictive rules operating on, the packing of distorted tetrahedra and octahedra, we could examine model packings in some detail to discover what kind of local super-unit aggregations of these two basic building blocks occur, and how these succeed or fail in embedding themselves in their surroundings in a spatially efficient way. We could try to identify the super-units themselves, and examine how they fit—or fail to fit—together in the overall densely-packed assembly. Bernal himself made some interesting relevant observations on the basis of the hard sphere model, when he pointed out [1] that tetrahedra could link together in extended, essentially linear chains, by sharing edges and faces. These units, which Bernal called "pseudonuclei" because of their denseness yet inability to crystallize, have helical geometry, and also are known as Boerdijk spirals (Fig. 4). Looking at an expanded ball and spoke model of his hard sphere liquid, we find that these linear structures branch, twist and turn as they apparently weave through the structure. In fact, there might well be a case for treating the tetrahedron as the basic building block, relegating the octahedron to the status of the defect—the octahedra, together with the larger holes, being formed at the interfaces between these pseudonuclei ribbons.

In an earlier work [6], we examined the kinds of $T + O$ units that could be found in the soft sphere packings, although this analysis was not carried through in sufficient detail to enumerate the local super-unit clusters that shared common edges (an operation that could with interest be resurrected?). Through consideration of correlation functions between tetrahedra and tetrahedra, tetrahedra and octahedra, and octahedra and octahedra, a picture could be built up of the kinds of linkings of pairs and triplets of polyhedra that are found in soft sphere models (see for example Fig. 5). Thus, some elements of the super-unit clusters of the individual tetrahedral and octahedral sub-units could be identified, although the kinds of searches made did not allow *complete* super-units to be enumerated. The results of this work do, however, underline the essential validity of the idea of building up the non-crystalline packing from the basic distorted tetrahedral and octahedral sub-units as building blocks.

One conclusion that can be drawn from this earlier work concerns the existence or otherwise of clusters of tetrahedra that might be identified as icosahedra. As the icosahedron can be thought of as 20 slightly distorted regular tetrahedra sharing a common vertex, this feature should be fairly

Fig. 4. An arrangement of face-shearing tetrahedra in a Boerdijk spiral (Bernal's "pseudonuclei").

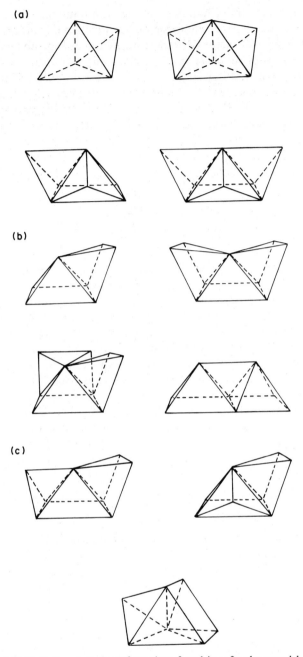

Fig. 5. Selected $T-T$, $O-O$ and $T-O$ configurations found in soft sphere models (from Ref. [6]).

easy to identify in a packing, and we have attempted to identify such local units [6]. These searches were instigated by attempts in the literature to discuss non-crystalline models in terms of relatively familiar polyhedral structures—if *crystalline* local units are not allowed, then perhaps familiar *non-crystalline* ones such as the icosahedron will serve as a basic unit of a non-crystalline packing instead? Despite extensive attempts, we were totally unsuccessful in identifying icosahedral units. This lack of success should not be surprising when we consider the inability of icosahedra to pack efficiently in space. Any attempt to pack them together would leave significant "holes" at the interfaces between them. In fact, an earlier "polytetrahedral" model of Sadoc [7] was constructed on the computer, using a building rule which biased the structure of the resulting assembly towards interpenetrating icosahedra. Even with this interpenetration, subsequent analysis [8] confirmed the model had a very low density probably because of the volume-expensive interfaces between adjacent icosahedral units. Thus, these empirical model studies show that, in both hard and soft sphere systems, tetrahedral sub-units do not in any significant number cluster into icosahedra. The message seems to be that the intermediate range order in idealized simple liquids is a much varied set of super-unit assemblages of distorted tetrahedra and octahedra than would be implied by a model based on imperfect packing of icosahedra. Even with the additional flexibility afforded by soft repulsions, such units do not appear in any significant numbers. This is perhaps not really surprising, as the coming together—or "unmixing"—of tetrahedra from more varied T/O environments to form complete icosahedral super-units would be entropically expensive, and possibly not greatly more favoured than the "unmixing" of tetrahedra and octahedra in the correct proportions to form crystal-related seeds. And the latter process we know from super-cooling is not exactly favoured. At strong undercooling, however, the icosahedral cluster might arguably be more favoured than the crystal-related ones, and hence it may perhaps act as an additional barrier to crystal nucleation.

10. SUMMARY

We have tried here to rationalize and contrast the structural principles of the idealized crystalline and liquid phases of spherical atoms (e.g. inert gases, simple metals) in a way which focuses attention on some interesting aspects of order and disorder. Classically, the crystal is thought of as an ordered, regular array of atoms, to which both local space groups and long-range translational symmetry operations can be applied to generate the crystal from an asymmetric unit. The idealized model of a liquid—deriving from Bernal's concept of random close packing of hard spheres—is, in contrast, a system in which these operations of symmetry do not appear to be particularly relevant. In fact, the model name—"random close packing of hard spheres"—focuses on the apparent randomness of the structure.

This is, however, to oversimplify matters: the liquid structure may have attributes of randomness in that it has no crystallinity, but it is certainly not lacking in some aspect of *order*. This *ordering* arises from the consequences of the "close packing" part of the concept: if the spherical atoms are close-packed, their organization *cannot* be totally random. The resulting structure is severely restricted by the constraints of volume exclusion—two atoms cannot overlap—and it is this "impenetrability", to use a term from *Alice Through the Looking Glass* coined in this context by Bernal, that controls the *restricted order* or *restricted randomness* (we could put it either way) that is the essence of liquid structure.

We have tried to present both crystal and liquid structures in terms of a "build-up" procedure which starts from the elementary local sub-units that are consistent with both the interatomic interactions (the "chemistry" of the assembly) and the necessarily high density of these condensed phases. The procedure moves upwards in the hierarchy to larger "super-unit" assemblages, which themselves require to be "embedded" without loss of overall density. In the crystal case, we identified tetrahedral and octahedral sub-units of four and six spheres, respectively; these can be operated upon to produce a unit cell which naturally embeds itself in an extended assemblage of identical unit cells. The ability of these two polyhedra to fill space in this way can be rationalized in terms of their dihedral angles: to fill space, we must be able to collect at a common edge an integer number of the polyhedra so that the sum of dihedral angles over all the common-edge-sharing polyhedra is 2π. Any less than that value and we would create a void; any greater would violate volume exclusion. Two tetrahedra and two octahedra is the only combination which fulfils this constraint, and this is the cluster—or "super-unit"—which we find in f.c.c and h.c.p. close-packed crystals.

Turning to the liquid, we considered the possibility of using the same two polyhedra in an attempt to "build up" the required non-crystalline structure we believe represents an ideal simple liquid. We were wary of abandoning the tetrahedron and octahedron as building blocks as they are (a) natural polyhedra consistent with low local potential energy and with volume exclusion constraints, and (b) the liquid is sufficiently dense that other possible polyhedral units would contain too much unoccupied volume. Clearly, however, as two tetrahedra plus two octahedra is the only combination capable of filling space locally, it would appear that retaining these building blocks must naturally lead to a crystal.

A way out of this apparent impasse is found by *relaxing the local symmetry*, and allowing the tetrahedral and octahedral sub-units to distort slightly. By how much is a matter for empirical study, but by considering distortions such that dihedral angles could vary by (i) $\pm 7°$, and (ii) $\pm 10.5°$, we found (i) four, and (ii) eight *different* ways of arranging our distorted tetrahedral and octahedral units with a common edge. Some of these clusters or "super-units"—e.g. five tetrahedra with a common edge—are highly symmetrical, and give interesting examples of a reduction of symmetry of the basic sub-unit allowing higher symmetry to be achieved in a super-unit aggregate of the sub-units. Referring to the two-dimensional model of Kawamura, we note that equilateral triangular and square units can be arranged to fill space around a vertex in four different ways, allowing the construction of both crystals and "liquids" in two dimensions. Similarly, the increased variability in the allowed kinds of three-dimensional local clusters or "super-units" of the basic tetrahedral and octahedral sub-units which is facilitated by this symmetry relaxation might again allow us to construct—by a "build-up" procedure similar in essence to that used in constructing a crystal from the two basic polyhedra—our non-crystalline model of a liquid.

We might note and stress here the nature of this build-up operation in both the crystal and liquid cases. In the former, the very limited number of possibilities almost by definition means the build-up operations will be standard symmetry operations operating upon regular polyhedra. In the liquid case, the procedure would not work without allowing some polyhedron distortion, and consequently the build-up procedures would not involve standard symmetry operations, but local ones which could perhaps be considered as "fuzzy", forgiving symmetry-ish operations at the local level. These operations would recognize the topology, but not refuse to operate because edge lengths and dihedral angles showed metric variations. We can also consider a *real* crystal in this context. Here, because of the thermal motions, the polyhedra would similarly not be perfect, and again, fuzzy and forgiving symmetry operations would be needed which took account of the connectivity of the polyhedral units but did not have to be fully consistent with the metric.

The problems, however, do not end here. Even if we can build up local space-filling super-unit clusters for the non-crystalline case, there is no guarantee we would be able to fit them together in a three-dimensional jigsaw-puzzle, without either violating volume exclusion by creating overlap, or leaving too large voids so that the density is unacceptably lowered. This is what we have called the "embedding problem" which is relevant to both crystals and liquids. In the crystal case, we have a formal solution in that only the 230 space groups are consistent with infinite lattices. In the case of liquids, we have yet to formulate a similar solution, which could be one of the results of the effective development of Bernal's proposed statistical geometry [1]. This would hopefully give us the rules for application of the relevant set of fuzzy symmetry operations: in the crystal, the rules are regular, in the liquid they are locally variable in a so far unknown way.

We have only begun to probe this part of the problem empirically by examining the polyhedron sub-unit—and in much less detail the super-unit—populations in both hard and soft sphere models. These initial studies have justified our way of looking at instantaneous liquid structure in terms of distorted tetrahedra and octahedra. They have also shown that, at least in the models examined, that other larger polyhedra exist as "defects" in the non-crystalline "polytetraoctahedral" structure. Although relatively small in number (around 7%), these larger polyhedra contribute 20–25% of the total volume (some—but probably not all—of which can be squeezed out by softening the repulsive atomic cores). Thus, the jigsaw-puzzle of even two *flexible* three-dimensional pieces does not quite fit together: the missing pieces are topologically different from, and enclose a larger volume than, the two basic sub-units.

Considering non-crystalline structures in terms of these basic—though distorted—chemically and physically reasonable building blocks is, we believe, a useful conceptual device. The symmetry of the units themselves—tetrahedra and octahedra—must be degraded to allow them to pack locally in any way other than those found in f.c.c and h.c.p. crystals. This is our first loss of symmetry. Secondly, the local "super-units" built up from these more elementary polyhedra will in general (though not in every case) be of low symmetry, unlike the two tetrahedron plus two octahedron combination found in close-packed crystals. And finally, to "embed" these units—i.e. to fit them together in a non-crystalline jigsaw-puzzle—appears not to be possible without a small number of larger, generally less regular polyhedra, that contribute 20–25% of the total volume of the system. The "build-up" procedures for crystal and liquid are essentially similar, though different in detail.

Fuzzy, forgiving symmetry operations that recognize topological rather than metrical identity are needed in the liquid case, together with their so far unknown rules of application.

This description in terms of varied super-unit clusters of distorted tetrahedra and octahedra also helps us to avoid assigning too much symmetry to non-crystalline structures. Although often proposed as having frequent occurrence in simple liquids, empirical evidence on models shows that highly ordered relatively large local clusters such as the icosahedron are essentially absent [6]. Models which have been designed to give a high occurrence of icosahedra exhibit unacceptably low densities, even when neighbouring icosahedra interpenetrate. The unit is essentially too locally ordered, and too difficult to embed, to be tolerated by a high entropy structure.

REFERENCES

1. J. D. Bernal, The structure of liquids. *Proc. R. Soc.* **A284**, 299–322 (1964).
2. J. L. Finney, The geometry of random close packing. *Proc. R. Soc.* **A319**, 479–493 (1970).
3. J. M. Ziman, *Models of Disorder*, p. 78. Cambridge Univ. Press, Cambridge (1979).
4. R. Collins, T. Ogawa and T. Ogawa, Information entropy and the structural geometry of two-dimensional liquids. *Prog. theor. Phys.* (Japan), **78**, 83–96 (1987).
5. H. Kawamura, A simple theory of hard disk transition. *Prog. theor. Phys.* **61**, 1584–1596 (1979).
6. J. L. Finney and J. Wallace, Interstice correlation functions: a new sensitive characterisation of non-crystalline packed structures. *J. Non-Cryst. Sol.* **43**, 165–187 (1981).
7. J. F. Sadoc, T. Dixmier and A. Guinier, Theoretical calculation of dense random packing of equal and non-equal hard spheres. Applications to amorphous metallic alloys. *J. Non-Cryst. Sol.* **12**, 46–60 (1973).
8. J. L. Finney, Modelling the structures of amorphous metals and alloys. *Nature* **266**, 309–314 (1977).

Computers Math. Applic. Vol. 17, No. 1–3, pp. 355–374, 1989
Printed in Great Britain. All rights reserved

0097-4943/89 $3.00 + 0.00

TOPOLOGICAL ASPECTS OF BENZENOIDS AND CORONOIDS, INCLUDING "SNOWFLAKES" AND "LACEFLOWERS"

S. J. Cyvin, J. Brunvoll and B. N. Cyvin

Division of Physical Chemistry, The University of Trondheim, N-7034 Trondheim-NTH, Norway

Abstract—A coronoid is a "benzenoid with a hole". Some aspects of the studies of topological properties of benzenoids and coronoids are reviewed. The survey includes the search for concealed non-Kekuléans, multiple zigzag chains and primitive coronoids. Enumerations and classifications of polyhexes (i.e. benzenoids and coronoids) are treated in some details. The original contributions of the present work are concentrated upon polyhexes with hexagonal symmetry.

Benzenoids with hexagonal symmetry are referred to as "snowflakes". The forms of all snowflakes with D_{6h} symmetry and h (the number of hexagons) $\leqslant 55$ are displayed as computer-generated drawings. In addition, the concealed non-Kekuléans with $h = 61$ and $h = 67$ are shown.

Coronoids with hexagonal symmetry represent another main subject of this work. Single coronoids in general are classified into regular including primitive, half essentially disconnected, essentially disconnected and non-Kekuléan systems. The smallest members of all these categories among coronoids with D_{6h} symmetry are displayed, viz. $h \leqslant 36$ for the primitive, $h \leqslant 54$ for the non-Kekuléan systems, and $h \leqslant 30$ for the other categories. Combinatorial formulas of K (the number of Kekulé structures) for some classes of coronoids with hexagonal symmetry are given, including a complete solution for the primitive systems belonging to D_{6h} or C_{6h}.

Finally the forms of the smallest ($h \leqslant 49$) multiple coronoids with D_{6h} symmetry ("laceflowers") are displayed.

1. INTRODUCTION

In the present paper some ideas about benzenoids and coronoids are expanded. Benzenoids were also treated in two papers [1, 2] of the first volume of *Symmetry: Unifying Human Understanding* [3], which may be consulted for precise definitions and for some key references in this topic. Benzenoid systems consist of identical regular hexagons in a plane, while coronoids (also touched upon in one of the cited papers [1]) are, loosely speaking, "benzenoids with a hole". By the term polyhexes we shall refer to the class of benzenoids and coronoids together. Benzenoid systems as mathematical (graph-theoretical) objects have obvious counterparts in benzenoid (polycyclic aromatic) hydrocarbons, which are very well-known in organic chemistry. More than 500 of such compounds have been described in the literature [4], and most of them have been synthesized. It is less-known that two cycloarenes, a new class of polycyclic aromatic hydrocarbons, also have been synthesized, viz. cyclo[*d.e.d.e.d.e.d.e.d.e*]dodekakisbenzene (kekulene), C_1 [5], and very recently cyclo[*d.e.d.e.e.d.e.d.e.e*]dekakisbenzene, C_2 [6]. These molecules correspond to coronoid systems and are depicted in Fig. 1.

The studies of topological properties of polyhexes have been intensivated during the last years. Even the literature which has appeared after the first volume of *Symmetry* [3] was published is too voluminous to be cited here. We only give the Refs [7–17] to a small portion of the existing relevant papers. All of them are from 1986 or 1987 and tend to represent as many authors as possible. Enumerations and classifications of polyhexes are substantial parts of the recent research in this area; a consolidated report by 14 authors on the present status has appeared [18]. To be precise, this report reflected the situation around the beginning of 1987. During a rapid development in the field a substantial amount of supplementary data is already available.

Enumerations and classifications of polyhexes are also the main subjects of the present paper. It contains some original contributions, which tend to be symmetry-oriented in the sense that emphasis is laid on the most symmetrical systems, i.e. those with hexagonal symmetry.

The numbers of Kekulé structures is an important aspect of polyhexes and will be considered in some details. This topic has been summarized and expanded in a recent book [19].

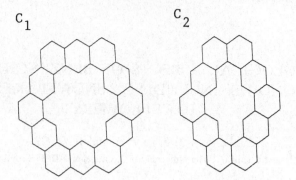

C_1 C_2

Fig. 1. Two primitive coronoids corresponding to synthesized cycloarenes.

2. EXAMPLES OF THE RESEARCH PROGRESS

Here we summarize some new results in the topological studies of polyhexes in relation to the two contributions [1, 2] in the first volume of *Symmetry* [3]. Some of the terms used in this section are defined more precisely in the subsequent sections. Here we only mention that the number of hexagons in a polyhex is denoted by h and the number of Kekulé structures by K.

Example 2.1

"The two smallest ($h = 11$) non-Kekuléan benzenoids ($K = 0$) with equal number of black and white vertices" [2] is a misleading statement. The category of benzenoids in question is also referred to as concealed non-Kekuléans [20]. It is true that no such systems with less than 11 hexagons exist, but there are more than two of them. The search for smallest concealed non-Kekuléans has continued since Gutman [21] in 1974 published the two frequently quoted forms (see, e.g. Ref. [2]). Finally, in 1986 eight such systems were identified [22]. Still later [23] it was definitely proved by computer enumeration that there exist exactly eight concealed non-Kekuléan systems with $h = 11$ (cf. Fig. 2); therefore any search for more of these systems is futile.

It was also found, according to a private communication from W. He and W. He (Shijiazhuang, China), that there are exactly 98 concealed non-Kekuléans with $h = 12$.

Example 2.2

In the paper [2] commented above it is also stated that the smallest benzenoid with C_{6h} symmetry occurs for $h = 19$, and that only one such system exists. The last part of this statement is wrong. There are exactly two benzenoids with $h = 19$ and C_{6h} symmetry [24]; cf. Fig. 3.

Example 2.3

In the other paper [1] considered here a concealed non-Kekuléan with $h = 43$ and D_{6h} symmetry was depicted (cf. Fig. 4). It was assumed there that this is the smallest system of this category. Later studies [24] have proved that this statement is true. It was deduced by computer enumeration that the system in question is unique, but in addition there are four concealed non-Kekuléans of the same size ($h = 43$) and C_{6h} symmetry; see Fig. 4.

Fig. 2. The eight existing smallest ($h = 11$) concealed non-Kekuléan benzenoids.

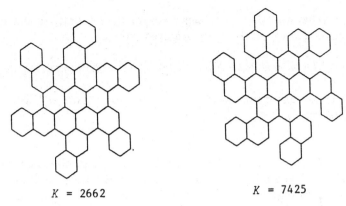

$$K = 2662 \qquad K = 7425$$

Fig. 3. The two existing smallest ($h = 19$) benzenoids belonging to C_{6h}. The numbers of Kekulé structures (K) are given.

Example 2.4

The benzenoid class of multiple zigzag chains, $A(m, n)$, was treated in some detail in one of the considered papers [2]. The exposition therein, as well as the modest extensions elsewhere [25] have been supplemented with further extensions and qualitatively new findings [16, 19].

For the number of Kekulé structures of $A(m, n)$ with fixed values of n the formulas have not only been extended to $n \leqslant 10$ with the last result:

$$K\{A(m, 10)\} = \frac{1}{3,628,800}(m + 1)(m + 2)(50,521m^8 + 606,252m^7 + 3,221,422m^6 + 9,895,860m^5$$
$$+ 19,220,389m^4 + 24,175,248m^3 + 19,240,308m^2 + 8,866,800m + 1,814,400). \quad (1)$$

Also a general formulation in terms of a determinant was produced, based on the John–Sachs theorem [26]. Instead of giving the general form it seems to be more interesting to give here the application to $n = 10$. Then the below determinant, when expanded properly, is equivalent to the polynomial form (1);

$$K\{A(m, 10)\} = \begin{vmatrix} \binom{m+2}{2} & \binom{m+3}{4} & \binom{m+4}{6} & \binom{m+5}{8} & \binom{m+6}{10} \\ 1 & \binom{m+2}{2} & \binom{m+3}{4} & \binom{m+4}{6} & \binom{m+5}{8} \\ 0 & 1 & \binom{m+2}{2} & \binom{m+3}{4} & \binom{m+4}{6} \\ 0 & 0 & 1 & \binom{m+2}{2} & \binom{m+3}{4} \\ 0 & 0 & 0 & 1 & \binom{m+2}{2} \end{vmatrix}. \quad (2)$$

Now we pass to the studies of K numbers for $A(m, n)$ with fixed values of m. The recursive properties of these quantities were attacked in a slightly different way from previously [2, 25]

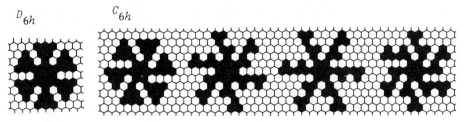

Fig. 4. The five smallest ($h = 43$) concealed non-Kekuléans with hexagonal symmetry.

inasmuch as the n value was allowed to step two units every time. In this new form a general formulation of the recurrence relation was achieved, viz.

$$K\{A(m, n + 2)\} = \sum_{j=0}^{m} (-1)^j \binom{m + j + 2}{2j + 2} K\{A(m, n - 2j)\}. \tag{3}$$

In spite of these new developments no explicit formula for $K\{A(m, n)\}$ with fixed values of m is known so far unless $m = 1$.

Example 2.5

A combinatorial K formula for a simple class of so-called primitive coronoids was given [1], kekulene (C_1 of Fig. 1) being one of the members. More extensive studies in this field have been performed, and some more general K formulas have been derived [27, 28].

Consider a cyclic single chain of hexagons with N corners (more precisely: N angularly annelated hexagons) and all the linear segments of equal length. This length is determined by a parameter j, defined so that $h = Nj$. Then

$$K(j, N) = 1 + (-1)^N + 2^{-N}[j + (j^2 + 4)^{1/2}]^N + 2^{-N}[j - (j^2 + 4)^{1/2}]^N. \tag{4}$$

For kekulene $N = 6$ and $j = 2$. With these parameters equation (4) reproduces correctly the number of Kekulé structures for kekulene, viz. $K = 200$.

Another class of coronoids is obtained by keeping the number of linear segments constantly six, but allowing for non-equidistant segments. Let their lengths be defined by six (not independent) parameters so that $h = a + b + c + d + e + f$. Then, according to an unpublished result by J. L. Bergan (Trondheim, Norway):

$$K(a, b, c, d, e, f) = abcdef + abcd + bcde + cdef + defa + efab + fabc$$
$$+ ab + bc + cd + de + ef + fa + ad + be + cf + 4. \tag{5}$$

On inserting $a = b = c = d = e = f = j$ for equidistant segments it is obtained

$$K(j) = (j^3 + 3j)^2 + 4 = (j^2 + 1)^2(j^2 + 4), \tag{6}$$

and finally for kekulene: $K(2) = 200$.

3. SOME DEFINITIONS AND THE INVARIANT Δ

The characteristic number (or invariant) Δ for a polyhex plays an important role in the topology of polyhexes. It is a positive integer or zero.

Throughout this paper we assume that a polyhex, P, is drawn (as usually) so that some of its edges are vertical. Let the n vertices of P be colored alternately black and white. The numbers of the colored vertices are $n^{(b)}$ and $n^{(w)}$, respectively. The peaks (p) and valleys (v) are the vertices on the circumference (perimeter) which point upwards and downwards, respectively. We adopt the convention that all peaks are white, and therefore all valleys black. Furthermore, if $n^{(b)} \neq n^{(w)}$, we assume as another convention $n^{(b)} > n^{(w)}$.

The invariant Δ is defined by

$$\Delta = n^{(b)} - n^{(w)}. \tag{7}$$

It has been pointed out [2] that also

$$\Delta = n_i^{(b)} - n_i^{(w)}, \tag{8}$$

where the difference is taken between the numbers of black and white internal vertices only. This feature is easily understood by the fact that the external vertices (which constitute the perimeter) consist of an equal number of black and white vertices. Hence they may be deleted as indicated in Fig. 5, whereby the counting for Δ is facilitated. A third way to determine Δ is often the easiest one, namely by taking the difference between the numbers of valleys and peaks, viz.

$$\Delta = n_v - n_p. \tag{9}$$

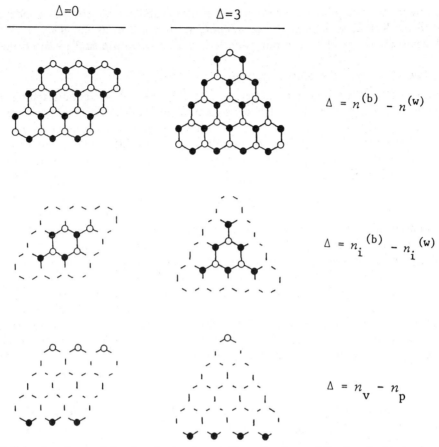

$$\Delta = n^{(b)} - n^{(w)}$$

$$\Delta = n_i^{(b)} - n_i^{(w)}$$

$$\Delta = n_v - n_p$$

Fig. 5. Colored vertices and Δ values for two benzenoids: anthanthrene (left column) with $\Delta = 0$, a Kekuléan system; triangulene (right column) with $\Delta = 2$, an obvious non-Kekuléan system.

Figure 5 explains this feature; in the bottom figures all vertices belonging to vertical edges are deleted, leaving only the peaks and valleys. Again it is clear that an equal number of black and white vertices was deleted. This proof of equation (9) is due to W. He and W. He according to a private communication. A more complicated proof was earlier conducted by Cyvin and Gutman [20]. It should be noted that the absolute numbers of peaks and valleys depend on the chosen orientation of P, but the difference (9) is invariant.

4. FURTHER DEFINITIONS AND THE "*neo*" CLASSIFICATION

4.1. Definitions

A polyhex is said to be Kekuléan when it possesses Kekulé structures ($K > 0$) and non-Kekuléan when it does not possess any Kekulé structure ($K = 0$). All Kekuléan systems are known to have $\Delta = 0$, but this is not a sufficient condition for being Kekuléan. Systems with $K = 0$ and $\Delta = 0$ exist and are referred to as concealed non-Kekuléan (cf. e.g., Figs 2 and 4). The other non-Kekuléans ($K = 0$, $\Delta > 0$) are called obvious.

Catacondensed polyhexes are those which have no vertex shared by three hexagons. When formulated in this way the definition applies to both benzenoids and coronoids. For benzenoids the catacondensed systems are often defined by the absence of internal vertices. All catacondensed polyhexes are known to be Kekuléan. Polyhexes which are not catacondensed are said to be pericondensed.

Some polyhexes among the pericondensed Kekuléan systems are designated essentially disconnected. They are defined by the presence of fixed bonds (i.e. single or double) in certain positions in all Kekulé structures. In essentially disconnected systems the total number of Kekulé structures is given by the product of K numbers for two or more subunits, which behave independently with regard to the Kekulé structures. These subunits are referred to as the effective units.

Kekuléan polyhexes which are not essentially disconnected are said to be normal.

4.2. Kekulé structures of normal benzenoids

The distinction between normal and essentially disconnected benzenoids among the Kekuléans is very important in the theory of Kekulé structure counts. There are infinitely many essentially disconnected benzenoids with $K = 9$ and any other possible (not arbitrary) value $K > 9$. In sharp contrast, the number of normal benzenoids with any $K > 1$ is limited. This statement is a corollary of a recently proved theorem [29]: For a normal benzenoid with h hexagons the minimum number of Kekulé structures is

$$K_{\min}(h) = h + 1. \tag{10}$$

This minimum K number is realized in exactly one system for a given h, namely the linear single chain (polyacene). The K numbers of normal benzenoids have also otherwise been studied extensively under the keywords "distribution of K" and "Kekulé structure statistics" [30–34]. In consequence of the above discussion it is sensible to ask: How many normal benzenoids have K (any integer $K > 1$) Kekulé structures? The answer is, for instance, 3, 12, 32, 54, 97, 150, 176, 256, 468 and 444 for $K = 10, 20, \ldots, 90$ and 100, respectively (from Ref. [32] and unpublished results). In addition to the lower limit (10) of K for normal benzenoids also the maximum values have been studied [30, 34, 35]. For $h = 1, 2, \ldots, 9$ and 10 one has $K_{\max}(h) = 2, 3, 5, 9, 14, 24, 41, 66, 110$ and 189, respectively. One of the results for an upper bound on K is:

$$K_{\max}(h) \leqslant 3^{h/2}; \quad h > 1. \tag{11}$$

This gives a constant upper bound for the quantity $(\ln K_{\max})/h$, which is related to the resonance energy of the benzenoid [2, 36]. We have found that this quantity also has a lower limit:

$$\tfrac{1}{3}[\ln(5 + 17^{1/2}) - \ln 2] < \frac{1}{h}\ln K_{\max}(h) \leqslant \tfrac{1}{2}\ln 3; \quad h > 1. \tag{12}$$

Here the sign of equality is valid for $h = 2$ and $h = 4$, only.

4.3. The "neo" classification

In conclusion of this chapter we define the *neo* classification, which takes account of all benzenoids. They are either normal (n), essentially disconnected (e) or non-Kekuléan (o).

Figure 3 shows two normal benzenoids with hexagonal symmetry. Some (concealed) non-Kekuléans are shown in Figs 2 and 4, the latter ones (Fig. 4) having hexagonal symmetry. Figure 5 contains a normal benzenoid (left) and an (obvious) non-Kekuléan (right). In supplement of these examples we show (Fig. 6) the two smallest essentially disconnected benzenoids with hexagonal symmetry.

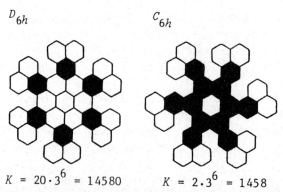

Fig. 6. The two smallest ($h = 25$) essentially disconnected benzenoids with hexagonal symmetry (D_{6h} and C_{6h}). The effective units are white. K numbers are given.

5. ENUMERATION AND CLASSIFICATION OF BENZENOIDS WITH HEXAGONAL SYMMETRY

The benzenoids with hexagonal symmetry (D_{6h} and C_{6h}) are very sparsely distributed among the totality of the systems. Among the 849,285 benzenoids with $h \leqslant 12$ there are only two systems with hexagonal symmetry, viz. benzene ($h = 1$) and coronene ($h = 7$); cf. Table 1 [18, 24, 37]. Therefore it would be hopeless to continue more extensive enumerations of the benzenoids with hexagonal symmetry as a subset of the total amount. Fortunately this is not necessary inasmuch as the systems of interest may be generated specifically by computer-aided procedures [24]. In Table 1 the listing for the two symmetry groups in question goes far beyond $h = 12$.

The benzenoids with hexagonal symmetry have been referred to as "snowflakes" [1, 38]. The reason for this is obvious from the treatise in another book on *Symmetry* [39]. However, the designation is most appropriate for the systems of D_{6h} symmetry (and not C_{6h}). Nevertheless it seems practical to use the term "snowflakes" for both symmetry groups. When we wish to distinguish between them we shall use the terms "proper" and "improper" snowflakes for D_{6h} and C_{6h}, respectively.

The values of h, the number of hexagons, for all snowflakes are discrete according to:

$$h = 6\eta + 1; \quad \eta = 0, 1, 2, \ldots. \tag{13}$$

Table 1. Benzenoids enumerated and classified according the the *neo* classification

h	Class†	D_{6h}	C_{6h}	Other symmetries	h	Class†	D_{6h}	C_{6h}
1	n	1	0	0	13	n	2	0
2	n	0	0	1	19	n	2	2
3	n	0	0	2	25	n	2	7
	o	0	0	1		e	1	1
4	n	0	0	6	31	n	5	24
	o	0	0	1		e	0	8
5	n	0	0	14	37	n	7	84
	e	0	0	1		e	1	44
	o	0	0	7	43	n	11	310
6	n	0	0	48		e	1	213
	e	0	0	3		o	1	4
	o	0	0	30	49	n	17	{2167
7	n	1	0	166		e	3	
	e	0	0	23		o	0	42
	o	0	0	141	55	n	30	{9158
8	n	0	0	643		e	4	
	e	0	0	121		o	1	312
	o	0	0	671	61	n	{59	
9	n	0	0	2531		e		
	e	0	0	692		o	1	
	o	0	0	3282	67	n	{100	
10	n	0	0	10375		e		
	e	0	0	3732		o	4	
	o	0	0	15979	73	n	{176	
11	n	0	0	42919		e		
	e	0	0	19960		o	7	
	o	0	0	78350				
12	n	0	0	{284918				
	e	0	0					
	o	0	0	384666				

†Abbreviations: *n*—normal; *e*—essentially disconnected; *o*—non-Kekuléan.
Data for $h \leqslant 11$ see Ref. [18] and the references cited therein. The entries for $h = 12$ are based on a private communication from W. He and W. He (Shijiazhuang, China). For the systems of hexagonal symmetries and $13 \leqslant h \leqslant 55$, see Refs [18, 24, 37]. The data for $h > 55$ are present results.

Here $\eta = 0$ pertains to benzene, a trivial snowflake. Otherwise $(\eta > 0)$ all snowflakes are pericondensed: coronene $(\eta = 1)$ and systems with coronene as a subunit. It is also an interesting fact that all snowflakes have $\Delta = 0$. Hence there are no obvious non-Kekuléans among them. Yet non-Kekuléan snowflakes exist (see, e.g. Fig. 4); they are all concealed non-Kekuléans.

All the existing snowflakes up to $h = 37$ have been depicted elsewhere [24], as well as selected examples for $h = 43$ [24]. An account of the non-Kekuléans with $h = 49$ is to be published, as well as selected non-Kekuléans with $h = 55$ [37].

In the present work we have pursued the enumeration of proper snowflakes, mainly because of the aestethical values. Here the D_{6h} systems are for the first time generated specifically. Figure 7 shows all the 90 systems with $h \leqslant 55$ in terms of computer-designed pictures. There are only two concealed non-Kekuléans among them. We have also identified the concealed non-Kekuléan proper snowflakes beyond $h = 55$, actually up to $h = 73$. The result is displayed in Fig. 8.

6. ANALYSING SINGLE CORONOIDS: THE *"rheo"* AND *"rio"* CLASSIFICATIONS

6.1. Introduction

The definitions of Section 4.1 apply to coronoids as well as benzenoids. We give at once (Fig. 9) two examples of essentially disconnected coronoids among the systems with hexagonal symmetry. Also the *neo* classification is applicable to coronoids. Table 2 [18, 40, 41] shows the results of enumeration of coronoids with special emphasis on those with hexagonal symmetry. They are classified according to *neo*.

It became soon clear that the K numbers "misbehave" within the class of normal coronoids. The root of this problem seems to be the lack of a specific generation procedure for normal polyhexes

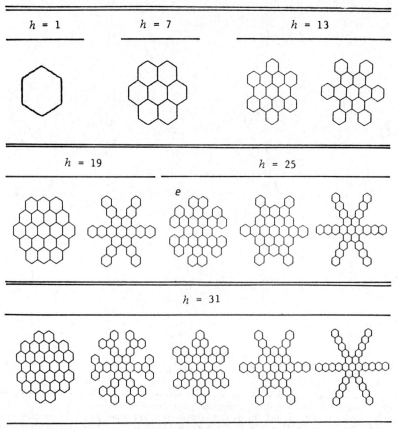

Fig. 7a

h = 37

h = 43

Fig. 7b

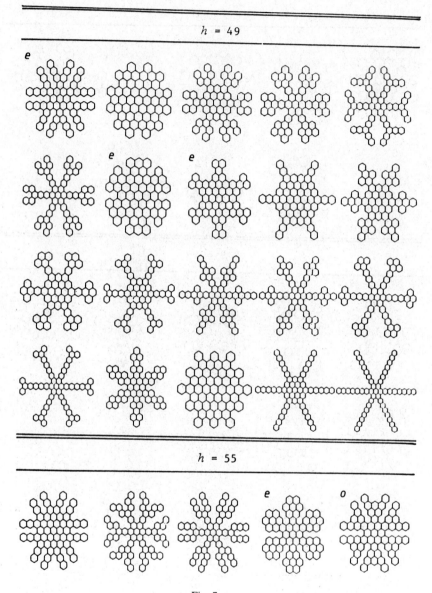

Fig. 7c

Fig. 7d

Fig. 7. Computer-generated "proper snowflakes" (benzenoids of symmetry D_{6h}) for $1 \leqslant h \leqslant 55$. Abbreviations: e essentially disconnected; o (concealed) non-Kekuléan. All others are normal (Kekuléan).

Fig. 8. All concealed non-Kekuléan proper snowflakes with $h \leqslant 73$: one system each with $h = 43, 55$ and 61, four systems with $h = 67$, and seven systems with $h = 73$.

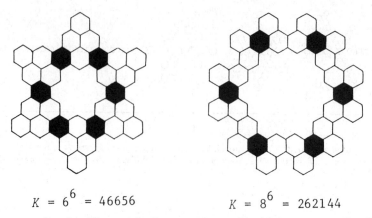

Fig. 9. The two smallest ($h = 30$) essentially disconnected coronoids with D_{6h} symmetry. The effective units are white. K numbers are given.

Table 2. Coronoids enumerated and classified according to the *neo* classification

h	Class[†]	D_{6h}	C_{6h}	Other symmetries	h	Class[†]	D_{6h}	C_{6h}
8	n	0	0	1	18	n	3	1
9	n	0	0	3	24	n	9	7
	o	0	0	2				
					30	n	18	41
10	n	0	0	24		e	2	3
	o	0	0	19				
					36	n	43	192
11	n	0	0	126		e	4	33
	e	0	0	2				
	o	0	0	155	42	n	$\left\{ 97 \right.$	$\left\{ 1113 \right.$
						e		
12	n	1	0	819				
	e	0	0	34	48	n	$\left\{ 221 \right.$	
	o	0	0	1100		e		
						o	1	
13	n	0	0	$\left\{ 5050 \right.$				
	e	0	0		54	n	$\left\{ 456 \right.$	
	o	0	0	7313		e		
						o	3	
					60	n	$\left\{ 1009 \right.$	
						e		
						o	20	

†See footnote to Table 1.
Data for $h \leqslant 12$, see Ref. [18] and references cited therein. The entries for $h \geqslant 13$ are from Refs [18, 40, 41], supplemented with present results.

(when coronoids are included). For normal benzenoids every system with $h + 1$ hexagons, say B_{h+1}, may be generated from a system with h hexagons, B_h, which also is a normal benzenoid. It is known that such a $B_h \to B_{h+1}$ addition (referred to as a normal addition [19, 41]) invariably makes the K number increase: $K\{B_{h+1}\} > K\{B_h\}$.

The distinguishing of "half essentially disconnected" (HED) coronoids [41] among the normal systems seems to give a remedy for the misbehaviour of K numbers as far as the single coronoids (with exactly one hole) are concerned. A HED coronoid possesses two schemes for its Kekulé structures; the set belonging to one scheme has edges with fixed bonds (single or double) in certain positions. An example is shown in Fig. 10. This characteristic feature, however, is not sufficient as a definition of the HED systems. So far this question has not been fully clarified. In fact only one short communication where the idea of HED coronoids was launched, has appeared. The best approach seems to be to start with a definition of "regular" polyhexes in terms of a generation procedure $P_h \to P_{h+1}$ (so-called regular addition [41]). When properly defined this procedure generates (a) all normal benzenoids and (b) a subset of normal coronoids which are not half essentially disconnected. The latter class (b) is referred to as "regular" coronoids. Hence the HED

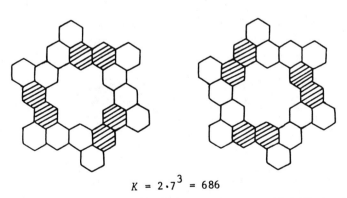

$$K = 2 \cdot 7^3 = 686$$

Fig. 10. The smallest ($h = 18$) HED coronoid with hexagonal symmetry (C_{6h}). The two (symmetrically equivalent) schemes for Kekulé structures are shown, and the K number is given.

coronoids could be defined as the Kekuléan systems which are neither regular or essentially disconnected.

The four categories regular (r), half essentially disconnected ($\frac{1}{2}e$), essentially disconnected (e) and non-Kekuléan (o) seem to cover all single coronoids, although no strict proof has been given to this effect. The *rheo* classification refers to these four categories. It is also practical to define the class of "irregular" coronoids as those which are Kekuléan and not regular. The *rio* classification refers to the categories regular (r), irregular (i) and non-Kekuléan (o).

A survey of the different classes of coronoids:

$$\text{normal } (n) \begin{cases} \text{regular } (r) \\ \left.\begin{array}{l} \text{half essentially disconnected } (\tfrac{1}{2}e) \\ \text{essentially disconnected } (e) \end{array}\right\} \text{ irregular } (i) \\ \text{non-Kekuléan } (o). \end{cases}$$

Here we shall not go further into more precise definitions of the coronoid classes (or categories), but give some examples from computer-aided enumerations of coronoids with hexagonal symmetry.

All single coronoids with the symmetries D_{6h} and C_{6h} have a hole of the shape of coronene or larger. The number of hexagons is

$$h = 6\eta; \quad \eta = 2, 3, 4, \ldots. \tag{14}$$

Here $\eta = 2$ pertains to kekulene (C_1 of Fig. 1). All these systems have $\Delta = 0$, as also is the case for the benzenoids with hexagonal symmetry (cf. Section 5).

The present examples are restricted to the coronoids of D_{6h} symmetry ("proper snowflakes with a hole"). Figure 11 shows systems of primitive coronoids; they are defined as consisting of one (circular) single chain of hexagons. All primitive coronoids are regular. The other (non-primitive) regular coronoids for some of the lowest h values are found in Fig. 12. The smallest HED coronoid

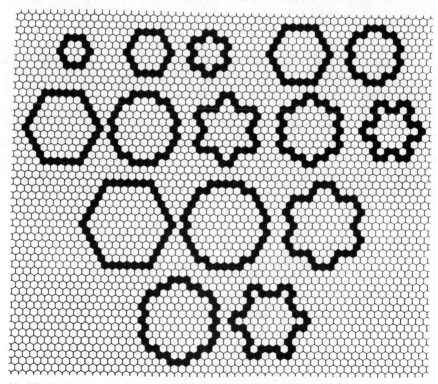

Fig. 11. All primitive coronoids of D_{6h} symmetry and $h \leqslant 36$: one system with $h = 12$, two each with $h = 18$ and 24, and five each with $h = 30$ and 36. The K numbers are: 200; 1300; 5780; 5780, 54,760; 19,604, 287,300, 571,540, [1,860,500, 1,860,500]; 54,760, 1,060,904, 7,761,800, [17,172,740, 17,172,740]. The bracketed figures pertain to isoarithmic systems.

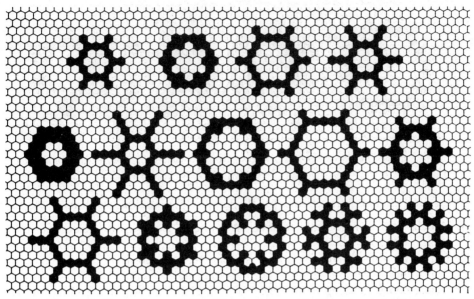

Fig. 12. All non-primitive regular coronoids of D_{6h} symmetry and $h \leqslant 30$: one system with $h = 18$, three with $h = 24$, and ten with $h = 30$. The K numbers are: 7776; 10,108, 63,536, 73,008; 27,508, 36,992, 73,008, 313,632, 497,664, 656,208, 1,060,904, 1,123,632, 2,592,000, 3,504,640.

of hexagonal symmetry belongs to C_{6h}; see Fig. 10. The smallest systems of this kind belonging to D_{6h} are shown in Fig. 13. We refer to Fig. 9 for the two smallest essentially disconnected coronoids belonging to D_{6h}. Finally we show some of the smallest non-Kekuléan coronoids of D_{6h} symmetry. They are all concealed non-Kekuléans. The reported numbers in Figs 9–14 are consistent with the data of Table 2.

7. COMBINATORIAL K FORMULAS FOR CLASSES OF SOME CORONOIDS WITH HEXAGONAL SYMMETRY

7.1. Primitive coronoids

Some primitive coronoids of hexagonal symmetry are members of the classes covered by equations (4) or (6) with respect to their numbers of Kekulé structures (K).

Here we give a quite general result for the K number of a primitive coronoid of hexagonal symmetry:

$$K = (u_0^2 + 2u_0 u_3 + u_3^2 + 1)^2 (u_0^2 + 2u_0 u_3 + u_3^2 + 4) = (u_0 + u_3)^6 + 6(u_0 + u_3)^4 + 9(u_0 + u_3)^2 + 4. \quad (15)$$

Fig. 13. The smallest HED coronoids belonging to D_{6h}: four systems with $h = 24$ and three with $h = 30$. The K numbers are: 2000, 4394, 4394, 11,664; 21,296, 148,176, 170,368.

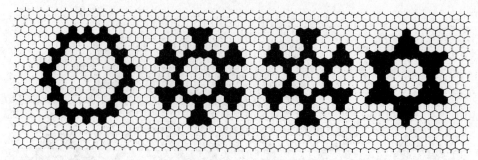

Fig. 14. The smallest concealed non-Kekuléan coronoids belonging to D_{6h}: one system with $h = 48$ and three with $h = 54$.

An algorithm for the quantities u_0 and u_3, along with U, u_1 and u_2, is given in the following:

1. Choose two symmetrically equivalent corners (A hexagons) generated by a rotation of 60°.
2. Depict the single chain from A to A inclusive, and call it U.
3. Delete the two end hexagons (A) to produce u_0.
4. Delete one end hexagon and the linear segment at the other end in the two different ways; call the fragments u_1 and u_2.
5. Delete both linear end segments to produce u_3.

Figure 15 illustrates this algorithm for two choices pertaining to the system with $K = 571,540$ in Fig. 11.

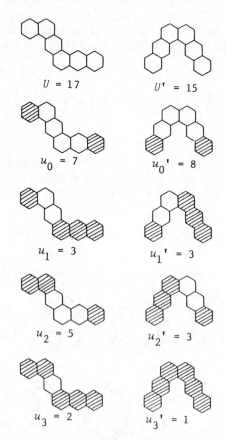

Fig. 15. Two catacondensed unbranched benzenoids, U (single chains), and their fragments u_i ($i = 0, 1, 2, 3$), represented by the unhatched parts. K numbers are given. The bottom-right drawing exemplifies the degenerate case of no hexagons with $K = u'_3 = 1$. Notice:

$$\begin{vmatrix} 7 & 3 \\ 5 & 2 \end{vmatrix} = \begin{vmatrix} 8 & 3 \\ 3 & 1 \end{vmatrix} = -1.$$

Let the K numbers of U and its fragments be denoted $K\{U\} = U$ and $K\{u_i\} = u_i$ $(i = 0, 1, 2, 3)$. One has

$$U = u_0 + u_1 + u_2 + u_3 \tag{16}$$

as a simple deduction from the method of fragmentation [42]. A more interesting theorem is (for numerical examples, see Fig. 15):

$$\begin{vmatrix} u_0 & u_1 \\ u_2 & u_3 \end{vmatrix} = u_0 u_3 - u_1 u_2 = -1. \tag{17}$$

The formula for K (15) was deduced by means of a symmetry-adapted method of fragmentation (SAMF) [38, 40, 43, 44]. In our example, notice that $u_0 + u_3 = u_0' + u_3' = 9$. On inserting this value into equation (15) one obtains correctly $K = 571,540$.

7.2. Classes of HED coronoids

Two classes of HED coronoids, each determined by two parameters (j, m), are depicted in Fig. 16. The example of Fig. 10 is a member of both classes. Combinatorial K formulas for the special cases with $j = 2$ of the two classes in question have been given previously [40], as determined by SAMF. On realizing that the members are HED coronoids we are able to derive such combinatorial formulas appreciably easier. In the present cases

$$K = 2E^3, \tag{18}$$

where E is the number of Kekulé structures for the pertinent effective unit (cf. Fig. 16), a single chain of three segments (for a general K formula pertaining to this unbranched single chain, see, e.g. Ref. [19]).

7.3. A class of regular coronoids

Finally in this section we give a two-parameter K formula for a class of regular coronoids; see Fig. 17. The special cases of $m = 2$ (j arbitrary) and $j = 2$ (m arbitrary) have been treated previously [38, 40]. The formula in Fig. 17, which was derived by SAMF, is a generalization of the previous findings.

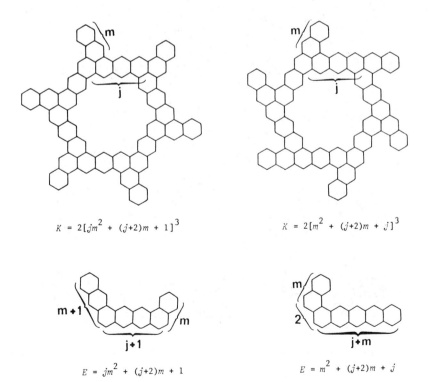

$$K = 2[jm^2 + (j+2)m + 1]^3 \qquad K = 2[m^2 + (j+2)m + j]^3$$

$$E = jm^2 + (j+2)m + 1 \qquad E = m^2 + (j+2)m + j$$

Fig. 16. Two classes of HED coronoids with the corresponding combinatorial K formulas. Also the K formulas for the effective units (E) are given.

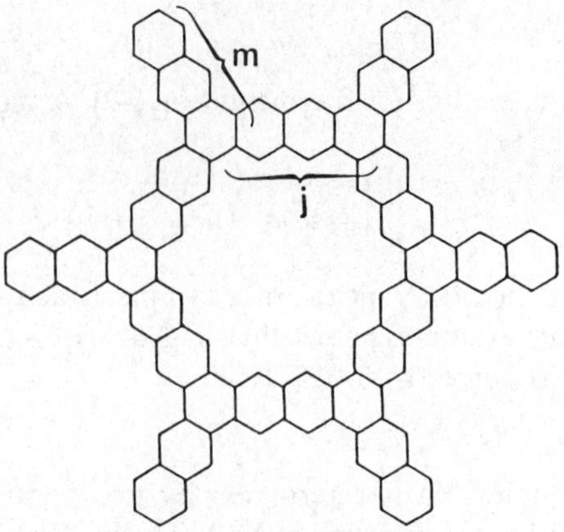

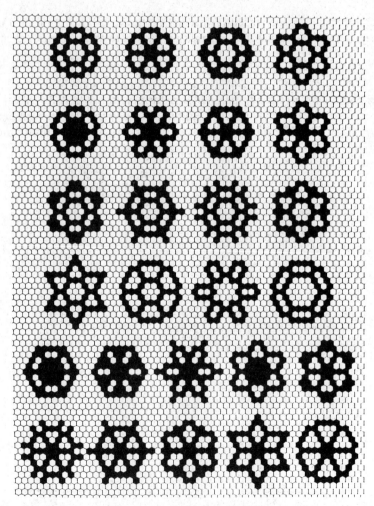

$$K = m^3(j^6 m^3 + 6j^4 m^2 + 9j^2 m + 4)$$

Fig. 17. A class of regular coronoids. The K formula is given.

Fig. 18. All laceflowers of D_{6h} symmetry and $h \leqslant 49$: one system with $h = 36$ and $h = 37$, two with $h = 42$, four with $h = 43$, eight with $h = 48$ and ten with $h = 49$.

8. MULTIPLE CORONOIDS WITH HEXAGONAL SYMMETRY

The classification of multiple coronoids (i.e. those with more than one hole) seems to be a large problem and has so far not been solved in a satisfactory way. Only as a fragmentary result Cyvin *et al.* [41] demonstrated four schemes for Kekulé structures of a double coronoid (two holes), which properly could be termed "one quarter essentially disconnected".

A multiple coronoid with hexagonal symmetry has at least six holes. We have coined the term "laceflowers" for these systems.

All the smallest ($h \leqslant 49$) laceflowers with regular hexagonal (D_{6h}) symmetry were first derived by hand. Later the completeness of this analysis was confirmed by generating exactly the same pretty systems (Fig. 18) with computer aid.

Acknowledgement—Financial support to BNC from the Norwegian Research Council for Science and the Humanities is gratefully acknowledged.

REFERENCES

1. H. Hosoya, Matching and symmetry of graphs. *Comput. Math. Applic.* **12B**, 271–290 (1986). Reprinted in *Symmetry: Unifying Human Understanding* (Ed. I. Hargittai). Pergamon Press, Oxford (1986).
2. S. J. Cyvin and I. Gutman, Kekulé structures and their symmetry properties. *Comput. Math. Applic.* **12B**, 859–876 (1986). Reprinted in *Symmetry: Unifying Human Understanding* (Ed. I. Hargittai). Pergamon Press, Oxford (1986).
3. I. Hargittai (Ed.) *Symmetry: Unifying Human Understanding.* Pergamon Press, Oxford (1986).
4. J. Cioslowski and M. Wala, Polycyclic benzenoid hydrocarbons—the primary data source. *Commun. math. chem. (match)* **21**, 195–258 (1986).
5. F. Diederich and H. A. Staab, Benzenoid versus annulenoid aromaticity: synthesis and properties of kekulene. *Angew. Chem.* **17**, 372–374 (1978).
6. D. J. H. Funhoff and H. A. Staab, Cycloarenes, a new class of aromatic compounds, Part 5. Cyclo-[d.e.e.d.e.e.d.e.e.]decakisbenzene, a new cycloarene. *Angew. Chem.* **25**, 742–744 (1986).
7. A. Graovac, D. Babić and M. Strunje, Enumeration of Kekulé structures in polymers. *Chem. Phys. Lett.* **123**, 433–436 (1986).
8. L. X. Su, Linear recursion relations and the enumeration of Kekulé structures of benzenoid hydrocarbons. *Commun. math. chem. (match)* **20**, 229–239 (1986).
9. P. Křivka, S. Nikolić and N. Trinajstić, Applications of the reduced graph model. Enumeration of Kekulé structures for certain classes of large benzenoid hydrocarbons. *Croat. chem. Acta* **59**, 659–668 (1986).
10. S. El-Basil, Gutman trees. Combinatorial-recursive relations of counting polynomials: data reduction using chemical graphs. *J. Chem. Soc. Faraday Trans. 2* **82**, 299–316 (1986).
11. J. V. Knop, W. R. Müller, K. Szymanski and N. Trinajstić, On the enumeration of 2-factors of polyhexes. *J. Comput. Chem.* **7**, 547–564 (1986).
12. F. J. Zhang, R. S. Chen, X. F. Guo and I. Gutman, An invariant of the Kekulé structures of benzenoid hydrocarbons. *J. Serb. Chem. Soc.* **51**, 537–543 (1986).
13. D. J. Klein, G. E. Hite, W. A. Seitz and T. G. Schmalz, Dimer coverings and Kekulé structures on honeycomb lattice strips. *Theor. Chim. Acta* **69**, 409–423 (1986).
14. W. He and W. He, One-to-one correspondence between Kekulé and sextet patterns. *Theor. Chim. Acta* **70**, 43–51 (1986).
15. J. R. Dias, A periodic table for polycyclic aromatic hydrocarbons. Part X. On the characteristic polynomial and other structural invariants. *J. molec. Struct. (Theochem)* **149**, 213–241 (1987).
16. S. J. Cyvin, B. N. Cyvin, J. Brunvoll and I. Gutman, Enumeration of Kekulé structures for multiple zigzag chains and related benzenoid hydrocarbons. *Z. Naturforsch.* **42a**, 722–730 (1987).
17. I. Gutman and H. Sachs, New approach to the Pauling bond order concept. *Z. phys. Chem.* **268**, 257–266 (1987).
18. A. T. Balaban, J. Brunvoll, J. Cioslowski, B. N. Cyvin, S. J. Cyvin, I. Gutman, W. C. He, W. J. He, J. V. Knop, M. Kovačević, W. R. Müller, K. Szymanski, R. Tošić and N. Trinajstić, Enumeration of benzenoid and coronoid hydrocarbons. *Z. Naturforsch.* **42a**, 863–870 (1987).
19. S. J. Cyvin and I. Gutman, *Kekulé Structures in Benzenoid Hydrocarbons. Lecture Notes in Chemistry.* Springer, Berlin (in press).
20. S. J. Cyvin and I. Gutman, Topological properties of benzenoid hydrocarbons. Part XLIV. Obvious and concealed non-Kekuléan benzenoids. *J. molec. Struct. (Theochem)* **150**, 157–169 (1987).
21. I. Gutman, Some topological properties of benzenoid systems. *Croat. chem. Acta* **46**, 209–215 (1974).
22. H. Hosoya, How to design non-Kekulé Polyhex Graphs? *Croat. chem. Acta* **59**, 583–590 (1986).
23. J. Brunvoll, S. J. Cyvin, B. N. Cyvin, I. Gutman, W. He and W. He, There are exactly eight concealed non-Kekuléan benzenoids with eleven hexagons. *Commun. math. chem. (match)* **22**, 105–109 (1987).
24. J. Brunvoll, B. N. Cyvin and S. J. Cyvin, Enumeration and classification of benzenoid hydrocarbons. Part II: Symmetry and regular hexagonal benzenoids. *J. Chem. Inf. Comput. Sci.* (in press).
25. I. Gutman and S. J. Cyvin, Topological properties of benzenoid systems. Part XXXV. Number of Kekulé structures of multiple-chain aromatics. *Mh. Chem.* **118**, 541–552 (1987).
26. P. John and H. Sachs, "Wegesysteme und Linearfaktoren in hexagonalen und quadratischen Systemen". In *Graphen in Forschung und Unterricht* (Ed. R. Bodeniek, H. Schumacher and G. Walter) pp. 85–101. Franzbecker, Bad Salzdetfurth (1985).
27. J. L. Bergan, S. J. Cyvin and B. N. Cyvin, Number of Kekulé structures of single chain corona-condensed benzenoids (cycloarenes). *Chem. Phys. Lett.* **125**, 218–220 (1986).

28. J. L. Bergan, B. N. Cyvin and S. J. Cyvin, The Fibonacci numbers and Kekulé structures of some corona-condensed benzenoids (corannulenes). *Acta chim. hung.* **124**, 299–314 (1987).

29. I. Gutman and S. J. Cyvin, Hexagonal systems with small number of perfect matchings (to be published).

30. S. J. Cyvin, Distribution of *K*, the number of Kekulé structures, in benzenoid hydrocarbons. Part I: Upper and lower bounds of *K*. *Commun. math. chem. (match)* **20**, 165–179 (1986).

31. B. N. Cyvin, J. Brunvoll, S. J. Cyvin and I. Gutman, Distribution of *K*, the number of Kekulé structures in benzenoid hydrocarbons. Part III: Kekulé structure statistics. *Commun. math. chem. (match)* **21**, 301–315 (1986).

32. S. J. Cyvin and I. Gutman, Number of Kekulé structures as a function of the number of hexagons in benzenoid hydrocarbons. *Z. Naturforsch.* **41a**, 1079–1086 (1986).

33. S. J. Cyvin, J. Brunvoll and B. N. Cyvin, Distribution of *K*, the number of Kekulé structures in benzenoid hydrocarbons. Part IV. Benzenoids with 10 and 11 hexagons. *Z. Naturforsch.* **41a**, 1429–1430 (1986).

34. R. S. Chen and S. J. Cyvin, Distribution of *K* the number of Kekulé structures, in benzenoid hydrocarbons. Part IA: Comments on upper bounds of *K*. *Commun. math. chem. (match)* **22**, 175–179 (1987).

35. I. Gutman, On Kekulé structure count of cata-condensed benzenoid hydrocarbons. *Commun. math. chem. (match)* **13**, 173–181 (1982).

36. R. Swinborne-Sheldrake, W. C. Herndon and I. Gutman, Kekulé structures and resonance energies of benzenoid hydrocarbons. *Tetrahedron Lett.* 755–758 (1975).

37. S. J. Cyvin, J. Brunvoll and B. N. Cyvin, Enumeration and classification of benzenoid hydrocarbons. Part IV: Concealed non-Kekuléans with hexagonal symmetry (to be published).

38. S. J. Cyvin, J. L. Bergan and B. N. Cyvin, Benzenoids and coronoids with hexagonal symmetry ("snowflakes"). *Acta chim. hung.* (in press).

39. I. Hargittai and M. Hargittai, *Symmetry Through the Eyes of a Chemist*, pp. 30–41. VCH, Weinheim (1986). Softcover edn: VCH, New York (1987).

40. S. J. Cyvin, B. N. Cyvin, J. Brunvoll and J. L. Bergan, Coronoid hydrocarbons with hexagonal symmetry. *Coll. Sci. Papers Fac. Sci. Kragujevac* **8**, 137–152 (1987).

41. S. J. Cyvin, B. N. Cyvin and J. Brunvoll, Half essentially disconnected coronoid hydrocarbons. *Chem. Phys. Lett.* **140**, 124–129 (1987).

42. M. Randić, Enumeration of the Kekulé structures in conjugated hydrocarbons. *J. Chem. Soc. Faraday Trans. 2* **72**, 232–243 (1976).

43. S. J. Cyvin, B. N. Cyvin and J. Brunvoll, Trigonal benzenoid hydrocarbons. *J. molec. Struct. (Theochem.)* **151**, 271–285 (1987).

44. B. N. Cyvin, S. J. Cyvin and J. Brunvoll, Number of Kekulé structures for circumkekulene and its homologs. *Mh. Chem.* (in press).

Computers Math. Applic. Vol. 17, No. 1–3, pp. 375–395, 1989
Printed in Great Britain. All rights reserved

A SET THEORETIC APPROACH TO THE SYMMETRY ANALYSIS OF HEXADECAHEDRANE†

E. R. KOUYOUMDJIAN

P.O. Box 60358, Palo Alto, CA 94306, U.S.A.

Abstract—The framework for a set theoretic approach to the symmetry analysis of hexadecahedrane is discussed.

A set theoretic formalism which incorporates sets of symmetry elements is proposed. This formalism relates the contributions of the local symmetry point sets (LPS) of the methine and dimethine units to the molecular symmetry point set (MPS) of hexadecahedrane. These contributions are through the intermediate contributions of the local symmetry point sets of the trimethine–methine/dimethine and the triquinacylene/peristylylene unit pairs.

The contributions of the LPS to the MPS, determine the contributions of the local symmetry point groups (LPG) of the constituent units to the molecular symmetry point group (MPG).

INTRODUCTION

The purpose of this study is to provide a framework for the accounting of molecular coordinate symmetry and structure in terms of the coordinate symmetry contributions of the constituent units and unit structures. This accounting in turn, affords an understanding of molecular coordinate symmetry and structure. The approach adopted illustrates the concept of symmetry as defined by "the relation of the parts to the whole" and as applied to molecular structure and coordinate symmetry, specifically in terms of:

1. The relation of unit and combinations of unit structures to the molecular structure.
2. The relation of the coordinate symmetry of unit and combinations of unit structures to the coordinate symmetry of the molecular structure.

Since the language used for the accounting of molecular coordinate symmetry is set theoretic in nature rather than group theoretic which is the conventional language of symmetry, certain terms must be defined, other terms redefined and the definition of still other terms extended. This is necessary in order to provide the framework within which this accounting is formalized. Furthermore, the parallel development between the proposed set theoretic formalism used for the accounting of the molecular coordinate symmetry and the corresponding group theoretical results will be emphasized.

This set theoretic approach for the accounting of molecular coordinate symmetry is specifically applied to the hexadecahedrane molecule ($C_{28}H_{28}$). Although hexadecahedrane has not yet been synthesized, the molecule is of theoretical interest, being the seventh member of the polyhedrane family ($C_{4n}H_{4n}$; $n = 1$–15). Each succeeding member past dodecahedrane ($C_{20}H_{20}$) is constituted of the 12 cyclopentylene units of the dodecahedrane molecule and two additional cyclohexylene units. The hexadecahedrane molecule belongs to the tetrahedral (Td) symmetry point group and has the highest degree of complexity relative to structurally simpler molecules such as methane (CH_4) tetrahedrane (C_4H_4), octahedrane or the truncated tetrahedrane ($C_{12}H_{12}$) and adamantane ($C_{10}H_{16}$) which also belong to the same symmetry point group.

---

†Dedictated to Grace W. Davis.

SYMMETRY ELEMENTS AND SYMMETRY OPERATIONS

Symmetry elements (axes, planes) have been conventionally used by several authors [1–20] for:

1. The classification and identification of symmetry point groups.
2. The identification of the symmetry of the equilibrium nuclear configuration of molecules within the Cartesian coordinate system to which symmetry point groups are assigned and referred to as the molecular coordinate symmetry.

It is interesting to note, that the elements of symmetry point groups are not symmetry elements, but are instead, symmetry operations which define symmetry point groups by satisfying the four criteria of mathematical groups.

The multiplicity of a symmetry element represents the number of times it must be successively applied in order to generate all the symmetry operations of its own kind, which may or may not constitute all the symmetry operations of the point group [21]. Therefore, the successive applications of a symmetry element yields the corresponding symmetry operations.

Furthermore, the interchangeable use between symmetry elements and symmetry operations which are inextricably related has lead to some ambiguity. In view of the above, a distinction seems to be appropriate. This distinction can be further emphasized by incorporating symmetry elements in well defined sets, while the corresponding symmetry operations are the elements of the well-defined symmetry point groups.

SOME SET THEORETIC CONCEPTS

Cantor (1845–1918) established and developed the theory of sets, which through the generality of its concepts has deeply influenced mathematics. He defined a set as "a collection into a whole, of definite, well distinguished objects (called the elements of the set) of our perception or of our thought" [22].

A set or collection of elements, also contains smaller sets known as subsets. A set "s" is a subset of a set "S" denoted by $s \subseteq S$, if every element "a" of "s" is also an element of "S", i.e. $a \in s$, implies $a \in S$.

Having defined a set and its subset, the two fundamental set theoretic operations [23] which are applicable to sets are considered.

1. The operation of intersection of sets

The result of the intersection of "n" sets, is the set of those elements which are common to every set, according to

$$(a, b, c) \cap (a, b, d) \cap (a, b, e) \cap \ldots \ldots \cap (a, b, z) = S_1 \cap S_2 \cap S_3 \cap \ldots \ldots \cap S_n = (a, b) = S, \quad (1)$$

where

      (i) $a, b, c \in S_1$; $a, b, d \in S_2$; $a, b, e \in S_3$; $a, b, z \in S_n$;
      (ii) $a, b \in S_1, S_2, S_3, \ldots, S_n$; $a, b \in S$,
      (iii) the symbol $\cap$ denotes the operation of intersection.

2. The operation of the union of sets

The result of the union of "n" sets, is the set of all elements of all the sets, according to

$$(a, b, c) \cup (a, b, d) \cup (a, b, e) \cup \ldots \ldots \cup (a, b, z) =$$

$$S_1 \cup S_2 \cup S_3 \cup \ldots \ldots \cup S_b = (a, b, c, d, e, \ldots z) = S, \quad (2)$$

where

      (i) $a, b, c, d, e, \ldots z \in S$;
      (ii) the symbol $\cup$ denotes the operation of union.

The operations of intersection and union, obey the following laws of algebra:

Commutation.

$$S_1 \cap S_2 = S_2 \cap S_1$$

$$S_1 \cup S_2 = S_2 \cup S_1. \tag{3a}$$

Association.

$$(S_1 \cap S_2) \cap S_3 = S_1 \cap (S_2 \cap S_3)$$

$$(S_1 \cup S_2) \cup S_3 = S_1 \cup (S_2 \cup S_3). \tag{3b}$$

Inclusion.

$$S_1 \cap S_2 = S_1, \tag{3c}$$

if and only if S_1 is included in S_2

$$S_1 \cup S_2 = S_2$$

if and only if S_2 is included in S_1.
Nullification.

$$S_1 \cap S_2 = \varnothing \tag{3d}$$

if $S_2 = \varnothing$ the empty set

$$S_1 \cup S_2 = S_1.$$

Distribution.

$$S_1 \cap (S_2 \cup S_3) = (S_1 \cap S_2) \cup (S_1 \cap S_3)$$

$$S_1 \cup (S_2 \cap S_3) = (S_1 \cup S_2) \cap (S_1 \cup S_3). \tag{3e}$$

These concepts will be incorporated into the framework and in the construction of the set theoretic formalism used for the accounting of the molecular coordinate symmetry.

SYMMETRY POINT SETS AND GENERATOR SETS

The symmetry point set (PS) associated with a given symmetry point group (PG) was previously defined [24] as the minimum set of symmetry element(s) (axes, planes), derived from all or any one of its subsets which are the generator sets (S_g). The generator set, contains a single symmetry element of each kind of symmetry elements of the symmetry point set. The relationship between the symmetry point set and the constituent generator sets for the non-cubic symmetry point groups was previously discussed [24].

The symmetry point set PS(Td) and any one of the 36 generator sets S_g(Td), for the cubic tetrahedral symmetry point group PG(Td), to which the hexadecahedrane molecule belongs, are shown in Fig. 1. The importance of the generator sets [24] resides in their contents which consist of the uniqueness of the symmetry elements as well as their operational nature. The symmetry elements of the generator set S_g(Td) could be labelled as the primary generating element (axis of highest order, i.e. the diagonally situated three-fold axis of proper rotation) and the auxilliary symmetry elements (two-fold axis and dihedral plane), one of which could be further labelled as the secondary generating element (two-fold axis). The information stored within the generator sets can be used for the construction of the symmetry point set.

The construction of the symmetry point set PS(Td)

This set can be obtained from the following:

1. All the generator sets. Since the PS (minimum set) is the set of symmetry elements of all the generator sets [24], the set theoretic concept of the union of sets as applied to generator sets,

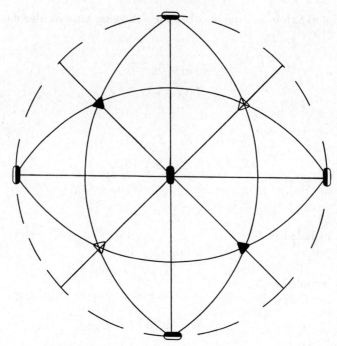

Fig. 1. The symmetry point set and generator set for the tetrahedral symmetry point group within the Cartesian coordinate system: $PS(Td) = (3C_2^1, 4C_3^1, 6\sigma_d)$; $S_g(Td) = (C_2^1, C_3^1, \sigma_d)$.

can be used according to

$$S_{g_1}(Td) \cup S_{g_2}(Td) \cup S_{g_3}(Td) \cup \ldots \cup S_{g_{36}}(Td) = PS(C_{3v})_1 \cup PS(C_{3v})_2 \cup$$

$$PS(C_{3v})_3 \cup PS(C_{3v})_4 \cup PS(D_2) = (3C_2^1, 4C_3^1, 6\sigma_d) = PS(Td), \quad (4)$$

where

(i) $S_{g_{1-36}}(Td)$ represent each of the 36 generator sets with their constituent symmetry elements.

(ii) $PS(C_{3v})_{1-4}$, $PS(D_2)$ are the sets of symmetry elements resulting from the decomposition of the generator sets into subsets, each containing a single symmetry element and their reorganization using the law of association. The symmetry elements of $PS(C_{3v})$ are the three-fold axis with three intersecting planes and are located throughout quadrants I–IV. The symmetry elements of $PS(D_2)$ are the two-fold axes which coincide with the Cartesian coordinate axes (Fig. 1).

(iii) $PS(Td)$ represents the comprehensive set of all the symmetry elements resulting from the union of all the subsets, which are the generator sets or the reorganized sets which are the symmetry point subsets.

2. *Any generator set.* In this instance, any one of the 36 generator sets can be used and expanded into the symmetry point set, through the sequential operational effect of the constituent symmetry elements upon each other. The operational effect of an *n*-fold axis of proper rotation which generates "*n*" *n*-fold operations of rotation, while still preserving its identity as a symmetry element (*n*-fold axis) is defined by

$$\overrightarrow{C_n^l} = C_n^1(C_n^1)^n = C_n^1(n\ e^{i\theta}) = C_n^1(n\ e^{i(360°/n)}), \quad (5)$$

where

(i) the symbol $\overset{\bullet\rightarrow}{}$ above the *n*-fold axis indicates its operational nature.

(ii) C_n^1 represents the *n*-fold axis after generating the "*n*" *n*-fold operations of rotation.

(iii) $(C_n^1)^n$ represents the "n" n-fold operations of rotation.

(iv) $n\,e^{i\theta}$ also represents the "n" successive operations of rotation as dictated by the multiplicity "n" of the n-fold axis.

(v) θ represents the angle of a single rotation and is dependent on the multiplicity "n" of the n-fold axis, according to $\theta = 360°/n$.

The operational effect of an n-fold axis of proper rotation is used in the expansion of the generator set into the larger symmetry point set through the generation of all the symmetry elements. This expansion, can be undertaken in two consecutive sequences of steps.

(a) The operational effect of the primary generating element which reproduces the auxilliary symmetry elements in a number of distinguishable positions, dictated by the multiplicity of the former element, according to

$$(\overset{\longleftrightarrow}{C_{2z}^1}, C_{3di_1}^1, \sigma_d) = ((3e^{i(120°)})C_{2z}^1, C_{3di_1}^1, (3\,e^{i(120°)})\sigma_d) = (C_{2(x,y,z)}^1, C_{3di_1}^1, 3\sigma_d) = (PS(D_2), PS(C_{3v})_1) \quad (6)$$

where

(i) $PS(C_{3v})_1$ represents the symmetry point set located in quadrant I, containing a three-fold axis and three intersecting planes.

(ii) $PS(D_2)$ represents the symmetry point set containing the three two-fold axes.

(b) The operational effect of the secondary generating symmetry elements, members of the symmetry point set $PS(D_2)$, which reproduce the symmetry elements of the symmetry point set $PS(C_{3v})_1$ in a number of distinguishable positions in quadrants III, II and IV, respectively and according to

$$(\overset{\longleftrightarrow}{PS(D_2)}, PS(C_{3v})_1) = (\overset{\longleftrightarrow}{C_{2(x,y)}^1}, \overset{\longleftrightarrow}{C_{2z}^1}, PS(C_{3v})_1))$$

$$= (\overset{\longleftrightarrow}{C_{2(x,y)}^1}, C_{2z}^1, ((2e^{i(180°)})PS(C_{3v})_1))$$

$$= (\overset{\longleftrightarrow}{C_{2(x,y)}^1}, C_{2z}^1, PS(C_{3v})_3 \cdot PS(C_{3v})_1)$$

$$= (C_{2z}^1, PS(C_{3v})_3, \overset{\longleftrightarrow}{C_{2x}^1}, (\overset{\longleftrightarrow}{C_{2y}^1}, PS(C_{3v})_1))$$

$$= (C_{2z}^1, PS(C_{3v})_3, \overset{\longleftrightarrow}{C_{2x}^1}, C_{2y}^1, PS(C_{3v})_2, PS(C_{3v})_1)$$

$$= (C_{2z}^1, PS(C_{3v})_3, C_{2y}^1, PS(C_{3v})_2, (\overset{\longleftrightarrow}{C_{2x}^1}, PS(C_{3v})_1))$$

$$= (C_{2z}^1, PS(C_{3v})_3, C_{2y}^1, PS(C_{3v})_2, C_{2x}^1, PS(C_{3v})_4, PS(C_{3v})_1)$$

$$= (C_{2(z,x,y)}^1, PS(C_{3v})_{1-4})$$

$$= (PS(D_2), PS(C_{3v})_{1-4})$$

$$= (C_{2(z,x,y)}^1, C_{3di_{1-4}}^1, 6\sigma_d)$$

$$= (3C_2^1, 4C_13, 6\sigma_d)$$

$$= PS(Td). \quad (7)$$

Equation (6) represents the expansion of the generator set into the set containing two subsets of the symmetry point set. Equation (7) represents the further expansion of the set of the above subsets into the symmetry point set $PS(Td)$ which is the set of the five subsets $PS(D_2)$ and $PS(C_{3v})_{1-4}$.

The symmetry point sets $PS(D_2)$ and $PS(C_{3v})$ as well as any one of the generator sets $S_g(D_2)$ and $S_g(C_{3v})$ for the symmetry point groups $PG(D_2)$ and $PG(C_{3v})$ are shown in Figs 2 and 3.

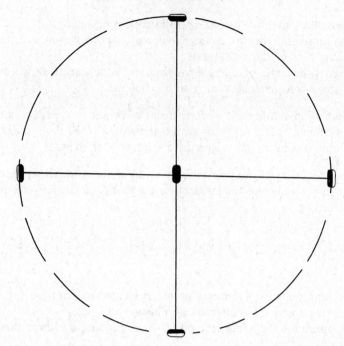

Fig. 2. The symmetry point set and generator set for the D_2 symmetry point group within the Cartesian coordinate system: $PS(D_2) = (C_{2z}^1, C_{2x}^1, C_{2y}^1)$; $S_g(D_2) = (C_{2z}^1, C_{2x}^1 \text{ or } C_{2y}^1)$.

The construction of the symmetry point set $PS(C_{3v})$

This set can similarly be constructed from the following:

1. All the generator sets. The union of all three generator sets yields the symmetry point set $PS(C_{3v})$ according to

$$S_{g_1}(C_{3v}) \cup S_{g_2}(C_{3v}) \cup S_{g_2}(C_{3v}) = (C_3^1, \sigma_{v_1}) \cup (C_3^1, \sigma_{v_2}) \cup (C_3^1, \sigma_{v_3}) = (C_3^1, 3\sigma_v) = PS(C_{3v}). \qquad (8)$$

2. Any generator set. In this case, any one of the three generator sets can be expanded into the symmetry point set through the operational effect of the generating element (three-fold axis of

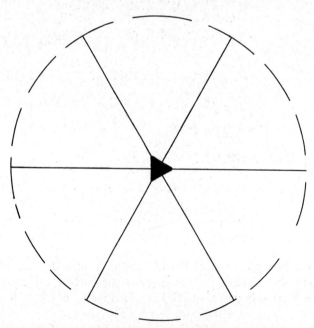

Fig. 3. The symmetry point set and generator set for the C_{3v} symmetry point group within the Cartesian coordinate system: $PS(C_{3v}) = (C_3^1, 3\sigma_v)$; $S_g(C_{3v}) = (C_3^1, \sigma_v)$.

proper rotation) upon the auxilliary element (vertical plane) according to

$$(\overrightarrow{C_3^1}, \sigma_{v_1}) = (C_3^1, (3\,e^{i(120^0)}\sigma_{v_1}) = (C_3^1, \sigma_{v_2}, \sigma_{v_3}, \sigma_{v_1}) = (C_3^1, 3\sigma_v) = PS(C_{3v}). \qquad (9)$$

Equations (8) and (9) represent the construction of the symmetry point set $PS(C_{3v})$ a subset of the symmetry point set $PS(Td)$ from all or any one of the generator sets $S_g(C_{3v})$. The constructions of the symmetry point sets $PS(Td)$ and $PS(C_{3v})$ from all or any one of the respective generator sets $S_g(Td)$ and $S_g(C_{3v})$ form the basis for the construction of the set theoretic formalisms relating the local symmetry point sets to the molecular symmetry point set for the accounting of the molecular coordinate symmetry.

MOLECULAR AND LOCAL SYMMETRY POINT SETS

The symmetry point set of a molecule is defined as the molecular symmetry point set (MPS), while the symmetry point set of a unit (atom, group of atoms) or of a combination of units is defined as the local symmetry point set (LPS).

The orthogonal projection of the molecular skeletal system of hexadecahedrane within the tetrahedral symmetry point set $PS(Td)$ (Fig. 4) shows the MPS as well as the LPS of the constituent units.

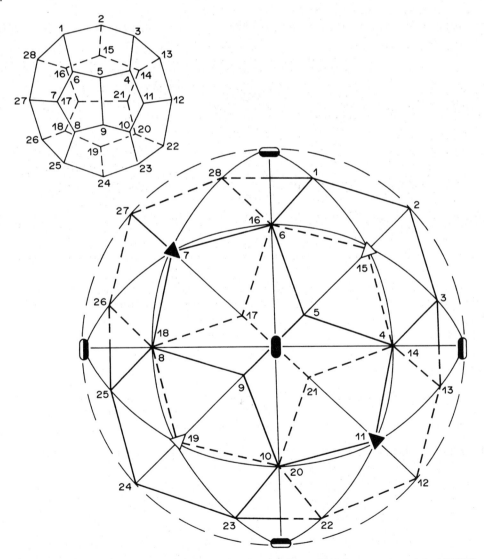

Fig. 4. The orthogonal projection of the hexadecahedrane skeleton within the symmetry point set $PS(Td)$.
(Solid and dashed lines represent the C–C bonds above and below the horizontal xy-plane.)

INTERSECTION OF LPS (INTERSECTION SETS)

The fact that a unit in a molecule has a given symmetry described and defined by the LPS does not necessarily imply that the symmetry of the unit is in its entirety or in parts, part of the molecular symmetry described and defined by the MPS. In other words, the local symmetry point set or its subset(s) may or may not be subsets of the molecular symmetry point set. This depends upon whether or not, the symmetry element(s) of the LPS pass through the center of mass of the molecule.

On the other hand, the intersection (intersection sets, i.e. sets of shared symmetry element(s)) of the local symmetry point sets of diametrically or diagonally situated units in a molecule are always subsets [24, 25] of the molecular symmetry point set since the shared symmetry element(s) pass through the center of mass of the molecule.

INTERCHANGE SETS

The symmetry element of any interchange set passes through the center of mass of the molecule, unit or combination of units irrespective of the presence or abscence of an atom at the center of mass. The interchange set is a subset of the MPS.

The operational effect of the symmetry element of the interchange set is responsible for the equivalent interchanges of:

1. The partially or totally shared symmetry element(s) of the intersection sets of the LPS of diametrically or diagonally situated units or combinations of units.
2. The units or combinations of units contributing to the intersection sets via the corresponding local symmetry point sets.

SYMMETRY POINT SETS AND SYMMETRY POINT GROUPS

Symmetry operations which constitute and define symmetry point groups satisfy the four criteria of mathematical groups:

1. Closure. The combination of two symmetry operations yields a unique result, which is another symmetry operation.

2. Associativity. The result of the combination of two symmetry operations with a third operation is the same as the combination of a symmetry operation with the result of the combination of the remaining two symmetry operations.

3. Reciprocality. Every symmetry operation has a reciprocal or inverse operation.

4. Identity. The combination of every symmetry operation with its inverse yields a unique result which is the identity operation.

The symmetry point group or collection of symmetry operations, also contains smaller groups known as subgroups. The symmetry operations of subgroups also satisfy the above criteria. Symmetry operations, elements of the symmetry point group and its subgroups, are:

1. Obtained through the applications and combinations of applications of the symmetry elements (axes, planes) of the symmetry point set or its subsets.
2. Represented by the defining relations [16], whereby each term is taken individually or in combinations with other terms. These relations are constructed from the symmetry elements(s) of any generator set, a subset of the symmetry point set.

The symmetry elements of the LPS and MPS, yield the symmetry operations of the corresponding LPG and MPG. The symmetry elements of the intersection and interchange sets yield the symmetry operations of the corresponding groups which are subgroups of the MPG.

APPLICATIONS OF SYMMETRY POINT SETS

The determination of site, intersite and site–intersite symmetry point groups

The use of symmetry point sets, i.e. LPS and the corresponding intersection sets, provide a method for the systematic determination not only of site (site–site), but also of intersite (intersite–intersite) and site–intersite symmetry point subgroups of the MPG. The concept of site symmetry point groups for a few selected molecular systems consisting of non-cyclic molecules with a central atom and a cyclic planar molecule without a central atom, as well as the uses of site symmetry point groups have been previously discussed [26–29]. However, the origin of these groups had not been mathematically formalized. Since the site symmetry point group (G_{Site}) must be a subgroup [26–29] of the MPG, the LPS which may or may not be a subgroup cannot be used as the site symmetry point group in spite of the fact that it represents the symmetry operations corresponding to the set of symmetry elements (LPS) of a unit in a molecule at a given site. Hence, a distinction must be made between the local and the site symmetry point groups.

On the other hand, the group of symmetry operations corresponding to the symmetry element(s) of the intersection set (a subset of the MPS) is always a subgroup of the MPG and represents in addition to site (site-site), intersite (intersite-intersite) and site–intersite symmetry point subgroups. This was clearly seen for the planar and the non-planar conformations of the lower cycloalkanes [25]. The LPG of the constituent methylene units is an invariant of the molecular symmetry point group, while the site, intersite and site–intersite symmetry point groups are subgroups which vary with the MPG of the different conformations for a given cycloalkane. In some instances, the MPG is instead a subgroup of the LPG as exemplified by cyclohexane and cyclopentane in the half-boat and envelope, twist conformations.

In view of the preceding, the definition of the site symmetry point group of a site in a molecule, expressed as "the group defined by the symmetry elements which pass through that site" [28, 29], must be expanded and formalized to include the following:

1. Site (site–site) symmetry point group. The group of symmetry operations corresponding to the symmetry element(s) of the intersection set of the LPS of adjacently, diametrically or diagonally situated units.

2. Intersite (intersite–intersite) symmetry point group. The group of symmetry operations corresponding to the symmetry element(s) of the intersection set of the LPS of diametrically or diagonally situated inter-units.

3. Site–intersite symmetry point group. The group of symmetry operations corresponding to the symmetry element(s) of the intersection set of the LPS of diametrically or diagonally situated units and inter-units.

The construction of the MPS

The LPS and MPS as well as the corresponding intersection and interchange sets are important. This importance resides in the relation of the symmetry elements of units and combinations of units incorporated in sets (LPS) to the symmetry elements of the molecular structure also incorporated in a set (MPS) via the intersection and interchange sets through a set theoretic formalism. The contributions of the LPS to the MPS in turn determine the contributions of the LPG to the MPG.

CONSTRUCTION OF THE MPS OF HEXADECAHEDRANE

The hexadecahedrane molecule is constituted of a fused system of 12 cyclopentylene ($\geqslant$CH—)$_5$ and four cyclohexylene ($\geqslant$CH—)$_6$ units, each unit in turn is constituted of methine ($\geqslant$CH) and dimethine ($\geqslant$CH—CH$\leqslant$) units. The symmetry contributions of all the methine and dimethine units are determined through the symmetry contributions of larger units (constituted of methine and dimethine units) which completely form the molecular structure.

In order to determine these contributions, use is made of the information gathered in the construction of the symmetry point set PS(Td) and its subset PS(C_{3v}) from all or any one of the respective generator sets S_g(Td) and S_g(C_{3v}).

1. The expansion of the generator set into the symmetry point set [equations (6), (7) and (9)] forms the basis for the selection of unit and combinations of unit structures which contribute to the molecular coordinate symmetry and structure through equivalent interchanges.

2. The construction of the symmetry point set from the union of all the generator sets [equations (4) and (8)] forms the basis for the construction of the set theoretic formalism used for the accounting of the molecular coordinate symmetry in terms of the symmetry contributions of the selected unit and combination of unit structures.

Therefore, the hexadecahedrane molecule can be visualized in terms of the four identical (A/B) unit pairs of diagonally situated A = triquinacylene HC(CH)$_3$(HCCH)$_3$ and B = heptacyclic peristylylene (HC(CH)$_3$)$_3$(HCCH)$_3$ units. Each triquinacylene unit is constituted of a fused system of three cyclopentylene units. Each heptacyclic peristylylene unit is constituted of a fused system of a central cyclohexylene unit surrounded by six cyclopentylene units. The hexadecahedrane molecule and the four identical (A = triquinacylene/B = peristylylene) unit pairs are shown in Fig. 5.

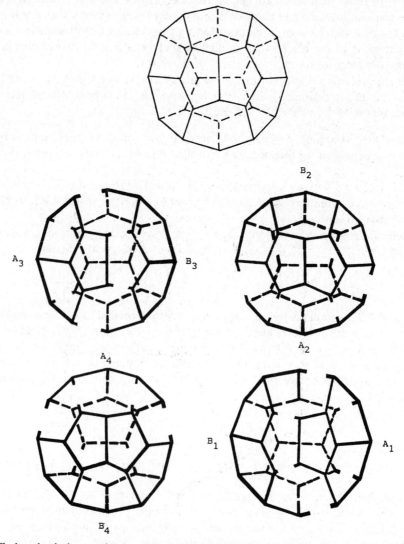

Fig. 5. The hexadecahedrane molecule and the four identical (A = triquinacylene/B = peristylylene)$_{1-4}$ unit pairs.

The orthogonal projection of the skeletal systems of the triquinacylene and peristylylene units within the symmetry point set PS(C_{3v}) to which these units belong are shown in Figs 6 and 7, respectively.

Each triquinacylene unit can also be visualized in terms of a central trimethine–methine unite and three peripheral dimethine units. Each peristylylene unit can be similarly visualized in terms of three diametrically situated trimethine–methine and dimethine units.

The orthogonal projection of the skeletal system of the trimethine–methine unit within the symmetry point set PS(C_{3v}) to which this unit belongs is shown in Fig. 8.

In view of the preceding, the MPS of hexadecahedrane will be expressed in terms of the contributions of:

    (i) The LPS of the triquinacylene and peristylylene units;
    (ii) The LPS of the trimethine–methine and dimethine units;
    (iii) The LPS of the methine and dimethine units.

The groups of symmetry operations corresponding to the MPS and LPS are the MPG and LPG of hexadecahedrane and the triquinacylene, peristylylene, trimethine–methine, dimethine, methine units shown in Table 1.

Construction of the MPS from the LPS of the triquinacylene and peristylylene units

Since the hexadecahedrane molecule can be inscribed in a cube, the sets of symmetry elements common to diagonally opposite triquinacylene/peristylylene unit pairs and the equivalent interchanges of the symmetry elements of these sets as well as of the contributing unit pairs are

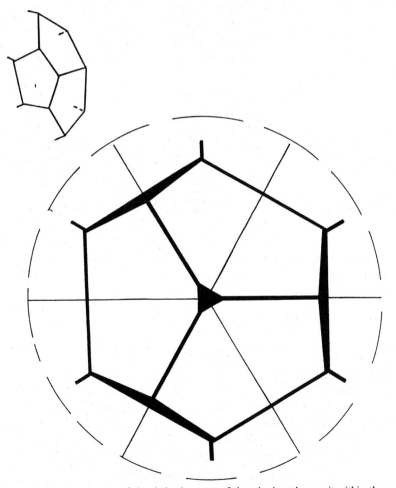

Fig. 6. The orthogonal projection of the skeletal system of the triquinacylene unit within the symmetry point set PS(C_{3v}).

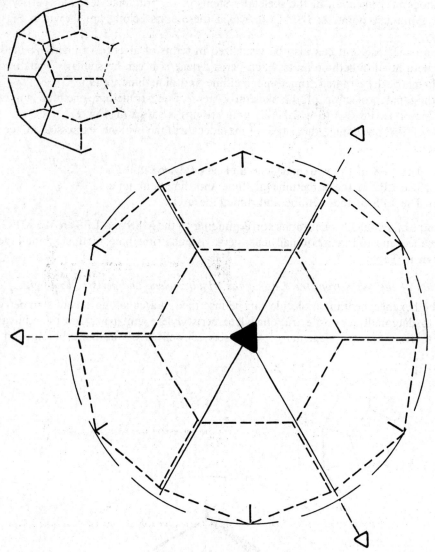

Fig. 7. The orthogonal projection of the skeletal system of the peristylylene unit within the symmetry point
set PS(C_{3v}).

considered in the construction of the MPS. There are four identical (A/B) unit pairs and four
resultant intersection sets (sets of shared symmetry elements) between the local symmetry point sets
according to:

$$(\text{LPS})_{A_{1-4}} \cap (\text{LPS})_{B_{1-4}} = (\text{LPS})_{(A\text{-}B)_{1-4}} = S_{\text{intersec.}} \tag{10}$$

Since the shared symmetry elements of the intersection sets pass through the center of mass of the
molecule, the intersection sets are subsets of the MPS. The corresponding groups of symmetry
operations constitute the molecular site–intersite symmetry point groups ($G_{\text{molecular site–intersite}}$) and are
subgroups of the MPG.

The symmetry elements of each pair of intersection sets as well as the contributing (A/B) unit
pairs are equivalently interchanged into the symmetry elements of the remaining pair of intersection
sets and their contributing (A/B) unit pairs through the operational effect of each of the three
two-fold axes of proper rotation which are the symmetry elements of the corresponding interchange
sets

$$(\text{LPS})_{(A\text{-}B)_{1,2:1,2:1,4}} S_{\text{interch.}} \overrightarrow{(C_2)}_{z:x:y} = (\text{LPS})_{(A\text{-}B)_{3,4:4,3:2,3}}, (\text{LPS})_{(A\text{-}B)_{1,2:1,2:1,4}}. \tag{11}$$

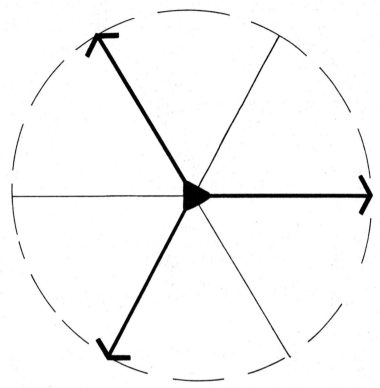

Fig. 8. The orthogonal projection of the skeletal system of the trimethine–methine unit within the symmetry point set $PS(C_{3v})$.

The interchange sets are subsets of the MPS as the symmetry element of each set passes through the center of mass of the molecule. The corresponding groups of symmetry operations represent

Table 1. Local and molecular symmetry point groups

Molecule	LPG$_{units}$		MPG
	Triquinacylene	Peristylylene	
	A_{1-4}	B_{1-4}	
Hexadecahedrane	C_{3v}	C_{3v}	
($C_{28}H_{28}$) or	C_{3v}	C_{3v}	
(HC(CH)$_3$)$_3$(HCCH)$_3$	C_{3v}	C_{3v}	Td
(HC(CH)$_3$)$_3$(HCCH)$_3$	C_{3v}	C_{3v}	

Units	LPG$_{units}$		LPG$_A$	LPG$_B$
Triquinacylene	Trimethine–methine	Dimethine		
($C_{10}H_{10}$) or		C_{2v}		
(HC(CH)$_3$)(HCCH)$_3$	C_{3v}	C_{2v}	C_{3v}	
		C_{2v}		
Peristylylene				
($C_{18}H_{18}$) or	C_{3v}	C_{2v}		
	C_{3v}	C_{2v}		C_{3v}
(HC(CH)$_3$)$_3$(HCCH)$_3$	C_{3v}	C_{2v}		

Units	LPG$_{units}$		LPG$_{(C_4H_4)}$
Trimethine–methine	Methine	Methine	
(C_4H_4) or	(central)	(adjacent)	
		C_{3v}	
(HC(CH))	C_{3v}	C_{3v}	C_{3v}
		C_{3v}	

Table 2. Local, molecular, intersection and interchange symmetry point sets

$LPS_{A_{1-4}}$	$LPS_{B_{1-4}}$	Intersection $(S_{\text{intersec.}})$	Interchange $(S_{\text{interch.}})$		MPS
$S(C_{3v})$	$S(C_{3v})$	$S(C_{3v})$	$S(C_2)_z$		$S(\text{Td}) = (3C_2^1, 4C_3^1, 6\sigma_d)$
$S(C_{3v})$	$S(C_{3v})$	$S(C_{3v})$	$S(C_2)_x$	$S(D_2)$	
$S(C_{3v})$	$S(C_{3v})$	$S(C_{3v})$	$S(C_2)_y$		
$S(C_{3v})$	$S(C_{3v})$	$S(C_{3v})$			

	LPS				LPS_A	LPS_B
Trimethine–methine	Dimethine					
	$S(C_{2v})$	$S(C_s)$			$S(C_{3v})_A = (C_3^1, 3\sigma_v)$	
$S(C_{3v})$	$S(C_{2v})$	$S(C_s)$	$S(C_3)$			
	$S(C_{2v})$	$S(C_s)$				
$S(C_{3v})$	$S(C_{2v})$	$S(C_s)$				
$S(C_{3v})$	$S(C_{2v})$	$S(C_s)$	$S(C_3)$			$S(C_{3v})_B = (C_3^1, 3\sigma_v)$
$S(C_{3v})$	$S(C_{2v})$	$S(C_s)$				

	LPS			$LPS_{(C_4H_4)}$	
Methine (central)	Methine (adjacent)				
	$S(C_{3v})$	$S(C_s)$			
$S(C_{3v})$	$S(C_{3v})$	$S(C_s)$	$S(C_3)$	$S(C_{3v}) = (C_3^1, 3\sigma_v)$	
	$S(C_{3v})$	$S(C_s)$			

the molecular site–intersite interchange point groups ($G_{\text{molec. site–intersite interch.}}$) which are subgroups of the MPG. The MPS is obtained by taking the union of all its subsets which are the intersection and corresponding interchange sets according to:

$$\text{MPS} = \left\{ \prod_{n=1}^{4} [(\text{LPS})_{A_n} \cap (\text{LPS})_{B_n}] \cup_n \right\} S_{\text{interch.}}(C_2)_z \cup S_{\text{interch.}}(C_2)_x \cup S_{\text{interch.}}(C_2)_y \tag{12}$$

and since

$$S_{\text{interch.}}(C_2)_z \cup S_{\text{interch.}}(C_2)_x \cup S_{\text{interch.}}(C_2)_y = (C_{2z}^1) \cup (C_{2x}^1) \cup (C_{2y}^1) = (C_{2z}^1, C_{2z}^1, C_{2y}^1) = S_{\text{interch.}}(D_2) \tag{13}$$

equation (7) reduces to

$$\text{MPS} = \left\{ \prod_{n=1}^{4} [(\text{LPS})_{A_n} \cap (\text{LPS})_{B_n}] \cup_n \right\} S_{\text{interch.}}(D_2) \tag{14a}$$

or

$$\text{MPS} = \left\{ \prod_{n=1}^{4} (\text{LPS})_{(A\text{-}B)_n} \cup_n \right\} S_{\text{interch.}}(D_2). \tag{14b}$$

Equations (14a, b) express the contributions of the LPS of the (A = triquinacylene/ B = peristylylene) unit pairs to the MPS of hexadecahedrane. The LPS and MPS as well as the intersection and interchange symmetry point sets are shown in Table 2. The resultant molecular site–intersite, site–intersite interchange as well as the MPG are shown in Table 3.

Construction of the MPS from the LPS of the trimethine–methine and dimethine units

In order to express the MPS in terms of the LPS of the trimethine–methine and dimethine units, the LPS of the triquinacylene and peristylylene units must be formalized.

Table 3. Local site, local site interchange, local site–intersite, local site–intersite interchange and local symmetry point groups. Molecular site–intersite, molecular site–intersite interchange and molecular symmetry point groups

Local site point group	Local site interchange point group	Local symmetry point group
$G_{\text{local site}}$	$G_{\text{local site interch.}}$	$LPG_{\text{trimethine–methine}}$
C_s	C_3	C_{3v}

Local site–intersite point group	Local site–intersite interchange point group	Local symmetry point groups
$G_{\text{local site–intersite}}$	$G_{\text{local site–intersite interch.}}$	LPG_A　　　LPG_B
C_s	C_3	C_{3v}　　　C_{3v}

Molecular site–intersite point group	Molecular site–intersite interchange point group	Molecular symmetry point group
$G_{\text{molec. site–intersite}}$	$G_{\text{molec. site–intersite interch.}}$	MPG
C_{3v}	D_2	Td

1. Construction of the LPS of the triquinacylene units from the LPS of the trimethine–methine and dimethine units. For each of the $A =$ triquinacylene units, there are three intersection sets between the local symmetry point set of the central trimethine–methine unit and the LPS of each of the three peripheral dimethine units according to:

$$(\text{LPS})_{(\text{trimethine–methine})_p} \cap (\text{LPS})_{(\text{dimethine})q_{1,2,3}} = (\text{LPS})_{(p-(q_1, q_2, q_3))} = S_{\text{intersec.}} \tag{15}$$

Each of the intersection sets is a subset of the LPS of the triquinacylene unit. The corresponding groups of symmetry operations constitute the local site–intersite symmetry point groups ($G_{\text{local site–intersite}}$) which are subgroups of the LPG of the triquinacylene unit.

The symmetry element of each of the three intersection sets as well as the contributing trimethine–methine, dimethine units are equivalently interchanged through the operational effect of each of the four three-fold axes of proper rotation, each an element of the corresponding interchange set according to:

$$(\text{LPS})_{(p-(q_1))_{1-4}} S_{\text{interch.}} (\overrightarrow{C_3})^{1-4} = (\text{LPS})_{(p-(q_2))_{1-4}}, (\text{LPS})_{(p-(q_3))_{1-4}}, (\text{LPS})_{(p-(q_1))_{1-4}} \tag{16}$$

where

(i) the subscripts 1–4 assigned to each of the three intersection sets refer to each of the four triquinacylene units.

(ii) the superscripts 1–4 assigned to each of the interchange sets, refer to each of the four three-fold axes of proper rotation located through quadrants I–IV.

Each interchange set is a subset of the LPS of each of the triquinacylene units. The corresponding groups of symmetry operations represent the local site–intersite interchange point groups ($G_{\text{local site–intersite interch.}}$) and are subgroups of the LPS of the triquinacylene units. The LPS for each of the triquinacylene units is obtained by taking the union of all its subsets which are the intersection sets and the corresponding interchange set

$$(\text{LPS})_{(A = \text{triquinacylene})_{1-4}} = \left\{ \prod_{g=1}^{3} \left[(\text{LPS})_{(\text{HC}\ll(\text{CH})_3)_p} \cap (\text{LPS})_{(\gg\text{CH}-\text{HC}\ll)q_g} \right]_{1-4} \cup_g \right\} S_{\text{interch.}} (C_3)^{1-4}, \tag{17a}$$

which reduces to

$$(\text{LPS})_{A_n} = \left\{ \prod_{g=1}^{3} [(\text{LPS})_{(p-q_g)}]_n \cup_g \right\} S_{\text{interch.}} (C_3)^n. \tag{17b}$$

Equations (17a, b) express the contributions of the LPS of the trimethine–methine and dimethine units to the local symmetry point sets of the triquinacylene units. The corresponding groups of symmetry operations constitute the LPG of the triquinacylene units. The local symmetry point sets of the trimethine–methine, dimethine and triquinacylene units as well as the intersection and interchange sets are shown in Table 2.

The local site–intersite, local site–intersite interchange and LPG of the triquinacylene units are shown in Table 3.

2. Construction of the LPS of the peristylylene units from the LPS of the trimethine–methine and dimethine units. For each of the $B =$ peristylylene units, there are three intersection sets between the local symmetry point sets of diametrically opposite trimethine–methine and dimethine units

$$(\text{LPS})_{(\text{trimethine–methine})s_{1-3}} \cap (\text{LPS})_{(\text{dimethine})t_{1-3}} = (\text{LPS})_{(s-t)_{1-3}} = S_{\text{intersec.}} \tag{18}$$

Each of the three intersection sets is a subset of the LPS of the peristylylene unit. The corresponding groups of symmetry operations constitute the local site–intersite symmetry point groups ($G_{\text{local site–intersite}}$) which are subgroups of the LPG of the peristylylene unit.

The symmetry element of each of the three intersection sets as well as the contributing trimethine–methine, dimethine units are equivalently interchanged through the operational effect of each of the four three-fold axes of proper rotation, each an element of the corresponding interchange set according to:

$$(\text{LPS})_{((s-t_1))_{1-4}} S_{\text{interch.}} (\overrightarrow{C_3})^{1-4} = (\text{LPS})_{(s-t_2)_{1-4}}, (\text{LPS})_{((s-t_3)_{1-4}}, (\text{LPS})_{((s-t_1)_{1-4}}, \tag{19}$$

where

(i) the subscripts 1–4 assigned to each of the three intersection sets refer to each of the four peristylylene units;

(ii) the superscripts 1–4 assigned to each of the interchange sets, refer to each of the four three-fold axes of proper rotation located through quadrants I–IV.

Each interchange set is a subset of the LPS of each of the peristylylene units. The corresponding groups of symmetry operations represent the local site–intersite interchange point groups ($G_{\text{local site–intersite interch.}}$) and are subgroups of the LPG of the peristylylene units. THe LPS for each of the peristylylene units is obtained by taking the union of all its subsets which are the intersection sets and the corresponding interchange set

$$(\text{LPS})_{(B=\text{peristylylene})_{1-4}} = \left\{ \prod_{g=1}^{3} \left[(\text{LPS})_{(\text{HC}\mathrel{<}(\text{CH})_3)s_g} \cap (\text{LPS})_{(\mathrel{>}\text{CH—HC}\mathrel{<})t_g} \right]_{1-4} \cup_g \right\} S_{\text{interch.}}(C_3)^{1-4}, \quad (20a)$$

which reduces to

$$(\text{LPS})_{B_n} = \left\{ \prod_{g=1}^{3} [(\text{LPS})_{(s-t)_g}]_n \cup_g \right\} S_{\text{interch.}}(C_3)^n. \quad (20b)$$

Equations (20a, b) express the contributions of the LPS of the trimethine–methine and dimethine units to the LPS of the peristylylene units. The corresponding groups of symmetry operations constitute the LPG of the peristylylene units. The LPS of the trimethine–methine, dimethine and peristylylene units as well as the intersection and interchange sets are shown in Table 2.

The local site–intersite, local site–intersite interchange and LPG of the peristylylene units are shown in Table 3.

Having constructed the LPS of the triquinacylene and peristylylene units from the LPS of the trimethine–methine and dimethine units, the MPS can now be expressed in terms of the LPS of the latter units.

Substituting the results expressed by equations (17a) and (20a) in equation (14a) we obtain,

$$\text{MPS} = \left\{ \prod_{n=1}^{4} \left[\left(\left\{ \prod_{g=1}^{3} \left[(\text{LPS})_{(\text{HC}\mathrel{<}(\text{CH})_3)p} \cap (\text{LPS})_{(\mathrel{>}\text{CH—HC}\mathrel{<})q_g} \right] \cup_g \right\} S_{\text{interch.}}(C_3)^n \right) \right. \right.$$

$$\left. \left. \cap \left(\left\{ \prod_{g=1}^{3} \left[(\text{LPS})_{(\text{HC}\mathrel{<}(\text{CH})_3)s_g} \cap (\text{LPS})_{(\mathrel{>}\text{CH—HC}\mathrel{<})t_g} \right]_n \cup_g \right\} S_{\text{interch.}}(C_3)^n \right) \right]_n \right\} S_{\text{interch.}}(D_2), \quad (21a)$$

which reduces to

$$\text{MPS} = \left\{ \prod_{n=1}^{4} \left[\left(\left\{ \prod_{g=1}^{3} [(\text{LPS})_{(p-q_g)}]_n \cup_g \right\} S_{\text{interch.}}(C_3)^n \right) \right. \right.$$

$$\left. \left. \cap \left(\left\{ \prod_{g=1}^{3} [(\text{LPS})_{(s-t)_g}]_n \cup_g \right\} S_{\text{interch.}}(C_3)^n \right) \right] \cup_n \right\} S_{\text{interch.}}(D_2) \quad (21b)$$

and since $(R \cup T) \cap (P \cup T) = (R \cap P) \cup T$ equation (21b) reduces to

$$\text{MPS} = \left[\prod_{n=1}^{4} \left[\left\{ \prod_{g=1}^{3} \left[(\text{LPS})_{(p-q_g)} \cap (\text{LPS})_{(s-t)g} \right]_n \cup_g \right\} S_{\text{interch.}}(C_3)^n \right] \cup_n \right] S_{\text{interch.}}(D_2). \quad (21c)$$

Equations (21a–c) express the contributions of the local symmetry point sets of the trimethine–methine and dimethine units to the molecular symmetry point set via the local symmetry point sets of the triquinacylene and peristylylene units.

Construction of the MPS from the LPS of the methine and dimethine units

In order to express the MPS in terms of the LPS of the methine and dimethine units, the LPS of the trimethine–methine unit must be formalized.

Construction of the LPS of the trimethine–methine unit from the LPS of the methine units

For each trimethine–methine unit, there are three intersection sets between the LPS of the central methine and each of the three adjacent methine units according to:

$$(\text{LPS})_{(\ni\text{CH}) \text{ central}} \cap (\text{LPS})_{(\ni\text{CH}) \text{ adjacent}_{1,2,3}} = (\text{LPS})_{(c-(a_1,a_2,a_3))} = S_{\text{intersec.}} \cdot \tag{22}$$

Each of three intersection sets is a subset of the LPS of the trimethine–methine unit. The corresponding groups of symmetry operations constitute the local site symmetry point groups ($G_{\text{local site}}$) which are subgroups of the LPG of the trimethine–methine unit.

The symmetry element of each of the three intersection sets as well as the contributing methine/methine unit pairs are equivalently interchanged through the operational effect of the three-fold axis of proper rotation an element of the interchange set,

$$(\text{LPS})_{(c-a_1)} S_{\text{interch.}}(\overset{\leftrightarrow}{C_3}) = (\text{LPS})_{(c-a_2)}, (\text{LPS})_{(c-a_3)}, (\text{LPS})_{(c-a_1)}. \tag{23}$$

The interchange set is a subset of the LPS of the trimethine–methine unit. The corresponding group of symmetry operations represent the local site interchange point group ($G_{\text{local site}}$) which is a subgroup of the LPG of the trimethine–methine unit.

As in the previous cases, the LPS of the trimethine–methine unit is obtained by taking the union of all its subsets which are the intersection sets and interchange set,

$$(\text{LPS})_{(\text{HC}\leqslant(\text{CH})_3)} = \left\{ \prod_{h=1}^{3} \left[(\text{LPS})_{(\ni\text{CH})c} \cap (\text{LPS})_{(\ni\text{CH})a_h} \right] \cup_h \right\} S_{\text{interch.}}(C_3), \tag{24a}$$

$$(\text{LPS})_{(\text{HC}\leqslant(\text{CH})_3)} = \left\{ \prod_{h=1}^{3} (\text{LPS})_{(c-a_h)} \cup_h \right\} S_{\text{interch.}}(C_3). \tag{24b}$$

Equations (24a, b) express the contributions of the LPS of the methine unit pairs to the local symmetry point set of the trimethine–methine unit. The corresponding group of symmetry operations constitute the LPG of the trimethine–methine unit.

The LPS of the methine, trimethine–methine units as well as the intersection and interchange sets are shown in Table 2.

The local site, local site interchange and LPG of the trimethine–methine unit are shown in Table 3.

Having constructed the LPS of the trimethine–methine unit from the LPS the methine units, the MPS can be expressed in terms of the LPS of the methine and dimethine units.

Substituting in equation (21a) for the LPS of the trimethine–methine units using the result expressed by equation (24a) we obtain

$$\text{MPS} = \left\{ \prod_{n=1}^{4} \left[\left(\left\{ \prod_{g=1}^{3} \left[\left(\left\{ \prod_{h=1}^{3} \left[(\text{LPS})_{(\ni\text{CH})c} \cap (\text{LPS})_{(\ni\text{CH})a_h} \right] p \cap_h \right\} S_{\text{interch.}}(C_3)^p \right) \right. \right.\right.\right.$$

$$\left. \cap (\text{LPS})_{(\ni\text{CH}-\text{HC}\leqslant)q_g} \right]_n \cup_g \right\} S_{\text{interch.}}(C_3)^n \bigg) \cap \left(\left\{ \prod_{h=1}^{3} \left[\left(\left\{ \prod_{k=1}^{3} \left[(\text{LPS})_{(\leqslant\text{CH})c} \cap (\text{LPS})_{(\ni\text{CH})a_h} \right] \cup_h \right\} \right.\right.\right.$$

$$\left. S_{\text{interch.}}(C_3)^{s_g} \right) \cap (\text{LPS})_{(\ni\text{CH}-\text{HC}\leqslant)t_g} \bigg]_n \cup_g \right\} S_{\text{interch.}}(C_3)^n \bigg) \bigg] \cup_n \right\} S_{\text{interch.}}(D_2). \tag{25a}$$

Equation (25a) expresses the contributions of the LPS of the methine and dimethine units to the MPS. These contributions are through the contributions of the LPS of the trimethine–methine/dimethine unit pairs and their contributions through the LPS of the triquinacylene/peristylylene unit pairs. Equation (25a) will not be further reduced in order to preserve the pattern of the symmetry contributions of the methine and dimethine units.

The MPS of hexadecahedrane has thus been constructed from the LPS of the constituent methine and dimethine units.

CONTRIBUTIONS OF THE LPG TO THE MPG OF HEXADECAHEDRANE

The contributions of the LPS to the MPS, in turn determine the contributions of the LPG to the MPG of hexadecahedrane.

The contributions of the LPG of the methine and dimethine units to the MPG is through three stages.

1. The relation of the local site symmetry point group to the LPG of the trimethine–methine units through the local site interchange point group.
2. The relation of the local site–intersite symmetry point group to the LPG of the triquin-acylene and peristylylene units through the local site–intersite interchange point group.
3. The relation of the molecular site–intersite symmetry point group to the MPG of hexadecahedrane through the molecular site–intersite interchange point group.

Furthermore, since groups can be factored into the corresponding subgroups, the point group can be expressed as the dot (direct, semi-direct, weak direct) product of the interchange and site symmetry subgroups [20, 27–29] according to

$$G = G_{\text{interch.}} \cdot G_{\text{site}} \tag{26}$$

which yields for:

1. The trimethine–methine units

$$\text{LPG}_{(\text{HC(CH)}_3)} = G_{\text{local site interch.}} \cdot G_{\text{local site}}$$

$$= C_3 \wedge C_s = C_{3v}. \tag{27}$$

2. The triquinacylene and peristylylene units

$$\text{LPG}_{(\text{HC(CH)}_3)(\text{HCCH})_3} = G_{\text{local site–intersite interch}} \cdot G_{\text{local site–intersite.}}$$

$$= C_3 \wedge C_s = C_{3v} = \text{LPG}_{(\text{HC(CH)}_3)_3(\text{HCCH})_3} \tag{28}$$

3. The hexadecahedrane molecule

$$\text{MPG}_{(C_{28}H_{28})} = G_{\text{molec. site–intersite interch.}} \cdot G_{\text{molec site–intersite.}}$$

$$= D_2 \wedge C_{3v} = \text{Td}. \tag{29}$$

Combining equations (27) and (29), the MPG can be expressed as the semi-direct product of its subgroups,

$$\text{MPG} = G_{\text{molec. site–intersite interch.}} \cdot G_{\text{local site interch.}} \cdot G_{\text{local site}}$$

$$= D_2 \wedge C_3 \wedge C_s = \text{Td}. \tag{30}$$

While combining equations (28) and (29), the MPG can also be expressed as

$$\text{MPG} = G_{\text{molec. site–intersite interch.}} \cdot G_{\text{local site–intersite interch.}} \cdot G_{\text{local site–intersite}}$$

$$= D_2 \wedge C_3 \wedge C_s = \text{Td} \tag{31}$$

The subgroup product structure is the same in equations (30) and (31), but the labelling of the subgroups is different depending upon the contributing units.

The above breakdown of the LPG and MPG (C_{3v}, Td) into the corresponding subgroup products (equations (27–29)) coincides with the previously expressed subgroup products [20, 27–29].
It should be noted that:

1. The local site symmetry point group is the group of symmetry operations common to the symmetry operations of the LPG of the methine/methine unit pairs. It is a subgroup of the LPG of the methine units as well as of the trimethine–methine units.
2. The local site–intersite symmetry point group is the group of symmetry operations common to the symmetry operations of the LPG of the trimethine–methine/dimethine unit pairs.

It is a subgroup of the corresponding LPG as well as a subgroup of the local symmetry point of the triquinacylene and peristylylene units.

3. The molecular site–intersite symmetry point group is the group of symmetry operations common to the symmetry operations of the LPG of the triquinacylene/peristylylene unit pairs. It is identical to the corresponding LPG and is a subgroup of the MPG.

The proposed set theoretic formalisms which account for the molecular and local symmetry point sets indicate the following:

1. Contributions of:

$$LPG_{(triquinacylene)} = C_{3v}$$
$$LPG_{(peristylylene)} = C_{3v}$$

via

$$G_{molec.\ site-intersite} = C_{3v},$$
$$G_{molec.\ site-intersite\ interch.} = D_2$$

to

$$MPG_{(hexadecahedrane)} = Td.$$

2. Contributions of:

$$LPG_{(trimethine-methine)} = C_{3v}$$
$$LPG_{(dimethine)} = C_{2v}$$

via

$$G_{local\ site-intersite} = C_s,$$
$$G_{local\ site-intersite\ interch.} = C_3$$

to

$$LPG_{(triquinacylene)} = LPG_{(peristylylene)} = C_{3v}.$$

3. Contributions of:

$$LPG_{(methine)} = C_{3v}$$

via

$$G_{local\ site} = C_s$$
$$G_{local\ site\ interch.} = C_3$$

to

$$LPG_{(trimethine-methine)} = C_{3v}.$$

The above outline gives the contributions of the various LPG.

The contributions of the LPG to the MPG cannot be determined reliably without the determination of the underlying contributions of the LPS to the MPS through the use of the set theoretic formalism.

CONCLUSIONS

In the present study, the molecular coordinate symmetry of hexadecahedrane has been accounted in terms of the symmetry contributions of the constituent units.

The previously proposed set theoretic formalism which accounted for the molecular coordinate symmetry for the conformations of the lower cycloalkanes [25] which belong to the non-cubic symmetry point sets and point groups is also applicable to hexadecahedrane. In this case, the proposed formalism is applied three times in succession. It is applied first, in determining the symmetry contributions of the triquinacylene/peristylylene unit pairs to the MPS. Second, in determining the symmetry contributions of the trimethine–methine/dimethine unit pairs to the local symmetry point sets of the triquinacylene and peristylylene units. Third, in determining the symmetry contributions of the methine unit pairs to the local symmetry point set of the trimethine–methine units. The sum total of these three consecutive applications of the formalism results in the determination of the symmetry contributions of the methine and dimethine units to the molecular symmetry point set.

The construction of the symmetry point sets PS(Td) and PS(C_{3v}) from all or any one of the generator sets S_g(Td) and S_g(C_{3v}) has been effected. This construction is important since it forms the basis for the selection of contributing unit structures and for the construction of the set theoretic formalims which accounts for the molecular coordinate symmetry. Therefore, the set theoretic analysis of hexadecahedrane is also applicable to structurally more complex or simpler molecules as methane, tetrahedrane, octahedrane (truncated tetrahedrane) and adamantane which also belong to the tetrahedral symmetry point set PS(Td) and point group PG(Td). The comparative differences reside in the nature of the (A/B) unit pairs, which in turn determine subsequent applications of the formalism. The importance of the generator set and its relation to the symmetry point set must be noted.

Furthermore, the proposed set theoretic formalism accounts for the origin of local site, local site–intersite and molecular site–intersite symmetry point groups and their relation to the LPG of the trimethine–methine, triquinacylene, peristylylene units and to the molecular symmetry point group through the corresponding interchange point groups. This set theoretic approach to the symmetry analysis of hexadecahedrane provides an accounting and consequently an understanding of the molecular coordinate symmetry through a hierarchy of well defined steps. This analysis yields a method for the selection of units and combinations of units which contribute to the molecular coordinate symmetry and provides in addition an illustration of the pattern of these contributions.

The proposed set theoretic formalism simply states that the symmetry shared by the parts and the symmetry responsible for the equivalent interchanges of the shared symmetry, constitute the symmetry of the whole.

The analysis of hexadecahedrane has provided an illustration and a formalization of repeated symmetries within symmetries through the fundamental language of set theory.

REFERENCES

1. N. J. Buerger, *Elementary Crystallography*. Wiley, New York (1956).
2. V. Heine, *Group Theory in Quantum Mechanics*. Pergamon Press, Oxford (1960).
3. N. Hamermesh, *Group Theory and its Applications to Physical Problems*. Addision-Wesley, Reading, Mass. (1962).
4. R. McWeeny, *Symmetry*. Pergamon Press, Oxford (1963).
5. S. L. Altmann, The crystallographic point groups as semi-direct products. *Phil. Trans. R. Soc.* **255A,** 216–240 (1963).
6. H. H. Jaffe and M. Orchin, *Symmetry in Chemistry*. Wiley, New York (1963).
7. D. S. Schonland, *Molecular Symmetry*. Van Nostrand-Reinhold, New York (1965).
8. G. G. Hall, *Applied Group Theory*. Elsevier, New York (1967).
9. J. R. Ferraro and J. S. Ziomek, *Introductory Group Theory*. New York (1969).
10. L. H. Hall, *Group Theory and Symmetry in Chemistry*. McGraw-Hill, New York (1969).
11. O. S. Urch, *Orbitals and Symmetry*. Penguin, Hamondsworth (1970).
12. F. A. Cotton, *Chemical Applications of Group Theory* (2nd edn). Wiley-Interscience, New York (1971).
13. D. N. Bishop, *Group Theory and Chemistry*. Clarendon Press, Oxford (1973).
14. A. D. Boardman, D. E. O'Connor and P. A. Young, *Symmetry and its Applications in Science*. Wiley, New York (1973).
15. D. B. Chesnut, *Finite Groups and Quantum Theory*. Wiley-Interscience, New York (1976).
16. C. D. W. Chisholm, *Group Theoretical Techniques in Quantum Chemistry*. Academic Press, London (1976).
17. S. L. Altmann, *Induced Representations in Crystals and Molecules*. Academic Press, London (1977).
18. L. L. Boyle and K. F. Green, The representation groups and projective representations of the point groups and their applications. *Phil. Trans. R. Soc.* **238 A,** 237–259 (1978).
19. P. R. Bunker, *Molecular Symmetry and Spectroscopy*. Academic Press, New York (1979).
20. R. L. Flurry Jr, *Symmetry Groups*. Prentice-Hall, Englewood-Cliffs, N. J. (1980).
21. E. R. Kouyoumdjian, On the number of generator sets for the non-cubic symmetry point groups. *J. chem. Educ.* **60,** 643 (1983).
22. F. Kamke, *The Theory of Sets*. Dover, New York (1950).
23. R. P. Hamos, *Naive Set Theory*. Van Nostrand Reinhold, New York (1960).

24. E. R. Kouyoumdjian, On the symmetry point sets for the non-cubic symmetry point groups. *Spectrosc. Int. J.* **4**, 193–206 (1985).
25. E. R. Kouyoumdjian, Contributions of local symmetry point groups to the molecular symmetry point group for the conformations of the lower cycloalkanes. *Spectrosc. Int. J.* **4**, 127–136 (1985).
26. R. L. Flurry Jr, Site symmetry in molecular point groups. *Int. J. Quantum Chem. Symp.* **6**, 455–458 (1972).
27. R. L. Flurry Jr, The uses of site symmetry in constructing symmetry adapted functions. *Theor. Chim. Acta* **31**, 221–230 (1973).
28. R. L. Flurry Jr, Site symmetry and hybridized orbitals. *J. chem. Educ.* **53**, 554–556 (1976).
29. R. L. Flurry Jr, Site symmetry and the framework group. *J. Am. Chem. Soc.* **103**, 2901–2902 (1981).

Computers Math. Applic. Vol. 17, No. 1–3, pp. 397–416, 1989
Printed in Great Britain. All rights reserved

CARBON AND ITS NETS

A. T. Balaban

Polytechnic Institute, Department of Organic Chemistry, Splaiul Independenţei 313,
77206 Bucharest, Romania

Abstract—The uniqueness of C—C bonds, which allows life to exist and gives birth to the infinite number of organic compounds, is analysed. A brief survey of the two known C allotropes, graphite and diamond, stresses how the different structure of the nets leads to such enormous differences in properties. The synthesis of diamond is surveyed. Other possible, hypothetical forms of elemental carbon are mathematically surveyed depending on the hybridization state of the C atoms, i.e. on the vertex degrees in the infinite lattice. Related nets with other atoms instead of carbon are mentioned. Then the symmetry of hydrocarbons having C skeletons that are fragments from the graphite net (benzenoid hydrocarbons) or from the diamond lattice (diamond hydrocarbons) is discussed, and the connection between symmetry and molecular stability is noted.

1. INTRODUCTION

For mankind, C should be the most precious element because all life is based upon it, and because coal and petroleum ensure most of the energy sources.

In the preceding volume, *Symmetry: Unifying Human Understanding* I mentioned the fascinating symmetries of the graphite and diamond lattices [1]. The present paper will review only aspects connected with this topic, and will attempt to survey other possible two- or three-dimensional carbon nets (lattices). Although none of these other nets has yet been prepared, it is of interest to discuss symmetries and relative energies or stabilities of such systems.

2. THE MOST PRECIOUS ATOMS IN THE UNIVERSE, OR WHY C?

We start from the well-known fact that all nuclides (isotopes) of the same element share the same electronic configuration. To simplify the discussion, therefore, we shall ignore the composition of the atomic nuclei for the time being, and shall treat the nucleus as a black box containing Z positive charges, around which the Z negatively-charged electrons occupy orbitals of various energies, according to the principles of quantum chemistry.

It took chemists more than 20 years to accept Staudinger's idea [2] that vinylic polymers are extremely long chains containing C—C covalent bonds. Likewise, cellulose and starch contain large numbers of six-membered pyranosic rings linked by covalent ethereal (C—O—C) bonds. Familiarity with association colloids such as soap micelles (loosely held bundles of smaller molecules or ions) had suggested that polymers possessed some kind of weaker intermolecular forces, but experiments by Staudinger, Mark, Meyer and Kuhn [3] provided evidence for the contrary, and confirmed that polymers had normal covalent bonds.

Why is there just one element, then, which forms the basis of life, and which is so ubiquitous in chemistry that more than nine-tenths of all known chemical compounds contain it? What is so unique about C—C bonds that a whole branch of chemistry (organic chemistry, the largest and most important one) is devoted to it, whereas all the remaining elements studied by inorganic chemists give rise to much less numerous and less important compounds? How will the disproportion between organic, inorganic and organo-elemental compounds evolve in the future?

These are important questions for chemists and philosophers of science alike.

Let us analyse first what is so unique about the element carbon.

Among the 100+ elements (i.e. types of electronic configurations around the nuclei) forming the building blocks of the whole universe, only two may form homoatomic chains whose stability is independent of the chain length. Therefore only these two elements, namely tetravalent carbon and divalent S, may give rise to "infinitely long" homoatomic chains. Though S also forms stable chains or rings of various lengths and shapes, all these substances are just allotropes of the element S;

they cannot give rise to a rich chemistry. When S has higher valency than two, the homoatomic S chains are no longer stable irrespective of the chain length. Chains of O, N, B, P, Si atoms are known but no stable homoatomic chains with more than 20 such atoms could be obtained; the longer the chain, the less stable the compounds.

Recently, it was discovered, however, that chains —$(SiMe_2—SiMePh)_n$— may also have kinetic stability; their extreme photolability in the near ultraviolet leads to applications for photoresists [4].

The reasons for the stability of homoatomic C chains are varied, and the most important factors are: small covalent radius, intermediate electronegativity, large bond energy due to strong overlap between bonding orbitals, lack of alternative bonding possibilities (such as ionic or covalent bonds). In most organic compounds, C is bonded mainly to H atoms (with matching intermediate electronegativity and high bond energy due to small covalent radius, hence to good orbital overlap); however, despite its high electronegativity, F also forms (with C chains) stable fluorocarbons and derivatives owing to its small covalent radius and strong bonds.

One should include a few words about silicates and about metallic clusters. Almost as stable as the long homoatomic C chains in organic polymers are the heteroatomic chains of silicates containing alternating Si and O atoms. Indeed, most of the earth's crust contains such chains; the abundance of elements on the earth's surface places O (49%) and Si (26%) on the first two places, with C occupying the 14th place (0.1%). The two elements, O and Si, which account for three quarters of the earth's crust (including the atmosphere and the hydrosphere) form with one another a much stronger bond than the O—O and Si—Si bonds. The tetravalency of Si can and, indeed, does give rise to a rich chemistry which will continue to develop. When, in addition to O and Si, other atoms such as C and H are present, a mixed or hybrid kingdom results (in addition to the mineral, vegetable and animal kingdoms): polysiloxanes (the so-called silicones), carbosilanes, etc., with numerous theoretically and practically significant applications.

During the last decades, metallic clusters have been amenable to detailed studies, mostly in organometallic systems. This area is also rapidly expanding and will undoubtedly cast light on many phenomena related to heterogeneous catalysis.

Neither of these two areas has, however, the potential of organic chemistry and neither is used in life phenomena, as we know them on earth. Nor has it been possible, so far, to discover such bonding in interstellar molecules. Cosmic abundances are largest for the lightest elements, H and He; O is third, while C and Si occupy the sixth and seventh places, respectively. Many "organic" molecules are known in interstellar clouds but they are quite different from the molecules of life; most of the cosmic molecules have triple bonds; on earth, a characteristic of molecules which are stable at high-temperatures ($>2000°C$) is that they also contain triple bonds (CO, N_2, HCN, NO, C_2H_2).

3. THE TWO KNOWN ALLOTROPIC FORMS OF C: GRAPHITE AND DIAMOND, OR THE SOFTEST AND HARDEST CRYSTALS

(a) Graphite

There exist two types of graphite: hexagonal and rhombohedral graphite [5–7]; they differ by the "vertical" correspondence between C atoms in the parallel planes: in the former this leads to ABAB...-type, in the latter to ABCABC...-type planes (Fig. 1).

The facile displacements of "molecular planes" relative to one another explain why graphite is as soft as talcum, the softest mineral, and why it writes in pencils. The Greeks must have used it for writing, hence its name [γραφειν (to write)], whereas in German pencil is called Bleistift, proving that lead, which is one of the softest metals, was used for writing, too.

Although diamond and graphite are elemental C, in real crystals of graphite or diamond, the relatively few peripheral atoms are connected to other atoms than C, probably H or O; one may call graphite an "honorary polycyclic benzenoid aromatic hydrocarbon" and diamond an "honorary polycyclic saturated hydrocarbon" or "honorary diamond hydrocarbon", meaning that they represent the limiting (asymptotic) cases of hydrocarbons with increasing numbers of C atoms and decreasing numbers of H.

(a)　　　　　　　　　　　　(b)

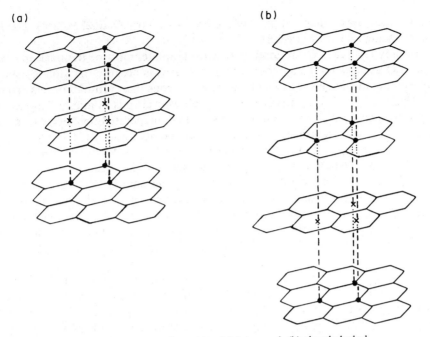

Fig. 1. The two types of graphite: (a) hexagonal; (b) rhombohedral.

The "honeycomb lattice" of carbon atoms in graphite is actually a huge planar molecule, wherein all atoms are connected by covalent bonds. The parallel planes are, however, loosely held by van-der-Waals-forces which are much weaker than covalent bonds, accounting for the softness of graphite, whose molecular planes may glide over one another. The strong anisotropy of graphite is also explained by its structure. The linear compressibility of graphite is 10^4–10^5 times larger on a direction perpendicular to the molecular planes than along directions within the molecular plane. The thermal and electrical conductivities in single crystals of graphite are about 200 times higher within the molecular planes than in the orthogonal direction. Also, the fact that graphite is black is due to light absorption in the huge aromatic chromophore.

Interatomic distances give evidence for the different kinds of bonding in graphite: its C–C covalent bond has a length of 141.5 pm, whereas the distance between the molecular planes is 335 pm. This fact leads to a low density (2270 $kg \cdot m^{-3}$ in the ideal case, usually less).

All C atoms in a planar graphite lattice have sp^2 hybridization, and all interatomic bond angles are 120°. The electronic delocalization within the molecular plane makes graphite the most stable allotropic form of C.

Of particular interest are the so-called intercalation compounds formed by graphite by including between the molecular planes various atoms such as alkaline metals, or molecules such as CrO_3, modifying thereby their reactivity, and extending slightly the inter-layer distance in the graphite net [7]. Most such intercalation compounds still conduct electricity.

The addition of F_2 to graphite leads to the white non-conducting "perfluorographite" $(CF)_n$ in which the C atoms with sp^3-hybridization cause buckling of the "molecular planes" (Fig. 2).

Since carbon has two stable nuclides, namely ^{12}C (98.9%) and ^{13}C (1.1%), and since the absorption cross sections for thermal neutron capture are quite low (0.0034 and 0.0009 barn,

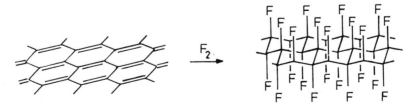

Fig. 2. Formation of "perfluorographite".

respectively), much lower than for protons (0.332 barn) or ^{17}O (0.235 barn), the first nuclear reactors were moderated with graphite.

Nowadays only few reactors are moderated with graphite; most of the nuclear power reactors use H_2O or D_2O as moderator and coolant systems. This is because the atomic mass makes C a less effective moderator in elastic interactions with neutrons than protium or deuterium, therefore the power density of the active zone is appreciably lower: U rods in graphite-moderated reactors must be spaced at larger distances than in H_2O- or D_2O-moderated reactors. Moreover, graphite may ignite at high temperature, and under neutron bombardment the C atoms are displaced accumulating strain energy (the Wigner effect responsible for the Windscale accident of the nuclear reactor in 1957). The Chernobyl tragedy proves that nuclear reactor accidents may become aggravated by the high temperatures caused by graphite combustion. A further drawback of graphite reactors is their positive void coefficient which by a feed-back loop may cause a catastrophe. In the US, the so-called N-reactor at Handford, Washington (which uses graphite as moderator and, like the Chernobyl-type reactors, lacks a large containment building and is used also for the manufacture of Pu) is reported to be designed so as to have a negative void coefficient [8].

A major use of graphite is nowadays for C fibers as reinforcement materials for composites. In oriented crystals, graphite can have a high modulus if the force is parallel to the molecular plane. The industrial C fibers used in high-technology composites are produced by pyrolysis of polymers in controlled conditions. The largest single U.S. producer of carbon fibers had sales of $64 million in 1985, despite the pace slackening of this market [9]. Such fibers are manufactured from pitch or polyacrylonitrile fibers; they are oxidized at 250° to lose much of the H, then carbonized to an extent of 98% at 800° in the absence of O_2, when cyclizations may occur, and finally graphitized at 1400–2500°C when loss of H_2, N_2, HCN, NCCN leaves a graphitic material with few N atoms.

Composites made from high-tech polymers (e.g. Kevlar, an aramid fiber with high tensile strength obtained from p-phenylenediamine and terephthaloyl chloride) and carbon fibers with high compressive strength, have many potential applications for replacing metals.

(b) Diamond

In a diamond single crystal (Fig. 3) all C atoms have sp^3-hybridization with tetrahedral bond angles and interatomic bond lengths 154 pm as in alkanes. The distance between parallel average planes is 140 pm, much smaller than in graphite, which is therefore less dense. Therefore the single crystal is a huge tridimensional molecule, explaining thus its hardness, transparency, high density (ideally $3510 \ kg \cdot m^{-3}$) and lack of thermal or electrical conductivity [5, 10].

Measurements of heats of combustion indicate that diamond is thermodynamically less stable than graphite under normal conditions by 0.2 kcal/mol. However, the diamond-to-graphite conversion has a high energy barrier: only on heating for a long time at 1500° under inert atmosphere at normal pressure is it possible to perform this conversion. Thus at 25° and 1 bar, diamond is metastable, i.e. it has a high kinetic barrier.

The reverse process, namely the production of artificial diamond from graphite, requires high pressure (Le Chatelier's principle accounts for the higher stability of the denser diamond at higher pressures). In the pressure vs temperature diagram (Fig. 4) the hatched area corresponds to the industrially applied process, which includes, in addition to C, a catalyst. Figures 5 and 6 represent the pressure/temperature/composition diagrams with Ni catalysts.

In manufacturing diamond it is essential to remember that at any given high pressure, the temperature must be high enough for the conversion to proceed at a convenient rate, but not higher

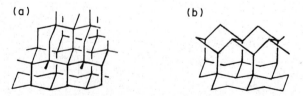

Fig. 3. The two types of diamond: (a) cubic (with the sphalerite lattice, having no eclipsed bonds); (b) hexagonal (with the wurtzite lattice).

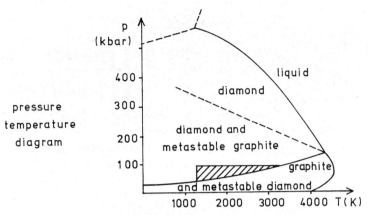

Fig. 4. The pressure–temperature diagram of C.

than the equilibrium indicates because always diamond is metastable relatively to graphite, and at the highest temperatures only graphite is stable. This is clearly seen in Fig. 5.

The diamond-producing apparatus, catalyst/solvent, and pressure–temperature conditions constitute a well-guarded industrial secret. From literature data [11–13] one gets general ideas, e.g. with two pistons one has the "Bridgman opposed anvils" (tetrahedral assemblies of four pistons may also be imagined or hexahedral ones with six pistons). Hydraulic rams act the pistons. Figure 7 illustrates the widely used "belt apparatus". The gasket material is a hydrated Al silicate (pyrophyllite). The high temperature (1700–1720 K) is obtained in the cell by passing electric current from the top to the bottom piston, after the onset of the required pressure (about 54 kbar, i.e. 5.4 GPa). The heat is evolved mainly in the graphite layers which are less conductive than nickel. Diamond forms at the interface by dissolution of graphite into the molten nickel and precipitation of the less soluble diamond, as crystals. After cooling, the nickel is dissolved in acid leaving the diamond crystals.

A different procedure for obtaining synthetic diamond involves the deposition of C atoms in vacuum on small diamond crystals. The C atoms are formed by pyrolysis of methane at 1000°C; the resulting larger diamond crystals are freed from graphite impurities by hydrogenation at 50 bar.

An interesting observation concerns tiny diamond crystals (typical diameters are 5 nm) in carbonaceous meteorites; the isotopic composition of accompanying Xe indicates that the meteorites did not originate in our solar system [14]. It is not certain how the initial diamond crystal resulted; further accretion then occurs as indicated above.

There exist two forms of diamond: the common (cubic) latticee (identical to the ZnS sphalerite lattice), where all C—C bonds are in staggered conformation, and the hexagonal (wurtzite) lattice which has Pitzer strain due to eclipsed C—C bonds. One can see adamantane subunits in the former lattice and iceane [15] subunits in the latter (Fig. 3).

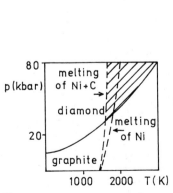
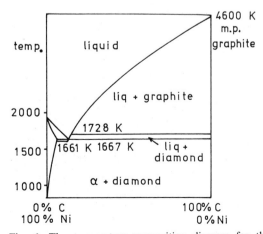

Fig. 5. The pressure–temperature diagram for the C—Ni system.

Fig. 6. The temperature–composition diagram for the C—Ni system at 54 kbar.

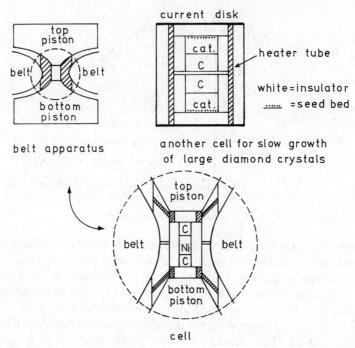

Fig. 7. The belt apparatus for synthetic diamond: at the bottom, an enlargement of the central zone is shown.

4. CARBON MOLECULES

We shall now discuss two types of relatively small C molecules C_n, with $n \leqslant 120$, which have been predicted to have some chemical stability.

(a) Buckminsterfullerene, C_{60}

A polyhedral C_{60} molecule with hexagonal and pentagonal faces, footballene or buckminsterfullerene was discussed [16]; it is still debatable if the experimental evidence is conclusive. In the spherical cavity, a metallic atom (e.g. Ln) may be incorporated. Figure 8 presents this molecule and a planar projection (Schlegel diagram).

A larger cluster of C atoms (C_{120}, archimedane) corresponding to a semiregular polyhedron with 12 decagonal, 20 hexagonal and 30 square faces, was discussed [17].

Considering that on replacing some of the six-membered rings in graphite by five-membered rings one induces a curvature in the plane of the lattice which converts it into a polyhedral quasi-spherical/system, one may speculate in abstract mathematical terms if an analogous procedure would be feasible for diamond by replacing some adamantane units in order to arrive at four-dimensional polytopes.

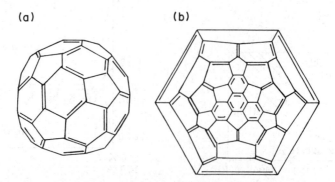

Fig. 8. (a) A view of buckminsterfullerene or footballene, C_{60}; (b) the same molecule in a Schlegel diagram showing all faces.

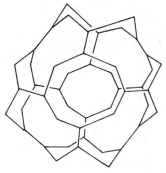

Fig. 9. A C_{48} cluster of C atoms.

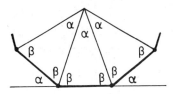

Fig. 10. Angles in a regular n-sided polygon, which is drawn with thick lines.

(b) A smaller sperhical C_{48} molecule

In a recent paper [18] on a three-connected C net (which will be presented in a subsequent paragraph), a C_{48} molecule was presented. It has six puckered eight-membered rings in a cubical arrangement, and eight nine-membered rings (Fig. 9). Its C atoms, unlike those of buckminsterfullerene, have sp- and sp^2-hybridization. Similarly to this molecule, it may accommodate a metal atom or ion inside the cavity.

(c) Cyclocarbyne (cyclokarbin)

The speculation on smaller molecules having a large ring of sp-hybridized C atoms with small steric strain was published in 1980 [20]. This molecule with either alternating single and triple bonds, cyclo—$(C\equiv C-)_{n/2}$, or with cumulenic double bonds, cyclo $=(=C=C=)=_{n/2}$, might be isolable if the angle strain is divided over a larger number of C atoms; e.g. for $n = 50$ the deviation from the bond angle of 180° becomes $\alpha = 7.2°$. In the general case one may easily see from Fig. 10 which shows the angles in a regular n-sided polygon that this deviation is $\alpha = 360°/n$: the deviation equals the angle with which the centre of the polygon "sees" each side of the polygon. The black dots in Fig. 10 symbolize the C atoms.

5. HYPOTHETICAL ALTERNATIVE C ALLOTROPES (INFINITE NETS)

The various possible nets will be systematically discussed according to the coordination number of C atoms: two-coordinated atoms will be denominated as sp-hybridized, three-coordinated ones as sp^2-hybridized, and four-coordinated ones as sp^3-hybridized (even when, owing to steric factors, the hybridization may no longer correspond to these pure states).

A. Systems Having sp-Hybridized C Atoms (Two-connected Nets)

(a) Polyyne C

The literature data on a linear macromolecular polyacetylenic (or poly-cumulenic) form of C (also called carbyne, Karbin or chaoite [21]) is still contradictory. It is obtained by oxidative polymerization of acetylene and appears to be different from diamond or graphite. It contains, probably, in addition to the linear chains, portionwise structures which link together several such chains such as those presented in Fig. 11 (of course in these portions the C atoms have sp^2- or sp^3-hybridizations).

(b) Other systems containing sp-hybridized C together with other types of C

In principle it is possible to introduce, into any two- or three-dimensional network, layers of sp-hybridized C atoms. This will cause rings in graphite to become elongated in one direction [Fig. 12(a)] or in several directions [Fig. 12(b)].

Similarly, a row of C—C bonds in diamond may be replaced by C—C≡C—C bonds as hypothesized by Melnichenko et al., affording the so-called polyyne-diamond [22] (Fig. 13).

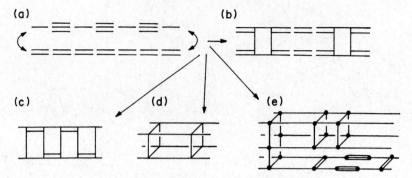

Fig. 11. Carbyne (a) and possible bridged structures (b–e).

B. Systems Having sp^2-Hybridized C Atoms (Three-connected Nets)

(a) Semiregular nets

Plannar lattices containing two or more types of regular polygons are called semiregular nets. The presence of angles far larger or smaller than 120° increases the energy of the lattice, causing it to be unstable. According to the generalized Hückel's rules in polycyclic systems, the presence of rings containing $4k$ atoms/electrons also causes destabilization ($k = 1, 2, 3, ...$) [23, 24].

The semiregular lattice $\{3, 12^2\}$ formed from triangles and dodecagons (bond angles 60°, 150°, 150°) and the lattice $\{4, 8^2\}$ formed from squares and octogons (bond angles 90°, 135°, 135°) are unfavourable on both above counts. The latter types of polygons may appear at fracture zones of graphite (Fig. 14). Such defects are annealed on heating [25]. Other possible defects with ring sizes ranging from 3 to 9 were enumerated by Dias [26]. The first systematic enumeration of semiregular C nets was published in 1968 [27].

One may estimate relative stabilities of three-connected nets (wherein each C atom is connected to three other C atoms) by a calculation [27] taking into account the atoms in the elementary cell. In graphite (Fig. 8) there are two C atoms in the elementary cell. In lattice $\{4, 8^2\}$ there are four C atoms in the elementary cell (Fig. 15). Following the procedure of Barriol and Metzger [28] the resonance energy of lattice $\{4, 8^2\}$ was calculated [27] to be $0.281\beta = -4.5$ kcal/mole. The strain energy was estimated to be 10 kcal/mole, leading to a total energy of $+5.5$ kcal/mole. In the same scale, graphite has a calculated resonance energy equal to the total energy (because of zero strain) of -6.1 kcal/mole. Thus, according to this approximate calculation, relatively to graphite, the net $\{4, 8^2\}$ is thermodynamically less stable by about $5.5 + 6.1 = 11.6$ kcal/mole. We recall that diamond is thermodynamically less stable than graphite under normal conditions by 0.2 kcal/mole. It is conceivable that the high C—C bond energy may lead to appreciable kinetic stability of less stable allotropic forms of C, making such forms able of existence.

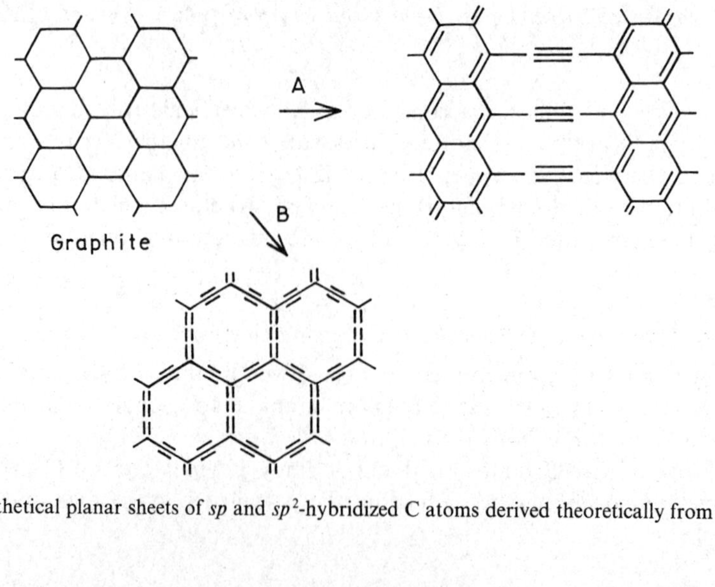

Graphite

Fig. 12. Hypothetical planar sheets of sp and sp^2-hybridized C atoms derived theoretically from graphite.

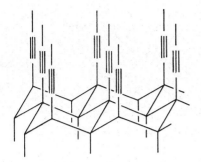

Fig. 13. Polyyne-diamond.

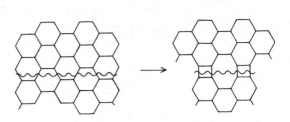

Fig. 14. Fracture zone in graphite leading, after gliding, to the formation of four- and eight-membered rings along the shear zone.

Newer calculations have been reported by Burdett and Lee [29] using the moments method for the same net $\{4, 8^2\}$ which is known to exist in the anionic part of $Ca[B_2C_2]$.

(b) Networks with pentagons and heptagons

Bathroom floor tilings corresponding to the previously described networks are fairly common. No tilings with pentagons and heptagons are known to the author, however. One may imagine two such lattices [27]; the first called $\{5^2, 7^2\}$ has eight C atoms in the elementary cell (Fig. 16); the second, called $\{5^3, 7^3\}$ has 12 C atoms in the elementary cell (Fig. 17).

A calculation similar to the preceding one [28] on the lattice $\{5^2, 7^2\}$ affords a resonance energy of -4.9 kcal/mole. Assuming the least strained geometry (angles in the regular pentagon are $\beta = 108°$; other angles are $\delta = 116°, \alpha = 144°, \epsilon = 136.5°$) one calculates a strain energy of $+3.8$ kcal/mole resulting in a total energy of -1.1 kcal/mole, i.e. conjugation energy prevails over strain energy; however, the total energy is higher than for graphite (namely by 5 kcal/mole), but less so than for lattice $\{4, 8^2\}$.

Burdett and Lee [29] also reported a calculation for the net represented in Fig. 16, which is known to exist in the anionic part of $Sc[B_2C_2]$. They also calculated energies for the least stable planar C net $\{3, 9^2\}$ having equilateral triangles and 9-gons (at each sp^2-hybridized carbon atom, one triangle and two eneagons meet).

(c) Other planar nets

The three-connected net presented in Fig. 18 contains decagons and hexagons (Hückel-type systems, hatched in the figure) and pentagons. The unit cell contains 12 atoms. The steric strain is small, so that this net is likely to present a favourable calculated total energy, similar to that of the net $\{5^2, 7^2\}$. Calculations are in progress.

(d) Tridimensional nets: a polyphenylene net reminiscent of zeolites

The first attempt to discuss a three-dimensional C net formed from sp^2-hybridized atoms was published in 1946 and 1950 by Riley and coworkers [30]. The net is formed by infinite chains of non-coplanar benzenoid rings linked as in ortho-tetraphenylene. The result is a three-dimensional net reminiscent of zeolites, including large holes and long channels.

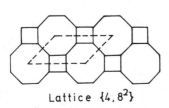

Lattice $\{4, 8^2\}$

Fig. 15. The planar lattice $\{4, 8^2\}$ with a unit cell (---).

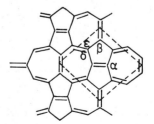

Fig. 16. The planar lattice $\{5^2, 7^2\}$ with a unit cell (---).

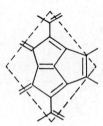

Fig. 17. The unit cell of the planar lattice $\{5^3, 7^3\}$.

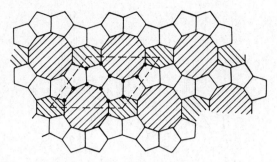

Fig. 18. A planar net with pentagons, hexagons and decagons; the unit cell, containing 12 C atoms (•), is marked with dashed lines.

(e) A hypothetical C allotrope with possible metallic character

Interesting three-dimensional networks formed from trigonal sp^2 C atoms were analyzed by Roald Hoffmann *et al.* [31]: the structure of $ThSi_2$, with C atoms instead of Th and Si. There are infinite polyene chains running along two dimensions with no conjugation along the third dimension. The density is calculated to be $2970 \, kg \cdot m^{-3}$, intermediate between that of diamond and graphite. The smallest rings are 10- and 12-membered. It was calculated that this form should have metallic conductivity, and that the C—C distances should be intermediate between those in graphite and diamond (the non-conjugated bonds should be longer than the polyenic ones). The calculated energy is higher than that of graphite by $0.74 \, eV/C$ atom. In order to make comparisons possible, one should keep in mind that $1 \, eV/atom$ corresponds to 23 kcal/mole, therefore the net of Fig. 19 is less stable than graphite by about 17 kcal/mole. The authors argue [32, 33] that this net may be able of existence owing to high kinetic barriers characteristic of C—C bonds.

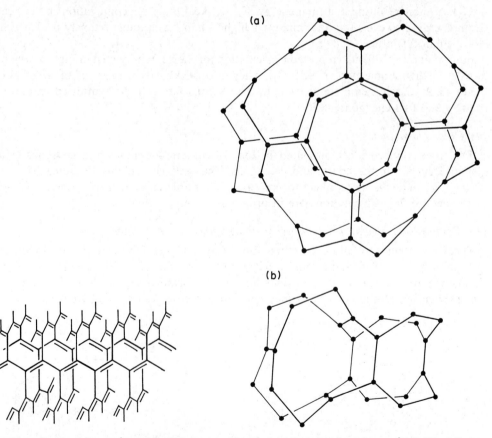

(a)

(b)

Fig. 19. A tridimensional net from sp^2-hybridized C atoms with possible metallic character.

Fig. 20. A tridimensional net from sp^2-hybridized C atoms with large cavities: (a) the larger cavity; (b) the smaller cavity.

(f) A tridimensional net with large cavities

An interesting tridimensional C net formed from sp^2-hybridized atoms with large quasi-spherical cavities and smaller quasi-cylindrical ones was discussed [18]. Its large cavities can be visualized by considering the semiregular polyhedron called truncated octahedron; in each of its 24 vertices three edges meet, belonging to two hexagons and a square. On replacing all edges that are common to a square and a hexagon by a sequence of two C—C bonds, i.e. by a C—C—C chain, one obtains the large cavity. It has six puckered (crown) octogons whose average planes are parallel to the faces of a cube, and eight puckered 9-gons (eneagons) whose average planes are parallel to the faces of an octahedron. Each octogonal face is surrounded by four 9-gons, and each 9-gon is surrounded by three 8-gons and three 9-gons.

In the net, adjacent large cavities share their 8-gonal faces. The "free spaces" between these cubically-packed quasi-spherical large cavities form the smaller cavities; these consist of two parallel puckered 9-gons (which are the bases of the quasi-cylinders) connected by three parallel C—C bonds.

The large and smaller cavities are shown in Fig. 20(a) and (b). The net is zeolite-like. The large cavities are isomorphic to the C_{48} cage molecules discussed above under Section 4.

The net has 8-gons as the smallest circuits; some of them are boat-shaped (three such 8-gons form a small cavity), other ones are crown-shaped (the faces common to two adjacent large cavities). Next, there are 9-gons, all of which have the same conformation, most easily seen in the representation of the smaller cavity [Fig. 20(b)].

C. Systems Having Only sp^3-Hybridized C Atoms (Four-connected Nets)

(i) A planar (square) lattice of C atoms with four-fold symmetry would have a high strain energy: all C atoms would be as strained as the central C in fenestrane ("planar C").

(ii) The tridimensional truncated octahedral lattice contains 12 C atoms in the unit cell (Fig. 21). The angle strain in the four-membered rings and the planar six-membered rings amounts to at least 12 kcal/mole of C atoms [27].

D. Systems with sp^2- and sp^3-Hybridized C (Three- and Four-connected Nets)

The paper published by the present author in 1968 [27] contained only a brief statement on the two lattices shown in Fig. 22 to the effect that the four-connected "planar" C atoms will lead to high strain. Other authors [33] discussed related topics.

A different approach was recently initiated, by imagining and calculating the energies of three- and four-connected nets possessing little strain; some of the nets may present metallic properties [34]. The densities of such systems (about 3000 kg/m^3) are intermediate between those of graphite and diamond.

Some of the atoms of Fig. 23(a), namely those which are shown without any double bonds, are forced to become tetracoordinated by forming an extra bond to a similar atom from a string of pentalene units in a parallel plane. For viewing this process, imagine that in a graphite-type stacking of molecular planes as in Fig. 23(a), half of the strings of pentalene units (black points) are translated midway between the molecular planes. Then the new bonds are formed, and the three- and four-connected lattice [Fig. 23(b)] results.

Fig. 21. The truncated octahedron.

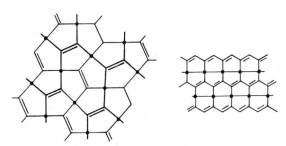

Fig. 22. C Nets with tetra- and tricoordinated C atoms; the former atoms are represented by black dots.

(a)

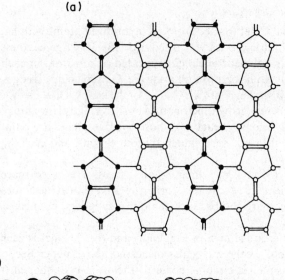

(b)

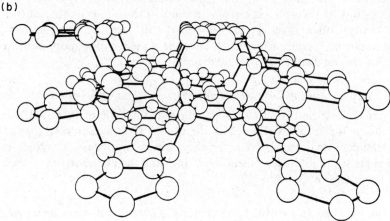

Fig. 23. (a) The starting planar arrangement of strings of pentalene units; (b) the 3,4-connected net.

The energy of this net may be calculated by means of the extended Hückel method in the tight binding approximation. The parameters used for C and the geometries are specified in Ref. [34]. The result is 1.19 eV C atom (27 kcal/mole) above the energy of graphite.

An energetically more favourable three- and four-connected net results by a similar procedure starting from the planar net containing five- and eight-membered rings [34] as seen in Figs 24 and 25. Again, the black atoms are translated with half the spacing between the molecular planes, and the atoms (black and white) which are shown without double bonds become tetracoordinated.

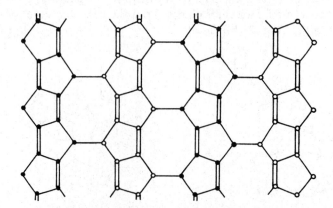

Fig. 24. The starting arrangement for another 3,4-connected net, similar to that of Fig. 23(a).

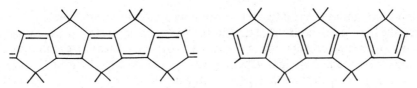

Fig. 25. Two types of delocalization in the strings of condensed five-membered rings in Fig. 24.

The energy of this net is calculated by the extended Hückel method to lie at 0.86–0.97 eV/C atom (20–22 kcal/mole) above the energy of graphite. The range of values is due to the fact that one may conceive a delocalized polyenic system, or two different types of localization (Fig. 25).

Finally, among the many imaginable nets of such type (discussed by Wells in a geometrical context) a higher-energy system depicted in analogous fashion contains eight- and four-membered rings [34], cf. Fig. 26.

The energy of this system is calculated to lie 1.27 eV/atom (29 kcal/mole) above the energy of graphite.

In concluding this chapter, mention should be made of strain-free oligoradicals [35] which contain eight-membered rings.

6. RELATED NETS WITH OTHER ATOMS INSTEAD OF C

Both the diamond and the graphite lattice may be encountered in compounds with other elements than C. Thus the diamond lattice (with lower cohesion because of the longer interatomic distances and smaller bond energies) occurs also in elemental Si (bond length 234 pm), Ge (bond distance 240 pm), and in gray tin (bond length 280 pm). On reaction with concentrated hydrochloric acid, this crystalline form of tin affords $SnCl_4 \cdot 5H_2O$, whereas the white polymorphic tin yields

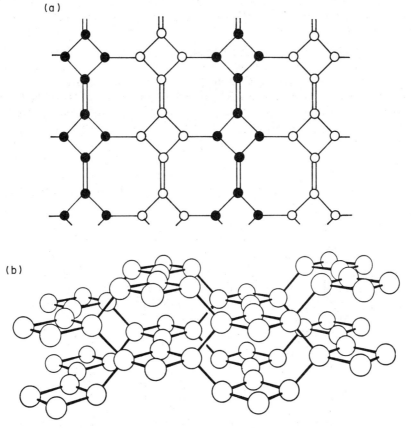

Fig. 26. The starting arrangement for a third 3,4-connected net (a); (b) the resulting net.

$SnCl_2 \cdot 2H_2O$ indicating that in the latter form with longer interatomic distances only two electrons in the valence shell are involved. In addition to these elemental lattices, the diamond structure is encountered also in cubic boron nitride, a compound whose hardness is second only to diamond. Since each pair of BN atoms has eight valence electrons exactly like a pair of C atoms, one might expect that under high temperatures and pressures the diamond-type lattice would result for $(BN)_x$; indeed, about a year after the successful diamond synthesis, Wentorf Jr performed this conversion [36] under conditions similar to those used for synthetic diamond, but using as catalysts/solvents alkali metal nitrides. The Knoop hardness of $(BN)_x$ is 4500 kg/mm^2 (compared to 9000 kg/mm^2 for diamond and 2100 kg/mm^2 for sapphire), therefore cubic $(BN)_x$ is used as an abrasive which, unlike diamond, may be heated in O_2 and does not react with Fe at high temperatures. This quality is useful in sharpening and shaping hard tool-steels and high-strength alloys.

On the other hand, hexagonal B nitride has a lattice which is similar to that of graphite, is thermally almost as stable and cleaves similarly to graphite; however, it is white, non-conducting, and does not afford intercalation compounds. It is used for high-temperature crucibles. Apparently, the π-electron delocalization is much smaller than in graphite, owing to the formal charges on quadrivalent B^- and N^+ atoms.

Passing now to the hypothetical nets discussed above, the trivalency of B and N would convert some infinite lattices into finite molecules. Alternatively, one may imagine new oxides of C; thus the large cavity of Fig. 9 or Fig. 20(a) would become $(CO)_{24}$ on replacing the vertices of degree two by O atoms, and the pairs of vertices of degree three by C=C double bonds.

7. SYMMETRY OF FRAGMENTS FROM THE GRAPHITE LATTICE

All benzenoid polycyclic aromatic hydrocarbons (PAH'S or benzenoids, for brevity) may be considered to represent fragments from the graphite lattice, whose peripheral C atoms are linked to H atoms.

Considerable interest is associated with these systems because of the proven carcinogenic activity of many compounds from this class possessing "bay-regions". Such a compound is benzo[a]pyrene, which results in many combustion processes and is present in cigarette smoke and in engine exhaust gases, cf. Fig. 27.

Benzenoids were classified into cata-condensed (catafusenes) and peri-condensed (perifusenes) depending on whether they do not contain or do contain, C atoms belonging to three adjacent six-membered rings (so-called internal C atoms). A newer definition is based upon the notion of dualist graphs [37], whose vertices are centres of hexagons and whose edges connect vertices corresponding to "condensed hexagons", sharing an edge and two C atoms as in naphthalene whose dualist graph has just one edge. One should note that unlike normal graphs where angles between edges are arbitrary, in dualist graphs the geometry is essential; for instance, the dualist graph of anthracene has two collinear edges (with angle 180°), whereas that of the isomeric phenanthrene has two kinked edges with an angle of 120°. Catafusenes have acyclic dualist graphs (trees), while perifusenes have three-membered rings in their dualist graphs. All catafusenes with the same number of vertices in their dualist graphs are isomeric. By this definition, a third category of benzenoids exists, namely corona-fused system (coronoids) whose dualist graphs possess larger rings than three-membered, which are not contours of aggregates of three-membered rings.

A "periodic system" of molecular formulas was proposed by Dias [38] for benzenoids C_xH_y on the basis of two parameters, N (the number of internal C atoms, i.e. zero in catafusenes, two in pyrene etc.) and d (the net disconnection, equal to one plus the number of tertiary C atoms minus the number of internal graph edges): $x - y = 2(N + d + 1)$. The number of tertiary C atoms is twice the number of hexagons minus two.

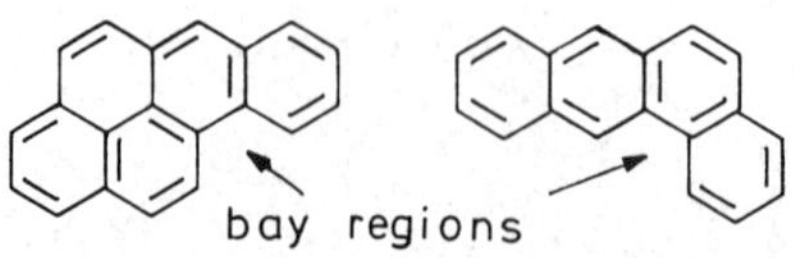

bay regions

Fig. 27. Two carcinogenic hydrocarbons with their bay regions: benzo[a]pyrene and benzanthracene. Neither has any symmetry.

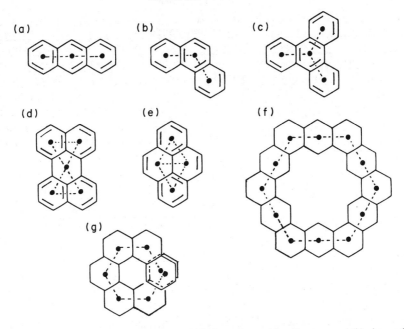

Fig. 28. The carbon skeletons of polycyclic benzenoid hydrocarbons: (a) anthracene and (b) phenanthrene are two isomeric ($C_{14}H_{10}$) non-branched catafusenes; (c) triphenylene ($C_{18}H_{12}$) is a branched catafusene; (d) perylene and (e) pyrene are perifusenes. In all cases a single Kekulé structure is shown; (f) kekulene, is a coronaphene; (g) is [7] helicene. In the last two cases double bonds are no longer included. The dualist graphs are indicated with broken lines.

Since annelation of catafusenes may be performed in three directions at a marginal benzenoid ring, the numbers C_h of non-branched catafusenes with h six-membered rings are easily calculated [36, 38]:

$$4C_h = 3^{h-2} + 4.3^{(h-3)/2} + 1, \quad \text{for odd } h;$$

$$4C_h = 3^{h-2} + 2.3^{(h-2)/2} + 1, \quad \text{for even } h.$$

Branched catafusenes can also be counted but complicated formulas result. In these formulas benzenoids include also helicenes, i.e. systems with superposed C atoms whose molecules cannot have a planar structure; helicenes cannot be considered as portions of the graphite lattice. If helicenes are left out from the benzenoid count (as we shall do henceforth), the enumeration becomes difficult even for non-branched catafusenes and only recursive computer algorithms can be employed for this purpose, up to given h values. So far, perifusenes have been counted only by computer algorithms for $h \leqslant 11$ [40, 41]. The enumeration of benzenoids is part of a larger unsolved problem in graph theory, namely the "cell growth problem" wherein a cell is a triangle, square or hexagon, and the "animal" consists of condensed cells, sharing an edge.

A coding system was devised for dualist graphs of catafusenes [37, 39] based upon the three orientations of edges in the graphite lattice, coded by digits 0, 1 and 2 for annelating a terminal benzenoid ring with angles 180°, 120° and 240°, respectively. The code rules state that coding starts from an endpoint of the dualist graph, that among all possible codes the minimal number is adopted on reading sequentially the digits, and that for branching one uses brackets.

Figure 28 illustrates catafusenes, perifusenes, coronoids and helicenes.

In turn, the symmetry of non-branched catafusenes can be used [37, 40] for systematic generation and enumeration. Table 1 presents the numbers of non-branched catafusenes with following symmetries: a = acenes (linear condensation); m = mirror symmetry; c = inversion centre; u = unsymmetrical, cf. Fig. 29.

Gordon, Cyvin, Gutman, Trinajstić, Knop and other authors [40–42] investigated in detail the numbers of Kekulé structures for benzenoids using various techniques. Perifusenes have been further classified on the basis of their Kekulé structures into normal ones (with at least one Kekulé structure), essentially disconnected ones having some single bonds in all Kekulé structures, and non-Kekuléan systems (radicals or polyradicals), i.e. systems without any Kekulé structure; in

Table 1. Numbers of benzenoids according to the type of condensation (cata, peri, corona), branching, symmetry and Kekuléan character

| | Catafusenes | | | | | | | Perifusenes | | Coronoids | | |
| | Non-branched† | | | | | Branched | Total | Kekuléan | | Kekuléan | | Total benzenoids |
h	a	m	c	u	Total			Yes	No	Yes	No	
1	1	0	0	0	1	0	1	—	—	—	—	1
2	1	0	0	0	1	0	1	—	—	—	—	1
3	1	1	0	0	2	0	2	0	1	—	—	3
4	1	1	1	1	4	1	5	1	1	—	—	7
5	1	4	1	4	10	2	12	3	7	—	—	22
6	1	3	4	16	24	12	36	15	30	—	—	81
7	1	12	44	50	67	51	118	72	141	—	—	331
8	1	10	13	158	182	229	411	353	671	1	0	1436
9	1	34	13	472	520	969	1489	1734	3282	3	2	6510
10	1	28	39	1406	1474	4098	5572	8535	15979	24	19	30129
11	1	97	39	4111	4248	16867	21115	41764	78350	128	155	141512
12	1	81	116	11998	12196	68925	81121	?	?	854	1100	?

†a = acenes (linearly condensed); m = mirror plane perpendicular to the molecular plane; c = centrosymmetric; u = unsymmetric.

graph-theoretical language, such systems are said to be non-decomposable into 1-factors, or to be devoid of perfect matchings. All systems with odd numbers of C atoms are non-Kekuléan; some peri-condensed systems with even numbers of C atoms are also non-Kekuléan, e.g. diradicals. All catafusenes are Kekuléan.

Calculations of π-electron energies in large peri-condensed benzenoid systems show [43] that such systems tend gradually towards graphite; these calculations allow to evidence how the topology of peripheral C atoms influences their properties.

Table 1 presents the numbers of benzenoids, excluding helicenes, according to [42].

8. SYMMETRY OF FRAGMENTS FROM THE DIAMOND LATTICE

Cyclic molecules strive to adopt low-energy strain-free conformations. For many molecules possessing saturated six-membered rings, such conformations are portions of the diamond lattice. Thus the chair form of cyclohexane, the two (cis and trans) isomers of decalin, and all the diamond hydrocarbons (adamantane, diamantane, and polymantanes) have carbon skeletons which are fragments of the diamond lattice. This was observed long ago by Wittig [44], Lukes [45] and others.

We shall pay special attention to the so-called "diamond hydrocarbons" or "polymantanes", consisting of adamantane units fused along a "face" (Fig. 30). Actually, adamantane is a tetrahedrane whose C—C bonds have been replaced by a 3C chain C—CH$_2$—C. The fusion involves joining two such tetrahedra along a face to yield a trigonal bipyramid.

In collaboration with von R. Schleyer [46], these diamond hydrocarbons have been enumerated and classified according to their symmetry, on the basis of their "tridimensional dualist graph". A coding system was devised based upon the four tetrahedral directions of the dualist graph, each denoted by one of the digits 1, 2, 3 or 4. This coding system was employed also for staggered rotamers of alkanes [39, 47] and was shown to possess a relationship with the three-digit coding of benzenoids [37]. As in the case of benzenoids, catamantanes and perimantanes have been defined according to whether the dualist graph is acyclic or cyclic, respectively. Catamantanes may be linear or branched, according to their dualist graphs. Whereas all catafusenes with the same number h of benzenoid rings are isomeric and have formula $C_{2+4h}H_{4+2h}$, only "regular catamantanes" are isomeric among themselves and have formula $C_{4n+6}H_{4n+12}$,

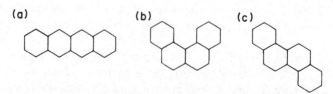

(a)　　　　　　　(b)　　　　　　　(c)

Fig. 29. Various symmetries of non-branched catafusenes for three isomeric $C_{18}H_{12}$ hydrocarbons: (a) acene; (b) mirror symmetry; (c) centric symmetry.

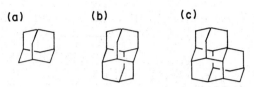

Fig. 30. Diamond hydrocarbons: (a) adamantane; (b) diamantane (earlier name—the emblem of the *19th IUPAC Congress* held in 1963 in London—was congressane); (c) triamantane.

where h, $n > 1$ are integers; they result on annulating a face of a polymantane on replacing three axial hydrogens by a trimethylenemethane group, resulting in a net addition of C_4H_4. The second kind of catamantanes called "irregular", are obtained by annulation at a face having less than three axial hydrogens, involving the net addition of C_xH_y, where $x, y < 4$. The smallest irregular catamantane is [1231]pentamantane, $C_{25}H_{30}$. A necessary and sufficient condition for a catamantane to be irregular is to have a code with the same digit separated by two other digits (e.g. the digit 1 in the above code). Table 2 presents all possible catamantanes with $n = 1$–6 adamantane units.

The coding starts from one end of the dualist graph and goes on registering the orientation of each bond; the orientations of the first two bonds are always 1 then 2. We adopt the convention of obtaining the minimal number when reading sequentially all digits, among all possible orientations of the dualist graph with respect to the tetrahedral coordinates; therefore the first two digits of the code will always be 12. When branching occurs, the branch is included in brackets; for geminal branching, the two branches are separated by a comma within the bracket. Thus code 121 designates the dualist graph shaped like the C-skeleton of *anti*-n-butane, code 123 designates its *syn*-rotamer, code 1(2)3 the C-skeleton of isobutane, and code 1(2, 3)4 the C-skeleton of neopentane.

In the tridimensional Euclidean space E_3, slight deviation from normal bond angles can accomodate such molecules as helicenes or helicene-like staggered C-rotamers of alkanes; however this is no longer possible for diamond hydrocarbons: when two vertices in the dualist graph of a polymantane are at the distance of one edge, these two vertices must be linked in the E_3 space, and the dualist graph must be cyclic. It is interesting to speculate about helicene-like polymantanes in a tetradimensional space E_4, where such ring-closure conditions for the dualist graph are not mandatory.

Table 2. All possible catamantanes with $n = 1$–6†‡

	Regular					Irregular					Total
	Linear			Branched		Linear			Branched		No.
n	Code	Sym.	Formula	Code	Sym.	Code	Sym.	Formula	Code	Sym.	
1	—	T_d	$C_{10}H_{16}$	—	—	—	—	—	—	—	1
2	1	D_{3d}	$C_{14}H_{20}$	—	—	—	—	—	—	—	1
3	12	C_{2v}	$C_{18}H_{24}$	—	—	—	—	—	—	—	1
4	121	C_{2h}	$C_{22}H_{28}$	1(2)3	C_{3v}	—	—	—	—	—	3
	123	C_2									
5	1212	C_{2v}		12(1)3	C_1	1231	C_s	$C_{25}H_{30}$	—	—	7
	1213	C_1	$C_{26}H_{32}$	12(3)4	C_s						
	1234	C_2		1(2, 3)4	T_d						
6	12121	C_{2h}		121(2)3	C_1						
	12123	C_1		12(1)32	C_1						
	12131	C_1		121(3)4	C_s	12132	C_1		12(1)31	C_s	
	12134	C_1		12(1)34	C_1	12314	C_1	$C_{29}H_{34}$	123(1)2	C_1	23
	12321	C_i	$C_{30}H_{36}$	12(1, 3)4	C_1				123(1)4	C_1	
	12324	C_2		12(3(12	C_s				12(3)41	C_s	
	12341	C_2		1(2)3(1)2	C_{2h}						
				1(2)314	C_1						
				12(3)14	C_1						
				1(2)3(1)4	C_2						

†In addition, one perimantane is possible for $n = 6$; it has code 12312, formula $C_{26}H_{30}$, and symmetry D_{3d}.
‡All systems with point groups C_1 or C_2 are chiral and give rise to two enantiomers.

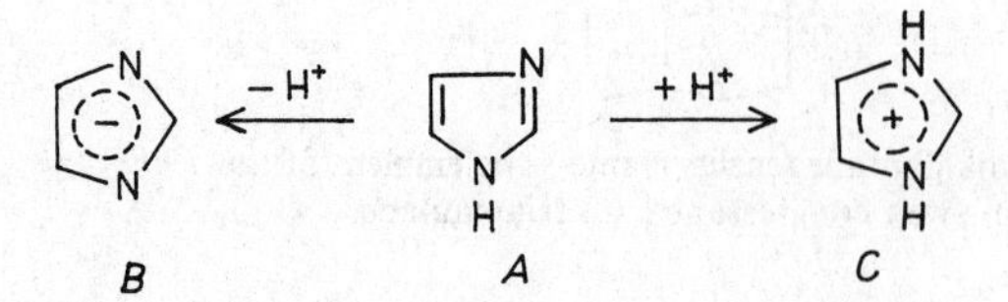

Fig. 31. Imidazole (*A*) has an unusually high basicity and high acidity because its anion (*B*) and cation (*C*) have higher symmetries than (*A*).

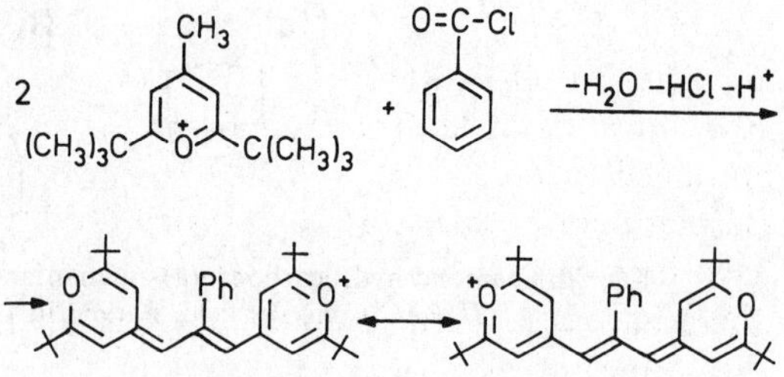

Fig. 32. Formation of a pyrylotrimethinecyanine in its two resonance forms; crosses indicate tert-butyl groups, C(CH$_3$)$_3$, and Ph phenyl groups.

Symmetry considerations have been helpful also in the enumeration of staggered conformers of alkanes [47]; with some restrictions and pruning, these staggered conformations are the dualist graphs of diamond hydrocarbons (polymantanes).

9. SYMMETRY AND STABILITY OF MOLECULES

In many cases, high molecular symmetry leads to enhanced stability. This is illustrated by many examples. Thus, isovalent conjugation always corresponds to higher stability than sacrificial conjugation, as illustrated by the high stability of cations and anions derived from azoles (pyrazole, imidazole), cf. Fig. 31. The acidity of the carboxyl group and the basicity of guanidine are due to the symmetry of the corresponding anion and cation, respectively. The spectacular bathochromic effects associated with the acid–base equilibria of polyarylmethane dyes such as phenolphthalein are also due to the formation of symmetrical ions whose delocalization becomes extended over a large array of atoms. The largest bathochromic shifts in dyestuffs (linearly depending on the number of =CH— groups) are present in symmetrical polymethine-cyanines with the general formula Ht=(CH)$_n$—Ht$^+$, where Ht is a heterocyclic system: pyridinium, (iso)quinolinium, pyrylium, cf. Fig. 32 [48]. With such cyanines (heptamethine-cyanines or longer cyanines) one may obtain electronic absorption in the infra-red, with obvious applications for photographic sensitization, etc.

It is interesting to observe that also in neutral molecules high molecular symmetry leads to enhanced stability. Schleyer demonstrated experimentally [49] that adamantane and diamantane, which are highly symmetrical molecules with the lowest energies among their valence isomers due to the lack of steric strain and of eclipsed interactions (Pitzer strain), are the final products in the AlCl$_3$-catalysed isomerizations of such valence isomers. Also triamantane and one tetramantane were obtained by such isomerizations [50]. Similarly, dodecahedrane with its very high symmetry, is formed by Rh-catalysed isomerization from an isomeric product named pagodane [51], as seen in Fig. 33. Symmetry, lack of strain and hence higher stability, are the reasons for the presence of adamantane and diamantane in crude oil [52].

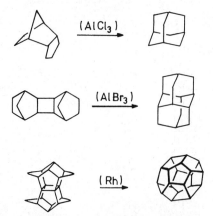

Fig. 33. Formation of highly symmetrical, less strained molecules, by isomerization: from top to bottom, adamantane, diamantane, dodecahedrane.

10. CONCLUSION

We have discussed the two known allotropes of elemental carbon (graphite and diamond), the presumed linear carbyne, and other hypothetical finite (molecules) and infinite C forms (allotropes). Despite the higher energy of all these forms than the energy of graphite, some of these nets may have high kinetic barriers for isomerization to graphite, so that they may be able of existence.

We have also examined related nets with other atoms instead of C.

Portions of graphite and diamond lattice, where the marginal C atoms are bonded to H, constitute the benzenoid and diamond hydrocarbons, respectively; these were discussed according to their symmetry. Of course, one could also discuss the hydrocarbons which result by considering portions of the other, hypothetical, nets. This has not been done because of space limitations.

It was pointed our briefly in the preceding section how important for the stability of molecules is their symmetry.

REFERENCES

1. A. T. Balaban, Symmetry in chemical structures and reactions. *Comput. Math. Applic.* **12B**, 999–1020 (1986). Reprinted in *Symmetry: Unifying Human Understanding* (Ed. I. Hargittai). Pergamon Press, Oxford (1986).
2. H. Staudinger, Über Polymerisation. *Ber. dt. chem. Ges.* **53**, 1073–1085 (1920).
3. V. M. Morawetz, Difficulties in the emergence of the polymer concept—an essay. *Angew. Chem. Int.* **26**, 93–97 (1987) and references cited therein.
4. L. D. David, The discovery of soluble polysilane polymers. *Chem. Br.* **23**, 553–556 (1987).
5. J. Donohue, *The Structure of the Elements*. Wiley, New York (1974).
6. B. T. Kelly, *Physics of Graphite*. Applied Science Publishers, New Jersey (1981); A. R. Ubbelohde and F. A. Lewis, *Graphite and Its Crystal Compounds*. Oxford Univ. Press, London (1960).
7. M. S. Dresselhaus, G. Dresselhaus, J. E. Fischer and M. J. Moran (Eds), *Intercalated Graphite, Materials Research Soc. Symp. Proc.*, Vol. 20. North-Holland, New York, and Elsevier, Amsterdam (1983); A. S. Whittingham and A. J. Jacobson (Eds), *Intercalation Chemistry*. Academic Press, New York (1982); M. E. Volpin and Yu. N. Novikov, Graphite as an aromatic ligand, *Topics in Nonbenzenoid Aromatic Chemistry* (Eds T. Nozoe *et al.*). Hirokawa Publ., Tokyo (1972).
8. D. Hanson, D. O. E.'s Hanford reactor to keep operating. *Chem. Engng News*, **65** (8), 23 (1987).
9. M. C. Reisch, High-performance fibers find expanding military, industrial uses. *Chem. Engng News* **65**, (5), 9–14 (1987); E. Fitzer (Ed.), *Carbon Fibres and Their Composites*, Springer, Berlin (1985).
10. J. E. Field (Ed.), *The Properties of Diamond*. Academic Press, New York (1979).
11. F. P. Bundy. Superhard materials. *Sci. Am.* **231**(2), 62–70 (1974); Diamond synthesis with nonconventional catalyst-solvent, *Nature, Lond.* **241**, 116–118 (1973); The p,T phase and reaction diagram for elemental carbon. *J. Geophys. Res.* **85**, 6930–6936 (1980); Direct conversion of graphite to diamond on static pressure apparatus. *J. Chem. Phys.* **38**, 631–643 (1963); F. P. Bundy, H. M. Strong and R. H. Wentorf Jr, Methods and mechanisms of synthetic diamond growth. *Chem. Phys. Carbon* **10**, 213–263 (1973); F. Bundy and J. S. Kasper, Hexagonal diamond, a new form of carbon. *J. Chem. Phys.* **46**, 3437–3446 (1967).
12. V. A. Zhorin *et al.*, Structural changes in graphite induced by the combined action of high pressures and shear deformation. *Zh. fiz. Khim.* **56**, 2486–2490 (1982); Formation of a sp^3 hybridized state in graphite by simultaneous exposure to high pressure and shear deformation, *Dokl. Akad. Nauk SSSR* **261**, 665–668 (1981).
13. R. H. Wentorf Jr, Solutions of carbon at high pressure. *Ber. Bunsenges. Phys. Chem.* **70**, 975–982 (1966); H. M. Strong and R. M. Chrenko, Diamond growth rates and physical properties of laboratory-made diamond. *J. Phys. Chem.* **75**, 1838–1843 (1971).
14. R. S. Lewis, T. Ming, J. F. Wacker, E. Anders and E. Steel, Interstellar diamonds in meteorites. *Nature, Lond.* **326**, 160–162 (1987).
15. C. A. Cupas and L. Hodakowski, Iceane. *J. Am. chem. Soc.* **96**, 4668–4669 (1974).
16. H. W. Kroto, J. R. Heath, S. C. O'Brien, R. F. Curl and R. E. Smalley, C_{60}. Buckminsterfullerene. *Nature, Lond.* **318**, 162–163 (1985); J. R. Heath, S. C. O'Brien, Q. Zhang, Y. Liu, R. F. Curl, H. W. Kroto, F. K. Tittel and R. E. Smalley, Lanthanum complexes of spheroidal carbon shells, *J. Am. chem. Soc.* **107**, 7779–7780 (1985); Q. L. Zhang, S. C. O'Brien, J. R. Heath, Y. Liu, R. F. Curl, H. W. Kroto and R. E. Smalley, Reactivity of carbon clusters: spheroidal carbon shells and their possible relevance to the formation and morphology of soot, *J. Phys. Chem.* **90**, 525–528 (1986).
17. A. D. J. Haymet, Archimedane. *Chem. Phys. Lett.* **122**, 421–424 (1985).
18. A. T. Balaban, On a 3-connected carbon net (infinite tridimensional lattice of sp^2-hybridized carbon atoms) and congeneric systems. *Rev. Roum. Chim.* **33**, 359–362 (1988).
19. D. J. Klein, W. A. Seitz and T. G. Schmalz, Icosahedral symmetry carbon cage molecules. *Nature, Lond.* **323**, 703–706 (1986).
20. A. T. Balaban, Is aromaticity outmoded? *Pure Appl. Chem.* **52**, 1409–1429 (1980).
21. A. M. Sladkov and Yu. P. Kudryavtsev, Polyacetylenes. *Usp. Khim.* **32**, 509–538 (1963); V. M. Melnichenko, A. M. Sladkov and Yu. M. Mikulin, Structure of polymeric carbon. *Usp. Khim.* **51**, 736–763 (1982).
22. V. M. Melnichenko, Yu. I. Nikulin and A. M. Sladkov, Layer-chain carbon. *Dokl. Akad. Nauk SSSR* **267**, 1150–1154 (1982).
23. M. Randić, Aromaticity and conjugation. *J. Am. chem. Soc.* **99**, 444–450 (1977).
24. N. Trinajstić, *Chemical Graph Theory*, Vol. 2, Chap. 3. CRC Press, Boca Raton, Fla. (1983).
25. J. R. Platt, Atomic arrangements and bonding across a twinning plane in graphite, *Z. Crystallogr.* **109**, 226–230 (1957).

26. J. R. Dias, Enumeration of graphite carbon-bond network defects having ring sizes ranging from 3 to 9. *Carbon* **22**, 107–114 (1984).

27. A. T. Balaban, C. C. Renţia and E. Ciupitu, Chemical graphs. Part 6. Estimation of the relative stability of several planar and tridimensional lattices for elementary carbon. *Rev. Roum. Chim.* **13**, 231–247 (1968); erratum, p. 1233.

28. J. Barriol and J. Metzger, Application de la méthode des orbitales moléculaires au réseau du graphite. *J. chim. phys.* **47**, 432–436 (1950).

29. J. K. Burdett and S. Lee, The moments method and elemental structures. *J. Am. chem. Soc.* **107**, 3063–3082 (1985); see also J. K. Burdett, Some structural problems examined using the method of moments. *Structure Bonding* **65**, 29–90 (1987); J. K. Burdett, Topological control of the structures of molecules and solids. *Acc. Chem. Res.* **21**, 189–194 (1988).

30. R. H. Riley, Chemical and crystallographic factors in carbon combustion. *J. Chim. phys.* **47**, 565–572 (1950).

31. R. Hoffmann, T. Hughbanks, M. Kertész and P. H. Bird, a Hypothetical metallic allotrope of carbon. *J. Am. chem. Soc.* **105**, 4831–4832 (1983).

32. M. Kertész and R. Hoffmann, The graphite to diamond transformation. *J. Solid State Chem.* **54**, 313–319 (1984).

33. I. V. Stankevich, M. V. Nikerov and D. A. Bochvar, Structural chemistry of crystalline carbon: geometry, stability, electronic spectrum. *Usp. Khim.* **53**, 1101–1024 (1984); *Russ. chem. Revs* **53**, 640 (1984); A. F. Wells, *The Third Dimension in Chemistry*. Clarendon Press, Oxford (1956); *Three-Dimensional Nets and Polyhedra*. Wiley-Interscience, New York (1977).

34. K. M. Merz Jr, R. Hoffmann and A. T. Balaban, 3,4-Connected carbon nets: through-space and through-bond interactions in the solid state. *J. Am. chem. Soc.* **109**, 6742–6751 (1987).

35. R. Hoffmann, O. Eisenstein and A. T. Balaban, Hypothetical strain-free oligoradical. *Proc. natn. Acad. Sci.* **77**, 5588–5592 (1980).

36. R. H. Wentorf Jr., Cubic form of boron nitride, *J. Chem. Phys.* **26**, 956 (1957).

37. A. T. Balaban and F. Harary, Chemical graphs. Part 5. Enumeration and proposed nomenclature of benzenoid *cata*-condensed polycyclic aromatic hydrocarbons, *Tetrahedron.* **24**, 2505–2516 (1968); A. T. Balaban, Chemical graphs. Part 7. Proposed nomenclature of branched *cata*-condensed benzenoid hydrocarbons. *Tetrahedron* **25**, 2949–2956 (1969); Challenging problems involving benzenoid polycyclics and related systems. *Pure appl. Chem.* **54**, 1075–1096 (1982).

38. J. R. Dias, A periodic table for polycyclic aromatic hydrocarbons, Part 1. Isomer enumeration of fused polycyclic aromatic hydrocarbons. *J. Chem. Inf. Comput. Sci.* **22**, 15–22 (1982); Part 2. Polycyclic aromatic hydrocarbons containing tetragonal, pentagonal, heptagonal and octagonal rings. *J. Chem. Inf. Comput. Sci.* **22**, 139–152 (1982); Part 3, Enumeration of all the polycyclic conjugated isomers of pyrene having ring sizes ranging from 3 to 9. *Math. Chem.* **14**, 83–138 (1983); *Handbook of Polycyclic Hydrocarbons. Part A. Benzenoid Hydrocarbons.* Elsevier, Amsterdam, 1987.

39. A. T. Balaban, Enumeration of catafusenes, diamondoid hydrocarbons and staggered alkane C-rotamers. *Math. Chem.* **2**, 51–61 (1976).

40. M. Gordon and W. H. T. Davison, Theory of resonance topology of fully aromatic hydrocarbons. I. *J. Chem. Phys.* **20**, 428–435 (1952); S. J. Cyvin and I. Gutman, Number of Kekulé structures as a function of the number of hexagons in benzenoid hydrocarbons. *Z. Naturf.* **41a**, 1079–1086 (1986); I. Gutman and O. E. Polansky, *Mathematical Concepts in Organic Chemistry*, p. 59 Springer, Berlin (1986).

41. J. V. Knop, W. R. Müller, K. Szymanski and N. Trinajstić, *Computer Generation of Certain Classes of Molecules*, p. 109. Kemija u industriji, Zagreb (1985).

42. A. T. Balaban and I. Tomescu, Algebraic expressions for the number of Kekulé structures in isoarithmic *cata*-condensed benzenoid polycyclic hydrocarbons. *Math. Chem.* **14**, 155–182 (1983); Chemical graphs. Part 41, Numbers of conjugated circuits and Kekulé structures for zig-zag catafusenes and (*j, k*)-hexes; generalized Fibonacci numbers. *Math. Chem.* **17**, 91–120 (1985); Chemical graphs. Part 40. Three relations between the Fibonacci sequence and the numbers of Kekulé structures for non-branched *cata*-condensed polycyclic aromatic hydrocarbons. *Croat. chem. Acta* **54**, 391–404 (1984); A. T. Balaban, C. Artemi and I. Tomescu, Algebraic expressions for Kekulé structure counts in non-branched regularly cata-condensed benzenoid hydrocarbons. *Math. Chem.* **22**, 77–100 (1987); A. T. Balaban, J. Brunvoll, J. Cioslowski, B. N. Cyvin, S. J. Cyvin, I. Gutman, He Wenchen, He Wenjie, J. V. Knop, M. Kovacević, W. R. Müller, K. Szymanski, R. Tosić and N. Trinajstić, Enumeration of benzenoid and coronoid hydrocarbons *Z. Naturf.* **42a**, 863–870   (1987);
A. T. Balaban, J. Brunvoll, B. N. Cyvin and S. J. Cyvin, Enumeration of branched cata-condensed benzenoid hydrocarbons and their numbers of Kekulé structures. *Tetrahedron* **44**, 221–228 (1988).

43. S. E. Stein and R. L. Brown, π-Electron properties of large condensed polyaromatic hydrocarbons. *J. Am. chem. Soc.* **109**, 3721–3729 (1987).

44. G. Wittig, *Stereochemie*, p. 148, Akademische Verlagsgesellschaft, Leipzig (1930).

45. R. Lukes, *Organická Chemie*, Vol. 1, p. 59. Naklad. Ceskoslovensk. Akad. Ved, Praha (1954).

46. A. T. Balaban and P. von R. Schleyer, Systematic classification and nomenclature of diamond hydrocarbons. I. Graph-theoretical enumeration of polymantanes. *Tetrahedron* **34**, 3599–3609 (1978).

47. A. T. Balaban, Chemical graphs. Part 27. Enumeration and codification of staggered conformation of alkanes. *Rev. Roum. Chim.* **21**, 1339–1343 (1976).

48. A. T. Balaban, Synthesis of pyrylotrimethinecyanines. *Tetrahedron Lett.* 599–600 (1978).

49. P. von R. Schleyer, A simple preparation of adamantane. *J. Am. chem. Soc.* **79**, 3292 (1957); C. Cupas, P. von R. Schleyer and D. J. Trecker, Congressane. *J. Am. chem. Soc.* **87**, 917–918 (1965).

50. V. Z. Williams Jr., P. von R. Schleyer, G. J. Gleicher and L. B. Rodewald, Triamantane. *J. Am. chem. Soc.* **88**, 3862–3863 (1966); W. Burns, M. A. McKervey, T. R. B. Mitchell and J. J. Rooney, A new approach to the construction of diamondoid hydrocarbons. Synthesis of *anti*-tetramantane. *J. Am. chem. Soc.* **100**, 906–911 (1978).

51. W. D. Fessner, B. A. R. C. Murty, J. Wörth, D. Hunkler, H. Fritz, H. Prinzbach, W. D. Roth, P. von R. Schleyer, A. B. McEwen and W. F. Maier, Dodecahedranes from [1.1.1.1] pagodanes. *Angew. chem. Int.* **26**, 452–454 (1987).

52. S. Hála, S. Landa and V. Hanus, Isolierung von Tetracyclo[6.3.1.0^{2,6}.0^{5,10}]dodecan und Pentacyclo-[7.3.1.1^{4,12}.0^{2,7}.0^{6,11}]tetradecan (Diamantan) aus Erdöl. *Angew. Chem.* **78**, 1060–1061 (1966).

Computers Math. Applic. Vol. 17, No. 1–3, pp. 417–423, 1989
Printed in Great Britain. All rights reserved

0097-4943/89 $3.00 + 0.00

C_{60}^B BUCKMINSTERFULLERENE, OTHER FULLERENES AND THE ICOSPIRAL SHELL

H. W. Kroto†

Department of Astronomy, University of California, Los Angeles, Calif., U.S.A.

Abstract—It seems remarkable that though carbon has been studied exhaustively for decades a new highly symmetric form, in addition to diamond and graphite, has recently been discovered. The C_{60}^B molecule is a cage with *t*-icosahedral symmetry analogous to that of a football, i.e. 12 pentagonal and 20 hexagonal faces. The molecule, buckminsterfullerene, is one of a family highly symmetric fullerene 5/6-face cage species. The spontaneous formation of C_{60}^B has led to the discovery of an icosahedral spiral (icospiral) network that explains the structure of carbonaceous particles such as soot. This new topological form solves the spheroidal wrapping problem using planar materials. Curiously giant fullerene structures can be generated as highly symmetric 3D cages by the introduction of only 12 pentagonal defects into a 2D hexagonal sheet network of any size.

INTRODUCTION

The discovery of the superstable C_{60} molecule in Sept. 1985 [1] has been rationalized on the basis of a new type of carbon compound: a graphitic closed cage in which the 60 atoms lie at the vertices of a truncated icosahedron, Fig. 1. The structure consists essentially of 12 five-membered rings separated by 20 benzenoid six-membered ones. Although historically this shape is listed as one of the semi-regular solids of Archimedes an interesting and highly appropriate historical reference structure is the hollow framework drawn by Leonardo da Vinci to illustrate the book, *De Divina Proportione*, by Pacioli [2].

Not only has the proposal that the species has this elegant symmetric structure given rise to much debate but so also has the name chosen [3–5]. The number of alternative names that have been proposed is large, however it is worth noting that in retrospect none has proven more appropriate than that chosen by us on the basis of the ideas of Buckminster Fuller [6] that led us to propose the structure in the first place [1]. Some favour the rather obvious (and somewhat dull) name footballene, but here there is a problem as in the U.S. this name applies to an ovoid structure and should be translated as soccerene. The appropriateness of the name, buckminsterfullerene (though perhaps a trifle long, does roll smoothly off the tongue—albeit an Anglo-Saxon one), will become apparent below and hopefully this will enable those allergic to the length to overcome their antipathy. Anyone afflicted by endemic pedantry might contemplate the IUPAC or *Chemical Abstract* names and if they can get them correct [7] so overcome the withdrawal symptoms. It is convenient to use the formula C_{60}^B to differentiate buckminsterfullerene from all other 60-carbon atom isomers, and perhaps prefix a symmetry symbol if necessary i.e. $(I_h)C_{60}^B$.

For the purpose of this article we are interested in that property which so often captivates the human imagination whether in childhood or as an adult (is there a difference?)—symmetry. There are few other molecules which fascinate, by their symmetry alone, more than does C_{60}^B. Organic chemists have long been driven to develop new synthetic techniques in order to produce new compounds with exotic structural symmetry. Perhaps the most famous recent example is the work of various groups that led to the elegant molecule dodecahedrane, a mammoth task, which was finally achieved by Paquette *et al.* [8]. As far as buckminsterfullerene is concerned its first appearance appears to have been in 1970 when Yoshida and Osawa [9, 10] in a highly perceptive piece of work on superaromaticity considered the possibility that such a molecule should be stable. They appear to have been influenced by the earlier very nice synthetic work of Barth and Lawton [11] who made corannulene, Fig. 2, and found that it was stable. C_{60}^B was also discussed subsequently by Bochvar and Gal'pern [12] and Davidson [13]. A rather remarkable idea which predates all these, however, was that of Jones (pseudonym—Daedalus of the *New Scientist*) who

---

†Permanent address: School of Chemistry and Molecular Sciences University of Sussex, Brighton BN1 9QJ, England.

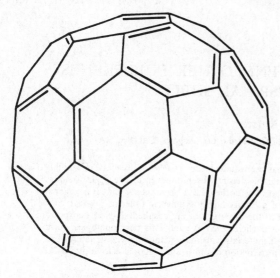

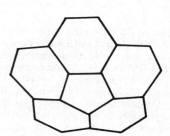

Fig. 1. The structure proposed for the superstable C_{60} carbon molecule discovered during laser vaporization of graphite [1]. The tetravalency of carbon is neatly accounted for as there are 12500 resonance structures.

Fig. 2. The corannulene structure presented in perspective to highlight the fact that it is a saucer shaped molecule [11].

contemplated, in a most imaginative way, the possibility that large spheroidal carbon networks might be formed by some high temperature graphite production process [14, 15]. These ideas led some chemists to consider the possibility of synthesising C_{60}^{B} by traditional organic chemistry approaches. However, the most important aspect of the C_{60}^{B} discovery is undoubtedly the fact that such an elegant molecule actually forms spontaneously. A study of this and other aspects of C_{60}^{B} has, as discussed below, shed new light on the structure of materials and revealed new structures with symmetries and shapes which do not appear to have been discussed previously.

C_{60}^{B} buckminsterfullerene, the truncated icosahedral carbon cage

The experimental evidence for a stable C_{60} species is now unequivocal. It was observed that when carbon is laser vapourized into a helium atmosphere and the resulting plasma allowed to nucleate certain magic numbers are observed in the mass spectrum of the resulting products. The original observation showed that a C_{60} species exhibited spectacular resistance to further growth and to a lesser extent so also did C_{70} [1]. Almost immediately we came to the conclusion that the closed t-icosahedral cage structure could readily explain the observation for C_{60} [1]. From the point of view of the closure of a hexagonal network the inclusion of pentagons is very important. Jones in his original proposal [14, 15] was aware of an integral requirement for the closure elegantly presented in the beautiful book of Thompson [16] who writes:

> "But here something strange comes to light, no system of hexagons can enclose space whether the hexagons be equal or unequal, regular or irregular, it is under all circumstances mathematically impossible. So we learn from Euler: the array of hexagons may be extended as far as you please, and over a surface either plane or curved, but it never closes in."

Euler's law states that for any polyhedron,

$$C(\text{No. of corners}) + F(\text{No. of faces}) - E(\text{No. of edges}) = 2.$$

For a system consisting entirely of hexagons $C + F - E = 0$ and so such a system cannot close. In general the closure of a three-connected network is governed by the general relation [17]:

$$12 = 3 \cdot n_3 + 2 \cdot n_4 + 1 \cdot n_5 + 0 \cdot n_6 - 1 \cdot n_7 - 2 \cdot n_8 \ldots.$$

In the 5/6 cases this formula indicates that closure can occur for any system which includes 12 pentagonal configurations and an unlimited number of hexagonal ones. The C_{60}^{B} structure elegantly fulfills this topological requirement with a satisfying, chemically stable structure in which each pentagonal ring is completely surrounded by benzenoid hexagonal ones [18, 19]. A most interesting further property of these 5/6 fullerene cages is the fact that C_{60}^{B} is the smallest cage that can contain completely isolated pentagons and this isolation cannot occur again until C_{70} [18, 19]. The observation of 70 as the second most prominent magic number lends further weight to the closed fullerene structure proposal in general and the buckminsterfullerene structure of C_{60} in particular.

From a symmetry point of view the C_{60}^B structure is extremely interesting in that it is *the* one which allows the maximum number of atoms to be distributed so that all are equivalent in terms that all positions are related by proper symmetry operations. From the point of view of chemical stability it is important to note that for extended carbon aggregates, hexagonal, followed by pentagonal, ring configurations are by far the most preferred structures. These factors endow C_{60}^B with unique geodesic *and* chemical structural stability. Such a structure is now accepted as also further stabilized by aromaticity and the fact that it will have a closed electronic shell configuration [20, 21]. For example this molecule has 12,500 resonance structures and on the basis of this criterion some degree of resonance stabilization is to be expected [22]. Perhaps more important than anything else is the fact that the molecule has no weakness, so from the point of view of chemical attack it should be as impregnable as the surface of graphite. It may be that this molecule is the most stable molecule that exists. Although it may be unstable relative to graphite in the bulk form there is now no doubt that it is stable relative to a flat sheet of hexagonal graphite with 60 atoms because such a sheet must have *ca* 22 dangling bonds on the edge which in the case of C_{60}^B are completely eliminated by intramolecular bond formation with the consequential gain in bond energy. A macroscopic piece of graphite (whatever that is!), may gain by having additional interlayer stabilization energy and this probably overcomes the destabilization due to unsatisfied valences at the edge when their numbers become insignificant in the bulk. The edges almost certainly will be cauterized by other atoms such as hydrogen in practice.

Other small fullerenes

In the original experiments it was already clear that 60 was not the only magic number, 70 was also relatively prominent. In addition, these two magic numbers were soon to be joined by some others [23]. If the closed cage C_{60}^B structural proposal were correct it seemed likely that these further observations should also be susceptible to explanation in terms of the closure principle. Such has turned out to be the case. It can be shown that varying degrees of relative geodesic/chemical stability are to be expected for certain numbered cages. The stability depends mainly on the how many pentagonal faces abut and the symmetry of disposition of these pentagonal faces [18, 19] and a careful study of the possible cage structures that are feasible leads rather satisfyingly to an extensive set of magic numbers: 24, 28, 32, 36, 50, 60 and 70 [18, 19] in excellent agreement with observation [23, 1]. We will call these structures, in general, fullerenes (neatly allowing us to retain buckminster—as a prefix for the C_{60}^B founder member of the family [4]). This is where the appropriateness of the name becomes evident as Buckminster Fuller's patents apply to a range of 5/6 ring structures and not just the 60 vertex one, in addition there is the possibility of giant fullerenes (see below), of the kind envisaged by Jones [14, 15], with many more than 60 atoms and there is actually some evidence that species with as many as 240 atoms form.

In the case of the small fullerenes it is worth noting that C_{20}^F with no benzenoid hexagonal rings at all is the smallest and is expected to be by far the least stable. By analogy with dodecahedrane it should presumably be called dodecahedrene. It is also worth noting that no C_{22}^F species can exist [24, 19] as the second fullerene must have at least one benzenoid ring and this can not occur until there are at least 24 atoms as in C_{24}^F. It is most gratifying that in at least one experimental observation the fullerene family appears to start fairly clearly at C_{24} [25]. During the two-and-a-half or so years that have elapsed since C_{60}^B was detected, occasionally the C_{28} cluster would also, under certain nucleation conditions, appear very prominent. It is thus most exciting that the simple cage theory predicts that the first high symmetry, relatively well-stabilized species that can be constructed is C_{28}^F with a rather neat tetrahedral structure, Fig. 3 [18]. This species has a most intriguing family relationship with Gomberg's triphenyl methyl, which was the fore-runner of free radical organic chemistry [26]. This result, if correct, suggests that the tetrahedral hydrocarbon $C_{28}H_4$, in which the carbon atoms at the centres of the four sets of trigonally abutting pentagons are hydrogenated, might be a relatively stable tetrahydrofullerene. Steps in stability are also to be expected for larger cages such as C_{32}^F and C_{36}^F [18]. There are other structures which are predicted to show some extra

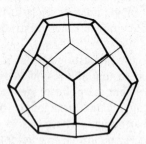

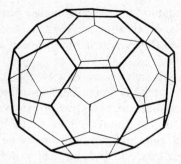

Fig. 3. After the first possible fullerene, C_{20}^F (dodecahedrene) which must be the most unstable, the first highly symmetric fullerene is the tetrahedral C_{28}^F cage shown here.

Fig. 4. One possible highly symmetric structure is depicted for a C_{50}^F cage which can avoid the involvement of fused triplets of pentagons.

stability, such as the rather elegant C_{50}^F structure depicted in Fig. 4 [18, 19]. The C_{50} signal is often seen to be prominent and it is most satisfying that the cage proposal is consistent with this observation also.

Probably as important as any other observation is result that C_{70}^F which is shown in Fig. 5, is the first cage, after C_{60}^F, which can be constructed without abutting pentagons [18, 19]. This cage is effectively two-halves of C_{60}^B separated by a ring of 10 carbon atoms and has a more-or-less ovoid shape. The observation that the C_{70} peak is second only to C_{60} in strength [1] can thus be readily explained and represents the most compelling further observation lending support to the fullerene closed cage structure proposal that is available at present.

THE ICOSPIRAL PARTICLE AND THE GIANT FULLERENES

The most intriguing aspect of the whole C_{60}^B story is the contention that such a symmetric molecule forms spontaneously. In order to explain this a straightforward scheme has been developed for nucleation leading to extended networks of sp^2 hybridized carbon atoms. It is proposed that the process is governed by two major factors:

(1) Energy driven nucleation, will lead to curved carbon network [27, 28]. (Note that this is in complete contradiction to the traditional assumption that flat graphitic microcrystals form spontaneously.)

(2) All further network growth takes place under epitaxial control [28].

The reason for factor (1) is straightforward as a flat hexagonal benzenoid C_{60} network will possess some 22 or so dangling bonds whereas a curved one will be able to eliminate these by bond formation, resulting in the most energetically favourable structure in the case of complete closure.

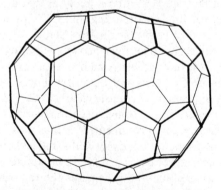

Fig. 5. It is curious that for fullerenes C_n^B with $n \leqslant 70$ the only ones that can be constructed without abutting pentagons are C_{60}^B itself and the C_{70}^F structure depicted here. The former has the shape of a football (or U.S. soccerball) whereas the latter has the shape of a rugby ball (or U.S. football). The above structure is readily produced by splitting C_{60}^B into two halves rotating by 36° and inserting a ring of 10 extra carbon atoms.

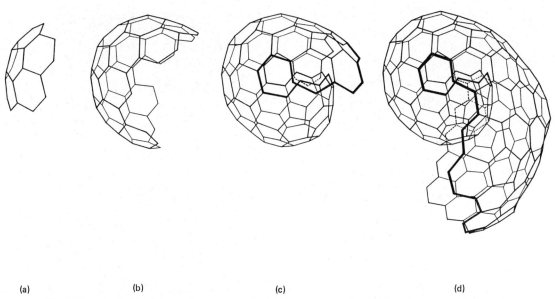

(a)   (b)   (c)   (d)

Fig. 6. Hypothetical initial growth sequence evolving from a corannulene carbon framework (a) through a broken egg-shell structure (b) to (c) a species in which edge by-pass occurs to an embryo in which the second shell is forming (d). The sequence leads in a natural way to both carbon particles and closed cages, fullerenes.

The network can curve towards closure by including five-membered rings. A typical nucleation scenario is depicted in Fig. 6 [28] in which we see that, in general, the mechanism results in an incompletely closed shell in which edge overlap occurs. To understand the structural implications of such a growth scenario the second premise (2) is required. As the spiral shell grows it is important to recognize that 12, and only 12, pentagonal configurations will occur in any 360° spheroidal section of the shell. One must also be aware that essentially all the curvature resides in these corannulene type cusps. It is at this point that a most interesting shape controlling effect becomes evident as the growing spiral evolves to form an object of ever increasing size. As fresh network forms it is likely to do so in such a way that the cusps in successive shells lie above each other, in other words the pentagons will be localized along radii emanating from the centre of the growing spiral. This is basically the epitaxial growth requirement.

As the object grows somewhat like a snowball accreting new surface, the resulting cluster should take on a shape consistent with that of a giant fullerene of commensurate size. Our study of the shape revealed a surprising result that as fullerenes grew in size the more-or-less round shape of C_{60}^B was effectively replaced by a polyhedral one [28]. This is depicted clearly in Fig. 7 where the change in shape on passing from the small to the symmetric giant fullerenes is shown. Here we see the remarkable result that as the cages get larger the object takes on an effectively icosahedral shape. The truncation still exists at the atomic dimensions but is lost on the macroscopic scale. In the closed object the 12 cusps in the surface are linked by a more-or-less smooth hexagonal net in the giant fullerenes. In fact the best way to think of this object is as a more-or-less flat flexible hexagonal net with 12 pentagonal defects which cause the net to fold and enclose space. As mentioned previously, such objects may actually form at least up to 240 atoms.

In general however the nucleation results in a spiral which grows to form a giant molecule/quasi-crystal consisting of concentric shells of quasi-icosahedral polyhedra. There is strong evidence in the elegant transmission electron microscope work of Iijima [29], that carbon particles do indeed possess such internal onion-like structure, which can further be explained in terms of the quasi-icosahedral spiral shell described here [28]. Of course in practice only approximate symmetry is to be expected though the basic polyhedral shape should be evident. This observation lends strong support not only to the nucleation mechanism in general but also to the C_{60}^B structure proposal in particular.

As far as the production of C_{60}^B itself is concerned, it is predicted that very occasionally, as the edges of the initial single shell embryos approach during growth, closure occurs and C_{60}^B or some

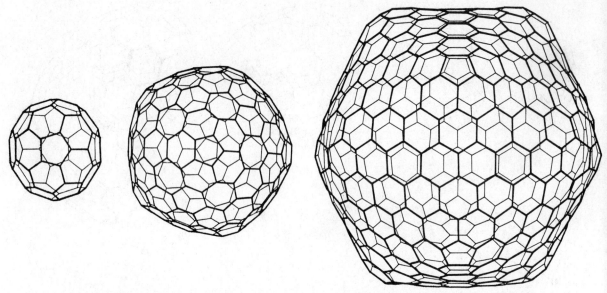

Fig. 7. A set of models for the closed symmetric fullerenes C_{60}, C_{240} and C_{540}. The diameters of the latter cages are two and three times that of C_{60}. The rapid shift towards quasi-icosahedral symmetry is striking.

other member of the fullerene family forms. There may be some evidence that there is an extra tendency to form the most favoured C_{60}^B cage in the fact that C_{62} peak is always weak relative to all other neighbouring ones. On closure no further growth can occur as the reactive edge has been eliminated. The nucleation scenario thus indicates that C_{60}^B is the *lone survivor* of the process and should *not* lie at the centre of a soot particle in general. The discovery that out of a chaotic system a single molecular species can remain when essentially all other carbon atoms have been absorbed into bulk material is a truly unique discovery. In this way it is most exciting to conjecture that C_{60}^B has been around since time immemorial and in particular is produced wherever carbon particles form both in the fireplace and in the outflows of carbon-rich red giant stars. In the latter case it may be detectable by some optical characteristic and as far as the ion C_{60}^{B+} is concerned it may even be responsible for the diffuse interstellar bands [30]. The point that it is the only molecule to survive out of chaos delineates C_{60}^B (and its positive ion) from all other contenders that have previously been proposed as carriers of the diffuse bands.

SUMMARY OF SYMMETRY ASPECTS

Some new and fundamental results on carbon particles have been described from the basis of their symmetry. It seems curious that although carbon has been the subject of so much study over so many years that the underlying symmetries addressed in this article have remained hidden for so long. Only now can such elegant structures be contemplated on the basis of the remarkable observation of the superstability of the C_{60} molecule which must be ubiquitously distributed both on earth and in space [30–32]. Nowhere is this more clear than in the work of Gerhardt *et al.* [33] who have shown that C_{60}^+ is a dominant ion in a sooting flame. This observation lends some of the strongest support for the nucleation mechanism we have proposed [27, 28] a key aspect of which is the prediction that C_{60}^B will be a by-product of carbon nucleation in general and soot formation in particular. It seems almost unbelievable that such a species could have remained unsuspected for so long. It is however the lone molecular survivor of the soot formation process and so probably lies hidden and difficult to separate and identify. Even so the intriguing aspect from the point of view of this article is that the work has led to molecules with most beautiful symmetries:

(1) C_{60}^B with a closed almost round *t*-icosahedral cage, Figs 1 and 2.
(2) Small fullerenes with many different cage symmetries, Figs 3–5.
(3) Giant fullerenes with quasi-icosahedral polyhedral shape, Fig. 7.
(4) The icospiral shell, Fig. 6.

It is interesting to conjecture that the icospiral shell may well be *the* primordial particle. It is highly likely that the first particle to form was carbonaceous and must have done so in the gas phase. If so our new observations point unequivocally to this structure. Indeed this icospiral structure may have even wider applicability in that it provides a way to form globular materials from sheets involving hexagonally packed entities by occasionally introducing defects. Another interesting possible rôle exists for this quasi-icosahedral, quasi-single-crystal particle, which is essentially a giant molecule; it is feasible that under isotropic pressure it will be an ideal embryo which with a minimal rearrangement serves as a growth nucleus for the diamond crystal.

The curious icospiral shell structure proposed here has C_s symmetry in its most symmetric form. The initial embryo is seen from the side in Fig. 6. It is effectively a spiral constructed from essentially triangular segments, more-or-less smoothly connecting the pentagonal cusps in such a way that the interlayer separation is as close to the graphitic distance of *ca* 3.4Å as possible. As successive segments form, the size must increase and in the case of the more symmetric structures a more or less icosahedral particle results. The spiral shell will have two helical edges which spiral away in opposite directions. It is intriguing to note that in one-and-the-same object one can have both of the highly bioemotive features, the helix and icosahedron.

Acknowledgements—I am very happy to acknowledge many discussions on this topic with: Bob Curl, Laurence Dunne, Patrick Fowler, Jim Heath, David Jones, Mike Jura, Ken McKay, Sean O'Brien, Alan Parsonage, Rick Smalley, David Wales and David Walton. I am also grateful to Mike Jura and the University of California, Los Angeles where this work was carried out.

REFERENCES

1. H. W. Kroto, J. R. Heath, S. C. O'Brien, R. F. Curl and R. E. Smalley, *Nature* **318**, 162–163 (1985); H. W. Kroto, *Science* (in press); R. F. Curl and R. E. Smalley, *Science* (in press).
2. F. Lucia Pacioli, *De Divina Proportione*. Ambrosiana (Diagrams by Leonardo da Vinci); *The Unknown Leonardo* (Ed. L. Reti), p. 71. McGraw-Hill, New York (1974).
3. P. J. Stewart, *Nature* **322**, 444 (1986).
4. H. W. Kroto, *Nature* **319**, 766 (1986).
5. A. Nickon and E. F. Silversmith, *Organic Chemistry—The Name Game. Modern Coined Terms and their Origins*. Pergamon Press, Oxford (1987).
6. R. Buckminster Fuller, *Inventions—The Patented Works of Buckminster Fuller*. St. Martin's Press, New York (1983); R. W. Marks, *The Dymaxion World of Buckminster Fuller*. Reinhold, New York (1960).
7. J. Castellis and F. Serratosa, *J. chem. Educ.* **83**, 630 (1986).
8. L. A. Paquette, R. J. Ternansky, D. W. Balogh and G. Kentgen, *J. Am. chem. Soc.* **105**, 5446 (1983).
9. E. Osawa, *Kagaku* (in Japanese) **25**, 854 (1970).
10. Z. Yoshida and E. Osawa, *Aromaticity* (in Japanese). Kagakudojin, Kyoto (1971).
11. W. E. Barth and R. G. Lawton, *J. Am chem. Soc.* **93**, 1730 (1971).
12. D. A. Bochvar and Gal'pern, *Dokl. Akad. Nauk SSSR* **203**, 610 (1973); I. V. Stankevich, M. V. Nikerov and D. A. Bochvar, *Russ. chem. Revs* **53**, 640 (1984).
13. R. A. Davidson, *Theoret. chim. Acta* **58**, 193 (1981).
14. D. E. H. Jones, *New Scient.* **3 Nov**, 245 (1966).
15. D. E. H. Jones, *The Inventions of Daedalus*, pp. 118–119. Freeman, Oxford (1982).
16. W. D'Arcy Thompson, *On Growth and Form*, Chap. IX, p. 737. Cambridge Univ. Press, London (1942).
17. B. Grunbaum, *Convex Polytopes*. Wiley Interscience, New York (1967).
18. H. W. Kroto, *Nature* **329**, 529–531 (1987).
19. T. G. Schmaltz, W. A. Seitz, D. J. Klein and G. E. Hite, *J. Am. chem. Soc.* **110**, 1113–1127 (1988).
20. P. W. Fowler and J. Woolrich, *Chem. Phys. Lett.* **127**, 78–83 (1986).
21. D. J. Klein, W. A. Seitz and T. G. Schmalz, *Nature* **323**, 703 (1986).
22. D. J. Klein, G. Schmaltz, G. E. Hite, W. A. Seitz, *J. Am. chem. Soc.* **108**, 1301–1302 (1986).
23. Y. Liu, S. C. O'Brien, Q. Zhang, J. R. Heath, F. K. Tittel, R. F. Curl, H. W. Kroto and R. E. Smalley, *Chem. Phys. Lett.* **126**, 215–217 (1986); S. C. O'Brien, Ph.D. Thesis, Rice University, Houston, Tex. (1988).
24. P. W. Fowler and J. I. Steer, to be published (see also Ref. 19).
25. D. M. Cox, K. C. Reichmann and A. Kaldor, *J. Chem. Phys.* **88**, 1588–1507, Fig. 6a (1988).
26. L. F. Fieser and M. Fieser, *Organic Chemistry*, pp. 551–555. Reinhold, New York (1956).
27. Q. L. Zhang, S. C. O'Brien, J. R. Heath, Y. Liu, R. F. Curl, H. W. Kroto and R. E. Smalley, *J. Phys. Chem.* **90**, 525–528 (1986).
28. H. W. Kroto and K. McKay, *Nature* **331**, 328–331 (1988).
29. S. Iijima, *J. Crystal Growth* **5**, 675–683 (1980).
30. H. W. Kroto, *Polycyclic Aromatic Hydrocarbons and Astrophysics* (Eds A. Leger *et al.*), pp. 197–206. Reidel, Holland (1987).
31. H. W. Kroto, *Carbon in the Galaxy* (Ed. S. Chang) *Proc. Conf*, NASA Ames Res Center, Nov. 5–6 (1987).
32. H. W. Kroto, *Phil. Trans. R. Soc.* **A325**, 405–421 (1988).
33. Ph. Gerhardt, S. Loffler and K. H. Homann, *Chem. Phys. Lett.* **137**, 306–309 (1987).

Computers Math. Applic. Vol. 17, No. 1–3, pp. 425–441, 1989
Printed in Great Britain. All rights reserved

ASYMMETRY THROUGH THE EYES OF A(NOTHER) CHEMIST†

M. Ács

Department of Organic Chemical Technology, Technical University of Budapest, Budapest,
P.O. Box 91, H-1521, Hungary

Abstract—Mechanism of chiral recognition is one of the most exciting problems in organic chemistry and biochemistry. Starting with a model of three-point interaction between two individual molecules, we come to a more complex recognition system composed of homostructural and homochiral molecules, mimicking the asymmetric surfaces catalyzing biochemical reactions in living organisms. These asymmetric surfaces can discriminate between the enantiomers of entirely different molecules, so they are able to separate the synthetically prepared optical isomer mixtures and this system is a good model for studying second order weak interactions playing an important role in enzyme catalyzed reactions. The asymmetrically situated binding points on the surfaces formed by the discriminating molecules require complementary positioning for interacting groups in the guest molecules, and this requirement cannot be completely fulfilled by both mirror image molecule. Compounds able to form such chiral surfaces are good and frequently applied resolving agents.

SYMMETRY AND CHIRALITY—PRINCIPLE OF APPROPRIATENESS

There are giant molecules among the building blocks of higher organized living organisms, such as the nitrogen–carbon skeleton of the porphyrine molecule in hemoglobin, which can be built up with the aid of manifold screw axes. The very complicated protein framework of hemoglobin is composed of four nearly identical protein chains [2], which are directed in such a way that the whole molecule has three (a true and two virtual) orthogonal twofold screw axes (Fig. 1). Compounds with fundamental importance for life are characterized by the lack of mirror symmetry.

In view of the encoding mechanism of giant asymmetric molecules for the construction of macroscopically symmetric living organisms composed of asymmetric molecules, a contradiction can be felt.

The external symmetry is a determining factor for our perception and judgement as well as for our physical activities. Symmetry is frequently related to the manifestation of beauty and harmony, although in some cases the embodiment of hideousness and fearfulness has also symmetrical patterns [3]. There are practical as well as rational reasons leading to the development of external biological symmetry. Let us consider humans. All activities in and against the external world can be grouped among the mechanical movements, and the probability for the success of a given activity depends on an equilibrium of forces. The simplest way to reach the equilibrium state is the use of pairs of forces. The pair in itself suggests bilateral symmetry. Obviously, the biological proportion of human bodies can not be derived so easily, but two interesting examples support the above reasoning.

(1) Nowadays, many babies come into the world with a partial weakness of the eye muscles causing the eyes to look in two different directions (squint or strabismus). This involves grave consequences. As a result of the non-parallel position of the eyes, pictures conveyed by the two eyes into the brain do not coincide, everything is seen in duplicate and movement becomes unbalanced. In each moment, the brain is forced to decide on choosing between the duplicate sets of information. This difficult task is solved by ignoring one of the pictures and after a while the eye conveying the "second" picture will be disconnected and will lose its ability to see [4].

(2) Dawkins, in his book entitled *The Selfish Gene*, sums up the replicator theories connected with the origins of life. The replicator is a spontaneously forming chemical system which is able to reproduce itself, for instance the DNA among the actual "polymolecules". Some replicators in the prebiotic soup had developed a defensive cover against other replicators and other dangers [5].

---

†Allusion is made to the book *Symmetry Through the Eyes of a Chemist* [1].

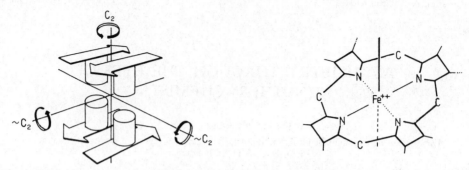

Fig. 1. Schematic representation for the four protein chains in hemoglobin and the carbon–nitrogen skeleton of the porphyrine molecule.

These covers must have been aspecific in their constitution and shape. The covering of a single polymolecule living in aqueous medium must have been either close-fitting, following the morphological shape of the replicator, or encapsulating water because of the dissolved substances necessary for the replicator's self-reproduction, they must have been shaped in the statically and geometrically most stable form: the sphere. At that moment, when this sphere was to get in touch with external partners, it had to be able to distinguish them from possible invaders. The less confusable marks on the extraneous body and the more specific the discerning system through which the newcomer had to pass, the more reliable the differentiation was.

As the building blocks for these systems are molecules and the symmetric arrangement of their atoms diminishes with their becoming more complicated, a level is reached where not only the entire molecular system but even the constituting units are composed of asymmetric molecules.

Like other asymmetric objects, asymmetric molecules have their mirror image pairs. This mirror image isomer (enantiomer, optical isomer, optical antipode)—in the case of chemical compounds—can be formed spontaneously, but can also be prepared synthetically. In our biosphere their spontaneous generation is rare. The reason for it is that biochemical reactions producing these asymmetric molecules, important for life, always take place on asymmetric surfaces. On these asymmetric surfaces only one of the mirror image isomers can be derived, always with the same configuration. Under achiral conditions both mirror image isomers form simultaneously in equal amounts, since there is no energy difference between them detectable by the actual earth-scale measuring techniques. Nevertheless, taking into consideration the parity-violating weak neutral current interactions, Mason pointed out that the perfect mirror image of an amino acid with L-configuration can only be an amino acid with D-configuration built from anti-matter [6]. Although the difference in energy between the two mirror-image terrestrial amino acids is $5.6 \cdot 10^{-19}$ eV, on human time-scale this difference is not enough for the enrichment of any of the enantiomers as an outcome of a given reaction pathway. At the same time, experimental results make it probable that the sense of the earth's rotation can affect the selectivity of reactions resulting in chiral products [7].

Evaluating these two pieces of information one may suppose that the preponderance of L-amino acids and D-sugars in nature could not have been incidental. As a consequence of the chiral discrimination, a slight excess of one of the enantiomers through billions of reaction steps has led to the complete suppression of its mirror image isomer.

Seelig [8] elaborated a relatively simple statistical model for illustrating a possible way of the generation of optical activity in nature. The model is related to a two-step reaction where the autocatalytic effect of the end-product predominates. During these reactions the prochiral substrate is transformed into a chiral intermediate, which is further transformed into the end product. The configuration of the intermediate (and, consequently, the end product) depends on the chirality of the autocatalyst. If there is a slight excess (say, $10^{-8}\%$) in favour of one of the enantiomers, after a finite number of repeated reactions, supposing a constant flow of the initial substrate, one of the optical isomers will prevail in the system (Fig. 2).

The combination of the above approaches gives a possible answer to the questions: "why just this one of the optical isomers?" and "how did it disappear?". Appropriateness has required barriers for the free cross-combinations of the optical isomers. Their random combinations would

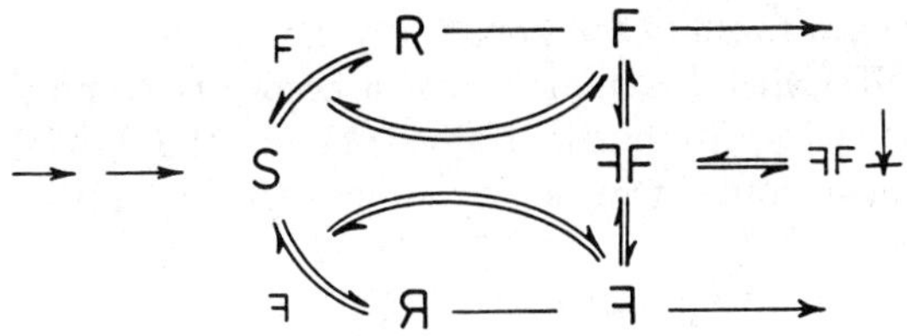

Fig. 2. How can a slight excess of one of the optical isomers induce the supression of its mirror image molecule in an autocatalytic two-stepreaction? The starting material (S) is achiral, the product (RЯ) and the intermediates (Fꟻ) are chiral. Slight excess generated randomly in any of the mirror image autocatalysts leads to its exclusive spreading.

have led to serious consequences. The number of stereoisomeric peptides built up from amino acids having only one asymmetric centre can be 2^n (n = the number of the asymmetric centres in the product molecule). For instance, the sweet taste of the dipeptide formed from L-aspartic acid and L-phenyl-alanine methyl-ester exceeds that of sucrose by 160 [9, 10]. Interestingly, the L-aspartic acid itself is tasteless, while the L-phenyl-alanine is bitter [11]. To taste the sweetness of the other stereoisomers (Fig. 3) we should have three other receptors, with different structures.

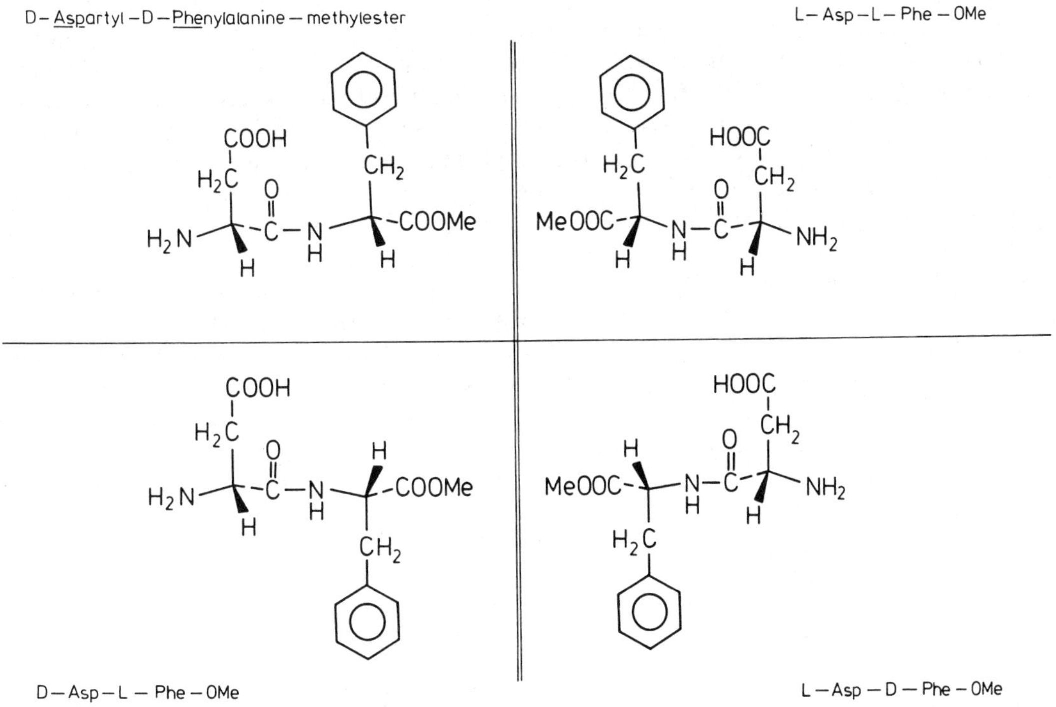

Fig. 3. Stereostructural variations for a dipeptide theme. L-aspartyl-L-phenyl-alanine methyl-ester marketed as ASPARTAME®.

Our reasoning then can lead to such a conclusion that symmetry is a requirement of stability and certainty, while asymmetry is the condition of unambiguous information storage and information transfer.

DOES NATURE TEND TOWARD SYMMETRY OR ASYMMETRY?

A large body of experimental evidence (in spectroscopy, crystallography, nuclear physics, chemical reactions etc.) [1] seems to confirm the existence and ruling of the law of symmetry. At the same time asymmetry is present [12], even in the world of elementary particles.

For its practical advantages, human preferences to symmetry are very strong—we are willing to complete an object with its mirror image for being able to apply (or extend the usage of) symmetrically-structured principles and rules.

There are other approaches to unify the apparently conflicting points of view. In Urmantsev's general system theory [GST(U)]: "...symmetry has a necessary complement and opposite that is a corresponding asymmetry". Within the framework of the GST(U), "...asymmetry is a system category indicating a noncoincidence of system (S) in features (F) with an account of modifications (M)" [13].

INTERACTIONS AMONG ASYMMETRIC (CHIRAL) MOLECULES AND CHIRAL SYSTEMS

One of the most exciting problems is the already mentioned unexplainable duality of the following "equation":

$$\text{asymmetric building blocks} \quad + \quad \text{asymmetric encoding system} \quad \Rightarrow \quad \text{symmetrical living organisms.}$$

Does the general information carrier, the multifold asymmetric double helix, really encode bilateral symmetry, which characterizes the major part of the animal kingdom? Bilateral symmetry in insects can be explained on the basis of circular asymmetries during early embryogenesis [14]. The individual cells—during ontogenesis—somehow "sense" their positions and this "positional" information is interpreted by embarking on position-specific developmental pathways. If the proper neighbouring cell(s) is (are) eliminated or somehow disappear(s), the remaining cell(group)s imitate or completely take its (their) function over. Only one important point is missing from the theory, i.e. how can the individual cells "sense" their relative positions in the developing system.

There are some well-known examples of completion or function take-over in nature, e.g. regeneration processes in nematodes or function take-over between the two hemispheres of encephalon, in cases where partial injuries have occurred [15]. Obviously, there should be some other relationship between asymmetry and symmetry than their antinomy.

Anet *et al.* [16] have shown that chiral objects cannot be split into achiral isometric objects, while it is obvious, that proper, even numbered combinations of isometric, asymmetric elements can have mirror symmetry. At molecular level, this means, that while achiral molecules have an infinite number of chiral conformations (Fig. 4), chiral molecules have no achiral conformations at all.

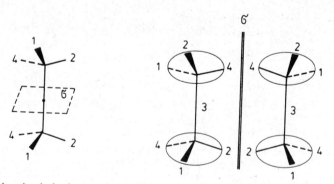

Fig. 4. Achiral molecule having a symmetry plane orthogonal to the main bond axis. All rotational isomers, but the eclipsed conformer, lack this symmetry element.

Partial or complete lack of symmetry is one of the non-negligible factors influencing reaction rate: e.g., during the reaction of two chiral molecules the activation enthalpies for the diastereo-isomeric transition states are different primarily because of difference in steric accessibility. The product enthalpy differences derive from the greater steric crowdedness of one of the diastereo-isomers; 5 kJ/mole difference in activation enthalpy at 25°C results in a ratio of 7.5 for the reaction rate of the favoured isomer, leading to an enantiomeric excess of 76%, under kinetic control [17]. It is easy to guess that in the reactions of molecules having chirally more complicated structures, the stereodifferentiating effects will dominate; the energy difference will increase with increasing chirality. For instance, in the case of enzymes the difference amounts to the energy domain necessary for the configurational inversion of the asymmetric centre (racemization). This value is generally in the range of bond-breaking energies.

All the above implies that both symmetry and asymmetry have relative meanings: any object (or system) can be treated as symmetric or asymmetric, if we choose proper observation points. Human beings, even nowadays, mostly choose those points where symmetry is more striking. For thousands of years human beings could have merely relied on external observations in cognition of nature leading to the consequences that the man-made or artistic objects are shaped and built up symmetrically. Thus the symmetric shapes and structures of our synthetic world may have their roots in copying natural objects and creatures.

As the human ability for abstractions has developed, applications of symmetry rules have become conscious, so the cognition processes have become simpler and easier to treat. This also holds for cases where only details or a few features of the model can be described as symmetrical.

However in nature asymmetry is at least as important as symmetry. In all important biochemical reactions, asymmetric molecules do play roles and so a number of "open" questions emerge from the field of chemistry and molecular biology.

MANIFESTATION OF ASYMMETRY IN MOLECULAR DIMENSIONS

As the two-dimensional, hexagonal benzene structure was amply supported by subsequent experimental work, Kekulé was tempted to extend his flatland structural chemistry to aliphatic and inorganic compounds [18]. Molecular optical activity, however, did not yield to any explanation based upon the two dimension approach.

Le Bel's pioneer work parallel to that of van't Hoff on the connection between optical activity in the fluid phase and enantiomorphous three-dimensional structures [19], gave organic stereochemistry its present form. The three-dimensional approach could provide the solution to a number of problems emerging in the field of organic chemistry, such as the above mentioned optical isomerism. Four different ligands attached to one carbon atom can be arranged in two mirror image (enantiomeric) ways (Fig. 5). Later on it turned out that optical isomerism is not exclusively bound

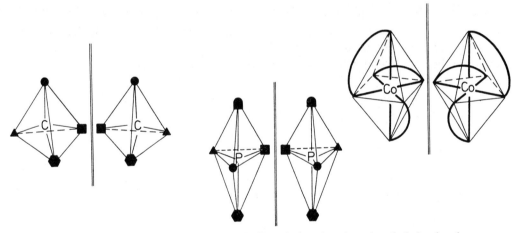

Fig. 5. Mirror image polyhedra of asymmetrically substituted carbon (tetrahedra), phosphorous (bipyramids) and cobalt (octahedra).

to the presence of asymmetrically-substituted tetravalent carbon atoms, since properly substituted tri- and tetravalent nitrogen, tri-, tetra- and pentavalent phosphorous and tricoordinated, bivalent sulphur atoms or asymmetric substitution along an axis, plane or helix may also lead to optical isomerism (Fig. 6) [20]. If the coordination number is greater than four, valencies of the central atom direct towards the vertices of a trigonal bipyramid (coordination number five), or of an octahedron (coordination number six) (Fig. 5).

The metal centred octahedral structure for hexacoordinated transition-metal complexes was proposed by Werner, who deduced the number and type of (geometric and optical) isomeric products expectable as a result of ligand replacement reactions of symmetrically substituted cobalt

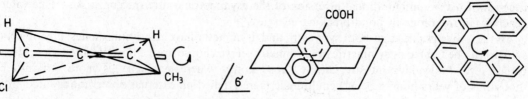

Fig. 6. Chiral arrangement along an axis, plane and helix.

compounds [21]. To support his theory, Werner had provided convincing experimental evidence, in separating first geometric, later on optical isomers of coordination complexes, thus confirming the unity of entire chemistry [22].

Natural substrates have their unambiguous natural pathways to react, but synthetic compounds may follow different and unpredictable ways in living organisms. This is particularly true for the enantiomers of chiral drugs. Dramatic differences in biological activity between the enantiomers may occur [23]. These differences manifest themselves both in therapeutic and in toxic side effects. Thus, e.g. the (−)-isopropyl-noradrenaline is an 800 times more effective bronchodilator than its

D-(−)-isopropyl-noradrenaline

(+)-isomer [24]. The racemate (1:1 mixture of mirror image isomers) can produce half of the therapeutic potency of the useful enantiomer. Enantiomers may convey opposite biological activity (antagonism) or reinforce the useful activity (synergism) of one another. In the latter case the racemate can have as strong activity as the enantiomer (Fig. 7) [25].

Fig. 7 (and Table 1). Configuration—antiallergic activity relationship in the series of pyrido [1, 2-a]pyrimidines [25]. The biological activities for the racemates are nearly identical to that of the S-(−)-isomers. If no interaction were among the enantiomers the racemates would show half equivalent biological potency.

Table 1. Antiallergic activity (i.v., μmole/kg)

X	Configuration		
	Racemate	S-(−)	R-(+)
H	0.6	0.3	54.8
4–OH	1.7	0.14	110.5
4–OEt	1.6	0.2	100.0
2–COOH	0.48	0.5	37.1

Differences in side effects may pose serious dangers. In the case of propranolol, the β-adrenergetic activity resides on S-(−)-isomer, while its mirror image has a contraceptive effect [26]. As both the racemic form and S-(−)-isomer are marketed [27], one can easily figure out how those young women (longing for babies) felt, when light was thrown on the real reason for their apparent sterility: taking the drug.

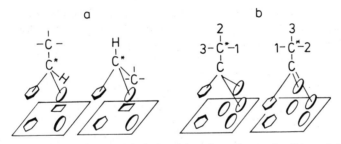

Propranolol   Thalidomide/ Contergan

The most shocking tragedy of our century caused by medicine was felt by those young women whom the racemic thalidomide was administered to when less than three months pregnant. Besides its minor tranquillizing effect, thalidomide also proved to be teratogenic. More than 15 years after its withdrawal from circulation, a systematic study of the separated enantiomers showed that the R-(+)-isomer is indifferent for teratogenity while the S-(−)-isomer was responsible for the dreadful catastrophes [28].

Enantiomers under natural conditions do not act uniformly. Since receptors are chiral surfaces, the differences in biological activity depend on the distance between the asymmetric (chiral) centre and the groups producing biological effect. The closer those groups, the greater the difference in activity is (Fig. 8).

Fig. 8. Configurational effect onto the biological activity—depending on the distance between centre of biological activity and the asymmetric centre (*asymmetric carbon atom). (a) The two centres are closely related, the enantiomers bind selectively; (b) no significant difference in biological activity.

A major part of the drugs acts through a receptor activating or inhibiting procedure. In both cases the drug has to show the characteristic features of the natural activators or substrates. One of the most significant ideas which helped elucidate what is happening when a substrate interacts with an enzyme was put forward by Fischer, who visualized the situation as a "lock and key" fit [29] as if the enzyme contained an indentation into which the substrate could fit very precisely. If the key fits well, the reaction takes place; the wrong key will not work.

This theory gives a relevant explanation for those cases, where the rest-state (conformation) of the receptor does not correspond to the form of the activated complex (Fig. 9). The activator/substrate bound to the receptor can induce an alteration in the receptor conformation. The alteration depends on structural features, i.e. shape, charge distribution, etc. of the activator/substrate molecule [30]. As the number of possible binding points on the receptor surfaces is greater than required by the natural substrate [31], an entirely different "key" may also induce the opening of the given lock. Otherwise the molecule bound irreversibly to the receptor may inhibit its functioning, as is the case, e.g. with the irreversible binding of carbon-monoxide or hydrogen-cyanide molecules to hemoglobin, instead of oxygen molecules [32] (Fig. 9).

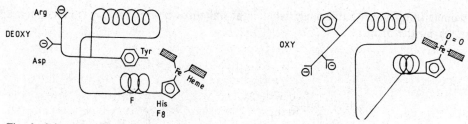

Fig. 9. Schematic representation of a receptor in the ground-state and in the activated complex (hemoglobin before and after oxygen take-up, after Ref. [29]).

The key–lock model suggests a rather static-mechanistic picture: a highly selective lock can only be opened with the proper key. However even these receptors (locks) can be fooled, compounds other than the key-molecules can also bind to them and activate, or inhibit, their functioning.

When elucidating the behaviour of chiral compounds, the combined model of the left and right hands and the fitting (not fitting) gloves is frequently used (Fig. 10).

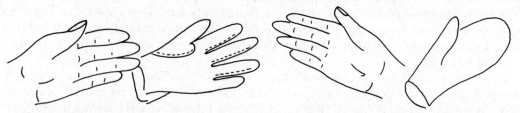

Fig. 10. Fitting of mirror image hands to left-handed gloves—the receptor is symbolized by the gloves, substrates by the hands. Even a mitten cannot be fitted to the mirror image hand, however it does fit smaller homochiral hands.

This picture can better describe such virtual anomalies as that the GABA (γ-amino-butyric acid, which itself is an achiral molecule) agonists and antagonists are chiral molecules, and only one of their mirror image isomers has an effect on the GABA-receptor (Fig. 11) [33].

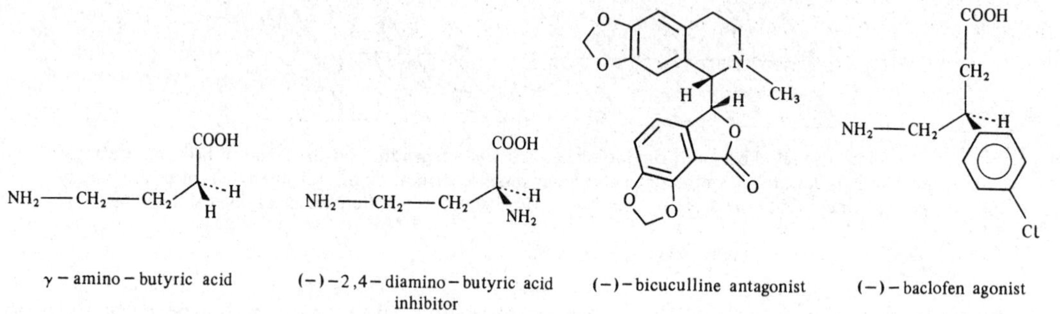

| γ – amino – butyric acid | (−)−2,4 – diamino – butyric acid inhibitor | (−)−bicuculline antagonist | (−)−baclofen agonist |

Fig. 11. GABA and various chiral inhibitor, antagonist and agonist molecules acting on GABA receptor.

The foregoing stresses the importance of the separation at least for pharmacological investigations. The separation procedures, called also optical resolutions, are based on the differences in mutual affinity of the chiral molecules or/and on the differences in physico-chemical properties of the produced diastereoisomers.

SEPARATION OF OPTICAL ISOMERS

The most efficient way to get optically pure isomers is the use of biological, i.e. microbial and enzymatic methods. Their efficiency is due to enzyme stereospecificity while the chemical specificity of enzymes limits their application. The physical separation methods are less efficient and rarely applied. The use of chiral chemical agents is the most widespread method for enantiomer separation [34]. Economic aspects can be decisive in selecting the proper reaction series for preparing a given optical isomer. Chiral effects can induce the selective generation of one of the enantiomers, this is called asymmetric synthesis [35]. The efficiency of the applied method depends upon the differences in strength and life-time of the contacts formed between the chiral "effect" and the substrate. As a result of the reaction taking place between chiral reactants, new formations (compounds, salts, complexes, adducts) come into being. The new substance incorporates all the starting materials, and consists of two well-defined and even achirally distinguishable groups of molecules. These groups are identical in their constitution and arrangement of covalently connected atoms, but their spatial arrangement is not enantiomeric any more—they are called diastereo-isomers. The physico-chemical properties of the diastereoisomers such as melting point, heat of fusion, solubility, spectroscopic characteristics are different. As a consequence of the different spatial arrangement in the two diastereoisomers, different interactions result in forming different second order contact. The question arises: what kind and what strength of structural differences may lead to the bulk separability of the diastereoisomers.

As an illustrative example for the distinctive mechanism and for the separation procedure let us take the optical resolution of racemic compounds via diastereoisomeric salt formation.

The method is described by a simple equilibrium equation showing only the starting and ideal final state of the optical resolution:

$$F\text{⅂} + RR \rightleftharpoons FR + \text{⅂}R$$

$$\begin{array}{l} F, \text{⅂} \text{ enantiomers} \\ R \text{ resolving agent.} \end{array}$$

The equation is not apt to demonstrate the interactions formed among chiral and achiral molecules determining the behaviour and final state of the given system. Without being aware of and understanding the nature of these types of interactions, one cannot evaluate and elucidate the results of an optical resolution, and there is no way for scientific design and prediction for the results of a given optical resolution.

In the reaction mixture of a resolution, solvent molecules have achiral effects on any of the chiral molecules. This achiral effect does not affect the relationship between the chiral molecules, only the chiral interactions are taken into account as a first step in studying this chiral system.

A selected molecule (F) can get in touch with homochiral (F) and heterochiral (⅂ and R) molecules. The resulting dimer associates (FF/F⅂ and ⅂⅂/F⅂) or salts (FR and ⅂R) are in diastereoisomeric relationship. In the simplest case with only dimer formation, the result of enantiomer separation is determined by the equilibrium and stability conditions of the chiral components (Fig. 12) [36]. Precipitation of the most stable and the least soluble dimer can be expected. In some cases the racemic dimer (F⅂) is so stable that under ordinary conditions (ambient temperature, neutral pH etc.) the starting racemate is recovered instead of one of the diastereo-isomeric salts [37].

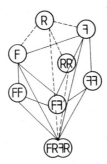

Fig. 12. Chiral (dimer) interactions possibilities in the reaction mixture of optical resolution via diastereoisomeric salt formation.

Diastereoisomeric dimers produced as a result of the homo- and heterochiral interactions can react in different ways. The question is: what kind and what strengths of interactions are needed to make diastereoisomers macroscopically distinguishable.

THE THREE-POINT INTERACTION MODEL OF
CHIRAL DISCRIMINATION

Having considered the overall behaviour of systems built up from chiral components significant differences were found already in the interactions of corresponding molecule pairs, that is to say differences and similarities between enantiomer associates or diastereoisomeric salts can already be discovered in comparing dimer formations.

A free one-point vertex interconnection between two tetrahedra can result in twice 10 diastereoisomeric dimer pairs for the homo- and heterochiral (racemic) interactions, while the enantiomers and resolving agent may form twice 16 diastereoisomeric salt pairs. If, in the interaction, more than one vertices are concerned and one of the possible contacts (the strongest) is fixed by, e.g. ionic bond, a hydrogen bond, etc., the number of possible dimers along an edge falls to 6 from 21 in the case of enantiomer associates, and from 36 to 9 for diastereoismeric salt pairs. Selection and fixation of one of the edges causes a further reduction in the number of possible contacts along the faces of the tetrahedra (Fig. 13).

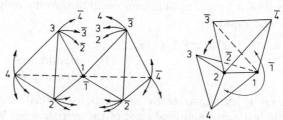

Fig. 13. The three-point interaction model for chiral discrimination. (a) fitting freedom for two-point interaction in a racemic dimer (one vertex-pair is fixed); (b) fitting freedom for three-point interaction in a racemic dimer (one edge-pair is fixed).

The first two interactions are equally probable in every corresponding diastereoisomer, but the third will be different, because of the mirror image relationship between the enantiomeric components of the diastereoisomers. Therefore, the diversity of properties leading to their separability is due to the differences in the third interactions both in the diastereoisomers [38] and in the non-racemic enantiomer mixtures [39].

The differential interaction energy between two fictive chiral tetrahedral molecules, such as two chiral ammonia, has been calculated by Salem et al. [40]. Using a six-centre forces approach, this was found to be 0.029 kJ/mole. Models based on two- or four-centre forces failed to give any chiral discrimination, i.e. no energy difference was found between the homochiral and racemic dimers. The "third", the differentiating interaction is rarely so well defined, e.g. hydrogen bond [41], charge-transfer complexes [42] (Fig. 14), that it could be identified by traditional, e.g. spectroscopic methods.

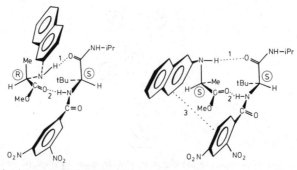

Fig. 14. A three-point interaction model showing the chiral discrimination between the enantiomers of N-(3, 5-Dinitorbenzoyl)leucine n-propylamide and methyl N-(2-naphtyl)alaninate. In the S–S-dimer three second order contacts are formed as indicated on the figure with dotted lines. In the S–R-dimer the π-donor-acceptor interaction can not be formed because of the spatial arrangement of the aromatic ligands. The presence and the lack of this π-π-interaction shows itself in the colour of the dissolved complexes: the S–S-dimer in solution is dark orange, while its diastereoisomer is pale yellow. (Based upon the data published in Ref. [42].)

In practice, these "pure" dimer interconnections do not exist, optical resolution itself is mostly linked to a phase transition (in our case to a liquid–solid phase transition). Characteristic features of the crystalline phase can be discovered even in solution, in the "last" moment before precipitation starts. Consequently, not only the dimer–contacts, but further dimer–dimer interactions should be taken into account. In this way the weak second order interactions, which are undetectable in single dimers, are multiplied, partly because of the formation of new types of interactions. As a result of this multiplication, macroscopic properties of diastereoisomers become distinguishable [43].

MACROSCOPIC DIFFERENCES AND WEAK SECOND ORDER INTERACTIONS

One of the routine processes of the optical resolutions via diastereoisomeric salt formation is the separation of N-methamphetamine enantiomers using natural 2R, 3R-tartaric acid, in anhydrous alcoholic medium [44]. The R-base-R, R-tartrate crystallizes because of its greater stability and poorer solubility (Table 2). The origin of these differences in physico-chemical

R–N–methamphetamine   R,R–tartaric acid

properties can be found among second order interactions. As the differences are amplified in solid phase, and the result of optical resolution occurs in solid state, single crystal X-ray diffraction is a proper method for their determination [44].

Table 2. Characteristic parameters for diastereoisomeric N-methamphetamine tartrates

Parameter	Diastereoisomer S	Base-2R, 3R-tartrate	R	Difference in favour of R–RR salt
Melting point (°C)	115		164	+49
Heat fusion (kJ/mole)	70		49	−21
Solubility (g/100g, EtOH)	4.2		0.16	×25
Hydrogen bonds				Symmetry
I.		N(2)–H(2A)...O(14)		1_{000}
II.		N(2)–H(2B)...O(12)		1_{010}
III.		O(16)–H(16)...O(18)		2_{0-12}
IV.		O(18)–H(18)...O(14)		2_{0-12}
V.		O(21)–H(21)...O(12)		1_{001}
C–H...O†				
1.	C(1)–H(1A)...O(20)		C(1)–H(1C)...O(20)	1_{00-1}
2.	C(1)–H(1C)...O(21)		C(1)–H(1B)...O(21)	2_{002}
3.	C(4)–H(4B)...O(14)			1_{000}
			C(5)–H(5A)...O(20)	1_{00-1}
4.	C(8)–H(8)...O(20)		C(8)–H(8)...O(20)	2_{102}
5.			C(9)–H(9)...O(16)	2_{102}

†Contacts between the cation and the tartrate anion with C...O distance less than 3.6 Å (Fig. 15).
Symmetry codes: 1: $x, y, z,$; 2: $2 - x, 1/2 + y, -z$.

The hydrogen bonding patterns for the two diastereoisomer salts are similar (Table 2) and the differences between the salts are caused by non-bonded, second order interactions (Table 2 and Fig. 15). In the less soluble R-base tartrate, an additional weak C–H...O contact [45] is found between the para hydrogen atom of the phenyl group and an oxygen atom of the alcoholic hydroxyl group attached to one of the neighbouring tartaric acid molecules.

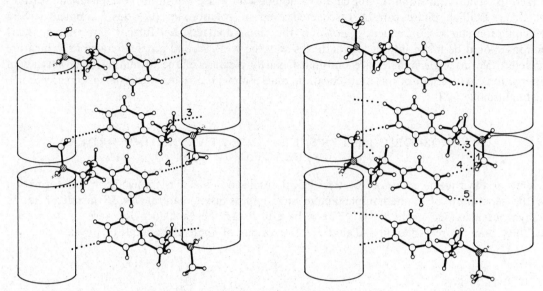

Fig. 15. Numbered dotted lines show the weak second order interactions in diastereoisomeric methamphetamine tartrates. For the hydrogen-bonding see Table 2. The tartrate anion chain is represented by the cylinders.

The close-contact of the above atomic groups discovered by X-ray crystallography has been confirmed as an attractive interaction by quantum-chemical calculations [46]. This special interaction, uncommon in organic chemistry has been proved to be so strongly bonding that the formation of crystalline structure in the R-metamphetamine derivative series is unimaginable without it. For instance the R-p-fluro-metamphetamine-R, R-tartrate cannot be crystallized from anhydrous solutions, and the crystalline solid can only form when encapsulating one molecule of water per salt unit. This water cannot be eliminated without decomposition [47]. In acidic tartrates the tartrate ion pattern showed a remarkable similarity. This similarity and the fact that tartaric acid is the most frequent resolving agent among chiral acids [48] merit our particular attention.

WHY IS TARTARIC ACID A GOOD RESOLVING AGENT?

The answer seems to be quite simple: the altogether four-membered carbon chain contains four groups capable of establishing hydrogen donor-acceptor bonds and two further groups forming hydrogen acceptor bonds. This large concentration of second order interacting groups renders highly selective chiral recognition and discrimination possible. However, the same possibility is given for the mirror image isomer of the racemate to be resolved—since the variability of the interconnections between the chiral base and tartaric acid makes it possible for the diastereoisomeric salts to have similar stability with obviously different structure.

This virtual contradiction is resolved by a special self-recognition mechanism of tartaric acid. The acid molecules form an infinite chain through a hydrogen-bonding system established between the carboxyl and carboxylate groups. The enantiomers of the chiral base approach the chain by Coulombic attraction. One base molecule is surrounded by four tartaric acids, members of the distinct infinite chains. The four chains linked by bases, by further hydrogen bridges [49], and by weak second order interactions, form a chiral channel (Fig. 16). The base properly fits and binds to the channel wall.

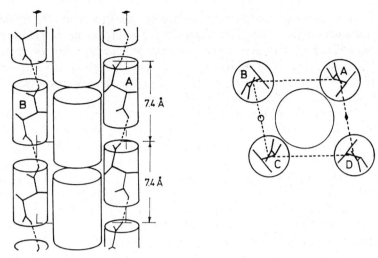

Fig. 16. Schematic arrangement of tartaric acid in its acidic salts. (After Ref. [45].)

Analysis of the six metamphetamine tartrate structures, and of ten randomly selected bitartrate structures available in the Cambridge Crystallographic Databank (116 structures by May, 1984) led to the conclusion that tartaric acid in its acidic structures forms a head-to-tail type bitartrate chain, independently of the cation quality [44].

Carström [50] pointed out that this pattern is characteristic not only for the acidic salts, but for the neutral salts, as well and even for the free acid.

The tartrate structure bears high order flexibility in spite of its repeating features or even due to them. Since the chains can move with respect to one another, the position of the binding points also changes, and the hydrogen bonding pattern can expand or contract according to the steric requirements of the cations. Thus its chiral recognition ability involves a large variety of chiral bases.

SYMMETRY VERSUS CHIRALITY AND THE PRINCIPLE OF COMPLEMENTARITY

The greater the self-accomodation ability of a given formation, the better its prospects for survival. If an organism proves to be capable of living, then its reproduction rate clearly depends on its degree of symmetry, under identical condition [51]. The more symmetric the "object", the faster its reproduction, and thus its occurrence in nature is more frequent than its less symmetric rivals (Fig. 17). Qualitative and quantitative direction of natural propagation, of how many, how

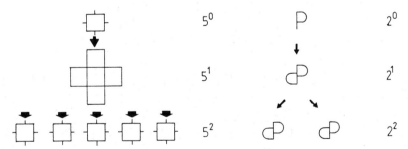

Fig. 17. Propagation possibilities for symmetric and asymmetric formations. Both formations are self-reproducing along the straight line. If every condition remains the same their propagation—and developing and splitting rate is identical then in the fourth generation the number of squares will exceed by more than 15 times the number of "P".

well structured and what sort of species do generate, is an issue in molecular dimensions. Most of the biochemical reactions such as e.g., protein synthesis take place on asymmetric surfaces. As it was formerly mentioned, adherence to the building instructions is of vital importance for accurate functioning. The more unambiguous the commands, the easier to adhere to technological discipline. If only this requirement were taken into account, it would be the postulation of the law of asymmetry. However, the end product of the processes depends not only on the "command" (or instructions) but on the subject and on its receptivity, too. The most frequent way of receiving a command is the temporary interconnection between messenger and addressee. The closer their connection, the more accurate the information transfer.

As the affinity forces are of electrostatic origin, a proper charge distribution is the condition of "contact formation", e.g. a complementary charge distribution, must characterize the substrate and the catalyst on the asymmetric surface in their reaction centres. The complementary charge distribution corresponds to a well defined chemical structure and vice versa. Complementary fitting is a well studied problem with nucleinic acids. In the double helix of DNA, a given purine base is always linked to the same pyrimidine base. The special hydrogen-bonding system connecting the two strands of the double helix requires proper arrangement of the base pairing (Fig. 18) [52]. For a better fitting of the amino acids to be transferred and reacted, a thymine–uracil change ($5H \rightarrow 5CH_3$) takes place in messenger—and transfer—RNA [53].

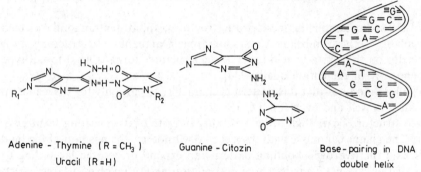

Adenine - Thymine (R = CH₃)
Uracil (R = H)   Guanine - Citozin   Base-pairing in DNA
                                                                                    double helix

Fig. 18. Complementary purine (A, G) and pyrimidine (T, U, C) bases linked by hydrogen bonds in the DNS strands.

The temporary close-fitting of the complementary surfaces leads to a temporary reduction of the asymmetry in the studied system (Fig. 19).

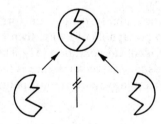

Fig. 19. Temporary close-fitting of asymmetric surfaces may reduce the "degree of asymmetry".

The complementarity can serve as a possible and plausible explanation of chiral discrimination or recognition.

An interesting example for complementary chiral discrimination can be observed in the crystal structures of strychnine and its methoxy derivative, brucine, which are frequently applied resolving

R = H :   strychnine
R = OCH₃ : brucine

agents. Both compounds contain several polar groups enabling them to form hydrogen bonds and second order interactions. A thorough study of their crystal structure was accomplished by Gould and Walkinshaw [54]. For both alkaloid molecules, a layered structure is characteristic. The geometry and conformation of the bases are identical independently of the anion. The distance between the layers varies according to the bulkness and structure of the guest molecules. The difference in their chiral discriminating ability originates from their different layer structures.

All the strychnine salts show a similar bilayer packing pattern. The strychnine molecules are packed back to back to allow the protonated amine nitrogen bond with the solvent/anion sheet. The hydrogen-bonding network tying together the bilayers can be formed by means of any molecule containing more than one polar group. The chiral recognition is provided by the chiral holes of the bilayers. The corresponding groups (e.g. phenyl) of the enantiomer can not be intercalated properly into these cavities. This close-fitting, or the lack thereof, allows a high order enantio-discrimination for strychnine (Fig. 20).

Strychnine   Brucine

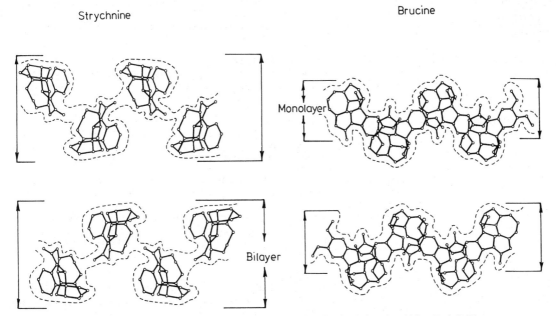

Fig. 20. The layered structure of strychnine and brucine in their salts. (After Ref. [54].)

The crystal structure of the brucine complexes is composed of monolayer sheets of brucine. The monolayers are separated and simultaneously tied by a channel comprising the anion and the molecules of crystal solvates. The brucine molecules are stacked with the plane of the indole ring perpendicular to the plane of the monolayer sheet, providing a corrugated chiral surface with many binding points for the guest molecules (Fig. 20).

Owing to the structural flexibility arising from the variable layer distances and the variability of the relative positions of the layer, these alkaloids are widespread resolving agents. Brucine has been applied as resolving agent for 175 racemic acids in 400 randomly selected resolution procedures) [48].

Although tartaric acid, brucine and strychnine are excellent resolving agents, they would not be proper information carriers because of the small activation energy differences between the chiral recognition/discrimination processes for two enantiomeric molecules. During the early stages in a number of optical resolutions the precipitating salt mixture comprises the two diastereoisomeric salts in equal amount, i.e. the racemate to be resolved is recovered in unchanged form. The diastereoisomeric ratio gradually shifts to the equilibrium value. The "messenger" cannot afford this time-consuming delay. Although reversible contact formations take place during biochemical reactions, it cannot mean an equilibrium competition between the enantiomers: interactions under natural conditions are practically forbidden between the receptor and the mirror image isomer of the natural substrate.

In enzyme-catalyzed biochemical reactions, an erroneous reaction occurs in every millionth turn-over. This value is quite characteristic of the order of the activation energy differences in such biochemical reactions.

A plausible explanation to this large difference is in the perfect fitting of complementary chiral surfaces in one case and the lack of it for its enantiomer. The complete complementarity implies morphological and charge distribution aspects, both having their origin in the chemical structure of the molecules.

REFERENCES

1. I. Hargittai and M. Hargittai, *Symmetry Through the Eyes of a Chemist.* VCH, Weinheim (1986). Softcover edn: VCH, New York (1987).
2. M. F. Perutz, Relation between structure and sequence of hemoglobin. *Nature* **194**, 914–917 (1962).
3. Beauty and ugliness or fright and disgust and abstractions closely related to our human biological, social and cultural heritage. Whatever symmetric patterns are beared by an object or a living creature if they are able to endanger us, it will exert a repulsing effect on human beings. Even within bilateral symmetry, we got used to certain proportions, and deviations may give us the impression of ugliness. (See some paintings of Hieronymus Bosch.)
4. See entries for the mechanism of sight and for strabismus in any encyclopaedia of biology, such as e.g., *Biológiai Lexikon* (Hungarian) (Ed. Straub F. Brúnó). Akadémiai Kiadó, Budapest (1978).
5. R. Dawkins, *The Selfish Gene.* OUP, Oxford (1975).
6. S. F. Mason, Origins of biomolecular handedness. *Nature* **311**, 19–23 (1984).
7. R. C. Dougherty, Asymmetric synthesis in a confined vortex: gravitational fields can cause asymmetric synthesis. *J. Am. chem. Soc.* **102**, 380–381 (1980); Critical evaluation of the paper: A. Peres, Asymmetric syntheses in spinning vessel. *J. Am. chem. Soc.* **102**, 7389–7390 (1980).
8. F. F. Seelig, System-theoretic model for the spontaneous formation of optical antipodes in strongly asymmetric field. *J. Theor. Biol.* **31**, 355–361 (1971).
9. J. M. Davey, A. H. Laird and J. S. Morley, Polypeptides, Part III. The synthesis of the C-terminal tetrapeptide sequences of gastrin, its optical isomers and acylated derivatives. *J. Chem. Soc.* 555–562 (1966).
10. R. H. Mazur, J. M. Schlatter and A. H. Goldkampf, Structure-taste relationships of some dipeptides. *J. Am. chem. Soc.* **91**, 2684–2691 (1969).
11. J. P. Greenstein and M. Winnitz, *Chemistry of Amino Acids.* Wiley-Interscience, New York (1961).
12. W. Heisenberg, *Physics and Beyonds (Encounters and Conversations)* [in the series of *World Perspectives* (Ed. R. H. Nanda)]. Harper and Row, New York (1971).
13. S. S. Urmantsev, Symmetry of system and system of symmetry. *Comput. Math. Applic.* **12B**, 379–405 (1986). Reprinted in *Symmetry: Unifying Human Understanding* (Ed. I. Hargittai). Pergamon Press, Oxford (1986).
14. K. Sander, Bilateral symmetry in insects: could it derive from circular asymmetries during early embryogenesis. *Comput. Math. Applic.* **12B**, 413–418 (1986). Reprinted in *Symmetry: Unifying Human Understanding* (Ed. I. Hargittai). Pergamon Press, Oxford (1986).
15. A. Ábrahám, *Anatómia.* (Hungarian) Medicina Könyvkiadó, Budapest (1970).
16. F. A. L. Anet, S. S. Miura, J. Siegel and K. Mislow, La coup du roi and its relevance to stereochemistry. Combination of two homochiral molecules to give an achiral product. *J. Am. chem. Soc.* **105**, 1419–1426 (1983).
17. E. Fogassy, M. Ács and L. Tőke, *Optikai izomerek elválasztása [Isolation of Optical Isomers* (Hungarian)]. Akadémia Kiadó, Budapest (1987).
18. S. F. Mason, *Molecular Optical Activity and the Chiral Discriminations.* Cambridge Univ. Press, Cambridge (1982).
19. J. A. Le Bel, Sur les relations qui existent entre les formules atomiques des corps organiques et le pouvoir rotatoire de leur dissolutions. *Bull. Soc. chim. Fr.* **22**, 337–347 (1874).
20. H. B. Kagan, *Organische Stereochemie.* Georg Thieme, Stuttgart (1977).
21. A. Werner, Sur l'activité optique des composés chimiques sans carbone. *Compt. rend.* **159**, 426–429 (1914).
22. I. Bernal and G. B. Kauffmann, The spontaneous resolution of cis-bis(ethylenediamine)dinitrocobalt(III) salts. *J. Chem. Educ.* **64**, 604–610 (1987).
23. M. Simonyi, On chiral drug action. *Med. Res. Rev.* **4**, 359–413 (1984).
24. F. Luduena, L. von Euler, B. F. Tullar and A. M. Lands, Effects of the optical isomers of sympathomimetic amines on the guinea pig bronchioles *Archs int. Pharmacodyn.* **109**, 392–400 (1957).
25. I. Hermecz, T. Breining, Z. Mészáros, Á. Horváth, L. Vasvári-Debreczi, F. Dessy, Ch. DeVos and L. Rodriguez, Nitrogen bridgehead compounds. 18. New antiallergic 4H-pyrido[1, 2-a]pyrimidine-4-ones. *J. Med. Chem.* **25**, 1140–1145 (1982).
26. R. M. Pearson, E. J. Ridgeway, A. Johnston and J. Vadukul, Concentration of D-propranolol in cervicovaginal mucus. *Lancet* **2**, 8417–8418 (1984); *Chem. Abstr.* **102**, 90222a (1984).
27. M. Negwer, *Organic-chemical Drugs and their Synonyms.* Akademie, Berlin (1987).
28. G. Blaschke, H. P. Kraft, K. Fickentscher and F. Köhler, Chromatographische Racemattrennung von Thalidomid und teratogene Wirkung. *Arzneimittel-Forsch.* (Drug. Res.) **29**, 1640–1642 (1979).
29. E. Fischer, Einfluss der Configuration auf die Wirkung der Enzyme. *Ber. dt. chem. Ges.* **27**, 2985–2993 (1894).
30. T. Nógrády, *Medicinal Chemistry. A Biochemical Approach.* OUP, Oxford (1985).
31. S. M. Rapoport, *Medizinische Biochemie.* VEB Verlag Volk und Gesundheit, Berlin (1977).
32. P. Bálint, *Orvosi Élettan.* [Medical Physiology (Hungarian)]. Medicina Könyvkiadó, Budapest (1981).
33. P. Krogsgaard-Larsen, γ-amino-butiric acid agonists and antagonists and uptake inhibitors design and therapeutic aspects. *J. Med. Chem.* **24**, 1377–1383 (1981); N. G. Bowery, D. R. Hill, A. L. Hudson, A. Doble, D. N. Middlemiss, J. Shaw and M. Turnbull, (−)-Baclofen decreases neurotransmitter release in the mamalian CNS by an action at a novel GABA-receptor. *Nature* **283**, 92–94 (1980).
34. S. H. Wilen, A. Collet and J. Jacques, Strategies in optical resolution *Tetrahedron* (Report No. 38) **33**, 2725–2736 (1977).

35. J. D. Morrison and H. S. Mosher, *Asymmetric Organic Reactions*. Prentice-Hall, Englewood Cliffs, N. J. (1971); M. Nógrády, *Stereoselective Syntheses*. Chemie, Weinheim (1986).

36. B. Zsadon, M. Ács, E. Fogassy, F. Faigl, Cs. Novák, Gy. Pokol and A. Újházy, Comparison of enantiomer separation by inclusion chromatography and by resolution via diastereoisomeric salt formation. *Reactive Polymers* **6**, 197–202 (1987).

37. During the optical resolution of 1-(3, 4, 5-trimethoxy-benzyl)-6, 7-dihydroxy-isoquinoline. HCl accomplished with natural 2R, 3R-tartaric acid, racemic hydrochlorid salt precipitates at ambient temperature. Successful enantiomer separation [precipitation of the (−)-isomer-tartrate] can only be achieved above 90°C.

38. E. Fogassy, F. Faigl and M. Ács, New method for designing optical resolutions and for determination of relative configuration. *Tetrahedron* **41**, 2837–2840 (1985).

39. E. Fogassy, F. Faigl and M. Ács, Selective reactions in solution of enantiomers. *Tetrahedron* **41**, 2841–2847 (1985).

40. L. Salem, X. Chapuisat, G. Segal, P. C. Hiberty, Ch. Minot, Cl. Leforestier and Ph. Sautet, Chirality forces. *J. Am. chem. Soc.* **109**, 2887–2894 (1987).

41. A. Dodashi, N. Saito, Y. Motoyama and S. Hada, Self-induced non-equivalence in the association of D- and L-amino acid derivatives. *J. Am. chem. Soc.* **108**, 307–308 (1986).

42. W. H. Pirkle and T. C. Pochapsky, Chiral molecular recognition in small biomolecular systems: a spectroscopic investigation into nature of diastereoisomeric complexes. *J. Am. chem. Soc.* **109**, 5975–5982 (1987).

43. J. Jacques, A. Collet and S. H. Wilen, *Enantiomers, Racemates and Resolutions*. Wiley, New York (1981).

44. E. Fogassy, M. Ács, F. Faigl, K. Simon, J. Rohonczy and Z. Ecsery, Pseudosymmetry and chiral discrimination in optical resolution via diastereoisomeric salt formation. *J. Chem. Soc. Perkin Trans* **2**, 1881–1886 (1986).

45. R. Taylor and O. Kennard, Crystallographic evidences for the existence of C–H...O, C–H...N and C–H...Cl hydrogen bonds. *J. Am. chem. Soc.* **104**, 5063–5070 (1982).

46. P. Laczik, G. Náray-Szabó and K. Simon, Molekula felismerés a borkösav és az *N*-metil-amfetamin elegykristályában (Hungarian). *Magy. kém. Foly.* **93**, 436–440 (1988).

47. Unpublished results.

48. Systematic survey of optical resolutions referred in *Chem. Abstr.* between 1910 and 1985.

49. M. Kuramoto, Y. Kushi and H. Yoneda, Structural study of optical resolution VII. The crystal structure of (−)$_{589}$-oxalatobis(ethylene-diamine)cobalt(III) hydrogen-*d*-tartrate dihydrate and comparison of the more and less-soluble diastereomers. *Bull. chem. Soc. Japan* **53**, 125–132 (1980).

50. D. Carström, The structure of the catechol amines IV. The crystal structure of (−)-adrenaline hydrogen tartrate. *Acta Crystallogr.* **B29**, 161–167 (1973).

51. M. Eigen and R. Winkler, *Das Spiel* (*Naturgesetze steuern den Zufall*). Piper, München (1975).

52. J. Rebek, B. Askew, P. Ballester, C. Buhr, S. Jones, D. Németh and K. Williams, Molecular recognition hydrogen bonding and stacking interactions stabilize a model for nucleic acid structure. *J. Am. chem. Soc.* **109**, 5033–5035 (1987).

53. R. D. Watson, *Molecular Biology of the Gene*. Benjamin, New York (1976).

54. R. O. Gould and D. Walkinshaw, Molecular recognition in model crystal complexes. The resolution of D- and L-amino acids. *J. Am. chem. Soc.* **106**, 7840–7842 (1984).

Computers Math. Applic. Vol. 17, No. 1–3, pp. 443–455, 1989
Printed in Great Britain. All rights reserved

CHEMICAL KINETICS AND THERMODYNAMICS

A HISTORY OF THEIR RELATIONSHIP

S. Lengyel

Central Research Institute for Chemistry, Hungarian Academy of Sciences,
Budapest, P.O. Box 17, H-1525, Hungary

Abstract—The history of the understanding of phenomenology and energetics of chemical reactions and their mutual relationship is given from a certain point of view. The differentiation of the ever growing knowledge on continua, where chemical reactions advance leading to the development of their thermodynamics and chemical kinetics as distinct science branches is considered. The integration of these two branches by a general variational principle of thermodynamics is carried out. The time evolution of chemical kinetic systems is studied in this integrated theory. The role of symmetry as an instrument of synthesis or integration is pointed out.

1. WHAT IS CHEMICAL AFFINITY?

Modern chemical kinetics

Modern chemical kinetics is a science describing and explaining chemical reactions as we understand them at present. In the beginning, it studied the dependence of the rates of the reactions proceeding in a system on the concentrations of the components then, also on other influences as temperature, electric field, radiation, etc. Later on, it turned to the molecular mechanism and special regimes, such as chain reactions, polymerization and degradation of polymers, flames, combustion, explosions, oscillation, atomic and molecular beam reactions, etc.

Modern thermodynamics

Modern thermodynamics is the thermodynamics of irreversible processes. It was developed on the basis of classical thermodynamics of equilibrium systems, called thermostatics by some authors. It is used in the study of transport processes as thermal conduction, diffusion, electric conduction, viscous flow, etc. Chemical reactions which are non-transporting irreversible processes are also investigated by methods of modern thermodynamics.

Chemical kinetics and (non-equilibrium or process) thermodynamics are, these days, two well-defined, distinct branches of science. Phenomenological theories of the reaction rates belong to the former, the study of energy dissipation or entropy production by chemical reactions, the dependence of the reaction rates on thermodynamic forces, etc. to the latter. Let us cast our thoughts back to the whole history of ideas and theories pertaining to chemical reactions. It is perhaps better to forget our present day's terminology and speak about "transmutations of matter". We realize that Hippocrates who is looked upon as "father of medicine" some 24 centuries ago assumed that it is similarity or "affinity" that drives bodies to combine. This assumption was disproved but the term survived and became inheritance of the nineteenth and twentieth centuries and serves now to denote the thermodynamic force causing chemical reactions to advance.

While in antiquity and later on up to the seventeenth century no experimental basis sufficient for setting up theories of chemical reactions existed, the second half of the seventeenth century was the first time scientific ideas about the causes of chemical reactions occurred. The epoch in the history of theoretical chemistry, starting then and lasting up to the middle of the nineteenth century, seems to have been devoted to searches for well-founded answers to questions: why some substances chemically react and why others do not? why some chemical reactions are slow, others fast? etc. We are not interested at present about going into details of the various theories proposed, with intention to systematize the experimental material that was currently at disposal. These theories played a transient role while the experimental material rapidly increased. However often the word affinity in this epoch was used, the first attempt to define it in terms of energy occurred no sooner than in the middle of the nineteenth century [1].

The modern thermodynamic definition of the affinity of a chemical reaction (van't Hoff [2], the first (1901) Nobel Laureate in chemistry) was based on Gibbs' [3] complete thermodynamic theory of chemical equilibrium states and affinity was considered as the force causing the reaction or as the maximal work performed by this force when the reaction advances. However, the constitutive (phenomenological) equation relating rate to affinity was still required.

2. HOW CHEMICAL REACTIONS PROCEED AND WHY?

Meanwhile, the basic equation of chemical kinetics, a differential equation describing the dependence of the reaction rate on the concentration, together with its solution was published for a first order reaction for the first time in the middle of the nineteenth century [4]. The general theory of the dependence of reaction rates on concentrations, in other words the fundamental equation of chemical kinetics, followed soon afterwards (Guldberg and Waage's kinetic mass action law [5]).

Thus, the understanding of chemical reactions has bifurcated: (a) phenomenological chemical kinetics investigated the rates of chemical reactions and evolution in time of chemical kinetic systems. A tremendous amount of experimental work was done, and the suitable differential equations were set up and solved, in a few decades around the turn of the century. (b) In chemical thermodynamics experimental material accumulated by measurement of the changes of thermodynamic functions (enthalpy, entropy, free energy, chemical potentials etc.) which resulted in systems by chemical reactions taking place along a path through equilibrium states. The thermodynamics of chemical equilibria was thoroughly studied in the cases of a considerable amount of reactions. The relationship between equilibrium constants and rate constants was analyzed.

Sixty years after the first publication of the basic chemical kinetic differential equation and its solution and 43 years after Guldberg and Waage's publication of the kinetic mass action law, a bridge was built between kinetics and classical thermodynamics of chemical reactions. A constitutive equation consistent with the Guldberg–Waage equation of chemical kinetics was heuristically found for ideal systems, i.e. for ideal gases and ideal dilute solutions [6, 7].

After bifurcation this was the first link found between chemical kinetics quite developed and non-equilibrium thermodynamics of chemical reactions in its prehistory. In terminology of the 1980s, for the elementary, reversible chemical reaction

$$\sum_{k=1}^{K} v_k' B_k \rightleftarrows \sum_{k=1}^{K} v_k'' B_k, \tag{1}$$

between the components $B_1, B_2, \ldots, B_K$ the Guldberg–Waage kinetic equation for the net reaction rate

$$\frac{d\xi}{dt} = \frac{d\vec{\xi}}{dt} - \frac{d\overleftarrow{\xi}}{dt}, \tag{2}$$

is written as

$$\frac{d\xi}{dt} = \vec{k} \prod_{k=1}^{K} c_k^{v_k'} - \overleftarrow{k} \prod_{k=1}^{K} c_k^{v_k''}, \tag{3}$$

where the vs are the stoichiometric coefficients, in elementary reactions equal to the numbers of colliding particles, i.e. zero, one or two.

The key of transformation is the equation relating the chemical potentials

$$\mu_k = \left(\frac{\partial g}{\partial c_k} \right)_{T, p, c_{k \neq j}}; \quad k = 1, 2, \ldots, K, \tag{4}$$

with the concentrations (c) of the components in an ideal system

$$\mu_k = \mu_k^\theta + RT \ln(c_k/c_k^\theta); \quad k = 1, 2, \ldots, K, \tag{5}$$

where the symbol θ labels deliberately chosen standard values and g is the density of the Gibbs functions.

Substitution of equation (5) into the kinetic equation directly leads to the Marcelin–Kohnstamm equation

$$\frac{d\xi}{dt} = \lambda(e^{A'/RT} - e^{A''/RT}) \tag{6}$$

or

$$\frac{d\vec{\xi}}{dt} = \vec{\lambda}e^{A'/RT}; \quad \frac{d\overleftarrow{\xi}}{dt} = \overleftarrow{\lambda}e^{A''/RT} \tag{7}$$

with

$$\vec{\lambda} = \exp\left(-\sum_{k=1}^{K} v'_k \mu_k^\theta/RT\right) \prod_{k=1}^{K} (c_k^\theta)^{v'_k}; \quad \overleftarrow{\lambda} = \exp\left(-\sum_{k=1}^{K} v''_k \mu_k^\theta/RT\right) \prod_{k=1}^{K} (c_k^\theta)^{v''_k} \tag{8}$$

$$A' = \sum_{k=1}^{K} v'_k \mu_k; \quad A'' = \sum_{k=1}^{K} v''_k \mu_k. \tag{9}$$

The explanation must be added here that reversibility of the chemical reaction was assumed [see double arrow of the chemical equation (1)] together with the condition that in equilibrium the rates of the forward reaction [left to right in equation (1)], and of the reverse reaction (right to left) are equal and also that the affinity $\bar{A}'$, causing the forward reaction to advance is equal with the backward affinity $\bar{A}''$ where the symbol $^-$ labels equilibrium values. These equilibrium conditions involve that $\vec{\lambda} = \overleftarrow{\lambda}$. Therefore, the arrows will be omitted over the symbol λ.

It should be remembered that the free energy of the system is changed by the reaction in unit time by the amount

$$-\sum_{k=1}^{K} v'_k \frac{d\xi}{dt} \mu_k + \sum_{k=1}^{K} v''_k \frac{d\xi}{dt} \mu_k = -A\frac{d\xi}{dt} \tag{10}$$

that is equal to the negative value of the maximal work the system can perform. The value

$$A = A' - A'' \tag{11}$$

is the affinity of the reaction as defined by van't Hoff.

Later on [8], the affinity could be related to Clausius' "non-compensated heat" a concept introduced in 1850 [9]. In present day's terminology the relation can be written as

$$\frac{\partial S}{\partial \xi} = \frac{A}{T}, \tag{12}$$

where T and S are temperature and entropy of the system where the reaction proceeds; ξ is the extent of the reaction.

In the meantime, experimental and theoretical kinetic study of chemical reactions continued in the 1910 and 1920s. Then, a further link with thermodynamics was not recognized but found by introducing the new concept of the activity coefficient [10], into the Guldberg–Waage equation in the case of chemical reactions in electrolyte solutions [11]. A thermodynamic definition of the activity coefficient starting from the Gibbs function, on the way through chemical potential and relative activity, leading to activity coefficient would have led to *recognition* of this link. However this did not happen, since, alas, the relation

$$a_k = y_k c_k / c_k^* \tag{13}$$

was used instead to define the activity coefficient y_k, a function of T, p and composition. Consequently, a direct replacement of the concentrations by the relative activities in the Guldberg–Waage equation (3) was refused.

3. WHAT IS PROCESS THERMODYNAMICS?

A few years later the foundation of "thermodynamics of irreversible processes" was laid down restricted, however, to linear constitutive equations. Reciprocal relations for the coefficients were derived by statistical study of adiabatically isolated, homogeneous "aged" systems fluctuating around equilibrium [12]. In other words, symmetry of the coefficient matrix of the constitutive equations relating fluxes with forces was proved. It was also shown that the reciprocal relations can be expressed in terms of a potential, and permit the formulation of a variational principle: the "principle of the least dissipation of energy". The deduction of the constitutive equation from this principle was carried out for heat conduction in a continuous medium.

After several preparatory contributions (Onsager's contribution included) a phenomenological theory of irreversible processes in continuous media was set up [13–15] and a second variational principle, the "principle of minimum production of entropy" recognized [16]. In the thermodynamics of irreversible processes in continua it was shown that by such processes in all volume elements entropy is produced or, what is equivalent, energy is dissipated, i.e. the transformation of the dissipated portion of the energy into useful work made impossible. The entropy production rate in unit volume is a bilinear expression, each term of which is a product of a Cartesian component of a flux and the conjugate force component. For example, in the process of viscous flow the viscous pressure tensor is the current density and the vector gradient of the mass center velocity divided by the temperature is the tensorial thermodynamic force; in heat conduction the density of the internal energy flow is the (vectorial) current density and the gradient of the reciprocal absolute temperature is the force. The current densities of the diffusion of the chemical components are the fluxes and the negative gradients of the chemical potentials divided by the temperature are the forces in diffusion processes. In chemical reactions both the current densities and the forces are scalars: the reaction rates are the fluxes and the affinities divided by the temperature are the forces. In the system of the constitutive equations all flux components may be functions of all force components and vice versa. The equations contain material constants. Fourier's law of heat conduction is an example for constitutive equations and heat conductivity for material constants. The knowledge of the constitutive equations is indispensable if concrete calculations are to be carried out.

The entropy production can always be expressed as

$$\sigma = \sum_{i=1}^{f} J_i X_i, \tag{14}$$

where the Js represent Cartesian components of the fluxes and the Xs the Cartesian components of the forces.

In the linear theory, from the linear constitutive equations

$$J_i = \sum_{j=1}^{f} L_{ij} X_j, \quad X_i = \sum_{j=1}^{f} R_{ij} J_j, \quad i = 1, 2, \ldots, f, \tag{15}$$

and the reciprocal relations

$$L_{ij} = L_{ji}, \quad R_{ij} = R_{ji}, \quad i, j = 1, 2, \ldots, f, \tag{16}$$

the existence of the flux potential

$$\psi = \frac{1}{2} \sum_{i,j=1}^{f} L_{ij} X_i X_j \tag{17}$$

and the force potential

$$\phi = \frac{1}{2} \sum_{i,j=1}^{f} R_{ij} J_i J_j, \tag{18}$$

follows from which the flux components and the force components can be respectively derived

$$J_i = \frac{\partial \psi}{\partial X_i}; \quad X_i = \frac{\partial \phi}{\partial J_i}; \quad i = 1, 2, \ldots, f. \tag{19}$$

Almost 35 years after that Onsager had recognized "the principle of the least dissipation of energy" and 20 years after the publication of Prigogine's "principle of minimum production of entropy", the most general variational principle of thermodynamics, the "governing principle of dissipative processes" was formulated [17]. According to this principle the time and space evolution of thermodynamic systems is governed in such a way that the function

$$\sigma - \psi - \phi, \tag{20}$$

keeps its extremum value in all volume elements at any time. Consequently, the variation of this function is zero

$$\delta(\sigma - \psi - \phi) = 0. \tag{21}$$

This is the local formulation of the governing principle of dissipative processes. Accordingly, for the whole system we have

$$\delta(\Sigma - \Psi - \Phi) = 0, \tag{22}$$

where the functions Σ, Ψ, Φ are obtained by integration over the whole volume of the system. Equation (22) is the global form of the principle. In equations (20) and (21) σ is the entropy production, ψ the flux potential depending on the forces and ϕ the force potential depending on the fluxes. All the three are densities.

Combination of equations (14), (17)–(19) and (21) is nothing more than application of the governing principle of dissipative processes in the linear theory. It leads to the equation

$$\sum_i^f \left(\left(J_i - \sum_{j=1}^i \frac{L_{ij} + L_{ji}}{2} X_j \right) \delta X_i + \left(X_i - \sum_{j=1}^f \frac{R_{ij} + R_{ji}}{2} J_j \right) \delta J_i \right) = 0, \tag{23}$$

where δX_i and δJ_i ($i = 1, 2, , \ldots, f$) are independent variations and can be arbitrarily chosen. Therefore, all the expressions in parentheses vanish, i.e. the linear constitutive equations are reobtained with symmetric coefficient matrices.

4. THE UNIFIED FIELD THEORY OF THERMODYNAMICS

The phenomenological theory of irreversible processes in continuous media elaborated in the 1940's, however, lacked the uniformity and completeness characteristic of the mechanics of continua and electrodynamics. The unified field theory of thermodynamics [18], was then published. In this theory the complete set of extensive state variables A_i ($i = 1, 2, \ldots, f$) is defined. Their local densities a_i moving with velocities v_i represent current densities $J_i = a_i v_i$ and, of course, are functions of time and position. The balance equations

$$\frac{\partial a_i}{\partial t} = -\operatorname{div} J_i + \sigma_i, \tag{24}$$

hold for the local values with the local source densities σ_i. The local density of the entropy

$$s = s(a_1, \ldots, a_i, \ldots, a_f) \tag{25}$$

is also a function of time and position. The intensive variables are defined by

$$\Gamma_i = \frac{\partial s}{\partial a_i}; \quad i = 1, 2, \ldots, f. \tag{26}$$

The local rate of entropy production in unit volume σ is obtained from the time derivative of s by combination with the balance equations and with the definition of the intensive variables

$$\sigma = \sum J_i \cdot \operatorname{grad} \Gamma_i + \sum \Gamma_i \sigma_i, \tag{27}$$

where summation is carried out over all values of the subscript. The first term on the right-hand side is the entropy production by the transport processes caused by the thermodynamic forces grad Γ_i. The second term is the entropy production involved with the inner generation of the extensive quantities, for example production of some components and consumption of some other components by chemical reactions.

5. LINEARLY INDEPENDENT CHEMICAL REACTIONS AND THEIR ENTROPY PRODUCTION

In the chemical kinetic system of S elementary reactions†

$$\sum_{k=1}^{K} v'_{kt}\mathbf{B}_k \rightarrow \sum_{k=1}^{K} v''_{kt}\mathbf{B}_k \quad t = 1, 2, \ldots, S, \tag{28}$$

and their reverses the concentration c_k of the component $\mathbf{B}_k$ will be changed by the reactions at a rate

$$\sigma_k = \sum_{t=1}^{S} (v''_{kt} - v'_{kt})\left(\frac{\mathrm{d}\vec{\xi}t}{\mathrm{d}t} - \frac{\mathrm{d}\overleftarrow{\xi}t}{\mathrm{d}t}\right) \quad k = 1, 2, \ldots, K, \tag{29}$$

while the intensive parameters are

$$\Gamma_k = \frac{\partial s}{\partial c_k} = -\frac{\mu_k}{T}; \quad k = 1, 2, \ldots, K. \tag{30}$$

Thus, the contribution of the complete system of all chemical reactions to the entropy production becomes

$$\sigma^{\text{chem}} = \sum_{t=1}^{S} \left(\frac{A'_t}{T} - \frac{A''_t}{T}\right)\left(\frac{\mathrm{d}\vec{\xi}_t}{\mathrm{d}t} - \frac{\mathrm{d}\overleftarrow{\xi}_t}{\mathrm{d}t}\right) = \sum_{t=1}^{S} \frac{A_t}{T}\frac{\mathrm{d}\xi_t}{\mathrm{d}t}, \tag{31}$$

where in conformity with equations (9) and (11) A'_t, A''_t and A_t ($t = 1, 2, \ldots, S$) are forward and backward affinities and affinities of the reactions, respectively and, in conformity with equations (2), (6) and (7), $\mathrm{d}\vec{\xi}_t/\mathrm{d}t$, $\mathrm{d}\overleftarrow{\xi}_t/\mathrm{d}t$ and $\mathrm{d}\xi_t\mathrm{d}t$ ($t = 1, 2, \ldots, S$) are rates of the forward and the backward reactions and the net reaction rates, respectively.

Among the collection of all reactions which, in general, are linearly dependent, sets of linearly independent reactions may be found. The maximum number of linearly independent reactions will be Q, if this is the rank of the $K \times S$ stoichiometric matrix $[v'_{kt}]$. A set of reactions is called stoichiometrically independent, if no chemical equation in this set can be obtained by linear combination of any subset of this set of chemical equations. The elements of the $K \times Q$ matrix $[v'_{ku}]$ of the independent set of reactions and the $K \times S$ matrix $[v'_{kt}]$ of the system of all reactions will be related by the equations

$$v'_{kt} = \sum_{u=1}^{Q} \gamma_{tu} v'_{ku}; \quad k = 1, 2, \ldots, K; \quad t = 1, \ldots, S, \tag{32}$$

with

$$\gamma_{tu} = \delta_{tu} + \beta_{tu}, \tag{33}$$

where β_{tu} ($u = 1, 2, \ldots, Q; t = 1, 2, \ldots, Q, Q+1, \ldots, S$) are the coefficients of linear combination and

$$\delta_{tu} = \begin{cases} 1 & u = t \\ 0 & u \neq t \end{cases} \quad \begin{array}{l} t = 1, 2, \ldots, Q, Q+1, \ldots, S, \\ u = 1, 2, \ldots, Q, \end{array} \tag{34}$$

is the Kronecker δ-symbol. Relations analogous with equation (32) hold also between elements of $[v''_{ku}]$ and $[v''_{kt}]$ with the same constant coefficients.

Substituting the equations which relate the stoichiometric matrices of the kinetic system of all reactions to those of a complete set of independent reactions into the definition of the affinities we obtain for the chemical entropy production

$$\sum_{t=1}^{S} \left(\frac{A'_t}{T} - \frac{A''_t}{T}\right)\left(\frac{\mathrm{d}\vec{\xi}_t}{\mathrm{d}t} - \frac{\mathrm{d}\overleftarrow{\xi}t}{\mathrm{d}t}\right) = \sum_{u=1}^{Q} \left(\frac{A'_u}{T} - \frac{A''_u}{T}\right)\frac{\mathrm{d}\xi^*_u}{\mathrm{d}t}, \tag{35}$$

with

$$\frac{\mathrm{d}\xi^*_u}{\mathrm{d}t} = \sum_{t=1}^{S} \gamma_{tu}\left(\frac{\mathrm{d}\vec{\xi}_t}{\mathrm{d}t} - \frac{\mathrm{d}\overleftarrow{\xi}_t}{\mathrm{d}t}\right); \quad u = 1, 2, \ldots, Q. \tag{36}$$

---

†The character S denotes here number of reactions in the kinetic system (28) and should not be confounded with the entropy.

6. EXTENSION BEYOND LOCAL EQUILIBRIUM

In the thermodynamic theories of irreversible processes in continua, in contrast to the non-equilibrium state of the whole system, locally cellular equilibrium was postulated. In other words, it was assumed that in each volume element (cell), the values of the densities of the independent state variables and state functions exist and that the latter are the same functions of the former as in equilibrium systems.

The first important extension beyond this basic assumption of local equilibrium was the derivation of an additional term in Fourier's second law of heat conduction [19]. This term which is linear in the second time derivative of the temperature was obtained from the kinetic gas theory and led to a telegraph-type, hyperbolic differential equation for the temperature as variable. This equation reflects the role of thermodynamic waves in heat propagation by conduction. The term added to the second Fourier law corresponds to a supplement to the force $(-\text{grad } T)$ and as was understood 30 years later [20], can be traced back to an entropy term quadratic in the current density of internal energy flow, added to the equilibrium entropy. Thus, Cattaneo's theory was an extension beyond local equilibrium.

Thirty years later, in the wave approach of thermodynamics [20] the local entropy density of the unified field theory of thermodynamics was extended beyond equilibrium by a term quadratic in the current densities of the purely dissipative transport processes. Thus, the extension of the thermodynamic forces grad Γ by terms linear in the first time derivatives of the current densities is involved. Linear constitutive equations are assumed between the current densities and the extended forces with the same coefficient matrix as in the unified field theory. Finally, by combination of these constitutive equations with the first and second derivatives of the local state equations and with the source-free balance equations, complete sets of telegraph-type hyperbolic equations have been derived both for the intensive variables Γ and the densities a of the extensive variables. These equations describe general thermodynamic waves, i.e. superimposed propagation by waves and conduction. Coefficients of these equations are relaxation times, diffusivities and conductivities.

For other extensions of the thermodynamics based on the assumption of local equilibrium see Ref. [21].

7. EXTENSION TO NON-LINEAR PHENOMENOLOGICAL EQUATIONS

In the first thermodynamic theories of irreversible processes in continua linear equations were postulated to relate fluxes to forces and vice versa. The proofs of the reciprocal relations and of the applicability of variational principles were given for the linear theory, only. Fourier's law of heat conduction and Fick's law of diffusion are linear constitutive equations and proved to be good approximations for over a century. Thus, the assumption of linear constitutive equations seemed to be justified, at least in the case of transport processes. However, it is obvious that a linear function will be a poor approximation to the exponential functions of the Marcelin–Kohnstamm equation. States near equilibrium will be no exceptions in this respect, since in equilibrium the affinities A' and A'' and the rates $d\vec{\xi}/dt$ and $d\overleftarrow{\xi}/dt$ may have high values while $\bar{A}' = \bar{A}''$ and $d\vec{\xi}/dt = d\overleftarrow{\xi}/dt$. Therefore, in

$$e^{A'/RT} = 1 + \frac{A'}{RT} + \frac{1}{2!}\left(\frac{A'}{RT}\right)^2 + \frac{1}{3!}\left(\frac{A'}{RT}\right)^3 + \cdots, \tag{37}$$

the terms of higher order are not negligible, and the same holds true for the reverse affinity A''.

The insufficiency of linear constitutive equations and the fact that for the formulation of variational principles in thermodynamics the existence of dissipation potentials is a precondition, made the generalization of the reciprocal relations for non-linear constitutive equations necessary [22, 23]. This generalization was postulated in the forms

$$\frac{\partial J_i}{\partial X_k} = \frac{\partial J_k}{\partial X_i}; \quad \frac{\partial X_i}{\partial J_k} = \frac{\partial X_k}{\partial J_i}; \quad i, k = 1, 2, \ldots, f, \tag{38}$$

where

$$J_i \equiv J_i(X_1, \ldots, X_k, \ldots, X_f); \quad X_i \equiv X_i(J_1, \ldots, J_k, \ldots, J_f) \tag{39}$$

are, in general, non-linear functions. Because of equations (38) the dissipation potentials

$$\psi = \psi(X_1, \ldots, X_i, \ldots, X_f) \tag{40}$$

$$\phi = \phi(J_1, \ldots, J_i, \ldots, J_f) \tag{41}$$

may exist for which

$$J_i = \frac{\partial \psi}{\partial X_i}; \quad i = 1, 2, \ldots, f, \tag{42}$$

and

$$X_i = \frac{\partial \phi}{\partial J_i}; \quad i = 1, 2, \ldots, f, \tag{43}$$

and equations (38) are nothing else than Maxwellian symmetries like

$$\frac{\partial J_i}{\partial X_k} = \frac{\partial}{\partial X_k}\left(\frac{\partial \psi}{\partial X_i}\right) = \frac{\partial}{\partial X_i}\left(\frac{\partial \psi}{\partial X_k}\right) = \frac{\partial J_k}{\partial X_i}, \tag{44}$$

$$\frac{\partial X_i}{\partial J_k} = \frac{\partial}{\partial J_k}\left(\frac{\partial \phi}{\partial J_i}\right) = \frac{\partial}{\partial J_i}\left(\frac{\partial \phi}{\partial J_k}\right) = \frac{\partial X_k}{\partial J_i}. \tag{45}$$

8. WHAT ARE THE FORCES DRIVING CHEMICAL REACTIONS TO ADVANCE? HOW IS THE EVOLUTION OF CHEMICAL KINETIC SYSTEMS GOVERNED?

During the 30 years which passed after the postulation of the generalized reciprocal relations several attempts were made to prove these in the case of a chemical kinetic system with the elementary chemical reactions (28) and their reverses. All attempts failed because of wrong choice of the thermodynamic forces.

Finally, it was shown that in the case of the chemical kinetic system (28) the generalized reciprocal relations will be satisfied, if we choose the collection

$$\frac{A_1'}{T}, \ldots, \frac{A_u'}{T}, \ldots, \frac{A_Q'}{T}; -\frac{A_1''}{T}, \ldots, -\frac{A_u''}{T}, \ldots, -\frac{A_Q''}{T} \tag{46}$$

[see right-hand side of equation (35)] as independent thermodynamic forces [24]. The flux conjugate to both A_u'/T and $-A_u''/T$ is $d\xi_u^*/dt$ as defined by equation (36). The constitutive equations are obtained from the Marcelin–Kohnstamm equations in the form

$$\frac{d\xi_u^*}{dt} = \sum_{t=1}^{S} \gamma_{tu} \lambda_t \left[\exp\left(\sum_{v=1}^{Q} \gamma_{tv} A_v'/RT \right) - \exp\left(-\sum_{v=1}^{Q} \gamma_{tv} A_v''/RT \right) \right]; \quad u = 1, 2, \ldots, Q. \tag{47}$$

It is easy to verify satisfaction of the generalized reciprocal relations, since the forces in the exponents are linearly independent. The inverse constitutive equations we obtain by direct inversion of equations (7) in the forms

$$\frac{A_t'}{T} = R \ln\left(\frac{d\vec{\xi}_t}{dt}\Big/\lambda_t\right); \quad \frac{A_t''}{T} = R \ln\left(\frac{d\overleftarrow{\xi}_t}{dt}\Big/\lambda_t\right); \quad t = 1, 2, \ldots, S. \tag{48}$$

Since, in contrast to the $2S$ forces, among which only sets of at most $2Q$ are independent, the collection of all rates $d\vec{\xi}/dt$; $d\overleftarrow{\xi}/dt$ ($t = 1, 2, \ldots, S$) are linearly independent, the matrix of the partial derivatives of the forces with respect to the fluxes is a diagonal matrix. Consequently, the inverse generalized reciprocal relations will be trivially satisfied.

Also the governing principle of dissipative processes was applied to systems of elementary reactions in homogenous continua [24, 25] and the kinetic mass action law derived from this principle.

The dissipation functions ψ and ϕ can be easily constructed and have the forms

$$\psi = 2R \sum_{t=1}^{S} \lambda_t \left[\exp\left(\sum_{v=1}^{Q} \gamma_{tv} A_v'/RT \right) + \exp\left(-\sum_{v=1}^{Q} \gamma_{tv} A_v''/RT \right) \right] \tag{49}$$

and

$$\phi = 2R \sum_{t=1}^{S} \left[\frac{d\vec{\xi}_t}{dt} \ln\left(\frac{d\vec{\xi}_t}{dt} \Big/ \lambda_t \right) + \frac{d\overleftarrow{\xi}_t}{dt} \ln\left(\frac{d\overleftarrow{\xi}_t}{dt} \Big/ \lambda_t \right) - \left(\frac{d\vec{\xi}_t}{dt} + \frac{d\overleftarrow{\xi}_t}{dt} \right) \right]. \tag{50}$$

The governing principle of dissipative processes pertaining to the complete system of the elementary chemical reactions is obtained then in the form of the following equation:

$$\sum_{u=1}^{Q} \left[\frac{d\xi_u^*}{dt} - 2 \sum_{t=1}^{S} \gamma_{tu} \lambda_t \exp\left(\sum_{v=1}^{Q} \gamma_{tv} A_v'/RT \right) \right] \delta\left(\frac{A_u'}{T} \right)$$

$$+ \sum_{u=1}^{Q} \left[\frac{d\xi_u^*}{dt} + 2 \sum_{t=1}^{S} \gamma_{tu} \lambda_t \exp\left(-\sum_{v=1}^{Q} \gamma_{tv} A_v''/RT \right) \right] \delta\left(\frac{A_u''}{T} \right)$$

$$+ \sum_{t=1}^{S} \left[\frac{A_t'}{T} - \frac{A_t''}{T} - 2R \ln\left(\frac{d\vec{\xi}_t}{dt} \Big/ \lambda_t \right) \right] \delta\left(\frac{d\vec{\xi}_t}{dt} \right)$$

$$+ \sum_{t=1}^{S} \left[-\frac{A_t'}{T} + \frac{A_t''}{T} - 2R \ln\left(\frac{d\overleftarrow{\xi}_t}{dt} \Big/ \lambda_t \right) \right] \delta\left(\frac{d\overleftarrow{\xi}_t}{dt} \right) = 0, \tag{51}$$

where the variation was carried out with respect to the complete set of independent variables. Therefore, according to equation (51), all expressions in brackets must be zero. Combination of the equations thus obtained gives us the constitutive equations (47) and (48), and, since the latter in ideal systems are equivalent with the Guldberg–Waage kinetic mass action law, as shown here, the basic equation of chemical kinetics was deduced from the governing principle of dissipative processes [24, 25]. It was also demonstrated [25] that for non-ideal systems the same deduction leads to the generalization of the kinetic mass action law in the form

$$\frac{d\xi_t}{dt} = \frac{d\vec{\xi}_t}{dt} - \frac{d\overleftarrow{\xi}_t}{dt} = \vec{k}_t \prod_{k=1}^{k} a_k^{v_{kt}'} - \overleftarrow{k}_t \prod_{k=1}^{K} a_k^{v_{kt}''}; \quad t = 1, 2, \ldots, S, \tag{52}$$

where

$$a_k = \exp\left(\frac{\mu_k - \mu_k^\theta}{RT} \right); \quad k = 1, 2, \ldots, K, \tag{53}$$

are the relative activities of the components in the system (not to be confounded with the as in equations (25) and (55), etc.). Chemical potentials and the density of the Gibbs function are defined in the unified field theory of thermodynamics as

$$\mu_k \equiv \left(\frac{\partial g}{\partial c_k} \right)_{T, p, c_{k \neq j}} \tag{54}$$

and

$$g \equiv g(r, t) = g[a_1(r, t), \ldots, a_i(r, t), \ldots, a_f(r, t)]. \tag{55}$$

As already mentioned, formula (52) was to date refused by some authors.

9. HOW DOES A CHEMICAL KINETIC SYSTEM EVOLVE IN TIME?

In chemical kinetics, concentrations of the components in the kinetic mass action law can be expressed in terms of the extent of the reaction provided that the system is closed. Thus, we arrive at a differential equation of one variable: the extent of the reaction. The solution of this equation is the extent of the reaction as a function of time.

In thermodynamics we have another possibility for the study of the evolution. By combination of the Marcelin–Kohnstamm equation (7) with De Donder's relation (12), also considering equation (11), we may obtain differential equations for the thermodynamic forces and the reaction rates whose solutions show us how the chemical reaction evolves in time.

Differentiation of the De Donder relation with respect to time leads to

$$\frac{d(A/T)}{dt} = s'' \frac{d\xi}{dt},$$
(56)

where the coefficient

$$s'' = \frac{\partial^2 s}{\partial \xi^2},$$
(57)

in general, depends on the extent of the reaction ξ and on time. In linear approximation s'' is a constant. Equation (56) can be split in two, one pertaining to the forward and the other to the reverse reaction:

$$\frac{d(A'/T)}{dt} = s'' \frac{d\vec{\xi}}{dt}$$
(56')

and

$$\frac{d(A''/T)}{dt} = s'' \frac{d\overleftarrow{\xi}}{dt}.$$
(56'')

These can be combined with the Marcelin–Kohnstamm equations

$$\frac{d\vec{\xi}}{dt} = \lambda \exp\left(\frac{A'}{T}\bigg/R\right)$$
(58')

and

$$\frac{d\overleftarrow{\xi}}{dt} = \lambda \exp\left(\frac{A''}{T}\bigg/R\right).$$
(58'')

We then obtain

$$\frac{d(A'/T)}{dt} = s''\lambda \exp\left(\frac{A'}{T}\bigg/R\right)$$
(59')

and

$$\frac{d(A''/T)}{dt} = s''\lambda \exp\left(\frac{A''}{T}\bigg/R\right)$$
(59'')

or

$$\frac{1}{\vec{r}^2} \frac{d\vec{r}}{dt} = \frac{s''}{R}$$
(60')

and

$$\frac{1}{\overleftarrow{r}^2} \frac{d\overleftarrow{r}}{dt} = \frac{s''}{R}$$
(60'')

according as we eliminate the reaction rates

$$\vec{r} = \frac{d\vec{\xi}}{dt} \quad \text{and} \quad \overleftarrow{r} = \frac{d\overleftarrow{\xi}}{dt}$$
(61)

or the forces A'/T and A''/T.

In linear approximation, i.e. assuming constancy of s'', we obtain the solutions

$$\frac{A'}{T} = \left(\frac{A'}{T}\right)_0 - R \ln\left[1 - t\frac{s''\lambda}{R} \exp\left[\left(\frac{A'}{T}\right)_0\bigg/R\right]\right]$$
(62')

$$\frac{A''}{T} = \left(\frac{A''}{T}\right)_0 - R \ln\left[1 - t\frac{s''\lambda}{R} \exp\left[\left(\frac{A''}{T}\right)_0\bigg/R\right]\right],$$
(62'')

of equations (59) and the solutions

$$\mathring{\mathbf{r}} = \mathring{\mathbf{r}}_0/(1 - ts''\mathring{\mathbf{r}}_0/R) \tag{63'}$$

$$\overleftarrow{\mathbf{r}} = \overleftarrow{\mathbf{r}}_0/(1 - ts''\overleftarrow{\mathbf{r}}_0/R) \tag{63''}$$

of equations (60) which describe the time evolution of the reversible elementary chemical reaction

$$\sum_{k=1}^{K} v_k \mathbf{B}_k \rightleftarrows \sum_{k=1}^{K} v_k \mathbf{B}_k \tag{64}$$

between the components $\mathbf{B}_1, \mathbf{B}_2, \ldots, \mathbf{B}_K$. Thus we may see how the affinities A' and A'' and the reaction rates $\mathrm{d}\vec{\xi}/\mathrm{d}t$ and $\mathrm{d}\overleftarrow{\xi}/\mathrm{d}t$ depend on time. In equations (62) and (63) the indices $_0$ label initial values.

By integration of the difference of equations (63') and (63'') we may go even one step further. We obtain

$$\xi = \int_0^t \left(\frac{\mathrm{d}\vec{\xi}}{\mathrm{d}t} - \frac{\mathrm{d}\overleftarrow{\xi}}{\mathrm{d}t}\right) \mathrm{d}t = \ln \frac{1 - (s''\mathring{\mathbf{r}}_0/R)t}{1 - (s''\overleftarrow{\mathbf{r}}_0/R)t}, \tag{65}$$

which is the solution of the kinetic mass action law of a reversible elementary reaction. The form of equation (65) does not depend on the kinetic orders of the forward and of the backward reaction. Comparison with the usual solution in classical chemical kinetics will give the significance of the constant s''. For example, in the case of the reversible elementary reaction

$$\mathbf{B}_1 + \mathbf{B}_2 \rightleftarrows \mathbf{B}_3 + \mathbf{B}_4 \tag{66}$$

the differential equation expressing the kinetic mass action law and its solution can be found on pp. 29–30 of Ref. [26], where the example represented by the gas phase decomposition of HI, $2\mathrm{HI} = \mathrm{H}_2 + \mathrm{I}_2$ is mentioned.

The method of combination of the De Donder relations with the Marcelin–Kohnstamm equations can be applied also to any chemical kinetic system of elementary reversible reactions, e.g. to the system represented by equations (28). In most cases, however,

$$s'' = \frac{\partial}{\partial \xi}\left(\frac{\partial s}{\partial \xi}\right) = \frac{\partial (A/\mathrm{T})}{\partial \xi}$$

is not a constant but a function of A/T and another function of ξ. Therefore, integration of equations (59) and (60) is only possible if we know these functions.

10. THE ROLE OF SYMMETRY IN THE STUDY OF IRREVERSIBLE PROCESSES

We have seen that in the case of constitutive (in other words phenomenological) equations symmetry plays an important role. The coefficient matrix

$$\begin{bmatrix} L_{11} & L_{12} & L_{13} & \cdots & L_{1f} \\ L_{21} & L_{22} & L_{23} & \cdots & L_{2f} \\ L_{31} & L_{32} & L_{33} & \cdots & L_{3f} \\ \cdot & \cdot & \cdot & & \cdot \\ \cdot & \cdot & \cdot & & \cdot \\ \cdot & \cdot & \cdot & & \cdot \\ L_{f1} & L_{f2} & L_{f3} & \cdots & L_{ff} \end{bmatrix} \tag{67}$$

is symmetric in any case. In the linear theory, the elements L_{ik} are constant coefficients in the linear constitutive equations. In the general non-linear theory, however, the elements of the above matrix are defined as

$$L_{ij} = \frac{\partial J_i}{\partial X_k}; \quad i, k = 1, 2, \ldots, f. \tag{68}$$

In both theories this symmetry was a condition for the existence of dissipation functions and thus opened the way for the deduction of the according constitutive equations from the general variational principle of thermodynamics. Constitutive equations are results in different branches of science, originally outside thermodynamics. Hooke's law (1660) of elasticity, Ohm's law (1827) of electric conduction, Fourier's law (1822) of heat conduction, the Navier–Stokes equations (1845) for viscous fluids, Fick's law (1855) of diffusion and the Guldberg–Waage kinetic mass action law (1867) of chemical kinetics should be mentioned here as examples of constitutive equations in different branches of science.

The general effect of the derivation of the constitutive equations of heat conduction [27–29], diffusion [30], viscous, laminar and plastic flow and turbulence [31–35], chemical kinetics [24, 25] etc. from a variational principle of thermodynamics is some kind of synthesis of the science of material and heat transfer, rheology, chemical kinetics, etc. with thermodynamics. We have seen that the study of chemical reactions once bifurcated into chemical thermodynamics and chemical kinetics. The symmetry manifested in the generalized reciprocal relations now makes possible the reattachment of the two science branches, treating chemical reactions by the methods of process thermodynamics and of chemical kinetics, respectively. Thus, we may consider symmetry as an instrument of synthesis, the synthesis of these two sciences. We may perhaps believe that it is a general property of symmetry that symmetry is an instrument of synthesis or integration.

REFERENCES

1. J. Thomsen, Die Grundzüge eines thermo-chemischen Systems. V. Über die Affinität mit besonderer Rücksicht auf die chemischen Zersetzungen. *Annln Phys.* 92, 34–57 (1854).
2. J. H. van't Hoff, *Études de Dynamique Chimique*. Frederik Muller, Amsterdam (1884).
3. J. W. Gibbs, *Trans. Conn. Acad. III* (1874–78).
4. L. Wilhelmy, Über das Gesetz, nach welchem die Einwirkung der Säuren auf den Rohrucker stattfindet. *Annln Phys.* 81, 413–428, 499–526 (1850).
5. C. M. Guldberg and P. Waage, *Études sur les Affinités Chimiques*. Brøgger et Christie, Christiania (1867).
6. R. Marcelin, Sur la mécanique des phénomenes irréversibles. *C.r. hebd. Séanc. Acad. Sci.* 105, 1052 (1910); Mécanique des phénomenes irréversibles a partir des données thermodynamiques. *J. Chim. phys.* 9, 399–415 (1911).
7. Ph. Kohnstamm, On osmotic temperatures and the kinetic signification of the thermodynamic potential. *Proc. Sect. Sci. K. ned. Akad. Wet.* 13, 778–788 (1911); Ph. Kohnstamm and F.E.C. Scheffer, Thermodynamic potential and velocities of reaction. *Proc. Sect. Sci. K. ned. Akad. Wet.* 13, 789–800 (1911).
8. Th.De Donder, L'Affinité. Applications aux gas parfaits. *Bull. Acad. r. Belg. Cl. Sci.* mai, 197–205 (1922).
9. R. Clausius, Über die bewegende Kraft der Wärme. *Annln Phys.* 79, 369–500 (1850).
10. G. N. Lewis, Outlines of a system of thermodynamic chemistry. *Proc. Am. Acad. Arts Sci.* 43, 259–293 (1907).
11. J. N. Brönsted, Die Bedeutung des Aktivitätsbegriffes für die chemische Reaktiongeschwindigkeit. *Z. phys. Chem.* 102, 169–207 (1922); *Z. phys. Chem.* 115, 337–364 (1925).
12. L. Onsager, Reciprocal relations in irreversible processes. *Phys. Rev.* 37, 405–426 (1931); *Phys. Rev.* 38, 2265–2279 (1931).
13. J. Meixner, Zur Thermodynamik der Thermodiffiusion. *Ann. Phys.* 39, 333–356 (1941).
14. I. Prigogine, *Études Thermodynamiques des Phénomenes Irreversibles*. Desoer, Liege (1947).
15. S. R. De Groot and P. Mazur, *Non-equilibrium Thermodynamics*. North-Holland, Amsterdam (1962).
16. I. Prigogine, *Bull. Acad. r. Belg. Cl. Sci.* 31, 600 (1945).
17. I. Gyarmati, On a general variational principle of non-equilibrium thermodynamics (in Russian). *Zh. fiz. Khim.* 39, 1489–1493 (1965); I. Gyarmati, On the "governing principle of dissipative processes" and its extension to non-linear problems. *Ann. Phys.* 23, 353–378 (1969).
18. I. Gyarmati, Non-equilibrium thermodynamics. Field theory and variational principles. Springer, Berlin (1970).
19. C. Cattaneo, Sulla conduzione del calore. *Atti del semin. matematico* e fisico dell'Università di Modena. Vol. III. pp. 83–101 (1948–49); where reference is made to a lecture held by the author at the University of Rome, Italy, 4 May (1946).
20. I. Gyarmati, On the wave approach of thermodynamics and some problems of non-linear theories. *J. Non-equilib. Thermodyn.* 2, 233–260 (1977).
21. J. Casas-Vázquez, D. Jou and G. Lebon (Eds), *Recent Developments in Non-equilibrium Thermodynamics*. Springer, Berlin (1984).
22. I Gyarmati, On the Principles of Thermodynamics. Dissertation, Budapest (1958); I. Gyarmati, On the phenomenological basis of irreversible thermodynamics II. *Periodica Polytech. Chem. Engng* 5, 321–339 (1961).
23. J. C. M. Li, Thermodynamics of nonisothermal systems. The classical formulation. *J. Chem. Phys.* 29, 747–754 (1958).
24. S. Lengyel, Deduction of the Guldberg–Waage mass action law from Gyarmati's governing principle of dissipative processes *J. Chem. Phys.* 88, 1617–1621 (1988).
25. S. Lengyel, On the relationship between thermodynamics and chemical kinetics. *Z. Phys. Chem.* (in press).
26. S. W. Benson, *The Foundations of Chemical Kinetics*. McGraw-Hill, New York (1960).
27. H. Farkas, The reformulation of the Gyarmati principle in a generalized "Γ" picture. *Z.phys.Chem.* 239, 124–132 (1968); On the phenomenological theory of heat conduction. *Int. J. Engng Sci.* 13, 1035–1053 (1975).
28. A. Stark, Approximation methods for the solution of heat conduction problems using Gyarmati's principle. *Ann. Phys.* [7] 31, 53–57 (1974).

29. P. Singh, The application of the governing principle of dissipative processes to Bénard convection. *Int. J. Heat Mass Transfer* **19**, 581–588 (1976); Non-equilibrium theory of unsteady heat conduction. *Acta phys. Hung.* **43**, 59–63 (1977); Formulation of Gyarmati's principle for heat conduction equation. *Wärme- Stoffübertragung* **13**, 39–45 (1980).

30. J. Sándor, Primenyenye variatsionarovo Printsipa Dyarmaty k Izotermicheskoi Diffuzii. *Zh. fiz. Khim.* **44**, 2727–2735 (1970); Application of the Gyarmati principle to multicomponent nonisothermal and chemically reacting diffusional systems. *Acta chim. Hung.* **67**, 303–320 (1971); The treatment of nonisothermal multicomponent diffusion in electrostatic fields on the basis of Gyarmati's variational principle. *Electrochim. Acta* **17**, 673–682 (1972).

31. Sz. Böröcz, An example for Gyarmati's integral principle of thermodynamics. *Z. phys. Chem.* **234**, 26–32 (1967); Derivation of the transport equations of non-isothermal hydrodynamics from the Gyarmati principle in different pictures. *Acta chim. Hung.* **69**, 329–340 (1971).

32. P. Singh, D. K. Bhattacharya, A new approximate method for laminar stagnation flow. *J. Non-equilib. Thermodyn.* **2**, 103–108 (1977); Application of Gyarmati's principle to boundary layer flow. *Acta mech.* **30**, 137–144 (1978).

33. D. K. Bhattacharya, Application of Gyarmati's variational principle to laminar stagnation flow problems. *Wärme- Stoffübertragung* **17**, 27–30 (1982).

34. J. Verhás, The construction of dissipation potentials for nonlinear problems and the application of Gyarmati's principle to plastic flow. *Z. phys. Chem.* **249**, 119–122 (1972).

35. Gy. Vincze, Deduction of the quasilinear transport equations of hydrothermodynamics from the Gyarmati principle. *Ann. Phys.* [7] **27**, 225–236 (1971); Deduction of the Reynolds equations of turbulence from the Gyarmati principle. *Acta chim. Hung.* **75**, 33–43 (1972); On the treatment of thermoelectrodynamical processes by Gyarmati's principle. *Ann. Phys.* [7] **30**, 55–61 (1973).

Computers Math. Applic. Vol. 17, No. 1–3, pp. 457–465, 1989
Printed in Great Britain. All rights reserved

SYMMETRIES AND THE NOTION OF ELEMENTARITY IN HIGH ENERGY PHYSICS

K. Gavroglu

Department of History of Science, Harvard University, Cambridge, MA 02138, U.S.A.

Abstract—The insights we have acquired about symmetries during the past 25 years have not only contributed to the construction of successful explanatory schemata in elementary particle physics, they have also modified the conceptual framework within which a series of philosophical and methodological issues of elementary physics are discussed.

The atomistic paradigm of high energy physics cannot any more be dismissed because the proposed elementary particles are too many (and, hence, it is claimed, they do not provide a simple account of nature) or because it is not possible to observe quarks in an isolated manner. The developments in particle physics have brought about radical changes to our notions of simplicity and observability, and in this paper we elaborate on these changes. It is as a result of these changes that the present situation in elementary particle physics justify us to claim that we have indeed reached a level of explanation where the constituent particles (quarks, leptons, gluons and intermediate bosons) used for the explanation of the various phenomena considered to be delineating a particular level in the descriptive framework of the physical phenomena and a specific stratum in the organization of nature, can be regarded as elementary.

1. INTRODUCTION

Does the present situation in elementary particle physics justify us to claim that we have reached a level of explanation where the constituent particles used for the explanation of the various phenomena can be regarded as elementary?

I will argue in what follows that we are presently in a position to systematically investigate this question and also provide an answer, in the affirmative, not because of any new experimental discoveries where *the* elementary particles have been observed, but only because of the insight we have acquired about symmetries during the past 25 years and which has brought about radical changes to the concepts of observability and simplicity which necessarily dominate any discussion about the philosophical and methodological aspects of elementary particle physics. The role of symmetries in constructing theories to account for the phenomena associated with particle physics has been analytically presented in the volume preceding the present one and will not be repeated here [1]. Nor will there be any examination of the series of problems arising out of the relationship of the notion of elementarity with much of our (Western) metaphysical tradition [2]. The implications of the problematique concerning "teleological" and "first cause" arguments to the concept of elementarity will interest us only indirectly, and we shall concentrate, not on the ontological status of the various entities considered to be elementary, but rather on the methodological role of these entities in the construction of theories.

What is, however, an elementary particle? What are the aims of elementary particle physics? An appropriate place to look for an answer, is the two thorough and detailed reports of the National Research Council (U.S.) since, if anything, they reflect a view compatible with the consensus of the high energy physics community.

> "We call a piece of matter an elementary particle when it has no other kinds of particles inside of it and no subparts that can be identified—we think of it as a point particle." [3]

And as for the subject itself it is stated that:

> "The nature and purposes of elementary particle physics concern both the discovery of new phenomena exhibited by matter (and other forms of energy) under extreme conditions and the understanding of known phenomena." [4]

It is, then, quite remarkable that given these "definitions", the survey of the literature on the whole [5], displays a truly paradoxical situation. Despite the fact that the accounts of the various developments conform absolutely with the definition of what an elementary particle is, nearly all the writers express reservations and doubts about whether the particles we would presently consider as elementary [leptons, quarks and intermediaries (the photon, the Ws, the Z, the gluons)] should really be given the status of the ultimate constituents of matter.

Two reasons are usually projected to justify nearly all the reservations expressed about the elementary character of all the structureless particles we know today and especially of the quarks. It is, firstly, remarked that not all of them have been seen and that all attempts to find free quarks have failed. And, secondly, that there are too many of those particles to consider them as the ultimate building blocks. In other words, despite the fact that leptons, quarks and intermediaries would be absolutely compatible with the "accepted definition" of what an elementary particle is, *further methodological criteria such as those of observability and simplicity are invoked in order to doubt the elementary status of the very same particles.*

The common conclusion, then, of most people who pass judgement on the present status of high energy physics, can be summarized as follows: granted that quarks, leptons and intermediaries are particles with no other kind of particles inside them, and they are in that respect point particles, we can neither observe in an isolated manner all of them, nor are they so few as to make up a convincing simple schema. Based, in effect, on this syllogism Schrader–Frechette [6] argues that the Kuhnian paradigm of the world being built up by elementary particles should be abandoned, and that atomism is in deep crisis. This particular claim has been convincingly rebutted by Cushing [7].

It may be argued that using explicitly stated methodological criteria for a further appraisal of physical theories is something to be encouraged. One still wonders, however, about the status and the degree of concensus reached for the criteria with respect to which various questions are to be appraised. Do the criteria for particles to be regarded as elementary express a consensus only good for the day-to-day activities of physicists? Yet when an overview of the developments is attempted there seems to be a shift to a new consensus this time about the *non*-elementarity of the very same entities which are regarded as elementary in the day-to-day activities!

2. SUCCESSFUL METHODS, BROKEN SYMMETRIES AND THE QUESTION OF SIMPLICITY

It is undoubtedly the case that reading the developments of high energy physics is necessarily influenced by one's metaphysical views, ontological beliefs and epistemological preferences. It should not however escape our attention, that more often than not, it is the impressive success of the methods employed to understand the phenomena and the ensuing confidence in the theory which becomes dominant in the evaluation of the developments rather than more sophisticated philosophical considerations. Therefore, a considerable amount of confusion can be dispelled, if, in such evaluations, the following are differentiated and kept separate:

1. The metaphysical beliefs and the ontological claims.
2. The successful methods.
3. The emerging picture of nature.

It is *then* a different question altogether, if as a result of our studies, we would decide to modify (1) because of (3) or choose to elevate (2) to a principle which seems to be the best for contributing to philosophical argumentation and so on. Let us take an extreme, yet especially characteristic case. The investigation of the problem of elementarity motivated by the success of S-matrix theory, and an analogous investigation motivated by the success of quantum field theory, will necessarily oblige the adoption of two different ontologies. In the case of the former, for example, the elementarity of particles is to be searched "in a reality" similar to that which emerged from the study of molecules and atoms and where complexity is expressed in terms of "excited states". The continuous subdivision of matter as a means of finding *the* elementary particles will be doubled since the question, "what does a particle consist of?", is meaningless for such an approach if the energies used to find the constituents are larger than the mass of the particle being searched into. Thus, the specification of elementarity becomes, in effect, synonymous to achieving a self-consistent derivation of any given particle by "everything else" [8]. The study of the same problem, motivated by the success of quantum field theory leads to quite a different situation. Here, the belief that it is possible to have a detailed space–time description of particles under extreme relativistic conditions by their fields is paramount. The "reality" where the elementary particles are to be sought is very similar to that which emerged from the study of electromagnetism with its "well-defined" procedures of translating the "tangible" particles into fields. In these examples, one can then see how the success of a particular method, under the influence of the "emerging reality" forces the adoption of a particular ontology [9].

It is interesting to note that Heisenberg's insistence to develop a way of viewing "elementarity" by extending arguments primarily used in atomic spectroscopy leads to the proposal that "what we have to look for are not fundamental particles, but fundamental symmetries" [10]. Such a proposal, however, does not provide us with an alternative framework to answer the questions we posed in the beginning. What is being sought is not the kinds of possible ontologies within which one could accommodate a notion of elementarity and be able to "read" consistently the various theories of particle physics. The opposite, in fact, is the case: the questions we posed presuppose a particular ontology—that of "the ultimate building *blocks*". What is sought is the contextual (and historical) character of "the ultimate".

We would like to argue that some of the problems mentioned above can be partly dispelled and a satisfactory answer be given to the questions we posed, (1) if elementarity is examined within a framework where the particles are to be regarded as elementary to the extent that they can be used to achieve a *unified* account of all phenomena, and (2) if it is realized that the development of elementary particle physics has brought about conceptual changes which have radically modified the admittedly controversial issues of *observability* and *simplicity*, and that, if anything, a case can be made that the criterion of observability cannot be identified with observing an entity in an isolated manner, nor that of simplicity with "fewness".

The emphasis on relating the study of the problem of elementarity to the question of seeking a set of laws providing a unified description of nature is not merely an attempt to take into account what has been the outstanding aim (and success) of high energy physics during the past years. What has been neglected, however, in the various discussions about elementary particles is an appraisal of the methodological status of the concept of elementarity within a context created by the attempts to construct a theory which provides a unified account of what seem to be different interactions giving rise to a class of phenomena. Finding (and I would say deciding) that a set of particles are elementary is meaningful only to the extent that it can be shown that these particles are sufficient for a unified account of as many phenomena which—intuitively, at least—we consider as delineating a particular level in the descriptive framework of the physical phenomena and a specific stratum in the organization of nature [11]. Each such level has a relatively autonomous status. It is this *relative* autonomy which is important here, since there is always something with respect to which autonomy is signified, and that no level is fully autonomous since some of the phenomena used to delineate each level do not unambiguously belong to a single level. Furthermore, it is the *relative* autonomy which allows reductionism from one level to the next, and yet the relative *autonomy* of each level is what confines the practice of constructionism to within each level.

Instead of, then, asking the question, what are the ultimate constituents of matter?, one should rather inquire about those (ultimate) constituents of matter which can be used to provide a unified description of phenomena and be, in turn, determined by this description. It is the latter that is historically meaningful, even though it was the first question that acquired a legitimacy on purely epistemological grounds. The (theoretical or experimental) search for the ultimate constituents of matter is then related with ways of "combining" them, proposing schemata by which we can build up the composite particles and the phenomena to which they give rise to, and understanding in a more fundamental manner those laws and regularities which have already provided an explanation for many phenomena and whose validity has been repeatedly tested. It is within such a framework that the status of the various constituents so far as elementarity is concerned, has to be appraised.

Simplicity as a criterion to be used for choosing among "competing" modes of explanation has been repeatedly invoked by physicists and philosophers alike, and it was usually the "more symmetric" mode that was eventually preferred. Its discussion is inherently difficult, especially if the aim is to reach a consensus on how the criterion of simplicity should be used in a consistent manner. Its only meaningful discussion seems to me is to argue about the relative merits of a particular criterion with respect to other such proposed criteria. The notion of the "ultimate constituents" of matter and the ways devised to reach a consensus about their identity has been inextricably related to the notion of simplicity. And it is within such a *problematique* that among the many modes whose explanatory power is roughly equivalent, the one with the fewer proposed (sub)particles if favored as having a chance of being "more fundamental". It is such a viewpoint which associates simplicity with fewness that is used to reject the quark model as providing a self-consistent account of phenomena in terms of *elementary* particles, because there are too many constituents which are particular to the quark model to be given the status of elementary.

Even though the adoption of such a particular criterion of simplicity cannot be comprehensively defended, there are some questions which can be posed independent of any specific criterion of simplicity. What happens, for example, if a particular criterion of simplicity appears to be violated in a systematic manner? How justified is one in using any such criterion, if one knows beforehand that it is bound to be violated? Can one talk of degrees of simplicity? Or, is there any meaning to the notion of approximate simplicity?

We will not attempt to answer these questions, but note that the developments in high energy physics seem to provide us the conceptual framework that allows their examination. Take, for example, the use of symmetries. It is no exaggeration to claim that symmetries have been regarded by physicists not only as principles of universal validity, but also as indications of the simplicity of nature at its deepest level. However, most symmetries are demanded from the theories, with the certainty that they are violated either by interactions which have not been taken into consideration or dynamically. Various techniques have been devised to calculate the contributions of these violations. In the unified theories of particle physics, the approximate character of the symmetries of the strong, electromagnetic and weak interactions, is possible to be explained as being a consequence of gauge invariance and renormalization. There are, really, only exact symmetries which govern all interactions and their approximate character is dynamically explainable [12]. These new insights we have gained into the structure of theories allow us to inaugurate a totally different approach to the question of simplicity, rather than being entangled in the deadlock brought about by the process of deciding how many is too many, or how few is not too many!

3. THE NOTION OF OBSERVABILITY

The developments in high energy physics, however, imply the possibility of a radical departure from a notion of observability so closely tied with the observation of entities in an isolated manner. Details of the quark model, on which some of the contents of this section are based, can be found in Lipkin [13] and Greenberg and Nelson [14].

The development of hadron physics has, until recently, consisted of a series of *ad hoc* rules, models and assumptions which were, at best, loosely connected to one another and even more tenuously related to an underlying dynamical theory. In the more recent past, however, one theory of hadrons has begun to emerge as something of a standard theory. This is the Yang–Mills theory of colored quark and gauge fields. Quantum chromodynamics (QCD) is a quantum field theory of the strong interactions with non-abelian gauge fields mediating the interactions between the quarks. The outstanding challenge posed by this theory is to learn to make reliable computations of hadron properties in a systematic fashion. Nothing comparable to the Feynman rules and the perturbation approximation series in quantum electrodynamics (QEC) exist for the bound state physics of the QCD. And there is no proof for the existence of a single bound state in any relativistic four-dimensional quantum field theory.

These difficulties notwithstanding, QCD has certain attractive features. It does not seem to be in conflict with any existing phenomenology of the strong interactions, and the symmetries that can be extracted from QCD are precisely the symmetries of the strong interactions and no more. Local gauge invariance of the color SU(3) and the formal existence of quarks transforming as the fundamental representation of this group are the only requirements and they seem to be sufficient to specify the theory.

Even though QCD has been constructed in close analogy with QED, the intermediaries of the strong force or the color charge quark, the gluons, have non-zero interactions (and self-interactions) among themselves. Exactly because gluons carry the strong color charge, it is possible for the color charge of a quark to be shared with the gluon cloud in addition to a color polarization phenomenon much like the charge screening of QED. Because the color charge is spread out rather than localized, the effective color charge will tend to appear larger at long distances and smaller at short distances.

The outcome of the competition between these two opposing tendencies depends on the number of gluon species that can share the color charge and on the number of quark types that can screen the color charge. If the color gauge group is SU(3), the net effect is one of antiscreening, that is, of a smaller effective charge at short distances. Extremely close to the quark, the effective color charge becomes vanishingly small, so that nearby quarks behave as if they are non-interacting free particles. This is the origin of the term asymptotic freedom.

Interestingly, asymptotic freedom does provide a partial, at least, justification of the parton model put forth to describe violent scattering processes: the measurable quantities are reproduced by assuming that the constituents of a proton are a swarm of non-interacting point entities.

Asymptotic freedom offers a qualitative explanation to the paradox of quasi-free quarks that are permanently confined. At the short distances probed in deep inelastic scattering, the effective color charge is weak, so the strong interactions between quarks can largely be neglected. As quarks are separated the effective color charge grows, so the strong interaction becomes more formidable. This is the property of confinement [15]. What confinement means in QCD is that all physical states are color SU(3) singlets. Confinement implies that the color degrees of freedom are in principle not observable in an isolated manner although they mediate the strong force. The quarks and gluons since they are not color singlets have no corresponding physical states.

The prediction of a new particle (and usually its discovery) is followed by a process of "elementarizing" it. There is firstly the assignment of quantum numbers (mass, charge, spin, strangeness etc.) and its assignment to one of the particle families (leptons, quarks, intermediaries). Particles are, thus, first labelled and classified. The process, however, of "elementarization" is not completed unless the procedures of observability are also specified. One of the reasons that quarks are not regarded as elementary is because these procedures of observability are taken to imply observing an entity in an isolated manner. This is, however, totally unwarranted since the procedures of observability can be specified in such a manner so as to dispel any reservations about the possibility of not recognizing in a unique manner what it is that is being observed. In case a newly discovered particle is not observed in an isolated manner, it can be claimed that fulfilling the following conditions specifies the particle uniquely:

(a) Account for already observed particles.
(b) Account for already observed interactions/decays.
(c) Account for already observed properties (e.g. magnetic moment).
(d) Account for any observed unexpected phenomenon.
(e) Predict particles/events *and* absence of events.
(f) Predict events unique to particular mode *because* of constraints involved.

This process of "elementarizing" a particle is just another way of utilizing the polymorphous role of the symmetry considerations in elementary particle physics.

These are procedures that do not allow for the possibility of either manipulating or intervening [16]. If, however, the impossibility to manipulate and intervene is stipulated by the theory itself, one is by no means justified in demanding that the only way a theory would be acceptable is if it responds positively to what *then* amounts to an externally brought-in criterion. If isolating a single quark is to be considered as the ultimate convincing evidence for the reality of the quarks and for accepting them as constituting elementary entities, is that not a way of negating, at least, the methodological implications of confinement which seems to be a *dynamical* property of gauge theories? Alternatively, the totality of the proposed steps that make up the procedures of observation seem to be consistent with these implications. A parallelism can be made with the quantum theory of atoms. *So far as quantum theory is concerned it is meaningless* to pose the question as to where an electron is after it "leaves" an outer orbit and before it "appears" at a lower one. If one wants to make a claim about the discreteness of space–time, this meaninglessness cannot be taken as an indication for any claim favoring the discreteness of space–time. And it is "doubly wrong", after taking the interpretation of the electron jump as giving indications of discreteness in the structure of space–time, to *then* criticize the theory because it uses continuous space–time parameters. The same circular argumentation seems to me is being used in the case of the quarks, when it is demanded, on the one hand, that they be freely observed, when, on the other hand, their role has been articulated through a theory where confinement is a property *derived* from those structural characteristics (gauge invariance and renormalization) which are at least a necessary (and for some a sufficient as well) condition for achieving a unified description of all interactions.

Let us consider some characteristic cases from the history of elementary particle physics which have forced us to rethink the whole question of observability.

At first sight, the law of energy conservation (and of linear and angular momentum) did not seem to hold in the weak-decay with an initial state composed of only a neutron and a final one composed of the two observable particles, the proton and the electron [17]. This situation prompted some physicists to question the validity of the law of conservation of energy when applied to individual microscopic processes. W. Pauli's suggestion first in 1930 and then in 1933 appeared at the time equally, if not more, preposterous. He proposed that a massless particle with zero charge and 1/2 spin and which because of its feeble interactions with surrounding matter escapes observation, is the carrier of the missing energy. This was something extremely bothersome since it was not like the other "unseen" particle, the photon, which could be accounted for as the quantum of the electromagnetic field by the then newly developed techniques of the second quantization. And, especially, after the demonstration of the particle-like behavior of the photon, the latter's status among the elementary particles of the period was hardly doubted. That was not, however, the case with the neutrino when it was first proposed. The change came after the proposal of a successful theory of weak interactions by E. Fermi in 1933–34, in analogy with QED. The subsequent corroborating evidence in favor of such a theory left no doubt about the "existence" of the neutrino long before its first observation in an isolated form in 1953.

The second example is somewhat more intriguing. Heisenberg's uncertainty principle allows for the law of energy conservation to be "violated" provided this violation occurs in processes whose duration and the extent of the violation are related by Planck's constant. One of the simplest implications of such a state of affairs is for an electron to emit a photon and in a little while to absorb the same photon. Since the details of the electron–photon interaction were among the best known quantities, the effects of this phenomenon should have been quite straightforward to calculate. The calculation was, indeed, quite straightforward, its results, however, turned out to be infinite: the charge of the electron, as a result of such an effect had to be modified by an infinite amount. This difficulty was resolved in the late forties by the work of S. Tomonaga, J. Schwinger, R. Feynman and F. Dyson, where the mathematical techniques used were followed by a new intepretation of the physical meaning of the parameters expressing mass and charge. The terms which were infinite expressed the various interactions of the "bare mass" m_0 and "bare charge" e_0, and which eventually gave the electron mass and charge their measured values. m_0 and e_0 would be the values of the electron mass and charge if all interactions were to be turned off—something impossible anyway and also devoid of any physical meaning and practical use. The way out of this difficulty was to put the physical mass and charge m_0 and e_0 plus the correction terms, whenever in the expressions there appeared the m and e. One now had two sources of infinities which "cancelled" each other: the one coming from the calculations of the various quantities when m and e are substituted by m_0 and e_0, and the other by the corrections [18].

The third example is related to the possibility provided by the "coloring" of the quarks to construct a fairly satisfactory schema for the strong interactions [19].

QCD, constructed in analogy with QED, and after a considerable amount of insight was gained about the gauge theories, possesses the quite remarkable property of asymptotic freedom. The closer the quarks are to each other inside the hadrons, the weaker the interaction among them is and they behave like "free" particles. If the potential between the quarks is of the form $\alpha(r)/r$, then asymptotic freedom follows from the structure of the theory. Because of quantum corrections the effective coupling constant of a quantum field theory depends on the distance scale r at which the coupling constant is measured. Thus, since $\alpha(r)$ tends to zero as r tends to zero it is, then, possible to use perturbation theory for small r. It should be strongly emphasized that the corresponding quantum mechanical effect in electromagnetism is vaccum polarization and what amounts to asymptotic freedom is achieved in large distances—a state of affairs which has influenced the formation of our "traditionally" held view about observing isolated entities. The confinement of quarks whose only proofs available are model dependent, seems to be quite indispensable for the only promising way of incorporating gravity into a unified description of all forces.

Recent developments introduce a different kind of "confinement" as well. The future success of the superstring theories is quite strongly dependent on devising a convincing method to show that the ten dimensions which are necessary to construct the theory can, in fact, be "compactified" to the four that make our space–time continuum [20]. The rest are there, but unobservable, all curled up, not having had a chance to unfold during the first instants of the big bang—allowing, in a variant of the inflationary universe, for the "existence" of (many) universes with different dimensionality [21].

The examples we mentioned display a move from (1) a situation where the unseen is *accounted* by a new theory and the procedures for its observation in an isolated form are explicitly and unambiguously stated, to (2) a situation where the proposed theory shows how to *tame* the catastrophes brought about by the unseen, proposing at the time procedures for observing manifestations of the unseen, to (3) a situation where remaining unseen is guaranteed by the theory itself modifying analogously the procedures of observation.

It may be remarked that there is no rigorous proof of confinement which is (relatively) model independent, and that such a situation cannot justify our placing so much emphasis on this concept. Such an argument, however, is quite irrelevant for what we attempt to do in this paper which is to answer the two questions we posed at the beginning. And one of the ways for providing an answer is to show that the developments in high energy physics seem to be establishing a framework which *legitimizes* the use of a set of concepts which should, at least, motivate us to question our beliefs about the observability of the ultimate building blocks. This is, obviously, not a claim for the correctness of the dominant theories in particle physics, but rather an appeal to realize that on a conceptual level, we are in a position to have theories, which allow for quite radical departures from a set of accepted procedures of observability.

Might not all these be a series of mathematical tricks to ensure that what is not observed stays unobserved, because basically it is not there to start with? After all, there is such a historical precedent. It is the ether, whose ever enriching "physical" attributes were postulated "as excuses for hiding evidence of it from experiment" [22]. The parallelism, however, cannot be sustained for one very crucial difference between the two. We now know that the main reason ether was introduced, was because of the prevailing prejudices in favor of the mechanistic outlook. It was impossible to imagine and accept the propagation of waves independent of a medium. One of the truly remarkable aspects of Einstein's 1905 paper, is that it shows that the ether was not a *necessary* notion for a consistent reading of both electromagnetism and mechanics [23]. His arguments convinced us that showing that something is unnecessary may have as tangible and measurable results as proving that something is right or wrong. For the case of ether every time there was a failure to observe an expected property, there was an enrichment of its physical attributes. The situation with the quark model is totally different. Every predicted property of the quark model has been corroborated, and further refinements were able to account for the observed deviations. In the case of the ether the additional physical attributes guaranteed that what was "expected" and looked for and not found, stays unseen. In the case of the quarks what was expected was found and the development of the theory gave rise to confinement. In the case of the ether, one had from the start an unsuccessful mode of explanation, whereas in the case of the quarks one had, right from the beginning, a successful model.

4. CONCLUSION

What I have attempted to do was to argue that appraising the developments in high energy physics within a context founded on an ontology of "a few, freely observable ultimate building blocks" as being *the* elementary particles is quite misleading, and does not really conform with the implications of the emerging conceptual framework of these recent developments. The insight we seem to have been gaining for the features of the "subnuclear level" is that a consistent and unified account of all phenomena of this realm can be satisfactorily built with quite a few particles taking part in gauge invariant and renormalizable interactions which necessarily confine some of the constituents. It is only in this sense that leptons, quarks and intermediaries can be regarded as elementary, and that the paradigm of elementary particle physics is in no crisis.

Does all this mean that there is no possibility for a further underlying structure to be discovered? Does this mean that leptons, quarks and the intermediaries will forever remain structureless, however much we try to find their structure? Nothing justifies any denial to such developments and if past experience is to have any guiding value new structures will almost certainly appear. One of my aims was to show that reaching the "smallest" and "structureless" constituents may in fact be a necessary condition in order to consider them as elementary, but it is by no means a sufficient condition. This latter requirement can only be fulfilled if these structureless constituents can actually provide a unified explanation of all the phenomena characteristic of a particular "realm", thus bringing forth the methodological significance of "elementarity" during each historical period. Concerning the developments of the last 25 years, nowhere is this significance more pronounced than in the changes brought to the process followed for elementarizing the particles, and especially in specifying their procedures of observability. The modifications to the notions of simplicity and observability have been precipitated as a result of our further understanding of the complex role of symmetries. They are not merely a convenient means for constructing theories, they also seem to be continually modifying the conceptual framework within which a series of philosophical and methodological issues of elementary particle physics are discussed.

Acknowledgements—I wish to thank Professor S. S. Schweber for his extremely helpful comments.

REFERENCES

1. P. D. Mannheim, Symmetry and spontaneously broken symmetry in physics of elementary particles. *Comput. Math. Applic.* **12B**, 169–183 (1986). Reprinted in *Symmetry: Unifying Human Understanding* (Ed. I. Hargittai). Pergamon Press, Oxford (1986).
2. W. A. Wallace, Elementarity and reality in particle physics. *Boston Studies in the Philosophy of Science* (Eds R. S. Cohen and M. Wartofsky), Vol. III. Reidel, Dordrecht (1968).
3. *Physics in Perspective*, p. 19. National Research Council (U.S.), Physics Survey Committee, Washington National Academy of Sciences (1972).
4. *Physics Through the 1990's*, p. 13. Elementary Particle Physics, National Academy of Sciences, U.S.A. (1986).
5. L. M. Brown and L. Hoddeson (Eds), *The Birth of Particle Physics*. Cambridge Univ. Press, Cambridge (1983); G. Buschor, Discovery of new elementary particles with unusual properties. *Universitas* **18**, 163–186 (1976); G. F. Chew, Elementary particles? *Physics today* **17**, 30–34 (1964); Cinquant' Anni di Fisica delle Interazioni Deboli (Fifty Years of Weak Interaction Physics). A cura di A. Bertin, R. A. Ricci, A. Vitale in occasione del cinquantesimo anniversario della formulatione della teoria di Fermi sul decadimento beta Bologna: Societa Italiana di Fisica; F. E. Close, *An Introduction to Quarks and Partons*. Academic Press, London (1978); M. Conversi (Ed.) *Evolution of Particle Physics*. Academic Press, New York (1970); G. Feinberg, *What is the World Made Of?* Anchor Books–Doubleday, New York (1977); E. Fermi and C. N. Yang, Are mesons elementary particles? *Phys. Rev.* **76**, 1739–43 (1949); W. Heisenberg, The nature of elementary particles. *Physics Today* **29** (March), 32–39 (1976); Y. Ne'eman and Y. Kirsh, *The Particle Hunters*. Cambridge Univ. Press, Cambridge (1983); R. Weingard, Grand unified gauge theories and the number of elementary particles. *Philosophy Sci.* **51**, 150–155 (1984).
6. K. Shrader-Frechette, Atomism in crisis: an analysis of the current high energy paradigm. *Philosophy Sci.* **44**, 409–440 (1977); K. Shrader-Frechette, High-energy models and the ontological status of the quark. *Synthèse* **42**, 173–189 (1979); K. Shrader-Frechette, Quark quantum numbers and the problem of microphysical observation. *Synthèse* **50**, 125–145 (1982).
7. J. T. Cushing, Models and methodologies in current theoretical high energy physics. *Synthèse* **50**, 5–101, 109–123, 544 (1982).
8. W. Heisenberg, The present situation in the theory of elementary particles. In *Two Lectures*, pp. 9–25. Cambridge Univ. Press, Cambridge, (1949); J. T. Cushing, The importance of Heisenberg's S-matrix program for the theoretical high energy physics of the 1950's. *Centaurus* **29**, 110–149 (1986).
9. M. Redhead, Some philosophical aspects of particle physics. *Stud. hist. phil. Sci.* **11**, 279–304 (1980); S. S. Schweber, Some philosophical reflections on the history of quantum field theory (to appear); J. Schwinger, *Particles, Sources and Fields*, Vol. I. Addison-Wesley, Reading, Mass. (1970); S. Weinberg, The search for unity: notes for a history of quantum field theory. *Dedalus* **107**, 17–35 (1977); S. Weinberg, The ultimate structure of matter, in *A Passion for Physics: Essays in Honor of Geoffrey Chew* (Ed. Carleton De Tar, J. Finkelstein, Chung-I Tem), pp. 114–127. World Scientific, Singapore (1985).
10. W. Heisenberg, Development of concepts in quantum theory, In *The Physicist's Conception of Nature* (Ed. J. Mehra), p. 273. Reidel, Dordrecht (1973).
11. P. W. Anderson, More is different. *Science* **177**, 393–396, 1972.
12. S. Weinberg, Conceptual foundations of the unified theory of weak and electromagnetic interactions. *Rev. mod. Phys.* **52**, 515–524 (1980).
13. H. J. Lipkin, Quarks for pedestrians. *Physics Rep.* **8**, 173–268 (1973).
14. O. W. Greenberg and C. A. Nelson, Color models of hadrons. *Physics Rep.* **32C**, 69–121 (1977).
15. M. Bander, Theories of quark confinement. *Physics Rep.* **75**, 205–286 (1981); J. Mandelstam, General introduction to confinement. *Physics Rep.* **67**, 109–121 (1980).
16. I. Hacking, *Representing and Intervening*. Cambridge Univ. Press, Cambridge (1983).
17. K. Gavroglu, Popper's tetradic schema, progressive research programs and the case of parity violation in elementary particle physics 1953–1958. *Z. allg. wiss.* **XVI**, 261–186 (1985); C. S. Wu and S. A. Moszkowski, *Beta Decay*. Interscience, London (1966).

18. S. S. Schweber, Shelter Island, Pocono and Olstone. The emergence of American quantum electrodynamics after World War II. *Osiris* **2**, 265–302 (1986).
19. W. Marciano and H. Pagels, Quantum chromodynamics. *Physics Rep.* **36C**, 137–276 (1978).
20. J. Schwarz and M. Green (Eds), *Superstrings: The First 15 Years of Superstring Theory*. World Scientific, Singapore (1986).
21. A. D. Linde, The inflationary universe. Notes, Loeb Lectures at Harvard Univ., Fall (1987).
22. S. Drell, Elementary particle physics. *Daedalus* **106**, 15–32 (1977).
23. G. Holton, *Thematic Origins of Scientific Thought: Kepler to Einstein*. Harvard Univ. Press, Cambridge, Mass. (1973); A. Miller, *Albert Einstein's Special Theory of relativity: Emergence (1905) and Early Interpretation (1905–1911)*. Addison-Wesley, Reading, Mass. (1981).

Computers Math. Applic. Vol. 17, No. 1–3, pp. 467–473, 1989
Printed in Great Britain. All rights reserved

0097-4943/89 $3.00 + 0.00

SYMMETRY-BREAKING PATTERN FORMATION IN SEMICONDUCTOR PHYSICS: SPATIO-TEMPORAL CURRENT STRUCTURES DURING AVALANCHE BREAKDOWN

J. Parisi, J. Peinke, B. Röhricht, K. M. Mayer and U. Rau

Physikalisches Institut, Lehrstuhl Experimentalphysik II, Universität Tübingen, Morgenstelle 14,
7400 Tübingen, B.R.D.

Abstract—Nonlinear current transport behavior during low-temperature avalanche breakdown in extrinsic germanium comprises the spontaneous symmetry-breaking evocation of spatial and temporal dissipative structures in the formerly homogeneous semiconductor. Representing a general feature of spatially distributed and compartmentalized reaction–diffusion systems, such kind of order–disorder transitions are discussed in terms of the interdisciplinary synergetics concept that embraces wide ranges of human activities both in the arts and in the sciences.

1. INTRODUCTION

The interdisciplinary science "synergetics" deals with complex systems that possess the fundamental property of spontaneous self-organization of their macroscopic behavior. In this sense, synergetics deals with the spontaneous emergence of order out of disorder. Far from equilibrium, the cooperation of a large number of systems may produce macroscopic spatial, temporal or functional structures. The processes involved are nonlinear, and in many cases stochastic [1]. Dramatic progress has been made in recent years in the understanding of such phenomena in quite different disciplines ranging from mathematics to physics and chemistry to biology and sociology.

Among the variety of nonlinear dynamical systems which can be studied experimentally, solid-state turbulence in semiconductors appears particularly interesting. Nonlinear current transport behavior during low-temperature avalanche breakdown of extrinsic germanium comprises the spontaneous symmetry-breaking evocation of both spatially inhomogeneous and temporally unstable dissipative structures in the formerly homogeneous semiconductor [2–8]. Such kind of non-equilibrium phase transition between different conducting states results from the autocatalytic nature of impurity impact ionization generating mobile charge carriers [9–11]. The simple and direct experimental accessibility via advanced electronic measurement techniques prefers semiconductors as a nearly ideal study object for complex nonlinear dynamics compared to other physical systems. Further representing a convenient model reaction–diffusion system that exhibits distinct universal features, the present semiconductor system may acquire general significance for many synergetic systems in nature. Finally, in view of the rapidly growing application of semiconductor technologies, the understanding, control and possible exploitation of sources of instability in these systems have considerable practical importance.

This paper provides a brief survey of our recent experimental investigations on the spatio-temporal nonlinear current flow in the post-breakdown regime of p-germanium at liquid-helium temperatures. Spatially, we report on filamentary flow patterns developing during avalanche breakdown. Temporally, we outline the typical universal scenarios demonstrated by means of the self-generated oscillatory behavior of distinct state variables. In particular, we concentrate on the relationship between the onset of low-dimensional chaotic dynamics and the break-up of spatial order during current filamentation. Such cooperative phenomena induced by the avalanche breakdown kinetics in semiconductors can be interpreted in terms of a phenomenological reaction–diffusion model known from chemical systems theory. The underlying symmetry concept may have the potential of bridging numerous branches of the sciences, the arts and other human activities.

2. SYSTEM

Our experimental studies were performed on single-crystalline p-type germanium material, having the typical dimensions of about $0.2 \times 2 \times 5$ mm^3 and an indium acceptor concentration of about

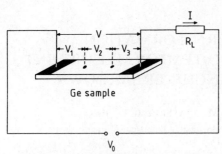

Fig. 1. Scheme of the experimental setup. The shaded areas on the germanium sample indicate the evaporated ohmic aluminum contacts.

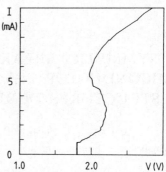

Fig. 2. Current–voltage characteristic obtained by applying the bias voltage V_0 to the series combination of the semiconductor sample and the load resistor ($R_L = 100\,\Omega$). The voltage V was measured along the sample. Further possible control parameters (magnetic field, electromagnetic or electron-beam irradiation) were not applied. The bath temperature was kept at 4.2 K.

$3 \times 10^{14}\,\text{cm}^{-3}$ (corresponding to a shallow impurity acceptor level of about 10 meV). The extrinsic germanium crystal carries properly arranged ohmic aluminum contacts evaporated on one of the two largest surfaces. The sample geometry and the electronic measuring configuration are sketched schematically in Fig. 1. To provide the outer ohmic contacts with an electric field, a d.c. bias voltage V_0 was applied to the series combination of the sample and the load resistor R_L (1 or 100 Ω). A d.c. magnetic field B perpendicular to the broad sample surfaces could also be applied by a superconducting solenoid surrounding the semiconductor sample. The resulting electric current I was found from the voltage drop at the load resistor. The voltage V was measured along the sample. The inner probe contacts (of about 0.2 mm diameter) served for monitoring independently the partial voltages V_i ($i = 1,2,3$) along the sample. Utilizing low-temperature scanning electron microscopy, two-dimensional images of current filament patterns were obtained by scanning the specimen surface with an electron beam and by recording the beam-induced current change in the voltage-biased specimen as a function of the beam coordinate (x, y). As described elsewhere [8, 12] in detail, the present imaging method combines a low-temperature stage with a scanning electron microscope. During the experiments, the semiconductor sample was always kept at liquid-helium temperatures (4.2 K or below) and carefully protected against external electromagnetic irradiation (visible, far-infrared).

Similar to the corresponding processes of structure formation in gaseous plasma discharges and atmospheric lightning, impact ionization of the shallow impurity acceptors can be achieved in the bulk of the homogeneously doped semiconductor at low temperatures. In the temperature regime of liquid helium most of the charge carriers are frozen out at the impurities. Since the ionization energy is only about 10 meV and electron–phonon scattering is strongly reduced, avalanche breakdown already takes place at electric fields of a few V/cm and persists until all impurities are ionized [13]. The underlying nonequilibrium phase transition from a low conducting state to a high conducting state is directly reflected in strongly nonlinear regions of negative differential resistivity [9, 14]. Under slight variation of distinct control parameters (electric field, magnetic field, temperature, electromagnetic irradiation and/or electron-beam irradiation) the resulting electric current flow displays a wide variety of spatial and temporal dissipative structures.

3. RESULTS

The autocatalytic process of impurity impact ionization is reflected in a strongly nonlinear curvature of the measured current–voltage characteristic in the post-breakdown regime. Figure 2 gives an example of a typical $I-V$ curve, clearly displaying negative differential resistance of S-shape behavior. According to the inherent multiplication of mobile charge carriers during avalanche breakdown, the resulting current flow drastically increases from typically a few nA in the pre-breakdown region up to a few mA in the post-breakdown region. Simultaneously, spontaneous

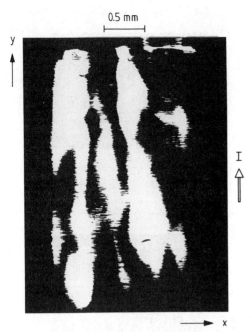

Fig. 3. Brightness-modulated image of the filamentary current flow in a homogeneously doped semiconductor during avalanche breakdown obtained by low-temperature scanning electron microscopy (load resistor $R_L = 1\,\Omega$, bias voltage $V_0 = 2.0\,V$, magnetic field $B = 0$ G, bath temperature $T = 4.2$ K, no electromagnetic irradiation). The dark regions correspond to the filament channels extending along the y-direction.

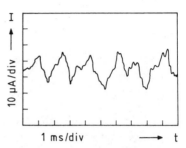

Fig. 4. Temporal structure of spontaneous current oscillations superimposed upon the steady d.c. current in the post-breakdown regime (load resistor $R_L = 1\,\Omega$, bias voltage $V_0 = 2.0$ V, magnetic field $B = 0$ G, bath temperature $T = 4.2$ K, no electromagnetic and no electron-beam irradiation).

emergence of both spatial and temporal dissipative structures in the electric carrier transport takes place.

The complex spatial behavior of our semiconductor system can be globally visualized by means of low-temperature scanning electron microscopy. Figure 3 shows a two-dimensional image of a typical current filament pattern developing in the nonlinear regime of the current–voltage characteristic. As reported elsewhere [8] in detail, the multifilamentary current flow becomes more and more homogeneous if the semiconductor system is driven further into its linear post-breakdown region at higher electric fields. Nucleation of additional filaments is often accompanied by abrupt changes between different stable filament configurations via noisy current instabilities.

The highly nonlinear current–voltage curve is further associated with the appearance of self-generated current and voltage oscillations. Both current I and partial voltages V_i ($i = 1,2,3$) display —superimposed upon the d.c. signals of typically a few mA and some hundred mV, respectively— temporal oscillations with a relative amplitude of about 10^{-3} in the frequency range 0.1–100 kHz. An example of spontaneous current oscillations is shown in Fig. 4. By means of slightly varying the distinct control parameters, the temporal behavior of the system variables V_1, V_2, V_3 and I changes dramatically, exhibiting the typical universal scenarios of chaotic nonlinear systems [2–7]. The observed dynamical phenomena include the intermittent switching between different oscillation modes (Pomeau–Manneville scenario), the period-doubling cascade to chaos (Feigenbaum–Grossmann scenario) and the quasiperiodic route to chaos (Ruelle–Takens–Newhouse scenario), as well as the suppression of quasiperiodicity through frequency locking. Moreover, we observed a transition from ordinary chaos to higher dimensional hyperchaos (Rössler scenario). As an example, we show in Fig. 5 the phase portraits obtained from different inner voltage drops V_i for different values of the control parameters V_0 or B, as indicated. From top to bottom, the sequence clearly demonstrates the structural change of the underlying attractor from a stable fixed point via a limit cycle and a quasiperiodic state to a chaotic state. The strange attractor then transforms from the ordinary chaotic state of low dimensionality (fractal dimension $d = 2.6$) into a strongly turbulent hyperchaotic state of higher dimensionality (fractal dimension $d = 4.3$).

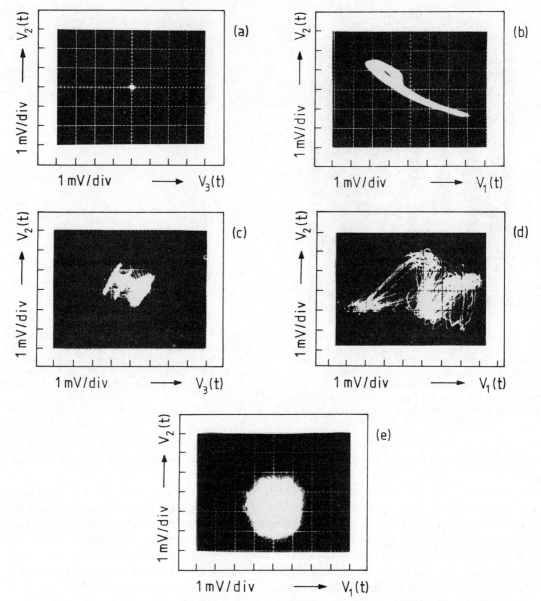

Fig. 5. Phase portraits of different dynamical states obtained by plotting two different partial voltages against one another for different values of the bias voltage V_0 and the magnetic field B (load resistor $R_L =$ 1 Ω, bath temperature $T = 4.2$ K, no electromagnetic and no electron-beam irradiation). (a) Fixed point ($V_0 = 2.220$ V, $B = -4.0$ G); (b) limit cycle ($V_0 = 2.184$ V, $B = 0$ G); (c) quasiperiodic state ($V_0 = 2.220$ V, $B = 2.1$ G); (d) chaotic state ($V_0 = 2.145$ V, $B = 31.5$ G); (e) hyperchaotic state ($V_0 = 2.145$ V, $B = 46.5$ G).

The phase plots of Fig. 5 provide information on both the temporal behavior and the spatial coupling of two spatially separated inner voltages. The apparent loss of spatial coherence between different sample parts indicates a break-up of the semiconducting system from strongly coupled into more independent subsystems (the partial voltages V_i become more independent). In this way, new actively participating degrees of freedom are gained, reflecting increasing dimensionality of the semiconducting system. Our experiments deal with a challenging example of a macroscopic system consisting of spatially separated and coupled subsystems which by themselves show nonlinear dynamical behavior. This model approach is further supported by additional experimental findings. First, the autocatalytic hot-carrier subsystems are diffusively coupled by long-range phonon propagation due to the extremely high lattice heat conductivity of germanium [6, 15]. Second, the spontaneous current and voltage oscillations are localized in the boundary region of current filaments [8, 15]. Finally, quasiperiodic current flow with two incommensurate intrinsic frequencies

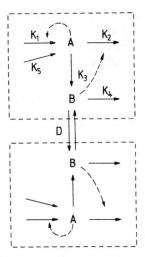

Fig. 6. Reaction scheme of a simple Rashevsky–Turing model system. Constant pools are omitted from the scheme, catalytic rate control is indicated by dashed arrows.

can be attributed to the simultaneous presence of two competing localized oscillation centers, arising spontaneously and interacting in a nonlinear way. Prior to the onset of low-dimensional chaos, the increasing strength of nonlinear coupling between the oscillators develops an increasing tendency to lock into commensurate frequencies. Now the spatially separated oscillation centers display temporally resonant response behavior [7, 16].

Following the above quasiperiodic approach to chaos via frequency locking in view of symmetry and structure formation processes, the spontaneous emergence of a highly symmetrical and well-ordered mode-locking structure developing from a more disordered state of quasiperiodicity represents a typical disorder–order transition, confining the system flow to only a few degrees of freedom. The still open question arises whether the complex temporal behavior of the latter chaotic state might be interpreted as a new degree of disorder (ensuing the quasiperiodic state) or even as a higher form of order (ensuing the mode-locked state), perhaps.

4. MODEL

The above spatio-temporal correlation phenomena can be qualitatively explained in terms of a simple dynamical model based on the generic Rashevsky–Turing theory of symmetry-breaking morphogenesis [17–20]. As the essential behavioral characteristic, breakdown of symmetry can in the simplest case be realized by a two-cellular symmetrical reaction–diffusion system consisting of cross-inhibitorily coupled, potentially oscillating two-variable subsystems (four-dimensional flow). The reaction scheme of the simplest version of Turing's model is sketched in Fig. 6. The two morphogens A are self-inhibiting via B and, to a lesser extent, cross-inhibiting each other via the symmetrical coupling between the two morphogens B. The excess of self-inhibition within each cell over the cross-inhibition generated by the other cell is compensated for, within either cell, by a path of self-activation (autocatalysis of A) which is not mediated through diffusion to the other side. D is the diffusion coefficient for the morphogen B. The effects of constant pools and reaction partners are comprised in the effective rate constants $K_1, \ldots, K_5$. The two-compartment structure of the four-dimensional flow in Fig. 6 is capable of eliciting spontaneous symmetry-breaking phase transitions and boiling-type turbulence [19, 20].

Experimental evidence of a variety of analogous spatio-temporal cooperative processes [2–8] suggests a reaction scheme for our semiconductor system qualitatively similar to that of the phenomenological reaction–diffusion model indicated in Fig. 6. Involving the concrete physical mechanism of impurity impact ionization coupled with energy relaxation and energy exchange of hot carriers through emitted phonons, the two morphogens of our model can be interpreted as the number density of moving charge carriers reflected in the electric current (activator) and the mean energy per carrier (inhibitor). The diffusive coupling is then governed by the energy exchange between different hot-carrier subsystems via long-range phonon propagation [11]. The dynamical possibilities of avalanche breakdown kinetics in semiconductors turn out to be directly projected onto the main behavioral characteristics of a generic Turing-type system, representing the most convenient prototype model for many different synergetic systems in nature.

5. CONCLUDING REMARKS

Of the variety of nonlinear dynamical systems that exhibit nonequilibrium phase transitions and deterministic chaos, semiconductors provide nearly ideal model systems for qualitative and quantitative investigation due to their simplicity both in preparation and experimental accessibility. In particular, recent advances in materials engineering and measurement techniques have opened up the possibility of a fruitful interaction between the theoretical understanding and experimental observations. Thus, a possible twofold impact of semiconductor physics on the life sciences can be elucidated. On the one hand, the powerful techniques developed for studying complex physical phenomena are very useful in the biological context. Equally useful, on the other hand, are certain physical concepts, such as symmetry and symmetry breaking, linear and nonlinear stability, frustration and constrained dynamics. Moreover, symmetry is not only one of the fundamental concepts in science, but also has the potential of bringing together the most diverse fields from mathematics to biology, the creative and performing arts, and most branches of the sciences.

It is emphasized, however, that biological systems are capable of feats of pattern formation (evolution, morphogenesis) which are far beyond the capability of any man-made systems. Nevertheless, the ultimate goal is a better understanding of the basic principles of spatio-temporal organization, in order to treat periodic diseases or other perturbations of "normal" dynamics in human oscillatory systems. Because biological systems have to produce and preserve a temporally and spatially ordered structure against the destructive forces of thermal noise and entropy, chaos seems to be of only negative influence in biology, sometimes considered as a pathological condition. However, one has to realize that the same mathematical model, which is responsible for the formation of stable patterns, is—at least principally—able to behave chaotically. Chaos is an inherent possibility, but is it, moreover, a necessary element in the formation and functioning of living organisms?

Acknowledgements—The authors thank R. P. Huebener, O. E. Rössler, S. Grossmann, H. Thomas, W. Prettl, W. Metzler and I. Gumowski for discussions.

REFERENCES

1. For a review see: H. Haken, *Synergetics* (1977); *Pattern Formation by Dynamic Systems and Pattern Recognition* (1979); *Dynamics of Synergetic Systems* (1980); *Chaos and Order in Nature* (1981); *Evolution of Order and Chaos* (1982); *Advanced Synergetics* (1983). Springer, Berlin.
2. J. Peinke, A. Mühlbach, R. P. Huebener and J. Parisi, Spontaneous oscillations and chaos in p-germanium. *Phys. Lett.* **108A**, 407–412 (1985).
3. J. Peinke, B. Röhricht, A. Mühlbach, J. Parisi, Ch. Nöldeke, R. P. Huebener and O. E. Rössler, Hyperchaos in the post-breakdown regime of p-germanium. *Z. Naturf.* **40a**, 562–566 (1985).
4. B. Röhricht, B. Wessely, J. Peinke, A. Mühlbach, J. Parisi and R. P. Huebener, Chaos and hyperchaos in the post-breakdown regime of p-germanium. *Physica* **134B**, 281–287 (1985).
5. J. Peinke, J. Parisi, A. Mühlbach and R. P. Huebener, Different types of current instabilities during low-temperature avalanche breakdown of p-germanium. *Z. Naturf.* **42a**, 441–443 (1987).
6. B. Röhricht, J. Parisi, J. Peinke and R. P. Huebener, Spontaneous resistance oscillations in p-germanium at low temperatures and their spatial correlation. *Z. Phys.* **B66**, 515–521 (1987).
7. J. Peinke, J. Parisi, B. Röhricht, B. Wessely and K. M. Mayer, Quasiperiodicity and mode locking of undriven spontaneous oscillations in germanium crystals. *Z. Naturf.* **42a**, 841–845 (1987).
8. K. M. Mayer, R. Gross, J. Parisi, J. Peinke and R. P. Huebener, Spatially resolved observation of current filament dynamics in semiconductors. *Solid St. Commun.* **63**, 55–59 (1987).
9. E. Schöll, Impact ionization mechanism for self-generated chaos in semiconductors. *Phys. Rev.* **B34**, 1395–1398 (1986); Instabilities in semiconductors: domains, filaments, chaos. In *Festkörperprobleme* (*Advances in Solid State Physics*), Vol. 26 (Ed. P. Grosse), pp. 309–333. Vieweg, Braunschweig (1986); E. Schöll, *Nonequilibrium Phase Transitions in Semiconductors*. Springer, Berlin (1987).
10. E. Schöll, J. Parisi, B. Röhricht, J. Peinke and R. P. Huebener, Spatial correlations of chaotic oscillations in the post-breakdown regime of p-Ge. *Phys. Lett.* **119A**, 419–424 (1987).
11. J. Parisi, J. Peinke, B. Röhricht, U. Rau, M. Klein and O. E. Rössler, Comparison between a generic reaction–diffusion model and a synergetic semiconductor system. *Z. Naturf.* **42a**, 655–656 (1987).

12. R. P. Huebener, Applications of low-temperature scanning electron microscopy. *Rep. Prog. Phys.* **47**, 175–220 (1984).

13. K. Seeger, *Semiconductor Physics*, 3rd edn. Springer, Berlin (1985).

14. J. Peinke, D. B. Schmid, B. Röhricht and J. Parisi, Positive and negative differential resistance in electrical conductors. *Z. Phys.* **B66**, 65–73 (1987).

15. R. P. Huebener, K. M. Mayer, J. Parisi, J. Peinke and B. Röhricht, Chaos in semiconductors. *Nucl. Phys. (Proc. Suppl.)* **B2**, 3–12 (1987).

16. U. Rau, J. Peinke, J. Parisi, R. P. Huebener and E. Schöll, Exemplary locking sequence during self-generated quasiperiodicity of extrinsic germanium. *Phys. Lett.* **124A**, 335–339 (1987).

17. N. Rashevsky, An approach to the mathematical biophysics of biological self-regulation and of cell polarity. *Bull. math. Biophys.* **2**, 15–25 (1940); Further contributions to the theory of cell polarity and self-regulation. *Bull. math. Biophys.* **2**, 65–67 (1940); Physicomathematical aspects of some problems of organic form. *Bull. math. Biophys.* **2**, 109–121 (1940).

18. A. M. Turing, The chemical basis of morphogenesis. *Phil. Trans. R. Soc.* **B237**, 37–72 (1952).

19. O. E. Rössler, Chemical turbulence: chaos in a simple reaction–diffusion system. *Z. Naturf.* **31a**, 1168–1172 (1976).

20. B. Röhricht, J. Parisi, J. Peinke and O. E. Rössler, A simple morphogenetic reaction–diffusion model describing nonlinear transport phenomena in semiconductors. *Z. Phys.* **B65**, 259–266 (1986).

SYMMETRY 2

Unifying Human Understanding
(Part 2)

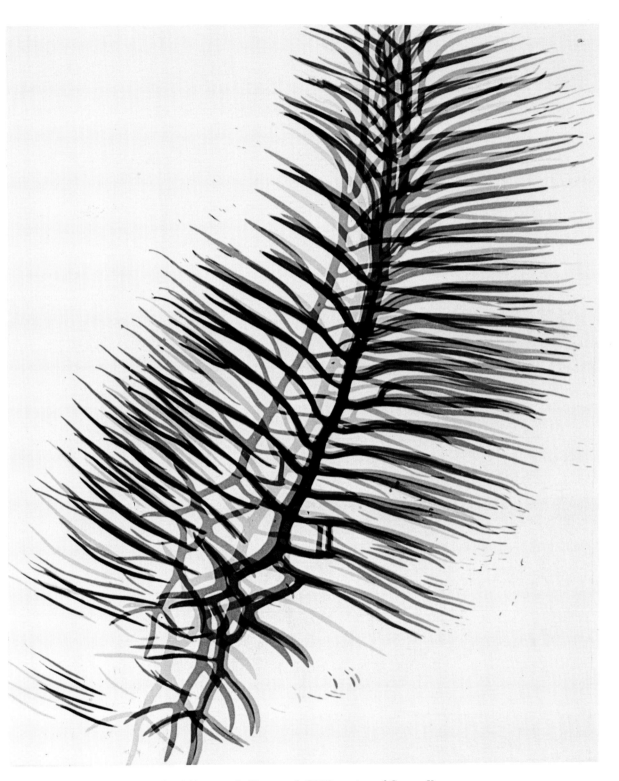

Dendritic network. Photograph (1962) courtesy of Gyorgy Kepes.

Computers Math. Applic. Vol. 17, No. 4–6, pp. 475–484, 1989
Printed in Great Britain. All rights reserved

0097-4943/89 $3.00 + 0.00

THE PERCEPTUAL VALUE OF SYMMETRY

P. LOCHER

Department of Psychology, Montclair State College, Upper Montclair, NJ 07043, U.S.A.

C. NODINE

Department of Educational Psychology, Temple University, Philadelphia, PA 19122, U.S.A.

Abstract—Everyday experiences demonstrate that symmetry affects the visual perception of form. This paper describes empirical evidence about the influence of static and dynamic symmetry of visual shapes and pictorial compositions on visual detection, attention, exploration and physiological arousal. Flash experiments demonstrate that symmetry is detected wholistically during the first glance. Viewers' eye fixations show that the axis of symmetry is used as a perceptual landmark for visual exploration. Dynamic symmetry enhances visual exploration; static symmetry restricts exploration. Finally, visual exploration of symmetrical compositions is shown to be linked to arousal and aesthetic judgments.

1. INTRODUCTION

Everyday experiences and much empirical evidence indicate that symmetry affects the visual perception of form. Research by experimental psychologists has repeatedly shown, for example, that the presence of symmetry in a pattern or visual composition can be detected more quickly than its absence, and that some types of symmetry, such as bilateral or vertical, are more readily detectable than others (see, for example, Ref. [1]). Art historians have noted the perceptual salience of symmetry in the visual arts and architecture. In fact, based upon his extensive examination of the composition and design of works of art, Bouleau [2] concluded that it is the symmetry contained in a composition, no matter how "hidden", which draws the viewer into that world of secret geometry which has governed the arts of design from classical antiquity. But what is it about symmetry that "catches the eye"? How is the perception of a visual composition influenced by the presence of symmetry? This paper will focus on possible answers to these questions. Specifically, the influence of symmetry on visual detection, attention, exploration and physiological arousal will be discussed.

Perceptual processes play a major role in Osborne's writings on aesthetics. He states that the "pleasure which many people experience in aesthetic perception derives from the successful exercise of enhanced perceptual activity upon an object adequate to arouse and sustain it at more than ordinary intensity" [3, p. 81]. According to Osborne, it is to things which produce enhancement of perception and arousal that we attribute aesthetic quality and call works of art. The aesthetic character of such an object is a function of its individual elements and its organic unity.

Berlyne's [4, 5] theory of psychoaesthetics, one of the first comprehensive psychological theories of visual aesthetics, will be used as a framework for considering evidence of the role that symmetry plays in perception. In Berlyne's view, the behaviors involved in the processing of a visual stimulus occur as part of a comprehensive "orientation reaction" to the information content of the stimulus pattern. This orientation reaction occurs during visual exploration and is signaled by a rise in arousal indicative of disorientation due to an encounter with a new or novel stimulus. The function of the orientation reaction is to arouse and sustain interest in the visual stimulus.

Thus, Berlyne identifies visual exploration as a primary indicator of visual interest and novelty which are directly linked to aesthetics through arousal. Exploration results from the viewer's state of curiosity which is aroused when he is confronted with a visual stimulus such as a picture or work or art. This curiosity about the information content of the stimulus can be satisfied only by examining the picture. Visual exploration calls into play basic perceptual and cognitive processes of attention, detection, discrimination and identification of pictorial information which ultimately reduce the viewer's curiosity state.

The structural complexity of a visual display (that is, the number of independent elements it contains) determines the intensity of the orientation reaction which results after the stimulus has begun to excite the sensory system. Exploration of a pattern and the resulting curiosity and arousal are, therefore, intrinsically bound up with the structural complexity of a stimulus, of which symmetry is a major component. In its simplest form, symmetry refers to balance, either static or dynamic, in the arrangement of similar elements similarly placed within a composition.

The exact duplication of structural elements about an axis of symmetry produces static or geometrical symmetry. This type of mirror-image arrangement of elements can be seen in the visual arts stimuli in Fig. 1. They consist of five original asymmetrical abstract paintings and their computer-generated single-axis and double-axis statically symmetrical transformations. The single-symmetry transformation for each original was generated by reflecting one-half of the original about the horizontal or vertical axis. The double-symmetry transformation was generated by reflecting the bottom left quarter about both the horizontal and vertical axes.

Dynamic symmetry, on the other hand, is achieved by differentially weighting and counter-weighting distributions of compositional elements about an imaginary axis of symmetry which serves as a fulcrum. Osborne [3] points out that the balance, or equality of weighting, about a medial axis achieved in this manner is frequently described by the term "harmony". He notes that Baroque or Rococo art may be considered symmetrical in this sense despite its deliberate asymmetry. Non-iconic abstract art is also frequently inspired by this form of dynamic balance or interplay of parts.

When static symmetry is introduced into an otherwise asymmetrical composition by transforming it, the number of non-symmetrical or non-concordant elements contained in the symmetrical transformation is reduced, as is irregularity. Therefore the transformed composition is less complex structurally and less informative than its asymmetrical counterpart. This change in information

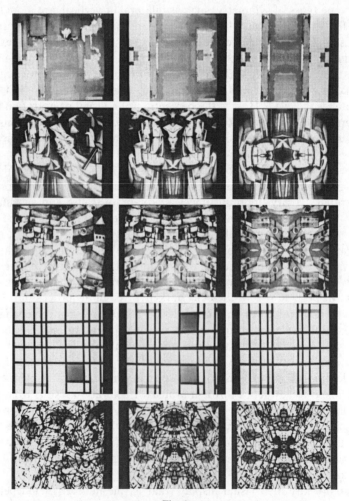

Fig. 1

content due to the introduction of symmetrical elements can be seen in the stimuli in Fig. 1. According to Berlyne's [4, 5] theory, the difference in structural complexity between the asymmetrical originals and their less complex symmetrical transformations should influence both visual exploration of the stimuli and their arousal potential.

In general, Berlyne's theory predicts that symmetrical compositions should receive less perceptual exploration and generate less arousal than asymmetrical compositions because symmetrical stimuli contain fewer unique elements. They are simpler informationwise. Osborne writes that "the symmetry of repeating patterns provides a very elementary aesthetic stimulus. It may serve to arouse attention, particularly if the repeating elements are unfamiliar or if they carry personal associations. But it cannot hold or enhance perceptual attention" [3, p. 81].

Dynamically symmetrical stimuli, on the other hand, have greater aesthetic potential because they enhance attention. As Osborne points out, observers must explore more of the presentational content of a dynamically symmetrical stimulus to perceive the interplay of elements located throughout the object which produce its "organic unity" or harmony. In contrast to the limited perceptual impact of statically symmetrical stimuli, dynamically symmetrical stimuli require increased visual exploration to reduce information uncertainty and curiosity.

The remainder of this paper describes empirical evidence about the influence of static and dynamic symmetry on the perceptual processing of visual stimuli.

2. GLOBAL DETECTION OF SYMMETRY

Symmetry is a property of a visual stimulus which catches the eye in the earliest stages of vision. The visual system seems capable of automatically processing symmetrically-organized stimuli without scrutinizing concordant features. Several studies have demonstrated that viewers can characterize shapes and patterns as symmetrical or asymmetrical on the basis of information contained in a single brief fixation. For example, Carmody et al. [6] tested subjects' ability to discriminate among three levels of symmetry in simple black random shapes which subtended a 9° visual field on the basis of a 100 ms flash presentation. Results indicated that detection of symmetry was highly accurate (93%). However, when shapes contained both symmetrical and asymmetrical components ("mixed" symmetry shapes), detection was less accurate (65%) presumably due to the presence of the conflicting aspects of symmetry within the same single shape.

More recently, Barlow and Reeves [7] demonstrated the accuracy and versatility of mirror symmetry detection under a wide range of conditions. Their stimuli consisted of random dot patterns which subtended approx. 2° at the subjects' eyes. Exposure durations for the various flash experiments ranged from 100 to 500 ms. Barlow and Reeves found that symmetry can be detected at horizontal, vertical and oblique axes, although each was not equally salient, and when the axis of symmetry was displaced to the right or left of midline. The perceptual mechanisms reponsible for the detection of symmetry showed a tolerance for "smeared" or imperfect symmetrical patterns which resulted from slight random fluctuations of pairs of dot elements about the axis of symmetry of the patterns.

In each of the above studies, stimuli lacked the size and number of feature dimensions (for example, line, shape, color, texture) of everyday visual stimuli such as those presented on T.V. screens. To determine whether global ("wholistic") detection of symmetry of such stimuli is also possible, we conducted an experiment in which subjects responded either "symmetrical" or "asymmetrical" following a 100 ms flash presentation of each composition included in Fig. 1. The compositions, which were more typical in size to everyday stimuli (35° visual), were presented randomly in the orientation in which they were generated and in a rotated 90°, 180° and 270° orientation. In addition, subjects either fixated the center of the composition or a point one-fourth pattern width off-center, either on or off the axis of symmetry.

Results showed that subjects were both accurate and confident in their ability to detect symmetry on and off axis in all of the transformations. Accuracy of detection for all conditions ranged from 91 to 99%. Vertical symmetry was as perceptually salient as horizontal symmetry. And differences in accuracy due to variations in local features among the compositions were small.

To determine whether the exact type of symmetry orientation of a multidimensional composition could be globally detected, we conducted a second investigation using the stimuli in Fig. 1. Subjects were asked to state after a 50 ms presentation of each stimulus whether it was asymmetrical, symmetrical about the vertical axis, symmetrical about the horizontal axis, or symmetrical about both axes. The percentages of correct detections for each type of composition were 79, 75, 64 and 74%, respectively. Thus, subjects were highly accurate in their perception of the specific type of symmetry present in these color compositions when they were presented for 50 ms.

The results of tachistoscopic viewing described above suggest that the redundant features which characterize symmetry are detected globally by the visual receptive field. Furthermore, they demonstrate that the mechanism for global detection of symmetry works efficiently for large complex color stimuli. It appears, then, that symmetry affects the way the viewer initially reacts to a visual stimulus. This reaction is immediate in that it seems to take place during the first glance.

Barlow and Reeves [7] have proposed a model of the mechanism responsible for symmetry detection which can account for the rapid simultaneous processing of large structurally complex stimuli. They suggest that the receptive field of the retina is divided into subregions comparable in size to the tolerance range for which the eye performs most efficiently. Presumably, feature detectors interact automatically to analyze subregions of the stimulus. With the onset of a pattern, the number of elemental units in each region is counted simultaneously. The presence of symmetry is signaled by a region comparator which determines the similarity of symmetrically placed subregions of the receptive field located about a putative axis of symmetry.

3. VISUAL EXPLORATION OF SYMMETRICAL STIMULI

As mentioned above, visual exploration is a primary component of the orientation reaction and an indicator of a viewer's level of curiosity. One of the best ways to monitor visual exploration is to record the viewer's eye movements as he explores a stimulus. The eyes move in spurts and pauses. For purposes of analysis, the pauses are most meaningful because they indicate which aspects of a stimulus are receiving visual attention. Thus, it is possible to superimpose a viewer's fixation pattern over a stimulus, align it to a calibration pattern, map out the location of fixations, and identify which pictorial features raise the viewer's interest and invite exploration. The map of the location of fixations provides a graphic record of *how* information is processed perceptually in order to satisfy curiosity.

Evidence about how the various types of symmetry influence visual exploration of stimuli is provided by Locher and Nodine [8–10] and Llewellyn-Thomas [11]. In an early study, Locher and Nodine [8] recorded the eye fixations of subjects performing a complexity rating task in which the stimuli consisted of large (36°) symmetrical and asymmetrical random shapes differing in complexity. It was found that subjects restricted their attention to one side of symmetrical shapes while the distribution of fixations for asymmetrical shapes was more complete. The difference in visual exploration is clearly seen in the eye fixation patterns in Fig. 2.

Note that the observer directed his gaze to the left side of the symmetrical shape and that almost none of the perimeter of the shape on the right-hand side of the axis of symmetry was examined. Informative features, defined as changes in angularity, along the entire perimeter of the asymmetrical shape, however, were fixated. Llewellyn-Thomas [11] also observed the presence of one-sided scanning strategies in the fixation patterns of subjects viewing the Rorschach inkblots. Presumably the global response to symmetrical shapes pre-programmed exploration to take advantage of the redundant nature of the relationship among elements in these shapes, resulting in the one-sided fixation patterns reported in these two investigations.

The stimuli used in these studies were simple unidimensional shapes. How is exploration affected when more complex multidimensional stimuli are transformed symmetrically to produce redundancy in artistic compositions? Recently, Locher and Nodine [10] showed how this form of static symmetry affected visual exploration. Subject's eye fixations were recorded while they rated each of the compositions in Fig. 1 for inclusion in a hypothetical art show. The visual compositions were somewhat larger than a standard sheet of paper (22 by 28 cm), subtending a visual angle of approx. 20°. Difference in the visual exploration of asymmetrical and symmetrical compositions are shown in the composite fixation patterns in Fig. 3 for the original Hoffman painting entitled "*The Golden Wall*" and its single- and double-axis transformations. The location of each fixation directed by all subjects to a composition is indicated by a change in direction of the line in the composite

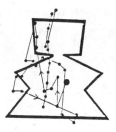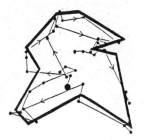

Fig. 2

pattern for that composition. Each composite in Fig. 3, therefore, depicts the density and spatial distribution of fixations used to explore pictorial information on these artworks.

An examination of the composite fixation patterns in Fig. 3 shows clearly that exploration of these compositions was influenced by symmetry. Contrast the fixation pattern for the single-axis

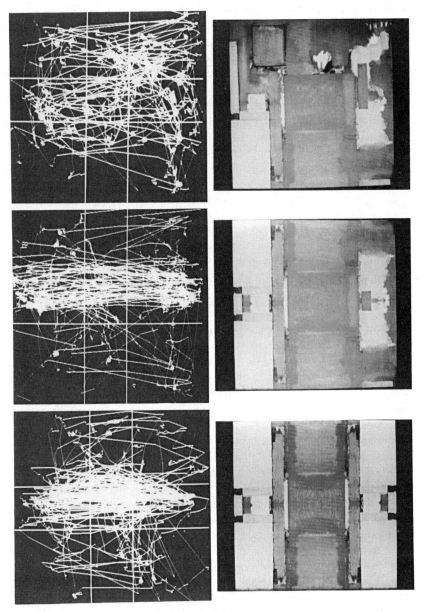

Fig. 3

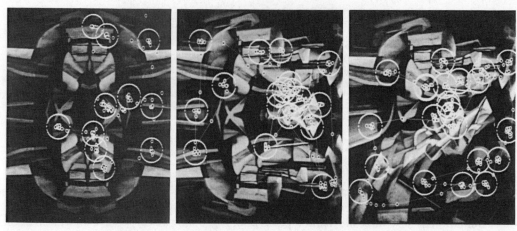

Fig. 4

transform with the picture of the transform. Note the features that stand out in the left and right half of the transform. In the color image which the subjects saw, the features on the left are blue surrounded by a homogeneous yellow area extending from the top to the bottom of the composition. The large "clear" area in the right half of the transformed artwork is a homogeneous yellow field with a small blue region in its center. The remainder of this half of the transform and the entire middle region of the composition are red.

As can be seen in the composite fixation pattern, attention focused on both of these areas of structural detail in the single-axis transform of Hoffman's painting. When individual scanning patterns were examined, it was found that this clustering of attention was due to the comparison of redundant details on either side of the horizontal axis of symmetry and to the comparison of details on one side of the composition with those on the other side. Attention was channeled to redundant details along the axis of symmetry rather than redundant details farther from the axis or than unpaired features. The concentration of fixations about the horizontal axis is more pronounced in the composite fixation pattern for the double-axis transformation of Hoffman's painting. In contrast to the "biased" exploration of the symmetrical stimuli, exploration of the original Hoffman was broadly distributed covering pictorial information throughout the entire composition.

Figure 4 shows the scanning pattern of a single subject viewing the original "*King and Queen Surrounded by Swift Nudes*" by Duchamp and its single-symmetry and double-symmetry transformations. Each circle indicates the area (percent) of the composition covered by one fixation. For the eye-movement recording system used in this research, each fixation covers 4% of the total composition which represents about as much information as a viewer can effectively take in with each glance. As with the composite fixation patterns, these typical individual fixation patterns show that a greater area of the original asymmetrical painting was explored than that of its less structurally complex single-symmetry transformed composition. And coverage of the double-symmetry transformed composition is considerably less than that for either the original or single-axis transformation.

The limited and biased fixation patterns shown in Figs 3 and 4 were also observed for the symmetrical transformations of artworks by Pollock and Klee. Overall, these patterns demonstrate the perceptual salience of the axis of symmetry of statically symmetrical multidimensional compositions. They also support Osborne's [3] view that the paired or concordant elements of statically symmetrical compositions provide only elementary aesthetic stimuli because they contain little structural complexity and are, therefore, of relatively little visual interest.

Dynamically symmetrical compositions, on the other hand, should require greater exploration in order to grasp the unity of the pictorial composition about a "felt" axis of symmetry. Osborne asserts that such compositions enhance perceptual activity and its resulting arousal. This is supported by a recent study conducted by Locher and Nodine [10]. They recorded subjects' eye fixations as they viewed pairs of representational artworks, one after the other, and decided which composition produced a "more pleasing expression of harmony and beauty". The effects of dynamic

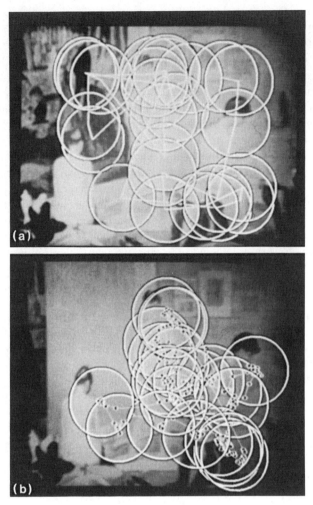

Fig. 5

symmetry on the spatial aspects of exploration was examined by determining how much area (percentage) of an artwork was covered by a subject's eye-fixation pattern. Figure 5 shows the composite fixation patterns of subjects viewing "*Les Poseuses*" by Seurat and a less dynamically symmetrical altered version of this painting. As in Fig. 4, each circle indicates the area of the display covered by one fixation.

Stimuli in Locher and Nodine's study consisted of six pairs of dynamically symmetrical paintings and an alternative, less symmetrical version of each. Two original paintings were altered by the researchers to disrupt dynamic symmetry ("*Les Poseuses*" and "*Le Chahut*", both by Seurat), and two paintings were altered to increase dynamic symmetry ("*Jour de Dieu*" by Gauguin; and "*Elinor, Jean and Anna*" by Bellows). Two additional artworks were selected because of the artists' deliberate use of dynamic symmetry ("*Grandes Baigneuses*" by Cezanne, and "*Composition*" 1921, by Mondrian). The artists of these paintings had produced their own alternative, less dynamically symmetrical versions ("*Baigneuses*" by Cezanne, and "*Composition with Red, Yellow and Blue*" 1922, by Mondrian).

Bouleau [2] has presented a detailed analysis of the interplay of compositional elements that create the symmetry of balance in Seurat's "*Les Poseuses*" (see Fig. 5). According to him, the model who stands facing us establishes the axis of the picture and divides it in half. Diagonals which extend from upper to lower corners of the painting provide the basic scaffolding used to position the pictorial elements in this dynamically symmetrical composition. Details of the painting of "*La Grande Jatte*" in the upper left corner "balance" details of the woman in the lower right corner. Similarly, an interplay of elements exists between details of the woman in the lower left surface and those of the pictures in the upper right of the composition. The researchers altered this painting to disrupt symmetry by removing most of the details of "*La Grande Jatte*" from the composition. This produced an asymmetrical imbalance in the weighting of elements in both the upper-left and lower-right corners of the painting.

Locher and Nodine found that the presence of dynamic symmetry in an artwork enhanced visual exploration of it. This perceptual enhancement is seen clearly in the fixation pattern in Fig. 5. Note that a larger area (52%) of the dynamically symmetrically "*Les Poseuses*" was explored when compared with its modified, more asymmetrical alterative composition (43%). Viewers attended to balancing areas of the original painting which, according to Bouleau's analysis, correlate with structural symmetry. Similar scanning patterns were found for the entire set of artworks.

Results of the eye-movement studies reported above demonstrate that both static and dynamic symmetry influence visual exploration. When static symmetry is present, exploration is restricted to one side of undimensional shapes and to paired features on either side of the axis of symmetry of multidimensional compositions. The biasing of exploration supports the view expressed above that statically symmetrical stimuli are characterized globally on first glance, that is, *before* the eye scans them. Otherwise, how, for example, would the observer whose fixation pattern is illustrated in Fig. 2 know what the half he failed to scan looked like? It was found [9] that the differences in fixation distributions between symmetrical and asymmetrical compositions illustrated in Fig. 3 were already present after only 3 s of viewing. For differences in the scanning strategies to be present so early during encoding, symmetry must have been detected soon after onset and used to direct exploration of local features of the compositions.

The global or wholistic grasp of the configurational properties of compositions seems to be a fundamental component of perception which occurs regardless of the type of task that the subjects were engaged in, namely, ratings of complexity [6] or judgments of aesthetic value [9]. It is the global response which presumably influences the orientation reaction proposed by Berlyne [4, 5] by pre-programming exploration to take advantage of the redundant nature of the relationship among elements in statically symmetrical compositions. Researchers have yet to determine whether the balance in dynamically symmetrical compositions can be detected globally.

4. AROUSAL POTENTIAL OF SYMMETRICAL COMPOSITIONS

Both Osborne [3] and Berlyne [4, 5] suggest that the pleasure which is experienced in aesthetic perception is the result of an arousal build-up or "boost" resulting from perceptual activity designed to encode the stimulus and thus reduce information uncertainty in the stimulus. The arousal potential of a visual stimulus is, therefore, directly related to its information content which provides the grist for processes of perceptiual and intellectual analysis. This is supported by much accumulated evidence that indicates that increasing the number of elements in geometric figures and patterns leads to a reliable increase in physiological arousal (see, for example, Ref. [12]). This rise in arousal is pleasurable provided the increase is not enough to drive arousal into an upper range which is aversive and unpleasant.

According to this view of psychoaesthetics, when static symmetry is introduced into a visual pattern, stimulus complexity of the transformation decreases, perceptual processing demands are reduced, and the arousal potential of the transformation is lowered. To test this hypothesis, Krupinski and Locher [13] performed the following experiment. Subjects examined a set of non-representational artworks which include the original works by Hoffman, Duchamp, Klee and Pollock and their single- and double-symmetry transformations depicted in Fig. 1. An additional 11 non-representational paintings such as Rothko's "*Untitled*", Vasarely's "*Our*" and Lissitsky's "*Proun ID*" were included in the stimulus set. Arousal was measured by skin conductance changes (GSR) elicited as subjects examined each composition for 9 s in order to judge its aesthetic potential.

Krupinski and Locher found that the average arousal increment decreased monitonically from the original asymmetrical paintings (32.7 units) to single-axis symmetrical transformations (18.4 units) to double-axis symmetrical transformations (5.7 units). The ordinate of the GSR recordings measured skin resistance changes in 100 unit increments. Differences in average arousal increment among the four different compositions within each level of static symmetry were small and non-significant suggesting that the stimuli in each level fell within the same range of subjective complexity.

Because of the greater variance in GSR among symmetry levels than among artistic compositions within a symmetry level, these findings demonstrate that changes in structural complexity caused by experimentally manipulating symmetry influenced arousal. Introduction of symmetrical details by transforming non-symmetrical original artworkers reduced the number and variety of unique elements and visual interest as reflected by eye fixation data. Correspondingly, perceptual activity caused by the less structurally complex symmetrical compositions produced less arousal.

Krupinski and Locher also provide evidence that symmetry influences the aesthetic judgments of the compositions. After viewing each compositions, subjects rated it on a five-point scale for inclusion in a hypothetical art show (1 = poor—definitely do not show, downright boring; 5 = excellent— shows exceptional creativity and originality, definitely show it). It was found that the more structurally complex asymmetrical originals were rated higher on average for inclusion in the art show (4.23) than the less complex single-symmery transformations (3.02), which were rated higher than the double-axis transformations (1.68).

These findings show that the degree of static symmetry in a picture is an important stimulus property which determines its arousal potential and how it is evaluated aesthetically. They support the psychoaesthetic models of Osborne [3] and Berlyne [5] which assert that the evaluative aspects associated with appreciation of art are directly related to the arousal potential of artworks to which structural complexity is a significant contributing factor.

5. ENCODING SYMMETRY

Artists and psychologists have long known that symmetry is a perceptually salient feature of visual stimuli that influences the way visual information is encoded. However, speculation about the nature of the perceptual mechanisms responsible for the encoding of symmetry has remained speculative. The research reported here has presented evidence that detection of symmetry, programming of visual exploration, and physiological arousal seem to be mediated by an initial global impression or gist of the stimulus.

The picture of perceptual encoding of symmetrical displays which emerges is the following. Viewers characterize a stimulus as being symmetrical globally on the basis of visual information picked up in the first (or first few) fixations. Having detected the presence of symmetry in an early stage of perceptual processing, visual exploration is then guided by attention mechanisms designed to take full advatage of the redundant nature of the information contained in statically symmetrical stimuli or the balance created by the interplay of elements in dynamically symmetrical compositions. The arousal experienced by the viewer as he attempts to satisfy his curiosity about how pictorial elements are related to the design of a composition is derived from the perceptual exploration carried out during the encoding of the visual stimulus.

6. CONCLUSION

The human eye–brain system seems to virtually resonate with symmetry. The rapid and accurate detection of static symmetry by the perceptual system is most likely a fundamental unlearned response. The perceptual value of symmetry was discovered by ancient artists and incorporated into the design of both their art and architecture. The invention of dynamic symmetry gave artists a powerful perceptual principle which allowed them to break away from the constraints dictated by strict symmetry and yet take full advantage of symmetry's potential in organizing a perceptually and aesthetically pleasing visual composition.

Both static and dynamic symmetry are detected globally and wholistically, and most perceptual theories assume that the eye–brain system uses the axis of symmetry as an anchoring point for visual exploration and analysis. Exploration is guided not only by the structural content of pictures but also by their emotional impact. Thus, exploration leads to physiological changes in arousal that are closely tied to aesthetic responses.

Finally, when it comes to the use of dynamic symmetry in art, one wonders whether it is the presence of symmetry or violations thereof that catches the eye and induces sensations of proportionality, balance and unity in a contemporary work of art.

REFERENCES

1. H. B. Bruce and M. J. Morgan, Violations of symmetry and repetition in visual patterns. *Perception* **4**, 239–249 (1975).
2. C. Bouleau, *The Painter's Secret Geometry*. Hacker Art Books, New York (1980).
3. H. Osborne, Symmetry as an aesthetic factor. *Comput. Math. Applic.* **12B**, 77–82 (1986). Reprinted in *Symmetry: Unifying Human Understanding* (Ed. I. Hargittai). Pergamon Press, Oxford (1986).
4. D. E. Berlyne, *Conflict, Arousal and Curiosity*. McGraw-Hill, New York (1960).
5. D. E. Berlyne, *Aesthetics and Psychobiology*. Appleton-Century-Crofts, New York (1973).
6. D. Carmody, C. Nodine and P. Locher, Global detection of symmetry. *Percept. Mot. Skills* **45**, 1267–1273 (1977).
7. H. B. Barlow and B. C. Reeves, The versatility and absolute efficiency of detecting mirror symmetry in random dot displays. *Vision Res.* **19**, 783–793 (1979).
8. P. J. Locher and C. F. Nodine, Influence of stimulus symmetry on visual scanning patterns. *Percept. Psychophys.* **13**, 408–412 (1973).
9. P. J. Locher and C. F. Nodine, Symmetry catches the eye. In *Eye Movements: From Physiology to Cognition* (Eds J. O'Regan and A. Levy-Schoen). North-Holland Press, Amsterdam (1987).
10. P. Locher and C. Nodine, The role of eye movements in the detection of perturbations of symmetry. *Fourth Eur. Conf. Eye Movements*, Goettingen, Germany, Sept. 22 (1987).
11. E. Llewellyn-Thomas, Eye movements and fixations during the initial viewing of Rorschach cards. *J. project. Tech. Person. Assess.* **27**, 345–353 (1963).
12. D. E. Berlyne, M. Craw, P. Salapatek and J. Lewis, Novelty, complexity, incongruity, extrinsic motivation and the GSR. *J. exp. Psychol.* **66**, 560–567 (1963).
13. E. Krupinski and P. Locher, Skin conductance and aesthetic evaluative responses to nonrepresentational works of art varying in symmetry. *Bull. Psychonomic Soc.* **26**, 355–358 (1988).

Computers Math. Applic. Vol. 17, No. 4–6, pp. 485–503, 1989
Printed in Great Britain. All rights reserved

MIND'S EYE

A. H. Lowrey

Laboratory for the Structure of Matter, U.S. Naval Research Laboratory, Washington,
DC 20375-5000, U.S.A.†

Abstract—The concept of symmetry is one of the great universal principles used to comprehend the enormous amounts of data encountered in both the worlds of natural phenomenon and of abstract knowledge. With the advent of computers, methodology has evolved to process and generate huge amounts of information. This information is often inconsistent and ambiguous and is similar to that encountered by human perception. This article develops some commonalities between applications of symmetry and applications of computer methodology to visual perception (robotic vision), to explore the impact of developing technology on general understandings about human knowledge. These commonalities suggest that advances in robotic vision will enlarge the study of symmetry, reveal astonishing new types of symmetry, and produce unexpected applications of philosophical interrelationships between abstract and perceptual knowledge.

INTRODUCTION

With this second book on symmetry, the editors confirm their belief that the symmetry concept is a basic principle that is useful to explain relationships between aspects of mathematics and physical, biological, and other natural phenomenon [1]. It is a concept whose *superimposition* on computational analysis is one approach for organizing vast amounts of information and computational techniques to produce understandable, accessible and useful results [2].

As these volumes illustrate, symmetry has relevance and meaning for an awe-inspiring range of human intellectual endeavor, ranging from the broad canvas of the arts to the unseeable realm of high energy physics. It has long been associated with beauty in both of the two cultures considered in C. P. Snow's discussions about disparities between the sciences and the arts. It is also discovered with astonishing variety in natural phenomenon and abstract knowledge. The scope of symmetry produces the hope that nature may possess an order that is accessible to the comprehension of the human mind [2].

Yet the meaning of symmetry is not precisely defined. Intuitive understanding of symmetry, derived from visual perceptions of simple geometric forms, appears straightforward. Developing understanding and applications most often proceed by formation of mathematical relations describing a transformation process. This process, in general, represents an implied action on spatial orientation to produce predictable perceptions, often that of an image indistinguishable from the original visual scene. Yet these two volumes illustrate that the symmetry concept encompasses far broader areas of human understanding than those involved with geometric transformation or abstract mathematical formulation. One of the many books dedicated to explaining and exploring the symmetry concept introduces a particular abstract concept of symmetry that is based on linguistic analogy. This clearly extends the study of symmetry to include the far reaches of poetic license [3]. The same book includes a description of symmetry as a disease. The numerous varieties of symmetry invoked in these current volumes imply that a precise definition cannot encompass the full meaning of the symmetry concept and that, in fact, it may not be possible to provide well-defined limitations.

This, then, results in the contention that a fundamental ambiguity is inherent in the concept of symmetry. One purpose of this article is to explore some of the specific ambiguity that arises in applications of the symmetry concept, even in supposedly well-defined areas. The importance of the presence of this ambiguity lies in its association with the nature of visual perception. This article will show, to some extent, that the attempts to develop robotic vision have also encountered

---

†Address provided for information only. The views presented here are those of the author and do not necessarily represent the Naval Research Laboratory.

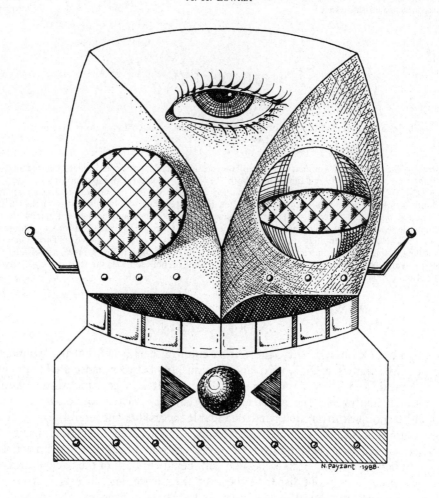

N. Payzant -1988-

ambiguity as a fundamental part of the attempts to simulate visual perception. This provides an essential conceptual commonality: ideas about symmetry and ideas about machine vision share the much-discussed limitations of human knowledge. They are both subject to the poorly understood, nonrational mental processing involved in creating the perceptions with which our reasoning and thinking begin.

Having established some of these inherent difficulties in defining and applying the symmetry concept and in applying computational technology to creating machine vision, this article suggests that methodologies being developed to deal with these difficulties have a similar basis and might well stimulate new approaches to both areas of research. In conclusion, there will be discussion of how the new computational technology will have a broad impact—possibly in areas of understanding which today appear quite remote from such machine-oriented reasoning.

VISION AS PROCESS: GENERAL CONSIDERATIONS

What does it mean to see? It can be argued that seeing is designed to know *what is where*. In this light, vision is the process of discovering what is present in the world and where is its relative location by means of some mental analysis of the stimulus present in a given visual scene. This description of vision as an *information-processing task* is an important starting place, both for the purposes of this article and for the myriad of activity prompted by the idea that computing technology has made possible the creation of a machine that could see. As will be discussed, the exploration of this information-processing task has provided profound insights into the nature of human perception and response to visual scenes with general philosophical implications about human knowledge and understanding. It is of interest here, as an introduction into some the many problems of robotic vision, to present some information about a particular class of human perception, namely that of optical illusion, and the visual information which prompts this type of perception. The purpose of this discussion is to illuminate some of the essential difficulties in creating a simple description of perception in terms of an elementary-information-processing task.

Machine vision has excited research since the early 1950s. In his general review of the developments in this technology, Ladd [4] presents the early realization that certain distinguishing features in a visual scene might be the fundamental elements for construction of visual perception. Edges, shading, texture and many of the other apparent components of visual information were quickly seized upon as the basis for attempts at machine algorithms for reproducing sight. It can be argued that this generally has been the basis for the development of image processing as a separate technical discipline. Certainly much effort using this type of visual classification continues today. However, it was soon realized that there were inherent limitations embodied in this approach to vision, most notably the inability to produce computational schemes that were applicable to generalized visual processing. Given an elaborate computer code and specialized scenes, some apparent discrimination was possible among, for example, shapes in a world of blocks or bagels in a bin [4]. However, such visual analysis was inadequate to reproduce general perception and, in fact, was itself subject to incompleteness and lack of definition. In the physical world, edges are fuzzy, colors or shades ill-defined and irregular, and textures nonuniform. Far more important was the realization that such analysis did not encompass much of the information necessary for perception and that the human mind could construct coherent visualizations almost independent of this type of visual information.

As an illustration of the above ideas, several types of optical illusions will be presented as a background for the discussion on vision and symmetry that follows. These exemplify the extent of the complexity of human perception that is being illuminated by research into robotic vision. Consider the subjective contours discussed by Kanizsa [5]. These are vivid illustrations of the unconscious visual processing from which human perception evolves. Figures 1(a)–(c) give the illusion of triangular figures whose outlines are clearly perceived but are not explicitly present in the visual information of the diagrams. The first figure illustrates a white triangle superimposed over another triangle in a planer arrangement; the white color of this triangle appears brighter than the background although they are in fact identical. The second figure suggests a curved arc and perhaps a three-dimensional arrangement. This effect was achieved merely by rotating the two top *PAC* Man-like figures 10° in a symmetrical fashion and slightly rearranging their location to align the perceived contour generated from the figure below. There are no curved lines involved in generating the illusion of the arc and, in fact, Fig. 1(c) shows that even the triangular cutouts from the circles are not necessary to create a bright undrawn figure.

In an article on the interpretation of visual illusions, Hoffman [6] discusses some of the factors which are likely to be incorporated in this visual processing. It is a fact that the eye is capable of perceiving a spherical shape given only the visual information generated by randomly placed lights located on a rotating sphere in a dark environment. In analyzing this phenomenon, Hoffman emphasizes the use of the learned laws of projection and the observation of rigid, generally smooth objects existing in the physical world. He cites a proof by Shimon Ullman that these two accepted beliefs can, in principle, provide a unique and correct solution to this problem. Hoffman discusses rules regarding perception of curvature in describing the intersection of objects. Hoffman and Richards proposed a principle of Transversality suggesting that the intersection of two surfaces is perceived as a concave discontinuity. They suggest that this provides some understanding about the famous goblet/face illusion of Rubin (Fig. 2). Here the illusion depends on the interpretation of object and background. When the object is considered to be a white goblet intersection with a black background, most of the curves are concave with normals of minimum curvature pointing into the black background. When the black faces are perceived as the objects, the concave intersections with the white background become the recognizable features such as nose, lips and chin whose normals of minimum curvature project into the white background. With respect to symmetry, this phenomenon is strikingly apparent in Fig. 3 (taken from Kanizsa [7]). The intersection of two figures is seen as junctions intruding into each figure as a background rather than the juxtaposition of two symmetric objects. Hoffman concludes that vision is an active inferential process exploiting regularities in the visual world and that mathematical investigation of this inferential power is a promising direction towards greater understanding of human vision.

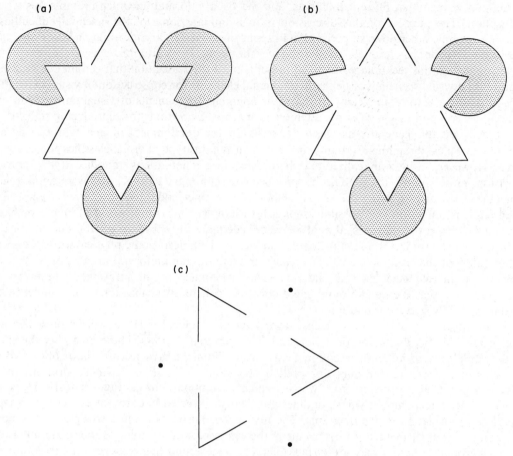

Figs 1(a)–(c). Kanizsa subjective contour.

The premises and implementation of these ideas are discussed extensively in this article; the illustrations and discussion of this section hopefully provide an introduction into the nature of the problems under consideration. It is noteworthy that these authors share common concerns about inferential, nonconscious visual processing. They emphasize the search for regularities and

Fig. 2. Goblet/face illusion after Rubin.

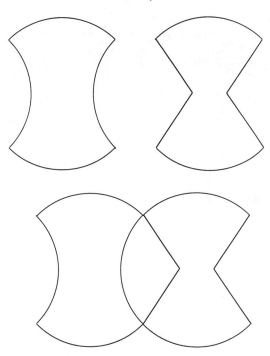

Fig. 3. Kanizsa figure showing perceptual transformation upon juxtaposition of two symmetric objects.

mathematical information, above and beyond that information present in the visual scene, for useful understanding of visual perception.

SYMMETRY: SOME GENERAL OBSERVATIONS

At first glance, symmetry is deceptively simple. First introductions to the symmetry concept often center on elementary transformations of planer figures. Rotations, reflections, inversions and the like become the first examples because it is easy to visualize the manner in which they recreate the original image. The transition to mathematics for concise descriptions then becomes obvious and even satisfying as some of the more elegant symmetries are developed. The mind often finds beauty in observations that are not readily accessible at first thought or apparent without some body of information in which they are embedded. The importance of this line of inquiry, directed towards developing sophisticated tools and language for succinct description of complex physical phenomena, is well illustrated by numerous essays in the first volume of this series [1]. For this article, it is useful to emphasize that inherent in this line of inquiry is a generalized concept of symmetry as a transformation *process*, and depends on mathematical reasoning and logical deduction to develop the language and descriptions for a scientific understanding of phenomenon in the physical world.

The dominance of this approach masks ambiguity about the role symmetry plays in human understanding and knowledge. A pointed example may be derived from considering a square cut from a child's modeling clay. Asked about transformations of this square, the response of a student trained in symmetry is predictable; the elementary symmetry operations are obvious and easily described by equations on paper. For a five-year-old child however, a similar question is likely to result in the square being squeezed into a three-dimensional lump and then rolled again into a square shape. Such a symmetry transformation is not mentioned in text books and is hard to describe by equations. Another example, more directly related to the purpose of this essay, is the presumed symmetry created by visual perception; " ... a strong tendency exists in thought to the extent that man may discover nature to be dominated by laws of symmetry even if, in many cases, one suspects he discovers what he himself has put there" [7]. Low-resolution visual information tends to create symmetric perceptions despite marked distortions at higher resolution [8]. The

purpose of these observations is to relate questions about symmetry to generalized questions about human knowledge and understanding.

One of the historical triumphs of the development of science has been the escape from a man-centered universe. Beginning with the Copernician revolution, developments up to the early twentieth century produced a series of laws and rules governing an orderly, predictable universe which appeared to operate in a manner quite independent of the investigator and the experiment. The presumption was clearly almost that of the clockwork natural world whose mechanisms, given enough time and attention, would be revealed and accessible to the human mind in a manner that would allow deterministic predictions of future actions and events. This view still has practical usefulness and seems to be the basis for Einstein's widely quoted awe at the amount of the workings of the physical world that are understandable to human intelligence. With the developments in this century, we have been forced to return, in part, to a mind-centered base for knowledge and understanding. The well-known limits on predictability and powers of observation and on the deterministic view of the physical world have created a new relationship between scientific problems, methods of experiment or analysis, and the nature of the solutions obtained. In a sense, this has created some new perspectives on the nature of scientific investigation and knowledge as exemplified by lines of inquiry relating the practice of physics to the disciplines of oriental religion [9].

In the context of this changing flux of ideas, this article attempts to relate the understanding of symmetry to these more generalized problems of human understanding. The efforts directed towards applying technology to human perception, robotic vision in this instance, have created methodologies that are clearly related to philosophical examinations of the nature of scientific thought. Some observations drawn from the work of Bertrand Russell provide a background for the further discussion of machine vision and the relationship between the symmetry concept and these generalized questions about scientific understanding. The following ideas are brief extractions from the book by Bertrand Russell entitled *Human Knowledge, Its Scope and Limitations* [10].

Russell's work deals with examination of the nature in which the human mind processes information. He posits certain first principles: of relevance to this discussion is the concept of a belief for which no further reason can be given, i.e. a belief that is a postulate based on a certain type of faith about the nature of the physical world and about the mechanisms of human understanding. In a scientific context, an example of such a belief is the sense that perception of a given experimental result repeated over a period of time is indicative of an observation relevant to the description of some portion of the physical world. This example is not completely representative of the complexity of the question posed by Russell since it is the very nature of perception and understanding that is being examined. However, this example raises the basic question of interpretation which is the main concept of importance for the current discussion. Russell suggests, for example, that such a type of scientific belief or principle is interpreted in the least questionable form. This interpretation itself is an example of a premise which, consciously or unconsciously is *assumed* in the reasonings of science. He also suggests that such questions about interpretation imply essential ambiguity when he posits that there are many statements about which we are more certain of their truth than of their meaning.

An important quality of these principles or beliefs, at least in ordinary thinking, is that they are the cause of other beliefs or concepts and are not derivative. This active character is important because these beliefs affect the process of reasoning or deduction. In the scientific context, Russell chooses to emphasize that these beliefs constitute the conscious or unconscious methodology of investigation and reasoning and are *assumed* rather than derived. He is trying to focus on assumptions that have created the great body of scientific knowledge and information. From a historical sense, it is obvious that many revolutions in science have resulted from challenges to the assumptions that governed the interpretation of observations and information in the physical world. It is this relation between assumptions and interpretations which concerns Russell when he says [10, p. 224]:

> "The question of interpretation has been unduly neglected. So long as we remain in the region of mathematical formulation, everything appears precise, but when we seek to interpret them, it turns out that the precision is partially illusionary. Until this matter has been cleared up, we cannot tell with any exactitude what any given science is asserting."

Interpretation then becomes a pivotal word that is central to the ideas of this article. The practical applications of the symmetry concept (both in the abstract and in the physical worlds) involve selective observations, assumptions about the process of transformations, and implicit conditions imposed by the nature of the solution or application desired.

The simple example of the model clay square shows this but the idea deserves further elaboration. It is easily visualized that interpretation of symmetry depends on the time scale of observations in much the same manner as it depends on spatial resolution. The human visual response time that makes television and films such convincing illusions provides a simple example. Time resolved pictures of the familiar rotating pinwheel reveal distinct symmetries that are not observable when it is spinning in the wind. In the extreme, a nonsymmetrical object in a static reference frame relative to the observer may acquire circular symmetry if rotated sufficiently rapidly about any arbitrary axis. Well-known examples of time-frame symmetry appear in experimental science. Molecular structure contains a classic example of phosphorous pentafluoride. When examined in a magnetic resonance experiment, the bond lengths are observed as equal while in an electron diffraction experiment distinct types of bond lengths are observed. The difference arises from the different duration of measurement; the electron diffraction experiment achieves essentially snapshots of molecules that are in static positions compared to the time scale of molecular vibrations and, despite the inherent averaging, resolution of different average bond lengths is possible. The magnetic resonance experiment is ten orders of magnitude slower than the electron diffraction measurements. Thus this experiment measures bond lengths for the molecule throughout the entire time span of a complex vibrational transformation known as psuedorotation. In this transformation, atoms exchange relative orientation and rearrangement of the bonding structure and the lengths are then all observed to be equal.

An important point of this discussion is that symmetry is in fact ambiguous; determination of symmetry depends on factors beyond the inherent properties being examined. Another observation, related to perception of symmetry but also serving as a general comment on the nature of visual perception, is that well-known factors of cultural conditioning influence the observations derived from visual information. Western culture, at least since the fifteenth century, has accepted perspective drawing as representative of three-dimensional (3-D) solid objects or depths of scene. However, there is considerable evidence that this perception of depth in 2-D representations is a learned experience. Dergowski[11] has reviewed experiments designed to test the nature of 3-D perception of culturally remote people. Specifically, with regards to an observation of symmetry, individuals not accustomed to depth perspective built model constructions designed to represent a 3-D perspective sketch of a cube as two squares with diagonal connections on a planer surface as opposed to creating a 3-D construct. It was similarly observed that groups of people from similar cultural background who apparently exhibited 3-D perception suffered confusion when trying to model the optical illusion of an "impossible" Penrose trident. In Fig. 4 the familiar 3-D view of the partial cube seems to have the correct perspective but it becomes flat for many people when rotated 45°; the extended drawing is the representation given by people not accustomed to perspective interpretation of pictures. It is apparent that perception of symmetry in a visual scene is thus ambiguous in this culturally dependant context. The 2-D interpretation is quite different from the 3-D understanding, and each interpretation is clearly a function of factors not present in the visual information at hand.

ROBOTIC VISION: ANALOGIES AND ANALYSIS

Because it appears possible to create machine vision, enormous human effort has been unleashed that encompasses the whole panoply of human knowledge, understanding and technology. It would be tempting to say that this development is being driven by technological advances. Indeed the unique expansion of computer power and information storage has certainly created a radically new set of tools with which to investigate this awesome concept. However, it is more generally true that these technological developments have primarily stimulated the continuing effort to understand human vision as a part of the understanding of human knowledge; a human activity that has existed at least since the time of recorded history (see author's note at end of article). The inherent involvement of these most general lines of inquiry is of particular significance because it demonstrates that advances in technology will influence the common understanding of these ancient problems. It is clear that this technology can provide challenges to the basic assumptions upon which our understanding is founded.

To create a manageable discussion, this section presents robotic vision in the perspective of analytical representations of human perception. It also discusses, in some small detail, the reasoned approach presented by Marr [12] of a single facet of this process, that portion of 3-D vision that creates illusions of depth from seemingly random patterns. As often occurs in scientific investigation, pathological cases (in this situation an optical illusion) allow isolation of specific phenomenon for insight into the understanding of the more general process. There is vigorous debate about the technical developments in this field; it is a broad and fertile research area for many

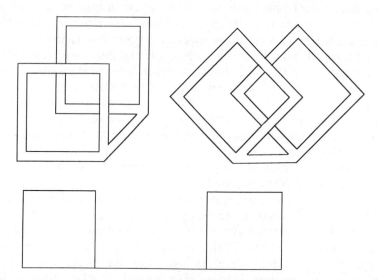

Fig. 4. Figures illustrating cultural relativity of depth perception. Top two representation of cube-like objects show change of depth perception upon orientation. Bottom extended figure is representation of 3-D construct made by people accustomed to 2-D perception.

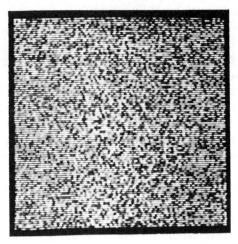 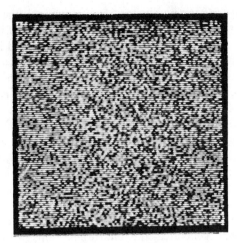

Fig. 5. Julesz stereogram illusion: left and right diagrams differ only by a uniform translation of a square section of random dot image in one image. From "Binocular depth perception of computer-generated patterns" by Bela Julesz from *The Bell System Technical Journal* **XXXIX**(5), 1129 (1960). Published by the American Telephone and Telegraph Company, Copyright 1960, used by permission.

ideas and methodologies [4]. The examples from the work of Marr are chosen as vivid examples of conceptual approaches to this subject, thus providing useful material for this discussion.

A brief and preliminary introduction of the specific example for the basis of our discussion is helpful in creating the context for the more extensive questions of interest. Consider the random dot stereogram experiment of Julesz [13] which is shown in Fig. 5. This consists of a pair of random dot images, identical except that a central square section of one has been uniformly translated a distance away from its original location. Perceived individually, these diagrams have the appearance of completely random images; unless exaggerated, translation of this segment of dots produces little noticeable change. Viewed stereoscopically, the junction of the two images produces the perception of this central square floating above the untranslated background image. The perceived image is a controllable function of the translation process. Depth perception can be related to the concept of stereoscopic fusion through the measure of the various amounts of displacement. Translation processes other than uniform displacment can create different effects on depth perception, and elaborate experiments are also possible with multiple correspondence. Detailed discussion of such extensive research is beyond the purpose of this article, but it is significant to realize the computational technology is creating new laboratories for examining age-old questions of vision and perception.

This phenomenon has been critically analyzed by Marr using a computational approach as a conceptual basis for machines that would create a similar interpretation from such images. It is important to stress that this discussion already raises a fundamental question about robotic vision: that of purpose. This section expands the question of purpose that is the guiding force behind the overall strategic development of robotic vision and is, in part, the driving force behind the development of the technology for accomplishing this end. According to Marr, the purpose of robotic vision is to build a machine that knows *what is where by looking*, that is to build a machine that could process images and discover what is present in the world and where it is located. It is important to realize, however, that such a machine must be able to create a computationally accessible and processable representation of this visual image in order to generate machine preception and in order to make this information useful in the general manner to which human beings are accustomed. This definition of functionality is particularly more elaborate than that for a television camera which can transmit only visual information, albeit transformed, distorted or "enhanced". Marr quotes Austin [14] emphasizing that viable robot vision should also at least roughly correspond to what an ordinary person knows to be true at first hand.

One reason for this emphasis is that visual processing may serve different needs or functions. This is shown by the analysis of the vision of a housefly presented by Reichardt and Poggio [15]. It is clear that about 60% of the fly's visual system is oriented toward processing a few simple relative motion parameters. For example, if the visual field of an object rapidly expands, say from a nearby looming surface, the motor mechanisms for landing are activated. The fly lands in the center of this visual field (inverting if the field is above) and when the feet touch the surface, the power system for the wings is switched off. Another feature of the system is apparently tuned to objects possessing both a small angular dimension in the field of view and a relative motion with respect to the background. This information is apparently delivered in such a manner that a fly a few feet away will be intercepted while an elephant at 100 yd does not register. The investigations propose a simple differential equation that apparently describes directional control. These studies clearly show that the purpose of vision in the fly is primarily to control motion by regulating the connections between the muscle systems upon which survival depends. It is not apparent that more complex information is an important aspect of the fly's visual processing. Lack of more complicated visual processing may well contribute to the speed of the fly's muscular reaction to visual stimulus, about 21 ms.

The purpose of human visual perception is obviously much more complex. It is this complexity that has generated the enormous range of investigations into human psychology and the generalized process known as perception as well as the studies concerning the suitable construct of logical circuitry and computational design that will reproduce in a manageable way the information about *what is where*.

It is important to recognize that Marr begins with the premise that vision is a computationally accessible process. This is not a trivial assumption. This assumption is an acknowledgement of visual perception as an intrinsic product of an unconscious, inferential process. This justifies a logical foundation for a methodological analysis and is patently necessary for any consideration of creating machine vision. From this premise, Marr proposes three levels of approach to dividing the problem into tractable forms: that of computational theory, that of developing representations and algorithms for expressing this theory, and that of hardware implementation for the physical and computational processing of the visual information. He makes a telling case that this is a necessary unified methodology for beginning to deal with the global problems mentioned above. The role of computational theory as the first step in this process cannot be underestimated. In the first place, it begins to formulate, in an explicit way, the nature of the information needed to produce visual perception from external optical stimulus. Secondly, this emphasis on theory provides a mechanism for analyzing and integrating the contraints on visual perception imposed by the active phenomenon of the real world, and for understanding how these constraints may be related to the mental processes of our imagination and thinking.

Computational theory is a broad term. Marr's belief is that the nature of the computations underlying visual perception depend more on the computational problems that need to be solved than on the mechanism with which they are performed. Thus, the housefly is built for a few types of computations based on visual stimulus and achieves rapid process time. Human perceptions encompass a larger variety of probems (and in fact are culturally conditioned and to some extent are dependent on previous experience) and require correspondingly longer response times that are clearly dependent on the nature of the problem encountered.

Representation and algorithms are the second level of approach and understanding of this problem. With a computationally theoretic understanding of the nature of the visual problems to be solved, it is feasible to search for the manner in which the visual stimulus will be encoded, either in the brain or in the machine. This then becomes the study of representations. Given an understanding of the representations involved, it is then possible to consider useful and/or efficient algorithms for processing the information in its represented form. At this level of investigation, much effort has been devoted to the psychology of perception. It is clear that a visual scene may produce several types of representations, and the nature of the interpretative problem to be solved will engage appropriate mechanisms to process appropriate information. Some exploration of the complexity of the visual process is given by Wolfe [16] who shows that the process of creating a visual representation involves several interactive processes which he calls hidden visual processes. Wolfe describes several experiments designed to probe the response of various physiological systems involved in visual perception. One example is an experiment designed to explore edge detection as a function of brightness and color. This experiment shows that the common features we associate with vision are not isolated phenomenon. In this case, the conclusion is that edge detection is not a color-sensitive visual function: isoluminescent edges created by distinct colors are difficult to perceive. This research concludes that the full range of human visual senses or processes has not yet been discovered and that perception is a complex interaction amongst many functions of the human brain and nervous system. This is a concrete example, for the present discussion, of a central assertion that visual perception involves information, assumptions and constraints above and beyond the visual stimulus present in the scene.

In this context, Marr emphasizes the importance and nature of the concept of process. Process is a vague word whose meaning is open to the choice of the user; considered use of the work perhaps creates the awareness of the delicate balance of Lewis Carroll's Humpty Dumpty who stoutly insisted that words were defined by the meaning he chose to give them. Marr chooses to use the word process as a unifying description of the application of the methodology discussed. At the abstract level of computational theory, process is examined in terms of what it accomplishes and why. What the process does is associated with the rules of theory which describe the process; the why inherent in the use of the word relates to the constraints imposed by the real world on the nature of the desired results. At the representation and algorithm level of visual perception, process is the relation between the representation and the problem-solving choices. This encompasses the information derived from the computational theory and the understanding of *real-world constraints on interpretation and desired results*. At the hardware or implementation level, process becomes the understanding of the transformations of the representations by the algorithms into the desired results and the choice of appropriate tools for accomplishing these tasks.

At this point, before presenting some of the detailed analysis of the Julesz illusion, return to the word *process* as a device for relating this discussion of robotic vision to symmetry. The language used here is seductively similar to discussions of symmetry as a transformation process [3]; there is also a clear correspondence with the language in the earlier discussions of symmetry presented here. It is tempting to dismiss this similarity as a simple-minded analogy. However, Marr's three-level methodology is applicable to problems in symmetry in a manner which is more than coincidence. The computational theory level, with its emphasis on what and why is essential in deciding the existence and usefulness of the symmetry concept for any particular question. As we have seen, symmetry is often in the eyes of the beholder, depending on time frame, inertial coordinate base, or even the vision of artistic license which creates symmetry in ways that are previously not found. Representation, transformation, and associated algorithms (or rule of relation) are the essence of symmetry as a process of implicate action. And the hardware level is the choice of tools with which to express symmetry in a manner accessible to others. Equally important, this methodology explicitly acknowledges the incorporation of the constraints imposed by the real world at all three levels of understanding. These constraints represent implicit extra information that is not present in the direct information being processed.

At this point, a brief survey of Marr's analysis of the Julesz illusion provides a concrete example of the implementation of this methodology. This survey is not intended to be comprehensive but rather is an illustration of specific concepts involved in robotic vision, concepts that bear on the relationship between symmetry and machine perception. This analysis begins with a description of stereopsis based on the binocular images created by the difference in spatial position of two eyes. The spatial positions in these images are characterized by a displacement called disparity. Marr chooses to restrict this term to mean the angular discrepancy the displacement creates with respect to the visual field of each eye. The Julesz illusion provides a concrete visualization of the problems involved in determining the generalized stereo disparity and using this information to create the subjective perception of depth. At the level of computational theory, subjective analysis of these two images may be described in terms of matching identical points on the image as seen by each eye. In this particular example, some points bear a one-to-one spatial relationship between images while the translated segment contains others that do not. Even for the identical portions of the diagram, the 2-D projection of the two images does not contain sufficient information for a unique solution for the correct matching of the points as perceived by each eye. There is a fundamental ambiguity, known as the false-target problem, that permits multiple matching of points if perception depends only on simple ray-trace analysis of images. The Julesz illusion provides an accessible example for understanding the questions implicit in the false-target problem.

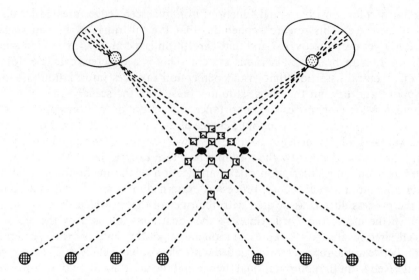

Fig. 6. Ray representation of false target problem central to stereo matching of left and right eye images:
after the work of Marr.

Consider the diagram in Fig. 6. Given the images of the sets of dots that relate to each eye, correct perception requires unique identification of the dots. This represents the true correspondence of the images in the physical world. There are 16 possible matches in the figure, of which only four represent the existing situation being perceived. The unique solution is provided by additional constraints, originating in the physical world, that allow the correct matching of identical points of positions on an object. Marr describes two simple facts known from ordinary experience in the physical world that restrict the solution of this problem and allow a unique image.

Deceptively simple, these facts are that a given point on a physical surface does in fact have a unique spatial position at a given time (with respect to human perception) and that surfaces are usually smooth and continuous. These two facts are noticeably similar to the learned information postulated by Hoffman in his analysis of illusions [6]. Marr formulates these constraints in terms of matching: if two potentially corresponding elements can have arisen from the same physical location, they can match. Only two elements are in fact the correct match, and the continuity of surfaces means that the angular disparity between points varies smoothly almost everywhere. In terms of the Julesz illusion, these matching constraints may be translated into rules: *compatibility*— black dots can only match black dots: *uniqueness*—almost always a black dot from one image can match only one black dot in the other image; and *continuity*—the angular disparity of the matches varies smoothly over the image. Boundaries and experientially less probable alignment of objects may create exceptions, but most perceptual experience is subject to these constraints, that are *derived from the nature of the physical world*.

The applicability of this analysis results from the fact that most visual scenes can be conveniently divided into segments or tokens of various levels of complexity. Reduction of a scene into such elements produces a representation of the image commonly known as a primal sketch. Marr suggests that the creation of stereopsis implies the existence of a buffer memory process within the brain using these primitive tokens. He labels the contents of this buffer memory the 2 1/2-D sketch. Such an idea corresponds with many of the ideas presented that consider visual perception as the result of unconscious nonlogical mental processing. This implicit presumption of distinct elements is an integral part of the premise that visual perception is a computationally accessible process and is a significant aspect of the fundamental relationship between the symmetry concept and visual processing. (This is discussed at greater length in the next section.) At the computational level of theory, the matching constraints derived from experience in the physical world create the fundamental assumptions of stereopsis: given a scene containing sufficient detail and fulfilling these matching constraints for the elements of the images, the correspondence is then assumed by the mind to be unique and a correct perception. Marr presents convincing arguments that this analysis is sufficient for understanding stereopsis.

A detailed discussion of the proposed algorithms for solution of the Julesz illusion is too extensive to be adequately covered here. However, a brief description of two proposed methods of solution for this problem will illustrate the implementation of these considerations. Marr first proposes a cooperative algorithm based on calculating the probability of matching of two points as a function of disparity. By dividing the problem into elements representing correspondence between spatial positions perceived by each eye, the probabilty that a particular element represents a correct solution to the false target problem is formulated as an iterative relationship between elements modulated by probabilities based on the matching constraints discussed above. The scene is modeled as several planer arrays of elements, each plane corresponding to a different angular disparity such that the intersection element represents the matching between points in each image related by a given angular disparity. The probabilities are calculated based on the constraints which are expressed as directional relationships between the elements in each plane. The uniqueness constraint inhibits probabiliity of matching along horizontal and vertical directions in the planes (these directions representing the lines of sight for each eye). This is because a given position for one eye implies that there is only one position for the other eye that represents a correct match. The smoothness or continuity constraint amplifies the probability of correct match for diagonal directions as matching for one point implies that matching for another is likely to occur with small but distinct displacments in each line of sight. An iterative problem is formulated that distinguishes the matching of points in the Julesz images as a function of the disparity and correctly identifies the translation that has been performed. A second algorithm is postulated in terms of multiple matching processes, starting with rough correspondence and converging to matching of fine detail. In this formulation, a buffer image or representation of the scene is a central site for multiple types of processing and interaction, both with the sensory data which is incoming (eye movements are postulated to be important thus implying the processing of multiple images created for each scene) and for interaction with implicit and possibly quite complex mental transformations of the information (thus perhaps providing a link for such well-known effects as those of past experience or presumed spatial relationships). This second formulation emphasizes the complexity and large amount of additional information necessary for creating perception from the sensory information about the visual scene.

In summary, the problem has been analyzed in terms of purpose. For the Julesz illusion this is easily understood as one of establishing the relationship between corresponding dots in the two images. Given this purpose, the possible visual relationships between the dots has been explored in terms of the information available from the scene. The inherent ambiguity has been analyzed and proposed mechanisms for determining unique solutions have been developed. This has been done by the use of additional information obtained from constraints imposed by physical world experience. By using these constraints, possible quantification of the visual information has been proposed that differentiates the dots in a manner likely to be accessible to the visual system. Computational algorithms are then proposd and implemented that in fact identify the differentiation of the dots in the illusion in a manner that reproduces the structural features perceived by human stereopsis. For this discussion, this process of analysis and the implications of its extension to other visual perception is most significant. This presentation of the problems inherent in robotic vision reveals questions related to generalized human perception. It thus follows that questions about robotic vision are inexorably intertwined with the broad and timeless questions about the nature of human knowledge, language and understanding.

A notable example is the use of the word process discussed above. If vision and perception are approached from a broader view, such that the human brain/mind taken as a whole plays the role of "hardware" as envisioned by Marr, the intrinsic processing and transformation of visual information into conceptual perception has long been recognized. If one accepts the visual orientation of descriptive language, the historical recognition of this phenomenon is manifest. The word "imagery" is used in fundamental ways that far transcend any limitations imposed by confines of visual stimulus or recollection. With the dawn of analytical psychology, enormous effort went into attempts to create unifying theories or ideas about the nature of implicit information processing creates the ordinary understanding of *what is where*. The work of Kanizsa reiterates, in another manner, some of the important relational concepts mentioned above and sets the stage for creating some examples to illustrate the indigenous relationship between symmetry and visual perception. In the collection of essays constituting a volume entitled *Organization in Vision* [7], Kanizsa deals with numerous aspects of this general problem of understanding human visual perception. This volume provides access to the many years of literature on the subject and provides some helpful perspective on the question considered here.

In one chapter, Kanizsa discusses the conflicting information about the influence of past experience on present visual perception. There is no question that such learned response plays a role in the subjective experience of present visual information. It is well-known that indistinct objects (such as those seen from afar) are often perceived as familiar forms drawn from conditioned images inherent in the mind [8]. Earlier discussion describes the well-known illusion for which the mind "fills in the blanks", creating familiar images in the presence of only partial visual stimulus. Past experience is often so dominant that visual perception is created in conflict with the stimulus present; impossible objects such as the Penrose drawing, or many of the works of M. C. Escher draw on the fact that the 2-D projection presents ambiguous information (as discussed above) and illusionary effects may be created drawing on the instinctive experience of depth. It is clear that the basis artistic representation of perspective is created by everyday experience of the 3-D world. It has also been shown that this visual experience is culturally dependent; some social structures exist for which 3-D interpretation of 2-D scenes is not present and for whom the familiar depth illusions have no meaning [11].

To emphasize the role past experience plays in perception, consider the fact that the human mind possesses an astounding ability to recognize other individual human faces. A commentary by Garfield [17] reviews this question and reiterates several significant points. First, it is apparent that the human mind can recognize a familiar face in a crowd of several hundred strangers in less time that would be required for complete processing of the visual information present in the scene. This ability to recognize familiar faces also seems to last over long periods of time; twenty years may lapse and yet people still recognize a face familiar from that previous period. It is clear that faces are complex visual stimuli, often encountered, which are familiar, and trigger complex emotional responses; all of which may contribute to the special perceptual ability with which they are associated. This is without a doubt, an example of experience-influenced perception.

However, a recent unusual technological development suggests that the nature of such experiential influence is not obvious. As suggested in 1973 by Howard Chernoff (a statistician at Harvard University) a cartoon face may be used to represent a surprising amount of information [18] (see Fig. 7). By using shape characteristics of various facial features, it is easy to postulate up to 10 distinctive identifiers for such a cartoon face. Shapes of the mouth, eyes, nose, etc. create a distinct facial expression that is recognizable at a glance. Given 10 settings for each feature, i.e., 10 gradations between happy smile and an angry frown, 10 billion expressions are combinatorially distinct and subtle differences between these combinations are accessible. Given training, these differences can create emotional/intellectual meaning that is useful for conveying large amounts of information. It is interesting to conjecture the role that perceived symmetry plays in this process. It is well-known that human faces are not symmetrical [19] and, given the experiential basis for face recognition, it is reasonable to assume that this lack of symmetry is one component of the identification process. Given the mind's tendency to "fill in blanks" and to impose symmetry, it is perhaps the deviations from symmetry that provide the clues to individual identity and character. Clearly the imposition of bilateral symmetry on the data face representation problem is a limitation of allowed expressions. It is intriguing to speculate that the differences from symmetrical shape and orientation may be more readily accessible and identifiable than the complete image itself; this reduced amount of information could account for the exceptionally rapid recognition of a friend in the crowd.

To return to Kanizsa's discussion of experience-influenced perceptions, his primary focus is on the difficulty of identifying in a clear manner the exact nature of this influence. Spatial arrangement illusions, identification illusions, and shape/transparency illusions are presented to communicate that the effect of experience does not produce consistent perception derived from a visual scene. It is clear that a purely empirical, experiential based analysis is not sufficient to satisfactorially identify the nature and extent of the necessary extra information used for visual interpretation.

Fig. 7. Simple cartoon representations of possible data face constructions. Elimination of bilateral symmetry increases possible information content.

The first essay in Kanizsa's book concerns interpretation of information in ways that go beyond the extent of available information. He considers perception to be the process that selects, discards, analyzes and integrates sensory data; his emphasis on the word process is to avoid a separation between vision and thinking. In his first chapter the familiar illusions of impossible objects, ambiguous interpretations and image filling of blank space are encountered to show that careful analysis of such pathological scenes raises questions whose consideration can extend the breadth of human knowledge and understanding. Of interest here is the conjuction of questions about what can be thought and what can be seen (or perceived). A primary conclusion is that the eye reasons in its own fashion, i.e. that visual perception is an exercise of human thinking and analysis which may follow different rules than reasoning and logical processing. Kanizsa thus strongly suggests that sufficient understanding of visual perception to create robotic vision well reveal and use computational processes that are not immediately obvious or deducible from other problems or experience.

It is important to recognize that this discussion is basically an attempt to organize, in the terminology of technology, ideas and understanding that have been part of the human knowledge base for a large part of history and have been important in many cultures and civilizations. Visual thinking and process is different from logical reasoning and analytical organization of information. The distinctive role of visual art and esthetics in history and culture is common knowledge. A contemporary discussion of the role of visual thinking has been organized by Georg Kepes in a volume entitled *Education of Vision* [20] which is part of a series based on ideas related to vision + value. In his introduction, Kepes stresses the impact visual experience has upon the psychological function of the human mind and in this context presents vision as a continuous creative process. This discussion and terminology are directly related to the ideas presented above. The first article in the volume is "Visual thinking" by the psychologist Rudolph Arnheim [21]. This article elaborates many of the characteristic aspects of the mental processing that are inherent in seeing. It is not surprising, therefore, to find that the psychological analysis arrives at the same concepts of distinct unconscious computation and analysis that are necessary for an adequate methodology for the creation of robotic vision. In this vein, a volume entitled *The Psychology of*

Perspective and Renaissance Art by Michael Kubovy [22] provides an artistic and humanistic analysis of the stereopsis problem discussed in computational terms above.

It is worthwhile remembering the complexity of the visual system postulated by Wolfe [16] and the apparent cooperative function of several physiological systems that create sight. This directly reinforces Marr's insistence of the importance of computational theory. It also suggests that new information, understanding and knowledge will certainly appear as the development of successful machine vision continues. New technology will allow creation of specialized experiments in processing visual information. These experiments will involve testing of controlled parameters, information processing and analysis of results in ways that are not currently accessible in biological systems. It follows, as will be discussed later, that our understanding of symmetry will also be affected; new perceptions form the basis for new symmetries.

GENERAL IMPLICATIONS: RELATION TO SCIENTIFIC PROCESS

The previous discussions about symmetry and vision can be viewed as a natural extension of the considerable debate over the nature of scientific thought and process. As we have seen from the work of Russell, the questions of ambiguity and the problems of forming "correct" perceptions are inherent in discussion and determination of scientific information. Because symmetry is most often invoked explicitly in scientific considerations, some further discussion of these problems in the scientific context are given to describe general relationships that are the premise of this article. Most of this discussion is drawn from a collection of essays by Michael Polanyi entitled *Knowing and Being* [23]. However, it is not difficult to realize that the questions considered here have long been discussed in many contexts (see author's note). It thus follows that pursuit of the quest of robotic vision and fuller comprehension of the symmetry concept will entail broad implications about the generalized nature of human knowledge and understanding. It is often stated that much progress in science is predicated on advances in technology. It is contention of this article that technological progress creates fundamental changes in human thinking, including areas that are far broader than those primarily encompassed by scientific considerations.

The first arguments for consideration appear in an essay entitled "The unaccountable element in science." [23] Polyani's focus is on the singular contribution of an ordinary scientist dealing with an ordinary scientific problem. He considers the contributions to scientific thought made by personal judgements that are distinct and irreplaceable by explicit reasoning. Central to these ideas is the considered premise that raw experience is devoid of all meaning and is made intelligible only by the powers of perception. This premise uses much the same concept as proposed by Kanizsa [7] or Kepes [20] in attributing a distinct type of human thought process to visual organization and perception that is quite different from reasoning and logical deduction. Many such ideas are explored by Marr [12] of which the buffer memory proposal for the 2 1/2-D sketch is only one. Much of the discussion concerns the question of ascribing formalized significance to information, i.e. what are the considerations about the probability that a particular set of data has valid or useful significance. Polanyi cites a remark attributed to Enrico Fermi that a miracle is an event whose chances of occurrence are less than one in 10. Another such rule, ascribed to Sir Robert Fisher, rejects patterns with probabilities of less than one in 20 as illusionary. These considered definitions of miracles differ only by numerical ratio. The point of these remarks is that the boundary between chance and pattern is arbitrary.

Polanyi then describes mathematics as only a formalized link between an intuitive surmise of significance and an arbitrary, informal decision to accept or reject on some basis of computed probability. In this light, randomness becomes conceivable only in relation to potential order and a determination of either is the result of an informal act of personal interpretation. Thus progress in science is conjectured to be a distinctly personal effort guided by the gift of perceiving a problem that is not observed by others, sensing a personal direction in the midst of apparent randomness, and eventually creating a solution that is a surprise to others. Polanyi postulates a similarity between perception and scientific intuition. Both are governed by rules that are perhaps unknowable but are certainly individual and distinct from other forms of human thinking. He concludes that scientific knowledge is accepted on the premise of hidden truths and thought processes which then becomes the motivation for further investigation. In this manner Polanyi equates scientific process with the process of perception. Both processes are motivated by the need to understand the foundations of seeing and perception and both processes are founded on the premise that we can make sense of experience because it "hangs together in itself". This discussion bears a remarkable similarity to Marr's analysis of vision. For Marr [12], robotic vision is clearly founded on the premise that perception is computationally accessible. It can be created only by using implicit information not present in the visual scene. Constraints derived from real world experiences of uniqueness and continuity are necessary to create the sensation and confidence that what we are perceiving "hangs together in itself". This is also similar to the hidden processes of Wolfe [16] and learned laws of Hoffman [6]. This similarity of analysis and even the use of language is not coincidence; it is a manifestation of the broad scope of the developments in machine vision.

A further elaboration of Polanyi's ideas is found in the essay entitled "The logic of tacit inference" found in the same volume [23]. Having concluded that understanding logical reasoning and deduction is not a sufficient basis for understanding the scientific process, Polanyi then looks for the area of human experience from which to draw information that will contribute to his analysis. He wants to find some human logic by which tacit or assumed mental processing (perhaps unconscious or unknowable) can achieve and uphold valid conclusions. Again, the example of perception becomes central to the discussion. He maintains that the capacity of science to perceive new and unique patterns differs from ordinary perception only because it has tools and training to integrate shapes that are not readily handled by ordinary perception. Trained perception is asserted to be the basis for all descriptive sciences. He concludes that there is no justification for separate approaches to scientific explanation, scientific discovery, learning and meaning; they are unified aspects of a general process of perception or understanding.

It thus seems reasonable to consider Marr's work (and in fact the whole effort directed towards robotic vision) as examples of generalized attempts to create representations and understanding of human thought and reasoning. It is significant that Marr's analysis of robotic vision encounters difficulties in creating perception from ambiguous information in a practical engineering context that is similar to those Polanyi encounters and describes in his philosophical endeavor for creating descriptions and understanding of scientific thought and process. It follows that the similarity of ideas and language result from treating inherently simlar questions and problems. From this, symmetry as a trained perception created and used for particular applications, is another example of such a process.

To this point, this essay is concerned primarily with the construction of ideas to show the similarity of the fundamental problems encountered in creating robotic vision and in the use and analysis of the symmetry concept. There appears to be a profound basis for this similarity in the phenomenon of the physical world. Polanyi boldly asserts the lack of fundamental meaning in raw experience; Marr deals with machine perception in terms of creating coherent pattern based on ambiguous information. From the discussions of Prigogine [24] concerning trajectory analysis and thermodynamics, a concept of a fundamental physical reality that corresponds to these ideas can be derived. From the extensive and profound presentation by Prigogine in the book *Order Out of Chaos* [24], the ideas relevant to this essay are summarized:

(i) The principles of thermodynamics necessitate that any given pattern or order will evolve in phase space in the course of forward time into chaotic equilibrium.

(ii) The second law of thermodynamics acts as a selection principle so that only those intial conditions which will lead to this chaotic equilibrium may occur for a given system.

(iii) It follows that any apparent pattern or order contains within it the same degree of chaos or randomness, in terms of its trajectory through phase space as is found in the final equilibrium.

It thus appears that ambiguity and disorder are inherent in the facets of the physical world that create the stimulus for our perceptions. It is resonable to believe that this uncertainty is rooted in all our attempts to deal with the environment in which we are inextricably immersed. Polanyi himself suggests that a Laplacean mind that computes future trajectories from present topography is merely transforming meaningless information into other meaningless information and not adding to the advancement of knowledge.

To conclude this section, it is interesting to note that Polanyi rejects a cybernetic interpretation of thought and behaviorism. This is counter to Marr's assumption that perception is computationally accessible. It appears likely that the debate between these two beliefs will focus not on the information present for mental or computational processing but rather on the nature and extent of the rules or extra information necessary for creating useful interpretations or patterns derived from the experiential information.

CONCLUSION: NEW WINE FROM NEW WINESKINS

The conclusion of the essay is expressed by the title "Mind's eye". The fundamental similarities between problems in robotic vision and those in symmetry arise from functioning of the central core of human perception, which is perhaps an almost trivial tautology. However, the many levels of understanding inherent and necessary for an adequate analysis of problems in either discipline give rise to rather more profound observations concerning future developments in both areas.

A central theme of this essay is the fundamental role of ambiguity inherent in the problems addressed by each area of research. In some sense, even the elementary mathematical analysis of textbook symmetry is a language for dealing with ambiguity. Transformation-related equivalence organizes vast amounts of perception into a concise and manageable representation. With the increasing sophistication of theories and scope associated with the symmetry concept, new computational technologies for processing information present in a visual scene and/or necessary for representing such information in an accessible manner will emerge. Conversely, new technologies for machine vision will suggest new manners of organizing information and will fuel development of complex ideas in symmetry.

It is also clear that technological developments in pursuit of robotic vision will generate new types of information. The developments of image-processing methods have created new information; false-color representation, image enhancement and the associated mathematics of pixel processing, and the basic efforts to identify fundamental elements of visual scenes already have created new types of information for which the technology associated with symmetry has been essential. The exponential growth of computational power, which has kept robotic vision as a viable carrot in the competition for useful applications, has simultaneously created the exploration of unforeseeable complexities in symmetry. It is also clear that the search for generalized robotic vision, driven, in Marr's terms, by the need to create a machine that knows *what is where*, will also spin off special-purpose technology designed to see *what* man cannot see and, certainly, *where* man cannot see. The type of perceptions created by such developments are bounded only by the limits of the imagination. It follows that new symmetries, new equivalences, and transformations, will result derived from specialized definitions of purpose.

With the acceptance of a boundless human imagination and the limitless vision of the "Mind's eye", the concluding assertion here is that the technological developments will irrevocably affect our understanding of human perception, knowledge and understanding. These timeless wonders are still at the center of revolutions in human ideas because the very assumptions and hidden processes behind them are subject to innovative challenge and examination.

Author's Note

At the request of the editor, I give here a few notes about the title of this article. Many readers will have seen this phrase or idea in many different contexts. The idea that human vision creates a perception of the world that is distinct from some more fundamental physical reality is found throughout recorded history. Philosophy and religion have been the natural areas of human culture for the propagation and survival of the ideas about this question which were part of ancient cultures. In a survey of the philosophies of India, Heinrich Zimmer [25] traces an origin of this concept to the Vedic religion of the Aryan culture which began in the frame of 2000–1000 B.C. He posits two fundamental lines of thinking which clearly address this concern; the first was the inquiry into the essential nature of the physical reality from which vision derives such changing perceptions; the second was into the basic functioning of the human self which controlled the essential creative processes of perception. A sense of this ancient understanding of this dilemma may be found in the following quote from the Brihadaranyaka Upanishad [26]

"He is the unseen seer, the unheard hearer, the unthought thinker, the ununderstood understander. No other seer than He is there, no other hearer than He, no other thinker than He, no other understander than He: He is the Self within you, the Inner Controller, the Immortal. What is other than He suffers."

The Platonic formulation of our perceptions being derived from images which are shadows of a true reality cast on the wall of the cave of our existence is a similar root in Western culture. The mystical pursuit of direct perception of reality is a transcultural phenomenon based on this recognition that the true eye lies within the mind.

To conclude the note, an extensive survey of contemporary exploration of this type of problem, as well as many others raised by the concept of artificial intelligence may be found in *The Mind's I* [27].

REFERENCES

1. I. Hargittai (Ed.), *Symmetry: Unifying Human Understanding*. Pergamon Press, Oxford (1986).
2. E. Y. Rodin, Foreword *Symmetry: Unifying Human Understanding* (Ed. I. Hargittai). Pergamon Press, Oxford (1986).
3. J. Rosen, *Symmetry Discovered*. Cambridge Univ. Press, Cambridge (1975).
4. S. Ladd, *The Computer and the Brain*. Red Feather Press: Bantam Books, Toronto (1986).
5. G. Kanizsa, Subjective contours. *Sci. Am.* April (1976). Also J. Walker, The amateur scientist. *Sci. Am.* Jan. (1988).
6. D. D. Hoffman, The interpretation of visual illusions. *Sci. Am.* Dec. (1983).
7. G. Kanizsa, *Organization in Vision*. Proger Scientific: Holt, Rinehart and Winston, CBS, New York (1979).
8. B. E. Lowrey, A space view of a symmetric object. *Comput. Math. Applic.* **12B**, 477–485 (1986). Reprinted in *Symmetry: Unifying Human Understanding* (Ed. I. Hargittai). Pergamon Press, Oxford (1986).
9. F. Capra, *The Tao of Physics*. Shambala, Boulder, Colo. (1975).
10. B. Russell, *Human Knowledge, Its Scope and Limits*. Simon and Shuster, New York (1948).
11. J. B. Dergowski, Pictorial perception and culture. *Sci. Am.* Nov. (1972).
12. D. Marr, *Vision*. Freeman, San Francisco (1982).
13. B. Julesz, Binocular depth perception of computer generated patterns. *Bell Syst. Tech. J.* **XXXIX**, 1125–1162 (1960).
14. J. L. Austin, *Sense and Sensibilia*. Clarendon Press, Oxford (1962).
15. W. Reichardt and T. Poggio, Visual control of orientation behavior in the fly: Part 1. A quantitative analysis. *Q. Rev. Biophys.* **9**, 311–375 (1976); Visual control in flies in recent theoretical developments in neurobiology (eds W. Reichardt, V. B. Mountcastle and T. Poggio) (1979); Visual control of the orientation behavior in the fly, Part II. Towards the underlying neural interactions. *Q. Rev. Biophys.* **9**, 377–438 (1976).
16. J. M. Wolfe, Hidden visual processes. *Sci. Am.* Feb. (1983). Some specific physiological mechanisms which contribute to such processes are discussed by M. S. Livingstone, Art, illusion and the visual system. *Sci. Am.* Jan. (1988).
17. E. Garfield, I never forget a face. *Scientist*, pp. 36–44, Feb. (1979).
18. I. Peterson, Pictures worth a thousand numbers. *Wash. Post Outlook* **C3**, 19 July (1987).
19. I. Hargittai, Limits of perfection. *Comput. Math. Applic.* **12B**, 1–17 (1986). Reprinted in *Symmetry: Unifying Human Understanding* (Ed. I. Hargittai). Pergamon Press, Oxford (1986).
20. G. Kepes, *Education of Vision*. Vision + value series. Brazillier, New York (1965).
21. R. Arnheim, Visual thinking, *Education of Vision*. Brazillier, New York (1965).
22. M. Kubovy, *The Psychology of Perspective and Renaissance Art*. Cambridge Univ. Press, New York (1986).
23. M. Polanyi, *Knowing and Being*. Routledge & Kegan Paul, London (1969).
24. I. Prigogine, *Order out of Chaos*. Bantam Books, Toronto (1984).
25. H. Zimmer, *Philosophies of India*. Bollinger Foundation Series No. 26. Pantheon, New York (1951).
26. R. C. Zaehner (Translations of Hindu scriptures) *Hindu Scriptures*. Dent, London (1966).
27. D. R. Hofstadter and D. C. Dennett, *The Mind's I*. Basic Books, New York (1981).

Computers Math. Applic. Vol. 17, No. 4–6, pp. 505–534, 1989
Printed in Great Britain

0097-4943/89 $3.00 + 0.00
Pergamon Press plc

NON-EUCLIDEAN GEOMETRIES AND ALGORITHMS OF LIVING BODIES†

S. V. Petukhov

Department of Biomechanics, Mechanical Engineering Research Institute, U.S.S.R. Academy of Sciences, Griboedov Str. 4, Moscow 101830, U.S.S.R.

Abstract—The importance is shown of non-Euclidean geometrical symmetric transformations and iterative algorithms for the structuring of supramolecular biological bodies. The variety of kinematics of biological movements is related with the "cyclomeric polymorphicity", or restructuring of the iterative algorithm in the biological structure. The author believes that the morphogenetic significance of iterative algorithms in biology is attributable to the mechanisms of interaction in biological layers of tissues and replication of supramolecular structures. The geometrical fundamentals of classical biomorphology need expansion and a generalized biomorphology has to be developed by replacing the conventional similarity symmetries by broader ranges of higher-order transformations (notably Möbius and projective). The progress of theoretical biology is today contingent on more extensive use of group-theoretic methods incorporating higher-order symmetries.

1. THE ERLANGER PROGRAM AND GEOMETRIZATION IN BIOLOGY

One hundred and fifteen years ago F. Klein, who then was 23 years old, put forward the famous Erlanger program which established an essentially new, group invariant view of geometry. Since then geometry has been the science of invariant groups of transformations and the Euclidean geometry, one of the possible geometries, each using its own set of transformations. These are similarity transformations in the Euclidean geometry, affine transformations in the affine geometry, Möbius, or circular, transformations in the conformal geometry and projective transformations in the projective geometry [e.g. 1, 2].

This approach restored to a significant degree the unity of geometry lost by the mid-nineteenth century because of the development of its new fields, clarified the interrelationships of different geometries and made it possible to construct new ones.

The revolution in geometry brought about by the Erlanger program was later extended to the physics and to philosophical views of space. The very fundamentals of human perception of space changed and ever since have been related to the concept of a mathematical group of transformations. H. Poincaré formulated this relationship in very simple terms, space is a group. The advent of the special theory of relativity gave birth to a new term, "geometrization of physics" which stood for the fact that formally this theory is a theory of invariants of some group of transformations (Poincaré–Lorentz group), or a geometry. The ideas of geometrization of physics, representation and description of its theories in the language of invariants of groups of transformations were extended to quantum mechanics, the theory of conservation laws, the theory of elementary particles and other physical fields. The group invariant approach and symmetry concepts become a cornerstone of today's group theoretical thinking [1].

This revolution entailed a revision of the philosophical fundamentals of physics; the view of what the initial laws of physics should be has changed. Since Newton the laws of nature had been formulated as differential equations; this changed completely. For basic physics the major initial laws are laws of symmetry. About this change Yu. B. Rumer and A. I. Fet, Soviet theoreticians, wrote

> "The development of physics in recent years has reversed, in a sense, the relation between equations of motion and symmetry groups. Now the symmetry group of a physical system is in the forefront, the concepts of this group and its subgroup contain the most fundamental information on the system. Consequently, groups become a primary, most profound element in a physical description of nature."
> [1, p. 4]

---

†The research reported in this article was partially presented in a motion picture *Cyclic Groups of Transformations in Biomechanics* (filmed by Soyuzvuzfilm with the author's script).

Also, in the words of H. Weyl [3], "the symmetry method is the guiding principle of today's mathematics and its applications".

"Nothing is more fruitful than cognition of oneself", wrote R. Descartes [4]. But what does cognition imply? To know basic biological phenomena in terms of mathematical natural sciences is to interpret these phenomena and their laws in a language of more profound concepts which are characteristic of mathematical natural sciences. The mainstream for biology to bridge the gap to exact sciences is penetration of group invariant concepts and methods into biology through study of biological symmetries. The characteristic headlines of papers on mathematical and theoretical biology are "The concept of a group and the perception theory" [5], "Biological similarity and the group theory" [6], "Research in non-Euclidean biomechanics" [7] etc. In other words, new daring attempts are made to build theoretical models of specific biological phenomena in the fields of morphogenesis, psychophysics etc. as formal theories of invariants of certain groups of transformations. In effect, the geometrization of physics goes hand-in-hand with attempts to geometrize biology. The theoretical biology of the future seems to be bound to become largely a group invariant biology. Natural sciences will then make a step to A. Eddington's ideal [8] of combining whatever we know of the physical world into one science whose laws could be expressed in geometric or quasigeometric terms.

For illustration of research along this line this article will describe the writer's research in higher symmetries and algorithms in self-structuring of living bodies. The research started with studies on the same kinds of symmetrical (Euclidean and non-Euclidean) algorithms for structuring the supramolecular organic forms in various lines and on various levels of the evolution of biological objects such as shells or molluscs and foraminifers, bodies of ring-shaped worms and miriapods, fish fins, bird features, flowers, leaves and offshoots. In the light of the Erlanger program, we will mean by the geometry of living bodies those geometrical groups of transformations whose properties dictate in supramolecular biological structures the standard integration of components into one whole.

Besides, symmetry of two figures implies that they can be made to coincide by some operation. Symmetry itself is a sum of two ideas, coincidence of figures and an operation which makes this coincidence possible. The former idea makes the concept of symmetry illustrative, whereas the latter is geometrically more meaningful, for from the group invariant, or geometrical, point of view far from every operation resulting in coincidence of some figures is admissible, only those are which belong to a "group of transformations", a stringent geometrical concept.

Studies of the laws and algorithms of organic shaping is a major direction in the biology of development which is expected by many to yield significant basic discoveries and important applications. Morphology has always been all important for biology but at the present time scientists in various disciplines give special attention to morphological self-organizations, properties of biological structures and their evolutionary transformations. Unlike, say, crystallography, mathematical biology does not rely on a generally acclaimed formal theory of morphogenesis, although numerous attempts have been made at creating such a theory with the aid of various initial control, engineering, diffusion reaction etc. models. Development of such a theory is difficult, largely because of a shortage of data on the common biological properties of morphogenesis which could be formalized and theoretically interpreted. R. Thom, a French researcher, was right when he said "...a geometrical attack of the morphogenesis problem is not merely justified, it is essential" [9]. The status of the mathematical biology of development is such that this science has yet to cover the evolution path from accumulation of knowledge on key morphological properties and adequate geometrical structures to the development of the desired theory, a path which has been covered by crystallography and other natural sciences dealing with objects much plainer than those of biology.

The very concept of symmetry emerged from ancient observations of the shapes of living bodies. We might witness this concept going full circle in a complex evolution of today's science of symmetry and taking up again its initial objects, living bodies and unraveling profound laws of living matter. This is all the more probable because, in the words of Weyl [3],

"symmetry in the broad or narrow sense is the idea using which man has for centuries tried to obtain an insight into and create order, beauty and perfection."

The reader will certainly agree with V. I. Vernadsky [10] who said

"Symmetry of living matter has not been thoroughly explored and its study remains a major task of the biologist.... This is the field of future tremendous fruitful endeavor."

2. THE HISTORY OF RESEARCH IN THE SYMMETRIES AND ALGORITHMS OF THE MORPHOLOGICAL SELF-ORGANIZATION IN BIOLOGICAL OBJECTS

"(Biological) structures are only in particular cases dictated by the functions they perform while in a general case they obey some mathematical laws of harmony. The variety of shapes has its own orderliness, independent of the function, its own system which manifests itself, in particular, in symmetry detection from stringent mathematical description." (S. V. Meyen et al. [11])

Symmetry in the forms of biological bodies has always attracted the attention of natural scientists as one of the most remarkable and mysterious natural phenomena. The references of this article mention only a small fraction of the literature on this subject. Macromolecular symmetries were discussed by a special *Nobel Symposium* [12]. School curricula in biology include numerous instances of rotational, translational and mirror symmetries, and also symmetries of similarities of scale in biological bodies such as the walking gear of animals, flowers and offshoots of vegetation etc. (Let us note once more that the group of similarity transformations is the core of the Euclidean geometry and so such symmetries may be referred to as Euclidean as opposed to non-Euclidean symmetries which represent transformations from non-Euclidean groups.) A deeper biological insight resulted in discovery of new facts about very different biological bodies of various size and sophistication obeying the symmetry principle (and algorithmicity, which is worthy of special attention).

Biological symmetry is embodied to a larger or smaller degree in numerous biological theories, some highly controversial: N. I. Vavilov's law of homological series; A. G. Gurvich's theory of the morphogenetic field; V. I. Vernadsky's theory of the non-Euclidean geometry of living matter; the biological significance of the diffusion reaction model of morphogenesis, developed by A. M. Turing, and self-organizing growing automata, whose theory is being developed by J. von Neumann's followers; morphogenetic mechanisms behind numerous psychophysical phenomena including the esthetic preference of the morphogenetically significant golden section which is expressed by Fibonacci numbers etc.

We should not overlook the fact that from the geometrical (i.e. group invariant) viewpoint the entire classical biomorphology is essentially an extension of the group of similarity transformations. This is the case for morphological studies and theories of mirror symmetry and asymmetry of biological bodies, multicomponent biological forms which embody similarity symmetry according to A. V. Shubnikov [13], scaled three-dimensional growth of biological objects, dwarfs and giants among organisms of the same species etc. On the other hand, there is not good reason to believe that the geometrical fundamentals of morphology are confined to the similarity group; Vernadsky's assumption [10] on the important biomorphological significance of non-Euclidean geometry has been awaiting checking for a long time. Section 3 will discuss specific matters in non-Euclidean biomorphology.

More light must be shed on the geometrical fundamentals of morphology because they will dictate the geometric specifics of morphological studies and facts and their interpretation. A change of geometrical fundamentals would entail changes in all superior levels of this science and dictate new requirements and approaches to the development of formal theories which should be consistent with these fundamentals. Biomorphology is related to many fields of biology such as the biomechanics of postures, biomechanics of growing and motoric movements, psychophysics of perception etc. The updating of its geometrical fundamentals is capable of giving rise to new research approaches and encourage discovery of new properties of biological self-organizations in these fields.

In the existing variety of geometrically legitimate organic forms we will concentrate on structures whose components are integrated into an entirety in compliance with certain rules or algorithms which are the same along various lines and on various levels of biological evolutions. These structures, which may be referred to as algorithmical, are of special interest in theoretical morphology and related sciences such as biomechanics, biotechnology, bionics, synergetics etc. What is important is that in addition to regularly shaped biological objects there are some in which the conjugation of components is less regular, if it exists at all. This paper will consider algorithmical supramolecular biostructures which are chains or manifolds decomposable ($S = S_1 \cup S_2 \cup S_3 \ldots \cup S_n \ldots$) into commensurable and regularly positioned elements (or motive units S_k). Figure 1 shows such manifolds discussed by the literature on biological symmetries. The general rule of representing decomposable manifolds is that the preceding motive unit is transferred

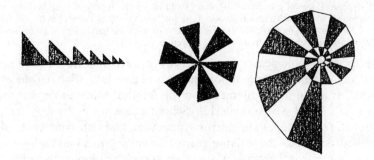

Fig. 1. Geometric examples of similarity cyclomeres or multiblock configurations with cyclic groups of similarity automorphisms (according to Ref. [13]).

into the succeeding one by a certain fixed similarity transformation g; in other words, the neighboring motive units S_k are mutually conjugated by an iterative algorithm

$$S_{k+1} = g * S_k. \tag{1}$$

Consequently, by reapplying the generating transformation g m times to a motive unit S_k a component S_{k+m} is obtained; mathematically speaking, in the set S_k a cyclic (semi-)group of transformations, G, is active which contains elements g^0, g^1, $g^2, \ldots, g^m, \ldots$ (a finite number of motive units in a biological object is neglected where necessary). In other words, this decomposition of the manifold thus organized includes a cyclic group of automorphisms and their motive units are aligned along the orbit of the appropriate cyclic group. For brevity, such configurations will be referred to as cyclomers, a term widespread in biology, no matter whether g is Euclidean or not in equation (1). Because Shubnikov [13] has made a significant contribution to research in similarity cyclomerism, the configurations shown in Fig. 1 are frequently referred to as Shubnikov symmetrical configurations. Section 3 will demonstrate the biological importance of non-Euclidean cyclomerism, which, thus far, has not been studied in biological bodies.

Similarity transformations in biomorphology are also known with reference to the scale of three-dimensional growth, which is fairly frequently observed in animals and vegetation over extensive periods of individual development and is accompanied with mutually coordinated growth behavior of small zones distributed in the volume of the body—a behavior which is geometrically described as a scaling transformation. With the transformation of as few as three points of the growing configuration known, transformation of the continuum of its points may be assessed. Note that the growth-related ability of living organisms of a variety of species to exist in morphologically identical modifications on various scales is probably the morphological property of living matter that man has been aware of for the longest time and has been reflected in scientific, mythological and fiction literature in which dwarf–giant relationships are variously described.

Do the Shubnikov similarity symmetry and the scale of the volume growth exhaust all geometrically legitimate kinds of mutual conjugation of parts in a structure and ontogenetic transformations in living bodies? Or do they act in biomorphology as very particular cases of the kinds which are built around non-Euclidean groups of transformations containing a similarity subgroup? The writer's research has provided a positive answer to this latter question. As S. Lie, a classical mathematical writer, noted [14], there are two basic ways to extend the similarity transformation group, either to Möbius transformations or to projective transformations (see the Appendix). Now let us proceed to new findings on non-Euclidean symmetries in mutual integrations of individual biological bodies.

3. NON-EUCLIDEAN GEOMETRIES AND ITERATIVE ALGORITHMS IN ORGANIC FORMS

"If we do not succeed in solving a mathematical problem, the reason frequently consists in our failure to recognize the more general standpoint from which the problem before us appears only as a single link in a chain of related problems. After finding this standpoint, not only is this problem frequently more accessible to our investigation, but at the same time we come into possession of a method which is applicable also to related problems." (D. Hilbert [15])

The biological value of similarity cyclomerism (Fig. 1) seems to be clear. What configurations form if the generating transformation g of equation (1) is a Möbius or affine transformation? And do such non-Euclidean cyclomerisms have biological analogs? Analysis reveals that the manifold of cyclomeric configurations noticeably expands in this case and includes, in addition to cylindrical, conic and helical forms associated with similarity cyclomerism, more complex configurations such as lyre-, sickle- bud-shaped etc. (Fig. 2). In these configurations the motive units may be different in shape and the variation of these units along the cyclomerism may be essentially non-monotone. What is important is that non-Euclidean cyclomerisms are as widespread in biological bodies along various lines and at various levels of evolution as the similarity cyclomerism, but in the science of biosymmetry they have not been studied. Remarkably, non-Euclidean and Euclidean cyclomerism are observed in analogous multicomponent biological bodies simultaneously.

In particular, the horns of numerous animals are helically or rectilinearly conical and thus can be described as similarity cyclomerism. In other animals the horns are essentially different and configured as non-Euclidean cyclomerism. Thus, Fig. 3 shows the horns of a *Pantholops hodgsoni* [16] which is described by a cyclomerism obtained by a Möbius generating transformation (of the so-called loxodromic type).

This example illustrates the general morphological procedure which reproduces the conventional procedure in which similarity cyclomerisms are analyzed. To begin with, the manifold of basic geometrical configurations is obtained, for instance by computer graphics. These configurations

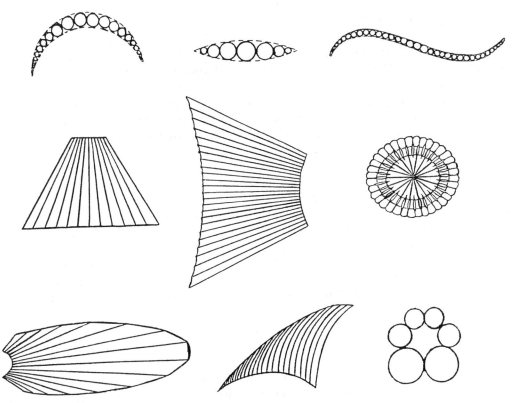

Fig. 2. Examples of Möbius, affine and projective cyclomeres.

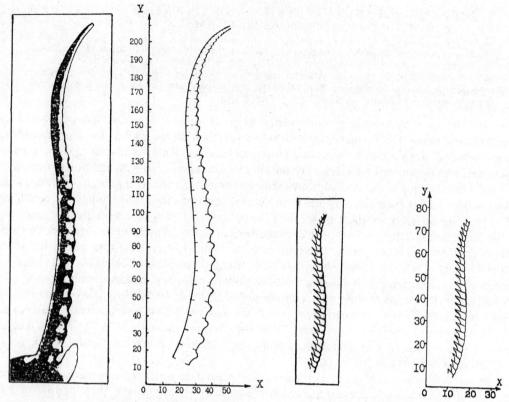

Fig. 3. A segmented horn of the *Pantholops hodgsoni*
orongo [16] and its model as a cyclomerism with a
Möbius generating transformation.

Fig. 4. A comb-like antenna of an insect and
its model as a Möbius cylcomere.

may be obtained by applying the iterative algorithm (1) to some motive unit, in particular, a point,
with g from the group G of, say, Möbius or projective transformations. These abstract
configurations and actual biological structures are compared. In establishing visual kinship the
coefficients of g for this biostructure are updated in the following procedure: the general analytical
form of transformation from G is written down (see the Appendix) and its coefficients are computed
by substituting in this general equation the coordinate of those associated points of motive
biological units which transform into one another by this transformation. Applying the transformation of g thus specified to the motive unit the required number of times the desired
cyclomerism is obtained which models the biostructure as a whole.

Non-Euclidean cyclomerisms are also observed in the antennas of insects (Fig. 4). Figure 5 shows
that the sequence of vertebral disks in the human dorsal vertebra is reproduced by Möbius
cyclomerisms, in contrast to the configuration of the spine of numerous animals, in particular
lizards, which is described by conventional similarity cyclomerism. By modifying the generating
transformation, and thus bending the "normal" configuration of the spinal cord cyclomerism,
models of morphogenetic anomalies of the spine can be obtained. This is consistent with our data
on Euclidean and non-Euclidean cyclomerisms, in man, animals and vegetation, both in normal
and pathological shapes. This also agrees with the formula of the morphology of plants that
ugliness is a version of the norm and that, as S. V. Meyen, a prominent Soviet researcher, wrote
in 1973,

> "The most important peculiarity of deviating forms from the structural viewpoint is in their approxi-
> mation, by means of the deviating character, to another (often closely related) taxon, where this character
> state is a normal one. This phenomenon was named by Krenke as 'the law of related deviations'." [18]

Möbius, as well as Euclidean, cyclomerisms are to be found in the configuration of bird feathers,
for in addition to straight or helical shapes there are lyre-shaped feathers, as in the case of
lyrebirds—so named because of the shape of their tails—or the tail of the Caucasian heathcock.

Möbius cyclomerisms of the lyre-shaped (or loxodromic) type are also observed in the structure of ambulacrum fields in sea urchins, discontinuous lateral generating lines in the helical shells of some molluscs, the long horn-like growth in the beak of the *Chasmorynchus niveus* bell-bird, sculptural lines in the body of the *Hippocampus guttulatus microstephamus* etc. Similarity cyclomerisms are most vivid in the shells of protozoa, but non-Euclidean cyclomerisms are also visible there (Fig. 6).

Both kinds of cyclomerisms are also observed in the structure and functioning of the vestibular organ and the eye muscles; the structure of the nervous system, bone tissue, vessels and muscles; positioning of biologically active points in the human body; cyclomeric buildings instinctively made by protozoa and social insects; paths of human and animal motions in normal and extraordinary conditions; formation of psychophysical delusions; the esthetics of proportions and shapes in architecture and art etc. They are manifest in unicellular as well as multicellular organisms; consequently, the cell is not a morphogenesis unit in a general case. Figures 7–14 show other cases of the variety of non-Euclidean cyclomerisms in living organisms.

Depending on the generating transformation, biological cyclomerisms may be explicitly discontinuous or practically continuous. Iterative algorithms are functionally useful in the generation of multicomponent biological structures because of the obvious convolution of information in the genetic encoding and morphogenetic implementation of such structures. In a living organism a broad range of cyclomerisms are usually present, in addition to morphological structures of a less regular nature. Furthermore, the cyclomeric structuring is reflected in the coloring series, weight parameters and numerous other characteristics of living organisms, as well as in the series of body units.

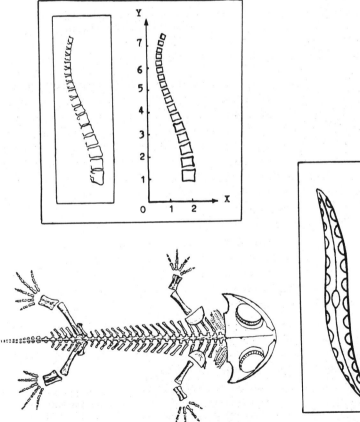

Fig. 5. Vertebra of the torso part of the human spine [17] and its model as a Möbius cyclomere. For comparison a similarity cylcomere in the structure of the torso part of the spine is shown [17].

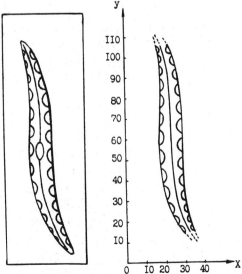

Fig. 6. The shell [19] and its model as a Möbius cyclomerism.

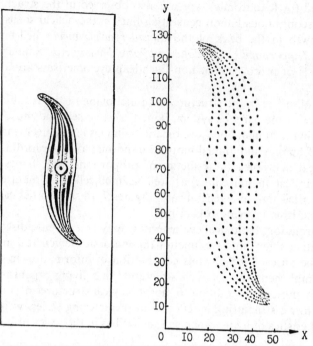

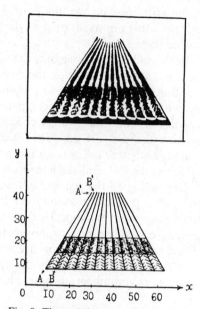

Fig. 7. Cyclomeric structures in unicellular organisms. *Left*: *Naviculla hippocampus* [17]. *Right*: a model cyclomeric structure whose lyre-shaped forming line is the orbit of a point in a Möbius loxodromic transformation; a sequence of the orbits makes a cyclomerism with other Möbius generating transformations.

Fig. 8. The multiblock structure of the scale of a bone fish [19] and its model as a cyclomerism with an affine generating transformation $X_{k+1} = X_k - 0.099\, Y_k + 5.24$; $Y_{k+1} = Y_k$. ABB′A′ is a specific cyclomerism.

The non-linearity of generating Möbius and projective transformations makes it possible for the size of individual cyclomerisms and their relations to vary in the same cyclomerism series; the geometrical invariants over the entire series are only those characteristics that are invariant with projective (or Möbius) transformations, e.g. the values of cross ratios or wurfs. Because Möbius

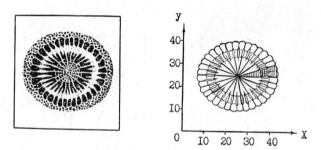

Fig. 9. A multiblock part of a four-ray *Tetracoralla* [19] and its geometrical model as a cyclomerism with a generating affine transformation $X_{k+1} = 0.94\, X_k - 0.4\, Y_k + 11.48$; $Y_{k+1} = 0.29\, X_k + 0.94\, Y_k - 5.82$; which transforms every radial segment (one of which is shaded for illustration) into its neighbor.

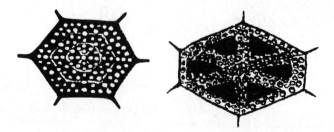

Fig. 10. Bone plates of the *Ostractones* [19] test with a configuration of affine cyclomerism.

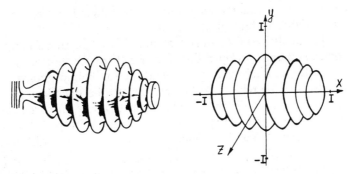

Fig. 11. An individual sexual capsule of *Eusertularia exserta* of the hydroid class [19] and its model as a cyclomerism with a projective generating transformation $X_{k+1} = (X_k + 0.258):(0.258\,X_k + 1)$; $Y_{k+1} = 0.966\,Y_k:(0.258\,X_k + 1)$; $Z_{k+1} = 0.966\,Z_k:(0.258\,X_k + 1)$.

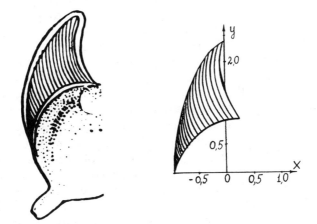

Fig. 12. The multiblock structure of the *Phylospadix scouleri* fruit [20] and its model as a cyclomerism with a projective generating transformation $X_{k+1} = -(X_k + 0.08):(0.08\,X_k + 1)$; $Y_{k+1} = 0.997\,Y_k:(0.08\,X_k + 1)$.

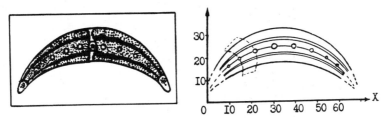

Fig. 13. The desmidean *Glosterium lebleinii* alga [20] and its geometrical model as a cyclomerism with a Möbius generating transformation of the hyperbolic type (the dashed square shows a cyclomerism variety).

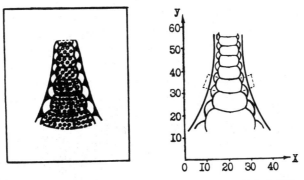

Fig. 14. A multiblock structure in the skeleton of the *Stichocapsinae radiolaria* [21] and its geometrical model as a cyclomerism with a Möbius generating transformation of the elliptical type (the dashed square shows a cyclomerism variety).

and projective transformations underlie the conformational and projective geometries, these abstract geometries materialize in biological structures.

Until recently the desirability and adequacy of non-Euclidean transformations in the morphology of living bodies remained an open question. Furthermore, numerous scientists believed that the Euclidean transformations were quite sufficient for biomorphology. In particular, J. Bernal, a well-known British researcher, and S. Carlyle used the Euclidean group of movements in their concept of generalized crystallography which was to cover, in particular, the symmetries of living bodies [22]. The complexity, importance and shortage of knowledge of the geometry of living bodies were summarized by Yu. A. Urmantsev, a Soviet student of biosymmetries, in 1971:

> "It would be no exaggeration to say that the geometrician who succeeds in developing the geometry not of the entire living natural but of a 'mere' flower will be worthy of rewards for both boldness and genius." [23].

The desire to use generalized geometric concepts in biology was felt as long ago as 1942, as in papers by D'Arcy W. Thompson [24], D. V. Nalivkin [25] and V. I. Vernadsky [10]. The research reported in this article has certain features which make it different. D'Arcy Thompson tried to find transformations which make one another into bodily shapes of various organisms such as perch and pike. He obtained, occasionally, very complicated curvilinear transformations which reflected, in particular, the relative autonomy of the morphological development of body organs; he did not use the Erlanger program as the basis in this comparative analysis. Nalivkin introduced curvilinear symmetries without reference to the Erlanger program; the specific rules that he proposed for construction of "symmetrical" transformations result in transformations which do not add up to groups of point transformations, because they disturb the group principle of the one-to-one correspondence of points and are not conventional in other fields of natural sciences. Vernadsky [10] did not make his views on the non-Euclidean geometry of living matter explicit and did not rely on the Erlanger program either. None of these papers considered special non-Euclidean iterative algorithms and the configurations they lead to with cyclic groups of non-Euclidean automorphisms. Unlike those papers, this article analyzes, above all, the rules of mutual conjugation of natural components in the individual biological body; this analysis proceeds along the lines of the Erlanger program, important for mathematical natural sciences, in terms of specific non-Euclidean groups of transformations which include the similarity subgroup; iterative algorithms which gave rise to discontinuous configurations with a cyclic (semi) group of non-Euclidean automorphisms are given special attention.

4. EUCLIDEAN AND NON-EUCLIDEAN ITERATIVE ALGORITHMS IN THE KINEMATICS OF BIOLOGICAL MOVEMENTS AND PHYSIOLOGICALLY NORMAL POSTURES

> "The movement is in many respects similar to an organ... seems to be a very useful idea, notably as far as a stable and universal movement such as locomotion is concerned." (N. A. Bernstein [26])

The above discussion dealt with the structure of static organic forms. But cyclomerisms also have a direct bearing on the kinematics of a broad range of biological movements which can, on numerous occasions, be interpreted as process in which cyclomerisms replace one another.

Growth transformations of numerous metamerically-structured biological bodies (larvae of insects, bodies of centipedes and worms etc.) in which metamery with one metameric step is taken over in certain ages by a metamery with another step is a trivial example of such transformations.

Less trivial cases are mutual replacements of cyclomerisms of different kinds, such as translational cyclomerism replaced by rotational cyclomerisms. This case is represented in Fig. 15 via the stages of colony development in phytomonadids (according to Ref. [27]), where, in addition to various rotational cyclomerisms, translational cyclomerisms are involved. In addition, the number of motive units changes with the age.

When non-Euclidean cyclomerisms are brought into the picture, a better insight is obtained into the relation of the development of organisms or their parts with the ability of the living matter to stay in different cyclomerism states. For example, in the development of the salamander fetus the sequence of somites which is described at an early stage by the similarity cyclomerism as a logarithmic helix changes into a lyre-shaped Möbius cyclomerism.

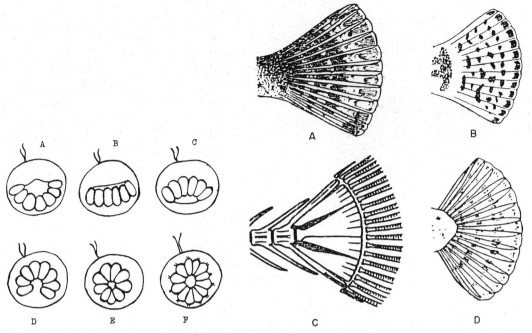

Fig. 15. Development of a phitomonadid colony [27]. A–E are excurvation stages in *Eudorina illinoisensis*.

Fig. 16. Caudal fins in *Scorpaena porcus* Linne (A), *Siniperca chua-tsi* (B), *Platessa platessa* (C) and *Rhombus maeoticus* (D) [28, 29] as cases of similarity cyclomerism.

In studying the biological cyclomerisms, the following theorem is useful. If a cyclomerism with a generating transformation g (such as similarity) is subjected to, say Möbius or projective, transformation S, the newly formed manifold with decomposition is again a cyclomerism with a generating transformation P:

$$P = S \cdot g \cdot S^{-1}. \tag{2}$$

By this theorem, with the generating transformation of the cyclomerism forerunner known, one can determine the generating transformation of the cyclomerism image which is obtained by an arbitrary transformation S. The form of the motive unit of the transformed cyclomerism is a transformed associated unit of the forerunner changed by S.

In analogy with the polymorphism of crystals whose lattices can, under certain conditions, be restructured with a change in their symmetry groups, the ability of living bodies to restructure their Euclidean and non-Euclidean cyclomerisms may be referred to as cyclomeric polymorphism. The concept of cyclomeric polymorphism sheds additional light on the fact that in individual development of multicomponent structures of the cyclomeric type, the transition from one cyclomerism to another proceeds as a relay race in the series of motive units (e.g. opening of cones as the scales ripen, of composite flowers such as daisies etc.) and may be referred to as the cyclomerism change wave. The natural morphogenetic movements in numerous multicomponent biological bodies such as the so-called excurvature in the development of the Volvox colony may be modeled as the simultaneous propagation of two or more cyclomeric change waves in the series with a fixed interval.

Euclidean and non-Euclidean cyclomerisms are involved in the apparatus of motoric movements in numerous animal organisms, in particular in the multineedle fish fins. Classical cases of similarity cyclomerisms in the tails of some fish are shown in Fig. 16; and cases of affine and projective cyclomerism in the tails of others, in Figs 17 and 18.

In the spirit of the above reasoning, numerous specifics of the structure of fins as multicomponent parts may be described and explained in terms of symmetric morphogenesis algorithms, leaving the locomotoric functions of these organs aside. Incidentally, the superficial but widespread view is that the fins are intended for swimming and that evolution made their structure optimal for this function and so the specifics of the structure may and must be derived from hydrodynamics. In

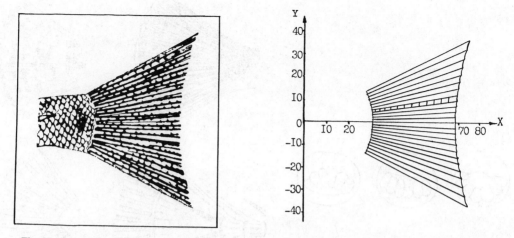

Fig. 17. The multiblock structure of the caudal fin in *Acanthurus* sp. [29] and its model as a cyclomere with a projective generating transformation $X_{k+1} = (0.998\ X_k - 0.087\ Y_k):(-0.0016\ Y_k + 1)$; $Y_{k+1} = (0.0347\ X_k + 0.9985\ Y_k):(-0.0061\ Y_k + 1)$ (one motive is shaded for illustration).

other words, knowledge of hydrodynamic equations is supposed to be sufficient for understanding the structure of fish fins. The above findings of group invariant analysis refute this view and draw attention to the multitude of fin functions, which include, in addition to the locomotoric function, the general biological function of contributing to the morphogenetic processes of inheriting the body form with algorithmical mutual conjugation of components.

This remark leads to a question, to what extent the structure of a living organism and the performance of some function in the environment may be taken by machine designers as a model to be imitated, bearing in mind the millions of years of evolution and natural selection. Studies of symmetry mechanisms in biological bodies reveal that far from everything in the structure of organs is dictated by optimal adaptation to specific functions in the environment by the laws of biological morphogenesis are very important. In effect, the machine designer can concentrate on optimal functioning in the environment and does not have to adapt to the needs of the internal requirements of the organism and purely biological requirements (volume growth in the individual development, inheritance of biological properties etc.). Still, the achievements of nature in the performance of functions must not be dismissed offhand. Nature itself has long solved numerous problems of functioning in the environment that designers face today. These solutions are not necessarily optimal in the usual sense but they are invaluable because they highlight the existence of challenges and demonstrate ways to tackle them; they stimulate human imagination and have acted as catalysts of technological progress in the entire course of human history.

Let us adopt a physiologically normal posture, a concept widely used in biology. In the entire set of postures or positions of its parts, such as the tail, the trunk etc., some positions of the body components *vis-à-vis* one another are inherited. They are instinctively adopted in a stereotype way in cases of fear, weariness, rest etc. They include the posture of rest of the starfish with symmetrical

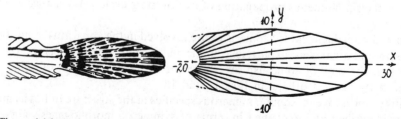

Fig. 18. The caudal fin in *Lumpenus lampetraeformis* [29] and its model as a cyclomerism with a projective generating transformation

$$X_{k+1} = \frac{0.83\ X_k + 1.27\ Y_k - 2.26}{-0.05\ Y_k + 1}; \qquad Y_{k+1} = \frac{-0.14\ X_k + 0.91\ Y_k - 3.26}{-0.05\ Y_k + 1}.$$

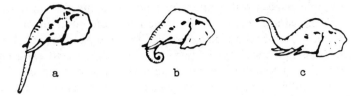

Fig. 19. Physiologically normal postures of the elephant trunk configured as similarity cyclomerisms (a, b) and a loxodromic Möbius cyclomerism (c).

positioning of the rays which are described as rotational cyclomerisms, the mutual positioning of components of the fin at rest, the metameric posture of a resting caterpillar etc.

Research into such in-born postures and positions of groups of mobile components ("segmentary" postures) of the support and motor apparatus is also important because they appear to be the references for construction of motoric movements and for the complex system of muscular drives with which genetically dictated characteristic conditioned reflexes are developed, the analyzers are coordinated etc. A better insight into their relations with the functioning of various system organisms such as neuromuscular, vestibular etc. will be useful in optimizing human postures and movements; reduction of muscular weariness in human operators; reduction of the detrimental effects of unfavorable factors such as vibrations and overloads; increasing the vestibular stability; choice of optimal postures in the cases of protracted immobility, for instance in treating broken extremities; forecasting and explaining human senso-motoric reactions under unusual conditions such as zero gravity in space missions; better coordination of space suits and exoskeletons with the specifics of the human support and motor apparatus; improving the movements of athletes; improving sporting equipment; and the development of senso-motoric systems in zoomorphous robots.

The author's research has revealed that the set of inherited postures includes, in addition to (segmentary) postures described by similarity cyclomerisms, postures which are described as non-Euclidean cyclomerisms, which significantly expands the data on the relationship of such postures with symmetrical morphogenesis algorithms. Besides, the kinematics of natural motoric movements (in analogy with the above morphogenetic movements) is often a transition from one cyclomerism type to another, or involves cyclomeric polymorphism.

In particular, the tubular *Glomeris romana*, whose segmented body is usually rectilinearly elongated and is metamerically configured is known to convolve into a segmented ring; this movement is described as the transition from one similarity cyclomerism to another.

Euclidean and non-Euclidean cyclomerisms describe the stereotype postures of the elephant's trunk, in which it is straight, coiled into a logarithmical helix or is lyre-shaped when the elephant blows it (Fig. 19). The lyre-shaped Möbius cyclomerism is characteristic of the attacking posture of the cobra whose body is obviously segmented and the configuration of the frightened rat's tail (Fig. 20). A kindred cyclomeric shape is also observed in stereotype postures of the multivertebrum necks of some birds such as flamingos and swans. The proboscis of many butterfly species, which

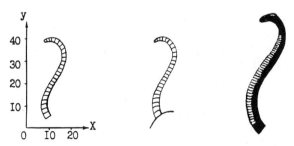

Fig. 20. Möbius cyclomeres in physiologically normal postures in animals. *Left*: a cyclomere with a Möbius generating transformation. *Middle*: posture of the tail in a frightened rat. *Right*: the posture of a cobra before attack (*photo*: S. I. Bortkevich).

is coiled into a logarithmic helix at rest, straightens when used and this movement is described in a natural way as the transition from one cyclomerism state to another. Similar symmetric algorithmical specifics of movements are also observed in some plants which are between animals and vegetation with respect to mobility. In particular, in response to exogenous irritation the branch of the sensitive mimosa, whose leaves are cyclomerically positioned, can fold its leaves along the branch but this new configuration of the sequence of leaves is also cyclomeric. The kinematics of various motoric movements such as walking and crawling in numerous animal organisms such as multipedea, caterpillars and ring-shaped worms can also be modeled as the transition from one cyclomerism to another, and waves of cyclomerism changes propagating along a multicomponent body.

The similarity of growth and motoric movements and their relationship with shaping in terms of cyclic groups of authomorphisms confirms that the organization of biological kinematics cannot generally be considered separately from the morphogenesis principles (which reflect the conditions in the internal environment of the organism and the self-structuring ability of this environment) or derived solely from the classical mechanics laws and the functional purpose of the movement, as some still try to do.

The specifics of self-conjugation of organic macrocomponents in the course of self-assembly in the Euclidean, Möbius and projective types of cyclomerism is worthy of special attention. As the literature noted long ago, supramolecular organic bodies are different from crystals, in particular, in the curvilinearity of their surface and in that their mutual conjugation into orderly ensembles does not generally occur by the crystallographic rule of the densest packing possible. In the biological self-assembly of individual bodies various reactive chemical groupings nonuniformly distributed over the surface of these bodies and generating surface energy are very important. The biological self-assembly is consistent with the well-known rule of stability in a minimal energy state, for the appropriate chemical groupings on conjugated surfaces react and release energy and the total surface energy of the system is less than the sum of the surface energies of individual organic surfaces.

What is important is that, by all accounts, a significant role in the mutual conjugation of curvilinear organic surfaces during self-assembly is played by surface lines having extreme mechanical and geometrical properties; it is along these lines that chemical groupings assemble (and entire ensembles of metabolic systems including groups of cells), which make possible the accretion of adjacent surfaces. To put it differently, biology, unlike conventional crystallography, may and must deal with the folding of conjugating adjacent curvilinear surfaces along special lines and zones which reflect, in particular, the properties of the surfaces' curvature and the tension distribution on the surface. For certain reasons surface curvature and asymptotic lines can be classified with the above special conjugation lines. Recall that a surface curve is referred to as an asymptotic line if its direction at every point coincides with that in which the normal curvature of the surface is equal to zero. A surface curve is referred to as a curvature line if its direction at every point coincides with that of the extremal value of normal curvature. Note that surface curvature (asymptotic) lines are invariants of Möbius (projective) transformations.

If mutual conjugation of adjacent organic surfaces proceeds along lines invariant with a certain group of transformations, then with a biological transformation (in particular, in the course of volume growth) of conjugated surfaces by a transformation from that group, new surfaces are obviously conjugated along the same types of surface lines and there is no need to move the chemical groupings which make the accretion possible to a significantly new relative position on every surface. This fact is attributable to the probable biological desirability of using certain groups of transformations in the mutual conjugation of numerous surfaces into one biological body along surface lines invariant with these transformations. From the proposed point of view, the case of Möbius transformations is associated with the positioning of surfaces with respect to one another along their curvature links; in the case of projective transformations, along asymptotic surface lines; in the case of similarity transformations, along curvature and asymptotic lines simultaneously (because the group of similarity transformations is an intersection of the groups of Möbius and projective transformations).

There is abundant evidence in favor of a morphogenetic significance of these lines. In particular, the surface curvature at a specific point is by no means significant morphogenetically. Curvature and asymptotic lines largely reflect the mechanics of hulls. For instance, in the case of thin (momentless) hulls for a broad range of surfaces (axial symmetrical surfaces with an axial symmetrical load) the surface curvature lines coincidence with the lines of their main tensions; in other words, curvature lines are identified by the extreme mechanical properties. In biology the latter fact justifies the positioning along these lines of the centers of chemical interaction between the organic surface and the environment, because the extreme mechanical tensions probably have an extreme impact on the opening of micropores; deformations of structural elements of chemical groupings on the surface; renaturation and denaturation of collagen molecules, morphogenetically important; biological rhythms in morphogenetic processes etc.

The author has developed a few models which represent the formation of Euclidean and non-Euclidean biological cyclomerisms with the assumption that replications of tissue layers are morphogenetically significant (valuable data on possible layer replication in various substances have been obtained by Weiss [30]).

5. THE KINEMATICS OF THE THREE-DIMENSIONAL GROWTH OF BIOLOGICAL BODIES WHOSE PARTS ARE NOT CYCLOMERICALLY POSITIONED

"Wisdom is rooted in watching with affection the way people grow." (Confucius, quoted from Ref. [31, p. 7]).

We have thus far considered the kinematics of three-dimensional biological growth with reference to cyclomeric bodies, in which it is described as the interchange of cyclomerisms (leaving out for a time the change in the number of motive units). Numerous biological bodies are, however, characterized by unorderly or complicated positioning of their parts and can be treated, at best, as ensembles of numerous cyclomerisms rather than a single cyclomerism. Nevertheless, these biological bodies are also capable of integral three-dimensional growth of geometrically consistent types.

The three-dimensional growth of living bodies is a challenging and mysterious case of the orderly cooperative behavior of numerous elements. This growth is essentially different from the surface growth of crystals, which occurs by accumulation of matter on the surface and does not involve the internal areas. The three-dimensional growth of living bodies entails change in the dimensions and frequently in the shape of the internal areas. The process is cooperative in the sense that by knowledge of the transformation of some points in the body figure, the transformation of the entire set of points in that figure may be determined.

This is the case for vegetation and animals (growth of leaves and flowers of some plants, larvae of insects, adult fish etc.) and is expressed in the proportional growth of all parts of the body; with the transformation of the position of three points in a body undergoing a known scaling transformation, the transformation of the entire set of points in the body can be determined.

The writer's research has revealed the existence of non-Euclidean kinds of three-dimensional growth not previously known, notably Möbius and affine. Let us take a closer look at the former.

If the three-dimensional growth is modeled using the concept of some active medium which grows through coordinated growth of every point-like zone of its volume, then the scaled three-dimensional growth occurs in this model if two conditions are observed: (1) scaling of every local zone of the body, or growth of every point-like zone with equal intensity in all directions (locally isotropic growth); (2) equality of the scales on which all local zones of the scaling body change. These local conditions may be regarded as postulates from which the specifics of scaling the "entire" body are derived.

But does nature not employ simpler types of growth changes which satisfy a simpler list of conditions, for instance containing only the first condition? Let us see what the corollaries would be of the local isotropic growth, a condition very natural physcially and one of the plainest local rules. Geometry states that with this type of local changes, in which every local zone is subjected to local transformations, the shape of the "entire" body may change dramatically from the layman's point of view because similarity "in the small" does not imply similarity "on the whole" in the general case (which is under consideration), where similarity transformations are not identical in different local zones. Transformations under which every local zone of the body

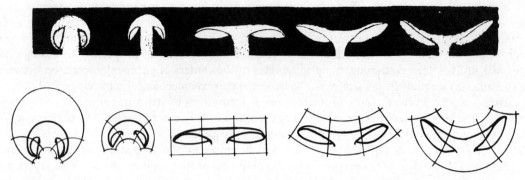

Fig. 21. Growth transformations in the cap of the fly-agaric *Amanita* and their modeling as Möbius transformations.

undergoes similarity transformations are referred to as conformal. (It was in terms of local similarity that Gauss developed a theory of conformal transformations.) For the case of a three-dimensional space, to which this discussion will be confined, all conformal transformations are known to be covered by the group of Möbius transformations.

Let us proceed to specific examples of adequate modeling of the three-dimensional growth of biological bodies in terms of Möbius transformations. As in the case of cyclomerism, Euclidean and non-Euclidean symmetrical transformations are carried out in analogous biological bodies. Thus, the caps of numerous kinds of mushrooms, such as *Lactarius vellereus* and *Morchella culenta*, scale up in their individual growth. On the other hand, in some mushrooms the cap transformation over an extensive period is described as Möbius. This case is illustrated in Fig. 21 for the fly-agaric Amanita which changes the cap shape. Similar changes in shape also occur in the caps of field mushrooms.

Möbius transformations are applicable also to pathological deformaties as well as normal growth. This also confirms that in normal and pathological self-formation of biological bodies kindred mechanisms and symmetry rules are involved.

Scaled and Möbius kinds of three-dimensional growth are also observed in the individual development of composite flowers, fruits etc. Because three-dimensional scaling also occurs in animals, we tried to describe nonlinear transformations occurring in the human skull with aging as Möbius (Fig. 22) and succeeded in obtaining a first approximation.

The above results of group invariant analysis legitimize the use of new tools of morphometric analysis, invariants of conformal and projective geometries, such as the wurfs mentioned above, in the research of biological structures. (Note that the founder of today's projective geometry J. Desargues, French architect and engineer (1593–1662), widely employed biological terminology, probably assuming a kinship between the projective structures and nature.) Some of these tools may be used in the construction of geometrical models of numerous organic bodies from multistaged symmetrical units, for the body is made of symmetrical (in the sense of, say, Möbius symmetry) blocks of the first order combined into symmetrical blocks of the second order etc. In particular, this principle manifests itself in the kinematics of the human body, where the mirror symmetry of the two halves of the body, which act as second-order blocks, is supplemented by an approximate Möbius† symmetry of the longitudinal proportions of the three-component kinematic blocks which comprise this structure, viz. the phalanxes of fingers, the three-membered extremities (shoulder–forearm–wrist and hip–shin–foot) and the three-membered body (in anthropology the body is subdivided into the upper, torso and lower parts). This is the conclusion obtained from the computation of the wurfs of these three-membered blocks from the anthropometric data of

---

†In the unidimensional case, the groups of Möbius and projective transformations coincide and so in this example the symmetry can of either kind and the concept of a wurf or a cross-ratio, which is an invariant of projective transformations, can be used [2].

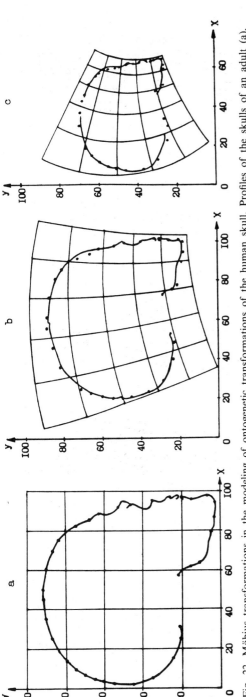

Fig. 22. Möbius transformations in the modeling of ontogenetic transformations of the human skull. Profiles of the skulls of an adult (a), a 5-year-old (b) and a newborn (c), taken from Ref. [32].

Table 1. The wurfs W of the human extremities and torso at different ages

	Shoulder A_1B_1	Forearm B_1C_1	Wrist C_1D_1	W	Hip A_2B_2	Shin B_2C_2	Foot C_2D_2	W	Upper part	Torso	Lower part	W
Embryo age (lunar months)												
4	2.37	1.90	1.55	1.33	2.76	2.66	1.84	1.26				
6	4.40	3.65	3.23	1.34	5.45	5.45	4.30	1.27				
8	6.17	5.05	4.53	1.35	7.40	7.60	5.80	1.27				
9	6.95	5.70	5.14	1.35	8.40	8.60	6.61	1.27				
Newborn	8.80	7.40	6.30	1.33	10.90	9.90	7.80	1.30	10.6	18.3	28.6	1.29
Age (years)												
1	12.5	10.4	8.7	1.33	15.8	14.4	11.5	1.30	15.6	25.0	41.7	1.32
4	16.7	13.7	11.2	1.33	22.4	20.2	15.8	1.30	18.9	31.4	58.4	1.32
7	20.7	16.5	13.1	1.33	28.0	24.8	18.9	1.30	21.0	35.3	71.7	1.33
10	23.2	18.4	14.4	1.32	32.1	27.8	20.7	1.30	22.5	38.0	80.6	1.34
13	26.3	20.7	16.0	1.32	37.7	32.2	23.6	1.30	24.5	42.2	93.5	1.34
17	30.5	23.9	18.3	1.32	44.4	37.3	26.9	1.29	26.9	47.8	108.6	1.33
20	32.3	24.5	18.8	1.33	45.4	37.5	27.0	1.30	25.3	51.8	109.9	1.29
Average and boundaries of ontogenetic fluctuations of W (%)	$1.33 \pm 1.5\% = 1.01P \pm 1.5\%$				$1.29 \pm 2\% = 0.98P \pm 2\%$				$1.32 \pm 2\% = 1.01P \pm 2\%$			

$P = 1.309\ldots$—golden wurf

The lengths (in cm) are taken from the work of Bunak [33].

Table 2. Wurfs W of the middle finger in man at different ages

Age (years)	AB	BC	CD	W
4	2.42	1.43	0.86	1.31
6	2.64	1.65	1.02	1.31
8	3.00	1.88	1.19	1.31
10	3.10	1.96	1.25	1.31
12	3.34	2.13	1.37	1.31
14	3.56	2.27	1.46	1.31
16	4.08	2.57	1.64	1.31
18	4.19	2.65	1.69	1.31
21	4.41	2.78	1.76	1.31

The lengths of phalanxes (in cm) are taken from the work of Rokhlin [34]. AB—the basal phalanx; BC—the medium phalanx; CD—the end phalanx.

V. V. Bunak [33] and D. G. Rokhlin [34]:

$$W = \frac{(C - A)(D - B)}{(C - B)(D - A)},$$
(3)

the expressions in parentheses being the lengths between the end dividing points A, B, C and D of the three members of every block. The values of these wurfs in all the blocks, at least during the entire individual post-natal development, group around the benchmark of $P \approx 1.31$ (see Tables 1 and 2). This is especially interesting because the growth of the human body is essentially nonlinear (Fig. 23); for instance, the upper part grows 2.4-fold, the torso 2.8-fold and the lower part 3.8-fold.

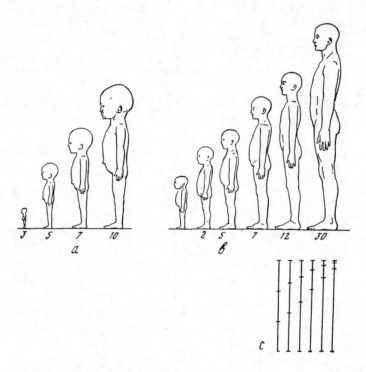

Fig. 23. Changes of the human body with age (according to Ref. [32]). (a, b) Antenatal and postnatal stages—(a) in lunar months (b) in years—the first on the left is a newborn. (c) Three-stretch parts whose wurfs are equal to 1.31.

A non-Euclidean analysis of the structure and growth reveals that all the three-membered blocks of the human kinematics in the straighted posture are practically Möbius equivalent and Möbius invariable during the lifetime. This is largely true of dwarfs and giants and a broad range of highly organized animals as well as normal human subjects. These findings are discussed in more detail elsewhere [35], where (see also the Appendix) it is shown that the value of wurfs, 1.31, in the body blocks is a function of the biological phyllotaxis laws of shaping; it coincides with the so-called golden wurf,

$$P = \frac{3 + \sqrt{5}}{2} = \frac{\Phi^2}{2} = 1.309\ldots,$$

conjugated with the Fibonacci numbers and the golden section,

$$\Phi = \frac{1 + \sqrt{5}}{2} = 1.618\ldots.$$

6. AUTONOMOUS AUTOMATA AND GROUP INVARIANT PROPERTIES OF LIVING ORGANISMS

"The basic feature of every new scientific idea is that it relates in some way two different series of facts." (M. Planck [36])

One of the promising ways to model the phenomena of biological shaping is application of the theory of growing automata to the modeling of morphogeneses [37]. This control engineering modeling is legitimized by the fact that in biological shaping a feedback control system is involved which is distributed throughout the developing body. Cellular automata, uniform arrays devised by von Neumann [38] and his followers, and the ideas of Lindenmayer's parallel grammars [39] are widely used for this purpose. In the light of the data presented in the preceding sections on the important morphological value of iterative algorithms, a sophisticated modeling approach can be developed where the concept of an autonomously growing automaton (or networks of such automata) would be useful.

Recall that a finite automaton is a dynamic system whose behavior at specified times (clock times), $1, 2, \ldots, p$, is described by the equation

$$x(p) = f[x(p-1), u(p-1)], \tag{4}$$

where $x(p)$ and $u(p)$ are variables which take on values from specified finite alphabets: $x(p)$ representing the internal state of the automaton at time p; $u(p-1)$, the state of the automaton input at the preceding time, which reflects the impact of the "environment" on the automaton. A special case is an automaton whose behavior does not depend on the environment at all. Such an automaton is referred to as autonomous; for such an automaton a change of state is obviously dictated (in a way similar to algorithms of biological cyclomerisms) by the iterative algorithm

$$x(p) = f[x(p-1)]. \tag{5}$$

But biological cyclomerisms are observed in relatively autonomous subsystems of the body such as horns, fins etc. In addition, the concept of autonomy is generally closely related with the genetic inheritance of shapes characteristic of the biological species independently of the environment. The relative autonomy of the subsystems in an organism is a prerequisite of its efficient functioning. All this confirms that our approach is correct in interpreting the biological cyclomerisms and their ensembles in terms of autonomous growing automata [equation (5)]. Other authors used growing automata, assuming timed control of the automaton state from outside while the morphogenetic significance and potential of autonomous automata were neglected. One should not overlook the well-known noise immunity of iterative algorithms which may make them especially desirable in living organisms.

Such modeling can be implemented, for instance, in a cellular automaton—which is usually a uniform array of numerous identical cells, each cell having several possible states and interacting only with a few neighboring cells. The idea of such an automaton is nearly as old as that of electronic computers. The research in this field was pioneered in the early 1950s by von Neumann. The "Life" game, devised in 1970 by J. Convay, and capable of simulating numerous aspects of biological development is the most widely known cellular automaton. Numerous problems in cellular automata are covered by the so-called "information mechanics". What is important is that von Neumann's automata include cells which are placed in the cells of the Cartesian network and this "Euclidean" disposition of cells, introduced from outside, is taken up in later papers. Breaking with this tradition, I used the data on the biological significance of non-Euclidean cyclomerisms to justify the use of cellular automata also based on networks with cyclic groups of non-Euclidean automorphisms in biological modeling. In the "non-Euclidean" cellular automata the timed change of state of the autonomous automaton is treated as attachment of a new motive unit of a non-Euclidean cyclomerism.

The research revealed new promising lines for the development of the theory of growing automata to be applied to biological morphogenesis. Firstly, the specific features of biological morphogenesis make it possible to develop a theory of autonomous automata which are hierarchically embedded into one another and functioning at every level with different clock time periods which are or are not multiples of those of the preceding or succeeding levels. The case of nonmultiple periods fits well the biological situation of forming, for instance, leave organs on a cylcomeric plant stalk, or structures which develop at a later stage relatively autonomously by their shaping algorithms.

Secondly, the idea of a succession of autonomy conditions f in equation (5) proves useful in interpreting the series of biological data on the dependence of the shape of cylcomeric structures on the specifics of the environment. Indeed, in a great number of cases a change of the habitation range for a species entails a change in their cyclomerisms, in that the kind of cyclomeric change which is easily interpretable in terms of changes of the autonomy conditions. This approach is equally applicable to other phenomena of cyclomeric polymorphism.

Thirdly, in addition to autonomous automata, the concept of a quasiautonomous automaton is useful in analysis of the body of morphological data, such an automaton is described by equation (4) with $u(p)$ making either little impact or only changing the generating transformation coefficients f rather than shape. In addition, variations of $u(p)$ with clock times may be stochastic as well as deterministic. This approach easily leads to the description of biological structures whose parts are interrelated in a random and irregular way. The resultant data on the biological significance of iterative algorithms force the writer to a conviction that for mathematical modeling of a broad range of biological development phenomena formalisms of discrete (or finits) mathematics are more adequate.

Furthermore, the idea arises that numerous organism subsystems function in the norm and in pathology as autonomous automata and their ensembles may prove important in medicine and biology and suggest treatment methods, professional selection and training procedures, ways to optimize the working conditions for operators of complex machinery etc.

The laws of interaction between parts in biostructures may be related with the concept of the biological space. In formal usage of physical terminology, the world of a biological body is a manifold of events where a system exists of events making an impact on other events which dictate its space structure. In that world the physical understanding of space and space–time is applicable, formulated by A. D. Aleksandrov (Member of the U.S.S.R. Academy of Science) in 1965,

> "Space–time is a set of various events in the world taken apart from all its properties other than those that are determined by the structure of the system of relations made by impacts of events on one another."
> [40, p. 55]

In our view, the whole issue of general interaction laws in the biological space must be treated as laws of interaction in the spaces of self-organizing ensembles of self-organizing growing automata. The geometrization of biology seems to be contingent on problem-oriented development of the theory of finite automata. In light of the preceding discussion, the following definition may prove useful: a biological space is space of self-organizing hierarchical ensembles of self-organizing growing aperiodic (in the general case) automata operating with different clock time periods. For brevity this ensemble may be referred to as a hyperautomaton.

In this formulation aperiodic implies rhythmically developing, in time, biological processes whose time structuralization may prove to be controlled by nontrivial iterative algorithms which will be treated in the following section.

7. APERIODIC ITERATIVE BIORHYTHMS

> "Whenever you have to deal with a structure endowed entity E try to determine its group of automorphisms, you can expect to gain a deep insight into the constitution of E in this way. (H. Weyl [3, p. 158])

Researchers in various countries have, for a long time, been studying time biorhythms and this field of natural sciences can claim its own traditions, terminology and challenging findings. Still, it has concentrated attention on periodic rhythms of physiological processes such as breathing and walking, repeating processes occurring simultaneously with periodic diurnal and seasonal changes etc. The reader of the literature on biorhythms may think that no biorhythms other than periodic are significant or possible. In point of fact, however, the range of biologically significant rhythms is broader and the periodic ones is a particular, albeit important, case.

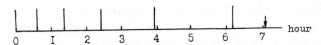

Fig. 24. The evacuation rhythms in *Arenicola marina* [42] described as time cyclomeres with a Möbius generating transformation $t_{k+1} = (1.292\,t_k + 0.593):(0.020\,t_k + 1)$. The arrow indicates the time of high tide when the observation terminates.

The periodic rhythms of any process may be interpreted as structuring by a transitive iterative algorithm (1) whose generating transformation is a parallel shift along the time axis by the period.

But are less trivial iterative algorithms, whose time structuring is dictated by more complicated generating transformations, not implemented in biological processes? This very important question, which is answered in the positive, draws attention to specific biorhythms and regularities in time structuring that are usually neglected but seem to be worthy of most serious attention and capable of significantly enriching the science of biological rhythms.

An illustrative example of an iterative algorithm whose generating transformation in scale similarity is the well-known moulting of crustacea [41]. The time between two moults is seen to increase monotonically with a constant scale during the entire lifetime over which moults occur. Not only is the total time between moults scaled but also, and with the same factors, the duration of every stage in the preparation for a moult. In addition, every moult is followed by a proportional increase in the size and mass of the organism. In other words, this biological transformation is a case of a remarkable organization of the most complicated biological processes in space–time which demonstrates the cyclomeric properties and adds legitimacy to the search for symmetrical structuring algorithms in time biorhythms that would be similar to such algorithms in biological series in space.

Aperiodic iterative algorithms with a Möbius (or projective, which is the same in a unidimensional case) generating transformation

$$t_{k+1} = \frac{1.292t_k + 0.593}{-0.02t_k + 1} \tag{6}$$

are demonstrated by, for instance, the *Arenicola marina*'s rhythmic evacuation (Fig. 24), which follows from our analysis of experimental data reported by P. R. Evans [42]. In laboratories these worm-like organisms demonstrate an automatic periodic rhythm of excretion with an interval of about 40 min. In natural conditions these creatures observe a quite different rhythm in which the intervals grow nonlinearly and follow algorithm (6). These organisms were observed in their natural milieu at low tide. The moment they are engulfed by high tide is shown in Fig. 24 by a vertical arrow.

Nontrivial Euclidean and non-Euclidean iterative algorithms are obviously at work in other rhythmic processes as well, such as the volley-like flapping of an excited *Bonasa umbellus* heath cock, some kinds of heart arhythmias etc. The analysis of non-Euclidean symmetries in biological processes such as complex structured communication signals between living organisms, rhythmical change of organism state parameters in normal and pathological functioning etc. must undoubtedly be continued and systematized. The search for such symmetries in morphogenetic processes such as moulting is especially important and challenging for general biology. The modeling of such processes could rely on the theory of aperiodic automata.

8. THE PROTOSYSTEM OF BIOLOGICAL REGULATION AND PSYCHOPHYSICAL MATTERS

"To take possession of space is the first gesture of the living, men and beasts, plants and clouds, the fundamental manifestation of equilibrium and permanence. The first proof of existence is to occupy space." (Le Corbusier [43, p. 30])
"I'm looking with tactile eyes. I'm feeling with a seeing hand." (J. W. Goethe)

The recognition of non-Euclidean symmetries in algorithmical mutual conjugation of parts of organic forms has dramatically extended the range of biological objects, the spatial behavior of whose parts may be quite legitimately and stringently described in geometrical terms. The existence in most diverse biological bodies of the same geometrical types of cooperative structure and behavior is seen as a major feature of biological evolution and suggests the existence and general biological significance of a morphogenetic regulation system which is responsible for this coordinated behavior of body parts in the growth and development of the organism.

This morphogenetic system of integrating regulation seems to exist concurrently with the nervous and humoral systems which also perform certain functions in integrating parts of the organism. The system seems to be a protosystem, above all in that it emerged earlier in the evolution (for the ability to grow is the most ancient property of living matter; morphogenesis and growth of geometrically regular and nontrivial forms is observed in protozoa, in particular, in unicellular organisms which have no nervous system). In other words, other systems are immersed in this one by their structural formations and also emerge in philo- and ontogenesis against its background.

Even in highly sophisticated organisms one can see at certain stages the secondary nature of incorporating the nervous system into the regulation of the behavior of organs and tissues in the course of individual development; before the essential elements of the nervous system emerge, parts of the organism act vigorously in a coordinated way similar to morphogenetic movements. For illustration let us recall two well-known biological facts [44]. (1) The first organ to move in a 4 mm long 3-week-old fetus is the heart; it starts beating for internal reasons because at that stage the heart has no nervous links; the heart contracts when there is no blood to pump. (2) The larvae of the *Tantogolabrus* fish on the first day of free swimming in a body of water do not respond to exogenous irritants or move without an effectual receptor system. The sensory system develops gradually in them to master the primary motion system for biologically efficient coordination of its in-born activity with the nature of exogenous irritation.

Note also that the three-dimensional growth, which is heavily dependent on the cyclomeric (or cyclogenetic) properties of the living matter, integrates the numerous parts of the body into a growing ensemble which acts to prevent the body transformations which occur with the aging from invalidating all the senso-motoric habits that were acquired at preceding ontogenesis stages when the body was "different". In this context the geometrical properties of three-dimensional growth and the biochemical media which make this growth possible have, in my view, a direct bearing on the formation in man of an inborn idea of the structure of his body, an idea which has for a long time been regarded as a major element in the spatial perception and coordination of movements. The morphogenetic regulation system which largely functions by cyclogenesis performs, among other things, the function of coordinated adjustment of numerous muscular, joint and other proprioceptors which contribute to man's awareness of the structure of his body. The morphogenetic regulation system which looks after three-dimensional growth acts for diverse proprioceptors and muscles as a distributed cooperation enforcing unit, a tuning fork. In other words, the various elements of the senso-motoric system act in unison, not only by virtue of their direct mutual links but also because they are immersed in an orderly growing environment which influences their operation and aids the brain in composing generalizing adequate images out of innumerable reports from the set of receptors. Without studying the morphogenetic regulation protosystem and its general biological significance one cannot understand to the full the more recent organism regulation systems, in particular the nervous system.

Now let us consider non-Euclidean symmetries in psychophysical phenomena. The idea of a close linkage between the specifics of spatial perception and the morphogenesis principles is deeply rooted. Since the times of Kepler the esthetic quality of the golden section has been related in the literature with its morphogenetic materialization in living bodies. Today's psychology has had much experience in using the higher geometry in similating the observed psychological phenomena.

Back in 1868, H. Helmholtz [45] tried to apply a certain mathematical concept related with the concept of groups of transformations to the psychology of perception. The desire to obtain an insight into the psychological phenomena in perception from the basic viewpoint of the concept of a group led H. Poincaré [46] to an interesting concept on the genesis of spatial ideas in an individual, and on the interrelationships between physical, logical and physiological fundamentals of geometry. Poincaré developed and consolidated this idea until he died. He believed that a group should be accepted as a key concept in the scientific cognition of the genesis of spatial ideas in an individual. He thought that the concept of a group, at least potentially, was in-born and existed before any individual experience.

With time the mathematical concept of a group won new ground in psychological research, notably in the so-called perception constance phenomena. In his enlightening paper, E. Cassirer [5] indicated that the analogy between the psychology of perception and geometry is, above all, in that for the perceived patterns (shapes) the specific properties are preserved even when their elements change. Thus, a musical tune does not change much if all its notes are shifted. This is also true of visual perception by dimensions, shape, color etc.

> "Indeed, what else is that 'identity' of the perceptual form but what, in a much higher degree of precision, we found to subsist in the domain of geometrical concepts? What we find in both cases are invariances with respect to variations undergone by the primitive elements out of which a form is constructed. The peculiar kind of 'identity' that is attributed to apparently altogether heterogeneous figures by virtue of their being transformable into one another by means of certain operations defining a group, is thus seen to exist also in the domain of perception." [5, p. 25].

Klein's Erlanger program suggests essentially that not every interpretation of a three-dimensional object is its geometrical characteristic. If we take an object and recognize only its individuality, its geometrical nature and significance would not manifest itself. According to Cassirer [5], a description of a spatial shape as such individually yields, at best, its geographical, or "topographical" rather than geometrical characteristic. To obtain the latter, a quite different approach is needed which was formulated by Klein as follows: the geometrical properties of any figures must be described by formulas which remain true when the system of coordinates changes; conversely, any formula which is in this sense invariant with the group of available transformations of coordinates represents a geometrical property. In compliance with this concept the properties which describe an object should not be defined for elements which add up to this object. This should only be done in the framework of the group with which the object is related. As soon as a group is replaced, quite new relationships emerge. What expressed "the same" geometrical concept may obtain a significance of its own whereas what appeared to be something quite different may prove to be identical in the new geometry. Thus, in terms of the metrical Euclidean geometry various conic sections (circumference, ellips, parabola and hyperbola) act as independent geometrical patterns having their own specific and distinct properties. This difference disappears when the same conic sections are regarded in terms of projective geometry because projective transformations may translate all the above conic sections into one another and thus deprive them of independent existence. In effect, the concepts of today's geometry acquire their inherent accuracy and become truly universal only insofar as the specific figures, suggested by intuition, are not regarded as predestined or rigid but rather as raw material which needs processing if it is to acquire a geometrical status. The mathematical definitiveness is not in the elements which the mathematicians treat as given quantities but in the rule whereby these elements are interrelated. Ever since the Erlanger program, nobody has asked whether two quadrangles are identical or whether two stretches are equally long without reference to a specific group of transformations within which the question must be answered. In other words, the definitiveness of a figure depends on the context where it is integrated and whose particular case it is.

Bertrand Russell who worked on geometrization of psychology wrote:

> "Il faut construire un pont en commençant à la fois par ses deux extrémités: c'est-à-dire, d'une part, en rapprochant les assomptions de la physique des données psychologiques et, de l'autre, en manipulant us données psychologiques de manieur à édifier des constructions logiques satisfaisant de plus prés aux axiomes de la géométrie physique." [47, p. VIII].

Are symmetrical principles of biomorphogenesis not used in the psychophysics of spatial perception? Does the organism not tend to perceive the world in terms of patterns of the same groups of transformations in compliance with whose principles it is shaped? Are the structuring principles that materialize in morphogenetic phenomena and psychological phenomena of spatial perception not akin?

These questions are especially natural because Helmholtz, Sechenov and Poincaré explored the idea of the leading role of the kinetic structure of the body (which is now known to be structured largely by morphogenetic principles with Möbius and other non-Euclidean symmetries) in the genesis of spatial concepts in an individual. In particular, this idea was the cornerstone of Poincaré's teaching on the physiological fundamentals of geometry and on the origin of spatial ideas in an individual [46]. (At this point it is important to remind the reader that the argument advanced in this paper on the essential importance of non-Euclidean geometry for this range of biological issues was not to be found in the writings of those authors.)

According to Poincaré, the very notion of space and geometry originates in an individual because of the activities of the body kinematics which incorporates the internal receptors of the attitudes and movements of body parts with respect to one another; i.e. there is in the kinematic arrangement something which precedes the notion of space. The specifics of the entire apparatus behind the kinematic activities of the body has been made consistent with the realities of the world by evolution. Consequently, every newborn organism masters adequate spatial notions not only through personal contact in the course of ontogenesis with objects of the environment, but also thanks to what preceding generations acquired and was consolidated in the course of phylogenesis in the body movement apparatus.

Poincaré believed that the body movements had a leading role to play in the genesis of the space concept. For an immobile being there would have been no space or geometry. If we had no measuring tool, Poincaré insisted, we could not build space; but we have and we use it instinctively—it is our own body.

> "The system of coordinate axes to which we refer all external objects is a system of axes invariably references to our body which we carry wherever we go." [46]

However, Poincaré and numerous other authors did not practically ask an important question, is it Euclidean or non-Euclidean groups of transformations and the related reference systems that are characteristics of the movement and space perception in man and animals? What we argue is that groups of non-Euclidean transformations and cyclogenetic principles contribute to the genesis of space ideas in the individual. The following discussion will advance more evidence in favor of this argument.

As far as Poincaré's ideas on a human body reference system of coordinates are concerned, all the elements of the human body change their Euclidean lengths as they grow and at the fastest rate in infancy when the organism finds its way in the world and when stable references of length are especially important. Although being deprived of a reference of the Euclidean measure, the body has invariably during its lifetime and in the life of the entire species references of non-Euclidean wurfs (Section 5), which are quite good as tools for geometric comparison and construction of coordinates.

Speaking of Möbius biosymmetries of kinematic blocks in the human body and the Poincaré theory, we have to recall the date of today's physiology on kinesthesia which are found to be closely related with those separating points whose positioning in the body is found to be Möbius. The kinesthesia mechanisms supply knowledge on the attitude and the movement of body elements, Poincaré believed that these ideas stemmed from the "muscular sensation". A modern physiological handbook states, however, that the sense of position and of movements of the joints is provided solely by appropriate receptors in the joints and there is no need in the enigmatic "muscular" sensation to explain the kinesthetic sensations [48].

Today's physiology also supplements Poincaré's teaching on the interrelationship of the body and spatial sensations by asserting that the ideas the individual has on his bodily shape are in-born. This assertion stems from studies of the so-called phantom sensations in invalids who feel the presence of a lost part of the body, not only in those whose extremities were amputated but also in those who were born without extremities. Consequently, the individual's idea of the structure of his body is in-born rather than dictated by his experience. This seems to be another argument in favor of the morphogenetic mechanisms contributing to the formation of spatial feelings in an individual.

The significance of the morphogenetic structures for spatial perception and active ranking of the environment by the organism is demonstrated by constructions made by insects and other organisms. Usually instinctive and following certain standards of shape, this activity is an enigma and a scientific problem. In numerous cases these construction works of individual organisms and joint efforts of numerous individuals appear to embody Euclidean cyclomeres. Zoopsychology (the stigmetry theory) knows that the construction shape is an active factor in instinctive construction activities and manages the work by itself.

The morphofunctional structure of the sensory systems in an organism embodies nontrivial morphogenetic symmetries and algorithms. Thus the human cochlea consists of three patterns, the ratios of whose lengths make the golden section Φ ($1:\Phi:\Phi^2$) and yield the golden wurf, 1.31 (Fig. 25), of Section 5, where other morphogenetic shapes, unrelated to the perception of sound were discussed. The cochlea shape is usually represented schematically as a logarithmic helix (which reproduces the classical similarity cyclomere) inside which lies Corti's organ of a kindred shape. In some animals such as cats, however, Corti's organ has a shape which is significantly different at the broad end of the cochlea from the logarithmic helix and tends to curl lyre-like in the inverse direction, and is described on the whole as a loxodromic Möbius cyclomerism.

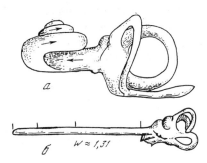

Fig. 25. The helical structure of the human ear cochlea and the golden wurf in it [49]. (a) Cochlea, coiled into a helix; (b) cochlea uncoiled into a straight line.

The non-Euclidean geometry is also applied to the description of visual perception. This kind of research was pioneered by R. Luneburg [50] who proved that the space of visual perception is described by the Lobachevsky geometry. These findings were followed by scores of papers in various countries where the idea of a non-Euclidean space of visual perception was extended and refined. A book by B. V. Rauschenbach [51], a Soviet researcher, is worthy of special attention. The Luneburg approach was thoroughly tested by G. Kienle [52] who arrived at essential conclusions on the importance of the Möbius geometry in structuring the space of visual perception.

In the main series of his experiments, where about 200 observers were involved, Kienle obtained about 1300 visual patterns of various kinds. In addition, 11 additional series yielded over 2000 measurements. The observers made judgments on the straightness of curves; the equality of angles, distances, segments and areas; the parallelism of straight lines; the existence of a common center of a family of circumferences etc. The experiments confirmed that the space of visual perception is described by the Lobachevsky (or hyperbolic) geometry; Kienle came to the conclusion that the well-known conformal (or Möbius) model developed by Poincaré was an adequate model of that geometry. He concluded his paper by writing:

"Poincaré's model of hyperbolic space, applied for the first time for a mapping of the visual space, shows a reasonably good agreement with experimental results." [52, p. 400].

Consequently, Möbius transformations, more specifically a subgroup which is associated with the Lobachevsky geometry, prove to be the prime factor in structuring the space of visual perception. Why it is that this particular group is so important remains an open question. Other non-Euclidean geometries may also be important for other kinds of spatial perception and for visual perception under conditions different from those used by Luneburg and Kienle.

In this light it is especially enlightening that Möbius symmetries and iterative algorithms are also to be found in microscopic movements of the eye apple, without which normal visual perception is known to be impossible. Ref. [53, p. 195] reports the path of a point moving along the eye retina against a background of microscopic apple tremor with respect to the retina receptors. The diameter of the central retina area is a mere 0.05 mm, the tremor frequency being up to 150 cycles/s and the amplitude approximately equal to half the diameter of a conecell. The nature of the image tremor is regular in that all five wave-shaped retina image drift lines with tremor show either 27 or 42 saw-like periods. Every tooth of the saw-like path is commensurable with neighboring salients and approximately orthogonal to the curvilinear draft path; in other words, the saw-like periods are integrated into a regular process. Our research into the structure of these curvilinear paths with a high frequency tremor has demonstrated that they are satisfactorily described as Möbius cyclomeres. The iterative algorithmical shape of the microtremor may be thought to be dictated by some autonomous automaton controlling the eye tremor.

Sensor perception, as in the above cases of biological kinematics, is largely structurally related with morphogenesis and may be treated as a structural extension of morphogenesis rather than something entirely new that nature devised at some evolutionary stage. In "mastering space" the organism seems to use the same structuring principles and algorithms, be it morphogenetic, kinematic or psychophysical mastering. This unity of structuring principles and algorithms for various systems and sophistication levels being coordinated and adding up to a living organism which "masters space" offers important advantages to living organisms. The writer is confident that the biological significance of iterative algorithms and non-Euclidean symmetry will be detected in the most diverse biological fields, in particular, in the encoding, notably in the nervous system, of biological information.

9. CONCLUSIONS

For biology and the mathematical modeling of biological phenomena to advance, the symmetrical and algorithmical properties of organic shaping need in-depth study. This article sheds light on previously unknown basic properties of morphological self-organization of living matter, observable in a broad range of supramolecular organic bodies of different evolutionary classes. These properties manifest themselves in that the structure and behavior of the set of individual parts in algorithmically organized biological structures are coordinated by finite parametrical groups (and semi-groups) of non-Euclidean transformations, above all Möbius and affine.

Shubnikov and Koptsik [54, p. 26] define symmetry as the broadest maximal group of object automorphisms. The author's own finding is that by disregarding non-Euclidean automorphisms biology overlooks symmetries and the very fact that quite a few living beings are symmetrically algorithmically structured. Bearing this in mind, morphology would have to dramatically expand its fundamentals by using non-Euclidean, rather than the conventional similarity, transformations if group invariant tools of modeling are to be efficiently applied and refined.

The discovery of non-Euclidean biological symmetries provides valuable data for the development of a formal theory of morphogenesis, the ability to explain the existence of these symmetries being an illustrative indicator of the adequacy of this theory. A reliable method has been found for the materialization of Vernadsky's idea [10] regarding adequate employment of the non-Euclidean geometry in biomorphology along the lines of the Erlanger program and one which makes it possible to develop biological theories in the spirit of the "geometrization of physics".

A recent paper [55] on nonbiological applications of iterative algorithms (unrelated with Euclidean, Möbius or projective transformations) describes a number of biological-like structures and notes that "these curious configurations demonstrated that the rules whereby the most complicated living tissues are structured may be plain to the point of absurdity".

The findings reported in this article go far beyond those results as far as relations with conventional biomorphology, non-Euclidean geometries, theory of automata, the kinematics of biological movements etc. are concerned.

Research in biosymmetries enhances the comprehension of the unity of nature, in the same vein that science discovers ever new general biological laws and mechanisms such as the genetic codes and bioenergy mechanisms. This article sheds light on and analyzes new, non-Euclidean, symmetrical and algorithmical properties of general biological phenomena in biological morphogenesis, the knowledge of which is indispensible for a broad range of theoretical and applied areas, including anthropological and zoomorphological robotics, ergonomics, biomechanics, biotechnology (above all, "morphoengineering", or directed morphogenesis control) etc. These results emphasize the morphological significance of the internal environment of the organism and suggest new approaches to understanding their relationship of morphogenesis and the mechanisms of genetic encoding and biochemical cycles [56]. The writer's findings are not only in favor of the argument that biology is a fertile field for the introduction of various symmetrical approaches, methods and tools of group-theoretical analysis [3, 54, 57 *inter alia*], but that the development of theoretical biology and biomechamics at this stage is largely dependent on vigorous utilization of group-theoretical methods with the use of non-Euclidean geometries.

Acknowledgement—The author is deeply grateful to K. V. Frolov (U.S.S.R. Academy of Sciences) for extensive help in this research.

REFERENCES

1. V. P. Vizgin, *The Erlanger Program and Physics* (in Russian). Nauka, Moscow (1975).
2. F. Klein, *Vorlesungen über nicht-Euklidische Geometrie*. Springer, Berlin (1928).
3. H. Weyl, *Symmetry*. Princeton Univ. Press, N.J. (1952).
4. P. Descartes, *Selected Works* (in Russian). Izdatelstvo inostrannoy literatury, Moscow (1959).
5. E. Cassirer, The concept of group and the theory of perception. *Phil. phenomenol. Res.* **5**(1), 1–36 (1944).
6. J. R. Derome, Biological similarity and group theory. *J. theor. Biol.* **65**(2), 366–378 (1977).
7. S. V. Petukhov, Studies in non-Euclidean biomechanics. In *The Biomechanics of Man–Machine Systems* (in Russian), pp. 38–39. Nauka, Moscow (1981).
8. A. Eddington, *The Philosophy of Physical Science*. Cambridge Univ. Press, Cambs. (1939).
9. R. Thom, *Structural Stability and Morphogenesis, an Outline of a General Theory of Model*. Benjamin, Boston, Mass. (1975).
10. V. I. Vernadsky, *The Chemical Structure of the Biosphere and its Environment* (in Russian). Nauka, Moscow (1965).
11. S. V. Meyen, B. S. Sokolov and Y. A. Shreider, The classical and non-classical biology: the Lyubishchev phenomenon. *Vest. Akad. Nauk SSSR* **10**, 112–124 (1977).
12. Symmetry and function of biological systems at the macromolecular level. In *Proc. 11th Nobel Symp.*, Stockholm (1969).
13. A. V. Shubnikov, The similarity symmetry. *Kristallografiya* **5**(4), 489–496 (1960).
14. S. Lie, *Theorie der Transformationsgruppen*, Ab. 3. Verlag, Leipzig (1983).
15. C. Reid, *Hilbert*. Springer, New York (1970).
16. L. A. Zenkevich (Chief Ed.), *The Animal Life* (in Russian). Prosveshcheniye, Moscow (1976).
17. E. Haeckel, *Naturlicheschopfungs Geschichte*. Verlag, Berlin (1898).
18. Meyen S. V., Plant morphology in its nomogenetical aspects. *Bot. Rev.* **39**(3), 205–260.
19. E. Haeckel, *Kunstformen der Natur*. Verlag, Leipzig (1904).
20. Al. A. Fyodorov (Chief Ed.), *The Vegetation Life* (in Russian). Prosveshcheniye, Moscow (1976).
21. M. G. Pertushevskaya, *Radiolaries of the Nassellaria order in the World's Oceans* (in Russian). Nauka, Leningrad (1981).
22. J. Bernal and S. Carlyle, Ranges of generalized crystallography. *Kristallografiya* **13**(5), 927–951 (1968).
23. Yu. A. Urmantsev, The specifics of space and time relations in living nature. In *Space, Time, Movement* (in Russian), pp. 215–241. Nauka, Moscow (1971).
24. D'Arcy W. Thompson, *On Growth and Form*, 2nd edn. Cambridge Univ. Press, New York (1942).
25. D. V. Nalivkin, Curvilinear symmetry. In *Crystallography* (in Russian), pp. 15–23. Metalurgizdat, Moscow (1951).
26. N. A. Bernstein, *Essays in the Physiology of Movements and the Physiology of Activity* (in Russian). Meditsina, Moscow (1966).
27. O. M. Ivanova-Kazas, *Comparative Embryology of Invertebrates* (in Russian). Nauka, Novosibirsk (1975).
28. L. S. Berg, *The USSR Food—Fish* (in Russian). Pishchepromizdat, Moscow (1949).
29. L. S. Berg, *Selected Works* (in Russian). Nauka, Moscow (1961).
30. A. Weiss, Replication and evolution in inorganic systems. *Angew. Chem.* **20**(10), 850–860 (1981).
31. W. W. Sawyer, *Prelude to Mathematics*. Penguin Books, Harmondsworth, Middx. (1955).
32. B. M. Petten, *The Human Embryology* (in Russian). Medgiz, Moscow (1959).
33. V. V. Bunak, Changes in the relative length of human extremity skeleton segments during the growth period. *Izv. Akad. Nauk RSFSR* **84**, 33–45 (1957).
34. D. G. Rokhlin, *X-ray Osteology and X-ray Anthropology* (in Russian). Biommedgiz, Moscow (1936).
35. S. V. Petukhov, *Biomechanics, Bionics and Symmetry* (in Russian). Nauka, Moscow (1981).
36. M. Planck, *Vortrage und Errinnerungen. Ursprung und Auswirkung Wissenschaftlichen Ideen*. Springer, Stuttgart (1949).
37. M. J. Apter, *Cybernetics and Development*. Pergamon Press, New York (1967).
38. J. von Neumann, *Theory of Re-producing Automata*. Univ. of Illinois Press, Urbana (1966).
39. A. Lindenmayer, Algorithms for plant morphogenesis. In *Theoretical Plant Morphology* (Ed. R. Sattler). Leiden Univ. Press, London (1978).
40. E. N. Andreyev, *The Microworld Space* (in Russia). Nauka, Moscow (1969).
41. T. H. Waterman (Ed.), *The Physiology of Crustacea*. Academic Press, New York (1960).
42. P. R. Evans, Adaptations shown by foraging shorebirds to cyclical variations in the activity and availability of their intertidal invertebrate prey. In *Cyclic Phenomena in Marine Plants and Animals* (Eds E. Maylor and R. Hartnoll), pp. 357–366. Pergamon Press, New York (1979).
43. Sh.-E. Le Corbusier, *The Modular*. Faber & Faber, London (1963).
44. J. M. R. Delgado, *Physical Control of the Mind, Toward a Psychocivilized Society*. Harper & Row, New York (1969).
45. H. Helmholtz, *Über die Tatsachen, die der Geometrie zu Grunde liegen*. Teubher, Berlin (1868).
46. H. Poincaré, *Oeuvres*, Vol. II. Gauthier-Villars, Paris (1916).
47. J. Nicod, *La Geometrie dans le Monde Sensible* (Pref. de B. Russell). Press Univ. France, Paris (1962).
48. J. E. Rose and V. B. Mountcastle, Touch and kinesthesis. In *Handbook of Physiology*, pp. 387–429. Am. Physiol. Soc., Washington, D.C. (1959).
49. T. A. Cook, *The Curves of Life*. Constable, London (1914).
50. R. K. Luneburg, The metric of binocular visual space. *J. opt. Soc. Am.* **40**(10), 627–642 (1950).
51. B. V. Rauschenbach, *Spatial Designs in Old Russian Paintings* (in Russian). Nauka, Moscow (1975).
52. G. Kienle, Experiments concerning the non-Euclidean structure of visual space. In *Bioastronautics*, pp. 386–400. Pergamon Press, New York (1964).

53. *Perception; Mechanisms and Models.* Freeman, San Francisco, Calif. (1972).
54. A. V. Shubnikov and V. A. Koptsik, *Symmetry in Science and Art* (in Russian). Nauka, Moscow (1972).
55. A. K. Dudeney, Wallpaper pattern generating programs. *V mire nauk.* **11,** 92–99 (1986).
56. L. V. Belousov, *The Biological Morphogenesis* (in Russian). Moscow Univ. Press, Moscow (1987).
57. B. K. Vainshtein, *Today's Crystallography: Symmetry of Crystals, Structural Crystallography Methods* (in Russian). Nauka, Moscow (1979).
58. S. P. Kurdyumov, Combustion eigenfunctions in a nonlinear environment and its structural laws. In *Contemporary Issues in Mathematical Physics and Computing Mathematics* (in Russian), pp. 217–243. Nauka, Moscow (1982).

APPENDIX

Mathematical Formalisms

Let us recall what the basic groups of transformations in this article, Fibonacci numbers, and iterative algorithms in various applied mathematical fields are.

For a three-dimensional space, the following interrelationship of finite groups of pointwise transformations hold, formulated by Lie back in 1893 [14]:

$$D \subset S \subset \begin{cases} M \\ LH \subset A \subset Pr, \end{cases}$$

where D is a 6-parametric group of Euclidean movements, S is a 7-parametric similarity group, M is a 10-parameter group of Möbius transformations, LH is an 11-parameter special linear inhomogeneous group, A is a 12-parameter group of affine transformations and Pr is a 15-parameter general projective group (the structure is very much like this one in the case of a two-dimensional space). Every group is associated with a certain geometry: S, Euclidean; M, conformal; Pr, projective. In effect, there are two possible ways to look for generalizations of similarity biosymmetries, conformal geometrical (M) and projective geometrical ($LH \subset A \subset Pr$). Both are described in the main text.

The group of similarity transformations which preserves the shape of figures includes transformations of parallel shift, rotation, mirror reflection and scaling. In a three-dimensional space, the general analytical expression of the transformation has the form

$$x'^i = K \cdot L_j^i x^j + b^i \tag{A.1}$$

(with an orthogonal matrix L_j^i, K being an arbitrary number and $i, j = 1, 2, 3$). The group of projective transformations consists of transformations which have, in Cartesian coordinates, the form

$$x_i' = \frac{c_{ij}x_j + b_i}{d_jx_j + 1}, \tag{A.2}$$

where c_{ij}, b_i and d_j are real coefficients which satisfy the condition that the determinant of the system (A.2) is nonzero, $i, j = 1, 2, 3$. These transformations change linear patterns into nonlinear ones and preserve the membership relations between them (Fig. A1).

The Möbius group has been used in mathematics and physics under different means, e.g. conformal, circular, analagmatic, inverse radii, Möbius, Liouville and Kelvin (after scientists who made significant contributions to studies of its mathematical

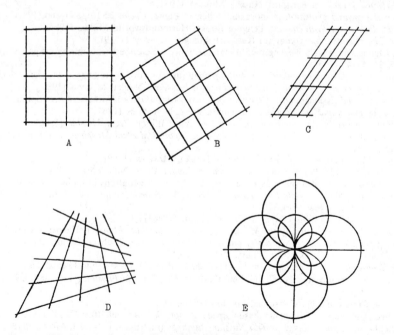

Fig. A1. Characteristic transformation of the Cartesian net (A) by various group transformations: Euclidean (B), affine (C), projective (D) and Möbius (E).

properties and the development of their physical applications). By definition, this is a group of pointwise transformations which preserve the values of angles and transform spheres into spheres. Möbius transformations are locally similar transformations, in that they maintain the shape "in the small" (by the Liouville theorem in the case of a three-dimensional space their group coincides with the group of conformal transformations; in the two-dimensional case the latter group is broader than the Möbius group). This implies that if a small vicinity about every point of a three-dimensional body is scaled, the vicinities about different points not necessarily being scaled equally, the entire body undergoes a stringently Möbius transformation (if the vicinities about all points are scaled to the same extent, the entire body undergoes a similarity transformation which is a particular case of the Möbius transformation). Möbius, or conformal transformations are distinguished from the other groups of transformations of a three-dimensional space by these properties. The local properties of conformal geometrical objects is a major field of differential geometry. Möbius transformations of three-dimensional patterns have a specific set of invariants, including surface curvature lines and their isogonals, ombilical points etc. These transformations divide the set of figures into closed classes of Möbius-equivalent figures that are somewhat broader (because of the additional three parameters of the group) than the conventional classes of equivalent figures in the similarity group. The totality of all Möbius transformations of a three-dimensional space add up to a 10-parameter group, whose analytical expression in canonical pentaspherical coordinates coincides with the group of Lorentz's transformations of 5 variables. Any Möbius transformation in an n-dimensional space may be represented as a superposition of one similarity transformation (A.1) and one inversion with respect to a sphere of unit radius:

$$x_i' = R^2 \frac{x_i - x_{0i}}{\Sigma(x_j - x_{0j})^2} + x_{0i}, \tag{A.3}$$

where R is the inversion (in this case unit) radius, x_{0i} are coordinates of the inversion center and $i, j = 1, 2, \ldots, n$. Any Möbius transformation in an n-dimensional space can also be represented as a product of no more than $n + 2$ common inversion transformations with respect to a sphere. With an infinite inversion radius, inversion with respect to a sphere changes into mirror reflection with respect to a plane.

In the infinitesimal form, Möbius transformations are

$$f^i = a^i + w^{ij}x^i + ax^i + v^jx^jx^i - \tfrac{1}{2}v^ix^jx^i,$$

$$L(x) = 2(a + v^jx^j),$$

where the parameters a^i determine infinitesimal shift transformations; w^{ij}, turn transformations; a, homothetic transformations; and v^i, Möbius proper transformations. Proceeding from this infinitesimal transformation f^i, a finite Möbius transformation is obtained by using Lie equations for determination of finite Lie transformations from given infinitesimal transformations:

$$\frac{\mathrm{d}x^i}{\mathrm{d}t} = f^i(x),$$

where t is a group parameter, $i = 1, 2, 3$. Solutions of these equations in this case are a similarity shift transformation $x'^i = x^i + A^i$, a turn transformation $x'^i = 0^{ij}x^j$, $0^{ij}0^{j1} = \delta^{i1}$, a homothetic transformation $x'^i = Ax^i$ and a Möbius proper inversion transformation (A.3).

Now let us consider the series of Fibonacci numbers, which is a recurrent sequence (with $n = 0, 1, 2, 3, \ldots$),

$$\{F_{n+2} = F_n + F_{n+1}\}: 0, 1, 1, 2, 3, 5, 8, 13, 21, \ldots. \tag{A.4}$$

These numbers have long since been known in biological phyllotaxis laws, whereby in numerous cases of symmetrical shaping of biological objects the numerical characteristics of their configurations are pairs of Fibonacci numbers from sequences of two types (parastichy and orthostichy):

$$\left\{ Q_n' = \frac{F_{n+1}}{F_n} \right\}: \frac{2}{1}, \frac{3}{1}, \frac{5}{3}, \frac{8}{5}, \ldots \to \Phi = \frac{1 + \sqrt{5}}{2} = 1.618\ldots \tag{A.5}$$

and

$$\left\{ Q_n'' = \frac{F_{n+2}}{F_n} \right\}: \frac{2}{1}, \frac{3}{1}, \frac{5}{2}, \frac{8}{3}, \frac{13}{5}, \frac{21}{8}, \ldots \to \Phi^2 = \frac{3 + \sqrt{5}}{2} = 2.618\ldots. \tag{A.6}$$

Various researchers have noted such regularities in the symmetry of numerous vegetable and animal objects (for a survey see Ref. [35, pp. 8–19]). Phyllotaxis has been discussed in a range of recent papers. A pioneering effort was made by Kepler who, in the seventeenth century, related in the laws of symmetrical morphogenesis with Fibonacci numbers and the golden section $\Phi = (1 + \sqrt{5})/2$ [see the sequence (A.5)], which has been known since ancient times and is so important in the esthetics of proportions.

Unlike sequences (A.5) and (A.6), the writer has studied elsewhere a sequence of cross-ratios [wurft, equation (3)] rather than affine or common ratios of neighboring number in the Fibonacci series. Three such neighboring numbers F_n, F_{n+1} and F_{n+2}, may be interpreted as lengths of three sequential stretches contained, in compliance with the recurrent property of the Fibonacci series (A.4), between four points F_{n+1}, F_{n+2}, F_{n+3} and F_{n+4}. The value of the wurf, equation (3), of such stretches is

$$W_n = \frac{(F_n + F_{n+1}) \cdot (F_{n+1} + F_{n+2})}{F_{n+1} \cdot (F_n + F_{n+1} + F_{n+2})}. \tag{A.7}$$

Consequently, the wurf series has the form

$$\{W_n\}: 1, \frac{3}{2}, \frac{5}{4}, \frac{8}{6}, \frac{13}{10}, \ldots, \frac{F_{n+2}}{2F_n}, \ldots \to P = \lim_{n \to \infty} \frac{F_{n+2}}{2F_n} = \frac{\Phi^2}{2} = \frac{3 + \sqrt{5}}{4} = 1.309\ldots. \tag{A.8}$$

In the same way that the limit of sequences of common ratios (A.5) of Fibonacci numbers is referred to as the golden section, the limit value P of the wurf sequence (A.8) obtained in this paper is referred to as the golden wurf. There are numerous convincing arguments in favor of the golden wurf $P \approx 1.31$ being the reference value of ontogenetically invariant wurf values of the three-membered kinematic blocks which comprise the human body (Section 5).

To conclude this appendix, iterative algorithms in a more or less explicit form are widely used in numerous applied mathematical fields, such as the theories of automata, iterations, optimal algorithms, multistep processes without "after-effects" in the theory of dynamic systems, Bellman's theory of dynamic programming, the ergodic theory, the theory of branching of solutions to nonlinear equations in cases of group invariance etc. Each of them has its own potential for interpreting the origin and biological significance of cyclomeric biostructures and for mathematization of biology.

Computers Math. Applic. Vol. 17, No. 4–6, pp. 535–538, 1989
Printed in Great Britain. All rights reserved

0097-4943/89 $3.00 + 0.00

SPIRAL PHYLLOTAXIS

R. Dixon

125 Cricklade Avenue, London SW2, England

Abstract—Three computer graphics are presented which have been inspired by the geometric theory of spiral phyllotaxis. The theory and its background are discussed, with special mention of Coxeter's paper on the subject. One of the graphics illustrates a planar corollary to helical phyllotaxis which the artist finds in Coxeter's presentation and suggests that we might best make sense of the helical pattern in plants through the idea of a compressible plane lattice.

"Phyllotaxis" literally means leaf-arrangement, but the patterns in plants that we study under that label may be formed equally well by any repetitive parts, such as florets, fruits, branches, petals and thorns, which bud and grow from a stem. Aided partly by the laws of physics, partly by evolved genetic instruction and partly by its own cybernetic intelligence, a plant optimizes its form in order to maximize its use of space, light and air. The plant must find an equitable organization as a solution to its space-filling task.

The most obvious patterns in plants are the point-symmetries, frequently found in the petal arrangements of flowers, as in the threefold iris, or the fivefold apple blossom. These patterns provide suitable arrangements for small numbers of like parts (petals) generated simultaneously at a growth point [1]. By contrast, spiral phyllotaxis occurs whenever like parts of a plant bud from a common stem *sequentially* in time, and may accommodate any number of parts.

Well-known examples of spiral phyllotaxis include pine cones and pineapples, the florets of composite flowers such as sunflowers or daisies, the leaves of a thistle or foxglove, the petals of a rose or lotus, the branching of a pear tree or cauliflower and many others. The great geometer of nature, Johannes Kepler seems to have been the first to notice that all of these spiral patterns share a curious law: the spirals occur only in *Fibonacci numbers*,

$$1, 1, 2, 3, 5, 8, 13, 21, 34, 55, 89, 144, \ldots.$$

So, for example, the fruit segments of a typical pineapple will be found to form two opposing sets of spirals, with 5 in one set and 8 in the other. In the case of pinecones the numbers may be 3 by 5. The florets on the head of a daisy or sunflower exhibit opposing spiral sets of 13 by 21, 21 by 34, 34 by 55, and even higher Fibonacci numbers. The branching of a pear sapling in spiral succession about the main stem shows a strong tendency to complete 3 revolutions every 5 branches, or 5 revolutions in 8 branches.

How can we explain this law of Fibonacci numbers in plant spirals?

For more than two centuries ever-increasing numbers of botanists and mathematicians have puzzled over this law [2] without much progress, until Coxeter [3] succeeded in relating the arithmetical pattern to a geometrical property. To begin with, the Fibonacci numbers reveal the presence of the *golden ratio*,

$$t = 1.618034 \ldots,$$

because 5/3, 8/5, 13/8, and so on, form the sequence of closest commensurable approximations of this incommensurable proportion. The golden ratio of two lengths is defined as

$$\text{large to small} = \text{sum to large}$$

giving

$$t : 1 = (t + 1) : t$$

and, therefore,

$$t^2 = t + 1.$$

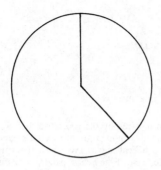

Fig. 1a. The golden section of a circle.

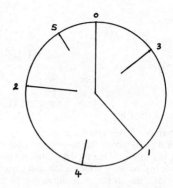

Fig. 1b. Successive additions of the golden angle.

The Greeks discovered that the golden ratio is present in the regular pentagon as the proportion of lengths in diagonal to side, and in the regular decagon as radius to side.

There is an extensive literature on the golden ratio testifying to its many properties in art and science, some of it reliable but much of it—alas—has passed unchallenged into a fairly widespread folklore riddled with superstition and error. Huntley [4] provides perhaps the best popular introduction.

So far as we know, the Greeks never considered a golden ratio of angular measure. This we can do in the following manner. Divide the circumference of a circle into two arcs bearing a golden ratio of lengths. The two arcs so made subtend angles of

$$222.492\ldots \deg = 2\pi/t \text{ rad}$$

and

$$137.507\ldots \deg = 2\pi/t^2 \text{ rad}.$$

Let either of these angles be called the *golden angle*. It makes no matter which, since an anticlockwise turn of one is equivalent to a clockwise turn of the other. A Fibonacci pattern of spiral phyllotaxis arises if the spiral or helical sequence of buddings on a stem occur at intervals of the golden angle.

Coxeter's explanation of the geometric properties of this arrangement involves the theory of principal convergents and continued fractions. It is therefore an essay in number theory, whose purpose is to show that the golden ratio is the only irrational to satisfy Tait's condition: having its sequence of convergents strictly alternating between above and below their irrational limit.

Coxeter's discussion introduces and illustrates Klein's geometric model of numbers—as directions through an infinite plane square lattice from any given lattice point as origin. Any number n is represented by a line whose gradient is n. Rationals such as 3/5 will be represented by lines which meet the next lattice point after travelling a distance of 3 by 5 squares. Irrationals are represented by lines through the lattice which never again cross another lattice point.

It follows that Coxeter is demonstrating (in his argument about a cylindrical lattice) the existence of a corollary condition to helical phyllotaxis in an infinite plane square lattice, namely the existence

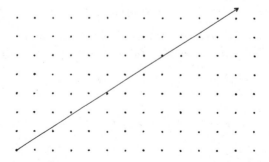

Fig. 2. The golden direction in an infinite plane square lattice.

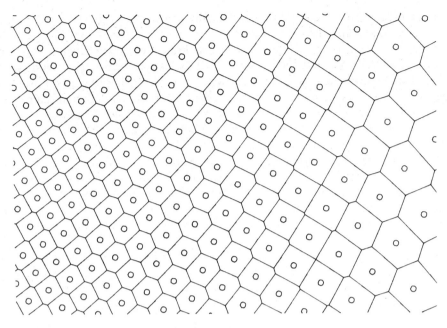

Fig. 3. The golden direction of an infinite square lattice.

of a *golden direction* in an infinite square lattice. The golden direction makes an angle to the initial lattice directions of arctan (t):

$$\arctan(t) = 58.2825\ldots \text{ deg.}$$

Only in the golden direction can a square lattice be continuously stretched or compressed without deviating far from regular hexagon or square packing or spacing. See the illustration of this (Fig. 3), here presented with Voronoi polygons—which enclose all points of the plane nearest to a given lattice point.

It is perhaps easier to imagine the connection between this property of a golden direction in an infinite plane square lattice and the spatial needs of plants. A plant form consisting of a sequence of like parts at staggered stages of their individual growths must grow as a whole, flexing its arrangement of parts while continuing to be spatially equitable. In this requirement, it is an important fact that the parts of plants do not grow uniformly [5]. A plant form is a staggered

Fig. 4. Sunflower. (Disc phyllotaxis of logistic growths.)

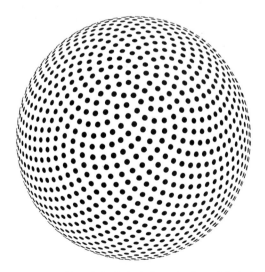

Fig. 5. Dandelion. (Spherical phyllotaxis of equal fruits.)

sequence of sigmoidal growths and it is this which finally presses the need for the degree of flexibility which golden ratio phyllotaxis best provides.

For example, the several hundred florets on a daisy head are neither equal in size nor do they form an exponential sequence, but form a sequence of sizes from centre (youngest) to periphery (oldest) somewhere between these two extremes—we are, of course, ignoring the naturally occurring broken regularities. This shows up in the smooth changing from one dominant pair of Fibonacci numbers to another as you count spirals at increasing distances from the centre.

As well as Klein's diagram, Coxeter presents the phyllotaxis diagram of the Bravais brothers, and shows it to be another geometric representation of numbers, namely as locations on a cylinder. Cylindrical latitude = value and cylindrical longitude = fractional part of the value. Multiples of any rational number form lines on the cylinder: multiples of any irrational form helices. In particular, multiples of t form the helix of spiral phyllotaxis. Its salient property is that continuous compression or stretching in the direction of the cylinder provides a continuously changing pattern of packing or spacing on the stem which is never far from either of the regular solutions of square or hexagon.

By a process of gentle distortion a cylinder may be turned into a cone, a cone into a disc and a disc into a sphere, thus transferring the argument to each of these forms. Using computer graphics [6] we can explore idealizations of spiral phyllotaxis in each of these forms with surprisingly simple algorithms for such subtle and demonstrable results.

REFERENCES

1. R. O. Erickson, The geometry of phyllotaxis. In *The Growth and Functioning of Leaves* (Eds J. E. Dale and F. L. Milthorpe). Cambridge Univ. Press, Cambs. (1983).
2. I. Adler, A model of contact pressure in phyllotaxis. *J. theor. Biol.* **45,** 1–79 (1974).
3. H. S. M. Coxeter, The role of intermediate convergents in Tait's explanation for phyllotaxis. *J. Algebra* **20,** 167–175 (1972).
4. H. E. Huntley, *The Divine Proportion*. Dover, New York (1970).
5. R. Dixon, The mathematics and computer graphics of spirals in plants. *Leonardo* **XVI-2,** 86–90.
6. R. Dixon, *Mathographics*. Blackwell, Oxford (1987).

Computers Math. Applic. Vol. 17, No. 4–6, pp. 539–594, 1989
Printed in Great Britain. All rights reserved

0097-4943/89 $3.00 + 0.00

A SYMMETRY-ORIENTED MATHEMATICAL MODEL OF CLASSICAL COUNTERPOINT AND RELATED NEUROPHYSIOLOGICAL INVESTIGATIONS BY DEPTH EEG

G. Mazzola,[1,2] H.-G. Wieser,[2] V. Brunner[3] and D. Muzzulini[4]

[1] Department of Mathematics, University of Zürich, Zürich, Switzerland
[2] Department of Neurology, University Hospital, Zürich, Switzerland
[3] Institute of Geophysics, ETH, Zürich, Switzerland
[4] Seminar of Applied Mathematics, ETH, Zürich, Switzerland

Abstract—This work presents (1) a mathematical model of classical counterpoint, based on distinguished symmetries between consonant and dissonant musical intervals and derived local symmetries, together with (2) an investigation of the electrical activity (depth EEG) of the human brain in relation with consonant and dissonant musical stimuli. Presenting a musical test program to 13 patients with electrodes implanted within different brain areas [hippocampal formations of both sides, planum temporale (near Heschl's gyrus) and/or placed epicortically at mediobasal limbic structures], we found that the reaction of depth EEG corresponds in a precise and quantified way to the postulates of mathematical counterpoint theory. The main results are: (1) The EEG of the hippocampus reflects the consonance–dissonance dichotomy for simultaneous intervals in a predominant way. (2) Within the right Heschl's gyrus, the EEG response to the distinguished symmetry between consonances and dissonances is significant. (3) The EEG of right hemispheric locations dominate the processing of music related to Gestalt perception in space–time (pitch/onset-time), in particular of successive intervals. (4) The geometrically distinguished pair of dichotomies (the consonance–dissonance dichotomy and dichotomy of proper tonal intervals from the major tonic) is reflected within the spectral density data of the classical θ-, α-, and β-frequency bands. These findings may help to understand the relation between music and emotion.

1. INTRODUCTION TO THE PROBLEM

There are but a few problems in musicology, on which there is so little consensus and so much dissension, as with the dichotomic concept of consonances and dissonances. Seemingly, the subject touches a fundamental function of music: to fulfill the human need for harmony, accord and peace within a balanced alternation to tension, aggression and movement.

On the other hand, history of music and musicology shows that the phenomenon of consonances and dissonances has always been an object of purely formal and even mathematical investigations. In modern times, it has been the famous sentence of Gottfried Wilhelm Leibniz, that listening to and understanding of music must be a secret calculation of our soul [1], which gave rise to a new and never ceasing belief that there must be a formal and quantitative access to the emotionally crucial category of consonances and dissonances.

Our present work follows this tradition in its broadest sense. After a detailed study of the symmetries within musical interval structures (Sections 2 and 3), we discuss a mathematical model of classical counterpoint, following the medieval tradition in its final development with Giovanni da Palestrina, as described by Johann Joseph Fux in his *Gradus ad Parnassum* [2] (Sections 4 and 5). The crucial statement of this theory is a deduction of the rule of forbidden parallels of fifths from a distinguished symmetry between the set of consonant intervals and the set of dissonant intervals. It will be seen that essentially, there are but two partitions of the set of intervals, distinguished by a particular symmetry, the second one being defined by the collection of proper intervals from the tonic, within the major tonality.

This model has been tested not only from the musicological point of view (Section 5), but also in the context of neurophysiological criteria, as they are traditionally developed from the fact of the electrical activity of the human brain, measured by the electroencephalogram (EEG), on the human scalp and—using recent developments—from intracranial depth electrodes.

To a large extent, the theoretical basement of the mathematical model of counterpoint seems to be confirmed (or, at least, not contradicted) on the level of electrophysiological responses of humans. We could observe strong arguments for a sophisticated processing of consonant and dissonant intervals within the left hemispherical limbic system, in particular the hippocampal formation, where a dominant link between emotional and memory functions has been postulated since the investigations of Papez [3] and MacLean [4, 5].

However, before a closer discussion of these neurophysiological methods can be undertaken, it is indispensable to recapitulate the complex problem of consonances and dissonances. Unfortunately, no history of the phenomenon has been written to date, but the subject appears in major treatises on the fundaments of music and its perception. Let us just mention some of the most influential ones: Fux [2], Euler [6], Stumpf [7], von Helmholtz [8]. An excellent modern treatise from the physical and physiological point of view is by Roederer [9], the psychological perspective of the problem is brilliantly exposed by de la Motte-Haber [10].

As quite generally, musical reality is composed of three levels: the symbolic, the physiological and the psychological one, the consonance–dissonance problem, too, is split into these perspectives, and historically, the most prominent theories of recent times are symbolic (Euler), physiological (von Helmholtz) and psychological (Stumpf). Let us briefly expose these approaches in order to make the problem more comprehensible.

As a mathematician, Leonhard Euler was interested in number theory. It is not astonishing that he picked up the Pythagorean theory, which starts from small prime numbers $p = 2, 3, 5, 7$, in order to describe the intervals as being frequency-quotients of small products of the first primes: octave $= 2:1$, fifth $= 3:2$, major third $= 5:4$, fourth $= 4:3$, etc. Observe that Euler's symbolic view perfectly matches with the Pythagorean metaphysical foundation of music by the mysterious tetractys-triangle $1:2:3:4$ [11].

Euler's classification of intervals uses a number theoretical function, the "gradus suavitatis" [6]. For a natural number a, consider its prime decomposition $a = p_1^{e_1} \ldots p_n^{e_n}$ as a product of powers of different primes $p_1, \ldots, p_n$. Setting

$$\Gamma(a) = 1 + \sum_{k=1}^{n} e_k(p_k - 1),$$

Euler defines $\Gamma(a/b) = \Gamma(a \cdot b)$ for a frequency quotient a/b, where a and b are relatively prime natural numbers. Hence the fifth $3/2$ has $\Gamma(3/2) = \Gamma(6) = 4$ and the tritonus $45/32$ yields $\Gamma(45/32) = \Gamma(1440) = 14$. In fact, Γ should be termed "gradus dissuavitatis" because the classical consonances, such as octave or fifth, have small Γ-values, whereas the tritonus and other dissonances have larger Γ-values. In order to get a picture for the "degree of pleasantness" defined by Γ, consider Fig. 1, where we took the value $10/\Gamma$ vs the intervals of the classical chromatic scale for just 2–3–5 tuning [12].

Observe that we do not recover any limit value for the transition from consonances to dissonances; in fact, Euler doesn't use the term of dissonant intervals. The gradus function only yields a partial ordering among the intervals.

Nonetheless, note that Γ is a formally reasonable affine function in the exponents $e_1, \ldots, e_n$ since it may be compared with the linear pitch function

$$\sum_{k=1}^{n} e_k \log(p_k),$$

from the mathematical point of view.

Comparing this symbolic classification of intervals with a psychological result shows a rather dramatic discrepancy between the two methods. We are referring to the famous work of Plomp and Levelt [13].

These authors offered pairs of sine-shaped (simultaneous) tones to musically non-trained persons to judge the psychological, verbally expressed degree of pleasantness of the different intervals. The choice of non-trained individuals was essential as the investigation was designed to avoid educational side-effects like recognition of the tritonus as being "diabolus in musica", whence a cultural and not a non-codified emotional response would have resulted. The experimental degree of pleasantness is shown in Fig. 2.

It is in conflict with Euler's curve (Fig. 1), in particular on the level of the tritonus. However, the conflict is only apparent, because the symbolic meaning needn't coincide with the psychological one. To be precise, the Γ-function primarily relates to frequency quotients which best can be realized within the physical production of sound. But the psychological valuation relates to a

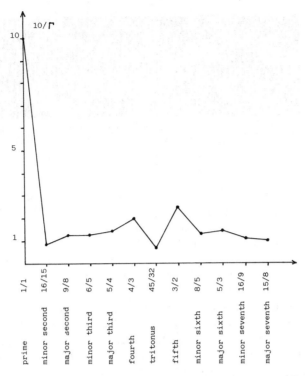

Fig. 1. This curve shows the classification of the intervals within an octave, as described by Euler [6]. To make evident the degree of pleasantness of an interval out of the just chromatic scale, we draw the values of $10/\Gamma$ assigned to the intervals, the factor 10 being a scaling number. Euler's Γ-function does not indicate any clear separation between consonances and dissonances, it purely induces a partial ordering among the twelve intervals.

complicated transformational process from sensorial perception to cognitive categories and finally ends with the motoric response in the verbal judgement. This multifold transformation has to be taken into account while comparing the symbolic with the psychological level of reality.

A third alternative in between the two precedent ones is proposed by the physiological approach of von Helmholtz [8]. It is a rather elaborate theory and has often been misunderstood. Helmholtz does not stick to the frequency or pitch of a tone, but considers its harmonics within the Fourier spectrum too. Hence, if the two tones of an interval have the respective frequencies f_1 and f_2, their sine-shaped harmonics will have frequencies $2f_1$, $3f_1, \ldots, nf_1, \ldots$ and $2f_2$, $3f_2, \ldots, mf_2, \ldots$, respectively. Following Helmholtz, the perception of dissonances is due to the presence of beats of nearby harmonics of the two tones which, for some mysterious reason, our ear (or better: what the brain makes out of the sensorial information from the ear) doesn't seem to like. Recall that the beat effect between two sine-waves $\sin(\alpha)$ and $\sin(\beta)$ results from the form

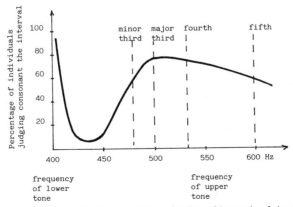

Fig. 2. The work of Plomp and Levelt [13] revealed a valuation of intervals of simultaneous sine-tones which differs from the quantities obtained by symbolic calculus like Euler's Γ-function.

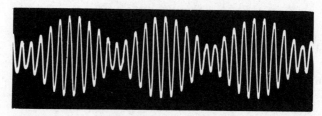

Fig. 3. For the linear superposition of two sine-waves with slightly different frequencies, beats appear as amplitude modulations. Following Helmholtz, the beats centred around 33 Hz among the harmonics of complex sounds are responsible for the sensation of dissonances.

$\sin(\alpha) + \sin(\beta) = 2\sin(\frac{1}{2}(\alpha + \beta))\cos(\frac{1}{2}(\alpha - \beta))$ of their linear superposition. This means that this latter will be a sine-wave with an amplitude modulation given by $\cos(\frac{1}{2}(\alpha - \beta))$. The shape of such a superposition of nearby sine-waves is shown in Fig. 3. Helmholtz considered beat frequencies centred around 33 Hz as being responsible for dissonance perception. His calculations of dissonance degrees for violin yield the descending chain: octave, fifth, fourth, major sixth, major third, natural seventh (7/4), minor third/minor sixth—a rather good coincidence with Euler's values. For a more detailed discussion, we refer to Ref. [14]. Notice however, that Helmholtz's calculus is strongly dependent on the sound colour of the two tones of the interval, which is heard. Furthermore, the beats physically disappear if the lower tone is presented to one ear and the upper tone is presented to the other ear separately (binaural listening). Nonetheless, the binaural experiments of Husmann [15] showed that consonances were also distinguished from dissonances as "points of light out of a muddy experience".

At this point, one would need to have precise knowledge on the physiological processing of sound-stimuli within the neuronal net. We shall shortly come back to this question. But still, Helmholtz's calculus is in contradiction to Plomp's and Levelt's findings. And there are fundamental postulates of musicology and music psychology putting into question Helmholtz's physiological perspective. In fact, Stumpf [7], as well as Albersheim [16] have pointed out that consonances and dissonances are cognitive entities showing a mutually polar position, a dichotomy of musical qualities [16, pp. 121, 261]. However, neither Euler's Γ-function nor Helmholtz's beat-calculus provide any qualitative criteria to distinguish between two classes of intervals. As it seems, the consonance–dissonance dichotomy is far from being a purely physiological phenomenon, it primarily resides on highly abstract, culturally codified characteristics.

Our discussion shows that the consonance–dissonance problem is mainly one of the interaction of three levels of musical reality: symbolic, physiological and psychological. From this point of view, the discrepancies in the different theories are only apparent contradictions, since one would have to know the transformations of information when changing the level of reality. The main difficulty to resolve these conflicts stems from the fact that one talks about *the* fifth, *the* major third and by doing so, one pretends to have one identical object to deal with, though we have a triply split phenomenology, i.e. three levels of reality.

To facilitate understanding of the following it seems necessary to give a synopsis of the complex way of sound from the moment it reaches the ear to the final processing within the auditory cortex. We refer to Figs 4–10 and to the respective legends for details. Here, we would simply stick to the main steps.

In outline, sound waves which reach the air column in the external acoustic meatus cause a comparable set of vibrations in the tympanic membrane, and thus in the chain of auditory ossicles. Similar vibrations occur at the foot plate of the stapes, but here the force per unit area of the oscillating surface is increased some twenty-fold. These are effective in overcoming the inertia of the perilymph thus producing pressure waves within it, which are conducted almost instantaneously to all parts of the basilar membrane (Figs 4 and 5). This pressure wave gives a configuration of resonance on the basilar membrane, Fig. 6. This configuration corresponds to an approximated Fourier decomposition of the complex sound in the sense that higher harmonics may interfere with each other and hence limit the analytical capacity to the first six or seven harmonics. The configuration is then transformed into electrical impulses from approx. 15,000 hair cells on the organ of Corti which is situated on the basilar membrane, Figs 7 and 8.

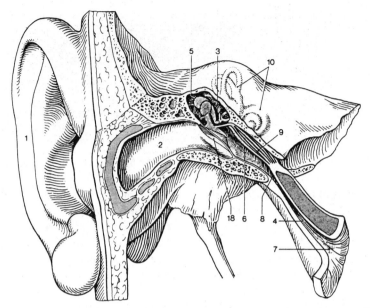

Fig. 4. When a sound wave arrives at the external ear [auricle (1); external acoustic meatus (2)] it arrives at the tympanic membrane (18). The latter transfers its deformations to the auditory ossicles of the middle ear (3, 5): the malleus, the incus and the stapes. The excitation then reaches the cochlea (inferior part of 10) [17].

It is, however, not known, in which way the information of the resonance configuration is processed to give a precise perception of pitch. This is above all due to the extremely complicated nervous path from the cochlea to the primary auditory cortex (Heschl's gyri) in the temporal lobe. We should stress that the pathway drawn in Fig. 9 is a strong simplification and that, in particular, there are different ipsilateral connections from the cochlea to the auditory cortex as well as some connections back to the cochlea.

The auditory system has at least five "relay stations" (see Fig. 9). This makes clear that by no means the approximated Fourier decomposition at the periphery can be transferred to the auditory cortex. For example, Fig. 10 shows that even the most elementary tonotopy on the auditory cortex is a multiple one.

The Fourier decomposition as it seems to work at the cochlear level, would be a reasonable foundation of Helmholtz's approach. However, it raises more problems than it tends to solve. One of the most elementary and still unsolved problems rises from the task to find a pitch from the frequencies of the given harmonics. To show how sophisticated a quantitative modelling of a central pitch processor may look, we just sketch the "template fitting process" proposed by Goldstein [20]. For a theoretical approach to pitch recognition using the analysis in the time domain, and being in

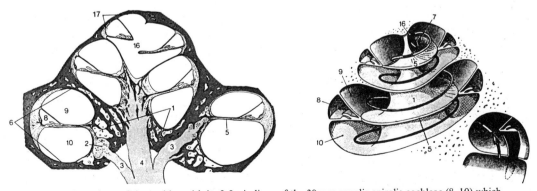

Fig. 5. Two views of the cochlea with its 2.5 windings of the 30 mm canalis spiralis cochleae (8–10) which is filled with lymph. The canal is divided into the scala vestibuli (9), the scala tympani (10) and the ductus cochlearis (8). Within the ductus cochlearis the spiral organ of Corti is situated on the basilar membrane (see Fig. 7). This organ generates the neuronal impulses which are sent to the cortex by the cochlear nerve (3, 4, left) [17].

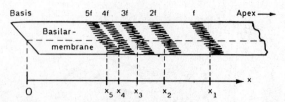

Fig. 6. The basilar membrane varies continuously in its width, mass and stiffness from the basal to the apical end of the cochlea. When exposed to a periodic oscillating pressure wave in the neighbouring peri-lymph, its behaviour varies with the spectrum of the oscillations. Here, the resonance tracts on the basilar membrane, which correspond to the spectral components, are shown. This "cochlear Fourier decomposition" may be seen as physiological correlate to Helmholtz's theory of consonances and dissonances.

contrast to the theories of Goldstein and Terhardt [21], we refer to Langner's report on the auditory system of the Guinea Fowl [22]. Goldstein's model is an important example for mathematical considerations concerning music perception, since our model of counterpoint will involve some elementary symmetry-considerations, which should be compared with the following calculation.

The hypothesis of Goldstein is as follows. Suppose for simplicity, that two sine-waves with frequencies f_a, f_b are given; they may be represented as resonance loci of a sound on the basilar membrane. Then we are looking for a basic frequency f and a natural number n, such that f_a is as near as possible to the nth partial nf of f and f_b is as near as possible to the $(n + 1)$st partial $(n + 1)f$ of f. In other words, we try to minimize the error-value

$$\sqrt{((nf - f_a)/nf)^2 + ((n + 1)f - f_b)/((n + 1)f)^2}. \tag{1}$$

This model fits rather well with experiments of Smoorenburg [23]. However, it resides on the complicated condition (1), which amounts to solve an extremal problem by vanishing of the partial derivative of condition (1) along the variable f, and then looking for an optimal n. . . . Furthermore, the special choice of n, $n + 1$ out of the general case n, m of relatively prime numbers is not justified. Presently, it is not evident, how such a problem of differential calculus should be solved by the "secrete calculations of the soul" in the sense of Leibniz.

Concluding, this means that the different theories on consonances and dissonances still have to

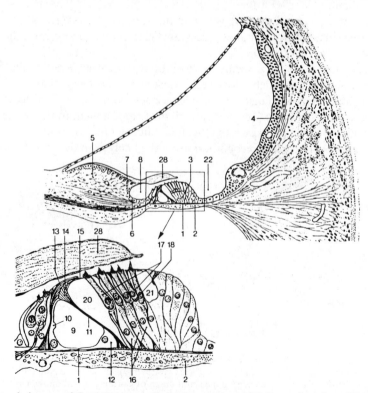

Fig. 7. The spiral organ of Corti (3) sits on the basilar membrane (1). It consists of roughly 15,000 hair cells (18) ordered in three to five rows along the ductus cochlearis [17].

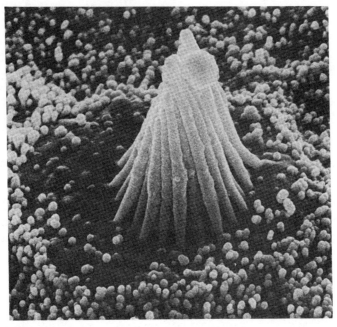

Fig. 8. (Scanning electron micrograph by R. A. Jacobs, California Institute of Technology of inner ear of bullfrog.) Each hair cell bears 30–50 hairs, the stereocilia and one sensible kinocilium. The hairs touch the tectorial membrane (see 28 in Fig. 7) and react to positional change of 3° which corresponds to the atomic diameter of 10^{-8} cm. The reaction time of a hair cell is some tens of μs, and a voltage change of ± 8–10 mV is produced. For details, see Hudspeth and Corey [18].

stay fairly uncorrelated to each other as soon as they belong to different levels of reality. However, there is a unique access to the dense chains of events in time, which are produced by music: the EEG taken from the scalp, but in particular that recorded directly from the relevant brain structures, reflect in an extremely complex way the electrical activity of the cortical neurons as they happen to react to musical stimuli. At present, only the recording permits a sufficient time

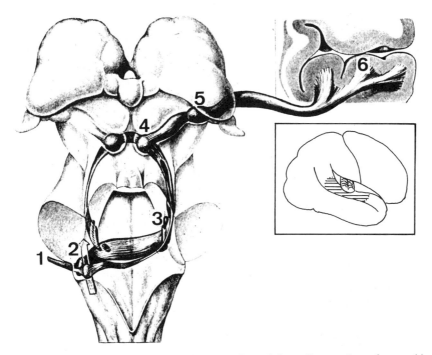

Fig. 9. This simplified picture shows the six relevant stations of the auditory pathway from cochlea to the Heschl's gyri of the auditory cortex. (1) Cochlear nerve, (2) cochlear nucleus, (3) superior olivary nuclei, (4) lower tectum (inferior colliculus), (5) medial geniculate body, (6) auditory cortex [19].

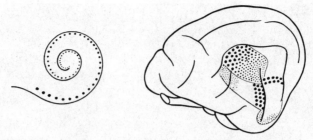

Fig. 10. Left is shown the tonotopy in the cochlear canal; on the right, we see the corresponding tonotopy on the auditory cortex of the cat. Large dots indicate high frequency, small dots indicate low frequency (according to Woolsey in Ref. [17]).

resolution for musical events with usual durations of 0.5 s or less. Other functional diagnostic means, such as positron emission tomography (PET), cannot follow a chain of say 10 musical events, each lasting 0.5 s, with a sample rate of roughly $100 \, s^{-1}$, which is necessary to capture the neuronal electric activity within 0–40 Hz, the range of most EEG waves. The technique of single-neuron-activity recordings measures much more locally the electric processes, and therefore their semantics of this technique may be different from those drawn from EEG recordings (see Creutzfeldt *et al.* [24]).

The study of the EEG, in particular of the depth EEG, is the only present-day possibility to inquire higher brain functions related to music in order to link the psychological and the physiological levels. Direct recordings from the unanesthetized human limbic brain are only accessible from candidates for surgical epilepsy therapy during the pre-operative evaluation. In some, carefully selected patients suffering from medically intractable complex partial seizures of suspected temporal lobe origin, stereotaxic depth recordings are still required for precise localization and delineation of the primary epileptogenic area.

Research work from surgically active epilepsy centres has provided interesting results concerning the limbic system and emotions. Except for those studies dealing with musical and auditory hallucinations accompanied by epileptic discharges or provoked by electrical stimulation [25–28], none of the studies which rely on depth EEG data deals with musical perception. However, in recent years, sophisticated surface EEG studies have dealt with these aspects; the most outstanding findings were reported by Petsche *et al.* [29]. Surface EEG studies dealing with mental performance were successful in relating interpretable EEG changes to tasks involving memorizing and reasoning [30, 31], listening to music, noise and voice, arithmetic and vision [32], visual-motor practice [33], hemispheric asymmetries [34] and musical recognition [35, 36]. From the above works one concludes that the two hemispheres participate in different ways in the processing of musical perception. In particular, the work of Giannitrapani [32] suggests that the left hemisphere should be particularly involved in processing musical stimuli. On the other hand, recent single neuron recordings on right and left temporal lobes of patients while listening to music [24] didn't show significant lateralization.

Despite these refined techniques, EEG recordings are burdened with two delicate problems, which still are far from being solved in practice. The *first* is the so-called inverse problem and refers to the fact that the EEG recording is just a projection of the very complex electric activity of dipole layers, consisting of large populations of active neurons, onto several determined electrodes. One should therefore always take care for the interpretation of EEG recordings in terms of localized and functionally specified brain activity, see also Gloor [37], Mazzola [38]. The *second* problem is the semantics of EEG. Up to now, it seems very difficult to surpass the most elementary meaning of EEG, like localization of epileptic foci. However, it is plausible that the different frequency bands, the potential variation evoked by well defined events, the distribution of EEG signals over time and space must reflect some semiotic information, though no semiosis has been successfully codified from the EEG expression into its content.

This being so, we have been involved in semiotic questions of the EEG, as related to music, for three reasons. First, it is well-known that the limbic system as well as music are related to emotions. Second, it seems that the sensation of consonances vs dissonances is a rather universal phenomenon, which has a good chance to be away from cultural variation like the mother tongue.

Third, we were interested in the neurophysiological foundations of counterpoint and its consonance–dissonance basis.

The point of our investigation was to choose the relevant distinguishing factors for consonances and dissonances. Of course, Euler's Γ-function would have been a prominent testing quantity. However, despite its admitted emotional value (different intervals being more or less pleasant), the partial ordering of intervals in Euler's (and Helmholtz's) sense is very unlikely to be of a reasonably stable character, i.e. it would certainly heavily depend on the context.

To us, the main obstacle against the testing of these ordering functions is the above mentioned marginality of such an ordering for musicological and psychological reasons. The classical counterpoint is built upon a sharp dichotomy:

Consonances: Perfect: prime, octave, fifth;
Imperfect: thirds, sixths;
Dissonances: Fourth (within the kernel of the classical counterpoint, the intervallic, two-part counterpoint, the fourth is undoubtedly dissonant [2]), seconds, tritonus, sevenths.

Even if we wanted to test the ordering of the intervals by some or another pleasantness function, we had to fight against the major obstacle of the just tuning which is the basis of all the pleasantness theories. In fact, these theories depend upon Euler's substitutional hypothesis: every non-just interval (i.e. its frequency ratio is out of the tuning lattice) has to be substituted by the nearest just one; and it is only the just substitute, to which a pleasantness function can be applied. However there is not the slightest evidence for the existence of a neurophysiological mechanism producing this shift from the non-just to the just intervals, and worse: even mathematically, this uniqueness of the nearest just interval will—in general—fail to exist, see Ref. [39], for a proof.

In the next section, we shall describe structures within the set of intervals, which allow an access for a mathematical treatment of the dichotomy aspect of consonances and dissonances.

2. THE GEOMETRY OF INTERVAL DICHOTOMIES

We will first examine the structures on the set of intervals, which yield a description of the dichotomic access to consonances vs dissonances, as it is applied in the classical counterpoint of Palestrina and Fux (till present!) and as it is put into evidence by the findings of music psychology following Stumpf [7] and Albersheim [16].

The first important question concerns the concept of the interval. On the one hand, this musicological concept is often vaguely defined because of mathematical confusions. Eimert [40], a prominent composer of serial and electronic music and an important theorist of music, seems to confuse the number of semitone steps of an interval with the boarding points. In Ref. [41], Wille rightly criticizes this confusion as a source of irreparable obscuration of musicological thinking.

On the other hand, the concept of interval is extremely context-dependent. Let us just mention five typical variants:

1. Within well-tempered tuning, the sum of three major thirds will yield an exact octave, whereas this sum in the just tuning yields a frequency ratio of 1.95, which corresponds to a b sharp, roughly 41 cents below the exact octave of c.
2. The direction from c to e (upwards) or from e to c (downwards) within the interval (c, e) of the major third may be specified.
3. Intervals may be thought of being pairs of simultaneous or else successive tones.
4. One may identify two intervals if the upper tone of the first yields the upper tone of the second by addition of one or several octaves.
5. Often, the tones of intervals are thought to be alterations of other tones. Hence a minor third (c, e flat) is not identical with an augmented major second (c, d sharp), even in well-tempered tuning.

Of course, the best competence for the intervallic thinking has to be attributed to the medieval tradition of counterpoint. We prefer this era of musical culture to the ancient Greek tradition since

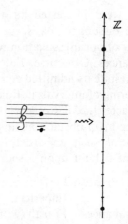

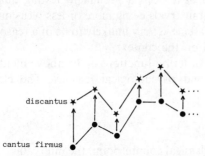

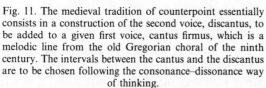

Fig. 11. The medieval tradition of counterpoint essentially consists in a construction of the second voice, discantus, to be added to a given first voice, cantus firmus, which is a melodic line from the old Gregorian choral of the ninth century. The intervals between the cantus and the discantus are to be chosen following the consonance–dissonance way of thinking.

Fig. 12. From the naive point of view, one would expect an interval to be a set of two notes, representable on the line $\mathbb{Z}$ of integers $\mathbb{Z} = \{0, \pm 1, \pm 2, \ldots, \pm n, \ldots\}$ for a well-tempered tuning, say. The counterpoint theory tells us that more information is needed to define an interval properly.

in the medieval tradition, the process of musical composition too (and not only the musical material) has been codified to a cultural process. The tradition of counterpoint shows the following image. Everything starts from a given melodic line, the cantus firmus (abbreviated c.f.), a feature from the old Gregorian choral, which will be a fixed prerequisite for our discussion.

The fundamental task of counterpoint is, to add one (and later: several) voice(s) to the c.f., the discant or organal voice, following determined rules (Fig. 11).

This terminology goes back to the eleventh and twelfth centuries and is some times referred to as the tradition of ars antiqua (see Apfel [42] for a thorough study of the medieval counterpoint and the dramatic history of its canonization, and Jeppesen [43] for a more concise history of the subject).

From the study of the contrapuntal tradition, one learns that an interval is not just a set of two tones on the line of pitch values, as would be expected (Fig. 12).

In reality, the concept of the interval contains three different informations:

(1) the c.f.-*tone* which may be seen as the "basis" of the interval;
(2) one of two possible *orientations*, upwards, resp. downwards, to indicate the direction of (vertical) displacement of the c.f. to the discantus. This corresponds to the *sweeping* (= +), resp. to the *hanging* (= −) counterpoint;
(3) the amount of distance between the two tones, the intervallic number or *dual number*. The term "dual number" stems from modern algebra, and we shall see that our terminology is a special case of the algebraic notion and not only a speaking expression (Fig. 13).

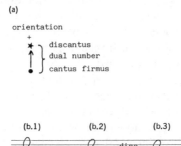

Fig. 13. (a) The notion of interval is composed of three separate informations: the c.f., the orientation and the dual number. (b.1) A third may be interpreted as (b.2) c.f. = g', sweeping, dual number = 4 semitone steps, whence discantus = h'; or (b.3) c.f. = h', hanging, dual number = 4, whence discantus = g'.

As we are concerned with the fundamental counterpoint structures, we will refrain from the time coordinates such as onset-time and duration. Of course, these coordinates are relevant on a more sophisticated level, but we wish to concentrate on the basic problem.

Furthermore, we shall restrict ourselves to the widespread context of well-tempered tuning modulo octave. One may as well interpret this as concentrating on the steps by semitones on the keyboard within one octave. The calculus modulo octave fits with many qualities of intervals in counterpoint, which are invariant under shift of multiples of octaves. For instance, a fifth ($=7$ semitones) is a perfect consonance together with the octave-augmented duodecime ($= 19$ semitones). Thus, we are given twelve tones $0, 1, 2, \ldots, 11$ which may be thought of as being counted from a fixed tone, c say, as the number of semitone steps. Number 12, the octave, would be identified with 0, 13 with 1, 14 with 2, etc.

In elementary number theory, this set is called the cyclic group of order 12 and denoted by $\mathbb{Z}_{12}$. The arithmetic on $\mathbb{Z}_{12}$ is very straightforward. Just add and multiply the numbers in $\mathbb{Z}_{12}$ as usual, but only retain from the result the non-negative remainder after division by 12. For example, you get $5.5 = 7.7. = 11.11 = 1$, and $1 + 11 = 2 + 10 = 3 + 9 = 4 + 8 = 5 + 7 = 6 + 6 = 0$. Note that $1, 5, 7, 11$ are the only invertible elements of $\mathbb{Z}_{12}$, i.e. those elements which admit a multiplicative inverse, see Fig. 14.

This restriction to $\mathbb{Z}_{12}$ does not mean that the intervallic theory of counterpoint completely reduces to $\mathbb{Z}_{12}$. We are just processing an abstraction from the total musicological reality. But abstraction is the dominant process for scientific progress, and even the very concept of symmetry would not be applicable if we hadn't the abstraction from contingencies at disposal.

It will be seen that the calculus still remains complicated enough within $\mathbb{Z}_{12}$, and that important results still persist on this level of abstraction. For a more geometric presentation of the relevant data of an interval, see Fig. 15.

From this representation, one sees that the interval is given by the triple $(\alpha_{\pm}, a, b)$. Usually, it is clear from the context, which orientation has to be chosen. Then, one simply writes the symbolic expression $a + \epsilon \cdot b$ to denote the interval. Hence we get c.f.$(a + \epsilon \cdot b) = a$, and $\alpha_{\pm}(a + \epsilon \cdot b) = a \pm b$, the discantus value.

For the mathematicians, we should remark that the symbol ϵ (epsilon) will effectively play the role of an infinitesimal quantity, one is accustomed with in differential calculus. This means that we really are interpreting an interval in counterpoint as an infinitesimal displacement of the c.f.: *The discantus is an infinitesimal variation of the c.f.*

We shall discuss this onset in extenso in the next section. Here, we would like to stick to the pure dual numbers b of the intervals $a + \epsilon \cdot b$, i.e., to quantities $b = 0, 1, 2, \ldots, 11$ out of $\mathbb{Z}_{12}$. We

```
a= c.f.
b= dual number
orientation  ∝ :D↦a+b
             +
             ∝ :D↦a-b
             -
∝ = sweeping counterpoint
+
∝ = hanging counterpoint
-
```

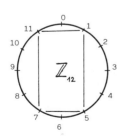

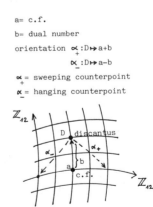

Fig. 14. The cyclic group $\mathbb{Z}_{12}$ represents the 12 semitone steps on an equally partitioned circle. The elements connected by a rectangle are the invertible elements, i.e. those x admitting an y with $xy = 1$.

Fig. 15. The correct geometric representation of an interval takes place in a two-dimensional space, whose coordinates are numbers in $\mathbb{Z}_{12}$. The first coordinate a designs the c.f. value, whereas the second designs the dual number b. The orientation then results as a diagonal (α_{+}) or codiagonal (α_{-}) projection of the discantus point D onto the first axis depending on whether one has a sweeping or a hanging counterpoint. The discantus value on $\mathbb{Z}_{12}$ is $\alpha_{\pm}(D) = a \pm b$.

want to term these numbers as usual: 0 = prime, 1 = minor second, 2 = major second, 3 = minor third, 4 = major third, 5 = fourth, 6 = tritonus, 7 = fifth, 8 = minor sixth, 9 = major sixth, 10 = minor seventh, 11 = major seventh. We shall come back to the total interval structure $a + \epsilon \cdot b$ in the next sections.

Our initial question concerns dichotomic structures on the set $\mathbb{Z}_{12}$ of dual numbers. The concept of a *dichotomy* is this: A dichotomy is a partition (X/Y) of $\mathbb{Z}_{12}$ into two disjoint subsets X, Y of dual numbers, each containing six elements. Hence, Y is the complement $\hat{X}$ of X. Note that the partition is not ordered, i.e. $(X/Y) = (Y/X) = (\hat{X}/X) = (X/\hat{X})$.

There are totally 462 dichotomies. We are primarily interested in the dichotomy $(K/D) = (0, 3, 4, 7, 8, 9/1, 2, 5, 6, 10, 11)$ (Fig. 16) of the six consonances K and the six dissonances D.

This (K/D) dichotomy is by no means the only historically reasonable possibility to define consonances (and dissonances). Especially the fourth often changed from the consonant side to the other and vice versa (we refer to Apfel [42] for this "tragedy of the homeless fourth"). However classical counterpoint of intervals clearly positions it within the dissonant intervals.

Further examples of dichotomies are canonically furnished by the proper intervals of the medieval modes as defined from their tonics. We denote by X_i the diatonic scale $X = \{0, 2, 4, 5, 7, 9, 11\}$ with tonic i, whence the following list of modal dichotomies:

X_0	Ionian	$(2, 4, 5, 7, 9, 11/0, 1, 3, 6, 8, 10)$
X_4	Phrygian	$(1, 3, 5, 7, 8, 10/0, 2, 4, 6, 9, 11)$
X_2	Dorian	$(2, 3, 5, 7, 9, 10/0, 1, 4, 6, 8, 11)$
X_9	Aeolian	$(2, 3, 5, 7, 8, 10/0, 1, 4, 6, 9, 11)$
X_7	Mixolydian	$(2, 4, 5, 7, 9, 10/0, 1, 3, 6, 8, 11)$
X_5	Lydian	$(2, 4, 6, 7, 9, 11/0, 1, 3, 5, 8, 10)$
X_{11}	Locrian	$(1, 3, 5, 6, 8, 10/0, 2, 4, 7, 9, 11)$.

Before bringing order into the "zoo" of dichotomies, let us explain a very natural reason for the existence of the (K/D) dichotomy—besides the artificial counterpoint theory.

To this end, look at the chromatic scale X with tonic c in the just tuning over the primes 2, 3, 5. We already encountered this configuration in Fig. 1, where we wrote down the ratios of the tones with respect to the tonic. Recall that the just tuning was a transcendental justification of the medieval music: the concrete music rather was an accidental attribute in the sense of Aristoteles. However, we are not very sure, whether the working musicians did really care very much about these speculative concerns.

If X is represented as a point set within the three-dimensional lattice spanned by log(2), log(3) and log(5), it looks like Fig. 17.

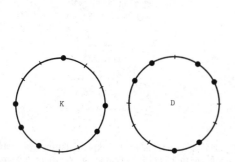

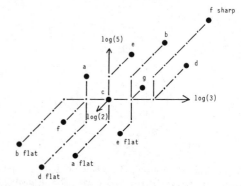

Fig. 16. The dichotomy of consonances $K = \{0, 3, 4, 7, 8, 9\}$ and dissonances $D = \{1, 2, 5, 6, 10, 11\}$ of classical counterpoint are but one of 462 possible partitions of $\mathbb{Z}_{12}$ into two disjoint subsets, each consisting of six elements. It is shown that it is one of the most important ones for topological and symmetry reasons.

Fig. 17. This configuration of 12 dots shows the chromatic scale for 2–3–5 just tuning. The 3-dimensional lattice is spanned by log(2), log(3) and log(5), three independent "vectors" which represent the octave, the fifth and the major third. Notwithstanding the irregular appearance, the configuration is highly symmetrical (see Fig. 18).

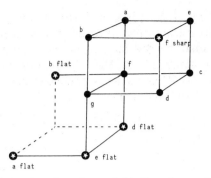

Fig. 18. The transformed chromatic scale under a suitable linear isomorphism reveals its full symmetry. Two parallel, vertical planes contain each six points: the front half and the back half. The names of the points refer to their original names before transformation. A uniquely determined inner symmetry A exchanges the lines through b, g, e flat and through a, f, d flat, then the lines through a flat, g, f sharp and through b flat, f, c and (hence) the points d and e. If transformed to well-tempered tuning, the symmetry becomes the autocomplementarity function of counterpoint.

In this position, it is difficult to believe that there might be any regularities on X. But after a linear symmetry transformation

$$S = \begin{pmatrix} 1 & 2 & 2 \\ -2 & -3 & -4 \\ 0 & 0 & 1 \end{pmatrix}, \quad \det(S) = 1,$$

SX appears as a very symmetrical object (Fig. 18).

In fact, SX is uniquely embedded in two parallel planes E_1, E_2 which stay vertically on our Fig. 18. This implies that X too, admits one and only one partition in two pieces which stay in two parallel planes:

$$X_1 = \{\text{e flat, g, b, d, a flat, f sharp}\};$$

$$X_2 = \{\text{d flat, f, a, e, b flat, c}\}.$$

Second, we recognize that X_1 and X_2 are isomorphic to each other under one and only one affine isomorphism $A = e^g \cdot A_0$, where e^g denotes the translation by $g = (-5, 2, 1)$ and where A_0 is the linear isomorphism

$$A_0 = \begin{pmatrix} 11 & 18 & 24 \\ -4 & -7 & -8 \\ -2 & -3 & -5 \end{pmatrix}, \quad \det(A_0) = 1.$$

For details, we refer to Mazzola [44]. If you write down the set X_1 in $\mathbb{Z}_{12}$ by simple transfer of the tone-names, c = 0, d flat = 1, etc., you precisely get the configuration of the discant tones of the consonant intervals as defined from the Leitton b in X. This defines a correspondence of the couple X_1, X_2 in just tuning, with the couple K, D in the well-tempered tuning. Observe that the transition from X to $\mathbb{Z}_{12}$ is a purely set-theoretic bijection by conservation of the names of the twelve tones! This means that, by no means, the unique symmetry between X_1 and X_2 has to be refound in terms of K and D. But, indeed a minor miracle happens:

Proposition 1

The unique symmetry A between X_1 and X_2 induces a unique symmetry $\tilde{A}$ from K to D, i.e. an affine isomorphism $\tilde{A} = e^2 \cdot 5$ transforming each consonance x in K to the dissonance $A(x) = 5x + 2$ of D. It is easily seen that $\tilde{A} \cdot \tilde{A} = 1$, and thus, $\tilde{A}$ is also a symmetry from D to K. The symmetry $\tilde{A}$ is called the autocomplementarity function of (K/D) (short: a.c. function).

Note that we are still dealing with dual numbers and not with tones. A musicological justification of symmetries on $\mathbb{Z}_{12}$ (=affine isomorphisms $e^t \cdot g : x \mapsto e^t \cdot g(x) = g \cdot x + t$, with invertible g and translation value t in $\mathbb{Z}_{12}$) will be given below.

At this point, we would like to show, how two dichotomies may be compared under symmetries on $\mathbb{Z}_{12}$. Suppose we are given two dichotomies $(X/\hat{X})$ and $(W/\hat{W})$. Then we say that $(X/\hat{X})$ is isomorphic to $(W/\hat{W})$ if and only if there is a symmetry f with $(fX/f\hat{X}) = (W/\hat{W})$, i.e. $fX = W$ or else $fX = \hat{W}$. Of course, being isomorphic to each other is an equivalence relation among the dichotomies, whence the partition of the collection of all 462 dichotomies into classes of isomorphic dichotomies. There are 26 dichotomy classes. The following list numbers these classes following the general numbering in Mazzola [45] and shows one representative dichotomy for each class (in fact the lexicographically first if ∗ precedes ∘):

63	∗∗∗∗∗∗∘∘∘∘∘∘	72	∗∗∗∗∘∘∗∘∗∘∘∘	81	∗∗∗∘∘∗∗∘∗∘∘∘
64	∗∗∗∗∗∘∗∘∘∘∘∘	73	∗∗∗∗∘∘∗∘∘∗∘∘	82	∗∗∗∘∘∗∗∘∘∗∘∘
65	∗∗∗∗∗∘∘∗∘∘∘∘	74	∗∗∗∗∘∘∘∗∗∘∘∘	83	∗∗∗∘∘∘∗∗∗∘∘∘
66	∗∗∗∗∗∘∘∘∗∘∘∘	75	∗∗∗∘∗∗∘∘∗∘∘∘	84	∗∗∘∗∗∘∗∘∘∗∘∘
67	∗∗∗∗∘∗∘∗∘∘∘∘	76	∗∗∗∘∗∗∘∘∘∗∘∘	85	∗∗∘∗∗∘∘∗∘∗∘∘
68	∗∗∗∗∘∗∘∘∗∘∘∘	77	∗∗∗∘∗∘∗∘∗∘∘∘	86	∗∗∘∗∘∘∗∗∘∗∘∘
69	∗∗∗∗∘∗∘∘∘∗∘∘	78	∗∗∗∘∗∘∗∘∘∘∗∘	87	∗∗∘∘∗∗∘∘∗∗∘∘
70	∗∗∗∗∘∗∘∘∘∘∗∘	79	∗∗∗∘∘∗∘∘∗∘∘∗∘	88	∗∘∗∘∗∘∗∘∗∘∗∘
71	∗∗∗∗∘∘∗∗∘∘∘∘	80	∗∗∗∘∗∘∘∘∗∗∘∘		

Observe just as a curiosity that the lexicographically first four classes are occupied by the above mentioned modal dichotomies. We have:

> No. 63—Lydian, Locrian;
> No. 64—Ionian, Phrygian;
> No. 65—Mixolydian, Aeolian;
> No. 66—Dorian.

At this point, it seems reasonable to give a musicological justification of these classes, i.e. of the equivalence relation defined by symmetries on $\mathbb{Z}_{12}$. One knows [39] that the group of symmetries on $\mathbb{Z}_{12}$ is generated by four particular symmetries:

$$C_1 = e^0 \cdot 11 : x \mapsto -x;$$
$$C_2 = e^0 \cdot 5 : x \mapsto 5x;$$
$$C_3 = e^3 \cdot 1 : x \mapsto x + 3;$$
$$C_4 = e^4 \cdot 1 : x \mapsto x + 4.$$

This (non-minimal) set of generators of the symmetry group has a nice topological property: call the number $d(x, y)$ the distance of two intervals x and y if $d(x, y)$ is the minimal length of a sequence of thirds to be added to x, in order to get y. For instance, $d(6, 7) = 2$ since $6 - 3 + 4 = 7$. This concept is important for the construction of chords in musicology. Now, it is straightforward that for all $i = 1, 2, 3, 4$, and for all x, y in $\mathbb{Z}_{12}$, $d(C_i(x), C_i(y)) = d(x, y)$, whence the distance of intervals is an invariant under the group of symmetries on $\mathbb{Z}_{12}$.

To view these facts geometrically, we mention that $\mathbb{Z}_{12}$ is isomorphic to the Cartesian product $\mathbb{Z}_3 \times \mathbb{Z}_4$ of the cyclic group $\mathbb{Z}_3$ and the cyclic group $\mathbb{Z}_4$ of order 3 and 4 (this is the Sylow decomposition of $\mathbb{Z}_{12}$). This permits us to draw the elements of $\mathbb{Z}_{12}$ as points of the torus as shown in Fig. 19.

In this geometry, the four generators C_1, C_2, C_3, C_4 appear as classical symmetries on the torus, see Figs 20(a–d).

They show a pregnant musicological interpretation. The symmetry C_1 is a transversal rotation of 180° and corresponds to the transition to the complementary interval. C_2 is a horizontal reflection and corresponds to the cycle of fourth, i.e. the semitone step is replaced by the fourth, a traditional way of counting "distances" in musicology. C_3 is a vertical rotation of 90° and corresponds to adding a minor third to an interval. C_4 is a tilting movement of 120° on the torus and corresponds to adding a major third to an interval. This means that every symmetry is musically significant in that it is composed of three fundamental types of symmetries: complementarity, cycle of fourth and stratification by thirds.

Coming back to the (K/D) dichotomy which is our final interest, why should this particular dichotomy have a special position among the 462 dichotomies? We noticed that (K/D) is given the autocomplementarity function $\tilde{A}$ transforming K and D into each other. Clearly, for each symmetry g, (gK/gD) still bears a uniquely determined symmetry between gK and gD, namely the conjugated

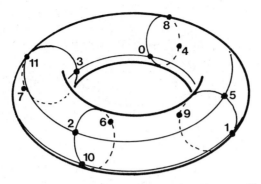

Fig. 19. The cyclic group $\mathbb{Z}_{12}$ is visualized as a set of 12 points on the torus. These are the crossing points of four small, vertical circles and of three larger, horizontal circles on the torus. The small circles correspond to the major third cycles, the large circles correspond to the minor third cycles on the chromatic scale. The topology of this configuration is musicologically significant.

symmetry $g \cdot \tilde{A} \cdot g^{-1}$, abbreviated $^{(g)}\tilde{A}$. More generally, if any dichotomy (X/Y) bears a uniquely determined symmetry f with $f(X) = Y$, this property is inherited by any other dichotomy isomorphic to (X/Y). We shall call such a dichotomy(class) a *strong* dichotomy(class).

Among the 26 dichotomy classes, besides the (K/D) class, there are five strong classes. The six classes are shown as configurations on the torus in Fig. 21.

On top, we have the unique strong modal dichotomy (No. 64): the dichotomy of the Ionian or major scale (!). At the bottom, we see the (K/D) dichotomy (No. 82), and, in between, the four remaining strong dichotomies Nos 68, 71, 75, 78 are seen. This spatial arrangement of the six strong dichotomies is by no means a casual one. The two polar antipodes, the major and the

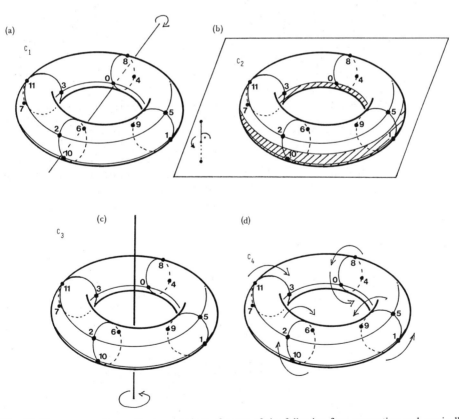

Fig. 20. Every symmetry on $\mathbb{Z}_{12}$ is a product of some of the following four generating and musically significant symmetries. (a) The symmetry C_1 is a horizontal rotation of 180° and signifies complementation of intervals. (b) The symmetry C_2 is a horizontal reflection and corresponds to the cycle of fourths. (c) The symmetry C_3 is a vertical rotation of 90° and corresponds to adding a minor third to an interval. (d) The symmetry C_4 is a tilting movement of 120° on the torus and corresponds to adding a major third to an interval.

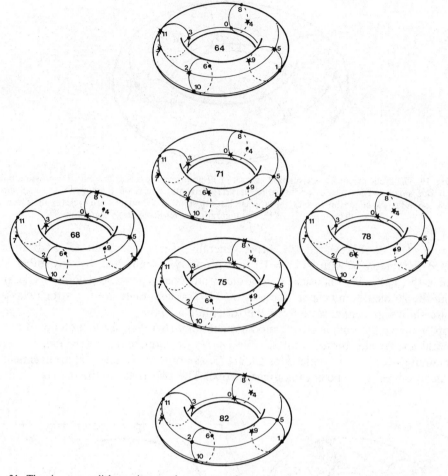

Fig. 21. The six strong dichotomies are shown as configurations (stars/dots) on the torus. The polar dichotomies No. 64, resp. No. 82 represent the six proper intervals on the diatonic scale from the major (or Ionian) tonic, resp. the consonances. The polar positions of these two dichotomies is motivated from the topology of the "toroidal representation" of the dichotomies.

consonance–dissonance dichotomies, are defined from topological properties based on the distance function $d(x, y)$ already discussed above. We just want to give two such topological properties defined on strong dichotomies.

  1. Call *diameter* of a strong dichotomy $(X/\hat{X})$ the number

$$d(X/\hat{X}) = \sum_{1 \leqslant i < j \leqslant 6} d(x_i, x_j),$$

where $x_1, \ldots, x_6$ is an arbitrary numbering of the elements of X. Evidently, $d(X/\hat{X})$ doesn't depend on this particular ordering and we have $d(X/\hat{X}) = d(\hat{X}/X)$. Observe that $d(X/\hat{X})$ is a class invariant.

  2. Call *span* of a strong dichotomy $(X/\hat{X})$ with a.c. function f the number

$$s(X/\hat{X}) = \sum_{i=1}^{6} d(x_i, f(x_i)).$$

Again, $s(X/\hat{X})$ is well-defined and is a class invariant of $(X/\hat{X})$. Figure 22 shows the positions of the strong dichotomies relative to diameter and span. We notice the polar positions of the major and the (K/D) dichotomies, as drawn in Fig. 22.

The (K/D) dichotomy shows the two halves being maximally separated on the torus, see Fig. 23. In polar position to (K/D), the major dichotomy shows maximal mixing of the two halves.

We would like to close this Section with a short comment on the historic role of the polar dichotomies Nos 64 and 82. As is well-known, the major–minor tonality replaced the dominant position of the counterpoint in theory and practice. The "disintegration" of the consonance–

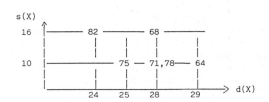

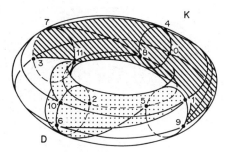

Fig. 22. This graphic shows the place of the strong dichotomies relatively to two coordinates: diameter $d(X/\hat{X})$ and span $s(X/\hat{X})$, which measure the mutual distribution of X and $\hat{X}$ on the torus. The diameter intuitively tells, how near the points of X are to each other, the span tells, how far the autocomplementarity function on X removes the points of X into the points of $\hat{X}$. The polar position of No. 64 and No. 82 becomes evident.

Fig. 23. The consonance–dissonance dichotomy is shown by two connected regions, K lined and D dotted. The minimal diameter of (K/D) is seen by a complete separation of the two halves of the dichotomy.

dissonance dichotomy as determinative musical category (remember Schönbergs emancipation of dissonances!) was accompanied by the development of the way of thinking based on the category of major (and minor) tonality. A possible role of the major dichotomy within this latter musicological way of thinking during the last three-hundred years should be challenging as a problem to the musicologist, since the antipodal position to the (K/D) dichotomy is given by a topological argument which has a determined musicological meaning.

3. COUNTERPOINT AS AN INFINITESIMAL VARIATION OF GREGORIAN CHORAL

After having recognized the outstanding position of the (K/D) dichotomy as a dichotomy of dual (interval) numbers, we are going to develop the counterpoint theory within a bipartite discussion. The first part deals with symmetries on intervals and their geometric/musicological meaning. The second part is devoted to a mathematical model of counterpoint leading to an explicit list of allowed and forbidden passages between two intervals and, in particular, to the rule of forbidden parallels of fifths.

As it has been shown in mathematical music theory [45], modulation may be expressed by a mathematical existence theorem, the existence of modulation quants, to be precise. Within mathematical counterpoint theory, there will appear a perfect parallelism: an existence theorem for allowed passages from interval to interval.

However, within counterpoint theory, the existence problem seems to be much more explicit than with modulation theory, as it is already stated from the musicological point of view, in contrary to the modulations. The problem is, to find a successor to a given interval by following certain rules.

It is quite astonishing that the existence problem has never been discussed in classical literature of counterpoint [2, 43, 46]. It seems that, to musicians and musicologists, existence relates to another ontological level as it does to the mathematician. It is the level of experience against that of axiomatic and logical deduction.

Let us just start with the computational part of the problem. As we have seen, an interval is given by an orientation (which is fixed by the context) and, within this fixed initial condition, by a new quantity $a + \epsilon \cdot b$ with a and b in $\mathbb{Z}_{12}$ being the cantus firmus and the dual number, respectively.

The mathematical meaning of the symbolic writing $a + \epsilon \cdot b$ is taken from the algebraic modelling of differential calculus in algebraic geometry. It intuitively means that we are given a point a on a surface and an infinitesimal tangent vector $\epsilon \cdot b$ [Fig. 24(a)].

In this terminology, to every cantus firmus point is attached its "tangent plane", i.e. the set of dual numbers $\epsilon \cdot b$, b in $\mathbb{Z}_{12}$, which is visualized in Fig. 24(b).

This geometric intuition is not far from reality if we recall that $\mathbb{Z}_{12}$ was represented as point set on a torus. The same is true for the set $\epsilon \cdot \mathbb{Z}_{12}$ of dual numbers, which are attached at every cantus firmus point. In other words, the above tangent plane is realized by a "small" tangential torus $\epsilon \cdot \mathbb{Z}_{12}$ at every point of $\mathbb{Z}_{12}$, as shown in Fig. 25.

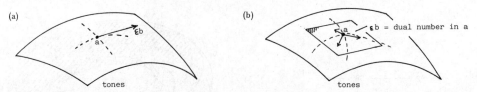

Fig. 24. (a) The symbolic writing $a + \epsilon \cdot b$ for an interval is visualized. The cantus firmus is shown as a point on a "surface of tones", the tangent vector $\epsilon \cdot b$ in a indicates the infinitesimal displacement of the discantus. (b) The set of intervals in a fixed cantus firmus a is shown as tangent plane to the "surface of tones".

We use the symbol $\mathbb{Z}_{12}[\epsilon]$ to denote the full set $\{a + \epsilon \cdot b; a, b$ in $\mathbb{Z}_{12}\}$ of intervals. From now on, we call the entities $a + \epsilon \cdot b$ dual numbers (and not only their infinitesimal parts $\epsilon \cdot b$). The consonant intervals will be the elements of the set $\tilde{K} = \mathbb{Z}_{12} + \epsilon \cdot K$, the dissonant intervals are the elements of $\tilde{D} = \mathbb{Z}_{12} + \epsilon \cdot D$, the complement of $\tilde{K}$ in $\mathbb{Z}_{12}[\epsilon]$. The tangent space $I_x = x + \epsilon \cdot \mathbb{Z}_{12}$ designs the intervals in cantus firmus x, the subset K_x of I_x denotes the consonances in x, the subset D_x the dissonances in x, i.e. $\mathbb{Z}_{12}[\epsilon]$ is the disjoint union of $\tilde{K}$ and of $\tilde{D}$ or of all I_x, whereas I_x is the disjoint union of K_x and of D_x, and finally, $\tilde{K}$ is the disjoint union of all K_x, whereas $\tilde{D}$ is the disjoint union of all D_x, see Fig. 26.

In mathematical music theory [45], a covering of a set of musical elements (tones) by subsets gives rise to a global structure: a so-called interpretation of this set of elements. Here, we have set up an interpretation of $\mathbb{Z}_{12}[\epsilon]$ by the K_x and the D_x. In particular, we shall write $I_x^{K/D}$ to denote the covering of I_x by K_x and by D_x. The *nerve* of an interpretation [45] in this case simply designs the fact that K_x and D_x are disjoint subsets of I_x and is a graph

$$\bullet\!\!-\!\!-\!\!-\!\!\bullet\!\!-\!\!-\!\!-\!\!\bullet$$
$$K_x \quad I_x \quad D_x$$

To end this mathematical excursion, let us explain, how arithmetic works on $\mathbb{Z}_{12}[\epsilon]$. You may add and multiply the symbolic expressions $a + \epsilon \cdot b$, $u + \epsilon \cdot v$ as usually in algebra, except that $\epsilon^2 = 0$;

Addition: $(a + \epsilon \cdot b) + (u + \epsilon \cdot v) = (a + u) + \epsilon \cdot (b + v)$

Multiplication: $(a + \epsilon \cdot b) \cdot (u + \epsilon \cdot v) = a \cdot u + \epsilon \cdot (a \cdot v + b \cdot u) \; [+\epsilon^2 \cdot b \cdot v = 0]$

What you have, is, what mathematicians call a commutative ring: addition and multiplication follow the usual rules of algebra, but division is not possible, generally. Within $\mathbb{Z}_{12}$, the invertible

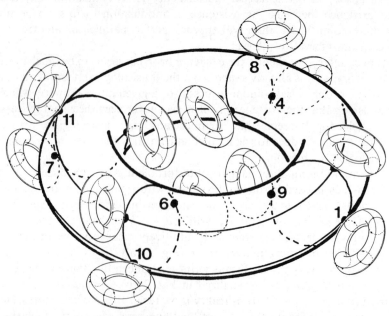

Fig. 25. The full set $\mathbb{Z}_{12}[\epsilon]$ of intervals is given by 12 small tori attached at a large one. The points on the large torus are the cantus firmus tones. In each cantus firmus, the 12 discantus points to be reached from the cantus firmus stay on a small tangential torus.

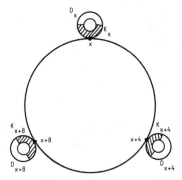

Fig. 26. Each tangent torus I_x is partitioned into the consonant half K_x (shaded) and the dissonant half. The picture shows a transversal section of $\mathbb{Z}_{12}[\epsilon]$ through a major third cycle.

elements are $1, 5, 7, 11$, whereas in $\mathbb{Z}_{12}[\epsilon]$, the invertible dual numbers are of the form $u + \epsilon \cdot v$, u invertible in $\mathbb{Z}_{12}$ and v arbitrary. We then have $u - \epsilon \cdot v$ as inverse of $u + \epsilon \cdot v$ since $(u + \epsilon \cdot v) \cdot (u - \epsilon \cdot v) = u^2 - \epsilon^2 \cdot v^2 = u^2 = 1$ in this case.

By these arithmetical operations, one introduces the symmetries on $\mathbb{Z}_{12}[\epsilon]$ as being the affine isomorphisms, i.e. the transformations of the form $g = e^{a + \epsilon \cdot b} \cdot (u + \epsilon \cdot v)$, $u + \epsilon \cdot v$ being invertible in $\mathbb{Z}_{12}[\epsilon]$. For any x in $\mathbb{Z}_{12}[\epsilon]$, we have (by definition, as already discussed above)

$$g(x) = a + \epsilon \cdot b + (u + \epsilon \cdot v) \cdot x.$$

These symmetries really are musically reasonable, see Fig. 27.

The point here really is, why we had to put $\epsilon^2 = 0$ in our context? Remember that ϵ announces the value which measures the displacement from the cantus firmus. In this view, higher powers of ϵ should measure "higher" displacements from the cantus firmus. Is this a musically reasonable onset? If one thinks of the history of counterpoint, it becomes clear that the three- or four-part composition didn't come up at once. It grew through several epochs from the Gregorian choral (for one voice only), became the ars antiqua for two voices as described by the old theorists (like Pseudo-Guido de Caroli loco, by whom we possess the oldest organum treatise) and developed to the three- and four-part composition as it is documented for the first time in a song of pilgrimage of Santiago de Compostela in the *Codex Calixtinus*. The counterpoint for several parts of the fourteenth and fifteenth century was built on the cantus firmus voice of the tenor, the bracket of the composed sound. But there are two techniques to be distinguished:

1. The technique of the English treatises as realized in masses by Machaut, for example. Apfel [42] calls this technique "vielstimmiger Verdoppelungsdiskant".
2. The technique of the continental treatises as realized in chansons by Dufay. Apfel calls it "freier vielstimmiger Diskant".

The English theory considers all voices exclusively in relation with the cantus firmus, the mutual

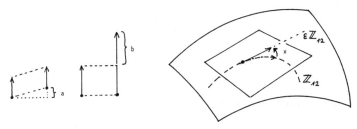

Fig. 27. The symmetries on $\mathbb{Z}_{12}[\epsilon]$ are musically reasonable. We consider the fundamental situations: (1) (left) e^a. = transposition by a; (2) (middle) $e^{\epsilon \cdot b}$. = augmentation by b; (3) (not visualized) u. = the "u-circle", i.e. an interval $a + \epsilon \cdot b$ with cantus firmus $= a$ and discantus $= a + b$ is transformed into $ua + \epsilon \cdot ub$, the interval with cantus firmus $= ua$ and discantus $= u(a + b)$, i.e. the u-circle is applied to both tones of the interval. For $u = 5$, the 5-circle is the well-known circle of fourths; (4) (right) the multiplication with ϵ means to view a tone x as being the discantus $\epsilon \cdot x$ of an interval relatively to the cantus firmus 0. Geometrically, this means to lift the "curve" through 0 and x to a tangent line $0 + \epsilon \cdot x$. This operation is the musically essential part of $1 + \epsilon$, which completes our list of generating symmetries.

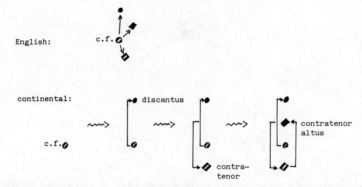

Fig. 28. The English and the continental technique of several parts counterpoint differ in so far as the first only considers the relations of the parts to the cantus firmus, whereas the second, following Monachus and Tinctoris, realizes a generative principle of mutual relationships among the four parts as shown.

positional relations of the voices were not relevant. Only the continental theory, as exposed by Tinctoris and Monachus, for example, took into consideration the higher relations (Fig. 28).

From this it follows that the contratenor not only has to consider the cantus firmus, but also the position of the discantus. Mathematically, this would result into a consideration of $a + \epsilon \cdot b + \epsilon^2 \cdot c$, c being the distance between the discantus and the contratenor, whereas the English reading would amount to consider the two intervals $a + \epsilon \cdot b$ and $a + \epsilon \cdot (b + c)$ from the cantus firmus to discantus and contratenor. This shows that it is consistent with the medieval musicology to distinguish different levels of interdependence among the voices. We shall, however, restrict ourselves to the basic two-part composition (or to the English technique, at best). This restriction is expressed with the above condition of the vanishing of the second power ϵ^2 of ϵ. We are really dealing with the "first approximation".

However, we have to return to the symmetries! We know that (K/D) is a strong dichotomy, the autocomplementarity function being $e^2 \cdot 5$. This fact translates to the $\tilde{K} - \tilde{D}$-interpretation of $\mathbb{Z}_{12}[\epsilon]$ as follows.

Proposition 2

(i) The symmetry $f = e^{\epsilon \cdot 2} \cdot 5$ on $\mathbb{Z}_{12}[\epsilon]$ exchanges $\tilde{K}$ with $\tilde{D}$, and every symmetry which has this exchange property, is of the form $e^w \cdot f$, w being in $\mathbb{Z}_{12}$.

(ii) Setting $^{(z)}f = e^z \cdot f \cdot e^{-z}$ for z in $\mathbb{Z}_{12}$, the only symmetry g which exchanges $\tilde{K}$ and $\tilde{D}$ and fixes the tangent space I_z in the cantus firmus z, is $g = {}^{(z)}f$.

For a proof, we refer to Ref. [46]. Instead, it is advantageous to realize the geometric situation described by the above proposition. To begin with, one checks that $^{(z)}f$ maps I_x onto I_{5x-4z} and exchanges K_x with D_{5x-4z}, resp. D_x with K_{5x-4z} according to the autocomplementarity function (see Fig. 29):

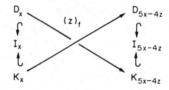

This means that we have a commutative diagram of interpretations:

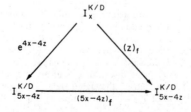

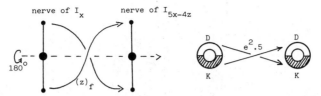

Fig. 29. On $\mathbb{Z}_{12}[\epsilon]$, the autocomplementarity function $^{(z)}f$ exchanges consonant and dissonant parts of the tangent spaces I_x and I_{5x-4z}. This is visualized by a screwing movement of 180° for the "nerve" of I_x.

The geometric meaning of this diagram is seen in Fig. 30. It tells us that $^{(z)}f$ looks like a screwing movement, which carries the space $I_x^{K/D}$ into the space $I_{5x-4z}^{K/D}$ and is obtained by a translation by $4x - 4z$ along the axis, followed by a rotation of 180°.

We are really confronted with what physicists in modern unification theories like to call *global symmetries* since, if you know, what happens under the autocomplementarity function in the fixed tangent space I_z, you know everything just by translating this situation to any other tangent space.

The global symmetry on $\mathbb{Z}_{12}[\epsilon]$ given by the autocomplementarity function, has properties which are decisive for the counterpoint way of thinking. In order to understand this, recall that we have to describe rules that tell us, which intervals can be reached from a fixed interval, and which ones are "forbidden". To be precise, we shall not deal with the last but one and with the next but one interval. Of course, this refined considerations would be of a certain relevance for deeper analysis, but we have to confront the elementary situation in the first place.

During the history of counterpoint, two *general principles* appeared.

1. Within the step for interval to interval (the so-called first species), a consonant interval has to be followed by another consonant interval, and the beginning has to be a perfect consonance, in other words: *only consonant intervals* are allowed.

2. *Independence of voices.* This second principle is different since it is purely negative. Observe that this is not the formal antithesis to the early parallel organum of the ninth century, in the sense that the discantus should be the pitch inversion of the cantus firmus melody. The countermovement of the discantus really has to be a contrast to the cantus firmus in some rather abstract sense.

Are there positive factors within this requirement of contrast and independency? The dominating structure of contrast, of course, is the consonance–dissonance dichotomy. In fact, the more sophisticated level of counterpoint (second species, cantus firmus vs two discant tones) suggests to put dissonances on the arsis (the non-accentuated time) as contrast to the consonances on the thesis. However, already within the first species, the passage from imperfect consonances to perfect ones (and vice versa) is the technique of contrast. This splitting among the consonant intervals reveals to be something like a transfer of the dissonance concept into the field of consonances. For, as we have seen in the discussion of the different theories of consonance and dissonance, the difference between perfect and imperfect consonances precisely stems from an increased dissonant ingredient within the imperfect consonances. This argument is valid either with Euler's gradus function, or with Helmholtz's theory.

So far, the situation seems to be paradoxical since one has to build sequences of consonant intervals (first principle) that should in some sense be alternating sequences of consonant and dissonant colour. In other words, not only dissonant intervals are to be separated from consonant

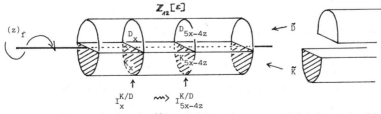

Fig. 30. The autocomplementarity function $^{(z)}f$ on $\mathbb{Z}_{12}[\epsilon]$ may be seen as a "global symmetry" in the sense of modern physics. The tangent space (a slice on our cylindrical visualization) I_x is transformed into I_{5x-4z} by a succession of a translation by $4x - 4z$ (along the axis), followed by rotation of 180°, i.e. the autocomplementarity function *on* I_{5x-4z}.

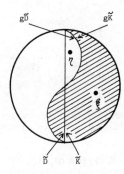

Fig. 31. For any couple of distinct intervals ξ and η, there is a suitable symmetry g on $\mathbb{Z}_{12}[\epsilon]$ such that the g-deformed *extended* dichotomy $(g\tilde{K}/g\tilde{D})$ separates ξ and η from each other, i.e. ξ sits in $g\tilde{K}$ and η sits in $g\tilde{D}$. This property of the consonances and dissonances is derived from the autocomplementarity function.

ones, but the same dichotomic onset should yield a separation among the consonant intervals, as if an imaginary (K/D) dichotomy was to exist.

However, in fact, the (K/D) dichotomy in the shape of the extended dichotomy $(\tilde{K}/\tilde{D})$ on $\mathbb{Z}_{12}[\epsilon]$ holds a property that fits with the above seemingly paradoxical requirement, and which is a consequence of the autocomplementarity symmetry, and more precisely of the fact that (K/D) is a strong dichotomy. The following theorem explicates this property. For a proof, we refer to Ref. [46].

Theorem 1

Let ξ and η be two different intervals (we write small Greek letters for intervals) in $\mathbb{Z}_{12}[\epsilon]$. Then, there is a symmetry g on $\mathbb{Z}_{12}[\epsilon]$ such that ξ lies in $g\tilde{K}$ and η lies in $g\tilde{D}$. Intuitively, this means that consonances and dissonances can be deformed in such a way as to separate any two different intervals, see Fig. 31.

This theorem is the basic feature for the following model of counterpoint. It should be read to the end that the consonance–dissonance dichotomy not only presents a fixed partition of the set of intervals, but it furnishes a means of global orientation among the intervals: any two intervals can be viewed as being either on the consonant or on the dissonant side after a suitable deformation of the fundamental dichotomy.

4. LOCAL SYMMETRIES IN COUNTERPOINT

The second part of this discussion deals with a mathematical model of counterpoint for the first species: note against note. We are given an arbitrary fixed consonant interval $\xi = x + \epsilon \cdot k$, k in K. The question is, which consonances η one is allowed to reach from ξ. We repeat that everything is intended to work within a fixed orientation, though this information is not explicitly used. But in calculating the finally permitted steps, the information will be relevant. According to the preceding discussion, we shall have to choose a symmetry g on $\mathbb{Z}_{12}[\epsilon]$ sharing certain properties of separation. We ask three properties from g:

(1) The consonance ξ has to be a *g-deformed consonance*, i.e. by definition, ξ is an element of $g\tilde{K}$.
(2) The deformed (extended) dichotomy $(g\tilde{K}/g\tilde{D})$ admits the autocomplementarity function $^{(x)}f$, leaving fixed the tangent space I_x.
(3) The symmetry g verifying properties (1) and (2) is such that $g\tilde{D}$ contains a maximal number of (real) consonances.

These requirements are quite natural. The first has already been discussed. The second claims that the deformed dichotomy $(g\tilde{K}/g\tilde{D})$ possesses the autocomplementarity function $^{(x)}f$ relative to the cantus firmus x of ξ. The third requirement simply claims a maximal freedom of choice, subjected to the conditions (1) and (2), a quantitative aspect of the independence principle. The main result, a counterpoint theorem runs as follows:

Theorem 2

Let ξ be a fixed consonance. Then

(i) there is a symmetry g satisfying conditions 1–3 above and such, that there is at least one consonance that is a g-deformed dissonance.

(ii) Let ξ and η be fifths. Then there is no g satisfying conditions 1–3 such that η is a g-deformed dissonance. This is only valid for fifths.

(iii) Let X_t be a mode with tonic t. Then, if both, the cantus firmus x and the discantus $\alpha_\pm(\xi)$ of ξ are in X_t, ther is a g satisfying conditions 1–3 and a g-deformed dissonance η with cantus firmus and discantus $\alpha_\pm(\eta)$ lying in X_t, η being a consonance.

For a proof, we refer to Ref. [46].

This theorem is read in the sense that the g-deformed dissonances η which are consonances, are the candidates for immediately succeeding ξ. Then the musicological statements are:

(i′) Within (well-tempered) chromatics, it is always possible to find a successor of a fixed consonant interval.

(ii′) There is exactly one general interdiction of parallels: that of the parallels of fifths.

(iii′) Within any mode, the successor equally exists.

Hence we are assured that there is no dead end when doing elementary counterpoint within our model. Observe that parallels of octaves and primes are not (generally) forbidden. However, this doesn't contradict the rules of counterpoint for a very simple reason. There are well-known examples, for example No. 51 in Ref. [47], where the passage from an octave to a prime is allowed. Older treatises even allow "killing" of voices, as the example from the *Loewener Traktat* [42]:

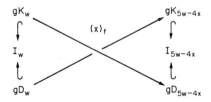

Now, these steps, when calculated modulo 12, become parallels of primes. For this reason, they should not be generally excluded and in fact aren't, as is readily seen from Table 1, where all possible steps relative to the sweeping counterpoint are listed in the sense that an asterisk * is put, whenever a passage from the interval of the top row to the interval of the right column corresponding to the star is allowed. This means that the step is allowed for at least one deformation symmetry g.

Before discussing some musicological comments on this model, we should like to make evident the geometric aspect of the counterpoint theorem.

A deformation g will again define an extended dichotomy $(g\tilde{K}/g\tilde{D})$ on $\mathbb{Z}_{12}[\epsilon]$ and hence a new interpretation $I_x^{g(K/D)}$ of every tangent space I_x by the dichotomy $(gK_x, gD_x) = (g\tilde{K} \cap I_x/g\tilde{D} \cap I_x)$. Now, the requirement (2) implies that under the autocomplementarity function $^{(x)}f$, $I_w^{g(K/D)}$ is mapped onto $I_{5w-4x}^{g(K/D)}$, i.e.

So far, the situation resembles the one for $\tilde{K}/\tilde{D}$, where $^{(x)}f$ is a global symmetry. But at a second sight, there is an important difference. Take, for example, the deformation symmetry $g = e^{\epsilon \cdot 8} \cdot (5 + \epsilon \cdot 4)$ for the consonance $\xi = \epsilon \cdot 9$. Then we have:

$$g\tilde{K} \overset{\bullet}{=} \mathbb{Z}_{12} \cdot (1 - \epsilon \cdot 4) + \epsilon \cdot e^8 \cdot 5 \cdot K,$$

Table 1

1. Direct motion of CF

```
CF:0 —→ 0   CF:2 —→ 2    CF:4 —→ 4     CF:5 —→ 5   CF:7—→ 7   CF:9—→ 9   CF:11—→11

0,4,7,9↴     0,3,7,9↴     0,3,7,8↴      0,4,7,9↴    0,4,7,9↴   0,3,7,8↴   0,3,8↴
  *  *0       *  *  *0      *  *  *0       *  *0       *  *0      *  *  *0    *  *0
*    *  *4    *     *  3    *     *  *3    *    *  *4   *    *  *4  *    *  *3  *    *3
*  *    *7    *  *    *7    *  *    *7     *  *    *7    *  *    *7  *  *    *7  *  *   8
*  *  *  9    *    *  9     *  *  *  8     *  *  *  9    *  *  *  9  *  *  *  9  *  *  8
```

2. Minor second upwards of CF

```
CF:4 —→ 5    CF:11 —→ 0

0,3,7,8       0,3,8
*  *  *  *0   *  *  *0
*  *  *  *4   *  *  *4
*  *    *7    *  *  *7
*  *  *  *9   *  *  *9
```

3. Minor second downwards of CF

```
CF:5 —→ 4    CF:0 —→ 11

0,4,7,9       0,4,7,9
*  *  *  *0   *  *  *  *0
*  *  *  *3   *  *  *  *3
*  *    *7    *  *  *  *8
*  *  *  *8
```

4. Major second upwards of CF

```
CF:0 —→ 2    CF:2 —→ 4    CF:5 —→ 7     CF:7 —→ 9     CF:9 —→ 11

0,4,7,9       0,3,7,9      0,4,7,9       0,4,7,9       0,3,7,8
*     *  *0   *  *  *  *0  *     *  *0   *     *  *0   *  *  *  *0
*  *  *  *3   *  *  *  *3  *     *  *4   *  *  *  *3   *  *  *  *3
*  *    *7    *  *    *7   *  *    *7    *  *    *7    *  *  *  *8
*  *  *  *9   *  *  *  *8  *  *  *  *9   *     *  *8
```

5. Major second downwards of CF

```
CF:2 —→ 0    CF:4 —→ 2    CF:7 —→ 5     CF:9 —→ 7     CF:11 —→ 9

0,3,7,9       0,3,7,8      0,4,7,9       0,3,7,8       0,3,8
*  *  *  *0   *  *  *  *0  *     *  *0   *  *  *  *0   *  *  *0
*  *  *  *4   *  *  *  *3  *     *  *4   *  *  *  *4   *  *  *3
*  *    *7    *  *    *7   *  *    *7    *  *    *7    *  *  *7
*  *  *  *9   *  *  *  *9  *  *  *  *9   *  *  *  *9   *  *  *8
```

6. Minor third upwards of CF

```
CF:2 —→ 5    CF:4 —→ 7    CF:9 —→ 0     CF:11 —→ 2

0,3,7,9       0,3,7,8      0,3,7,8       0,3,8
*  *  *  *0   *  *  *  *0  *  *  *  *0   *  *  *0
*  *  *  *4   *  *  *  *4  *  *  *  *4   *     *3
*  *    *7    *  *    *7   *  *    *7    *  *  *7
*     *  9    *     *  *9  *     *  *9   *     *9
```

7. Minor third downwards of CF

```
CF:5 —→ 2    CF:7 —→ 4    CF:0 —→ 9     CF:2 —→ 11

0,4,7,9       0,4,7,9      0,4,7,9       0,3,7,9
*  *  *  *0   *  *  *  *0  *  *  *  *0   *  *  *  *0
*  *  *  3    *  *  *  3   *  *  *  3    *     *  *3
*  *    *7    *  *    *7   *  *    *7    *  *  *  *8
*  *  *  9    *  *  *  *8  *  *  *  *8
```

8. Major third upwards of CF

```
CF:0 —→ 4    CF:5 —→ 9    CF:7 —→ 11

0,4,7,9       0,4,7,9      0,4,7,9
*     *  *0   *     *  *0  *     *  *0
*  *  *  *3   *  *  *  *3  *  *  *  *3
*  *    *7    *  *    *7   *     *  *8
*     *  *8   *     *  *8
```

—continued opposite

Table 1—*continued*

9. Major Third downwards of CF

CF:4 ⟶ 0   CF:9 ⟶ 5   CF:11 ⟶ 7

```
0,3,7,8          0,3,7,8          0,3,8
* * * *0         * * * *0         * * *0
* * * *4         * * * *4         * * *4
* *   *7         * *   *7         * * *7
* * * *9         * * * *9         * * *9
```

10. Fourth upwards of CF

CF:0 ⟶ 5   CF:2 ⟶ 7   CF:4 ⟶ 9   CF:7 ⟶ 0   CF:9 ⟶ 2   CF:11⟶ 4

```
0,4,7,9      0,3,7,9      0,3,7,8      0,4,7,9      0,3,7,8      0,3,8
* * * *0     * * * *0     * * * *0     * * * *0     * * * *0     * * *0
* * * *4     * * * *4     * * * *3     * * * *4     * * * *3     * * *3
* *   *7     * *   *7     * *   *7     * *   *7     * *   *7     * * *7
* * * *9     * * * *9     * * * *8     * * * *9     * * * *9     * * *8
```

11. Fourth downwards of CF

CF:5 ⟶ 0   CF:7 ⟶ 2   CF:9 ⟶ 4   CF:0 ⟶ 7   CF:2 ⟶ 9   CF:4 ⟶ 11

```
0,4,7,9      0,4,7,9      0,3,7,8      0,4,7,9      0,3,7,9      0,3,7,8
* * * *0     * * * *0     * * * *0     * * * *0     * * * *0     * * * *0
* * * *4     * * * *3     * * * *3     * * * *4     * * * *3     * * * *3
* *   *7     * *   *7     * *   *7     * *   *7     * *   *7     * * * *8
* * * *9     * * * *9     * * * *8     * * * *9     * * * *8
```

12. Tritonus of CF

CF:5 ⟶ 11   CF:11 ⟶ 5

```
0,4,7,9          0,3,8
    * *0          * *0
* * *  3          * * *4
*   * *8          * * *7
                  *   *9
```

i.e.

$$gK_w = w + \epsilon \cdot (5 \cdot K + 8 - 4w) \quad \text{and} \quad gK_{5w} = 5w + \epsilon \cdot (5 \cdot K + 8 + 4w),$$

whereas

$$K_w = w + \epsilon \cdot K \quad \text{and} \quad K_{5w} = 5w + \epsilon \cdot K.$$

This shows that the g-deformed consonances in w and in $5w$ do not come from one another by a translation of $4w$ within the cantus firmus part, as it was the case with (real) consonances in w and in $5w$. Geometrically, this means that $^{(x)}f$ is no more a global symmetry, but a local symmetry. Figure 32 shows the details.

This means that one cannot view $^{(x)}f$ as composition of translation plus rotation since the tangent spaces are subjected to deformations which depend on the cantus firmus (CF in Table 1) parameter, see Fig. 33 for a visualization of this local situation.

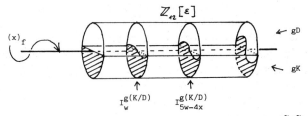

Fig. 32. In contrast to the extended consonance–dissonance dichotomy $(\tilde{K}/\tilde{D})$, the g-deformed dichotomies $(g\tilde{K}/g\tilde{D})$ no longer possess a global symmetry. Their autocomplementarity functions $^{(x)}f$ are "local symmetries", since, in general, the tangent space I_w, covered by the g-deformed consonances gK_w and the g-deformed dissonances gD_w, is no longer transformed into I_{5w-4x} via succession of a translation and a rotation (see Fig. 30). In fact, the g-deformed consonances in $5w - 4x$ do not coincide with the translation of the g-deformed consonances in w, by $4w - 4z$, in general.

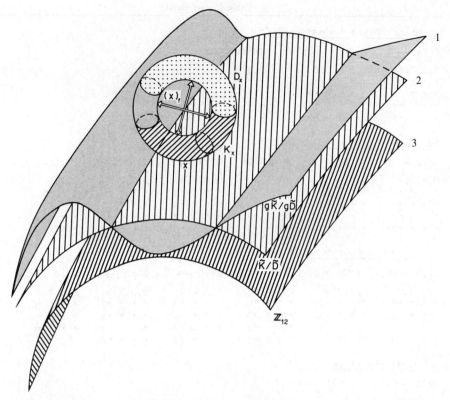

Fig. 33. The local character of the g-deformed dichotomies $(g\tilde{K}/g\tilde{D})$ (1) in contrast to the global character of (K/D) (2) is to say that the surface 1 splits the tangent space at x into two pieces gK_x and gD_x, which depend on the cantus firmus x, in contrast to the global splitting by the (undeformed) consonances and dissonances: their surface 2 stays at constant distance from the cantus firmus torus (3).

In physics, local symmetries are used to explain the appearance of forces. We recognize a nice analogy to physics in our model of counterpoint: Here too, the contrastful or, better, the strong steps from one interval to the next one, appear through local symmetries as a consequence of the autocomplementarity function. Note that the modulation model in Ref. [45] too, yields a close analogy to instances of modern physics (namely the principle of forces being transmitted through quanta like photons, gluons, bosons and gravitons).

Though this analogy is pure coincidence, it is clear that the mathematical model is a common structural principle to both: music and nature, at least in the sense of Hanslick [48], who maintained that music emerges from the brute noise by means of mathematics.

5. SOME MUSICOLOGICAL COMMENTS

We conclude the theoretical part of this paper by some musicological comments on the preceding model of elementary counterpoint. In general, it cannot be expected that the counterpoint theorem covers all the classical rules of counterpoint (two voices, first species), because we only considered the situation within $\mathbb{Z}_{12}$ and concerning the immediate successor interval. Nonetheless, we can observe an extensive congruence of the mathematically calculated possibilities with those from music theory. The following remarks can be made to illustrate these observations.

A. Examples from Fux's Gradus ad Parnassum

We refer to the book by Tittel [47], where examples of Fux are discussed. Except for the following three examples, everything within the total ten examples for different modes is in congruence with our Table 1.

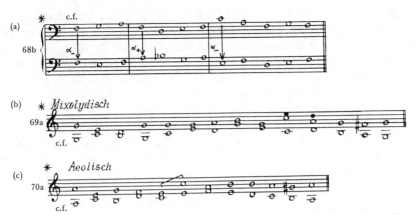

Fig. 34. Three examples from Fux [2] showing that the mathematical model is in good accordance with the musicological point of view. (a) Two changes of orientation, from hanging to sweeping counterpoint, and vice versa, make it necessary to divide this example into three parts: bars 1–3, 4–7, 8–12. (b) The two black notes design an error of the pupil of Fux, an "ottava battuta". This is also forbidden by the mathematical model. (c) The bad motion from bar 5 to bar 6 is forbidden by the mathematical model, Fux tolerates it as a limit case.

A.1. No. 68b, Lydian in Ref. [47]. [See Fig. 34(a)]

This hanging counterpoint shows one single step which is not allowed in our table: from bar 3 to bar 4. In fact, we observe a crossing of voices there, lasting till bar 7. However, our model does not cover changes of orientation! This means that we have to split this example into three pieces: bars 1–3, bars 4–7, bars 8–12, with three orientations: hanging, sweeping, hanging. Then everything works.

Remark. If you want to apply the table of allowed steps to the hanging counterpoint in order to decide, whether a step $x + \epsilon \cdot y$ to $u + \epsilon \cdot v$ is allowed, just apply the symmetry of pitch inversion $x \mapsto 12 - x$ and $u \mapsto 12 - u$ to the cantus firmus and consult the table for $12 - x + \epsilon \cdot y$ going to $12 - u + \epsilon \cdot v$.

A.2. No. 69a, Mixolydian in Ref. [47]. [See Fig. 34(b)]

Here everything works except the step from bar 10 to bar 11. There, we encounter a mistake of Fux's pupil, and Fux corrects it as to fit with our table, too. The pupil didn't observe the Italian rule of the "ottava battuta".

A.3. No. 70a, Aeolian in Ref. [47]. [See Fig. 34(c)]

Following our table, only the passage from bar 5 to bar 6 is not allowed. Tittel too, positions it as the limit of the permitted cases; Fux tolerates it.

B. Symmetries

The aspect of symmetries related to (K/D) is completely present within musicological considerations, though the autocomplementarity function does not seem to have been noticed. But the subsets $K_p = \{0, 7\}$ and $K_i = \{3, 4, 8, 9\}$ of perfect and imperfect consonances have been related to symmetries, see Fig. 35.

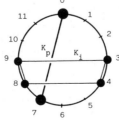

Fig. 35. Each subset, the perfect consonances K_p and the imperfect consonances K_i has inner symmetries. Both admit an inversion, K_i is even stable under a cycle of fourths. However, all of K doesn't admit any symmetry. Mizler [2] tried to make consonant the fourth in order to extend the symmetry of complementation on K_p to all of K.

The "principle of octave completion" is valid for K_i, it tells that to an imperfect consonance, the complement, i.e. the completion to the octave, is also an imperfect consonance. This obviously refers to the symmetry C_1 of complementarity (see Section 2). On the other hand, K_p is not stable under C_1, because the fourth and the fifth should be in C_1-correspondence within K. It is remarkable that L. Chr. Mizler in his translation of Fux's *Gradus* remarks [2, p. 38] that the fourth should also be consonant, because it is part of the triad c, g, c'; and this is the octave completion of the fifth. However, even together with the fourth, no cycle of fourth would fix the consonances, although, from the "acoustical" point of view, i.e. from the simplicity of the ratio of frequencies, the fourth should be nearly as consonant as the fifth. The fact that in counterpoint this position is not accepted, shows clearly that the medieval way of thinking the intervals is far from a purely acoustical point of view. It really seems to reflect a structural need for compositional purposes.

C. Rules of Counterpoint

The first species of counterpoint essentially resides on one rule: don't move into a perfect consonance by direct motion, i.e. when both voices simultaneously ascend, or descend, respectively. In particular, "covered" (parallels of) fifths and octaves are forbidden. In the mathematical model, covered fifths are allowed without exception.

However, if the cantus firmus moves $\pm$ a major second or a major third, covered octaves (or primes) are forbidden (in the mathematical model), whereas parallel octaves are allowed. Furthermore, steps such that the material sits within a diminished seventh chord are forbidden, which is very reasonable from the musicological point of view (tritoni!). Further, even within chromatics, steps within the augmented triad are not allowed in the mathematical model, a positive property in the sense that tonality cannot be broken up in direction of Messiaen's first mode [49]. But the musicological comparison is not yet fully developed, and more work to understand the two transitions from the mathematical model in $\mathbb{Z}_{12}$ to music theory and vice versa has to be done.

6. METHODS, TEST MATERIAL AND ANALYSIS OF THE NEUROPHYSIOLOGICAL INVESTIGATIONS

The present investigations are based upon a music test program presented to 13 patients at the University Hospital of Zürich, from 1984 to 1987. Every patient has been confronted twice with a tape program of $\frac{1}{4}$ h duration, and perceived through monophonic earphones. The patients were suffering from medically intractable complex partial seizures of suspected mediobasal temporal lobe origin and underwent presurgical evaluation with a view towards surgical epilepsy therapy. For an extensive EEG analysis, besides the scalp electrodes (their placement was according to the Hess system), intracerebral stereotaxic depth electrodes or foramen ovale electrodes have been implanted for precise localization of the primary epileptogenic area.

Two patients had stereo EEG explorations by means of several multichannel electrodes allowing for refined bipolar recording techniques (see Fig. 36). The other patients were examined by use of one- and four-contact foramen ovale electrodes placed epicortically at the mediobasal aspects of

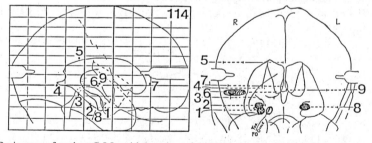

Fig. 36. Brain map of patient C.J-L. with location of depth electrodes. Bipolar derivations referred to in this article are: right hippocampus (RCA = 2/1–3), left hippocampus (LCA = 8/1–3) and right planum temporale close to Heschl's gyrus (RH = 6/5–6 and 6/8–9). The ten contacts of the multicontact probes are numbered from the inside out. Contact length and intercontact spacing is 1.5 mm; the diameter of the hollow-core electrode is 0.7 mm. In addition a "foramen ovale electrode" (FO) was inserted with its tip at the right uncus hippocampi. Since sintered AgAgCl was used for this epicortical electrode, DC recording was possible. (For details see Wieser *et al.* [50].)

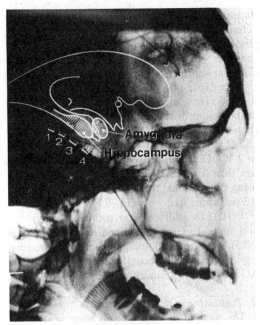
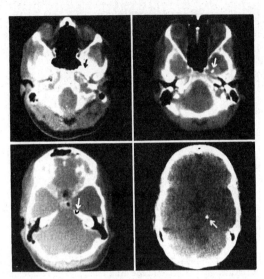

Fig. 37. Left: X-ray (lateral view) showing the foramen ovale electrode within the cannula together with the implanted depth probes. The contours of ventricular system and hippocampus are superimposed according to their positions in the previous neuroradiological examination. Numbers 1-4 indicate the contact position of the multiple-contact electrode. Right: Computertomogram of a patient with a right FO electrode (arrows). As can be seen, the tip of the electrode is in the ambient cistern.

the temporal lobes, close to the region of amygdala and hippocampus (see Fig. 37). See also Table 2 for more specifications.

None of the patients considered the (voluntary) test as being disagreeable.

It can be argued that epileptics probably are not comparable to normal humans with respect to their EEG profile. As a general observation, this is true, but these objections should not be overestimated for several reasons. First, all music tests were performed in interictal periods, and, epileptiform potentials are rather easily distinguished from those generated by musical stimuli by the experienced epileptologist. Second, the precise localization of the epileptogenic area from electrophysiological, histological and postoperative data, allows for a good estimation of possible influences of the epileptic disorder. Third, we should stress that to date, except for depth recordings in medically intractable epileptics, no possibility exists to observe states of the brain with the dense time resolution necessary for tests of musical structures.

A. The Music Program

The attempt to measure the performance of the brain during musical perception provides major methodological problems. Musical events are highly complex since, first, the sounds are situated in high-dimensional parameter spaces; second, the time density of the sound events is large; third it is not clear at all, what sort of information out of the totality of the perceived music is selected for further cognitive and/or emotional processing; last, many of the most elementary musicological concepts (like "interval", as we have seen) are not clearly defined—which makes even the discussion of experimental results quite difficult.

Many of the existing tests related to music deal with extreme situations from the musicological point of view. On one hand, click-tests are purely acoustical experiments, since the most elementary musical parameter of pitch is missing. On the other hand, some tests dealing with uninterrupted time windows of 16 s [32], and others, where some dozens of tones or even entire pieces of music are presented, cannot give a sufficiently detailed information about time density of musical syntaxis and semantic because many parameters, like rhythm, sound colour, harmony, melodic line, etc. are presented at once in complex pieces by Mozart and Tchaikovsky. These "realistic" tests probably are too complex since many partial effects to the brain will interfere and even possibly cancel each other. Therefore, we tried to stick to an extremely simple and clear (as we hope) semiology of

Table 2

PATIENT	SEX	BIRTH YEAR	SOCIAL POSITION	MUSICAL EDUCATION	PREFERRED MUSICAL STYLES	MEDICAL DATA	
C.L.	f	64	secretary	plays piano	Haydn, Tchaikowsky	1-bilateral limbic epilepsy with strong involvement of limbic midbrain area	FO 35 (14-07-87)
C.J-L.	m	50	academic	none	swiss folklore	2-unilateral right mesiobasal limbic epilepsy sucessfully operated (= seizurefree since op.)	FO 2; SEEG 114 (11-04-84) AHE R (Nr.94) (06-11-84)
F.R.	f	44	housewife	none	operettas	3-bilateral TLE;only moderate amelioration since palliative op.	FO 24 (24-12-86) AHE R (Nr.158) 12-02-87)
H.H.	f	62	ladie's taylor	none	none	4-left mesiobasal limbic epilepsy with secondary focus right temporal; schizophrenia-like personality disorder	FO 34 (06-07-87) (waiting for op.)
H.G.	m	53	farmer	none	popular songs	5-partial epilepsy with complex partial and generalized seizures;no clear-cut focal epileptogenic area detectable; inoperabel	FO 30 (01-06-87)
K.M.	f	46	civil servant	none	popular songs	6-right mediobasal limbic epilepsy	FO 29 (04-05-87) (waiting for op.)
L.M.	m	68	student	plays violin	Mozart, Britten	7-right mediobasal limbic epilepsy no clear-cut seizure originating substratum detectable; further examinations with stereotactically implanted electrodes necessary	FO 32 (16-06-87)
M.E.	f	49	hairdresser	none	classics & popular songs	8-bilateral limbic epilepsy (seizure origin right; auras left). Further examinations (sel.TL-Amobarbital-Memory test and PET studies) necessary	FO 33 (25-06-87)
M.M.	f	67	cook	none	popular songs	9-right mesiobasal limbic epilepsy; seizurefree since op.	FO 25 (26-02-87) AHE R (Nr.160) (25-03-87)
N.D.	m	76	schoolboy	plays piano	Mozart, Elton John	10-left mesiobasal limbic epilepsy seizure free since op. (amygdalo-hippocampectomy and removal of an oligodendroglioma)	FO 22 (11-11-86) AHE L (Nr.154) (13-01-87)
R.R.	m	57	technical merchant	music amateur no instruments	Rock & Pop	11-right mesiobasal limbic epilepsy seizurefree since op.	FO 26 (02-04-87) AHE R (Nr.166) (21-07-87)
S.M.	m	54	farmer	none	German folklore	12-left mesiobasal limbic epilepsy; with exception of 1 seizure, seizurefree since op.	SEEG 134 (26-01-87) AHE L (Nr.161) (31-03-87)
W.W.	m	45	typesetter	plays guitar	classics	13-left mesiobasal limbic epilepsy seizurefree since op.(amygdalo-hippocampectomy, removal of an hamartoma)	FO 23 (02-12-87) AHE L (Nr.167) (20-02-87)

musical structures. To be precise, we have chosen the well-tempered tuning because the average European is best accustomed to this tuning and because we are testing a counterpoint theory which strongly resides on this tuning. In later investigations we shall extend the music test to just and more exotic tunings as well.

It is known that musical signs (in the sense of semiology) are easily loaded with connotational meaning. For instance, listening to a violin timbre may evoke your memory to associate this sound with your hated grandmother, who miserably played the violin, etc., i.e. some strong emotion may heavily disturb the musical structures you are hearing. Thus, we strictly avoided natural instruments and the LP-recordings of well-known musical pieces and interprets. We also had to do so because the recordings within the limbic system, which is thought to represent the "emotional brain", could easily have provoked emotional artifacts.

For this reasons, we made use of the music computer, developed at the Department of Mathematics of the University of Zürich, see Fig. 38 and Ref. [51].

Fig. 38. The music computer $M(2, Z)Z^2$-o-scope of the University of Zürich consists of a monitor (top) to visualize tones as points by pitch and onset-time on the plane; (2) a periphery-unit (bottom) for input of musical and mathematical data; (3) a sound generator Yamaha TX7 (behind periphery unit) together with a music sound computer Yamaha CX5M (below monitor); (4) a terminal (below CX5M) and a CPU (not visible). This machine was designed for analytical purposes of mathematical music theory. All affine transformations on the point sets of parametrized tones are possible (for details, see Ref. [51]). For scientific music tests, in particular for Gestalt-tests, this computer is indispensable to produce objectively reproducible musical events.

Our music computer allows a complete control of the parameters, in particular the modelling of the instrumental character, the performance of all affine transformations on the two-dimensional space of pitch and onset-time, a central task for contrapuntal research and Gestalt investigations related to music.

With these prerequisites, we restricted ourselves to clear and simple problems concerning the structure of counterpoint, such as simultaneous and successive intervals, contextual tests, questions regarding the autocomplementarity function for the (K/D) dichotomy. We used the following instrumental colours, synthesized by a Yamaha TX7 digital sound generator through the algorithms for six operators (see Fig. 39):

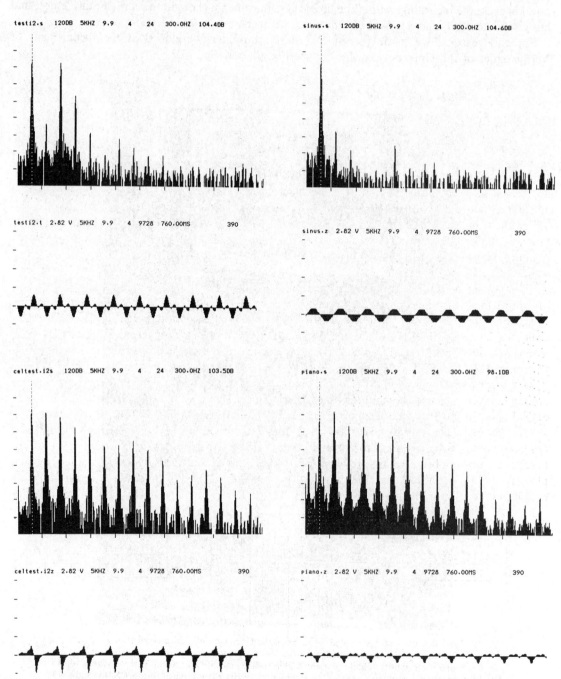

Fig. 39. Four timbres have been used in our investigation. Their spectra and the wave shapes are shown. (1) TEST (left, top), (2) SINUS (right, top), (3) CELTEST (left, bottom), (4) PIANO (right, bottom). They were produced by the Yamaha CX5M and TX7 with the Yamaha-algorithms using six coupled sine-waves or generators.

1. TEST

This is a neutral, rather clear voice, like an organ, but less natural in sound, the envelope being a smoothened rectangle [Fig. 39(1) for the spectrum].

2. SINUS

The envelope of this nearly pure sine wave is that of TEST [Fig. 39(2) for the spectrum].

3. CELTEST

This is the cello sound on the TX7 sound generator, without the vibrato, which would have been to complex for connotational reasons [Fig. 39(3) for the spectrum].

4. PIANO

This is the piano sound of the sound generator, not a very Steinway-like one, to say the truth.

The music program itself included several subtests of different orientations. Each candidate was confronted with totally about 700 sound events, their durations ranging from 0.1–3.0 s. As to the program for the first candidate, C.J-L. who was explored by stereo-EEG, we refer to Fig. 40. This first program was designed to make a general test of what could be of further interest, and we learned very much from this, as will be seen below. As a consequence, we could concentrate on a second program with more elaborate tests for some particularly important questions issued from the first test. In particular, parts 3 and 5, regarding isolated intervals were varied in sound colour and from the combinatorial point of view. As a consequence, we had to omit the repetitions (27 and 31) of subtests 25 and 29, as well as the subtests 9, 13, 15, 17, 21 (see Fig. 40). In addition, we varied the sound colours of subtests 11 and 19. The new test lasted 3 min longer (17′08″) than the first one, and this unfortunately is the maximal time for such a strenuous listening task to the patient. The patients were asked to close their eyes in order to avoid "noise" from visual inputs.

We now discuss the important subtests.

1. The RANDOM-sequence (see Partition 1) is played with TEST-sound and is a sequence of optimally distributed, successive consonances and dissonances. It is derived from a rare dodecaphonic series where not only all chromatic tones but also all intervals appear, and this is done in such a way, that consonant and dissonant intervals appear in strict alternation. This series is repeated three times to yield the RANDOM-sequence. This series is taken from the classification of Eimert [52], No. 421 in his list.

This means that, with respect to pitch, frequency of appearance of intervals and distribution among consonances and dissonances, one is given an optimal distribution. The sequence has been played accelerando from the beginning to the end, from duration $= 1.4$ s/tone to 0.1 s/tone. The RANDOM-sequence was played three times during the program (1, 7, 23; see Fig. 40) and was designed to test whether and how the different recorded brain structures would react to the described strategy of interval distribution. Further, we hoped that the RANDOM-sequence would play a role of neutralizer between the more specialized subtests in case it would show an equilibrated response on the EEG recordings. We preferred this RANDOM-subtest to the often

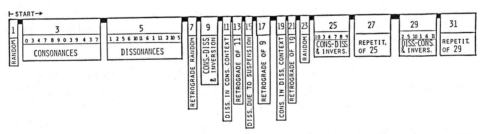

Fig. 40. Sequences of subset presentation constituting the music interval test for the first patient C.J-L. Its entire duration is 14 min. The relative duration of the subsets is preserved in this figure. Black segments indicate pauses. Large numerals indicate the subsets, as described. Small numerals in subset tests 3 and 5 refer to the musical intervals as determined by the 12 semi-tones in the interval-size continuum. In subsets 25 and 29, small numerals refer to the consonant (dissonant) interval which was confronted successively with every dissonant (consonant) interval.

Partition 1

used white noise for statistical references, since white noise is an event outside the category of musical events where pitch is an elementary constituent which white noise doesn't have.

2. Isolated consonances and dissonances. These intervals were played as successive and as simultaneous pairs of tones, and corresponding to size, complementarity and using the TEST-sound, see Partition 2.

Partition 2

Except for the first patient C.J.-L., the different intervals were also played for the SINUS, CELTEST and PIANO sound, and in a different ordering corresponding to the RANDOM-series, see Partition 3.

3. Consonances and dissonances in different contexts. To all patients, we presented a counterpoint note against note from Fux's book. However we replaced one consonance by a unprepared dissonance. In the upper row of Partition 4 we have the test for patient C.J.-L. which was presented with the TEST-sound, in the lower row, we have the test for the other patients; the colours TEST and CELTEST were used.

Then, every candidate was confronted with a sequence of dissonant intervals derived from the above examples of Fux by alterations. However, this time we kept only one interval to remain consonant. Again the first row shows the test for C.J.-L., and the second the one for the other candidates, see Partition 5. The colours were the same as in the preceding subtest.

TEST, CELTEST, SINUS, PIANO

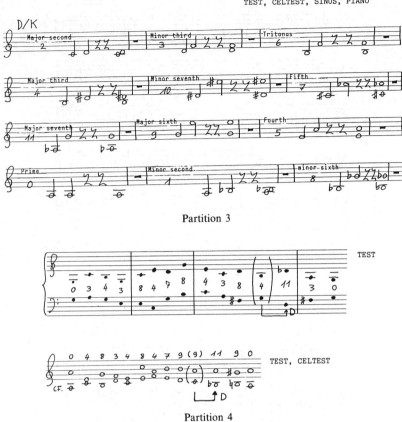

Partition 3

TEST

TEST, CELTEST

Partition 4

TEST

TEST, CELTEST

Partition 5

The first candidate could also hear the retrograde of the two examples. Every interval had a duration of 0.6 s. The first candidate was also confronted with a prepared dissonance with the character of a suspension ("Wechsel-4-6-Dissonanz" in German, see de la Motte [53, p. 48]: Partition 6). The TEST-sound was applied.

TEST

Partition 6

4. Test of the autocomplementarity function. The idea to test this symmetry is as follows: every consonant interval is confronted with every dissonant interval, and vice versa, see Partitions 7 and 8.

Partition 7

Partition 8

In this situation, one looks at the EEG of every dissonance confronted with a fixed consonance x and compares these six EEG recordings with each other. One hopes that the value $\tilde{A}(x) = 5x + 2$ will show a particular shape within the EEG. The idea may be compared to a criminalistic confrontation of a series of candidates for the crime, with a witness. Each interval has a duration of 0.5 s, the sound is TEST.

B. The Analysis

In principle, every sound event is analyzed by the same procedure based on the standard Fourier power spectrum [54]. But we equally checked the EEG outputs visually.

A complete analysis has been accomplished for the first candidate. The EEG was recorded on PCM tape (476 samples/s), then demultiplexed and finally digitized (256 samples/s) in order to make the calculations on the CDC of ETH Zürich, using programs for Fourier analysis from the Institute of Geophysics, ETH Zürich. The data of the other patients are directly processed in digital form. These data are actually involved in an automatic analysis program. An automatic processing of a dozen tests has been unavoidable since every candidate was confronted twice with about 700 events recorded in 16 to 32 different EEG channels. Thus roughly ten million numerical data had to be analyzed. The complete analysis of the first candidate comprised a processing of 10'260 numerical data from the spectral analysis, and referred to three bipolar channels: RCA (= Right Cornu Ammonis), LCA (= Left Cornu Ammonis) and RH (= Right Heschl), see Fig. 36.

The quantitative analysis of EEG waves is a problem which is intimately related to the expected semantics. *A priori*, it is not known, what sort of numerical data should be extracted from the EEG to get a reasonable semantic for music perception, if such data can be found at all. To begin with, we looked at data extracted from the traditional frequency bands, and this choice was rather successful. As Giannitrapani rightly remarks [32], one should attack the semantical problem of the EEG from the topographical distribution of the activities within the different frequency bands.

As we didn't observe relevant activities above 40 Hz, we limited the analysis to the four classical bands $\delta = 0–4\,\mathrm{Hz}$, $\theta = 4–8\,\mathrm{Hz}$, $\alpha = 8–14\,\mathrm{Hz}$ and $\beta = 14–40\,\mathrm{Hz}$.

If E denotes an event (a tone, an interval of a determined duration), we deduce the power $P(E)$, the δ-power $P_\delta(E)$, the θ-power $P_\theta(E)$, the α-power $P_\alpha(E)$ and the β-power $P_\beta(E)$. From this the *spectral participation vector*

$$S(E) = (P(E), P(E)/P_\delta(E), P(E)/P_\theta(E), P(E)/P_\alpha(E), P(E)/P_\beta(E)$$

is derived. The coordinates $S_\delta(E) = P(S)/P_\delta(E)$ etc. are called the *participation coefficients*. This formal terminology is inspired by the well-known fact that S_α is vigilance-dependent and that S_β seems to be coupled with "active intelligence" (Giannitrapani), i.e. S_β decreases if the candidate intensely tries to find out the meaning of some proposed configuration. (Muscular artifacts seem to be excluded in this finding.)

We should draw our attention to the remark of Giannitrapani [32] that for his music test (Tchaikovsky's Marche Miniature), the two segments of each 16 s with 40 s in between showed a low reliability (Spearman's r^2), in particular for the β-band (see Fig. 41), such that a tone-by-tone analysis is preconized for (β-) participation coefficients, or for the participation vector.

To obtain a reasonable mean-value of participation, we introduced the sum

$$v(E) = S_\theta(E) + S_\alpha(E) + S_\beta(E),$$

which is called the *participation* (value) of E. We omitted S_δ here, because it could easily be affected by noise. As every other numerical value, the participation is at most a plausible mathematical construction. Its semantic qualification only justifies the use of the concept. Intuitively, $v(E)$ is something like the length of the vector $(S_\theta, S_\alpha, S_\beta)$, see also Fig. 42.

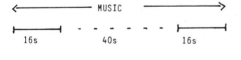

Fig. 41. The music test of Giannitrapani [32] presents Tchaikovsky's Marche Miniature. Two sequences of each 16 s, with 40 s space were compared with each other to calculate the reliability. This was much lower than the corresponding reliability while resting, in particular for β-frequencies.

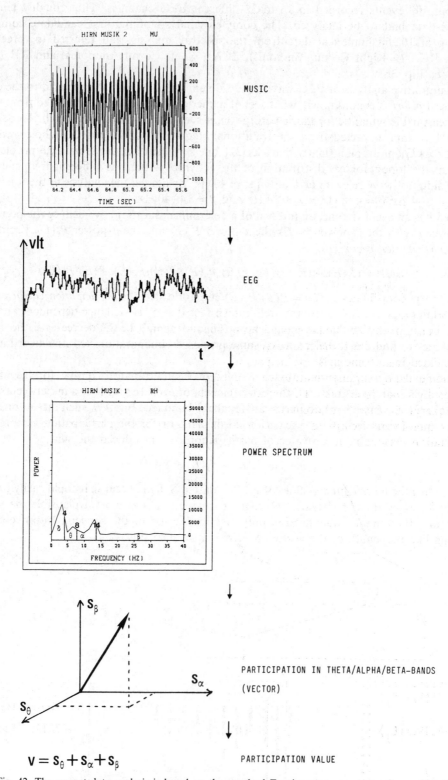

Fig. 42. The present data analysis is based on the standard Fourier power spectrum, from which the δ-, θ-, α- and β-frequency bands are extracted to define a spectral participation vector, whose three coordinates (of five) S_0 = total power/θ-power, etc. are shown, as well as the "average" participation value.

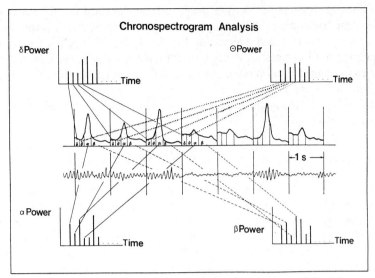

Fig. 43. In this chronospectrogram analysis, the successive time windows of 1 s define a sequence of four-dimensional vectors whose coordinates are the δ-, θ-, α- and β-powers.

If one starts from the well-known chronospectral representation (Fig. 43), the spectral participation is essentially the vectorial representation of the power distribution, with the inverse coordinates. Hence the participation vector and the vector of the different partial powers are related by a transformation which is essentially the Cremona birational transformation known to algebraic geometers. This means that the chronospectral representation associated with a sequence $E_1, E_2, \ldots$, of events, is linked to the participational representation, as sequence of three-dimensional vectors as shown in Fig. 44, by a classical Cremonese-like algebraic transformation.

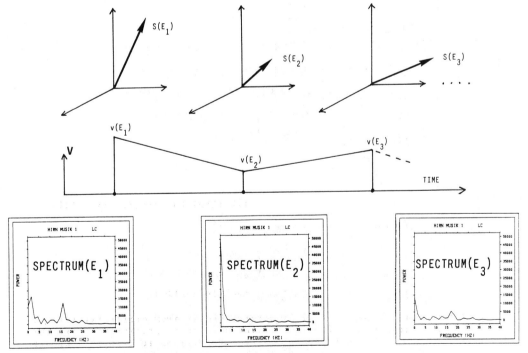

Fig. 44. The sequence of events $E_1, E_2, \ldots$, yields a sequence of spectra and from there a sequence of spectral participation vectors, resp. participation values. The formalism of chronospectrograms and the formalism of spectral participation are related by an algebraic Cremona transformation and (hence) are essentially equivalent.

Note that it is not reasonable to analyze events whose duration is too short, since the power spectrum may become too coarse. For example, extremely short musical events of 1/20 s duration yield only two values for $P(E)$ and everything breaks down. This alludes to something like an uncertainty principle for EEG tests to perception of music. However, the minimal durations of the events which we analyzed, is 0.5 s, which seems to be at the limit of tolerance for the given frequency bands, in particular as to the narrow δ- and θ-bands.

For candidate C.J-L., all participation vectors have been calculated, resulting in 10,260 numerical data, as mentioned above.

7. GENERAL OBSERVATIONS OF EVOKED EEG RESPONSES

To begin with, let us look at some examples which show that musical stimuli may give rise to evoked potentials or, vice versa, suppression of EEG patterns with the onset of music.

A first series of observations by visual examination is shown on Figs 45–47. They indicate marked "evoked potentials" at the beginning of the musical event, within the first half second (Figs 45 and 46). However, in Fig. 47, the reaction to four units of six successive pairs of intervals (test of the autocomplementarity function) seems to show increasing latency (up to nearly 1 s), as the unit is presented the third and the fourth time.

The next figures (Figs 48–50) illustrate three remarkable findings. The example in Fig. 48 relates to seemingly opposite reactions of the locally recorded EEG. In the upper example (a) slow focal activity is suppressed at the onset of the musical stimulus, a single tone, whereas the second example (b) refers to the start of a slow focal activity of same frequency (3/2 Hz) at the onset of the musical stimulus. Figure 49 shows a rather long periodicity (6 s) of the EEG, which seems to be driven by the 1/3 s periodicity of the melodic sequence. This suggests that not only single evoked potentials at the beginning of a musical stimulus are reflected within EEG activities, but the reaction has to be considered along the entire unit of music, like an uninterrupted chain of tones as shown in Fig. 49. Figure 50 shows that even for somebody who, as it is the case here, is not interested in music, the reaction to a rather long musical sequence (7 s) may persist during the whole presentation.

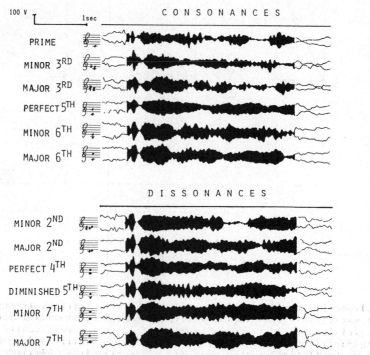

Fig. 45. Evoked responses in the right auditory cortex to consonant (top) and dissonant (bottom) musical intervals. The area between the two corresponding EEG channels (patient C.J-L., upper = 6/5–6, lower = 6/8–9) is blackened during the time the interval was presented.

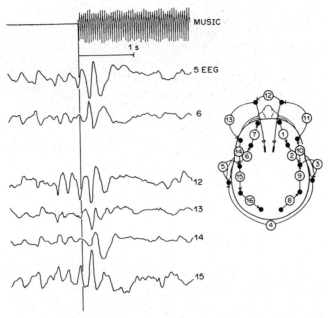

Fig. 46. EEG (patient C.L.) before and during presentation of an isolated simultaneous musical interval (major second). The microphone signal, as written by the EEG machine, is at the top of the figure, the EEG channels 5, 6, 12–15 represent selected bipolar recordings as indicated. Note that channels 12 and 13 record from the symmetrically inserted "foramen ovale" electrodes. In contrast to the right-sided EEG derivations, where no gross changes are visible, the illustrated left EEG derivations show a marked "evoked potential", with the main negative component of about 180 ms left temporal.

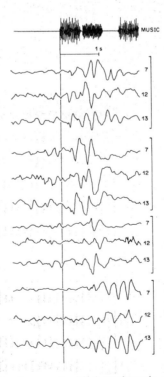

Fig. 47. Patient C.L.: four EEG sections, each with three channels, are illustrated underneath the other. They are selected with reference to the onset of the 12 segments (each segment represents the confrontation of one particular consonance with all dissonances or vice versa; segment duration 11 s) of the autocomplementarity function test (see text). Topography of the EEG derivations is the same as in Fig. 46. Note the marked EEG changes, which predominate left, during the first 2 s following the start of the segment. They occur earlier at the onset of this particular subtest (upper part) and with an increasing delay as the subtest proceeds (lower part of the figure). As the EEG channel 7 derivates from left frontal scalp, the illustrated EEG reactions might represent a phenomenon similar to K-complexes.

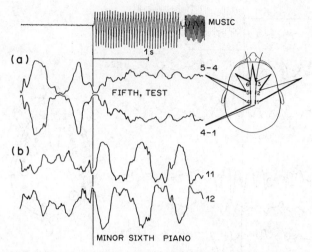

Fig. 48. (a) Patient M.E.: bilateral, actually *left* predominant limbic temporal lobe epilepsy: bipolar EEG recording from the *left* parahippocampal gyrus (multicontact "foramen ovale" electrodes) showing marked desynchronization of the slow focal rhythmic activity at the onset of the first tone of the successive fifth interval (TEST-sound). (b) In contrast to (a), in this patient (L.M. with suspected *right* temporal lobe epilepsy), slow high voltage rhythmic activity of about the same frequency as in (a) was recorded from the *right* uncus hippocampi as soon as the first tone of the minor sixth (PIANO-sound) was heard.

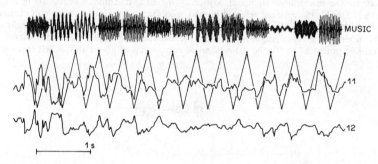

Fig. 49. Patient M.M.: right bipolar "foramen ovale" electrode recording from parahippocampal gyrus during listening to RANDOM-sequence (see text). The frequency of the tone succession of 2.2/s seems to drive the underlying slow rhythmic EEG activity (during 6 s). The inserted zigzag should help to visualize this synchronization.

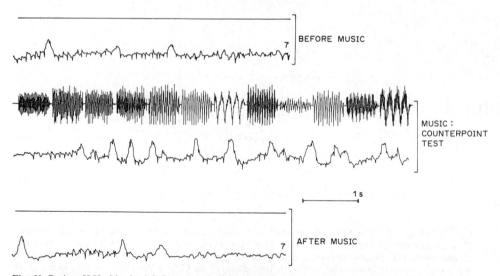

Fig. 50. Patient H.H.: bipolar left frontal scalp EEG derivation before (above), during (middle) and after (below) the first COUNTERPOINT-subtest. In this patient with no musical interests and no gross *visible* scalp EEG changes to the music test, only during this particular subtest a certain bilateral frontal EEG reaction was seen.

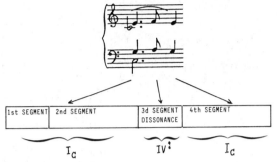

Fig. 51. A prepared fourth dissonance of suspension-like character is presented (3d segment) to patient C.J-L. Three segments are taken for the analysis, the 3d is compared with its predecessor and successor. The segment lengths indicate the relative duration of the segments. The duration of the dissonant 3d segment is 1.1 s.

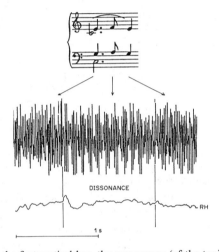

Fig. 52. Patient C.J-L.: At the first vertical bar, the consonance (of the tonic major triad) changed to a subdominant chord containing the dissonant fourth. (Above: EEG machine-deformed microphone signal of music.) Coincident with this harmonic change, the EEG at right Heschl's gyrus (RH) shows a marked evoked potential.

Some more interesting findings emerged for the first patient. The prepared dissonance of the fourth (suspension) shows that dissonances are perceived within codified contexts. This test was performed twice. We measured four segments (Fig. 51) of which the third represents the dissonant part. This dissonance is clearly marked by an evoked potential as shown in Fig. 52.

If the power is measured in all channels (RCA, RH, LCA), one notices that it yields a very strong maximum in every channel, from the second to the third and fourth segment, on the dissonant *third* segment (Fig. 53).

Observe that the left hippocampus nearly doubles its power as the prepared dissonance appears, the mean increase being 69% above the mean of the preceding and the succeeding consonance. The distinguished position of the left hippocampus for recognition of simultaneous intervals will be corroborated by the results of the next sections.

The RANDOM-subtest for the first patient was presented five times. The total sequence of 13.5 s duration was cut into three equal pieces of each 4.5 s duration for our analysis. Since the RANDOM-sequence moves accelerando with the first patient, the first segment contains 4 tones, the second contains 10 and the last contains 23 tones. The data are visualized in Fig. 54.

One recognizes that with the exception of P and S_β of RH, all values could be supposed to be constant within the standard deviation (indicated). Hence, the limbic EEG seems to be insensitive for accelerated motion, and the equal distribution of the tones, intervals and K–D representatives is reflected by the invariance of the participation vectors against the choice of 4, 10 or 23 tones out of the total sequence.

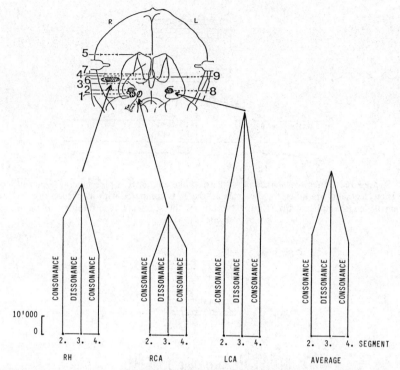

Fig. 53. The increase of spectral power during the prepared dissonance (3d segment) with respect to the preceding and succeeding consonances (major tonic triad) is shown. The scale is in units of the program of the Department of Geophysics, ETH, Zürich. Observe the leading role of the left hippocampus.

It is interesting to observe the decrease of P and of S_β in RH, the right auditory cortex. This means that the acceleration of the motion is related to a strong increase of (relative) β-power. If one considers the sensitivity of the right hemisphere for Gestalt-qualities, the increase of temporal density in the melodic ambit during the accelerando of the RANDOM-subtest probably demands an increased active intelligence within the auditory cortex in order to understand the melodic Gestalt, an interpretation which confirms Giannitrapani's hypothesis on the role of the β-band.

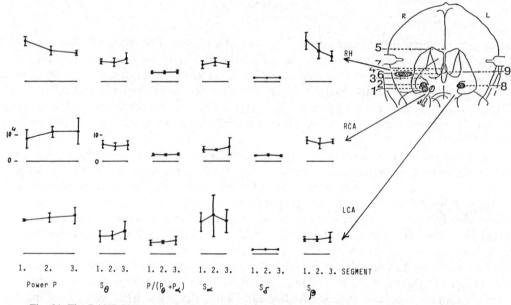

Fig. 54. The RANDOM-sequence of 13.5 s was presented five times to C.J-L. Each sequence was split into three 4.5 s portions which were analyzed as shown. Observe the increase of relative β-power for the right Heschl, which is parallel to the increase of musical tempo of the RANDOM-sequence. The limbic recordings do not show any significant changings along the three segments.

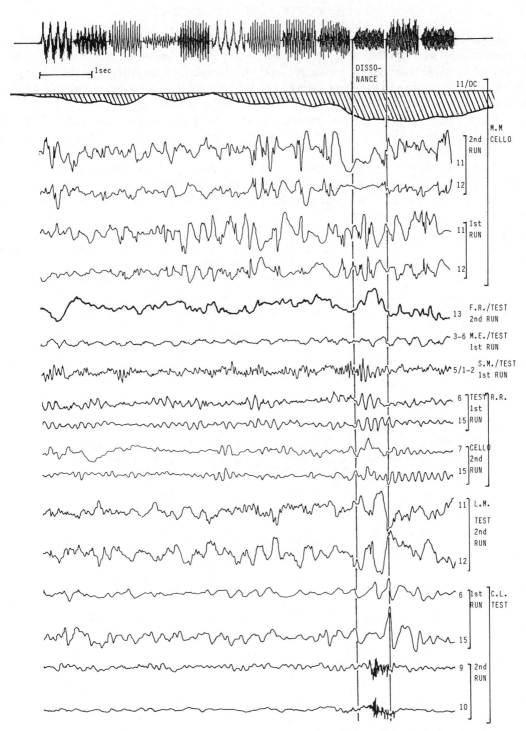

Fig. 55. Synopsis of electrophysiologically traced EEG changes from different patients at various locations during the first COUNTERPOINT-subtest. Remember that this subtest consists of a defect counterpoint arrangement note against note from Fux's *Gradus ad Parnassum*. Among the 12 consonant intervals, we changed the 10th into a dissonant major seventh (marked between vertical bars). Note that every patient is represented with a variable number of channels (according to the square brackets on the right). With the exception of patients M.E. and S.M., the identification of the EEG derivations follows the schema shown in Fig. 46. Patient M.E.'s EEG location is depicted in Fig. 48(a). Patient S.M. had stereotactically implanted depth electrodes, the shown EEG derivation (= 5/1–2) recorded from left anterior cingulate gyrus. Note (a) the uppermost channel, with the DC recorded EEG (right against left foramen ovale Ag/AgCl-electrode), which shows a right positivity (or left negativity) in anticipation of the dissonant interval (which the patient heard the third time); and (b) the muscle activity in the two lowermost channels.

8. DEFECTIVE COUNTERPOINT

This subtest of a concrete counterpoint note against note concerns a dissonance (major seventh) which we implanted in this classical contrapuntal motion. Most of our patients showed clear EEG changes at the appearance of the dissonance after nine well-composed consonances. Figure 55 shows 18 examples of EEG changes which were easily recognized by visual inspection.

Contrary to this result, most of the patients did not show any particular change of EEG activity for the second part or this counterpoint subtest, i.e. when dissonances were presented with only one consonant exception.

This suggests that to our patients, who are "normal Europeans" as to their musical preferences and habitudes, the consonant context really is not equivalent to the dissonant context, as far as the vigilance for a change of intervals into the complementary category of the (K/D) dichotomy is concerned.

Only in two patients (C.L. and L.M. were about 20-years old, musically well-trained and very sensitive personalities) visible changes at the appearence of the consonant interval within the dissonant "counterpoint" have been observed, see Fig. 56.

It is interesting to notice that most of these examples refer to left hemispheric EEG channels. This may be related to left hemispheric dominances in processing of music, as has been suggested by Giannitrapani (Fig. 57) and as it can be deduced from our results reported in the following sections.

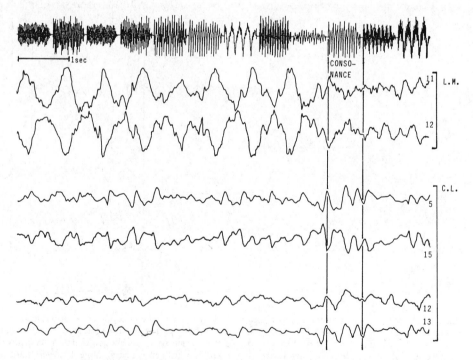

Fig. 56. EEG reactions of two patients (L.M., C.L.) to the second part of the COUNTERPOINT-subtest. This consists of a sequence of dissonances derived by alterations from the example of Fux, used in the first part of the COUNTERPOINT-subtest, except for the 10th consonance which is unaltered. In general, in the examined population, a reaction to the single consonance (marked between bars) was not detected. These two illustrated patients are the only exceptions and differ from the other patients by their high standard musical education, interests and activities. Patient L.M.: bipolar right foramen ovale electrode recordings, for identification see Fig. 48(b). Patient C.L.: bipolar left temporal basal scalp electrodes $(= 5, 15)$ and left foramen ovale electrode recording $(= 12, 13)$; for identification see Fig. 46.

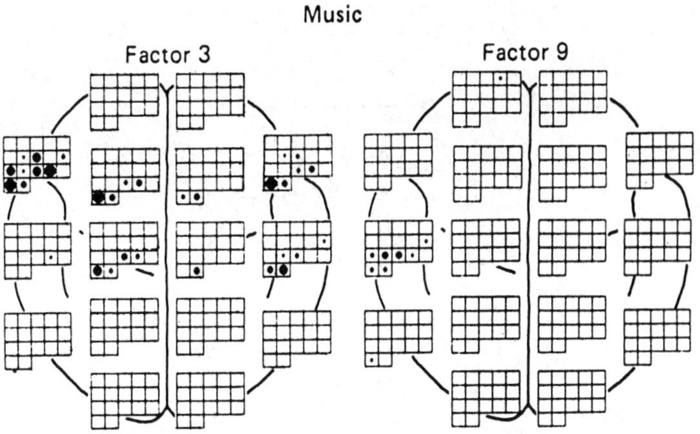

Music

Factor 3   Factor 9

Fig. 57. The result of a factor analysis of a music test (Tchaikovsky's Marche Miniature) by Giannitrapani [32] is shown. The square fields correspond to 17·2 Hz wide frequency bands (increasing from left/top to right/bottom) from 0 Hz to 34 Hz. The size of the dots indicates the loading of the factor. The analysis refers to 8 s scalp EEG during presentation of music. Observe the left dominance in high frequencies for factor 9 (out of 10 factors).

9. SPECTRAL DATA FROM ISOLATED CONSONANCES AND DISSONANCES

The subtest of isolated intervals has been analyzed by computer for the first candidate and furnished a series of remarkable results which confirm the theory of dichotomies in view of their importance for music perception. Every interval was presented four times with the TEST-sound; every time first lower tone, then upper tone, then both together, see Fig. 58. For every interval, we analyzed five time-windows of 1.5 s each, A, B, M_1, M_2, T. Here, A and B stands for the two isolated tones of the interval, M_1 and M_2 denotes the first and second half of the simultaneous interval, and T stands for a segment 0.5 s after the interval, within the tacet of 3 s between two intervals.

For the successive intervals, the participation coefficients $S_\theta, S_\alpha, S_\beta$ were calculated for A and B, in RCA, RH and LCA. From this, the quotients

$$S_\theta(A)/S_\theta(B), \; S_\alpha(A)/S_\alpha(B), \; S_\beta(A)/S_\beta(B),$$

were built in order to get the relative change of the participation coefficients from the first to the second tone of the interval. The graphical representation [Fig. 59(a)] shows the corresponding values. We see that the increase of the relative powers $1/S_\theta, 1/S_\alpha, 1/S_\beta$ is valid for all channels and all frequencies, the maximum being 73.5% for (RH, β), see Fig. 59(b). The increase for the α- and β-bands (35% and 36%) clearly exceeds the increase for the θ-band (15%). Equally, RH (37%) and LCA (28%) exceed RCA (21%).

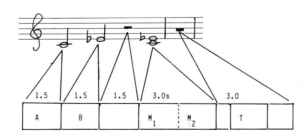

Fig. 58. Patient C.J-L.: the analysis of isolated intervals (successive and simultaneous) was carried out for five segments A, B, M_1, M_2, T of each 1.5 s duration, as indicated.

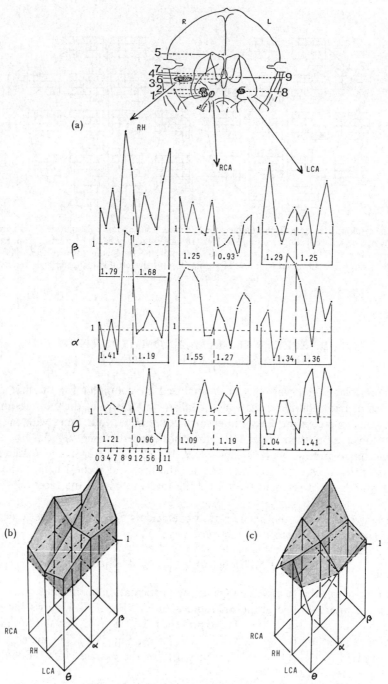

Fig. 59. Patient C.J-L.: the successive intervals (segments A, B) were analyzed by comparison of the participations via $S_\theta(A)/S_\theta(B)$, $S_\alpha(A)/S_\alpha(B)$, $S_\beta(A)/S_\beta(B)$. Figure 59(a) shows the participations for all bands, locations and intervals. Figure 59(b) shows the average quotients over all intervals, as functions of band and location. Figure 59(c) shows the average quotients for consonances divided by the average quotients for dissonances. The strong lateralization is recognized.

If the values of the quotients $S_\theta(A)/S_\theta(B)$ etc. for the consonances are compared to those for the dissonances, we observe a strong lateralization. The values for the ratio $S_\theta(A)/S_\theta(B)[K]$: $S_\theta(A)/S_\theta(B)[D]$, etc. ($K$—average; D—average) are seen in Fig. 59(c). This K–D polarization value for successive intervals is pronounced within the right hemispheric channels RCA, RH, a fact that matches with the Gestalt perception of successive intervals instead of a harmonic perception of simultaneous intervals.

The three following segments M_1, M_2, T have been tested relatively to the participation coefficients $S_\theta, S_\alpha, S_\beta$ and to the participation $v = S_\theta + S_\alpha + S_\beta$. However, we were not interested in the numerical values themselves, but in the ordering that is generated by $S_\theta, S_\alpha, S_\beta$ and v among the 12 intervals. This was done in view of the central dichotomy (K/D) structure from music psychology, musicology and mathematical music theory. Consonances and dissonances had to be tested for their antipodal, polar behaviour within the EEG data. Hence the intervals were given twelve ranks $6, 5, 4, , 3, 2, 1, -1, -2, -3, -4, -5, -6$ according to their position in the ordering defined by the participation data, for example: RCA, β, fourth presentation of M_2:

S:	15.45 >	14.99 >	9.46 >	9.34 >	8.28 >	7.67 >	7.5 >	7.45 >	7.0 >	5.14 >	4.34 >	3.39
Interval:	4	10	7	0	8	2	9	11	5	3	6	1
Rank:	6	5	4	3	2	1	-1	-2	-3	-4	-5	-6

The ranks are normed such that their sum is zero. For each of their four presentations, and for every datum $S_\theta, S_\alpha, S_\beta$, RCA, RH, LCA, the consonant rank $r(K) = \Sigma_{x_i \text{ in } K} r(x_i)$ was calculated, and average for the four presentations was built for every segment M_1, M_2, T. This gives three 4×3 matrices

	RCA	RH	LCA		RCA	RH	LCA		RCA	RH	LCA
	-2.5	3	-3.5		2	6.75	2.75		-3.25	0.25	-3.25
	1.5	6.25	-7.75		0.5	0.75	2.25		-2.5	3	-2.5
	8.25	-1.75	-7		1.75	2.75	4		1.25	-6	3.5
	4.25	-0.25	-9.5		-1.25	4	4.75		-1.25	-1	-1.25
		M_1				M_2				T	

which is graphically shown in Fig. 60(a).

There is a clear difference between the segment T without music and the two segments with music. For the average, the participation value v, all channels show a very weak polarization. Both K and D are equally distributed among the 12 ranks. This in contrast to the segments M_1 and M_2 which show pregnant polarization between K and D. If one compares the absolute value $r(K)$ of the consonant rank sum with the maximal possible value of $r(K) = 21 = 1 + 2 + \cdots + 6$, the percentage of polarization, 100%. $r(K)/21$ is seen in Fig. 60(b).

Further, the distribution of the consonant ranks in segment T is very symmetric, in contrast to the situation for M_1 and M_2, i.e. the tacet does not show any lateralization in contrast to the activity during the interval perception. Like in previous findings for simultaneous intervals, polarization is strongest for the left hippocampus LCA, which seems to be dominant for dichotomic separation of consonances vs dissonances.

Several observations of rank-distributions have occasioned us to give a closer look at the upper six and lower six intervals within the ranking. The example on Fig. 61 shows the segment M_2 for LCA. We recognize a maximal polarization. Every dissonance is higher ranked than every consonance, we have an average $v_D = 35.6$ for D vs an average $v_K = 21.6$, i.e. $v_D/v_K = 1.68$. But this small miracle increased when we tested the RCA of the same segment: the dichotomy from the positive and the negative ranks was

$$(9, 0, 1, 11, 3, 2/10, 6, 7, 8, 5, 4),$$

whose class is the antipodal major dichotomy No. 64 (see Section 2). In fact,

$$e^3 \cdot 5(9, 0, 1, 11, 3, 2/10, 6, 7, 8, 5, 4) = (0, 1, 3, 6, 8, 10/2, 4, 5, 7, 9, 11).$$

Therefore we were interested in the appearance of strong dichotomies defined by the upper and lower halves, i.e. positive and negative ranks. The calculus of the dichotomy classes for the three segments M_1, M_2, T, the three channels RCA, RH, LCA and the four runs of the test shows the following distribution, see Table 3.

Here, the underlined numbers refer to strong dichotomies. In order to interpret this result, observe that we have the following probabilities:

(1) the probability to hit a representative of a strong dichotomy is 31.2%;
(2) the probability to hit a particular strong dichotomy is 5.2%;
(3) the probability to hit a particular couple of different strong dichotomies is 1.53%.

This implies that, within M_1, the strong classes appear 40.35% above probability; within M_2, they appear 6.8% above probability and within T, they appear 46.6% *below* probability. Polarization in the sense of strong dichotomies seems to be significantly above probability for M_1 and M_2, whereas, when intervals are no longer perceived, the appearance of strong dichotomies in T is significantly below pure probability.

Furthermore, the appearance of No. 82 in M_1, M_2, T is 60.2%, 220%, 60.2% above probability, whereas the appearance of No. 64 in M_1, M_2 is 380.8%, 220% above probability, and No. 64 doesn't appear in T.

Finally, the couple $\{64, 82\}$ appears with 16.35-fold probability in M_1, with 32.7-fold probability in M_2 and it does not appear in T. The couple $\{64, 75\}$ also appears twice in M_1, but No. 75 doesn't appear anywhere else.

This means that the antipodal couple $\{64, 82\}$ appears extremely often during the music segments M_1, M_2 and disappears in segment T (tacet).

In order to really understand what this could mean and whether these results statistically persist, more investigations are needed.

Table 3

		1st	2nd	3d	4th	RUN
M_1	RCA	75	69	68	(64)	
	RH	67	72	71	(82)	
	LCA	(64)	(64)	69	75	
M_2	RCA	(64)	81	67	72	
	RH	77	(82)	72	73	
	LCA	(82)	(64)	73	69	
T	RCA	85	69	70	70	
	RH	(82)	65	76	72	
	LCA	68	65	83	77	

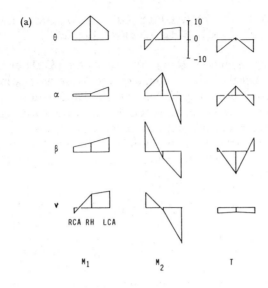

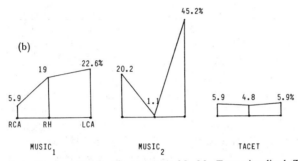

Fig. 60. (a) The polarization matrices for the segments M_1, M_2, T are visualized. The indicated values correspond to the consonance ranks (=sum of the ranks of the consonances) in the respective cases (see text), which measures the degree of separation between consonances and dissonances by the linear ordering defined through the participation coefficients. (b) This graphic indicates the percentage of polarization relative to the maximal (=100%) polarization.

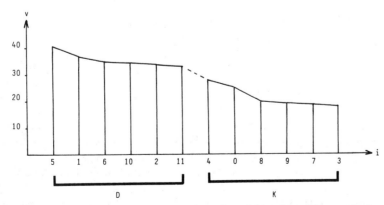

Fig. 61. Patient C.J-L., segment M_2, LCA, first presentation of this subtest. Here maximal polarization is achieved. The participation value of every consonance is lower than the participation value of every dissonance.

10. TESTING THE SYMMETRY OF AUTOCOMPLEMENTARITY ON THE INTERVAL SET

The autocomplementarity function $\tilde{A}(x) = 5x + 2$ for the (K/D) dichotomy is a fundamental symmetry for the mathematical model of counterpoint, as we have seen in Section 4. Therefore, the question, whether this function may be correlated to EEG data is important for discussion of perception and neuronal processing of consonances and dissonances.

We have tested all transitions from a fixed consonance x to the six dissonances (and vice versa), each test being played four times. From this we calculated the frequencies of hitting the correct value $\tilde{A}(x)$ and compared them with the corresponding probabilities. For example, consider the transition $1 \rightarrow ?$:

Partition 9

Then, choose a frequency band, a channel, a test number, for example, RCA, fourth run, and rank the six consonant intervals such that their participation S_θ decreases, the ranks going from 1 to 6:

Consonance:	3	0	8	7	4	9
S:	6.34	6.26	5.11	2.99	2.64	2.53
Rank:	1	2	3	4	5	6

From these relative positions of the six target intervals, we choose the lowest participation, i.e. the highest relative power, here within the θ-band. We shall consider the ranks 5 and 6 and we allow an "uncertainty" of the brain regarding the octave, within which the tones of the interval are perceived. To be precise, together with y in D, $12 - y$ too, if it sits in D, will be looked at, and the same for consonances. For example, 2 and 10 are dissonances, whereas 5 is dissonant, but 7 is consonant.

If we have chosen an interval x, the number $I = 1, 2, 3, 4$ of the test, the frequency band (θ, α, β) and the channel (RCA, RH, LCA), we write 1 if $\tilde{A}(x)$ or $12 - \tilde{A}(x)$ appears among ranks 5 or 6, and we write 0 in the other cases. This defines a 3×3 matrix $M(x, i)$ with coefficients 0 or 1, e.g.

$$M(1, 4) = \begin{array}{c} \\ \text{RCA} \\ \text{RH} \\ \text{LCA} \end{array} \begin{array}{ccc} \theta & \alpha & \beta \\ \left[\begin{array}{ccc} 0 & 0 & 1 \\ 1 & 0 & 1 \\ 0 & 1 & 1 \end{array} \right] \end{array}.$$

Hence, each $x = 0, 1, 2, \ldots, 11$ produces a sequence of four matrices $M(x, 1), M(x, 2), M(x, 3), M(x, 4)$. Then, the average $M(x) = 1/4 \cdot (\Sigma_{i=1}^4 M(x, i))$ is calculated, e.g.

$$M(1) = \begin{pmatrix} 0.25 & 0.25 & 0 \\ 0 & 0.5 & 0.25 \\ 0.75 & 0.25 & 0.25 \end{pmatrix}.$$

These coefficients are the average direct hits. They have to be compared with the probability matrices for the desired ranks and intervals modulo complements. Let x be a fixed consonance.

First case. If $y = \tilde{A}(x)$, we have $y = 6$, or else $12 - y$ is not in D. The probability to hit ranks 5 or 6 is $1/3$.

Second case. With $y = A(x)$, $12 - y$ too is in D, and $y \neq 6$. Up to arrangement, the pair $(y, 12 - y)$ can be distributed in 15 ways among the six ranks. Up to arrangement, one and the same pair can be distributed in six ways among the ranks 1–4. Hence the probability to find y or

$12 - y$ among the ranks 5 or 6 is $(15 - 6)/15 = 3/5$. Hence the average matrix $M(x)$ has to be compared to

$$M_1 = \begin{pmatrix} 1/3 & 1/3 & 1/3 \\ 1/3 & 1/3 & 1/3 \\ 1/3 & 1/3 & 1/3 \end{pmatrix}, \text{ in the first case, and to,}$$

$$M_2 = \begin{pmatrix} 3/5 & 3/5 & 3/5 \\ 3/5 & 3/5 & 3/5 \\ 3/5 & 3/5 & 3/5 \end{pmatrix} \text{ in the second case.}$$

Figure 62 shows the frequencies of direct hits for $A:K \to D$ and for $A:D \to K$, averages taken over all consonances, resp. all dissonances. The meaning of the graphics is, that the probability level is drawn as horizontal plane (100%), and the excess appears as a darkened relief.

The result gives a refined image on the topographical and spectral frequency of direct hits for the autocomplementarity function. It is recognized that the frequency to hit in the direction $K \to D$ is somewhat higher than for the reverse direction $D \to K$. Probably this is due to the fundamental role of K as basis of the interval theories, for instance for the second species of counterpoint, dissonances are only allowed on the non accentuated arsis time. The average frequency for both directions $K \leftrightarrows D$ is:

	θ	α	β
RCA	107.45	109.4	103.35
RH	106.0	92.48	138.5
LCA	87.88	113.35	112.45

This means that the RH for the β-band shows a very good hitting frequency, 38.5% above probability. This again confirms the hypothesis of Giannitrapani [32], that active intelligence could be related to high β-power, and with the theory of Gestalt-processing within the right hemisphere. The β-bands for the three channels have an average of 15.1% above probability, they are higher qualified than the θ- and the α-bands.

Also RH has an average of 14.17% above probability, thus being better qualified than the limbic RCA and LCA for this task. Apparently, the space–time processing within this counterpoint context is done by the right auditory cortex and for the β-band.

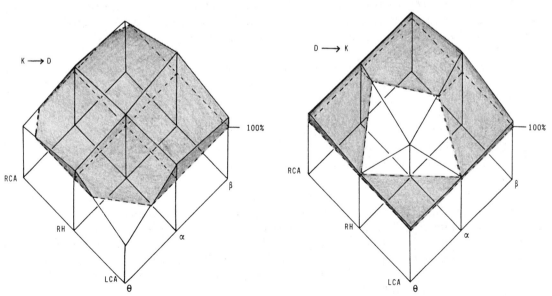

Fig. 62. Test of the autocomplementarity function (or symmetry) for patient C.J-L. Left graphic: direct hit frequencies from K to D, average over all consonances. Right graphic: direct hit frequencies from D to K, average over all dissonances. The level 100% indicates the *a priori* probability to hit the correct functional value $\tilde{A}(x)$ for an interval x. Observe the significant result for RH and the β-band.

Finally, it should be remarked that the six dissonances that appear as target intervals for $\bar{A}$-values, show an average rank distribution over all frequency bands, which is very close to the mean rank 3.5; RCA:3.493, RH:3.493, LCA:3.500. The same situation is true for the singular frequency bands. This means that the hitting frequencies are not due to the unequilibrated appearance of some particular target intervals.

11. DISCUSSION

Undoubtedly, listening to music has much to do with emotion. And it is not less sure that the elementary, but fundamental dichotomy of consonances and dissonances is an extremely important factor for the evocation or suppression of emotional response to music. It is known that the musicologist Kirnberger [55] has attributed a single quality to every interval: major second = anxious or very sad, major seventh = tender, etc. On the other hand, since the work of Papez [3] and MacLean [4, 5], the limbic system with the hippocampal formation has been interpreted as the "emotional brain", and more recent experiments of Aggleton and Mishkin [56] with monkeys clearly show that the amygdala (which has strong connections to the hypothalamic nuclei) is one of the most interesting relays in this respect.

The results of mathematical music theory made it possible to give precise accounts to the classical postulates concerning the dichotomic character of consonances vs dissonances, and to make them accessible to quantitative analysis. Our findings concerning the quantitative correlations between EEG and (K/D) suggest the following.

1. The elementary structure of the counterpoint, the (K/D) dichotomy, *is processed within the limbic system*, predominantly within the left hemisphere, as fas as *simultaneous* intervals are concerned (harmonic processing).

2. The correlation in terms of the *participation vectors* derived from spectral density coefficients are *semantically valuable for perception of music*.

3. The *right hemisphere* dominates the processual efficiency as soon as *patterns within space–time* have to be analyzed.

4. *Simultaneous* intervals are processed with a dominance of the left *limbic* system, whereas *successive* intervals and related structures of counterpoint (autocomplementarity function) are dominantly processed within the *right neocortical Heschl's gyri*, as soon as the (K/D) dichotomy is concerned.

5. The *strong dichotomies* may be related to understanding of musical intervals. The general situation, however, is not clear at present. The *very frequent appearance of the antipodal couple of the major dichotomy* (No. 64) *and the (K/D) dichotomy* (No. 82) are reasonable from the musicological point of view.

6. The elementary *autocomplementarity symmetry* between consonances and dissonances, an interesting structure for counterpoint, is significantly *"recognized" within the β-EEG of auditory cortex*. (Of course, this symmetry is much simpler to handle than the differential quotients within the template fitting model of Goldstein for example, it is by no means an impossible task for the "unconscious calculations of the soul" in the sense of Leibniz.)

From these findings it follows that the limbic system may have good chances to make *topographically precise the connection between emotions and music* (i.e. consonances and dissonances).

Certainly, one should avoid interpreting the correlation between the EEG of the limbic system and musical structures too tightly. It is not clear from the theory of the EEG, how far EEG recordings may directly reflect neuronal activities of circumscribed areas of the limbic system. To solve this problem, it would be necessary to make precise modellings and experiments on the electrical activities of the hippocampal formations. In particular, since the electric field of the amygdala is a closed one, in contrast to the field of the hippocampus, conclusions from epicortically recorded EEG are rather difficult, as far as the amygdala is concerned.

Finally, the semantic potential of musical structures with respect to emotions should be investigated with care. Indeed, if perception of music is correlated to the (EEG of the) limbic system, this does not automatically imply that emotions are directly produced by musical stimuli.

Very probably, the hypothesis of Winson and Abzug [57], concerning the gating of information within the hippocampus, is applicable in the sense that music should not produce emotions, but only allow a neuronal gate to the emotional memory to be opened in order to become effective by its own.

However our findings are only the very beginning of a detailed investigation of the electrophysiology of perception of music. We hope that we shall know more about this subject after an extensive analysis of the presently accumulated material.

Acknowledgement—This work was supported by Schweizerischer Nationalfonds (3.087-084).

REFERENCES

1. G. W. Leibniz, *Epistolae ad Diversos*, Vol. I. Leipzig (1734).
2. J. J. Fux, *Gradus ad Parnassum*. Mizler, Leipzig (1742).
3. J. Papez, A proposed mechanism of emotion. *Arch. Neurol. Psychiat.* **38**, 725–743 (1937).
4. P. MacLean, Psychosomatic disease and the "visceral brain": recent developments bearing on the Papez theory of emotion. *Psychosom. Med.* **11**, 338–353 (1949).
5. P. MacLean, The triune brain, emotion, and scientific bias. In *The Neurosciences: Second Study Program* (Ed. F. Schmitt), pp. 336–348. Rockefeller Univ. Press, New York (1970).
6. L. Euler, Tentamen novae theoriae musicae (. . .). op. omn. (Ed. A. Speiser) Ser. III, Vol. XI, Zürich (1960).
7. C. Stumpf, *Tonpsychologie*. Leipzig (1883–90).
8. H. von Helmholtz, *Lehre von den Tonempfindungen als physiologische Grundlage der Musik*. Nachdruck, Darmstadt (1968).
9. J. Roederer, *Introduction to Physics and Psychophysics of Music*. Springer, New York (1973).
10. H. de la Motte-Haber, *Musikpsychologie*. Laaber, Laaber (1984).
11. G. Mazzola, Die Rolle des Symmetriedenkens für die Entwicklungsgeschichte der europäischen Musik. In *Symmetrie*, Vol. 1, pp. 405–416. Institut Mathildenhöhe, Darmstadt (1986).
12. M. Vogel, *Die Lehre von den Tonbeziehungen*. Syst. Musikwissenschaft, Bonn-Bad Godesberg (1975).
13. R. Plomp and W. Levelt, Tonal consonance and critical bandwidth. *J. acoust. Soc. Am.* **38**, 548 (1965).
14. D. Muzzulini, *Physikalische und mathematische Ansätze in der Musiktheorie: Helmholtz–Oettingen–Mazzola*. Musikwiss. Semin. Univ. Zürich (1987).
15. H. Husmann, *Vom Wesen der Konsonanz*, Heidelberg (1953).
16. G. Albersheim, *Zur Musikpsychologie*. Heinrichshofen, Wilhelmshaven (1974).
17. W. Kahle, *Taschenatlas der Anatomie*, Vol. 3. Thieme, Stuttgart (1979).
18. A. J. Hudspeth and D. P. Corey, Sensitivity, polarity, and conductance change in the response of vertebrate hair cells to controlled mechanical stimuli. *Proc. natn. Acad. Sci. Am.* **74**(6), 2407–2411 (1977).
19. H.-G. Wieser, Musik und Gehirn. *Revue Suisse Méd.* **7**, 153–162 (1987).
20. J. L. Goldstein, An optimum processor theory for the central formation of the pitch of complex tones. *J. acoust. Soc. Am.* **54**, 1496 (1973).
21. E. Terhardt, Zur Tonhöhenwahrnehmung von Klängen. II. Ein Funktionsschema. *Acustica* **26**, 187–199 (1972).
22. G. Langner, Evidence for neuronal periodicity detection in the auditory system of the Guinea Fowl: implications for pitch analysis in the time domain. *Exp. Brain Res.* **52**, 333–355 (1983).
23. G. F. Smoorenburg, Pitch perception of two-frequency stimuli. *J. acoust. Soc. Am.* **48**, 924 (1970).
24. O. Creutzfeldt, G. Ojemann and E. Lettich, Single neuron activity in the right and left human temporal lobe during listening and speaking. In *Fundamental Mechanisms of Human Brain Function* (Ed. J. Engel Jr, G. A. Ojemann, H. O. Lüders and P. D. Williamson), pp. 69–81. Raven Press, New York (1987).
25. W. Penfield and P. Perot, The brain's record of auditory and visual experience. *Brain* **86**, 596–696 (1963).
26. E. Halgren, R. D. Walter, D. G. Cherlow and P. H. Crandall, Mental phenomena evoked by electrical stimulation of the human hippocampal formation and amygdala. *Brain* **101**, 83–117 (1978).
27. P. Gloor, A. Olivier and L. F. Quesney, The role of the amygdala in the expression of psychic phenomena in temporal lobe seizures. In *The Amygdaloid Complex* (*INSERM Symp. No. 20*) (Ed. by Y. Ben-Ari), pp. 489–498 (1981).
28. H.-G. Wieser, *Electroclinical Features of the Psychomotor Seizure*. Fischer–Butterworths, Stuttgart–London (1983).
29. H. Petsche, H. Pockberger and P. Rappelsberger, EEG studies in musical perception and performances, In *Musik in der Medizin* (Eds R. Spintge and R. Droh), pp. 53–79. Springer, Heidelberg (1987).
30. B. Stigsby, J. Risberg and D. H. Ingvar, Electroencephalographic changes in the dominant hemisphere during memorizing and reasoning. *Electroenceph. clin. Neurophysiol.* **42**, 665–675 (1977).
31. G. Dolce and H. Waldeier, Spectral and multivariate analysis of EEG changes during mental activity in man. *Electroenceph. clin. Neurophysiol.* **36**, 577–584 (1974).
32. D. Giannitrapani, *The Electrophysiology of Intellectual Functions*. Karger, Basel (1985).
33. J. Busk and G. C. Galbreith, EEG correlates of visual-motor practice in man. *Electroenceph. clin. Neurophysiol.* **38**, 415–422 (1975).
34. M. Hirshkowitz, J. Earle and E. Paley, EEG asymmetry in musicians and nonmusicians: a study of hemispheric specialization. *Neuropsychologia* **16**, 125–128 (1978).
35. G. McKee, B. Humphrey and D. H. McAdam, Scaled lateralization of alpha activity during linguistic and musical tasks. *Psychophysiology* **10**, 441–443 (1973).
36. J. L. Walker, Alpha EEG correlates of performance on a musical recognition task. *Physiol. Psychol.* **8**, 417–420 (1980).
37. P. Gloor, Volume conductor principles: Their application to the surface and depth electroencephalogram. In *Presurgical Evaluation of Epileptics* (Ed. H.-G. Wieser and C. E. Elger), pp. 59–68. Springer, Berlin (1987).

38. G. Mazzola, *Computer aided reference-free modelling of electrical brain activity from surface and depth EEG-recordings*. Preprint, Univ. Zürich (1987).

39. G. Mazzola, *Mathematische Betrachtungen in der Musik*. Lecture Notes, Univ. Zürich (1987).

40. H. Eimert, *Lehrbuch der Zwölftontechnik*. Breitkopf und Härtel, Wiesbaden (1952).

41. R. Wille, *Mathematische Sprache in der Musiktheorie*. J.buch Überblicke Mathematik. Bibliographisches Institut, Mannheim (1980).

42. E. Apfel, *Diskant und Kontrapunkt in der Musiktheorie des 12. bis 15.* Jahrhunderts. Heinrichshofen, Wilhelmshaven (1982).

43. K. Jeppesen, *Kontrapunkt*. Breitkopf und Härtel, Wiesbaden (1980).

44. G. Mazzola, Konsonanz-Dissonanz und verborgene Symmetrien. In *Mikrotöne II* (Ed. F. R. Herf), pp. 95–104. Helbling edn, Innsbruck (1988).

45. G. Mazzola, *Gruppen und Kategorien in der Musik*. Heldermann, Berlin (1985).

46. G. Mazzola and D. Muzzulini, *Der Kontrapunkt und die (K/D)-Dichotomie*. Preprint, Univ. Zürich (1987).

47. E. Tittel, *Der neue Gradus*. Doblinger (1959).

48. E. Hanslick, *Vom Musikalisch Schönen*. Breitkopf und Härtel, Wiesbaden (1980).

49. O. Messiaen, *Technik meiner musikalischen Sprache*. Leduc, Paris (1944).

50. H.-G. Wieser, C. E. Elger and S. R. G. Stodieck, The "Foramen Ovale Electrode": A new recording method for preoperative evaluation of patients suffering from mesiobasal limbic temporal lobe epilepsy. *Electroenceph. clin. Neurophysiol.* **61**, 314–322 (1985).

51. G. Mazzola, Das $M(2, Z)/Z^2$-o-scope. In *Musik und Mathematik, Salzburger Musikgespräch 1984* (Ed. H. Götze and R. Wille), pp. 92–94. Springer, Berlin (1985).

52. H. Eimert, *Grundlagen der Musikalischen Reihentechnik*. Universaledition, Wien (1964).

53. D. de la Motte, *Harmonielehre*. Bärenreiter, Kassel (1976).

54. G. M. Jenkins and D. G. Watts, *Spectral Analysis*. Holden-Day, Oakland (1968).

55. J. P. Kirnberger, *Die Kunst des reinen Satzes in der Musik*. Berlin and Königsberg (1771).

56. P. Aggleton and M. Mishkin, The amygdala: sensory gateway to the emotions, In *Emotions: Theory, Research, and Experience* (Ed. R. Plutchik and H. Kellerman), Vol. 3. Academic Press, New York (1985).

57. J. Winson and C. Abzug, Gating of neuronal transmission in the hippocampus: efficiency of transmission varies with behavioral state. *Science* **196**, 1223 (1977).

Computers Math. Applic. Vol. 17, No. 4–6, pp. 595–611, 1989
Printed in Great Britain. All rights reserved

SQUARE SPIRALS, DIMENSIONALITY AND BIOPOLYMERS

J. F. Liebman, J. M. Richter, M. J. Bienlein and S. S. Kulkarni

Department of Chemistry, University of Maryland, Baltimore County Campus, Baltimore,
MD 21228, U.S.A.

Abstract—This paper illustrates the diversity and esthetic beauty of the spatial patterns produced by a class of heuristic procedures describable by a one-parameter algorithm. While the patterns may provide insight into the structures of biopolymers and other macromolecules, the main focus of this work is the demostration of how two-dimensional order may arise from one-dimensional processes.

We begin this article with a discussion of an organized one-dimensional arrangement of points. These points are so chosen to correspond to a one-dimensional lattice and so we consider only those points that are an integer distance from the origin. It may appear more correct to consider all the points, both a positive and negative distance, than just the positive even if one immediately acknowledges the computational expediency of our simplifying assumption. However counterintuitive be the assumption,† the set of all points corresponding to 0 and the positive integers has the same number of points as the seemingly larger set of the points that correspond to all of the integers. Nothing is thus conceptually lost by our decision to just consider points that are a positive integer from the origin. Our selection of points is equivalent to the elementary mathematical function $f(x) = 0$ for all non-integer x and $f(x) = 1$ for all positive integers x. Our one-dimensional lattice is also equivalent to an infinite string which has been knotted every cm.

This paper presents a simple—perhaps far too simple—picture for the emergence of higher-dimensional order from "merely" the one-dimensional order of a one parameter paradigm. In particular, we consider the transformation of a one- to a two-dimensional lattice, and with this so consider evolution from one- to two-dimensional order. We opt to take our (half-) infinite one-dimensional lattice and "proceed on a [square] spiral starting at the origin" [2]. Proceeding in a rather related way to that of this just cited Ref. [2], one may identify each lattice point of the half-infinite line with a lattice point in the plane.‡ That is, the "1-point" $x = 0$ corresponds to the "2-point" $(x, y) = (0, 0)$, $x = 1$ to $(x, y) = (0, 1)$, $x = 2$ to $(x, y) = (-1, 1)$, $x = 3$ to $(x, y) = (-1, 0)$, $x = 4$ to $(x, y) = (-1, -1)$, etc. We now define an n-spiral by "decorating" the original line every nth lattice point. This decoration can be described in various ways. For example, we may say that our earlier function $f(x)$ is non-zero only when x is an integer multiple on n. We may opt to tie a bow instead of "merely" a knot every n cm.

The one-dimensional lattice is seemingly analogous to an arbitrarily long polymer composed of one type of monomeric subunit. However, many such "real" polymers twist and turn upon themsleves and thus have secondary and tertiary structure that are ordered or organized in all three spatial dimensions. Simple examples include the α-helices that are formed by the "polypeptide that would be a protein" poly-aaaa [3]. Of course, the majority of the distinct chemical species called proteins are ever so more involved, and thereby interesting, because they contain more than one type of amino acid monomeric subunit and the arrangement of the subunits is manifestly

---

†This assertion, and many others that are even less intuitively reasonable, may be rigorously proven in terms of the analysis given by G. Cantor [1].

‡Strictly speaking, our spiral is left-handed while that in Ref. [2] is right-handed. As such, there is a symmetric relation between our spiral and that earlier given, and the spiral handedness may "merely" subconsciously reflect the fact that the coauthor responsible for programming the spiral (JMR) is left-handed and another author (JFL) has often considered himself to be "ambi-levorous", i.e. equally bad with both hands. The reader should also note that because of the nature of our computer code and printer, our square spirals look rectangular. S/he should also be assured that they are mathematically, if not pictorially, square and, again for computatorial reasons have 66 characters across and 66 lines down.

aperiodic.† Indeed, it has been argued that one reason why catalytically active proteins, more commonly called enzymes, are as large as they are is to allow for the proper three-dimensional arrangement, flexibility, orientation and general organization of active sites. Two quotes from recent reviews are worth citing. From Rebek [5] we cite:

> "It would be most desirable to have access to systems in which the distance between convergent functional groups could be specified in increments of say, even 0.5 Å. It is unlikely that access to such structures will be possible without a tremendous synthetic investment, and we have already suggested [elsewhere, 6] that enzyme structure may be a response to the problems involved with such fine-tuning of small molecules."

Taking a somewhat different tack, Kell [7] suggested: "Enzymes [may be] so big in order . . . to act as channellers of thermal energy to their active sites."

Whether either analysis has validity or not, the design of enzymes constitutes a rather major "engineering feat" because there are some rather tight tolerances for atomic sizes, bond lengths and both bond and torsion angles and because proteins/enzymes are, by definition, all composed of the universal —NHCHRCO— building block. However obvious it is to the biochemist, it should be nonetheless be noted that there are even greater structural restrictions on polypeptides than that just implied. This building block is directional, i.e. they are only joined "head-to-tail" to form the larger "dimeric" units —NHCHRCONHCHR′CO— and seemingly never do either of the alternative "head-to-head" —NHCHRCOCOCHR′NH— or "tail-to-tail" dimers —COCHRNHNHCHR′CO— appear. Furthermore, while except for R = H, all of the building blocks are chiral and so "should" appear equally often in both mirror-image forms. However, real proteins are composed of but one type of handed amino acids, and all of their handedness is the same. Admittedly, we have omitted here any appended coenzymes, i.e. small, non-protein species that contain the catalytically active sites of many enzymes. The presence of coenzymes adds considerable diversity to both structure and activity.

This is all rather complicated. Nonetheless, in a severely simplified form, the protein metaphor may be used if one prefers a polypeptide chain wherein every nth amino acid is a recurring special one that is somehow functionalized and the rest are featureless, say for $n = 5$ corresponding to, say a 1-aspartyltetraglycyl [i.e. {—NHCH(CH$_2$COOH)CO[NHCH$_2$CO]$_4$—} or asp-gly-gly-gly-gly] repeat unit. Of course, the analogy is somewhat belied in that our square spiral is imposed on the one-dimensional "species" while the real molecule that is such a copolymer of aspartic acid and glycine would have its own and certainly different secondary structure. As such, our model is rather artificial and so the term "special" rather than "catalytically active" was chosen as to not to alienate any biochemist of confuse anyone else.

With the above caveats, most of the remainder of this article is dedicated to pictures of the special sites generated by n-spirals. In all that follows a * in the center denotes the beginning of the spiral while # is used for all of the other special sites. The 1-spiral is of course every lattice point. The 2-spiral looks like a chessboard or two-dimensional NaCl lattice. This, too, is rather obvious— given a point described by the integer coordinates (x, y). It is a special site when the sum of the coordinates $x + y$ is divisible by 2. This occurs only when both x and y are even or both odd. As such, "alternate" points in the two-dimensional lattice are decorated, and hence the observed pattern. What occurs for the higher n-spirals? Figures 1–14 reproduce our computer generated n-spirals: n was chosen somewhat artificially as the prime numbers 1, 2, 3, 5 and 7 and their lower powers. Numerous two-dimensional patterns seem to arise, although "explanations" for them analogous to that of the chessboard still evade us. Nonetheless, they demonstrate how one-dimensional order can be transformed into two-dimensional order by a simple geometric wrapping and mapping, and so give inferences as to how the one-dimensional primary structures of proteins are transformed through chemical interactions into their much more complicated, though also much more beautiful, three-dimensional secondary and tertiary structures and with this, into the highly efficient and selective species that are usually called enzymes.

---

†Aperiodicity is important in biomolecules. Another example of what we are referring to are polynucleotides wherein helical structures arise from chains of just one type of base instead of all four bases normally found in DNA or the even greater number of bases normally found in RNA. Indeed quoting Schrödinger: "We believe a gene—or perhaps the whole chromosome fibre—to be an aperiodic solid." [4]

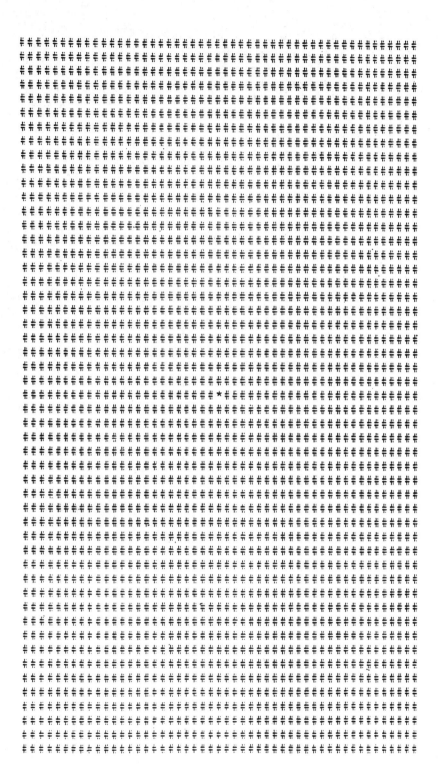

Fig. 1. Performing an interval of 1 spacing; finish printing 1 spiral.

Fig. 2. Performing an interval of 2 spacing; finish printing 2 spiral.

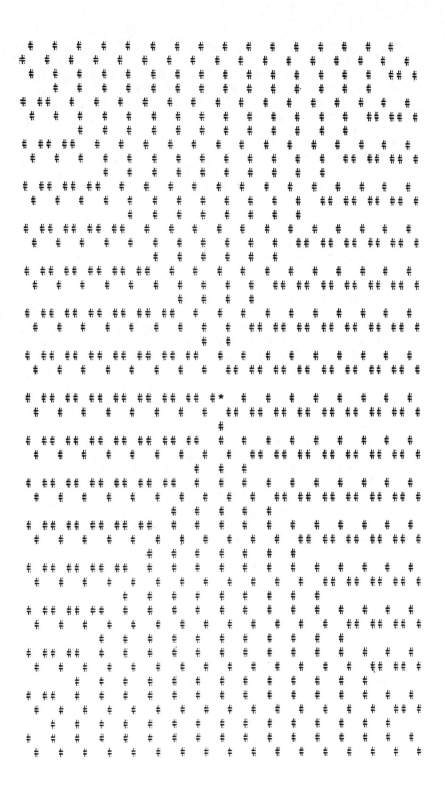

Fig. 3. Performing an interval of 3 spacing; finish printing 3 spiral.

Fig. 4. Performing an interval of 4 spacing; finish printing 4 spiral.

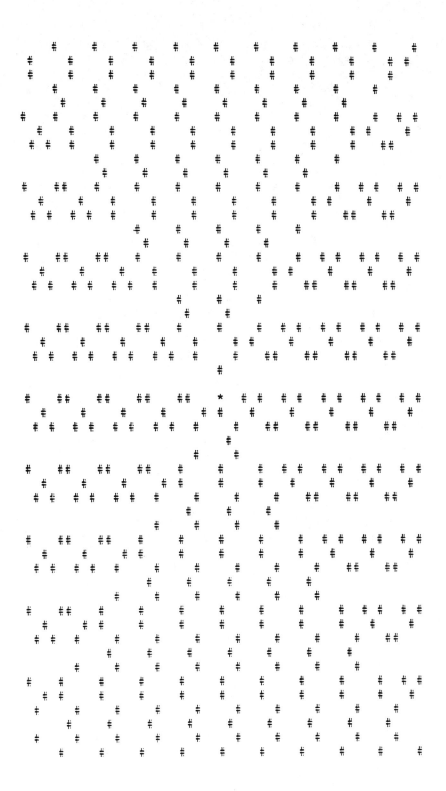

Fig. 5. Performing an interval of 5 spacing; finish printing 5 spiral.

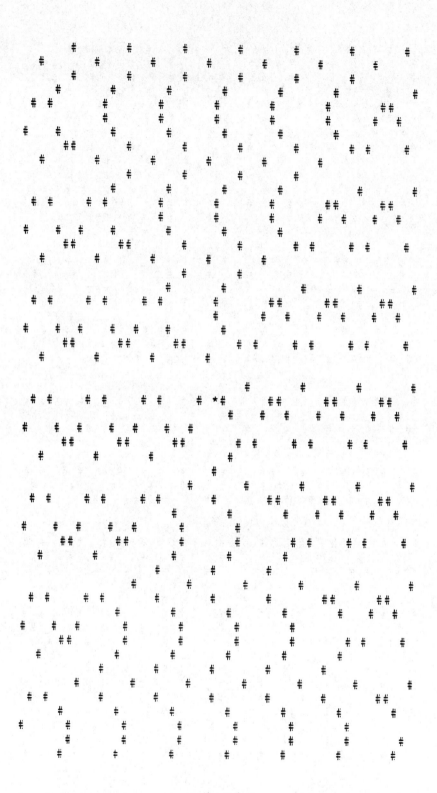

Fig. 6. Performing an interval of 7 spacing; finish printing 7 spiral.

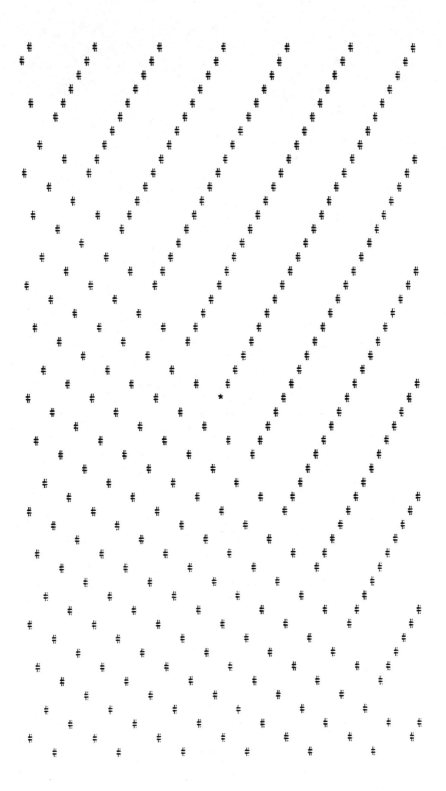

Fig. 7. Performing an interval of 8 spacing; finish printing 8 spiral.

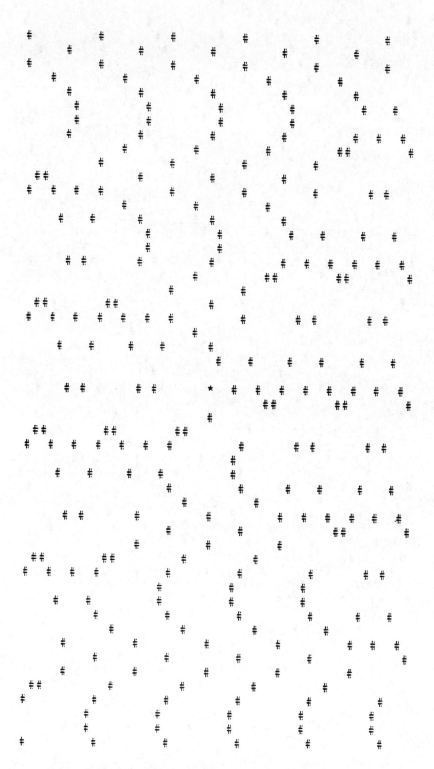

Fig. 8. Performing an interval of 9 spacing; finish printing 9 spiral.

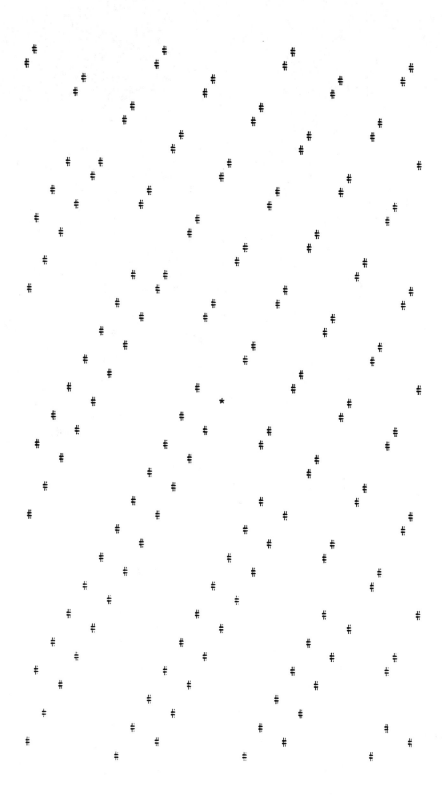

Fig. 9. Performing an interval of 16 spacing; finish printing 16 spiral.

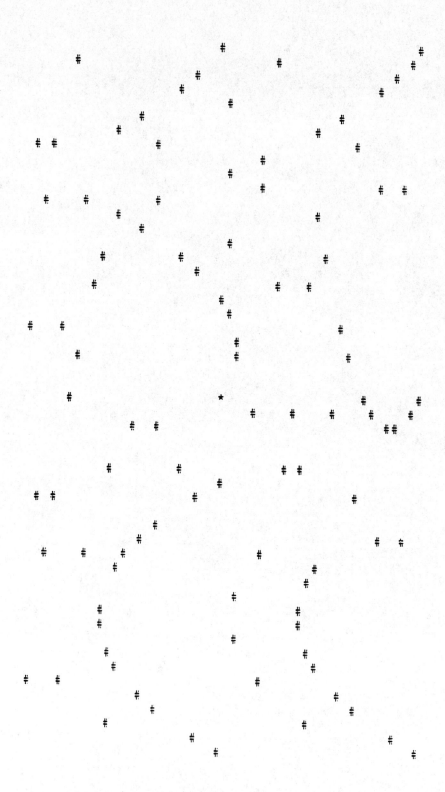

Fig. 10. Performing an interval of 25 spacing; finish printing 25 spiral.

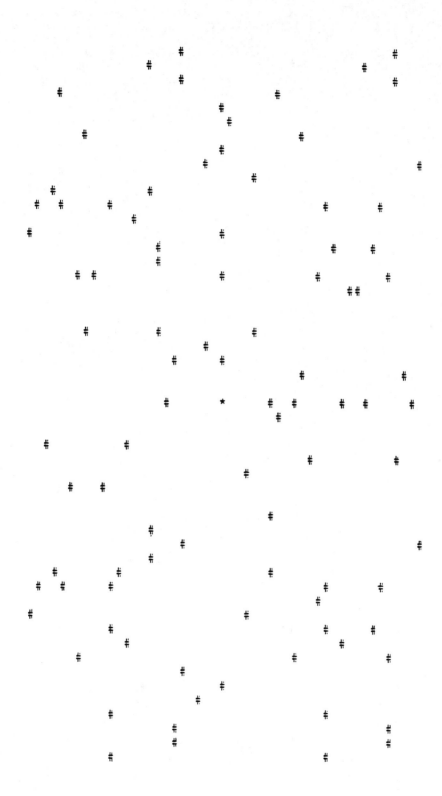

Fig. 11. Performing an interval of 27 spacing; finish printing 27 spiral.

Fig. 12. Performing an interval of 32 spacing; finish printing 32 spiral.

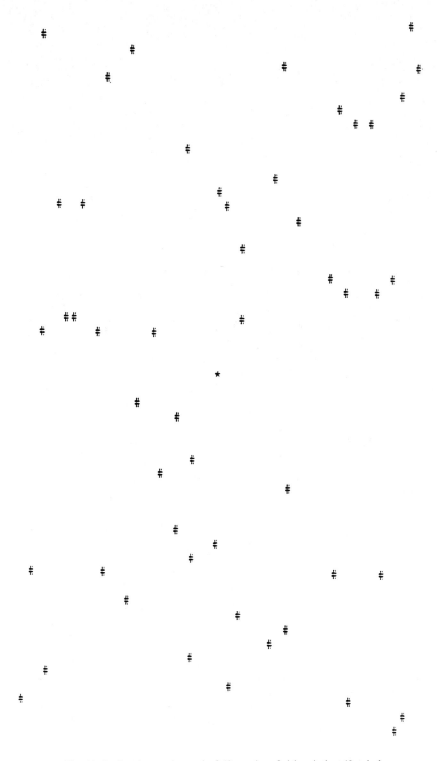

Fig. 13. Performing an interval of 49 spacing; finish printing 49 spiral.

Fig. 14. Performing an interval of 64 spacing; finish printing 64 spiral.

Acknowledgements—The authors wish to thank the University of Maryland Baltimore County Campus Computer Center for the gift of computer time, also Louise M. Garone, Fred Gornick, Austin P. Platt, Robert F. Steiner, Courtlandt C. Van Vechten Jr, Deborah Van Vechten and James S. Vincent for their comments and advice.

REFERENCES

1. G. Cantor, *Contributions to the Founding of the Theory of Transfinite Numbers* (translated, with accompanying introduction and notes, by P. E. B. Jourdain). Dover, New York (1955).
2. R. Stølevik, J. Brunvoll, B. Cyvin and S. Cyvin, Strange patterns generated from simple rules with possible relevance to crystallatization and other phenomena. *Comput. Math. Applic.* **12B,** 789 (1986). Reprinted in *Symmetry: Unifying Human Understanding* (Ed. I. Hargittai). Pergamon Press, Oxford (1986).
3. C. Chothia, Principles that determine the structure of proteins. *Ann. Rev. Biochem.* **53,** 537 (1984).
4. E. Schrödinger, *What is Life* (published as combined volume with *Mind and Matter*), p. 64. Cambridge Univ. Press, Cambridge (1967).
5. J. Rebek Jr, Progress in molecular recognition. In *Environmental Influences and Recognition in Enzyme Chemistry* (Eds J. F. Liebman and A. Greenberg). VCH Publishers, New York (in press).
6. J. Rebek Jr, L. Marshall, R. Wolak, K. Parris, M. Killoran, B. Askew, D. Nemeth and N. Islam, Convergent functional groups: synthesis and structural studies. *J. Am. chem. Soc.* **107,** 7476 (1985).
7. D. B. Kell, Enzymes as energy "funnels". *Trends Biochem. Sci.* (*TIBS*) **7,** 349 (1987).

Computers Math. Applic. Vol. 17, No. 4–6, pp. 613–638, 1989
Printed in Great Britain

0097-4943/89 $3.00 + 0.00
Pergamon Press plc

THE GOLDEN SECTION IN THE MEASUREMENT THEORY

A. P. Stakhov

Vinnitsa Polytechnic Institute, Vinnitsa 286021, U.S.S.R.

Abstract—The main ideas and results of the algorithmic measurement theory, i.e. the constructive trend in the mathemitic measurement theory are discussed. The theory in question dates back to the problem of choosing the best system of scale weights (Fibonacci, thirteenth century). The theory uses the asymmetry principle of measurement and is connected with the brilliant mathematical achievements, viz, the golden section, the Fibonacci numbers and Pascal's triangle. New methods of number representation, i.e. the Fibonacci and the golden ratio codes have been developed. They are used in computer engineering.

1. INTRODUCTION

The main event in the history of mathematics nowadays is the displacement of the mathematical researchs towards the extreme and optimal problems. This process is reflected in the mathematical measurement theory in the form of the algorithmic measurement theory [1–3]. Arising in the early seventies from the practical application of the analog-to-digital conversion [1] the theory exceeded the limits of the technical problems and found unexpected connections with the well-known mathematical discoveries.

The algorithmic measurement theory lies on the crossing of the continuous and discrete mathematics and solves uniquely the problem of representing the continuous values and real numbers. Besides, the algorithmic measurement theory is closely connected with the notion of a real number and plays a fundamental part in the theory of notations [4].

It should be noted that the new measurement theory is deeply connected with the Fibonacci numbers and the golden ratio. The purpose of the article is to give the popular presentation of the new measurement theory based on the golden section.

2. THE GOLDEN SECTION

Johannes Kepler said that geometry has two treasures—the Pythagorean theorem and the golden section. The former can be compared with pure gold, the latter with a precious stone.

The golden section arose from the division of the line segment AB with the point C in the extreme and mean ratio (Fig. 1), that is,

$$\frac{AB}{CB} = \frac{CB}{AC}. \tag{1}$$

It is reduced to the equation

$$x^2 = x + 1. \tag{2}$$

The positive root of the equation $\alpha = (1 + \sqrt{5})/2$ is called the golden ratio and the division of the line segment in the ratio (1) is called the golden section.

Being the root of the equation the golden ratio has the property

$$\alpha^2 = \alpha + 1. \tag{3}$$

Expression (3) can be rewritten as

$$\alpha = 1 + \frac{1}{\alpha}; \quad \alpha - 1 = \frac{1}{\alpha}. \tag{4}$$

Hence when subtracting 1 from $\alpha = 1.618$ it varies inversely as $1/\alpha = 0.618$. It is proved that the golden ratio is the only positive integer having this property. It should be noted that 1.618 and 0.618 are supposed to express the ratio of the golden section.

Substituting many times the expression $\alpha = 1 + (1/\alpha)$ for α in the right-hand side of equation (4) we find the representation of the golden ratio in the form of the continued fraction

$$\alpha = 1 + \cfrac{1}{1 + \cfrac{1}{1 + \cfrac{1}{1 + \cdots}}}.$$

The simplicity of this representation emphasizes the fundamental character of the golden ratio. The representation of the golden ratio

$$\alpha \sqrt{1 + \alpha}$$

gives one more representation of α:

$$\alpha = \sqrt{1 + \sqrt{1 + \sqrt{1 + \sqrt{1 + }}}} \cdots$$

Multiplying or dividing the left- and right-hand sides of equation (3) by α, we obtain the mathematical expression combining the powers of the golden ratio:

$$\alpha^n = \alpha^{n-1} + \alpha^{n-2}, \tag{5}$$

where n is an integer.

From equation (5) it follows that the geometric progression (with the radix α)

$$\alpha^n = \alpha \cdot \alpha^{n-1} \tag{6}$$

has both the multiplicity (6) and the additivity (5) properties, i.e. the powers of the golden ratio have the properties of the geometric and arithmetic progression.

The golden section is widely used in geometry. It is proved that $\alpha = (1 + \sqrt{5})/2 = 2\cos 36°$.

Using this equation we can show that in the pentagram $ABCDE$ the cross points of the diagonals F, G, H, K, L, divide them in the golden section and form the pentagram $FGHKL$ (Fig. 2). The pentagram aroused admiration with the Pythagoreans. It was a symbol of health and a landmark.

The pentagram comprises a set of wonderful figures which are widely used in the works of art. In ancient Egypt and classical Greece the law of the "golden cup" was well-known. It was used by architects and goldsmiths.

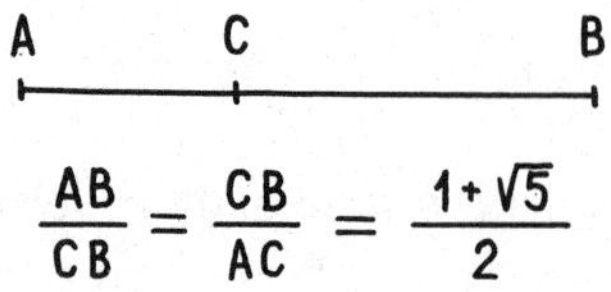

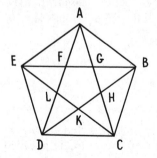

A   C   B

$$\frac{AB}{CB} = \frac{CB}{AC} = \frac{1 + \sqrt{5}}{2}$$

Fig. 1. Golden section. According to Johannes Kepler, Geometry has two treasures—the Pythagorean theorem and the golden section. The former can be compared with pure gold, the latter with a precious stone. The golden section was known to the ancient Babylonians and Egyptians. It forms the basis for Pythagoras' teaching of the number harmony of the world. The term "the golden section" was introduced by Leonardo da Vinci.

Fig. 2. Pentagram. The intersection points of the diagonals divide them in the golden section and form a new pentagram. It aroused admiration with the Pythagoreans. It was a symbol of health and a landmark.

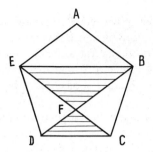

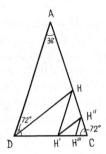

Fig. 3. Golden cup. In ancient Egypt and classical Greece "the law of the golden cup" was well-known. It was widely used by architects and goldsmiths. The majority of the Greek vases satisfy the golden cup proportions.

Fig. 4. Golden triangle. The fact that the bisectrix DH of the angle D coincides with the diagonal of the pentagrams and divides the side AC by the point H in the golden section intrigued the Pythagoreans. The golden ratio was widely used by artists of the Italian Renaissance. Leonardo da Vinci created his "*Mona Lisa*" using the golden triangles.

If we draw the diagonals BE, BD and EC (Fig. 3) in the pentagram $ABCDE$ the dashed part has a form of the "golden cup" which is characterized by the ratios

$$\frac{EB}{DC} = \frac{EF}{FC} = \frac{BF}{FD} = \frac{1+\sqrt{5}}{2}.$$

One more remarkable figure involved in the pentagram is the golden triangle, for example, ADC (Fig. 4), whose base is the side of the pentagram. This triangle has the vertex angle measuring 36° and the base angles measuring 72° each.

The Pythagoreans were greatly interested in the fact that the bisectrix DH of the angle D coincides with the diagonal DB of the pentagram and divides the side AC by the point H in the golden section. So, the new small golden triangle DCH is generated. If we draw the bisectrix of the angle H to the point H on the side AC and continue this procedure endlessly, we get an infinite sequence of the golden triangles.

The same recurrent property is inherent in the golden rectangle $ABCD$ (Fig. 5) in which the ratio of the sides is the golden section:

$$\frac{AB}{AD} = \alpha.$$

Removing the square $AEFD$ from the rectangle $ABCD$ we have the new golden rectangle $EBCF$ where $FE/EB = \alpha$.

If we continue this procedure endlessly we get an infinite sequence of the golden rectangles. The pentagrams are closely connected with the dodecahedron (Fig. 6) which symbolized the harmony of the world in the Pythagorean teaching.

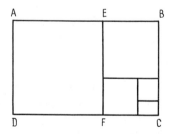

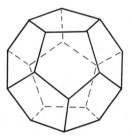

Fig. 5. Golden rectangle. The golden rectangle is drawn with its sides in the ratio

$$\frac{AB}{AD} = \frac{1+\sqrt{5}}{2}.$$

The law of the golden rectangle is widely used in architecture. In particular, the proportions of the Parthenon satisfy the law.

Fig. 6. Dodecahedron. The dodecahedron was the main geometric figure of the Pythagoreans and symbolized the harmony of the universe. The sides of the dodecahedron are the pentagrams built according to the golden ratio.

According to the well-known mathematician B. L. van der Waerden, Pythagoras might have borrowed the idea of the golden section from the ancient Egyptians and Babylonians during his long travellings throughout the ancient East. The golden section penetrates the antique culture of Greece. "According to Plato and to the ancient cosmology in general," writes the distinguished Soviet scholar A. F. Losev, "the world is a certain proportional entity subjected to the law of the harmonic division, that is the golden section..." [5, p. 56]

In the epoch of Italian Renaissance the golden section is again raised to the main aesthetic principle. Leonardo da Vinci called it "Sectio aurea" from which the terms the "golden section" or the "golden number" came from. Under the influence of Leonardo da Vinci the famous mathematician Luca Pacioli published in 1509 the first book on the golden ratio under the title "De Divina Proportione", i.e. "On the Divine Proportions". The golden section was applied to many paintings of Leonardo da Vinci (the brilliant "Mona Lisa" in particular), Titian, Raphael, Dürer. During that period the golden section penetrates into music. The Netherlandish composer Jacob Obrecht (1430–1505) used the golden section in his compositions which may be compared to "a cathedral created by a brilliant architect".

After the epoch of the Renaissance the golden section fell into oblivion for almost two centuries. In the middle of the nineteenth century the German scientist Zeising made an attempt to formulate the universal proportionality law discovering once again the golden section.

At the end of the nineteenth century the German psychologist Gustav Theodor Eechner carried out a series of psychological experiments aimed to define the aesthetic impression made by the rectangles with different side dimensions. The experiments turned out favourable for the golden rectangle.

In the twentieth century the idea of the golden section was again greatly concerned. In the first half of the century the music theorist L. Sabaneev formed the general rythmic equilibrium law and grounded the golden section as the aesthetic norm of a musical composition. The distinguished French architect and artist Le Corbusier made his Modulor based on the golden section. The famous Soviet producer Eisenstein used the golden section in making his films (e.g. "The Battleship Potemkin"). The golden section is found in Pushkin's verses. The academician G. V. Tsereteli proved that the poem "The Knight in the Tiger's Fell" was written by Shota Rustaveli with the help of the brilliant ratio. The well-known Hungarian composer Bela Bartok widely used the golden ratio in his music compositions.

In the twentieth century the golden section also penetrates into biology, physics and mathematics (brain rhythms, cardiovascular activity of mammals, the theory of measurement and encoding).

3. HOW DO RABBITS MULTIPLY?

The golden section is closely connected with the so called Fibonacci numbers [6], discovered in the thirteenth century by the famous Italian mathematician Leonardo of Pisa (Fibonacci).

The Fibonacci numbers constitute the sequence

$$1, 2, 3, 5, 8, 13, 21 \ldots ,$$

where any number equals the sum of the two preceding numbers. Fibonacci discovered this number sequence while solving the well-known problem of rabbits' multiplying which was presented in his book "Liber Abaci" published in 1202.

The essence of the problem is this: suppose there is one pair of rabbits in an enclosure on the first day of January; this pair will produce another pair of rabbits on February first and on the first day of every month thereafter, each new pair will mature for one month and then produce a new pair on the first day of the third month of their life and on the first day of every month thereafter. The problem is to find the number of pairs of rabbits after n months since the first birth.

The solution of this problem was a great contribution to the combination theory. While solving the problem Fibonacci discovered the first recurrence formula

$$\varphi(n) = \varphi(n-1) + \varphi(n-2) \tag{7}$$

generating the Fibonacci sequence. By this discovery he anticipated the method of recurrent relations regarded as the most appropriate for solving combinatorial problems.

The problem of rabbits multiplying generated the mathematic theory of biological populations [7].

The absolute numbers evaluated by the recurrence formula [7] depend on $\varphi(0)$ and $\varphi(1)$. For example, if we are given $\varphi(0) = 0$, $\varphi(1) = 1$ then the formula [7] generates the number sequence

$$0,\ 1,\ 1,\ 2,\ 3,\ 5,\ 8,\ 13,\ 21,\ 34,\ \ldots. \tag{8}$$

Fibonacci didn't continue his research of the properties of his number sequence. However, at the beginning of the nineteenth century the works dedicated to the special properties of the Fibonacci numbers began to "multiply like the Fibonacci rabbits". The French mathematician Lucas introduced the generalized Fibonacci numbers. They start with any two positive integers and any term beginning with the third one is equal to the sum of the two previous terms. Lucas introduced an interesting number sequence (the Lucas sequence) which is evaluated by the similar recurrence formula

$$L(n) = L(n-1) + L(n-2), \tag{9}$$

with the initial terms $L(0) = 2$ and $L(1) = 1$.

The recurrence formula (9) generates the Lucas sequence

$$2,\ 1,\ 3,\ 4,\ 7,\ 11,\ 18,\ 29,\ 47,\ \ldots. \tag{10}$$

Let's define some properties of the Fibonacci and Lucas numbers. First we form the number sequence, the so called ratios of the neighbouring Fibonacci numbers.

$$\frac{1}{2},\ \frac{3}{1},\ \frac{4}{3},\ \frac{7}{4},\ \frac{11}{7},\ \frac{18}{11},\ \frac{29}{18},\ \frac{47}{29},\ \ldots$$

What do these sequences tend to? They tend to the limit which is the golden section, i.e.

$$\lim_{n \to \infty} \frac{\varphi(n)}{\varphi(n-1)} = \lim_{n \to \infty} \frac{L(n)}{L(n-1)} = \frac{1+\sqrt{5}}{2}.$$

Let's state the connection between the golden ratio powers and the Fibonacci and Lucas numbers. So, we extend the sequence (8) and (10) to the left taking into account that the terms after the second may be found as the sum of the two preceding terms in the line. The sequences formed are given in Tables 1 and 2.

Using formula (5) we can get the analytical expressions for the powers of the golden ratio (see Table 3). Comparing the Tables 1–3 we see that any power of the golden ratio can be written as

$$\alpha^n = \frac{L(n) + \varphi(n)\sqrt{5}}{2}.$$

Table 1

n	0	1	2	3	4	5	6	7	8	9
$\varphi(n)$	0	1	1	2	3	5	8	13	21	34
$\varphi(-n)$		1	−1	2	−3	5	−8	13	−21	34

Table 2

n	0	1	2	3	4	5	6	7	8	9
$L(n)$	2	1	3	4	7	11	18	29	47	76
$L(-n)$		−1	3	−4	7	−11	18	−29	47	−76

Table 3

n	0	1	2	3	4	5	6	7	8	9
α^n	$\dfrac{2+0\sqrt{5}}{2}$	$\dfrac{1+1\sqrt{5}}{2}$	$\dfrac{3+1\sqrt{5}}{2}$	$\dfrac{4+2\sqrt{5}}{2}$	$\dfrac{7+3\sqrt{5}}{2}$	$\dfrac{11+5\sqrt{5}}{2}$	$\dfrac{18+8\sqrt{5}}{2}$	$\dfrac{29+13\sqrt{5}}{2}$	$\dfrac{47+21\sqrt{5}}{2}$	$\dfrac{76+34\sqrt{5}}{2}$
α^{-n}	$\dfrac{2+0\sqrt{5}}{2}$	$\dfrac{-1+1\sqrt{5}}{2}$	$\dfrac{3-1\sqrt{5}}{2}$	$\dfrac{-4+2\sqrt{5}}{2}$	$\dfrac{7-3\sqrt{5}}{2}$	$\dfrac{-11+5\sqrt{5}}{2}$	$\dfrac{18-8\sqrt{5}}{2}$	$\dfrac{-29+13\sqrt{5}}{2}$	$\dfrac{47-21\sqrt{5}}{2}$	$\dfrac{-76+34\sqrt{5}}{2}$

Two more important formulas based on the Tables 1–3 are formed connecting the Fibonacci and Lucas numbers with the golden ratio:

$$\varphi(n) = \begin{cases} \dfrac{\alpha^n + \alpha^{-n}}{\sqrt{5}}, & \text{for odd } n; \\[2ex] \dfrac{\alpha^n - \alpha^{-n}}{\sqrt{5}}, & \text{for even } n. \end{cases} \tag{11}$$

$$L(n) = \begin{cases} \alpha^n + \alpha^{-n}, & \text{for even } n; \\[1ex] \alpha^n - \alpha^{-n}, & \text{for odd } n. \end{cases} \tag{12}$$

Expression (11) is known as a Binet formula. The Fibonacci and Lucas sequences may be found in different branches of learning. The most surprising is the botanical phenomenon of phyllotaxis (8) which was at first singled out by J. Kepler. The best known manifestation of phyllotaxis is the arrangement of the florets of a sunflower or of the scales of a pine cone, in spiral or helical whorls rotating clockwise and counterclockwise. These spirals are called the parastichy pairs.

If we count the clockwise and the counterclockwise parastichy pairs we see that the ratios are constant and specific for different plants. The pine cones have

$$\frac{5}{3}, \frac{8}{5}, \frac{13}{8}, \frac{21}{13}, \frac{34}{21},$$

the majority of sunflowers have

$$\frac{55}{34}, \frac{89}{55}.$$

The numerators and denominators represent the Fibonacci numbers and the fractions tend to the golden ratio.

In the twentieth century the Fibonacci numbers aroused great interest. In the middle of the century the Fibonacci Association was established in the U.S.A. Since 1963 it has been publishing the mathematic magazine "*The Fibonacci Quarterly*". Two "*International Conferences on Fibonacci Numbers and Their Applications*" took place in the U.S.A.

4. SOME FIBONACCI TRIGONOMETRY

It is well-known that the trigonometric equations of the Lobachevski geometry are expressed by the hyperbolic functions and the geometry is called hyperbolic.

According to R. Luneburg, the solid geometry of the human visual perception is the Lobachevski geometry. V. Vernadski's hypothesis says that the non-Euclidean geometry is of great importance for the studying of living matter. The expressions for the hyperbolic sine and cosine are of the form

$$\sinh x = \frac{e^x - e^{-x}}{2} \tag{13}$$

$$\cosh x = \frac{e^x + e^{-x}}{2} \tag{14}$$

Comparing these formulas with expressions (11) and (12) we see the connection between the Fibonacci and Lucas numbers and the hyperbolic functions. To state this connection we divide the representation of the Fibonacci and Lucas numbers into two pairs of forms:

$$\varphi(2k) = \frac{\alpha^{2k} - \alpha^{-2k}}{\sqrt{5}}, \qquad \text{with } n = 2k, \tag{15}$$

$$\varphi(2k+1) = \frac{\alpha^{2k+1} + \alpha^{-(2k+1)}}{\sqrt{5}}, \qquad \text{with } n = 2k+1, \tag{16}$$

$$L(2k) = \alpha^{2k} + \alpha^{-2k}, \qquad \text{with } n = 2k, \tag{17}$$

$$L(2k+1) = \alpha^{2k+1} - \alpha^{-(2k+1)}, \qquad \text{with } n = 2k+1. \tag{18}$$

Some similarity between the forms and the hyperbolic functions ensures the introduction of the Fibonacci and Lucas trigonometric functions.

The Fibonacci sine and cosine are the continuous functions corresponding to expressions (17) and (18) defined for any real number,

$$\operatorname{sinL} x = \alpha^{2x+1} - \alpha^{-(2x+1)}$$

$$\operatorname{cosL} x = \alpha^{2x} + \alpha^{-2x}.$$

Thus, the connection between the hyperbolic and the Fibonacci trigonometry is expressed in the form

$$\sinh x = \frac{\sqrt{5}}{2} \operatorname{sinf}\left(\frac{x}{2 \ln \alpha}\right);$$

$$\cosh x = \frac{\sqrt{5}}{2} \operatorname{cosf}\left(\frac{x - \ln\alpha}{2 \ln \alpha}\right);$$

$$\operatorname{sinf} x = \frac{2}{\sqrt{5}} \sinh[(2 \ln \alpha)x];$$

$$\operatorname{cosf} x = \frac{2}{\sqrt{5}} \cosh[(2x + 1) \ln \alpha].$$

Some identities of the Fibonacci trigonometry are given without proof:

$$\operatorname{sinf} x + \operatorname{cosf} x = \operatorname{sinf}(x + 1);$$

$$\operatorname{sinL} x + \operatorname{cosL} x = \operatorname{cosL}(x + 1);$$

$$\operatorname{sinf} x \cdot \operatorname{cosf} x = \frac{1}{\sqrt{5}} [\operatorname{sinL}(x - 1)];$$

$$\operatorname{sinf} y \cdot \operatorname{cosf} x = \frac{1}{\sqrt{5}} [\operatorname{sinL}(x + y) - \operatorname{sinL}(x - y)];$$

$$\operatorname{sinf}^2 x + \operatorname{cosf}^2 x = \operatorname{cosf} 2x.$$

The connection between the golden ratio and the hyperbolic functions which is given by the Fibonacci trigonometry shows that the hyperbolic functions represent the "secret" expression of the golden section in mathematics. This idea generates some view on the theory of the hyperbolic functions and the hyperbolic Lobachevski geometry via the golden section.

5. PASCAL'S TRIANGLE OR ANOTHER WAY OF RABBITS MULTIPLYING

Let's consider the array of the binomial coefficients which is called Pascal's triangle. It is known that

$$(a + b)^n = c_n^0 \cdot a^n \cdot b^0 + c_n^1 \cdot a^{n-1} \cdot b^1 + \cdots + c_n^k \cdot a^{n-k} \cdot b^k + \cdots + c_n^n \cdot a^0 \cdot b^n,$$

where the numbers c_n^k are called the binomial coefficients. The above formula is the binomial one. This name involves a historical injustice because the formula was well known to the Central Asian mathematicians in the eleventh century. In Western Europe the formula was known to Pascal who proposed the method of evaluating the binomial coefficients via their arrangement in the array which is called Pascal's triangle. Let's consider the so-called Pascal's triangle:

$$
\begin{array}{ccccccccc}
c_0^0 & c_1^0 & c_2^0 & c_3^0 & c_4^0 & c_5^0 & \cdots & c_n^0 \\
 & c_1^1 & c_2^1 & c_3^1 & c_4^1 & c_5^1 & \cdots & c_n^1 \\
 & & c_2^2 & c_3^2 & c_4^2 & c_5^2 & \cdots & c_n^2 \\
 & & & c_3^3 & c_4^3 & c_5^3 & \cdots & c_n^3 \\
 & & & & c_4^4 & c_5^4 & \cdots & c_n^4 \\
 & & & & & c_5^5 & \cdots & c_n^5 \\
 & & & & & & & \vdots \\
 & & & & & & & c_n^n
\end{array}
$$

i.e.

1	1	1	1	1	1	1	1
	1	2	3	4	5	6	7
		1	3	6	10	15	21
			1	4	10	20	35
				1	5	15	35
					1	6	21
						1	7
							1

1	2	4	8	16	32	64	128

The rows of Pascal's triangle are numbered from the top but the upper row with the binomial coefficients

$$c_0^0 = c_1^0 = c_2^0 = \cdots = c_n^0 = 1$$

is called the zero row.

The K-row begins with the binomial coefficient $c_K^k = 1$. The columns are numbered from the left to right; the leftward extreme column consisting of the only one (C_0^0) is called the zero column. The nth column involves the binomial coefficients

$$c_n^0, c_n^1, \ldots, c_n^k, \ldots, c_n^{n-k}, \ldots, c_n^n,$$

where $c_n^k = c_n^{n-1}$. Pascal's triangle is based on the recurrent relation

$$c_n^k = c_{n-1}^k + c_n^{k-1} - 1.$$

Therefore, the above array of the binomial coefficients will be termed as Pascal's 0-triangle. The binomial coefficients and Pascal's triangle have many applications in different branches of mathematics. "This array has a set of wonderful properties," wrote J. Bernoulli. "Just now we've shown that it conceals the essence of the theory of connections, but those who are closer to geometry know that it hides a lot of fundamental secrets from other branches of mathematics."

We consider some of the secrets. It should be noted that the binomial coefficients are elegantly expressed by the factorials:

$$m! = 1 \cdot 2 \cdot \ldots \cdot m.$$

It is proved that

$$c_n^k = \frac{n!}{K!(n-k)!}.$$

We have considered the binary sequence $1, 2, 4, 8 \ldots 2^n$ and the Fibonacci sequence $1, 1, 2, 3, 5, 8, \ldots, \varphi(n)$. These number sequences prove to be closely connected with the binomial coefficients and Pascal's triangle.

Let's sum up the binomial coefficients in the nth column of Pascal's 0-triangle. As a result we have a binary digit 2^n, i.e.

$$2^n = c_n^0 + c_n^1 + \cdots + c_n^n. \tag{19}$$

This formula is of great importance for the theory of binary codes.

As is generally known, the number of all n-binary digit combinations is 2^n. If we remove the subset of the code combinations comprising k ones and $n - k$ zeros (with $k = 0, 1, \ldots, n$) from the set of n code combinations, the number of code combinations equals $c_n^k = c_n^{n-k}$. This is represented in formula (19) giving the general number of n code combinations as a sum of the binomial coefficients.

To state the connection between Pascal's 0-triangle and the Fibonacci numbers we shift each row of Pascal's 0-triangle one column rightwards. So, we have the array of the binomial coefficients which is called Pascal's 1-triangle.

1	1	1	1	1	1	1	1	1
	1	2	3	4	5	6	7	
		1	3	6	10	15		
			1	4	10			
				1				
1	1	2	3	5	8	13	21	34

If we sum up the binomial coefficients in the columns of Pascal's 1-triangle the obtained number sequence consists of the Fibonacci numbers!

We can also shift each row of Pascal's 0-triangle 2, 3, ..., p-columns rightwards (p is a positive integer). Thus, we have the modifications of Pascal's 0-triangle which are called Pascal's 2-, 3-, ..., p-triangles, respectively:

$p = 2$

1	1	1	1	1	1	1	1	1	1	1	1
	1	2	3	4	5	6	7	8	9		
		1	3	6	10	15	21				
			1	4	10						
1	1	1	2	3	4	6	9	13	19	28	41

$p = 3$

1	1	1	1	1	1	1	1	1	1	1	1
	1	2	3	4	5	6	7	8			
		1	3	6	10						
1	1	1	1	2	3	4	5	7	10	14	19

Considering the number sequences obtained when summing up the binomial coefficients arranged in Pascal's 2-, 3-, ..., p-triangles we note that the number sequences have some strict relationship. With $p = 2$ the nth term of the sequence is formed from the preceding terms by the recurrence formula

$$\varphi_2(n) = \varphi_2(n-1) + \varphi_2(n-3).$$

with $p = 3$ the recurrence formula is of the form

$$\varphi_3(n) = \varphi_3(n-1) + \varphi_3(n-4).$$

For the arbitrary p the new number sequence taken from Pascal's p-triangle is given by the general recurrence formula:

$$\varphi_p(n) = \begin{cases} 0, & \text{with } n < 0, \\ 1, & \text{with } n = 0, \\ \varphi_p(n-1) + \varphi_p(n-p-1), & n > 0. \end{cases} \tag{20}$$

Thus, we have discovered the infinite set of the number sequences consisting of the Fibonacci p-numbers. These number sequences comprise the binary sequence (for $p = 0$) and the classical Fibonacci sequence (for $p = 1$).

Let's go back to the problem of the rabbits multiplying. Fibonacci stated that each new pair of rabbits would mature in one month. Suppose that rabbits mature in p months, where $p \in \{0, 1, 2, \ldots, \infty\}$. We can solve this problem for the two cases, i.e.

(1) the reproduction begins with a "baby pair" of rabbits;
(2) the reproduction begins with an "adult pair" of rabbits.

Let $\varphi_p(n)$ denote the number of pairs of rabbits for n months after the births have taken place. The law of reproduction depends on the maturity of the original pair of rabbits for $0 \leqslant n \leqslant p$. As for the original "baby pair" of rabbits, they mature for the first p months, so no increase can be observed, i.e.

$$\varphi_p(0) = \varphi_p(1) = \cdots = \varphi_p(p) = 1. \tag{21}$$

If reproduction starts with an adult pair of rabbits the pair will produce a monthly increase of a pair of rabbits for the first p months, i.e.

$$\varphi_p(0) = 1, \varphi_p(1) = 2, \ldots, \varphi_p(p) = p + 1. \tag{22}$$

Let $n \geqslant p + 1$. Then in the nth month $\varphi_p(n - p - 1)$ baby pairs of rabbits produced by all the pairs of rabbits living $(p + 1)$ months ago will be added to $\varphi_p(n - 1)$ pairs of rabbits already living in the $(n - 1)$th month. This gives the following recurrent relation for $\varphi_p(n)$, with $n \geqslant p + 1$, i.e.

$$\varphi_p(n) = \varphi_p(n - 1) + \varphi_p(n - p - 1). \tag{23}$$

Let's consider the cases of the generalized Fibonacci numbers given by the recurrent relation (23) under the initial conditions (21) and (22).

(1) *For $p = 0$.* The essence of this condition is that the rabbits mature "in a moment", i.e. the "baby rabbits" mature just after birth. It is easy to see that in the case the solution is the binary sequence $1, 2, 4, 8, \ldots, 2^n$.

(2) *For $p = 1$.* We have the classical variant of the Fibonacci problem.

(3) *For $p = 2$.* The solution comprises of two number sequences under the initial conditions (21) and (22):

$$1, 1, 1, 2, 3, 4, 6, 9, 13, 19, \ldots$$

$$1, 2, 3, 4, 6, 9, 13, 19, 28, 41, \ldots$$

(4) *For $p = \infty$.* The essence of the condition is that the rabbits mature endlessly. It is clear that if the reproduction starts with a "baby pair" of rabbits, no increase is observed, i.e. for any n $\varphi_p(n) = 1$.

If the reproduction starts with an adult pair of rabbits, for $p = \infty$ the law of reproduction is given by $\varphi_p(n) = n + 1$ which generates the sequence of natural numbers.

6. HOW MANY GOLDEN SECTIONS CAN BE FOUND?

It was stated above that the ratio of the Fibonacci neighbouring numbers tends to the golden ratio. But what does the limit of the ratio of the neighbouring Fibonacci p-numbers equal to?

Let x denote the limit, i.e.

$$\lim_{n \to \infty} \frac{\varphi_p(n)}{\varphi_p(n-1)} = x.$$

Then $\varphi_p(n)/\varphi_p(n-1)$ is represented in the form

$$\frac{\varphi_p(n)}{\varphi_p(n-1)} = \frac{\varphi_p(n-1) + \varphi_p(n-p-1)}{\varphi_p(n-1)} = 1 + \frac{1}{\frac{\varphi_p(n-1)}{\varphi_p(n-2)} \cdot \frac{\varphi_p(n-2)}{\varphi_p(n-3)} \cdots \frac{\varphi_p(n-p)}{\varphi_p(n-p-1)}}. \tag{24}$$

For $n \to \infty$ all the ratios $\varphi_p(n-j)/\varphi_p(n-j-1)$ in the right-hand side of formula (24) tend to x and the aim is to find the real root of the equation

$$x = 1 + \frac{1}{x^p}$$

or

$$x^{p+1} = x^p + 1. \tag{25}$$

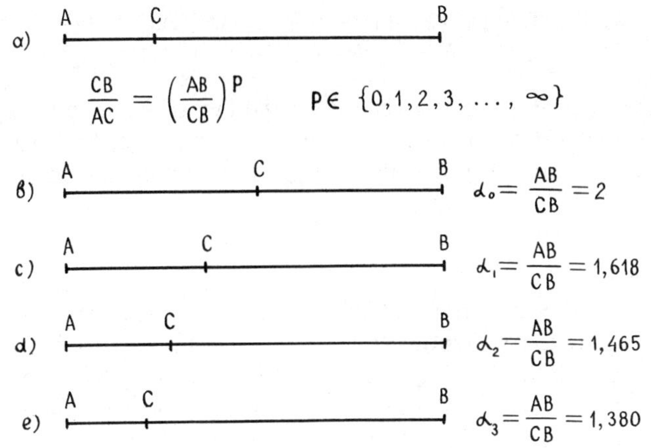

Fig. 7. Golden p-section. This geometric ratio is the generalized classical problem of the golden section. The main conclusion following this problem proves the existence of the infinite set of p-sections fundamentally connected with the Fibonacci p-numbers and the array of the binomial coefficients called Pascal's triangle.

We obtain the same equation while solving the problem of division the line segment AB at C (Fig. 7) in the ratio as

$$\frac{CB}{AC} = \left(\frac{AB}{CB}\right)^p,$$ (26)

where $p \in \{0, 1, 2, \ldots, \infty\}$.

For $p = 0$ the division of the line segment in the ratio (26) is reduced to the classical dichotomy [Fig. 7(b)] and to the classical section for $p = 1$ [Fig. 7(c)]. Thus, the division of the line segment in ratio (26) is called the generalized golden segment or the golden p-section and α_p, the real root of equation (25), is called the golden p-ratio.

The approximate values of the golden p-ratios for the initial p are given in Table 4.

So, we have proved that there exists an infinite set of the golden p-sections, each of them being the limit to which the ratio of the Fibonacci neighbouring p-numbers tends.

Let's state some properties of the golden p-ratio α_p.

From equation (25) we have

$$\alpha_p^n = \alpha_p^{n-1} + \alpha_p^{n-p-1}$$ (27)

on the other hand,

$$\alpha_p^n = \alpha_p \cdot \alpha_p^{n-1}.$$

It means that some powers of the golden p-ratio have both the properties of geometric and arithmetic progression.

It is easy to show that the powers of the golden p-ratio α_p^n and the Fibonacci p-numbers given by the recurrence relation (20) are connected by the formula

$$\alpha_p^n = \varphi_p(n-1) \cdot \alpha_p + \sum_{j=1}^{p} \varphi_p(n-j-1) \cdot \alpha_p^{-(p-j)}.$$

For $p = 0$ this formula is of the form

$$2^n = 2^{n-1} \cdot 2$$

and for $p = 1$

$$\alpha_1^n = \varphi_1(n-1) \cdot \alpha_1 + \varphi_1(n-2).$$

Table 4

p	0	1	2	3	4	5
α_p	2	1.618	1.465	1.380	1.324	1.285

7. HOW THE FIRST CRISIS IN THE FOUNDATIONS OF MATHEMATICS WAS OVERCOME

The early Pythagorean mathematics was based on the law of commensurability. According to this law, any two values Q and V have some common measure, i.e. both values are divisible by it. Thus, they can be written as a ratio of the comprime numbers m and n:

$$\frac{Q}{V} = \frac{m}{n}.$$

Considering the ratio of the diagonal and the side of the square denoted by $\sqrt{2}$ (Fig. 8) the Pythagoreans encountered a contradiction. Indeed, suppose that $\sqrt{2} = m/n$, where m and n are comprime numbers. Then $m^2 = 2n^2$. Hence, m^2 is even and m is even, respectively. However, n is odd; and if m is even, m^2 is divisible by 4, and hence, n^2 is even. Thus, n is also even. But n can't be even and odd simuntaneously! This contradiction shows that the premise of the commensurability of the diagonal and the side of the square is wrong and therefore $\sqrt{2}$ is irrational.

The discovery of the incommensurability startled the Pythagoreans and caused the first crisis in the foundations of mathematics. The discovery of the irrationals generated a notion of the complex mathematical abstraction which is not based on human experience.

According to the legend, Pythagoras committed a "hecatomb", i.e. sacrificed one hundred oxen to the gods. The discovery was worthy of the sacrifice since it was "a turning point" in mathematics. It ruined the former system created by the Pythagoreans and generated a lot of new and celebrated theories.

The importance of the discovery may be compared with the discovery of the non-Euclidean geometry or the theory of relativity at the beginning of the twentieth century. Along with these theories the problem of the incommensurable line segments was well-known to educated people. Plato and Aristotle often discussed the problems of the "incommensurability" [9, pp. 72–73].

To overcome the crisis in mathematics the famous geometer Eudoxus developed his "exhaustion" method and created the theory of values. It is one of the greatest achievements in mathematics throughout its history. Eudoxus' theory of the incommensurability (see *"Principles of Euclid"*, Book 5) coincides in general with the modern theory of irrationals suggested by Dedekind in 1872.

The measurement theory of geometric values [10] is based on the group of continuity axioms which comprises either two axioms, viz. the axiom of Eudoxus–Archimedes and the Cantor axiom or the axiom of Dedekind.

The Eudoxus–Archimedes axiom (axiom of measurement)

For any two line segments A and B a positive integer n can be found, so that $nB > A$ (Fig. 9).

The Cantor Axiom (of the "contracted" line segments)

If there is an infinite sequence of the "enclosed" line segments $A_0, B_0, A_1, B_1, \ldots, A_n B_n, \ldots$ (Fig. 10), i.e. each line segment is part of the preceding one, there exists at least one point of intersection C for all the line segments.

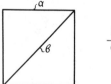

$$\frac{\beta}{\alpha} = \sqrt{2}$$

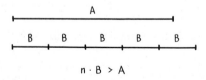

$$n \cdot B > A$$

Fig. 8. Incommensurable line segments. According to the legend, Pythagoras discovered the incommensurable line segments while examining the ratio of the diagonal to the side of the square. The discovery brought about the crisis in the foundations of mathematics. As for the influence on science it may be compared with the discovery of the non-Euclidean geometry in the nineteenth century and the theory of relativity at the beginning of the twentieth century.

Fig. 9. Eudoxus–Archimedes axiom. The axiom was suggested by the famous geometer Eudoxus aiming to overcome the crisis in the foundations of mathematics connected with the discovery of incommensurability. It is the reflection of the "exhaustion" method with which Eudoxus created the theory of values.

The principal result of the measurement theory of geometrical values is a proof of the existence and uniqueness of the solution q for the basic equation of measurement:

$$Q = qV, \tag{28}$$

where V is the unit of measurement; Q is the measurable value and q is the result of measurement.

The proof of equation (28) is that the sequence of the contracted line segments is formed from the unit of measurement V by the measurement axiom and the so-called measurement algorithm. This process tends to infinity and the sequence converges to the measured line segment. The real number q appears as a result of measurement.

It is difficult to imagine that the formulation of the axioms and the creation of the mathematical theory of measurement was the result of more than a 2000 year period in the development of mathematics. The axioms and the mathematical theory of measurement comprise a set of great mathematical ideas influencing the formation and the development of different branches of mathematics.

It should be noted that the measurement axiom is the reflection of Eudoxus' "exhaustion" method in modern mathematics. The axiom involves a 1000 year experience of man in measuring distances, areas and time intervals. It is a brief representation of the easiest algorithm of measuring the line segment A by the line segment B which is less than A, consisting in the successive applying B to A and counting the number of B's put on A (the counting algorithm).

The above measurement algorithm is the basis of various fundamental notions of arithmetic and the number theory, viz. of the notion of the natural number ($n' = n + 1$), the prime and the composite number, multiplication, division, etc. In this connection the Euclidean definitions of the prime (the "first") and the composite number ("the first number is measured by one", "the composite number is measured by some number") are of great interest.

The measurement axiom generates the division theorem which plays a fundamental part in the arithmetic of integers. The theory of divisibility and the theory of comparison are based on the theorem.

It should be noted that the subject of arithmetic, i.e. the research of the "general properties of the positive integers 1, 2, 3, ..." arises from the counting algorithm which generates both the natural numbers and the theories connected with them.

Cantor's theorem introduced in 1872 involves one more unusual phenomenon of mathematical thinking, i.e. the abstraction of the completed infinity. To clear out this notion we compare the Cantor axiom with the Eudoxus–Archimedes axiom. As for the Cantor axiom, the infinite set of "contracted" line segments together with the connecting point is considered to be given by all the objects simultaneously. The "completed" infinite sets represent the most distinctive feature of Cantor's set-theoretic mathematical style.

Though of an empirical origin the Eudoxus–Archimedes axiom is constructive and implicitly rests upon the more "simple" abstraction of the infinity which is called the abstraction of potential feasibility.

According to the Eudoxus–Archimedes axiom, the number of measurement steps and the involved segments is always finite $nB > A$, but it is potentially unlimited which makes the constructive sense of the infinity as a potential, forming category.

In this connection we note the intrinsic discrepancy of the set-theoretic measurement theory and arisen from it the theory of real number admitting in the initial conditions (the continuity axioms) of the coexistence of the dialectical contradictory ideas of the infinite (the actual, "static" and complete infinity as in Cantor's axiom and that of Dedekind and the potential, the so called "forming", incomplete infinity as in the Archimedes axiom).

However, the abstraction of the actual infinity plays a fundamental part in the classical mathematic measurement theory because it generates the "possibility" of the infinite duration of measurement with the inevitable occurrence of the irrational. . . .[10, p. 24]

8. CAN THE MEASUREMENT BE COMPLETED FOR THE INFINITE PERIOD OF TIME OR THE CONSTRUCTIVE APPROACH TO MEASUREMENT?

The idea of measurement as a process completed during the infinite period of time finds, on the one hand, a deep gap between the "experimental data of the natural research" [11, p. 20]; on the other hand, according to A. A. Markov, "to think of the infinite, i.e. incompleted process as of a completed one is impossible without committing violence upon the mind rejecting such a contradictory fantasy" [12, p. 41].

The paradoxes and the contradictions in Cantor's theory of sets discovered at the beginning of the twentieth century shook considerably the foundations of mathematics and brought about a new crisis (the third one since the discovery of the incommensurable line segments). To overcome this crisis different efforts were made. The most radical is the constructive approach which excludes completely the abstraction of the actual infinity and prefers the abstraction of the potential infinity.

What will happen to the mathematical measurement theory if the abstraction of the actual infinity is excluded?

It means that the mathematical measurement theory should be constructed on the idea of the measurement infinity according to which any measurement is performed in the finite number of steps and also on the constructive idea of the potential feasibility according to which we abstract ourselves from the limited possibilities of choosing the number of steps for a given measurement.

The stated methodological basis results in the unavoidable measurement error due to the finite property of the measurement act, viz. the quantizing error and in a new interpretation of the problems of the mathematical measurement theory.

In proving equation (28) it is important to find the proper measurement algorithm by which the sequence of the "contracted" line segments is formed from V. When the number of steps is infinite the measurement algorithm has no influence on q. It ensures only the representation of q in different notations.

For a given number of n measurement steps there appears a difference between the n-step measurement algorithms concerning the "accuracy" of measurement defined by the quantizing error. So, the second constructive idea of the effective measurement algorithm in terms of the quantizing error is very important and the synthesis of the optimal measurement algorithms is considered to be the fundamental problem of the algorithmic measurement theory.

9. GEOMETRIC PROBLEM PUT FORWARD BY N. I. LOBACHEVSKI

Lobachevski was one of the first mathematicians who paid attention to the non-constructive character of the set-theoretic measurement theory. After the futile efforts to win the recognition of the official academic scientists in Russia of that time Lobachevski published in 1840 his brilliant mathematic work "Geometric research on the theory of parallel lines" in German. Lobachevski begins his work as follows: "In geometry I have found some imperfections which I consider to be the reason of the fact that this science hasn't exceeded the limits of the state in which it came to us from Euclid. I refer to the imperfections and vagueness of the first notions of the geometric values and, in the end, the main gap in the theory of parallel lines." [10, p. 10]

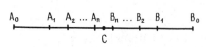

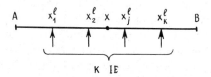

Fig. 10. Cantor axiom. The axiom suggested by Cantor in 1872 comprises an unusual creation of the mathematical thinking, i.e. the abstraction of the completed infinity which was assumed by Cantor as a basis of his theory of infinite sets. The abstraction was criticized by the constructive mathematics usimg the more "simple" idea of the infinity which was called the abstraction of potential feasibility.

Fig. 11. Mathematical pattern of measurement. The pattern is based on the notion of the indicator element. The comparison is performed with respect to other homogeneous values (measures) due to the notion of the indicator element.

By the following the celebrated geometer put forward a programme of geometric research including three problems:

1. The foundations of the initial notions in geometry.
2. The measurement algorithms.
. 3. The problem of parallel lines.

Later the geometry developed along the path defined by Lobachevski. The solution of the problem of parallel lines resulted in the non-Euclidean geometry stirred the geometric science and brought about the fundamental study of the intial geometric notions. Riemann and Helmholtz tore geometry away from the real substrate with which it had been connected for a 1000 years. They gave a new interpretation of such notions as "point", "distance", "straight line", "plane" and "space".

The solution of the problem of the measurement algorithms caused a certain bewilderment. What did Lobachevski mean by that? Of what importance were the researches in that direction for science?

The problem of measurement can be stated in terms of geometry. Suppose, we have the line segment AB with some point X (Fig. 11). The problem is to find the length of the line segment AX. It is performed via K indicator elements (IE). After application of the jth IE ($j = 1, 2, \ldots, K$) to some point $X_j^l \in AB$ the line segments AX and AX_j^l are compared, i.e. the relations "less" ($AX < AX_j^l$) and "greater than or equal" ($AX \geqslant AX_j^l$) are defined. The relations "less" and "greater than or equal" are coded by the binary digits 0 and 1 generated by IE. The problem of measuring the line segment AX by IE is reduced to decreasing the aperture of x according to IE "indications". The procedure of measurement is as follows: in the first step IE are applied to some points of the line segment AB and the aperture is decreased to the line segment A_1B_1 from AB according to IE "indications"; in the next step IE are applied to the points of the aperture defined in the preceding step.

The restrictions s are imposed on the measurement process. The system of the formal rules strictly defining in each n-step for each line segment AX the set of points of IE application depending on the "indications" in the preceding steps under the resctrictions s imposed on the measurement is called the (n, k, s) measurement algorithm.

How many different (n, k, s) algorithms are to be found? Let's take a simple case, for example, $k = 1$ and there are no restrictions ($s \equiv 0$) as for the moving of IE along the line segment AB. Let it be necessary to measure AX to the "accuracy" of N [Fig. 12(a)]. It means that the set of points to which IE can be applied is rigorously restricted and defined by the points $X_1, X_2, \ldots, X_{N-1}$.

Let's denote the number of different measurement algorithms by P_N to the accuracy of N. It is obvious that for. $N = 2$ there exists the only measurement algorithm which consists in the

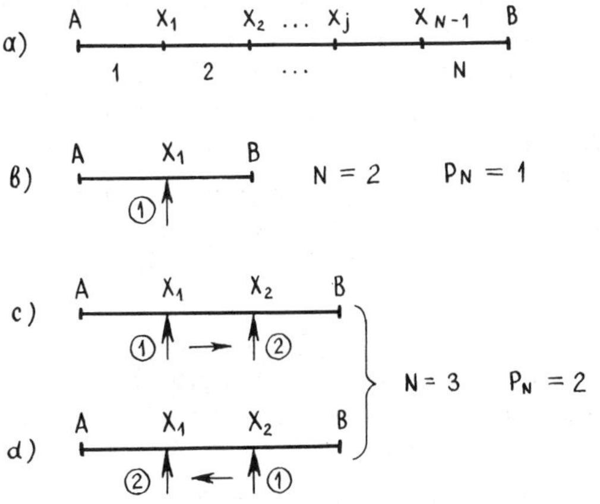

Fig. 12. On the estimate of the number of different measurement algorithms. The number of different measurement algorithms to the "accuracy" of N is rather large. For example, for $N = 256$ the number constitutes 1.13×2^{497}; it exceeds considerably the estimate of the number of electrons in the universe (136×2^{256}).

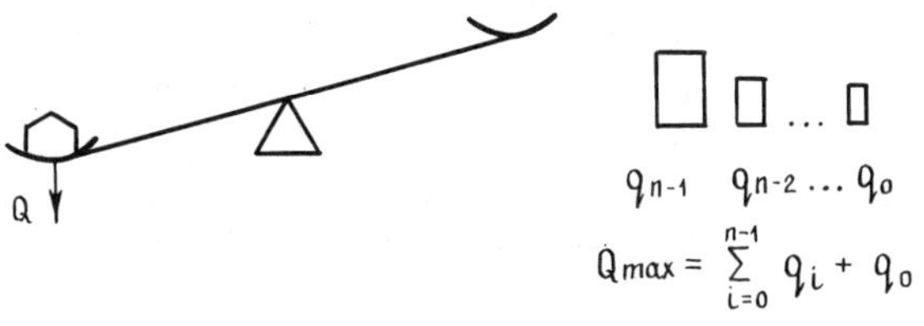

Fig. 13. Problem of choosing the best system of scale weights. The problem was first stated and solved by the Italian mathematician Leonardo of Pisa (Fibonacci) in the thirteenth century. Fibonacci proves that the binary system of scale weights is the best one and also discovers the binary notation underlying modern computer engineering.

Table 5

N	2	3	4	5	6	7	8	9	10
P_N	1	2	5	14	42	132	229	1430	4862

application of IE to the point X_1 [Fig. 12(b)]; for $N = 3$ there are two algorithms, viz. the first one consists in the application of IE to the point X_1 and then to the point X_2 [Fig. 12(c)]; the second consists in the application of IE to the point X_2 and then to the point X_1 [Fig. 12(d)].

In general the measurement algorithm to the accuracy of N can be represented as a process of the successive binary partitions of some ordered finite set of N elements into separate ones. It is shown [13] that the number of the partitions, i.e. the number of different measurement algorithms is equal to

$$P_N = \frac{1}{N} \cdot C_{2(N-1)}^{N-1} \tag{29}$$

where C_m^n is a number of combinations of m by n.

The values of P_N for the initial N are given in Table 5.

With increasing N, P_N increases greatly and the precise evaluation of P_N by equation (29) is impossible. Using the Stirling formula we have the approximate expression for P_N:

$$P_N \approx \frac{2^{2(N-1)}}{N \cdot \sqrt{\pi \cdot (N-1)}}. \tag{30}$$

P_N values evaluated by expression (30) are shown in Table 6; it follows that for large N the number of different measurement algorithms is rather great. Remember that the Eddington's estimation of the electron quantity in the universe is no more than $136 \cdot 2^{256}$ which is less than the number of different measurement algorithms for $N = 256$.

The above examples show the complexity of the Lobachevski problem concerning the measurement algorithms. It is clear that there is no need to analyse all possible measurement algorithms. It is sufficient to build the best one. The aim of the algorithmic measurement is to solve these problems.

10. THE PROBLEM OF CHOOSING THE BEST SYSTEM OF SCALE WEIGHTS OR HOW THE BINARY NOTATION WAS INVENTED

In the thirteenth century Leonardo of Pisa (Fibonacci) set forward a problem choosing the best system of the scale weights and included it in his book "*Liber Abaci*".

The essence of this problem is in the following (Fig. 13): what system of scale weights taken separately is used for weighing the Q loads from 0 through Q_{max} of the error no more than q_0? The system of scale weights ensuring the maximal meaning of the greatest weighed load Q_{max} under the equal conditions is called optimal. In Russian historic and mathematic literature [14] this problem is also called the Bachet–Mendeleyev problem in honour of the French mathematician of the seventeenth century Bachet de Mesiriaque and the founder of Russian metrology D. Mendeleyev.

There exist two variants of this problem. In the first variant the weights are put only on the free scale, in the second variant they are put on both [15].

Let's consider the first variant. It is clear that the weight of the greatest load which may be weighed using a certain system of scale weights $\{q_{n-1}, q_{n-2}, \ldots, q_0\}$ an error q_0 is equal to the sum of all the scale weights adding q_0. However, the scale weights should be arranged so that they compose any load from 0 through Q_{max} divisible by q_0. It is clear that with given q_0 and the system of the scale weights n the weight of the greatest load is a function of n, i.e. $Q_{max} = \varphi(n)$. The problem is to find the system of scale weight which with given q_0 and n yields

$$Q_{max} = \max_{\{q_{n-1}, \ldots, q_0\}} Q_{max} .$$

Table 6

N	16	64	256
P_N	$1.13 \cdot 2^{23}$	$1.13 \cdot 2^{116}$	$1.13 \cdot 2^{497}$

The example reveals the essence of the scale weights problem. Let it be necessary to find the system of four scale weights among $\{10, 5, 3, 1\}, \{8, 4, 2, 1\}, (5, 3, 2, 1\}$.

The fact is that the first variant isn't satisfactory since it doesn't allow to compose the loads of 7, 12, 17. The second and the third variants meet the requirements but the second is preferable, since $Q_{max}^{(2)} = 16$ that is greater than $Q_{max}^{(3)} = 12$.

The first variant of the scale weights problem was solved by Fibonacci. He proved that the binary system of scale weights $\{1, 2, 4, 8, \ldots, 2^{n-1}\}$ was optimal. He discovered the "binary" measurement algorithm which is widely used in modern analog-to-digital converters and also the method of the binary representation of numbers

$$Q = \sum_{i=0}^{n-1} a_i \cdot 2^i, \tag{31}$$

which defines the information arithmetic and circuitry fundamentals of modern computer engineering.

The binary digits $a_i \in \{0, 1\}$ in equation (31) encode one of the two positions of the scales; thus, $a_i = 1$ if the scales are in the initial position after adding the next weight to the free scale and $a_i = 0$ otherwise.

It should be noted that the binary notation shows a deep connection between the measurement algorithm and the methods of number representations (notations).

11. NEUGEBAUER'S HYPOTHESIS

The culmination of the historical development of the notations is the decimal notation. Despite the seeming simplicity it comprises a deep mathematical idea.

There are two concepts as for the origin of the positional notations, i.e. the counting and the metrological one. The concept of the "finger count" explains easily the origin of the decimal, quinary and at last icosahedral notation but finds it difficult to explain the origin of the notations with the "non-finger" bases, i.e. the binary, the duodecimal and the Babylonian sexagesimal notations. To overcome this difficulty the American historian of mathematics O. Neugebauer in 1927 put forward a hypothesis of a metrological origin of the positional notations [16].

Neugebauer's main idea is that each positional notation was preceded by the long historical stage in the development of the system of weights and money. For example, the notation with the base 60 is a result of the sexagesimal notation appeared in ancient Babylon. According to Neugebauer, "the emergence of the notation from the original distinct system of weights is closely connected with the positional notation. The positional notation is nothing but the methodical refusal of indicating measure units in writing." [16, p. 124]

Neugebauer's hypothesis says that the new ideas concerning the positional notations should be looked for in the measurement theory. Neugebauer's hypothesis is confirmed by the binary notation created in European mathematics. It was developed by the author while solving the problem of choosing the best system of scale weights. The fact is that the methods of positional representation of numbers are treated as the measurement algorithms. So, the research of the measurement algorithms can be regarded as a study of creating of the new notations.

12. ASYMMETRY PRINCIPLE OF MEASUREMENT

The finiteness and potential feasibility principles of measurement making the basis of the constructive measurement theory are "external" with respect to measurement; they are so general that the constructive measurement theory can be reduced to a certain trivial result (e.g. to the proof of the "optimality" of the Fibonacci "binary" measurement algorithm).

To get the non-trivial results a certain general principle ensured by the essence of measurement should be added to the basis of the constructive algorithmic measurement theory. The principle is based on the studying of the weighing procedure.

The analysis of the above mentioned "binary" measurement algorithm gives a general property for any measurement reduced to the weighing on the weigh-scales. In the first step of the "binary"

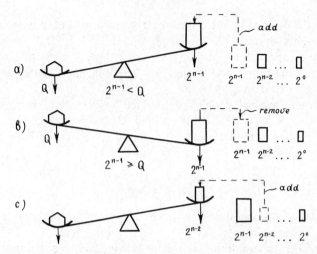

Fig. 14. Asymmetry principle of measurement. The principle underlying the algorithmic measurement theory has put the two well-known problems stated by Fibonacci, viz. the problem of scale weights and the problem of "rabbits multiplying" on the same mathematical basis (the Fibonacci p-numbers). The deep inner connection between the "rabbit mutiplying" and the procedure of weighing on the weigh-scales with the "sluggishness" p has also been discovered.

weighing algorithm the largest 2^n weight is put on the free scale [Fig. 14(a)]. The cases $2^{n-1} < Q$ [Fig. 14(a)] and $2^{n-1} \geqslant Q$ [Fig. 14(b)] result at that. In the former case [Fig. 14(a)] the next step consists of adding the next large weight to the free scale. In the latter case the "weigher" should perform two operations, i.e. remove the previous weight [Fig. 14(b)] so that the weigh-scales are returned to the initial position [Fig. 14(b)]. Then the next weight is put on the free scale. The defined property of measurement constitutes the asymmetry principle of measurement.

In the latter case the "weigher's" actions are defined by the two factors. At first he has to remove the proper weight and then take into account the time spent for returning the weigh-scales to the intial position.

The main idea of the algorithmic measurement theory based on the asymmetry principle of measurement consists in introducing the relaxation period of time into the mathematic measurement pattern and the measurement algorithm.

Owing to their physical or biological sluggishness the artificial and natural threshold and switches confirm the idea of the "switch relaxation" and the asymmetry of comparison. For example, the asymmetry property of the biological neuron consists in the refractory period following the impulse generation; the neuron loses its sensitivity and can't be agitated irrespective of the input power.

We introduce the discovered property into the problem of scale weights. Let's consider measurement as a process running during the discrete periods of time; let the operation "add the weight" be performed for the unit of the discrete time and the operation "remove the weight" (which is followed by the returning of the weigh-scales to the initial position) be performed for p units of the discrete time with $p \in \{0, 1, 2, \ldots, \infty\}$. In this case p defines the "sluggishness" of the weigh-scales. The case for $p = 0$ corresponds to the ideal case when we neglect the "sluggishness" of the weigh-scales. The case corresponds to the classical problem of scale weights.

Further generalization of the problem of scale weights consists in increasing the number of the weigh-scales from 1 to K, the same Q load being put on the left-hand scale of the weigh-scales (the "parallel" measurement of the Q load by means of K weigh-scales in n-steps).

The n-step algorithm of measuring the Q load on K weigh-scales with the "sluggishness" p is required.

13. THE FUNDAMENTAL RESULTS OF THE ALGORITHMIC MEASUREMENT THEORY

To solve the above stated problem it is necessary to define the notion of the optimal measurement algorithm. We refer to the above measurement pattern based on the indicating elements. It is clear that the indicating element corresponds to the weigh-scales in the problem of scale weights. The pattern of the indicating element reduces the measurement problem to the problem of searching the point X on the line segment AB in n-steps by means of K IE. The above approach allows to use the methods of one-dimensional search for synthesizing the optimal measurement algorithms [17]. Let us consider the (n, k, s) algorithm for any points $X \in AB$ and get a partition of the line segment AB into N line segments Δ_i, where $AB = \Delta_1 + \Delta_2 + \Delta_3 + \cdots + \Delta_N$. Then from this partition we choose the largest line segment $\Delta_{\max}$ and define the effectiveness of the (n, k, s) algorithm by means of the length of this line segment or rather by the pure number $T = AB/\Delta_{\max}$ which we call the (n, k) accuracy of the algorithm. By the optimal (n, k, s) algorithm we mean the (n, k, s) algorithm ensuring the maximal (n, k) accuracy.

The case is widely used when the (n, k, s) algorithm ensures the partition of the line segment AB into N equal parts of Δ. In this case $AB = N\Delta$, the (n, k) accuracy $T = N$, i.e. for that class of the (n, k, s) algorithms the criterion of optimality coincides with the number of N quantization levels provided by the (n, k, s) algorithm. The restriction s imposed on the measurement algorithm are brought about by the "sluggishness" of the p weigh-scales. With given p the (n, k) accuracy of the (n, k, s) optimal algorithm is a certain function of the number of n-steps and K indicator elements, i.e.

$$N = \varphi_p(n, k).$$

The strict mathematical solution of the generalized problem of scale weights (with regard for the asymmetry principle of measurement) was obtained by the author together with Viten'ko in 1970 [1]. To explain the essence of the solution we refer to the above constructed mathematical pattern of measurement. We stimulate the "sluggishness" of the weigh-scales by means of the IE position. Before the lth step of the (n, k, s) algorithm the jth IE $(l = 1, 2, \ldots, n; j = 1, 2, \ldots, k)$ is in the position of $p_j(l)$, where

$$0 \le p_j(l) < p$$

$$p_j(l + 1) = p_j(l) - 1.$$

Since the weigh-scales in real measurement correspond to IE in the mathematical measurement pattern the physical sense of the function $p_j(l)$ is as follows.

If $p_j(l) = 0$ it means that the weigh-scales are put in the initial position [Fig. 14(a)]. If $p_j(l) = p$ it means that the weigh-scales are placed to the contrary position in the lth step [Fig. 14(b)]. After removing the weight the weigh-scales are put in the initial position in p-steps (Fig. 15). In every step IE position is decreased by one unit, i.e.

$$p_j(l + 1) = p - 1;$$

$$p_j(l + 2) = p_j(l + 1) - 1 = p - 2;$$

$$\cdots\cdots\cdots\cdots\cdots\cdots\cdots$$

$$p_j(l + p) = 0.$$

Before the lth step we number IEs so that we put them in order according to the rule:

$$p_k(l) \ge p_{k-1}(l) \ge \cdots \ge p_2(l) \ge p_1(l), \tag{32}$$

where $p_j(l)$ is the position of the jth IE in the lth step. We denote IE positions by $p_1, p_2, \ldots, p_k$ in the first step of the (n, k, s) algorithm ordered in accordance with condition (32). It is clear that the (n, k) accuracy of the (n, k, s) optimal algorithm depends not only on n and k but on the initial positions of IE denoted by $p_1, p_2, \ldots, p_k$. The relation is given by $\varphi_p(n, k) = \varphi_p(n; p_1, p_2, \ldots, p_k)$.

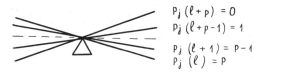

$$P_j(\ell + p) = 0$$
$$P_j(\ell + p - 1) = 1$$
$$P_j(\ell + 1) = P - 1$$
$$P_j(\ell) = P$$

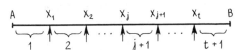

Fig. 15. "Sluggishness" of the weigh-scales. Under the "sluggishness" of the weigh-scales we mean a number of the discrete periods of time p, during which the weigh-scales are returned to the initial position after removing the proper weight from the left scale.

Fig. 16. Synthesis of the optimal measurement algorithm. Under the optimal measurement algorithm we mean the n-step measurement algorithm ensuring the division of the initial aperture into the maximum number of units.

Let the initial positions of IE ordered according to condition (32) be so that $p_1 = p_2 = \cdots = p_t = 0$ and $p_{t+1} > 0$. This means that only the first ts of IE are applied to the points of the line segment AB in the first step of the (n, k, s) algorithm.

Suppose there exists an optimal (n, k, s) algorithm which under the above restrictions, divides the line segment AB into

$$\varphi_p(n; \underbrace{0, 0, \ldots, 0}_{t}, p_{t+1}, p_{t+2}, \ldots, p_k)$$

equal parts. Let the first step of the algorithm ensure the application of ts of IE to the points $X_1, X_2, \ldots, X_t$ (Fig. 16). Depending on the position of x on the line segment AB we can write $t + 1$ cases based on the indications of IE after the first step of the (n, k, s) algorithm:

$$X \in [AX_1]; X \in [X_1 X_2]; \ldots; X \in [X_t B].$$

Let $X \in [AX_1]$. In this case all ts of IE have reached the position of p and the rest $(k - t)$ of IE are decreased by one unit. Putting in order IE position according to condition (32) we have a set of IE positions in the second step of the (n, k, s) algorithm for the line segment AX_1:

$$p_{t+1} - 1, p_{t+2} - 1, \ldots, p_k - 1, \underbrace{p, p, \ldots, p}_{t}.$$

Since the (n, k, s) algorithm has only $n - 1$ steps available, applying the optimal $(n - 1, k, s)$ algorithm to the line segment AX_1 we can divide it by induction into

$$\varphi_p(n - 1; p_{t+1} - 1, \ldots, p_k - 1, \underbrace{p, \ldots, p}_{t})$$

equal parts.

After the first step of the algorithm we obtain the line segment $[X_j, X_{j+1}]$, i.e. $X \in [X_j, X_{j+1}]$. In this case j of IE remained in the zero position, $(t - j)$ of IE reached the position of p and the rest of $(k - t)$ of IE are decreased by one unit. Putting in order the position of IE in accordance with condition (32) we get a set of IE positions in the second step of the (n, k, s) algorithm for the line segment $[X_j, X_{j+1}]$:

$$\underbrace{0, 0, \ldots, 0}_{j}, p_{t+1} - 1, \ldots, p_k - 1, \underbrace{p, p, \ldots, p}_{t-j}.$$

Applying the optimal $(n - 1, k, s)$ algorithms to the line segment $[X_j, X_{j+1}]$ we can divide $[X_j, X_{j+1}]$ by induction into

$$\varphi_p(n - 1; \underbrace{0, 0, \ldots, 0}_{j}, p_{t+1} - 1, \ldots, p_k - 1, \underbrace{p, \ldots, p}_{t-j})$$

equal parts.

From this follows the recurrence formula for evaluating the function $\varphi_p(n, k)$ for the optimal (n, k, s) algorithm:

$$\varphi_p(n, k) = \varphi_p(n; \underbrace{0, \ldots, 0}, p_{t+1}, \ldots, p_k) = \sum_{j=0}^{t} \varphi_p(n - 1; \underbrace{0, \ldots, 0}_{j}, p_{t+1} - 1, \ldots, \underbrace{p, \ldots, p}_{t-j}). \qquad (33)$$

To define the initial conditions for evaluating $\varphi_p(n, k)$ one must take into account that for $n = 1$ and $p_1 = p_2 = \cdots = p_t = 0$, $p_{t+1} > 0$ the optimal $(1, k, s)$ algorithm consists in the partition of the line segment AB into $t + 1$ equal parts, i.e.

$$\varphi_p(1; 0, \ldots, p_{t+1}, \ldots, p_k) = t + 1. \qquad (34)$$

Let's introduce the definition

$$\varphi_p(n, k) = \begin{cases} 0, & \text{for } n < 0, \\ 1, & \text{for } n = 0. \end{cases} \qquad (35)$$

The "physical sense" of the function $\varphi_p(n; p_1, p_2, \ldots, p_k)$ implies for $n \geqslant p_1 > 0$ the equality

$$\varphi_p(n; p_1, p_2, \ldots, p_k) = \varphi_p(n - p_1; 0, p_2 - p_1, \ldots, p_k - p_1). \tag{36}$$

Under the initial conditions (34) and (35) the recurrence formula (33) is the fundamental mathematic result of the algorithmic measurement theory, via the result the theoretically infinite set of the optimal measurement algorithms (for defined n, k, and p) is given in the generalized form.

14. UNEXPECTED RELATIONS

The recurrence relation (33) proves to comprise a set of formulas of the discrete mathematics under the initial conditions (34) and (35).

Let $p = 0$. In this case the recurrence formula (33) and the initial condition (34) are of the form:

$$\varphi_p(n; k) = \varphi_p(n; \underbrace{0, 0, \ldots, 0}_{k}) = (k + 1) \cdot \varphi_p(n - 1; \underbrace{0, 0, \ldots, 0}_{k}) \quad \varphi_p(1; \underbrace{0, 0, \ldots, 0}_{k}) = k + 1.$$

The solution of the above recurrent relation is the combinatorial formula $(k + 1)^n$, i.e.

$$\varphi_p(n, k) = (k + 1)^n.$$

So, the n-step algorithm of measurement consists in the successive partition of the initial aperture AB (and all the other apertures) into the $k + 1$ equal parts and generates the method of the positional $(k + 1)$th number representation, i.e. the "decimal" method for $k = 9$ and the "binary" method for $k = 1$. Let $p = \infty$ and $p_1 = p_2 = \cdots = p_k = 0$. This means that the measurement is performed by the weigh-scales with the infinite sluggishness, i.e. in the process of measurement the weigh-scales reach the contrary position only once.

Examine the function

$$\varphi_p(n; \underbrace{0, 0, \ldots, 0}_{j}, \underbrace{p, p, \ldots, p}_{k}). \tag{37}$$

The physical sense of the problem implies that for the $(k - j)$ IE in the position p with n can't affect the function (37) which depends only on the number of steps of the algorithm n and the number of j IE in the zero position, i.e.

$$\varphi_p(n; \underbrace{0, \ldots, 0}_{j}, \underbrace{p, \ldots, p}_{k - j}) = \varphi(n, j). \tag{38}$$

With regard to function (38) the recurrence formula (33) for the function

$$\varphi_p(n; \underbrace{0, \ldots, 0}_{k})$$

and the initial condition (34) are expressed as

$$\varphi(n, k) = \varphi(n - 1, 0) + \varphi(n - 1, 1) + \cdots + \varphi(n - 1, k) = \varphi(n, k - 1) + \varphi(n - 1, k); \tag{39}$$

$$\varphi(1, k) = k + 1. \tag{40}$$

It is easy to show that the relations (39) and (40) are given by the binomial coefficients, i.e.

$$\varphi(n, k) = C_{n+k}^k = C_{n+k}^n,$$

and the corresponding measurement algorithm given by Pascal's triangle generates the method of numbering the positive integers as a sum of the binomial coefficients.

For $k = 1$ and $p_1 = 0$ the recurrent relation (33) is of the form

$$\varphi_p(n; 0) = \varphi_p(n - 1; 0) + \varphi_p(n - 1; p).$$

Let us consider the function $\varphi_p(n - 1; p)$. From function (37) we have

$$\varphi_p(n - 1; p) = \varphi_p(n - p - 1; 0)$$

and therefore the recurrent relation (37) is of the form

$$\varphi_p(n; 0) = \varphi_p(n - 1; 0) + \varphi_p(n - p - 1; 0).$$

Using for simplicity $\varphi_p(n) = \varphi_p(n; 0)$ we get the recurrent relation $\varphi_p(n) = \varphi_p(n-1) + \varphi_p(n-p-1)$, which coincides with the recurrent relation giving the Fibonacci p-numbers.

The corresponding measurement algorithms form the class of the Fibonacci measurement algorithms where the system of scale weights is given by the Fibonacci p-numbers. The Fibonacci measurement algorithms generate the Fibonacci p-codes considered below.

The fundamental result of the algorithmic measurement theory along with the other unexpected results is demonstrated in Fig. 17. "Unexpectedness" of the result obtained consists in stating the deep mathematical connection between the algorithmic measurement theory and some principle mathematic achievements, i.e, the Fibonacci p-numbers and binomial coefficients.

The result obtained is of great interest both for the combinatories and the theory of numbers. However, it entails the principle methodological conclusions if it is allowed that "in its origin the notion of number which became the basis of arithmetic was not only of a concrete nature but was inseparable from the notion of measurement which later underlay geometry. In the further development of mathematics the notions are differentiated and are combined at each higher stage." [18, p. 16]

15. FIBONACCI CODES

The proved above possibility of measuring the values by means of the Fibonacci p-numbers is equivalent to the expression of any positive integer N in the form

$$N = a_{n-1} \cdot \varphi_p(n - 1) + a_{n-2} \cdot \varphi_p(n - 2) + \cdots + a_0 \cdot \varphi_p(0), \qquad (41)$$

where $a_i \in \{0, 1\}$ is the binary digit in the ith bit of code (41); $\varphi_p(i)$ is the weight of the ith bit evaluated by the recurrence formula (20):

$$n \text{ is the number of digits in the Fibonacci } p\text{-code.}$$

The representation of the positive integer in the form of code (41) is called the Fibonacci p-code of the number N. We write the Fibonacci p-code as

$$N = a_{n-1} a_{n-2} \cdots a_0.$$

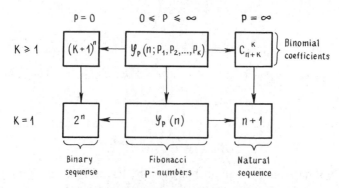

Fig. 17. The fundamental result of the algorithmic measurement theory. The recurrent relation $\varphi_p(n; p_1, p_2, \ldots, p_n)$ describing the set of the optimal measurement algorithms comprise a number of well-known combinatorial formulas, viz. the formula for the binary and natural number sequence, for the Fibonacci p-numbers and the formula for the binomial coefficients.

Since for $p = 0$, $\varphi_0(i) = 2^i$ and for $p = \infty$, $\varphi_p(i) = 1$ for any i it implies that for $p = 0$ the Fibonacci p-code (41) coincides with the classical binary code; for $p = \infty$ it coincides with the trivial representation of N as a sum:

$$N = \underbrace{1 + 1 + \cdots + 1}_{N},$$

which is called the "unitary" code. Thus, the Fibonacci p-codes (41) for distinct $p \geqslant 0$ sort of fill up the gap between the classical binary code and the "unitary" code regarding them as special cases for $p = 0$ and $p = \infty$.

We can prove that for given $p \geqslant 0$ and $n \geqslant p$ for any positive integer N there exists the only representation of N in the form of

$$N = \varphi_p(n) + \zeta, \tag{42}$$

where

$$0 \leqslant \zeta < \varphi_p(n - p). \tag{43}$$

Note that for $p = 0$ the expressions (42) and (43), are of the form

$$N = 2^n + \zeta; \quad 0 \leqslant \zeta < 2^n.$$

As opposed to the classical binary code the Fibonacci p-codes for $p > 0$ are the redundant codes, i.e. some representations of the form (41) correspond to one and the same positive integer N. The so called normal or minimal form of N in the Fibonacci p-code is of great importance.

The minimal form of N in code (41) is obtained from expansion of N and the remainders r by formulas (42) and (43) until the remainder is equal to zero. In the minimal form each unit bit $a_i = 1$ is followed by at last p zero bits $a_{i-1} = a_{i-2} = \cdots = a_{i-p} = 0$.

The minimal forms of 15 in the Fibonacci 1 and 2 codes are given below.

$$p = 1$$

	21	13	8	5	3	2	1	1
$15 =$	0	1	0	0	0	1	0	0

$$p = 2$$

	13	9	6	4	3	2	1	1	1
$15 =$	1	0	0	0	0	1	0	0	0

Different representations of the same N in code (41) can be obtained one from another by means of convolution and development which are based on (20) connecting the bit weights in code (41). The operations are illustrated by the examples:

$$p = 1$$

$$p = 2$$

$$15 = \begin{cases}
\begin{array}{ccccccccc}
13 & 9 & 6 & 4 & 3 & 2 & 1 & 1 & 1 \\
1 & 0 & 0 & 0 & 0 & 1 & 0 & 0 & 0 \\
0 & 1 & 0 & 1 & 0 & 0 & 1 & 0 & 1 \\
0 & 1 & 0 & 1 & 0 & 1 & 0 & 0 & 0 \\
0 & 1 & 1 & 0 & 0 & 0 & 0 & 0 & 0
\end{array}
\end{cases}$$

The developments are denoted by └──↑↑ and the convolutions are denoted by ↑└──┘ . It should be noted that while performing convolution and development in the code of N the value of the represented is not changed.

16. THE GOLDEN p-RATIO CODES

Let us weigh the load A by means of the system of scale weights α_p^i, where i takes its value from the set of the integers and α_p is the golden p-ratio. The representation of the real number A satisfies the measurement algorithm:

$$A = \sum_{i=-\infty}^{+\infty} a_i \alpha_p^i, \tag{44}$$

where $a_i \in \{0, 1\}$ is the binary digit in the ith bit of code (41); α_p^i is the weight of the ith bit; α_p is the radix of the notation.

We call the representation of A in the form of equation (44) the golden p-ratio code of A. Equation (44) gives theoretically an endless number of methods of positional representation of real numbers since each p has its own notation.

The radix in equation (44) is the golden p-ratio α_p which is the irrational for $p > 0$; thus, the golden p-ratio codes are the notations with irrational radices of the α_p type.

First the notations were introduced in 1957 by George Bergman who considered the special case of the notation (44) corresponding to $p = 1$.

Note that for $p = 0$, $\alpha_0 = 2$ the golden p-ratio codes (44) are reduced to the classical binary notation which is the only exception, i.e. the radix is the natural 2. As for their mathematical structure the golden p-ratio codes (44) are similar to the Fibonacci p-codes (41). This similarity is defined by the fact that the weight of any ith bit is equal to the weight of the $(i-1)$th and the $(i-p-1)$th bits.

As in the Fibonacci p-codes the main feature of the notation (44) is the multivalued representation of numbers. Different code representations of the same number in equation (44) can be obtained by means of the convolution and development based on the relation (27). Here is the binary representation of the positive integers in the golden 1-ratio code:

$$0 = \quad 0 \quad 0 \quad 0 \quad 0, \quad 0 \quad 0 \quad 0 \quad 0$$

$$1 = \begin{cases}
0 \quad 0 \quad 0 \quad 1, \quad 0 \quad 0 \quad 0 \quad 0 \\
0 \quad 0 \quad 0 \quad 0, \quad 1 \quad 1 \quad 0 \quad 0
\end{cases}$$

$$2 = \begin{cases}
0 \quad 0 \quad 0 \quad 1, \quad 1 \quad 1 \quad 0 \quad 0 \\
0 \quad 0 \quad 1 \quad 0, \quad 0 \quad 1 \quad 0 \quad 0
\end{cases}$$

$$3 = \begin{cases}
0 \quad 0 \quad 1 \quad 1, \quad 0 \quad 1 \quad 0 \quad 0 \\
0 \quad 1 \quad 0 \quad 0, \quad 0 \quad 1 \quad 0 \quad 0
\end{cases}$$

$$4 = \begin{cases} 0 \quad 1 \quad 0 \quad 1, \quad 0 \quad 1 \quad 0 \quad 0 \\ 0 \quad 1 \quad 0 \quad 1, \quad 0 \quad 0 \quad 1 \quad 1 \\ 0 \quad 1 \quad 0 \quad 0, \quad 1 \quad 1 \quad 1 \quad 1 \end{cases} \tag{45}$$

$$5 = \begin{cases} 0 \quad 1 \quad 0 \quad 1, \quad 1 \quad 1 \quad 1 \quad 1 \\ 0 \quad 1 \quad 1 \quad 0, \quad 0 \quad 1 \quad 1 \quad 1 \\ 1 \quad 0 \quad 0 \quad 0, \quad 1 \quad 0 \quad 0 \quad 1 \end{cases}$$

$$6 = \begin{cases} 1 \quad 0 \quad 0 \quad 1, \quad 1 \quad 0 \quad 0 \quad 1 \\ 1 \quad 0 \quad 1 \quad 0, \quad 0 \quad 0 \quad 0 \quad 1 \end{cases}$$

As follows from equations (44) the golden p-ratio code of A is divided by a decimal point into the integer and the fractional parts; moreover, for $p > 0$ the notions of the integer and the fraction in the golden p-ratio code don't coincide with the similar notions in the traditional notations. For example, the integer 4 is expressed in the golden 1-ratio code as the mixed number

$$4 = 101, 01 .$$

On the other hand, the irrational

$$\alpha_1^3 = \left(\frac{1 + \sqrt{5}}{2}\right)^3$$

is expressed as the integer

$$\alpha_1^3 = 1000 .$$

Specifically, the irrational α_p, i.e. the ratio (44) for any $p > 0$ is represented as the radix of the traditional notation:

$$\alpha_p = 10 .$$

The main peculiarity of the golden p-ratio codes in comparison with the Fibonacci p-codes is that the left- or right-shift of the code has a strict mathematical sense of multiplication or division of the represented number by the radix α_p. This ensures representation of numbers with the floating point and simplifies considerably multiplication in the golden p-ratio code.

For example, the integer 4 is represented in the golden 1-ratio code as

$$4 = 101.01 = (10.101)\cdot\alpha_1^1 = (1.0101)\cdot\alpha_1^2 = (0.10101)\cdot\alpha_1^3 .$$

Any real number is represented in the golden p-ratio code in the form

$$A = \alpha_p^n + \zeta ,$$

where

$$0 \leqslant \zeta < \alpha_p^{n-p} .$$

Using this formula we represent the real number A in the normal or minimal form where each unit bit is followed by at least p zero bits.

17. CONCLUSIONS

Two problems set in ancient times played an important role in the development of science. Those were the measurement and the number harmony of the world. Later these problems combined by the central Phythagorean philosophical thesis, i.e. "Numbers define the essence of all things" developed separately. The first problem connected with the discovery of the incommensurable line segments played an important role in the development of mathematics; the second problem connected with the golden section influenced art and aesthetics.

The discovery of the golden section and the Fibonacci numbers in the new (algorithmic) measurement theory is indicative of their generality.

The golden section and the Fibonacci numbers are expressed in the algorithmic measurement theory in the generalized form, i.e. the Fibonacci p-numbers and the golden p-sections.

The fundamental character of the Fibonacci p-numbers and the golden p-sections are expressed by their simple mathematical connection with the binomial coefficients and Pascal's triangle, the so-called universal generator. The idea of this infinite golden p-sections gave a new impulse to the forming of modern ideas of the systems' harmony [19]. The same relationship (the Fibonacci p-numbers and the golden p-sections) in the optimal measurement algorithms and optimal (harmonic) structures of self-organized systems reflects a deep inner connection between the algorithmic measurement theory and the theory of self-organized systems.

The Italian mathematician Leonardo of Pisa (Fibonacci) became famous for two mathematical discoveries, viz. the problem of choosing the best system of scale weights when he invented the binary notation and the problem of the rabbits multiplying which gave the Fibonacci numbers.

The asymmetry principle of measurement combined both Fibonacci problems and showed the inner connection between the "rabbits multiplying" and the weighing on the scales with the sluggishness p. This similarity can generate a good idea for the mathematical theory of biological populations.

The constructive idea of the infinity (the abstraction of the potential feasibility) underlies the new measurement theory. The new measurement algorithms and methods of number representation (first of all notations with irrational radices) are of great interest to the constructive theory of numbers.

The binary notation which underlies the computer engineering nowadays doesn't meet the requirements of high performance and reliability. This notation is characterized by a lack of redundancy which hampers to create the self-checking integrated circuits for the computer. So, the Fibonacci and the golden p-ratio codes are of great practical interest since they aid to build new information, arithmetic and circuit engineering fundamentals and develop the Fibonacci computers with higher reliability. Moreover, the unexpected "intrusion" of the Fibonacci numbers and the golden section into the measurement theory and via it into the encoding theory and computer arithmetic should be regarded as a reflection of the "fibonaccisation" of science nowadays.

REFERENCES

1. I. V. Viten'ko and A. P. Stakhov, Teoriya optimalnikh algoritmov analogo-tsifrowogo preobrasowaniya. *Pribory Sist. Awtom.*, wyp.11, izd-vo Kharkovs-kogo universiteta, Kharkov (1970).
2. A. P. Stakhov, *Vvedeniye v Algoritmicheskuyu Teoriyu Izmereniya*. Sov. Radio, Moskva (1977).
3. A. P. Stakhov, *Algoritmicheskaya Teoriya Izmereniya*. Znaniye, Moskva (1979).
4. A. P. Stakhov, *Kody Zolotoy Proportsiyi*. Radio i Svyaz', Moskva (1984).
5. A. F. Losev, Isotoriya filosofiyi kak shkola mysli. *Kommunist* 11(56), 55–56 (1981).
6. N. N. Vorobyov, *Chisla Fibonacci*. Nauka, Moskva (1978).
7. A. A. Himelfarb, L. R. Ginsburg and P. A. Poluentov i dr, *Dinamicheskaya Teoriya Biologicheskikh Populyatsiy*. Nauka, Moskva (1974).
8. Y. A. Urmantsev, Zolotoye secheniye. *Priroda* **17**, 33–40 (1968).
9. *Istoriya Matematiki*, Vol. 3, Chap. Tomakh. Nauka, Moskva (1970).
10. Y. S. Dubnov, *Izmereniye Otrezkov*. Fizmatgiz, Moskva (1962).
11. N. A. Shanin, *O Rekursivnom Matematicheskom Analize i Ischisleniyi Arifmeticheskikh Ravenstv R. L. Gudsteina. Vstupitelnaya Statya k Knige R. L. Gudsteina, "Rekursivny Matematichesky Analiz"*. Nauka, Moskva [R. L. Goodstein, *"Recursive Mathematical Analysis"*] (1970).
12. A. A. Markov, *O Logike Konstruktivnoy Matematiki*. Znaniye, Moskva (1972).
13. N. Y. Vilenkin, *Kombinatorika*. Nauka, Moskva (1959).
14. I. Y. Depman, *Istoriya Arifmetiki*. Uchpedgiz, Moskva (1959).
15. V. F. Gartz, *Luchshaya Sistema dlya Vesovykh Gir'*. Sankt-Peterburg (1910).
16. O. Neugebauer, *Lektsiyi po Istorii Antichnyk Matematicheskikh Nauk, T. I. Dogrecheskaya Matematika*. ONTI NKTP SSSR, Moskva-Leningrad (1937). [O. Neugebauer, Vorlesungen über Geschichte der antiken mathematischen Wissenschaften, Erster Band. Vorgriechische Matematik. Springer, Berlin (1934)].
17. G. Wilde, *Metody Poiska Ekstremuma*. Nauka, Moskva (1967). [D. J. Wilde, *Optimum Seeking Methods*. Prentice-Hall, Englewood Cliffs, N. J. (1964)].
18. E. Kol'man, *Istoriya Matematiki v Drevnosti*. Fizmatgiz, Moskva (1961).
19. E. M. Soroko, *Strukturnaya Harmoniya Sistem*. Nauka i Teknika, Minsk (1984).

Computers Math. Applic. Vol. 17, No. 4–6, pp. 639–652, 1989
Printed in Great Britain. All rights reserved

0097-4943/89 $3.00 + 0.00
Copyright © 1989 Pergamon Press plc

BUCKLING PATTERNS OF SHELLS AND SPHERICAL HONEYCOMB STRUCTURES

T. Tarnai

Hungarian Institute for Building Science, Dávid F.u.6, Budapest-1113, Hungary

Abstract—Analogy between the post-buckling equilibrium form of complete spherical shells with the mandrel inside and the form of living spherical honeycomb structures is investigated. The primary aim of this paper is to describe the typical topological–geometrical properties of the multi-dimple buckling pattern of a complete spherical shell on the basis of this analogy. It was found that, although the sphere itself is the most symmetrical form, the buckling pattern on it (consisting of pentagons, hexagons and heptagons) is asymmetric.

1. INTRODUCTION

In stability research on thin elastic shells, circular cylindrical shells [Fig. 1(a)] are the most intensively studied structures and almost every important detail of their buckling has been cleared up [1, 2]. So, the buckling shape of a cylindrical shell under different external loads is also known [3].

A thin-walled circular cylindrical shell subjected to uniform axial compression, in the advanced phases of the buckling process, buckles in a diamond pattern. If the cylinder is short the end supports admit local buckles only in a narrow circumferential strip in the middle of the cylinder [Fig. 1(b)], but if the cylinder is long a buckling pattern close to the Yoshimura pattern can develop. The Yoshimura pattern [Fig. 1(c)] in fact is a polyhedron composed of equal n-gonal antiprisms consisting of equal isosceles triangles. The Yoshimura pattern is an inextensional buckling pattern, i.e. it is obtained from the cylindrical surface by isometry. This isometry, however, is not a continuous transformation, so the cylindrical surface can reach the Yoshimura pattern only through extensional deformation with a snap-through phenomenon. The Yoshimura pattern is valid for shells with zero wall thickness [4]. In reality, shells have finite wall thickness and consequently finite bending stiffness also, and so the buckling pattern has no sharp edges but a curvature with a small but finite radius along the edges.

The main feature of this kind of buckling pattern of circular cylindrical shells is a high order of symmetry that has been confirmed by several experiments [3].

Another common shell form is the sphere. Buckling of thin-walled complete spherical shells subjected to uniform external pressure has also been investigated [4; cf. 1, 2] but scanty information is available on the buckling pattern especially on that at an advanced stage of the buckling process. Data available on the buckling pattern of complete spherical shells are quite meagre since researchers have mostly dealt with local phenomena and with the buckling shape in the early phases of the buckling process with the formation of a single dimple. It is not known whether there exists an inextensional buckling pattern of complete spherical shells similar to the Yoshimura pattern of cylindrical shells, which would be useful in theoretical analyses. It is not known either whether the buckling pattern of the sphere, where the buckled state occurs with the formation of a large number of dimples, has certain kinds of symmetry.

As a conjecture, an interesting theoretical argumentation on the development of the buckling pattern, which is tacitly supposed to be a rounded "polyhedron" with icosahedral symmetry, can be found in Ref. [2]. To our knowledge, the only series of experiments, which was executed also in the advanced phases of the buckling process and could produce a large number of dimples forming a honeycomb on the complete spherical surface, was performed at Stanford University by R. L. Carlson *et al.* [5].

The aim of this paper is to try to discover the main tendencies in the formation of the buckling pattern of complete spherical shells by means of the Stanford experiments [5] and of analogies with spherical honeycomb structures in nature.

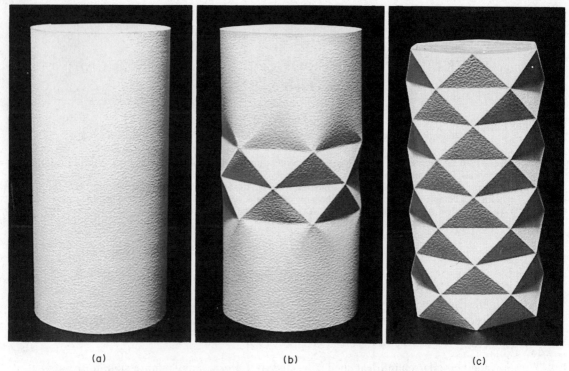

Fig. 1. Paperboard models for representation of buckling of a circular cylindrical shell: (a) the cylinder prior to buckling; (b) local buckling; (c) the Yoshimura pattern for $n = 6$.

2. CIRCLES ON A SPHERE

For spherical shells there exists an inextensional local buckling form which is a circular dimple on the spherical surface obtained by reflection of a spherical cap in the plane of the base circle of the cap into the sphere, as shown in a meridional section in Fig. 2. The spherical cap, however, can reach its inverted position only through extensional deformation with a snap-through phenomenon. If several local buckles of this kind develop, a packing of circles on the sphere, similar to that in Fig. 16(a), is obtained. If the diameter of the circles is increased the packing system becomes a covering system. However, deformation of the dimples fitted to the covering system—or more correctly, to the Dirichlet cells of the circles—cannot be inextensional any longer. (The Dirichlet cell is the central projection of a face of the convex polyhedron determined by the planes of the circles onto the surface of the sphere.)

It is of interest to mention that an inextensional buckling pattern fitted to a covering system of circles can be obtained for the discrete version of the buckling problem, i.e. where a triangulation is made on the sphere and the spherical triangles are replaced by plane triangles, and so the spherical caps are replaced by pyramids. If any three of the circles have at most one common point then the vertices of the Dirichlet cells of the circles, which are the vertices of the bases of the pyramids, will be coplanar. Thus, every pyramid can be inverted by reflection in the plane of its base (Fig. 3). This is known as "dimpling" [6] which considerably increases the stiffness of a spherical grid [7].

It is well-known that the densest packing of equal circles in the plane is arranged so that the centres of the circles are at the vertices of a regular triangular tessellation. The same arrangement is obtained for the thinnest covering of the plane with equal circles [8]. (On the spherical surface, however, there are only finitely many numbers of circles for which the solution of the problem of the densest packing and that of the thinnest covering result in the same arrangement of the centres of the circles [9].) In the plane, the Dirichlet cells of the circles in both cases are regular hexagons. One can think that an economical arrangement of circles would be obtained by producing a hexagonal tessellation on the sphere also. However, it is known as a consequence of the Euler theorem that it is impossible to enclose a simply connected domain of the three-dimensional space by a polyhedron bounded merely by hexagons [10]. (In spite of this fact, a Hungarian graphic artist

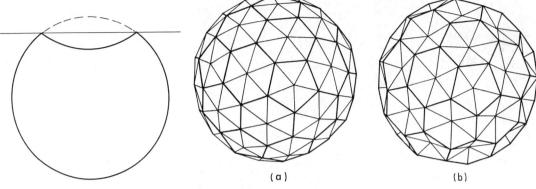

Fig. 2. Inextensional single-dimple buckling of a spherical shell. A meridian
section.

Fig. 3. A triangle polyhedron inscribed into a sphere: (a)
prior to buckling; (b) in the state of post-buckling with
an inextensional buckling pattern.

(a)   (b)

once tried it. The result can be seen in Fig. 4. Maybe the artist would have been in a trouble if he had have to draw the honeycomb also on the other side of the sphere.) A trivalent polyhedron (i.e. a polyhedron in which three edges meet at each vertex) mainly composed of hexagons always need polygons bounded by less than six sides.

The Euler theorem also involves that in a trivalent polyhedron bounded by pentagons and hexagons the number of pentagons is always 12. But Grünbaum and Motzkin [11] have proved that for every non-negative integer m satisfying $m \neq 1$, there exist trivalent convex polyhedra having f_k k-gonal faces such that $f_5 = 12$, $f_6 = m$, $f_k = 0$ for $k \neq 5, 6$. Goldberg [12] has shown that if $m = 10(T - 1)$, where $T = b^2 + bc + c^2$, b and c are non-negative integers, not both zero, then there exists at least one trivalent polyhedron, having 12 pentagonal faces and m hexagonal faces, with icosahedral rotational symmetry. The geometrical meaning of b and c can be seen in Fig. 5 where the triangle composed of dashed lines is a face of the icosahedron. Following Coxeter [13] we denote such a tessellation by the symbol $\{5+, 3\}_{b,c}$ in which $5+$ and 3 mean that the tessellation consists of polygons having 5 and more than 5 (i.e. 6) sides and 3 edges meet at each vertex. This tessellation is known as the topological dual of the triangular tessellation denoted by $\{3, 5+\}_{b,c}$, which has great importance in virus research [14, 15]. It is easy to see, if $bc(b - c) = 0$ then the tessellations can also have a plane of symmetry in addition to the icosahedral rotational symmetry. For instance, the triangle polyhedron in Fig. 3(a) is constructed in system $\{3, 5+\}_{2,2}$ but the pentagonal–hexagonal pattern in Fig. 3(b), obtained from it by buckling, represents a tessellation $\{5+, 3\}_{2,0}$.

Both dense spherical circle packing and thin spherical circle covering can be constructed in icosahedral symmetry. In the packing problem the tessellation $\{3, 5+\}_{b,c}$ is useful [16] but in the covering problem usage of the tessellation $\{5+, 3\}_{b,c}$ has the advantage [17], as a consequence of the fact that the packing and covering problems are dual counterparts of each other.

To illustrate covering the sphere with circles some models with icosahedral symmetry in rotation are presented in Figs 6–8. The arrangements in Fig. 6 have got a plane of symmetry but those in Figs 7 and 8 have not. The honeycomb pattern on the geodesic dome shown in Fig. 9 is also a result of a spherical circle covering [18].

We mention here that, in molecular biology, a clathrin cage of small coated vesicles, in general, forms a trivalent polyhedron with equal edges, bounded by pentagons and hexagons [19]. The geometry of these clathrin cages is in a certain relation with spherical circle covering. For instance, the "soccerball" polyhedron in Fig. 6(b) has been identified as a clathrin lattice. (In many cases, however, the pentagons and hexagons in the clathrin cage are not planar, similar to the cells of the equal-edged honeycomb in Fig. 10.) On the other hand, since the clathrin cages have an economical form, considering the pentagons and hexagons of clathrin cages as approximations of Dirichlet cells we could cover the sphere with 16 and 20 equal circles in arrangements better than the previous conjectures [20].

In spherical circle packings it frequently occurs, and it can be seen in Fig. 16(a) also, that circles cyclically joined to each other are centred at vertices of quadrangles and pentagons. In such cases, a relatively large space, not occupied by the circles, is formed inside the quadrangles and pentagons,

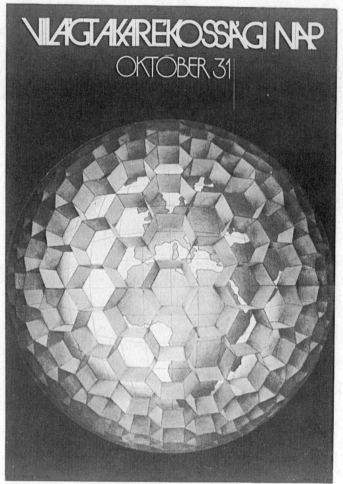

Fig. 4. World-day of saving. (A poster in Hungary.)

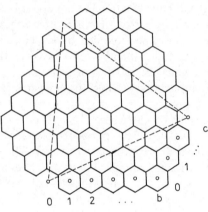

Fig. 5. Meaning of the Coxeter symbol $\{5+,3\}_{b,c}$.

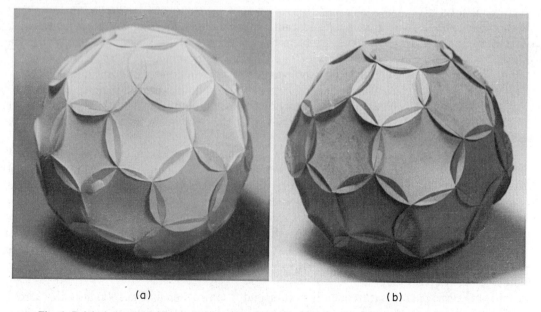

(a)   (b)

Fig. 6. Polyhedron model of covering the sphere with 32 circles in system $\{5+,3\}_{1,1}$: (a) covering with equal circles; (b) covering with two kinds of circles. (Courtesy of Dr M. J. Wenninger.)

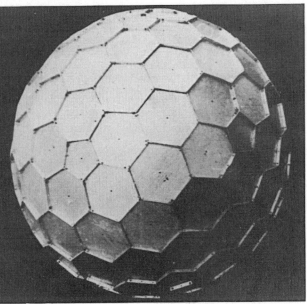

Fig. 7. Polyhedron model of covering the sphere with 72 circles in system $\{5+,3\}_{2,1}$, using two kinds of circles.

Fig. 8. Polyhedron model of covering the sphere with 132 circles in system $\{5+,3\}_{3,1}$ using three kinds of circles. An aluminium model designed by G. N. Pavlov for a statue set up in Gagarin, U.S.S.R. (Courtesy of Dr G. N. Pavlov.)

especially if these polygons are regular. Moreover, in the latter case the Dirichlet cells of the circles form tetravalent and pentavalent vertices which are not characteristic of economical cell arrangements. These facts enable us to think that the real buckling pattern of a spherical shell is closer to the system of Dirichlet cells of circle covering than to that of circle packing. (However, it should be noted that in nature spherical circle packings [21] do not always stand out in sharp contrast to spherical circle coverings because of unknown constraints.)

Fig. 9. Geodesic dome near Kirov, U.S.S.R., erected in 1976. (Courtesy of Dr G. N. Pavlov.)

Fig. 10. Model of a spherical "hexagonal" net with equal edges in system $\{5+, 3\}_{6,6}$.

3. BUCKLING OF COMPLETE SPHERICAL SHELLS

Based on the preceding geometrical argumentation and on the mechanical idea of Gioncu and Ivan [2] one can imagine the full buckling pattern of the complete spherical shell as a honeycomb composed of pentagons and hexagons corresponding to a covering system of not necessarily equal circles on the sphere in icosahedral symmetry. Reality, however, shows a different picture.

During the buckling process of complete thin-walled spherical shells subjected to uniform external pressure, in general, only one dimple appears which gradually deepens and so leads to collapse of the sphere. This kind of buckling behaviour, usually, is due to initial imperfections. If increase in the dimension of the dimple is restricted then it enables additional dimples to develop under an increasing external pressure.

This principle was applied in the Stanford experiments [5], using mandrel-restricted specimens. Each shell specimen was made by electroplating on a hollow wax mandrel. After the plated mandrel was removed from the heated plating bath, the mandrel contracted more upon cooling than the metal shell, so a gap was produced between the mandrel and the shell. The specimen was gradually evacuated and under external air pressure the shell buckled. The buckling process was described in Ref. [5] as follows:

> "During air tests conducted with the mandrel inside, it was generally found that buckling at a low pressure (less than 50 percent of theoretical) appeared as a single dimple whose size depended on a gap dimension. The dimple was approximately circular . . . If the test was continued, additional dimples were observed to 'pop in', usually singly, and at higher pressures . . . At the higher buckling pressures, a different type of response was observed (initial buckling at pressures greater than 50 percent of the theoretical value). In these instances, the shell was more highly stressed, and the transition to the buckled state occurred with the formation of a large number of dimples. The transition was very rapid and it was not possible to detect a dimpling sequence by visual observation, i.e., the dimples appeared to form simultaneously."

The resulting buckling pattern of one of the specimens is shown in Fig. 11(a). The shell did not collapse as it was supported inside by the wax mandrel. Simultaneous formation of a large number of dimples shows that the specimen was close to a "balanced" shell structure, where there are no large differences in the local resistances to buckling.

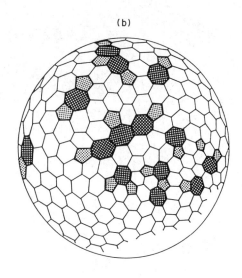

Fig. 11. Buckling pattern of a complete spherical shell. (a) Air-system test with wax mandrel inside the specimen (courtesy of Professor N. J. Hoff). (b) Pentagons and heptagons in the buckling pattern.

Apart from a region along the contour of the sphere we have identified the buckling pattern of the specimen in Fig. 11(a) and drawn it in Fig. 11(b). Examination of Fig. 11(b) reveals that the buckling pattern contains not only pentagons and hexagons, as expected, but heptagons also, and the buckling pattern has no symmetry at all. It is of interest to observe that many of the dimples in Fig. 11(a) contain circular swelling in the middle as small "anti-dimples". (It may be noted that an "anti-dimple" as an inextensional form can also be produced on a dimple in Fig. 2 by reflection in a plane intersecting the inverted spherical cap in a complete circle; and also on a dimple of the polyhedron in Fig. 3(b) by reflection of the apex of the inverted pyramid in a plane intersecting all the lateral edges of the inverted pyramid.)

It is not clear from this experiment whether the honeycomb shape of the buckling pattern with its irregularities is accidental or it represents certain rules generally valid for spherical honeycombs in nature. Since this example of buckling pattern is the only one available to us, to detect the tendencies we have considered analogies with other spherical honeycomb structures in nature.

4. ANALOGIES IN NATURE

Figure 12(a) shows the quasi-hexagonal pattern of the skeleton of the radiolarian *Aulonia hexagona* depicted by Haeckel [22] and analysed from a morphological point of view by D'Arcy Thompson [10]. It should be noted here that the accuracy of Haeckel's drawings was questioned after D'Arcy Thompson's opinion was published [23]. Namely D'Arcy Thompson, in his letter of 9 March 1947 to H.S.M. Coxeter, wrote:

"As to Haeckel, I wouldn't trust him round the corner, and I have the gravest doubt whether his pentagonal dodecahedron and various others ever existed outside his fertile fancy. I believe I may safely

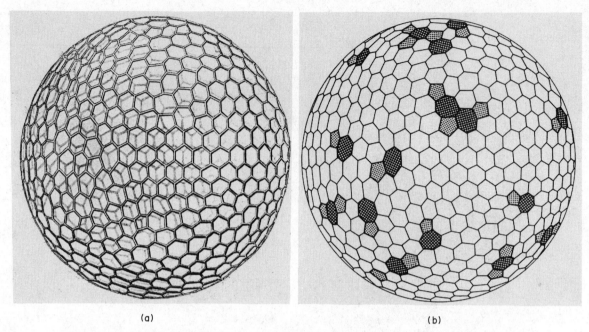

(a)   (b)

Fig. 12. (a) Aulonia hexagona Hkl (from Ref. [22]) and (b) pentagons and heptagons in its network.

say that no type-specimens of these exist in the British Museum, or anywhere else. He was an artist, a pattern-designer, a skilled draughtsman. He had a minute professorial salary in a small University. The *Challenger* paid eight guineas apiece for as many plates as he chose to draw, and he kept on drawing them, and lived on the proceeds (so they used to say) till the end of his life. He represents a thoroughly bad period in Natural Science."

However, not every natural scientist agrees with D'Arcy Thompson. For instance, S.A. Kling, an outstanding expert in radiolaria [24] in his letter of 20 October 1980 to the author, wrote:

"As to Haeckel's illustrations and D'Arcy Thompson's comments (of which I was unaware), I believe that the large majority of the pictures are accurate. It was my understanding that most or all of the drawings published by Haeckel in the Challenger Reports were actually made by his illustrator, A. Giltsch. Although some of the drawings may be somewhat idealized or even the products of imagination, the plates as a whole do not, I believe, exaggerate the diversity of forms in the oceans (taking into consideration the fact that some of his species were actually extinct forms derived from older seafloor sediments). Occasionally I have had the experience of discovering a specimen of one of Haeckel's more bizarre forms that I thought didn't exist. So I think that D'Arcy Thompson's view is quite extreme and puzzling. Haeckel was highly regarded as a biologist both in his own time and today. His system turned out to be quite unnatural, but his large volume of descriptive work, produced in a relatively short time, remains the foundation of radiolarian taxonomy."

Figure 13(a) confirms Kling's views. This figure shows a scanning electron micrograph of a spherical radiolarian similar to Haeckel's *Cenosphaera vesparia* [22]. Considering the radiolaria in Figs 12 and 13 and the tendencies of distribution of pentagons and heptagons in their spherical honeycomb pattern, Fig. 13(b) gives nearly the same picture as Fig. 12(b). This convinces us that Haeckel's drawing presented here in Fig. 12(a) is correct.

Comparing the line drawing of the buckling pattern of the spherical metal shell [Fig. 11(b)] and the line drawings of the structural networks of the radiolarian shells [Figs 12(b) and 13(b)] it can be ascertained that:

(1) The honeycombs are asymmetric.
(2) The honeycombs are trivalent "polyhedra" bounded by pentagons, hexagons and heptagons.
(3) It is typical that a heptagon has a pentagon neighbour joined to it by a whole side. [Only one exception is seen in Fig. 13(b) where there is an individual heptagon, i.e. completely surrounded by hexagons.]
(4) It is typical that a pentagon has a heptagon neighbour joined to it by a whole side. [Only one exception is seen in Fig. 11(b) where there is an individual pentagon, i.e. completely surrounded by hexagons.]

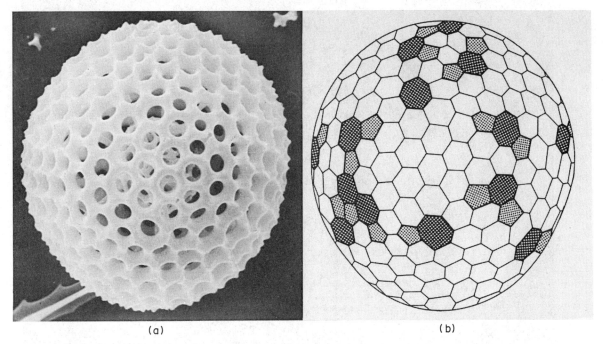

Fig. 13. (a) Scanning electron micrograph of a spherical radiolarian (courtesy of Dr S. A. Kling) and (b)
pentagons and heptagons in its network.

One can find three kinds of typical morphological modules in these honeycombs: hexagons,
pentagon–heptagon juxtapositions and pentagon–heptagon–pentagon juxtapositions. Occurrence
of individual pentagons and heptagons is not frequent, so they may be considered as exceptions.

There is a surprising similarity between Fig. 11(a) and Fig. 13(a), namely the circular openings
in the radiolarian shell seem to have the same morphological properties as the circular "anti-
dimples" on the buckled metal shell.

Several other examples of spherical honeycombs and quasi-hexagonal patterns can be found
in nature, e.g. certain pollen grains [25], coated pits and coated vesicles [19, 26]. Though not
spherical, the Benard convection is also worth mentioning: when a shallow fluid layer is heated
uniformly from below and cooled from above hexagonal convection cells appear in the fluid
layer at a certain temperature gradient [27]. If surface tension dominates only a weak hexagonal
pattern develops [28]. Such a formation is shown in Fig. 14(a). Because of the boundary conditions,
surface tension and imperfections there are relatively many individual pentagons and heptagons
in the pattern [Fig. 14(b)]. Moreover, a hexavalent vertex also takes place. Haken [29] believes that
there is a close relation between the "hexagonal" buckling pattern of thin shells and the
"hexagonal" pattern of the Benard convection, because both develop due to instability: the shell
buckling is due to elastic instability, the Benard thermal convection is due to hydrodynamic
instability.

Returning to the typical morphological modules we may say that the *pentagon–heptagon
juxtapositions* are well-known as dislocations in crystals of hexagonal close-packed structure.
Such a dislocation is shown in Fig. 15 which is obtained from a soap bubble layer forming
a hexagonal arrangement of equal circles in the plane by drawing the Dirichlet cells
(Wigner–Seitz cells) of the circles. A pentagon–heptagon pair introduces an additional row
of hexagons into the hexagonal pattern, or conversely, it contracts three rows of hexagons
into two rows. (The rows are inclined at 30° to the vertical direction.) Pentagon–heptagon
juxtapositions were discovered in the equal-edged hexagonal basketwork of large coated pits
and coated vesicles by Heuser and Evans [26], who noted that identical dislocations occur in
beehives and diatoms, too. Looking at Fig. 15 it is apparent that, as a consequence of close packing
of equal circles in the plane, two of the sides of the heptagon are very short compared to the others.
This property of the heptagons more or less remains valid also in packing of equal circles on the
sphere.

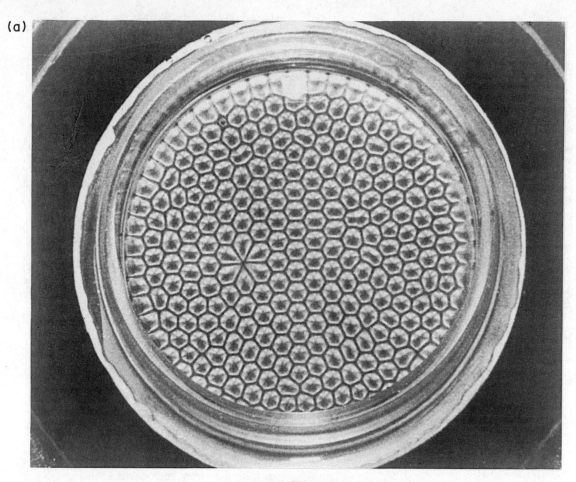

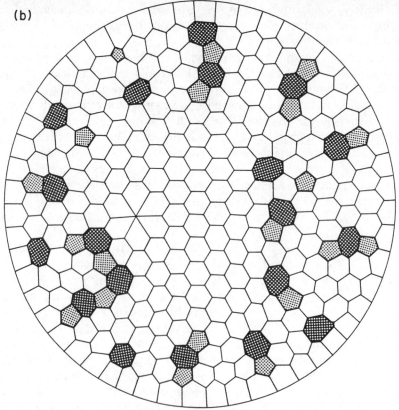

Fig. 14. (a) Benard convection cells (courtesy of Professor E. L. Koschmieder) and (b) pentagons and heptagons in their pattern.

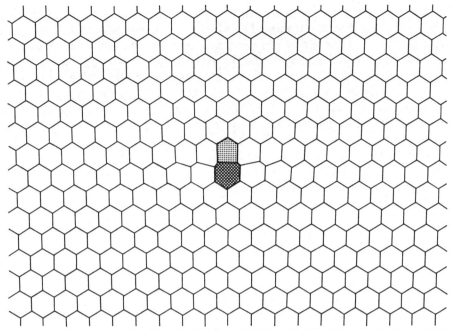

Fig. 15. Dislocation in a two-dimensional bubble layer. Dirichlet cells of bubbles. (On the basis of a figure from Ref. [30].)

Figure 16(a) shows a spherical sculpture composed of 485 equal truncated conical shells made of stainless steel. It is, in fact, a random packing of equal circles on the sphere. Drawing the Dirichlet cells of the circles [Fig. 16(b)] we obtain also pentagons and heptagons. It is seen that the shape of the pentagon–heptagon pairs, in may cases, is quite close to that in Fig. 15. Our feeling

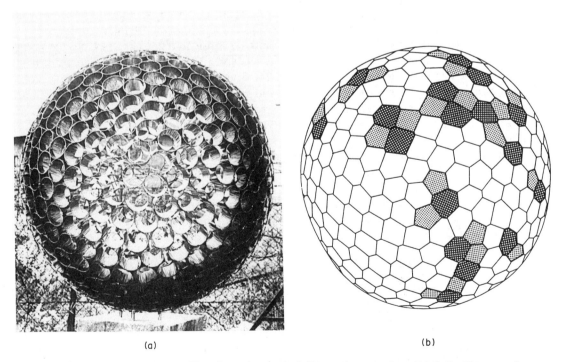

(a)   (b)

Fig. 16. (a) A sculpture of Bálint Józsa, composed of 485 equal truncated conical shells. (Courtesy of the artist.) (b) Pentagons and heptagons in the system of Dirichlet cells of it.

is that this spherical honeycomb is full of dislocations. The polygons (pentagons and hexagons included) in many cases look like truncated quadrangles (squares). Our feeling is supported also by the fact that a circle on the sphere can never be touched by six others (at most by five) if the circles are equal, consequently regular hexagons circumscribed about the respective circles never occur among the Dirichlet cells. This is one of the reasons why we believe that the buckling pattern of the spherical shell is obtained by circle covering rather than by circle packing. Moreover, as mentioned previously, in special cases pentavalent vertices can be formed by the Dirichlet cells. [Near to the contour of the sphere in Fig. 16(a) towards the northwest there are five circles in a nearly regular arrangement. The two hexagons, two pentagons and one heptagon in Fig. 16(b) are very near to becoming five elongated pentagons with a common pentavalent vertex.] The morphological properties discovered in parts (b) of Figs 11–13 are also valid here and *pentagon–heptagon–pentagon juxtapositions* can be seen, too. Two pentagons can be connected to a heptagon side by side, topologically, in three different ways. Considering Fig. 16(b) and also Figs 11(b), 12(b) and 13(b), however, we found only two of them. These are depicted in Fig. 17. We did not find a pentagon–heptagon–pentagon juxtaposition in which the two pentagons have a common side. In the case of two pentagons, both neighbours of a heptagon, joined side by side, we could always decompose the picture into pentagon–heptagon juxtapositions and such a pentagon–heptagon–pentagon juxtaposition where the two pentagons had no common side.

If there are pentagons, hexagons and heptagons in a trivalent polyhedron then, using the notation in Section 2, the Euler theorem involves that $f_5 - f_7 = 12$. In the case where there are no individual pentagons and heptagons [this is the case in Fig. 16(b)], since the number of pentagon–heptagon pairs does not affect this equality, it follows that the number of pentagon–heptagon–pentagon triplets is equal to 12. (However, as seen previously, individual pentagons and heptagons can occur—this number should be considered only as an expected value.) Thus, it seems that the pentagon–heptagon–pentagon juxtapositions in these spherical honeycombs play a role similar to that of pentagons in spherical honeycombs consisting of pentagons and hexagons.

5. CONCLUSIONS

The morphological resemblance between the buckling pattern of the spherical metal shell in Fig. 11(a) and the shape of the radiolarian shell in Fig. 13(a) (and other spherical honeycombs in nature) is startling. However, resemblance should be considered nothing more than simple analogy, and conclusions derived from this analogy should only be considered conjectures from the point of view of prediction of the buckling pattern of complete spherical shells with the mandrel inside.

Care is necessary in this field since a buckled shell and, for example, radiolaria are very distant from each other in scale, microstructure, function and circumstances by way of the fact that one of them is an artificial object and the other is a living structure. Moreover, formation of the skeleton of radiolarians is very poorly understood [24] (D'Arcy Thompson tried to explain it by a soap-bubble-aggregate model [10]). In spite of the fact that the formation of buckling of a complete spherical shell is much simpler than development of the skeleton of a radiolarian, because of the highly non-linear character of the buckling problem we do not know the exact buckling pattern either.

In nature (on a mechanical level) the principle of the minimum potential energy is valid and things intend to have a shape in which the potential energy is a minimum. In reality, however, there are always several random disturbances and imperfections which are not known and in many cases

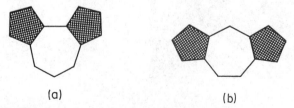

(a)   (b)

Fig. 17. Pentagon–heptagon–pentagon juxtapositions: (a) pentagons on the first and third sides of the heptagon; (b) pentagons on the first and fourth sides of the heptagon.

constraints or accessory conditions are not known either under which the minimum of the potential energy is looked for. This is why theory is able only to approximate reality. A good example of this is Kelvin's space-filling minimal tetrakaidecahedron. Although Kelvin's curved-faced symmetrical polyhedron mathematically is conjectured to be the minimal energy equivolume space-filling cell shape, it very rarely occurs in nature and only in a very distorted form [31]. In the case of a complete spherical shell, however, even the mathematical solution (or conjecture) of the buckling problem of the perfect shell with several dimples has not been found.

Besides analogies we used certain geometrical argumentations, though we are aware of the fact that a difficult mechanical problem (and what is more, a biological problem) cannot be reduced to a geometrical problem and buckling of spherical shells cannot be explained by the problems of the densest spherical circle packing and of the thinnest spherical circle covering, but we think that knowing these geometrical problems we can get closer to the understanding of the several-dimple buckling of complete spherical shells.

Although the shape of the buckled sphere in Fig. 11(a) has been influenced by random imperfections we believe that the shape itself is not accidental but some of its details show certain tendencies and in this sense the buckling pattern is typical. The following main features of the buckling pattern were ascertained:

(a) Although the sphere, as a whole, is the most symmetrical shape, the buckling pattern on it has no symmetry at all.

(b) The buckling pattern forms a trivalent polyhedron bounded by pentagons, hexagons and heptagons.

(c) In the buckling pattern a pentagon, in general, has a heptagon neighbour, and vice versa; individual pentagons and heptagons completely surrounded by hexagons rarely occur.

Acknowledgements—The research reported here was supported by OTKA Grant No. 744 awarded by the Hungarian Scientific Research Foundation.

Note added in proof

After the manuscript went to press, it came to our knowledge that Hutchinson [32] was the first to predict theoretically a hexagonal buckling pattern on a part of a spherical shell, but when he wrote his paper [32] he was not aware that the pattern predicted was a pattern of hexagons. Shortly afterwards, however, he did notice this and communicated the observation to Koiter who then referenced the fact in his fundamental paper [33].

REFERENCES

1. L. Kollár and E. Dulácska, *Buckling of Shells for Engineers*. Akadémiai Kiadó, Budapest/Wiley, Chichester (1984).
2. V. Gioncu and M. Ivan, *Buckling of Shell Structures* (in Romanian). Editura Academiei, Bucharest (1978).
3. N. Yamaki, *Elastic Stability of Circular Cylindrical Shells*. North-Holland, Amsterdam (1984).
4. N. J. Hoff, Selected papers in *Monocoque, Sandwich, and Composite Aerospace Structures*. Technomic, Lancaster, Basel (1986).
5. R. L. Carlson, R. L. Sendelbeck and N. J. Hoff, Experimental studies of the buckling of complete spherical shells. *Exp. Mech.* **7**(7), 281–288 (1967).
6. R. B. Fuller, *Synergetics. Explorations in the Geometry of Thinking*. Macmillan, New York (1975).
7. C. J. Kitrick, Nonlinear analysis of normal and inverted geodesic domes under the action of concentrated loads. M.S. Thesis, Univ. of Cincinnati, Ohio (1983).
8. L. Fejes Tóth, *Lagerungen in der Ebene auf der Kugel und im Raum*, 2nd edn. Springer, Berlin (1972).
9. G. Fejes Tóth and L. Fejes Tóth, Dictators on a planet. *Studia Sci. math. Hung.* **15**, 313–316 (1980).
10. W. D'Arcy Thompson, *On Growth and Form*, 2nd edn. Reprinted, Cambridge Univ. Press, Cambs. (1963).
11. B. Grünbaum and T. S. Motzkin, The number of hexagons and the simplicity of geodesics on certain polyhedra. *Can. J. Math.* **15**, 744–751 (1963).
12. M. Goldberg, A class of multi-symmetric polyhedra. *Tôhoku Math. J.* **43**, 104–108 (1937).
13. H. S. M. Coxeter, Virus macromolecules and geodesic domes. In *A Spectrum of Mathematics* (Ed. J. C. Butcher), pp. 98–107. Auckland Univ. Press, New Zealand/OUP, Oxford (1972).
14. D. L. D. Caspar and A. Klug, Physical principles in the construction of regular viruses. *Cold Spring Harb. Symp. quant. Biol.* **27**, 1–24 (1962).
15. A. S. Koch and T. Tarnai, The aesthetics of viruses. *Leonardo* **21**, 161–166 (1988).
16. T. Tarnai and Zs. Gáspár, Multi-symmetric close packings of equal spheres on the spherical surface. *Acta crystallogr.* **A43**, 612–616 (1987).
17. T. Tarnai and M. J. Wenninger, Spherical circle-coverings and geodesic domes. *Struct. Topol.* (in press).
18. G. N. Pavlov, Compositional form-shaping of crystal domes and shells. In *Spherical Grid Structures* (Ed. T. Tarnai), pp. 9–124. Hungarian Inst. Building Science, Budapest (1987).

19. B. M. F. Pearse and R. A. Crowther, Structure and assembly of coated vesicles. *A. Rev. Biophys. biophys. Chem.* **16,** 49–68 (1987).
20. T. Tarnai and Zs. Gáspár, Covering the sphere with equal circles. Lecture presented at *Colloquium on Intuitive Geometry*. Balatonszéplak, Hungary (1985).
21. T. Tarnai, Spherical circle-packing in nature, practice and theory. *Struct. Topol.* **9,** 39–58 (1984).
22. E. Haeckel, *Report on the Radiolaria*. Report on the scientific results of the voyage of *H.M.S. Challenger* during the years 1873–1876. *Zoology (Edinb.)* **18** (1887).
23. H. S. M. Coxeter, Review on symmetry, *In Symmetry* by H. Weyl. Princeton Univ. Press, N.J. (1952). *Am. math. Mon.* **60,** 136–139 (1953).
24. S. A. Kling, Radiolaria. In *Introduction to Marine Micropaleontology*, Chap. 9 (Ed. B. U. Haq and A. Boersma). Elsevier, New York (1978).
25. M. R. Bolick, Mechanics as an aid to interpreting pollen structure and function. *Rev. Palaeobot. Palynol.* **35,** 61–79 (1981).
26. J. Heuser and L. Evans, Three-dimensional visualization of coated vesicle formation in fibroblasts. *J. Cell Biol.* **84,** 560–583 (1980).
27. L. A. Segel, Non-linear hydrodynamic stability theory and its applications to thermal convection and curved flows. In *Non-equilibrium Thermodynamics Variational Techniques and Stability*; *Proc. Symp.* Univ. of Chicago, Ill., 1965. Univ. of Chicago Press, Ill. (1966).
28. E. L. Koschmieder, Surface tension driven Benard convection. Presented at *16th Int. Congr. of Theoretical and Applied Mechanics*, Lyngby, Denmark, Poster No. 348 (1984).
29. H. Haken, *Synergetics: an Introduction*, 2nd edn. Springer, Berlin (1978).
30. Ch. Kittel, *Introduction to Solid State Physics*, 3rd edn. Wiley, New York (1967).
31. P. Pearce, *Structure in Nature is a Strategy for Design*. MIT Press, Cambridge, Mass. (1978).
32. J. W. Hutchinson, Imperfection sensitivity of externally pressurized spherical shells. *J. Appl. Mech.* **34,** 49–55 (1967).
33. W. T. Koiter, The nonlinear buckling problem of a complete spherical shell under uniform external pressure, I–IV. *Proc. Kon. Nederl. Akad. Wet.* **B72,** 40–123 (1969).

Computers Math. Applic. Vol. 17, No. 4–6, pp. 653–669, 1989
Printed in Great Britain. All rights reserved

SYMMETRY IN FREE MARKETS

B. P. Fabricand†

Pratt Institute, Brooklyn, NY 11205, U.S.A.

Abstract—Free markets are shown to be highly symmetric under an interchange operation involving market participants. The symmetry operation leads to a conservation law of information: there exists no information that permits one person to have a greater expectation than any other. Symmetry violations arising from privileged and late-breaking information may occur and lead to better-than-average performance. The role of government in symmetric markets is briefly explored.

INTRODUCTION

The concept of symmetry has been crucial to the development of modern physics. For casting symmetry in such a pivotal role, we are indebted to Albert Einstein and his 1905 paper introducing special relativity [1].

Before then, the connection between symmetry and the laws of physics was recognized (if it was recognized at all) only peripherally. The main interest was in the conservation laws. If a system under observation remains unchanged under some symmetry operation—technically, if the system Hamiltonian stays constant—then some observable describing the system remains unchanged. That observable is then said to be invariant or conserved. Thus, a system unaffected by a spatial translation conserves momentum. A system unaffected by a temporal translation conserves energy. A system unaffected by a spatial rotation conserves angular momentum. A system unaffected by an interchange of neutrons for protons or protons for neutrons conserves isotopic spin. These are just some of the great conservation laws that prove so useful in predicting the future behavior of real-world systems.

Einstein deepened the idea of symmetry. He introduced the daringly speculative postulate that the basic laws of physics themselves be symmetric under what is known as a Lorentz transformation —they must remain invariant under a rotation in four-dimensional space-time. In other words, two observers moving with constant velocity relative to each other must experience the same laws of physics even though their temporal and spatial measurements on some commonly observed event will differ. Time and space are no longer absolute constructs but depend on the observer. Einstein's hypothesis proved so successful that Lorentz invariance is now the *sine qua non* of all modern theories.

Whether or not symmetry will occupy so pivotal a role in the social sciences is not known at present. Some limited applications with far-reaching implications can be and are made in this article to free markets. These furnish particularly suitable systems for study. Not only are they of significance to all societies, but they are readily handled with probability theory and statistics, at least in their more tractable forms. Risk and reward, which accompany all human activity (or the lack of it), and the expectation of each market participant are the observables of interest. Because their calculation can become enormously complicated, we shall begin our analysis with some simple games of chance, what we may term money markets. In such markets, the participants risk money to make more money. This approach offers several advantages: (1) the risks are easily calculated; (2) the rewards are well-known and uncomplicated by considerations of value received for goods and services; (3) all possible future outcomes can be foreseen and (4) they are realized in a short time. Some or all of these benefits must be given up as our study broadens to include the more complex public markets in which we as civilized human beings participate. The pari-mutuel market, the stock market and the market-organized society will undergo examination as to the nature and consequences of their symmetries. Finally, the problems faced by both individuals and governments in enhancing free market expectations are briefly explored.

---

†Present address: Box 1107, New Milford, CT 06776, U.S.A.

A GAME OF ROULETTE

As a first application of symmetry to free markets, we consider the casino game of roulette. The mechanical details should be well-known. A small white ball is dropped onto a slotted, rotating wheel. After rolling around the rim of the wheel several times, the ball loses momentum and drops into a numbered slot that may be colored black, red or green. In American roulette, there are 18 red slots, 18 black slots, and 2 green slots labeled "0" and "00". The roulette player seeking to augment his wealth must predict where the ball will fall. For him, the present leads to not one but many possible futures, each with its own probability of occurrence. In this sense, games of chance make good models of real-world happenings and are widely used as such in scientific theories. The enormous popularity of gambling may stem from this correspondence.

Suppose our gambler decides to bet on one of the 18 red numbers. If the wheel is unbiased, the ball is equally likely to land on any of the 38 numbers. Out of these 38 possible outcomes, there are 18 ways of winning and 20 ways of losing. We say that the player has 18 chances in 38 of winning, or a win probability of 18/38. In other words, he can expect to win on average 18 times for every 38 spins of the wheel. This then is his risk. The reward for winning, the player's profit, is set by the casino operator at $1 for every $1 bet. The reward for losing is, of course, loss of the stake.

In any given game, a gambler risking $1 on such a bet may win $1 or lose $1. However, he is more likely to lose than win. The question to be asked is this: over many games, how much will our gambler win or lose for each dollar bet? The answer is called his "expectation" and it may be found as follows: on average, a bet on a red number wins 18 times in every 38 games. In 20 games, the stake is lost. For $1 bets, he wins $18 and losses $20 over 38 games, on average. His net return is a loss of $2 for every $38 bet. The expectation is just the net return divided by the total amount bet, or $-$2/38. The minus sign indicates a loss or negative expectation; a plus sign would indicate a profit or positive expectation. In pennies, the expectation is $-5.3¢$. We must understand that this is an average, a result to be expected over a large number of games for an unbiased wheel. Whenever he plays roulette, a person can expect to lose 5.3¢ per dollar bet. But there may be winning streaks producing a profit and losing streaks producing greater-than-expected losses. Some players may even break the bank on very rare occasions. Appendix A exhibits a general formula from probability theory for calculating the expectation.

When the expectation for all players is the same, the returns per dollar bet are normally distributed and centered at the expectation, as in Fig. 1. That means that more people realize a return per dollar bet of $-5.3¢$ than any other return. As we move to the right or left along the curve, fewer people realize greater or lesser returns. Note the symmetry of the bell-shaped curve. As many roulette players appear in the right wing of the curve as the left wing.

THE ROULETTE MARKET

The roulette market, like all others, involves transactions between people. There must be both a buyer and a seller. The roulette player buys the chance to make more money. The casino operator sells him that chance. The casino's expectation is $+5.3¢$, the exact opposite of the players. That means that the casino hopes to generate revenues of 5.3¢ for every dollar bet by the players. From this money, the operator must cover the costs of running the marketplace (the casino), the interest on any debt outstanding, amortization of the debt, taxes and at the same time secure a reasonable

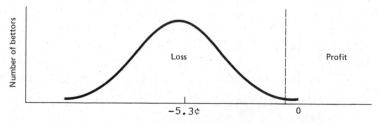

Fig. 1. Distribution of returns from red–black bets in roulette.

return on his investment (his cost of capital). We shall discuss the seller's overall expectation in the section on consumer markets.

From the buyer's perspective, the roulette market is an excellent example of a free market. Everybody is welcome to play regardless of race, color, religion, sex, national origin or political belief. Nobody is excluded. The rules of play are established and enforced by the casino, and they are the same for all. Every player can act as he wishes within the terms set forth in the rules of play. The casino neither influences the outcomes nor interferes with the players' betting methods. Neither does it make any attempt to confiscate, renege on or redistribute the players' winnings. Finally, all players have equal access to any and all information that may aid them in attaining their goals.

Except where governments limit competition, the casino operator also conducts his business in a free market. He must compete with all other casinos in attracting customers. To do this, he provides suitable ambience, offers quality entertainment and serves good food and drink at reasonable cost. Most importantly, he fixes the players' expectation at a level that permits reasonable profits for the casino and at the same time rewards attractive enough to entice people to the roulette table. Competitive conditions in the American roulette market have produced an equilibrium expectation for the player of $-5.3\not c$. At this figure, both buyer and seller seem to derive a maximum of satisfaction from their market transactions. The casino operates on a reasonable profit margin and the players pay a fair price for the delights offered. Such a market has been called "efficient" [2]. We shall see that it is also symmetric. In such markets, all participants know all there is to know in setting the terms of transactions.

SYMMETRY IN THE ROULETTE MARKET

Regarding the roulette market as our system under observation, we can now ask if there are symmetry operations that leave it unchanged. And, if so, what is conserved? We shall dispense with the obvious symmetry that the laws of physics, which govern the behavior of all people and roulette apparatuses, are the same wherever and whenever the game is played. Of significance to us is the constancy of the roulette market when any person whosoever is interchanged with any other person. No matter who is playing and no matter who the substitution is, the probabilities, rewards and expectation are exactly the same. A person placing his bets at random can expect to do as well as anybody else. As far as the roulette market is concerned, one person looks just like any other. It makes no difference whether the roulette gambler is an Einstein or a village idiot, a king or a beggar. We can say that the roulette market is symmetric under the operation of player interchange.

The conserved quantity is information. There exists no valid information from any source that enables one person to do better than another at roulette. Betting systems, mathematical knowledge, intuition, intellect, clairvoyance, crystal balls, astrology, witchcraft—the painful losses endured by all serious roulette gamblers offer mute testimony that none of these can change the expectation one iota. Everything known about roulette is built into the game. We have here a market of equal opportunity; every player has the same expectation. That is not to say there is equality of reward. Over a number of games, the people sitting around the roulette table realize very different profits and losses, as indicated in Fig. 1. Some will do average, some better and some worse. Furthermore, there is no predicting beforehand who will do what. A player's past record gives no hint of his future performance.

Achieving better-than-average performance on a consistent basis in the roulette market is equivalent to violating interchange symmetry. To accomplish this, a player must acquire valid information not generally known to either the other players or the casino operators and exploit it. From what has been said so far, such information may seem impossible to get. But, maybe not. Reality is much more complex than the rather simple considerations discussed above might lead us to believe, and we must delve more deeply into the validity of interchange symmetry.

In our roulette model, we assumed the existence of an unbiased wheel and took 18/38 as the "true" probability of the ball falling into a red slot. When plugged into the theory of probability, this number yields a negative expectation. Suppose, however, that the actual probabilities for each slot are not known, as they are not for a real rotating roulette wheel in a casino. All mechanical devices are to some degree imperfectly constructed, and it is very likely that the wheel is somewhat biased and the probabilities for the 38 slots are not all equal. If this be the case, no amount of logic and mathematics in themselves can describe the behavior of this real-world object. Probability theory can say something about the future only when the right probabilities are fed in. Failing this, a positive expectation in roulette becomes a possibility.

We may divide roulette aficionados into two groups. People in the first group are proponents of the biased wheel model of roulette. They insist that the deviations from equal probabilities are great enough to produce a positive expectation. To see the effect of a wheel imbalance, suppose the casino operator is unaware that the probability of "red" is 20/38 instead of 18/38. The expectation of a bet on "red" would then be $+5.3¢$ rather than $-5.3¢$. A player knowing this could reap an average profit of $5.3¢$ for every dollar bet on red.

People in the second group are proponents of the unbiased wheel model. Included in this group are the casino operators. The latter, from their knowledge of the roulette mechanism, have risked their money on equal probability for all slots. They claim that any deviations of the unknown probabilities from equality are too small to measure. Therefore, nobody possesses or can possess any information that allows symmetry violation and its consequence, better-than-average performance on a consistent basis.

Now the argument is joined on the proper grounds. We have a one-on-one confrontation between two groups of participants in the roulette market. Whose conception of reality is more correct? Is the market symmetric? Suffice it to say here that casinos have been offering roulette to all comers for hundreds of years without going bankrupt. They have not bothered to modify the rules of play or the payoffs, and they have not excluded potential customers. The market has shown not the slightest sign of disruption from a symmetry violation. Still, an argument such as this can be resolved only by careful observation and measurement by competent and reliable people. There is no recourse in the marketplace to "true" probabilities or "higher" authority.

Interchange symmetry and its related conservation of information, it should be clear, arise solely in free markets. Controlled markets, by contrast, have by definition one or more of the following features: arbitrary exclusion of players, censorship of information, interference with playing methods, or tampering with earnings, any one of which destroys symmetry through its effect on the expectation. The converse is not the case. Free markets may also exhibit symmetry violations allowing some people to do consistently better than average.

A SYMMETRY VIOLATION IN THE BLACKJACK MARKET

Blackjack or "21" is one of the most popular of all gambling games and one of the most lucrative for the casinos. We need not go into the details except to say that like all such games, it held out a negative monetary expectation to the player until the early 1960's when a symmetry violation severely disrupted the blackjack market. The cause was newly discovered information in the form of a profitable system of blackjack play devised by mathematician Edward O. Thorp [3]. Thorp had uncovered a market inefficiency, a faulty assignment of probabilities to the future outcomes in blackjack. The model used by the casinos to calculate those probabilities assumed a constant deck of 52 cards and made no allowance for the diminution of the deck as the game progressed. With probabilities calculated for a diminishing deck, Thorp was able to establish a profitable method of play, one that yielded a positive expectation.

Armed with his newfound knowledge, Thorp proceeded to put his theories to the test. Card dealers, who at first scoffed at the lucky newcomer, became by degrees increasingly bewildered and frustrated. The consternation of casino operators mounted as funds drained from their coffers, until, finally, Thorp was barred from playing. Disguises and changes of venue no longer sufficed once dealers caught on to the style of play. The respite was short-lived. With the publication of Thorp's book, hordes of expert blackjack players descended on the gambling palaces seeking quick fortunes. Was there any way out of the quandary? By now the casino experts understood that Thorp's strategy had exploited a deficiency in their knowledge of blackjack, and they frantically sought a remedy. Their very existence jeopardized, the casinos changed the rules of play to ensure a negative expectation for the players based on the new probabilities. But one problem led to another. Under the altered rules, play at the blackjack tables fell off sharply, obliging a return to the old ones.

Multiple decks of cards from a shoe were next introduced, making it harder for players to "case" the deck, a prerequisite for profitable operations. More frequent shuffling made the Thorp system more difficult to apply, although it slowed the game down considerably. Winning players were vigorously excluded or taken care of by nimble-fingered dealers. Needless to say, this modern day gold rush was effectively detoured. Now, matters have come full cycle. The original rules are still in force and blackjack is again played in many places with one deck dealt and shuffled by hand. Adjustment to the new market conditions has been completed and equilibrium prevails. Casino profits are increasing, blackjack is more popular than ever, yet Thorp's system remains as valid now as then. When is good information not good information? When the market is not free!

New information has the potential for severely disrupting a free market. It may be an invention, a technical advance, a superior product and even a revolutionary theory such as Newton's System of the World. It may burst on the scene like an earthquake and it may sneak in so surreptitiously that not many people appreciate its significance and act on it. How a free market adapts to it depends on its structure. In free consumer markets, we shall see that such symmetry violations result not only in benefits to the innovator but in greater expectations for all.

THE *PARI MUTUEL* MARKET

We turn to a much more complex market, one in which the public as a whole must calculate the probabilities of future outcomes and their associated rewards. That the *pari mutuel* market is an almost ideal laboratory for the study of market mechanisms under institutional conditions has been largely ignored until recently. Yet, here we have a microcosm of society struggling to combat as best it can its own unique set of problems and uncertainties. Here are thousands of diverse individuals with easy access to enormous amounts of data trying to see what the future holds in store. Their problem is everybody's: to determine the probabilities and rewards of the possible future outcomes of an event, in this instance a sporting event. How each fares depends on the symmetry of the market. If interchange symmetry applies, there is conservation of information and all participants have the same expectation. If not, those with access to relevant information of limited availability can expect to de better than average.

The popular sport of horse racing serves to illustrate the operation of the *pari mutuel* market. A person wishing to bet on a horse (we consider only bets to win) buys a ticket from the race track which identifies the horse chosen, the amount bet, the date and the number of the race. The race track collects all the money wagered to win by all bettors into a win pool, from which it deducts a certain amount known as the "take". Part of the take goes to the track for operating costs and part to the government as a tax. When the race is over, the track redeems the winning tickets with the money remaining in the pool, usually 80–85% of the original.

Like the casino, the race track plays a passive role in the wagering. It establishes and enforces the rules of play, but neither influences the outcomes of the races nor interferes with the players' betting methods. However, a major difference between the two arises in their handling of the win probabilities and rewards. In the roulette market, the casino operator must estimate the win probabilities and set the rewards so as to ensure a negative expectation for the roulette player and a positive expectation for himself. The *pari mutuel* market operator, on the other hand, takes no interest in the win probabilities. And he affects the players' rewards only indirectly through the size of the take. He does not in any way compete with the player for the money remaining in the pool after removal of the take.

Who then fixes the expectation of each horse bettor? Against whom does he compete for the money at stake? Remarkably, his competition is none other than all his fellow racegoers taken as a body. Their collective opinion of the winning probabilities of each horse and the take determine the expectation of the individual gambler. This relationship is shown in the following formula, which is derived in Appendix B:

$$E_i = \frac{P_i}{p_i}(1 - f) - 1,$$

E_i is the expectation of a player betting on one of the horses i in the race. P_i is the true (and unknown)

Table 1. Probabilities and returns in the *pari mutuel* market (1955–1962)

Dollar odds	Public's probability (%)	True probability (%)	Number of horses and winners	Expected number of winners ±2 s.d.'s	Return dollar bet (¢)
0.40–0.55	56.9	71.3	129—92	73 ± 11	+3.4
0.60–0.75	50.2	55.3	295—163	148 ± 18	−7.1
0.80–0.95	44.9	51.3	470—241	211 ± 22	−3.8
1.00–1.15	40.6	47.0	615—289	250 ± 25	−2.4
1.20–1.35	37.1	40.3	789—318	293 ± 27	−8.1
1.40–1.55	34.1	37.9	874—331	298 ± 28	−6.1
1.60–1.75	31.5	35.5	954—339	301 ± 29	−4.8
1.80–1.95	29.3	30.9	1051—325	308 ± 30	−10.5
2.00–2.45	26.3	28.9	3223—933	848 ± 50	−6.5
2.50–2.95	22.8	23.0	3623—835	826 ± 50	−13.5
3.00–3.45	20.1	20.9	3807—797	765 ± 50	−11.0
3.50–3.95	18.0	18.6	3652—679	657 ± 46	−11.6
4.00–4.45	16.2	16.1	3296—532	534 ± 42	−15.3
4.50–4.95	14.8	15.5	3129—486	463 ± 40	−10.6
5.00–5.95	13.2	12.3	5586—686	737 ± 50	−20.1
6.00–6.95	11.4	11.0	5154—565	588 ± 46	−18.0
7.00–7.95	10.0	9.9	4665—460	467 ± 41	−16.4
8.00–8.95	9.0	8.2	3990—328	359 ± 38	−21.8
9.00–9.95	8.1	8.2	3617—295	293 ± 33	−14.7
10.00–14.95	6.5	6.0	12,007—717	780 ± 54	−20.7
15.00–19.95	4.7	4.0	7041—284	331 ± 35	−26.4
20.00–99.95	2.5	1.4	25,044—340	626 ± 50	−54.0
			93,011—10,035†		

†The number of winners exceeds 10,000 because of dead heats.

winning probability of horse i, and p_i is the winning probability of horse i as estimated by the betting public as a whole. f Is the fraction of the win pool removed as the take.

It is easily seen that the higher the take the less the expectation of the gambler, no matter which horse he bets on. It is the same for all. For purposes of symmetry analysis, however, the probability ratio P_i/p_i is critical. If, for example, the public's estimate of the win probability p_i equals the true probability P_i for all horses, then $E_i = -f$. That is, all bettors have exactly the same expectation independent of the choice of horse. Just as in the roulette market, one person would be "pari mutuelly" indistinguishable from any other person. Symmetry holds and information is conserved.

An abundance of data from the past 40 years of horse racing permits us to test the validity of substitution symmetry. One of the first and most complete compilations is reproduced in Table 1 [4]. It shows the results of 10,000 races run at American thoroughbred tracks from 1955 to 1962. The 93,011 horses in these 10,000 races are grouped according to the "dollar odds," which denote a bettor's profit for every dollar risked should his horse win. The odds are related to the horse's win probability as estimated by the betting public (see Appendix B). The lower the odds, the greater the public's win probability for the horse. The horse starting at the lowest odds has the greatest win probability and is known as the "favorite". High-odds horses have small win probabilities and are known as longshots.

The "true probability" for each group is taken equal to the win percentage, which is found by dividing the number of winners by the total number of horses (column 4). For all groups, the win percentage is an adequate statistical representation of the "true probability" for our purposes.

Column 5 indicates the expected number of winners for the group when the public's probabilities are assumed correct and the anticipated range of statistical fluctuation (two standard deviations). Finally, the last column presents a bettor's return for each group, the amount won or lost per dollar bet if one dollar were staked on each horse.

It should be noted that the dollar odds fluctuate during the betting period before the race according to the money wagered on each horse. The race track displays this information on the totalizator board or "tote," usually after each minute of betting. In Table 1, the final odds at the close of betting are used. These odds determine the rewards.

SYMMETRY IN THE *PARI MUTUEL* MARKET

Table 1 reveals a remarkably close correlation between the public's estimates of the win probabilities and the true probabilities for most of the odds groups. Indeed, these two probabilities are equal within statistical fluctuations for all horses starting at dollar odds between \$2.50 and \$9.95. Discrepancies do arise, however, for horse starting at odds outside this range. The public underestimates the win probabilities at low (short) odds and overestimates them at high (long) odds. Returns vary from a *profit* of 3¢ per dollar bet for the shortest odds horses (there are very few of these) to a loss of over 50¢ per dollar bet for the longer odds horses. An expectation of $-16¢$ over the entire odds range would have obtained had the public's win probabilities been correct.

The data point to a *pari mutuel* market that is quasi-symmetric under the interchange operation. For all groups other than favorites and longshots, interchange symmetry holds. All bettors on horses in these groups have the same expectation—their returns are indistinguishable (within expected fluctuations) as far as the *pari mutuel* market is concerned. A person betting at random does as well as anybody else. There are departures from perfect symmetry for favorites and longshots which allow better-than-average performance for favorite players and worse-than-average performance by longshot players. The market thus distinguishes the latter two groups from all others.

Such a violation of interchange symmetry implies the existence of valid information not generally known to the betting public. It is, therefore, not adequately factored into the wagering on favorites and longshots. Because of its limited availability, the public underbets the favorites and overbets the longshots. Favorite bettors know something the rest of the public does not. And those people betting on the horse favored by the public near the close of betting are able to utilize this information to realize better-than-average returns. They will not win, but they will lose less than expected.

What is the nature of this information? How does it arise? We can understand this symmetry violation by examining the forecasts of those expert public handicappers as published in the racing journals, notably the *Daily Racing Form* in America. These brave gentlemen make their selections the day before the races take place. They have access to the past performances of the horses, the conditions of the next day's races, and any other information such as workouts available up to their deadlines. We shall refer to such information as "historical" data, in contrast to "late-breaking" information that becomes available afterwards. Late-breaking information may include the trainers' assessments of how their horses feel the morning of the race, unpublicized workouts, jockey changes, weight changes and the track condition. The experts must base their selections on historical data.

How do they do? Table 2 contrasts their returns per dollar bet with that of the post favorites, the horses having the most money bet on them at the close of betting for four different 5000-race samples. The take at the time averaged to 16%, and it is clear that the experts' returns per dollar bet fluctuate narrowly about this figure. Their average is also 16%. This is just what we expect if the public determines winning probabilities accurately. We can conclude, therefore, that on the basis of historical data, interchange symmetry exists in the *pari mutuel* market. There is no historical information available that allows one person to do better than any other in the *pari mutuel* market.

Now look at the post favorites. Although not profitable, the returns from bets on the crowd's choices are consistently better than those of the experts. By simply betting on these horses, a person will do better than average. His expectation is greater than that of other bettors. In light of this finding, we are forced to admit the existence of information that violates market symmetry, that is, late-breaking news. This information is not known to the general public when they set the winning probabilities of the horses, and it leads to a skewness in the returns. These are *not* normally distributed as would be the case if everybody had the same expectation. More people realize better-than-average returns than less-than-average returns (because favorite players are most numerous), and the normal curve is skewed toward higher returns.

The *pari mutuel* market serves as a powerful first example of how a free group of people wholeheartedly acting in their own selfish interests utilize historical information in the best possible

Table 2. Experts' return per dollar bet in four 5000-race samples

	¢			
Public favorites	−10.8	−6.6	−9.2	−7.5
Reigh count	−16.5	−16.0	−14.6	−12.1
Armstrong	−14.3	−14.9	−12.9	−14.7
Sharpshooter	−18.6	−17.6	−14.7	−13.2
Hermis	−15.0	−15.5	−18.7	−14.8
Handicap	−15.4	−18.2	−20.9	−17.3
Sweep	−15.9	−18.5	−19.0	−16.4
Trackman	−16.4	−15.7	−18.0	−12.8

way to make the best possible predictions of future happenings. It is an open arena where the thousands of diverse opinions of thousands of diverse individuals clash, interact and modify one another in such a way that the final consensus is a closer approximation to the truth than that of any single person. Not only does the public determine winning probabilities accurately, but it does so in a manner whereby the horses voted most likely to succeed do best. A remarkable performance!

For scientific readers, late-breaking information removes a degeneracy in the *pari mutuel* market. Instead of one expectation for all bettors on the basis of historical information, there is a spectrum of expectations, the greatest being for favorite players and the smallest for longshot players.

THE STOCK MARKET

The invention of money introduced a flexible medium of exchange that made feasible the deferred transfer of goods and services. In effect, money became a call on the future output of producers. And it became also a convenient method of storage, an effective means for saving. When advancing technology rendered practicable projects of a scope beyond the means and abilities of a lone entrepreneur, it was but a step to pooling the monetary savings of individuals and channeling them into commercial and industrial ventures beyond the means of any one person. Venice, Genoa and the other Italian city-states made wondrous use of this device. Their merchants and craftmen together with other small savers built tall ships, stocked them with expensive cargoes of Italian manufacture, and sent astute captains abroad to trade their goods for foreign merchandize and gold on a favorable basis. So prosperous did the city-states become that they were able to initiate and finance the great cultural flowering of the Renaissance. Northern Europe saw a similar development in the cities of the Hanseatic league.

The savings pool evolved in time into a modern capital market. Shares denoting fractional ownership of industrial and commercial organizations engaged in new and innovative ventures were parceled out to people willing to risk their present savings for possible future rewards. Investors received, in other words, a piece of the action proportional to their investment. Soon, those shares became transferable from one person to another, and stock markets sprung up to facilitate the exchange and allocate funds to new and old businesses. We have here the rudiments of the modern corporation with shares of common stock designating multiple ownership of the business. The Muscovy Company (chartered in 1555) and the Dutch East India Company (chartered in 1602) are two early examples of the genre.

The stock market's importance in a capitalist society can hardly be overemphasized. It is the preferred channel for directing flows of individual savings into the creation and expansion of the means of production. It functions as a ready market in which corporations, both established and start-up, may raise money by selling stock to new fractional owners. How much money a company can raise depends on its prospects, as judged by the investing public. Innovative and efficient businesses find little difficulty in tapping the savings pool to finance their expansion plans and the development of new products and services. Inefficient, slothful companies, on the other hand, that produce obsolescent and inferior goods find it difficult to sell stock. In this manner, the stock market allocates the public's savings to businesses in accordance with the public's wishes. The great advantage to a company is that it gets the money free and clear of debt restrictions.

The stock market offers to the individual investor a savings alternative. A person buying the stock of some company thinks that the return on his money, from price appreciation and dividends (cash payments out of profits made to stockholders), may be greater than the return available elsewhere. A person selling a stock thinks he can do better either in another stock or in some other investment instrument. In the market place for stocks, the opinions of these two groups of people continually clash, interact and modify one another until a price level for a stock is reached whereat there are about as many shares being bid for as being offered. What the investing public is trying to do is to decide on a price for a stock that adequately reflects the risks involved in securing a good return on the investment. With each passing moment, the risks are evaluated and reevaluated in the light of any new information that arises, and the market price of each stock fluctuates with each revelation. The thoughts of the investing public concerning the prospects of the many businesses upon which our society relies are mirrored in the market quotations on the financial pages of the daily newspapers. These prices represent the collective wisdom of millions of buy, sell, don't buy and don't sell decisions. Each of these numbers takes into account all possible futures and their related probabilities, as seen by the investing public.

Like other free markets, the stock exchanges set and enforce the rules of investing. They place no restrictions on who may play, they do not interfere with investment strategies, they do not try to reallocate profits and losses, and they do not influence stock prices. Finally, they encourage the fullest dissemination of financial information.

SYMMETRY IN THE STOCK MARKET

The calculation of an investor's expectation in the stock market is more difficult than in the simple games of chance considered above because investment outcomes are contingent on the length of time stocks are held. Holding times must be standardized in order to compare the returns from stocks owned for varying time intervals. For this purpose, the compound interest formula serves (see Appendix C). We shall reduce the profit and loss on all stock transactions to a percentage rate of return compounded yearly (denoted by r in what follows) as given by

$$A = P(1 + r)^n,$$

where P is the initial amount invested and A is what the investment is worth n years afterward. An investor's expectation is then the average of the percentage rates of return over a large number of investments.

Initially, we shall determine an investor's expectation when buying stocks at random. That is, stocks will be selected blindly with no regard to a corporation's past performance, future prospects, economic conditions or any other information. In other words, we assume interchange symmetry holds and no one investor is superior to any other in selecting stocks for profit. Once this is done, claims of better performance will be analyzed for possible symmetry violations.

We proceed as follows: (1) pick a stock at random; (2) pick a purchase data at random and note the price; (3) pick a later sales date at random and note the price; (4) using the compound interest formula, compute the interest rate r; (5) repeat the same procedure over and over again; (6) average the r s to obtain the expectation.

Carrying out this procedure for all stocks and all time periods over stock market history would be an endless task. However, a good random sampling of the total stock population has been published using 1715 stocks listed on the New York Stock Exchange during the interval from January 1926 through November 1960 [5]. The results are shown in Fig. 2. For the 1715 stocks, 56,557,538 holding periods, varying from one month to the full 35 years covered by the study, went into the graph. The expectation calculated from the data is +9% per year compounded annually. An extension of the research to 1976 revealed no significant alterations of the results [6].

These measurements of the returns on common stocks suggest the possible outcomes available to an investor who bought blindly at any time between 1926 and 1976. Most likely, he would have chosen stocks and holding periods giving returns near 9% per year compounded annually.

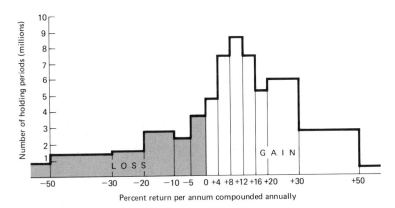

Fig. 2. Distribution of returns from random investments in stocks listed on the New York Stock Exchange (1926–1960).

Sometimes, much higher yields would have been obtained—some made fortunes—and sometimes much lower, even to a complete loss of capital. The worst losses took place during the Great Depression; almost a 50% loss was taken by the average investor for stocks owned outright if he bought in September 1929 and sold late in 1932. However, overall, the rates of return on common stocks far exceeded those of any other investment instrument.

The distribution of returns is near normal, which is what we expect if every investor has the same expectation. A deviation from the normal distribution occurs in the right shoulder of the curve. There are too many returns between 12 and 50%. This skewness is reminiscent of the *pari mutuel* market, where late-breaking information seemed to be the cause.

Let us examine the claims of the experts with regard to better than random performance. One hint comes from an item in *The New York Times*, 16 August 1967:

> "A member of the Senate Banking Committee sought to prove today that it is possible to pick a portfolio of stocks that would do better than most mutual funds simply by throwing darts at the New York Stock Exchange list. Senator Thomas J. McIntyre, Democrat of New Hampshire, reported to the committee that he had done just that, and gotten better investment results than the average of even the most growth-oriented mutual funds. A hypothetical $10,000 investment made 10 years ago in the senator's dart-selected stocks would be worth $25,300 now."

A mutual fund is an organization that collects money from people and invests it for the purpose of capital growth from appreciation and dividends. The fund managers spend full time investigating and analyzing investment opportunities and so, presumably, should be able to do better than average. The senator told the committee that he tried the experiment after fund managers disputed testimony by Paul Samuelson, Nobel Laureate in economics, that random stock selection yields investment results as good or better than those achieved by the funds.

Expert performance is exposed more definitively in the following summary covering the years from 1927 to 1960 [7]:

1927–1935

> "It can, then, be concluded with considerable assurance that the entire group of management investment companies proper (as opposed to the sample here studied) failed to perform better than an index of leading common stocks and probably performed somewhat worse than the index . . . ."

1934–1939

The overall gain in asset value of the six largest companies averaged out to 53.7% including dividends paid out. Standard & Poor's 420 stock index showed a 66.6% increase and only one of the funds exceeded this figure by achieving 70.6%.

1940–1949

The same six companies had an average gain of 129% compared to 97% for the Standard & Poor index.

1951–1960

The performance of 58 companies over this ten-year period averaged to a gain of 221%, more than tripling the initial capital. But Standard & Poor's Composite Index of 500 stocks rose 322%, almost 50% better than the average of the funds. This figure was exceeded by only three of the funds.

Quoting Graham, Dodd and Cottle:

> "These results do not appear to us to be as satisfactory as they should be. They suggest that the investment companies as a whole—and practicing security analysts as a whole—might well examine their basic approaches to both the selections of common stocks for purchase and the decision to sell 'less satisfactory' holdings."

Matters have not improved since. From 1958 to 1967, the average return per annum compounded annually for 146 funds was 10%, the same as for each of the Dow–Jones averages of industrial, rail and utility stocks. In the 10 years ending 31 March 1976, the net asset value of the average mutual fund was up 46%, far behind the 60% gain in the Dow–Jones industrial average. [4].

From *U.S. News and World Report*, 31 August 1987:

> "The blue-chip bull has created an ironic twist: Over the past five years, the average equity fund—a mutual fund that invests in a broad selection of stocks—hasn't bettered or even kept pace with either the soaring Dow–Jones Industrial Average or the Standard & Poor's 500 Index. You might have done better investing in an index fund . . . ."

An index fund invests in stocks found in an index, usually the Standard & Poor's 500, so that its portfolio mirrors the index.

Further evidence comes from years of testing of the random walk model of the stock market [8]. Price changes in stocks satisfy random walk criteria and are, presumptively, a series of random numbers. If such be the case, there is no way of predicting price changes and conservation of information exists for the stock market. Evidence to the contrary is lacking, a very unusual situation in economics.

SYMMETRY VIOLATION IN THE STOCK MARKET

The analysis of the preceding section points to a symmetric stock market, one in which every investor has the same expectation. The skewness in the distribution of returns from common stock investments, however, suggests the possibility of a quasi-symmetric stock market similar in nature to the *pari mutuel* market. Does the stock market adjust instantaneously to late-breaking information or is such information discounted over time? If the discounting process is of sufficient duration, can astute investors acting on the late-breaking information perform better than average?

The effect of late-breaking information is analyzed most thoroughly in *"The Science of Winning"* [4]. There it is shown that the investing public's response to surprising quarterly earnings reports —earnings widely divergent from estimates by financial analysts—is *not* by any means reflected instantaneously in stock prices, but is a process that may go on for months. During this period of adjustment, there are opportunities for better-than-average performance.

The response to quarterly earnings reports was tested as follows:

(1) Quarterly earnings projections for some 1500 stocks were obtained from investment services and brokerage houses. For each company, the highest estimate was compared with the actual earnings reported by the corporation in its quarterly report in the *Wall Street Journal*. The percentage change from the estimate was calculated. Companies were then separated into six categories: (i) those showing better-than-expected earnings of 20% or more; (ii) those with better-than-expected earnings between 10 and 19%; (iii) 0–9%; (iv) those showing worse-than-expected earnings of 1–9%; (v) 10–19% and (vi) 20% or more worse than expected.

(2) The price of each stock was recorded as of the opening of the stock market on the second business day following the announcement of quarterly earnings and again, approx. 3 months later, at the opening of the stock market on the second business day following the succeeding earnings announcement. For each stock, the percentage change in price over the time interval between earnings reports was calculated.

Tables 3 and 4 display the results for two different time intervals. The "expectation" row gives the range of results to be expected if a person blindly picks a sample of this group size from the total sample and observes the number doing better than average.

Clearly, there is a symmetry violation on the basis of late-breaking information on earnings reports. Those stocks with better-than-expected earnings not only perform better than the others, but they do so on a statistically significant scale. Further statistical tests are applied in Ref. [4] and point to the same conclusion.

Table 3. Price performance of stocks by group, 1 August 1969–1 September 1971

Group	I	II	III	IV	V	VI	Total
Number of stocks	981	732	1907	1440	960	2260	8280
Mean price change	+5.9%	+2.9%	+1.4%	−0.2%	−1.2%	−2.5%	+0.4%
Number of stocks performing better than the market	605 (62%)	484 (66%)	1157 (61%)	829 (58%)	536 (56%)	945 (42%)	4556 (55%)
Number of stocks performing equal to or worse than the market	376 (38%)	248 (34%)	750 (39%)	611 (42%)	424 (44%)	1315 (58%)	3724 (45%)
Expectation	540 ± 28	403 ± 24	1049 ± 36	792 ± 34	528 ± 30	1243 ± 40	

Table 4. Price performance of stocks by group, 1 September 1971–15 May 1973

Group	I	II	III	IV	V	VI	Total
Number of stocks	1193	683	2158	1682	1161	2524	9401
Mean price change	+3.7%	+2.4%	+0.6%	−1.4%	−2.3%	−3.5%	−0.7%
Number of stocks performing better than the market	680 (57%)	394 (58%)	1189 (55%)	842 (50%)	551 (47%)	1001 (40%)	4657 (50%)
Number of stocks performing equal to or worse than the market	513 (43%)	289 (42%)	969 (45%)	840 (50%)	610 (53%)	1523 (60%)	4744 (50%)
Expectation	597 ± 32	342 ± 26	1079 ± 40	841 ± 36	581 ± 32	1262 ± 44	

On the basis of historical information, the stock market, like the *pari mutuel* market, is symmetric and there is conservation of information. Late-breaking information causes a symmetry violation and permits some investors to perform better than average. We have another example of a market in which each investor uses all available information to pursue his own best interests and, by so doing, creates a market with a high expectation for all. In the absence of evidence to the contrary, we must conclude that this free-market expectation is the best possible at present.

SYMMETRY IN CONSUMER MARKETS

We turn now from money markets to consumer markets. In so doing, we face a new and difficult challenge in evaluating the expectation. What is the value of goods and services received for each dollar spent? In money markets, the expectation is simply calculated by dividing the net profit or loss by the amount risked. To make such a calculation in consumer markets, however, we must assign monetary values to the sensual and psychological utility generated by the purchase of goods and services. Needless to say, there exists no generally accepted model of human behavior that permits such a calculation, and obtaining the expectation in this manner can lead only to ambiguity and controversy. In lieu of the expectation, therefore, we shall focus on conservation of information to demonstrate symmetry in free consummer markets.

To start, let us look at a typical market, that of automobiles in America. There are many manufacturers, both foreign and domestic offering a wide variety of makes and models in all price ranges. In order to stay in business, an automaker must compete with all the others for the consumer's dollar. The risks are enormous, the constraints tight. A fickle public demands cars possessing quality, style and extra features. Dealer networks for sales and service are mandatory. Research and development to keep abreast of product and manufacturing innovations are essential lest the competition or some market newcomer introduce a superior automobile and gain market share.

All these activities require a huge money inflow from car sales, not only to cover operating costs but to reap an adequate return on the invested capital (or else, why not liquidate and invest the money elsewhere?). To reach the necessary sales level, our automaker must negotiate a fair price for his cars with the consumer. That is vitally important. Should his prices be too high, he may sell too few cars and lose money. Should they be too low, he may sell more cars and still lose money. Usually, there is an optimal price leading to optimal sales and profits. The optimal price does, of course, vary with economic conditions. Since nobody can accurately predict what these might be like months hence, our manufacturer must continually adjust prices on the basis of current sales figures. Such adjustments may take the form of sales incentives such as rebates and low-interest loans. During recessions, the optimal price may even lead to an operating loss.

On the other side of the equation, Mr Average Consumer is faced with choosing a car from a bewildering array of makes and models. He may decide on anything from a high-priced luxury item like a Rolls Royce to a low-priced utilitarian Volkswagen Golf. Expensive cars offer more in quality, style and social status, but even the inexpensive models perform well their basic function of getting people from one place to another.

This brief introduction to a consumer market brings us to a consideration of its state of knowledge. Firstly, with regard to the producers, we ask: does every successful automobile manufacturer possess all the information necessary to secure his niche in the car market? If the answer is yes, then we have conservation of information. That is, no one producer knows anything that enables him to drive his competition out of existence and keep out newcomers. Certainly, there will be times when one manufacturer gets a momentary edge by inventing a new and advanced component, or by developing lower cost manufacturing techniques, or by a new styling, or by some innovation anywhere along the intricate chain of producing and selling an automobile. When something like that happens, the rest of the market must and usually does adjust in the direction of greater efficiency. The Japanese automakers, for example, broke into the world market with new and lower cost cars and now compete very successfully with the established producers. At other times, companies die because they are unable to adapt to changed conditions. Studebaker and American Motors, among many others, have passed from the scene. Sometimes, markets expand and allow increased market share for newly born and mature producers. And sometimes markets contract driving even some of the most efficient producers into extinction. These are market fluctuations that may arise anywhere and anytime because of changes in technical, economic and political conditions. But, unless one producer gobbles up market share to the point of extinguishing his rivals, we have to conclude that information is conserved. The constancy of market share among the major producers over the years argues strongly for this view.

Conservation of information implies equality of expectation for all producers. We can measure the expectation in the same way as for the financial markets: divide the net profit or loss by the amount risked and average over some convenient time period. For businesses, the figure of interest is the per cent return on net worth. It is calculated by dividing the company's net earnings by the company's net assets (the value of what it owns less any liabilities). This is the number that tells us how much money a company makes or loses on its investment, how successful or unsuccessful it is. In a similar manner, the bank interest rate on savings tells a depositor how much he profits from his deposit. Table 5 shows the per cent return on net worth for seven automobile manufacturers for each year during the ten year period 1978–1987. Negative returns arise when a company loses money for the year.

The returns on net worth vary considerably, just as we should now expect for any risk venture. Each company's fortunes ebb and flow on management decisions concerning production levels, model and style changes, quality control, debt load and advertising budgets. All companies are hostage to the political and economic uncertainties that obscure the consumer's purchasing power. The three American companies, General Motors, Ford and Chrysler, suffered considerably in the early 1980's from Japanese competition and a strong dollar. Chrysler would have gone bankrupt except for a government loan bailout. Nowadays, Chrysler and Ford have come back strongly while General Motors languishes with an inadequate return on investment and the foreign manufacturers suffer from a weak dollar.

By averaging the returns on investment, we obtain a reasonable value for the expectation of an automobile manufacturer. Omitting the years 1980–1982 for Chrysler, which are distorted by government intervention, the expectation is 16.9%. Given the statistical smallness of the sample, the returns on investment are not inconsistent with an equal expectation for all manufactures. In the absence of evidence to the contrary, we conclude that interchange symmetry exists for manufacturers in free markets.

Interchange symmetry applies only to the expectation. Just as people in the roulette market differ in an infinity of ways other than their expectation, manufacturers differ in the products offered, management and production techniques and many other ways. However, all look the same to the car market in terms of expectation.

A 16.9% return is a good deal higher than those available from savings instruments such as government bonds, and apparently furnishes an adequate reward to business owners for business risks and headaches. It also seems low enough to keep newcomers out of the market. Such an

Table 5. Percent return on net worth for automobile manufacturers, 1978–1987

	1978	1979	1980	1981	1982	1983	1984	1985	1986	1987 (estimated)
Chrysler	−8.3	−71.4	−561.2	−97.3	−31.0	38.5	45.3	38.8	26.3	21.0
Jaguar	—	—	—	−41.9	4.6	67.6	27.1	67.0	24.8	25.0
Ford	16.4	11.2	−18.1	−14.4	−10.8	24.7	29.6	20.5	22.1	28.0
General Motors	20.2	15.2	−4.5	1.9	5.3	18.0	18.8	13.6	9.7	8.0
Honda	7.6	12.3	27.5	17.2	15.0	16.7	17.7	19.2	11.1	10.0
Subaru	42.4	39.9	41.7	39.4	38.5	33.8	30.1	28.9	29.8	−15.0
Volvo	9.4	13.0	7.3	11.2	7.3	15.2	23.3	21.0	13.8	13.5

Table 6. Percent return on net worth for petroleum companies, 1978–1987

	1978	1979	1980	1981	1982	1983	1984	1985	1986	1987 (estimated)
Atlantic Richfield	14.6	19.1	22.2	19.3	16.6	14.5	14.1	27.0	11.7	19.5
Amoco	15.1	18.0	20.4	18.0	16.5	15.4	17.0	16.9	8.2	10.5
British Petroleum	11.8	32.5	24.2	13.9	8.3	9.0	12.1	16.1	8.2	13.5
Chevron	13.4	19.2	21.7	18.7	10.4	12.5	11.5	9.3	5.8	6.0
Exxon	13.7	19.1	22.2	19.5	14.6	16.9	19.2	18.6	15.6	13.5
Mobil	12.6	19.1	21.5	16.6	9.4	10.8	10.1	11.4	9.6	8.5
Royal Dutch	13.6	29.3	18.5	13.6	11.8	14.1	14.3	14.4	10.2	10.5
Texaco	9.0	16.5	17.9	16.8	10.0	8.4	7.6	8.2	4.4	4.0

expectation furnishes bountiful rewards to Mr Average Consumer, who revels in the quantity, richness, variety and affordability of automobiles offered for sale. Again, the market appears efficient, with both producer and consumer deriving a maximum of utility from their transactions.

Table 6 exhibits the percentage returns on investment for some of the larger companies in the petroleum industry. The products sold in this market are not as subject to consumer vagaries as automobiles and the returns on investment do not vary as much. The expectation is 14.6%. And again, the returns for all companies over the years are consistent with the same expectation for all.

The expectation in a free market is self-regulated. Too high an expectation invites competition from newcomers and old competitors anxious to share in the high profits. Too low an expectation stimulates corrective action, which can take many forms: liquidating the business, buying new businesses with better expectations, installation of new officers, takeovers by others who think they can utilize the company's resources to get a better return on investment. Too low an expectation in a whole industry may also be a symptom of serious illness. It may arise from many sources, such as high labor costs, high taxes, government control of prices or excessive government regulation. Whatever the reason, low profits impede the replacement of aging equipment and reduce the levels of research and development spending. The ensuing deterioration of the industry may induce newcomers, both foreign and domestic, to seize the opportunity to establish themselves. In general, it appears that free market expectations are set with due regard for both the long-term and short-term health of the industry.

What about symmetry on the consumer's side? Mr Average Consumer must weigh the advantages and disadvantages of the cars offered for sale and decide on what is best for him. The question is this: does every consumer have the same expectation as all others? Or, equivalently, does the buyer get what he pays for?

Just as in all other free markets, the consumer has at his disposal a vast amount of information. There are the advertising claims put out by the various manufacturers, the results of road tests by various testing organizations, and word-of-mouth evaluations by relatives, friends and neighbors. Finally, our consumer can test-drive the car himself to see if it suits him. There would appear to be no information available to one consumer and not to another that enables him to say one car is a better buy than another. Subjective differences of opinion always arise. However, for every person saying car A offers better value dollar for dollar than car B, somebody else can be found who says the opposite. Again, in the absence of evidence to the contrary, we assign to all consumers the same expectation.

It should be evident that the expectation in a consumer market may vary with changing economic conditions. A sluggish car or petroleum market, for example, may result in a buyer's market that enhances the consumer's expectation through lower prices. Only an equal expectation for all exists at any given time. This does not mean each and every consumer gets the same return per dollar spent. Some people may pick up bargains and others may overpay. Many random factors can enter into the final price determination, and the expectation is only the central value of all these possibilities. As with the other free markets we have considered, this expectation appears to be the best possible.

THE ROLE OF GOVERNMENT IN SYMMETRIC MARKETS

We assume that all governments have as one of their primary goals the enhancement of each and every citizen's economic expectation. To achieve such an end, a government must establish the conditions necessary for expanding a nation's production of goods and services to the maximum possible extent. And it must foster an equitable distribution of those necessities and niceties of life consistent with the highest standard of living for all. These are noble intentions. They usually are. However, fulfilling noble intentions in the real world is fraught with difficulty and danger. Reality is never as simple-minded as so many people make it out to be. Economic, sociological and psychological models of reality designed to cope with the real world are woefully inadequate and cannot reliably predict the future consequences of present actions. All too often, "the road to hell is paved with good intentions". With little scientific knowledge to go on, nations can choose their paths to economic greatness guided only by their philosophies of government. Of these, the two most important are those of Adam Smith ("the least government is the best government") and Karl Marx ("centralize all instruments of production in the hands of the State"). We shall briefly summarize the arguments of these diametrically opposed views and assess them from the standpoint of market symmetry.

Adam Smith's "invisible hand" is the basis of capitalism. This principle may be stated as follows: the greatest prosperity for all comes about, as if guided by an invisible hand, when and where each and every individual has a maximum of freedom to decide upon and pursue his own best interests. Many implications follow, both economic and political. Here we have the premise that the collective wisdom of all the people, which is brought into focus by the marketplace and the voting booth, can handle the massive problems besetting all of us more ably than any king, dictator or ruling elite. Here we have the notion that a surging sea of random decisions by countless millions of people generates the fads and fashions of the times, the great discoveries and inventions, the finest art, the most abundant production and the most widespread and equitable distribution of goods and services. Here we have the thought that profit motivates people to act far more often for the betterment of mankind than to its detriment. On the political side, it becomes the function of government to maximize the freedom of all its citizens. Adam Smith's very sophisticated idea appears to be rooted in what is known in physics and mathematics as the random walk. Instead of random steps leading to a normally distributed random variable, there are random decisions leading to a normal distribution of possible prosperities, with maximum prosperity most likely. In this century, the random walk has been responsible for a deeper understanding of natural phenomena ranging from the behavior of the fundamental particles of the universe to practical human affairs.

Marxism, on the other hand, promotes the idea that the production of goods and services under capitalism can never be sufficiently large to reward the proletariat for its labor. Maximum production can be achieved only by taking the means of production out of private hands and placing them under State control. Marx apparently did not subscribe to the notion, so in vogue today, that a Robin Hood type redistribution of wealth from rich to poor in itself could materially benefit the poor. His goal, a bountiful and equal slice of the pie for each and every individual, was predicated on much larger production. Politically, there is much for government to do. It must take over private property and nationalize industry. It must hire people to run the means of production and distribution and plan the nation's economic future. It must control wages and prices and direct the course of research and development. Under Marx's philosophy, public servants assume a role akin to Plato's philosopher-king.

Looking at these two philosophies in terms of market symmetry, there is a striking difference in their treatment of information. We have presented cogent evidence that free markets are symmetric to a high degree. The people participating in these markets, whether as producers or consumers, strive to do the best they can for themselves and, in so doing, utilize all available information to make the market as efficient as possible. Conservation of information thereby becomes a feature of the marketplace. At the same time, symmetry violations can take place. When they do, the marketplace affords ample opportunity and reward for those rare individuals who are able to perceive and correct market inefficiencies. These are unpredictable achievements by gifted people fortunate to be in the right place at the right time. Inventions or new ways of doing things or new systems of thought that reveal the future more clearly than before mark their appearance. Often, these innovations spark a vigorous exploitation of ideas and resources that enhances a nation's quality of life through the introduction of new and improved products and services.

In contrast, Marxism is an asymmetric, elitist philosophy. All information is vested, not in the people, but in the State, which means in those people running the country, the Communist Party.

It is their job to determine what should be produced, how much, by whom and for whom. No competition from private sources is tolerated. Party rulers are responsible for originating and promoting new products and services and improving the quality of life. Their decisions are final. The people as a whole are excluded and have little input.

The rationale behind government intervention into free markets is the possession by public servants of information superior to that possessed by the marketplace. Such exclusive information presumably qualifies them to try their hand at improving free market efficiency. They may, for example, fix the size of businesses through anti-trust policies, regulate wages and prices, hand out subsidies to troubled industries, control the money supply through central banking activities, back projects they deem worthy, and redistribute income through progressive tax policies.

However, what if free markets are indeed symmetric? What can public servants hope to add to the store of knowledge if all valid information is utilized? Their "exclusive" information is then nothing more than noble intentions. They are merely processing notoriously untested and unreliable social and economic data in reality models that have trouble predicting tomorrow's date. The effect on market efficiency is more apt to be harmful, even destructive, than beneficial. Recall the inferior performance of the experts associated with the *pari mutuel* and stock markets. These markets are inherently much simpler than national economic systems and, one might think, more susceptible to expert management. Yet, the public as a whole, with each person pursuing his own best interests in the tradition of Adam Smith's invisible hand, consistently outperforms the experts.

In symmetric markets, it would appear, the role of government should be sharply delimited. Maximizing the freedom of its citizens, maintaining law and order, protecting the nation from internal and external tyranny, negotiating trade treaties with foreign powers, establishing a system of weights and measures—these are functions that should provide benefits to free markets over and above the costs of government. Any other government intrusions into free markets should be taken with extreme caution.

Again, evidence to the contrary is lacking. The greatest Western powers, Rome, Great Britain and the United States of America, all became great when their governments played an insignificant role in business affairs. A comparison of government spending relative to private and business spending illustrates this point very clearly. When government grew to sizable proportions, both the Roman and British Empires entered into protracted declines. And, even in America, the growth of government has slowed economic expansion to a crawl. Finally, it may be noted that those countries starting out with massive government interference in the marketplace never achieve a standard of living for their citizens on a par with free market societies.

REFERENCES

1. A. Einstein, *Annls Phys.* **17**, 891–911 (1905).
2. E. Fama, *J. Finance* **25**, 383–417 (1970).
3. E. O. Thorp, *Beat the Dealer*. Vintage, New York (1966).
4. B. P. Fabricand, *The Science of Winning*. Van Nostrand Reinhold, New York (1979); Paperback edn: Whitlock, New York (1983).
5. L. Fisher, *J. Business Univ. Chicago* **XXXVIII**(2), 149–161 (1965).
6. R. Ibbotson and R. Sinquefield, Stocks, bonds, bills and inflation: the past (1926–1976) and the future (1977–2000). Financial Analysts Research Foundation, University of Virginia, Charlottesville, Va (1977).
7. B. Graham, J. Dodd and S. Cottle, *Security Analysis* (4th edn). McGraw-Hill, New York (1962).
8. R. A. Brealey, *An Introduction to Risk and Return from Common Stocks*. MIT Press, Cambridge, Mass. (1969).

APPENDIX A

Expectation in the Roulette Market

The concept of average performance is defined concisely by the expectation, which is given by a formula from probability theory:

$$E = P_1 o_1 + P_2 o_2 + \ldots,$$

E is the expectation, the Ps are the probabilities of all possible outcomes, and the os are the rewards attached to each outcome. Using this formula to calculate the expectation of "red" bets in roulette, we have

$$E = 18/38 \, (\$1) + 20/38 \, (-\$1) = -\$2/38.$$

APPENDIX B

Expectation in the Pari Mutuel Market

The expectation of a horse bettor, using the formula from Appendix A, is

$$E_i = P_i o_i + (1 - P_i)(-\$1),$$

E_i is the expectation of a gambler betting on horse i, P_i the true (and unknown) probability of winning for horse i, o_i the reward if it wins, $(1 - P_i)$ the probability that horse i loses, and \$1 the stake. The term o_i is identical with the odds to a dollar against horse i winning the race, or the dollar odds, for short. For example, if horse i is at odds of two to one (2/1), its winning will net its backers a profit of two dollars for every dollar bet. In terms of money bet,

$$o_i = \frac{M - m_i - f M}{m_i},$$

where M is the total amount of money wagered on all the horses in the race, m_i the money wagered on horse i, f the "take", and $f M$ the total deducted from the win pool for the track and state. The numerator is the money in the win pool available to be paid out as a profit to the people who bet on horse i.

The odds may be related to the probability of winning p_i as assigned by the public to horse i using

$$p_i = \frac{m_i}{M}.$$

This is just the fraction of money in the win pool bet on horse i. Combining the three formulas in this appendix, we get the expectation for bettors on horse i,

$$E_i = \frac{P_i}{p_i}(1 - f) - 1.$$

APPENDIX C

Expectation in the Stock Market

Expectation in the stock market is given by the yearly rate of interest compounded yearly based on the initial investment. Thus, P dollars invested for n years at an interest rate of $r\%$ compounded yearly will amount to A dollars as given by the compound interest formula,

$$A = P(1 + r)^n.$$

As an illustration, suppose we buy 100 shares of General Motors stock at \$55 per share and sell it 1 year later for \$60 per share. What is our return per dollar invested? The original investment is \$5600 (\$5500 for the stock plus \$100 commission). On the sale of the stock 1 year later, we receive \$6000 less \$100 in commissions plus \$380 in dividends paid by the company, or a net of \$6280. Substituting in the formula,

$$6280 = 5600 (1 + r)^1$$

and $r = 0.12$ or 12%. Every dollar invested in General Motors was worth \$1.12 1 year later. Averaging the rs for all transactions then gives the expectation.

Computers Math. Applic. Vol. 17, No. 4–6, pp. 671–695, 1989
Printed in Great Britain. All rights reserved

0097-4943/89 $3.00 + 0.00

SYMMETRIES IN MUSIC TEACHING

M. Apagyi

Apáczai Csere János Education Centre, Art School, P.O. Box 197, Pécs, H-7601, Hungary

Abstract—The study presents some simple possibilities of symmetrical structures of the same principle either in small or large pieces of music.

In part–whole relationships of musical manifestations there are symmetries in dynamics, rhythm, melody, harmony, tonality and form. It is one of the crucial tasks of music pedagogy to recognize and practice them in a creative way.

SYMMETRIES IN MUSIC TEACHING

The present paper touches upon some problems of symmetries in music, employing experiences obtained in teaching.

The most frequent types of symmetry in music:

- (a) bilateral (essentially, it is a reflection across a vertical or horizontal axis);
- (b) translational (it can involve shifting, reversing, or translocation which is motion implying repetition of a certain fundamental unit). Practically, it is the symmetry of recurrence resulting from shifting, the consequence of which is that the fundamental unit overlaps itself time and time again.
- (c) rotational (it means turning around across an axis with the end being the starting position).

Before pointing out symmetries in actual compositions, let us outline the possibilities of symmetrical construction of musical elements.

By musical element we mean tone as the base of every kind of sound. Tones are determined by their intensity, duration, pitch and timbre. All of them may constitute different symmetries as well.

The parameters mentioned above interrelated, and together they characterize tones. However, here we are discussing the possible symmetries formed by intensity, duration and pitch separately.

SYMMETRY OF INTENSITY

Intensity of tone is called dynamics, which has a wide range from the lowest to the loudest. It is influenced considerably by spatial situations, consequently a gradual dynamical increase makes us feel as if we are coming nearer, while a decrease makes us feel as if we are moving away.

(a) In Fig. 1 there is a gradual strengthening and softening, respectively, at continuous sounding of the same pitch. This process represents a reflection across a vertical axis, which, practically, results in bilateral symmetry.

(b) Bilateral symmetry is obtained by uniformly interrupted repetition (staccato) of tone of the same pitch, however, with different dynamics (Fig. 2).

(c) Gradual strengthening of tones of the same pitch and repetition of the units result in translational symmetry (Fig. 3). In this case, the three tones of intensifying dynamics make the fundamental unit.

(d) Multiple repetition of action denoted under (a) leads partly to translation, and partly to rotation (Fig. 4). The same can be illustrated in another figure to make rotation unambiguous (Fig. 5).

SYMMETRY OF DURATION—SYMMETRY OF RHYTHM

Duration of tones are denoted by their values, by which we can distinguish tones sounding for longer or shorter times. Depending on values of tones different rhythms can be formed.

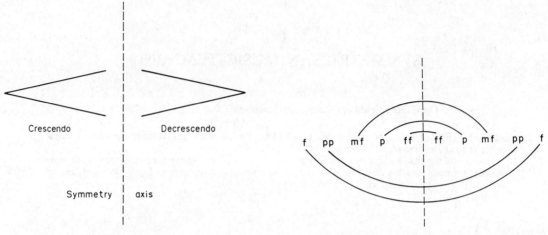

Fig. 1. Crescendo–decrescendo symmetry.   Fig. 2. Bilateral symmetry of different dynamics.

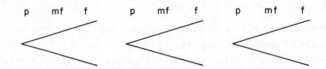

Fig. 3. Translational symmetry of gradual strengthening.

Fig. 4. Rotation of crescendo–decrescendo.

Symmetry can refer to both duration and rhythm.

(a) Tones of the same pitch and intensity with regular intervals can result in translation, where one tone makes up the fundamental unit (Fig. 6).

(b) Gradual acceleration and deceleration of tones of the same pitch and intensity creates bilateral symmetry. Spatial effect is unambiguous in this case (Fig. 7).

(c) Bilateral reflection of different rhythms (Fig. 8).

The example in Fig. 8 is taken from Minuet in C major by Mozart (namely its first four bars, which we will return to later). Mozart's minuet makes it obvious that, for example, ostinato (permanently repeated) rhythm means a translational symmetry.

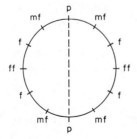

Fig. 5. Rotation of strengthening and softening illustrated by a ring.

Fig. 6. Translation of tones of same duration.

Fig. 7. Gradual acceleration and deceleration as bilateral symmetry.

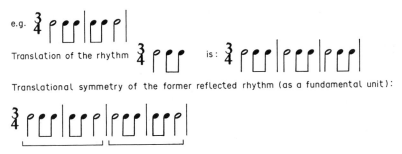

Fig. 8. Examples of bilateral and translational symmetries in rhythm.

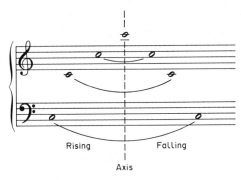

Fig. 9. Different pitches of the same tones as bilateral symmetry across a vertical axis.

PITCH SYMMETRY—MELODY SYMMETRY

Pitch is characteristic of tones, e.g. the order in pitch of a certain tone refers to octaves. Tones of different pitch make up melody. Pitch also denotes spatial relations, e.g. the terms high and low tone.

(a) Different pitches of the same tone as bilateral symmetry across a vertical axis are shown in Fig. 9. The reflection is formed by the retrograde-part and built on succession in time.

(b) Different pitches of the same tone as bilateral symmetry across a horizontal axis appear in Fig. 10. Here the point is spatial simultaneity of rising and falling; however, it can appear as a succession in time as well.

(c) Translation of different pitches of the same tone is illustrated in Fig. 11.

(d) Figure 12 shows rotation of different pitches of the same tone. In this case translation occurs as well. It can be observed that translation and rotation are often associated.

As an example of symmetries of intensity, rhythm and pitch we can use Prelude and Waltz in C from the series "*Games*" by György Kurtág (Fig. 13). Even this quite simple piece can prove that symmetric constructions manifest themselves in various ways, thus creating all of the musical form.

Fig. 10. Different pitches of the same tone as bilateral symmetry across a horizontal axis.

Fig. 11. Translation of different pitches of the
same tone.

Fig. 12. Rotation of different pitches of the same tone.

In Prelude (Libero) the symmetry of pitch is rather concealed, and can be traced back to the basic situation (Fig. 14).

The symmetry of Waltz (Guisto) is built on the basic situation introduced at Prelude, i.e. C′ is the centre of symmetry, which is emphasized by the composer at the very end. Starting from C′ and corresponding to each other, the octaves of contrary direction succeed each other gradually. The rhythm of Waltz has translational symmetry (Fig. 15) while its dynamics have a bilateral one (Fig. 16).

Fig. 13. Note of Prelude and Waltz in C by G. Kurtág.

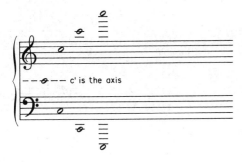

Fig. 14. Reflection of C tones across a horizontal axis.

Fig. 15. Crotchets—translational symmetry.

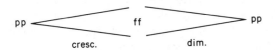

Fig. 16. Crescendo–diminution—bilateral symmetry.

Fig. 17. Reflection of tones of different pitch across a vertical axis.

(e) Reflection of tones of different pitch across a vertical axis (Fig. 17). The tones reflected results in a so-called "retrograde-part" and succession in time can be seen. It can be illustrated by the sixth piece of "Six unisono melodies" from volume one of "*Microcosmos*" by Bartók (Fig. 18). The order of the tones is reflected in a retrograde-like way from C″ which is denoted as the axis. We can see clearly how symmetry and form are related within one work.

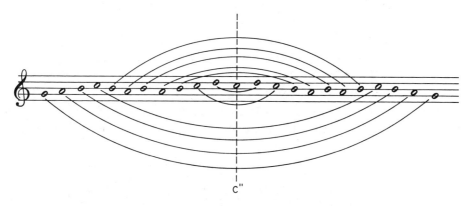

Fig. 18. Note of the sixth piece of "Six unisono melodies" from volume one of "*Microcosmos*" by Bartók and its figure of explanation.

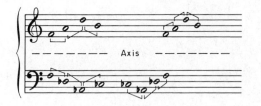

Fig. 19. Reflection of tones of different pitch across a
horizontal axis.

Fig. 20. Translation of tones of different pitch.

(f) Reflection of tones of different pitch across a horizontal axis (Fig. 19). The intervals of
the low part coincide with those of the high part, however, they are of opposite direction. Both
spatial simultaneity and succession in time are equally possible.

(g) Translation of tones of different pitch (Fig. 20). In both cases the question of how
translational symmetry is formed related to rotation emerges. In musical structures we can refer
to translational symmetry not only in the case of sequentia but also in the case of imitation, canon
and fugue as well. Apart from these cases, we should mention here variational genres—where
translation results from slight changes—and transposition—where the key of a piece of music is
changed. However, no detailed discussion of this subject can be given at this time.

(h) Rotation of tones of different pitch (Fig. 21). Here too, rotation and translation intertwine.

Now let us have a look at some pieces of music where reflection across a horizontal axis can
be found. The first is "*Leggiero*" by Lajos Papp. Its tonal system is a 1:5 scale model built on
D. Here the distance between the tones is measured in halves (Fig. 22).

The next example is taken from "*Games*" by Kurtág ("Palm exercise"). Instead of single tones,
here we find groups of tones. The place of the horizontal axis is obvious, as well as the central
reflexion beginning at *pp*: the right hand goes from white keys to black keys, the left hand proceeds
from black keys to white keys (♮0—white keys, #0—black keys). Here the rhythm becomes
inverted as well (Figs 23 and 24). In the same way, if we have a look at the note of "Hommage
á Paganini" (from "*Games*" by Kurtág) both the reflection across a horizontal axis and the
correspondence of parts of contrary direction at the beginning and the end of the composition can
be seen (Figs 25 and 26).

Examples of translational symmetry

First, Concerto (after Vivaldi) by Bach, a section from part three. Sequence is the characteristic
structural principle of the work: the first two bars are repeated dominantly, a new sequence is
started beginning from bar seven, and another one from bar 22 (Fig. 27).

In a section of part one of the same Bach Concerto symmetry is again created by sequences
(Fig. 28). Another example can be taken from a passage of "Toccata in D major" by Bach (Figs
29 and 30).

In the following examples the sequence of repetitions and shiftings, namely rotation and
translation, respectively, are intertwined:

Bach: Concerto (after Telemann), movement III, passage (Fig. 31)
Bach: Concerto (after unknown composer),
      movement II, passage (Fig. 32)
Bach: Toccata in D major, passage (Fig. 33).

(i) Reflection of tones of different pitch across horizontal and vertical axes. This combined
form of reflection is characteristic of dodecaphonic compositions. They are built on multiple

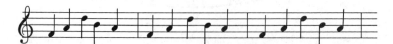

Fig. 21. Rotational tones of different pitch.

Fig. 22. Note of "*Leggiero*" by Lajos Papp.

Fig. 23. Short–long, long–short rhythm-figure of explanation.

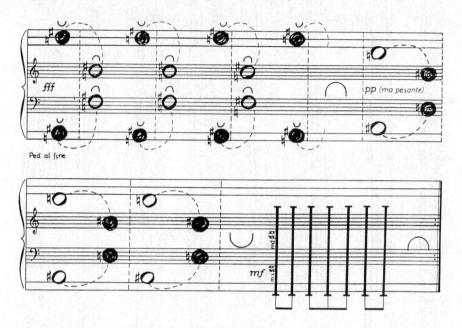

Fig. 24. Note of "Palm exercise" by G. Kurtág.

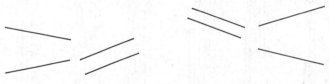

Fig. 25. Denotion of directions.

reflection of 12 scale Reihe, so we can see bilateral, central and translational symmetries as well. See Webern Variations for piano (Op. 27), movement I. Relations are denoted by numbers (Fig. 34).

Some pieces of the series for piano "Gradus" by Samuel Adler are also built on the Reihe technique. He created a table, to serve as a standby, in which he made each of the retrograde, inversion and retrograde inversion built on each tone of the basic line denoted as prime. In the first piece (8a) we can find the tones of the first prime line starting from F and those of the twelfth inversion-line built on C (Figs 35 and 36).

SYMMETRY OF HARMONY

Up to now, we have been discussing symmetries coming from tones as musical elements, which are related to intensity, duration, rhythm, pitch and melody.

However, one can often find a symmetric arrangement of harmonies in pieces of music. One of the most frequent pair in this respect is symmetric linkage of the tonic (T) and the dominant (D), which, in this case, means the I–V degrees and one of the inversions of V degree (Fig. 37).

Such an arrangement of harmonies can be observed in a great number of pieces of music. For example Sonata in C major (K 279), in G major (K 283), in D major (K 311) by Mozart; also in his *Fantasy* in D minor, or Minuet in C major; here we can enumerate *Ecossuise* in G major

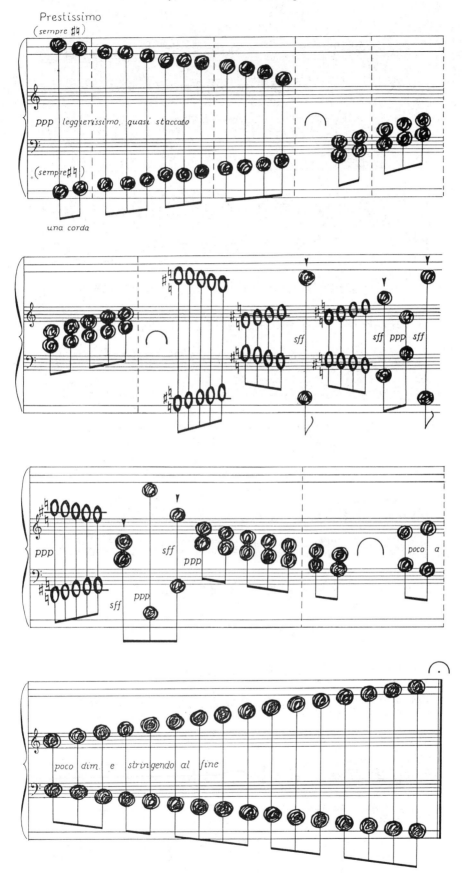

Fig. 26. Note of "Homage á Paganini" by Kurtág.

Fig. 27. Bach: Concerto, movement III, passage-note.

Fig. 28. Bach: Concerto, movement I, passage-note.

Fig. 29. Bach: Toccata, passage-note.

Translation of rhythmic units of the high part is transferred into the low part in bar 140, then in bar 151 it returns to the high part:

The rhythmic unit of the low part changes its place in the same way:

Fig. 30. Explanation with two short examples of note.

Fig. 31. Bach: Concerto, movement III, passage-note.

by Beethoven and *Ländler* in D major, *Ländler* in E flat major by Schubert. Of the above mentioned pieces, the last contains T–D functions only, and in its middle part one can find T D D T linkage (Fig. 38).

The harmony scheme of the piece is shown in Fig. 39. So, in this case, symmetry of harmony is in close relationship with symmetry of form.

SYMMETRY OF TONALITY

In tonal pieces of music, in which defined harmony functions (tonic, dominant, subdominant are connected, the starting mode often changes in the course of the piece, then it comes back. Schubert's latter *Ländler* starts from E flat major, then is modulated into B flat major, and again to the original key of E flat major (Fig. 40).

Modulation is possible not only into major key but into minor key as well; e.g. into the parallel minor. Schubert's *Letzte Walzer* in F major is modulated into the parallel minor: D minor. Its harmony scheme is shown in Fig. 41.

Fig. 32. Bach: Concerto, movement II, passage-note.

Fig. 33. Bach: Toccata, passage-note.

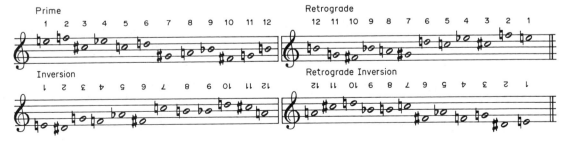

Fig. 34. Webern Variations for piano, Op. 27, movement I, first lines.

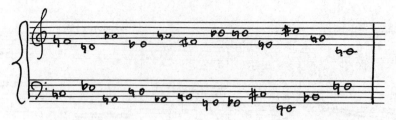

Fig. 35. Tones of the first prime line starting from F and tones of the inversion-line built on C.

8

	Prime →	1	2	3	4	5	6	7	8	9	10	11	12	← Retrograde	
Inversion ↓	1	F♮	D♭	A♭	E♭	A♮	F#	B♭	B♮	E♮	C#	G♮	C♮	1	
	2	A♭	F♮	B♮	F#	C♮	A♮	D♭	D♮	G♮	E♮	B♭	E♭	2	
	3	D♮	B♮	F♮	C♮	F#	D#	G♮	G#	C#	B♭	E♮	A♮	3	
	4	G♮	E♮	B♭	F♮	B♮	G#	C♮	C#	F#	D#	A♮	D♮	4	
	5	D♭	B♭	E♮	B♮	F♮	D♮	G♭	G♮	C♮	A♮	E♭	A♭	5	
	6	E♮	D♭	G♮	D♮	A♭	F♮	A♮	B♭	E♭	C♮	F#	B♮	6	
	7	C♮	A♮	E♭	B♭	E♮	D♭	F♮	G♭	B♮	A♭	D♮	G♮	7	
	8	B♮	A♭	D♮	A♮	E♭	C♮	E♮	F♮	B♭	G♮	D♭	G♭	8	
	9	F#	D#	A♮	E♮	B♭	G♮	B♮	C♮	F♮	D♮	G#	C#	9	
	10	A♮	F#	C♮	G♮	C#	B♭	D♮	D#	G#	F♮	B♮	E♮	10	
	11	E♭	C♮	G♭	D♭	G♮	E♮	A♭	A♮	D♮	B♮	F♮	B♭	11	
	12	B♭	G♮	D♭	A♮	D♮	B♮	E♭	E♮	A♮	G♭	C♮	F#	12 ↑	
		1	2	3	4	5	6	7	8	9	10	11	12	Retrograde Inversion	

Preliminary exercise

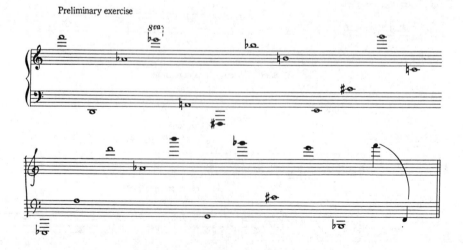

Fig. 36. Piece eight of "Gradus" by S. Adler—table, note of the work 8(a).

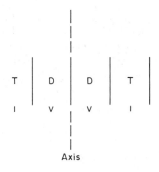

Fig. 37. Symmetry of harmony of tonic-dominant.

Fig. 38. Note of *Ländler* by Schubert.

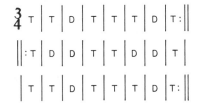

Fig. 39. Harmony scheme of *Ländler* by Schubert.

E flat major	B flat major	E flat major
Basic key	Dominant key	Basic key
8 bars	8 bars	8 bars

Fig. 40. Tonality scheme of *Ländler* by Schubert (example of tonality symmetry).

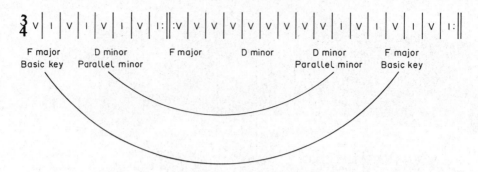

Fig. 41. Harmonies and tonality scheme of *Letzte Walzer* by Schubert.

Fig. 42. Note of *Letzte Walzer* by Schubert.

Fig. 43. Bach: Minuet in D minor its note and tonality scheme.

Fig. 44. Notes of two folk-songs.

The piece is characterized by translation symmetry as well, which emphasizes the middle part consisting of fifth (Fig. 42). In the following example we can see minor key which is modulated into its parallel major in the middle part Bach: Minuet (Fig. 43).

All these examples show that symmetries of tonality are in close relationship with the form of the pieces. In each case, however, we have focused on symmetry of tonality only.

SYMMETRY OF FORM

Four-line form

Symmetries are easy to recognize in the four-line Hungarian folk-songs. In particular, two of them are very common: A A^5 A^5 A and A B B A forms.

Translational symmetry is also characteristic of both folk-songs (Fig. 44).

One-unit form

Some pieces consist of a single period, thus creating a one-unit form often divided into a question and an answer. We can take as an example Unisono melody, the first part of *Microcosmos* by

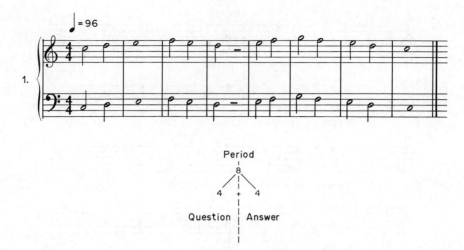

Fig. 45. Note of the piece one of Unisono melody by Bartók and denotion of symmetry of form.

Bartók. The complete piece is a period of eight bars, in which the four-bar question is answered by the other four bars. In this way equilibrium is created. Bartók called it "symmetric balanced structure" (Fig. 45).

Two-unit form

It consists of two mainly repeated periods. From the point of view of symmetry this is the most typical version. It is well illustrated by Mozart's Minuet in C major. Sixteen bars make up the pieces with eight bars in each of the two units. Both units can be divided into $2 + 2 + 4$ bars; in this respect the piece is asymmetric. Symmetry and asymmetry generally occur jointly, only their proportion differs. In some cases symmetry is more obvious; some other cases show asymmetry first of all (Fig. 46).

Fig. 46. Note of Minuet C major by Mozart.

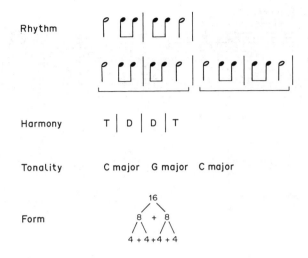

Fig. 47. Table of symmetries in Minuet in C major by Mozart.

Some of the symmetries in this piece relate only to certain details, others relate to the whole piece. Structure of the same principle, both in details and the whole, is demonstrated by the following example (Fig. 47).

Three-unit form

It consists of three sections; the third is either a repeat with no change or a variation. For example, the structure of *Ecossuise* in G major by Beethoven (Fig. 48).

Trio form

It is a multiple, recapitulational structure in which each unit is compound.

Movement *Menuetto al Rovescio* from Sonata No. 41 by Haydn serves as an example (Fig. 49).

The movement is characterized by different types of symmetry. Both in Menuetto and Trio the second period represents retrograde inversion of the first one. Bars corresponding to each other are denoted by numbers. Naturally, the melody is associated by symmetry of harmony, and symmetry characterizes the form as well (Fig. 50).

Rondo form

It consists of theme and interludes with a formula (Fig. 51).

We can mention two couperin-pieces: "*The Reeds*" and "*The Reapers*".

Movements of pieces with more than one movement can be arranged in a symmetrical way. A well-known example is Bach's "*Musikalisches Opfer*", in which the parts are arranged in a symmetry as follows: RICERCAR FIVE CANONS TRIOSONATA TRIOSONATA FIVE CANONS RICERCAR. (Separate study should be devoted to examination and analysis of Bach's

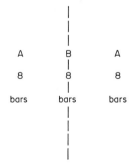

Fig. 48. Symmetry of forms in *Ecossuise* by Beethoven.

Fig. 49. Note of *Menuetto al Rovescio* by Haydn.

mastery to structure canons and fugues from the point of view of symmetry. Great variety of different symmetries can be found in his works and acme can be observed in the cycle "*Die Kunst der Fuge*".)

We can take an example of multimovement symmetry from the twentieth century. It is *Canticum Sacrum* by Stravinsky, which was first performed in St Marks cathedral in Venice. It has five movements, and some of the criticians found its structure the same as that of the cathedral having five domes, since the last movement of the work is nearly an exact retrograde-version of the first one. So its scheme is like the one shown in Fig. 52.

GOLDEN SECTION

When we speak of symmetry we must make mention of the golden section. It is one of the most typical ratios according to which the whole is to the major part as the major part to the minor one. This ratio in itself is asymmetric, however if positive and negative golden cuts are present together—being equalized—they make up a symmetry (Fig. 53).

Fig. 50. Figure of symmetry of form of the Haydn Minuet.

Fig. 51. Rondo form.

Fig. 52. Scheme of a cyclic piece. (Taken from the book *Stravinsky* by White.)

Fig. 53. Golden section—figure of explanation.

Fig. 54. Golden cuts of piece one of *Microcosmos* by Bartók.

Fig. 55. Note of Prelude 1 by Chopin.

In piece one of *Microcosmos* by Bartók the peak tones represent the intersections of golden cut, which can be found in bar three and after bar five, respectively (Fig. 54).

The peak tone in question is the F″ which is the intersection of negative golden cut, while that of the answer is the G″ which is the intersection of positive golden cut throughout the piece.

Another symmetry in golden section is that regularity is constantly repeated. For example in the case of Fibonacci-numbers: 1, 2, 3, 5, 8, 13, 21, 34, 55 . . . any new number in the sequence is made up of the preceeding two numbers: $1 + 2 = 3$; $2 + 3 = 5$; $3 + 5 = 8$ etc.

The next piece is Prelude 1 by Chopin. Here the golden section can be detected in dynamics and melody, based upon the number of bars. It has 34 bars starting with Mezzoforte; this continuously increases up to bar 21. Dynamics fortissimo in bar 21 represents the peak—here we find the highest note as well. From this point on both melody and dynamics are decreased (Figs 55 and 56).

The positive golden cut in this case represents an asymmetric ratio formally, as it stands alone. At the same time a translational symmetry can be found throughout the Prelude.

Another example of simultaneous presence of symmetry and asymmetry is taken from *Dance of Kites* in Vol. III of *Microcosmos* by Bartók (Fig. 57).

In all practicality it can be divided into three parts each consisting of eight bars. In the first eight bars four tones are reflected across a horizontal axis between two G tones (Fig. 58).

In the second unit the two constantly sounding tones are E′ and g sharp with five tones reflected (Fig. 59).

In bars 14–15 we can find the positive golden cut preparing for part three. Here, for a short time, a reflection across the vertical axis can be observed. Then in the last four bars reflection across the horizontal axis returns.

In the third unit the organ point is represented first by G″ and b, then by E″ and D″. At the end the two outside G tones (Fig. 60).

The relationship of the parts consisting of constantly sounding (pedal point) and changing tones in the three units (Fig. 61).

The starting position of the first unit reappears in the second one, however, in a contrary way. While the third unit unifies both versions on the pattern of thesis–antithesis–synthesis.

However, this unification is created not only by a simple repetition of the former two units but by means of elaboration as consummation, therefore ending in perfection of the musical material.

Supposingly, this rich variety of musical symmetries is not by chance, since all the tempored tonal system is built up symmetrically.

Lajos Bardos discussed this issue in details in one of his essays published in the periodal *Parlando* (February, 1979). (Figures are taken from his essay.)

White keys of a tempered piano make up the diatonic heptatonic scale, which can be presented as a perfect fifth as well:

F C G D A E H

Black keys make up the pentatonality:

F sharp/G flat C sharp/D flat G sharp/A flat
D sharp/E flat A sharp/B flat.

The two scales symbolized in circle of fifths make up the tempered dodecaphony (Fig. 62). Tones D and G sharp, opposite each other, denote two centres of symmetry.
Tone D is centre of symmetry of the diatonic scale (Fig. 63).
Tone G sharp (A flat) is the centre of symmetry of the pentatonic scale (Fig. 64).

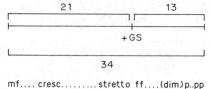

Fig. 56. Figure of explanation of Prelude 1 by Chopin.

Fig. 57. Bartók: *Dance of Kites*—note.

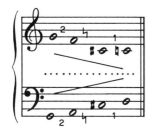

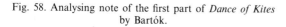

Fig. 58. Analysing note of the first part of *Dance of Kites* by Bartók.

Fig. 59. Analysing note of the second part of *Dance of Kites* by Bartók.

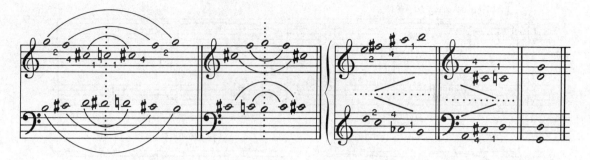

Fig. 60. Analysing note of the third part of *Dance of Kites* by Bartók.

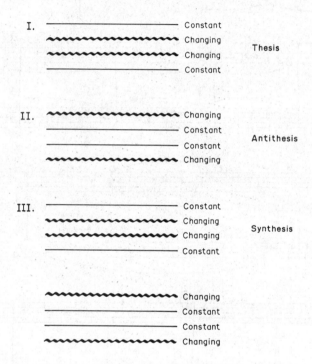

Fig. 61. Figure of explanation of *Dance of Kites* by Bartók.

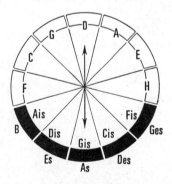

Fig. 62. Illustration in circle of fifths.

Fig. 63. Symmetry of diatonic scale—note.

Fig. 64. Symmetry of pentatonic scale—note.

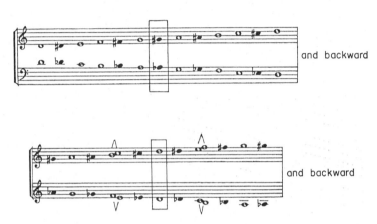

and backward

and backward

Fig. 65. Symmetry of chromatic scale—two notes.

From these, we may conclude that the chromatic scale is also symmetric in case of D and G sharp (Fig. 65).

The symmetry centre of the white keys is the tone G sharp, while D is that of the black keys.

The essay we are referring to does not deal with different types of symmetries of time and space. However, it takes conditions of time and space of music occurrences preconditioning each other. Within it, time symmetry is generally characterized by succession and reflection through a vertical axis; while simultaneity and reflection through a horizontal axis characterize space symmetry.

We could use numerous other examples of symmetries in music but the types presented above are the simplest occurrences.

Its importance is indicated by the fact that from the very first moment of education and training it occurs and reoccurs time by time as a basic problem in different ways, and examples can be found in pieces by each composer.

However it is very important that pupils, in the process of teaching, should understand and recognize the different structures including those other than symmetry. With this aim in mind, e.g. when learning types of symmetry it is very reasonable to take the possibilities enumerated in the study as a starting point and exercise them in improvisations and composing. The experience is that such kind of work and activity deeply influence pupils from emotional point of view.

Thus improvising, composing and interpreting jointly promote the progress in learning the tongue and bulk of music.

Computers Math. Applic. Vol. 17, No. 4–6, pp. 697–708, 1989
Printed in Great Britain. All rights reserved

UNCERTAINTY PRINCIPLE AND SYMMETRY IN METAPHORS

L. Bencze

Department of Modern Hungarian, Eötvös Loránd University, P.O. Box 105, Budapest, H-1364, Hungary

Abstract—Analogies taken from the humanities are common epistemological means in teaching and explaining phenomena in sciences. This paper, however, follows a more or less opposite approach by taking formal and functional analogies from the sciences (uncertainty principle and symmetry) in order to illustrate the phenomenon of metaphor in language in an attempt to get closer to its nature, behavior and function.

Interdisciplinarity characterizes current trends in sciences and humanities. This is not just a matter of fashion. Aristotle was still able to command all the sciences, and like a shepherd he could gather them as one flock into one sheepfold—philosophy. With Christianity and from the early Middle Ages onwards, scientific knowledge disintegrated into different branches of science and humanities in an accelerating way. This process has achieved an immensely high degree of specialization in the twentieth century. This trend has been accompanied by a stronger and stronger need for integration within each science as well as among all sciences [1, pp. 381–413]. This began as early as the twelfth century with the foundation of universities and was expressed in the name itself as the etymology shows [unus (—one) + verto (—to turn): universus (—turned into one, combined into a whole)].

One of the unifying principles of an integrated system of knowledge of man and his world seems to be the concept of symmetry [2].

For more than 20 years I have walked under Norman and Gothic vaults and arches, which day after day imposed the ideas of reflectional, rotational and translational symmetries on my mind (Figs 1 and 2) [3]. Tension and dynamism in my wife's designs and tapestries (and perhaps in our marriage) may be the manifestation of a sophisticated balance of dominating reflectional, translational and color symmetries and of their absence as well (Figs 3 and 4) [4, 5]. Translational symmetry was probably responsible for the boredom of musical training in my childhood as well as for the nice melodies I cannot forget [6]. For almost 30 years I have been studying and teaching problems of rhythm, rhyme [7] and metaphor in literature and language and have tried to find the reason for their aesthetic functions. Yet, it was only recently that some works reminded me that a common core and connecting principle of all the above may be symmetry [2, 8–10].

In relation to human life the phenomenon of symmetry can be compared to health. If one is healthy he is not interested in it, does not take care of it and does not feel it. The presence of health is something unnoticed in life and the absence of it, i.e. illness, is something which is noticed and which directs all our attention, aims and wishes to health. Symmetry in nature and art surrounds us permanently, more or less unconscious in us and unnoticed by us, while the lack of it calls our attention to its repletive omnipresence. Such is the linearity of time in human life, this unidirectional and irreversible passing and lethal asymmetry of the two endpoints in our lives—birth and death. This personal and irreversible process of the individual pushes the human mind towards the idea of a bilateral reflectional symmetry, i.e. to the hope and belief of another life to come with its antisymmetrical items of reward and punishment. According to this there was a paradise (lost) at the beginning of time and there will be a paradise (regained) at the end of time and the axis of reflection is a Messiah in the middle of time. His person is again a symmetry in itself, the incarnation of divinity (Deus → homo) and the deification of man (homo → Deus) [11].

If a linguist approaches these phenomena of symmetry exclusively from a linguistic point of view and deals with the meanings of expressions he says they are metaphors. A metaphor comprehends a "real life", of which the traditional name is *literal sense*, and an "uncertain life to come", called *figurative sense* [12, p. 3; 13, p. 552; 14].

Fig. 1. The nave of St Martin's Basilica at the Archabbey of Pannonhalma, Hungary. (*Photo*: A. Alapfy.)

Skipping a vast number of definitions for metaphor [15, pp. 300–331; 16, 17], I quote here a few examples beginning with Aristotle:

"Metaphor is the transference of a strange name",

in Bywater's translation,

"Metaphor consists in giving the thing a name that belongs to something else" [18, 19];

and another one from I. A. Richards, a great twentieth century scholar of metaphor research:

"In the simplest formulation when we use a metaphor we have two thoughts of different things active together and supported by a single word, or phrase, whose meaning is a resultant of their interaction" [20, p. 93];

and two from two encyclopedias:

"Metaphor. A condensed verbal relation in which an idea, image, or symbol may, by the presence of one or more other ideas, images, or symbols, be enhanced in vividness, complexity or breadth of implication" [21]; "By common definition and by etymology a metaphor is a transfer of meaning both in intension and extension" [22].

Consequently, the term *metaphor* in its etymological and wider sense means every transference in language, i.e. tropes (Cicero: verborum immutatio—changing of words) [13, pp. 552, 553, 556] while traditionally its meaning has also been restricted to transference based on similitude, e.g. "he

Fig. 2. The splendid vaulting of St Benedict's Chapel at the Archabbey of Pannonhalma, Hungary. (*Photo*: A. Alapfy.)

is a lion", which means he is as brave, as strong etc. as a lion (Cicero: verbum translatum—transference of words) [13, pp. 552, 553, 556].

One of the most common and ridiculous superstitions of the Western civilization has been that metaphors (or tropes in general) are to adorn style [20, p. 90; 23, p. 359]. It is a tradition which may be traced back to a superficial reading of Quintilian and a fatal misunderstanding as a consequence of it: "... *there are some* (i.e. tropes) which are intended solely to the purpose of embellishment" (*eruntque quidam* tantum ad speciem accomodati) [24, p. VII. VI. 5] (present author's italics). Several times he makes clear distinctions between tropes, which are to express meaning(!) and those which are to ornament speech: "... some are for the sake of meaning, others for the sake of decoration" (...quosdam gratia significationis, quosdam decoris assumi) [24, p. VIII. VI. 2]; "... tropes employed to express our meaning involve ornament as well, though

Fig. 3. *Struggle for Light*; tapestry, wool, 150 × 150 cm. Designed and woven by S. Örsi, 1978.

Fig. 4. *The Seventh Door*; tapestry, wool, 122 × 172 cm. Designed and woven by S. Örsi, 1985.

the converse is not the case" (... qui significandi gratia adhibentur esse et ornatum, sed non idem accidet contra); "... to make our meaning clearer ... or to produce decorative effect..." (quia significantius est aut quia decentius) [24, p. VIII. VI. 6].

Though Quintilian was interested in the type which decorates speech, he was still aware of the other type as well. In doing so he relied on Aristotle, for whom the information-giving nature of metaphor was of vital importance [25]:

> "We will begin by remarking that we all naturally find it agreeable to get hold of new ideas easily: words express ideas, and therefore those words are the most agreeable that enable us to get hold of new ideas. Now strange words simply puzzle us; ordinary words convey only what we know already; it is from metaphor that we can best get hold of something fresh." [26]

In any case metaphor has been a "disturbing enigma" for scholars [12, p. 13] since Aristotle. There are at least three reasons for this:

(1) nature of metaphor;
(2) birth and behavior of metaphorical relations in speech/text;
(3) effect, function and relation of metaphor concerning human personality (thinking, emotions, instincts etc.).

According to different approaches of various disciplines to metaphor, theories have been labeled as comparison (Quintilian), analogy (Aristotle), interaction (Black, Richards), improper usage (Locke, Wittgenstein), opposition, logical absurdity (Beardsley), matter of emotion (Carnap), intuition (Wheelwright), substitution, similarity, juxtaposition, identity, tension, collision, fusion, deviance, anomaly, mistake etc. [12, p. 3; 15, pp. 300–331; 22, 27]

This simple list of the technical terms of theories refers to the essence of metaphor: "the literal meaning does not disappear" [12, p. 13] but goes hand in hand with the figurative meaning in every moment (cf. comparison *of two*, interaction *between two*, tension *between two* etc.). The result is a tension in our mind, an oscillation [15, p. 313]. Is it a matter of empirical falsity or semantic anomaly [34], of relation of denotata and/or of significata [29, p. 50], of intension and/or of extension [22]? When one says "Peter is a lion", he sets up a contradiction according to the rules of traditional logic, as he transgresses logical categories: "Peter is a man", "A lion is a lower animal", consequently "A man is a lower animal"! Yet we have no problem in understanding the sentence "Peter is a lion". Suddenly we recognize common features in *Peter* and in the *lion* (tertium comparationis) [19, pp. 90–92; 30, pp. 27; 38; 153]; (ground) [20, p. 93; 28, 35, pp. 25–47, 33]. We do not care about all the potential features of the two meanings, but simply make some of the virtual features actual (*brave, strong* etc.) [35, p. 44]. At this stage another problem arises: which features are actualized? *Brave, strong* or some others or both? Do we delete all of the potential and some of the virtual features? [15, pp. 302, 314]. At once we realize that the interpretation of metaphor is not totally uncertain, yet it is not as certain as the sentence "Peter is six feet tall" can be true. In addition we feel it is not just a matter of intuition vs reasonable thinking or experience. We know that our interpretation of a metaphor depends on context, situation, culture and education, personal age and historical period, just like everyday classifying (see below) [36]. That is why for speakers of European languages "my ducky", for example, may or may not be a nice thing to say to a woman but "my little elephant" certainly is not (cf. the meanings of *head* in English, French, German and Hungarian) [37].

The phenomenon of transgression of existing categories is common both in poetic metaphors as well as in new inventions in sciences. If one says "A wolf is an animal" he will classify *animal* as genus and *wolf* as a species somewhere under the genus *animal*. But if he says "A man is a wolf" he will not follow everyday or scientific biological experience and we could say again that he has violated the logical rules of classifying or a kind of tabula Porphyriana based on it [38, pp. 202–250, 229]. Everyday and scientific classifying themselves may also be totally different without being really illogical. Everyday classifying, not unlike metaphorical classifying, may depend on age, education, culture and social status, period, genre and mental condition (see above).

When my 3-year-old daughter consistently called every animal, even a fish on the kitchen table, *bow-vow* and every plant and flower *kertyschoo*, the naming was funny yet the classifying perfect, i.e. animal kingdom, vegetable kingdom [15, pp. 300–331; 39–45]. Similarly, when in Hebrew both an eagle and a bee were called *oph*, that categorizing seems naive to us at first glance because we translate *oph* as *bird* in our system of categories. The ancient Hebrew system was unlike ours however. The meaning of *oph* was "flying being with wings" and thus, the classifying was again perfect [15, p. 317]. The same can be found in the distinction of *meat* and *fish* (*caro* vs *piscis*) in Catholic moral theology from the ancient times throughout the Middle Ages up until modern times. From the viewpoint of fast (Lent), common sense and even local tradition were authoritative in deciding what was meat and what was fish [46]. Thus, every animal that lived and breathed on the surface of the earth was looked at as *meat*—mammals, birds etc. Every animal that lived in and around water was *fish*. Therefore, during Lent people were allowed to eat frogs, turtles, cockles, shellfish, crayfish, beavers, wild ducks, crakes, seagulls etc. This distinction, which included even mammals in the term *fish* had nothing to do with science. Yet, there is a clear logic in it: the place of life [1, p. 402]. Something similar happens in verbal jokes as well [47].

As far as metaphorical language is concerned it provides a more holistic view of man and his ideas in everyday life even though it is not precise in a scientific way, not to mention the fact that it was the unusual associations and unexpected relations which helped develop natural sciences as well, as pointed out in the story of Newton's apple. Every invention or new idea in science is a kind of rebellion against and a violation of existing categories just like the language of children, poets and early civilization myths. In each case there is something in common: the introduction of new categories as a result of a new system of classifying. Therefore, a metaphor is much more than simply a break in the semantic isotopy of a text [15, p. 312; 48–50]. "No advance was possible in the intellectual life of man without metaphor." [51]

The importance and the problems of metaphor which I have tried to sketch above may be responsible for the two extreme views on interpretation: one says metaphors make language totally obscure, while the other says they can be analyzed as exactly as facts in sciences [15, p. 323]. In an attempt to reach a more differentiated understanding of the problem of metaphor I introduced the uncertainty principle to the question in 1981 [52].

Then I suggested that every manifestation of a human being, first of all an artistic one, is *potentia* which is actualized, i.e. interpreted, by the recipient, another human being (in case of language). Consequently, what a man says is always polyvalent. It is the circumstances (who, what, to whom, where, why etc.) that make his utterance more or less definite. Let us try to compare metaphorical language with scientific language. The latter tends to be accurate, so that it expresses extremely little with one word and thus achieves exactness and totally excludes ambiguity. Metaphors, on the other hand, tell us a lot, but inexplicitly, ambiguously and in an undefined manner. Abstract language is subsidiary and artificial. Metaphors are not simply there to decorate language but to achieve conciseness, serve totality in cognition and reduce the number of signs. For instance, the wording of the label on a bottle of Tokay wine "Rex vinorum et vinum regum" (Wine of kings and king of wines) could be scientifically explained, at least in theory, but the process would fill books. Even in chemistry and physics approximations are often the only practicable way. Thus, for example, to make an exact calculation and provide a description of all possible wave functions of an iron atom would fill an entire library and that of a uranium atom would require more paper and ink than there is matter in the solar system. Or take, for example, the term "electron cloud" in the description of atomic structures which is one of the most telling metaphors among the many used in natural sciences [53]. It is important in my analogies here that metaphors are always "cloudy" in common experience [35, p. 47].

There is an uncertainty in the question of metaphor as well. We know exactly the two "names" (*man–wolf*), yet we cannot link them adequately (Fig. 5). Therefore, a literary analysis or any everyday interpretation cannot come near anything like a mathematical or scientific one (except in the case of the uncertainty principle of physics), which totally eliminates ambiguity and establishes equivalences and equations. Any interpretation must retain some degree of uncertainty. It sets only the limits within which ambiguity may exist and alternatives (fierce, hostile etc.) may be chosen. It offers end values (*man* and *wolf*) and the oscillation of mind can spring into existence in any channel (fierce, hostile etc.) between them. Consequently, the number of intepretations is not unlimited [37]. Ambiguity and tension remain within certain limits. For example in the famous metaphorical saying "Homo homini lupus"—"Man is to man a wolf" Plautus: Trinummus II.4 [54, 38, pp. 202–250, 229], the channel can be wild, hostile, inimical, fierce etc., but obviously not

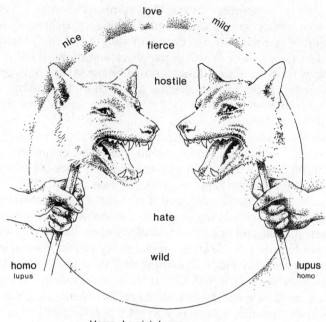

Homo homini lupus.
Homo homini aut deus, aut lupus.

Fig. 5. (Drawing: M. Barabás.)

mild, friendly, serving, adoring etc. Otherwise it would perhaps say: "Homo homini deus"—"Man is to man a god", as Erasmus quotes it as a proverb: "Homo homini aut deus, aut lupus". The oscillation of mind here seems to correspond to connections between different parts of the brain. A metaphor makes the whole man react. In metaphorical thinking, the scholastic principle can be altered in the following way: "Nihil est in intellectu quod simul non sit in anima, corpore et sensibus"—Nothing is in the mind that is not simultaneously in the soul, body and senses (see below also).

This can be illustrated by an analogy from chemistry, too. In a molecule, which is constructed of atoms, we know that there are certain electrons which are shared between atoms and that there are others which belong to individual atoms, at least to a good degree of approximation. In a metaphor, we also know which items of meaning can be common (hostile, fierce etc.) and which cannot (mild, nice etc.), but we can never be certain of the actual common one or ones (hostile and/or fierce etc. and/or both and/or others). This depends on the individual, his age, education etc. [38, p. 219]. So we must take into consideration both bonding and nonbonding electrons, using the molecular language vs features of meaning [10]. It seems much easier and certainly more clear-cut to describe the affinity and repulsion of electron pairs with one another than those in the meaning of a metaphorical expression. Wolves may be fierce both toward other animals and toward each other, may quarrel over prey and with each other etc. These relations are also defined much more by education, common sense etc. than biology. The two meanings in the metaphor remain the same, just as the two nuclei remain unchanged in the chemical bond, whereas the relationship between the two meanings is subject to various conditions, just as the bonding between the two atoms can be described, at best, by a probability distribution function of the electrons, and is more sensitive to changes on various conditions than the nuclear positions. While we know the number of chemical bonds in a molecule, we do not even know the number of features in a metaphor. We can speak of positions of greater and smaller symmetries and this is why, for example, "my little elephant" can be nice in one language and culture and insulting in another. The problem of certainty and uncertainty in the nature of metaphor can be revealed more closely if we take the concept of symmetry into consideration in a more detailed way.

It is all the more necessary that we do this because metaphor was discovered as an example of symmetry as early as, at least, Shubnikov and Koptsik, though they share the common European

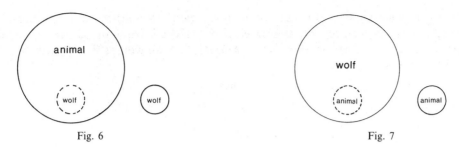

Fig. 6   Fig. 7

opinion of a narrow interpretation of the use of metaphor and restrict it to poetic language (cf. the remarks above on Quintilian):

"A specific feature of poetry such as the metaphoric content of its language develops within a unified scheme of groups of projective transformations. Writing the Aristotelian metaphor in the form of ratio

$$\frac{\text{What age is}}{\text{for life}} \quad \frac{\text{so evening is}}{\text{for day}}$$

we find that other tropes (poetic comparisons and contrasts) are formed in an analogous manner." [23, p. 359]

The concept of antisymmetry in grammatical metaphors as the unusual and parallel use of Hungarian plural suffixes was raised by Fónagy [15, p. 310].

Scientific investigations of metaphor go back to Aristotle, just as the concept of symmetry goes back to Greek thinking. The word ἡ συμμετρία had various meanings, the most important of which were: commensurability, due proportion, symmetry, one of the characteristics of beauty and goodness, fixed proportion, suitable relation, convenient size and harmony of life [55].

In today's literature of symmetry a kind of wider sense corresponds to the Greek meanings: "A broader interpretation allows us to talk about degrees of symmetry, to say that something is more symmetrical than something else" [10]. In this sense we "call objects equal in relation to some particular feature if both objects possess this feature" [23, p. 1] and thus "we introduce the idea of two objects being more or less equal" [23, p. 1]. This type of symmetry is called "material symmetry" [9] and applies to metaphor while the so-called "geometrical symmetry", "as a special kind of geometric law" [23, pp. 4, 2] and geometric regularity does not [4, 10, 56]. I use the terms bilateral, rotational, translational and color symmetries as well as asymmetry in the sense in which they are generally accepted and established in the literature [2, 10, 23, p. 359; 57–60].

If we consider the symmetry of the sentence "A wolf is an animal" from a logical point of view, i.e. we categorize the *animal* as genus, the *wolf* as a species of it, we shall find reflectional symmetry and asymmetry in their extensional relation as the set of animals includes the set of wolves (Fig. 6). A similar symmetry and asymmetry can be found in their intensional relation but in a reciprocal way, as the set of features of a wolf includes a set of animal features (Fig. 7). This symmetry and asymmetry is also present in a simple sketch of the dichotomic structure or binary oppositions of the so-called Tabula Porphyriana (or Arbor Porphyrianus):

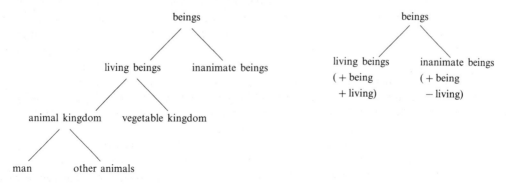

The symmetry is in the feature *being* which exists in both; the asymmetry, or at best antisymmetry (traditionally called *differentia specifica*), is in the features *living–inanimate*, with the feature *living* existing in one and missing in the other etc. This is why it could also be looked at as antisymmetry.

Now let us see what happens if we say "A man is a wolf". We disregard the system of Tabula Porphyriana in this sentence, regardless of whether our tabula is based on science or is determined by culture, language etc. (see the examples of *fish*, *bird* etc. above), as if *wolf* were a genus and *man* a species of it:

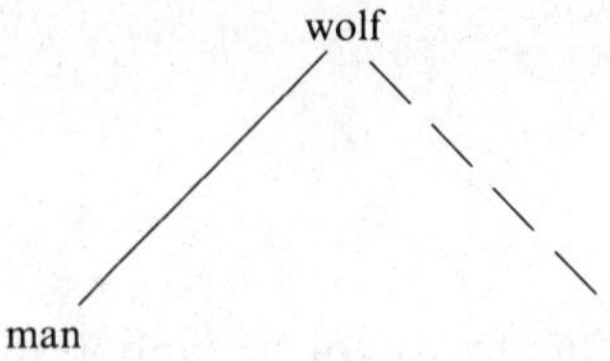

while according to the common system, *man* and *wolf* are both species somewhere under different generi, as can be seen quite clearly even in a mere sketch:

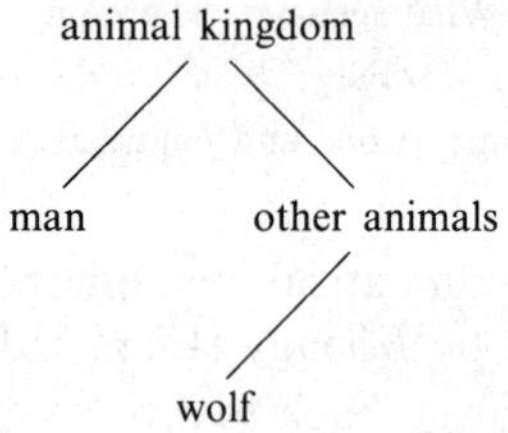

Yet, we cannot say that the sentence "A man is a wolf" is simply illogical and violates the rules of thinking and of language because we—in a European culture—can understand it without difficulty. The solution to the problem, I think, is in the existence of the above-mentioned uncertainty in the symmetry and asymmetry of the meanings in the sentence, not to mention that it is symmetrical and asymmetrical even in its acoustic or written form:

A man is a wolf

where the copula *is* is the axis of reflection or the border of translation. Metaphorical relation is a kind of mirror which is polished semantically, socially and culturally (cf. the example "little elephant" above).

Now let us suppose that the two meanings (*man–wolf*) are two different and amorphous objects opposite each other. They can both be turned in space separately in every direction. Both have certain parts on their surfaces which are identical or similar either in shape or color (fierce, hostile etc.). Consequently, if we turn them continuously there will be certain moments and stages when identical or similar parts face each other (fierce–fierce etc.) or are back to back. The two objects (man–wolf) in and of themselves are asymmetric, but in the above-mentioned moments and stages symmetries arise between parts. This is when we discover common features of *man* and *wolf* (hostile, fierce etc.). The probability of such turnings depends on the number of identical or similar parts, on the speed and direction of the turnings and the cleverness of the person who is turning them (cf. the problem of the Rubik cube, and see Fig. 8). If one is clever, the probability of symmetrical stages in an interval grows and the frequency and probability of the returning of such stages will be more or less stabilized (cf. the problem of certainty and uncertainty of metaphor as discussed above). This means that a clever person can identify more common features of *man* and *wolf* more quickly than others. We can make this play more complicated and refined and put a lens between the objects. In this case the work of the lens will also depend on the distance of the objects from the lens and on the nature of lens itself. Such lenses can be culture, education, age etc., as above. Therefore, let us add omnidirectional motion in space to the earlier rotation of the two objects. Then theoretically there will be more possible stages of symmetry as the lens can enlarge or reduce shapes, i.e. a square of this size □, for example, can be symmetrically adequate with this size □ etc. (see Fig. 9). If the lens in question happens to be a fish-eye then a shape ○ can be symmetrical with ⬭. That is why "my little elephant" can be nice—depending on the cultural lens.

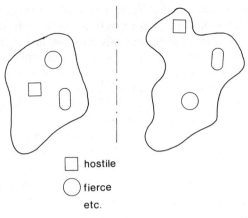

□ hostile

○ fierce

etc.

Fig. 8

The two meanings in metaphor are amorphous in comparison with one another. Yet, there are more or less symmetrical items (fierce, hostile etc.) in them while others are asymmetrical. The existence of items belong to the possibility of the objective, external world (cf. *potentia* above). Whether they appear or are thought or recognized as symmetries (cf. *actual*ization above) depends on two main groups of conditions: one is the rotational and omnidirectional motion, which is the brightness and cleverness of the speaker; the other is the mirror or lens/lenses and the objects, the type of lens, the type of amorphous object—which is the individual's language, culture, education, age, historical period, social status, context, situation etc. (cf. the "circumstantiae" in classical rhetoric) [13, pp. 91, 139, 377, 399 etc.]. These result various values which can be subdivided as aesthetics, knowledge-invention, emotional tension, humor, poeticity etc. The two main factors (motions and lens + object) define the borders of metaphor–nonmetaphor in an exact and precise way. Within these borders one stage, one moment (*fierce* or *hostile* or *wild* etc.), i.e. the appearance of metaphor/symmetry as metaphor/symmetry, is more or less uncertain.

By means of this model the nature, behavior and effect of metaphor can be described—I am convinced—more precisely and exactly than ever. It comprehends several different characteristics of metaphor which have been emphasized separately by various approaches, and provides an explanation of the tension, oscillation, cognitive and emotional procedures, invention and poeticity, and the role of science and art in our mind and life. It also alludes to the relationship between the external world and the world in language (cf. Petőfi's TeSWest theory: text structure—world structure) [61], which is a kind of symmetry as complicated as symmetry in metaphor. Besides, it reminds us of the external, formal symmetry of the shape of neocortex, which is the material source of the metaphorical operations discussed above. There may also be a functional connection between metaphoric thinking and the structure and working of the brain.

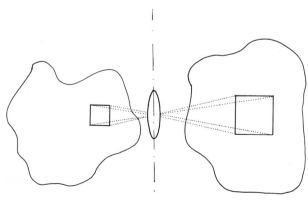

Fig. 9

Since Sperry's experiments the research of left and right hemispheres has become fashionable [15, p. 324; 62, 63]. It is widely known that the left hemisphere is the center of abstraction and speech, while the right hemisphere is the center of music and iconic thinking. It is also well-known that there has been a contradiction in localizing certain functions in the brain. Lashley's experiments seemed to prove there is no special place of memory in the brain, while Penfield succeeded in connecting certain senses to certain neurons [62, pp. 201–204]. If certain parts of the neocortex are cut out, other parts may take over their functions—primarily in childhood. This contradiction and uncertainty of definite localization and the possibility of the changing place of functions seems to correspond to the unpredictability in metaphorizing which is, at the same time, predictable within certain limits in its special symmetry, as we have seen above. The brain as a whole (including the limbic system, the center of emotions and other parts) is the same in every man, though not regarding the level of neurons and their connections. A certain connection is established within one man's brain but not in another's, though the neural possibilities are more or less equally given (cf. determination by age, education, culture etc. above). If the neural connection is produced, it is not sure that the neural route is the same. The length and number of intermediate connections on it can be different and can exist in parallel [64]. This also seems to correspond to the special uncertainty and symmetry of semantic features in metaphor which can be produced by various rotations etc. in our model. Consequently, both the nature of the brain and the nature of the metaphorizing mind include a kind of risk, not unlike human life itself [65, pp. 1, 9], which is not limitless in either case (see the introductory remarks to this paper). Degrees of risk in finding symmetry in language and art consist of a wide range of chances (cf. Figs 1–4). The use of metaphor in language—as I hope I have proven—is a "carefully calculated risk" [65, p. 94] as metaphor connects known and unknown as a means of requiring a knowledge of unknown via known [26, 38, p. 217], art and inventions are also adventures into an unknown part of the world with the purpose of making it known. In the symmetry of metaphor—one could say—the unknown is reflected by the known. At the same time the concept and existence of symmetry itself ensures the limits of risk. This anxiety and hope is expressed by Black and Boyd:

"No doubt metaphors are dangerous—and perhaps especially so in philosophy. But a prohibition against their use would be a willful and harmful restriction upon our power of inquiry." [35, p. 47];
"The use of metaphor is one of many devices available to the scientific community to accomplish the task of accommodation of language to the casual structure of the world" [66].

This accommodation by metaphors belongs to the most routine of human acts, so much so that it is taken to extremes even when used without similarity, e.g. in the physics of quarks where there are technical terms like "naked charm state" and "naked bottom state" [1, pp. 381–413; 67]. That is why I think Petőfi is perfectly right in saying that "normal and figurative messages can be handled in the same way" (see also the everyday and scientific classifying above) [68]. I also hope that the introduction of the correlation of the uncertainty principle and symmetry in metaphor may be fruitful and stimulating for modeling facts in natural sciences. The risk of certainty and uncertainty, the struggle to reduce it and keep it to a limited extent, to evaluate and ramify it in metaphor has removed the concept of metaphorical symmetry from the geometrical one. Yet, the key to the enigma of metaphor may also be the uncertainty principle in symmetry.

REFERENCES

1. L. Bencze, Conscious tradition, unconscious construction or subconscious metaphors? Certain levels of text cohesion and coherence. In *Text Connexity, Text Coherence. Aspects, Methods, Results* (Ed. E. Sözer). Buske, Hamburg (1985).
2. I. Hargittai, *Szimmetria egy Kémikus Szemével (Symmetry Through the Eyes of a Chemist)*. Akadémiai Kiadó, Budapest (1983).
3. L. A. Cummings, A recurring geometrical pattern in the early renaissance imagination. *Comput. Math. Applic.* **12B,** 981–997. Reprinted in *Symmetry: Unifying Human Understanding* (Ed. I. Hargittai). Pergamon Press, Oxford (1986).
4. Gy. Dóczi, Seen and unseen symmetries: a picture essay. *Comput. Math. Applic.* **12B,** 39–62. Reprinted in *Symmetry: Unifying Human Understanding* (Ed. I. Hargittai). Pergamon Press, Oxford (1986).
5. V. Molnár and F. Molnár, Symmetry-making and breaking in visual art. *Comput. Math. Applic.* **12B,** 291–301. Reprinted in *Symmetry: Unifying Human Understanding* (Ed. I. Hargittai). Pergamon Press, Oxford (1986).
6. R. Donnini, The visualization of music: symmetry and asymmetry. *Comput. Math. Applic.* **12B,** 435–463. Reprinted in *Symmetry: Unifying Human Understanding* (Ed. I. Hargittai). Pergamon Press, Oxford (1986).
7. B. Pavlović, On symmetry and asymmetry in literature. *Comput. Math. Applic.* **12B,** 197–227. Reprinted in *Symmetry: Unifying Human Understanding* (Ed. I. Hargittai). Pergamon Press, Oxford (1986).
8. I. Hargittai, Degas' dancers: an illustration for rotational isomers. *J. chem. Educ.* **60**(2), 94 (1983).

9. I. Hargittai (Ed.), Limits of perfection. *Comput. Math. Applic.* **12B**, 1–17. Reprinted in *Symmetry: Unifying Human Understanding*. Pergamon Press, Oxford (1986).
10. I. Hargittai and M. Hargittai, *Symmetry Through the Eyes of a Chemist*. VCH, Weinheim (1986).
11. Pope Leo the Great, 1. Sermo in Nativitate Domini. In *Breviarium Congregationis Sancti Mauri O.S.B. Pars Hyemalis*, pp. 230–231. Vindobonae, Ex Typogr. Mechitaristarum (1842).
12. J. J. A. Mooij, *A Study of Metaphor. On the Nature of Metaphorical Expressions, with Special Reference to Their Reference*. North-Holland, Amsterdam (1976).
13. H. Lausberg, *Handbuch der literarischen Rhetorik. Eine Grundlegung der Literaturwissenschaft*, Band I–II. *Zweite, durch einen Nachtrag vermehrte Auflage*. Hueber, München (1973).
14. J. P. van Noppen (Ed), *Metaphor and Religion*. Vrije Univ., Brussels (1983).
15. I. Fónagy, Metafora. In *Világirodalmi Lexikon*, Vol. 8. Akadémiai Kiadó, Budapest (1982).
16. W. A. Shibles, *Metaphor: an Annotated Bibliography and History*. The Language Press, Whitewater, Wisc. (1971).
17. J. P. van Noppen (Ed.), *Metaphor: a Bibliography of Post-1970 Publications*. Benjamin, Amsterdam (1985).
18. I. Bywater, Aristotle, *De Poetica*. In *The Works of Aristotle Translated into English under the Editorship of W. D. Ross*, p. 1457b. Clarendon Press, Oxford (1946/1952).
19. *Aristotle, The Poetics. With an English Translation by W. H. Fyfe*, p. 1457b. Harvard Univ. Press, Cambridge, Mass./Heinemann, London (1960).
20. I. A. Richards, *The Philosophy of Rhetoric*. Oxford Univ. Press, New York (1936/1950).
21. A. Preminger (Ed.), *Encyclopedia of Poetry and Poetics*, p. 490. Princeton Univ. Press, N.J. (1965).
22. P. Edwards (Chief Ed.), *The Encyclopedia of Philosophy*, Vol. 5, pp. 284–289. Macmillan, New York/The Free Press, London (1967).
23. A. V. Shubnikov and V. A. Koptsik, *Symmetry in Science and Art*. Plenum Press, New York (1974).
24. Quintilian, *The Institutio Oratoria of Quintilian. With an English Translation by M. E. Butler*, Vols I–II. Harvard Univ. Press, Cambridge, Mass/Heinemann, London (1953/1960).
25. L. Bencze, A metafora meghatározása és helye az arisztotelészi fogalomrendszerben (On the place and definition of metaphor in Aristotle's conceptual system). *Filol. Közl.* **29**, 1–18, 273–284 (1983).
26. W. R. Roberts, Aristotle, *Rhetorica*. In *The Works of Aristotle Translated into English under the Editorship of W. D. Ross*, p. 1410b. The Clarendon Press, Oxford (1946/1952).
27. M. Black, Metaphor. *Proc. Aristotelian Soc. New Ser.* **55**, 273–294 (1954–1955).
28. W. Abraham, *A Linguistic Approach to Metaphor*. Peter de Ridder Press, Lisse (1975).
29. H. Kubczak, *Die Metapher*. Carl Winter Univ. Heidelberg (1978).
30. S. R. Levin, *The Semantics of Metaphor*. The Johns Hopkins Univ. Press, Baltimore, Md (1977).
31. J. R. Searle, Metaphor. In *Expression and Meaning. Studies in the Theory of Speech Acts*, pp. 77–116. Cambridge Univ. Press, Cambs. (1979).
32. A. Ortony (Ed.), *Metaphor and Thought*. Cambridge Univ. Press, Cambs. (1979).
33. P. Recoeur, *The Rule of Metaphor*. Routledge & Kegan Paul, London (1978).
34. R. E. Sanders, Aspects of figurative language. *Linguistics* **96**, 56–100, 60 (1973).
35. M. Black, Metaphor. In *Models and Metaphors. Studies in Language and Philosophy*. Cornell Univ. Press, Ithaca, N.Y. (1962).
36. I. Fónagy, Reported speech in French and Hungarian. In *Direct and Indirect Speech* (Ed. F. Coulmas), pp. 255–309. de Gruyter, Berlin (1986).
37. I. Fónagy, Nyelvek a nyelvben (Languages in language). *Ált. Nyelv. Tanulm.* **12**, 61–105 (1978).
38. G. A. Miller, Images and models, similes and metaphors. In *Metaphor and Thought* (Ed. A. Ortony). Cambridge Univ. Press, Cambs. (1979).
39. B. Leondar, Metaphor and infant cognition. *Poetics* **4**, 273–278 (1975).
40. M. C. Beardsley, *Aesthetics. Problems in the Philosophy of Criticism*. Harcourt-Brace, New York (1958).
41. J. Bernicot, L'étude expérimentale des métaphores. *Année psychol.* **81**, 465–484 (1981).
42. R. Brown, *Words and Things*. The Free Press, Glencoe, Ill. (1958).
43. R. Brown, *A First Language: The Early Stages*. Harvard Univ. Press, Cambridge, Mass. (1973).
44. J. Piaget, *Six Psychological Studies* (Translated by A. Tenzer). Random House, New York (1967).
45. B. Kaplan, Radical metaphor. Aesthetic and the origin of language. *Rev. exist. Psychol. Psychiat.* **2**, 75–84 (1962).
46. H. Noldin, *Summa Theologiae Moralis iuxta Codicem Iuris Canonici*, XXVII edn (Recognovit et emendavit A. Schmitt); Vol II, *De praeceptis Dei et Ecclesiae. Oeniponte–Lipsiae*, p. III.I.II.I. 677b (1941).
47. I. Fónagy, He is only joking: joke, metaphor and language development. In *Hungarian General Linguistics* (Ed. F. Kiefer), pp. 31–108. Benjamin, Amsterdam (1982).
48. A. J. Greimas, *Sémantique Structurale*. Larousse, Paris (1966).
49. J. Dubois, Fr. Edeline, Ph. Minguet, J. M. Klinkenberg, Fr. Pire and H. Trinon, *Rhétorique Générale*, p. 220. Larousse, Paris (1970).
50. J. Dubois, Fr. Edeline, J. M. Klinkenberg and Ph. Minguet, *Rhétorique de la Poésie*, p. 88. Ed. Complexes, Brussels (1977).
51. Müller (Cited by W. A. Shibles), *An Analysis of Metaphor in the Light of W. M. Urban's Theories*, p. 127. Mouton, The Hague (1971).
52. L. Bencze, On the metaphorical animal. *Annls Univ. Scient. bpest. Rolando Eötvös nominatae. Sectio ling.* **12**, 215–222 (1981).
53. E. R. MacCormac, Meaning variance and metaphor. *Br. J. phil. Sci.* **22**, 145–159 (1971).
54. M. Black, More about metaphor. In *Metaphor and Thought* (Ed. A. Ontony). Cambridge Univ. Press, Cambs. (1979).
55. H. G. Liddell and R. Scott, *A Greek–English Lexicon*. Clarendon Press, Oxford (1953).
56. A. L. Mackay, But what is symmetry? *Comput. Math. Applic.* **12B**, 19–20. Reprinted in *Symmetry: Unifying Human Understanding* (Ed. I. Hargittai). Pergamon Press, Oxford (1986).
57. J. Brandmüller and R. Claus, Symmetry. Its significance in science and art. *Interdiscipl. Sci. Rev.* **7**, 296–308 (1982).
58. J. Rosen, *Symmetry Discovered. Concepts and Applications in Nature and Science*. Cambridge Univ. Press, Cambs. (1975).

59. N. Shejkov, *Leben und Symmetrie*. Urania, Leipzig (1982). Hungarian translation: *Élet és Szimmetria*. Gondolat, Budapest (1987).]
60. H. Weyl, *Symmetry*. Princeton Univ. Press, N.J. (1952).
61. J. S. Petőfi, Thematisierung der Rezeption metaphorischer Texte in einer Texttheorie. *Poetics* **15,** 289–310 (1975).
62. St Rose, *The Conscious Brain*. Weidenfeld & Nicolson, London (1973).
63. C. Sagan, *The Dragons of Eden. Speculations on the Evolution of Human Intelligence*. Random House, New York (1977).
64. B. Gulyás (Ed.), The Brain–Mind Problem., *Philosophical and Neurophysiological Approaches*. Leuven Univ. Press, Leuven (1987).
65. N. Rescher, *Risk, A Philosophical Introduction to the Theory of Risk Evaluation and Management*. Univ. Press of America, Lanham, Md (1983).
66. R. Boyd, Metaphor and theory change. In *Metaphor and Thought* (Ed. A. Ortony), pp. 356–408, 358–359. Cambridge Univ. Press, Cambs. (1979).
67. E. R. MacCormac, Religious metaphors: linguistic expressions of cognitive processes. In *Metaphor and Religion* (Ed. J. P. van Noppen), pp. 47–70, 61. Vrije Univ., Brussels (1983).
68. J. S. Petőfi, Metaphors in everyday communication, in scientific, biblical, and literary texts. In *Metaphor and Religion* (Ed. J. P. van Noppen), pp. 149–179, 289. Vrije Univ., Brussels (1983).

Computers Math. Applic. Vol. 17, No. 4–6, pp. 709–713, 1989
Printed in Great Britain. All rights reserved

MODERN SYMMETRY

A. Hill

Department of Mathematics, University College London, Gower Street, London WC1E 6BT, England

Abstract—In previous studies [1, 2], the author—writing primarily as a geometric abstract artist—has attempted an approach to symmetry/asymmetry from a phenomenological point of view taking mathematics to be regarded as "...the theoretical phenomenology of structure" [3].

The main ideas of what is accepted as the mathematical treatment of symmetry have an extremely long history. It is not even clear that the study originates with the ancient Greeks; suffice to say it can be loosely regarded as a part of geometry and as such is therefore one of the earliest forms of "science" we can find. What we can say is that so far the earliest example of a "system" remains the books of Euclid, and it is generally accepted that the priority here was the methodology rather than the individual theorems. However, although research may change this, "Euclid" is a cornerstone in Western mathematics, and the "elements" include topics not restricted to quantitative and "metrical" geometry.

As a system or branch of mathematics the part we call plane geometry is closed, there are no more theorems to be discovered therein. However, this is not true of many topics we can regard as initiated by "Euclid" such as projective geometry and polytope theory. If one is interested in symmetry, the implications of the "classical concept", we find that mathematicians and also physical scientists—particularly chemists—were initiating a more abstract concept of symmetry which nevertheless could be seen as the result of contemplating fundamental features of very simple "structures". It is tempting to call this concept topological symmetry, but the term has not gained any currency despite the fact that the argument, if naive, is neither illogical nor a solecism. The notion will suggest that what is called symmetry and asymmetry can exist as features of a connected structure which remain invariant under certain simple deformations such that the feature then can be regarded as strictly qualitative and independent of quantitative considerations, thus belonging to the elastic geometry—"elastic lines" as in "rubber sheet" geometry, as topology is often described in the literature of popularizing (science and mathematics).

What are known as the five Platonic solids were generally conceived as literal "solids", the forms known as the tetrahedron, cube, octahedron, dodecahedron, icosahedron, most generally conceived as crystallographic or volumic "sculptural" forms. Essentially of course they can be seen as another set of forms, and perhaps the earliest well-known examples are the drawings of Leonardo da Vinci in which they appear as "skeletal" forms. These representations were the step which led to the geometry of the solids being represented "schematically" so that the prism-shaped "limbs" ("edges") of Leonardo's "closed lattices" could be replaced by the lines ("wires" in the case of a model) or the pencil lines on paper of the linear models. The final stage in this development had to wait for the spirit of the modern/abstract way of conceiving structures—we finally arrive at the Schlegel diagram in which the lengths of the lines and the area of the faces no longer carry over the "symmetry" of the figures represented. Finally it becomes immaterial whether the "lines" follow any regular feature, i.e. they need not be "straight", and in drawing them with "curves" these curved lines may in fact be uniquely different. We have "joined up" or connected a set of points just as we please, and what is drawn can still be regarded as pertaining to some form of physical structure; the wayward paths or "connectives" can be "seen" as elastic strings each of which seems to have undergone a unique deformation as if the "elasticity" of each line were intrinsically different or unique.

It is often pointed out as strange that the ancient Greeks did not notice the fundamental qualitative law which holds between the relation of the dimensional elements of a polyhedral structure, now referred to as points (zero-dimensional)—"corners"; lines (one-dimensional)—"edges"; and polygons (two-dimensional) or "faces". It seems Descartes was almost able to grasp the notion but it was Euler by whose name the famous theorem is known, not surprisingly since Euler was a founding father of topology and with his theorem a vast edifice of theorems having to do with connectivity was initiated.

If we look at the newer form of geometry as the study of structural features of "amorphous" or informal linear structures, schematic diagrams of degrees of connectivity, what we then say of the five Platonic structures is that each of the respective sets of elements is unidentifiable, interchangeable; we have a structure of utmost regularity—and redundancy—belonging to the set of regular coverings in two-dimensional space: plane tessellations either infinite—the three lattices made respectively from three-, four- and six-sided cells or polygons—or closed, as in the five closed systems which exhaust the possibilities for a closed system. When we ask: what else is there? the answer is that we can exhibit structures all of whose respective elements are distinguishable and permit no interchanging. Such structures are described as asymmetric. Finally we show that a structure may have some of its respective elements interchangeable while some remain identifiable; we don't call these "both symmetric and asymmetric" but say that they are symmetric by virtue of exhibiting some symmetries.

It was only in the last century that the daunting task of enumerating all possible polyhedral structures attracted the attention of mathematicians. While much has been learnt between the first efforts and today, no-one is very confident that the problem is going to suddenly become easy and in due course solved. From the point of view of symmetry we discover that when we ask about the possibilities the answer is that polyhedra with any symmetry at all fade out of the "catalogue" as the "size" gets greater, i.e. as the number of vertices (points or corners), edges (or lines) and faces increases. So, nearly all polyhedra are asymmetric! There are then three sets of symmetric polyhedra: the famous Platonics, a mere five; the no less famous Aristotelian solids exhibiting an almost equally high degree of symmetry, of which there are 13, another set revered, and rightly so, by the ancient greeks; and lastly an infinite but diminishing set which struggle for existence, as one might put it, exhibiting various "degrees of symmetry".

By contrast we can construct (or "exhibit") a family of linear structures which, conversely, are linear all symmetric. This family did not "exist" until it was "invented" at the turn of the century. The family or form is known as a *tree* and we can confidently add that it was there all the time but had not been looked at mathematically.

A tree, like the Schlegel diagram and maps, is a topologically linear structure, one-dimensional since it consists of lines and points, but so connected that there are no closed areas—loops, polygons, circuits, faces . . . The enumeration of classes or families of trees has been solved although the overall problem continued to look rather intractable until recently. Amongst symmetric trees there is one special family or species which can be defined such that, however large, they will always be symmetric. Such trees, while being quite simple structures, have as yet no simple short name and are known as *homeomorphically irreducible*. The instruction goes like this: your tree must not contain points of degree two, which means that each point connects at least three lines or only one line, the latter being called the terminal points of the tree. For example, some capital letters of the Roman alphabet are trees of this kind—T, Y, X, G, H, K—while others are not, having points of degree two—A, E, F, L, M, N, V, W, Z.

Let us look at the following question of "pattern making"—it can be looked upon as purely mathematical or as belonging to gestalt theory or even aesthetics: we wish to partition circles with homeomorphically irreducible trees—we will call them *hitrees*. If we take each hitree in turn and

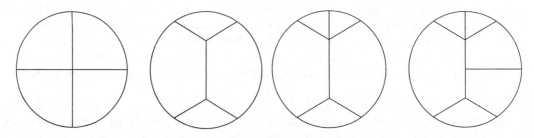

Fig. 1

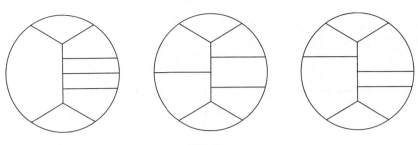

Fig. 2

use it as a form of partition we soon discover that we can obtain several distinguishable partitions from the same hitree. In mathematical terms we say that a tree may take a distinct number of embeddings and this is a feature of its symmetry. Thus a tree may permit only one embedding, as shown in Fig. 1 (and here we keep to hitrees). While others will clearly allow more than one (see Fig. 2).

If we were to exhaust all such patterns or partitions, up to say hitrees with 20 lines, we would soon discover that an increasing number of them would be asymmetric despite the fact that all of the hitrees employed and all that we may choose are symmetric.

Now it should come as no surprise that each pattern (or partition) is in fact a representation of a polyhedron, if we replace the circle by a polygon so that the terminal points of the hitress are joined by straight lines or edges we have changed nothing but the new "diagram" or "figure" can be taken as a Schlegel diagram. Thus, there exists a species of polyhedron generated in this manner: mathematically they would be described as having a *homeomorphically irreducible spanning tree*.

Returning to our plane patterns or partitions (or maps) we could ask that the circle be replaced by a rectangle—for example a square; let us also specify that the lines of the tree partitioning our square are to be parallel with its sides. Our pattern consists of horizontal and vertical lines only, thus the cells or areas of the partition will be "rectangles". In this special family of partitions the recognizable (but not topologically) distinct partitions of the circle with "orthogonally embedded" trees produces many more possibilities and these are of course increased when the circle is replaced by the square (see Fig. 3).

Whereas we can find only two for one of our trees when the surround is a circle, as soon as the circle is replaced by the square it becomes obvious that the number is more than double and the reader can see that it is not hard to reproduce the other four.

Elsewhere I have discussed the fact that these orthogonal partitions constitute the most characteristic compositional schemes of the abstract paintings of Piet Mondrian, certainly between the years 1918–44. The computer scientist Frieder Nake [4] and I were able to propose how to enumerate all possible "Mondrians" in the rectangular format. This can be extended to deal with the lozengical format which Mondrian frequently adopted, and this in turn comes up with some surprising results.

The example on the right-hand side is in fact the scheme chosen for what must be one of Mondrian's most strikingly simple compositions as the painting consists of just the two intersecting black lines (or in the hitree) on a white lozengical format. Of course what the viewer is confronted

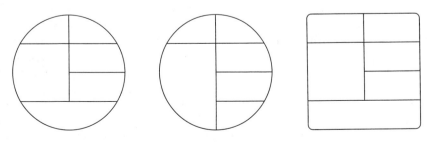

Fig. 3

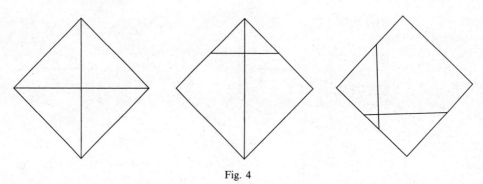

Fig. 4

with contains many other features, the lozenge appears not to be a perfect square, the lines are not of exactly the same thickness, and when we examine the resulting polygons it is clear that even if the lines were of the same thickness the arrangement has no metrical symmetry, the two areas adjacent to the two sides of the triangular area are not of the same size, and it follows that the remaining area is not "symmetric". Bisecting the triangle and extending the "axis" helps one see this to be the case. One need hardly add that depending where the point of intersection of the lines (the four degree node of the tree) is placed the artist could choose a great number of possibilities by which he could in this manner partition the lozenge.

Despite his importance for modern art as one of the most respected "geometric" abstract painters and a founder of abstract art, Mondrian took no interest or inspiration in any aspect of mathematics, not the time-honoured golden section nor any other formula; the idea of calculation was inimical to him; his works, although extremely rigorous and perfectionist, resemble more the free toccata than the canonical fugue. It is often stressed in art historical exegesis that he belonged to the mystical and spiritual stream; whatever truth there may be in that his essential importance is that of a radical plastician, perhaps the very last *painter*, a constructive painter no longer relating to *La Belle Peinture* and the continuous rhetoric of painting as it had been from Lascaux to Van Gogh. His ideas of space, time, surface, structure came out of cubism and took painting—as with other great innovators of the time—to the position of the tableau object, the autonomous plastic art work. For the initiated modern artist there is no turning back from this arrival point: it is the watershed.

To the artists who followed in the wake of neo-plasticism (Mondrian and the group known as De Stijl)—pioneer constructivism, and the less messianic formalism inaugurated by Jakobson and his school, it is indeed the sciences and mathematics, although not exclusively, which provide a continuing inspiration and link with the scientific ethos as opposed to movements which seek for an identity in such areas a "automatism", the mystic, the unconscious, all somehow part of the modernist thrust along with the ubiquitous expressionism, not to mention the stereotypic "humanism" which is set to optimize the conventional image and icon which characterize the work of the work-a-day artist.

Essentially modernism is not wholly identified with the formalist direction, it indeed recognizes its complement viz. the irrational, the subversive and anarchistic as best demonstrated by the iconoclastic dadaists and the montage pieces by the constructivists. It sees an end to the slothful conventions whereby art is to be equated with what is generally accepted as being such if only because it is done in artists' studios, old conventions jacked up by the inclusion of some modern terms of plastic grammar and syntax, the attempt to project a modern art in the terms of the old—all of this can safely be abandoned. The modern artist has within his grasp modern science, modern mathematics, modern concepts due to various other disciplines, and by relating to these things—although this ensures no guarantee—he continues, paradoxically if you like, the tradition whereby the artist while being an individual (perhaps even a solipsist) works in a context a large part of which parallels and reflects the thinking of the new age. No longer is he the servant of the Church, nor need he replace Church by State (Marxism), but equally no longer is he trapped in the labyrinth of egotistic romanticism which leads to excesses such as self-expressionism, nor of course need he mock the artists of the past by peddling an uncountable variety of "modernistic" formulations of the secular traditions in art; portraiture, "landscape", "still life"—all destined to be a decadent charade which all too easily finds admirers.

Some abstract artists have favoured symmetry and made great works which espouse the notion; one thinks of Brancusi's *Endless Column*. Others, like Mondrian, have strenuously avoided symmetry. They may not have known of the words of the celebrated French biologist Claude Bernard which can be rendered as: "It is the asymmetric that creates life."

The idea of asymmetry has too often been relegated to the areas of the non-important, the non-beautiful, even if it is understood that in nature perfect symmetry is never to be found, only some form of approximation to the mathematical ideal.

Clearly the artist is free to choose—there is both symmetry and asymmetry. However, let us end by stating that in asymmetry the idea of an approximation is rather meaningless. To put it technically, if a connected structure has an automorphism group characterized as being the identity class this means that it has the feature that all its elements are distinguishable; this is a special and most fundamental condition—certainly if regarded from the point of view of information theory, it represents an "absolute" as soon as this can be demonstrated. What is fascinating is that no algorithms exist for determining whether a given structure is asymmetric or not. In the case of large structures it becomes necessary to painfully apply the procedures to determine the automorphism number. Significantly it is in chemistry that such information may prove crucial, and much effort has gone into the research, also undertaken by mathematicians. To that extent the problem remains one of a series that have yet to be solved; the answer will have practical rewards and no less in mathematics a most profound step will also have been taken.

Which subsumes which, the demonstrable metric concept of symmetry/asymmetry or the "abstract" concept of the automorphism group? The latter is equally demonstrable and in order to furnish a proof one is forced to state a piecemeal account of the "neighbourhood situation" of every point and this has to be done by a sequence of simple observations; it is a giant piece of "micro-checking", as one might put it, although one is not dealing in the real micro-world, just the zero-dimensional implications of a strucutre of at most three dimensions. *Generalization* being one of the key strategies in mathematics—"generalize it up to the next dimension"—one can quickly grasp that the concept of symmetry/asymmetry poses many difficult questions. Some we may require for solving recondite problems, others remain more like nightmare chess problems. It is unlikely that art can contribute to this daunting area, as it once did in the Renaissance, but there is no "logical" reason why not since the notions of symmetry/asymmetry belong, in a sense, to both science and art.

REFERENCES

1. A. Hill, Art and Mathesis—Mondrian's structures. *Leonardo* **1,** 3 (1968).
2. A. Hill, *Program:Paragram:Structure DATA—Directions in Art, Theory and Aesthetics.* Faber & Faber, London (1968).
3. P. Bernays, Comments on Wittgenstein's philosophy of mathematics. Ratio II No. I (1959).
4. F. Nake, *Asthetik als Informations-Verarbeitung. 6.3. Mondrian Strukturell-Topologisch.* Springer, New York (1974).

Computers Math. Applic. Vol. 17, No. 4–6, pp. 715–730, 1989
Printed in Great Britain. All rights reserved

0097-4943/89 $3.00 + 0.00
Copyright © 1989 Pergamon Press plc

SYMMETRY AND TECHNOLOGY IN ORNAMENTAL ART OF OLD HUNGARIANS AND AVAR-ONOGURIANS FROM THE ARCHAEOLOGICAL FINDS OF THE CARPATHIAN BASIN, SEVENTH TO TENTH CENTURY A.D.

Sz. Bérczi

Department of General Technics, Eötvös Loránd University, Rákóczi ut 5, Budapest, H-1088, Hungary

Abstract—Earlier investigations [1] have revealed that mirror double frieze types were used in the ornamental art of the Old Hungarians, on archaeological finds from the tenth century A.D. On archaeological finds from the Avar-Onogurian times (seventh to ninth century A.D.) a rich variation set of double frieze patterns has been discovered by the author. Over mirror reflection, half turn and simple translation also occur as doubling operations of ornamental friezes on belt buckles, the main finds of the Avar-Onogurian archaeology. After the elaboration of the mathematical background of all kinds of double friezes built from a simple element, the possible technological roots of these ornamental inventions have been discussed and their relations to the plane symmetry patterns summarized. The suggestion, that double friezes were cut out of plane symmetry patterns made mainly by weaving makes the technological origin of these ornamental structures probable.

The structural classification of ornamental adornations has been used in a comparison of different cultural communities. Both the frequency of occurrence of different frieze, double frieze and plane symmetry structures and the inventive developments of basic (natural) patterns are characteristic to cultural communities. Examples of the survival of structures and the distribution of m–g structure from the supposed Pontusian Greeko-Scythian source region have been shown.

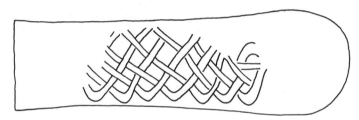

Weaving pattern on a belt buckle from Avar-Onogurian archaeology. Courtesy of Èva Garam, Hungarian National Museum Budapest. (Tiszafüred-Majoros, 577 grave.)

INTRODUCTION

Ornaments from different parts of the world show that inventions were always present with man. In the organization of the ornament we can find old discoveries about the structure of arrangements on the plane, but these discoveries are hidden in the colors and rich forms of folk art or in sculpturings and carvings on metal, bone or pottery relics of ancient peoples among the findings in archaeology.

A tool is necesssary to decipher these discoveries: the concepts and language of geometry, the oldest branch of mathematics. Although the constructors did not understand mathematics, they did do it. How?

Ornaments are built up from repeating discrete congruent elements in a regular way. There is a strong concept to determine and classify the variable appearance of structures generally called regular. This concept is *Symmetry*. The concept had been developed originally by the Greeks with the meaning: proportional, commensurable (the whole to its parts). Today symmetry is an operation which can be carried out on a regular pattern without causing any changes in the pattern. Symmetry may rearrange some or all of the building congruent elements, but keeps the pattern unchanged as a whole. So symmetry preserves the pattern in its state, leaves the structure invariant. Therefore symmetry belongs to the pattern, it is a characteristic of the structure. The best known symmetry is the external mirror symmetry of many animals and man. This symmetry frequently characterizes folk art arrangements [2] like that in Fig. 1, which is a life-tree scene, preserved from

Mirror-symmetric molecule from the
Hungarian folk art

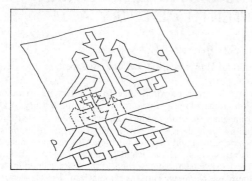

Congruency by overturning (or by
mirror reflection)

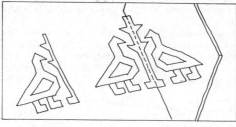

Reproduction from its half part by
mirror reflection (or by overturning)

Repeating of the mirror-symmetric
molecule anywhere on the plane by a
mirror reflection (or by overturning)

Fig. 1. Mirror-symmetric structure from Hungarian folk art and three different characteristics of this
structure with respect to the meaning of the symmetry operations.

old myths all over the world. Not only one symmetry belongs to a regular pattern, but a family
of them, which is called a group; the symmetry group of the pattern. The structure of the
ornamental pattern can be given by its symmetry group in which different symmetries of the pattern
are inter-related. But the symmetry group can be built up by only some of the symmetries of the
group, because of these inter-relations. The generators of different symmetry groups on the plane
may be translations, rotations, reflections and their combinations. In the case of the generation
of discrete frieze groups (which is the name of regular discrete arrangements of elements along the
line), translation, mirror reflection, glide reflection, half turn and some of their combinations
determine seven different types. They are given in Fig. 2 with their generators, simple patterns,
natural patterns visible in everyday phenomena and with an ornament of the Old Hungarians from
the Carpathian Basin, ninth to tenth century A.D.

OLD HUNGARIAN ORNAMENTAL ART

I began structural studies using the symmetry principle on the archaeological finds from the Old
Hungarians. They are characterized by palmette motifs on the level of form-elements. On the level

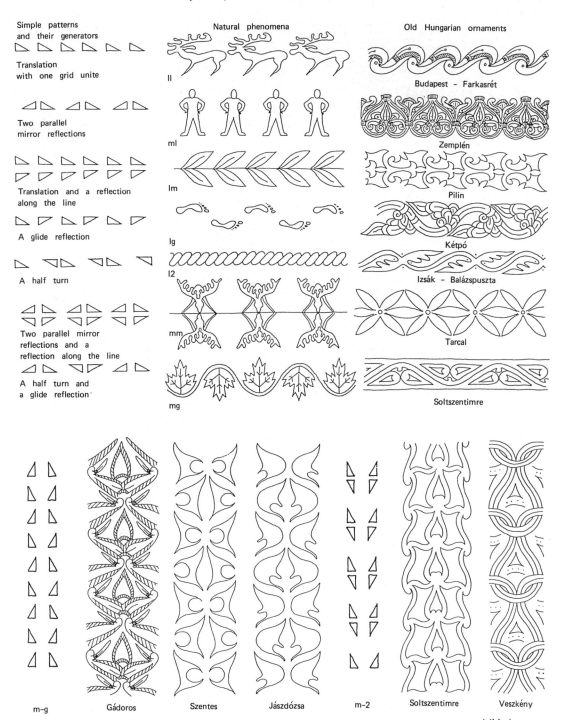

Fig. 2. *Upper half:* classical frieze groups given by their simple patterns, generators, patterns visible in everyday phenomena and their representation in Old Hungarian ornamental art (the archaeological sites are also given). *Lower half:* the double friezes discovered in Old Hungarian art are given by their simple patterns and some representatives (the archaeological sites are also given).

of structures, all seven symmetry types appeared in ornaments but I recognized more complex patterns, too. If I consider the patterns of friezes as threads, these more complex patterns are double threads. This was the point in the investigation when I began to suspect intuitive mathematical developments in ornamental art. Because the seven frieze patterns appear in nature, but the double frieze patterns do not. Double friezes stand halfway between friezes and plane symmetry groups (which are sometimes called wallpaper groups after the regular patterns of tapestries). Mathematics

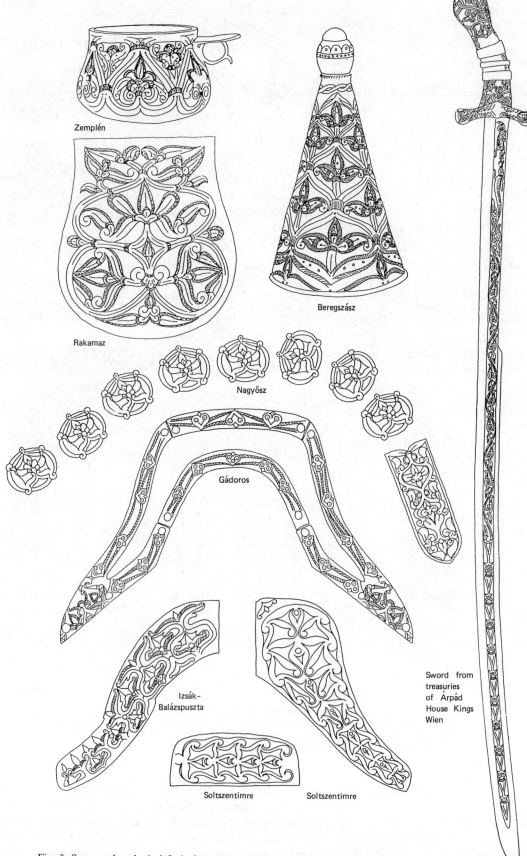

Zemplén

Rakamaz

Beregszász

Nagyősz

Gádoros

Izsák–
Balázspuszta

Soltszentimre   Soltszentimre

Sword from
treasuries
of Árpád
House Kings
Wien

Fig. 3. Some archaeological finds from the Hungarian Conquest, ninth to tenth century A.D., in the
Carpathian Basin showing the rich ornamental art of the Old Hungarians.

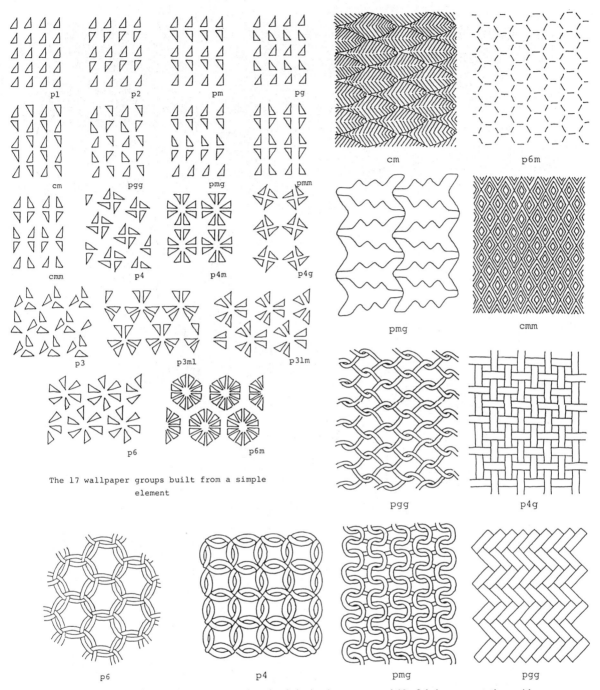

The 17 wallpaper groups built from a simple element

Fig. 4. The 17 plane symmetry groups given by their simple patterns and 10 of their representations with patterns from natural phenomena (*cm, p6m, cmm*) and from technologies (fur-fitting—*pmg*—from the Ethnographical Museum of Budapest, the curtain is from Hanti art, Western Siberia; woven structures— *pgg, p4g, pmg, pgg*; and chained ring structures—*p6, p4*). Compare them with the double friezes in Fig. 7.

also gives the whole set of different regular arrangements of the plane on the level of invariance by symmetry transformations [3]. These 17 plane symmetry groups are given in Fig. 3 by their simple patterns. Some of them can be found in nature, e.g. the structure of feathers on birds (*cm* type), the structure of a regular honeycomb (*p6m* type) or the structure of the skin of reptiles (*cmm* type), but more of them can be discovered in technologies. For example, optimal fitting of furs results in a *pmg* pattern, so it is only indirectly present in nature; its discovery is forced by conditions, the necessity to find extreme solutions in combinations of elements [4]. This is the case for different weaving and thread, or chain-connecting, technologies, as shown in Fig. 4.

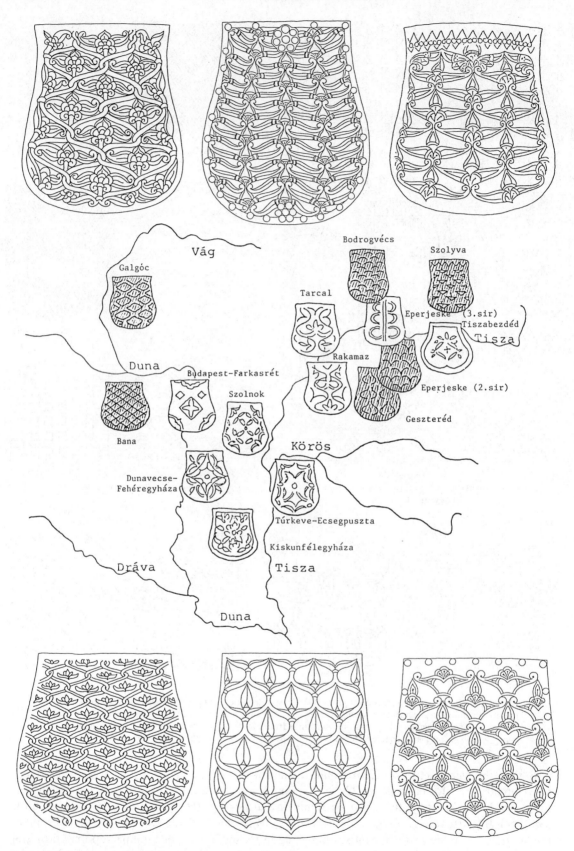

Fig. 5. The archaeological sites of the sabretache plates found in the Carpathian Basin from the Hungarian Conquest, ninth to tenth century A.D. The six sabretache plates with *cm*-type ornamental structures are shown with their fine adornations.

The most frequently occurring structure of plane symmetry types is the *cm* one among the finds of Old Hungarians' archaeological relics. This structure has been found on saddles, on a silver cup, but mainly on sabretache plates. Sabretache plates were decimeter-sized metallic (silver with gold covering) plates, they were the coverings for the holders of fire instruments. The sabretache plates of leaders were adorned with rich ornamentation built up from palmettes, frequently in a life-tree structure. The very rich ornamental design is so characteristic of these plates that this period of art of conquering Hungarians is called the Art of Sabretache plates in art history of the Carpathian Basin (Fig. 5).

Ornamental structures can be used in sketching the development of Old Hungarian ornamental art. Old Hungarians preferred a "family" of structures which can be deduced from each other (Fig. 6). They are constructed from two kinds of "molecules" of form-elements—the mirror symmetric palmette bunch and the palmette thread with 1*g* frieze structure—unifying them in the mirror double frieze. From here two directions of development were followed: (1) they enlarged the double frieze by repeating it infinitely on the plane by two parallel mirrors, so forming a *cm*-type pattern; or (2) they individualized palmette bunches taking them out of the palmette-net, but preserving its mirror symmetry.

What are the roots of Old Hungarian ornamental art considering only its structural side? It is necessary to study the cultural relations of the community to answer this question. The Old Hungarians came to the Carpathian Basin from the Pontusian steppe; on the other hand, remnants of Avars and Avar-Onogurians have been found in the Carpathian Basin. Structure of ornaments seems to be a surviving deep layer of the ornament, it is a kind of heritage of a community. Therefore I began to study the ornamental art of the Avar-Onogurian people, who were the last great migration into the Carpathian Basin (in the seventh century A.D.) before the Hungarian Conquest [5].

AVAR-ONOGURIAN ORNAMENTAL ART

Belt buckles are the most characteristic findings in the Avar-Onogurian archaeology from the seventh to the tenth century A.D. A far richer variety of double friezes has been found on them in my studies than in the Old Hungarian period. The description and classification of the ornamental structures of Avar-Onogurian double friezes needs a slightly different approach to frieze patterns from that which concerns the seven frieze groups. We turn our attention to the congruency operations which can generate the frieze pattern from one element. Earlier we saw that four simple congruencies can be generators of operators of repetition in frieze structure building: the translation (with grid unit), the half turn, the mirror reflection and the glide reflection. As a fifth operation, a composite one can also generate regular repetition of a simple element along the line: it can be composed from half turn and mirror reflection (for example). The symbols of these five generators and their patterns are, respectively: *t*, 2, *m*, *g* and *mg*. To these five basic frieze patterns the frieze groups of 11, 12, *m*1, 1*g* and *mg* can be related. But the patterns of the sixth and seventh frieze groups 1*m* and *mm* can not be fitted to these five basic patterns, because in them the repeating simple elements are double perpendicular to the direction of the line of the frieze pattern (by mirror reflection), so they exhibit a double frieze structure.

Studying Avar-Onogurian frieze ornaments gave me the idea of deducing them by doubling the five basic frieze pattern types [6]. This doubling can be carried out by the four simple frieze-building congruencies: *t*, 2, *m* and *g*. Using them as element generators perpendicular to the five (ready) basic frieze patterns (i.e. perpendicular to *t*, 2, *m*, *g* and *mg* patterns) I could obtain 20 different double frieze patterns, as shown in Fig. 7 (left side). The elements generated by the doubling operation—individually element by element—form a frize pattern which is a duplication (a congruent pair) of the initial pattern and parallel to it. (Metaphorically it is a kind of "DNA-doubling" on the plane along the line, with different kinds of doubling operations.) For comparison the Avar-Onogurian belt buckles have been tabulated in a similar order. (In Fig. 7 a supplementary first row was given to double friezes: the five basic friezes and their counterparts among the belt buckles from Avar-Onogurian art.)

The operation of doubling a frieze pattern can be continued in the direction in which it was begun (i.e. perpendicular to the initial given frieze pattern) and so newer and newer copies of the starting

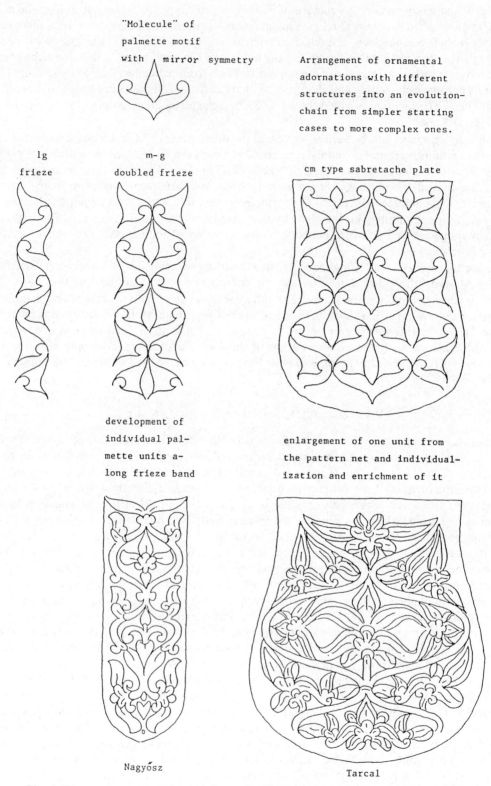

"Molecule" of
palmette motif
with mirror symmetry

Arrangement of ornamental
adornations with different
structures into an evolution-
chain from simpler starting
cases to more complex ones.

lg
frieze

m-g
doubled frieze

cm type sabretache plate

development of
individual pal-
mette units a-
long frieze band

enlargement of one unit from
the pattern net and individualization and enrichment of it

Nagyósz

Tarcal

Fig. 6. We can use ornamental structures to sketch the developments in ornamental art. The Old Hungarians preferred a "family" of structures which can be deduced from each other. They can be constructed from two form-"molecules" by operations which double a frieze by a mirror and then stretch the frieze over the whole plane with two parallel reflections. Further developments were the individualization of a plamette bunch taken from the net.

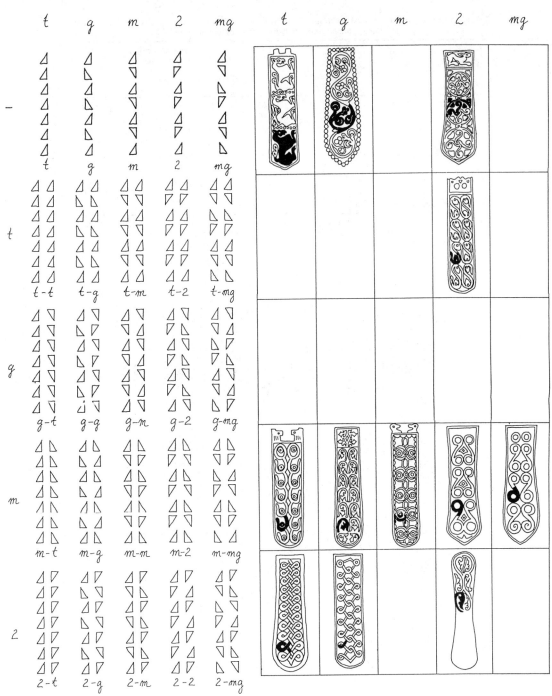

Fig. 7. Classification of Avar-Onogurian double friezes according to their pattern-forming operations. Columns mark the order vertically, rows mark the order in the horizontal direction. The archaeological sites of the belt buckles are: *t*, Nagymágocs-Ótompa; *g*, Mártély (Csongrád c.); *2*, Klárafalva (Csongrád c.); *t* − *2*, Tiszafüred-Majoros; *m* − *t*, Csuny (Moson c.); *m* − *g*, Győr; *m* − *m*, Nemesvölgy (Moson c.); *m* − *2*, Óföldeák (Csanád c.); *m* − *mg*, Keszthely; *2* − *t*, Regöly (Tolna c.); *2* − *g*, Tiszafüred-Majoros; *2* − *2*, Szeged-Kundomb.

frieze pattern can be added to the earlier ones, according to the order of the doubling operation. This is the point where plane symmetric structure forming technologies may appear as associations, as generators of ideas as to the origin of or intellectual background for the intuitive mathematical developments of Avar-Onogurian ornaments, the double friezes.

It is a common condition for regular structure building technologies that during the construction repeating congruent elements must be fitted in at least two main directions. If the fitting (the

Table 1. Matrix of connections and generating relations between 5 simple frieze patterns, 20 double friezes and 9 plane symmetry groups (with second order rotations only)

Double frieze	Plane symmetry group of the enlarged double frieze				
	t	g	m	2	mg
t (translation with a grid unit)	$t-t$ $p1$	$t-g$ pg	$t-m$ pm	$t-2$ $p2$	$t-mg$ pmg
g (glide reflection)	$g-t$ pg	$g-g$ pgg	$g-m$ cm	$g-2$ pgg	$g-mg$ $g \times mg$
m (mirror reflection—two parallel mirrors)	$m-t$ pm	$m-g$ cm	$m-m$ pmm	$m-2$ pmg	$m-mg$ cmm
2 (half turn)	$2-t$ $p2$	$2-g$ pgg	$2-m$ pmg	$2-2$ $p2$	$2-mg$ $2 \times mg$
mg (half turn and mirror reflection)	pmg	$g \times mg$	cmm	$2 \times mg$	$mg \times mg$

The symbolization of the matrix elements $g \times mg$, $2 \times mg$ and $mg \times mg$ mean, that the symmetry group of the structure characterizes not the element arrangement of the pattern, but the arrangement of "molecules" built from elements of the generator frieze patterns (see Fig. 8).

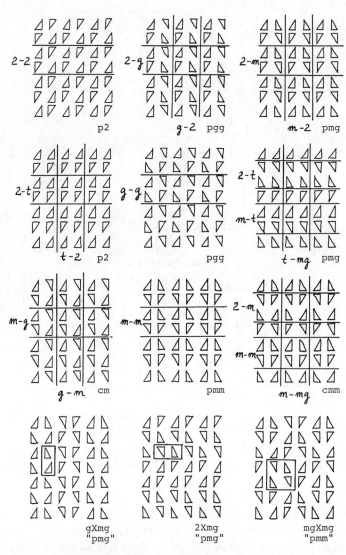

Fig. 8. Plane symmetry groups and their different patterns (according to the woven structure in Table 1) with examples to show the redundancy when double friezes are cut out of them. In the fourth row the three "molecular" symmetry patterns from Table 1 are shown and their "molecular" blocks are marked with a frame.

Győr	Szeged	Nagyősz	Pécs	Esztergom
7.–9. C.A.D.	9. C.A.D.	9. C.A.D.	11. C.A.D.	11. C.A.D.

Gate of the village church at Őriszentpéter, twelfth century

Ornamental frieze from Tarnaszentmária, eleventh century

Fig. 9. The survival of ornamental structures in the Carpathian Basin from the Árpád House Kings period (archaeological sites and ages are given).

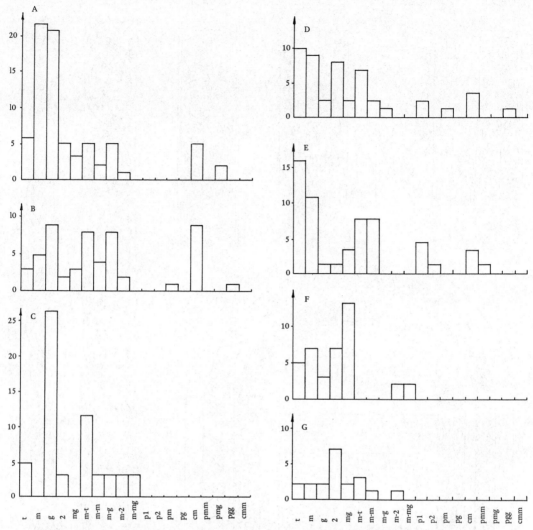

Fig. 10. Frequency of occurrence of freize, double frieze and plane symmetry patterns (not higher than second order rotations). (A) In Romanic architecture during the Árpád House Kings from Hungary (Gerevich Collection). (B) In ornamental art of the Conquesting Hungarians (Bérczi Collection). (C) Avar-Onogurian ornamental art from seventh to ninth century A.D. (Hampel Collection). (D) Pontusian-Scythian ornamental art (Artamonov-Forman Collection). (E) Altai-Scythian ornamental art (Rudenko Collection). (F) Celtic ornamental art (Déchelette Collection). (G) Greek ornamental art (Schuchhardt Collection).

attaching) of neighbouring elements is uniform in a direction, then to this attaching operation one of the four simple frieze generator congruencies can be corresponded in the geometry. Four of them appear in Fig. 7. Their continuation means repetition of the same type of congruency. The fifth type, which is a composite one, can be used as an alternately changing operation of two different generator congruencies: with the earlier example this composite *mg* fitting can be the alternation of half turn and mirror reflection.

The continuation of the doubling operation results in the double friezes growing into patterns which are enlarged on the plane. So our matrix of double friezes in Fig. 7 transforms into a matrix where plane symmetry patterns appear in the cross-places of rows and columns. Completing this matrix with the fifth enlarging generator operation—the composite *mg* alternating operation—a fifth row is added to it and so the whole matrix becomes symmetric. In the matrix in Table 1 the matrix elements are given by the symbols of the patterns of the plane symmetry groups, respectively. If we compare the Table 1 matrix to Fig. 4 we can see that the first 9 plane symmetry groups of the whole set of 17 appear in Table 1. The common characteristics of these nine types is that they all have rotational symmetry not higher than second order. The nine types are the following: *p*1, *p*2, *pm pg*, *cm pmg*, *pgg*, *pmm* and *cmm*.

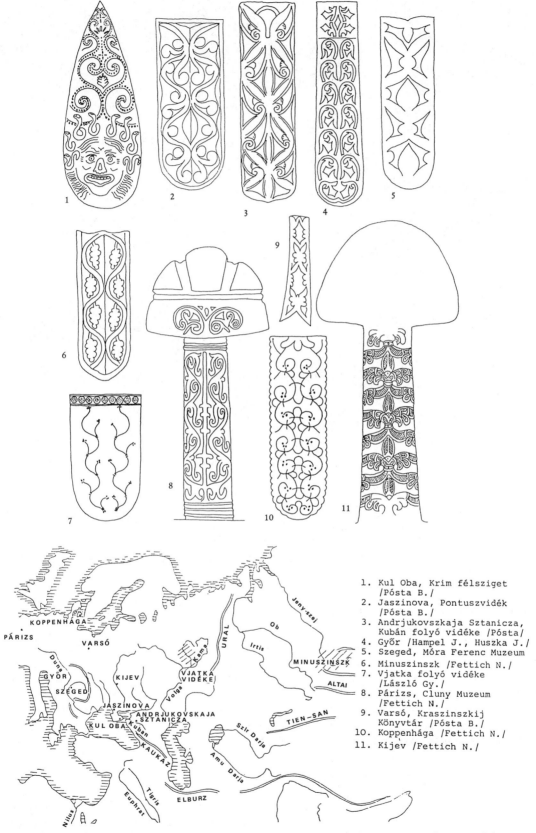

1. Kul Oba, Krim félsziget /Pósta B./
2. Jaszinova, Pontuszvidék /Pósta B./
3. Andrjukovszkaja Sztanicza, Kubán folyó vidéke /Pósta/
4. Győr /Hampel J., Huszka J./
5. Szeged, Móra Ferenc Muzeum
6. Minuszinszk /Fettich N./
7. Vjatka folyó vidéke /László Gy./
8. Párizs, Cluny Muzeum /Fettich N./
9. Varsó, Kraszinszkij Könyvtár /Pósta B./
10. Koppenhága /Fettich N./
11. Kijev /Fettich N./

Fig. 11. Occurrences of the $m - g$ structure among archaeological finds from the Eurasian steppe belt.

Fig. 12. The characteristic pattern of a Pontusian-Scythian gold dish from Kul Oba Kurgan (Krim Peninsula) shows the oldest to date occurrence of the $m - g$ double frieze structure, in the fifth century

Comparison of Table 1 with Fig. 7 shows the redundancy of the plane symmetry groups with respect to the given classification of double frieze types. This redundancy is more obvious in Fig. 8, where double friezes are considered as bands cut out from different plane symmetry patterns. The redundancy of plane symmetry patterns means that a double frieze pattern can be cut out from different plane patterns, and, on the other hand, more than one double frieze band can be cut out from one plane symmetry pattern. This means that one kind of pattern discovered from technology could serve as the sample pattern for more than one double frieze type. Therefore, the classification of double friezes given earlier in the paper is more exact in distinguishing double frieze patterns than plane symmetry patterns.

THE SYMMETRY METHOD IN ARCHAEOLOGY

The structure identification method can be used in two different approaches in archaeology. But the two approaches have common roots. The universality of basic structures, both in their natural appearance and their mathematical background. In more detail, the universality of structures summarized by the symmetry approach means that from natural phenomena and technological activities all kinds of structural basic types could have been discovered and used in ornamental adornations in all communities. However, the development beyond this basic set is a result of intellectual work and so may be characteristic of a cultural community. From the universal set of basic structures—on the other hand—communities might have selected, and so distinguished and preferred, different types [by cultural customs, traditional representations (e.g. Fig. 9) etc.], so the frequency of occurrence of different basic patterns may also be characteristic of a cultural community. Therefore, two kinds of measurement have been carried out in my archaeological research. I have measured extra developments (in a different sense in Hanti art [1])—and now especially in Avar-Onogurian and Old Hungarian archaeology—and the frequency of occurrence (Fig. 10) of different ornamental structures in great collections of archaeological materials from cultural communities of the Old Hungarians, Avar-Onogurians [7], Pontusian Scythians [8] and Altai-Scythians [9], from Celtic [10] and Greek [11] collections and, finally, architecture in the Carpathian Basin during the Árpád House Kings of Hungary [12]. Concentrating on the ornamental structures of peoples, stratifying each on the others in time (communities on earlier communities), I have observed the survival of characteristic structures. Those of the $1g—m−g—cm$ structural series are especially prominent. At the same time, differences between different ages might have been caused by the appearance (migration) of new communities with new customs. (Among other archaeological finds, changes in motifs discussed in art history also prove changes in cultural life occurred during the same period.)

In a third branch of research I have searched for the earliest occurrences of the characteristic $m−g$ structure. This structure occurs all over the Eurasian steppe (Fig. 11) from the Great Migration Period, scattered by steppe peoples, and can be found mainly on belt buckles. The earliest occurrence (to date) of this structure is from Pontusian Scythia (Iron Age, fifth century B.C.) from the Krim Peninsula (Fig. 12). The structure appears in a 12-droplet form on a gold dish among the relics from a Kurgan excavation at Kul Oba [13]. Judging by the head of Madusa which adorns the structure, it had a Greeko-Scythian master. This structure did not occur in the Celtic, Greek and Altai-Scythian collections from the same period.

A more comprehensive comparison of ornamental structures can be carried out when the structural method, the identification and classification of ornamental structures, becomes an accepted, standard part of the description of archaeological finds. Until that time, individual comparative works may help in the development of the structural method in archaeology.

SUMMARY

In my work I have shown possible technological roots of ornamental structures from archaeological finds, mainly from the Carpathian Basin. I have also suggested different methods to use from symmetry investigations on ornamental art from archaeological finds. I have presented the first results based on these methods: inventive developments of double frieze patterns in Old Hungarian and Avar-Onogurian ornamental art; frequency of occurrence of frieze, double frieze and plane symmetry structures in collections of cultural communities; and examples of survival and scattering of characteristic structures.

Both the mathematical background of the rich set of double friezes and the adaptation of the structural method in archaeology promise further rewards from using the symmetry principle for investigations on structure in different disciplines.

Acknowledgements—Valuable discussions with Emil Molnár and the possibility of publishing within an interdisciplinary approach, by István Hargittai, are gratefully acknowledged. I am grateful for the support to this work given by the Catholic University of Leuven, Belgium, the Collegium Hungaricum of Leuven, Stephen Muselay, Hermann Roelants and Balázs Gulyás.

REFERENCES

1. Sz. Bérczi, Escherian and non-Escherian developments of new frieze groups in Hanti and Old Hungarian communal art. In *Proc. Interdisciplinary Congress on M. C. Escher, University of Rome, 1985.* North-Holland, Amsterdam (1986).
2. I. Hargittai and Gy. Lengyel, The seven one-dimensional space-group symmetries illustrated by Hungarian folk needlework; and The seventeen two-dimensional space-group symmetries in Hungarian needlework. *J. Chem. Educ.* **61**, 1033–1034 (1984); **62**, 35–36 (1985).
3. H. S. M. Coxeter, *Introduction to Geometry.* Wiley, New York (1961).

4. Sz. Bérczi, Szimmetriajegyek a honfoglalás kori palmettás és az avar kori griffes-indás diszitõmüvészetben. *Cumania* **10,** 9–60 (1986).
5. Gy. László, *Études Archéologiques sur l'Historie de la Societé des Avars* Vol. XXXIV. Archaeol. Hung, Budapest (1985).
6. Sz. Bérczi, New double frieze types in the Avarian communal art. Presented at the *Symmetry Symp. Exhib.* Univ. of Darmstadt, F.R.G. (1986).
7. J. Hampel, *A Honfoglalási Kor Hazai Emlékei.* Budapest (1900).
8. Artamonov and Forman, *Goldschatz der Skythen.* Prague (1970).
9. S. I. Rudenko, *Kultura Naseleniya Gornava Altaja v Skifskom Vremeni.* Moscow (1953).
10. J. Déchelette, *Manuel D'Archéologie Préhistorique Celtique et Gallo-Romaine. IV. Second Age du Fer on Époque de La Téne.* Picard, Paris (1927).
11. K. Schuchhardt, *Schliemann Ásatásai.* Hungarian Acad. Science, Budapest (1892).
12. T. Gerevich, *Magyarország Románkori Emlékei.* Egyetemi Nyomda, Budapest (1938).
13. I. B. Brasinszkij, *Szkita Kincsek Nyomában.* Helikon, Budapest (1985).

Computers Math. Applic. Vol. 17, No. 4–6, pp. 731–749, 1989
Printed in Great Britain. All rights reserved

0097-4943/89 $3.00 + 0.00

THE GEOMETRY OF DECORATION ON PREHISTORIC PUEBLO POTTERY FROM STARKWEATHER RUIN

P. J. CAMPBELL

Mathematics and Computing, Beloit College, Beloit, WI 53511, U.S.A.

Abstract—This paper contributes further data and analysis to a growing body of literature that use mathematics to enhance interpretation of a culture from styles of its artifacts.

The mathematics employed is the classification of repeating patterns. The artifacts whose patterns are analyzed are specimens of prehistoric Pueblo pottery from Starkweather Ruin in New Mexico. The vessels are housed in the Logan Museum of Anthropology at Beloit College.

The present paper provides:

    (a) mathematical background on pattern analysis;

    (b) a survey of literature employing these techniques in an anthropological context;

    (c) data from observations on the complete corpus of decorated pottery vessels from Starkweather Ruin, a Mogollon-Pueblo site near Reserve, New Mexico;

    (d) commentary and analysis.

MATHEMATICAL BACKGROUND

Native and modern peoples throughout the world, from the ancient Egyptians to the twentieth-century Dutch artist Maurits Escher, have experimented in various media with *repeating patterns*. They have investigated the different ways of systematically repeating a basic discrete design element, or *motif*, in patterns along a strip or on a surface. Being systematic means reproducing the basic design motif in the same size by following a uniform rule on where to place each, based on considerations of symmetry. Creative artisans in each culture must have come up against the strong limitations that, as we will see, are imposed by being systematic in this sense. [In the definitive technical terminology of Grünbaum and Shephard [1, p. 165; 2, pp. 204–205], we are restricting consideration to *discrete non-trivial (monomotif) patterns*.]

The commonly accepted notions of symmetry in two dimensions are susceptible to analysis into four basic geometrical elements:

    (1) *reflection* of the motif across a straight line, producing a mirror-image, characteristic of so-called bilateral symmetry;

    (2) *translation*, or repitition at regular intervals, of the motif in a straight line;

    (3) *rotation* of the motif about a center, so that it repeats at regular angular intervals;

    (4) *glide reflection*, a combination of translation followed by reflection, best illustrated by the pattern produced by a person's footprints.

These basic elements are called *symmetries*. Synonymous mathematical terms are *rigid motions*, *congruence transformations*, *distance-preserving maps* and *isometries* (from the Greek for "same size").

We say that a pattern possesses a particular symmetry if the motion of that symmetry, when applied to the pattern as a whole, takes every exemplar of the motif into another one exactly. The collection of symmetries that a pattern possesses is called its *symmetry group*, where "group" is used in the technical mathematical sense to refer to fundamental ways in which the symmetries interact algebraically. For example, performing one symmetry motion, followed by a second one, must result in a combined motion that is a symmetry already in the collection.

Patterns may be classified by their symmetry groups. If two symmetry groups are the same (*isomorphic*), then the patterns they represent are of the same (*symmetry*) class. Grünbaum and Shephard [1, p. 164; 2, pp. 38–40] give the technical mathematical details: two symmetry groups are of the *same class* if one can be transformed into the other via conjugation by an affine transformation of the plane. For monochromatic patterns this condition is equivalent to just the existence of a group isomorphism between the symmetry groups, as Schwarzenberger [3, pp. 12–13] notes.

The investigation below is based upon symmetry group classification. Finer classifications, by *henomeric type* and by *diffeomeric type*, are introduced by Grünbaum and Shephard [1, pp. 17; 2, p. 220]. The quantity of data available to us does not warrant our using a finer classification, although archaeologists should be aware that Grünbaum and Shephard's investigation have revolutionized the field of pattern analysis by providing clarification and systematic classification of concepts.

We first consider *monochromatic* patterns, those in which all exemplars of the motif are executed in a single color against a background of a different color.

Patterns can be classified by the number of directions in which they admit translation.

Finite patterns allow no translations (nor glide reflections) and hence are limited in their symmetry elements to rotations about a single point, plus possibly reflections as well. Their symmetry groups of finite patterns are called *point groups*. There are two families: the *dihedral* groups, which have reflections, and which characterize the symmetries of flowers with bilaterally symmetric petals; and the *cyclic* groups, with no reflection symmetries, which characterize the symmetries of pinwheels or flowers whose petals do not have bilateral symmetry. The notation we will use for these are *dn* for the former and *cn* for the latter, where *n* is the number of petals. For example, *d*4 and *c*4 are the groups of symmetries of a four-leaf clover and a swastika, respectively. There is no restriction on the conceivable number of petals or rotation arms, so that there is an infinite number of groups in each of the two families. Thus, even after the motif and the colors for figure and background have been selected, there is still very wide (if monotonous) latitude for the artisan in choosing a pattern to execute the work.

One-dimensional patterns, those in which translations are allowed along a single axis, are variously referred to as *frieze patterns*, *strip patterns*, or *band patterns*. Here a very stark limitation asserts itself. Mathematical analysis [2, p. 218] shows that *there are only seven classes of strip patterns*! That is, once the basic motif is selected, and the colors of figure and background specified, there are seven ways to repeat the motif systematically. Several notations have been devised for these seven symmetry groups; with a special concern for easy adaptability of the notation to patterns with more colors, we have elected to use the notation of the *International Crystallographic Union* [4]. The left part of Fig. 1 shows samples of each monochromatic pattern, together with the notation. The basic motif for all the sample patterns is an asymmetric right triangle. Crowe and Washburn [5] give a similar table that cross-references notations of several authors.

The international notation succinctly summarizes the symmetries of the pattern. The full notation for a monochromatic strip pattern is made up of four symbols:

(a) The first is always a *p* (for "primitive"), indicating that every symmetry moves every motif exemplar.

(b) The second symbol is an *m* (for "mirror") if the pattern has vertical reflection lines, that is, reflection symmetries perpendicular to the direction of the pattern. The symbol is a *1* if no such symmetry is present.

(c) The third symbol is an *m* if the central axis along the length of the pattern is a reflection line, and an *a* if it is a line along which glide reflection takes place without mirror reflection being present. Again, a *1* symbolizes lack of symmetry.

(d) The fourth symbol is a *2* if the pattern had two-fold (i.e., half turn, or 180°) rotation symmetries. Otherwise the symbol is a *1*.

Crowe [6] and Zaslow [7] give further practical details on how to classify strip patterns.

Patterns that repeat in more than one direction on a two-dimensional surface are called *repeating plane patterns*, *periodic patterns*, or *wallpaper patterns*. There are only 17 classes of periodic patterns, shown in Fig. 2. Since these patterns figure only occasionally in the Pueblo pottery to be discussed later, we refer the reader to Zaslow [7] and Schattschneider [8] for discussion of notation, classification, and recognition of these pattern classes.

Three-dimensional symmetry groups describe placement of atoms in crystals, and crystallography was the motivation for their enumeration in the late nineteenth century. There are 230 such groups, called *Fedorov groups*. Mathematicians and crystallographers have also enumerated the symmetry groups in four and higher dimensions—see Schwarzenberger [3, pp. 132–135] for historical remarks.

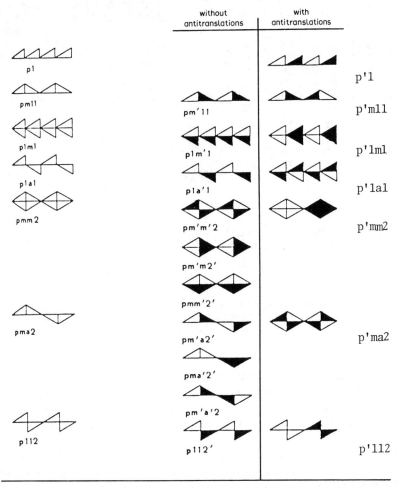

Fig. 1. The seven monochromatic and 17 bichromatic strip pattern classes, with international notation.
(Adapted from Shubnikov and Koptsik [34, Fig. 208, p. 271].)

For patterns in which the motif is executed in more than one color, the scheme of color repetition must be systematic; and this means that the symmetries of the uncolored pattern systematically permute the colors. Of interest in connection with Pueblo pottery are the *bichromatic* patterns, in which the motif appears in two different colors, or in two colors against a neutral background that is either uncolored (such as an unpainted pot surface) or of a third color. (Note that Lockwood and Macmillan [9] use the variant term dichromatic to embrace all of: uncolored patterns, "particoloured" (our bichromatic) patterns, and "grey" patterns.)

It is important to note that so-called "black-on-white" pottery will generally have a monochromatic pattern, because the motif occurs only in black; while some "polychrome" pottery, such as red-and-black-on-cream, will have a bichromatic pattern, based on the two colors in which the motif appears. In counting the number of colors involved in the symmetry, it is necessary to decide whether all of the colors on the object embody the motif, whether the background itself changes color, or whether one color functions as a neutral background for the others.

Some Starkweather "black-on-white" pots feature a cross-hatching that we have decided to consider as a distinct color, so that the two colors black and hatched appear against a white background. The patterns are consequently analyzed as bichromatic.

Each of the monochromatic symmetry groups discussed above gives rise to one or more bichromatic symmetry groups, depending on how many different ways the color change can interact

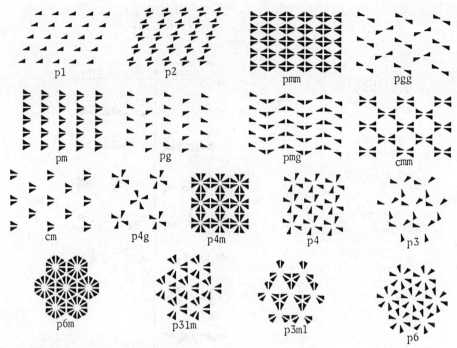

Fig. 2. The 17 monochromatic periodic pattern classes. (Adapted from Shubnikov and Koptsik [34, Fig. 150, p. 157], which in turn was adapted from M. J. Buerger, *Elementary Crystallography* (rev. edn). Wiley, New York (1963).)

with the symmetries of the uncolored pattern. The rotation symmetry of the cyclic *point groups* can either preserve color (giving back the original monochromatic pattern), or else change color, thereby producing one new infinite class of patterns. (The color change only "works out" if n is even.) The dihedral point groups offer color the opportunity of interacting with either or both of the rotation and reflection symmetries, resulting in two new infinite classes patterns.

There are a total of 17 *bichromatic strip groups*, shown with their notation on the right of Fig.1. Translations that reverse color are called *antitranslations*. For these we agree with Crowe and Washburn [5] in deviating slightly from the international notation of Fig. 1, replacing p_a and $p_{a'}$ by p' for typographical reasons. Crowe and Washburn further describe the pattern classes and give a guide and flowchart for their recognition.

At variance with this result, both Washburn [10] and Macdonald and Street [11] arrive at a total of 21 bichromatic strip symmetry patterns. Not surprisingly, their analyses are based on a different criterion for which patterns should be considered distinct. Jarratt and Schwarzenberger [12] explain: ". . . [Macdonald and Street] begin by selecting a representative pattern corresponding to each of the seven uncoloured frieze groups and then colour the patterns. Resulting coloured frieze patterns are considered equivalent if they can be superimposed on one another (modulo permutations of the colours)." As in Jarratt and Schwarzenberger [12] and Crowe and Washburn [5], the analysis of this paper is based on the criterion of symmetry class and the resulting 17 pattern classes. The differences between the two criteria are discussed with examples in a later section of this paper. examples in a later section of this paper.

The 17 two-dimensional symmetry groups produce 46 *bichromatic periodic groups*—see Lockwood and Macmillan [9, pp. 63–66, 198–202] and Shubnikov *et al.* [13, Fig. 144 opposite p. 220]—and the 230 three-dimensional Fedorov groups produce 1651 bichromatic groups in three dimensions, called *Shubnikov groups*—see Shubnikov *et al.* [13, pp. 175–201].

Lockwood and Macmillan [9, pp. 67–70] remark that a polychromatic period pattern having rotation can have only three, four or six colors; and they illustrate all 11 such patterns. Again, their observation is peculiar to their definitions, which are based on a very restricted view of color symmetry. Jarratt and Schwarzenberger [12], Senechal [14] and Wieting [15] use the more-conventional definitions to arrive at 96 four-colored groups (Wieting illustrates them all), and a total of more than 900 groups for two to 15 colors. Other authors suggesting a more restricted view are Macdonald and Street [11, 16] and van der Waerden and Burckhardt [17]. Loeb [18] imposes restrictions springing from crystallographic considerations.

Schwarzenberger [19] gives a definitive overview of color symmetry which dispels most of the confusion that has beset the subject. All investigators agree completely in the case of two-color patterns.

Despite the above remarks, it may still be surprising that beyond bichromatic coloring, increasing the number of colors does *not* vastly increase the number of pattern classes of *strip* patterns. Jarrett and Schwarzenberger [12] show that the number of strip pattern classes with N colors is 7 when N is odd, 17 when N is divisible by 2 but not by 4, and 19 when N is divisible by 4. (Lockwood and Macmillan [9, p. 68] claim that polychromatic strip patterns can be based only on the uncolored classes $p111$, $p1m1$ and $p1a1$; their observation is again based on a narrower definition of a pattern.)

LITERATURE SURVEY

Washburn [10, pp. 11–12] enumerates some of the few papers that have used symmetry patterns to classify and analyze designs. Several others have appeared, however, and we find it useful to survey chronologically all of the contributions, especially since some omitted by Washburn concern artifacts produced by native peoples of the Americas.

Speiser [20] enumerated the monochromatic point, strip and periodic pattern classes and offered illustrations from ancient Egyptian ornamentation as collected by Jones [21]. Woods [22] presented a mathematical enumeration of the point, strip and periodic patterns classes, including all uncolored and bichromatic symmetry groups. Buerger and Lukesh [23] correlated the common knowledge of chemists about symmetry groups in crystallography with ornamentation in two dimensions. Their article may be the origin of the name "wallpaper groups".

Stafford [24] analyzed the colored repeating patterns on the amazingly beautiful embroideries produced around 200 B.C. by the Paracas culture of Peru. It appears she was unaware of the mathematical analysis of symmetries. Although the initial work on uncolored patterns had been done in the late nineteenth century by crystallographers, had been exposited in Woods [22], and had found its way into mathematical books such as Birkhoff [25] and Hilbert and Cohn-Vossen [26, pp. 56–88], it was certainly not common knowledge, even among mathematicians. Color symmetry was first investigated only in the 1940s and 1950s. The Paracas embroideries feature mainly translations and glide reflections with color changes. Irregularities, instances of "broken" or unbalanced symmetry, are common.

Brainerd [27] suggested analyzing ceramic designs according to symmetry properties. Mueller [28, 29] applied the theory of one- and two-dimensional monochromatic symmetry groups to analyze the Moorish patterned ornamentation on the walls of the Alhambra in Spain. Shepard [30] used the seven monochromatic strip groups to classify strip patterns on pottery from the American Southwest. Garrido [31] used the monochromatic strip and period groups to classify ornaments on monuments of ancient Mexico. He noted relative percentages of occurrence of each pattern class, pooling together monuments from the wide diversity of Mexican cultures.

MacGillavry [32] presented examples from the work of the Dutch graphic artist M.C. Escher (1898–1971) to illustrate each of the 17 wallpaper pattern classes, as well as some of the color pattern classes. Escher made several prints specifically for the volume, in order to fill out the catalogue. It is interesting to observe that Escher attributed his initial interest in his intertwining figures to a visit to the Alhambra in 1935–1936; he felt that the patterns of the Moors could be very greatly enriched by using animate figures for motifs, a practice forbidden in Islam. Also worthy of note is that Escher's half-brother was a chemist, who no doubt was aware of the use of symmetry groups in chemistry.

Rappoport [33] was referred to by Shubnikov and Koptsik [34], but we have not seen the former work.

In 1971 Donald W. Crowe began a series of articles [6, 35–38] analyzing monochromatic symmetry patterns in African artifacts, including Bakuba decoration (Zaïre), Benin bronzes (Nigeria), and Begho pipes (Ghana). This work has been furthered by Zaslavsky [39–41].

Plate 1. 22399 Reserve pitcher. Neck has six elements, motif 19 and pattern *p112′*. Body has seven elements, motif 16 and pattern *p112′*. Body design continues across handle.

Plate 2. 22427 Reserve pitcher. Neck has six elements, a bar motif and pattern *pma2*. Body has five elements, a cross and bar motif and pattern *pmm2*. Handle has five elements, a wavy line motif and pattern *pm11*.

Plate 3. 22403 Reserve pitcher. Neck has four elements, motif 36R and pattern *p112*. Body has five elements, motif 36 and pattern 36. Handle has two elements, motif 73 and pattern *c2*. Body and neck patterns are mirror images.

Plate 4. 22409 Reserve eccentric olla (pitcher). Neck has five elements, cross motif and no pattern. Lower neck has six elements, motifs 6 and 41 and no pattern. Body has six elements, motifs 19, 46R and 47R, and pattern *p112* (with deviation from exact symmetry).

Plate 5. 22423 Tularosa bowl. Interior has four elements, motif 19 and pattern *p112′* (with deviation from exact symmetry).

Recent years have seen notable further attempts to arouse interest among anthropologists in applying symmetry techniques to analysis of patterns on artifacts. Zaslow and Dittert [42–44] focused on decorations on ceramics of the Hohokam culture, largely from the site at Snaketown, Arizona. Many examples of this pottery bear two-dimensional patterns. A major feature of Zaslow and Dittert [43] is an effort to detect an evolutionary development in Hohokam design. The inferred development parallels established chronology and even suggests some refinements of the sequence. The authors maintain that "cultural continuity is implied by the continuity in pattern development"

Plate 6. 22387 Tularosa pitcher. Neck has seven plus elements, shaded bars and pattern *p112* (with deviation from exact symmetry). Body has five elements, motif 16 and pattern *p112′* (with deviation from exact symmetry). Missing doube (?) handle, right-angle jump in border, "probably Starkweather" (Nesbitt).

Plate 7. 22395 Jar. Neck has five elements, motif 71, and pattern *pma2* (with deviation from exact symmetry). Upper body has four elements, motif 37 and pattern *p111*. Lower body has seven elements, motifs 53 and 73 and pattern *c2*. Additional symmetry element of curved vs jagged.

Plate 8. Representative ceramics from Starkweather Ruin.

Plate 9. 22398 Tularosa pitcher. Rim has 17 elements, motif 25, and pattern *pm11* (with deviation from exact symmetry). Body has three elements, motifs 52 and 53, and pattern *p′112* (with deviation from exact symmetry). Handle has three elements, wavy line motif, and pattern *p112*.

Plate 10. 22435 Jar. Part of neck and most of body missing. Body has motif 16 and pattern *p2′*.

(p. 25) and tentatively suggest "a connection between social factors and the pattern class selected for ceramic decoration" (p. 26).

In addition, Zaslow, a chemist, offered in Ref. [7] a copiously illustrated guide to employing one- and two-dimension monochromatic symmetry groups in analyzing ceramic decorations. Worthy of special note is a section on building two-dimensional patterns by placing horizontal strips adjacent to one another (pp. 27–36). In Ref. [45] he concentrated on handedness ("mirror orientation") of motif elements as an important component of pattern development and indicator of chronology.

Zaslow [46] identified "shared pattern features" in Hohokam ceramics and Oaxaca Valley mosaic panels, concluding that close similarities preclude independent development. Zaslow and Lindauer [47] used geometric patterns on ceramics to infer an Anasazi influence on Hohokam ceramics.

The papers confined themselves to monochromatic patterns, as had all previous authors except Woods [22] and Stafford [24]. As a result, Fig. 11 (p. 25) of Zaslow [7], classified $p2$ there, should be the bichromatic pattern $p_b'2$, in the notation of Lockwood and Macmillan [9]; and Fig. 14 (p. 30), classified $p4gm$, may be viewed as $p'4gm$ ($p_c'4gm$, in the notation of Crowe and Washburn [5]) if one considers the shadings as one color, and as a much more complicated color group if the shadings are considered two different colors.

The major feature of Washburn [10] was consideration of bichromatic patterns. Washburn classified designs on 152 whole vessels, 80% from a single site at Mariana Mesa, and the others from sites within 5 miles of it. Following Woods [22], she referred to reversal of the two colors under a symmetry as the process of *counterchange*. Washburn made a finer distinction for two-dimensional patterns, so that she regarded as different two designs whose appearance changes according to alignment of the design vertically or horizontally [10, pp. 26–27].

This gave her a total of 23 monochromatic two-dimensional patterns based on a fixed orientation. We might call them the 23 *oriented* wallpaper patterns, because the differences in appearance trace to differences in orientation of the motif. Washburn distinguished between designs with the same pattern but aligned at 90° to each other. This suggests the further possible refinement of also distinguishing between designs with the same pattern but different *handedness* of the motif, or aligned at 180° ("upside down") or 270° to each other.

Such distinctions cannot apply to monochromatic strip patterns, so she still had seven of those. However, in enumerating bichromatic strip patterns she arrived at a total of 21 (instead of 17). Here the expanded number is due not to fixed orientation of the strip, but to grouping of motif copies into "units." Though her illustrations [10, Fig. 19, p. 26] suggest to the reader that the grouping is based on proximity of motifs, in fact the basis for grouping is not explained. However, Macdonald and Street [16] also arrived at the number 21.

Examples are easily constructed that support Washburn's contention that one *should* distinguish two designs whose symmetry groups are the same, and which at the same time defy the ability of her system of classification to do so. In our Figs 3(a) and (b) we use thick dark lines to represent one color, thin light ones to represent the other. The strips depicted have the same symmetry group $p'mm2$ and the same motif, a line segment at 45° to the horizontal. One strip appears to be made up of Xs and the other of lozenges.

Washburn's analysis was based on choosing a basic "unit" of the pattern, possibly larger than the fundamental motif. If we regard a single-color X and a single-color lozenge as the respective "units" of the two patterns, we classify both as $1–2_2 11_2$, in Washburn's notation. On the other hand, if we regard a bicolor X and a bicolor lozenge as the respective "units," then we arrive at her different class $1–2^2 11^2$ for what are precisely the same strips as before.

The impact of the argument above is made stronger by coloring parts of the patterns, producing one strip made of single-color bow ties—or is it made of two-color diamonds [Fig. 3(c)]?—and another made of two-color bow ties—or is it made of single-color diamonds [Fig. 3(d)]?

Washburn's distinctions also led her to a correspondingly large number for the "counterchanged" wallpaper patterns than the traditional 46 bichromatic ones.

Apart from the incorporation of counterchange, the great contribution of Washburn's book was the theoretical background (pp. 3–10) she offered for the establishment of symmetry analysis as an important tool of the anthropologist. She has continued to use symmetry considerations in studying ceramics [48, 49].

Schattschneider [8] offered examples of Chinese lattices in 14 of the 17 monochromatic pattern classes, while Niman and Norman [50] recounted a classroom activity based on identifying pattern classes in Islamic art. Rose and Stafford [51] outlined a possible course in the mathematics of symmetry, providing identification algorithms for pattern classes. Crowe and Washburn [52] provided flowcharts for identifying the monochromatic and bichromatic pattern classes. Ascher and Ascher [53] noted a correspondence between the symmetries of designs on Inca pottery and the organization of the Inca quipu recording system. Crowe and Washburn [5] analyzed the bichromatic patterns present on nineteenth-century pottery of San Ildefonso Pueblo, a village on the Rio Grande between

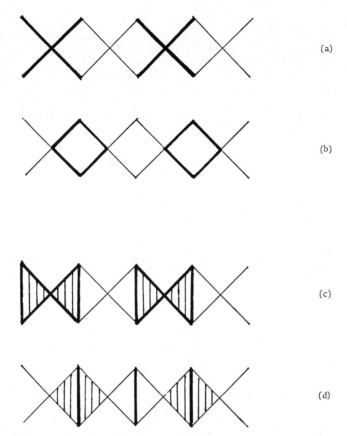

Fig. 3. Patterns of the same symmetry class distinguished [(a) from (b), (c) from (d)] under the classification systems of Washburn [10] and MacDonald and Street [11].

Los Alamos and Santa Fe, New Mexico. This pottery represented a revival for sale to tourists and exhibits a Spanish floral influence. Crowe and Washburn worked from illustrations in Chapman [54]. They showed that 14 of the 17 bichromatic patterns occur, and they offered comparisons with older San Ildefonso pottery styles, concluding that changes in pottery patterns exhibit cultural change.

A comprehensive treatment of analysis of plane patterns on cultural artifacts was given by Washburn and Crowe [55] and a tutorial module was published by Crowe [72].

THE STARKWEATHER SITE

We turn now to a description of Starkweather Ruin and analysis of the symmetries on the pots recovered from it. Excavation at the site was undertaken during the seasons of 1935 and 1936 by Paul H. Nesbitt, professor of anthropology at Beloit College, in Beloit, Wisconsin. Nesbitt was accompanied by undergraduates from the College, thereby continuing a strong tradition of involving students in research. Beloit students have participated in excavations in many parts of the world; and the Logan Museum on the campus contains many artifacts from those expeditions, in displays created by students.

Readers desiring a concise introduction to the relevant peoples, cultures, and periods of the North American Southwest may consult Willey [56] for both general background and specific terminology.

Relevant to our discussion below are a few background remarks. Archaeologists distinguish four major prehistoric "cultural subareas" in the Southwest, which "correspond in large degree to natural environmental conditions" [56, p. 179]: Mogollon, Anasazi, Hohokam and Patayan (see Fig. 4). Present-day descendants are known for the last three groups (respectively, the Hopi and Zuñi

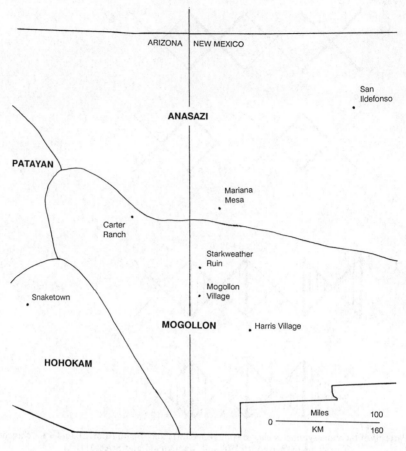

Fig. 4. Map of cultural subareas and sites referred to in text. (Adapted from *Emil W. Haury's Prehistory of the American Southwest* [Eds J. J. Reid and D. E. Doyel]. Univ. of Arizona Press, Tucson, Ariz. (1986).)

Pueblo Indians, the Pimas and Papagos, and Yuma-speaking Indians). It is unknown, however, what happened to the Mogollon people.

At the time of the excavation of Starkweather Ruin, the Mogollon culture had only just been identified by Emil Haury. In 1933 Haury had excavated Mogollon Village (named after the nearby Mogollon Mountains), a site close to Starkweather. It had yielded only pithouses, spanning the period from 550 to 900 A.D. [57]. The following year he had excavated a second site, Harris Village, in the Mimbres Valley some distance away, where he had found a similar sequence of remains spanning the same time range. Haury had found the sequence to be "fundamentally different from the Basketmaker-Pueblo (Anasazi) sequence to the north" [58, p. 298].

It was in order to obtain more information on the Mogollon culture that Nesbitt undertook to excavate at Starkweather. Nesbitt does not discuss in Ref. [59] how he chose that particular site, or even the origin of the name of the ruin [59, pp. 9, 10, 35–36, 78]:

> "Starkweather Ruin is situated . . . 3.5 miles west of Reserve, Catron County, New Mexico, in Section 3 Range 19 west, Township 7 south. The site occupies the west section of a small mesa which towers approximately eighty meters above the San Francisco river situated three miles to the northeast, and forty meters above Starkweather canyon directly to the south. The mesa is about 1850 meters above sea level. . . . The top of the mesa . . . measures approximately 400 meters east and west and 150 meters along the north–south axis. . . . Thirty-two houses were excavated, twelve pueblo surface houses and twenty pitrooms. . . . When the pueblo stone house builders came to Starkweather, the early pit dwellers had already moved elsewhere, and this later culture is not to be thought of as having evolved from that of the earlier occupants. Certain ceramic types do seem however to be derived from the pithouse period. . . . The ceramic collections from Starkweather village consist of the following materials:
>
> a. Approximately 12,000 sherds from the pitrooms and 5 complete and incomplete vessels.
> b. 302 complete specimens and several thousand sherds from the pueblo rooms and outlying burial ground."

Nesbitt classified the earlier pitroom culture as Mogollon, with two specimens of wood from one of the pithouses both giving bark dates of 927 A.D. He cautiously suggested that Mogollon occupation of the site may have lasted from 650 to 1100 A.D. He found the pithouse pottery as a whole to be "identical to that reported by Haury from the Mogollon and Harris village sites" [59, p. 114). His main conclusion was that the Mogollon culture is similar to the Basketmaker-Pueblo, with some Hohokam influence, in opposition to Haury, who had postulated the Mogollon as a third major cultural province. Nesbitt dated the first Mogollon settlements in New Mexico to 700–900 A.D., later than Haury. He also concluded that the Mogollon, Hohokam and Pueblo cultures derived certain ceramic types from the same source but that there was no evidence that the Mogollon was derived from any specific eastern group. For early reaction to Nesbitt's report, see Colton [60] and Kidder [61].

In the later 1930s tree-ring dating (and subsequently, radiocarbon dating) confirmed Haury's view of the Mogollon as a separate cultural area. Nesbitt, however, was right in arguing for cultural continuity rather than people moving in and out [62].

Our concern is not with the Mogollon but with the later Pueblo pottery found at Starkweather, which—unlike the browm Mogollon pottery—features painted designs. There are two such types: Tularosa black-on-white, dating from the Pueblo III horizon (1050–1300 A.D.) and the "earlier related type, probably ancestral" [59, p. 93] Reserve black-on-white from Pueblo II horizon (900–1100 A.D.).

COMMENTARY AND ANALYSIS

Table 1 summarizes observations made on the Starkweather pottery. The motif numbers given are keyed to Fig. 5.

Table 2 cross-classifies the data of Table 1 by type of pottery and symmetry of design. No expert was available to classify pots not otherwise identified; however, all but two of the untyped pots are manifestly either Reserve or Tularosa black-on-white.

The data of Washburn [10] included only one Reserve pot design but 69 Tularosa pot designs, compared with the 18 Tularosa, 36 Reserve, and 26 untyped designs of Starkweather. She distinguished between *symmetry of structure* and *symmetry of whole design* (pp. 18–19). Using the former appears to be close to the practice adopted by the present author. This author recorded (see Table 3) 12 designs with an asterisk, denoting that the pattern was slightly marred, and 11 more as possessing no symmetry.

The Tularosa-producing Starkweather site and the Mariana Mesa sites investigated by Washburn are contemporaneous Pueblo III sites of the eleventh and twelth centuries, roughly 60 miles apart. Tularosa black-on-white occurs at both locations. Other types of decorated pottery occur at Mariana Mesa but not at Starkweather, and the "probably ancestral" Reserve pottery of Starkweather is all but completely absent at Mariana Mesa. Washburn concluded (p. 172) that the distributions of pattern classes from the several sites were similar. The small number of Tularosa vessels at Starkweather, together with the differences in classification practices, does not afford close comparison with her data. However, the Starkweather data as a whole bear out the same themes, while offering some individuality.

The predominant symmetry pattern class at Starkweather is $p112$, whose characteristic element is a half-turn. Only two of the 17 bichromatic strip patterns occur, $p'112$ and $p112'$, and both of these prominently feature the half-turn. Three of the monocolor strip patterns, $p1a1$, $p1m1$, and $pmm2$ scarcely occur at all. *Pueblo art at Starkweather strongly preferred the symmetry of the half-turn; made some use of vertical mirror lines; and almost entirely avoided horizontal mirror lines and glide reflections.* Although Washburn's data included a larger proportion $pmm2$ patterns, all of these "vanished" under her stricter classification by design symmetry. She appears to have observed a greater proportion of two-dimensional patterns, including some bichromatic ones. The avoidance of bilateral symmetry was a consistent feature of Western Hemisphere pottery; such symmetry was prominent only in Cochle, Panama [63].

The preference for half-turns and avoidance of glide reflection would be easy to explain if the

Table 1. Data on potter

Vessel	Form	Type	Plate	Neck (rim)			Second neck (or rim)		
				Elements	Pattern	Motif	Elements	Pattern	Motif
21000	Bowl		—	7	pm11	30	33	pm11	30
22385	Eccentric pitcher	R	31B	9	p112	13			
22386	Pitcher	T	33G	4	p112	10			
22387	Pitcher	T	—	7+	p112*	Other			
22388	Bowl		—	5?	Can't tell				
22389	Bowl		—	4	Can't tell				
22391	Bowl		—	4	p112*	36,46			
22392	Bowl		—	2	c2	52			
22394	Bowl	R	29C	10	p112	52	7	p112	52
22395	Jar		—	5	pma2*	71	4	p111	37
22396	Pitcher	R	31F	5	pm11	5			
22397	Pitcher	R	31I	4	p111	36			
22398	Pitcher	T	33F	17	pm11*	25			
22399	Pitcher	R	30B	6	p112′	19			
22401	Bowl	R	29E	2	c2	7,53			
22401	Pitcher	R	31H	18	p112	tick			
22401/2	Bowl	T	32B	4	p112′*	18			
22403	Bowl	T	32A	4	p111	Other			
22403	Pitcher	R	31C	4	p112	36R			
22409	Eccentric olla	R	31G	5	None	Cross	6	None	6,41
22411	Pitcher		—	3	p112*	73			
22411/3	Pitcher	T	33D	10	pmm2	11	4 rows of 41	cmm	39,10
22417	Ladle		—	9	p2	10			
22418	Pitcher	T	33A	5	p112	36R			
22419	Hanging vessel		—						
22422	Pitcher		—	?	pma2	Line			
22423	Bowl	T	32D	4	p112′*	19			
22424	Bowl	R	29B	4	p112	other			
22424/2	Bowl	R	29F	7	p111	other	5	p111	Other
22425	Pitcher	R	31E	4	pm11	39			
22425	Pitcher	R	30C	6	p112	36			
22425/5	Jar		—	5	None				
22427	Pitcher	R	30A	6	pma2	Line			
22434/3	Pitcher		33B		None				
22435	Jar		—		Can't tell				
22475	Pitcher	R	30D	6 rows of 48	cmm	39			
22476	Ladle	R	29D	2	c2	33,67			
22479	Bowl	Red-on-gray			Can't tell				
22479/1	Bowl	Red-on-buff		?	p112	52			

Not available for direct examination were the Starkweather vessels of Nesbitt's Plates 31A, 31D, 32C, 32E, 33C and 33E. Plate 32E. 22417, 22418 (2 items) and 22426.

Vessel—Logan Museum catalog accession number, which may cover several objects of the same provenience. Form—determined by author's by Nesbitt plate or Logan Museum catalog: left blank if neither of these classifies. T for Tularosa black-on-white; R for reserved instances of the basic motif. Pattern—An asterisk denotes deviation from exact symmetry. c2, c4: point symmetry pattern classes. p111, pattern classes (see Fig. 2). p2′: bichromatic periodic pattern class. Motif—Correlated to illustration numbers in Fig. 263, p. 167 of for bowls are entered in the data table in the same columns as for necks for pitchers. Patterns that covered the entire inside surface of

potter had executed part of the design, then turned the pot over, and repeated the same design while turning the pot in the same direction as before. The sense of motion, conveyed from the motion of the pot to the dynamic of the design, would favor half-turns and tend to work against the employment of horizontal mirror lines. The conjectured method, however, would not absolutely preclude horizontal mirror reflection, which would be easy to execute; so its absence may also be attributable to Pueblo esthetic taste.

The slight, perhaps deliberate, marring of the symmetry of some designs may be a manifestation of a cultural norm that artifacts should be left incomplete or imperfect, because completion is a bad omen, possibly associated with death [63]. A similar norm affected other American cultures and is

Starkweather Ruin

...ments	Body Pattern	Body Motif	Elements	Handle Pattern	Handle Motif	Remarks
22?	*pma2* None	Other	18	*pm11*	30	"Starkweather?" 3-bulb shape
5	*p112*	6		None		
5	*p112'**	16				Missing double (?) handle; right-angle jump in border; "probably Starkweather"
						Restored
						Restored
						Both patterns the same
7	*c2*	53,73				Additional symmetry element of curved vs jagged; no handle
3	*p112*	36,36R		None		
4	*p112*	Other	6	None	Lines	
3	*p'112**	52,53	3	*p112*	Line	
7	*p112'*	16				Body design continues across handle
7	*p1m1**	Lozenge, 73R	6	None	Line	
5	*p112*	36	2	*c2*	73	Body, Neck patterns reverses of each other
5	*p112**	19,46R, 47R				Nesbitt groups it in plate with pitchers
3	*p112'*	16				Handle missing
6	*p112**	5,23,46	4	*p111*	49,50R	Double handle, crossed over; checkerboard on second neck
3	*p111*	52,53		None		Figure handle (see Nesbitt p. 95).
5	*p112*	56,73				Most of top missing; lug on one side, other side broken away; frets on design
4	*p112*	Line, 29	?	*pma2*		
						Both patterns the same
bands: 14, 16, 4, 11	*p112*	13				
bands of 6	*p112*	56,64	4	*c4*	51	
4	None	Spiral				Repaired; called "vase" in card catalog
5	*pmm2*	Cross, line	5	*pm11*	Line	
50	*p2*	61				
	p2'	16				Most of bottom missing; called "olla" in card catalog
7	*p112'*	16				Coiled handle; checker-board on neck
						20% fragment
						20% fragment

...32E depicts the item with catalog number 22393; the other plates appear to correspond, not necessarily respectively, to item 22415,

...istent subjective classification. Agrees with Nesbitt plates and Logan Museum catalog card, except where noted. Type—as determined ...k-on-white. Plate—in Nesbitt [59]. Elements—number as counted by a naive observer, so that an element may consist of one or more ..., *pma2*, *p1m1*, *pmm2*: strip pattern classes (see Fig. 1). *p'112*, *p112'*: bichromatic strip pattern classes (see Fig. 1). *cmm*, *p2*: periodic ...hburn [10], reproduced here as Fig. 5. An R indicates the motif occurs in the opposite handedness than shown there. Strip patterns ...owl are entered in the data table under the heading for body.

a marked feature of Islamic art ("only Allah is perfect"—see [39, pp. 137–151] for an illustration of deliberate defects in the context of magic squares and their symmetries).

The presence at Starkweather of the (presumably) older Reserve pottery, together with the presence of Tularosa pottery at both Starkweather and Mariana Mesa, suggests the possibility of diffusion of style proceeding from one location to the other. We are mindful, however, of the cautions of Deetz and Detlefsen [64], who demonstrated the "Doppler effect" in archaeology: inferred rates of diffusion of styles can vary greatly with directionality of site sampling. We offer no speculation about connections between Mariana Mesa and Starkweather Ruin, or indeed any assertions about origins or spread of the design styles found at Starkweather. We know only that Pueblo people came

Table 2. Starkweather data on pattern classes tabulated by pottery type and location of pattern, with number of vessels analyzed indicated

	Tularosa			N = 8	Reserve			N = 16	Untyped			N = 15	All Combined			N = 39
	Neck	Body	Handle	Total	Neck	Body	Handle	Total	Neck	Body	Handle	Total	Neck	Body	Handle	Total
c2					2		1	3	1	1		2	3	1	1	5
c4							1	1							1	1
p111	1	1	1	3	3			3	1			1	5	1	1	7
p1a1														0		
pm11	1			1	2		1	3	2		1	3	5		2	7
p112	3	2	1	6	7	6		13	3	2		5	13	10	1	24
pma2						1		1	2	1	1	4	3	1	1	5
p1m1						1		1						1		1
pmm2	1			1		1		1					1	1		2
p'112		1		1										1		1
p112'	2	1		3	1	2		3		1		1	3	4		7
p2									1	1		2	1	1		2
cmm	1			1	1			1					2			2
p2'										1		1		1		1
None			2	2	2	1	3	6	2	1		3	4	2	5	11
Can't tell									4			4	4			4
Totals	9	5	4	18	19	11	6	36	16	8	2	26	44	24	12	80

Fig. 5. Design elements. (Reproduced from Washburn [11, Fig. 263, p. 167].)

to live at Starkweather, their descendants dwelt there, and after a long occupation the inhabitants left and the site was abandoned.

We are not able to provide any sequencing of the pottery and designs found there, except to infer with Nesbitt that the cruder Reserve black-on-white preceded the Tularosa black-on-white.

Painstakingly-thorough analysis might permit the drawing of further conclusions about social organization of the inhabitants of the site. Longacre [65, 66] conducted a detailed "design element analysis" of 175 design elements on 6,000 sherds from a Pueblo settlement at the Carter Ranch Site, 80 miles from, and contemporaneous with, Starkweather. Since he worked from sherds, Longacre was not in a position to analyze symmetries of patterns of complete vessels, as we have done here. He hypothesized that:

> "If there were a system of localized matrilineal descent groups in the village, then ceramic manufacture and decoration would be learned and passed down within the lineage frame, it being presumed that the potters were female as they are today among the western Pueblos. Nonrandom preference for design attributes would reflect this social pattern." [65, p. 1454]

This hypothesis is still controversial. Longacre showed that design elements were associated with one of three groupings of rooms plus a kiva; and that "different types of pottery were associated with different rooms, and a set of stylistically distinct vessels was associated with each kiva and associated burials."

He concluded that the hypothesis of localized matrilineal groups was supported, that the household was the basic local unit, and that "communities made up of from one to three localized matrilineages were united through the mechanism of centralized ritual."

A study by Deetz [67] of the dynamics of stylistics change in the ceramics of the Arikara in South Dakota reached similar conclusions about that society, based on a detailed analysis of all features of the pottery and the same hypothesis:

> "Under a matrilocal rule of residence, reinforced by matrilineal descent, one might well expect a large degree of consistent patterning of design attributes, since the behavior patterns which produce these configurations would be passed from mothers to daughters [among the Arikara, most likely from grandmothers to granddaughters—see p. 97], and preserved by continuous manufacture in the same household. Furthermore, these attribute configurations would have a degree of mutual exclusion in a community, since each group of women would be responsible for a certain set of patterns differing more or less from those held by other similar groups." [67, pp. 2, 97]

Potters in market societies are known to be extremely conservative in style, because of their economic dependence on the market [68]. Longacre and Deetz suggested that matrilineal societies foster distinct pottery styles whose character is conserved not by market pressure but through tradition of the matrilineage. If they are right, then it should in theory be possible, with the help of symmetry analysis in addition to study of other ceramic attributes, to characterize the style or styles peculiar to each of the many Pueblo sites for which a large corpus of pottery is available.

One prevailing feature of the Starkweather collection points to the source of practical difficulty in even a limited version of such an undertaking: no two of the pots from Starkweather have the same design. Individual creativity may vary so much within a local style that the features of the style become impossible to discern accurately. The great variation among the Starkweather pots can be explained in part by features of Pueblo society. Many of the pots were found in burials, and the pot buried with a person in some sense must have partaken of that person's individuality. Moreover, a buried pot cannot be imitated (except from memory). Also, pots were made largely for family use, quite possibly one at a time as need arose. Another factor promoting individualization of designs within the familial tradition may have been need for a sense of achievement [69].

However, to the extent that the hypothesis of Longacre and Deetz holds true, the diffusion of similarities across Pueblo ceramics from widespread locations needs to be accounted for.

In this paper our purpose is more limited. Despite the differences noted above between Starkweather and Mariana Mesa ceramics, our major conclusion is that pattern analysis provides an easy comparison and confirmation of the similarities of the two sites, thus supporting Washburn's assertion:

> "It would appear that the use of symmetry classes to measure the similarities in design structure is a very consistent, objective procedure that can yield accurate, reproducible, and comparable results. The method clearly demonstrates the high degree of similarity of design structures among the Upper Gila area inhabitants and in this way confirms the postulated existence of an interacting community of potters." [10, p. 173]

Table 3. Comparison of pots by design symmetry class

	Washburn (Tularosa)		Starkweather			
	Design symmetry	Structure symmetry	Tularosa	Reserve	Untyped	Total
c1	1	1	0	0	0	0
c2	2	3	0	3	2	5
c4	0	1	0	1	0	1
d4	1	0	0	0	0	0
?	1	0	0	0	0	0
		5	0	4	0	6
p111	29	12	3	3	1	7
p1a1	0	0	0	0	0	0
pm11	3	7	1	3	3	7
p112	14	21	9	16	6	32
pma2	0	4	2	1	4	5
p1m1	1	0	0	1	0	1
pmm2	0	9	1	1	0	2
		53	16	25	14	54
p′112	0	0	1	0	0	1
p112′	6	(included above under p112)	3	3	1	7
p1		1	0	0	0	0
p2		5	0	0	2	2
pgg		3	0	0	0	0
pmg		2	0	0	0	0
cmm		1	0	1	0	2
p2′		0	0	0	1	1
		11	1	1	3	5
None		0	1	6	3	11
		0	1	6	3	11
Can't tell		0	0	0	4	4
		0	0	0	4	4
		69	18	36	26	80

THE VALUE AND MEANING OF SYMMETRY ANALYSIS

Of what significance is the work of this paper? For mathematicians, Pueblo pottery designs serve as realizations of abstract symmetry groups and offer their students practice in identification of symmetry elements and patterns. For anthropologists, pattern analysis represents a *new tool for identification and differentiation of patterned artifacts from closely-related cultures*. Enumeration of patterns employed by a culture, and comparison with those not employed, may yield insight into the esthetic sense and design process of the artisan. Consideration of the "geometric coercion" [63] imposed by the limited number of pattern classes casts a different light on the unlikelihood of the same pattern appearing on artifacts of two cultures, such as Valdivia in Ecuador and Middle Jomon in Japan [70]. For anyone, knowledge of the patterns affords an additional mode of appreciation of the artwork.

Does this style of analysis have anything to offer to the Native American who may be a descendant of the artisan? Are we uncovering or discerning Native American mathematics, or merely viewing sacred cultural remains through the eyes of a foreign and secular technology? Was the mathematics in the mind of the potter, or are we imposing it? Undoubtedly the symmetries are present in the pottery designs. The psychological difficulty in recognizing the devising of pottery designs as a form of mathematics is the preconception that there is only one true style of geometry: the written deduction of Euclid, subsequently "purified" by Hilbert to an axiomatic system completely devoid of figures and diagrams. The relevant analogue of this conception is the framework of abstract group theory in which it can be proved (with some difficulty!) that there are exactly seven symmetry groups of strip patterns, 17 of periodic patterns, and so on. Consider, though, the first investigators who tried to enumerate the 230 monocolor space groups. They did not proceed by rigorous logical group-theoretic reasoning; on the contrary, each of several individuals came up with incomplete enumeration in which some groups were listed twice. Comparisons and painful correction eventually led to a complete and correct list [3, pp. 132–133]. Were these crystallographers doing mathematics? Yes, they were, of a mixed inductive–deductive sort. So too were the Pueblo potters, though we have no narrative, only the pots, to stand as the record of their reasoning.

As a more current example of inductive mathematics we may cite the continuing investigation of what types of convex pentagons can tile the plane—that is, cover the plane with same-size, same-shape replicas without gaps or overlaps. One mathematician pronounced the subject closed and announced there were only eight types. A number of years later the topic was written up in a popular science magazine, and the article provoked one amateur to discover a ninth type; and a second amateur—a housewife with no formal education in mathematics beyond high school "general mathematics" 36 years earlier—then found five more over the next two years. A decade has passed, and still no one knows if the list is complete. The full fascinating story is related in Schattschneider [71]. She notes (p.166):

> "The mind and spirit are the forte of all such amateurs—the intense spirit of inquiry and the keen perception of all they encounter. No formal education provides these gifts. Mere lack of a mathematical degree separates these 'amateurs' from the 'professional'. Yet their dauntless curiosity and ingenious methods make them true mathematicians."

CONCLUSION

This paper has followed the practice of Washburn [10] in examining actual pottery vessels. This practice is important, because a photograph of one side of a vessel may be misleading in tending to indicate a degree of symmetry not borne out by the other side of the vessel. We have, however, concurred with Crowe and Washburn [5] in our scheme for classifying bichromatic patterns, rather than following the counterchange scheme of Washburn. The materials examined are related to those studied by Washburn, and much different from the ones studied by Zaslow and Dittert [43] (Hohokam culture, characterized by two-dimensional patterns), by Crowe and Washburn [52] (nineteenth century San Ildefonso culture, characterized by great variety and abundance of bichromatic patterns), and by Shepard [30] (who concentrated on identifying monochromatic band patterns on assorted Pueblo pottery).

Acknowledgements—I wish to dedicate this paper to Druscilla Freeman (deceased 9 February 1988), Curator of the Logan Museum at Beloit College, who suggested examining the Starkweather materials and provided ready access to them. I also wish to thank Dan Shea, Professor of Anthropology at Beloit College, who suggested pertinent literature and offered helpful comments.

REFERENCES

1. B. Grünbaum and G. C. Shephard, A hierarchy of classification methods for patterns. *Z. Kristallogr.* **154**, 163–187 (1981).
2. B. Grünbaum and G. C. Shephard, *Tilings and Patterns*. Freeman, New York (1987).
3. R. L. E. Schwarzenberger, *N-dimensional Crystallography*. Pitman, San Francisco (1980).
4. *International Tables for X-Ray Crystallography* (3rd edn.), Vol. 1. Kynoch Press, Birmingham (1969).
5. D. W. Crowe and D. K. Washburn, Groups and geometry in the ceramic art of San Ildefonso. *J. Algebra Groups Geometr.* **3**, 263–277 (1985).
6. D. W. Crowe, The geometry of African art: II. A catalog of Benin patterns. *Hist. Math.* **2**, 253–271 (1975).
7. B. Zaslow, A guide to analyzing prehistoric ceramic decorations by symmetry and pattern mathematics. In Pattern Mathematics and Archaeology. Arizona State University Anthropological Research Papers No. 2. Department of Anthropology, Arizona State Univ., Tempe, Arizona (1977).
8. D. Schattschneider, The plane symmetry groups: their recognition and notation. *Am. math. Mon.* **85**, 439–450 (1978).
9. E. H. Lockwood and R. H. Macmillan, *Geometric Symmetry*. Cambridge Univ. Press, New York (1978).
10. D. K. Washburn, A symmetry analysis of upper Gila area ceramic design. Papers of the Peabody Museum of Archaeology and Ethnology. Vol. 68. Harvard Univ., Cambridge, Massachusetts (1977).
11. S. O. Macdonald and A. P. Street, On crystallographic colour groups, In *Combinatorial Mathematics IV* (Eds L. R. A. Casse and W. D. Wallis), pp. 149–157. Springer, New York (1976).
12. J. D. Jarratt and R. L. E. Schwarzenberger, Coloured plane groups. *Acta Crystallogr.* **A36**, 884–888 (1980).
13. A. V. Shubnikov, N. V. Belov, *et al.*, *Colored Symmetry*. Pergamon, New York (1964).
14. M. Senechal, Color groups. *Discrete appl. Math.* **1**, 51–73 (1979).
15. T. W. Wieting, *The Mathematical Theory of Chromatic Plane Ornaments*. Dekker, New York (1982).
16. S. O. Macdonald and A. P. Street, The seven friezes and how to colour them. *Util. Math.* **13**, 271–292 (1978).
17. B. L. van der Waerden and J. J. Burckhardt, Farbgruppen. *Z. Krystallogr.* **115**, 231–234 (1961).
18. A. L. Loeb, *Color and Symmetry*. Wiley, New York (1971).
19. R. L. E. Schwarzenberger, Colour symmetry. *Bull. Lond. Math. Soc.* **16**, 209–240 (1984).
20. A. Speiser, *Die Theorie der Gruppen von endlicher Ordnung* (4th rev. edn). Birkhäuser, Basel (1956).
21. O. Jones, *The Grammar of Ornament*. Quartich, London (1856). Reprinted, Van Nostrand Reinhold, New York (1974).
22. H. J. Woods. The geometrical basis of pattern design. Part I: Point and line symmetry in simple figures and borders.

Part II: Nets and sateens. Part III: Geometrical symmetry in plane Patterns. Part IV: Counterchange symmetry in plane patterns. *J. Text. Inst.* **26**, 197–210, 293–308, 341–357 (1935); **27**, 305–320 (1936).

23. J. J. Buerger and T. Lukesh, Wallpaper and atoms. *Technology Rev.* **39**, 338–342, 370 (1937).
24. C. E. Stafford, *Paracas Embroideries: A study of Repeated Patterns.* Augustin, New York (1941).
25. G. D. Birkhoff, *Aesthetic Measure.* Harvard Univ. Press, Cambridge, Mass. (1933).
26. D. Hilbert and S. Cohn-Vossen, *Anschauliche Geometrie* (1932). Translated as *Geometry and the Imagination.* Chelsea, New York (1952).
27. G. W. Brainerd, Symmetry in primitive conventional design. *Am. Antiq.* **8**(2), 164–166 (1942).
28. E. Mueller, Gruppentheoretische und strukturanalytische Untersuchung de Maurischen Ornamente aud der Alhambra in Granada. Thesis, Univ. of Zurich (1944). Reviewed in *Math. Rev.* **12**, 478 (1951).
29. E. Mueller, El estudio de ornamentos como aplicación de la teoría de los grupos de orden finito. *Euclides, Madr.* **6**, (59), 42–50 + 2 plates (1946).
30. A. O. Shepard. The symmetry of abstract design with special reference to ceramic decoration. *Contr. Am. Anthrop. Hist.* **47**, 211–293 (1948).
31. J. Garrido, Les groupes de symétrie des ornements employés par les anciennes civilisations du Mexique. *C.r. hebd. Séanc. Acad. Sci. Paris* **235**, 1184–1186 (1952).
32. C. H. MacGillavry, *Symmetry Aspects of M. C. Escher's Periodic Drawings.* International Union of Crystallography by A. Oosthoek's Uitgeversmaatschappij NV, Utrecht (1965). Reissued as *Fantasy and Symmetry: The Periodic Drawings of M. C. Escher.* Abrams, New York (1976).
33. S. Kh. Rappoport, *Nondescriptive Forms in Decorative Art* (in Russian). Izd. Sovietskii Khudozhnik, Moscow.
34. A. V. Shubnikov and V. A. Koptsik, *Symmetry in Science and Art* (2nd rev. edn). Plenum Press, New York (1974).
35. D. W. Crowe, The geometry of African art: I. Bakuba art. *J. Geometry* **1**, 169–183 (1971).
36. D. W. Crowe, Geometric form and pattern in art, In *Africa Counts: Number and Pattern in African Culture* by Claudia Zaslavsky, pp. 172–189. Prindle, Weber & Schmidt, Boston (1973). Reprinted, Lawrence Hill, Westport, Conn. (1979).
37. D. W. Crowe, The geometry of African art: III. The smoking pipes of Begho, In *The Geometric Vein: The Coxeter Festschrift* (Eds Chandler Davis *et al.*), pp. 177–189. Springer, New York (1981).
38. D. W. Crowe, Symmetry in African art. *Ba Shiru: J. Afr. Lang. Lit.* **11**(1), 57–71 (1982).
39. C. Zaslavsky, *Africa Counts: Number and Pattern in African Culture.* Prindle, Weber & Schmidt, Boston (1973). Reprinted, Lawrence Hill, Westport, Conn. (1979).
40. C. Zaslavsky, African patterns. *Maths Teacher* **70**, 386 (1977).
41. C. Zaslavsky, Symmetry along with other mathematical concepts and applications in African life, In *Applications in School Mathematics* (Eds. S. Sharron and R. E. Reys), pp. 82–97. National Council of Teachers of Mathematics 1979 Yearbook. National Council of Teachers of Mathematics, Reston, Va. (1979).
42. B. Zaslow and A. E. Dittert Jr, The symmetry and pattern analysis displayed in Hohokam ceramics painting. *J. Ariz. Acad. Sci.* **11** (*Proc. Suppl.*) 9 (1976).
43. B. Zaslow and A. E. Dittert Jr, The pattern technology of the Hohokam, in Pattern Mathematics and Archaeology. Arizona State Univ. Anthropological Research Papers No. 2. Department of Anthropology, Arizona State Univ., Tempe, Ariz. (1977).
44. B. Zaslow and A. E. Dittert Jr, Pattern theory used as an archaeological tool: A preliminary statement. *S West. Lore* **43**(1), 18–24 (1977).
45. B. Zaslow, Mirror orientation in Hohokam designs and the chronology of early Hohokam phases. *Kiva* **45**, 211–225 (1980).
46. B. Zaslow, Pattern dissemination in the prehistoric southwest and Mesoamerica: a comparison of Hohokam decorative patterns with patterns from the upper Gila area and from the valley of the Oaxaca. Arizona State Univ. Anthropological Research Papers No. 25. Department of Anthropology, Arizona State Univ. Tempe, Ariz. (1981).
47. B. Zaslow and O. Lindauer, Anasazi influence on post-Sacaton Hohokam decorations. Paper read at the *1985 Meeting of the Soc. Am. Archeology, Symp. Organization of the Classic Period of Hohokam Society,* Denver, Colo. (1985).
48. D. K. Washburn, The Mexican connection: cylinder jars from the Valley of Oaxaca. *Trans. Ill. St. Acad. Sci.* **72**, 70–84 (1980).
49. D. K. Washburn, Symmetry analysis of ceramic design: two tests of the method on Neolithic material from Greece and the Aegean, In *Structure and Cognition in Art* (Ed. D. K. Washburn). Cambridge Univ. Press, Cambridge (1983).
50. J. Niman and J. Norman, Mathematics and Islamic art. *Am. math. Mon.* **85**, 489–490 (1978).
51. B. Rose and R. D. Stafford, An elementary course in mathematical symmetry. *Am. math. Mon.* **88**, 59–64 (1981).
52. D. W. Crowe and D. K. Washburn, Flowcharts as an aid to the symmetry classification of patterned design. *Int. Congr. Antropology,* Vancouver, B.C. (1983).
53. M. Ascher and R. Ascher, *Code of the Quipu.* Univ. Michigan Press, Ann Arbor, Mich. (1981).
54. K. M. Chapman, *The Pottery of San Ildefonso Pueblo.* School of American Research, Monograph No. 28. University of New Mexico Press, Albuquerque, N.M. (1970).
55. D. K. Washburn and D. W. Crowe, *Symmetries of Culture: Theory and Practice of Plane Pattern Analysis.* Univ Washington Press, Seattle, Washington (1988).
56. G. R. Willey, *An Introduction to American Archaeology. Vol. 1: North and Middle America.* Prentice-Hall, Englewood Cliffs, N.J. (1966).
57. E. W. Haury, *The Mogollon Culture of Southwestern New Mexico.* Medallion Papers 19. Gila Pueblo, Globe, Ariz. (1936).
58. S. A. LeBlanc, Development of archaeological though on the Mimbres Mogollon, In *Emil W. Haury's Prehistory of the American Southwest* (Eds. J. J. Reid and D. E. Doyel). Univ. Arizona Press, Tucson, Ariz. (1986).
59. P. H. Nesbitt, *Starkweather Ruin: A Mogollon-Pueblo Site in the Upper Gila Area of New Mexico, and Affiliative Aspects of Mogollon Culture.* Logan Museum Publications in Anthropology, Bulletin No. 6. Logan Museum. Beloit College, Beloit, Wis. (1938).
60. H. S. Colton, Review of Nesbitt (1938). *Am. Antiq.* **5**, 352–354 (1940).
61. A. V. Kidder, Review of Nesbitt (1938). *Am. Anthrop.* **41**, 314–316 (1939).
62. D. Shea, Oral communication (Jan. 1988).
63. D. Shea, Oral communication (1981).

64. J. Deetz and E. Detlefsen, The Doppler effect and archaeology: a consideration of the spatial aspects of seriation. *S West. J. Anthrop.* **21,** 196–206 (1965).
65. W. A. Longacre, Archaeology as anthropology: a case study. *Science* **144,** 1454–1455 (1964).
66. W. A. Longacre, *Archaeology as Anthropology: A Case Study.* Anthropological Papers of the University of Arizona No. 17 Univ. Arizona Press, Tucson, Ariz. (1970).
67. J. D. F. Deetz, *The Dynamics of Stylistic Change in Arikara Ceramics.* Illinois Studies in Anthropology No. 4. Univ. Illinois Press, Urbana, Ill. (1965).
68. G. M. Foster, The sociology of pottery: questions and hypotheses arising from contemporary Mexican work, In *Ceramics and Man* (Ed. F. R. Matson), pp. 43–61. Aldine, Chicago, Ill. (1965).
69. W. T. Levey, Early Teotihuacan: an achieving society. In *Mesoamerican Notes 7–8: Teotihuacan and After: Four Essays* (Ed. J. Poddock), pp. 25–68. Univ. the Americas, Mexico (1966).
70. E. Estrada *et al.,* Possible transpacific contact on the coast of Ecuador. *Science* **135,** 371–372 (1962).
71. D. Schattschneider, In praise of amateurs, In *The Mathematical Gardner* (Ed. D. A. Klarner), pp. 140–166 + 5 color plates. Prindle, Weber & Schmidt, Boston, Mass. (1981).
72. D. W. Crowe, Symmetry, rigid motions, and patterns. HiMAP Teacher Training Module 4. COMAP, Inc., Arlington, Mass. (1986). Reprinted in *UMAP Jl* **8**(3), 207–236 (1987).

Computers Math. Applic. Vol. 17, No. 4–6, pp. 751–789, 1989
Printed in Great Britain. All rights reserved

0097-4943/89 $3.00 + 0.00

IN THE TOWER OF BABEL:
BEYOND SYMMETRY IN ISLAMIC DESIGN

W. K. Chorbachi

Harvard University, Dudley House, Cambridge, MA 02138, U.S.A.

Abstract—A personal account of an interdisciplinary inquiry into the study of Islamic geometric design and architectural decoration touching on the fields of History, History of Science, Scientific Theory of Symmetry and History of Art. The study stresses the necessity of the use of a common scientific language of Symmetry Notation in order to discuss and communicate in a precise manner about Islamic geometric pattern. To understand Islamic geometric design, it is necessary to move beyond the symmetry issues, to the step-by-step process of design. This is based on primary sources of scientific manuscripts of practical geometry written specifically for the Muslim artisans. The research demonstrates not only a direct meeting but a collaborative work between science and art in Islamic civilization.

The story of arrogant men building the Tower of Babel (Genesis 11) reads as follows:

> "Now the whole earth had one language and few words. And as men migrated from the east, they found a plain in the land of Shīnär and settled there. And they said to one another, 'Come, let us make bricks, and burn them thoroughly.' And they had brick for stone, and bitumen for mortar. Then they said, 'Come, let us build ourselves a city, and a tower with its top in the heavens, and let us make a name for ourselves, lest we be scattered abroad upon the face of the whole earth.' And the Lord came down to see the city and the tower, which the sons of men had built. And the Lord said, 'Behold, they are one people, and they have all one language; and this is only the beginning of what they will do; and nothing that they propose to do will now be impossible for them. Come, let us go down, and there confuse their language, that they may not understand one another's speech.' So the Lord scattered them abroad from there over the face of all the earth, and they left off builidng the city. Therefore its name was called Bābel, because there the Lord confused the language of all the earth; and from there the Lord scattered them abroad over the face of all the earth."

This paper examines the relationship between the fields of "History, History of Science, Scientific Theory and the Process of Islamic Geometric Design". The theme of the Tower of Babel and the curse of the multiplicity of tongues runs throughout my discussion of these topics because of the lack of a common language, either in the study of Islamic geometric pattern or in interdisciplinary discourse. If the curse of Babel has beset us in the field of Islamic Art, we need not despair, for there is hope that is can be rectified. In the Biblical analogy, the curse from the Old Testament is finally, removed in the New Testament. It is only through the grace of God and the true demonstration of love by man that the curse of multiplicity of language with the resulting confusion of tongues is to be undone and people will be able to understand each other, as on the day of Pentecost (Acts 2:7).

My approach is frankly etiological. Just as the etiological passages from the Tower of Babel in the Old Testament explain how things came to be in the world (Fig. 1), this paper explains how things came to be in my attempt to study and document the direct meeting of science and art in Islamic civilization. It also takes up the background and origins of the methodological and interpretive trends that have characterized the study of the history of Islamic art and Islamic geometric pattern and ornament. I have been studying this material for 17 years, beginning in 1970. This account will span the first 7 years of my research in this area, from 1970 to 1977. The material is presented chronologically, and in a somewhat personal manner.

I begin with the process through which I originally came upon the material because of the crucial importance of tools and methodology to this kind of work. Information of this kind is rarely revealed publicly, nor is it often published. It is from these starting points, too often ignored, that one can learn most, for they involve more than tools and methodology; they involve the logical process of interdisciplinary research.

My account begins in 1971, with a meeting of a small group of people interested in Islamic art and architectural decoration. We were all looking at the same monument and the same portion of its decoration. In each case, our descriptions, analyses and even the naming of parts and shapes of

THE CONFUSION OF TONGUES

And they said, Go to, let us build us a city and a tower, whose top may reach
unto heaven . . . So the Lord scattered them abroad from thence upon the face of all
the earth . . . (Genesis 11: 4,8)

Fig. 1. The Tower of Babel: the Confusion of Tongues. Illustrated by Gustave Doré in *The Doré Bible Illustration* (Dover, New York).

that decoration were totally different from those of the person sitting next to us. We all saw what we saw, and we each spoke in our own terms and in our own language. We walked out as if we had not been together, we had not communicated, we had not understood each other. We were all speaking different languages. From that very early moment, I realized that there was something amiss in the study of Islamic architectural decoration. We lacked the proper tools, and proper or common language. We would never be able to truly classify, analyze and comprehend the material if we continue in this way.

Shortly thereafter, on a memorable afternoon, as I skimmed through books on Islamic architecture, I made the simple observation that, since the tenth century, an increasing number of geometric figures were used with a parallel increase in sophistication of the patterns of geometric design. These observations led to an obvious question: whoever had created these elaborate geometric designs must have mastered a knowledge of practical geometry that enabled him to have achieved the resulting structures or geometric patterns. If the Muslim artists, artisans, architects, builders, designers, carpenters and craftsmen knew geometry, they could not have acquired it spontaneously. They must have learned it, and therefore they must have been taught. But how were they taught? What knowledge of geometry was available for teaching? Who was teaching, and with what books or manuals? If such textbooks or manuscripts existed, then we should look for them, study their nature, clarify the problems that they resolved, distinguish what they considered as problematic in their own materials and find the geometric methods of construction they used to achieve the designs and patterns that are now recognized as artistic masterpieces. Such an approach would bring us closer both to an objective comprehension of the methods of design these artisans used and to understand the step-by-step process of Islamic geometric design.

In the summer of 1971, as I started my search for a thesis topic, the geometry and architectural decoration was still on my mind. I cannot help recall the initiation of this now rewarding project. I explained my observations about the development of Islamic geometric designs to my advisor and I expressed my wish to find a textbook of geometry or manuscript that was written specifically for artisans to teach them how to design and to study the manuscript in order to come to an objective understanding of Islamic geometric design and architectural decoration. The immediate response was that there was no such thing. To this I responded that I would look for it, and only if I don't find it would I be able to say there is no such thing. Thus, I was faced with a most emphatic assertion that my proposal and preliminary conclusion would never be confirmed by finding the material evidence of manuscripts written for the artisans, and the whole project was doomed to failure.

Such quick and categorical negation was typical of a widespread assumption on the part of those in the field that no such manuscripts or written documentation ever existed, an assumption that proved to be invalid. I was asked to broaden this topic in anticipation of failure at my original questions. I therefore included other relevant issues, such as a survey of those aspects of Islamic civilization that seem to reflect the widespread interest in geometry. This was to guard against failure, and possibly to serve as a stepping stone for documenting the influence of geometry in both the arts and society in general. The list of questions began to expand as I set about the task of finding the artisans' textbooks. For example, can one show that the interest in science or geometry was part of the average cultured person's background in the ninth or tenth century? What practical geometry had been developed by the tenth century? What caused the growth of this phenomenon? Geographically, where did it begin and in what directions did it spread?

My ultimate aim remained, however, to find out what kind of geometry was taught to artisans; what they knew; what problems they faced in their design; and, if geometric theory was available to artisans, how long it took before it was no longer the exclusive property of the scientific community. When did it filter down to the craftsmen and architects? Did science and are not only meet, but actively collaborate in Islamic civilization?

I tackled these questions from the standpoint of a strong background in art and in Islamic studies and with a general background in history, historiography and research methodology. The last three were due to two outstanding professors, Constantine Zuraiq of the American University of Beirut and George Maqdisi, then at Harvard. Their training provided the tools that are necessary to take advantage of the extensive resources of the Harvard library system. I read through catalogues and indices of manuscript collections available in libraries throughout the world. By the end of the week I had a large pile of index cards referring to geometry manuscripts. I sorted them to see which were edited, which had known authors, which had known contents, which were in what library or city, etc. And as I was going through the cities card stack, it struck me that there were several manuscripts in the Khudabakhsh Library in Patna, India. Another look at the index cards revealed that a number of the Patna manscripts were copied in the city of Mosul in northern Mesopotamia, very early in the thirteenth century (632 A.H./1234 A.D.). Two questions came to mind simultaneously: (1) How did those manuscripts get to Patna? (2) Someone must have been studying or working on geometry in Mosul ca 632 A.H./1234 A.D. Who was that?

Research on the history of Muslim schools and teaching in the thirteenth century revealed that there were two very famous scholars in Mosul at the time: Kamāl al-Dīn Yūnis bin Manʿa and Athīr al-Dīn al-Abharī. The former was recognized as the most outstanding teacher at the Muslim main school of the time in Mosul (the school was subsequently named after him "al-Madrasah al-Kamālīyah").

I reported the results of this first week of research—how I was led to Mosul and the teaching of Kamāl al-Dīn Yūnis bin Man'a whose tracks and findings I proposed to follow. To my surprise, this plan was immediately dismissed on the grounds that what I had identified was due to the Nestorian Church and activities related to its revival. I was dismayed, especially as the name of the teacher in Mosul seemed very Muslim. I continued my search for information about Kamāl al-Dīn Yūnis bin Man'a, and immediately discovered that an incredible amount of information was available. Nearly every historical source of that period that I consulted had an entry on him, and these included: Ibn Khallikān's history, *Obituaries of the Eminent Men and Histories of the Leading Contemporaries*; Ibn abī Uṣaybi'a's history, *The Choicest News of the Generations of Physicans and Scholars*; Ibn al-Fūwaṭī's work on thirteenth century history, *Comprehensive Occurrences and Beneficial Experiences of the Seventh Century of Hijra*; and, most incredibly, the history of the Muslim scholars of the *Shāfi'ī* theological school, *The Great Classes of Shāfi'ī Scholars* of al-Subkī [1]. This last source clearly classified Ibn Man'a among the outstanding Muslim scholars of the *Shāfi'ī* school. Brockelmann's *History of Arabic Literature* [2] listed his writings and revealed that there was at least one manuscript available of his works that could be of great interest to this research. To my amazement, a transliterated title appeared on Brockelmann's pages: *Risāla fīmā yaḥtāju ilayhī al-ṣāni'u min a'māl al-handasa*, on which Kamāl al-Dīn Yūnis bin Man'a had written a commentary entitled *Sharḥ al-a'māl al-handasiyya*. The title of the main work, or the subject of Ibn Man'a's commentary, means literally "A treatise on what the artisan needs of geometric problems" while the title of the commentary is "Commentary on the geometric problems". This main title corresponded exactly to the contents of the manuscript geometry textbooks I had visualized, yet I had never dreamt that it would be the actual title of a manuscript.

The work turned out to be that of Abū'l-Wafā' al-Būzjānī, a well-known scientist and mathematician who lived in Baghdad from approximately 945 A.D. until his death about 987 A.D. The geometry text was singled out as being an important document and of specific interest to the historians of Islamic art in 1855 by the Austrian historian of science, F. Woepcke, and there were a number of recensions of the manuscript available in various libraries across the world [3]. Thus by the third week I had located an example of the kind of manuscript I was looking for and identified its author.

At this point, several observations are in order. First, simple but correct reasoning and logic are the basis for most research. In general one should not deny the existence of something without first looking for it. Second, the emphatic response to my topic proposal, "there is no such thing", is quite revealing, and seems to reflect the prejudices of Western scholars after World War I. To them, Islamic civilization could not have been intellectual. They assumed that Muslim craftsmen were people of minimal knowledge and education, capable only of minimal creative expression, whose genius, if they were uncommonly intelligent, consisted exclusively in committing two or three patterns to memory. Their lifetimes were spent merely reproducing those two or three patterns. Such opinions also proliferated as a result of anecdotal stories told by earlier English travelers, for example, the following quotation from Archibald Christi:

> "Oriental workers carry intricate patterns in their heads and reproduce them easily without notes or guides. There is a story that tells of an English observer, seeing a most elaborate design painted directly on a ceiling by a young craftsman, [the observer] sought the artist's father to congratulate him on his son's ability, but the father replied that he regarded the boy as a dolt for he knew only one pattern, but his brother was indeed a genius—he knew three!" [4]

Ignorant Islamic craftsmen supposedly knew only the number of their ten fingers and this indicated the limit of their intelligence or education. So imbedded was this ridiculous assumption that in the summer of 1971 inquiry into the craftsmen's knowledge of geometry seemed absurd. No one inquired as to who designed those sophisticated patterns and how.

The third point is the dismissal of the possibility that observable scientific activity in the thirteenth century was Islamic, and its attribution to the Nestorian Eastern Church or its revival, i.e. to Christian civilization. The assumption that Islam and Islamic civilization brouht nothing new, that the Umayyad period simply inherited from the Byzantine and merely reproduced it in distorted ways, is an overwhelming barrier to progress in the field. Islamic civilization was never given the slightest chance of being scientific or pragmatic and of having intellectual activities. Thus, Kamāl al-Dīn Yūnis bin Manʿa was dismissed as a Christian from the Nestorian Church. Scientific geometry textbooks written for the artisans/architects were non-existent.

Three weeks after beginning my search, I located the artisans's geometry textbooks. Before departing for Europe, I found at the Harvard Fogg Library a reference in a book on ornament from 1910 to 1911, indicating the presence of a collection of architect's drawings, the "Mirza Akbar Collection" at the Victoria and Albert Museum in London [5]. There, it took the staff five days to locate the collection in storage. I was astounded by the number of drawings it contained and by the size of the collection. The staff of that section shared my surprise and gathered around the table in amazement at these drawings, which are of an architect's workshop from the eighteenth and nineteenth centuries. In 1981, I examined similar material in two Arab towns; it is still in the hands of the artisans today. These scrolls (Fig. 2) were not only the basic reference manual, but also served as a design book from which artisans chose the appropriate pattern to be used in architectural decoration or in the workshop.

My next stop was Paris to examine a Persian translation of the manuscript of Abū al-Wafā, al-Būzjānī. I had taken with me a shopping list, as George Maqdisi used to call it, of other items of possible interest in that library. The shopping list for the Bibliotheque Nationale included a mansuscript without title or author, mentioned in the catalog only as "a manuscript of geometry problems with geometric figures" [6]. My first glance at the folios of this manuscript made it clear that this was a much more important finding than the manuscript of Abū al-Wafā'. Here, the complex geometric patterns of design were recognizable in the drawings of the repeat units, which were distinctively illustrated. Also, in contrast to the simple shapes and polygons of Abū al-Wafā's manuscript, the complex geometric shapes in this manuscript, "On interlocking similar and congruent figures", indicated a much higher and later stage of development.

By the time I returned to Cambridge, I had located a range of written material, in the history of Islamic science and geometric design from the tenth century to the mid-nineteenth century, lying in library and museum storage rooms all over the world. In point of fact, my material turned out to be so convincing that it is now being used and propagated even by those who demonstrated such a strong skeptical attitude towards it at the beginning. Though locating the manuscripts took only two months, acquiring microfilms and/or photocopies of these documents without any backing or support took several years. Meanwhile, I was struggling to decipher the material, and to find an appropriate language in which to discuss it and describe the geometrical patterns with which it dealt.

At a very early point in my research, I was aware of the existence of an appropriate scientific language, namely that of Group Theory and Crystallography. As early as 1944, Edith Müller had written a thesis on Group Theory and Symmetry Notation of the patterns of Moorish ornament in the al-Hambra Palace [7]. Earlier, Andreas Speiser had called particular attention to Islamic art in his chapter on ornament in *Die Theorie der Gruppen von endlicher Ordnung* as early as 1927 [8]. It was only in 1935 that the scientific findings of Point Group Theory were fixed in the International Tables of Crystallography. These notations became the most widely used language by the chemists. It is possible that E. Müller had followed the line of inquiry that A. Speiser had suggested. She deserves considerable credit for systematically articulating the connection on a monument. Her thesis work takes into account the scientific theoretical finding and reveals the value of these scientific theories to the understanding and classification of Islamic geometric patterns. This information was neglected by all historians of Islamic art until the early eighties, and no one has even attempted to bring up the issue within the field. It is possible that the difficulty of her scientific languange makes her material and research inaccessible to the historian of art. As I tried to read that book, I realized that it was too complicated and too scientific for either students of Islamic art history or general art historians. Thus E. Müller's study remained the only one to apply Group Theory and Symmetry Notation to the study of Islamic geometric pattern.

To understand her book, I needed to understand scientific theory, and I spent endless days at the chemistry library studying the basics of Group Theory and the theory of symmetry and its notational systems. I found that the language was difficult and did not completely serve my purposes. There were so many different notation systems for the symmetry groups that it was confusing for an outsider to the field of chemistry to attempt to evaluate or select one. This was especially true

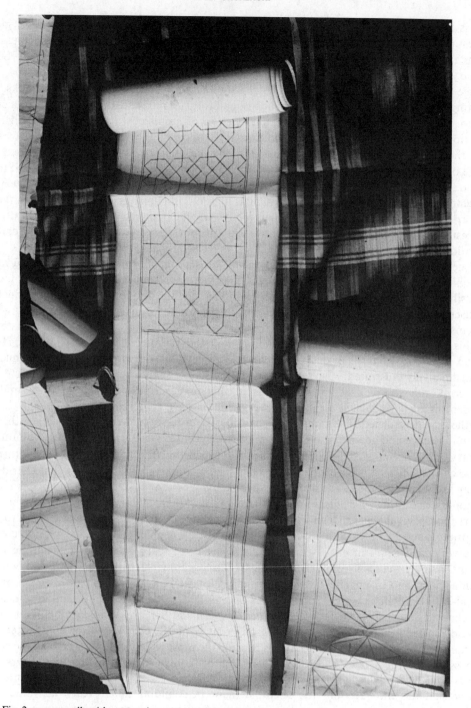

Fig. 2. paper scrolls with geometric patterns and drawings from the master-artisan's workshop in an Arab town 1982.

because I needed this notation as a tool to help classify the material that I was studying: I did not intend to add to the existing confusion in the field or Art History, specifically, the chaotic state of loss into which the study of pattern and ornament had fallen after an incredible blossoming at the end of the nineteenth century.

By the mid-seventies, the field of Islamic Art experienced an overwhelming surge of publications and revived interest. The number of books published increased tremendously relative to the previous few decades. Yet the standard of the publications was academically at its nadir, especially the study of geometry and ornament. This chaotic situation involves three concerns:

1. The preoccupation in the field of Islamic Art with issues of patronage—Royal, Princely and Mystical.

2. The proliferation of visual description and the psychology of visual perception as approaches to the studies.
3. The popularity of Linguistics, and later Semiotics, as the "sciences" within the arts whose languages were most developed, and that could be used as tools to attain a more scientific level of understanding of geometric pattern and art.

The shortcomings of the above concerns all point to the need for a scientific language and methodology with which to understand and systematically categorize Islamic geometric pattern.

1. Preoccupation in the Field of Islamic Art with Issues of Royal–Princely and Mystical Patronage

In the mid-seventies, there was a general precoccupation in much of the field with issues of patronage, both Royal–Princely and Mystical. I will not go into the wide range of problems created by the concentration on the Royal–Princely patronage or how and why it gained the main grounds in the field. Perhaps it was no coincidence that this was the period when oil money flowed and art historians attempted to lure these moneys to their field. In those years, the moneys of the imperial courts of Persia, in particular, played a very active role in the Islamic art scene. Exhibitions, conferences and publications multiplied.

A very well-known, active group of international mystics supported specific publications and pushed certain Islamic mystical ideas. Their main doctrine was that of the "Principle of the Unity of Being". They tried to show that, ultimately, all differences in outward manifestations are "inwardly united at the Center. They are the bridge from the periphery to the Center, from the relative to the Absolute, from the finite to the infinite, from multiplicity to Unity", quoting Seyyed Hossein Nasr [9]. The introductions and forewords of a number of books permeating such mystical themes were written by either Seyyed Hossein Nasr or Titus Burckhardt. These books include: *The Sense of Unity: the Sufi Tradition in Persian Architecture* by Nader Ardalan and Laleh Bakhtiar; *Sufi: Expression of the Mystic Quest* by Laleh Bakhtiar; *Islamic Patterns* by Keith Critchlow; and *Geometric Concepts in Islamic Art* by Issam el-Said and Ayse Parman. Frequently S. H. Nasr's main ideas are illustrated in geometric drawings in these books or are used in long quotes. A typical example is the circle with its center (Fig. 3) from N. Ardalan's *The Sense of Unity*, where these two drawings of the (*Ẓāhir*), the manifest of apparent, and the (*Bāṭin*), the hidden or internal, are represented as the center of the circle, stand for body and soul, respectively:

> "The Manifest (*Ẓāhir*): The Consideration of God as Hidden and the Manifested pertain to 'space'—to 'qualified' and 'sacred' space . . . Taken as Manifested, God becomes the reality that englobes all, that 'covers' and encompasses the cosmos. In this view, physical manifestation may be regarded as the innermost circle of a set of five concentric circles, followd by the other states of being respectively, with the outermost circle symbolizing the Divine Essence. . . ." [S. H. Nasr, *Science and Civilization in Islam*, p. 93.]
>
> "The Hidden (*Bāṭin*): [This] can be regarded as symbol of the microcosm, of man, in whom the physical is the most outwardly manifested aspect and his spiritual nature the most hidden. . . ." [S. H. Nasr, *Science and Civilization in Islam*, p. 94.] [10]

The "Principle of the Unity of Being" pervades the content of these works, sometimes even their titles. On occasion, it is pushed to a point of scientific fallacy such as the claim that all geometric patterns of Islamic art are derivable through a single method of construction based on the subdivision of the circle, in order to declare this art work an example of the "Unity of Being". This

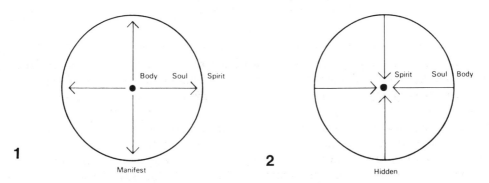

Figs 3.1 and 3.2. Mystical symbolism applied to the circle and its center in N. Ardalan's *The Sense of Unity* [10].

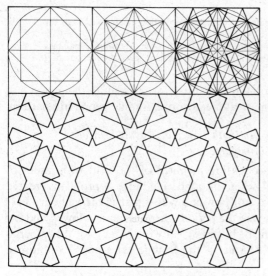

Fig. 4.1. The stress upon subdivision of the circle in deriving all geometric patterns. From I. El-Said, *Geometric
Concepts in Islamic Art* [12].

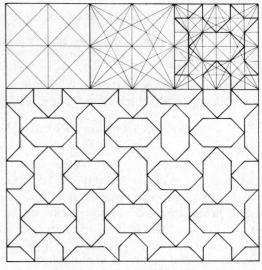

Fig. 4.2. The circle does not appear in deriving this
geometric pattern. From I. El-Said, *Geometric Concepts
in Islamic Art* [13].

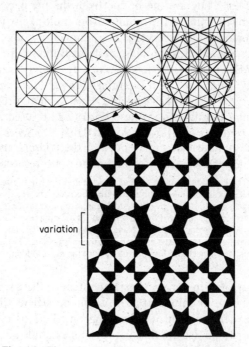

Fig. 4.3. The scheme of the circle does not fit the
long rectangular unit, a variation zone is marked. From
I. El-Said, *Geometric Concepts in Islamic Art* [14].

argument appears in *Geometric Concepts in Islamic Art* by I. El-Said. In his introduction to it, Titus
Burckhardt states that all the geometric patterns are derived by the same "method of deriving all
the vital proportions of a building (or a pattern) from the harmonious division of a circle . . . which
is no more than a symbolic way of expressing Unity (*Tawḥīd*), which is the metaphysical doctrine
of Divine Unity as the source and culmination of all diversity" [11]. This is illustrated in the mesh
of the subdivisions of the circle from which the pattern is developed (Fig. 4.1) [12]. In some cases,
however, the authors neglected to draw in the circle, ironically revealing how unfundamental its
existence is to the alleged "unique way" or "only way" of deriving all patterns (Fig. 4.2) [13]. And
finally, there are a few cases of designs where it was absolutely impossible to hide the fact that the
analytical method did not hold. These are illustrated (Fig. 4.3) as containing a non-standard zone

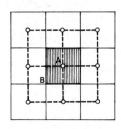

Fig. 5.1. Mystical interpretation given to the two distinct 4-fold symmetry centers of brickwork by L. Bakhtiar in *Sufi: Expressions of the Mystic Quest* [15].

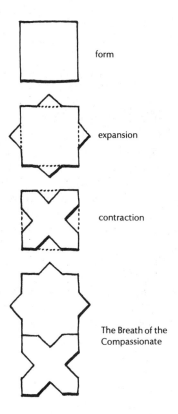

form

expansion

contraction

The Breath of the Compassionate

Fig. 5.2. Mystical symbolism applied to the 8-pointed star and cross pattern by L. Bakhtiar in *Sufi: Expressions of the Mystic Quest* [16].

Fig. 5.3. Very popular ceramic pattern with an 8-pointed star and cross pattern.

labeled as a "variation"! [14] The elongated rectangular area obviously belongs to a 2-fold symmetry group and cannot be masked by the generalization of the "one way" that is overwhelmingly represented in the square of a 4-fold symmetry group.

We know from scientific theory that there are 17 different and distinct groups of two-dimensional patterns that are periodic in two independent directions. The laws of symmetry for these 17 patterns have been established and recognized by the international scientific community of crystallographers since the mid-thirties. Yet here we are in the mid-seventies not acknowledging this fact and declaring that there is only one way to draw all the patterns. One cannot help but point out that the theme of the book *Geometric Concepts in Islamic Art* involves a scientific fallacy, in order to meet the demands of the desired mystical interpretation.

As to the last book, L. Bakhtiar's *Sufi: Expressions of the Mystic Quest*, a few remarks will suffice. In the right-hand image of the brickwork (Fig. 5.1) the points A and B, which are the distinct roto-centers for a 4-fold rotational symmetry operation 4 and 4′, are declared to be "*shuhūd*, presence [conscious witnessing of God's presence] (A), and *ghabat*, absence [unconscious witnessing of God's presence] (B), in a brickwork pattern at the Masjid-i-Hakīm Isfahān, Iran" [15]. The most popular pattern in Islamic geometric design, the interlocking 8-pointed star and cross (Fig. 5.2), becomes "form, expansion, contraction, the Breath of the Compassionate"! [16] This pattern exists in simple brickwork, in tiles, in wood and in solid gold! I wonder if the artisan who made this design thought of it as form, expansion, contraction and the Breath of the Compassionate God? Is this not a simple geometric design that involves a 4-point rotational symmetry? (Fig. 5.3) It is one thing to believe in mysticism and to follow in its practices and experience its positive effects. But it is a totally different matter when a new set of interpretations and symbols is created and propagated under the guise of historical truth. The symbolic mystical interpretations that have proliferated in these books on Islamic geometric design, pattern and ornament are based on a modern understanding of Islamic literature. There is no documented evidence that such interpretations were given to the art forms when they were created hundreds of years ago.

Earlier writings of S. H. Nasr reveal the core of a program to educate modern man to understand the language of symbolism in order to revitalize traditional sciences. He declares in *Man and Nature: the Spiritual Crisis of Modern Man* that:

> "Such a revitalization of the traditional sciences, however, requires a re-discovery of the true meaning of symbolism and the education of modern man to understand the language of symbolism in the same way that he is taught to master the languages of logic or mathematics."

The general public unfortunately remains unaware of this. If in these books, that are now readily available on the market, their authors had made clear that the presented views were modern understandings of old forms, turning them into symbols, there would be no reason to object. The problem lies in presenting these modern mystical views as historical truths, as if these symbols were the meanings at the time the art forms were created. The non-Islamicist who is exposed to these books will anachronistically assume that a modern interpretation is the historical truth. Where does one draw the line between true historical research and the creation of and attribution of symbolic meaning to forms from the past? How can we redeem the geometric shapes, forms and patterns from the shrouds of mystical interpretations in order to see the precise scientific design at their basis?

2. Visual Description, Perception and Arnheim

Meanwhile, a sudden surge in the popularity of several approaches in the theoretical field of Art History overwhelmed Islamic art. This situation has created an urgent need to extract the field from superficial analyses that do not contribute to the understanding of geometry and pattern.

The most prominent of these approaches was that of visual perception and visual description, popularized by Rudolf Arnheim, whose book had just been published and who was teaching at Harvard [18]. This art-historical approach stressed the process of selective vision, and its main approach to the study and understanding of art was based on visual description and on the psychological interpretation of perception, including such elements as balance, movement and tension. The emphasis was on what one sees and the gradual mechanics of what and how one sees. Thus, looking at the graphic drawings of a brick wall from the tomb of Muḥammad Makkī al-Zangānī in Kharraqān (486 A.H./1093 A.D.) through the detailed sequence in Figs 6.1–6.3 the viewer attempts to depict the possible selective mechanics of visual perception and description of the pattern. A person might first notice the V-forms, then the vertical X-forms or the horizontal X-forms, and only then realize that there are dots or circles, and need to balance them, connect them and see a square relationship or a bilateral relationship in the grouping. Similarly, the viewer starts perceiving a vertical or horizontal orientation, rhythm and repetition. As long as a person is looking, and whenever there is a stop or visual pause, there is an immediate process of selection and new perception. In all this, we remain on the surface level of description.

3. Linguistics and Semiotics as Fashion

While looking at these small geometric shapes (Figs 6.2 and 6.3) or similar ones, I often heard people proclaim loudly in discussions that this is a phoneme, this is a moneme or this is a signifier,

Fig. 6.1. General view of the Kharraqan tomb tower 1093 A.D. Photo: S. P. and H. N. Seherr-Thoss, *Design and Color in Islamic Architecture*, p. 53. Smithsonian Institution, Washington, D.C. (1968).

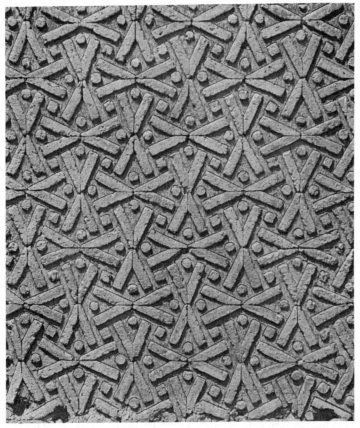

Fig. 6.2. Brickwork geometric pattern from the Kharraqan tomb tower 1093 A.D. Photo: S. P. and H. N. Seherr-Thoss, *Design and Color in Islamic Architecture*, p. 61. Smithsonian Institution, Washington, D.C. (1968).

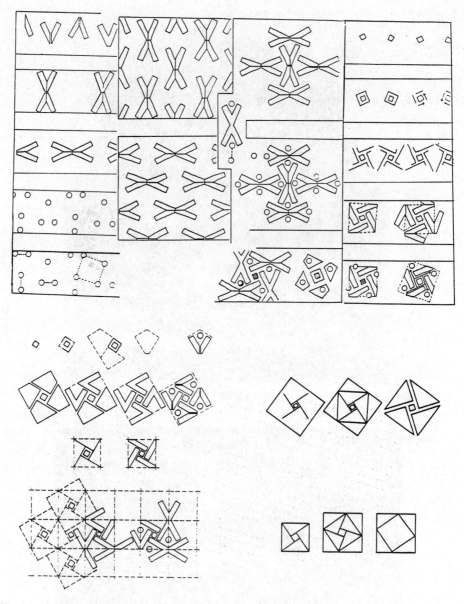

Fig. 6.3. Visual description and a sequential depiction of the underlying geometric structure of the Kharraqan brickwork pattern.

and I wondered what was being signified. This is the second influence, the fashion of linguistics, and its popularity was high at that time and persisted into the eighties. In the article "Symbols and signs in Islamic architecture", Oleg Grabar continues to use such terms of moneme, phoneme and semiotics: "a theme such as the muqarnas which involves nearly all morphemes of decoration..." [19]. What does that signify, I ask? If we are going to use linguistics, semiotics or appropriate another language, let us at least use one that can tell us something about the structure of this geometry. If we are going to keep picking up small elements, such as morphemes and larger elements such as monemes, what are we comprehending about the pattern or the design that is really there? There is also here a strong atomistic tendency that appears rooted in a predisposition by a certain political tendency among sociologists in particular and writers of contemporary Arab history to describe Islamic culture and civilization as "atomistic" and the society as "a mosaic". This maintained that there was no relationship between one part of that aggregate to the other. It was an aggregate at random. There was no structure. There were elements; everything was reduced to those elements. The art historians further legitimized this as a view by the way they looked at these Islamic

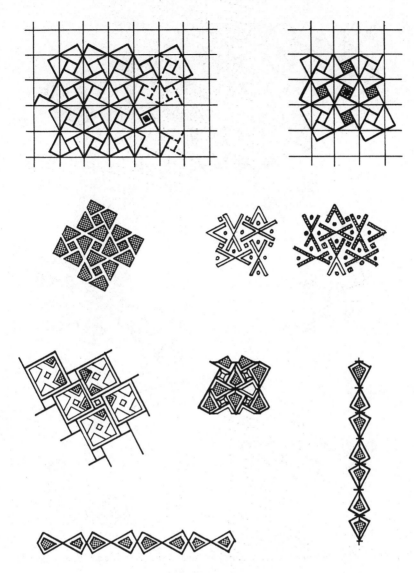

Fig. 6.4. Development of different patterns from the same underlying geometric structure of the Kharraqan brickwork pattern.

geometric patterns and mosaics. Yet if we continue to look at the pattern that we are describing, slowly and systematically, we cannot help seeing that there are some actual relationships that are taking place between these brick shapes; there is some expected order within the brick pattern; and it appears if we stop long enough to take a deeper look. Those geometric elements and brick shapes do not occur at random. "A whole is not the same as a simple juxtaposition of previously available elements", Jean Piaget insists in his *Structuralism* [20]. For if we now look at the circles that form a square (Fig. 6.3), we will see in their midst a real square and around the square we will see four lozenges in a 4-fold rotational pattern. If we return to where we started, we must ask these questions: Where is the morpheme? Where is the moneme? What is being signified? Where is the semiotic relationship here? It is obviously clear that this kind of terminology, borrowed from the linguistic and semiotic languages, is not going to get us far; for it cannot even tell us that there is a strong geometrical structure underlying this visual pattern of bricks; nor can it tell us that this tapestry of brickwork has a meaning or symbolizes and signifies a specific concept. This is not to say that the sophisticated language and methodology of the science of Linguistics and Semiotics cannot be of use as an analytical tool in other fields, particularly in literary analysis; but in the case of geometric patterns, we have already a precise scientific language developed for this purpose.

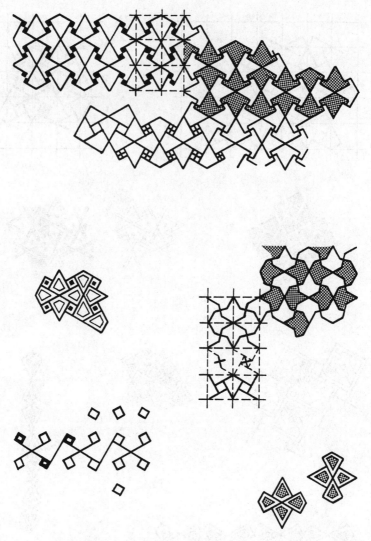

Fig. 6.5. Development of different patterns from the same underlying geometric structure of the Kharraqan
brickwork pattern.

THEORETIC RELATIONSHIPS AND DESIGN STRUCTURE

Moreover, regarding the underlying structure of this pattern (Figs 6.2 and 6.3) and judging
by the evidence found in the Paris manuscript "On interlocking similar and congruent figures"
(*Fī tadākhul al-ashkāl al-mutashābiha aw al-mutawāfiqa*) that I had been working on (Fig. 7), four
or five clear geometric construction steps can lead us to the underlying basic structure: if we use
symmetry operations to operate on, or moved around the elements of the structure, then different
relationships appear (Figs 6.4 and 6.5). In this way we have the development of different patterns
from the same underlying elements. As J. Piaget had seen, there is a strong relationship among those
elements that creates the process of composition, and together they create a whole. Compositions
have laws that regulate them. By means of such reasoning, we come very close to group structure
and Group Theory.

The final shape in the fifth step of the construction from the Paris manuscript (Fig. 7) is one of
the most often used designs in Islamic art. It is most popular in woodwork and in ceramics. A very
early example of its use in wood is in the door of the Mosque of Imam Ibrahim in Mosul (498 A.H./
1104 A.D.) (Fig. 8.). It is used in ceramic in the side wall of the Iwan of Masjid-i-Jami in Isfahan,
originally built around 515 A.H./1122 A.D. and redecorated later in 1112 A.H./1800 A.D. (Figs 9.1
and 9.2). This design is related to a very important problem in geometry, and has clear theoretical

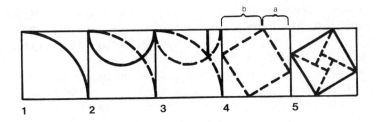

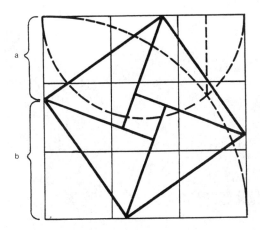

Fig. 7. Five-step construction of the underlying structure of the pattern as given in the Paris manuscript
Ancien fond Persan No. 169.

or scientific ramifications. In the Islamic tradition of practical geometry, as early as the time of Abū'l
Wafā' in the tenth century, there was a concentrated interest in the division of geometric areas such
as a triangle and a square. The square gained specific attention in two respects: (1) how to divide
it into a given number of squares (two, three or more), when the length of the side of one was known;
and (2) how to construct a square equal in size to the sum of the areas of two, three or more given
squares (Fig. 10). As we shall see, Classical Greek Geometry and the Pythagorian Theory deals with
a specific case of these problems. The Greek method of proof for the Pythagorian Theory is given
in Euclid's *Elements*, Book I, Proposition 47 (Fig. 11.1), which relies on a long proof of similar
triangles and application of areas, while the method in the Islamic manuscripts is closer, in its
dependence on dissection, to the Indian proof (Fig. 11.2) of Bhāskara (born 1114 A.D.), given
by Sir Thomas Heath in his commentary on *Euclid's Elements* [21]. The Islamic approach for
the artisans depends upon a practical method of proof (Fig. 12.1), in which the second square b
is bisected into two equal rectangles; the rectangles are then cut through their diagonal into two
triangles. The resulting four triangles are then placed around the square c, with their hypotenuses
adjacent to or coinciding on the sides of the square c. In the middle a square area is left, the smallest
square a is placed in it so as to fit the whole area. The visual clarity in this method allows the artisan
to dispense with the need for logical proof of the relationship: $a^2 + b^2 = c^2$. This does not mean
that the Muslim artisan and the scientist–geometrician, such as Abū'l Wafā', did not distinguish
between the necessity for logical proof or for following methods of construction that the scientist
had examined through correct proof, in contrast to the informal trial-and-error method. On the
contrary, in his chapter "On the division of the squares and their combination", he stresses that the
artisans should be aware that the reliance on trial and error in their constructions is not the route
to correctness, even though the drawing might seem to appear as visually correct. On the contrary,
the methods that the geometricians have demonstrated as correct through logical proof are the
methods that artisans should follow because, upon repetition, these methods will always prove
to be correct, unlike methods based on trial-and-error repetition or visual approximation. For a
geometrician, once a problem is correctly proved, visual appearance is of no further importance,

even if the drawing appeared visually incorrect or correct. Abū'l Wafā' recounts that in an assembly of scientist–geometricians and artisans the two groups arrived at different methodologies for the construction of one square from the sum of three squares. The artisans wanted to use the method of dissection of the squares and adding the cut parts together to construct the larger square. They also brought several other methods, some of which could be proved, others not, although those that could not be proved correct still appeared visually correct to the eye or through the visual imagination of the viewer. He shows some of the incorrect usage, he says, in order to make the artisans aware of both the correct and the incorrect methods, and so that they would clearly know enough to not accept the incorrect methods. Ultimately, the clever and dextrous artisan would only depend upon the method of proof and not on that of trial and error [22].

Fig. 8. Wooden door of the Mosque of Imam Ibrahim in Mosul dated 498 A.H./1104 A.D.

Fig. 9.1. The north-west Iwan of Masjid-i-Jami in Isfahan showing the use of the pattern in large scale in four places on the facade and inside the vault of the Iwan. Photo: S. P. and H. N. Seherr-Thoss, *Design and Color in Islamic Architecture*, p. 187. Smithsonian Institution, Washington, D.C. (1968).

This type of geometric algebra made available a number of mathematical and algebraic problems in drawn geometric illustrations that we see proliferating in these popular Islamic art designs, some of which are shown here (Figs 6.1–6.5, 8, 9.1 and 9.2), and which will appear again in the geometric problem from the Paris manuscript that will be discussed later (Figs 19.1–19.20). In rare instances, the architect–artisan seems to declare silently through visual evidence his precise knowledge of this geometric fact by placing his design in a prominent position of the decoration, as is the case with the treatment of the facade of the Iwan in the Isfahan, Masjid-i-Jami' (Fig. 9.1), or by actually having the area at the heart of the central square of the design carry his name or signature "This is the work of Muhammad Ibn Mu'min Muhammad Amin..." (Fig. 9.2); obviously in this placement he is declaring to posterity that he "knows and knows that he knows...", as the Arabic saying

Fig. 9.2. The ceramic panel showing the pattern in large scale and in the center of the small square carrying the name of the designer–artisan who made it. Photo: S. P. and H. N. Seherr-Thoss, *Design and Color in Islamic Architecture*, p. 189. Smithsonian Institution, Washington, D.C. (1968).

goes. The sides of the square (Fig. 12.2) indicate its division into two segments a and b where the sum of the side is equal to $a + b$ and $(a + b)^2 = a^2 + 2ab + b^2$. Notice that in the case of the design of the Isfahan wall ceramic (Fig. 9.2), the size of length a is half of b, but does not necessarily have to be so. It is possible that this specific case of the ratios $(a : b = 1 : 2)$ is used by the artisan because it simplifies the task of measuring and cutting. Figures 12.1 and 12.2 follow this specific convention, while Fig. 7 shows the more general form of the theorem where a does not have a direct proportion to b, or where the side of the small inner square is $b - a$. Thus:

$$(b - a)^2 + 4\left(\frac{ab}{2}\right) = c^2,$$

$$(b - a)^2 + 2ab = c^2,$$

$$b^2 - 2ab + a^2 + 2ab = c^2$$

and

$$b^2 + a^2 = c^2 \quad \text{or} \quad a^2 + b^2 = c^2.$$

In these geometric designs, the Muslim artisan has demonstrated two different important relationships: the Pythagorian Theorem and the expansion of the second degree binomial.

Here it is appropriate to return to the mystical interpretation to discuss this specific design. K. Critchlow includes it in his book on *Islamic Patterns* (Fig. 13); he bases his analysis on the dodecagon and the square and remarks:

> "This coincidence of the twelve and four suggests a zodiacal symbolism controlling or embracing the four axial kite shapes which can be taken to symbolize the four seasons, the four elements and the four qualities of heat and cold, moist and dry; the central sphere symbolizes the quintessence as a reflection of the bounding square." [23]

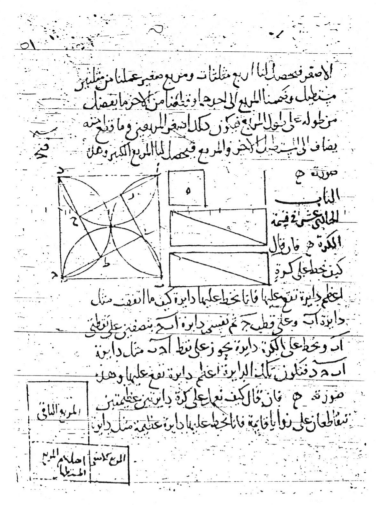

Fig. 10. Illustration showing one of the problems treating the construction of a large square. From the manuscript of Abū'l-Wafā' al-Būzjānī, Cairo, Dar al-Kutub.

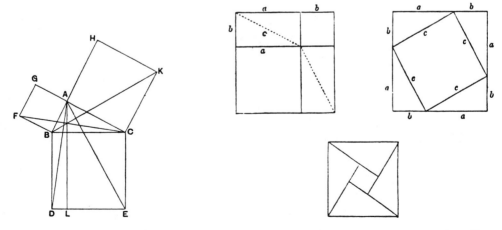

Fig. 11.1. The Greek method of proof of Euclid's Proposition 47. From Sir Thomas Heath, *Euclid's Elements* [21].

Fig. 11.2. The Indian method of Bhāskara. From Sir Thomas Heath, *Euclid's Elements* [21].

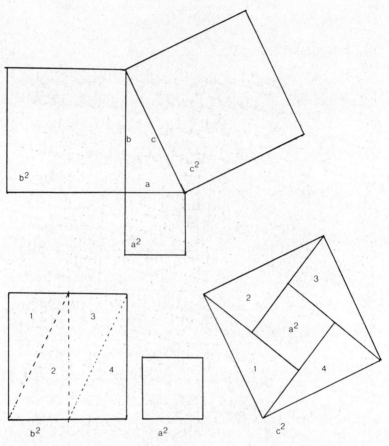

Fig. 12.1. The Islamic geometric design revealing the Islamic method for the visual presentation of the relationship $a^2 + b^2 = c^2$.

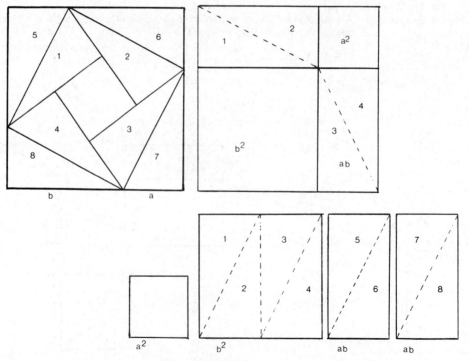

Fig. 12.2. The Islamic geometric design revealing the Islamic method for the visual presentation of the relationship $(a + b)^2 = a^2 + 2ab + b^2$.

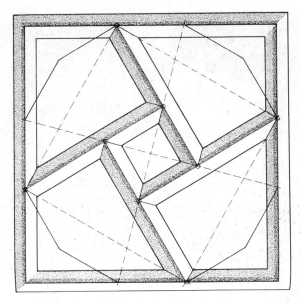

Fig. 13. The same Islamic design as given by K. Critchlow in *Islamic Patterns* [23].

I continue to question the attribution of these contemporary interpretations to the old traditional Islamic geometric forms. I wonder how it happended that, throughout the hundreds of folios of manuscripts I have examined dealing with Islamic geometric design I did not come across any such remarks or interpretation. Why did I not even find a marginal note, from later times, to this effect as we see in the illustration (Fig. 14) where there is plenty of space available for any one to subsequently add such a comment. As historians, our task is to come as close as possible to the original truth, depending on documents that are historically verifiable. The only comments in these manuscripts, that are non-scientific and that occur in these manuscripts, usually at the end of the given text and say: "...and Allah knows best". This is the only religious utterance that these Muslim scientists repeated time and again (Fig. 19; see Fig. 19.2), and there is no trace of any mystical or cosmological implication in this phrase. Rather, it reflects a very prominent Muslim belief in the humility of man before the knowledge of his Creator. Although the scientist is sure that his construction methods are correct because the geometric patterns produced are exact, still, even in this case of certainty, he humbly refrains from saying this is certain truth, but rather that true knowledge lies only with his God: Allah, the all knowing. This humble attitude of the scientist conforms to the general attitude or norm of Muslim beliefs.

THE NEED FOR A SCIENTIFIC LANGUAGE AND METHODOLOGY TO UNDERSTAND AND SYSTEMATICALLY CATEGORIZE AND DESCRIBE ISLAMIC GEOMETRIC PATTERNS

By the mid-seventies (1974–1976), I was fully concentrating on the Paris manuscript No. 169, "On the interlocking of similar or congruent figures", and working out in detail each problem found in it. In January 1977 something happened in my research. One day, the latest issue of *Scientific American*, January 1977, was brought to my attention because of an article in it on tiles and pattern that I could be interested in. As I looked at the cover, I recognized one of the geometric shapes that I was working on. This seemed unusual since the article was announcing contemporary scientific findings of a new mathematical relationship and a new geometric shape. I was sure, however, that I knew that geometric shape. Later, I took the magazine and some pages from my work and from the Islamic manuscript that I had been studying to two of my professors. It was a painful surprise that there was absolutely no recognition on their part of what I had seen nor of its significance. What was announced in *Scientific American*, by Martin Gardner in "Theory of tiles; extraordinary non-periodic tiling that enriches the theory of tiles", as a new finding by Roger Penrose of two shapes that can tessellate the space to infinity in a non-discrete group pattern (that extends to infinity

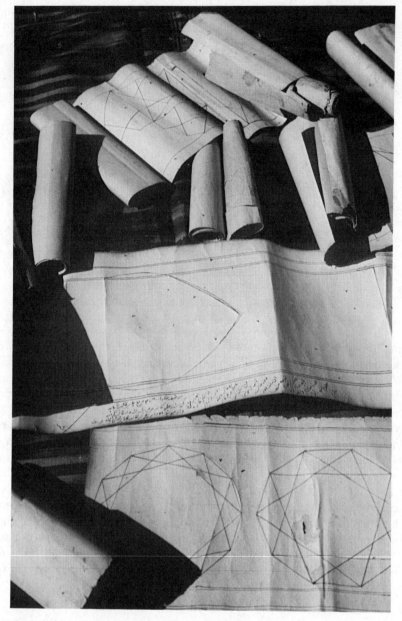

Fig. 14. Islamic artisan's scrolls with geometric drawings showing how marginal notes could be added.

but does not repeat) was not new! (Fig. 15) [24] Though the theory of non-periodic tiling does not appear, this geometric configuration of the shapes and pattern was in the Paris manuscript and it was described as a relationship derived from the decagon and a pentagonal star, named in the manuscript the "pentagonal seal" (Fig. 16). These so-called new shapes and pattern that are now even used by material scientists to analyze the structure of new materials (quasicrystals, schechtmanite) [25] and are now called Penrose tiling, have a geometric form or shape that has been known for hundreds of years by Muslim scientist–designers. Only the method of drawing it and the relationships it involves are different: while the modern Penrose method depends on the golden ratios and golden triangles, stressing the value of the angles, the old Islamic method uses the center angle for the decagon but stresses the ratio and relationship of the length of the radii of the decagon and its sides, also stressing the proportions of the lengths of these lines. This will be discussed in more detail in a future publication.†

---

†The details of this problem will be published in a joint paper with Arthur Loeb in the near future.

SCIENTIFIC AMERICAN

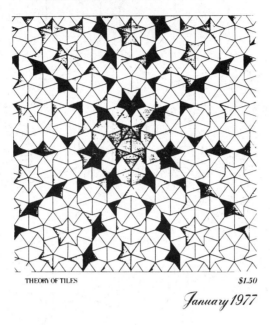

THEORY OF TILES $1.50

January 1977

Fig. 15. The cover of *Scientific American*, with decagon
and pentagonal star of the Penrose tiles [24].

At this time I was still plowing through chemistry books, and one day by chance I came across one whose title attracted me: *Color and Symmetry* by Arthur Loeb, who taught at Harvard [26]. I attended his first class meeting of the second semester of the academic year. I showed him the material I was working on and the article by Gardner which he not only knew but had discussed in his last class. The wisdom in an Arabic saying declares:

"He who knows not and knows not that he knows not, shun him. And he who knows not and knows that he knows not, awaken him. And he who knows and knows that he knows, follow him."

Quickly I found that A. Loeb had the language I was seeking, and that he had clarified and made more accessible to art students the complicated language of various symmetry notation systems and even the International Tables of Crystallography (Fig. 17). My study of symmetry had led to many books. Most pertinent to the arts was Hermann Weyl's classic, *Symmetry* (1952), HSM. Coxeter's text, *Introduction to Geometry* (1961) and A. V. Shubnikov and V. A. Koptsik's *Symmetry in Science and Art* (1972; translated into English by G. D. Archard, 1974) [27]. None of these were designed to serve the specific needs of art and design. Rather, they expanded on the discussion of symmetry; the second used the notation of the International Tables of Crystallography, and the third included immensely detailed and exhaustive enumerations that go far beyond what an art historian needs.

A good example of the multiplicity of languages and the confusion of tongues is the comparison table showing seven of the more popular notation systems for the Plane Symmetry Groups (Fig. 18), included within a very short article, "The Plane Symmetry Group: their recognition and notation" by Doris Schattschneider in *The American Mathematical Monthly* (June–July 1978) [28]. In contrast to these, A. Leob's notation system clarified each and every center point of the symmetry group; gave the symmetry value at each of the symmetry roto-centers (which is of great use to the art student and to the designer artist or architect); explained the role of the interaction of circles in space and revealed the properties of space, fundamental regions and unit cells, reflections and enantio-morphy, and distinguished between mirrors and glides by underlining the symmetry value number for the centers that fall on mirror lines and using an inverted v($^\wedge$) over the symmetry number of glide point centers. Moreover, his system could be easily taught to students of art, in a very short time,

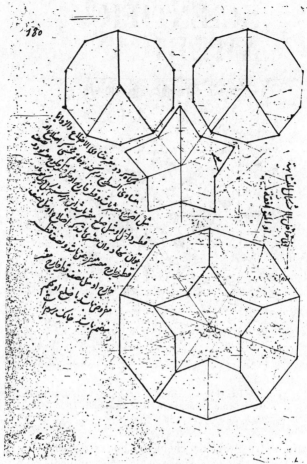

Fig. 16. Folio 180a from the Paris Islamic geometric
manuscript "On the interlocking of similar or congruent
figures", showing the design with the decagon and pentagonal star similar to the Penrose tiles.

for in a sense a large part of it is geared to them. Most essential is that he also recognizes that the problem we face in the field is again one of a multitude of languages, that this is causing a confusion of tongues, and that, ultimately, we have to choose one language in which to communicate with each other.

The multiplicity of different notational systems has also already crept into the computer world. By 1987 at least half a dozen software graphics programs based on symmetry became available on the market. Each used its own notational system and codes or code names to the symmetry patterns of Group Theory that the programs produced. In some cases, an approximate word is used that describes an image, such as flower, snow flake, or another fancy word far less precise than the original language developed by the crystallographers in 1935. The books on symmetry do not do any better and they tend to add to the confusion caused by a multiplicity of tongues. The academic community concerned with this subject must decide how to re-establish the use of a consistent and specific set of notation as a standard reference for this limited number of patterns.

The importance of Group Theory and its notational system for Islamic art lies in the fact that it provides a tool for exact cataloging of the infinite number of geometric designs used in Islamic art. It is also helpful as an analytical tool in recognizing the symmetry used within a design. Moreover, it provides a precise language and terminology by which those who are interested in these patterns can communicate precisely with each other about these patterns. All this might seem redundant to the scientists who have been involved in the study of symmetry yet, for the art historians, it is still an unacknowledged tool.

Table 3   Exhaustive list of configurations of symmetry elements in the plane

Number of distinct rotocenters with finite symmetry value	Combination of symmetry values	Configurations	I.T. notation	Illustrated by Figure number
0	1	1		32
		$1m$		
	∞	∞		8
		$\infty mm'$		36
		∞m		35
		∞g		34
	$1\infty\infty$	$1\infty\infty'$	$p1$	20
		$1\infty\infty'mm'$	pm	37
		$1\infty\infty'mg$	cm	39
		$1\infty\infty'gg'$	pg	38
1	k	k	C_k	7b
		km	D_k	33
2	$22'\infty$	$22'\infty$		12b
		$\underline{22}'\infty$		41
		$2\hat{2}\infty$		40
3	236	236	$p6$	15, 18
		$\underline{236}$	$p6m$	42
	244	$244'$	$p4$	17
		$\underline{244}'$	$p4m$	44
		$\underline{244}$	$p4g$	43
	333	$33'3''$	$p3$	16
		$\underline{33}'3''$	$p3m1$	46
		$3\hat{3}3'$	$p31m$	45
4	2222	$22'2''2''''$	$p2$	23
		$\underline{22}'2''\underline{2}'''$	pmm	52
		$2\hat{2}2'2''$	cmm	51
		$2\hat{2}2'\hat{2}'g/g'$	pgg	48
		$2\hat{2}2'\hat{2}'m/g$	pmg	49

Fig. 17. The Plane Symmetry Group notation table of A. Loeb [26].

COMPARISON OF NOTATION FOR THE PLANE SYMMETRY GROUPS

Internat'l (short)	Pólya; Guggenheimer	Niggli	Speiser	Fejes Tóth; Cadwell	Shubnikov-Koptsik	Wells Bell & Fletcher
$p1$	C_1	C_1^{I}	C_1, Abb. 17	W_1	$(b/a)1$	1
$p2$	C_2	C_2^{I}	C_2, Abb. 18	W_2	$(b/a):2$	2
pm	D_1kk	C_s^{I}	C_s^{I}, Abb. 19	W_1^2	$(b:a):m$	3
pg	D_1gg	C_s^{II}	C_s^{II}, Abb. 20	W_1^3	$(b:a):\tilde{b}$	4
cm	D_1kg	C_s^{III}	C_s^{III}, Abb. 21	W_1^1	$(a/a)/m$	8
pmm	D_2kkkk	C_{2v}^{I}	C_{2v}^{I}, Abb. 22	W_2^2	$(b:a):2\cdot m$	5
pmg	D_2kkgg	C_{2v}^{III}	C_{2v}^{III}, Abb. 24	W_2^3	$(b:a):m:\tilde{a}$	6
pgg	D_2gggg	C_{2v}^{II}	C_{2v}^{II}, Abb. 23	W_2^4	$(b:a):\tilde{b}:\tilde{a}$	7
cmm	D_2kgkg	C_{2v}^{IV}	C_{2v}^{IV}, Abb. 25	W_2^1	$(a/a):2\cdot m$	9
$p4$	C_4	C_4^{I}	C_4, Abb. 26	W_4	$(a:a):4$	10
$p4m$	D_4^{*}	C_{4v}^{I}	C_{4v}^{I}, Abb. 27	W_4^1	$(a:a):4\cdot m$	11
$p4g$	D_4°	C_{4v}^{II}	C_{4v}^{II}, Abb. 28	W_4^2	$(a:a):4\odot\tilde{a}$	12
$p3$	C_3	C_3^{I}	C_3, Abb. 29	W_3	$(a/a):3$	13
$p3m1$	D_3^{*}	C_{3v}^{I}	C_{3v}^{II}, Abb. 31	W_3^1	$(a/a):m\cdot 3$	15
$p31m$	D_3°	C_{3v}^{II}	C_{3v}^{I}, Abb. 30	W_3^2	$(a/a)\cdot m\cdot 3$	14
$p6$	C_6	C_6^{I}	C_6, Abb. 32	W_6	$(a/a):6$	16
$p6m$	D_6	C_{6v}^{I}	C_{6v}, Abb. 33	W_6^1	$(a/a):m\cdot 6$ (some alternatives exist)	17

CHART 6.   Sources referred to in the table are listed in the References. The groups are listed in consecutive order as they appear in the International Tables of X-ray Crystallography, [13]. Note that Speiser interchanges the Niggli notations of C_{3v}^{I} and C_{3v}^{II} (figure numbers in the Speiser column are for the 2nd, 3rd, and 4th editions of his book).

Fig. 18. D. Schattschneider's comparative table of notation for Plane Symmetry Groups [28].

AN EXAMPLE OF THE PROCESS OF ISLAMIC GEOMETRIC DESIGN

Stepping beyond the symmetries of Islamic geometric art will lead us to the mechanism of the process of Islamic geometric design. It is only when we follow the step-by-step procedure of constructing the geometric design that we come to understand it in full. The manuscript evidence and documentation thus becomes of primary importance, for it alone can lead us to the heart of this matter. It is not enough to give a quick translation or even an edition of the text in the original languages: neither can by itself lead to comprehension of the process of design. This is why it is essential to study in detail the scientific manuscripts and old documents that are now available to us, for they can bring us closest to the true historical procedure of the Islamic science of design. It is also essential to understand their scientific significance, because only through this can we place them in the larger context and recognize their importance in the science of geometric design. For instance, the following example from the Paris manuscript folio 192b turns out to be a fascinating one in its use of a strict algorithm with irrational numbers. Though based on a very strict algorithm and set of proportion, as we shall see, this geometric method of design and its forms is not a closed, dead-end system. It is rather one whose strength lies in its simplicity and strictness of derivation, for these two characteristics give it the open power for generating an infinite number of design variations from a single simple set of proportions.

In terms of working with primary sources, folio 192b is a problematic one since it has four illustrations and only three texts (Fig. 19). One assumes that three illustrations have a text while one illustration does not, and that the text closest to each illustration belongs to it. After translating the texts and comparing it to the illustrations, I found that text and illustration did not seem to correspond. I therefore proceeded to analyze the geometry of the illustration and reconstruct the design. Time and again, I came back to the texts and the folio that had now become exactly like

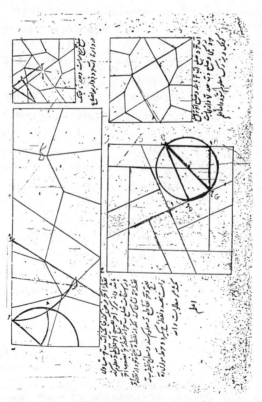

Fig. 19. A sample geometric problem from the Paris manuscript "On the interlocking of similar or congruent figures", folio 192b.

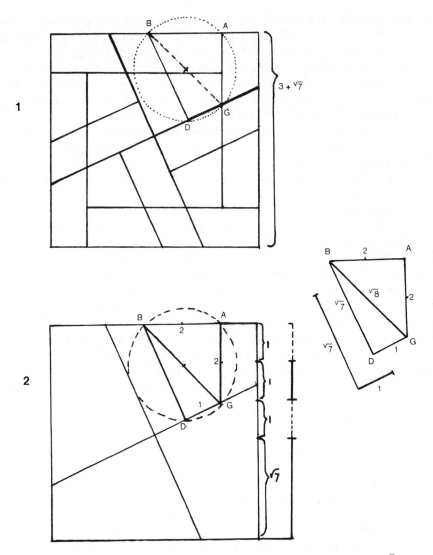

Fig. 19.1. The given geometric problem designed with a square unit: $3 + \sqrt{7}$.

Fig. 19.2. The given quadrangle ABGD with proportions $1:2:2:\sqrt{7}$ inscribed with a circle.

a puzzle. While I was analyzing the geometry of the illustration in the upper right corner, some specific numbers and irrational numbers emerged. One day as I began reading the text in the upper left corner, the numbers looked familiar. Suddenly, I realized that these numbers related in values to the asymmetric quadrangle of the illustration in the upper right side of the folio. The pieces of the puzzle fell together. The two upper texts should be linked to make one text, and this one text belonged not to the illustrations next to which it was copied, but to the illustration in the middle right side of the folio. Clearly this was a copyist's error, which in turn tells us something else about this unique manuscript: that it is copied and thus there must have been another manuscript like it. My main hope here is that the original manuscript has survived somewhere and will be recognized or discovered one day. Using another manuscript, one can resolve some of these textual problems.

The art historian Midhat S. Bulatov dealt with this problem in his book *Geometric Harmonization in the Architecture of Central Asia of the 9th–15th Centuries* (*Geometricheskaiia Garmonizatisiia Arkhitektury Srednei Azii IX–XV vv.*), published in Moscow in 1978 (the text of the original Persian manuscript was translated into Russian by Vil'danova). I differ with M. S. Bulatov on several points of reconstruction regarding this problem which he treats very briefly. He improperly concludes that this problem was not accompanied by an explanatory text. He states that "the following description of the construction (in the original text of the manuscript) does not correspond to the drawing, which is decoded in the following way . . ." [29]. In other words, the problem is solved independently from the text of the manuscript even though Vil'danova had translated the first two texts that belong to this geometric problem (illustrated in the middle right side of the folio).

When dealing with original documents, it is permissible to take some liberty in proposing the solutions to the problems if there is no text available for the drawn geometric illustrations. However, in these solutions would only be approximations to the process of the original Islamic method. After all, there could be as many different ways of handling a problem as there are contemporary persons interested in treating it. One should treat these documents with the same precision and respect which any historical document is given. Every detail of the original has to be worked out in full, to attempt to approximate as closely as possible the original truth, just as any historian would document an episode in the past. After one has verified it, then there are interpretations and implications that could be suggested or given.

In the case of treating geometric drawings or illustrations in manuscripts one has to examine the original physical manuscript. This is a vital issue, for many construction marks were only done by the sharp needle end of the copyist's metal divider to lightly scratch the paper surface marking these points of construction that were not inked. This fact means that photographs and microfilms are inevitably incomplete documents for they cannot depict these uninked marks and the original must be checked for any complete investigation. In general, the Soviet teams of scholars are limited to work on microfilms of manuscripts and without being able to examine the original manuscripts. They should be commended for the efforts that they have invested in returning to these primary sources which the rest of the Western scholars have demonstrated a skeptical attitude toward, a lack of interest in, as well as an incapability of retrieving and handling them. Second, one would wish that they had more access to some of the main issues and specific scientific findings on the international scene that are of great aid in recognizing the global significance of these original documents in the history of Islamic sciences.

The following segment gives the steps of the process of this design and their implications as they relate to other issues of interest or conern to geometry and design:

Fig. 19.1. In the texts the length of the side of the square is given as: $3 + \sqrt{7}$.

Fig. 19.2. The quadrangle ABGD inscribed inside the circle has as the length of its sides: side AB = 2 units, side AG = 2 units and side DB = $\sqrt{7}$. And the following comment is made: "And from here the small and large proportions are determined, and Allāh knows best."

Let us look in detail at the mathematical specifics of this quadrangle and see how it is generated in a very simple but strict algorithm.

Fig. 19.3. This shows an isosceles right-angle triangle whose sides are each given the length of 2 units, making its hypotenuse BG = $\sqrt{8}$.

Fig. 19.4. The midpoint of the hypotenuse BG is taken as the center of a circle and an arc of this circle is drawn such that angle A is on its circumference, such that all three vertices of the triangle lie on the circumference. The length of 1 unit compass opening is marked as D from G on the circumference, where GD = 1. Point D is connected to B by line DB. Line DB = $\sqrt{7}$, it follows that angle D is a right angle.

Fig. 19.5. This shows the second right-angle triangle with proportions $1:\sqrt{7}:\sqrt{8}$.

Fig. 19.6. These two triangles together make the kite shape ABGD.

Fig. 19.7. All vertices of the kite-shaped asymmetric polygon ABGD, with side lengths of 1, 2, 2, $\sqrt{7}$, lie on the circumference of the circle.

Fig. 19.8. When mirror-reflected on the side BD of length $\sqrt{7}$, the quadrangle kite-shape creates a semiregular pentagon that has all of its sides equal (2 units), while its two opposite angles A and A′ are both right angles.

Fig. 19.9. This shows how a cruciform area of a width equal to 1 unit is created when a gnomon with the width = 1 added from points A (Fig. 19.9b) and D (Fig. 19.9c), on both sides AG and BD, moving A to A′, B to B′, G to G′ and D to D′. A third strip unit with width equal to 1 unit is measured from point A on AG and drawn parallel to AB (Fig. 19.9d). Adding the first two gnomons of 1 unit to the quadrangle ABGD makes it retain its original proportions $1:2:2:\sqrt{7}$ in the larger asymmetric quadrangle kite shape A′B′C′D′ (Fig. 19.9e), which is then rotated four times at its right angles D′ to form the large square unit (Fig. 19.9f).

Fig. 19.10. This shows the side equal to 1 unit measured three times, and $\sqrt{7}$ adding to $3 + \sqrt{7}$.

The 1-unit strip (as a result of the third strip seen in Fig. 19.9) is shown running around the borders of the square.

Fig. 19.11. The quadrangle is rotated within the square, showing all the added lines as a result of the addition of the gnomon; also, the different segments of proportions $1:2:\sqrt{7}$ are depicted on the side, as the rotated quadrangle generates the larger square unit of side length $3+\sqrt{7}$.

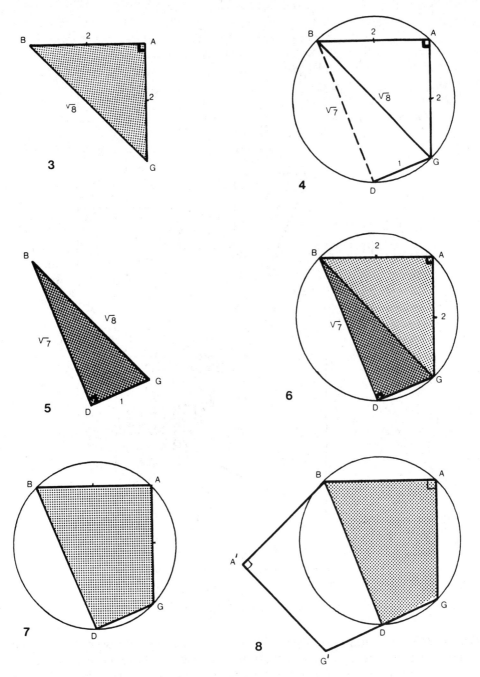

Fig. 19.3. The first triangle is a right-angled isosceles with proportions $2:2:\sqrt{8}$.

Fig. 19.4. The first triangle is circumscribed and a unit of length 1 is marked on the circumference.

Fig. 19.5. The second triangle DBG, proportions $1:\sqrt{7}:\sqrt{8}$.

Fig. 19.6. The two triangles combined to form the asymmetric quadrangle ABGD.

Fig. 19.7. The asymmetric quadrangle ABGD with sides 1, 2, 2, $\sqrt{7}$.

Fig. 19.8. The semiregular pentagon with side lengths of 2 units.

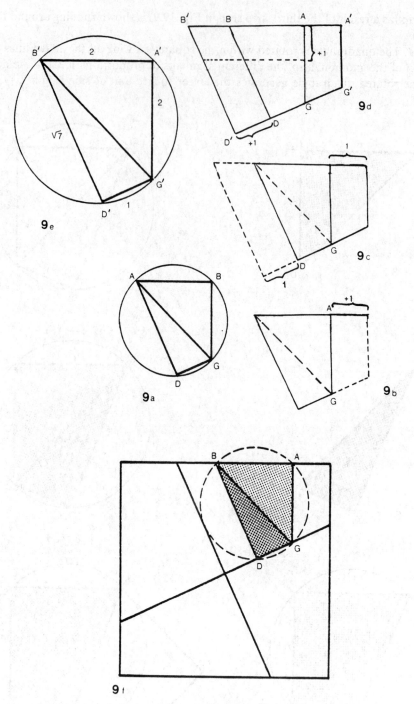

Fig. 19.9. The addition of gnomon strips to enlarge the quadrangle but retain the original proportions.

Fig. 19.12. The final drawing of the 4-fold rotation around a point is seen. This is made possible due to the characteristic proportions and opposite right angles of the symmetric quadrangle AGBD.

Fig. 19.13. For easier visual reading, the quadrangles are colored. The sides of the square indicate its division into two segments (a) and (b) where: the sum of the side is $= a + b$; and $(a + b)^2 = a^2 + 2ab + b^2$.

Fig. 19.14. When we apply the symmetry operations of mirror reflection and glides to the whole unit a pattern will develop to tessellate the plane, in a $2\hat{4}\hat{4}$ or $p4g$. It has a 4-fold roto-center through which two perpendicular glide lines pass, to carry them to the next enantiomorphic 4-fold symmetry center. The 2-fold roto-centers are located at the corners of the square, lying on two mirrors inter-

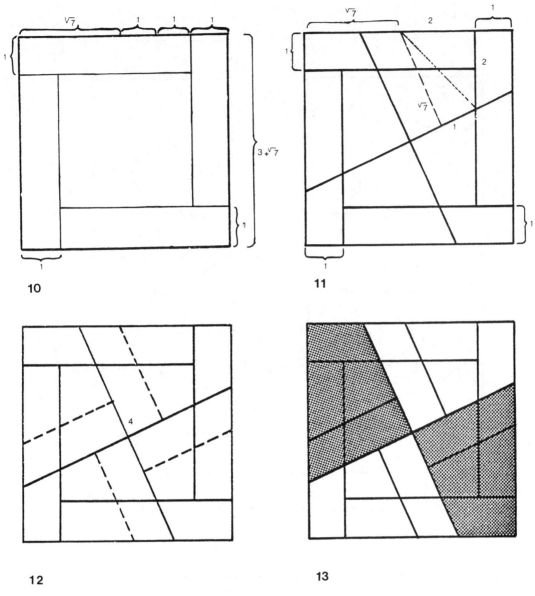

Fig. 19.10. The large square unit of side length $3 + \sqrt{7}$.

Fig. 19.11. The different proportions of the quadrangle in 4-fold rotation to generate the large square unit of side length $3 + \sqrt{7}$.

Fig. 19.12. Center for the 4-fold rotation of the quadrangle ABGD with all the line subdivisions.

Fig. 19.13. The quadrangles shaded for easier visual identification, dividing the side into two segments.

secting at right angles. So the 2-fold centers lie on mirrors, one 4-fold rotational symmetry center is a glide image of the other 4-fold rotational symmetry center.

Fig. 19.15. This shows the pattern colored in the simplest manner to reveal four small kite shapes rotating around the center of each square. The coloring also reveals how this pattern would be a very good subject for ceramic or woodwork, requiring only three different shapes: a small symmetric kite, a lozenge and a quadrangle.

Fig. 19.16a. When the quadrangle is taken without the subdivisions and is repeated through the symmetry operations, we can clearly see the semiregular pentagons and how they tessellate the plane in 4-fold symmetry $24\hat{4}$. The two opposite right angles of the semiregular pentagon allow for the 4-fold rotation.

$\underline{2}\,4\,\overset{\wedge}{4}$ = P4g

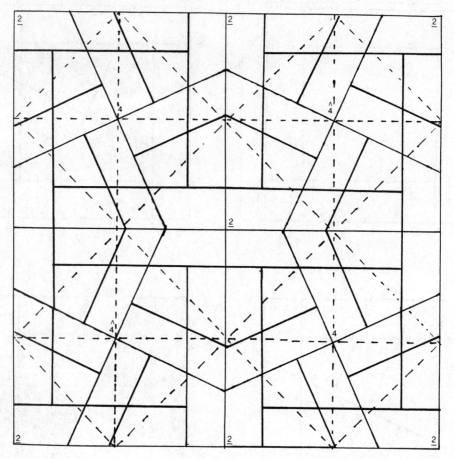

Fig. 19.14. The quadrangle repeated in a $\underline{2}4\overset{\wedge}{4}$ symmetry operation; the 2-fold centers are on perpendicular intersecting mirror lines, while the two 4-fold centers are on glide lines.

Fig. 19.15. A simple shading for the pattern.

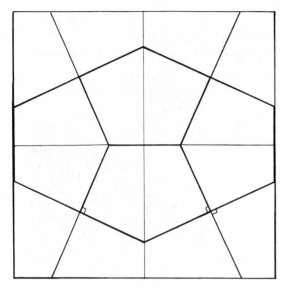

Fig. 19.16a. The repeated quadrangle without the subdivisions creates $\underline{24}4$ tessellation with semiregular pentagons.

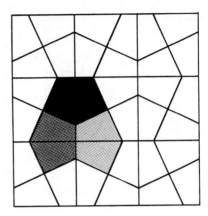

Fig. 19.16b. The pentagons are shaded for easier visual identification.

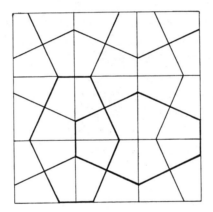

16_1

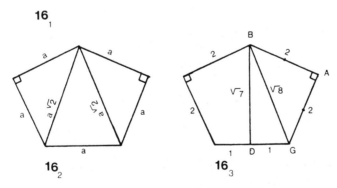

16_2

16_3

Fig. 19.16c$_{1-3}$. The pattern of a "favorite street tiling in Cairo" of a tessellation with pentagons and a comparison of the Islamic and Western derivation of the semiregular pentagons.

Fig. 19.16b. Here, the semiregular pentagons are colored in three shades to make them easier to see.

Fig. 19.16c$_{2-3}$. Two different semiregular pentagons are drawn at the bottom of the page. On the right side is the Islamic pentagon, where $\sqrt{7}$ is the critical value in the design. On the left is the Western one given by J. A. Dunn in an article on "Tessellations with pentagons" [30]. Dunn's

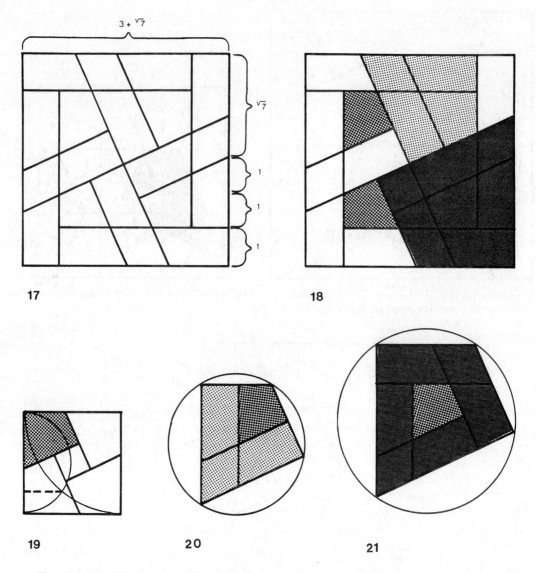

Fig. 19.17. The 4-fold rotation of the asymmetric kite-shaped quadrangle ABGD leaves no empty space
at the center of the larger square unit $(3 + \sqrt{7})$.

Fig. 19.18. The pattern is shaded to show three similar but different sized symmetric kite shapes.

Fig. 19.19. The small symmetric kite shapes in 4-fold rotation around a point leave a square at the center.

Fig. 19.20. The symmetric kite shape in the medium size.

Fig. 19.21. The symmetric kite shape in the larger size.

pentagon has an isosceles pentagon triangle that has the critical length of $a\sqrt{2}$ for the two equal
sides while the third side is a or any given length.

This tiling (Fig. 19.16c_1) is referred to as the "favorite street tiling in Cairo". In it, the tessellation
is considered hexagonal, each hexagon being a combination of four semiregular pentagons.
However, this tessellation is based on the 4-fold rotation of the semiregular pentagon, with sides
equal to two units and two opposite right angles. The latter characteristic permits the 4-fold rotation
of symmetry group $2\hat{4}4$ or p4g.

Fig. 19.17. The addition of the gnomons of 1-unit width on sides AG and BD of the quadrangle
kite shape of side proportions $2:2:1:\sqrt{7}$ allows it to retain the original set of proportions in the
larger form, as was seen in Fig. 19.9. When this larger asymmetric kite-shaped quadrangle is rotated
four times around a point, it creates a larger square (the square of sides $3 + \sqrt{7}$ given in the text
of the manuscript) leaving no empty area at the center.

Fig. 19.18–19–21. Within the square we see that there are three symmetric kite-shaped quadrangles of different sizes that have mirror symmetry, making each of the two sides equal to the other (note Figs 19.9 and 19.11). The kite shapes are similar in proportions but have three different sizes.

Fig. 19.19. When these symmetric kite shapes are rotated four times around a point, they form a larger square and leave a smaller square at the center. We can now see how this pattern relates to the design on the wooden door of Imam Ibrahim, Mosul, and to the design from the Isfahan Masjid-i-Jami (Figs 7–8). The sides of the square indicate its division into segments a and b where the sum of the sides is = a + b, and $(a + b)^2 = a^2 + 2ab + b^2$. This is exactly like the first set that had the theoretical concentration, presented in the visual description and the structural analysis from the brick wall of the Kharraqan Towers.

The main lesson that can be learned from this detailed example of documented manuscript evidence of the process of Islamic geometric design is that there is a specific way to design and construct each geometric design. Clearly, one can generalize that each problem of geometric design that is documented has its own steps of construction and that other than the Diophantine equation [31],

$$\frac{1}{k} + \frac{1}{l} + \frac{1}{m} = 1,$$

contrary to the claims of the adherents of the mystical interpretation in the seventies, there is no other single formula for deriving all the geometric patterns in Islamic art. The so-called unique method of construction is based upon the subdivision of the circle; it is not a replacement of a scientific formula and it cannot be propagated as the unique way to derive and construct all Islamic geometric patterns. The science of symmetry of patterns tell us that there are 17 different periodic two-dimensional groups and 7 groups periodic in a single direction (string or ribbon), also that each of these groups could have an infinite number of different designs. As seen, these Islamic geometric manuscripts give us samples of the infinite design variations of the basic 17 periodic groups; these documented geometric problems or examples in turn could be the basis for developing many new sets of design.

This last group of illustrated designs was an exercise in which I tried to explore the generative and creative power that, I believe, is at our fingertips once we understand this rich tradition of Islamic geometric design. This special quadrangle of side proportions $1:2:2:\sqrt{7}$ is carried through various symmetry operations to produce over 80 designs. A small sample of them are presented here to show the design potential of this geometric form:

Fig. 20.1. A 2-fold symmetry operation rotating it 180° on each side in 22'2"2''' point groups that generates a pattern of these asymmetric kites that tessellate the plane.

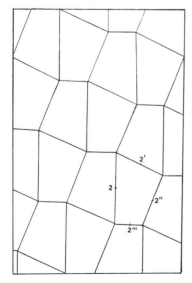

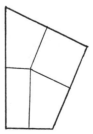

Fig. 20.1. The asymmetric quadrangle ABGD in a 22'2"2''' symmetry pattern.

Fig. 20.2. The asymmetric quadrangle ABGD subdived with perpendicular bisectors.

Fig. 20.3. A 22′2″2‴ tessellation with the subdivided quadrangle ABGD.

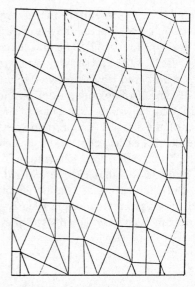

Fig. 20.4. The diagonal of the smaller quadrangles are drawn.

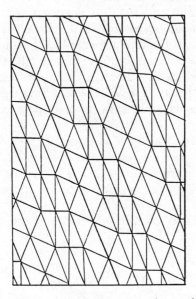

Fig. 20.5. The diagonals of the squares and rectangles are drawn.

Fig. 20.6. The second diagonals for the square and retangles are drawn.

Fig. 20.2. When the perpendicular bisectors of all the sides are drawn, the quadrangle is divided into four areas, generating four small quadrangles:

1. A small square of a 1-unit side length.
2. A rectange with side lengths equal to 1 and 1/2.
3. A small asymmetric kite-shaped quadrangle.
4. A small asymmetric kite-shaped quadrangle.

The two small quadrangles, that are the result of the subdivision, are both similar to each other and to the original quadrangle, retaining the original side proportions $1:2:2:\sqrt{7}$.

Fig. 20.3. The subdivided quadrangle is used to tessellate the plane in a 22′2″2‴ pattern.

Fig. 20.4. The subdivision is increased by drawing diagonal lines for the smaller quadrangle.

Fig. 20.5. The subdivision is increased by drawing the diagonal lines for the smaller square unit and the smaller rectangle.

Figs 21.1.–21.12. Colored examples (unfortunately reproduced in black and white) of patterns generated from the asymmetric quadrangle ABGD.

Fig. 20.6. The second diagonals are drawn for the smaller square and smaller rectangle.

Figs 21.1–21.12. The patterns resulting from 2-fold and 4-fold rotations of the asymmetric kite-shaped quadrangle ABGD with proportions $1:2:2:\sqrt{7}$ generate a number of patterns. These patterns in turn are multiplied through further subdivisions and selective symmetric coloring of the shapes to develop an infinite number of patterns, some examples of which are seen in these photographs.

I chose this special quadrangle because I felt that its geometry was very strict and its algorithm unique. In fact it is so strict and yet somehow so simple that, at first glance, the asymmetric quadrangle or kite shape it produces looks boring or not visually interesting. However, when used through different symmetry operations and coloring it showed the potential to generate an infinite number of patterns. In this regard, I would like to quote A. Loeb from his article "Algorithms, structures and models":

> "We have observed that our perception of an apparently complex configuration is altered when, instead of attempting a complete description of the object, we generate the configuration from a small number of relatively simple modules together with an algorithm for assembling them....
>
> "Generally, we do not know the modules and algorithms which would generate a given complex configuration. The role and process of science would seem to consist of a search for appropriate modules and algorithms which generate models whose behavior resembles adequately that of the complex configuration studied. The analogy between model and observed configuration is limited and quite subjective, depending on the observer, and the purpose, context and background of the experiment.
>
> "In design, the algorithmic approach generates with simple means a rich repertoire of patterns transcending the repertoire of the 'naked eye'. In addition the conceptual component of such a generated pattern has an esthetic appeal of its own, and constitutes an important link between art and science." [32]

If to A. Loeb "the role of science is to search for appropriate modules and algorithms which generate models whose behavior resembles" the simple algorithm and shape we have seen in this Islamic design, then art too has to search for the proper scientific languages and tools to generate new forms and expressions.

CONCLUSION

Over 10 years ago, I stood firm against attempts to direct me away from the core contents of the Islamic manuscripts of practical geometry in order to deal with the issues of Princely and Royal patronage, the circles of the intelligentsia, or the intellectual activities of tenth century Baghdad and its role in the writing of the first manual of practical geometry that we have. At that time I asked these questions: Where does the key to the existence and creation of these texts lie? Is it in the specific achievements and readiness of science at that time and in that place? Is it in the patronage of the royal intellect? Is it in the interests and practically of the age? Is it in the person of a particular scientist–geometrician and his interest in and play with the scientific materials that were available to him? Or is it in the intrinsic needs of the arts, artisans and architects, which in turn ultimately directed scientists to create these texts. Is that not their true *raison d'être*?

The true patron of the scientists who wrote these ancient manuscripts was art. It was the artisans and the architects who called for the services of science and scientists to assist them in solving the design problems that they were facing. And as in the case of Islamic art in the past, science must come to the service of the arts, whether we are talking today of Islamic art, of Western art or of art generally, today more than ever before, for otherwise I cannot imagine how the arts can move into the twenty-first century.

In the mid-seventies people completely shunned attempts to return to Islamic tradition. Today, turning toward the tradition is, for some, the fashionable thing to do and a fashionable subject to focus international conferences on. For others, it is something frightful to be avoided, for it implies a reactionary conservative return to medieval times, and is feared as being the source of militant radicalism. Throughout Islamic history, there has always been a voice declaring that the Muslim tradition is relevant to contemporary times. I have tried, visually, to show here that returning to the study of medieval Islamic tradition does not necessarily mean to advocate a move from the present century backwards to medieval times, whether in the field of science and geometric design as in the case here, or in other fields. Rather, the opposite may be the case, for Islamic tradition is so strong that, if we are in touch with the language of the present time and ground ourselves in this strong old tradition, we can arrive at an expression that is not only contemporary but that could be meaningful and valid in the coming century. (The design examples presented here are already more than 10 years old.)

> "And it was revealed unto Muhammad, peace be upon Him to 'Say: Are those who know equal with those who know not?' But only men of understanding will pay heed." [33]

Acknowledgements—This paper is based on two public lectures: "In the Tower of Babel: history, history of science, scientific theory and the process of Islamic geometric design", presented at MIT on 3 and 7 November 1987. A personal note of appreciation should be added here for the support given by Dr Arthur Loeb and Dr Hazem Sayed for the generous support they gave in the making of the lectures and this article.

REFERENCES

1. Ibn Khallikān, Aḥmad ibn Muḥammad (d.681 A.H./128 A.D.), *Wafayāt al-A'yān*, Vol. 4 (Ed. Muḥhammad Muḥyiddīn 'Abd al-Ḥamīd), pp. 396–401. Maktabat al-Nahaḍa, Cairo (1948); Ibn abī Uṣaybi'a, Aḥmad ibn al-Qāsim (d.669 A.H./1258 A.D.), *Kitāb 'Uyūn al-Anbā' fī Ṭabaqāt al-' Aṭibā'*, Vol. 1 (Ed. A. Müller), pp. 306–308. Königsberg: al-Matba' a al-Wahabiyya, Cairo (1882–1884); Ibn al-Fūwaṭī' al-Baghdadi, Kamāl al-Din 'Abdul Razzāq Abū'l-Faḍl (d.723 A.H./1323 A.D.), *al-Ḥawādith al-Jāmi'a wa-' l Tajārib al-Nāfi'a fī l-Mi' a al-Sābi'a* (Introduced by Musṭafā Jawād), pp. 149–150. Nu'mān al-Aḍamī Press, Baghdad (1932); al-Subkī, Tāj al-Dīn 'Abd al-Wahhāb ibn Alī (d.722 A.H./1370 A.D.), *Ṭabaqāt al-Shāfi'iyya al-Kubrā*, Vol. V, pp. 158–162. Al Maṭba'a al-Ḥusayniya, Cairo (1905–1906).
2. C. Brockelmann, *History of Arabic Literature/Geschichte der arabischen Litterature*, Supp., p. 859. Brill, Leiden (1937, 1943–1949).
3. F. Woepcke, Analyse et extrait d'un recueil de construction géométrique par Aboûl Wefâ. *J. Asiatique*, 5th Ser. **5**, 218–256, 309–359 (1985).
4. A. H. Christie, *Pattern Design: an Introduction to the Study of Formal Ornament*, p. 255. Clarendon Press, Oxford (1929). [First edition (1910) under the title *Traditional Methods of Pattern Designing*.]
5. A. H. Christie, *Pattern Design: an Introduction to the Study of Formal Ornament*, pp. 43, 258, 264. Clarendon Press, Oxford (1929). [First edition (1910) under the title *Traditional Methods of Pattern Designing*.]
6. E. Blochet, *Catalogue des Manuscrits Persans de la Bibliothèque Nationale*, Tome Deuxième Nos. 721–1160; 47, No. 23. Imprimerie Nationale, Paris (1912).
7. E. Müller, *Gruppentheoretische und Strukturanalytische Untersuchungen der Maurischen Ornamente aus der Alhambra in Granada* (Inaugural-Dessertation for a Doctorate from the University of Zürich). Buchdruckerei Baublatt, Rüschlikon (1944).
8. A. Speiser, *Die Theorie der Gruppen von endilcher Ordnung*, pp. 91–95. Springer, Berlin (1937).
9. S. H. Nasr, Foreword. In K. Critchlow, *Islamic Patterns: an Analytical Cosmological Approach*. Thames & Hudson, London (1976).
10. N. Ardalan and L. Bakhtiar, *The Sense of Unity: the Sufi Tradition in Persian Architecture*, p. 13. Univ. of Chicago Press, Ill. (1973).
11. T. Burckhardt, Introduction. In I. El-Said and A. Parman, *Geometric Concepts in Islamic Art*. World of Islam Festival, London (1976).
12. I. El-Said and A. Parman, *Geometric Concepts in Islamic Art*, p. 15. World of Islam Festival, London (1976).
13. I. El-Said and A. Parman, *Geometric Concepts in Islamic Art*, p. 38. World of Islam Festival, London (1976).
14. I. El-Said and A. Parman, *Geometric Concepts in Islamic Art*, p. 45. World of Islam Festival, London (1976).
15. L. Bakhtiar, *Sufi: Expressions of the Mystic Quest*, p. 102. Avon Books, New York, (1976).
16. L. Bakhtiar, *Sufi: Expressions of the Mystic Quest*, p. 16. Avon Books, New York (1976).
17. S. H. Nasr, *Man and Nature: the Spiritual Crisis of Modern Man*, pp. 130–131. A Mandala Book, London (1968).
18. R. Arnheim, *Visual Thinking* (1971); *Art and Visual Perception: a Psychology of the Creative Eye* (1954, reprinted 1971). Univ. of California Press, Berkeley.
19. O. Grabar, Symbols and signs in Islamic architecture. In *Architecture and Community Building in the Islamic World Today: the Aga Khan Award for Architecture*, pp. 25–34. Aperture/Islamic Publications, New York (1983).
20. J. Piaget, *Structuralism*, p. 8. Harper & Row, New York (1970, first published in French 1968).
21. Sir Thomas L. Heath, *Euclid's Elements*, Vol. I, Proposition 47, pp. 354–355. Dover, New York (1956).
22. Abū al-Wazfā' al-Buzjānī (945–987 A.D.), *Risāla fīmā yaḥtāju ilayhī al-ṣāni'u min a'māl al-handasa* ("A treatise on what the artisan needs of geometric problems"). Manuscript.
23. K. Critchlow, *Islamic Patterns: an Analytical Cosmological Approach*, pp. 72–73. Thames & Hudson, London (1976).
24. M. Gardner, Theory of tiles: extraordinary non-periodic tiling that enriches the theory of tiles. *Scient. Am.* **Jan.,** 110–121 (1977).
25. D. R. Nelson, Quasicrystals. *Scient. Am.* **Aug.,** 42–51 (1986).
26. A. L. Loeb, *Color and Symmetry*. Wiley, New York (1971).
27. H. Weyl, *Symmetry*. Princeton Univ. Press, N.J. (1952); H. S. M. Coxeter, *Introduction to Geometry*. Wiley, New York (1961); A. V. Shubnikov and V. A. Koptsik, *Symmetry in Science and Art* (1972) (translated from Russian into English by G. D. Archard). Plenum Press, New York (1974).
28. D. Schattschneider, The Plane Symmetry Group: their recognition and notation. *Am. math. Mon.* **June–July,** 439–450 (1978).
29. M. S. Bulatov (translator Vil'danova) (*Geometricheskaiia Garmonizatsiia Arkhitektury Srednei Azii IX–XV vv.*) *Geometric Harmonization in the Architecture of Central Asia of the 9th–15th Centuries*, p. 344. "NOUKA", Moscow (1978).
30. J. A. Dunn, Tessellations with pentagons. *Math. Gaz.* **394,** 366–369 (1971).
31. A. L. Loeb, *Color and Symmetry*, pp. 16–24. Wiley, New York (1971).
32. A. L. Loeb, Algorithms, structures and models. In *Hypergraphics: Visualizing Complex Relationships in Art, Science and Technology* (Ed. D. W. Brisson), p. 61. American Association for the Advancement of Science (1978).
33. Qur'an: XXXIX, 9.

Computers Math. Applic. Vol. 17, No. 4–6, pp. 791–813, 1989
Printed in Great Britain

0097-4943/89 $3.00 + 0.00
Pergamon Press plc

RECONSTRUCTION AND EXTENSION OF LOST SYMMETRIES: EXAMPLES FROM THE TAMIL OF SOUTH INDIA

P. Gerdes

Faculty of Mathematical Sciences, Eduardo Mondlane University, C.P. 257, Maputo, Mozambique

Abstract—Many traditional Tamil threshold designs are made out of one closed, smooth line. An investigation of Tamil patterns which do not conform to their cultural standard as they are composed of two, three or more superimposed closed paths, shows that these designs are "degraded" versions of originally more symmetrical patterns made out of single "never-ending" lines. The probably original versions are reconstructed and their mathematical potential is explored.

INTRODUCTION: TAMIL THRESHOLD DESIGNS

During the harvest month of Margali (mid-December to mid-January), the Tamil women in South India used to draw designs in front of the thresholds of their houses every morning. Margali is the month in which all kinds of epidemics were supposed to occur. Their designs serve the purpose of appeasing the god Siva who presides over Margali.

In order to prepare their drawings, the women sweep a small patch of about a yard square and sprinkle it with water or smear it with cow-dung. On the clean, damp surface they set out a rectangular reference frame of equidistant dots. Then the curve(s) forming the design is (are) made by holding rice-flour between the fingers and, by a slight movement of them, letting it fall out in a *closed, smooth* line, as the hand is moved in the desired directions. The curves are drawn in such a way that they surround the dots without touching them. "The ideal design is composed of a *single continuous line*" [1, p. 123; present author's italics]. Figure 1 shows examples of threshold designs which satisfy the Tamil *ideal*. These drawings display one [Fig. 1(a)], two [Figs 1(b, c)] or four [Fig. 1(d)] bilateral symmetries. However, there exist other traditional threshold designs which do not conform to the Tamil standard as they are composed of two, three or more superimposed closed paths. Figure 2(a) shows an example, made out of three separate closed lines [Figs 2(b–d)]. According to Layard, these designs "definitely represent a spirit of artistic development, ... but are *technologically degraded*" [1, p. 149; present author's italics]. They are "imitations"; they only may give the impression of being composed of one "never-ending" curve. Does the degradation really lie in a failure to create more intricate, single closed line designs?

In this paper an alternative hypothesis will be presented: most of the designs formed of a "plurality of never-ending lines", analysed by Layard, are degraded versions of originally single closed path figures.

ANALYSIS AND RECONSTRUCTION OF A *PAVITRAM*-DESIGN

Names given to designs formed of a single "never-ending" line are normally *pavitram*, meaning "ring" and *Brahma-mudi* or "Brahma's knot". "The object of the *pavitram* is to scare away giants, evil spirits, or devils" [1, p. 138]. Is it not strange that the design in Fig. 2(a), although composed of three superimposed closed paths, is nevertheless called *pavitram*? Is it possible to construct a design rather similar to Fig. 2(a) but made out of only one line?

On the one hand, the outer part of the design displays a rotational symmetry of 90° [see Fig. 3(a)], but, on the other hand, the inner part displays only a rotational symmetry of 180° [see Fig. 3(b)]. How may we remove this inconsistency between inner and outer parts?

Vertically two "sinoidal" curve-segments pass around the centre [compare Fig. 2(a) with Fig. 4(a)]. If horizontally the same happens [see Fig. 4(b)], then it turns out to be possible to complete a design [Fig. 4(c)] that not only displays a rotational symmetry of 90° (looking from

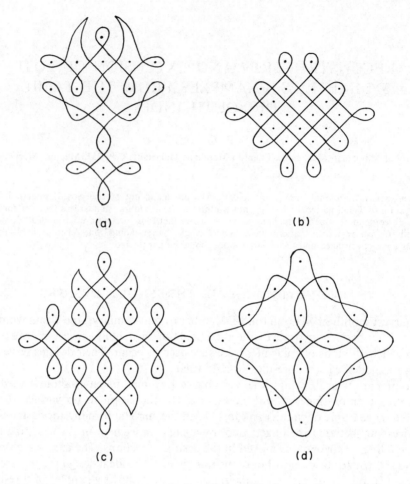

Fig. 1. Tamil threshold designs composed of one "never-ending" curve. *Source*: Ref. [1, p. 137].

the left or the right, from the bottom or the top, the figure remains always the same) but is also similar to the *pavitram* [Fig. 2(a)]. Moreover, the design [Fig. 4(c)] turns out to be composed of only one closed, smooth path!

The two designs, the one reported by Layard [Fig. 2(a)] and the one reconstructed by the author [Fig. 4(c)], differ only in two details, as illustrated in Fig. 5. When in Fig. 4(c) the curve-segments come so close in P and Q, that they almost touch, they may create the false impression that P and Q constitute points of intersection, as in Fig. 5(a).

The single closed line design of Fig. 4(c) is probably the *original* design. The reported plural closed line pattern [Fig. 2(a)] is a degradation of Fig. 4(c), a consequence of deficient transmission from one generation to another, caused, for example, by unclear drawing or unprecise memorization. As "no attempt is made to preserve the patterns. They are trodden upon almost immediately after they have been made and are soon obliterated" [1, p. 123], transmission becomes easily deficient.

A SECOND EXAMPLE

The unnamed threshold design, showed in Fig. 6(a), composed of five closed curves [Figs 6(b–d)], displays a rotational symmetry of 90°. One may suppose that the original of this pattern was formed of a single closed line and displayed the same rotational symmetry. By eliminating on each side one "false" junction, like in the previous example (junctions P and Q), four designs turn out to be the possible original pattern (see Fig. 7).

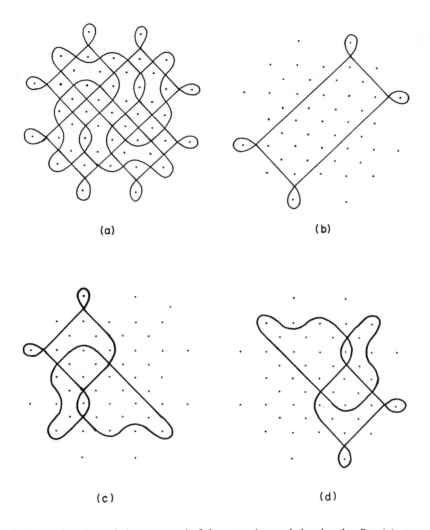

(a)

(b)

(c)

(d)

Fig. 2. Reported *pavitram*-design composed of three superimposed closed paths. *Part* (*a*), *source*: Ref. [1, p. 132].

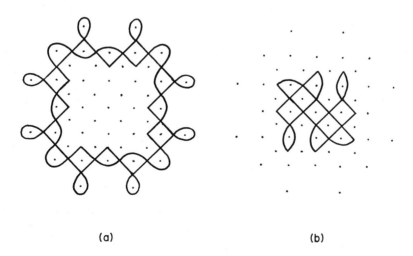

(a)

(b)

Fig. 3. The outer and inner parts of the reported *pavitram*-design.

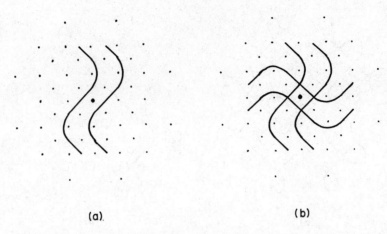

(a)   (b)

(c)

Fig. 4. Reconstructed *pavitram*-design.

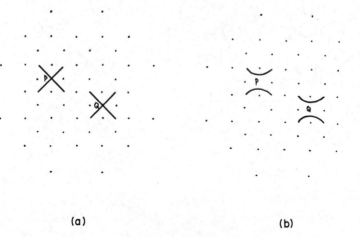

(a)   (b)

Fig. 5. Distinction.

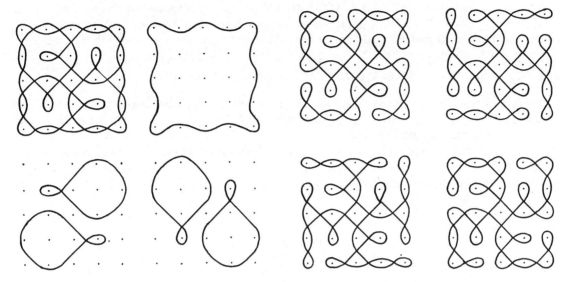

Fig. 6. Unnamed threshold design composed of five closed curves. *Source*: Ref. [1, p. 141].

Fig. 7. Four possible original patterns.

TRANSFORMATION RULES

Before advancing with the reconstruction of other traditional Tamil threshold designs, it seems appropriate to analyse what happens with the number of closed lines when one introduces or eliminates a junction, passing from one pattern to another. As a junction may be introduced by joining two distinct curves or by joining two parts of the same line, two situations have to be distinguished. In the first case, the number of lines decreases by one. In the other case, this number increases by one. As a junction may be formed of two crossing parts of the same curve or of two distinct intersecting curves, two more (inverse) situations arise, when one also considers the elimination of junctions. Now the number of lines increases by one in the first case, but decreases in the latter case. Table 1 summarizes the transformation rules.

In the example of the reconstruction of a *pavitram*-design, we applied rule 4 twice. The same rule was used four times in the second example, decreasing the total number of lines from five to one.

RECONSTRUCTION OF THE "ROSEWATER SPRINKLER" AND "SWINGING BOARD" DESIGNS

We give now two applications of transformation rule 1. The "rosewater sprinkler" design [see Fig. 8(a)] is composed of four closed curves [Figs 8(b–e)]. By introducing junctions in P, Q and R on the vertical axis, one obtains a very similar design (Fig. 9), composed of only one closed path.

Table 1

Rule No.	Nature of transformation	Situation before transformation	Situation after transformation	No. of lines after transformation
1	Introduction of a junction	)(	✕	−1
2	Introduction of a junction	)(	✕	+1
3	Elimination of a junction	✕	)(	+1
4	Elimination of a junction	✕	)(	−1

The "swinging board" threshold design illustrated in Fig. 10, is formed of three superimposed "never-ending" lines. When one now creates junctions in S and T on the horizontal symmetry axis, the probably original pattern made out of a single closed curve, appears (see Fig. 11).

THE *PAVITRAM*-DESIGN REVISITED

We transformed the unnamed threshold design, shown in Fig. 6(a), into a single closed line pattern, eliminating four junctions by application of rule 4. The "inverse" process is also realizable: four new junctions may be introduced with rule 1. As Figs 13 and 14 illustrate, there are two possibilities of doing this in such a way that the final design displays the same rotational symmetry of 90° as Fig. 6(a) and is formed of only one "never-ending" line. The closed path in Fig. 14(b) constitutes another possible original for this unnamed threshold design [Fig. 6(a)], as both patterns

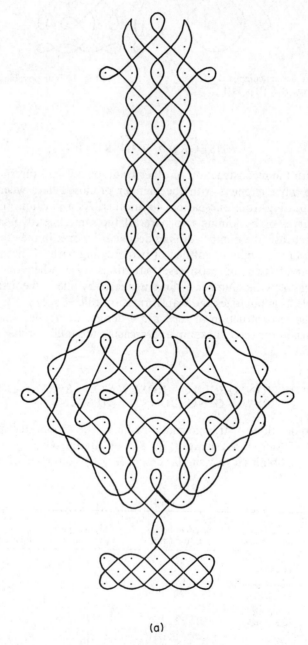

(a)

Fig. 8—*continued opposite*

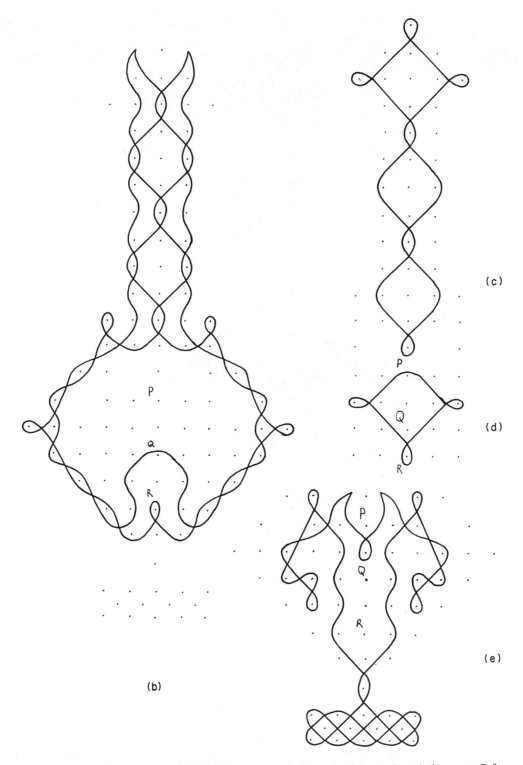

Fig. 8. Reported "rosewater sprinkler" design, composed of four closed curves. *Parts (a, b), source*: Ref. [1, p. 147].

are still rather similar. Surprisingly Fig. 13(b) is identical with our reconstruction [Fig. 4(c)] of the *pavitram*-design [Fig. 2(a)], but without the border ornamentation. In this form, the pattern can be easily extended, as will be shown after a brief reference to the notion of extension inherent to the Tamil design tradition.

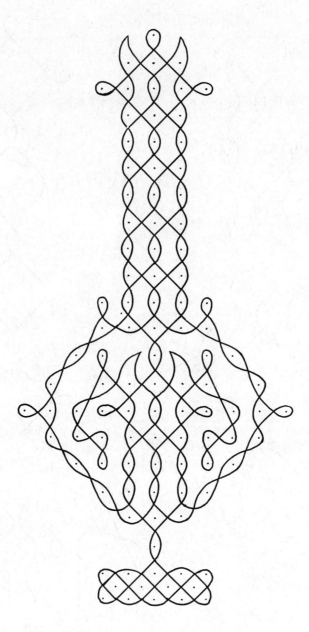

Fig. 9. Reconstructed "rosewater sprinkler" design.

EXTENSIONS IN THRESHOLD DESIGNS

The Tamil threshold designs shown in Fig. 15, made out of single closed curves, are rather similar in structure. Both may be considered as "rectangular" patterns (see Fig. 16) with "circular" ornamentation at those border dots that are not vertices of the rectangle. In this sense, Fig. 15(b) may be called a possible *extension* or *generalization* of Fig. 15(a). Figure 17 constitutes the next step.

Figure 18(a) is another possible extension of the Tamil design shown in Fig. 15(a). In turn, this pattern has been used as an element in the building up of the "cradle" threshold design, seen in Fig. 18(b). The elaborate "swinging board" pattern we analysed before (Figs 10–12) may be considered as a juxtaposition and partial superimposition—and therefore as an extension—of the single closed line elements shown in Fig. 19.

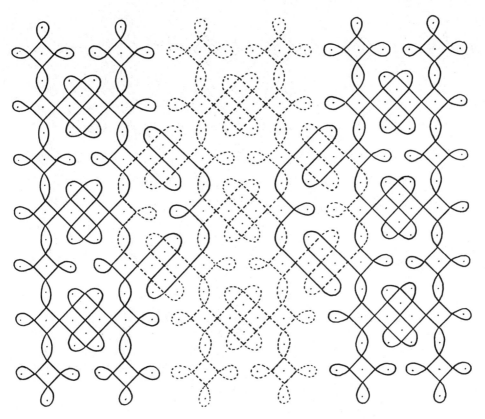

Fig. 10. Reported "swinging board" design. *Source*: Ref. [1, p. 145].

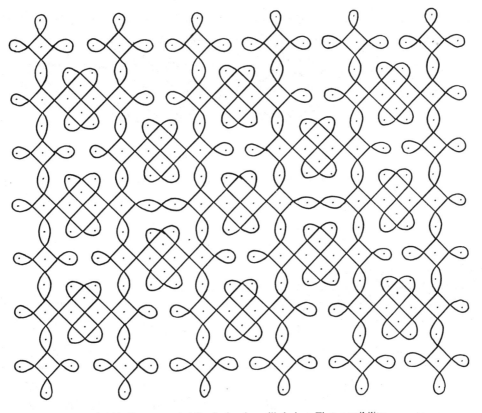

Fig. 11. Reconstructed "swinging board" design. First possibility.

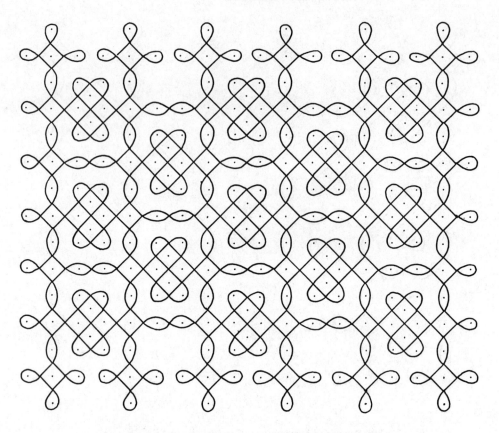

Fig. 12. Reconstructed "swinging board" design. Second possibility.

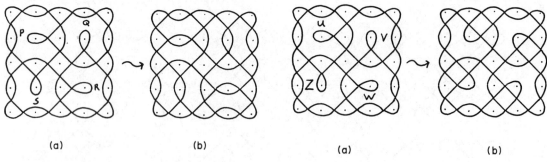

(a)   (b)   (a)   (b)

Fig. 13. Transformation by introduction of junctions in P, Q, R and S.

Fig. 14. Transformation by introduction of junctions in U, V, W and Z.

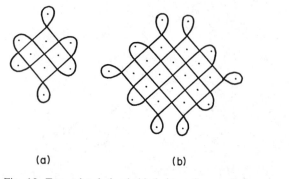

(a)   (b)

Fig. 15. Two related threshold designs. *Source*: Ref. [1, p. 137].

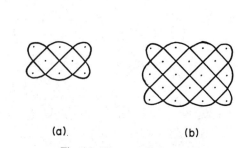

(a)   (b)

Fig. 16. "Rectangular" patterns.

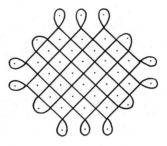

Fig. 17. Next step in the extension.

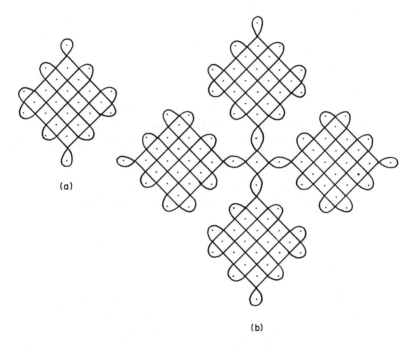

(a)

(b)

Fig. 18. Building up of the "cradle" threshold design.
Part (b), source: Ref. [1, p. 140].

Fig. 19. Single closed line elements.

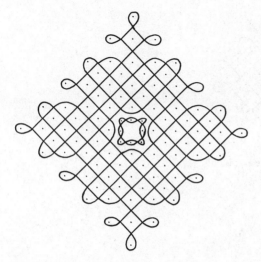

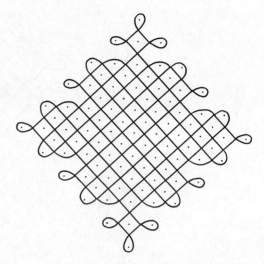

Fig. 20. Reported Brahma's knot. *Source*: Ref. [1, p. 143].

Fig. 21. Reconstructed Brahma's knot.

The original of the Brahma's knot shown in Fig. 20, formed of four closed smooth curves, may be easily reconstructed. As illustrated in Fig. 21, it is sufficient to elongate the curve-segments near its centre. Our hypothesis becomes still more credible if we compare this reconstructed Brahma's knot with the traditional Tamil threshold design shown in Fig. 22. Both "obey" the same geometrical algorithm. Figure 23(b) shows a further extension of these two patterns, this time with eight points on each square side. These closed curves are built up out of four "branches" in such a way that two consecutive branches are symmetrical in relation to the horizontal or vertical axis.

EXTENSIONS OF THE RECONSTRUCTED *PAVITRAM*-DESIGN

The pattern of the reduced, reconstructed *pavitram* [Fig. 13(b)] may be easily extended, as illustrated in Fig. 24 for the cases $p = 9$ and $p = 17$, where p denotes the number of dots on the side of the square reference frame. The drawing of the first quarter of the design in the case $p = 13$, as illustrated in Fig. 25, shows very clearly that a relatively simple *geometrical algorithm* lies at the basis of a, at first sight, rather intricate design. After completing each quarter, one turns around the corner and repeats the drawing. This explains the rotational symmetry of 90° that the final designs [Figs 13(b) and 24(a,b)] display. Not only extended square patterns are possible. Figure 26 shows the non-square extension 17×9, where 17 denotes the number of dots in the first row and 9 the number of dots in the first column. Generally, the design may be extended to all rectangular reference frames $(4m + 1) \times (4n + 1)$, where m and n represent arbitrary natural

Fig. 22. Another traditional Tamil design.

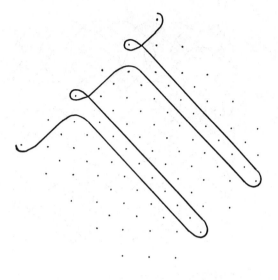

(a)

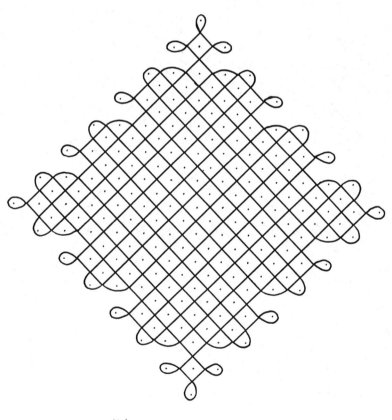

(b)

Fig. 23. (a) Illustration of the geometrical algorithm.
(b) Further extension.

numbers. All extended versions may be considered as built up out of partially overlapping *"cells"*. The cells are the reconstructed *pavitram*-design without border decoration [Fig. 13(b)]. Partially overlapping in the sense that two neighbouring cells have a common border row (or column) of dots. Figure 27 shows the 9 × 9 extension of the reconstructed *pavitram*-design, complete with border ornamentation.

(a)

(b)

Fig. 24. (a) 9 × 9 extension. (b) 17 × 17 extension.

EXAMINATION OF A BRAHMA'S KNOT

"Brahma's knot" is generally a name only given to certain designs formed of a single "never-ending" line. The Brahma's knot shown in Fig. 28, however, is made out of five superimposed closed curves (see Fig. 29). As, moreover, two of these five pass through the dots

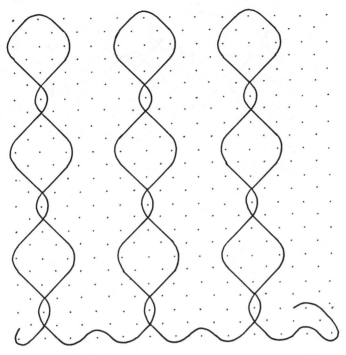

Fig. 25. Geometrical algorithm illustrated in the case 13×13.

P and Q instead of surrounding them, it becomes highly probable that we do not deal here with the original pattern. The clue to the rediscovery of the original Tamil design seems to lie in the peripheral closed curve [Fig. 29(a)]. Is it possible to modify this curve in such a way that it also will "fill up" the interior of the point square grid?

On each "side" of the peripheral line, the same "loop" (see Fig. 30) repeats itself three times. What will happen if there are only one or two such loops on each "side"? In the first case (Fig. 31), the single "never-ending" curve "fills up" the reference square. Although this does not happen in the second case [Fig. 32(a)], the square grid may be "filled up" [Fig. 32(b)] by introducing four times an element rather similar to the loop and that we have met already in Figs 6(c, d). Now, if one applies transformation rule 1 in S, T, U and V, one obtains the design with the horizontal and vertical symmetry axes shown in Fig. 32(d), composed of only one smooth closed path. The

Fig. 26. 17×9 extension.

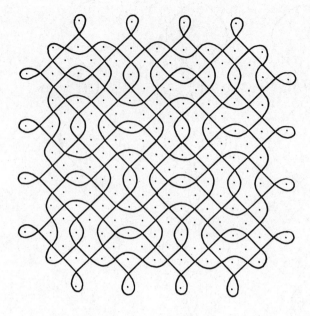

Fig. 27. Extension of the reconstructed *pavitram*-design.

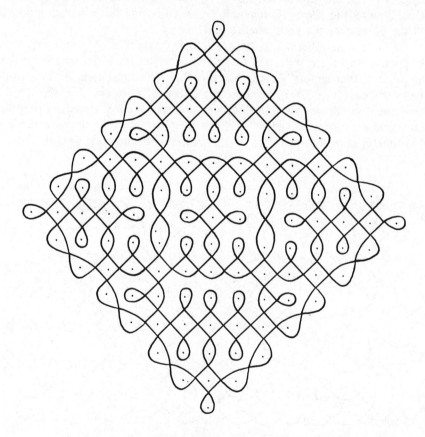

Fig. 28. Reported Brahma's knot. *Source*: Ref. [1, p. 144].

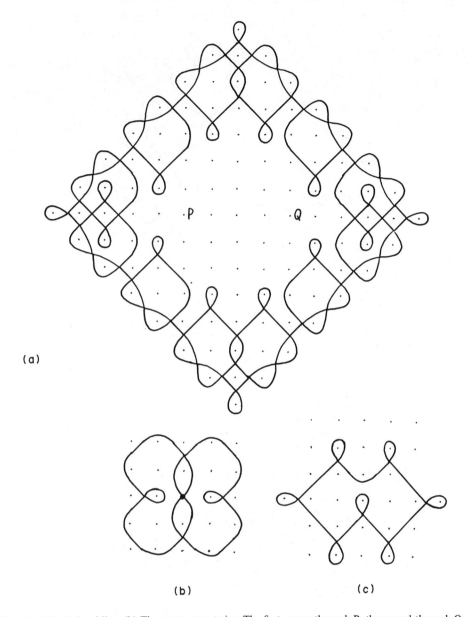

(a)

Fig. 29. (a) Peripheral line. (b) The same curve twice. The first passes through P, the second through Q.
(c) The same curve twice, at the top and at the bottom of the design.

Fig. 30. A loop.   Fig. 31. One loop on each "side".

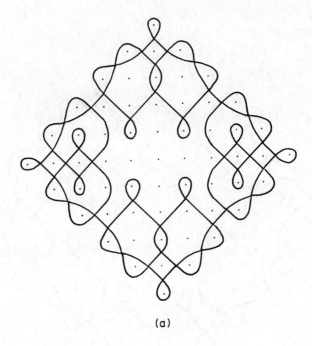

(a)

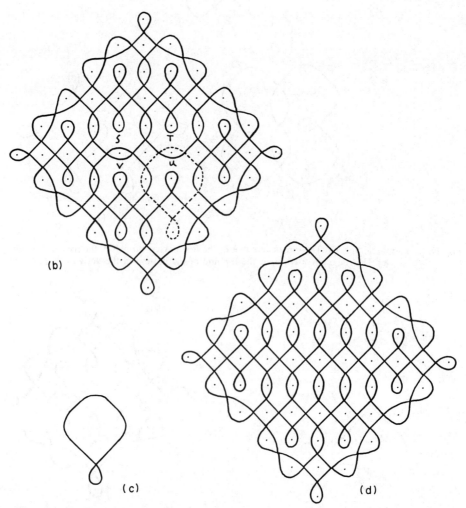

(b)

(c)

(d)

Fig. 32—*continued opposite*

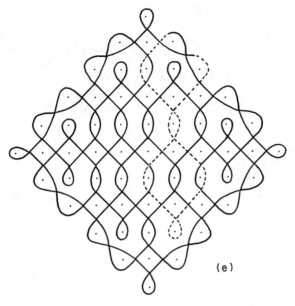

Fig. 32. "Filling up" the reference square grid in the case of two loops on each "side".

branches [Fig. 32(e)] that were introduced by these junctions, are not strange to the Tamil tradition as we met them before in the "rosewater sprinkler" design (Fig. 8) and they also appear in the reconstructed and extended *pavitram*-pattern (see, in particular, Fig. 25). Using the same geometrical algorithm in the case of three loops on each side, one gets the double symmetrical pattern of Fig. 33, which is probably the original Brahma's knot.

Fig. 33. Reconstructed Brahma's knot.

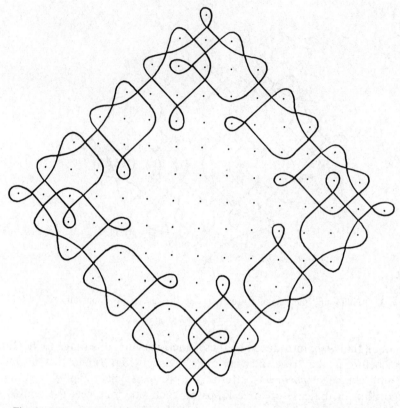

Fig. 34. New peripheral line that displays a rotational symmetry of 90°.

SOURCE OF INSPIRATION

The peripheral closed curve of Fig. 29(a) has vertical and horizontal symmetry axes. It is possible to change the orientation of the loops on two opposite sides in such a way that a new peripheral line appears (Fig. 34). This one displays a rotational symmetry of 90°. Figure 35 shows what happens in the case of one loop on each side. Figures 36 and 37 illustrate the extensions that may

Fig. 35. One loop on each "side".

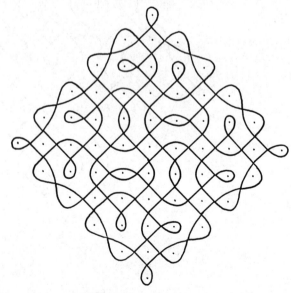

Fig. 36. First extension.

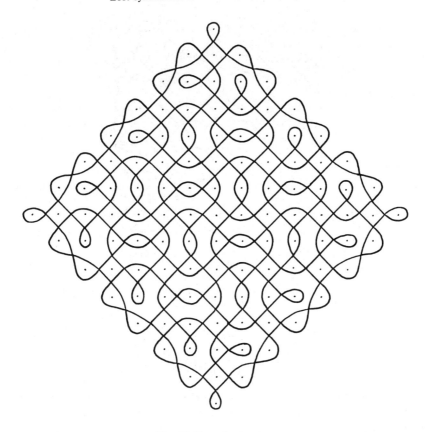

Fig. 37. Second extension.

be obtained when one "fills up" the interior in the cases of two and three loops on each side. All these curves are smooth, single "never-ending" lines.

Let us return to Fig. 14(b), and denote this design by A. By a reflection of this pattern around its vertical symmetry axis, one obtains the design shown in Fig. 38(b) (representation $-A$). If one now joins two A and two $-A$ elements according to the scheme

$$\begin{array}{cc} -A & A \\ A & -A \end{array}'$$

in such a way that two neighbouring "cells" have a common border row (or column) of dots, one discovers the interesting single closed path design shown in Fig. 39 with horizontal and vertical axes of symmetry. If one "isolates" its 5×5 centre, one gets another single "never-ending" line pattern [Fig. 40(a)]. Let B denote this design. By a reflection of B around one of its diagonals, one obtains Fig. 40(b) (representation B^T). By constructing analogously a design according to the

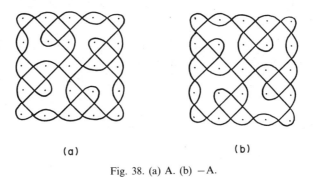

(a)   (b)

Fig. 38. (a) A. (b) $-A$.

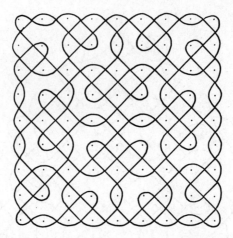

Fig. 39. Single closed curve constructed according to the
scheme

$$\begin{array}{cc} -A & A \\ A & -A \end{array}$$

scheme

$$\begin{array}{cc} B^T & B \\ B & B^{T} \end{array}$$

one discovers the single line pattern in Fig. 41. Isolation of its 5×5 centre leads us once again back to pattern A.

Many other variations are possible.

CONCLUDING REMARKS

Most of the threshold designs on which Layard's study was based, were published in Madras in 1923. At that time the Tamil tradition to draw patterns in front of the thresholds of their houses during the Margali month was in decline. The designs we analysed, formed of a plurality of "never-ending" curves, are "degraded" versions of originally single closed line patterns. These

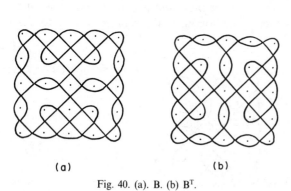

(a)   (b)

Fig. 40. (a). B. (b) B^T.

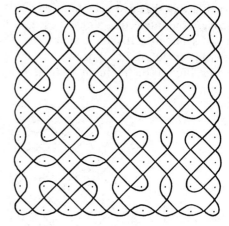

Fig. 41. Single closed curve constructed according to the
scheme

$$\begin{array}{cc} B^T & B \\ B & B^{T} \end{array}$$

patterns did not "fall from heaven"; they were not "occasionally" discovered. On the contrary. Systematic analysis led to their invention. Transformation rules as well as geometrical algorithms had been discovered. Notions of bilateral and rotational symmetry became well-anchored. Their inventors developed (or disposed of) clear ideas of extension and generalization. They constructed designs that had to satisfy certain chosen criteria like "singleness", continuity, smoothness and symmetry.

Our analysis reveals the high *mathematical potential* [2–4] of the tradition that led to the Tamil threshold designs. Further research on its origin and possible influence on the (maybe early or ancient) mathematical (and artistic) development in India (or elsewhere) seems necessary. Also the social, educational and technical factors that contributed to the aforementioned "degradation", i.e. to the development of designs which do not conform to the cultural standard of single closed line patterns, deserve further study. In some cases [e.g. Fig. 6(a)] the "degradation" may have facilitated the memorization.

The use of the Tamil threshold designs in the mathematics classroom may be suggested (cf. our study on the possible uses of traditional Angolan sand drawings in the mathematics classroom [5]).

REFERENCES

1. J. Layard, Labyrinth ritual in South India: threshold and tattoo designs. *Folk-Lore* **48,** 115–182 (1937).
2. P. Gerdes, Zum erwachenden geometrischen Denken. Eduardo Mondlane Univ., Maputo (1985).
3. P. Gerdes, Reconstruction and extension of lost symmetries: examples from the Tchokwe of North-East Angola. Eduardo Mondlane Univ., Maputo (1988).
4. P. Gerdes, *Ethnogeometrie: kulturanthropologische Beiträge zur Genese und Didaktik der Geometrie*. Franzbecker Verlag, Bad Salzdetfurth (1988).
5. P. Gerdes, On possible uses of traditional Angolan sand drawings in the mathematics classroom. *Educ. Stud. Math.* **19,** 3–22 (1988).

Computers Math. Applic. Vol. 17, No. 4–6, pp. 815–826, 1989
Printed in Great Britain. All rights reserved

THE ROLE OF SYMMETRY IN JAVANESE BATIK PATTERNS

A. Haake

Institut für Kristallographie und Mineralogie der Johann Wolfgang Goethe-Universität,
Senckenberganlage 30, D 6000 Frankfurt am Main, B.R.D.

Abstract—Batik has a long tradition in Java, Indonesia. The ancient periodic patterns contain meaningful symbols some of which have a meaning for the Javanese people even today. Only certain plane groups are used frequently while others rarely appear in a batik pattern. There is a strong relation between symmetry groups present and the symbolic content of a pattern.

INTRODUCTION

Basic textiles as weavings, knitwear etc. are the result of periodic intersections of threads forming a homogeneous texture. These textures can become decorated in different ways.

Embroideries often follow the natural symmetry of the basic textiles. *Prints* are bound to the printing tools—blocks or machine—which ask for periodic repetition of the unit. However, there is no practical necessity in doing *paintings* or *drawings* symmetrically.

What does "batik" mean?

Batik is an ancient method of textile decoration which has been practiced in many places all over Asia since prehistoric times.

In Java, Indonesia the technique was developed more than anywhere else. From here it spread out to European countries during the last 100 years. The names of the tools and even the name "batik" were adopted from the Indonesian language. "Batik" means "drawing with wax" (Fig. 1). Batik is a dyeing process: melted wax is applied on the cloth with a special pen called "canting". It reserves parts of the cloth which shall remain white (or in the present colour) from the hydrous solution of the dyestuff. After dyeing and fixation the wax is removed by boiling. Repetitions of these steps lead to multicoloured patterns.

The vegetable dyes which were the only ones available in the past needed a long time to penetrate the cloth. Under these circumstances direct painting or printing would not give sharp contours

Fig. 1. Applying of wax with the "canting".

because the dye solution would run in all directions without control. Therefore, the resist dye was the only possible way to get colourfast fabrics with distinct patterns [1].

For whom and for what purpose were batiks produced?

In Central Java a limited range of colours were used to produce loin cloths with patterns of highly symbolic content for the royal courts of Surakarta (Solo) and Yogyakarta. The traditional colours for batiks here are dark blue from indigo and all shades of brown/yellow/red from native plants.

The symbols on these cloths (size $\approx 1 \times 2.40$ m) and their relative positions were regarded as a protection against evil influences. In the eighteenth century a law was enacted which prohibited the use of certain patterns for the public; this law was followed strictly until the Second World War [2].

SYMBOLIC CONTENT AND SYMMETRY OF TRADITIONAL PATTERNS

A Systematics of Traditional Javanese Batik Patterns [3]

I. So-called "geometric patterns"

1. "Banji": crossing bands; Bronze Age pattern, maybe, of chinese origin; (Banji means Chinese "10000"), see Fig. 2.

2. Stencil patterns (stylized flower and fruit profiles, sometimes containing creatures) [Figs 3(a)–(f)].

> (a) "Ceplok": rosettes, stars, squares etc. (Ceplok means metal ornament) [Figs 3(a)–(c)].
> (b) "Ganggong": special form of "Ceplok" named after cryptocoryne ciliata, a water plant with long seed hairs [Fig. 3(d)].
> (c) "Kawung": intersecting circles (Kawung means fruit of arenga saccharifera) [Figs 3(e), (f)].

3. Inclining borders "Lereng" or "Garis Miring" (including the most popular varieties "Parang"; Parang means dagger, knife) [Figs 4(a), (b)].

4. "Nitik" and "Anyaman": imitation of weaving patterns (Nitik from "tik" = dot; Anyaman means wickerwork); the shape of the ornaments of this group is similar to "Ceplok" or "Garis Miring" types [Figs 5(a), (b)].

Fig. 2. "Banji" patterns (B. = Chinese: 10000) intersecting bands, like a weaving, the basic pattern is in use from prehistoric times; it is the symbol for lucky life. Symmetry p4gm. (a) "Kerton" from Central Java, $a_0 = 11$ cm; (b) the pure Chinese form from the coast, $a_0 = 15$ cm.

Fig. 3. Stencil designs. (a) "Slobok" (means wiggle), a pattern which is also well-known in the Pacific archipelagos. Symmetry: p4mg; $a_0 = 7.5$ cm. (b) Ceplok "Peksi Kreno" (Peksi means bird, Kreno means many fruits). Symmetry: conceptional c1m1; $a_0 = b_0 = 9$ cm. (c) Ceplok "Sekar Arum" (Sekar means flower, Arum means smell, taste). Symmetry: p2gg; $a_0 = b_0 = 10$ cm. (d) Ganggong "Satryo Wibawa" (Satryo means knight, Wibawa means mighty). Symmetry: p4mm disturbed by a five-leaf flower in one of the four-fold axes; from this centre the name giving seed hairs originate (compare with systematics [I 2(b)]), $a_0 = 10.5$ cm. (e) "Kawung Picis" (Kawung means fruit of a sugar palm, Picis means small). Symmetry: p4mm, $a_0 = 3$ cm. (f) "Kawung Sen" (Sen means Dutch coin). Symmetry: p4mm, $a_0 = 7$ cm.

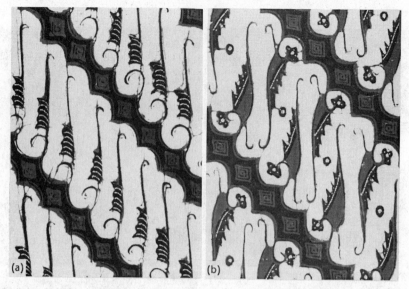

Fig. 4. Inclining borders. (a) "Parang Ukel" (Parang means dagger, Ukel means curl). $a_0 = 2$ cm, $b_0 = 10$ cm. (b) "Parang Plentong" (Plentong means round). $a_0 = 6$ cm, $b_0 = 13$ cm.

II. So-called "non-geometric patterns"

1. "Semen"-patterns: arrangements of plants, creatures and Hindu symbols without regarding the background structure; (semen from "semi" means sprout).

  (a) Patterns of stylized plants only (see Fig. 10).
  (b) Patterns of stylized plants and creatures.
  (c) Landscape-like patterns containing Hindu symbols [Figs 6(a), (b)].

2. "Buketan": influenced by European paintings, not really traditional.

III. Samplers

1. "Tambal": (means patchwork); triangles, squares or other shapes are filled with various patterns from group I and/or II (Fig. 7) [4].

2. "Sample-piece": a catalogue of available patterns of a batik painter (the names of the patterns are mentioned in each field).

3. Compositions: from patterns of group I and II new designs originate—often with ostensible symmetry (Fig. 8).

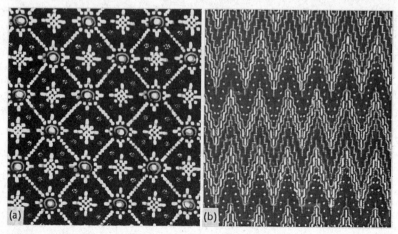

Fig. 5. Weaving imitations. (a) "Nitik Cinde" (Nitik from "tik" meaning dot; Cinde means silk patola originating from India, which were saved as family heirlooms). Symmetry: c2mm; $a_0 = 5.5$ cm, $b_0 = 5.4$ cm. (b) "Tirtateja" (meaning glittering light on water). Symmetry: p2mg; $a_0 = 4$ cm, $b_0 = 12$ cm.

Fig. 6. "Semen" patterns with Hindu symbols. (a) "Semen Sido Asih" (meaning happy life in love); the central motifs are "pohon hayat" (tree of life) and "meru" (mountain of gods) flanked by Garuda wings, peacocks, and deers (very tiny). Symmetry: p1m1; $a_0 = 42$ cm, $b_0 = 38$ cm. (b) "Semen Gurda" (Gurda meaning Garuda or banyam tree; see also Fig. 14). Symmetry: c1m1; $a_0 = 39$ cm, $b_0 = 61$ cm, ($a_0' = 38$ cm).

Fig. 7. "Tambal" (meaning patchwork); each patch contains its own pattern from I and II of the systematics [4]. (By courtesy of Galerie Smend, Köln.)

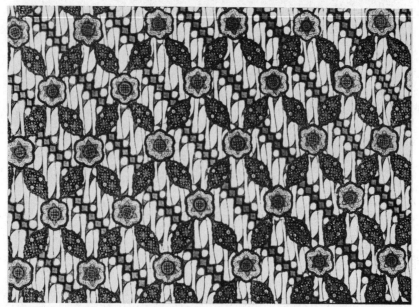

Fig. 8. "Ceplok Prabu Anom" variety (Prabu means prince, Anom means second), a composition with patterns of group I and II of the systematics. Usually the filling patterns lower the ostensible symmetry of the composition. $a_0 = 22$ cm, $b_0 = 23$ cm.

IV. "Pasisiran"-batiks (Pasisiran means coast, beach)

All coastal patterns show foreign influence during the last centuries [Figs 9(a), (b)] [5].

The symmetry in these patterns, especially in the old court patterns, bears much information about the Javanese people—their beliefs and the origin of their philosophy [3].

The magic power of a pattern—caused by the order of symbols, for example—played an important role in the life of a Javanese. It is said that some people checked a cosmic calendar every morning and compared it with their personal horoscope. By choosing a particular piece of batik to wear, which was supposed to have a corresponding magic power, they tried to diminish negative and intensify positive effects on their projects for the day.

Fig. 9. "Pasisiran" batiks (Pasisiran = coast). (a) "Lengko-lengko" (meaning zig-zag) from Pekalongan in bright blue/white. This type often has a "kepala" (= head) and narrow borders at the edges, (height 100 cm). (b) "Peksi Naga Liman" (bird, snake, elephant) a court pattern from Cirebon, where family relations between the Islamic sultanate and the Chinese imperial Ming dynasty (fifteenth century) caused a strong Chinese influence on the arts. Due to the size of the patterns there is no plane symmetry, $a_0 = 90$ cm [5].

Some patterns were reserved for special people (nobility) or special occasions (weddings, circumcisions, cremations, for example, see above) [6]. For centuries all this accurate work was done by drawing without knowledge of the mathematical laws of plane symmetry. The perfect realization of a design was one of the conditions for the magic power of a cloth. However, sometimes "mistakes" were built in intentionally to disturb the perfect symmetry. The reason may be the Islamic belief that only Allah can be faultless [7].

Most of the patterns of Central Java have been in use for generations—and already a long time before the "cap" (a copper printing block for the wax resist) had been developed about 1820 A.D. *The symmetry was not caused by this tool, but the tool was created to produce the batik more efficiently* to cover the increasing demand for batik clothing for exports and for everyday use.

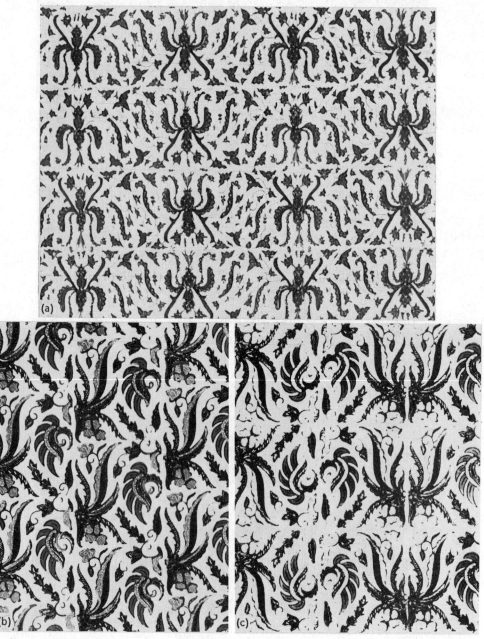

Fig. 10. "Pisang Bali" (meaning turned banana). The name of the pattern contains a two-fold rotation: Alternating rows of upward and downward growing plants. (a) A handdrawn example of the Central Javanese pattern. Symmetry: p2mg; $a_0 = 57$ cm, $b_0 = 21$ cm. (b) Part of a stamped batik cloth from Jakarta; the mirror-related pair of stamps was not used to construct the complete design. Symmetry: p1. (c) The same pattern, as it should have been done.

The "cap" work is interesting for symmetry reflections in another respect: for good results the resistant wax has to be applied on both sides of the cloth; and that asks for a set of blocks with patterns related by a mirror line. For these patterns such a pair of caps generates a full design. The man who uses these tools must know the intended pattern exactly. Otherwise the lack of a certain symmetry in a classic pattern will easily show that the caps were "borrowed".

A good example for the latter is an old pattern: "Pisang Bali". (Pisang means banana, Bali means balik which means (see Manuser) "turned over"). The name already contains the rotational symmetry. The original pattern has always p2mg symmetry, but the copy often has pm or p1 only [Figs 10(a–c)].

Batiks are also produced and worn outside of Central Java, but the patterns do not have the protective value as those of this area. The coastal regions of Java—influenced by various foreign cultures—developed a quite different style. Large motifs—mostly of Chinese or European origin mixed with Javanese symbols—which are degraded to be just decorations without any magic content. Due to the size of the motifs those cloths show one-dimensional symmetry only and are of lesser interest for plane symmetry—in spite of their evident beauty [see Figs 9(a), (b)].

SYMMETRY ELEMENTS AND TRADITION

Hundreds of old and new *drawn* Javanese batiks have been investigated referring to symbolism and plane symmetry groups.

It seems that symmetry elements reveal the principles of Javanese philosophy (Fig. 11) [3].

1. Translation

It is striking to notice that even the symmetry of patterns with a tiny unit is carried out thoroughly over the whole cloth ($\approx 1 \times 2.4$ m). Meditation is very common all over Asia since ancient times. The concentrating on the steady and neat repetition of a motif had a meditative effect on the creating person: the order of the pattern was transmitted to her spirit (Batik making is a female domain). Another reason for the repetition of a motif was the wish to multiply its power and transmit it to the wearer of the batik cloth [7].

2. Rotation

In old court designs one will not discover a hexagonal lattice. However, there are plenty of examples for the square, the rectangular (primitive or centred), and the oblique lattice. A great number of the stencil patterns as "Ceplok", "Ganggong", "Kawung" and nearly all of the "Nitik" belong to space group p4mm (compare with above systematics (I), Figs 3(a)–(f) and 5).

Some of the first three types and the "Banji" (Fig. 2)—a prehistoric motif which equals a basic weaving—have symmetry p4gm. The majority of the "inclining border" patterns follows symmetry

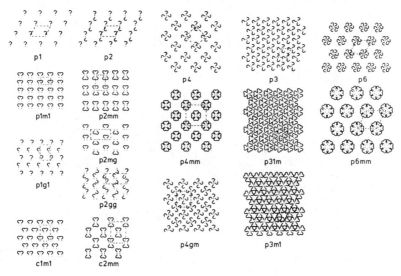

Fig. 11. The 17 plane groups [3].

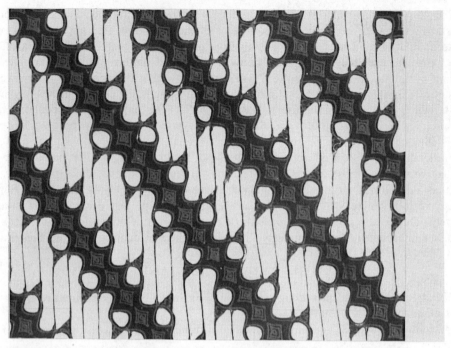

Fig. 12. "Parang Barong", a "Parang Rusak" variety. (Parang means dagger, Rusak means destroying or destroyed, Barong means giant). In this size it was reserved for the sultan, his first wife, and his crown prince only. Symmetry: p2; $a_0 = 10$ cm, $b_0 = 24$ cm.

p2, but there are also examples for p1 and even one-dimensional repeat. The main royal design "Parang rusak" (means destroying dagger) and countless variations of it represent this group (Fig. 12).

The predominance of two- and four-fold axes is a matter of the ancient Asian philosophy models "Mancapat" (Mandala) and "Dualism". The latter has entered batik patterns in various forms beside the two-fold rotation axis (see section 3, mirror-line). The model of Mancapat is a relic of the Hindu-Javanese era which was ended by the entry of Islam $\approx$ 1580 A.D. However, it was still kept as an heirloom in the palaces and huts. It represents a compass; its directions and centre are synonymous with Hindu gods, colours, days etc. The centre has always the highest rank,

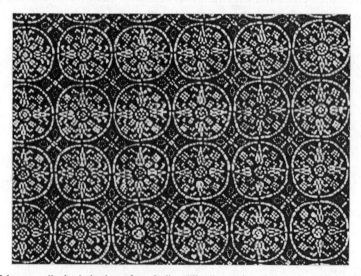

Fig. 13. "Jelamprang", the imitation of an Indian "ikat" weaving (reservation of the threads before weaving). It represents "Mancapat", the compass model of Eastern philosophy, but it is interpreted in two ways: "Wishnu's Weapon" (Hindu) and "Nine Holy Saints of Java" (Islam). Symmetry: p4mm; $a_0 = 6$ cm.

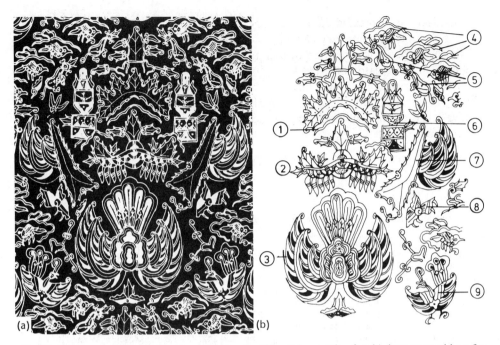

Fig. 14. Main symbolic figures in semen patterns. 1—"Meru" (mountain of gods), here crowned by a fire symbol and stylized plant; 2—"Pohon Hayat" (tree of life); 3—"Sawat Garuda" (Sawat means double wings with tail). These main centre motifs are flanked by: 4—Merus with small birds; 5—"Binatang" (land animal, deer); 6—"Bangunan" (building, temple); 7—"Pusaka", (heirloom, weapon); 8—"Kupu-kupu" (insect); 9—"Burung" (bird, here peacock) [see also Figs. 8(a), (b)].

e.g. "Bathara Guru" (Shiva), "multicoloured" and "Kliwon" the highest day of the Javanese five-days week, the four directions mean the other days, colours, gods, respectively (Figs 13) [6].

Considering the fact that more than 90% of the Javanese population have been followers of Islam for 400 years it is astonishing that there are very few examples of three- or six-fold rotational symmetry in batiks; and these rare examples are of later origin than the mentioned symbolic patterns.

Some historians wrote about an inner opposition against the entry of Islam in the courts of Central Java. Indeed, there are many residues of Hinduism and Animism which Javanese practice in their religion today. Therefore the lack of three- and six-fold symmetries, which are very common in Islamic art of the past and of today, could be another point in favour of this theory [3].

3. Mirror-line

This fundamental symmetry element is very common in "Semen" patterns. The main Hindu symbols—as "Meru" means mountain of gods, "Pohon Hayat" means tree of life, or "Garuda" means mythical bird, the symbol for sun—contain their own mirror line [Figs 14(a), (b)]. Pairs of other motifs are arranged on both sides of the line. "Dualism" means roughly "coexistence of opposites". The mirror is generating such coexistence. Many other examples of this principle can be found in a Semen batik: dark/light, eagle/snake symbolizing upper and lower world, or heaven and sea, e.g. (compare with the above section rotation).

CONCLUSIONS

In traditional Javanese batiks symmetry is a matter of the ancient Asian philosophy models "mancapat" (Mandala, windrose) and "dualism" (coexistence of opposites).

The entry of Islam ($\approx$ 1580 A.D.) could not banish these principles from the royal courts of Central Java.

—Three- and six-fold axes (common in Islamic patterns) are rarely found in Java.

—"Mancapat" is present in form of the four-fold axis—mainly as p4mm.

—"Dualism" is expressed in the change of colour, the combination of Hindu symbols and the symmetry elements two-fold axis and mirror line.

—The steady periodic repetition of a unit (translation) is proved to have meditative effects on the creator of a batik.

—Batik patterns which were subject to foreign influences during the last century (see "Pasisiran"-batiks) have no symbolic content. Instead of plane symmetry they have at most band-symmetry. In Javanese batiks loss of symmetry means also loss of symbolic content.

Please note: The pronunciation of Indonesian "c" in English is "ch".

REFERENCES

1. J. E. Jasper and M. Pirngadie, *De Inlandsche Kunstnijverheid in Nederlandsch Indie*, Vol. 3: De Batikkunst. Mouton, s'Gravenhage (1916).
2. A. Veldhuisen-Djajasoebrata, On the origin and nature of Larangan: forbidden batik patterns from the central Javanese principalities. In *Irene Emery Round Table on Museum Textiles, 1979 Proc.* (Ed. M. Gittinger) pp. 201–221. The Textile Museum, Washington, D.C. (1980).
3. A. Haake, *Javanische Batik–Methode, Symbolik, Geschichte*. Schaper, Hannover (1984).
4. A. Haake, Tambal–javanisches Patchwork in Batiktechnik. *Textilkunst* 12(3), 117–119 (1984).
5. N. S. Djoemena, *Ungkapan Sehelai Batik* (*Batik—its Mystery and Meaning*). Penerbit Djambatan, Jakarta (1986).
6. A. Veldhuisen-Djajasoebrata, *Bloemen van het Heelal*. Sijthoff, Amsterdam (1984).
7. K. R. T. Hardjonagoro, The place of batik in the history and philosophy of Javanese textiles. In *Irene Emery Round Table on Museum Textiles, 1979 Proc.* (Ed. M. Gittinger), pp. 223–242. The Textile Museum, Washington, D.C. (1980).

Computers Math. Applic. Vol. 17, No. 4–6, pp. 827–835, 1989
Printed in Great Britain. All rights reserved

0097-4943/89 $3.00 + 0.00
Copyright © 1989 Pergamon Press plc

SYMMETRY IN THE MOVEMENTS OF T'AI CHI CHUAN

J. G. HARRIES

Research Centre for Movement Notation, Faculty of Visual and Performing Arts, Tel Aviv University,
Ramat Aviv 69978, Tel Aviv, Israel

Abstract—The inherent reflective symmetry of the human body allows the orientation of positions and movements relative to its median plane. In addition, the body may be oriented symmetrically in space, relative to defined directions. Such orientations are easily definable in EW movement notation. A number of types of symmetry are distinguished, and these are demonstrated through notation, pictures and verbal descriptions. These relatively simple symmetries are applied to the Chinese martial art of T'ai Chi Chuan, with the aim of obtaining a clearer idea of "how symmetrical" it is; a contribution to the comparison of styles. A similar treatment could be employed in relation to other kinds of human movement, including dance. A more sophisticated analysis would result from the fuller exploitation of the concepts of EW movement notation.

1. INTRODUCTION

Reflective symmetry is an inherent condition of the physical structure of the human body and its motor behaviour. The body is capable of adopting countless positions in which its corresponding paired members (arms, legs and their parts) are distributed in bilateral symmetry about the median plane (the vertical plane which passes through the centre of the trunk and divides the body into the right and left halves). In simple cases, this basic symmetry of structure allows for definition of the orientation of the parts of the body in relation to the median plane. In addition, the body may be perceived as oriented symmetrically in space, in relation to specified axes and directions.

Both of these orientative descriptions can be simply and efficiently achieved in EW (Eshkol–Wachman) movement notation [1]. In the approach which underlies this notation, positions of the body are analysed as deviations from a quasi-arbitrary symmetrical stance: upright standing, arms downward with palms facing the body, feet parallel. This *zero position* is symmetrical about a plane from which, together with the horizontal plane, all directions are derived and measured. The system of reference based on this conception is a system of spherical coordinates; the joint of each limb is regarded as the centre of a sphere; all directions in space are defined according to the spherical coordinate system. Furthermore, all of these imaginary spheres are reproductions of *a single spherical system* inasmuch as they are all oriented alike—i.e. "zero directions" are parallel, and (therefore) all directions identified by the same numerical label are also parallel.

Symmetrical positions of the body are among the phenomena which can be expressed in the most simple and concise manner in terms of EW notation. Movements of paired limbs in which their trajectories are symmetrical with respect to the body's median plane, are given symbolic representation with equal ease. Because EW notation relates these events to time as well as to space, it is also well-suited for describing sequences of movement in which symmetrical positions of paired limbs are not assumed simultaneously but in sequence. Similarly, a passage of movement may be described which occurs symmetrically in successive repetitions. These are the most simple cases expressible in the symbols of the notation.

The types of symmetry referred to in this paper are as follows:

(1) *Bilateral symmetry of positions.* Positions are distinguished in which the axes of the limbs are symmetrical relative to the median plane: the amount of horizontal deviation of paired limbs from the median plane is equal and in opposite senses, vertical deviation remains constant and equal, and the parts are rotated equally and oppositely about their axes. (Unequal rotated states in corresponding paired members can be defined; the overall position would then be partially symmetrical.)

(2) *Bilateral symmetry of movements.* The trajectories of the movements of corresponding paired limbs are symmetrical relative to the median plane. Symmetrical movements always imply symmetrical positions. (The converse is not true, however, since symmetrical positions can be reached by means of asymmetrical movements.)

(3) *Sequential bilateral symmetry of positions.* As for type (1), except that the positions of the paired limbs are reached at a different moment in time.

(4) *Sequential bilateral symmetry of movements.* A movement or series of movements occurs symmetrically, first in one defined relation to the median plane and then in the bilaterally symmetrical relation to it and in the corresponding paired limbs.

(5) *Sequential symmetry of directions.* The same movements are performed, or positions reached, in different directions in symmetrical relation to a given axis on the horizontal plane.

(6) *Translation.* The same series of movements is performed at different locations, at different moments in time, as a result of transport of the body (i.e. by means of steps etc.).

In most movements of the body there is at least some symmetry. In the present context, the term *partial symmetry* is used when most parts of the body move symmetrically or are in symmetrical positions, with only minor exceptions.

Finer distinctions could be made, by using other concepts implicit in EW notation. For example, movements of limbs not of the same pair may be characterized as symmetrical movements with respect to the angular definition of their trajectories. Again, the plane of symmetry may be distorted by physical curving or torsion resulting from (definable) movements of the trunk, without this preventing the use of the concept and symbolization of symmetry in relation to the trunk itself.

The relatively simple symmetries described above will be applied to a short form of the Chinese martial art of T'ai Chi Chuan [2]. This constitutes a style of movement of non-aggressive character compared to the so-called "harder" martial arts; the movements are usually performed slowly, and no striking or other physical contact is involved. It is of combative origin, and is not primarily or deliberately aesthetic. It does not immediately appear preponderantly symmetrical, but has obviously symmetrical elements. The form is considered in the light of the criteria outlined above, and it may be possible to obtain thereby an idea of "how symmetrical" it is, in a non-colloquial sense. This would be a contribution to the comparison of styles. It appears that longer forms of T'ai Chi Chuan contain more repetition, which necessarily implies either identities patterned in time, or sequential symmetries.

2. TYPES OF SYMMETRY

The different types of symmetry enumerated above, will be demonstrated through notation, sketches and verbal descriptions, in examples taken from the T'ai Chi form. The form will then be examined, and the successive units of action ("postures") will be related to these types of symmetry.

EW movement notation provides a more exact, unambiguous and concise description of the positions and movements than can be done using words and pictures. Use of the concept of symmetry (denoted in the notation by the symbol $)$, not only enhances the conciseness of the score, but also brings out a specific definable formal aspect of the subject matter of the notated record: relations between parts of the body, and of the ways in which these are changed. See Fig. 1.

Fig. 1. The rotational symmetry of the T'ai Chi T'u diagram (shown in its clockwise and counterclockwise forms) signifies the continuous balance of increase and decrease of *yin* and *yang*: as *yin* increases, *yang* decreases by the same amount, and vice versa [3]. The EW symbol on the right is used in notating both bilaterally symmetrical positions and movements. Their coincidental resemblance may help us to bear in mind that both symbols are related to events developing in time as well as to static phenomena.

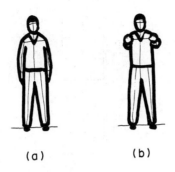

Fig. 2. (a) Bilateral symmetry of positions. (b) Bilateral symmetry of movement.

The first two examples are from the initial movements of the T'ai Chi "Standard Short Form", and are the least complex of the whole sequence.

(1) Bilateral symmetry of positions

The position [Fig. 2(a)] would be described in an instruction book of T'ai Chi in terms such as the following:

"Standing erect, facing forward. The head is up. The arms and hands hang at the sides. The chest is not held up or forced out. The left foot is the distance of a step to the left, the foot pointing straight ahead; the right foot is parallel to it."

Below, the position is shown in EW notation; beside each limb group a verbal indication is given of the information contained in Scheme 1:

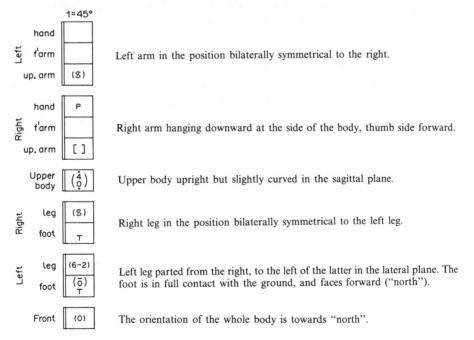

Left arm in the position bilaterally symmetrical to the right.

Right arm hanging downward at the side of the body, thumb side forward.

Upper body upright but slightly curved in the sagittal plane.

Right leg in the position bilaterally symmetrical to the left leg.

Left leg parted from the right, to the left of the latter in the lateral plane. The foot is in full contact with the ground, and faces forward ("north").

The orientation of the whole body is towards "north".

Scheme 1

Note. "North" in T'ai Chi parlance is a direction chosen as the front to which all other directions (south, east etc.) are referred. The same function is served in EW notation by "absolute (0)".

(2) Bilateral symmetry of movements

See Figs 2(a, b). The movements follow the position described in the preceding example. In textbook terms, this would be described somewhat as follows:

"Allow your arms to rise in front of you to a point about shoulder height, as if your wrists were being drawn upward by invisible strings. One hand is in front of each shoulder. Keep the hands loosely as you raise the arms. Elbows are slightly bent. When the wrists reach shoulder height, the palms straighten."

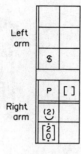

Scheme 2

The two columns of EW notation (Scheme 2) show how the arms are raised forward, rotating equally and in opposite senses about their longitudinal axes, until the axis of the unstretched arm is straight forward from the shoulders, in bilateral symmetry relative to the body's median plane and the direction in which the body is oriented ("north"). The hands are shown to hang passively during this movement, and only afterwards raised to continue the line formed by the arm.

(3) Sequential bilateral symmetry of positions

In a notation which consists of the description of a sequence of positions only, such as the positions arrived at, at the close of "postures" in T'ai Chi Chuan (cf. the sequence of positions in Ref. [4, pp. 14–20]), an overall position of the body which is bilaterally symmetrical to the *preceding* overall position can be denoted by a symmetry symbol applied to all parts of the body simultaneously. In the example (illustrated in Fig. 3), the second column of the notation (Scheme 3) indicates the bilateral symmetry of the positions fully notated in the first column; and the third column indicates the bilateral symmetry of the positions in the second—i.e. identity with the first column.

(4) Sequential bilateral symmetry of movements

The movements written out *in full* in the score show the passage from the position shown on the left of Fig. 4 (see preceding notation examples), to a position (fourth from left) in which all paired parts of the body are in bilateral symmetry to the first, relative to the median plane and

Fig. 3. Sequential bilateral symmetry of positions.

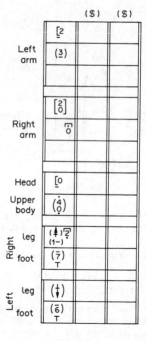

Scheme 3

Fig. 4. Sequential bilateral symmetry of movements.

also in relation to "west" [in EW notation: (6)]. The intervening movements by no means represent the shortest possible path between these two states; the actual movements include the following.

Bending of the right leg and straightening of the left (causing a shift of weight onto the right foot), so that the ball of the foot can be raised from the ground and then returned with the foot now pointing "southwest"—notated (5); at the same time the upper body is turned to face in the same direction; the leg movements return the weight to the left foot. The right foot is then released and brought near the left, the ball lightly touching the ground near it before stepping out to the "northwest", notated (7), contact with the ground being heel-first, as the body begins to turn to the right. During the whole of the foregoing, the arms are brought into a formation "as if holding a large ball", with the right hand in front of the lower abdomen and the left above it, in front of the chest, their palms facing each other. Finally, as the right foot is brought into full contact with the ground, pointing "west", the right leg flexes and the left straightens, the upper body completes its turn towards "west", and the arms are brought into positions bilaterally symmetrical to those of the starting position (see Scheme 3) the trajectories of the hands being such as to allow the palms to pass close to one another.

All of this is notated in detail in Scheme 4.

The passage is enclosed between extended double barlines, the first marked with two points, indicating repetition of the whole from that moment; the symmetry symbol at the end of the marked passage indicates that in executing the repeat, all movements of each side are now performed by the corresponding paired limbs of the other, in bilateral symmetry to the original (explicitly written) movements. Compare the last four stages sketched in Fig. 4 with the first four.

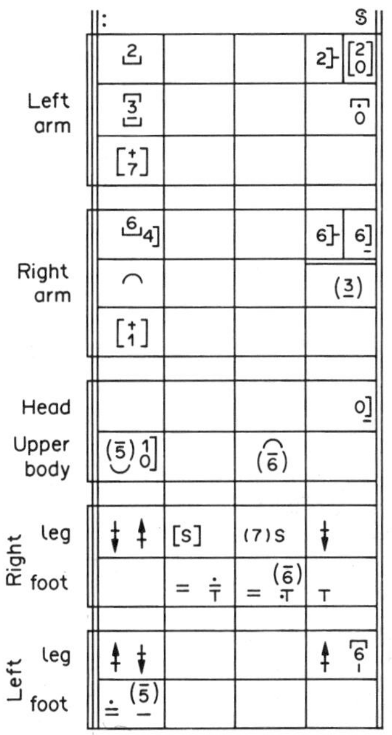

Scheme 4

Ordinary forward walking is an everyday example of this type of symmetry. The shorthand sign for a step in EW notation is S (without points); a number in parentheses indicates the direction on the horizontal plane.

Note. A practised reader of movement notation would easily recognize a symmetrical passage in a score written in full detail—i.e. one in which use was not made of the symmetry symbol. The use of the symbol is appropriate either for conciseness, or when it is wished to emphasize this aspect of the movement material.

(5) Sequential symmetry of fronts

If a passage of movement is repeated in such a way that in its second occurrence the body faces in a new direction, this will be an instance of rotational symmetry on the horizontal plane. In the notation, this will be seen as some integral rotation of front, together with a repetition of the passage, indicated by its enclosure within extended double barlines marked with double points. (The movement sequence may itself be either symmetrical or asymmetrical.)

(6) Translation

If, as is very frequently the case, the simple repetition of a passage of movement results in the second occurrence of the movements taking place in a new region of the floor space, this progression will be an instance of translatory symmetry, regardless of whether the movement sequence itself is symmetrical as in the two forward steps in this notation (Scheme 5): right, then left; or asymmetrical as in sideways stepping: open, close (as occurs in the tenth posture of the Standard Short Form). It is expressed in the notation as simple repetition, indicated by double bar lines enclosing the passage to be repeated and marked with double points.

Some parts of the body may move symmetrically while others do not; for instance, the leg movements may be in bilateral symmetry while the arms execute new movements during each cycle. Or, two overall positions of the body may be symmetrical, but with some variation. Such cases of *partial symmetry* would be shown by the symmetry symbols for the appropriate limbs, while the new material is written explicitly. The example (Scheme 6), taken from Cheng Man-Ching's short form [5, 6], shows two positions in which the second is bilaterally symmetrical with the first in all respects, except that the palm of the right hand is turned upward instead of downward.

It can be baffling to search for logic among the actions of T'ai Chi Chuan (when proficiently performed), as they melt one into another in a single, almost uninterruptedly smooth stream of movement. This evanescence is, however, captured in the symbols of movement notation, and

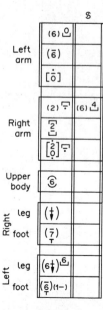

Scheme 5   Scheme 6

correspondences and contrasts can then be detected among the graphic symbols spread over a few pages of manuscript. A practised reader of EW movement notation relies upon the possibility of scanning such a score and noting the formal interrelations. When instances of a specific formal aspect (such as symmetry) are searched for, the data can be extracted from the notated record, and summed up in a table such as the one shown in the Appendix.

Following this procedure in the case of the T'ai Chi Standard Short Form, 53 instances of symmetry were identified, 28 of them complete and 25 partial. Thirteen of the symmetries were translation. As has been noted, bilaterally symmetrical movements always entail bilaterally symmetrical positions, and this explains why all 10 cases of sequential bilateral symmetry of movements are associated with sequential bilateral symmetry of positions. It is noteworthy that there are no instances here of the converse: bilaterally symmetrical positions reached by means of asymmetrical movements.

Positions with non-sequential bilateral symmetry are found to be reached by means of non-sequential (concurrent) bilateral symmetry of movements in almost all of the seven instances identified. The exception is the first posture ("Preparation"), which begins and ends with bilateral symmetry but passes from one to the other by asymmetrical movement.

The eighth posture, "Grasp bird's tail", is regarded by proponents as the basis of T'ai Chi Chuan [5, p. 118]. It is the only posture in which symmetries of all six types are detectable, albeit partially.

In 6 out of the 26 postures, no symmetry was discovered—neither within the posture itself, nor in relation to any other posture of the form.

A more thorough study of the form would include an examination of the leg movements alone, and of the movements of the upper part of the body alone (torso, head and arms), with regard to their symmetry or lack of it. Other styles of T'ai Chi Chuan would be studied in a similar way, and could then be compared.

Notated records of three forms of T'ai Chi Chuan have been made as part of a research project at the Research Centre for Movement Notation at Tel Aviv University [2, 4].

3. CONCLUDING REMARKS

A treatment such as the foregoing could be employed in relation to other fields of human movement, including the art of movement, dance. As already indicated, a more sophisticated and rigorous analysis could result from the fuller exploitation of the EW spherical system of reference with its multiple axes of symmetry, rotated states of limbs and symmetry about a non-rectilinear line of symmetry. With the aim of presenting the approach in as simple a manner as possible, this paper has been limited to a small number of types of symmetry, and applying them to a single sub-style of one area of human movement. However, even in the light of this relatively superficial and limited survey, it is evident that the concepts of movement underlying EW movement notation can serve as a powerful tool in the study of symmetry aspects of movement.

Acknowledgements—The author is indebted to Professor N. Eshkol, head of the Research Centre for Movement Notation at Tel Aviv University, for advice and help on vital points in this paper. The author owes his acquaintance with the T'ai Chi Chuan Standard Short Form to Tew Bunnag and his student Ingrid Flint, but bears sole responsibility for any inaccuracies which may be evidenced here.

REFERENCES

1. N. Eshkol and A. Wachman, *Movement Notation*. Weidenfeld & Nicolson, London (1958).
2. N. Eshkol, T. Sapir, M. Shoshani and J. Harries, The quest for T'ai Chi Chuan—2. Tel Aviv Univ., Israel (in preparation).
3. T. H. Jou, *The Tao of Tai-Chi Chuan*. Tuttle, Rutland, Vt (1980).
4. N. Eshkol, J. Harries, R. Sella and T. Sapir, The quest for T'ai Chi Chuan. Tel Aviv Univ., Israel (1986).
5. M.-C. Cheng and R. W. Smith, *T'ai Chi*. Tuttle, Rutland, Vt (1983).
6. M.-C. Cheng, *Cheng Tzu's Thirteen Treatises on T'ai Chi Chuan*. (Translated by B. P. J. Lo and M. Inn). North Atlantic Books, Berkeley, Calif. (1985).

APPENDIX

A Table of Symmetrical Correspondences Between the Postures of the T'ai Chi Chuan Standard Short Form

The postures	Bil. sym. of movts	Bil. sym. of posns	Seq. bil. sym. of movt	Seq. bil. sym. of posn	Seq. sym. of fronts	Translation	Number of Symmetries
1. PREPARATION	.	*	.	.	.	*26	2
2. BEGINNING	*	*	.	.	.	.	2
3. WILD HORSE PARTS MANES	.	.	*	*	.	*	3
4. WHITE STORK SPREADS WINGS	.	.	.	.	.	.	0
5. BRUSH KNEE & TWIST STEP	.	.	*	*	.	*	3
6. PLAY FIDDLE	.	.	.	.	.	.	0
7. REPULSE THE MONKEY	.	.	*	*	.	*	3
8. GRASP BIRD'S TAIL	[*]	[*]	[*]	[*]	[*]	.	5
9. SINGLE WHIP	.	.	.	.	.	*11	1
10. WAVE HANDS LIKE CLOUDS	.	.	[*]	[*]	.	*	3
11. SINGLE WHIP	.	.	.	.	.	[*]	1
12. HIGH PAT ON HORSE	.	.	.	.	.	.	0
13. KICK WITH RIGHT FOOT	.	.	[*15]	[*15]	[*15]	[*15]	4
14. STRIKE EARS WITH FISTS	[*]	[*]	.	.	.	.	2
15. KICK WITH LEFT FOOT	.	.	[*13]	[*13]	[*13]	[*13]	4
16. CREEP DOWN ⎫ 17. COCK STANDS ON ONE LEG ⎬	.	.	*18+19	*18+19	*18+19	*18+19	4
18. CREEP DOWN ⎫ 19. COCK STANDS ON ONE LEG ⎬	.	.	*18+19	*18+19	*18+19	*18+19	4
20. FAIR LADY AT SHUTTLES	*	.	[*]	[*]	[*]	*	5
21. NEEDLE AT SEA BOTTOM	.	.	.	.	.	.	0
22. FAN THROUGH BACK	.	.	.	.	.	.	0
23. TURN, STRIKE, PARRY, PUNCH	.	.	.	.	.	.	0
24. APPARENT CLOSE UP	.	.	[*]	[*]	.	.	2
25. CROSS HANDS	.	.	[*]	[*]	.	.	2
26. CONCLUSION	.	.	*	*	.	*2	3
	6	7	10	10	6	13	=53

It is customary among teachers of T'ai Chi Chuan to divide the form into short sequences of movement referred to as "postures". The picturesque names of all postures in the Standard Short Form are given in order on the left of the table, and numbered from 1 to 26. The types of symmetry attributed to each are indicated on the same line, by asterisks * placed in the appropriate columns allotted to the types of symmetry indicated at the head of the table. Thus, only bilateral symmetry of positions is attributed to posture No. 1, "Preparation"; whereas posture No. 2, "Beginning", has bilateral symmetry of movements as well. The first posture is the only case here in which bilateral symmetry of positions is not linked with bilateral symmetry of movements; i.e. the positions were reached by means of asymmetrical movement.

The whole sequence of postures was notated (as in the schemes given in the main text). Inspection of the resulting score permits easy identification of the different types of symmetry posited. The divisions according to the postures have mostly been adhered to, both "horizontally", i.e. not dividing the sequences of movement at other moments in time; nor "vertically", e.g. not treating movement of the upper part of the body separately from the leg movements. Divisions could be made so as to permit more complete correspondences; here, wherever the symmetry is incomplete, the related postures are noted as "partial symmetries", indicated in the table by the use of square brackets [*]. In one case, by regarding two postures as a single sequence, full symmetry was established (postures 16–19). This is the only point at which the limits of a posture have been ignored.

Sequential symmetries may be in immediate succession, or may be obtained between postures separated by other postures. In the latter case the number of the related posture is given following the asterisk. For example, posture No. 9 ("Single whip"), is related by translation to No. 11. This is shown as *11 in the column allotted to translation.

The numbers in the extreme right-hand column indicate the number of different symmetries attributable to each posture; and the bottom line shows how many instances of each type of symmetry were distinguished over the whole form.

Computers Math. Applic. Vol. 17, No. 4–6, pp. 837–885, 1989
Printed in Great Britain. All rights reserved

0097-4943/89 $3.00 + 0.00
Copyright © 1989 Pergamon Press plc

SYMMETRY ASPECTS OF BOOKBINDINGS

M. ROZSONDAI

Department of Manuscripts and Rare Books, Library of the Hungarian Academy of Sciences,
P.O. Box 7, Budapest, H-1361, Hungary

B. ROZSONDAI

Structural Chemistry Research Group of the Hungarian Academy of Sciences, Eötvös University,
P.O. Box 117, Budapest, H-1431, Hungary

Abstract—Geometric and other relations of decorated leather bookbindings are analysed. Symmetry properties of the ornamentation, the symmetries of motifs and of the layout, and some correlations of symbols and ideas, furthermore the occurrence of one- and two-dimensional space groups and interlace designs are demonstrated by Romanesque and Gothic bindings, Hungarian, Italian, French and German Renaissance, as well as by Baroque and rococo bindings, and finally, by some pieces of modern bookbinding art.

INTRODUCTION

"Numero pondere et mensura Deus omnia condidit"—"God created everything by number, weight and measure." Isaac Newton dedicated these words (Fig. 1), his "tessera", to a Hungarian student, Ferenc Páriz Pápai Jr, the possessor of the album held by the Department of Manuscripts and Rare Books of the Library of the Hungarian Academy of Sciences [1]. The Academy, now embracing all branches of knowledge from arts and humanities through natural sciences to applied sciences, was

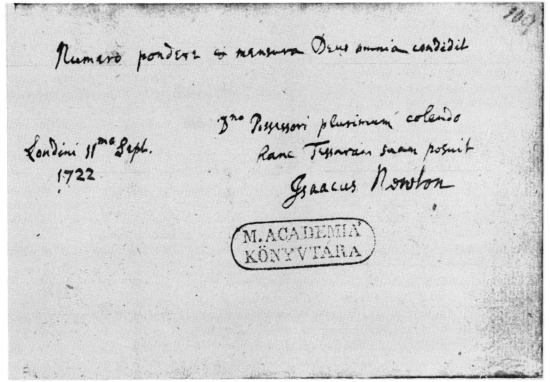

Fig. 1. Isaac Newton's autograph in the Library of the Hungarian Academy of Sciences. [Shelf-number: Tört. naplók, kis 8° 6.] Reproduced by permission. The authors are grateful for the kind permission to reproduce material from the holdings of the Library of the Hungarian Academy of Sciences (abbreviated hereafter in figure captions as Bibl. Acad. Budapest; no special mention of the permission will be made).

founded in 1825 as a "Learned Society" to promote above all Hungarian language and literature. Thus, in addition to scientific books and periodicals, the Library contains literary monuments and bequests, manuscripts and rare books. Most of our examples of bookbinding decorations will be taken from there.

Jan Amos Komenský (Comenius), the Czech educator, who founded modern visual teaching by his most renowned work, *Orbis Sensualium Pictus* (*The Visible World in Pictures*, Nuremberg 1658; London 1659) [2], set up the ideal of pansophia, a unified science reflecting the indivisibility of nature. His book, which he drafted during his stay (1648–1654) in Sárospatak, Hungary, is also a realization of these ideas in teaching Latin and other foreign languages. It contains a passage on "Bibiliopegus, the Book-binder", as well as others on "Printing, the Book-sellars Shop, a Book, and a School".

Recalling Newton's maxim, one of the striking manifestations of measure $= \tau\grave{o}\ \mu\acute{e}\tau\rho o\nu$ is symmetry as it appears in nature and in man's work. It may also be a bridge, as the present and a previous special issue of this journal exemplify, between different fields of artistic and scientific human activities.

In the following discussion we try to investigate geometric and analogous relations throughout the history of bookbinding decoration. No attempt has been made to give a complete representation of all periods and styles; the selection of examples was influenced by our personal interest and the accessibility of material.

One of the conspicuous geometrical relations is symmetry. Point groups and space groups will be used here to classify symmetry properties of ornaments but no knowledge of the theory is assumed. A point group (with the properties of a mathematical group) is a set of symmetry operations, which leave at least one point of the object fixed in space. For the planar figures to be discussed, the symmetry operations include reflection through a plane (mirror plane, m) and rotation about an axis (two-, three-, four-fold etc. axis, 2, 3, 4,... or, in other notation, C_2, C_3, C_4,...). Mirror planes and rotation axes are perpendicular to the plane of the figure. Conventional notations of symmetry groups indicate the basic symmetry operations of the group. In the case of space groups, additional symmetry operations are translation and glide reflection, i.e. a reflection combined with a translation. A space group applies to an infinite lattice or periodic pattern, and when we speak of the one- or two-dimensional space-group symmetry of a decorative pattern, we regard it as a section of an infinite structure. For that matter, imposing geometrical relations such as congruence, symmetry or similarity on real objects is more or less an approximation. Deviations from exact relations are more apparent on handicraft products—historical bookbindings typically belong to this category—and the extent of deviations depends on several factors such as technique, craftsmanship, tools and material used. Sometimes a deviation can be intentional.

Stone, wood, clay tablets, parchment or papyrus scrolls bore written records in ancient times before the book in its present form appeared. Its precursor was the diptych, a pair of ivory tablets or wooden boards, possibly decorated, protecting the inscription in wax inside. Some folded parchment sheets were then placed, later also fastened, between the panels—and thus the book was born! All the essential elements that constitute a book today had been brought together by the dawn of the Middle Ages. A handwritten and illuminated book was itself a great asset; ivory or precious metal covers and jewels added to its value and impressiveness [3]. Even in recent centuries jewelled or ivory bindings have been occasionally applied to special or ceremonial books.

Yet the bookbinder's craftsmanship finds its proper expression in the preparation and embellishment of leather bookbindings. Leather had been a favourite material for covering books from the early Coptic bindings until the age of large-scale book production when the cheaper binding materials of cloth and paper replaced it, at least partially.

Leather bindings were most commonly decorated by means of finishing tools, each having an engraved design on its face and producing a corresponding blind or gold-tooled impression on the leather. A pallet or fillet was used to impress a line or parallel lines, a smaller stamp to have a unit motif. The roll, a tool with a brass wheel, with an engraved pattern on its circumference, and the panel, a larger block of metal, were in general use from the fifteenth to the nineteenth century. The tool itself and its impression are denoted by the same words, viz. a stamp, a roll, a panel. A simple technique of reproducing an impressed motif or design of a bookbinding is to put a piece of paper on it and rub it with various soft lead pencils.

Even after the invention of printing in the 1450s, when hundreds and hundreds of identical copies of an edition were produced, hand bookbinding transformed each copy into a unique entity. The purchaser and user had the book bound in most cases. Bookbinding research, which grew out of bibliophilic interests in the last century, goes into the details of the "when, where, by whom, for whom" etc. of a book and its binding, and, relying on the exact identification of tools used for decoration, on owners' notes in the book, on fragmental pages used as auxiliary material in binding, on archival sources etc., has become a complex field of study. Decorated bookbindings reveal a lot about the books and the culture of a given age.

BOOKBINDING IN THE MIDDLE AGES

Some Coptic leather bindings from Egypt have come down to us [4] from the sixth to the eighth centuries but such bindings existed in earlier centuries. In Europe the earliest leather bindings date from the Carolingian age (ninth and tenth centuries). These are followed by the books bound in Romanesque style (twelfth and thirteenth centuries). 139 such bindings are recorded today [5], and they represent a fully developed art of book decoration.

It must be remembered that at the time of the *Romanesque* bindings, *Gothic* art was flourishing and dominating in the architecture of Europe [6], and the same intellectual trend, Scholasticism, influenced bookbindings and architecture alike. Applied arts—including bookbinding—are in general characterized by a certain delay in relation to fine arts.

Romanesque bindings, i.e. their layout (Fig. 2), are related rather to Coptic or Islamic bindings than to those of the Carolingian and Ottonian ages. The transition from Romanesque to Gothic bindings is, however, quite continuous. Motifs are similar, and even the recutting of some Romanesque tools, especially palmetto and dragon stamps has been noted [7].

The most frequent Romanesque and Gothic stamps (Fig. 3) represent lily (fleur-de-lis), palmetto, foliation, rosette (four, five and sixpetalous), birds, a pair of birds, deer, dragon, eagle, double-headed eagle, griffin, the Holy Lamb, mermaid, monkey, pelican (the Christ-symbol), unicorn, and the symbols of the four Evangelists (Matthew, Mark, Luke and John: angel, lion, ox and eagle). On Romanesque bindings usually there are many stamps, certainly more than on Gothic bindings. Only eight kinds of stamps (Fig. 5) but altogether more than 600 (!) impressions of them figure on the upper and lower cover of the early Gothic leather binding of a parchment codex from the fourteenth century (Fig. 4). The proper Gothic binding shows a looser layout and a much smaller number of repetitions of the stamps.

Bilateral symmetry of motifs (Fig. 3) (point group m or C_s) and rotations with mirror planes (point groups 3m, 4mm, 5mm and 6mm or C_{3v}, C_{4v}, C_{5v} and C_{6v}, subscript v for vertical) are common, but pure rotational symmetry (point groups 2, 3, 4, . . . or C_2, C_3, C_4, . . .) is rarely applied. A playful collection of symmetries appears on a roll (Fig. 6), including point group C_3 with

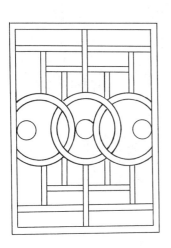

Fig. 2. Layout of three Romanesque bindings after [5], Nos 1, 7, 13.

Fig. 3. Stamps from some Gothic bindings of the Bibl. Acad. Budapest.

Fig. 4. Early Gothic leather binding, upper cover. National Széchényi Library, Budapest: Clmae 79.
Reproduced by permission.

Fig. 5. Stamps on the binding in Fig. 4: pair of birds, doubleheaded eagle, the Holy Lamb, pelican, deer,
dog, monkey, bird.

Fig. 6. A roll from a Gothic binding made in Augsburg. The motifs have C_{4v}, C_{4v}, C_3, C_{3v}, C_{6v}, C_{4v}, C_{6v}, C_{4v}, C_{5v} and C_{4v} point-group symmetry, respectively. Bibl. Acad. Budapest: Ráth 1058. This roll is identical with that in Ref. [8, Plate 171, 6].

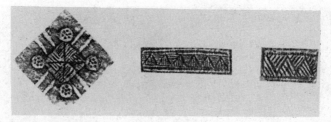

Fig. 7. Hatched motifs from Gothic bindings. Bibl. Acad. Budapest: Inc. 192, Inc. 668.

a three-fold rotation axis. Asymmetric motifs are often confined to a symmetric planar figure like a circle, a lozenge, or a square, and the encircling line is also shown.

Hatched motifs (Fig. 7) can be considered to possess colour symmetry, which means a combination of a geometrical symmetry operation (reflection, rotation, translation etc.) with a simultaneous permutation of colours. The use of hatching to indicate colours in heraldry explains the name "fer azuré" (azured tool) of the hatched stamps introduced in the sixteenth century on French Renaissance bindings.

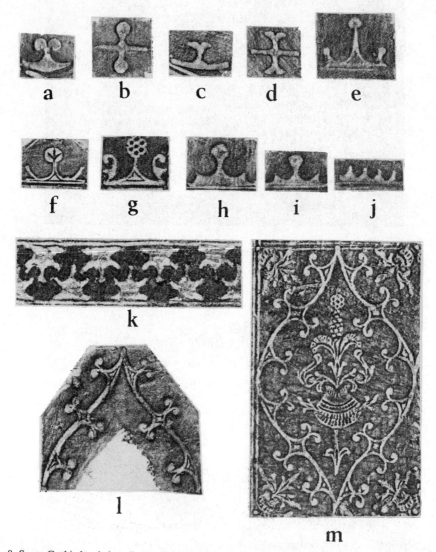

Fig. 8. Some Gothic headed outline stamps (a–k) and two blocks (l, m) derived from them. Bibl. Acad. Budapest. The complemental space between stamps k gives the popular "cloud" pattern [9].

Fig. 9. Gothic binding from a workshop in Herzogenburg (Ref. [8, Plate 27]). Bibl. Acad. Budapest: Inc. 534.

The headed outline or cusped edge stamps (Kopfstempel in German) played an important role in the decoration of leather bindings (Fig. 8). The central field of the upper cover, within the borders, was decorated with curved branches, with a repeated vine-like tooling, especially in the southern parts of Germany (Fig. 9). This pattern became popular in Austria, Bohemia and Hungary, and it is a good example of technical progress as demanded by the increasing book production. In the 1470s a whole curved diamond-shaped compartment (Fig. 9) was circumscribed in twelve steps by impressing eight double-headed cusped edge stamps [Fig. 8(a)] and four stamps with two "heads" on the opposite sides [Fig. 8(b)]. A large number of operations was needed to fill the central panel. To accelerate the working process, a half curved branch was engraved in the tool [Fig. 8.(l)], and the bookbinder obtained a whole curved lozenge in two actions. Soon after the panel with all the details engraved was introduced [Fig. 8(m)]. The half curved branches began a new life on some Renaissance bindings in the early sixteenth century, and they appeared as ogee branches (Fig. 10).

Fig. 10. Early Renaissance binding with ogee pattern, lower cover, from a secular workshop in Buda [10].
Bibl. Acad. Budapest: Inc. 877.

The cuir ciselé, i.e. the cut-leather bindings make up a special group of Gothic bindings [11]. The outline of the pattern is cut into the dampened leather and is emphasized by punching (stippling) the background by a pointed tool (Fig. 11). This embellishment required skilful masters. A similar appearance could be reached more simply by the ingenious use of the cusped edge stamp. The master had to choose the most suitable stamps to get an indented outline of an oak leaf (Fig. 12), or he had to think over carefully the proportions and the distances between the headed outline tools (Fig. 13). Thus, he achieved a turnover of foreground and background, and a good appearance of the oak leaf or of a four-leaved fleuron. The impressed curved lines mark the nervure of the leaf.

The upper cover of Gothic bindings shows a more abundant ornamentation than the lower cover (Fig. 14). One or two borders, marked out by fillets, surround a central field, which is then

Fig. 11. Cut-leather binding, Germany, fifteenth century [11, item 62]. National Széchényi Library, Budapest: Clmae 74. Reproduced by permission.

subdivided by oblique straight lines to form a diamond pattern (Fig. 15), or by an arched vine-like network into curvilinear compartments (Fig. 9). While rectangular or rather square, semicircular and quadrant subfields were preferred on Romanesque bindings (see Fig. 2), resembling structures of Romanesque architecture, perpendicular crossings of lines were avoided in the central field of Gothic bindings. The central rectangle is subdivided only by its main diagonals (Fig. 12), or, more often, by additional lines parallel to them (Fig. 16). On the binding in Fig. 15, the short sides of the rectangle are divided into two sections, the long sides into three sections by intersecting lines, while the main diagonals do not appear in the pattern. Using the latter and, consequently, an equal number of divisions on the sides of the rectangle, would have produced too slim diamonds. In all these cases, the original C_{2v} point-group symmetry of the rectangle, with two mirror planes and a two-fold rotation axis perpendicular to the plane of the figure, is retained if we disregard the pattern within the subfields. An interesting skewly oriented quasi-diamond tiling is shown in Fig. 13. The minor discrepancy between the two sides of the rhomboid is emphasized by the alignment of one and two cusped edge stamps along them, respectively.

The symmetries of a field and of the motifs which fill it often disagree. A pentamerous rosette can sit in the centre of a diamond (Fig. 15). The half diamonds along the sides of the central rectangle contain different motifs. The foliage on the binding in Fig. 14(b) breaks through the rigid frames of the rectangle and has four-fold rotational symmetry, point group C_4. Figure 17 shows a naturally simple arrangement of alternating drop-shape figures, all pointing downwards in the central field with the antlers in this unnatural orientation, and outgrowing the corner boxes. Adjoining single stamps in the inner border take the shape of a tracery so characteristic of Gothic architecture and decorative art. David's six-pointed star (Solomon's seal) in the outer corners encloses a pentapetalous flower. Bearing in mind that a rosette is the Virgin Mary's symbol, is this a hint at her line of descent?

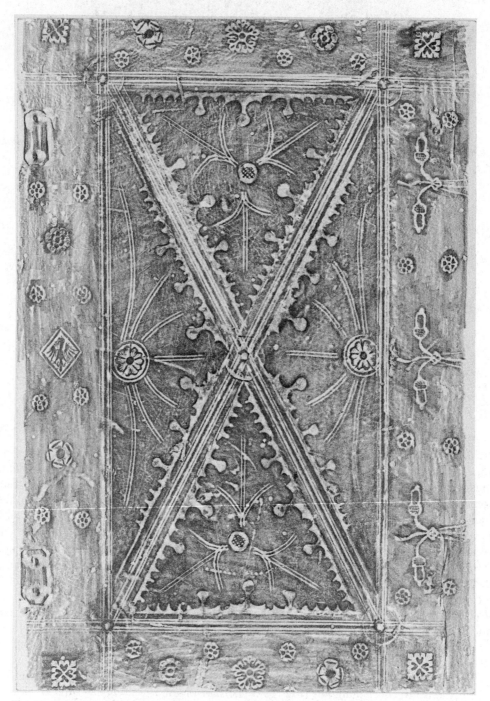

Fig. 12. German Gothic binding with leaf relief obtained by the headed outline tools in Fig. 8(h–j). Monastery bindery, Aldersbach, Bayern. Bibl. Acad. Budapest: Inc. 746.

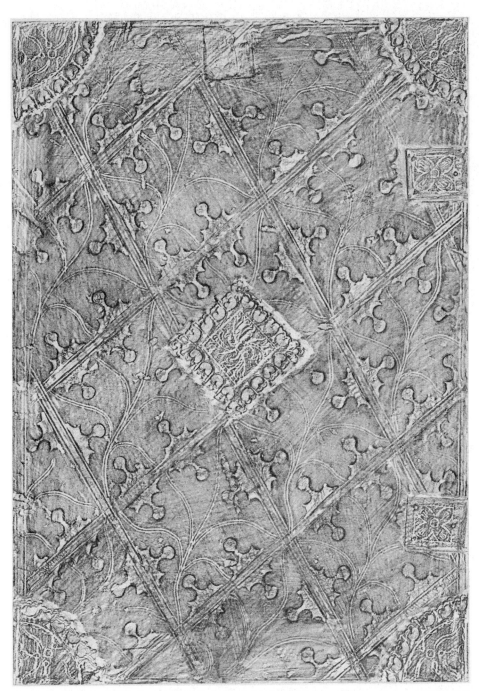

Fig. 13. Gothic leaf-relief binding. Probably Ingolstadt. National Széchényi Library, Budapest: Clmae 223. Reproduced by permission.

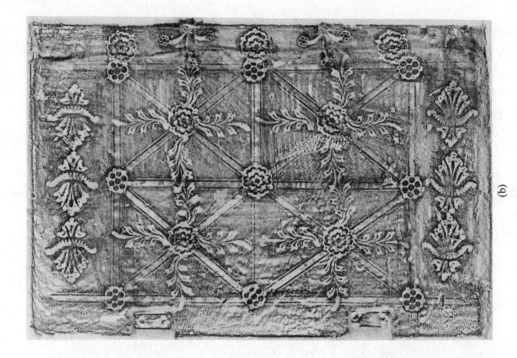

(b)

(a)

Fig. 14. Upper (a) and lower cover (b) of a Gothic binding. Near Linz. Bibl. Acad. Budapest: Inc. 400.

Fig. 15. Monastery binding from Vienna [12]. Bibl. Acad. Budapest: Inc. 148.

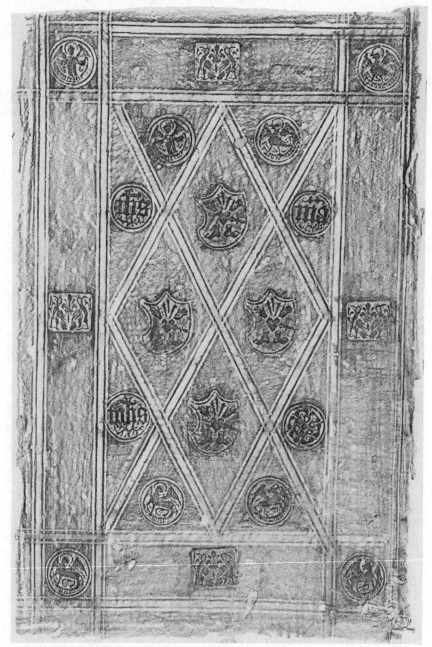

Fig. 16. Monastery binding from Weddern near Dülmen [13]. Pelbartus de Themeswar: *Pomerium sermonum de sanctis.* H. Gran, Hagenau (1520). Bibl. Acad. Budapest: RM III 93.

A symmetry of ideas and persons is represented on a simply decorated lower cover (Fig. 16) by the images of Christ's suffering in the shields (two hands, two feet, three nails, a heart and spear), the names of Jesus, his mother Mary and her symbol the rosette beneath, and John, the disciple whom he loved and who stood with Mary by his cross, and the symbols in the corners and half diamonds of the four Evangelists who told the story of the Passion. A binding decorated with the same tools is kept in the British Library [14].

The edges of a binding, and sometimes its central field too, are decorated by "frieze" patterns. The seven possible symmetries of infinitely repeating patterns, the one-dimensional space groups have been nicely illustrated by Hungarian needlework [15]. Similarly, border patterns from bookbindings are shown in Fig. 18, and classified according to symmetry groups. (See e.g. Ref. [17] for notation and explanation.) Such patterns were produced either by repeated impressions of single

Fig. 17. Gothic binding from Ulm [13]. Bibl. Acad. Budapest: Ráth F 1065.

stamps [as most of the patterns in Fig. 18 and the "cloud" pattern in Fig. 8(k)] or, especially later, on Renaissance bindings, by a roll. The period of the design, i.e. the shortest distance at which the motif recurs, shows the dimensions, the perimeter of the roll (Fig. 19), or, rather with geometric ornaments, the same motif is repeatedly engraved in the periphery, and it is difficult to find out the true dimensions of the tool. The floral curls in Fig. 18(b) fit roughly in space group 1g, while there exist actually at least four variants of flowers and birds, and translation remains the only symmetry operation.

While searching for and selecting from examples of space groups of border designs occurring on bookbindings, one may reflect upon how and why symmetries of ornaments were chosen, preferred or neglected. The conventions of the given style, its stock of forms and motifs seem to be decisive.

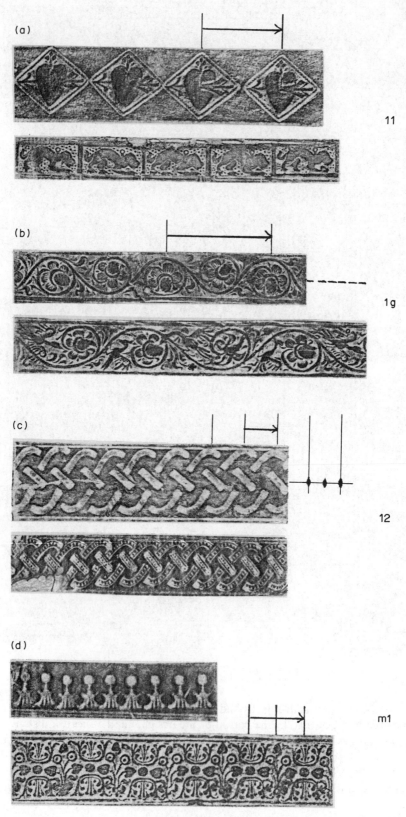

(a)

11

(b)

1g

(c)

12

(d)

m1

Fig. 18(a)–(d).

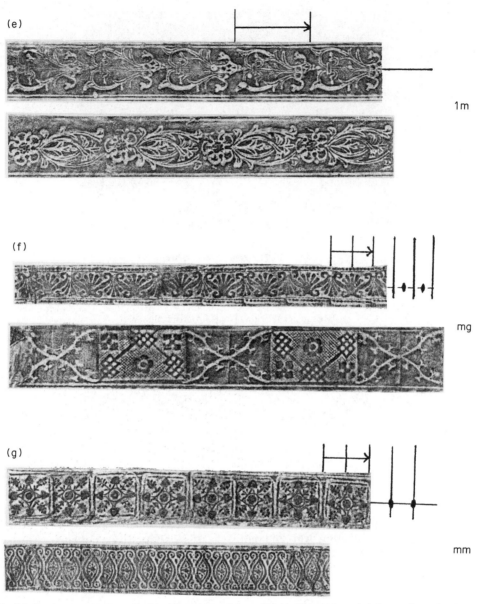

Fig. 18. Border patterns from Gothic (G) and Renaissance (R) bindings [16], and their one-dimensional space groups. → translation vector, — reflection plane, - - - - glide reflection plane, ◗ two-fold rotation axis. (a) Monastery bindery Buda (G)–Dominican bindery Vienna (G); (b) both from Augsburg (G); (c) both from Vienna (G); (d) Vienna (G)–Vienna (R); (e) Buda (R)–Venice (R); (f) Florence (R)–Minden, Germany (G); (g) Bamberg (G)–Spanish (R). Bibl. Acad. Budapest.

Gothic figures of animals, birds etc. or a hunting scene (Fig. 19), all viewed from the side, present no symmetry at all, and the simple repetition of these motifs in a strip leads to space group 11 [Figs 18(a) and 19]. Other arrangements with further symmetry elements are not likely to occur, because it rarely happens that an ordinary animal is represented upside down; a nice dragon, however, may be an exception (Fig. 20)! Bookbinders of the Renaissance, however, rejected such restrictions when they impressed a roll horizontally along the edges of the cover with portraits or mythological and allegorical figures. We have not met with a head-to-head tail-to-tail alignment of animals making up symmetry group ml. At any rate, a mirror plane or a glide reflection plane in an arrangement of asymmetric motifs, when realized by single stamps, would require both "enantiomers" (mirror image copies) of the tool. Flowers, foliage, vine or geometrical forms offer a wider variety of symmetries of either the motif itself or the pattern it comprises.

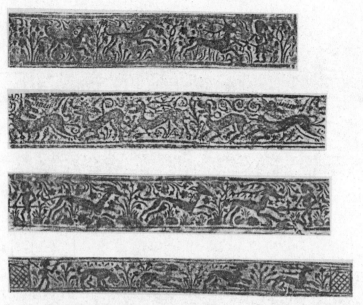

Fig. 19. Rolls with hunting scenes on late Gothic bindings from Augsburg (the upper three) and Memmingen (below) [16]. Bibl. Acad. Budapest.

Fig. 20. Dragon stamps forming a border design, space group 12. Drawing by Éva Kovács-Rozsondai after Ref. [18].

Masters of Romanesque and Gothic bookbindings obviously did not speculate about symmetry, they just applied it intuitively. The spirit of their age, the object of their artistic endeavour, i.e. the book cover, and the nature of their materials and tools determined their work. The symmetries of general layout, patterns covering fields and motifs making up patterns do not necessarily harmonize, and thus the resulting complete decoration may have a lower symmetry than its components, or even no symmetry at all.

RENAISSANCE BOOKBINDINGS

While the Gothic style is international and its general criteria are the same all over Europe, the Renaissance has national marks, and these are characteristic of the country whose "make" the given binding is. The bookbindings to the south of the Alps differ considerably from those of the Transalpine area.

Everyone who is fond of beautiful books must have heard of the Corvinus bindings. The once famous library of the Hungarian king Matthias Corvinus (1458–1490) held about 2000 codices [19]. Approximately one-tenth of the stock has survived, scattered throughout 44 towns in 14 countries. There are 47 Corvinus manuscripts and 3 incunabula† today in Hungary. Some of the books have their original silk, velvet or coloured gold-tooled leather bindings. The upper and lower covers of the decorated leather bindings are identical, apart from the title or the author's name, impressed at the top of the lower cover. Italian Renaissance and oriental influences are mixed with local decorating tradition on these typically Hungarian products (Fig. 21). The floral motifs: rosette, calyx-flower, leaves, peltate, palmettos and tulips are in general gilded, the cablework is blind, the

---

†Books printed in the fifteenth century.

(c)

(b)

(a)

Fig. 21—*continued overleaf*

(d)

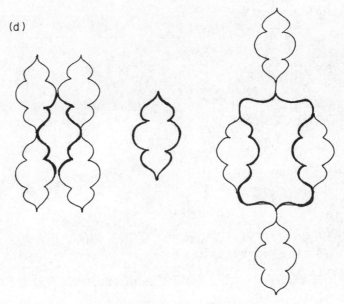

Fig. 21. Gold-tooled Corvinus bindings. (a) Österreichische Nationalbibliothek (ÖNB): Cod. Lat. 1037. (b)
National Széchényi Library, Budapest: Clmae 422. (c) ÖNB: Cod. Lat. 1391. Reproduced by permission.
(d) Schemes of the central designs after Ref. [20].

punch dotting coloured. Either the royal coat of arms or Matthias' raven (corvus in Latin, hence
his epithet) is placed in the centre. Among the extant copies one cannot find two identically
decorated bindings. The overall symmetry is quite simple, C_{2v}, except for some details. An
interesting feature of the large number of emphasized central panels has been noted [20], a leap of
a form into its negative, an interconversion of figure and its background. All these forms can be
imagined as enclosures shaped by surrounding copies of one and the same motif [Fig. 21(d)], which
by itself also appears as a central piece. The figure/background effect is even more pronounced in
the repeated pattern of Fig. 22, which was obviously inspired by a certain type of oriental carpet
(Fig. 23).

Some tools of the Corvinus bindings, first of all the flower-cup, were recut and used in other
binderies in Buda in the first three decades of the sixteenth century. In addition, new stamps and
rolls were produced (Fig. 24), and among them the different interlaced knotwork motifs and the
rolls combining palmettos with leaves of the acanthus became extremely popular. A number of
their variants existed [Fig. 25(a)] on Hungarian Renaissance bindings. The acanthus leaves were a
common ornamentation in Greek architecture in the fifth century B.C. and revived in the fifteenth
and sixteenth centuries all over Europe [Fig. 25(b), (c)]. The interlace or strapwork also occurs in
architecture (Fig. 26).

Italian Renaissance bindings are lightly decorated and pleasantly spaced out. A large scale of
knotwork stamps, arabesque centrepieces, linked arabesque circles as borders are their most
prominent stylistic features (Figs 27, 28). Coptic and Islamic motifs as well as Persian elements like
the peas-trailer or clasper can be recognized on them (Fig. 29).

Different types of Renaissance bindings developed in France in the sixteenth century. A number
of the books bound for Jean Grolier, the bibliophile, were decorated by complicated interlace work
(Fig. 30). Another trend of book ornamentation, which was connected with the French royal court
and had remained in practice well into the seventeenth century, created elegant masterpieces by
simpler geometrical means. These semé bindings (semer = to sow) provide an opportunity to
demonstrate two-dimensional space groups.

Only some of the 17 possible two-dimensional space groups occur on bookbindings. The continuity
of the pattern is broken not only by field boundaries but also by variant or extrinsic elements of
decoration, and the number of repetitions of the basic motif or tile is too small to speak of an
"infinitely repeating" pattern. On a semé binding, the lattice points, in which the motifs are placed,

Fig. 22. Gold-tooled Corvinus binding with repeated pattern. Österreichische Nationalbibliothek: Cod. Lat. 2271. Reproduced by permission.

Fig. 23. Holbein-carpet. Anatolia, sixteenth century. 150 × 220 cm. Museum of Applied Arts, Budapest: Inv. 14785. Reproduced by permission.

Fig. 24. Hungarian Renaissance binding [21] of the so-called Virginia Codex, a Hungarian linguistic record.
Franciscan monastery, Buda, 1510–1522. Bibl. Acad. Budapest: K 40.

(a)

(b)

(c)

Fig. 25.(a) Rolls of palmettos and acanthus leaves from Hungarian Renaissance bindings. (b) Frieze from the Erechtheion, Acropolis, Athens. (c) Pilaster head from king Matthias' palace, Buda.

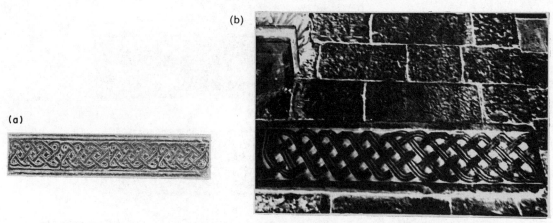

Fig. 26.(a) Knotwork border design composed of single stamps from the Hungarian Renaissance binding: Bibl. Acad. Budapest: Inc. 1018. (b) A five-strand knotwork frieze in the town-hall yard of Trogir, Yugoslavia.

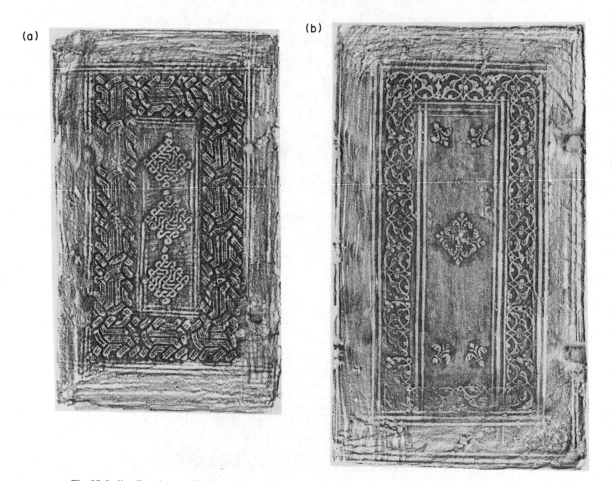

Fig. 27. Italian Renaissance bindings, sixteenth century, with different types of knotwork (a), and arabesque circle border (b). Bibl. Acad. Budapest: K 394, Ant. 195.

Fig. 28. Arabesque centre and corner pieces. Bibl. Acad. Budapest: Ant. 899, RM IV F 164.

Fig. 29. Gold-tooled Renaissance binding. Venice [22, 23]. National Széchényi Library, Budapest: Clmae 313. Reproduced by permission.

Fig. 30. French Renaissance binding for Jean Grolier [23, Plate 116]. Österreichische Nationalbibliothek: 73.E.20 (ES 191). Reproduced by permission.

form a rectangular (space group pmm, see Ref. [17] for notation) or a diamond tiling (cmm) (Fig. 31). The symmetry of the pattern is then determined by the symmetry of the underlying motif and the lattice. The fleur-de-lis motifs in a lozenge-type arrangement (Fig. 32) form a pattern with symmetry cm (Fig. 31). A system of alternating motifs is a superposition of two or more lattices (Fig. 33). Although this array has a low geometrical symmetry, it possesses further symmetry elements that include permutations of the motifs. This is then a case of colour symmetry, with the motifs representing the different colours. The Gothic diamond tiling (Figs 15 and 16) and the analogous vine-like diaper (Fig. 9) also belong to space group cmm (Fig. 31). The binding from the Corvinian Library (Fig. 22) has a higher tetragonal symmetry p4m (Fig. 34) if the details and the layering of the interlace work are disregarded. A much wider variety of space groups and colour groups occurs on decorated papers used as book covers or lining [24] or as wallpaper.

The symmetry of an interlace pattern can be studied at different levels. First, we may regard the pattern as a composition of lines and figures in the plane they decorate, exactly as it appears, disregarding its three-dimensional appearance. The two interlaced square frames in Fig. 35 thus possess only an eight-fold rotation axis but no mirror planes (point group $\mathbf{C}_8$). This figure can also

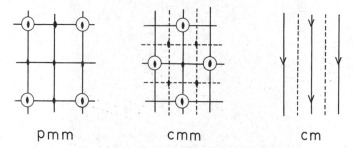

pmm   cmm   cm

Fig. 31. Two-dimensional space groups of the rectangular (pmm) and the lozenge (cmm) lattice, and a lozenge-type array (cm) of motifs with $\mathbf{C}_s$ point-group symmetry. ○ and V motifs, — reflection plane, ---- glide reflection plane, ◖ two-fold rotation axis.

Fig. 32. Semé binding, first half of the seventeenth century, probably from the Netherlands. Museum of Applied Arts, Budapest: Inv. 51.86. Reproduced by permission.

be regarded as an eight-pointed star, with "hidden" parts of the strips added. Now it has eight reflection planes in addition to the eight-fold axis (point group $\mathbf{C}_{8v}$). A third way of looking at the figure is to consider its two or more layers and introduce symmetry operations such as a combination of reflection with a permutation of layers. Geometric and such combined symmetry operations make up the layer groups. The double-square motif (Fig. 35) can be regarded as lying in a two-sided plane, or it can simply be characterized as a three-dimensional object of point-group

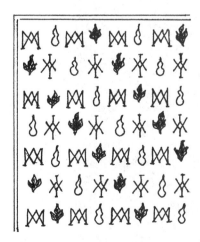

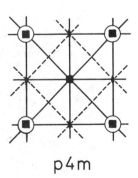

p4m

Fig. 33. Pattern, space group p1, composed of double letters M and Y, tears and flames on a semé binding. Sketch after Ref. [4, Plate 93].

Fig. 34. Symmetry elements of the two-dimensional space group p4m. See Fig. 31 for notation, and ■ four-fold rotation axis.

Fig. 35. Interlace motif from a Renaissance binding. Venice. Bibl. Acad. Budapest: RM IV F 164.

Fig. 36. Italian-type Renaissance binding, middle of the sixteenth century. Bibl. Acad. Budapest: Ant. 833.

Fig. 37. Portraits of Luther and Melanchthon on the central panels of upper and lower cover, respectively, of the binding by Thomas Krüger, Wittenberg, 1571 [25]. *Novum Testamentum.* Interpreted by Th. Beza. H. Stephanus, Genevae (1567). Bibl. Acad. Budapest: 542.338.

symmetry D_8, with an eight-fold axis and eight two-fold axes perpendicular to it. Further examples of layered motifs and patterns are shown in Figs 18(c), 27(a), 30, 36 and 54. On bookbindings, a braid of continuous strands or an interlace of loops and strands is often simulated by single stamps [Fig. 26(a)], and imperfections at the linkages may make an exact evaluation of the structure difficult. The four-strand knotwork in Fig. 18(c) (upper) is obtained by two single stamps: one straight, one curved. Note the different slopes of the ascending and descending branches, giving serrated rather than symmetric wave lines.

✳   ✳   ✳

In the countries north of the Alps panels and rolls dominate on Renaissance bookbindings. Both have mainly figural decorations, a portrait or a scene from the Holy Bible etc. The upper and lower covers are almost the same, except for the central panels, which, however, are related conceptually. Thus, if leaders of the Protestant Reformation are shown, Luther is accompanied by Melanchthon (Fig. 37), Calvin by Bèze, and if a "Reformer roll" is used we can nearly always see the following four portraits: Martin Luther, Johann Hus, Erasmus Rotterdamus and Philipp Melanchthon (=Marti–Iohan–Eras R–Phi Me: Fig. 38).

Personified virtues such as Justitia, Fortuna (Fig. 39), Fides (faith), Spes (hope), Caritas (Christian love), Patientia, Prudentia (prudence and providence), Fortitudo (strength of mind, courage) and Temperantia (moderation) are also favourite figures of the panels and rolls (Fig. 40) on Renaissance bindings of the German type. The counterpart of Justitia is Fortuna, sometimes Lucretia or Judith. The thematic symmetry of the panels on the upper and lower cover exists here, too.

A classical element was revived with the Janus-faced Prudence on a leather binding (Fig. 38). The inscription of the panel: "Seek advice from me—I who am called Prudentia—if you wish for counsel in your affairs." Janus, the Roman god with two opposite faces looking forward and backward, gives good advice and is considerate and provident like Prudentia on our panel. Janus is the god of all beginnings, and it is advisable to begin everything with consideration and circumspection, i.e. prudence.

The virtues as well as the Muses (Fig. 40) and the seven liberal arts (Fig. 41) are as a rule represented in every field of Renaissance art, hence also in book illustrations (Fig. 42) and on bookbindings. Note the analogous depiction of arts and virtues in Fig. 41.

If we see the panel of the Judgement of Solomon (1 Kings 3:16–28) on the upper cover then we see another scene from the Old Testament, viz. Samson with the lion and—in the background—he

Fig. 38. German Renaissance binding with a Reformer roll and a Prudentia panel. Bound by M W (meaning Melchior Wagner, Leipzig?) after 1567 [25]. Bibl. Acad. Budapest: RM IV 427.

Fig. 39. Justitia and Fortuna on the upper and lower cover, respectively, of a German Renaissance binding by H W (meaning Hans Welcker, Nuremberg?) [25]. Daniel Wintzenberger: *Warhafftige Geschichte und gedenckwirdiger Händel* . . . Dresden (1583). The inscription under Justitia: "Suum cuique juste tribuo" is a saying attributed to the Roman jurist Ulpianus; under the "ambiguous" Fortuna: a variant from Tristia 5.8.15 by Ovid. Bibl. Acad. Budapest: RM IV 428.

is taking away the gates of the city of Gaza (Judges 14:6; 16:3) (Fig. 43). Or on another binding: The most outstanding heroine in the Old Testament is Judith (the book named after her is one of the Apocrypha), who beheaded Holofernes, the general of the Assyrian troops of Nebuchadnezzar, and thus saved Israel from its enemies. Judith inspired many artists to depict her and her feat in paintings, in sculptures and in metal engravings. In topical symmetry parallel to Judith, we see either Justitia holding a sword and a pair of scales or—more often—Jael, who inflicted something very similar on another enemy, Sisera (Judges 4:21) (Fig. 44).

The strongest testimony of faith is Abraham's readiness to offer his only son Isaac (Genesis 22). "No one is so great as Abraham! Who is capable of understanding him?"—asks Søren Kierkegaard (1813–1855), whose influence is larger today than it was in his own time, and for whom the story of Abraham was the greatest paradox of faith. And yet he included a "Panegyric upon Abraham", "the knight of faith", in his book [26]. The example of Abraham's faith also fascinated customers who had their books bound in the sixteenth century. On a panel decorating the upper cover of a book (Fig. 45) we see Abraham with a sword in one hand, the other hand lying on the head of his son, who is bending forward, and below, to the right, is the sacrificial fire in a pot, while in the centre of the picture, in the background, Abraham is climbing Mount Moriah with Isaac and, finally, in the upper right corner, is the angel of the Lord in the clouds, seizing Abraham's sword. The inscription reads "Abraham credidit deo"—Abraham has believed in God. (Present perfect, as it has already been demonstrated!) The middle panel of the lower cover is decorated with the arms of the Dukes of Württemberg with the initials of their motto (Fig. 45): V D M I E—Verbum Domini manet in [a]eternum (a variant of Psalm 118 = 119:89). The word of the Lord remains for ever. We think that the panel of Abraham's sacrifice was chosen deliberately to emphasize the motto and the arms, and this is again a case of conceptual symmetry.

Let us return to Kierkegaard's vision and interpretation. He sketches four variants of what had happened and how the last act of this sacrifice was reached [26]. Each of them is fearful and shocking but what is more amazing is that he draws a parallel between the deed of Abraham and the weaning of a child from its mother. Kierkegaard comes to optimistic conclusions only in the cases of the mother and child. Father and son, mother and child are presented in analogous situations—as regards a kind of separation—and in four aspects. Intellectual symmetry is mostly indirect, hidden symmetry, what is more, the symbolism itself is inherently symmetrical. The author's life reveals a

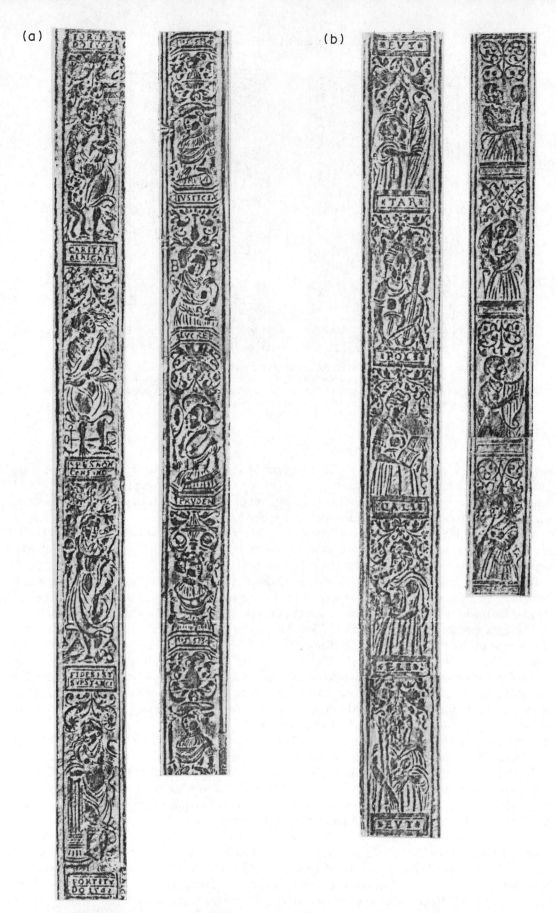

Fig. 40. Rolls with representations of the virtues (a), and the Muses (b). Bibl. Acad. Budapest: RM IV 177a, RM III F 175b, 542.169, RM III 92.

Fig. 41. The seven arts and the virtues on panels. Inscriptions (above) Grammatica–Dialectica–Redorica [!]–Arithmetica –Musica (two figures)—1574 (digit 4 reversed!)—Geometria–Astronomia; (below) Justicia [!]–Prudencia [!]–Fortitudo–Temperanci [!]–Fides–Spes–Charitas–Paciencia [!] [25]. Bibl. Acad. Budapest: 542.495, Ant. 944.

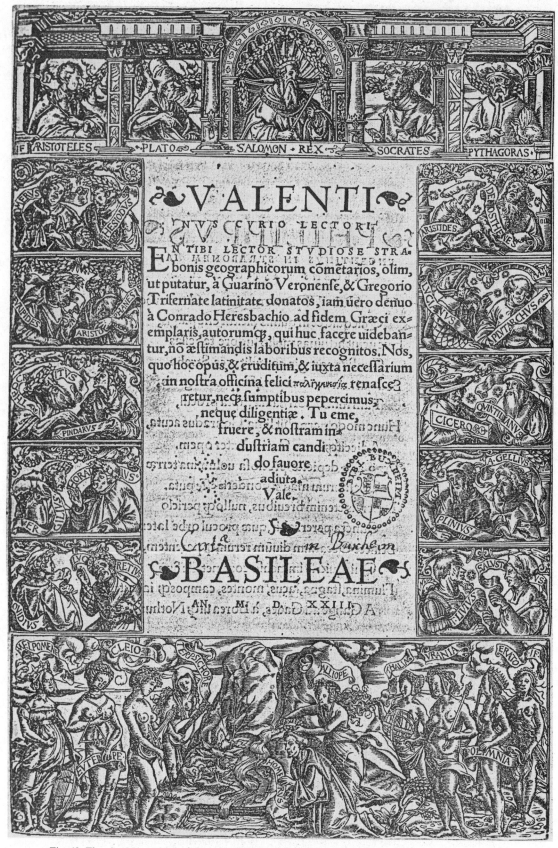

Fig. 42. The nine Muses on the title page of Strabon: *En tibi Strabonis geographicorum commentarios a C. Heresbachio recognitos.* Valentenus Curio, Basileae (1523).

Fig. 43. The Judgement of Solomon and Samson with the lion on a binding by Caspar Kraft [25]. Bibl. Acad. Budapest: 542.150.

Fig. 44. Judith with Holofernes and Jael with Sisera on a binding. Inscription under Judith from Psalm 144 = 145: 19; under Jael: Judges 5: 31. Bibl. Acad. Budapest: 542.496.

Fig. 45. Abraham's sacrifice and the coat of arms of the Dukes of Württemberg on a binding. *Kurtze auszlegung über ... Euangelia ... in Crobatischer Sprach ...* (Ed. Primus Truber). Tübingen (1562). Bibl. Acad. Budapest: Ráth 1772.

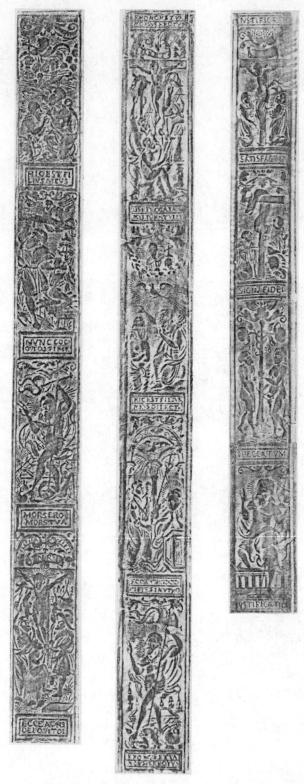

Fig. 46. Biblical scenes on German Renaissance rolls. Metropolitan Szabó Ervin Library, Budapest: Bq
0941/69; Bibl. Acad. Budapest: RM IV F 596, Ráth 1069–70. Reproduced by permission.

(a)

(b)

Fig. 47. (a) A Justification panel. Pietro Martire Vermigli, In *Epistolam S. Pauli Apostoli ad Romanos Commentarii*. Perna, Basileae (1568). Bibl. Acad. Budapest: 543.024. (b) The Justification represented on two panels. Jean Calvin: *Institutio Christianae religionis*. Rebulius, Genevae (1561). Bibl. Acad. Budapest: 542.080.

further dimension of the structure. The retold story of Abraham and the parallel cases of mother and child are reflections of Kierkegaard's emotional crisis after his engagement and tragic rupture with Regina, subtle allusions to the waves in his soul of hope and despair, faith and final resignation.

The scene of Abraham's sacrifice is often engraved on rolls, too, together with other scenes from the Bible. In general, such a roll consists of four little pictures. The Crucifixion and the Resurrection are represented on all three rolls shown (Fig. 46). These two scenes occur on panels in another connection [27]. The four scenes on the panel in Fig. 47(a) are divided by the tree of life: to the left the Fall (Adam and Eve under the tree of knowledge of good and evil), underneath Damnation and Moses with the tablets of stone; this side of the tree is dead. The right-hand side of it is in leaf, since this half of the panel comprises the Crucifixion and the Resurrection.

The complex theme and the method of representation come from the paintings of the Allegory of the Fall and the Redemption or Justification by Lucas Cranach Sr and his workshop [28, 29]. These paintings and other similar ones, e.g. that in the Budapest Museum of Fine Arts (Fig. 48) reflect the idea—the exegesis of Justification—of the great Reformer Martin Luther and his circle. In the painting in Budapest we can observe Mount Sinai, where the Lord God gave Moses the two tablets of stone, the Fall, the serpent of brass and Death. In the middle Adam or Everyman is sitting between a Prophet and John the Baptist, both of them pointing to the Crucifix. Mary is kneeling on the top of a mount—opposite Moses—accepting a small child gliding down on golden rays; on the two sides of the Crucifix we see the Holy Lamb and the Bethlehem scene with the Angels, the shepherds, and the Holy Family in the stable; and at last Christ triumphant over Death. All this can hardly be accommodated on one panel [cf. Fig. 47(a)]. The picture of Justification, divided symmetrically into two parts along the tree, occurs commonly on two panels [Fig. 47(b)] on the upper and lower cover of the given leather binding. Sin and Justification, Law and Redemption, Death and Resurrection; grave-stone, skeleton, objects, persons, gestures and ideas are positioned in a wonderful antisymmetry, in other words implying less geometrical rigour, in a counterpoint.

Even if figural representations appear on a Gothic or Renaissance binding, no definite connection to the contents of the book can be recognized. (See the captions to the figures named here.) The book of sermons by Pelbartus de Themeswar is decorated by the insignia of the Passion (Fig. 16). The covers of a New Testament which was interpreted by Bèze and printed in Geneva were nevertheless embellished by the portraits of Luther and Melanchthon (Fig. 37). It is of course thus, since the book was bound in Wittenberg, the stronghold of the Lutheran Reformation. Similarly, the Lutheran dogma of Justification is represented on Calvin's work [Fig. 47(b)]. Justitia appears on a

Fig. 48. *Fall and Redemption* (*Law and Grace*). Painting, German master, middle of the sixteenth century [30]. 82 × 59.5 cm. Deposited at the Budapest Museum of Fine Arts. Reproduced by permission.

book of historical events (Fig. 39), and Abraham's sacrifice on a collection of Gospel commentaries (Fig. 45).

FROM BAROQUE TO MODERN BINDINGS

Baroque leather bindings catch the eye by their rich gold tooling (Fig. 49). Their characteristic ornamental elements are the elongated leafy spiral, the curl, borrowed from late Renaissance "fanfare" style bindings, voluted C- and S-shaped figures, sprays and floral motifs (Fig. 50). The asymmetrical spirals are skilfully combined to give heart-shaped and other symmetrical constructions, which in turn form the diamond or marquise-shaped central panel and the enchantingly harmonizing corner pieces. These masters must have had therefore a pair of stamps to impress both mirror image forms of the asymmetrical spiral motifs. Two or three borders run along the edges. One of these rolls (Fig. 50) reminds us of an arabesque circle border pattern used 200 years before [Fig. 27(b)].

The brightest period of Hungarian Baroque bindings in the eighteenth century is associated with the Jesuits in Nagyszombat (today Trnava, Czechoslovakia), who maintained their Academy and printing house there (predecessors of Eötvös University and the University Press, Budapest). These bindings bear Jesus' monogram, IHS, the Jesuits' sign in the middle of the upper cover, and correspondingly the letters MARIA on the lower cover (Fig. 50).

There is no remarkable borderline between Baroque and rococo bindings. The frequent occurrence of shellwork gave the name to the style (rocaille = shell, shellwork in French). The central piece is often omitted, and only the borders of the cover(s) are decorated (Fig. 51). Books were the essential ornaments of the splendid library halls of the Baroque and later periods, and, very often, only the spines of these books were embellished.

A distinct period of Hungarian binding began in the first half of the eighteenth century and lasted for about a hundred years. The overwhelming majority of these popular coloured parchment bindings was produced in the town of wealthy urbanized peasants, Debrecen (Fig. 52). Their layout follows the interlace outlines of French Renaissance bindings, and the uncoloured strips surround brightly coloured fields with guilt or coloured fleurons and the curls known from Baroque bindings. The master of the binding in Fig. 52(a) even turns the flower-pot upside down for the sake of preserving the symmetry of the rectangle (C_{2v}). Colouring seems to have been applied instinctively to enhance brilliance but no conscious colour symmetry appears.

A curiosity of book construction, and an example of a three-dimensional point-group symmetry is the dos-à-dos binding (Fig. 53). Two, sometimes more books, related by their contents and usage, are bound together "back to back" with one common board and their spines on opposite sides.

(a)

Fig. 49—*continued overleaf*

(b)

Fig. 49. Hungarian Baroque bindings, eighteenth century. (a) Bibl. Acad. Budapest: 542.987. (b) Metropolitan Szabó Ervin Library, Budapest: Bf 0910/234 (MS). Reproduced by permission.

Fig. 50. Ornamental elements of Baroque bindings. The upper roll is from the binding in Fig. 49(a); the second roll is from the binding in Fig. 27(b).

(b)

(a)

Fig. 51. Rococo bindings, eighteenth century. (a) Bibl. Acad. Budapest: 542.989. (b) Library of the Transtibiscan Reformed Church District, Debrecen: G 6113. Reproduced by permission.

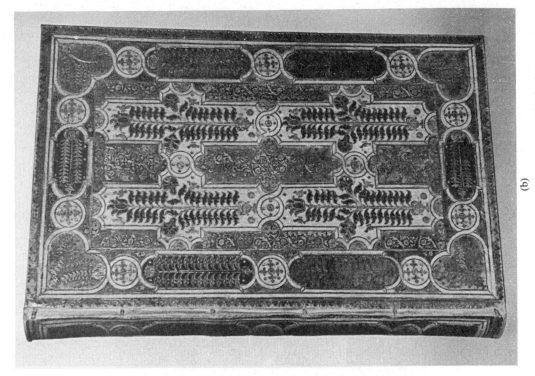

(b)

(a)

Fig. 52. Hungarian coloured parchment bindings from Debrecen, eighteenth century. (a) National Széchényi Library, Budapest: RK 89. (b) Library of the Transtibiscan Reformed Church District, Debrecen: A 48. Reproduced by permission.

Fig. 53. Hungarian dos-á-dos binding, eighteenth century. Eötvös Loránd University Library, Budapest:
RMK I 423a, RMK I 435 I–II. Reproduced by permission.

(Another example and references can be found in Ref. [4, item 91].) This structure possesses a two-fold rotation axis parallel to the spines and a reflection plane perpendicular to it (point group C_{2h}). How practical this form proves to be for a two-way pocket travel dictionary! In some modern pocket dictionaries the two integrated parts have rather one common spine and no board in the middle between the "running" and the "reversed" pages. The symmetry of this book is again trivial, point group C_{2v}, with the two-fold axis perpendicular to the spine. It is left to the reader to meditate on combinations of symmetry operations with an inversion of, say, a Hungarian–Italian dictionary to its Italian–Hungarian counterpart.

Some modern artists have created three-dimensional bookbindings [31] that resemble rather a sculpture or space construction, and which have lost their practical purpose of protecting and decorating a book in use.

Though masters of Baroque and rococo bindings retained much of the elements of preceding late Renaissance (curls, arabesque circles, interlace outlines, layout) or even late Gothic bindings [bird-in-vine [32], and cf. Fig. 18(b)], they succeeded in creating a new type of bookbinding decoration, distinguished by its appearance and effect. The wells of innovative power seem, however, to have been exhausted by the nineteenth century. At the same time when in architecture the different "neo" styles follow, bookbinders cannot but historicize, and the period is marked by phrases like Etruscan style, Cathedral style, and flourishing species of neo-Renaissance. It is in such nostalgic works that the technical perfection of hand bookbinding reaches a level unknown before (Fig. 54). Para-

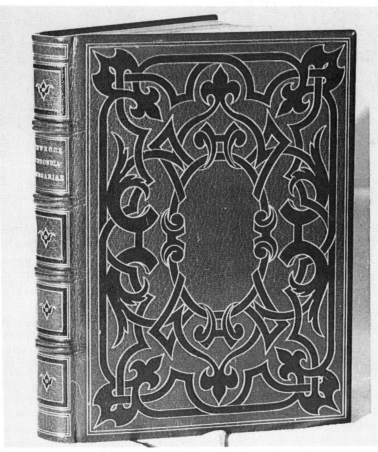

Fig. 54. Neo-Renaissance binding by Marius Michel for György Ráth, bibliophile, director of the Museum of Applied Arts, Budapest. End of the nineteenth century. Bibl. Acad. Budapest: Ráth 1044.

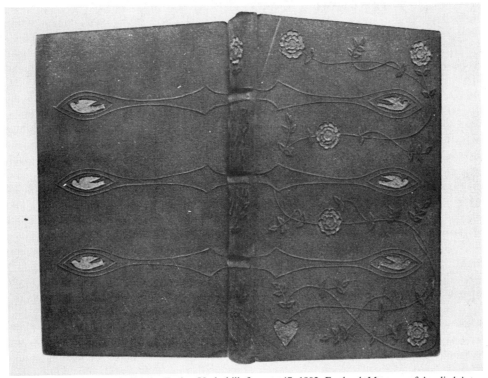

Fig. 55. Art nouveau binding by Evelyn Underhill, fec. op. 47, 1902, England. Museum of Applied Arts, Budapest: Inv. 6965. Reproduced by permission.

Fig. 56. Representation of the labyrinth on the external wall of the Cathedral in Lucca, Italy. Photograph by and courtesy of István Orosz, Budapest.

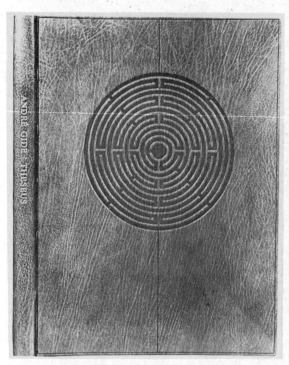

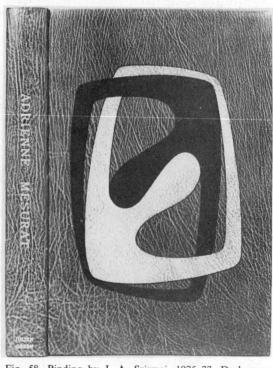

Fig. 57. Binding by J. A. Szirmai, 1983–84. Terra-cotta goatskin (Niger), on front cover blind impression by the use of a linocut. 32 × 25 cm. Cat. No. 45 [35]. André Gide: *Theseus*. Officina Bodoni, Verona (1949). Photograph by J. A. Szirmai. Reproduced by permission.

Fig. 58. Binding by J. A. Szirmai, 1975–77. Dark grey goatskin (oasis), on front cover onlays in black and light grey goatskin. 29 × 20 cm. Cat. No. 25 [35]. Julien Green: *Adrienne Mesurat*. Société Les Exemplaires, Paris (1929). Photograph by J. A. Szirmai. Reproduced by permission.

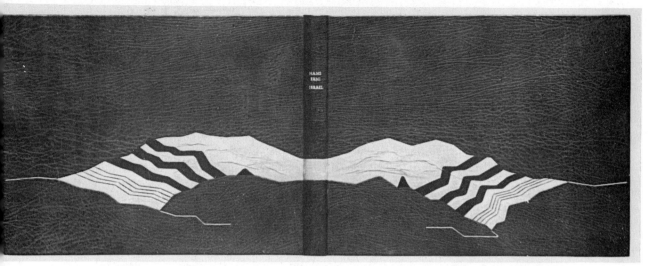

Fig. 59. Binding by J. A. Szirmai, 1972. Terra-cotta goatskin (oasis); onlays on front and back cover, partly crumpled, in yellow ochre, dark brown and green. 23.5 × 31.5 cm. Cat. No. 3 [35]. Hans Erni: *Israel. Ein Skizzenbuch.* Scheidegger, Zurich (1971). Photograph by J. A. Szirmai. Reproduced by permission.

doxically, at the same time, bookbinding as a handicraft loses ground and declines with the advent of machine binding.

The turn of the century is a landmark in the *art* of bookbinding. The modern bookbinder claims to be an artist who creates a work of art and not simply a decorated product of handicraft; he or she re-creates the contents of the book, expresses his or her impression of the work to be bound. Principles and forms, artistic endeavour and materials of modern fine art are reincarnated in modern bookbindings. A few examples have been selected here to show how apparently simple structures imply in the most intricate and sophisticated manner the arsenal of "symmetrology" [33]. Elements are consciously composed into a unit, and parts of the book, upper and lower cover, are often treated as a whole (Fig. 55).

The Labyrinth, originally the Minoan palace in Crete, has always provided an exciting adventure for the human mind, a source of secret, a many-folded symbol, among others, of the nether world and, at the same time, of redemption from death [34]. What it meant for Comenius is concisely expressed in the title of his work *Labyrinth of the World and the Paradise of the Heart.* It was represented in architecture (Fig. 56), arts and literature, as with, for example, the maze in the splendid book, *Three Men in a Boat,* by J. K. Jerome. Professor Szirmai's bookbinding (Fig. 57) is based on a circular symmetry, which is modified by a pseudo-symmetry of four-fold rotation and reflection. Similarity is expressed by the set of concentric circles, and catamorphy, the lowest category of geometrical relationships [33], by the decreasing number of radial passages and dead ends along the inner circles. On another binding by Szirmai (Fig. 58), the double figure is again a brilliant and ingenious superposition of two-fold rotational symmetry and deviation from it, symmetry and antisymmetry (two-colour symmetry of the figure on a "neutral" background), and, above all, the transfiguration of a phenomenon taken from the contents of the book. We must apologize to the reader for disclosing the keyword: it is schizophrenia. Figure 59 represents symmetry in a broad sense, so to say, a "topological symmetry". Design and generalized symmetry extends over front and back cover.

P. L. Martin and M. Jeagle use rotations and reflections on their bindings (Figs 60 and 61) combined with affine transformation, reflection with "blowing up" (homothetic reflection [33]), and colour transformation.

CONCLUSIONS

Throughout the history of bookbinding decoration, the rectangular form of the book cover has been decisive. The general layout of decorated leather bindings most often conforms to the two

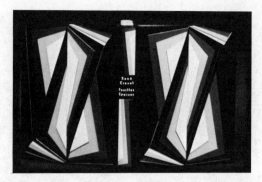

Fig. 60. Binding by Pierre Lucien Martin, 1966. Black box-calf, onlays in various shades of gray calf, doublures red peau de suède. René Crevel: *Feuilles eparses*. Original illustrations by Bellmer, Arp, Miro, Ernst. Paris (1965).

Fig. 61. Binding by Martin Jaegle, 1973. Dark blue oasis goatskin, onlays in white, red and dark red, line tooling in white. Voltaire: *Candide*.

mirror planes of the rectangle. One or more border designs along the sides leave a smaller, again rectangular (decorated) area in the middle. Details of Romanesque and Gothic figures, animals, floral motifs, religious symbols, or subdivision of fields disturb this symmetry. The orientation of such motifs is influenced by the conventional direction of view, i.e. the distinction between top and bottom of the book cover. The symmetries of border patterns do not indicate a preference of their outer or inner edges. A unified global composition, the emphasis on the central design, harmonizing border decoration and centre and corner pieces, and the symmetry of details begin with Renaissance bindings, either with geometric or floral patterns or with portraits and scenes. Interlace motifs and patterns become again popular. Colour symmetry occurs only sparsely.

Bookbinding decoration in each age retains some elements from its previous periods and follows with a certain delay the main styles of the arts. Decoration requires imaginative power and skill, above all in the case of cutleather bindings. Some nice parallels of geometry and symbols, depicted scenes, persons and ideas can be recognized, nevertheless, the representation has hardly any relation to the contents of the book.

The modern art of bookbinding breaks with tradition. It treats one or both book covers and even the complete book as a whole. The necessity of border designs is eliminated. Symmetry and the different kinds and levels of geometric relations are consciously applied or just abandoned. An allusion, albeit indirect, to the message in the book is intended.

Acknowledgements—Our thanks are due to Professor and Mrs Aladár and Éva Sarbu, Ms Éva Pröhle and Mr Bob Dent, who read the manuscript and made valuable suggestions to amend its English. We gratefully acknowledge the permission and photographs for illustrations from institutions and persons named in the figure captions.

REFERENCES

1. P. Gergely, *Pápai Páriz-album a Magyar Tudományos Akadémia Könyvtárában* (The Pápai Páriz Album in the Library of the Hungarian Academy of Sciences). Publicationes Bibliothecae Academiae Scientiarum Hungaricae 21, Budapest (1961).
2. J. A. Comenius, *Orbis Sensualium Pictus*. Facsimile of the third London edn (1672) (Introduction J. Bowen). Sydney University Press, Sydney (1967).
3. F. Steenbock, *Der kirchliche Prachteinband im frühen Mittelalter von den Anfängen bis zum Beginn der Gotik*. Deutscher Verlag für Kunstwissenschaft, Berlin (1965).
4. P. Needham, *Twelve Centuries of Bookbindings 400–1600*. The Pierpont Morgan Library—OUP, Oxford (1979).
5. F. A. Schmidt-Künsemüller, *Die abendländischen romanischen Blindstempeleinbände*. Hiersemann, Stuttgart (1985).
6. E. Panofsky, *Gothic Architecture and Scholasticism*. Archabbey Press, Latrobe, Pa. (1956).
7. G. D. Hobson, Further notes on Romanesque bindings. *Library* **15**, 161–211 (1934–35); Some early bindings and binders' tools. *Library* **19**, 202–249 (1938–39).
8. E. Kyriss, *Verzierte Gotische Einbände im alten deutschen Sprachgebiet*. Textband. Tafelband 1–3. Max Hettler, Stuttgart (1951–58).
9. O. Mazal, Gotische Einbände mit Kopfstempeldekoration aus der Inkunabelsammlung der Österreichischen National-bibliothek. *Gutenberg Jb* 473–481 (1962).
10. É. Sz. Koroknay, *Magyar reneszansz könyvkötések* (Hungarian Renaissance Bookbindings). Plate 36. Cahiers d'histoire de l'art 6. Akadémiai Kiadó, Budapest (1973).
11. F. A. Schmidt-Künsemüller, *Corpus der gotischen Lederschnitteinbände im deutschen Sprachgebiet*. Hiersemann, Stuttgart (1980).

12. M. Rozsondai, Wiener Dominikanereinbände in der Bibliothek der Ungarischen Akademie der Wissenschaften. *Gutenberg Jb* 234–244 (1981).
13. M. Rozsondai, The popularity of Pelbartus de Themeswar in Europe as demonstrated by bookbindings (in Hungarian). *Magy. Könyvszle* **100**, 300–319 (1984).
14. M. M. Foot, *The Henry Davis Gift: A Collection of Bookbindings*. Vol. 2, item 327. The British Library, London (1983).
15. I. Hargittai and Gy. Lengyel, The seven one-dimensional space-group symmetries illustrated by Hungarian folk needlework. *J. Chem Educ.* **61**, 1033–1034 (1984).
16. M. Rozsondai, Historical fine bindings in the Library of the Hungarian Academy of Sciences and their connection to book history (in Hungarian). Thesis, Budapest (1985).
17. D. Schattschneider, In black and white: how to create perfectly colored symmetric patterns. *Comput. Math. Applic.* **12B**, 673–695 (1986). Reprinted in *Symmetry: Unifying Human Understanding* (Ed. I. Hargittai). Pergamon Press, Oxford (1986).
18. M. M. Foot-Romme, Influences from the Netherlands on bookbinding in England during the late fifteenth and early sixteenth centuries. *Act. XIe Congr. Int. Bibl.* pp. 39–64, Bruxelles (1979).
19. Cs. Csapodi, *The Corvinian Library: History and Stock*. Akadémiai Kiadó, Budapest (1973).
20. É. Sz. Koroknay, Oriental influences in Hungarian Renaissance bookbinding (in Hungarian). *Müv. tört. Ertesitö* (1–2), 1–17 (1968).
21. É. Sz. Koroknay, Eine ungarische Renaissance-Einbandgruppe vom Anfang des 16. Jahrhunderts. *Gutenberg Jb* 361–371 (1966).
22. I. Schunke, Venezianische Renaissanceeinbände: Ihre Entwicklung und ihre Werkstätten. *Studi di bibliographia e di storia in onore di Tammaro de Marinis*. Vol. IV, pp. 123–200, Plate XXV, Verona (1964).
23. O. Mazal, *Europäische Einbandkunst aus Mittelalter und Neuzeit: 270 Einbände der Österreichischen Nationalbibliothek*. Plate 99. Akad. Druck- u. Verlagsanstalt, Graz (1970).
24. M. M. Foot, The Olga Hirsch collection of decorated papers. *Br. Libr. J.* **7** (1), 12–38 (1981).
25. M. Rozsondai, Signierte Renaissance-Einbände deutschen Typs aus dem 16. Jahrhundert. *Gutenberg Jb* 290–339 (1988).
26. S. Kierkegaard, *Fear and Trembling* ... (Translated, introduction and notes by W. Lowrie). Princeton Univ. Press, Princeton, N.J. (1968).
27. K. von Rabenau, Reformation und Humanismus im Spiegel der Wittenberger Bucheinbände des 16. Jahrhunderts. *Von der Macht der Bilder*. Beiträge des C.I.H.A.–Kolloquiums "Kunst und Reformation" (Ed. E. Ullmann) pp. 319–328. Leipzig (1983).
28. Lucas Cranach d.Ä., *Verdammnis und Erlösung* (1529). Öltempera auf Lindenholz. 80 × 115 cm. Gotha, Museen der Stadt, Schlossmuseum. Inv. 722/676.
29. Lucas Cranach d.Ä. (Werkstatt), *Sündenfall und Erlösung*. (Um 1530). Nicht bez. Öltempera auf Holz. 51 × 99 cm. Weimar, Kunstsammlungen, Galerie im Schloss.
30. Zs. Urbach, The allegory of the Fall and Redemption (in Hungarian). *Diakónia: Evangélikus Szemle* **5**(2), 19–29 (1983).
31. *Modern British Bookbinding*. Bibliotheca Wittockiana, Bruxelles, Koninklijke Bibliotheek, 's Gravenhage 1985. Exhibition Catalogue. Designer Bookbinders, London (1985).
32. M. M. Foot, *The Henry Davis Gift: A Collection of Bookbindings*. Vol. 1, 2nd impression, Plate IV.19.A. The British Library, London (1982).
33. E. Makovicky, Symmetrology of art: coloured and generalized symmetries. *Comput. Math. Applic.* **12B**, 949–980 (1986). Reprinted in *Symmetry: Unifying Human Understanding* (Ed. I. Hargittai). Pergamon Press, Oxford (1986).
34. K. Kerényi, *Labyrinth-Studien. Labyrinthos als Linienreflex einer mythologischen Idee*. 2. erweiterte Aufl. Albae Vigiliae 10. Rhein, Zürich (1950).
35. J. A. Szirmai, *Boekbandkunst*. Catalogus van boekbanden door J. A. Szirmai ... in de Universiteitsbibliotheek. Amsterdam (1984).

Computers Math. Applic. Vol. 17, No. 4–6, pp. 887–897, 1989
Printed in Great Britain. All rights reserved

515—A SYMMETRIC NUMBER IN DANTE

L. de Freitas

Rua Ribeiro Sanches 24-3°, 1200 Lisbon, Portugal

Abstract—The author compares the well-known mythological figure of Narcissus contemplating his own image in the water with palindromes and "symmetric" numbers, that allow themselves to be read both ways with equal result, and points out the kabbalistic relevance of such "mirrorings" in the context of the "anagrammatic genius" of the hebrew sacred language. After stressing the fact that the paradigm of the mirror is common to oriental as well as western traditional metaphysical and religious speculation, he comes to mention the mysterious use of the number 515 by means of which Dante Alighieri, in his *Divine Comedy*, prophesies the advent of God's Envoy. Critical ingenuity has tried to explain why Dante chose this numerical cipher, with no great success. Yet, the author proves that the mystery of that "palindrome-number" becomes clear if it is considered in the light of the mirroring analogy, often practised by mystics, philosophers and poets. In fact, it results from the properties of the Pythagorean "sacred triangle" 3·4·5, if taken as a "Narcissic system", as well as from the numbers expressing the angles of the "head" of a regular pentagon, again if taken after the mirror-paradigm. 515 proves to be the commutating factor between divine perfection above and its reflexion "in the waters". The text concludes by choosing other instances of the use of this symmetric number, with analogous meaning, although rendered cryptic due to dangers of heresy and fear of the Inquisition, in works by Albrecht Dürer and in a Portuguese painting of the sixteenth century, attributed to Master Jorge Afonso.

Narcissus lost in the contemplation of his own image reflected in the water is generally taken as a mythological warning against the dreadful risks of self-adoration. Yet the Narcissus legend holds more than one meaning, and we can easily perceive in its setting a type of perfect symmetry foreboding the final unhappy end. According to yet another viewpoint, the reflection of Narcissus looking back at him from the mirror of the waters, while he lovingly admires himself, reminds us of the palindromes used in ancient epigraphy, such as the one inscribed in the pavement of the church of San Giovanni that, according to some scholars,† inspired Dante and led him to write in the very last verses of the *Paradiso*:

---

†See Pézard's note [1] on verse 144 of the *Paradiso* (XXXIII). Pézard believes that Dante found his basic inspiration for the image of the wheel in Jeremiah (XVIII: 1–6).

"... as a wheel moves smoothly, free from jars,
My will and my desire were turned by love,
The love that moves the sun and the other stars."†

The palindrome in question is written around a circle representing the sun and reads:

EN GIRO TORTE SOL CICLOS ET ROTOR IGNE,

which freely translated means 'I am the sun, I am that wheel moved by fire whose spinning sets the spheres in round motion". Of course this "fire" is love that moves the sun and the other stars. The interesting point, however, according to our "Narcissistic" perspective, is that one can as well read a palindrome from end to beginning as from beginning to end. (Another palindrome somewhat resembling the one above is *"In girum imus nocte et consumimur igni"*.)

If we consider the notion of a palindrome made of numbers instead of letters, we find at once that it is possible to conceive of an infinite number of such arrangements, say 12321, 1221, 212, 11, etc. It is noteworthy that such numerical arrangements are referred to in everyday speech in all the Iberian languages by a specific word, *capicua*, presumably from the Catalan (or the Latin) for "head" and "tail", and that to *capicuas*, to this day, a beneficial and talismanic connotation is traditionally attached: they bring luck. Why? It is a mystery. However, I incline to see in it a trace of the kabbalistic tradition: maybe not too far-fetched a guess when we consider the millenary presence of Jews in the Iberian Peninsula and the intimate cultural exchanges between Hebrew and Christian learned classes through the centuries.

Indeed, "symmetric" or "mirror" numbers are quite relevant to Kabbalists. The possibility of reading those numbers indifferently from left to right or vice-versa with equal results derives its importance from the anagrammatic nature of the sacred language of Canaan, of the "linguistic genius" of which the reversing of syllables constitutes a fundamental aspect, used to prevent common ears from understanding certain secret matters, namely those biblical writings and prophetic visions related to the Reign of the Millenium.‡ The quality of symmetry displayed by numbers such as the mysterious and famous "Number of the Beast", 666, mentioned in the *Apocalypse* draws our attention to the paradigm of the *mirror*. Mirrors either natural, like the calm surface of waters, or man-made, have always fascinated philosophers, poets, artists and visionaries. There is a subtle and poignant enigma in the mirror's reflection, for it is "alive" while devoid of "reality", at the same time reversing the reflected subject by placing its right side at the left, its left side at the right or, when horizontal, what is on top below and vice-versa.

We find innumerable instances of *reflection* and speculation (from lat. *speculum*, "mirror") in different civilizations and cultures, specifically in our own, where the mirror has been extensively used by theologians, mystics, magicians, alchemists, poets and artists. St Paul states that we must learn to contemplate the original Light of God through its imperfect reflections "as in a glass mirror". Before him Moses descended from Mount Sinai after seeing Jehovah in a mirror, since the direct vision of God is an unbearable experience for mortals; his skin, after that, shone like polished metal, reverberating with Jehovah's glory in such a dazzling way that no son of Israel could look at him directly. At the root of this tradition there is, most probably, the similarity of the Hebrew words for "vision" and "mirror" [4]. Seeing the divinity with the help of mirrors is mentioned in *Numbers*. The sun itself was but a reflection of God's light, a mirror hung in the sky according to Plutarch, made of glass according to Philolaos. Saïs, the goddess holding a mirror, is connected with the Mosaic tradition. Equivalent traditions can be found in Oriental lore and religion, along with a wealth of legends and beliefs linked to the subject. In India whole areas of metaphysical meditation have been centred on the notion of the image in the mirror, which turns out to be the correct parallel of the nature of the individual ego, as compared to Brahman or the Atman. The true Self, as true Being, reflects Himself in the possibilities of being, "as the world-picture painted by the Self on the canvas of the Self" (Śaṅkarâcárya) [5]. *Ábhása*, a sanskrit word, in Śilpa usage means "painting"

---

†Canto XXXIII, 143–145. I follow the English translation of Dante's *Paradise*.
‡See Ref. [3]. The *Apocalypse* of St John has been intensely searched for mystic clues through the centuries, in the context of the prophetic thought of Christianity, especially since Gioachino di Fiore saw in it the message of the *Vangelo Eterno*, the evangel of the coming Age of the Holy Ghost. He unwittingly started the millenaristic upheavals that shook all Western Europe in the late Middle Ages and the beginning of the Renaissance.

and, together with *citra*, "art", designates fundamentally the *image*, as Coomaraswamy explains, "owing such reality as may be theirs to That whose image they reflect" [5]. The same distinguished art historian and thinker brings to our attention a beautiful description of the creation found in *Pañcaviṃśa Brāhmaṇa* (VII, 8, 1), according to which the waters being ripe unto conception, Váyu, the wind or *prâna*, moved over their surface, "wherefrom came into being a lovely thing (that is, the world-picture), there in the waters Mitra-Varuna beheld-themselves-reflected (*paryapaśyat*)".

Now we understand that the legend of Narcissus tells also a story of creation as image-in-the-waters, where the Creator beholds himself in his own image, which is the man in his insubstantial yet divine nature as "reflection". The divine or true Self "dies" in his longing to unite with himself, and such "death" becomes the everlasting manifestation and transformation into space and time of the "world-picture", that is to say, of all possibilities of being. Such is, in Genesis (I. 2), the spirit of God moving over the waters and creating Light, without which the image is not visible. The created universe where men are born, live and vanish, in the same manner as worlds are created and pass away, is but the reflected image of the divine Self in his perfection, with no begining nor end. This is a metaphysical and religious concept that inspired traditional art and literature throughout the centuries, in the East as well as in the West. Coomaraswamy has shown in his masterly studies, especially those he dedicated to the fourteenth century German mystic Meister Eckhart, whose genius enabled him to subsume the spiritual essence of medieval Europe, that the community of thought behind Christian art of the Middle Ages and that of India and China exists beyond doubt; see his vast bibliography, especially Refs [6, 7].

The metaphysical roots and spiritual aims of another thinker and poet of the Middle Ages, Dante, show again a close link with the East, specially with the Islamic tradition. The *Divine Comedy* is the outstanding monument of Dante's genius, not alone for its pure literary beauty but also as a treasure of encyclopedic knowledge ablaze with the light of a profusion of marvellous symbols. Dante sums up and inspiredly shapes in diamantine form many of the fundamental ideas of a tradition that springs from a Platonic philosophy† combined with an elaborate exegesis of biblical texts, namely Ezekiel's visions (the *Maaseh Merkabah*, or the "*Work of the Chariot*", which together with the study of Genesis constitutes the fundamental base of *Hokhmath Hakabbalah*, the esoteric wisdom of the Kabbalistic tradition) [9] and the Johannine gospel, as well as from a confirmed Sufi inspiration [10] and possibly from teachings of the Kabbalah, as the vision of the car of Beatrice (*Purgatorio*, XXIX, 91–114), so like the first vision of Ezekiel, whose name he invokes, seems to validate.‡ Many of the symbols in the *Divine Comedy* have been studied and interpreted by modern erudition, revealing the close familiarity of the poet with a vast spectrum of traditional motives, along with his masterly capacity to put them to subtle use as intensifiers of understanding and feeling. Some of Dante's symbolic motives have remained quite clear through the ages, some others have been obscured by the change of cultural references which took place since the fourteenth century. One of the most baffling mysteries concerns the numerical symbol 515 used by the poet in the *Purgatorio* (XXXIII, 43); Beatrice prophesies the coming of one who shall avenge the wrong done to Christendom:

> "Surely I see, and so foretell, not far
> Ahead now, stars that shall not need to strive,
> But shall bring in, secure from let or bar,
> The times when a Five-hundred-ten-and-five,
> God-sent, shall smite the thief, and smite the giant
> That sins with her, and leave them not alive." [12]

Dorothy L. Sayers, author of this translation, comments as follows: "*a Five-hundred-ten-and-five*": the person thus mysteriously designated is undoubtedly the same as the "Greyhound" of *Inferno*

---

†Since Bruno Nardi's studies it became clear that Dante's ideas though starting from an Augustinian source, derived mainly from the Neo-Platonic approach, as expressed by Arabic philosophers like Avicenna, Alfarrabi, etc. In the perspective of the "mirror-paradigm" it is relevant to remark, with Nardi, when he sums up the involved Platonic conceptions, that the first of these concerns God, conceived as Light; the "intelligences" of the celestial spheres come next, which, *like mirrors*, reflect the rays of His light in order to impress forms into matter. See Ref. [8].

‡Contemporary tradition reports that Dante sustained a long friendship with Immanuel ben Salomon ben Jekuthiel (1270–1330), a learned person and a poet himself. See Ref. [11].

(I, 101), who is also alluded to in *Purgatorio* (XX, 15). Critical ingenuity has exhausted itself over this numerical cipher, with no great success. Some have (with a good deal of juggling) extracted from it a rebus of Henry VII. If they are right, this passage must have been written before Henry's death in 1313, and also before *Purgatorio* (VII, 96), which laments the failure of Henry's expedition into Italy. Since, however, Dante continued to the end to hope for the advent of an Emperor of the right sort, it is probably better to take the prophecy in a general sense. What the 515 was intended to mean, we shall now probably never know" [13]. The identity of the person designated as a "Greyhound" (*Veltro*) in the *Inferno*, and the one designated as "515" in the *Purgatorio*, is accepted by most scholars, and generally taken as the image of "some hoped-for political saviour who should establish the just World-Empire" (Sayers), at the same time as the probable symbol of the reign of the Holy Ghost on earth, the visible kingdom of God. This "God-sent" person seems to announce the prophetic advent of Jesus Christ at the end of the cycle and embodies, aside from any other circumstances of place, time and identity, a divine figure made in the shape of a man, very close to the embodiment of Shaddai, the God of the Armies [as in Exodus (VI: 1–3)], traditionally inscribed in the "shield of David" (*Maghen David*) or "seal of Solomon", and also related to the mysterious Angel of the Covenant, Metatron. I cannot develop here the meaningful aspects of this relationship 515—Shaddai-Metatron, nor the interesting avenues of enquiry opened by Dante's mention of the "*Veltro*".† It is necessary, now, to make clear the solution of the "problem" of 515, as I found it a couple of years ago. However, before I disclose that solution, I would like to remind the reader that in 1921 Professor Rodolfo Benini published a paper on Dante's *Inferno* in which he reported his discovery of three couples of numbers, 3 and 9, 7 and 22, 515 and 666, charged with an undoubtedly important meaning, for they recur again and again throughout the poem. Benini found that 666 verses separate the prophecy of Ciacco from the prophecy of Virgil, 515 verses separate the prophecy of Farinata from that of Ciacco; again 666 verses are counted between the prophecy of Brunetto Latini and that of Farinata; and again 515 between that of Nicolas III and that of Messire Brunetto. This is quite remarkable. As we know, 666 is the "Number of the Beast" mentioned in the *Apocalypse*, that is to say, the number of the Antichrist; it is consequently logical to suppose that 515 cannot but represent, in some manner, Christ as opposed to the evil of the Beast. This led Benini to assume that 515 could perhaps be transformed into the Latin notation DXV meaning "*Dante Veltro di Cristo*" ("Dante the Greyhound of Christ") [15], a rather ludicrous suggestion in view of the humble Christian feelings of the poet. René Guénon rightly excluded this interpretation of Benini and proposed a slight change in the order of the letters, making it DVX, the latin word (*Dux*) for "leader" or "conductor" [16]. There is still, to use Sayer's words, some deal of juggling in Guénon's approach.

Now we shall take the second couple of numbers recurring in the *Inferno*, 7 and 22. Both are charged with great "magic" power. The 7, probably the most magic of numbers, is related by Dante to the divisions of the *Purgatorio*; 22, among other qualifications, is the number of letters of the sacred Hebrew alphabet, and also the number of regular polygons with an integer for their angle to the centre that can be inscribed in a circle.‡

We cannot engage, however, our enquiry in the many developments that could be started from these magical and theurgic properties, for they would take too long and distract us from our main purpose. What is really relevant here is the fact that the couple formed by 7 and 22 is linked to the symbolism of the circle by a simple relation, well known to the "operative" masons and builders since the most ancient times, establishing the approximate value of the number π: $\frac{22}{7} = 3.14 \ldots$.

It becomes most interesting to learn that 314 is the symbolic number of Shaddai according to the numerical value attributed to letters by the Kabbalists, exactly the same as the three "mother" letters of the Hebrew alphabet, *aleph*, *mem* and *shin*; the value of the words El-Shaddai ('God-Almighty') being 345. The mysterious Angel Metatron, "'the garment', or visible manifestation of the Ain Soph" that "governs the visible world, preserves the unity, harmony and revolution of the spheres, planets and heavenly bodies", the "Commander, under the Will of the Deity, of all the myriads of

---

†I treat these and many other related questions in my book [14].
‡The relation between the concerned polygons and the number of letters of the Hebrew alphabet was first disclosed by Abellio [17].

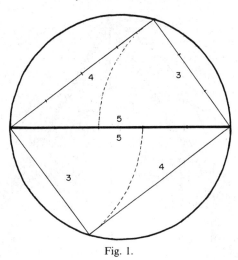

Fig. 1.

the angelic hosts, of the next or Ye'tzeer-atic World" [18], corresponds also to the number 314 through the value of the letters of his name. Shaddai and Metatron, both expressed by 314, are the manifestations of the power of Almighty God, El-Shaddai, expressed by 345, which concerns also the nature of "the most beautiful triangle" of the Egyptians, as Plutarch calls it in his book on Isis and Osiris: that is the famous triangle of Pythagoras. The old Egyptians saw in its vertical and horizontal cathetuses the representation of the male and female aspects of the deity, the hypotenuse symbolising Horus, the divine son. There is something similar to the Egyptian geometric statement of the triple nature of God in the Sufi tradition. In a text on the letters *alif*, *lam* and *mim*, Kamaloh din Abdouazâq Kâshâni, the Persian commentator and exegete of Ibn 'Arabi, states that these letters correspond to the numbers 1, 3 and 4, *alif* or 1 corresponding to the divine Self or the *initium* of being, *lam* or 3 referring to Gabriel or the angel of Knowledge and Revelation that dwells in the centre of being, and *mim* or 4 corresponding to the Prophet, the closing up of being [19].

It is common knowledge that the theorem of Pythagoras deals basically with the once highly sacred triangle 345, possessing the rare property expressed in the theorem, that is, the squares of its cathetuses equal the square of its hypotenuse: $3^2 + 4^2 = 5^2$. When we inscribe this triangle in a circle, we find that there is place for another one, but only one. Once inscribed in the circle the two triangles have in common, superposed, their hypotenuses; this allows us to write $5 + 5 = 10$, or 1. We have, then, twice, 314 where we had once 345. The number 1 places itself in the diameter of the circle, corresponding to the "surface of the waters", and we have on both sides of the parting line the values $3^2 + 4^2$, that is 25, of which $\sqrt{25} = 5$. Suddenly we are confronted with the 5–1–5 structure, so mysteriously pronounced by Beatrice in Canto XXXIII of the *Purgatorio*: the "Messo di Dio" is named after the symbolism of the sacred triangle of Egyptians, Kabbalists, Sufis and Temple-builders of the Christian Middle Ages, and his numerical symbol corresponds to the symmetry of his double nature, participating of the world "above" and of the world of the "reflection in the waters"!

And now we remember that Dante invokes Ezekiel, when the glory of the chariot shines on the other side of the river that flows in the Earthly Paradise. He invokes then the *Maaseh Merkabah*, the 'Work of the Chariot', so fundamental for traditional esoteric wisdom. Following the suggestion of Dante we open the book of Ezekiel to the first chapter, and are at once jolted by a revelation (never noticed by scholarly exegetes): the first words of the Prophet, just before the vision of the Chariot, are these: "In the thirtieth year, on the fifth day of the fourth month, as I was among the exiles by the river Chebar, heaven opened and I saw visions from God". (Ezekiel I. 1: *The New Jerusalem Bible*.) This confirms beyond doubt the correctness of our reasoning. We fully realise that 515 is a very appropriate and supremely intelligent symbol, universal by its geometrical base and its mystical resonance in many religious and metaphysical realms, fit to receive embodiment at any moment of history and expressing, with the maximum of clarity possible, the hypostatic union of the two natures, the divine and its manifestation or reflected "world-image" (as confirmed by the appearance of the Gryphon in Canto XXIX,

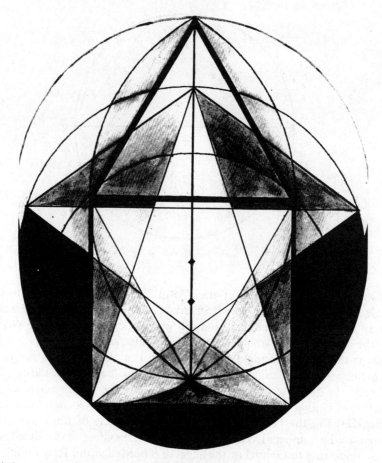

Fig. 2. Relations uniting the "luminous delta" with the pentagon, the "Vesica Piscis" and "philosopher's
stone" symbol.

"...A triumph-car, on two wheels travelling,
And at the shoulders of a Gryphon drawn†).

There is another way leading to the full understanding of the cipher 515 (in fact it was the first
one for me, when I suddenly came upon the solution of the mystery). It starts with the consideration
of the regular pentagon, which vast anthropomorphic symbolism is impossible to summarize, even
in the simplest sense. I must confine myself, in this brief notice, to the basic statement, taken for
granted, that the triangle uniting three successive vertices of the regular pentagon, called the
"luminous delta" by certain followers of Masonic traditions, had a great deal of importance in the
metaphysical speculations of the inner circles of the Templars and related esoteric groups. One must
keep in mind that Dante seems to have been one of the chiefs of an association called *Fede Santa*,
a Third Order affiliated to the Templar Order [16]. The "luminous delta", taken as the "head" of
the Anthropos, or divine man, and as the visible manifestation of the Godhead, expresses the
Self in the sense of pure ideality, outside space or time. In traditional terms the intellect of humans,
considered in relation to the "head" of the pentagon as divine knowledge and awareness, is no more
than a shadowy and inverse reflexion of the pure Light above, seen upon the dark waters of Chaos.
Inverse—that is, not only upside down but, like a palindrome (or a "*capicua*"), reading from right
to left what is left to right. If we apply this reverse mirroring property of our paradigm to the

†Translated by Dorothy L. Sayers. See Ref. [12]. The Gryphon is a combination of the fore-part of an eagle with the hind-part
 of a lion. It combines the divine (the bird), being of gold incorruptible, and the animal (the human), being mingled of
 red and white; white and red are the colours assigned by Dante to the Old and the New Testaments, respectively.
 Dorothy L. Sayers remarks: "here again [Dante] emphasizes the meeting of the two Dispensations in the Incarnation".

Fig. 3. Albrecht Dürer, *Melencolia I*, 1514.

numbers expressing in degrees the angles of the so-called "luminous delta", which are 108, 36 and 36, we obtain, "on the surface of the water", the numbers 801, 63 and 63. Their addition gives the total 927, the number of the absurd "image" mirroring the right total of 180° of the true triangle. We can conceive of a symbolic system where these "impossible" numbers express our decayed condition after the Fall and reflect, in a despairing way, the lost presence of the divinity. Such was the vision of the religious man of the Middle Ages, and such is still the vision of many philosophers, mystics and poets, to this day. "The things of this world," wrote Baltasar Gracián, "have to be contemplated upside down to be seen in their correct position." Dante was a deeply religious man, unhappy about the idea of a world immersed in corruption, atrocities and spiritual as well as material misery; but a man who cherished the hope of a drastic change and the establishment of true religion, true brotherhood and true love among humans. He favoured the Franciscan movement, the ideas of Joachim of Fiore, the aims of the Templars. He wished the dismal reflexion

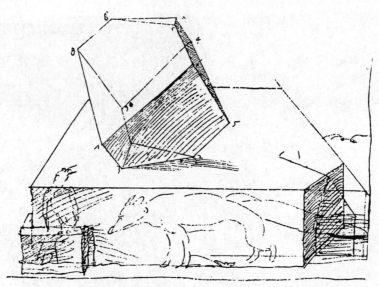

Fig. 4. Dürer: The Dresden study for the *"Melencolia"* polyhedron. *Apud* E. Panofsky and F. Saxl, *Dürer's Melencolia I*. Leipzig, Berlin, 1923.

of the Godhead on the "rivers of Babylon" to be restored to its perfection, true to the eternal archetype. But how to convert the absurd 927 into the purity of 180, the perfect triangular number? The answer is 515. In fact, 5.15 is the converter of 927 into 180:

$$\frac{927}{180} = 5.15.$$

This is a remarkable fact, and may seem too strange and rare a way of asserting the "lost" symbol of the Paraclete, hardly understandable in Dante's epoch. However, it is not. And I will conclude these lines with only two other facts that bring confirmation to our findings.

The first concerns Albrecht Dürer. It is possible to prove that he knew the meaning of the cipher 515, which he disguised in the representation of a strange polyhedron composed of triangles and pentagons, in the famous engraving *"Melencolia"*, dated 1514. The polyhedron in question stands above a greyhound, near to a sphere! Its angles give opportunity to some numerical extrapolations that prove quite near the "complex" of 515. However another engraving, representing the vision of St Eustace probably of 1501, has it in a much simpler and clearer form; it shows, along with five greyhounds in the foreground, the miraculous stag with the crucifix between the horns, each horn having symmetrically five points. This five-pointed structure of the stag's horns, with the crucified *Unigenitus* in the centre, following the 5–1–5 disposition, is selfconsistently repeated by Dürer in the St Eustace painting, on the right-hand side of the Paumgärtner triptych (in the *Alte Pinakothek* of Munich). We have to omit, for lack of space, still other works by the Master of Nüremberg that assert his clear perception and repeated use of the Dantean cipher for the Envoy of God, on which he probably received information during his voyages to Italy. Again, Dürer was undoubtedly linked to Templar esoteric circles, as Dante had been before him.

The second confirmation I want to mention here is given by a majestic painting of the Portuguese school, probably a work of Master Jorge Afonso, representing the "Apparition of Christ to the Virgin". It was commissioned by a queen for a women's convent near Lisbon. It exhibits at a very conspicuous place in the composition, with strange emphasis, what all scholars take for the date, that is, the number 1515. This date, however, cannot be accepted uncritically,† and may turn out

---

†In fact it was refused by Professor Reynaldo dos Santos, who attributed, to the "Apparition" under discussion, a date in the twenties of the sixteenth century. There is a strong suspicion that the first 1 in the "date" 1515 is apocryphal, a question to be settled, we hope, by a proper scientific examination of the paint. Despite the mentioned suspicions, exclusively inspired by historical, chronological and style-analysis concerns, my interpretation of the cipher or "date" in the painting is absolutely new.

Fig. 5. Albrecht Dürer, *Vision of St Eustach*, probably 1501.

to be simply the cipher 515 after some far from innocent repainting. In fact, the panel was the object of an unexplained amputation at some moment in its obscure history, being deprived of exactly the planks where the cipher was painted; the amputation took place, most certainly, during the most active period of the Inquisition, and specially after the application (in Portugal, after 1565) of the decisions relative to art taken during the Council of Trent. The missing planks were found in this century and restored to their original position in 1940.

It is certainly very intriguing to find the cipher 515, disguised as a "date", placed in so prominent a place in the painting, on a shield hold by a child angel; the idea that the "date" enacts the same role as Dante's cipher for the Paraclete is unavoidable. More so when attention is paid to a group of self-evident facts, among which are: (a) the fact that the subject of the Apparition to the Virgin is not scriptural and was viewed with strong hostility after the Council of Trent; (b) the fact that the

Fig. 6. *Apparition of Christ to the Virgin*, Portuguese School (attributed to Master Jorge Afonso) 1515(?).
Museu de Arte Antiga, Lisbon.

Templar Order survived in Portugal until the middle of the sixteenth century, under the name of
the Order of the Knights of Christ, permeating the whole Portuguese history from its beginnings
until the reign of João III (1536–1557); (c) the fact that King Manuel was the Master of the Order
of Christ, and the painting concerned was completed during his reign and was executed, according
to recent criticism, by the King's painter and (d) the fact that the cult of the Holy Ghost, quite in
the line of the ideas of Joachim of Fiore, was and still is† extremely strong among the Portuguese;
they believed, with greatest fervour during Manuel's reign, that they had been chosen by God to
prepare the coming of the Paraclete (Dante's "Messo di Dio") and the opening of the Millenium,
designated as "The Fifth Empire" or "The Reign of the Holy Spirit".

   All this is very much in the spirit of Dante. I think, for the reasons above and because of several
other clues, that the 'Apparition of Christ to the Virgin', the masterpiece of (probably) Master Jorge
Afonso now in the *Janelas Verdes* Museum of Lisbon, commissioned by Queen Leonor in the first

---

†The cult of the Holy Ghost is most lively nowadays in the Azores, in Brasil, and among many groups of Portuguese
   emigrants in the U.S.A. There is a vast bibliography on the subject in Portuguese, but practically none abroad. In French,
   I advise Ref. [20]. In English there is no bibliography, to my knowledge, except two papers by Mari Lyn Salvador [21].

quarter of the sixteenth century, exhibits the cipher 515, with the same intention and meaning that Dante gave to it two centuries before. This belief of mine grew stronger when I realised that, at the entrance of the church of the Templars in Tomar—the centre of the Templar Order in Portugal and one of the most important monuments in Europe—the Master who built the porch engraved in the stone—once again!—the cipher 515.

The Messenger of the 515, Lima de Freitas, 1984. Private collection.

REFERENCES

1. A. Pézard, *Dante—Oeuvres complètes*, col. Bibl. de la Pléiade, NRF-Gallimard, Paris (1979). Also comments by G. Villani, *Cron.* **I,** 60.
2. D. L. Sayers and B. Reynolds, *Paradise*. Penguin Books (1962).
3. O. V. de L. Milosz, *Oeuvres Complètes* (Ed. A. Silvaire) Vol. VII, Paris (1961). See also M. Eliade, *Cosmos and History: The Myth of the Eternal Return*. New York (1954). R. C. Cohn, *The Pursuit of the Millenium*. New York (1970). M. E. Reeves, *The Influence of Prophecy in the Later Middle Ages: A Study of Joachimism*. Oxford (1969). And *Joachim of Fiore and the Prophetic Future*. London (1976).
4. J. Baltrusaitis, *Le Miroir*. Elmayan-Le Seuil, Paris (1978).
5. A. K. Coomaraswamy, *The Transformation of Nature in Art*. Dover, New York (1956).
6. M. Eckhart, *The Transformation of Nature in Art* (1st edn) (1934).
7. A. K. Coomaraswamy, *Christian and Oriental Philosophy of Art*. New York (1956).
8. B. Nardi, *Sigieri di Brabante nella Div. Com. e le fonti della filosof. di Dante*. Rivista di fil. neoscolastica (1911–1912); also *Intorno al tomismo di Dante e alla quistione di Sigieri*. Giornale Dantesco XXII, 50.
9. L. Schaya, *The Universal Meaning of the Kabbalah*. George Allen & Unwin, London (1971).
10. A. Palacio's polemic studies comparing the *Divine Comedy* to Ibn'Arabi's *Futuhat*, specially *La Escatologia Musulmana en la Divina Comedia* (2nd edn) Madrid-Granada (1943).
11. R. Guénon, *L'ésotérisme de Dante*, Gallimard, Paris (1974).
12. D. L. Sayers, The Penguin Classics (1st Publ). Penguin, Harmondsworth (1955).
13. D. L. Sayers, The Penguin Classics. Note at the end of Canto (XXXIII, 336). Penguin, Harmondsworth (1955).
14. L. de Freitas, *515—Le Paradigme du Miroir* (in press).
15. R. Benini, Per la restituzion della Cantica dell'Inferno alla sua forma primitiva, In *Nuovo Patto* (Sept.–Nov. 1921).
16. R. Guénon, *L'ésotérisme de Dante*. Gallimard, Paris (1974).
17. R. Abellio, *La Bible—Document chiffré* (8th edn). Gallimard, Paris (1950).
18. Avicebron, *Qabbalah, or The philosophical Writings of Solomon Ben Yehudah Ibn Gebirol or Avicebron* (translated from the *Zohar* by I. Myer, LL.B.) first published in Philadelphia (1888); reprint by Robinson & Watkins, London (1970) and (1972).
19. H. Corbin, L'idée de Paraclet en philosophie iranienne, In *Rivista dell. Accademia Nazionale dei Lincei*, Rome (1971).
20. G. Durand, Iconographie et symbolique du Saint Esprit and G. Gusdorf, Les empires du Saint Esprit: Mythistoire et idéologie, both papers published in the collective volume *Os Impérios do Espírito Santo na simbólica do Império*, Edn of the Instituto Histórico da Ilha Terceira, Angra do Heroismo (Azores) (1985).
21. M. L. Salvador, Symbolism and ephemeral art: an analysis of the aesthetic aspects of the Festas do Divino Espírito Santo. In *Os Impérios do Espírito Santo na simbólica do Império*, see Ref. [20]; and *Festas Açoreanas: Portuguese Celebrations in the Azores and California*. The Oakland Museum (1981).

Computers Math. Applic. Vol. 17, No. 4–6, pp. 899–906, 1989
Printed in Great Britain. All rights reserved

SYMMETRY IN CHRISTIAN TIME AND SPACE

S. Kerr

4209 Avenue F, Austin, TX 78751, U.S.A.

Abstract—The Christian church strives to conform its shaping of time and space to the life of Christ, the archetype of the relationship of man to God. The liturgical year and sacred space have evolved from Jewish and pagan roots according to the changing perception of this archetype. Through liturgy and architecture, the church mediates the individual human experience of time and space. The liturgical year is symmetrical with the Christian structure of time, neither linear nor cyclic, but spiraling towards union with God. Similarly, Christian architecture reflects the community's experience of the sacred. A clearly defined threshold, path, central focus and altar articulate the spiritual reality which the church seeks to express and worship.

Let us describe symmetry in Christian time and space by delineating the fundamental symmetry from which all others flow. Symmetry in the Christian shaping of time and space is a question of the congruence of these edifices with the realm of spiritual reality. Bear in mind, then, that I will be discussing not a physical symmetry—I will even bypass the symmetry of the Latin cross, which not only reflects the bilateral symmetry of man (all crosses do that) but man's verticality, his being rooted on earth but thrust towards heaven. Nor will we be concerned here with the relationship between symmetry in Christianity and in other major world religions, in Islam, Buddhism, Hinduism, but leave these parallels to be observed by others, while we focus on the spirituality of the tradition which believes in a triune God, for it is in the Holy Trinity, that asymmetrical configuration, that we find symmetry when we add the stabilizing fourth element, the believer.

We begin, as did the Hebrew people, with the Father, understood as creator, ultimate reality, ground of all being. From the Father is begotten the Son, the incarnation of the Father, the long-yearned for Messiah who delivered Israel from an illusory salvation in a political and historical context, from a parochial identification of their interests with the almighty Yahweh's, and from self-hatred, often projected onto other people. The dialogue between the Father and the Son is the Holy Spirit, at first seen only momentarily, as at Jesus' baptism and at the theophany on Mt Tabor, but later, after Pentecost, understood to be sustained in his mediation between the Son and the Father.

Now as man begins to appropriate for himself Jesus' experience, to enter into this archetype, this way of apprehending in faith his experience as the relationship of son to father, so is established the foundational symmetry for Christianity, most succinctly described by Hans Urs von Balthasar:

> "Christ is the archetypal relationship of man to God, a relationship measured only by itself, and he is this as a true historical man.... The measure of being included in the original Christ-experience is at the same time the measure of the apostolic mission: the more a person participates (*teilnimmt*), the more must he (and can he), in turn, communicate (*teilgeben*), not out of a gratuitous generosity, but in virtue of the intrinsic teleology of the experience. In this way there arises a lively unity of descent and ascent..." [1].

Thus, the symmetrical relationships which we will describe are rooted in this foundation, that the believing community, in its long meditation on the meaning of the reality of the incarnation of the triune God, has created structures which are symmetrical with that divine experience. Behind the various maps for the ordinary life of faith and for the mystical life charted by the saints through the centuries—the desert fathers, monastic spirituality, the medieval mystics, Aquinas, Francis, Bernard, Dominic, the analytical giants of the counter-reformation, Teresa, John and Ignatius, and the modern saints of simplicity, Therese, Charles de Foucauld, Dorothy Day, Padre Pio—are religious buildings configured by the requirements of worship and the sacraments, and a structure for organizing time known as the liturgical year.

The Christian liturgical year has its roots in the Hebrew festal calendar which, like the first calendar cycles of most primitive religions, was lunar; only Egypt, Mexico and Peru had solar calendars [2]. The Jewish lunar months were divided by sabbatical weeks; the Mosaic sabbath (Exod. 20:8–11) is based on seven as a perfect number, expressing completion and fullness. The sabbath was a deliberate interruption of work and productivity, a sacrifice of one's labor to God. It created a psychological climate for a change of mood, but it did not break through a cyclical conception of time. That the seven-day week is congenial to man is seen in the fact that the Jewish sabbatical week and the Chaldean planetary week evolved independently, and were adopted by the Greco-Roman world which had no week, although it had a more accurate lunar calendar.

The early Christians abolished the Jewish sabbath and instituted a Lord's day, *dominica*, to celebrate the resurrection. Since they were a persecuted minority with no means of effecting social changes, this was not a day of rest, but a day when the work hours were framed by worship, by morning or evening liturgies, in which the paschal mystery of the Lord was celebrated. This *dominica* was called an eighth day, an octave, and was understood to escape from the cyclic time of the pagans and their Jewish ancestors, and to establish a creative tension between the recurring structure of the week and the eternity of God shared by redeemed humanity. (We will encounter a similar interpretation of the number eight in the early Christians' choice of the octagon for their baptistries.)

Thus, the Easter celebration evolved from the Jewish liturgical cycle which had gradually replaced nature rituals with festivals commemorating Israel's history of salvation. The most important of these was Pesach. This feast contains nomadic roots, for it is celebrated at the first full moon after the vernal equinox when a young male lamb is sacrificed, roasted and eaten. But its central focus is the reenactment of the deliverance of Israel from the hands of Egypt. When Jesus transformed this feast by identifying himself with the sacrificial lamb ["Take, eat, this is my body." (Matt. 26:26–28)], he lifted this feast from its local origin and universalized it into the feast of the salvation for all human beings of every time.

In retaining the link with nature (Easter is celebrated on the first Sunday after the first full moon after the vernal equinox), the early Christians thus kept all the pagan and primitive meanings of a spring festival of rebirth, incarnate in a season of green, of bud and flower, of fresh beginnings. But the cyclic observation of the vernal equinox is lifted first into the hope of renewal in human history recalled in the Passover festival, and then universalized and made utterly transcendent in the final eschatological hope of Easter.

Thus, in Easter we see worked out the symmetry of the archetype of the Christ experience, moving simultaneously towards man, with the divine content of salvation, and towards the creator, with the earth's seasonal rhythm and human history.

To commemorate this feast each Sunday, then, keeps the believer steadily in touch with the Paschal mystery. This weekly rhythm echoes a cycle innate to the human creature. It is interesting to note that totalitarian governments, in opposition to the week's Christian origin, have tried to substitute another cycle; the French republican government tried to abolish the seven-day week and replace it with a ten-day unit, while the Soviets experimented with a five-day week. They were unable to uproot this pattern which, far from being of human contrivance, has its origin in a divinely given rhythm. As Josef Pieper has acutely stated,

> "While man can make the celebration, he cannot make the festive occasion and the cause for celebrating. The happiness of being created, the existential goodness of things, the participation in the life of God, the overcoming of death—all these occasions of the great traditional festivals are pure gift. But because no one can confer a gift on himself, something that is entirely a human institution cannot be a real festival." [3]

In the creative struggle for symmetry with the divine archetype towards which the Christian calendar evolves we find the balance necessary for true celebration:

> "Many authoritarian fests do not realize anything more than the dissolving of the rationality of everyday. They amount to exercises of loosening in the context of having no context. Such exercises at most make the tensions of everyday tolerable, but they do not create a playful and festive countertension." [4]

The celebration of Easter, which was eventually established as a yearly occasion around the end of the first and the beginning of the second century, was originally a weekly feast. The church calendar continued to evolve from the inspiration of this model: the celebration of Christmas began in the fourth century as a logical development of the church's meditation on the life of the God-man. Having commemorated his death, they desired to honor his birth. Here too we find the interweaving of nature rituals and history which characterized the dating of Easter. Jesus is closely associated with sun symbolism (we will see this association influence the use of light in church design), so his birth is observed at the winter solstice; his conception near the spring equinox (25 March). Since according to Luke 1:26, John the Baptist was conceived six months before Jesus, his conception is observed at the autumnal equinox, and his birth at the summer solstice (24 June). Thus, the two equinoxes and the two solstices are brought into the Christian world view and transformed from cyclic rituals to feasts of the birth of life and light into this world's darkness. The meaning of John the Baptist's feasts is likewise associated with the sun's activity, for they were seen in the fourth century as cosmic confirmation of John's own words "He must increase; but I must decrease" (John 3:30), for after the summer solstice, the days grow shorter [5].

Once again, spiritual meaning is rooted in earth's reality; and it required centuries for the church to evolve this understanding: the feast of Christmas began in the fourth century when conflict with the Arians focused attention on Jesus' humanity; the feast of the Annunciation was introduced in the seventh century when the church turned attention to the contemplative dimension of the inner life.

Since Easter had a period of 40 days of preparation, Christmas by analogy acquired a preparatory period, although several centuries passed before it stabilized in the eleventh century at four Sunday's duration. From the earliest celebration though, the tension between the incarnation of the Lord in the historical Jesus and the eschatological return of the risen Lord kept this feast from being either cyclic or abstract. The analogy between Lent and Advent is clear in that both have the same liturgical color of purple; both are penitential in character; but Advent is not simply a repetition of Lent. As a later development, it has a subtle character, combining elements of waiting and anticipation, joy and wakeful attention on the threefold birth of the Lord into individual hearts, into history and at the parousia.

The liturgical year is divided into three seasons: the Advent–Christmas–Epiphany season, which is tied to a solar origin, the Lent–Easter–Ascension–Pentecost season with a lunar determination, and ordinary time. But ordinary time does not mean a lack of celebration, for the church made not only the incarnation and resurrection of the Lord the object of festivity, but they also celebrated events in the life of Mary, memorials of martyrs and saints, and festivals of doctrine such as Trinity Sunday and Corpus Christi. These also evolved through analogy: Mary, the martyrs and saints were celebrated as those who had most thoroughly imitated Christ. The tendency in the evolution of the Christian year is to express the latent eternal festival of universal assent to the world as a whole [6].

The Marian feasts are the fruit of the church's understanding of the human spirit which is receptive to the Lord. The apocryphal accounts of Mary develop her story in the same pattern as Jesus'; thus Giotto's paintings in the Arena chapel in Padua depict a cycle of Mary's life which includes an annunciation to Anna, a nativity of Mary (with a bath like that in the scene of Jesus' nativity) and a presentation of Mary in the temple. The two story cycles diverge in mid-life, but they mirror each other at death, for the tradition of Mary's assumption follows the paradigm of Jesus' ascension.

The symmetry between Jesus' and his followers' deaths is the basis for the early veneration of the martyrs. Even as early as the pages of the new testament, the church finds in a man's death a basis for honoring him. Just as Jesus had followed authenticity undeterred by the cost, so did Stephen, Paul, Peter and the early martyrs.

Although the first to be venerated were all martyrs, the church came to understand that the death to self necessary to a life of holiness and virtue was as real as the death of the whole man; thus by the fourth century, the first non-martyrs were honored as saints—Anthony, Basil and Gregory. We find in the cult of the saint the same fruitful tension between the archetype of Christ and the specific world of the particular saint which keeps believers from empty repetition. (One early desert father believed he'd been commanded by God to throw himself down a well in imitation of Abraham's sacrifice of Isaac. He was fished out by his disgusted brothers, but they had a hard time forgiving him his error.)

From the desert fathers to the middle ages extends a tradition directing the believer to make his path symmetrical to divine, not human, truth: as one guide, his name unfortunately lost, vividly urges:

> "Make sure you are no ape! That is to say, see that your impulses to silence or speech, fasting or eating, solitude or company, come from within out of an abundant love and devotion of spirit, and not from outside through the windows of your bodily sense, your ears and your eyes." [7]

While the church came dangerously close to love of death in the excesses of bridal mysticism with its language of victim and sacrifice, a biblical love of life ultimately counterbalanced that tendency. Death had to be the consequence of love of life, not sought for its own sake, for the church to find sanctity in death.

The contemporary church also insists that devotion to the saints be based on historical truth. Gone from the liturgical calendar are the likes of Valentine and Christopher, for the church wants to emphasize the work of the body of Christ throughout history, in diverse cultures and situations. The saint's feast day is located as closely as possible to his death, which is regarded as his birthday into heaven, *dies natale*, and it is his death which is his gateway into the church calendar.

Let us linger for a moment to look at a saint who followed the pattern of the Lord's life in most concrete ways: St Francis of Assisi is credited with beginning devotional practices of a highly analogous nature: purportedly he created the first nativity scene in a cave near Assisi, so that he could see a reenactment of the Lord's birth. He promoted the devotion of the 14 stations of the way of the cross, so that he could walk again the *via dolorosa* which led his Lord to Golgotha, and he bore in his body the stigmata, a reflection of Christ's five wounds. Where Francis imitated Christ in a corporal way, other saints follow him no less devotedly into other realms of human experience so that the phrase "the body of Christ" used to describe the church universal means that no nuance or aspect of human reality remains untouched by the Lord. His archetype is universal, but always rooted in lived human experience. In their diversity, the saints help bring the believer into harmony with the will of God.

Since the liturgical year is interwoven with the civil calendar, let us observe some of the differences and points of intersection, with an eye to noting how the Christian perspective of time shapes and diverges from the secular one. Both systems use the death of Christ as the pivot around which all time is organized; the year is either B.C. or A.D. (There is a movement to use the abbreviations b.c.e. for "before christian era" and c.e. for "christian era" to avoid the Christian root of this practice.) The civil year begins on 1 January, a cyclic day, and is divided into months. The Christian year begins with the first Sunday in Advent, having concluded with the feast of the Kingship of Christ. Thus, the end of the church year prepares for its beginning.

In the liturgical year, months have no significance. It is true that local devotions have arisen which depend on months, for example, the month of May is given to Mary's honor in England, but the church has always refrained from incorporating such devotions into her liturgical year in order to preserve the universal character of that year. For Christians in the southern hemisphere, religious occasions must not be linked to any particular season.

In the Christian organization of time, the relentless movement of historical time, the *chronos*, is held in tension with the time of *kairos*, of God's activity, of his transcendence made manifest. Because of this creative tension, the believer can locate his personal experience in a universal context. In the symmetry between his life experience and the liturgical year, he discovers the meaning of his own existence. This may happen on several levels:

1. In the rhythm of the day, the church sees a reflection of the human lifespan and the experiences of Jesus' life. Infancy is the dawn, called prime, the first hour in the divine office. Youth is lauds, the morning's praise; middle age, the afternoon, and old age, vespers, the evening sacrifice. Thus, as the church prays the liturgy of the hours, the sun's rising is associated with Jesus' resurrection, the third hour, Terce, with the outpouring of the Holy Spirit, the sixth hour, Sext, with the nailing to the cross, and None, the ninth hour, with his death [8].
2. In identifying certain spiritual stages, the Christian can describe his own experience as "lenten" or as "ordinary time" or as "festal glory". He may see his struggle embodied in the life of a particular saint.
3. In the patient journey up the spiral which includes both repetition and ascent, the weekly and annual renewal of Easter, the return of favored saints' days, along the continuum of his personal lifespan, the Christian can map his own story.
4. In holding a balance between communal and subjective experience, the liturgical year allows the believer to find personal meaning in these corporate rhythms. He

is never encouraged to efface his own inner reality for the sake of the community's schedule, but to locate himself in the universal context. His own experience is thus not the measure for his self-interpretation.

Symmetry in sacred space grows from the same correspondence between the archetype of Christ and our human engagement with him which has given us the liturgical year. Christianity is a religion of incarnation, "the God who is seen, heard, and touched in the man Jesus is, at the same time, the Man who sees, hears, and touches God." [9] Thus, we find an active identity between subject and object, between the church building and the spiritual reality of the believer, an identity observed by St Paul when he wrote the Corinthians "Do you not know that you are God's temple and that God's Spirit dwells in you?" (I Cor. 3:16). Just as when an individual achieves integration, he produces the symmetrical mandala shape, so does the church express the union of heaven and earth in integrated structures.

Church architecture is not primarily an unconscious or personal process, although elements of individual unconscious projection undoubtedly have a role to play in how buildings are shaped, but religious architecture expresses the group's self-understanding, its aspirations, theology and mission (see Figs 1–3). Again it is St Paul who articulates this truth:

"So then you are no longer strangers and sojourners, but you are fellow citizens with the saints and members of the household of God, built upon the foundation of the apostles and prophets, Christ Jesus himself being the cornerstone, in whom the whole structure is joined together and grows into a holy temple in the Lord; in whom you also are built into it for a dwelling place of God in the Spirit." (Eph. 2:19–22)

Since architecture is a collective enterprise, the buildings in which Christians embody their experience of the sacred also express their relationship between profane and sacred materials, profane and sacred space. For the Christian, the transcendent realm can never be isolated from his mundane experience.

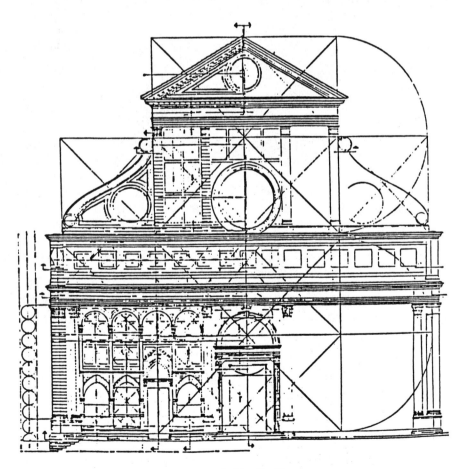

Fig. 1. Sta Maria Novella, Florence.

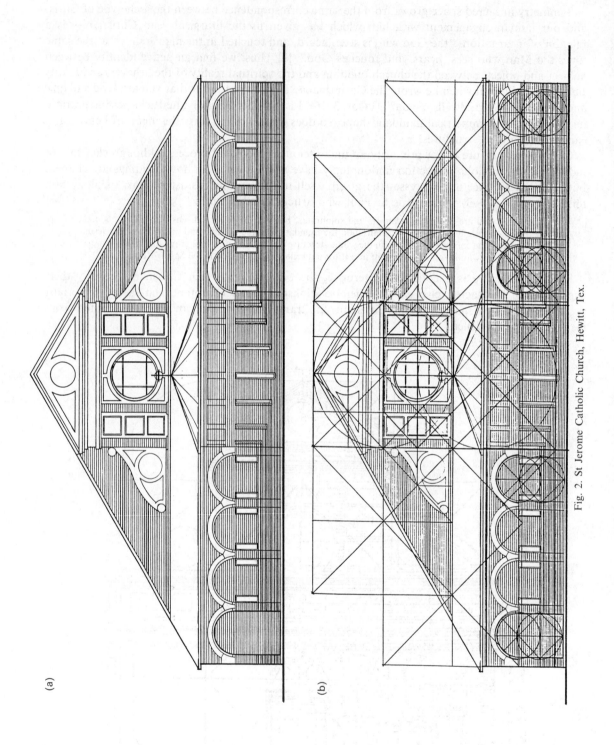

(a)

(b)

Fig. 2. St Jerome Catholic Church, Hewitt, Tex.

That the identity between subject and object is active means not only are our churches determined by our experience of the sacred, but they then have a role in shaping that experience. One could study the history of church architecture to learn how worship in an early Christian basilica prepares the way for monastic worship etc., but I want to focus on four elements common to all Christian churches in the west, whether Romanesque or Gothic, Baroque or modern: a strong threshold, a delineated path of forward progression through the building, an altar and a clearly defined center.

Let us consider these elements in sequence, observing the symmetrical relationship between the architecture and the inner experience of the spiritual world.

As he makes the transition from profane to sacred space, the believer is ritually crossing all the various thresholds he encounters each day, from sleeping to being awake, from home to work, from intimate to public interactions. Thus, the doors on a church are heavy and large, so that one has to enter purposefully. Church doors have no half measures, such as screen doors in domestic buildings, which permit a degree of privacy to be combined with a degree of openness. When entering a sacred space, one is either in or out, and what occurs inside is thus protected from inquisitive and casual attention. It is appropriate to the intentions of commerce that the doors of shops open without our touch; but the doors to the sacred must be pushed open physically.

The sense of crossing a threshold is extended by the use of the holy water font. This is a ritual remembrance of the sacrament of baptism through which one enters Christian life. Thus, with each entry to the church, one physically reenacts his spiritual entry into the divine milieu. The sign of the cross, the holy water, the genuflection or bow—all ritually express the crossing over into sacred space.

Many elements in church architecture are designed to create continuity and directionality—both necessary for a path. Floor patterns, the juncture of walls and ceiling, mosaics, windows and aisles often contribute to the sense of a physical journey to be made through the church. The path is an ubiquitous image in the Christian cosmos; beginning with Adam's expulsion from the garden, the Israelites' exodus in the desert, continuing throughout scripture and Christian literature. Since the space calls for action, once the believer has entered the church, he begins his journey forward. His imagination leads the way, as the interior world of the sacred acquires visibility.

There is a group of religious buildings which do not have a strong sense of path, but their distinctive character derives from their different purpose, to commemorate the beginning and the end of the Christian journey. From the beginning of Christian architecture, baptistries, mausoleums and shrines were often circular, suggesting no movement but concentrating energy in a sense of rest. Baptistries were sometimes octagonal, because the number eight was understood, as we saw in the discussion of the week, to refer to Sunday, the eighth day when Jesus was resurrected from the dead. So in baptism, one enters the realm of eternal life, as does the martyr in death.

The strong sense of center which these shrines convey is also evident in the churches where a path is articulated, for the center gives purpose and intelligibility to the path. One journeys not aimlessly but towards the center, a point which is usually defined by an apse, and/or a dome and/or an altar. In Christian architecture, the center is spiritually interpreted as heaven. Apses and domes create a sense of center with no precise locus; rather than defining it themselves, they surround the space which then becomes the center.

The altar gives an emotionally and visually satisfying focus to the journey, for it is there the eucharist is celebrated and witnessed, there the believers both observe the ritual and participate in it. Then the altar becomes a point of departure for a reversed journey out into the world again, so the building dynamically structures the rhythm of an encounter with the transcendent, redirecting the believer to the ordinary. Whereas pagan temples were built for the gods only, Christian churches were designed to make a home for God and His people.

Thus, the threshold, the path, the center and the altar transmit the existential meanings of Christianity to the environment as a whole [10]. The interconnectedness of these four elements is inherent in their meaning: crossing the threshold, one enters a sacred space which evokes a response of forward movement, a spiritual journey towards communion, celebration and the gathering together of all one's experience into an integral meaning, which, to be transformed, must be offered up at the altar, and then directed out into the world again.

As we regard the symmetry between the Christian shaping of sacred space and liturgical time, let us not fail to see that this endeavor to reflect the divine has been a font of creativity for the Western church. When living contact with the archetype of Christ has waned, when the church stumbles into empty repetition and decadent copying, the ashes of decline are transformed into the nest for the phoenix to rise again, so that space and time are once more fit structures for man to converse with the triune God.

REFERENCES

1. H. U. von Balthasar, *The Glory of the Lord*, pp. 305–306. Ignatius Press, San Francisco, Calif. (1982).
2. The following discussion of the liturgical year owes much to: N. M. D. Boulet, *The Christian Calendar*. Burns & Oates, London (1960).
3. J. Pieper, *In Tune with the World: a Theory of Festivity*, p. 46. Harcourt-Brace, New York (1963).
4. G. M. Martin, *Fest: the Transformation of Everyday*, p. 45. Fortress Press, Philadelphia, Pa (1976).
5. For an extended discussion of these feasts, as well as a readily available history of the liturgical year, see: A. Adam, *The Liturgical Year*. Pueblo Press, New York (1981).
6. J. Pieper, *In Tune with the World: a Theory of Festivity*, p. 23. Harcourt-Brace, New York (1963).
7. Anon., *A Study of Wisdom: Three Tracts by the Author of The Cloud of Unknowing*, p. 32. SLG Press, Oxford (1980).
8. O. Casel, *The Mystery of Christian Worship*, p. 89. Newman Press, London (1932). For further associations of the hours with events in Christ's life see: A.-M. Roguet, O. P., *The Liturgy of the Hours*, pp. 107–108. Liturgical Press, Collegeville, Minn. (1971).
9. von Balthasar, *The Glory of the Lord*, pp. 324–325. Ignatius Press, San Francisco, Calif. (1982).
10. G. Davies, *Temples, Churches, and Mosques*, p. 96. Pilgrim Press, New York (1982).
11. C. Norberg-Schulz, *Meaning in Western Architecture*, p. 112. New York (1980).

Computers Math. Applic. Vol. 17, No. 4–6, pp. 907–911, 1989
Printed in Great Britain. All rights reserved

URBAN SYMMETRY

M. Di Cerbo

33 Howard Street, New York, NY 10013, U.S.A.

Abstract—The author reveals his source of inspiration, the urban grandeur of the American city. He discusses his interpretation of this subject matter, and the influences other artists have had on his work. He describes the mediums he works in, drawings, prints and paintings; their particular qualities and how each process effects his personal expression.

I was born in Paterson, N.J., which is located 20 miles northwest of New York City. My family made frequent trips to New York, and as far back as I can remember, I felt awestruck when approaching the George Washington Bridge, one of the numerous monumental structures that dominate the Manhattan skyline. I have always had a great interest in architecture. I attended college in New York and started painting my urban images in my last year of graduate school.

I began by sketching the buildings of lower Manhattan. Shortly afterwards I stopped working from life and my compositions, although based on the architecture of Manhattan, became my own urban fantasies. Like all artists, I was also affected by other works of art. I was particularly influenced by a group of painters, who came to be known as the "Precisionists". Charles Sheeler, and in a more romantic mode Edward Hopper, were of prime importance. Their themes were often extracted from the urban environment and their styles ranged from a photographic realism to a Cubistic and almost Futuristic abstraction. This style first appeared in the 1920s. These artists were applying the lessons of European modernism, particularly the analytical methodology of Cubism to the American subject.

My work is certainly Americana, but there is no influence of Cubism. While the Precisionists started from reality, my roots were in non-objective painting. The Precisionist painting process entailed constant editing. I create linear blocks that, with added details, become skyscrapers revealed in their most primal state. My work is Classical in approach. I start out, by challenging myself with technical problems involving the use of perspective and the use of color. The concrete, steel and glass of the American city are my sources of inspiration. I try to capture my sense of urban elegance, by composing a geometry of forms with patterns of intense light and lyrical colors, that allude to the soaring architecture of skyscrapers. I present the city from the perspective of both an ant and eagle; buildings reaching endlessly upward or falling to infinite chasms below.

I begin with a small sketch, that has three or more vanishing points. I violate the laws of modern perspective. I am interested in drama not accuracy. I strive for a vision of pure energy and raw forces. I change and rework this sketch, playing with angles and shapes, until I arrive at a satisfactory composition of basic forms. This sketch, which is usually done on tracing paper, is transposed on to rag paper. I then add the windows and other architectural details and the buildings come to life. To achieve the dark tonal areas, graphite pencil is applied in a rubbing fashion, similar to the way children use crayons. The gray areas are obtained by applying a light coat of pencil, then rubbing this area with a shading pencil stump until the desired effect is achieved.

The light areas are achieved, by applying a light coat of pencil, then erasing most of the pencil with a kneaded eraser. Several coats of these techniques are usually needed; a workable fixitive is applied between coats. The last step is applying a permanent sealer. In *Metope* (Fig. 1), the viewer is at a higher point than any of the buildings pictured, and if he or she is looking out of a window, they would have to lean out in order to gain this viewpoint. There is a strong directional pull from the bottom left of the picture to the rear, a vertical pull form the bottom to the top and a more subtle horizontal tug from right to left. The vanishing points are extreme; in this composition as well as my lastest works, there are no side vanishing points. This helps to reinforce the feeling of power that large structures impart. Your eyes are drawn into the composition, then arrested by the groundplane and the block of buildings that arise from it. Your eyes move up these buildings to the rooftops which then pull you towards the background and top of the drawing.

Fig. 1. *Metope*; drawing, graphite on paper, $15\frac{1}{2} \times 12\frac{1}{2}''$, 1986. (*Collection*: The Next City Corporation, New York.) In this composition, the observer is leaning out of a window.

The best of these "finished drawings" become paintings and or prints. All of my prints are intaglios. The processes of etching, aquatint and drypoint are included in each print. First I aquatint the plate and achieve a range of tonal areas. Then I reinforce these areas with line etching. The subtle effects are produced with drypoint tools.

In *Uptown* (Fig. 2), one is drawn into the composition from the right side. Because of the way we read in the Western world, we tend to view things from left to right. A vanishing point on the left and below the composition, reinforced with strong architectural elements, forces the viewer to enter the picture from the right. The most dominant directional pull, however, is a vertical thrust upwards. We are not drawn into or repelled by this composition and it is a more natural viewpoint.

In contrast, in *Nightide* (Fig. 3), we are pulled into and then dropped into the picture. This is a bird's eye view, and its almost like being in a plane thats flying into a canyon of large buildings. There is deep depth of field and an awareness of movement in this intaglio. We tend to think of skyscrapers as static structures, but they are so large, and the elements so repetitive, that we have to move our head or eyes in order to see them. I have exaggerated this sense of movement, and in the process have created a feeling of mystery.

My paintings are dramatically different than my drawings or prints. Their size and the use of a wide range of colors account for this difference. Their size (in the 4×6 ft range or larger, I have done several mural commissions) tends to intimidate the viewer, and color always introduces varied visceral reactions.

Fig. 2. *Uptown*; intaglio, etching, aquantint and drypoint, $21\frac{1}{2} \times 16\frac{1}{2}''$, 1979. (*Collection*: The Brooklyn Museum, Brooklyn, N.Y.) In this print, I force the viewer to enter from the bottom right.

Fig. 3. *Nightide*; intaglio, etching, aquatint and drypoint, 36 × 28″, 1982. (*Collection*: The Victoria and Albert Museum, London.) There are over 10,000 cross-hatched lines in this print.

Fig. 4. *Infernus*; painting, acrylic on canvas, 54 × 68″, 1984. (*Collection*: Refco Corporation, Chicago, Ill.)
In this painting, I use a variety of earth colors combined with the amorphous light of night.

The paintings are painted on a horizontal work surface. With a finished drawing as a model, I lay in the basic shapes of the buildings. I then tape off each building and fill in these masked areas with the appropriate color. When this is done, I use masking tape to tape off the windows and other architectural elements. Changes are made in composition, color etc. until I arrive at the finished product.

Infernus (Fig. 4), is one of my more complex compositions. There are multiple vanishing points and the underlinging composition is two diagonal lines from corner to corner. You are pulled in every direction and the infinite variety of angles, architectural details and varied size of the buildings reflect the seemingly chaotic but functioning American metropolis. The colors I use are a combination of the laws of color interaction, together with the colors of visual reality. In this painting I use a variety of earth colors combined with the amorphous light of night. My color is always secondary to my use of perspective. It is an additive subordinate to a strong linear structure. The lights in each building are all on or all off. I do this to reinforce the directional pull, and to heighten the sense of order.

My work is about the gigantic man-made forms that dominate my surroundings. The human form is never depicted, but we are always aware of man's presence. The subject matter is man-made and the observer is always conscious of his or her vantage point, just outside the limits of the composition. Skyscrapers reflect the twentieth century in the same way that, Gothic cathedrals symbolize the middle ages. In my work, I try to capture the power of these monumental geometric structures.

The surfaces of my paintings are pristine, and rigorous order defines all of my work, in this respect my work is Classical. But I change familiar forms into ubiquitous symbols of twentieth century man, and this aspect of my work, I consider romantic.

Computers Math. Applic. Vol. 17, No. 4–6, pp. 913–919, 1989
Printed in Great Britain. All rights reserved

SYMMETRY AND WAYWARD NATURE

A. SENIOR

31 Bellevue Rd, Ealing, London W13 8DF, England

Abstract—The author, as a visual artist, stresses the perceptual aspects of symmetry. He discusses the particular qualities of natural patterns and how they relate to our notion of order. Concerning his work, three main lines of approach are detailed: the early landscape paintings of China and the philosophy underlying them, recent developments in our understanding of natural patterns (Mandelbrot), and the use of mirror image reflections. A selection of works are discussed in detail.

INTRODUCTION

The images in this essay represent the outcome of several lines of enquiry—all of which pursue the perceptual aspects of symmetry. I shall concentrate on the early landscape painting of China; recent developments in our understanding of natural patterns [1] and certain aspects of mirror image reflections in an attempt to trace the diverse influences which symmetry has had on my work.

SYMMETRY AND THE ARTIST

As a visual artist, I wish to stress the perceptual aspects of symmetry—how we learn to see before we learn to speak. Our initial encounter with the world is aesthetic: language and mathematics are nets which enable us to capture and objectify experience.

Symmetry is a net that guides our understanding of pattern. It also concerns the discernibility of difference—on a beach of pebbles it is the symmetry of the sea urchin that catches our attention.

In the sensory world, formal symmetry can be broken by the addition of another quality: one half of a symmetrical object can be coloured and this difference would not exist in an abstract formulation of the object.

Symmetry is closely related to our notions of order, so that when we encounter a patch of weeds, flowing water or clouds, our sense of order is challenged, and there is a tendency to describe natural patterns as chaotic or at least asymmetrical.

I would argue that we are encountering a richer kind of order here. This point is taken up by David Bohm in his book: "*Wholeness and the Implicate Order*" [2]. We have tended to define order in terms of predictability, but this may be true only of simple levels of order. Randomness (unpredictability) may be the vital ingredient in the order of complex natural systems.

I arrived at this viewpoint intuitively. To analyse this process is extremely difficult because it involves tacit forms of knowing, which as yet we have no means of formalizing. We recognize patterns whole, not merely as an assemblage of parts. These subjective experiences form the raw material of an artists approach to reality. As the Russian film director Andrei Tarkovsky said:

"The artist breaks down reality in the prism of his perception and uses a foreshortening technique of his own to show different sides of reality." [3]

Aspects of Symmetry

(1) Chinese landscape painting

My interest in the painting of China and Japan grew out of my encounters with Zen Buddhism and Taoism—of particular interest to me is the period of Sung and Yuan Dynasties (960–1368 A.D.). Painting in this period was particularly influenced by Taoism and Zen Buddhism.

Both philosophies recognized a distinction between human notions of order and natural order—the latter came to be known as Li. Li refers to the randomized order found in flowing water, clouds, trees .... In fact, the whole conception of nature was radically different in the East. The Chinese term for nature is Tzu-Jan, which means "of itself so", a process which happens

Fig. 1. *Sung Landscape.*

spontaneously, not one that is driven by outside forces. Here, then, was the philosophical basis for a type of painting that embraced the waywardness of nature as representing a deeper level of order akin to that proposed earlier.

In the paintings, themselves, we discover that the transient effects of time and place are sacrificed to reveal underlying formative principles. Natural pattern is seen as a manifestation of Tao–the Way; or in more modern terms an aspect of the global properties of the universe. This landscape painting became the major vehicle for the communication of these views. How was this achieved? Colour was largely eliminated, and form and space came to dominate the images. Individual forms were built up from simple abstract units: brush-strokes, dots and washes of black ink. Larger units, such as mountains, we built up from aggregates of small forms in a molecular fashion.

Chinese painters were the first to make systematic use of controlled accidents: ink splashes, ink spreading on wet paper or silk, and the random qualities of dry brush. These techniques provided a direct means of bringing Li (natural order) into their work.

Figure 1 represents my attempt to recreate a Northern Sung landscape in order to experience this way of seeing from the inside. It is, of course, a Westernized approach, since I am not a master of Oriental brushwork. It was painted on raw silk, using Chinese ink and a few washes of earth colour. It contains the classical elements: trees, waterfalls, mountains and a temple. It is painted in the method outlined above.

(2) New developments in the study of patterns in nature

The study of natural pattern in the West has been fragmentary until fairly recently. There are a few classics [4]; but much of the material was scattered in journals. The development of computers, and especially computer graphics, has created renewed interest. Abstract mathematical ideas could now be given visual form.

Figure 2 *Space Seed* grew out of my collaboration with Robert Dixon. We were both fascinated by patterns such as the sunflower and ripples on water. Robert Dixon, having solved the basic geometry, went on to realize a number of these patterns as computer graphics. I was more

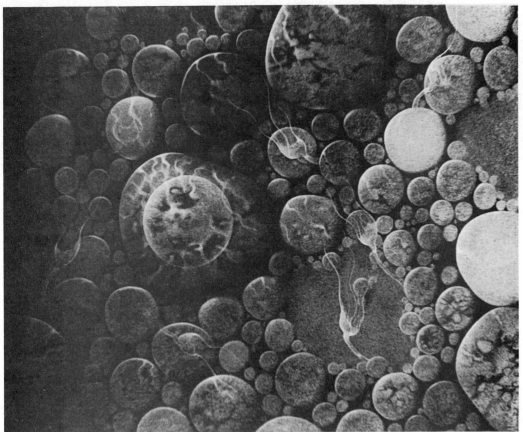

Fig. 3. *Moonstone.*

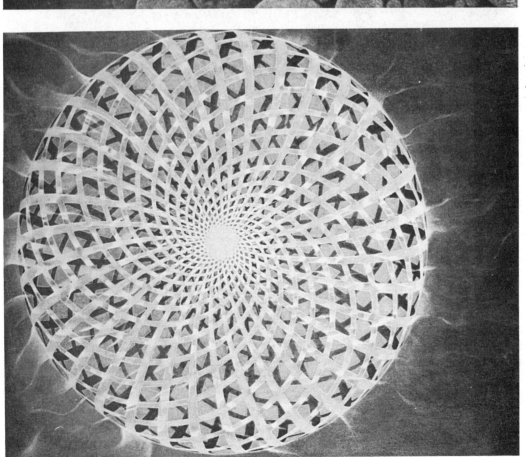

Fig. 2. *Space Seed.*

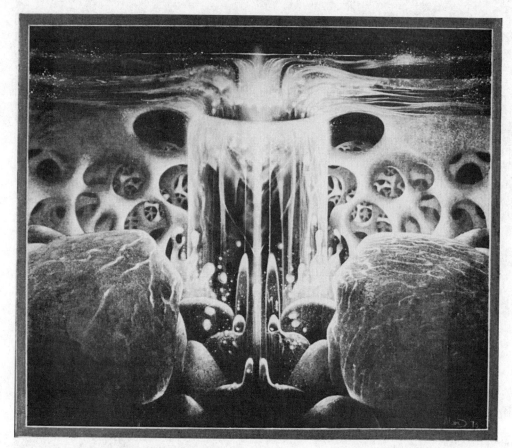

Fig. 4. *Still Point*.

concerned with relating these experiments to the more wayward examples to be found in nature (real sunflowers, real organisms). I took one of these basic patterns and treated the image imaginatively as if it were a micro-organism, injecting random effects of texture and light. It became an improbable diatom.

Figure 3 *Moonstone* was inspired by an image of Mandelbrot called an *Apollonian Gasket*—a two-dimensional image formed by the packing of scaled circles. I introduced broken symmetries, spaces, and then created a sense of texture by the multiple applications of randomized paint. The original image contains many interacting colours—cool at the top, moving to warm at the bottom. The image grew into an imaginative rockpool with a reflected moon—partly real, partly abstract.

(3) Mirror-image symmetry

Several of the paintings illustrated contain mirror-image reflections. My interest in such obvious examples of symmetry relates to my interest in pattern generally. Images with a dominant central axis have a particularly powerful effect on the observer; it challenges the widely held notion that key elements be placed off centre in a composition. The viewers intent is focussed by the sense of completion, of other worldliness. It looks real but it cannot be, the real world is open and changing.

Figure 4 *Still Point* contains a mirror image. The title refers to the state experienced in meditation—when the mind ceases to chatter and a magical stillness ensues. The image uses the metaphor of a water splash suspended in time to represent this state.

Each side of the image has slightly different qualities of light and texture. It stands on the borderline between the real and the abstract.

Figure 5 *Mind and Nature II*. The tree is placed at the centre like an icon. It contains a mirror image formed by reflecting the asymmetrical parts of the tree as a silhouette. I wanted the tree to take on some of the aspects of one of those many armed Hindu figures—an image of the power

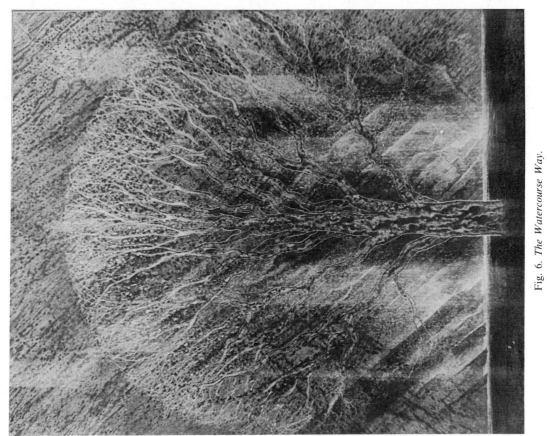

Fig. 6. *The Watercourse Way.*

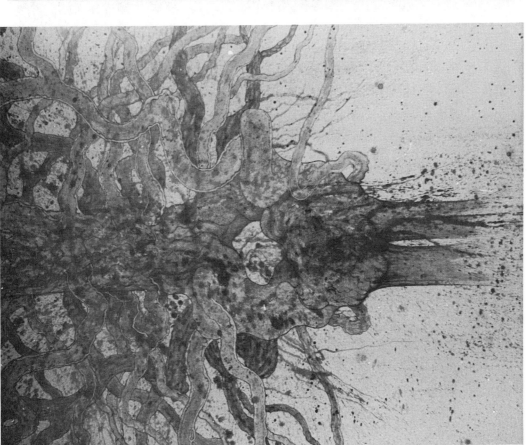

Fig. 5. *Mind and Nature II.*

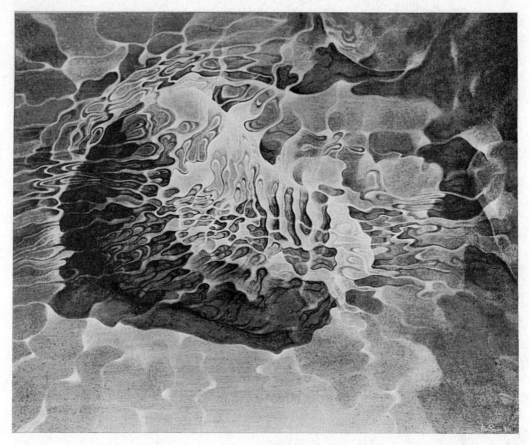

Fig. 7. *Insight II.*

of nature. The painting is carried out in a loose manner in contradiction to the ordered drawing, the stillness of the drawing fragmented by the energy of the paint.

ATTEMPTING TO INTEGRATE VARIOUS APPROACHES

Many of my painting are a fusion of the approaches outlined above.

Figure 6 *The Watercourse Way*. The title derived from the Taoist conception of nature as following its own way—Tzu-Jan. The image grows from the tacit recognition of self-similarity: in nature the branching of trees resembles the branching of water. The tree is an imaginary creation of branching limited by a circle which completes it. A waterfall cascades down behind. The whole image is imbued with energy—energy which is mutable; it moves through everything appearing here as a tree or there as water.

Figure 7 *Insight II*. The magical patterns of light in moving water, a small corner of nature which points to a wider truth: everything is process. What we take to be solid, immutable is also changing —we are like ripples in a stream. In observing natural patterns we never encounter aesthetic mistakes—it and just as "of itself so", it is its own measure.

Figure 8 *Memory Traces*. Within a small corner of the world: a group of pebbles, a leaf, a twig and a breeze—interactive patterns emerge. Patterns of flow permeate everything—frozen in the pebbles or moving in the breeze.

Fig. 8. *Memory Traces.*

REFERENCES

1. Mandelbrot, *The Fractal Geometry of Nature*. Freeman, San Francisco, Calif. (1983).
2. D. Bohm, *Wholeness & The Implicate Order*. Routledge, Kegan & Paul, London (1980).
3. A. Tarkovsky, "*Sculpting in Time*". Bodley Head (1986).
4. T. D'Arcy W. Thompson, *On Growth and Form* (2 Vols). Cambridge Univ. Press, Cambs. (1942).

Computers Math. Applic. Vol. 17, No. 4–6, pp. 921–930, 1989
Printed in Great Britain. All rights reserved

SYMMETRY IN PICTURES BY YOUNG CHINESE CHILDREN

Y.-M. Ha

Education Department, Nanjing Normal University, Nanjing, People's Republic of China

Abstract—Symmetry often appears in young children's drawings. Various kinds of symmetry in drawings are identified and their order of development studied. The reasons why young Chinese children frame their scribbles in symmetry and why the symmetry is mainly in the conformations and not in the colours of the picture are also discussed.

1. INTRODUCTION

When pictures by young children are exhibited, they transport us into a world where we are reminded of our own childhood and experience many pleasant feelings.

In pictures by young Chinese children, we can see traces of a happy and interesting life of their own. Besides, we often find them framed in the form of symmetry, which well deserves study. This way of constructing a picture cannot only be seen in drawings of geometric patterns, flowers, trees, the "beauty" of nature, transport vehicles and crafts, styles of buildings and animals, but also very often in pictures dealing with sorts of human forms. On the part of the children, the younger they are, the more they draw their pictures in this way. As a matter of fact, no exceptions would be found among groups of children, no matter what country they live in and what nation they belong to. They draw pictures the way children in China do.

How many styles of symmetry are used in young children's drawings? In what sequence do they develop? What is the cause of the symmetry? It was questions such as these that prompted our research.

2. TYPES OF SYMMETRY IN CHINESE CHILDREN'S DRAWINGS

Mainly, there are four styles. The first is a kind of mirror-reflection symmetry, which comprises the main part of all symmetry. A picture drawn in this way is like the curves formed by two figures of the same size and shape which are put above and below, or on opposite sides, of a real or an imaginary axis. The curves come about in very much the same way as a figure before a mirror. The figure and its reflection in the mirror make the curves. Either the outline of the figure on the left is the same as on the right (the upper part may differ from the lower) or the upper outline is the same as the lower (it may be that the left differs from the right). The inaccuracy in symmetry of this kind by young children, which causes differences between the two symmetrical parts, as this study reveals, is due to the limitation of the growth of both their brains and bodies. Thus, the concept of symmetry is discussed here in only a general manner.

Mirror-reflection appears most frequently in children's drawings of human shapes, and takes the various forms shown in Fig. 1.

Shape in Fig. 1(a)

This shape is most often seen in drawings of "tadpoles"—a human with only a head and with the legs going directly down from the jaw. There are no arms at all. On the head, there are, in most cases, a pair of symmetric eyes and a mouth on the vertical axis dividing the head equally into two. Sometimes, if included, ears and nose would also appear in the same way. With regard to the ears, one of the following two situations always applies: (1) they are not drawn; (2) if the

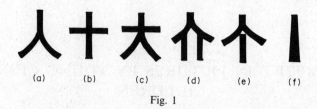

(a)   (b)   (c)   (d)   (e)   (f)

Fig. 1

child includes them you always see a pair. You almost never see a situation in which only one ear is in view. See Figs 2, 3 and 5.

Shape in Fig. 1(b)

In pictures of this type, human beings have begun to enjoy the use of the body trunk, arms and legs. The most eye-catching characteristic of this symmetry is that the arms are stretched horizontally. Also, on each of the hands there are now fingers. However, because of their weak sense of numbers, the number of fingers on each hand are not equal. The feet are often represented by two corresponding circles. Still there are occasions where the two feet point in the same direction (the tips of the two feet point either towards the left or the right). See Figs 4, 6 and 13.

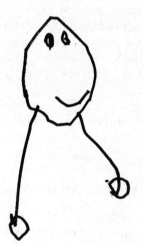

Fig. 2. Drawn by a four-year-old boy.

Fig. 3. Drawn by a five-year-old boy.

Fig. 4. Drawn by a five-year-old girl.

Fig. 5. Drawn by a four-year-old boy.

Fig. 6. Drawn by a five-year-old boy.

Fig. 7. Drawn by a six-year-old girl.

Fig. 8. Drawn by a six-year-old girl.

Shape in Fig. 1(c)

The human beings in drawings of this kind have horizontally-stretched arms, the legs standing apart with the toes pointing towards the reader, who most probably thinks that this tadpole human is doing physical exercise to music. See Fig. 9.

Shape in Fig. 1(d)

This shape exhibits an obvious difference from the shape in Fig. 1(c). Instead of outstretched arms, they hang downwards with a gap between them and the trunk. The legs stands apart with their toes pointing forward. The hair, hairstyle, decorations and things in the hands are all made to look symmetrical. See Figs 7, 8 and 10.

Shape in Fig. 1(e)

Using this symmetry, children generally like to hang the arms away from the trunk. The legs often appear close together. There are also occasions when the legs are lost (see Fig. 12).

Fig. 9. Drawn by a six-year-old boy.

Fig. 10. Drawn by a six-year-old girl.

Fig. 11. Drawn by a six-year-old girl.

Fig. 12. Drawn by a seven-year-old girl.

Shape in Fig. 1(f)

In drawings of this kind, one can see that the arms are attached parallel to the trunk (Fig. 11 shows this exactly).

Some animals drawn by young children are not exempt from symmetry either, as shown in Fig. 14. In this drawing the tortoise's head sticks out, as does the tail; with each of the four feet pointing away from the body. All these indicate the presence of the up-and-down symmetry. In Fig. 15, the butterfly looks like its real-life subject because of the application of symmetry.

When drawing buses, children will finish them in such a way that the front part looks the same as the back (see Fig. 16).

We also find in flower pictures that on the two sides of the trunk (the axis), flowers (including the petals), as well as the leaves, are equally allotted both in number and shape. In addition, in Fig. 17 there are even two symmetrically located vases with flowers in them.

Figures 18 and 19 show a spring willow and a tree from which the leaves are falling. The symmetry in these two pictures lies in the equally-numbered branches (or almost so), leaves and twigs at the tree ends, all with the trunk as the axis.

Fig. 13. Drawn by a seven-year-old girl.

Fig. 14. Drawn by a five-year-old boy.

Fig. 15. Drawn by a five-year-old girl.

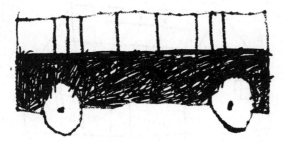

Fig. 16. Drawn by a six-year-old boy.

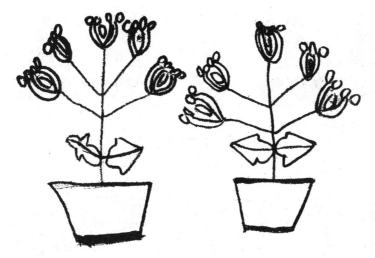

Fig. 17. Drawn by a six-year-old girl.

Figure 20 is a piece of house drawing. Generally speaking, usually only one side-profile is seen but this has been changed here with the result that both side views are seen.

Figures 21 and 22 are pictures of people in certain settings. In Fig. 21, two boats of the same structure, carrying the same number of rowers, display horizontal symmetry in their position. The same applies to the rowers, handling exactly the same oars in exactly the same way, with each pair of partners back-to-back. All this shows that the picture exhibits a very high level of symmetry.

Fig. 18. Drawn by a six-year-old boy.

Fig. 19. Drawn by a six-year-old boy.

Fig. 20. Drawn by a six-year-old boy.

Fig. 21. Drawn by a six-year-old boy.

Fig. 22. Drawn by a seven-year-old girl.

Figure 22 shows a scene in which children are walking along a road in lines. Fourteen children in parallel vertical lines are moving ahead in step. The group leader is made the only exception, walking ahead to the right of the lines. Notice also the equal number of trees on the two sides of the road. There are three on each side, and each of the three stands slanting towards the other side, overhanging the road. From each trunk grow two branches, on which another two twigs are drawn, and each twig has one leaf on the end. Careful study of this picture gives the observer the impression that symmetry works all the way through and that balance is felt everywhere and at every level.

The second type of symmetry is a sort of rough symmetry, which means that at first glance the picture makes you think it's symmetrical, but actually the right does not really correspond to the left and nor does the upper part with the lower part. Figure 23 belongs to this group. It shows a child having a meal. He sits at the table with his back to the observer. There is a tablecloth on

Fig. 23. Drawn by a six-year-old girl.

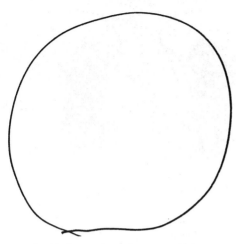

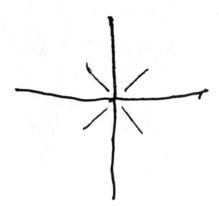

Fig. 24. Drawn by a three-year-old boy.   Fig. 25. Drawn by a three-year-old boy.

the table which is drawn using the "unfolding technique". There are some plates on the table cloth and various numbers of chairs of different shapes at the four sides. A casual look may give an overall impression that the picture possesses symmetry. But a closer look will prove otherwise. In some places things are different, e.g. in the upper part of the picture, the number of chairs on the two sides differs (three on one side, only two on the other), and each of the chairs has a distinctive appearance. Also, on the right the tablecloth has two corners, whereas it remains normal on the left. Apart from this, the positions of the plates contradict the concept of symmetry and as does the difference in the number of fingers between the child's right and left hand. But as far as the whole picture is concerned, it is in a state of approximate symmetry.

The third sort of symmetry is a radiation pattern. No correspondence is made horizontally or vertically. In Fig. 24, every point on the curved line is the same (or nearly the same) distance from the centre of the circle. In children's drawings, this sort of symmetry appears first. Figure 25 is an example of the type of symmetry shown in Fig. 1(b). Based on the cross conformation, lines radiate at all angles from the intersection. The shining sun, Fig. 26, is a favourite subject in the scribblings and drawings of children all over the world. First, they draw a circle or sometimes fully scribble the space which they regard as their sun. The second step is to draw the sunlight radiating from the sun heart. Furthermore, they might add to the sun a crested rim, just like the radiating petals of a sunflower. This symmetry of radiation is often called "Mandala" type. Figure 27 is a handkerchief design, showing no differences between the sizes of the four corners nor between the

Fig. 26. Drawn by a four-year-old boy.   Fig. 27. Drawn by a four-year-old girl.

Fig. 28. Drawn by a six-year-old girl.

lengths of the four sides. In the centre is the sun. A good balance is maintained in pictures with this symmetry.

The fourth type of symmetry is two-direction symmetry, which refers to the regular recurrence of a figure in two opposite directions. That is, in the two directions, a model figure appears repeatedly at equidistant intervals. This type of symmetry shows a strong rhythm. Examples: trees in Fig. 22; grass in Fig. 26; Fig. 28, "the skipping children"; Fig. 29, the pandas.

3. ORDER OF DEVELOPMENT OF THE VARIOUS KINDS OF SYMMETRY IN CHINESE CHILDREN'S PICTURES

According to our observations, the earliest symmetric outline in children's scribbles is circles. We see that they start with and go on repeating circles throughout their childhood drawings. Why is this? In fact, this is not because they are simply abstract geometric figures, but they are things they want to describe from their own world. Circles are their immediate expression of images in their life, because, at a certain stage of physical development, their limbs will grow to be more flexible, helping a lot in the movement of the elbows and wrists, which would naturally cause circles to come out in their pictures. But why is it that square shapes can only develop in children's minds after the presence of circles? The reason is simple: the four sides of a square are of the same length and its four angles are all right angles, which young children cannot draw—it is still beyond their ability. This explains why circles appear ahead of other geometric figures.

Based on their ability to draw circles, children then develop to draw "tadpoles" and the sun, followed by symmetric clothes and hairstyles, and on to decorative articles and stage props. Up to this point, in step with each other there appear symmetric drawings of flowers, grass, trees, animals, buildings and transport vehicles. On the whole, only two-direction symmetry comes a bit later, since education is a necessity before children arrive at this level.

4. WHAT CAUSES CHINESE CHILDREN TO DEVELOP THESE FORMS OF SYMMETRY IN THEIR DRAWINGS?

Study of children's pictures shows that the average characteristic is that each pair of symmetrical parts, having the same shape, combine to form the whole. The simplification is obvious. The figures,

Fig. 29. Drawn by a six-year-old boy.

both simple in conformation and relationship, make it easier for a child to give the concept and speciality of a thing he or she wants to have in the picture. For this reason, in the earlier drawings of children occasionally we see symmetry.

According to the theory of Gestalt psychologists, Gestalt concerns the organs and structures. These organs are in stratifications and, therefore, they produce various kinds of perceptions which are the result of the brain receiving stimulations from outside and organizing them, not what one imagines. All the Gestalt pictures that give pleasant stimulations, are simple to, in the main, the most easily understood and always take the form of the best and most regular symmetry in harmony.

According to the findings of some psychologists, for stimuli to visual perception which have some common qualities, such as the same shape, similar colours, the same distance and the same size, the alike parts will form a whole in the course of perception. This is the process of organization performed by the perceptive system on the stimuli to visual perception—action of simplification. However, when irregular stimuli come into sight, the child's perceptive system very often automatically develops a stong tendency to change them. On the one hand, regularity and symmetry are emphasized by amplifying, unfolding and displaying the relevant parts; and on the other hand, they will cancel or change the complicated, irregular and non-symmetric parts which are obstructing their way of constructing a brief and regular drawing of symmetry. So, shapes having slight differences between their parts are often taken as symmetric in their drawings.

Pursuit of a simple structure through "simplification" in perception is due to an organic desire for the internal balance, consequently producing a spontaneity of self-regulation which is an "inner requirement". The internal balance is broken when some visual stimuli come into sight that have the property of imbalance. Then the organic activity is to alter these stimuli to an extent where balance is achieved through rearranging the parts of the stimuli by likeness similarity and continuity. That is, reducing the complexity—i.e. the principle of simplification. This will surely result in the appearance of simple and symmetric shapes in the centre of the field of vision, and the symmetry will fully satisfy the "inner requirement".

The above explains why children always draw a man's head in the shape of a sphere that enjoys symmetry. It is the principle of simplification that makes children draw spheres for all round-like things, not the explanation given by a previous theory that: a child does that just because he first looks at the head, generalizes it to be a sphere and then uses some abstract geometric figures to draw it. Obviously, this explanation exaggerates children's abilities—the simple reason being that they do not yet possess a very developed thinking system. Hence, the theory cannot be accepted.

What gives rise to the appearance of symmetry in the shapes in Figs 1(c–e)? Apart from the simplification principle that plays its rôle in children's drawing of two-sided symmetry, there appears to be a certain psychological power that drives them to pursue the completeness of what they want to draw. When drawing the figure of a man, children stretch its two arms horizontally

Fig. 30. Drawn by a seven-year-old girl.

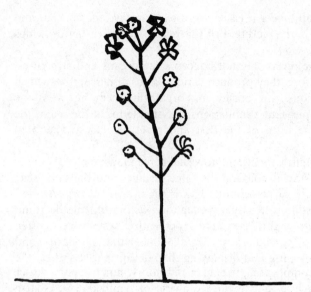

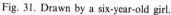

Fig. 31. Drawn by a six-year-old girl.

Fig. 32. Drawn by a six-year-old boy.

and separate the two legs. This is not an expression of their feelings, but an avoidance of obscurity—the arms are hard to distinguish from the body trunk—so both the arms and legs should be more clearly displayed. Once I designed a test, suggesting a child add turnip leaves to cover a rabbit's ears—the motive being for the picture to be made more complete—but there was immediate opposition from the child. He argued: "Um, the ears will be lost, people will think it is a cat!" From this we can see that a child takes the ears of a rabbit as one of its most recognizable features, which is something he will not allow to be changed. Hence, the purpose of putting the arms outstretched is just a need for them to catch the eye.

In Fig. 20, the drawing of both sides of the house is also an expression of the eagerness for completeness. As far as he's concerned, the child has the fact in his mind that a house has two sides. Then, why miss out any of them? As a result, this kind of symmetry is used.

To close, my last point is that my findings show that symmetry in children's drawings is mainly used in combining and permutating shapes. While in colouring, they often try to make each part different, particularly when doing human figures, in which symmetry is a rare occurrence (see Figs 17, 21, 22, 28, 30–32). Why children do this is still unknown at the present stage of this research. I hope to have the answer in the near future.

Acknowledgement—Thanks are due to F. Shao for the translation of this paper into English.

Computers Math. Applic. Vol. 17, No. 4–6, pp. 931–953, 1989
Printed in Great Britain. All rights reserved

FROM GEOMETRICAL RIGOR TO VISUAL EXPERIENCE

F. Lantos

Art School of Apaczai Education Center, P.O. Box 17, Pécs, H-7601, Hungary

Abstract—Simple elements and basic operations are used to create increasingly complex geometrical systems. Before the procedure and the vast number of variations would get out of hand, however, the artist's intuition and bias step in to establish a uniquely artistic visual experience.

Art has always been either directly or indirectly related to nature. Nature has provided inspiration for different possibilities of representation either by its extremely rich realm of sight or by the specialities of its structure and the problems of different depth and complexity of its visual and spacial logic. Sometimes both together played an important role.

In this present paper an attempt is made for a simultaneous observation of nature's constructional logic together with a set of visual operations which may look very simple but can hold a great number of possibilities as regards its visual appearence.

Starting from the phenomena of nature a lot of examples show that the circular form is directly or indirectly decisively determinant regarding either the arrangement of different forms or the logic of movement of their function.

As a matter of fact it is almost obvious that the circle form preconditions its counter form, the square. These two forms as counter complementary pair forms hold such profound meanings and correlations that can serve as a common basis for endless phenomena and matter.

I think it can be rightfully declared that all the form-creating and modelling processes inferrible from the "circle–square"—a concentrated form—can refer back to those phenomena and objects that are somehow related to them. Thus, a seemingly abstract constructive procedure can include all the moments of a nature-principled concept, even more it can be directly related to the realm of sight under some circumstances.

The present paper cannot of course aim to present the whole treasury of the above mentioned relationships. It can only undertake to flash a visual mentality through a logically followed visual sets of operations based on natural experience and observation. This is done from two points of view; the ramifying openness of the shape formation (originating from the same root) on the one hand, and the varied possibilities of the receptive "ego", on the other.

I hope there is no contradiction between the following set of figures and the contemporary scientific results, so it can cast a light on some possible connections of the scientific concept and the artistic creative work. It can also reveal the crucial points in pedagogy that include the practical possibilities of an integrated education.

The circle that possesses infinite-fold rotational symmetry and the same number of symmetry planes preconditions the square, which in turn has a four-fold rotational symmetry and four symmetry planes.

The square form also preconditions the circle form. Nevertheless the two forms together have only a four-fold rotational symmetry and four symmetry planes (Fig. 1).

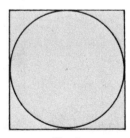

Fig. 1

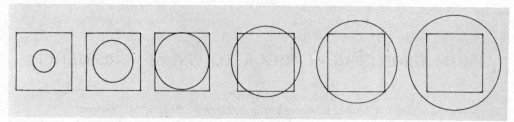

Fig. 2

Some proportions of the circle and square forms are shown in Fig. 2. The square is a more adequate form of enframing. Furthermore the circle form, enclosed in a square, is considered to be a basic element, a pair form holding contradictory characteristics.

If the concentric circles that are enclosed in a square are complemented with the characteristic lines of the quadrangle, the division over the median line arises automatically. The result of this division is a "quarter-cycle" that has got only one-fold rotational symmetry and one symmetry plane (Fig. 3). In this case the elemental angle is 360°, and a complete rotation leads to congruency.

The larger the number of symmetric elements of a form the less it is adequate to originate different new visual effects by multiplication. The circle together with the square is such a form. In a line diagram variant for example it has only one possibility (Fig. 4).

Arising from its single symmetry element, the "quarter-cycle" form makes possible a great number of variations of arrangements (Fig. 5).

If the components of this element (quarter-cycle) are put in a combination frame and each component is referred to all the others, a derived set of elements, is obtained, containing all the mirror counterparts of the asymmetric elements as well (Fig. 6).

Bilaterally symmetrical elements are shown in Fig. 7. Asymmetric elements and their mirror counterparts appear in Fig. 8. The result is $4 \times 14 = 56$ mirror analogous elements.

A set of 63 elements is obtained by adding the six symmetric elements and the "original", bilaterally symmetrical "quarter-cycle" element to the previous ones.

The components in line drawing can be, of course, filled in or coloured in different ways. This results in further variations. An example of this is the next element chosen at random showing the different possibilities of filling in Fig. 9.

The number of variants is 24 if the element is filled in with four colours, this time white, black, blue and red. A further six variations can be obtained if filled in by three tints (Fig. 10).

Filling in with two tones means that some divisions remain blank. This is valid, in fact, within the same family of elements only for certain components. Thus five further variations can be obtained as shown in Fig. 11.

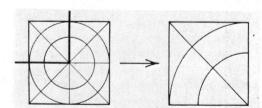

Fig. 3

Fig. 4

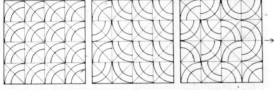

Fig. 5

Fig. 6

Fig. 7

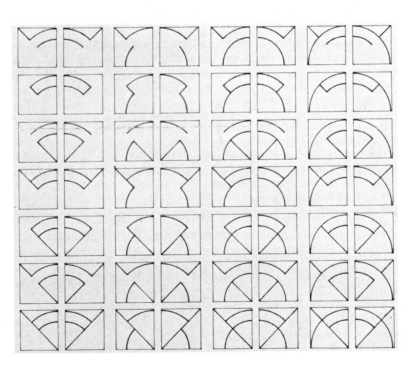

Fig. 8

Fig. 9

Fig. 10

Fig. 11

This example illustrates the great number of possibilities of forming new visual effects or designs, even within a single family of elements either by mere repetition of components or by the multiplication of the components and their different ways of filling in.

If different symmetry components have a definite role in combining the elements, they also remain an important factor in systematical arrangements of different elements. Thus we could hold possession of possibilities that inspired a succession of operations of different character and complexity. These operations, regarding their visual programmes, multifold notional and compositional meanings, specifically belong to the field of art. Besides, at the same time, they reveal such exact and subjective creative possibilities as well, that provide creative tasks organically derived from each other, taking the given circumstances into consideration on the long run.

From among the alternatives of operations those that refer to the different connections among the elements and patterns are chosen. The connections mainly result from the rotation of two or more superposed layers as compared to each other, and from the different forms and patterns derived from this rotation.

Rotation in this case means that one of the elements or patterns stays still while the other one rotates around a well-defined rotation centre in a predetermined direction and degree.

Examples

See Figs 12. Leftward rotation round the tip (Fig. 12.1). Rightward rotation round the tip (Fig. 12.2). Leftward rotation round the midpoint (Fig. 12.3). Rightward rotation round the

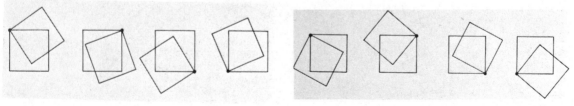

Fig. 12.1   Fig. 12.2

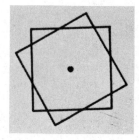

Fig. 12.3

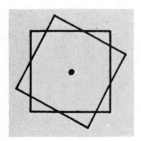

Fig. 12.4

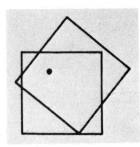

Fig. 12.5

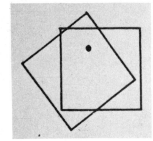

Fig. 12.6

midpoint (Fig. 12.4). Leftward as well as rightward rotation round an optional point of the quadrangle (Figs 12.5 and 12.6). Each type of rotation and the degree of rotation within one type give different results.

Let's follow the procedure on two, superposed identical elements and their symmetric pairs, first after a 90° rightward rotation round the midpoint (Fig. 13).

Rotation of an identical form pair of element, round the left top point in a degree determined by the apex and the median point of the quadrangle (26.5°, 45°, 63.5°) in Fig. 14.

The same type of rotation with the symmetric counter element in Fig. 15.

Fig. 13

Fig. 14

Fig. 15

Some phases of the previous rotation are taken out. First the second phase from the last line of Fig. 15. New results of the interaction of the two elements are obtained through the different ways of colouring and the resulting possibilities of other operations. (The negatives of the blank and filled in parts can be split, for example). This is first followed on a "comprehensive" figure then on the phases taken out from it (Fig. 16).

The same can be done by leaving out the constructional lines in the last phase of the second line in Fig. 14, as seen in Fig. 17.

The result is different if the whole set of superposed patterns, made up of more elements, is rotated. So it is not a rotation per element. At the same time the possibility of rotation per element by 90° round the midpoint within a pattern is not excluded. This means that each new superposed basic pattern, even in the case of identical components, will be different (Fig. 18).

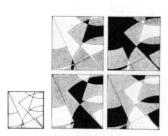

Fig. 16

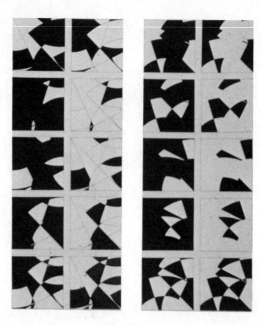

Fig. 17

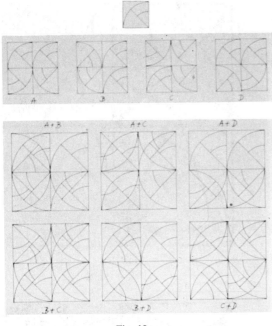

Fig. 18

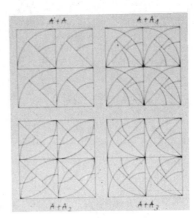

Fig. 19

The question becomes somewhat simpler if the superposed patterns are made up of elements of equal number, same form and same position, i.e. the patterns are totally identical. So the changes in the visual characters, resulting from the set of operations, can be more easily followed (Fig. 19).

The pattern bearing mark "A" is in basic position, "A_1" rotated by 90°, A_2" by 180°, "A_3" by 270° to the right round the midpoint.

If the above mentioned basic positions are rotated to the left round the left top point of the pattern by one unit (one unit of the quadratic division), and the newly rotated pattern is extended over the whole surface of stationary pattern a new configuration is obtained. (For the sake of better understanding one element of the rotated element is maintained in each figure), as given in Fig. 20.

Obviously the pattern-character is dependent on the number of the component parts. The result is different under the same logic of operation and number of components if the grid has a 2×2, 3×3 or 5×5 division, as illustrated by Fig. 21.

Fig. 20.1

Fig. 20.2

Fig. 20.3

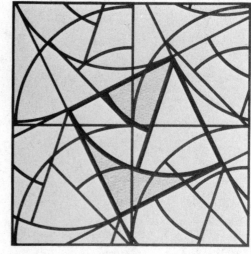

Fig. 20.4

Fig. 21.1

Fig. 21.2

Fig. 21.3

Fig. 21.4

Fig. 21.5

Fig. 21.6

Fig. 22

Further figures try to follow, of course without the claim of completeness, the interaction of a 4 × 4 grid division pattern pair. The "quarter-cycle" element (Fig. 3) is drawn in each square in the same position. The resulting two identical patterns are superposed and rotated by 180° in relation to the starting position (Fig. 22).

In further steps the basic pattern, already rotated by 180° round the left top point, is rotated again in phases of the quadratic units. Thus together with the basic position and the 90° rotation, as the last phase the result is nine different patterns of Fig. 23.

If some elements, specified by means of filling in, are put in this pattern, the differences in character derived from the different components and from the rotation can be clearly seen. Neglecting the two extreme phases the result is 3 × 7 different form-variation (see Fig. 24.1–7).

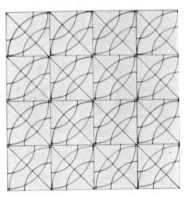

Fig. 23.1

Fig. 23.2

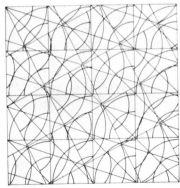

Fig. 23.3

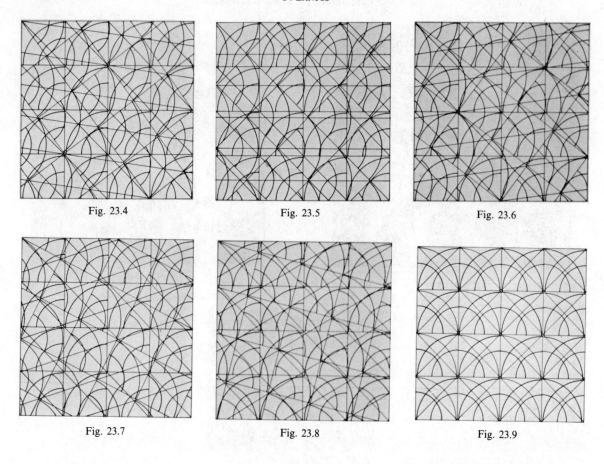

Fig. 23.4   Fig. 23.5   Fig. 23.6

Fig. 23.7   Fig. 23.8   Fig. 23.9

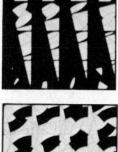

 The element of the top line (Fig. 24 top).

 The element of the bottom line (Fig. 24 bottom).

Fig. 24.1

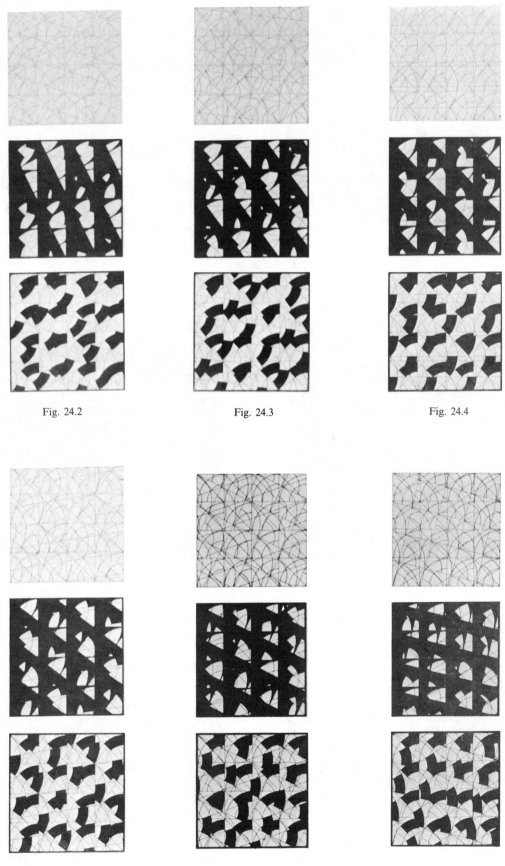

Fig. 24.2   Fig. 24.3   Fig. 24.4

Fig. 24.5   Fig. 24.6   Fig. 24.7

If the fifth rotation phase of Fig. 23 is taken out the next two pair patterns (Fig. 25) can be obtained by means of filling in the elements. Pattern pair No. 1—A, A_1. Pattern pair No. 2—B, B_1.

The various ways of these pattern connections are shown in Fig. 26 (Fig. 26):

If the tones of the starting and the rotated patterns are different various tonics can be obtained even within the same pattern (Fig. 27).

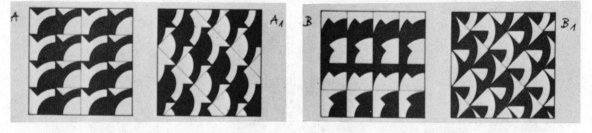

Fig. 25

Fig. 26.1. $A + A_1$.

Fig. 26.2. $B + B_1$.

Fig. 26.3. $A + B_1$.

Fig. 26.4. $B + A_1$.

Fig. 27.1

Fig. 27.2

Fig. 27.3

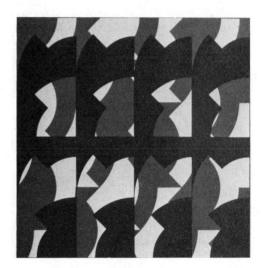

Fig. 27.4

The visual effect becomes more and more complex, and the spatial effect more and more complicated if more than two patterns are superposed.

Some examples are shown in Fig. 28.

The operation can be carried on, if similarly to Figs 16 and 17, where only one pair of elements was used, the final pattern is again taken into pieces. The $B + B_1$ arrangement is taken as an example to illustrate this, together with its negative variation (Fig. 29).

If the filled in and blank parts (Fig. 30.1) are split in the positive pattern, these together with their negative variations result in further six patterns (Fig. 30.2–7).

The same technique can be applied to the $A + A_1$ set, this time more freely in a repeated superposition and rotation, as seen in Fig. 31.

If the number of the components is more and more increased, the results become more and more complex (Fig. 32). The basic structure of the following patterns is made up of $8 \times 8 = 64$ elements.

Figure 33 shows one part of the above pattern doubled and superposed in different ways (rotated and slid along).

In the previous examples all the sets of operations though in rough outlines, could be logically followed.

Fig. 28.1

Fig. 28.2

Fig. 28.3

Fig. 28.4

Fig. 28.5

Fig. 28.6

Fig. 29.1

Fig. 29.2

Fig. 30.1

Fig. 30.2

Fig. 30.3

Fig. 30.4

Fig. 30.5

Fig. 30.6

Fig. 30.7

Fig. 31.1

Fig. 31.2

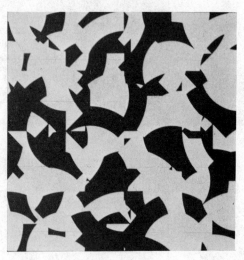

Fig. 31.3

Fig. 31.4

Fig. 31.5

Fig. 31.6

Fig. 31.7

Fig. 31.8

Fig. 31.9

Fig. 31.10

Fig. 31.11

Fig. 31.12

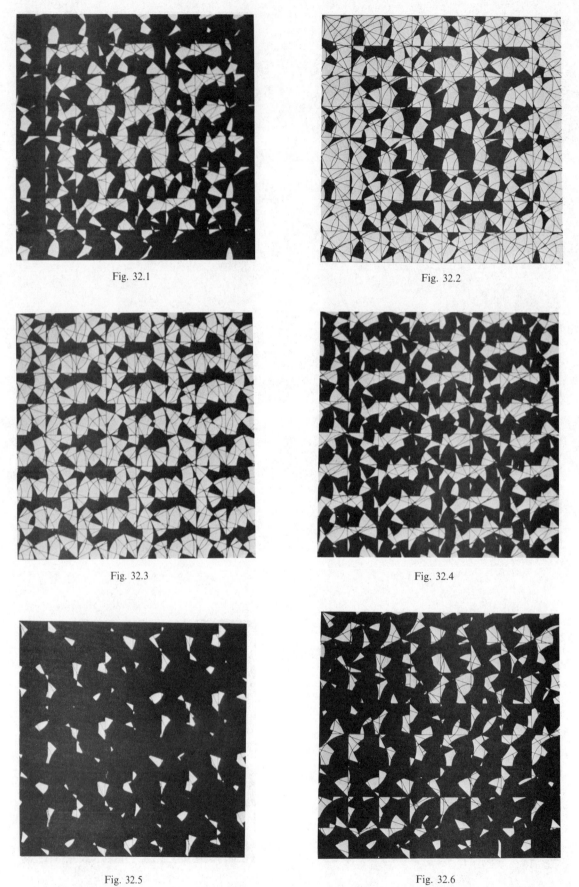

Fig. 32.1

Fig. 32.2

Fig. 32.3

Fig. 32.4

Fig. 32.5

Fig. 32.6

Fig. 33.1

Fig. 33.2

Fig. 33.3

Fig. 33.4

Finally I would like to point out how the results of objective operations can be resolved and made more complex in a way that is hard, or almost impossible, to be logically followed. In other words: how objective compositions, including symmetric basic situations, through the succession of superposed operations—maintaining their inner compositional order—can turn into specifically individual creative programmes. Figures 34–37 illustrate this.

Fig. 34.1

Fig. 34.2

Fig. 34.3

Fig. 34.4

Fig. 34.5

Fig. 34.6

Fig. 35.1

Fig. 35.2

Fig. 36.1

Fig. 36.2

Fig. 36.3

Fig. 36.4

Fig. 36.5

Fig. 36.6

Fig. 37

Computers Math. Applic. Vol. 17, No. 4–6, pp. 955–999, 1989
Printed in Great Britain. All rights reserved

0097-4943/89 $3.00 + 0.00
Copyright © 1989 Pergamon Press plc

ORNAMENTAL BRICKWORK

THEORETICAL AND APPLIED SYMMETROLOGY AND CLASSIFICATION OF PATTERNS

E. Makovicky

University of Copenhagen, Institute of Mineralogy, Østervoldgade 10, DK-1350 Copenhagen, Denmark

Abstract—In the present study historical brickwork ornaments are used as a vehicle for testing existing pattern classifications (Wyckoff sets/pattern types) and development of new classifications. Patterns can form progressive series in which incremental changes produce a series of new patterns spanning two selected end-member patterns. The following types of series have been defined and their symmetrological aspects discussed: expansion–reduction series including the simple element-expansion (-reduction) cases; element-propagation (i.e. accretional) series; variable-fit series; intercalation series; catamorphic (i.e. complication/simplification) series which embrace multiplicative catamorphy, pattern metamorphosis and element-additive catamorphy as well as the omission series; element-substitution series; pattern-reduction and partial overlap series; as well as affine series and the series of freely-hinged patterns. Applicability of these concepts was successfully tested on four large families of ornamental patterns as well as on the comprehensive pattern sets from the Great Mosque of Cordoba (Spain) and from the village of Jork (North Germany). In addition, the role of simple stacking notation and the results of selective colouring applied to simple brickwork patterns were studied. Detailed analysis of order–disorder phenomena in ornamental patterns is given as well.

1. INTRODUCTION

After having worked with complex patterns and symmetries [1–3], a question arose in the author's mind about which spectrum of patterns and symmetries was produced using one of the simplest means of ornamental expression—brick masonry. Study revealed on the one hand the extraordinary richness of brickwork patterns and on the other hand the limited spectrum of principles and of groups of symmetry used by the artisans. They became a vehicle for the development of alternative systems of symmetrological/structural analysis of patterns. Some of these are of general validity, applicable to all series of patterns, from brickwork through mosaics to the families of crystal structures. The present effort is parallel to the developments taking place for complex crystal structures, e.g. various polytype nomenclatures as well as papers [4, 5] and Ref. [6].

These studies led from plain masonry towards more and more complicated ornamental brickwork patterns. However, an attempt was made to limit the present study to patterns composed of real, simply shaped bricks (or corresponding stone elements), in most cases stopping short of polygonal or curvilinear ceramic tiles prepared in special molds. Furthermore, large-scale patterns of glazed and multicoloured bricks (like those on the domes and walls of Iranian and central Asian mosques), in which the role of individual brick all but disappears, were omitted as well.

2. ROSETTES, FRIEZES AND PLANE PATTERNS

Zero- and one-dimensional periodic patterns are widely used on buildings as a means of adorning plain brickwork. Both purely axial (rotational) as well as combined rotational-reflectional groups of symmetry occur in *zero-dimensional* brick patterns (rosettes). The latter symmetry implies that all bricks that meet at an angle have either been ground to conform with reflection planes or the square elements have been used on these planes. If this is not the case, individual bricks meet in a herringbone fashion and, on the level of individual bricks, the pattern will violate the potential (or actual on the large scale) reflection planes.

Such patterns correspond to the situation often found at unit-cell level in twinning of crystal structures and, consequently, they can be described as simple or complex twins in two dimensions,

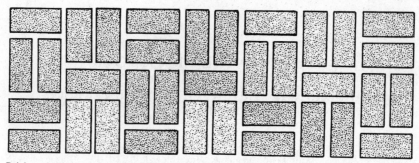

Fig. 1. Brickwork frieze from Gamle Kongevej, Copenhagen (1906 A.D.). Band group *pmg*2. This frieze
is not extendable into a uniquely defined two-dimensional pattern.

according to the number of composition planes. Pattern portions between two adjacent composition planes represent fragments of a two-dimensional periodic pattern.

   One-dimensional friezes were equally widely used (Figs 1 and 2). All frieze patterns can be divided
into two large groups. The first group combines friezes which are of typically one-dimensional
character [Figs 2(c) and (d)]. They are "self-contained" within distinct parallel boundaries. The
second type of friezes are those which represent obvious cut-outs from two-dimensionally periodic
patterns and their parallel limits often represent lines traceable in the same cut-outs in other
directions (i.e. perpendicular or oblique to infinite extension) [Figs 2(a) and (b)]. In the majority
of cases the two-dimensionally periodic versions can be unambiguously reconstructed [e.g. from
Figs 2(a) and (b) or in Fig. 24]. Sometimes, the two-dimensional patterns and their one-dimensional
cut-outs appear on the same building [Figs 2(b) and 35]. In other cases one-dimensional cut-outs
were not extended sideways beyond limits which could reveal their order–disorder properties
(Fig. 52). Finally, for some patterns multiple extension possibilities exist so that no definite idea
about the possible two-dimensional precursor can be deduced (Fig. 1). Again, bricks were ground
to shapes, or square elements inserted in many simple brickwork friezes if lengthwise mirror planes
were desired in stead of glide-reflection planes.

   Friezes representing cut-outs of two-dimensional patterns will often contain elements of *local
symmetry* of higher order than those allowed by the seven frieze groups. Such symmetry elements
will not fit into the overall frieze group either by position or their order [e.g. the local four-fold
axes and diagonal mirror and glide-reflection planes in Figs 2(a) and (b)]. They might be used by
the artisan when frieze pattern is "rounding a corner" to become a frame.

   In the *two-dimensionally infinite* periodic patterns cases of higher local symmetry can occur as
well. The point group symmetry of any brick in the wall face is 2*mm* and in many cases only some

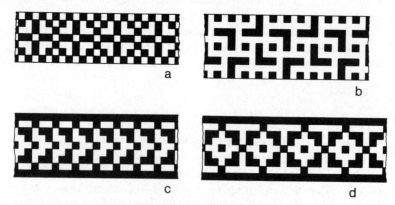

Fig. 2. One-dimensional patterns from eastern gates of the Great Mosque of Cordoba (built under Al
Mansur, 987 A.D.). (a) Band group *pmm*2. A cut-out from a two-dimensional pattern *p*4*mm*. (b) band
group *p*112. A cut-out from a two-dimensional pattern in Fig. 35. (a) and (b) contain numerous local
symmetry elements preserved from the plane patterns. (c) Band group *p*1*m*1; reduction version of pattern
in Fig. 2(d). (d) Band group *pmm*2. This pattern also represents unit-cell twinning of the pattern in
Fig. 2(c).

and not all of these symmetry elements become valid for the entire, one- or two-dimensionally infinite pattern. In many cases the local high symmetry of a brick or of a brick course (layer) results in order–disorder phenomena to which a special section has been devoted.

With respect to various symmetry elements of the relevant frieze or plane group, the pattern elements (motifs) can assume two distinct types of positions: (1) *general positions* which do not lie on any symmetry operators and (2) *special positions* which lie on one or on an intersection of several symmetry operators. Such elements must themselves display these local symmetry conditions if the symmetry conditions if the symmetry group is not to be violated. These distinctions are important for certain pattern classifications [7] and for brick colouring to obtain dichroic symmetry groups. Another group of phenomena important for understanding brickwork patterns are super- and subperiodicities, pseudosymmetry as well as superpositions (mixing) of symmetries, summarized in Ref. [3].

3. CLASSIFICATION OF PATTERNS

Pure unit-mesh and plane-group characterization yields only partial picture of pattern variability. Therefore, further classification of patterns can only be achieved by studying their individual elements as well as their mutual relationships and their positions in respect to symmetry elements.

Attempts to create such classifications have been perfected by Grünbaum and Shephard. Their series of contributions has been summarized in Ref. [7]. Parallel efforts for a rather specialized topic, the crystallographic point configurations have been perfected and summarized by Fischer and Koch [8].

For all 17 plane groups, there are 72 types of element (motif) configurations in both special and general sites (Wyckoff positions). Some of these configurations differ from each other only by their relative position and orientation in the unit mesh and not in the site symmetry and the configuration itself. Lumping such relevant cases together we obtain 51 types of *Wyckoff sets* which were singled out by crystallographers [9] and geometricians [7]. For point motifs such lumping yields only 30 distinct cases, the lattice complexes of crystallography; reduction to point motifs makes them non-applicable to the majority of plane patterns.

Two types of notation were devised for Wyckoff sets: (1) an explicit crystallographic one with a wealth of information but of corresponding complexity [8] and (2) a tabulation of Wyckoff sets by Grünbaum and Shephard [7] as so-called *pattern types*. In their refined classification of 52 pattern types a consecutive numbering is used. As in the case of lattice complexes, pattern types can be used not only for the plane group they were derived for but also for a variable number of lower plane groups which are motif-transitive for a given pattern type.

The efficiency of these classifications for real patterns, often with problems and ambiguities in selection of elements (motifs), with several Wyckoff sets (pattern types) of the same or of different kind in one pattern, nesting or engulfment of motifs, their very uneven sizes or with interpretational problems due to the differences between the ideal design and its technical realization, will be tested here. It should be stressed, however, that already the authors of these classifications perceived that without sizable quantitative data or without detailed special classifications for certain pattern types (e.g. the homeometric straight-line-segment or circular-disc patterns of Ref. [7]) all notations give only a limited picture of patterns examined. As is the case for crystal structures, the efficiency of these notations can be high for simple cases but its power becomes progressively limited with increasing pattern (structure) complexity.

As was stressed by Ref. [3], the choice of elements in any complex pattern (except for diaper patterns with a neutral background), can often be done in many ways. However, the patterns we study here were created by humans and meant for perception by humans. Certain brick elements were used to serve as clearly defined, eye-arresting elements and can be quite naturally used for the pattern classification. In the simplest brickwork patterns there is a natural element (elements)—a brick or bricks in several orientations and sizes. This convenient element (or elements) already recedes into the background for patterns of moderate complexity for which larger elements have to be defined. In all cases, the geometrical and symmetrological relationships of (diverse) elements, their adjacency conditions [7], must be studied as completely and many-sidedly as possible.

There exists another way of classifying patterns, a way which does not necessarily require *a priori* definition of elements and explicit analysis of adjacency conditions or pattern types. In their entirety, patterns can form progressive (or incremental) series in which one end-member evolves into another one via a finite or infinite number of intermediate patterns. This might be in many cases a practical answer to the otherwise formidable classification problems raised, e.g. by Ref. [10]. Moreover, this was often the way the artisans worked, creating new patterns by modifying the existing ones. The *incremental* or *progressive series* can be of various types according to the type of changes involved. The following categories apply to the subject of this study:

The *expansion–reduction series* in which certain elements of the pattern expand in area and, as a consequence, in importance at the expense of other elements which become reduced in area and significance. This process may or may not entail reduction of symmetry (Fig. 3). The reduction in symmetry and/or change of unit mesh may take place when *selective expansion–reduction* occurs in which only certain parts or some elements of the pattern are expanded or reduced whereas others remain unchanged. Symmetry and Wyckoff positions do not change in these series before end members are reached. In spite of this, more complex series can pass in their development through singular points in which some adjacency conditions change (e.g. Fig. 4.7.7, pattern series LPP 42-15, 42-13 and 42-16 in Ref. [7]). A profound change in symmetry may take place when end members are reached so that the last vestiges of certain elements (motifs) become eliminated (Fig. 3). The choice of elements in a particular pattern of this category is neither subjective nor arbitrary but it follows from the comparison of adjacent patterns in the pattern progression. Often, *element-expansion series* (or, in the opposite progression, *element-reduction series*) can be defined, in which all elements remain unchanged except for one (or a group of more) which expands (or is reduced) in the consecutive members of the series (Fig. 4).

The *element-propagation* or *accretional series* in which certain elements of the pattern propagate in number from the simplest end-member onwards, towards the potentially infinitely extensive end-member. The individual elements do not change their shapes and sizes in the process, they just accrete progressively into larger and larger aggregates. Some patterns form series which can grow

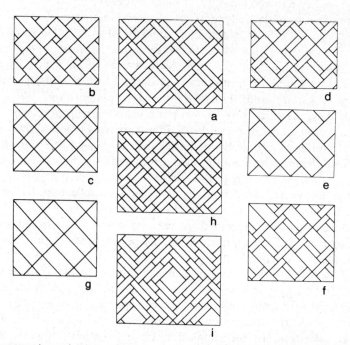

Fig. 3. Three expansion–reduction series (*b–c*, *d–e* and *f–g*) derived from the brick pattern (a) (Art Nouveau, the Central Railway Station, Copenhagen and depicted in Persian illustrated manuscripts, sixteenth century). Two members of an intercalation series based on the same pattern (h: Art Nouveau, Blegdamsvej, Copenhagen, i: from various incomplete friezes).

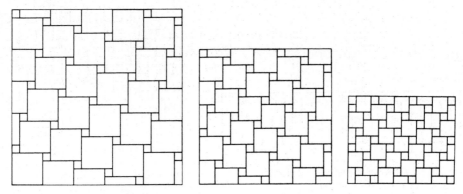

Fig. 4. Three members of element-expansion series based on two sets of squares. Plane group $\underline{p}4$, Wyckoff positions $\underline{a}$ and $\underline{b}$ (pattern type PP32^2).

without limits (Fig. 5). Other patterns (Fig. 6) can only form limited series with only several members because a qualitative change takes place in them at a certain Nth homologue, caused by the geometry and relative size of elements involved. Patterns of this series can either be understood on the basis of the element-propagation principle or by applying the expansion–reduction principle and taking the expanding block as a whole. The accretional (or element-propagation) series is the

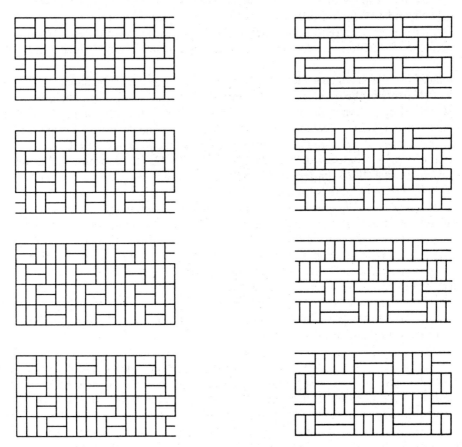

Fig. 5. Element-propagation (accretion) series with un-limited increments in one direction. Plane group $\underline{p}2$, constant elements surround positions $\underline{a}$, incremental elements are in $\underline{e}$ and for odd members also in $\underline{b}$. Pattern type PP7 and $7^m + 8^n$; $\underline{m}$ = integer part of $N/2$; $\underline{n} = 2$ ($N/2$-integer part of $N/2$).

Fig. 6. A limited (four-term) accretion series connected with adjustments in adjacency conditions. Plane group $\underline{cmm}$, un-changing elements are in positions $\underline{d}$, incremental elements in $\underline{e}$ and for odd members also in $\underline{b}$ (pattern types PP19 and $19^m + 20^n$; $\underline{m}$ and $\underline{n}$ as in Fig. 5).

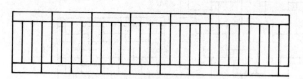

Fig. 7. Variable fit of horizontal and vertical elements (1:4 and 1:3, respectively).

Fig. 8. Ornamental brick pattern from the vault of cara-vanserai Ribat-i-Mahi [Khurasan, Iran; eleventh/twelfth century (?1154)]. The small-scale square design (follow the vertical bricks), ideally _p4mm_, is worked out in bricks as _cmm_.

principal type of series which is encompassed by the term _homologous series_ in chemical crystallography.

Symmetry groups do not vary throughout such series: all the elements involved in incremental accretion assume the same Wyckoff positions (pattern type) except for the central ones in the odd members of the series: these are in sites with higher local symmetry and lower multiplicity (Figs 5 and 6). This also implies that all non-central incremental elements have this, for their sites excessive, own symmetry. Accretional series can be complicated by making two adjacent, regularly alternating intervals with accretion into unequal homologues (i.e. with different numbers, N_1 and N_2, of incremental elements). An appropriate change in unit mesh and symmetry will follow as a consequence.

The _variable-fit series_ in which _m_ elements of the first kind progressively match with an increasing number _n_ of elements of the other kind. Alternatively, only one element may be present and edges of the first kind of _m_ such elements will match with the edges of the other kind of _n_ such elements. In these series (progressions), the sizes and shapes (ratios of edge lengths) of elements necessarily change by smaller or larger increments. Simple progressions of matches (1:1, 1:2, 1:3, etc.) can occur but also complex matches such as 25:30, 25:31, 25:32, etc., are possible as well (Fig. 7). In some instances (e.g. block patterns) this principle is combined with the accretional principle whereas others represent pure variable-fit cases. This case represents the other type of series encompassed by the "homologous series" in crystallography. Symmetry elements of pure variable-fit patterns represent intersection of plane groups of the two components.

The _intercalation series_ in which additional elements are introduced between the original elements of the pattern. In this way, the pattern is expanded without expanding or reducing the original elements and the number of element types in it increases. The elements intercalated in the consecutive members of these series must differ in shape and/or size. The boundaries to the element-propagation series or to the element-expansion series may be subtle if the elements intercalated are not sufficiently different from the existing ones. Examples of intercalation series are represented by the pair in Figs 8 and 9.

The _complication_ or _simplification_ (or _catamorphic_) _series_ represent the cases in which certain basic motifs are being progressively more and more (or less and less) complicated (i.e. undergo catamorphic changes [3, 11] by changing their shapes while preserving their position in the pattern and sometimes also preserving their role and/or fundamental outlines. _Multiplicative catamorphy_ is the case when the shapes change in a quantitatively expressible way, e.g. in the number of their vertices [11] or morphological subunits (central elements in Figs 8 and 9). _Metamorphic catamorphy_ or pattern metamorphosis is the case when the shapes do not change in an ennumerable or otherwise quantitative way. It represents "baroquization" (or debaroquization) of the pattern by complication or simplification of lines, attached ornamental elements, etc. The Zapotec pattern in Fig. 10 represents baroquization product of simple zig-zag stripe pattern with symmetry _pgg_. Boundaries towards the intercalation series may be subtle in some cases, when one has to decide

Fig. 9. Flat brickwork pattern (*p4mm*) from the minaret of the Malik Mosque, Kirman, Iran, late eleventh century. Horizontally oriented bricks between pattern elements (compare Fig. 8) have been omitted. The same pattern, accentuated in turquoise tiles, is on the minaret of the Jami Mosque, Niger, Iran (beginning of the thirteenth century). A number of complication–simplification, intercalation, omission as well as multiplication derivatives exist in arid Islamic lands.

Fig. 10. Hook-and-step key ornament in relief, constructed from shaped stone elements. Zapotec religious centre at Mitla, Central Mexico (about 1200 A.D.). The raised and recessed elements are symmetry-equivalent, giving a dichroic plane group *pgg′2′*.

what are newly introduced elements and what should be considered as parts of the progressively altered old elements. Intercalation itself also represents a deep transformation of the pattern; in fact it is a special case of (*element-*) *additive catamorphy* (Fig. 11) on the level of entire pattern. Symmetry and unit mesh (as well as pattern types) may remain preserved or may be variously altered in catamorphic processes, leading both to isomorphic, subgroup- or even supergroup symmetries.

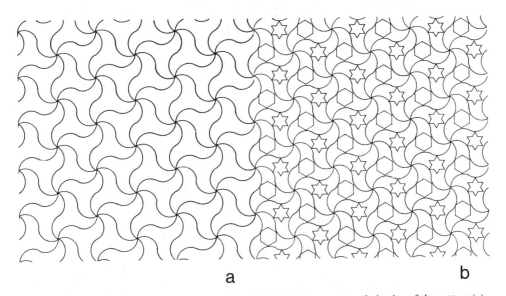

a   b

Fig. 11. Tile mosaic (b) from Alhambra, Granada (Spain) (*p3*) represents a derivative of the pattern (a) from Alcazar, Sevilla (Spain) (*p6*). Example of element-additive catamorphy.

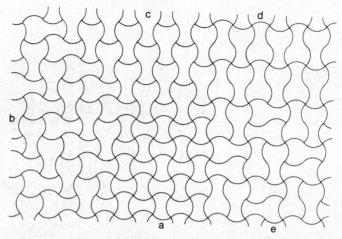

Fig. 12. Fundamental pattern (a) of arcuated tiles, *p4gm*, and its omission ("fish-tile") derivatives: (b) the "herringbone" pattern *pg*, (c) the parallel arrangement *cm*, (d) the antiparallel arrangement *pmg* and (e) the square arrangement *p4*. Pattern (a) is an omission derivative of a pattern of partly overlapping circles, *p4mm*. Modern Danish garden tiles.

A special case of catamorphic series is represented by the *omission series* in which certain parts (in the most obvious cases line segments) of the pattern are progressively omitted from it, leading to new patterns (i.e. omission derivatives) which cannot always be characterized as simpler or more complicated than the starting pattern. Changes in symmetry may or may not take place during the derivation by element omission. Combination of omission with other derivation mechanisms is very common. Examples are in Figs 12 and 13.

Fig. 13. Pattern of glazed bricks from the gates of the Khudoyar-Khans palace at Kokanda, central Asia, 1871. Plane group *p4*, PP32 (+41) or, for recessed portions, PP31, 41. An omission derivative of a square grid.

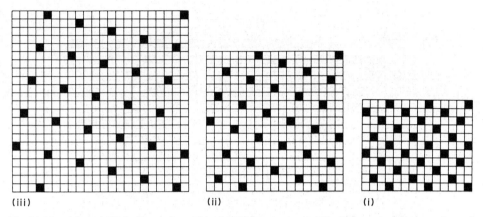

Fig. 14. The element-substitution series with a changing percentage of black elements. Three members of the series are shown; superstructures with plane group symmetry $p4$ were selected.

An *element-substitution series* results when the pattern represents a regular array of identical elements, a steadily increasing percentage of which are replaced by another element. This process can also be visualized as colouring of selected elements. The substituted (i.e. coloured) elements are distributed regularly according to chosen selection rules and define *supermesh* (supercells) of the basic mesh. If the metric of supercells corresponds to that of basic mesh, tesselation series or progressions are formed the mesh of which can progressively rotate against the axes of the basic mesh as its size increases (i.e. the number of substituted elements decreases) (Fig. 14). Such cases are connected with symmetry reduction.

Pattern reduction or *partial overlap series* represent progressions of patterns in which adjacent portions of the pattern overlap to a variable extent on periodically spaced "reduction lines". The spacing of reduction lines and/or the length and/or even the direction (in respect to the orientation of reduction lines) of the vector of overlap (reduction) will vary throughout the series. Reduction brings elements into adjacency which otherwise were separated from each other by other elements. As a rule, some or all of the lines (elements, areas) of the pattern will be omitted in the zones of overlap, thus creating a new pattern, an omission derivative of the original overlapped patterns (Figs 15 and 16). This case in crystallography is known as crystallographic shear. Another type of partial overlap series can be described as preserving the same mesh and the same shape of all elements but the latter change progressively their size(s) and, as a corollary, the degree of their overlap increases or decreases throughout the series (Fig. 17 [10]).

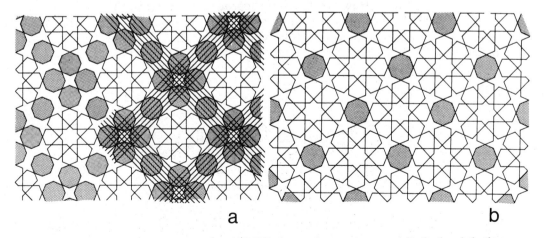

Fig. 15. Two Arabic mosaics from Bourgoin (1973). Pattern *b* represents a partial-reduction derivative of pattern *a*. Zones of reduction are indicated in *a*.

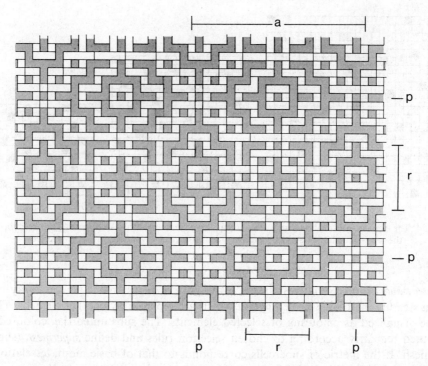

Fig. 16. Relief brickwork from caravanserai Ribat-i-Sharaf, Iran, which was constructed about 1120 and restored about 1154. A reduction derivative of a large-scale square pattern akin to that from Damghan and Simnan, Iran. Plane group _p4mm_ with unit mesh oriented at 45° to that of the parent pattern. _r_ denotes pattern stripes which were eliminated by reduction on planes _p_.

Affine series are obtained by progressive affine deformations of an initial pattern. Examples of affinely deformed patterns derived from the patterns of much higher (tetragonal) symmetry are in Figs 18 and 56. Related to these are pattern series in which only one or more elements undergo affine deformations whereas other elements remain unchanged, except for their rotation and shifts connected with the affine changes of the former elements. These are *"freely-hinged patterns"* with the same linkage pattern (topology) of sturdy elements and their "hinge points" are preserved throughout the series. More complicated types of freely-hinged patterns allow more complex continuous transformations of selected elements, which can be classified under pattern metamorphosis (e.g. from hexagons via "ditrigons" into triangles).

All these series can be defined in the strictest geometric sense, with only the relevant elements present throughout each series. However, it is more appropriate to describe these series in practice by allowing for small additional ornaments, diverse practical solutions (e.g. openwork vs flat, or incised brickwork) or for minor faults and deformations whenever it is obvious that the group of studied patterns was derived from the same underlying principles. This approach will be followed in the ensuing pattern description.

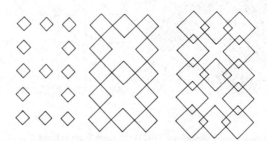

Fig. 17. Simple partial overlap series based on squares (_p4mm_).

Fig. 18. Relief brickwork from the minaret of the Jami' Mosque, Gurgan, Iran (twelfth century). Conscious affine transformation of a tetragonal pattern with the plane group p4; present symmetry is p2.

4. PLAIN BRICKWORK

Stacking Notation vs Pattern Type

Plain brickwork represents the basis from which the ornamental brick-laying developed. At the same time the plain brickwork—alongside with weaving and basketry—represents one of the basic subconscious imprints of symmetry which become fixed in everybody's mind during early childhood [3].

In the plain brick patterns (called *bonds* among professionals), bricks are laid in horizontal sequences (*courses*). A typical *brick* represents a rectangular prism with the edge ratio 1:2:4. Other ratios have been used as well, e.g. 1:2:3 in the Etruscan and Islamic buildings [12]. The flat-lying brick with the longest edge horizontal in the wall face is called a *stretcher*; that with the longest dimension perpendicular to the wall face is a *header* [13]. In all functional brickwork definite geometric relationships exist between the consecutive bricks in the brickwork: headers can either cap a stretcher or they can cap a joint between two stretchers. The 1:2:4 stretchers in two consecutive courses must be offset by 1/2 or 1/4 (or 3/4) of their length. In the 1:2:3 brickwork, they can be offset by 1/3 or 2/3, the header being exchanged by a closer.

Because of these relationships, a short-hand notation can be devised for brick stacking, somewhat analogous to that used in crystallography for the description of close packing of atoms in crystals (i.e. for the stacking of close-packed planes of atoms). Similar to crystallographic situations, it is complementary to the description by means of symmetry groups and relevant Wyckoff positions or by means of pattern types (Table 1). If stretchers are denoted by capital letters, headers can be denoted by lower-case letters and other fractional widths of bricks by using subscripts. All bricks centered in one vertical column will be denoted by the same letter of alphabet, e.g. $\underline{A}$ and $\underline{a}$. Their translation analogues then can be primed, e.g. $\underline{A}$, $\underline{A}'$, $\underline{A}''$, etc.

The stretchers displaced by 1/2 of their length from the $\underline{A}$ position (i.e. centred on the joint between $\underline{A}$ and $\underline{A}'$) will be denoted by $\underline{B}$. Those displaced by 1/4 and 3/4 of the stretcher length are represented by $\underline{C}$ and $\underline{D}$ depending on the displacement direction considered as positive (Figs 19–22). Stretchers displaced by 1/3 and 2/3 of stretcher length can be ascribed letters E and F, etc. Headers will have the same notation (in lower-case letters) as the stretchers (or in some cases imaginary stretchers) under which they are symmetrically placed.

This notation must be applied to bricks in one course as well. Starting with the stretcher $\underline{A}$, the stretcher after one intervening header is shifted by 1/2 of the full stretcher width against the hypothetical position $\underline{A}'$. Thus, it represents the position $\underline{B}$ (primed if taken exactly). The intervening header itself lies symmetrically above the stretcher which must be denoted as $\underline{D}$ from what was said above. Thus, the entire sequence is $A\&d\&B\&c\&A'$, etc. [Fig. 20(c)]. Some more complicated cases were observed in which the simple notation (without primes) does not distinguish between stacking variants composed of the same course type. Exact usage of primed notation should suffice to remedy this situation. However, it can be replaced by indicating the symmetry and unit mesh of the pattern.

The simplest *bonds*, the English bond and the Dutch bond, have stretchers and headers on sites of high symmetry (intersections of mirror planes) (Table 1, Fig. 19). In modern times, when the

Table 1. Plain brickwork

Name	Plane group of symmetry	Occupied Wyckoff positions	Pattern type†	Stacking sequence
English bond	*pmm*	*a, c, d*	PP16³	*A–a&b–A–a&b*
Dutch bond	*cmm*	*a, d*	PP20, 19	*A–a&b–B–a&b–A*
Plumb bond	*cmm*	*a*	PP20	*A–B–A–B*
Oblique veneer	*p2*	*a*	PP8	*A–C–B–D–A*
Stretcher veneer	*pmg*	*c*	PP13	*A–C–A–C*
Stretcher veneer	*pmg*	*a, c*	PP12, 13	*A–C–B–A*
Monk bond	*pmg*	*c³*	PP13³	*2A&d&2B&c– 2C&a&2D&b*
Monk bond	*p2*	*a, e*	PP7, 8	*2A&d&2B&c– 2C&a&2D&b*
English garden style	*cmm*	*a, d⁶*	PP19⁶, 20	*a&b–A–B–A–B–A–a&b–B*

†Grünbaum and Shephard's [7] pattern types. Superscript indicates number of distinct cases this position is occupied.

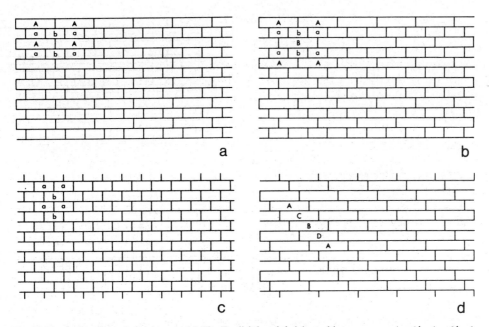

Fig. 19. Examples of simple brickwork. (a) The English bond, brick stacking sequence $A-a\&b-A-a\&b-A$, plane group *pmm*. (b) The Dutch bond, brick stacking sequence $A-a\&b-B-a\&b-A$, plane group *cmm*. (c) The Header bond, stacking sequence $a-b-a-b$, plane group *cmm*. (d) An oblique brick stacking sequence of stretchers $A-C-B-D-A'$, plane group *p2*.

brickwork became a veneer of constructions made of other materials, brickwork composed of only stretchers assumed importance. The Plumb bond (Table 1) with the stacking sequence $A-B-A-B$ occurs also in a variant composed of only headers, the Header bond [Fig. 19(c)] $a-b-a-b$. The other cases have stretchers displaced by 1/4 of their width, either in a monoclinic pattern (*p2*) or in orthorhombic sequences $A-C-A-C$ or $A-C-B-C-A$ (both *pgm*) [Figs 19(d) and 20(a)–(b)].

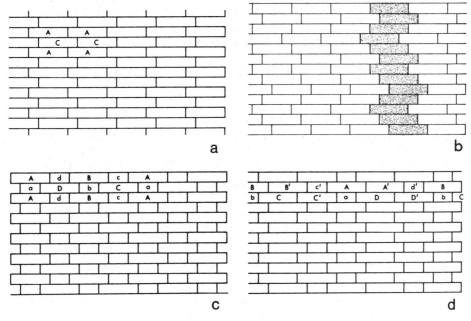

Fig. 20. Further examples of simple brickwork. (a) An orthorhombic stacking sequence of stretchers, $A-C-A$, plane group *pgm*. (b) The Stretcher bond with the sequence $A-C-B-C-A$ (*pgm*) and a stacking error. (c) The Flemish bond, brick stack sequence $A\&d\&B\&c-a\&D\&b\&C$, plane group *cmm*. (d) The Monk bond, the brick stacking sequence $2A\&d\&2B\&c-2C\&a\&2D\&b$, plane group *pgm*.

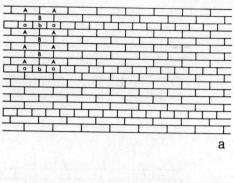

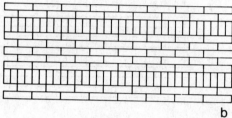

Fig. 21. An oblique version of the Monk bond, *2A&d&2B&c–2C&a&2D&b*, plane group *p*2. Headers are accentuated by shading. Copenhagen, Nørre Allé, modern.

Fig. 22. (a) English garden style bond, brick stacking sequence *a&b–A–B–A–B–A–a&b*, plane group *cmm*. (b) Headers–on–edge bond, plane group *cmm*.

More complicated bonds, e.g. the Flemish bond [Table 1, Fig. 20(c)], contain both stretchers and headers in the same course. The more complicated sequences of two stretchers and one header, the Monk bond, with the orthorhombic sequence *pgm* (Fig. 20) and a monoclinic alternative (*p*2 (Fig. 21) can both be coded as *2A&d&2B&c–2C&a&2D&b*.

In new world cities, such as Toronto and Montreal, the more complex "English garden style" with several (usually five) courses of stretchers intercalated by a simple row of headers (Table 1, Fig. 22), sometimes made more ornamental by arranging the headers vertically, is frequent. This does not complete the gamut of possible combinations, e.g. single courses *A&d&B&c* can intervene between several courses of pure stretchers, etc.

5. ORNAMENTAL BRICKWORK TECHNIQUES

Ornamental brickwork is mostly used as veneer for buildings (façades), filling of spaces and as infilling of interspaces between the beams of half-timbered houses (so-called *nogging*). In most cases the bearing requirements are relaxed and the match not allowed in functional brickwork becomes possible. Bricks can be layed perpendicular to each other, horizontally, vertically or diagonally in the wall face. Joints can be accentuated in white or they may be left empty, i.e. dark in the rays of desert sun. Thick mortar can assume an active role in the creation of the pattern. White plaster plugs between bricks (common in Islamic architecture), marble, limestone or merely white-washed elements can be used in a number of ways. In the Islamic desert countries, blue-glazed bricks were used to contrast with the drab brown-yellow bulk of the walls. In newer times red and yellow (buff) bricks are used to create similar colour effects.

A progression of more and more sophisticated and ornate techniques of pattern construction can be defined:

(1) Flat ornamental brickwork based on unmodified bricks: individual bricks can be prominent or they may coalesce into larger elements.
(2) Flat ornamental brickwork with (a part of) the bricks modified (clipped, rubbed, "gauged", ground or prepared in special molds).
(3) Flat ornamental brickwork accentuated by partly filled joints (Fig. 23).
(4) Relief brickwork and the rare brick openwork (Fig. 24).
(5) Two-coloured brick patterns; stone slabs may be used to create the colour effect (Fig. 25).

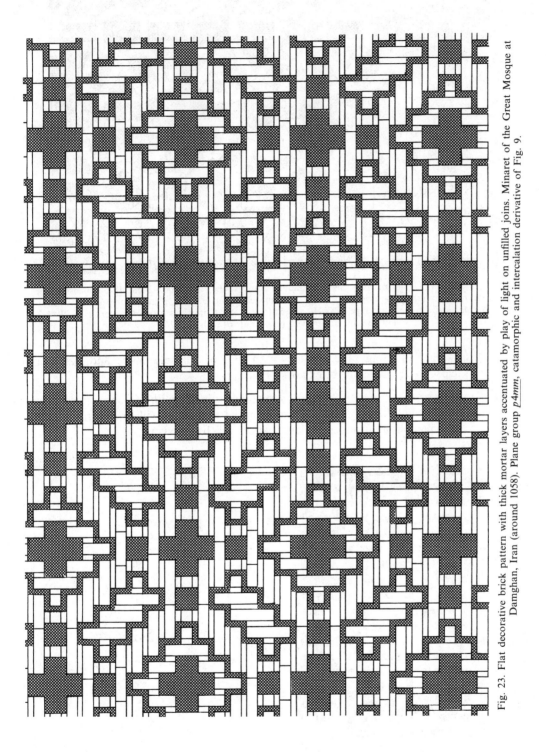

Fig. 23. Flat decorative brick pattern with thick mortar layers accentuated by play of light on unfilled joins. Minaret of the Great Mosque at Damghan, Iran (around 1058). Plane group _p4mm_, catamorphic and intercalation derivative of Fig. 9.

Fig. 24. Relief brick pattern from the Kharraquan tomb tower, Iran 1067–1068. Two-dimensional extension of a rare hexagonal brickwork frieze. Archimedean tiling (3.4.6.4), plane group *p6m*; Wyckoff positions *c*, *b* and *a*, i.e. pattern type PP49, 50 and 51 for polygons, positions *e* and *d* (pattern type PP47 and 48B for linear elements. Compare with Fig. 26.

(6) Cases transitional to tilings and mosaics with tiles of variable shape and size: periodic stonework close in character to brickwork (Fig. 26).

Classification categories outlined in Section 3 sometimes coincide with and sometimes transcend these pattern types. In the following, several prominent pattern series will be illustrated and their symmetry, pattern-type combinations and classification will be discussed.

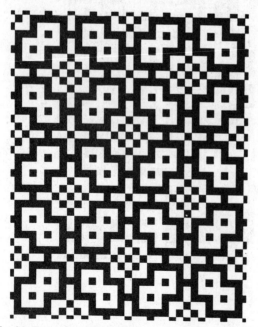

Fig. 25. Two-coloured brick-and-marble pattern from the Great Mosque of Cordoba (gates built by Al Mansur, 987 A.D.). Plane group *p4mm*; catamorphic relationships to patterns in Fig. 23.

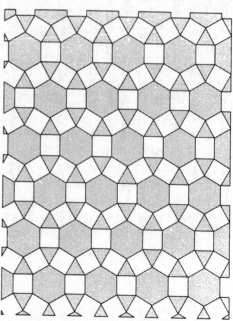

Fig. 26. Relief in raised blue faience square tiles. 'Ala ad-Din tomb tower in Varamin, Iran, 1289 A.D. (Mongol period). Symmetrological facts as in Fig. 24.

Fig. 27. The herringbone bond, plane group _pgg_. The first member of an element-propagation series (see Fig. 28).

Fig. 28. The second member of the herringbone-pattern series (plane group _pgg_). Replica of a historical house from sixteenth century, Hamburg. Peterstrasse, Carl-Toepfer Foundation; Art Nouveau, Main Railway Station, Copenhagen.

6. SELECTED PATTERN FAMILIES

The herringbone pattern

The herringbone pattern, _opus spicatum_, belongs to very old and widespread brick or stone patterns. It is found on Roman buildings and brick floors, in the old Zimbabwe ruins, in the walls of Romanesque buildings throughout Europe, Islandic turf walls or in the mud-cake walls of Slovak or Hungarian utilitarian village architecture.

In its pure form, it displays symmetry _pgg_ with all bricks in the same special position _a_ on two-fold axes (Fig. 27). Thus, the higher symmetry of bricks, 2_mm_, is excessive in this pattern and those modern pavements with sinusoidal bricks (brick symmetry 2) have the same plane group _pgg_ as the classical brickwork (all represent pattern type PP10).

Element-propagation homologous series, with the same symmetry _pgg_, will be generated when individual bricks are replaced by stacks of two, three or more elements (bricks) (Fig. 28). The critical combination of bricks is achieved when the thickness of these stacks is equal to the brick width. This situation results in a block pattern with symmetry _p4gm_. If stacking continues, increase in the number of stacked elements reverts the pattern to the symmetry _pgg_.

The stacked elements represent in general cases Grünbaum and Shephard's PP9 (eventually repeated several times) which for odd numbers of the series combines with PP10.

The herringbone pattern can be altered into a mirror-twinned one (_pmg_, PP12) when the interlocking ends of bricks are replaced by square elements (Fig. 29). On the other hand, its symmetry can be reduced either by making the two originally glide-related sides unequal (_p_2, PP8^2) or by reducing the symmetry of a brick to _m_ as in the case of the fish-tile pattern [Fig. 12(b), _pg_, PP6]. A very interesting herringbone pattern with two brick sizes (Fig. 24, right-hand side) from Jork, Germany, has the space group unchanged, _pgg_, but one axis doubled. Consequently, two sets of special positions (_a_ and _b_) on two-fold axes are occupied (PP10^2). Colouring of herringbone pattern will be treated separately.

The herringbone pattern, which itself represents "unit-mesh twinning" of a plain brick pattern, can be easily twinned (Fig. 55) and in such a form it represents the fundamental element for creation of large-scale ornaments and inscriptions composed of multicoloured glazed bricks which are found on Iranian and Central Asian religious buildings and mausolea.

The "interweaving pattern"

The interweaving pattern with symmetry _p4gm_ is extremely widespread in ornamental art, from ancient Egypt and Crete, through Greece, Islamic countries to half-timbered architecture of Northern Europe and North-Europe Art Nouveau.

Basically, all its forms are composed of straight to curved segments, perpendicular to each other, and "vortices" based on four-fold axes. Two extreme versions exist: (1) The simple "interweaving"

Fig. 29. Triangular fields contain herringbone patterns with plane group symmetries _pmg_ and _pgg_, respectively. Central Jork, south of Hamburg, Germany.

pattern with the straight portions (PP36) accentuated whereas the special positions _a_ (PP34) on four-fold axes are reduced to simple squares (PP41). In this case, the pattern resembles plain weaving and plaiting patterns (Fig. 30). (2) Patterns with the vortex motifs worked out in detail as swastikas (PP32) or crosses (PP41). The straight elements and the elements on four-fold axes may be worked out as mutually isolated motifs or they may merely represent distinct portions of the same sinuous lines. Straight portions can be split into elements lying on and those parallel to _m_ and even those lying in special positions _b_ (on intersections 2_mm_) can be separated from the rest of the pattern (Figs 31 and 32).

Metamorphoses between the extremes of this pattern type can be considered continuous. They involve a number of possibilities: accretional homologous series for the "linear" elements, metamorphic or multiplicative catamorphy for those elements that are in special portions on four-fold axes, expansion–reduction series as well as intercalation series.

The simple interweaving pattern (Fig. 30) has bricks on _m_ planes (i.e. 2_mm_ intersections; PP36) and simple squares on four-fold positions (PP41 in lieu of PP34). An accretional homologous series with the single bricks multiplied into a stack of two, three, etc. parallel bricks exists (Fig. 33). This series at the same time represents an element-reduction series for the square elements in respect to the expanding rectangular ones. The limiting case is a block pattern with the square elements eliminated.

The version with visually accentuated four-fold rotational symmetry dates back to painted ornament in old Egypt (Fig. 48 in Ref. [3]). The straight-line intervals preferably do not coalesce

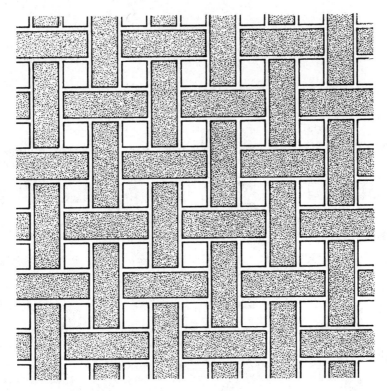

Fig. 30. The simplest member of the family of "interweaved" brick patterns from Jork, near Hamburg, West Germany; red and white bricks with whitewashed joins. Plane group *p4gm*. (Also on historical to pseudohistorical and Art Nouveau buildings, e.g. City Hall, Copenhagen.)

(PP35). They lie parallel to one system of reflection planes, straddling the other *m* system (Figs 31–33). Additional ornamental elements may be introduced (centred on *2mm* intersections) and the existing ones baroquized leading to very complicated results (Figs 32 and 34). Selective reduction of the original elements yields patterns as in Fig. 35. A rare dichroic form of such pattern is in Ref. [3], Fig. 13. These two patterns are related to each other via only minor selective expansion/reduction of some portions although they came from the opposite poles of the Islamic world and are separated by several centuries.

Fig. 31. Marble-and-brick ornament from the portals of El Hakam (961 A.D.), The Great Mosque of Cordoba, Spain. Plane group *p4gm*. The foliation pattern on the triangular limestone blocks was omitted. Catamorphic relationships to the ornaments in Fig. 30 and following figures.

Fig. 32. A relief brick pattern constructed from simple and gauged bricks from the gate of the caravanserai Ribat-i-Sharaf, Iran (Khurasan) from mid-twelfth century A.D. Only the V-shaped recessed elements are in general positions, the other sets are in various special positions. Plane group *p4gm*.

Fig. 33. The second member of the homologous series of "interweaved patterns" from the half-timbered Portau House (and elsewhere) at Jork, near Hamburg (seventeenth century). For given brick dimensions third and fourth (final) members exist. Plane group *p4gm*.

Another dichroic example of this pattern (*p4'gm'*) with dichroic swastikas was designed by Woods (1935–1936) [14, Fig. 1]. The bar-and-cross pattern in Fig. 36 is intermediate between the two extremes of this catamorphic series, with both elements playing a balanced visual role.

Given the popularity of "interweaving patterns", the universal strict adherence to the original plane group *p4gm* is remarkable. What cases of symmetry reduction have we observed? They are just a few: a relief brickwork pattern from eleventh century Iran (Fig. 37), very similar in details to the pattern from Cordoba (Fig. 31) has symmetry *pgg* (not observed by Ref. [10] for Islamic patterns!). It was produced by conscious omission of horizontal arms of all swastikas in the pattern. It is an omission derivative (PP34 altered into PP18). For technical reasons patterns like the one in Fig. 36 become easily affinely deformed into unintentional *pgg* versions. A conscious version of this type is shown in Fig. 56 and in carved or painted forms it profusely occurs in Chinese, Japanese and Celtic patterns. Finally, selective element expansion and reduction led to the pattern in Fig. 38.

The rotated-block pattern

This is one of the simplest and most universal nogging patterns. A combined, variable-fit and accretional series exists with the homologues differing in the brick thickness to length ratio; homologues with 2, 3, 4 or more bricks per one square block exist. Symmetry is *p4gm*. In the odd members, the central brick lies on *2mm* (PP36) while all the other bricks straddle one system of *m* planes only (PP35). There are no elements positioned on four-fold axes (Fig. 39).

The block pattern also represents the common end-member of the multiple-stack herringbone series (Fig. 28) and of the interweaving series (Fig. 33) as well as of selective expansion–reduction series [Fig. 3(d,e)]. Separate white elements may be introduced on *2mm* intersections.

Symmetry reduction of block patterns to *p4* will occur for sinusoidal or fish-bone bricks [Fig. 12(e)]. Symmetry-reduced versions with normal bricks are rare. In Fig. 40, bricks of length insufficient to construct four-layer square blocks were supplemented by ornamental, glide-related square elements. Unfortunately, the masons failed to produce periodic nogging patterns so that the *p2* pattern reproduced here had to be idealized from their aperiodic creations. Patterns such

Fig. 34. Raised decorative brick pattern from the minaret of the Kalan Mosque in Bukhara (1127, the Seljuk era). Plane group *p4gm*. One of the common patterns in Islamic art (e.g. the Büjük Madrasa, Konya, Turkey, 1251, as a stone relief) or as a more complicated derivative, e.g. on Tamerlan's Palace Ak-sarai in Shahrisabz, Central Asia, 1380–1405.

as the last example in Fig. 6, can be considered as rectangular block designs (symmetry *cmm*) similar to the present square category.

Large-scale square patterns

Patterns of this family were widely used in various catamorphic varieties in many parts of Islamic world. They exist in two basic forms: (a) as a flat brickwork, eventually adorned with some openings to enhance selected lines and points; (b) patterns of raised bricks, oriented in all directions on recessed background or two-coloured, marble and brick patterns.

Fig. 35. Marble-and-brickwork pattern from the eastern gates of the Great Mosque of Cordoba built under Al Mansur (987 A.D.). Plane group *p4gm*.

The flat-brick versions, apparently the original form of these patterns, represent in principle herringbone pattern twinned on two systems of planes perpendicular to each other (Figs 8 and 9). These patterns only rarely come out with square symmetry (Figs 9 and 23); most of them are orthorhombic, *cmm* (Fig. 8) or even *pmm*. In these, the two sets of twin (mirror) planes are applied with somewhat different frequencies. Increasing this difference, the obvious relationship of such patterns to perfect square patterns can disappear (Fig. 55).

The pattern in Fig. 9 allows definition of fundamental elements for this category in its ideal, *p4mm* form. Wyckoff positions *a* (0, 0) are occupied by brick crosses engulfed by serrated

Fig. 36. Light-and-shadow accentuated brick pattern from the façade of the ruined Il Khamid Mosque at Haftshuya, early fourteenth century, plane group *p4gm*. Catamorphic relationships to the patterns in Fig. 30 and the following.

Fig. 37. Relief brickwork pattern from an eleventh century Seljuk tomb at Quarraquan, Iran. Plane group *pgg* as an omission derivative (in the swastikas) of the pattern with symmetry *p4gm* (Fig. 31).

Fig. 38. Selective expansion–reduction variant of the pattern from Fig. 30; plane group symmetry reduced to *pmg*. Portau House, Jork, near Hamburg, West Germany.

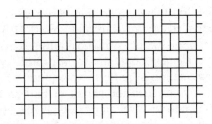

Fig. 39. Ornamental nogging, infilling and pavement block pattern with plane group *p4gm*. End-member of the expansion–reduction series of interweaved and stacked herringbone patterns.

brickwork squares, positions (b) (1/2, 1/2) by crosses, whereas positions $\underline{c}$ (1/2, 0) by extended "knots". The background stripes, paired around planes (10) placed at $x = 1/2$ of square mesh are in Fig. 23 decomposed into segments on $\underline{d}$ $(x, 0)$ and decorative openings on $\underline{f}$ (x, x).

In the other category, the same elements can be discerned although the freedom in their application and relationships naturally is much greater. Closest to the above "original" configuration is the configuration which can be traced in the "original" square (its edge indicated by $\underline{a}$) in the reduction-derivative pattern in Fig. 16. In Fig. 25, besides altered shapes of some elements, the previously important boundary elements positioned on Wyckoff positions $\underline{d}$ are reduced to zig-zag lines. "Knots" on $\underline{c}$ are replaced by their *duals*, i.e. by strings of corner-sharing squares in Fig. 25 and partly also in Fig. 16.

This family represents a typical catamorphic tree of relationships, so varied that only some of them can be mentioned here. In spite of all catamorphic changes, the kinship of all these patterns remains obvious. If the departures from square symmetry *p4mm* in the outstanding example are

Fig. 40. Block pattern with symmetry $\underline{p}2$. Idealized nogging from a guild house (1577 A.D.) from Goslar, West Germany.

ignored, the lowermost member of the *complication series* consists of only cross-and-square elements positioned on $\underline{a}$ and divided by "background zones" (10) of square mesh (at $x = 1/2$), i.e. PP41^2 (and 40). Introduction of "knots" on positions $\underline{c}$ entails creation of crosses on $\underline{b}$ (Fig. 9) with additional optional elements on f and it also produces splitting of zones (10) into elements on $\underline{d}$ (Fig. 23). The combination of pattern types present become PP41^{2+1}, 40, 38$^{(2)} \pm$ 39, i.e. simple intercalation has greatly enriched the gamut of pattern types discernible.

The appearance of duals to the knots in Wyckoff position $\underline{c}$, reduction of stripes on $\underline{d}$ into zig-zag lines, coalescence of elements on f with those on $\underline{b}$ (Fig. 25), change and engulfment of the latter by new elements as well as varying size ratios of crosses and serrated squares on $\underline{a}$—all represent examples of *pattern metamorphosis*. *Multiplicative catamorphy* in these patterns derives from the multiplicity of brick stacks (or of imaginary stacks of squares) that form individual elements of these patterns. This multiplicity varies independently from the type or number of intercalated elements present in the pattern. For example, the squares on position $\underline{a}$ have their edge-serration composed of 2, 3, 4 or 5 elements or the "knots" on $\underline{c}$ have 3 to 5 subunits.

Finally, at least the raised and two-coloured patterns can be understood as *omission derivatives* of multiple square patterns $p4mm$ with nested squares of several sizes positioned in Wyckoff positions $\underline{a}$ and $\underline{b}$; the process of line omission is combined for two-coloured patterns with the use of antiphase boundaries (Fig. 25). The number of squares, ratios of their sizes as well as ratios of their edge lengths to the unit mesh size vary between patterns. All these variations take place on a background of a small-square grid which underlies all of these patterns.

7. COLOURED PATTERNS

The beauty and distinctiveness of ornamental brick patterns can be heightened by the introduction of two sets of differently coloured bricks, usually red and yellow or red and white. Perfect colouring [7] is rarely achieved—and rarely intended in brick patterns. The reasons are manifold: the patterns may be technically difficult, they may require more intellectual effort, bricks of more than two colours or also of more complex shapes. Often artists created entire families of splendid patterns from bricks of two colours without using, and without thinking in terms of, *antisymmetry*. Finally, in the simple brickwork (plane groups cmm, pmm, pmg and pgg) all bricks, headers and stretchers, lie only in special positions and preclude those colouring schemes which require the presence of elements in general positions.

In the Dutch bond [Fig. 19(b)] stretchers lie in the special positions $\underline{a}$ on intersections $2mm$ (PP20) whereas headers lie in $\underline{e}$ on m and glide planes (PP19). The only *translationengleiche* dichroic group possible is $p_{C}mm$ (Fig. 41). Should black and white stretchers alternate in every course, antisymmetry groups $p_{b'}2$ and $p_{b'}mg$ will result. All groups with antireflections are excluded.

Another frequent approach to the colouring of Dutch bond involves picking out every nth course of stretchers (or headers) by colouring it differently from the rest of the masonry. Colourings with two or more coloured courses will not be examined here.

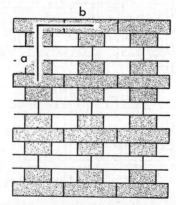

Fig. 41. Two-coloured Dutch bond with plane group of anti-symmetry $p_{C}mm$.

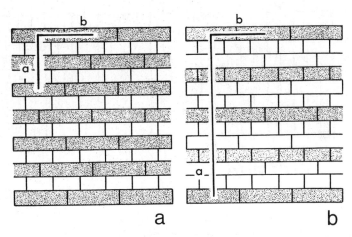

b   b

a   b

Fig. 42. Stripe-coloured Dutch bond from various buildings in Copenhagen and elsewhere. New unit mesh is indicated. Plane groups (a) *cmm*, (b) *cmm*.

Colouring every second course [Fig. 42(a)] will leave the symmetry unchanged, *cmm*. Alternation of two uncoloured and one coloured course leads to *cmm* with a large unit cell [Fig. 42(b)]. Leaving three uncoloured courses results in *pmm*, unless the adjacent headers are appropriately coloured as well (Fig. 41); four uncoloured courses produce a still larger unit cell with *cmm*, etc. To summarize, when the colouration has period *na'* equal to 1 (mod 2) *a'*, where *a'* is the periodicity of the underlying Dutch bond, single-stripe colouring yields centred cells with $a = 2na'$. If the colouration has a period 2 (mod 2), primitive cells result with $a = na'$. Finally, if the colouration period $n = 1.5$ (mod 1), centred cells with $a = 4na'$ are formed in which alternatively the stretcher and the header courses become coloured.

The resulting *uncoloured* symmetry represents the intersection of symmetry and translation elements of the stripe pattern with the symmetry group *smm* (line pattern PC4 in Ref. [7]) which has incrementally increasing periodicity *a*, with those of the fundamental brick pattern (*cmm*). These patterns are best treated as stripe patterns, forgetting the underlying brickwork. Undoubtedly this was the way they were understood by their builders. Cases with two courses thick coloured stripes lack internal mirror symmetry in each stripe (they consist of one header and one stretcher course). They potentially produce two different stripe groups, *s*1*m* and *smm* in our orientation.

Herringbone pattern has underlying symmetry *pgg* with all bricks in special positions on two-fold axes, precluding such colouring schemes as *pgg'*. Figure 43 represents *pg'g'* whereas Fig. 44(a) represents *pgg*, a maximal isomorphic subgroup of lowest index of the original plane group, with the tripled *a* axis. An attempt to colour the herringbone pattern using doubled *a* axis yields $p_{a'}g$

Fig. 43. Two-coloured herringbone pattern, plane group of antisymmetry *pg'g'*. Tile floor at Villach, Austria.

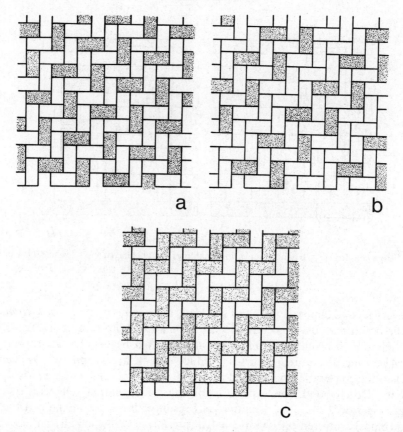

Fig. 44. Coloured herringbone tile patterns. Plane groups (a) _pgg_ (subway station in Toronto), (b) _pg_, (c) $\underline{p_{a'}g}$.

[Fig. 44(c)] from which an omission derivative _pgg_ with quadrupled _a_ axis as well as further patterns with symmetry _p2_ can be derived. Again, it is relatively easy to obtain uncoloured supercells and subgroups but only a limited number of dichroic groups.

Colouring of the two-brick block patterns (Fig. 39) with symmetry _p4gm_ leads to dichroic plane groups $\underline{p_C4}$ and _pgg′_ (Fig. 45). Colouring _p4′mg′_ reminds one of a simple two-coloured chess-board pattern.

Colour accentuation of North German patterns by white square elements in special positions (Figs 30 and 46) usually does not change the original plane group of symmetry. However, in Fig. 47, the colouring of clustered squares (in general positions) leads to the dichroic plane group _p4′m′m_. The other elements of this pattern, positioned on _m′_ and 4′ represent "grey" special positions.

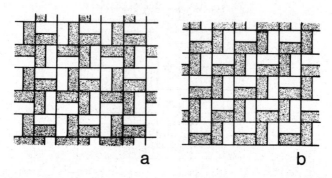

Fig. 45. Coloured block patterns. (a) $\underline{p_C4}$ from Tivoli Gardens, Copenhagen; (b) _pgg′_.

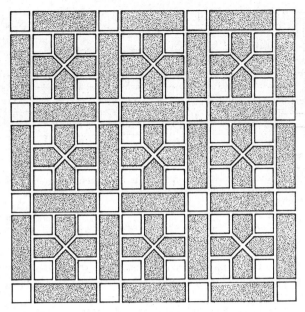

Fig. 46. Red-and-white decorative brick pattern from a nogging panel of the Portau House, Jork, near Hamburg, West Germany (seventeenth century). Plane group *p4mm*.

Colouring of brickwork often produces typical element-substitution series. If we outline the primitive pseudosquare (diamond) cell in Fig. 48, its edges can be enlarged two, three or more times by omitting the intervening dark crosses. Similar series can be based on English bond (Fig. 49). Expanding only the vertical axis by its integer multiples, *similar* centred patterns can be obtained, whereas other multiples give only primitive patterns.

Fig. 47. Ornamental brickwork in red and white with whitewashed joins. Portau House, Jork near Hamburg, seventeenth century. Dichroic plane group $p4'm'm$.

Fig. 48. Pseudotetragonal pattern (plane group *cmm*) obtained by colouring of the Flemish bond. Solvgade, Copenhagen (modern).

Fig. 49. Pseudohexagonal pattern (plane group *cmm*) obtained by colouring of the English bond. A magnificent double-brick form adorns the Kalta Minar minaret (1850) in Khiva, central Asia. It has been applied in several alternating dark-on-light and light-on-dark versions.

The nicest corpus of two-coloured patterns consists of the red and white brick-and-marble ornaments of the Great Mosque of Cordoba, Spain. This magnificent collection deserves a separate treatment.

8. ORDER–DISORDER (OD) PHENOMENA AND TWINNING

If local adjacency or neighbourhood conditions between one element (usually a band) of the pattern and the adjacent such element (band) are well defined but they fail to determine unambiguously the position of the third (and, as a consequence, also of the fourth, fifth, etc.) element (band) in respect to the first one, a disordered pattern can result instead of a strictly ordered one. This general definition and its implications can best be explained using the pair of brick patterns in Fig. 50.

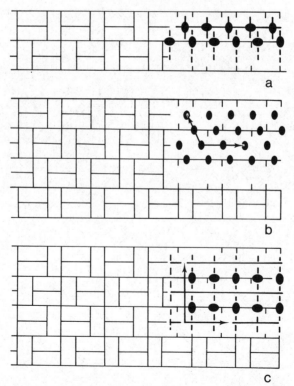

Fig. 50. OD in banded patterns. (a) Band group *pmm*2 of an individual band and the interband symmetry elements (local two-fold axes and glide planes). (b, c) Two maximally ordered patterns produced from such band pairs: (b) *p*2 and (c) *pmg* (Art Nouveau, Vognmagergade, Copenhagen).

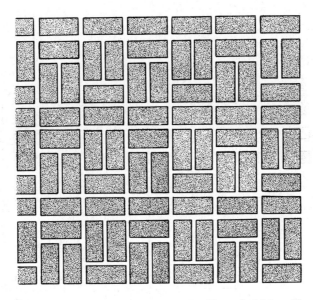

Fig. 51. Second member of the *pmg* series. Red bricks with whitewashed joints. Vicarage at Notmark, Als, Denmark (about 1688).

These patterns can be divided into bands two bricks high, each with band symmetry group *pmm*2. This high symmetry is not reflected in the band-to-band relationship: besides the translation with the horizontal component of only 1/3 of horizontal periodicity, two adjacent bands are related by two-fold rotation axes and vertical glide planes [Fig. 50(a)]. The presence of vertical reflection planes in each band allows two equivalent positions for the vertically standing headers to the right [Fig. 50(b)] or to the left of such a header immediately above them. It means that two fully eqivalent positions are allowed for each consecutive band in respect to its immediate neighbour.

This situation allows construction of completely disordered patterns, with no periodicity in band stacking. Such patterns would not be aesthetically appealing and masons always choose the periodical, ordered variants instead.

The two ordered versions with the shortest periodicities of band stacking are called *maximally ordered patterns* [Figs 50 (b) and (c)]. The case sub (b) represents the stacking version which can be described as the right–right–... shifts (or left–left–... shifts) whereas sub (c) contains a right–left–right–left sequence instead. Comparison of Figs 50(a)–(c) shows that the version (b), plane group *p*2, preserves only a certain selection of symmetry elements from the band group and from the interband relationship whereas sub (c), *pmg*, preserves another selection of these elements. In both cases the preserved elements change from band-group or local elements into generally valid ones. An infinite number of ordered stacking sequences with more complicated right–left shift sequences are possible besides the two maximally ordered ones.

In the case presented here, both maximally ordered patterns each give rise to a homologous series. For the second members of these series we have real examples: the *p*2 version adorns brick-paved sidewalks in the old city of Aachen [Fig. 5(b)], whereas the *pgm* version is found as ornamental nogging on the walls of the old vicarage at Notmark on the peninsula of Als, Southern Jutland, Denmark (Fig. 51). Both of them are found in the brick-paved paths of the University of Western Australia in Perth.

The pattern in Fig. 52 again has horizontal bands with symmetry *pmm*2. They are limited by two horizontal uninterrupted courses of bricks. The local four-fold axes of Greek crosses are not utilized in band or plane symmetry although they definitely strike the eye of the beholder.

Interband translations and symmetry elements are the same as in the previous example as are also the plane groups, *p*2 and *pmg*, of the two resulting maximally ordered patterns. The original frieze was only two band widths high, missing the opportunity to develop fully the potential OD relationships.

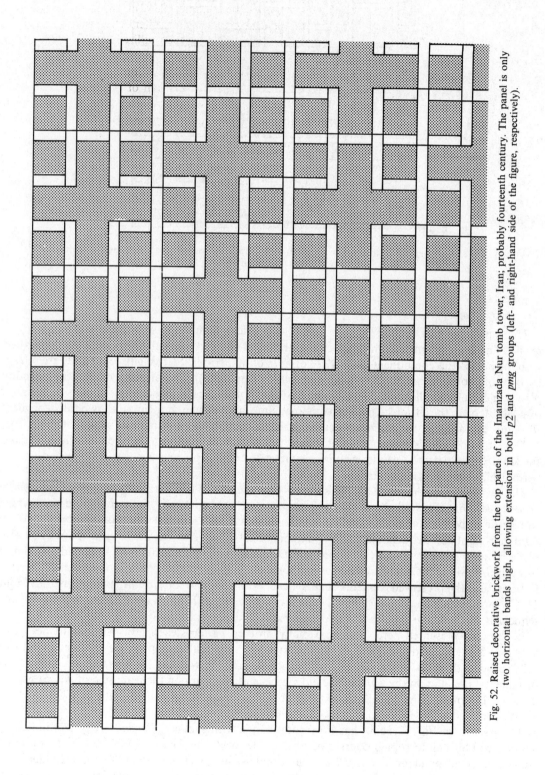

Fig. 52. Raised decorative brickwork from the top panel of the Imamzada Nur tomb tower, Iran; probably fourteenth century. The panel is only two horizontal bands high, allowing extension in both _p2_ and _pmg_ groups (left- and right-hand side of the figure, respectively).

In Fig. 53, the raised "embroidery" pattern consists of strips which run diagonally and display band symmetry _pm_11. The only relationships between two adjacent stripes are (1) the translation with the parallel component equal to 1/4 of band periodicity and (2) partial glide planes perpendicular to it. If we select the filled rectangle as the reference element, the "tail" in the next stripe can be attached below its left (Fig. 53) or right-hand lower corner. The resulting stacking of rectangles is right–right–... (Fig. 53) or right–left–right–... (not utilized by the artists) with respective plane groups _p_1 and _pg_.

OD phenomena for the pair of simple bonds, the English and Dutch bond [Figs 19(a) and (b)], result from the fact that the periodicity of the course of headers is half of that of the courses of stretchers above and below it. Thus, either a stretcher or a join of two stretchers can lie with equal probability over any of the headers. Two maximally ordered versions are simply the above two bonds (Table 1). All the symmetry elements of a single stretcher course, band group _pmm_2, become incorporated into these plain groups whereas different selections of symmetry elements of the header course, with the same band group, are employed in the two cases.

Contrary to this example, the Monk bond (Fig. 20) and its oblique version (Fig. 21) as well as the Stretcher bond pair based on the stretcher offset of 1/4 length [Figs 19(d) and 20(a)], are based on the same position-ambiguity principle as for the preceeding complex ornamental patterns. A more complex ordered pattern from a potential OD family comes from the medieval fortifications of Tangermünde, East Germany (Fig. 54). It envelopes the entire tower at the city gate without a stacking fault! However, the lower parts of the tower and the gate are enveloped by the oblique (i.e. spiralling) version of the same pattern, _p_2.

Because of the simple length:width:thickness ratios for bricks and the simple ratios of brick dimensions to the magnitudes of brick displacement, brick patterns are eminently suited for _twinning_, i.e. reflection or glide-reflection of the original pattern on newly inserted planes of symmetry. In this process of reflection (or glide-reflection) some pattern directions become altered and the reflected pattern meets the original one along certain composition (or contact) lines.

Another twinning mechanism is represented by rotation of the pattern around a single _n_-fold axis and joining of rotated segments on composition planes. The two types of twinning are often combined, yielding complex, multiply-twinned patterns with point-group symmetry, such as _mm_2 or _4mm_. As in crystal structures, the reflection plane often is in such cases substituted by a glide reflection plane; the point group of the twin then represents a large-scale approximation of the true, detailed situation.

If the twin elements, especially the reflection or glide-reflection planes, are applied periodically at short intervals, a new pattern with periodicity longer than before is created by the so-called _twinning on unit-mesh scale_. In Fig. 55, the first glide-plane twinning applied as frequently as possible, creates a herringbone pattern. This pattern is in turn twinned on regularly spaced vertical mirror planes. In the upper portion of the minaret this pattern is again twinned on periodical horizontal mirror planes, situated five horizontal brick courses apart, resulting in a pattern of vertically stacked lozenges.

Often a close relationship exists between twinning on unit-cell scale and the presence of stacking OD. For example, the zig-zag pattern in Fig. 54 and the spiral pattern from the same tower can be interpreted either as two maximally ordered stacking sequences or the first one (_pmg_) represents a mirror-twinned version of the latter one (_p_2) with twin planes six courses apart.

9. AFFINE AND OTHER TRANSFORMATIONS

The exact metrics of brick patterns will always depend on the dimensions of bricks and the thickness of the mortar-filled joins. Thus, a tetragonal pattern designed for certain ratios of brick dimensions will be _affinely_ deformed to orthorhombic if other type of bricks (or thickness of mortar layers) is used. For example, the tetragonal pattern, _p_4_gm_ in Fig. 36 was abstracted from such a slight deformation. Naturally, from such experiences arose the idea to consciously exploit such transformations (Figs 18 and 56).

Projective-perspective transformations can be achieved only by special prefabrication of bricks and notable examples appear to be absent. Therefore, it is tantalizing to find such red-and-white brick patterns, worked out in all variants, from complete lack of understanding of perspective,

Fig. 53. Lace-like relief brickwork from the Kalan minaret in Bukhara, Central Asia (1127–1129 A.D.). Plane group p1, a potential OD pattern.

Fig. 54. Zig-zag brickwork pattern accentuated by dark-glazed headers. Plane group *pmg*, stacking notation *A&d&B'* with the first element undergoing a sequence *A–C–B–D–A'–C'–B'*–reversal to *A*. A member of a potential OD family. Tangermünde (Germany), Neustädter Tor, about 1440.

through false perspective in which the periodicity in the receding direction is not reduced, to perfect perspective, in the lintel "key-stones" of the gates in the youngest portions of the Great Mosque of Cordoba. Were the latter well-understood examples made by the original Arabic artists or were they inserted only by the nineteenth century (or earlier) restorers?

10. CASE STUDY 1

The Great Mosque of Cordoba, Spain

The Great Mosque of Cordoba (La Mezquita Aljama de Córdoba) was built in four stages by the founder of the Iberian Omayyad dynasty and his successors. The first part, built by Abdu'r Rahman I in 785, was doubled by Abdu'r Rahman II in 848, and tripled by El Hakam II in 961, extending further and further in the SW direction. In 987 Al Mansur nearly doubled the area, adding several new aisles on the NE side of the previous mosque.

The red-and-white brick-and-marble ornaments which concern us here are primarily found in the magnificent portals of the gates constructed under El Hakam II in the SW corner of the mosque and under Al Mansur along the NE front of the new portions built during his rule. Several portals in the older parts of the mosque contain these ornaments as well. Especially the older portals have been rebuilt several times, the last time during the Late Gothic and Plateresque periods. The majority of portals were carefully restored by V. Bosco and M. Inurria during a number of years prior to 1914 [15, 16]. Although very interesting as examples of band groups, similarity and catamorphy, the omnipresent carved limestone ornaments are outside the scope of this paper.

The white-and-red, marble-and-brick ornament of the Mosque can be divided into three groups:

(1) panels with two-dimensional patterns above gates and windows (Fig. 57);
(2) straight framing bands (Fig. 2);
(3) patterns in the "key stones" of the flat lintels above the gates (Fig. 58).

Fig. 55. Flat ornamental brick patterns from the Firuzabad minaret, Iran, early twelfth century (slightly simplified). The herringbone pattern in the bottom part is twinned into a zig-zag pattern (plane group *pgm*) and is twinned again in the upper part into a diaper pattern *cmm*. Shading of horizontal bricks was added by us.

In category (1) a small number of eye-attracting elements heavily predominate: swastikas, "centred squares" and the coalesced pairs of such squares, the other elements being used less frequently. The combined swastika and centred-square ornaments are constructed on a square submesh which is filled by them in different sequences (Figs 59–61). Among these *substitution variants* both the chequerboard and the simple striped filling schemes are used. On the other hand, all these patterns represent members of one *element-propagation series*, in which the large square elements become separated by one, two or three black lines (stripes) interspaced with white stripe segments and a corresponding number of small white square elements. In the best ornaments, some

Fig. 56. Affine transformation of the ideal interweaved pattern from Fig. 36 (*p4gm* → *pgg*) together with expansion of recessed interspaces. Designed by J. Tomás.

parts of the black lines have been omitted in order to conform with the symmetry of inserted elements (Figs 60 and 61). They represent *omission derivatives* of respective homologues with complete square lattices (like the one in Fig. 59). No reliable cases with two opposing orientations of swastikas in the same pattern of this category have been found and the existing cases apparently represent errors (Fig. 59; original to Fig. 61).

In this family, the grid with only centred squares present displays the highest symmetry, *p4mm*. The symmetry of the swastika and centred-square patterns with the checkerboard arrangement is reduced to *p4* (Figs 60 and 61). Patterns with swastikas and squares arranged in rows have symmetry lowered to *p2* although they keep their square metrics. The coalesced centred squares (Fig. 57) reduce the pattern symmetry to rectangular, *cmm*.

992 E. MAKOVICKY

Fig. 57. Double square ornament based on square grid; symmetry *cmm*. Portal built under Al Mansur (987). The window openwork (*p4gm*) is related to patterns in Figs 31 and 35.

In agreement with their high-symmetry origins, the checkerboard square elements represent PP41; the small white squares and their simple interconnecting segments in the original grid without omissions represent PP41 and PP40 but with shorter translations between them. In orientationally well-defined cases, swastikas represent PP32. The entire patterns are easier to classify in terms of

Fig. 58. Perspective-projective transformation of frieze patterns in ornamental "key-stones". Gates of Al Mansur (987).

Fig. 59. Marble-and-brick patterns from the portal of Al Mansur. The left-hand field represents second member of element-propagation series of tetragonal square-and-swastika ornaments, without omissions on intercalated lines. Plane group *p*4, with errors.

white elements: PP41, 32 and 30 for Fig. 60 and PP41, 32, 30^2 and 38 for Fig. 61. Thus, a number of elements have their own symmetries higher than prescribed by their site symmetries in the plane group *p*4.

Several other patterns are based on the same visual elements as the previous family. The large-scale square pattern (Fig. 25), the lowest member of the multiplicative catamorphic series described in Section 6 has been used repeatedly on the wall panels of the Mosque (Fig. 59). Figure 62 represents a fine example of pattern-reduction process. It is based on a simple square grid filled by centred squares (*p*4*mm*). Planes of partial overlap and reduction represent planes (110) of square mesh; they are spaced $3\underline{d}$ (110) apart and the overlap vector in each plane is $\underline{d}$ (110). Minor omissions were used to enhance the resulting configurations, plane group *pmm*.

Fig. 60. Marble-and-brick pattern (plane group *p*4) from the portal of El Hakam II. An omission variant of the lowest member of the element-propagation series of tetragonal square-and-swastika ornaments from this mosque.

Fig. 61. Marble-and-brick ornament form the portal of El Hakam II. Plane group *p*4. An omission variant of the third member of the element-propagation series of square-and-swastika ornaments (Figs 59 and 60). Similarity to the Art Nouveau brick ornament in Fig. 3(a) suggests possible sources of inspiration. Original panel contains periodicity faults.

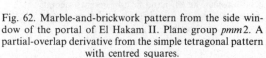

Fig. 62. Marble-and-brickwork pattern from the side window of the portal of El Hakam II. Plane group *pmm*2. A partial-overlap derivative from the simple tetragonal pattern with centred squares.

Fig. 63. Brick-and-marble pattern from the inner tympanum of small gates in the southern corner of Abd'al Rahman II's part of the Great Mosque. Plane group *p*4.

Three distinct modifications of "interweaving" patterns (*p4gm*) occur, spanning nearly the entire spectrum of the patterns discussed in Section 6 (Figs 30, 31 and 35). A magnificent intercalation derivative of a simple, two-dimensionally interconnected meander pattern in Fig. 63 has plane group symmetry *p*4 with symmetry-excessive nested squares on the one set of four-fold positions (pattern types PP32, 41^5, 31 and 30^2; in red only: PP32, 41^2).

The *cmm* pattern in Fig. 64 allows several interpretations, for example as an omission derivative of rows of paired squares or as a "freely hinged" pattern akin to that in Fig. 8. Only one panel of specially molded elements (tiles) was recorded (*p4mm*), composed of four- and eight-fold stars. The recently published sketches of Inurria [16] suggest that a number of patterns were lost forever. Only one of them, *p4gm*, was found in a fragment large enough to allow full reconstruction (Fig. 36 in Ref. [16]).

In spite of their lively two-coloured appearance, all patterns in the Great Mosque at Cordoba represent uncoloured plane groups. The only recorded exception is a simple serrate (i.e. zig-zag) motif, with white and red stripes alternating, giving the dichroic group $p_{a'}gm$.

Fig. 64. Marble-and-brickwork pattern from the gate of San Miguel (possibly installed when rebuilt by El Hakam II). Plane group *cmm*.

Table 2. Approximate statistics for plane groups used in selected brickwork and mosaics

Type, locality or region	Space group incidence (in rounded-off percents)																	
	p1	p2	pm	pg	cm	pmm	pmg	pgg	cmm	p4	p4mm	p4gm	p3	p3m1	p31m	p6	p6mm	Obs
Plain brickwork (Western)		20				10	30		40									10
Grand Mosque of Cordoba		5				11	5		15	21	21	21						19
Village of Jork, Germany		10					19	14	10		19	29						21
Islamic mosaics (the Bourgoin collection [1])		0.5				6	1		15		32	3			1	3	39	200

"Obs" describes number of distinct patterns studied.

Among the one-dimensional ornaments, many can be classified as stripes cut out from two-dimensional ornaments [e.g. Figs 2(a) and (b)]. Especially around the gates of El Hakam II, diaper ornaments of red and white bricks based on square symmetry *p4gm* are typical. They have been altered into rectangular ones, *cmm*, and even *pmm* by minor modifications. Two modifications often occur in the same frame, in its horizontal and vertical portions. Untransformed or properly perspectively transformed (Fig. 58) they occur in the keystones of the lintels, accompanied by other ornaments.

One-dimensional ornament in Fig. 2(d) is based on "centred" squares which were used profusely in the two-dimensional ornaments. The version in Fig. 2(c) can be obtained by its reduction to one half on regularly spaced planes of overlap. Restoration to Fig. 2(d) can proceed by unit-cell twining of the reduced version on mirror planes perpendicular to the axis of the pattern.

The ornamental "keystones" of the lintels display a variety of patterns. A part of them represent clear cut-outs from two-dimensional ornaments whereas others are clear one-dimensional (frieze) patterns, as in Figs 2(c) and 58. Their variable treatment has been discussed in Section 9.

In general in the red-and-white brick ornaments from the Great Mosque of Cordoba:

(1) The choice of plane groups of symmetry is quite limited (Table 2). Often, for lower symmetries pseudotetragonal metrics are used. All the patterns are described by uncoloured plane groups; only one serrate striped pattern has dichroic symmetry.

(2) In the majority of cases the patterns are easily legible, with clearly discernible eye-attracting elements positioned on axes of symmetry. The point-group symmetry of these elements complies with, or exceeds, the local symmetry of the pattern. Only rarely [Figs 2(b) and 35] is the viewer forced to analyse the pattern. Selection of elements is limited to several basic types in the majority of patterns. Most of the patterns belong to only several series and are interrelated in them by accretion, substitution, intercalation, expansion–reduction, complication and omission.

(3) Combination of axial patterns (e.g. *p4*) with the reflection-based ones (e.g. *p4mm*) in one set of panels above a single gate, or of such elements in a single pattern (e.g. swastikas with the point-group symmetry 4 and "centred" squares, point group symmetry *4mm*) causes an internal tension in them, which is augmented by their colour play as well as by the contrast of their strict rectilinear character with the curvilinearity of the floral motifs on the intervening, buff-grey carved panels.

11. CASE STUDY 2

Village of Jork (District of Stade, the "Altes Land" SW of Hamburg, Germany)

The village of Jork (pronounced York) was inhabited since the twelfth century. It grew at the junction of important roads on the south banks of the Elbe. In 1221 "Majork" was given to a newly founded Andreaskloster. Since then it has prospered from fruit production.

The village is laid out along the E–W and N–S roads. The multistoried houses have ornamental gables and ornamental entrances oriented towards the road. They are half-timbered and about a third of them display ornamental nogging. The colour effect is that of whitewashed beams, red bricks with whitewashed joins and green ornamental doors.

Fig. 65. Gauged-brick patterns with symmetries _p4mm_ and _cmm_ as well as interweaved patterns, _p4gm_.
Triangular fields have herringbone pattern nogging. Central Jork, Germany.

The majority of houses of interest, "Ziegelfachwerke", have been built prior to 1800. However, the building and rebuilding has continued until the present time. For example, the same patterns were used on the houses built in 1950's as on the historical Portausches Haus (seventeenth century). Similarly, house No. 45 in Osterjork was built about 1750, but the original uninteresting brickwork was replaced by the splendid ornamental nogging only in 1936. Thus, the patterns have been in use over an extended period of time, the old ones repaired and perhaps altered to some extent during the inevitable repairs of the half-timbered houses, and much of them was lost during modern alterations.

Central panels of half-timbered facades in Jork are usually filled by more fanciful patterns whereas the lateral, often triangular fields represent twinned or untwinned herringbone patterns _pgg_ or _pgm_ (Figs 27, 29 and 65). Alternation of herringbone patterns with more elaborate, usually square patterns is also common. Twinning or point-group designs may assume considerable importance on modern houses. Framing of diagonally oriented square designs is common (Fig. 66). Gauged-brick patterns nearly always obey symmetry _p4mm_ or its dichroic derivatives (Figs 46, 47, 65 and 66). Besides these, the principal families and homologous series are those mentioned in Section 6: various herringbone patterns (Figs 27 and 28), interweaved patterns (Figs 29, 30, 33, 38 as well as Figs 65 and 66), simple or decorated block patterns and patterns close to those in both halves of Fig. 55 and in Fig. 8. A rare modern example of _p2_ derived by substitution from a pattern akin to that in Fig. 3(h) was seen as well. Because of the limited size of the timber mesh available for nogging, the unit mesh size is always small compared to the brick dimensions. The only exceptions are patterns such as in Fig. 66 (lower part) from which only fragments too small for unambiguous interpretation often occur. Contradictory styles, techniques, symmetries and orientations (mutual rotation by 45°) render a lively and dynamic character to the facades (Fig. 65) that is further heightened by white joins or whitewashed elements although these are not compulsory.

Again, rectangular and square plane groups describe the bulk of patterns employed; they are supplemented only by rare modern examples of _p2_. In Jork, stress is on groups which contain prominent glide-reflection planes whereas _p4_, prominent in the Islamic or Art Nouveau examples, was not found (Table 2). Thus, in North German brickwork in Altes Land, the means used for achieving dynamism of plane patterns were somewhat different from those used by the Moors of Cordoba. Symmetries _m_, _2mm_, and less frequently also _4mm_ represent the bulk of the point group designs and twinned patterns at Jork.

Fig. 66. Rectangular panels with ornamental nogging, plane groups *p4mm* (above) and *pmg* (below).
Flanked by interweaved patterns, second homologue.

12. EPILOGUE

Symmetrological studies and pattern analysis represent a powerful method for exact analysis of archeological and art–historical problems. On the one hand they allow more exact descriptions of art schools and traditions, on the other hand they reveal in clearly definable terms some of the ways human creativity develops within an individual, a group or through generations of artists.

Several ways of symmetrological/structural classification have been used in this study for the same problems. It is hoped that the reader can see the advantage of using several complementary ways of classification. Each of them reveals a different aspect of the pattern and its relationships to other pattern groups. These relationships are not only of academic interest: some of them, however complex (e.g. pattern-reduction or omission), were definitely exploited by the ancient ornament designers although we are sometimes unable to decide whether they were developed at the locality or copied from pattern books artisan groups undoubtedly carried with them. The stacking notation which at first seems to belong to the realm of recreational science proved very handy as a shorthand replacement of drawings whenever the author found a new pattern.

The efficiency of fundamental symmetrological descriptions (e.g. Wyckoff positions or pattern types) is greatest for simple patterns: it decreases with increasing complexity of patterns chiefly because our powers of image synthesis from symbol strings have firm limitations. Classification into series of different kind is more powerful precisely in this realm of complex patterns. Some series (e.g. accretional or element-substitution series) can be described by short mathematical expressions whereas other types (e.g. various catamorphic series) elude such description. However, they should not be dismissed on these grounds: (1) we are dealing with art where free improvization and further ornamentation of basic design is a rule, not an exception, and (2) in the realm of physical science catamorphic changes relate some crystal structures to other structures not usually considered as their kin.

Our study shows how the universality of artistic medium, a brick, brings about universality, recurrency or at least similarity of ornamental solutions. The same fundamental families occur again and again, notwithstanding the distances and time covered. Of course, differences and local fashions exist even in plain brick styles. One of the most important influences appears to be the size of area available for the pattern and the viewing distance. Their increase goes hand in hand with increase in unit mesh size and increasing complication of the motif. The large-scale ornaments on minarets and facades of large public buildings became very complicated, almost mannerist in nature. They appear to avoid boredom produced by easiness of viewer's intuitive analysis and to satisfy the "horror vacui" that the architects shared with other cultures. It perhaps is remarkable that these large patterns in the majority of cases preserve close relationships to their much smaller and simpler predecessors.

One aspect of brickwork patterns, clearly determined by their medium, is the limited choice of plane groups of symmetry we observe for them. Hexagonal and trigonal cases are exceptional; they occur either as rare openwork or require special bricks (tiles). The bulk of cases belong to rectangular or square space groups; oblique solutions are much less frequent. More exact analysis of plane group statistics faces a number of problems. Statistics over patterns and pattern varieties found on an architectural object or in an area appears to be the simplest solution. However, it is often a problem to define which patterns and especially which varieties are sufficiently different from the similar ones to warrant an independent entry in the list. On the other hand this approach does not reveal the frequency of occurrence of individual patterns on the object. For example, some *p4* and *p4mm* patterns occur repeatedly on a number of panels on the Great Mosque of Cordoba, whereas other *p4*, *cmm*, etc. patterns have been used only once or twice. Although we do not meet here the problem of interlacing ribbon vs line patterns, common for the analysis of Arabic mosaics, dichroic patterns occasionally occur and have to be reduced to plain ones.

The sources of material to this study have been manifold. Besides the author's own observations, drawings and photographs, a great number of books and exhibitions were scanned for patterns. They almost exclusively contain photographs from which patterns had to be reconstructed and drafted. The following are considered best among the sources consulted: Castejón and de Arizala [15], Du Ry [17], Hrbas and Knobloch [18], Hunt and Harrow [19], Mitchell [12], Montes Ruiz [16] and Wade [20].

Literature which pays attention to ornamental nogging is rare in spite of a number of books which deal with half-timbered houses; the great majority of patterns illustrated come from our own observations. Hansen [21], Benzon [22] and Konst [23] have been consulted, among others. Zapotec ornaments can be seen, for example, in Coe [24].

Acknowledgements—Patterns in this paper have been drawn by Mrs R. Larsen, Mr J. Tomás, Mr M. Rageth, Mr E. Säkäjärvi, Mr L. Caceres and Miss J. Marker as well as by the author himself. Mrs Merete Lykke Johansen and Mrs Una Koester have typed and Dr J. Bailey has reviewed the manuscript. Interest and encouragement by my family, crystallographic colleagues from various countries and the editor, Professor I. Hargittai, as well as the assistance of Professor A. Arribas (Salamanca), Dr G. Roper (Itzehoe) and Dr C. Otero-Diaz (Madrid) in obtaining unavailable references are gratefully acknowledged.

REFERENCES

1. E. Makovicky and M. Makovicky, Arabic geometrical patterns—a treasure for crystallographic teaching. *Neues Jb. Miner. Mh.* 58–68 (1977).
2. E. Makovicky, The crystallographic art of Hans Hinterreiter. *Z. Kristallogr.* **150**, 13–21 (1979).
3. E. Makovicky, Symmetrology of art: coloured and generalized symmetries. *Comput. Math. Applic.* **12B**, 949–980 (1986). Reprinted in *Symmetry: Unifying Human Understanding* (Ed. I. Hargittai). Pergamon Press, Oxford (1986).
4. S. Andersson, Eine Beschreibung komplexer anorganischer Kristallstrukturen. *Angew. Chem.* **95**, 67–80 (1983).
5. E. Makovicky, The building principles and classification of bismuth–lead sulphosalts and related compounds. *Fortschr. Miner.* **59**, 137–190 (1981).
6. E. Makovicky, The building principles and classification of sulphosalts based on the SnS archetype. *Fortschr. Miner.* **63**, 45–89 (1985).
7. B. Grünbaum and G. C. Shephard, *Tilings and Patterns*. Freeman, New York (1987).
8. W. Fischer and E. Koch, Lattice complexes. *Int. Tabl. Crystallogr.* **A**, 820–848 (1983).
9. H. Burzlaff, W. Fischer and E. Hellner, Die Gitterkomplexe der Ebenengruppen. *Acta Crystallogr.* **A24**, 57–67 (1968).

10. B. Grünbaum, Z. Grünbaum and G. C. Shephard, Symmetry in Moorish and other ornaments. *Comput. Math. Applic.* **12B**, 641–653 (1986). Reprinted in *Symmetry: Unifying Human Understanding* (Ed. I. Hargittai). Pergamon Press, Oxford (1986).
11. K. L. Wolf and R. Wolff, *Symmetrie*. Böhlau, Münster-Köln (1956).
12. G. Mitchell (Ed.) *Architecture of the Islamic World*. Thames & Hudson, London (1978).
13. J. Flemming, *The Penguin Dictionary of Architecture*. Penguin, Hammondsworth (1987).
14. D. W. Crowe, The mosaic patterns of H. J. Woods. *Comput. Math. Applic.* **12B**, 407–411 (1986). Reprinted in *Symmetry: Unifying Human Understanding* (Ed. I. Hargittai). Pergamon Press, Oxford (1986).
15. R. Castejón and M. de Arizala, *La Mezquita Aljama de Cordoba*. Editorial Everest, Leon (1971).
16. R. Montes Ruiz, Mateo Inurria: Dibujos de la Mezquita y de Medina—Azahara en el Museo de Bellas Artes de Cordoba. *Apotheca Rev. Dep. Hist. Arte Univ. Cordoba* **4**, 61–107 (1984).
17. C. J. Du Ry, *Art of Islam*. Abrahams, New York (1970).
18. M. Hrbas and E. Knobloch, *Umění Středni Asie*. State Publishing House for Letters and Arts, Prague (1965).
19. A. Hunt and L. Harrow, *Islamic Architecture—Iran I*. Scorpion, London (1977).
20. D. Wade, *Pattern in Islamic Art*. The Overlook Press, Woodstock, N.Y. (1976).
21. H. J. Hansen (Ed) *Architecture in Wood*. Faber & Faber, London (1971).
22. G. Benzon, *Gammelt dansk bindingsvaerk*. Det Benzonske Forlag, Copenhagen (1984).
23. K. Konst, *Altländer Ziegelmuster und Altländer Giebel- und Torinschriften*. Druckerei Konst, Veersen (1984).
24. M. D. Coe, *Mexico*. Thames & Hudson, New York (1986).

Computers Math. Applic. Vol. 17, No. 4–6, pp. 1001–1008, 1989
Printed in Great Britain. All rights reserved

INFLUENCES OF THE IDEAS OF JAY HAMBIDGE ON ART AND DESIGN

H. J. McWhinnie

Department of Design, University of Maryland, College Park, MD 20742, U.S.A.

Abstract—The system of dynamic symmetry, as an approach to decision-making in design, by Jay Hambidge in the early 1920s had a strong influence on industrial design as well as upon the work of painters.

This paper will review the design theories of Jay Hambidge and dynamic symmetry in terms of the use of proportions in design which are humanistically based upon the proportions of the human body. Recent research in the general area known as *the golden section hypothesis* will be reviewed as a means of justification not only for Hambidge's theory of proportions in design but as a validation for recent ideas in design, especially those of Robert Venturi. The paper will demonstrate the utility of studies from the behaviorial sciences for questions of design history as well as for present-day design theory and practice.

The work of contemporary artists and designers will be shown to demonstrate that these ideas, the golden section and dynamic symmetry, while popular in the 1920s, have a renewed sense of relevance to current ideas and approaches in design theory. The influence upon industrial design, while it may not be clear as say the influence of dynamic symmetry or the work of Bellows or Rothko, is nevertheless also quite clear. This influence is most evident in the work of Jay Walter Teague senior and in the well-known book *Design This Day*. The Hambidge design system of dynamic symmetry was used by designers at Tiffany's in the early 1920s and Jay Hambidge in his writings argued for an industrial design based upon the proportions of the golden section. The Hambidge system is employed today at the design department of Ohio State University and by Professor Lee Payne in the industrial design program at Georgia Tech in Atlanta.

INTRODUCTION

My intention when I first began the literature search upon which this paper is based, was to prepare a single review of the published material that would focus upon the uses of the golden section in recent twentieth century art and design, as reflected in the recent art literature as presented by a computer data file, *Art Modern (1972–1983)*. My original intention was not to move beyond that specific data base, since it represented the more recent shows and articles relative to this topic. I have tried to keep with those basic intentions in the materials that I have developed for this paper.

The volume of published material caused me to decide to prepare two separate reviews of this material. I have tried to group the materials for the first review in terms of articles on the visual art, painting, sculpture, architecture that used the golden section and the principles of dynamic symmetry as a direct result of the work of Jay Hambidge [1].

The second review, the subject of this presentation, focuses upon a wider collection of materials, aesthetic concerns and a greater range of media used; all however, are still within the general topic of the use of the golden section and dynamic symmetry in works of twentieth century art and design.

That the golden section has become an obsession for many creative artists throughout history is a matter of record. This writer is no exception! There is something about these proportions which has fascinated many of us. This review will only cover some of the more recent experiments. We leave it to others to go back in time and into history.

If one begins, as I have now learned to do, to go about an art museum and look for the golden rectangle or any of the five basic root rectangles outlined by Jay Hambidge (1867–1924) [2] in his system of dynamic symmetry; one will find to their surprise that they will come to see golden rectangles almost everywhere. I recently toured an exhibition of nineteenth and early twentieth century small French paintings at the National Gallery of Art, Washington, D.C. Almost all of the paintings show influences of the use of one of the root rectangles in their design and composition. One of the laws of pictorial composition that the artist of the nineteenth century learned, was to, within the plane or rectangle of the picture surface itself, locate and find the square of that area and its leftover segment which Hambidge called "the recopical". In almost every work in that exhibition, which is really quite a good random cross section of nineteenth century French painting, important compositional events occur at the critical point which defines the square [3].

PRINCIPLES OF DYNAMIC SYMMETRY

Who was Jay Hambidge? He was an obscure American illustrator who lived in the first two decades of the twentieth century and who formulated a theory of symmetry that was based upon his own studies of Greek vases and architecture.

He conducted research on the measurement of the human figure at the Harvard Medical School in 1918–1919 and many years before Le Corbusier wrote about his more famous *Modular*, Hambidge developed a clear theory about the relationships of order and proportion in art and design with the human figure itself.

What is the golden section or the divine proportion? Notions of ratio and proportion are among the fundamentals of mathematics and technology. The former is defined as the quantitative comparison between two things or aggregates belonging to the same kind of species, and latter as the equality of two ratios (A is to B and C is to D).

Of considerable interest, for centuries, has been a fundamental ratio variously called the ratio of Phidias, the golden section, the golden rectangle, the divine proportion, or simply Φ (phi) from the initial letter of the name of the Greek sculptor, Phidias. It is expressed mathematically as the ratio 1:1.618, derived from $\underline{1 + 5}$ (Jordan) [4].

In addition to its application in mathematics, Φ also appears in the structure of plants and shells and in the proportions of the human body; in Egyptian, Greek, Roman and Gothic architecture; in the Renaissance masterpieces of Leonardo da Vinci, Botticelli, Hals, Turner and other artists; in the composition of music, delightfully in the interval of the minor sixth; and as recently as March 1984, at the University of Maine, in choreography, in an exercise described as an experiment in spatial relationships.

What has been described by Professor Jordan is a joint workshop that he did with Fran Merriet of Deer Isle in which students acted out the basic root rectangles of the Hambidge system as dance or body movement. The music measurements of the rectangles of the golden section were placed upon a dance studio floor and the workshop participants acted out these dimensions in the movements of their bodies.

Jordan argues that psychological studies attest to the aesthetic acceptance of the divine proportion as a ratio based in nature. Thus, both technologists and humanists, it is contended, are ruled by a dynamic symmetry according to a pre-existing system of ratio and proportion. The humanist readily acknowledges his debt to science, and the technologist concedes but for (such) harmony beautiful to contemplate, science would not be worth following [5]. The ubiquity of Φ, and its roles as a bridge linking technology, art and the humanities, is a part of the theme of this study on the influence of Jay Hambidge.

Many psychologists as well as artists have become involved in golden section research. D. D. Stuhl [6] has reported some recent developments in terms of the development of visual language. Research at the University of Kansas, Department of Design, is using classical symmetry theory to explore whether a formal framework can be established for the study of the relationships present in visual and artistic forms. Such a framework could be extensively applied and might bring about the formation and conventionalization of a visual language. It was always Jay's intentions to apply dynamic symmetry to a wide range of artistic problems and while he used analysis of classical art as examples to demonstrate his theories; he really was not that concerned with the analysis and study of Greek art.

Many artists have conceptualized the golden section as a means to establish order. But the events of the twentieth century artistic forms have taught us that order can come in many forms, assume many shapes and have different looks. Many artists have explored the golden section within a cubistic context. Part of its versatility as a compositional as well as design construct, lies in the fact that the golden section and the other root rectangles in the Hambidge system, may be employed in a wide range of art from realistic to abstract.

One of Jay Hambidge's associates, the painter George Bellows used the system of dynamic symmetry to compose his action-filled paintings of the fight scenes in New York City in the 1920s.

Fig. 1. George Bellows, *Dempsey and Firpo*. Oil, $51 \times 63\frac{1}{4}''$, 1924. (*Collection*: Whitney Museum of American Art, New York, Acq. No. 31.95.) One of the late paintings of Bellows based upon the Hambidge system of dynamic symmetry. Almost every line and slope in the composition falls at a precise joint in a Hambidge root five rectangle.

This painting (Fig. 1) is only one of many examples which used the system to create dramatic tensions of movement in the composition.

THE GOLDEN SECTION AND HUMAN MOVEMENT

I am not at all sure the properties of the square have been well-understood. The square and not the rectangle is the basic formal property of much of contemporary art. Most painters use rectangles as their picture surfaces, since the square is felt to be so static a shape; but it is still the basis of compositional formats and forces even within the traditional rectangles upon which most of us paint.

Some of the more interesting and recent work in the use of the golden section has been achieved in multi-media presentations, in dance and in human movement works. Massine in an interview speaks of his 1917 work parade [7]. That work introduced cubism into the theatre. Cubism for Massine held the key to the manipulation of groups of dancers. He orchestrated them as though their movements were harmonies, their postures, chords. He built a symmetry on counterpoint and was always composing in the third dimension. Classic forms have their own basic vocabulary and symmetry and structure, as in cubism. The use of the golden section undergirded their effects.

It has been very interesting for me to note that one does not find too much discussion of the golden section in the literature in reference to sculpture or other three-dimensional art forms but one does find it as a vital concept in movement and dance. This is certainly an area for which one could do a considerable amount of additional research. Why has the golden section and dynamic symmetry been so limited to work of two-dimensional art? An account of the performance entitled

"787" staged by students of the advanced design course at the Department of Art, California State University at Humbold, is important to consider [8]. In this respect, the primary aim of the performance was to produce multi-media presentations through which the golden section might be experienced directly. It was conceived in terms of a ritual of about 16 min in a darkened room. This performance started slowly and simply and moving with increasing speed and complexity to a climax. In this paper, the golden section is first analyzed for its mathematical values. The planning of the project and the performance of the event are described in this article in some detail. From the discussion it was not possible to establish the extent of the viewers' grasp of the aesthetic or other significance of the golden section as a result of the performance. But the conception of using the section as a basis for stimulation of the imagination was very useful. One only wished that there was a follow-up to this study and some way to demonstrate that this method was, in fact, a useful learning experience within the design education. Much additional research could be conducted which explores the golden section in terms of human movement relationship.

I have already noted in this paper that more recently Fran Merritt of Deer Isle, Me, in the spring of 1984 conducted some workshops in dynamic symmetry in which he had his students act out in human movement, the proportions of the various root rectangles including the golden section. His was an attempt for them to understand the basic proportions used in dynamic symmetry and to practice these through the use of kinesthetic cues as well as through purely visual exercises [9].

THE MODULAR

One cannot discuss the golden section in twentieth century art without considering the work of the French artist and architect, Le Corbusier. A recent study of Le Corbusier's method of developing a pictorial idea through several versions with particular reference to the tapestry *La Femme et Le Moneau*. This work is one of a series produced in the period 1946–1964. Its geometrical framework is based on the square and golden section and derived from the juxtaposition of found objects. This study showed how pictorial motifs in the tapestry can be traced back to his earliest work as a painter when he still used his family name, Jeanneret. At that time he was working in the purist style with particular reference to his development of an interlocked glazes motif. His use of a limited number of pictorial words is shown to have been carried through to his architectural designs.

It is, of course, Le Corbusier's development of his *Modular*, the use of the golden section as a comprehensive system of proportions entered into the history of twentieth century architecture [10]. Unfortunately for Jay Hambidge, when one discusses dynamic symmetry in art today, it is Le Corbusier and not Jay's own efforts that are remembered and noted. We have seen, I hope in this short review, that the golden section is a far more universal idea and it cannot really be ascribed to one's discovery. Many have been refreshed and nourished by these conceptions in their own creative efforts, not only in our century but throughout art history as well.

To put is another way, the golden section becomes the image itself and is used for the symbolic power it seems to possess regardless of any subject matter. I hope in these reviews of the work of other artists who have used dynamic symmetry in their works, to show that it can be used within the full range of artistic style, from the representative to the abstract. (In July 1985 the Hambidge Center hosted a symposium at Rabun Gap, Ga, on dynamic symmetry, the golden section and the work of Jay Hambidge.)

The modern American artist Randall Craig also uses the principles of symmetry in the composition of the proportions of his sculptures. Even the formulas themselves are alluded to in the works as metaphors of his system of design.

It is almost impossible to consider the golden section without getting into questions of design as a basic visual language. The utility of the golden section has been demonstrated as a powerful candidate for the basic structure of visual language. If we look at St Uhr's article, he takes up the theme of William Huff's study *An Argument for Basic Design* [11]. In that book, the activity of planning is an activity of structuring, and that consequently the study of structure is in fact a definition of basic design. This critic aims to bring Huff's study up to date with the addition of new items and terms within the definition of design as it is understood today. By selecting individual

Fig. 2. Walter Burt Adams, *Evanston Today*. Oil, 20 × 24″, 1970. (*Collection*: Joseph Levy Jr, Chicago, Ill. *Photo*: Willard Nickerson, Evanston, Ill.) Walter Burt Adams published a book on dynamic symmetry and has made use of these methods for many years in his own work. (Reproduced by permission of the artist.)

terms such as perspective, symmetry and the golden section; he goes on to analyze what terms signify in current design methodology and praxis with reference to a number of distinct fields of art. Jay Hambidge back in 1922 in a lecture at the Art Institute of Chigago, Ill., postulated that, in fact, his system of dynamic symmetry could, in fact, serve as a basic structure for much work in design [12]. We have, I feel, almost come full circle in our voyage of discovery of visual structure and artistic language. Did it not all begin so far as Western art is concerned with the ancient Greeks?

Lee Payne of the Industrial Design Department, Georgia Tech in Atlanta, recently completed a Master's Thesis on the general subjects of the implications of Jay Hambidge, dynamic symmetry and golden section upon American industrial design including 25 years of his own work. Payne's research focused upon the nature of and importance of the human frame of reference in matters of design and proportion [13]. Many of his insights were based upon the work of a Russian, Boursalavitch, whose book *The Golden Number* (1958) opened up important insights from the general field of vision and perceptual research [14]. It may be that the field of vision itself may be the vital factor, the missing link.

He found that:

    (a) the human visual limit comes to 1.5661 (which is close to the golden rectangle);
    (b) the field of color vision comes to 1.61018 (golden mean)
    (c) proportion of human binocular vision (golden mean).

Stone and Collins showed that in general the human field of binocular vision is 1.665 and that our general field of vision is delimited by the golden section.

Payne, as a part of his research, surveyed 60 industrial design firms to assess any possible influences from golden section research [13]. Out of the 60 questionnaires, 18 responded and indicated in their forms they could document no such influence. Only four design educators responded positively in Payne's study. These were from:

Ohio State University, Columbus, Ohio.
Syracuse University, N.Y.
Georgia Tech in Atlanta, Ga.
Western Washington State College, Bellingham, Wash.

Payne also noted and documented some golden section influences upon:

Archibald Weldon whose 1938 design for Revere Ware is in perfect proportion to the golden section.
DNA Standards for Paper and Type Design of Black & Decker Products.

Fig. 3. Dorothea Rockburne, *Narcissus*. Three panels, oil on gessoed linen, $97 \times 123 \times 6\frac{3}{8}''$, 1984. (*Gallery*: Xavier Fourcade, Inc., New York. *Photo*: John Ferrari.) Rockburne uses the motifs of the methods of dynamic symmetry directly as construction guidlines in her work. (Reproduced by permission of the Fourcade Gallery, New York.)

Fig. 4. Randall J. Craig, *Dynamic Circle*. Resin and fiberglass, 42″ dia × 1½″, 1984. (Collection of artist.) Randall Craig uses the construction lines of dynamic symmetry as background. In this work, the basic structure can be seen applied to a circle as well as to a square. (Reproduced by permission of the artist.)

Payne observed that the modern anthroprometric charts that industrial designers now employ use more recent and scientific methods in the determination of proportions and yet most of those measurements come close to the golden mean.

CONCLUSIONS

Some artists such as Dorothea Rockburne use the structural proportions of dynamic symmetry as the subjects of their paintings (Figs 3 and 4).

The use of a system of symmetry seems to provide in the work of art an essential order which is necessary for the human appreciation and expression. It provides the artist with a structure of reality or with a model of the world that can be transformed into a personal artistic expression and yet can also communicate a general and more universal expression.

REFERENCES

1. H. J. McWhinnie, Review of research in the golden section. *Leonardo* **19**(1) (1986).
2. J. Hambidge, *Principles of Dynamic Symmetry*. Dover, New York (Reprint 1968).
3. E. Walter, Jay Hambidge and the development of the theory of dynamic symmetry. Unpublished Ph.D. Study, Univ. of Georgia, Athens (1976).
4. R. Jordan, Personal communication (1984).
5. R. Jordan, Unpublished paper (1984).
6. D. D. Stuhl, A language of form. *Vis. Lang.* **13**(4), 379–409 (1979).
7. D. Sylvester, Massine in conversation. *Studio Int.* **198**(968), 52–54 (1974).
8. J. B. Havill, The multi-media performance 987 based on the golden ratio. *Leonardo* **9**(2), 130–132 (1976).
9. F. Merritt, Personal communication (Deer Isle, Me) (1984–1985).
10. Le Corbusier, *The Modular*. Paris (1947).

11. W. Huff, *An Argument for Basic Design*. Hastings House, New York (1965).
12. J. Hambidge, Unpublished notes on his 1921 Chicago lecture among the Hambidge Papers at the Archives for American Art.
13. L. Payne, Dynamic symmetry and industrial design. Unpublished MFA thesis, Univ. of Georgia, Athens (1983).
14. Boursalavitch, *The Golden Number* (1958).

Computers Math. Applic. Vol. 17, No. 4–6, pp. 1009–1026, 1989
Printed in Great Britain. All rights reserved

0097-4943/89 $3.00 + 0.00
Copyright © 1989 Pergamon Press plc

SYMMETRY IN CHINESE ARTS AND CRAFTS

L.-X. Yu

Department of Fine Arts, Nanjing Normal University, Nanjing, People's Republic of China

Abstract—Chinese art is very rich in symmetry because symmetry is deeply rooted in the minds of the Chinese people. It is a kind of world outlook as well as a symbol of the ideal and hope.

1. INTRODUCTION

Chinese art has a long long history and there are various symmetrical forms in it. Symmetry, it is believed in China, not only can be found in the treasure of Chinese art as one of the most important aesthetic rules for the formal beauty, but also is deeply rooted in the mind of the Chinese people as a kind of world outlook and a symbol of the ideal and hope. Symmetry has long been closely related with the life of the Chinese people.

2. SYMMETRICAL CHARACTERISTICS OF CHINESE ARTS AND CRAFTS AS SEEN IN THE TAI JI PATTERN

Figure 1 is a very interesting design. The circle is divided into two, showing a symmetrical state of inversing, black–white and real–imaginary, contrasts. It has an impression of a dynamic motion which reminds one of two fish running after one another merrily. Though the black and white are clearly two parts, one can find no clear-cut boundary between them, thus making an integrated circle. This differential and integral treatment of symmetry form is ingenious indeed and gives people a sense of beauty.

Is the Tai ji pattern only a picture for appreciating? Of course not. Are there any symbolic meanings in it? Yes, the Chinese believe the Tai ji pattern symbolizes the "positive" and "negative" nature of the universe.

What do "positive" and "negative" mean? In ancient Chinese philosophy, "positive" and "negative" were a pair of concepts used to account for the nature of universe. The original meanings for "positive" and "negative" were that the part facing the sun was the "positive" and the opposite part was called "negative". The philosophers in the Zhou Dynasty (1122–770 B.C.), however, took over the concepts in explaining the universe and "positive", "negative" became their philosophical

Fig. 1. Tai ji pattern.

Fig. 2. Bronze mirror (Tang Dynasty 618–907 A.D.).

term. Everything was attributed to two aspects, i.e. "positive" and "negative". Philosophers used them to explain the two opposing but independent forces in nature. For example, Bo Yangfu, a philosopher in the last years of the West Zhou Dynasty, explained an earthquake by this "positive" and "negative" theory. He said: "earthquake is due to the 'positive' being forced under within and the 'negative' being forced not to coming out." In the book of *Laotse* there was the theory that everything in the world consisted of the "positive" and the "negative". It was pointed out that the contradiction between the positive and negative is the intrinsic quality of everything. The author of *Yi Zuan* expounded further: "positive" and "negative" following one another is the principle of nature.

This "positive" and "negative" concept has a deep influence in traditional Chinese medicine as well. Since the Zhou Dynasty, the "positive" and "negative" has been one of the basic theories of traditional Chinese medicine. Therefore, the Chinese believe that the Tai ji pattern represent not only a kind of beauty in form but also the world outlook. We will talk more about this in Section 4. This is the Chinese symmetry concept in Chinese arts and crafts and the symmetry characteristics in terms of Chinese arts and crafts forms (Fig. 1).

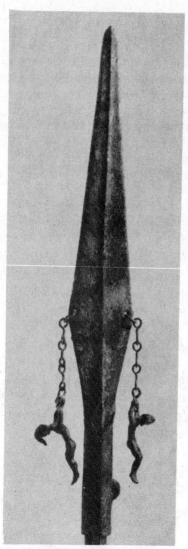

Fig. 3. Bronze spear-head (Han Dynasty 206–24 B.C.).

Fig. 4. Decorative pattern in architecture (Qing Dynasty 1644–1911 A.D.).

Fig. 5. Stone-cut pattern (Northern Wei Dynasty 386–534 A.D.).

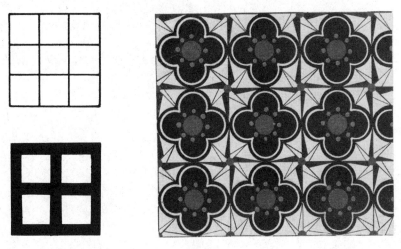

Fig. 6. Cotton prints.

Fig. 7. Fabric pattern (developed from "田" pattern).

Fig. 8. Caisson ceiling pattern (Dun-huang caves), Tang Dynasty.

Fig. 9. Bronze mirror (Tang Dynasty).

3. SYMMETRY FORMS IN CHINESE ARTS AND CRAFTS

Chinese arts and crafts have a long history as well as various kinds. The typical symmetry forms are as follows:

(i) Two-side (up–down and or left–right) symmetry

Two identical, similar or different patterns are arranged on the two sides of the middle line as a left–right and up–down pairs. We call it mirror symmetry if the pattern on both sides is the same (Fig. 2). If the patterns on both sides are well balanced but are different or similar textures (Figs 3 and 4). They are called non-mirror-image forms. From the viewpoint of the Chinese, symmetry on both sides conveys a feeling of solemnity of dignity.

(ii) Translational symmetry

This refers to the forms produced by the up–down or left–right movement of the pattern. The popular Er Fang Lian Xu pattern belongs to this one (Fig. 5).

(iii) Jiu Gong Ge pattern and ⊞ pattern symmetry

This symmetry has the square shaped ⊞ as its base, with patterns arranged in the square (Figs 6 and 7).

(iv) ∗ type symmetry (radiant symmetry)

This kind of symmetry is shown by the patterns radiating from the point in all directions (Figs 8 and 9).

(v) Symmetry in the Tai ji pattern

The Tai ji pattern is formed by dividing the picture with an S usually with patterns arranged in the S frames. Sometimes the S shape is repeated, linearized or modified to form patterns.

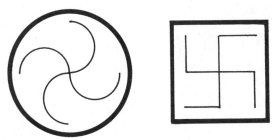

Fig. 10. The repeated, linearized or modified S-type patterns.

Fig. 11. Earthenware wheel (about 5000 years ago).

Fig. 12. Lacquer tray (warring states, 475–221 B.C.).

Fig. 13. Lacquer tray (Western Han Dynasty 206–24 B.C.).

Fig. 14. Patterns on outer coffin (made by silver, Tang Dynasty).

Fig 15. Veined porcelain (Five Dynasties 907–960 A.D.).

Fig 16. Stone-cut pattern (two phoenix, Sung Dynasty 960–1279 A.D.).

Fig. 17. Stone-cut pattern (two phoenix, Yuan Dynasty 1279–1368 A.D.).

Fig. 18. Stone-cut pattern (Ming Dynasty 1368–1644 A.D.).

Fig. 19. Colour pattern in fine architecture (Qing Dynasty 1644–1911 A.D.).

Fig. 20. Blue cotton prints with two phoenix.

Fig. 21. Chinese paper-cut.

The list of Figs 10–12 show different Tai ji patterns in different periods since the new stone age. It tells how the tradition is continued and how the Chinese love this type of symmetry (Figs 11–21).

4. HOW ARE THE CHINESE SYMMETRICAL PATTERNS DEVELOPED IN ARTS AND CRAFTS

There are many reasons why the Chinese people have a special interest in symmetry.

(i) Inspiration from nature

In nature there are numerous things in symmetrical forms such as butterflies, snowflakes, flowers, leaves and mineral crystallines. They give people aesthetic delight. The human body is symmetrical, too. The symmetry form are well-balanced which make the physical activities in life easier. Then comes harmony and harmony means beautiful. The deep love for nature and harmony inspired the Chinese people to create many artistic symmetrical forms.

Fig. 22   Fig. 23

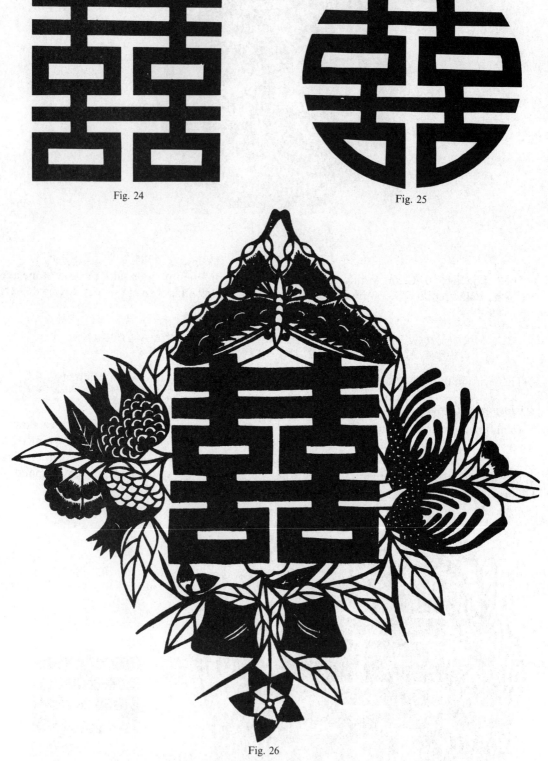

Fig. 24   Fig. 25

Fig. 26

Figs 22–26. Chinese character "double happiness".

(ii) Symmetry expresses one's ideal, wishes and emotion

(1) Symmetrical patterns. The symmetrical patterns of Figs 12–21 is called **Xi-Xiang-Feng** (meeting happily). It is developed from the pattern of Fig. 11 (earthenware wheel). The subjects

Fig. 27. Paper-cut with mandarin duck pairs.

of these types are often butterflies, mandarin ducks, phoenixes, parrots, fish or flowers in pairs which are arranged in a head to tail position as if they are playing, merrily running one after another.

This impresses one with a sense of endless motion and expresses the wish of lasting prosperity and lasting love between wife and husband.

(2) Symbolic meaning of double happiness 囍 . 囍 can not be found in any Chinese dictionary, yet everybody in China knows it's a Chinese character meaning double happiness.
喜 means happiness, two 喜 put together in symmetry emphasizes the degree of happiness, therefore, 囍 is the symbol of happiness. When a couple get married, as a custom 囍 is put on the door and window for decoration.

Fig. 28. Bronze ting (cooking vessel) (Shang sixteenth century B.C.).

(a)

(b)

Fig. 29—*continued opposite*

(c)

(d)

Fig. 29 (a)–(d). Door gods and couplet pairs on the door of a peasant family.

(a)

Fig. 30—*continued opposite*

(b)

Fig. 30 (a) and (b). The door gods.

There is a story about the original meaning of 囍 :

In the Ming Dynasty (1340–1661 A.D.), there was a scholar called Fang Mingqiu, who was once going to Peking for an exam. Passing by Suzhou city he saw a notice posted in the street by a rich man, saying if anyone could offer a second line of the couplet with the first one given, he would be his son-in-law.

Unable to offer the second line the scholar left remembering the given part of the couplet.

In Beijing he was asked by the examiner to compose the first line of a couplet with the second half given. It so happened that the line he had learned by heart in Suzhou was the best line to match this second line. He gave the answer immediately which surprised the examiner greatly. In this way he passed the Imperial exam with great honour and became a provincial officer.

Going back he returned to Suzhou again. Seeing no man had offered the correct answer, he tried the line learned in Beijing. He became the son-in-law of the rich man. The success in the exam and getting a wife made him so excited and happy that he wrote two (happiness) on the wall to celebrate his wedding.

In modern China, wherever there is a wedding, people still use 囍 as the decoration, though the meaning of it is slightly different from the original. Now it means happiest time and best wishes.

(3) The meaning of mandarin ducks in paper-cut. Like 囍, the mandarin duck paper-cut is often used on wedding occasions. Mandarin ducks in a pair represent never parting partners, which are the symbols of ever lasting love between wife and husband. The most important thing is that the ducks must be in pairs.

It's almost a curse if one duck paper-cut is presented for wedding celebration, for it means the love of the couple can't last long, one of them will be short lived. This is something that the Chinese people will avoid as taboo.

(4) Symbolic meaning of an ancient bronze tripot. The shape of the ancient bronze tripot in China is mirror-image symmetrical. On it are bold, vigorous, mysterious and symmetrical patterns which convey the feeling of dignity and solemnness. The pot itself was used by the slave owners as the symbol of power about 3000 years ago.

(iii) The impact from literature

Symmetry has an important role in Chinese literature. For example, the couplets written on scrolls usually posted on doors and pillars are symmetrical both in content and length of lines. In Chinese literature, there is a five-character (or seven) and eight-line poetry. This kind of poetry requires not that only the length of every line in the poem is the same, but also that the content of the first-half and the second-half of the second and third lines is symmetrical.

The following is a poem by Wang Wei:

山居秋暝　　（唐）　王维

空山新雨后，　　天气晚来秋。

明月松间照，　　清泉石上流。

竹喧归浣女，　　莲动下渔舟。

随意春芳歇，　　王孙自可留。

The title is *A Village in Autumn Dusk*. Let's examine the third and fourth lines to see how strict its symmetry is, the third line is "Bright moon shines on the pines", the fourth is "Clear springwater flows over the stones". Bright moon and clear water, flow and shine, pine and stone are pairs. The third line describes a static state, appealing to one's eyes, while the fourth line describes a dynamic state, appealing to one's ears. Not only the parts of speech but also the tones of the characters are symmetrical. The strict symmetry of poetry has a great impact on the development of symmetrical forms in arts and crafts. One can obtain a clear idea from the pairs of dragon and phoenix in Fig. 19, Qing Subao and Yuchi Jingde (door gods) in Fig. 30 with 日 (sun) up left and 月 (moon) up right.

Fig. 31. Old or young (turn it upside down and see).

Fig. 32. Wood cut pattern.

Since the Ming Dynasty, there was a song spreading among the people of the villages in Yi County of Hebei Province, which is very interesting in the sense of its symmetrical music and words

| 1 3 5 | 5 3 1 | 1 5 3 | 3 1 5 | 6 4 2 |
| 歌好听 | 歌好唱 | 啦啦啦 | 啦啦啦 | 哥教妹 |

| 2 4 6 | 5 1 3 | 3 5 1 | 1 3 5 | 5 3 1 |
| 妹教哥 | 啦啦啦 | 啦啦啦 | 唱好歌 | 听好歌 |

The meaning of its words is that the song is beautiful to hear and pleasant to sing, La-la-la, La-la-la, the boy is teaching his girlfriend to sing a song, the girl is teaching the boy to sing another song, La-la-la, La-la-la, let's sing a pleasant song and listen to a beautiful song. So we can find symmetry in singing this song in the normal order as well as in the inverted order. Now please look

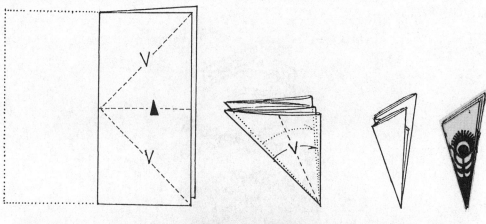

Fig. 33. Tuan Hua paper-cut (by author).

at Fig. 32, and see whether the images are the same when putting the picture in a normal way and putting it upside down.

There is also symmetry in the art of paper-cutting, especially in "Tuan Hua" paper-cutting which is very popular and welcomed among the Chinese people, because it is made by cutting pieces of folded paper with a pair of scissors and a special knife in very simple ways and with results of great variety. Tuan Hua paper-cutting is a very popular art activity in schools and in the countryside of China. It is predicted that this art form will contribute with satisfactory result to the development of symmetrology.

Acknowledgements—Translated by Zhang LianYing, Fu Jun and Wang Jia Liang.

Computers Math. Applic. Vol. 17, No. 4–6, pp. 1027–1040, 1989
Printed in Great Britain. All rights reserved

0097-4943/89 $3.00 + 0.00
Copyright © 1989 Pergamon Press plc

SPACE ⇌ COLOR ⇌ SYMMETRY

O. Zouni

Vrilission 22, 15236 P. Pendeli, Greece

Abstract—The author refers to the direct or indirect use of symmetry in her art. She uses straight lines, all of which begin from the same point, in order to transmit in design and color a beam of light. She works colors in their positive and negative aspects. In other works she interrupts her planes and divests the cut points, thus, giving the impression of a refraction which, in the similar elements, is like two successive reflections. Her compositions with the linear perspective are either developed symmetrically in a vertical axis or in parallel removal of this axis.

Most of the elements in my work arise from the observation of nature, the environment and visual phenomena. In nature a symmetrical structure is observed, which, however, is not always absolute; for example, the leaves, the flowers and the number of elements is often odd. Nature as well as art breaks the rule. Art cannot easily comply with the laws of strict geometry, which is why when it starts to be codified, to repeat itself, it loses all its spontaneity. I use symmetry in my work as a means of realizing an idea, because by mind is preoccupied by an idea and not by a theory. I set myself a geometrical problem in order to materialize my thoughts.

Since 1961 I have been investigating problems of space and geometry. I have been working in stages. The geometrical method helps me to materialize my idea–conception. The basis of my work is a linear design. My main problem is to organize the two-dimensional surfaces and three-dimensional spaces, so that I can express my visual experience.

I do my paintings or sculptures first on a small scale. The preparation of a model or notes allows me to vary form and color, exploring all possibilities, before moving to the final scale.

STUDY OF A BEAM OF LIGHT

Light is a phenomenon that I have persistently observed and it is something that has greatly influenced my work. I study a light beam on many successive levels, or how it develops on various surfaces. This study is based on the observation of light on vertical planes. Firstly, the final result is translated into a linear design, secondly into colors and, therefore, concrete visual illusions are created. For example, a straight line looks like a curved one (Fig. 1) or, likewise, a plane looks curved. I started with constructions of two vertical planes that at certain points are increased and form boxes (Fig. 2). The top of the beam starts from one of the levels and is developed on the other planes according to the angle which it makes (Fig. 3). This study of beam induces me to create sculptures (Fig. 3), as well as environments with many elements in positive and negative colors (Fig. 4). In these first works, I drew the beam with a symmetrical structure which sometimes divides the work into two equal parts and at other times one of these two parts progresses further than the other (Fig. 5). Progressing from there, I reached the point where I could work with two elements which reflected one another but only within the design (Figs 1 and 4). The colors in the symmetry axis are reversed, and thus a work is created with twin elements in a positive and negative synthesis (Fig. 6). Sometimes these elements are set free from their reflected positions, appearing more and more on the basis of being divided into a grid according to the proportion of their sides (Fig. 7).

Sometimes I place my sculptures on a base so that parts of them are reflected on it. These bases could be of stainless steel or water, or simply a mirror (Fig. 8). The most recent, and still developing idea, is one of a multireflection—with a mirror placed not only on the base but also on the upper part of the sculpture.

This multireflection, this multisymmetry, creates a feeling of vastness, which I also pursue in other ways in my work. In the works of this group that are drawn symmetrically there is only one position for the observer to see this symmetry, in any other position it is obscured (Figs 2 and 9). This is something that is natural, due to its volume.

Fig. 1. Twin sculptures, positive–negative, each 215 × 50 × 45 cm, 1978.

Fig. 3. Sculpture, 196 × 90 × 43 cm, 1973.

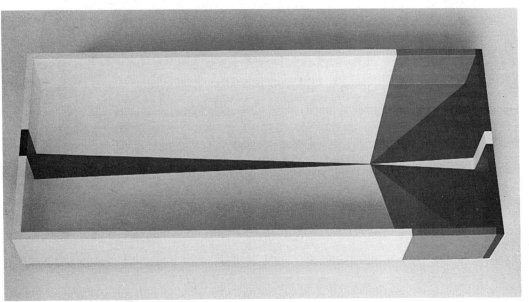

Fig. 2. Open box, construction in wood, 122 × 50 × 11 cm, 1972.

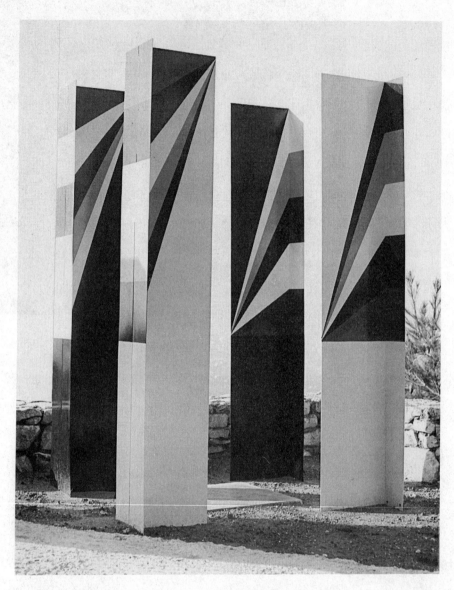

Fig. 4. Environment with four elements, positives–negatives, 250 × 160 × 140 cm, 1976.

Fig. 6. Twin sculptures, positive–negative, each 82 × 22 × 14 cm, 1975.

Fig. 5. Twin sculptures, positive–negative, each 182 × 62 × 32 cm, 1974.

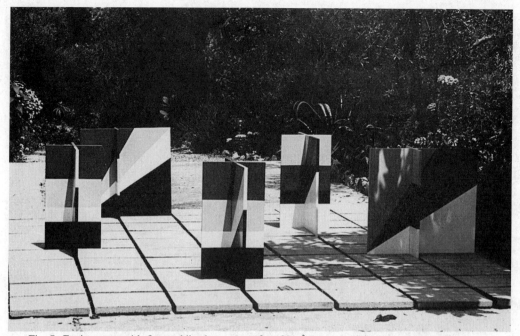

Fig. 7. Environment with five mobile elements, surface 20 m², concrete slabs 50 × 50 cm, each element 88 × 84 × 43 cm, 1974–1980.

Fig. 8. Sculpture, 134 × 82 × 52 cm, 1983.

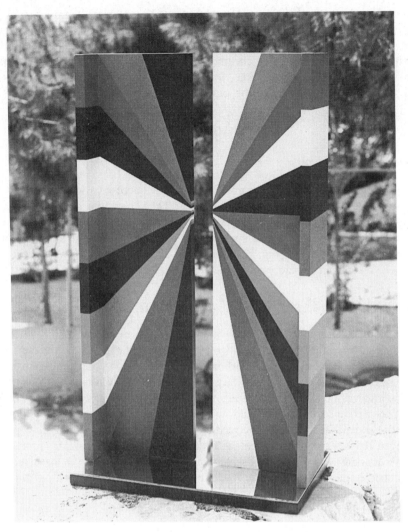

Fig. 9. Twin sculptures, positive–negative, each 72 × 19 × 14 cm, 1974.

Fig. 10. Twin construction in wood, positive–negative, each 130 × 130 cm, 1979.

A general conclusion from this group of my works is that starting from a visual phenomenon, an observation of nature, I finally reach a linear geometrical design, with an intellectual development of colors. Concerning the colors, what interests me, is whether there is a balance in intensity or not. Usually, I work with chromatic contrasts and often develop the same theme in both its positive and negative aspect (Fig. 10). Color is a basic element in my work. I work with basic colors and the tones I use serve the purpose of projecting the basics.

CUTTING OF PLANES

I have sometimes developed my designs by dividing the surface into 8 × 8 or 16 × 16 modules and I have tried to take advantage of the 5/8 of the surface that is equal to the golden section. The final surface, a result of the superimposition of many successive surfaces, is cut and the parts that are created, are kept apart, side by side (Fig. 11). Thus, the cut forms—parallelograms, squares, triangles, circles, diagonals—upon removal give the impression of refraction. A refraction which is caused by the interference of a surface that is revealed with the simultaneous removal of the cut sections of the first picture (Fig. 12). Therefore, according to the laws of refraction in similar materials, it is a result of two successive reflections and of two successive symmetries.

LINEAR PERSPECTIVE—IMPRESSION OF VACANCY

In my work I have been using the linear perspective more and more. Perspective has been a common problem for painters throughout history who want to express the third dimension in their work. That is, they try to transmit the depth of the real world on a particular surface level. I have been using perspective as a way of penetrating the feeling of emptiness and thus I create a deserted space of chaos and vastness that is stressed more by the human absence (Fig. 13). Many times I interrupt the perspective by opening a window on the picture, a second plane. This window usually "looks" on to the wall (Fig. 14), which either stays as it is, creates a new painting surface (Fig. 15) or even reflects the environment with the interference of a mirror (Fig. 16). The lack of a human figure or some other element of scale reference, permits the eye to lose itself in the vastness and the geometrical space takes the size that is identified with emotionally at that very moment

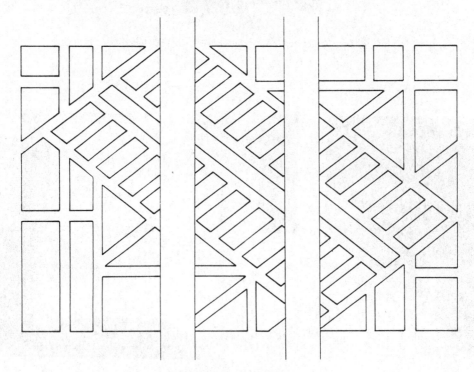

Fig. 11. Linear design, 1968.

Fig. 12. Construction in wood, 150 × 130 cm, 1972.

Fig. 13. *Horizon*, 115 × 290 cm, 1986.

Fig. 15. *Nature Through Illusion*, construction in wood, 250 × 160 cm, 1985.

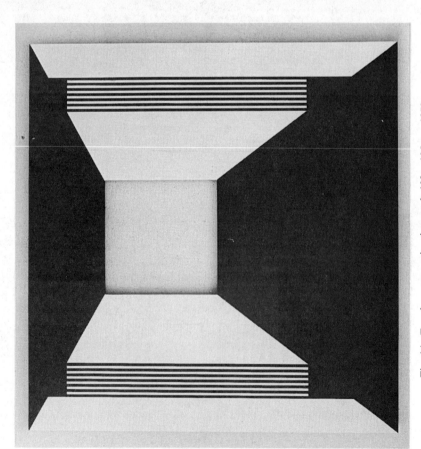

Fig. 14. *Temple*, construction in wood, 109 × 105 cm, 1979.

Fig. 16. Construction in carton with mirrors, 101 × 131 cm, 1985.

by the spectator. There is a series of my works, whose composition is absolutely symmetrical to the vertical axis that divides the picture (Fig. 17).

This is something that is observed in the history of art. For example, *The Lord's Supper* by Leonardo da Vinci, where the architectural environment of the work is developed in a central perspective, with absolute symmetry, and where the visual point is on Christ's head—a symbolic and obvious intention of the painter.

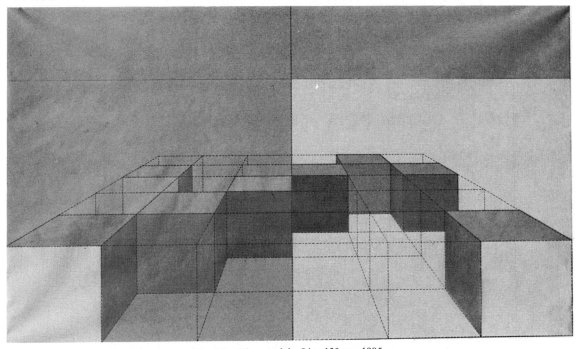

Fig. 17. *Cubes-module*, 94 × 130 cm, 1985.

Fig. 18. *Stairs in Perspective*, construction in carton, 71 × 71 cm, 1979.

In my works, the position of the observer, where this symmetry can be seen, is the frontal one (Fig. 18). In some of these works there is an additional small element that disturbs their symmetrical composition, this element also varies in size according to the observer's position, because it is on a different plane (Fig. 19). In another series of my works, bigger than the previous one, this axis of development of the composition is transposed towards one side of the picture (Fig. 20). The same thing, naturally, may happen with the transposition of the horizon. This, of course, is not a personal practice of mine, but maybe it reveals a symmetry of compositions. With this research

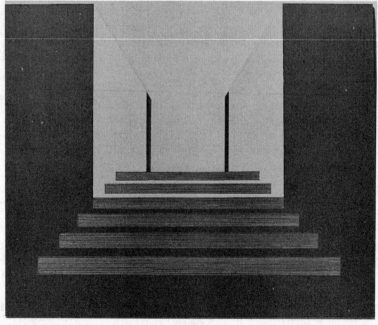

Fig. 19. *Stairs in Perspective*, construction in carton, 71 × 85 cm, 1979.

Fig. 20. *Altar*, construction in wood, 170 × 250 cm, 1985.

Fig. 21. *Double Space*, construction in wood, 120 × 250 cm, 1979.

into placing the vertical and horizontal axis of development of the composition within the picture I was led to works with double twin spaces, on which the horizon is identified but the vertical axis is in such a relation that the two spaces become independent instead of being united (Fig. 21).

STRIATIONS IN A LOGARITHMIC SCALE

In some parts of my compositions, which I started approximately in 1976, I created stripes with a progressive increase in their thickness. Each time they are studied according to the geometrical problem within which I place them. These stripes increase the impression of depth and, according to the mathematical succession that they follow—e.g. logarithmical—they give the level surface the impression of a curved one (Figs 13, 15, 20 and 21). This occurs because the mathematical succession is in fact the projection by means of straight lines from a certain point of a curved line, which comes from the distortion of another symmetrical parabola or catenary by the parallel transposition of the axis of its symmetry.

These stripes are also another way of using plastics and geometrical means in order to create depth, the optical illusion. I feel that illusion exists in art. Therefore, in order to stress illusion, I use perspective, symmetry with mirrors or the reflection of water, as well as other geometrical means.

Computers Math. Applic. Vol. 17, No. 4–6, pp. 1041–1066, 1989
Printed in Great Britain. All rights reserved

AD INFINITUM

J. ZVILNA

408 Keewatin Avenue, Toronto, Ontario M4P 2A5, Canada

Abstract—In the continuous (and obsessive?) search for an *understanding of the genesis of visual form*, a radical (L. *radix*, a root) form-generating system of relationships, interdisciplinary language, is in the process of being designed.

In this work, the emphasis is on the genesis of form in spacetime: through transformations, growth and abstractions, an infinite number of elements generate an infinite number of forms, all having one and the same intrinsic trihedral structure.

Symmetries emerge as if planned.

GLOSSARY

A doctrine of one. The core of IDL; simultaneous restraint, wonder and surprise.

Element. Intangible, visible emptiness, defined by planes, containing memory and predictable behaviour.

Experience. The way to know; the source of behaviour (applies to the living and non-living).

Form. Shape having an intrinsic structure.

Formative. Communicating development in ft or st.

ft. A world of two spatial dimensions plus time; fieldtime.

hc. Helical configuration.

IDL (I-DE-AL). Interdisciplinary language, a visual all-embracing system of patterns and forms in ft and st; the objective of the IDL—contact with the real, what is, tangible and intangible; the essence of IDL—the potential in emptiness.

Ignorance. A filter over the mind, that allows one to think and dream the impossible, e.g. an interdisciplinary language; cooperation with real time; form, structure and meaning of emptiness, etc.

Memory. The storehouse of experiences.

Natural order. Decoded nature's order; nature plus man.

Nature's order. An order not decoded as yet.

Pattern. The way elements are arranged in ft or st.

Process. The generating of pattern or form.

st. A world of three spatial dimensionsl plus time; space time.

Structure. The intrinsic order of pattern or form.

To understand. To relive; to go back in ft or st to the source and return step by step with memories—inscribed in mind or matter—all the way up to the present.

SCOPE OF THE SEARCH

How do forms in ft and st emerge? This is a question that has haunted me for some time now. Is there a single answer? So far I see a solution unfolding in short essays.

(1) "Rotation and form" [1]

The paper describes a rotational–translational movement as a form-generating process and its application:

> "Many events in nature—the sunset, the migration of birds, the hibernation of animals, the eclipse of the sun, etc. have been found as the result of ROTATION. Without the earth's rotation about its axis and around the sun there would be no day–night, no summer–winter, no consciousness–unconsciousness."

(2) "Colored symmetries in space–time" [33]

In this essay I managed to extract the organizing principles—economics, ecology and ecoethics—and also to *make* the building block—*one* type of element that grew in an infinite number of forms, all having *one* structure. The understanding of the genesis of forms in ft and the genesis of the element in its pre-phenomenological stage had to wait until the third paper.

(3) "Ad infinitum"

The work that follows continues to elaborate the concept of *restraint*, using the same structure as in the second paper. The essence of this form-generating system in ft and st—emptiness—is emerging. The discovered *infinite* number of fundamental elements are used to develop an *infinite* number of forms. The organizing principles, important as they are, are not repeated from the second paper. Also, for the properties of the element—memory, predictable behaviour, etc.—and the making of the element, the reader is referred to the second paper.

Fig. 1. Exhibition poster, 43 × 43 cm.

PREFACE

After my lecture "Rotation and form" at a 1976 Montreal conference [1], one of the questions asked was, "Why are you so obsessed with helical movement in spacetime?" For years, I myself have been puzzled about this. Helical configurations have been emerging not only in my visual exercises, but even where no visual signs could suggest them. For example, in Fig. 2 the two arrangements of bars are made with rotational and translational movements. I believe that the ordering process is reflected somehow in the internal structure. But how?

Returning to the question about helices—how embarrassed I was! A speaker should know his subject and all the answers to all the questions! I hesitated for a minute, searching furiously in my chambers of limited knowledge and intrinsic memory bank. To my astonishment—suddenly I knew the answer. I remembered an event so clearly that I was able to give a very convincing answer! I will return to this happening later, but first some background information on my life and the beginning of my search.

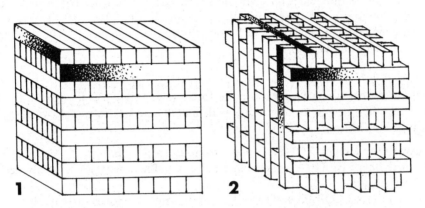

Fig. 2. Is there a hidden internal structure?

INTRODUCTION

During the 1930s I studied architecture. In the latter part of the 1940s I made toys and games. In the 1950s I returned to architecture, to work. I also wanted to study architecture my own way—going back to the ultimate sources [2]. But to do that I felt it necessary first to investigate the genesis of visual form. In the 1960s the beginning of the chosen direction was shown in several exhibitions.

During the 1970s at the School of Architecture, University of Waterloo, I taught (learned) a course about form as an interdisciplinary language. In 1981 emerged "A doctrine of one"—a single post to dance around. The search for the

FORM IN ARCHITECTURE

had become

ARCHITECTURE OF FORM:

an all-embracing questioning of form, content and values in the broadest possible context [3–18].

METHOD

The architecture of visual form searched for and described in this work, including its genesis, is based on a single method: movement in ft and st [19–23].

A movement involves space, time, mind (how, why), hand (energy, direction, orientation) and matter (what). Some of the ingredients are parts of nature's order. Others I include to add human content—without it the final product would be without signature. Thus another description of the method used in this inquiry, in addition to the one above, is COOPERATION WITH NATURE. "Cooperation" means to think and act as nature does, hoping to contact the real. To achieve that contact, reality itself is invited to participate in the formative processes.

Real time is a phenomenon I have always been puzzled about [24–32]. It takes part in all events in this work or experimental studies. To know more about time I have been inviting it to PARTICIPATE in my investigations. This strategy generated, in 1981, "A doctrine of one": single type elements, made by the rotation in st, resulting in an infinite number of forms. Later it was discovered that there are an INFINITE number of fundamental elements that could be coordinated in an INFINITE number of forms, all having ONE intrinsic structure—the central issue of this work.

TECHNIQUES

To live on flat land is different from living in space. Growth in space happens in all directions; on a flat field growth is in all directions also, but only in one plane. In the proposed system, form-generating methods and strategies in ft and st have the same nature. The differences dictate the use of different techniques.

1. Technique used in ft

I am living on the earth's surface, exposed to forces (e.g. gravity) which govern my being here. Since I have been destined to communicate with myself and others mainly visually, I read the patterns in Fig. 3 as follows:

Fig. 3.1. Growth on a HORIZONTAL plane from a centre point.

Fig. 3.2. Growth on a VERTICAL plane like a pattern on the inner wall of a container—the result of rising liquid.

Fig. 3.3. Growth on a SLOPED surface. The spiral pattern indicates that matter has been in a continuous motion, exposed to a linear force, gravity.

The above diagrams are drawn. They do not display true growth in ft—SIMULTANEOUS development in all directions.

A new technique is needed!

The technique chosen for the studies in ft is based on observation of the interaction between a substance and the forces working on it. A glass sheet, covered with a dense layer of carbon black, becomes a field for the interplay of forces.

In Fig. 4.1 oil is allowed to flow between the prepared glass sheet (15 × 20 cm) and a plastic one 2 mm above, through three holes in the plastic sheet. Slight vibrations of the horizontal plates generate lines—revealing stages in growth (Fig. 4.2). To understand the formative process, the photographic enlargement of the growth pattern should be exhibited horizontally (Fig. 4.3).

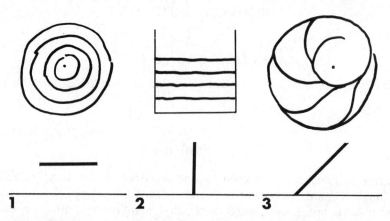

Fig. 3. Diagrammatic growth patterns on horizontal, vertical and sloped planes.

Fig. 4.1. Growth on the horizontal plane.

Figure 5.1 shows the growth of a spiral, when the carbon-coated sheet of glass (18 × 26 cm) is set vertically in continuous rotation in a vessel with rising lacquer thinner. To understand the process which generated the image, a sketch is included (Fig. 5.2). It shows the pattern of growth on an imaginary rotating plate. Every 6° the level of liquid marks a straight line on the sheet. The sheet rotates 540°—until the whole surface is covered with the pattern.

In the actual process (Fig. 5.1), as the sheet rotates, the introduced vibrations generate lines: the liquid moving upwards raises the carbon particles, permanently solidifying them at the highest position of the miniature wave in the form of a straight line—a tangent of the growing spiral. The same event is repeated rapidly and the curve of the inwardly moving spiral becomes visible. It continues along its path until the whole carbon-coated glass sheet is covered with the spiral-forming straight lines. The dynamic and irregular lines in the centre are the result of violent vibrations introduced on the surface of the liquid. It should be understood that the patterns generated during the process will not be washed off, either by the rising liquid or the vibrations on the surface. It is the nature of this technique that the patterns, once generated, are permanent. The same applies when oil, instead of lacquer thinner, is used.

The pattern in Fig. 6.1 is generated as in Fig. 4.2, except that the oil is fed through only one hole in the upper plate and the connected plates are in continuous rotation. Since the amount of oil in this case is limited, the final result is a closed system.

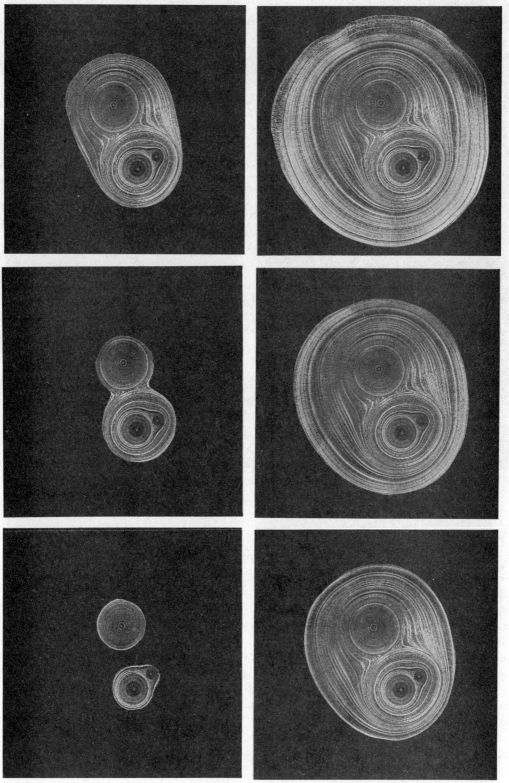

Fig. 4.2. Stages of growth on the horizontal plane.

Fig. 4.3. Exhibition at Harvard University (1965).

2. Technique used in st

The elements used in formative processes in ft—carbon black particles and oils, I borrowed from carbon technology. The elements in st are generated as follows:

    (a) a square bar (Fig. 8.1) is cut at 45° (Fig. 8.2);
    (b) the right side of the bar is displaced by simultaneous rotational and translational movement (Fig. 8.3);
    (c) the empty space between the left piece and the right piece of the bar is filled with a substance (Fig. 8.4);
    (d) the resulting form is removed.

Applying the same movement as in Fig. 8.2 the plane BEPS is moved, fixing the location of the plane every 9° (Fig. 9). Thus, Fig. 9.1 is the situation after 90° rotation, Fig. 9.2—after 180° etc. A sketch (Fig. 10) illustrates the movement 0°–90°. Connecting the points in space, shown in Fig. 9.4A, elements are achieved. The manufacturing of the elements is described in the previous paper [33].

By changing the cross-section of the bar and also its orientation (horizontal or vertical) during the cutting, an infinite number of different elements is achieved. From the infinite number of different types, seven have been selected for further investigations (Fig. 12).

The genesis of the element, as described above, brings forth the following conclusions:

    1. The nature of the pre-phenomenological stage of the element (Fig. 8.3) is EMPTINESS [34–37].

Fig. 5.1. Growth on the vertical plane.

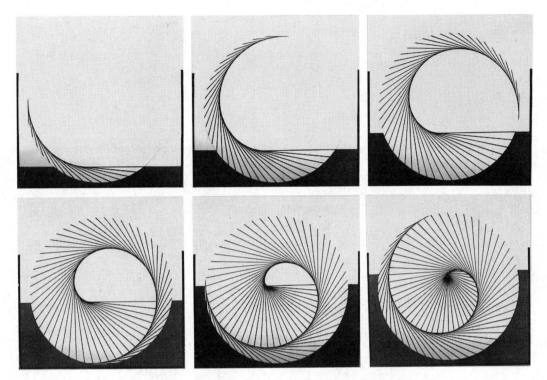

Fig. 5.2. Diagrammatic growth pattern on the vertical plane.

2. The emptiness is informed emptiness—the space between the two planes, not yet named as "an element" has EXPERIENCED helical movement.

3. The memory of the helical movement is forcing one to expect in further developments a display of PREDICTABLE BEHAVIOUR—a growth in helical configurations.

4. Since the origin of the element, visible but intangible, is empty (Fig. 8.4), the next part of this work is ILLUSION.

AD INFINITUM

After a period of believing, experimenting and writing about the possibility of a universal building block [33], I experienced serendipity—a happening one is not even searching for! I discovered that there is an infinite number of basic elements, as shown in Fig. 12.

The summary of the genesis of the element, as described in the last few pages, and the directions that will be taken in the last part of this work, are as follows:

1. The materialized elements, originating from emptiness, have been generated by ONE strategy—cooperation with nature.

2. They all are made by ONE tactic—rotation in spacetime.

3. In each type, a...h...p...z, there are four elements A, B, C and D, the same size but of different orientation. The nature of the elements is selective: only members of the same type are attracted to each other and will ignore elements of other types. At the same time "h"-type elements marked as "h I" (vertical) and "h—" (horizontal) are again the same; orientation in the process of connection makes the difference. The same applies to all types—an axis of symmetry exists through the type "a".

4. The elements will be connected in ONE way only, leading to an INFINITE number of forms, all having ONE structure—trihedral.

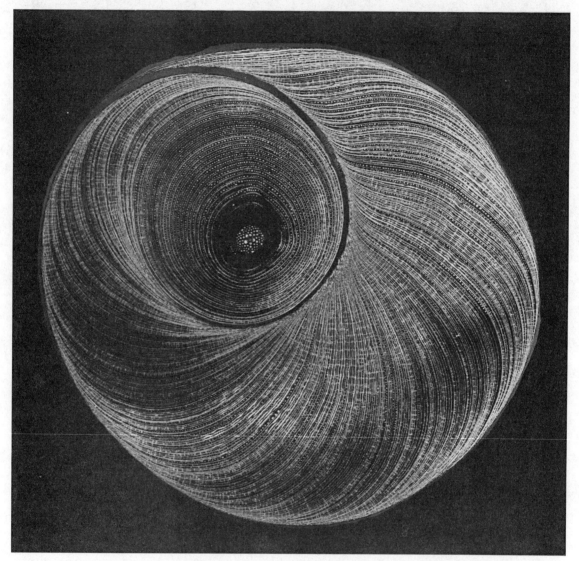

Fig. 6.1. Growth on the sloped plane in a continuous rotation around the centre of the plane.

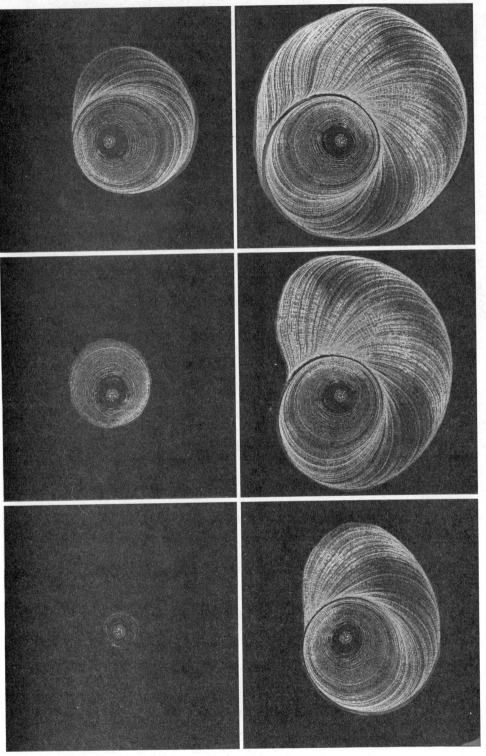

Fig. 6.2. Stages of growth on the sloped plane.

Fig. 7. Exhibition at MIT (1965).

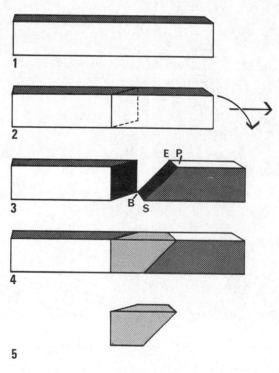

Fig. 8. Genesis of an element ("building block").

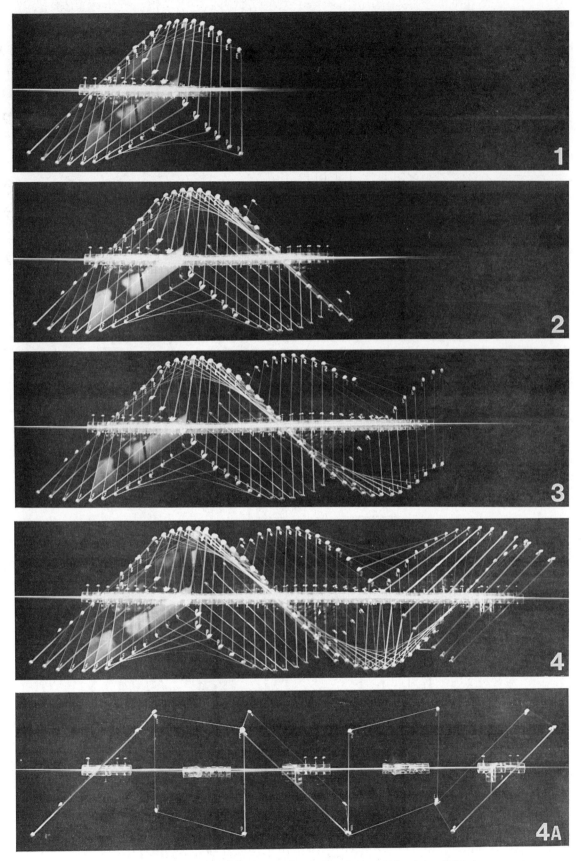

Fig. 9. Time–motion study: a rectangular plane in a continuous rotational and translational movement, photographed every 9°, **1–4**; **4A**, every 90°.

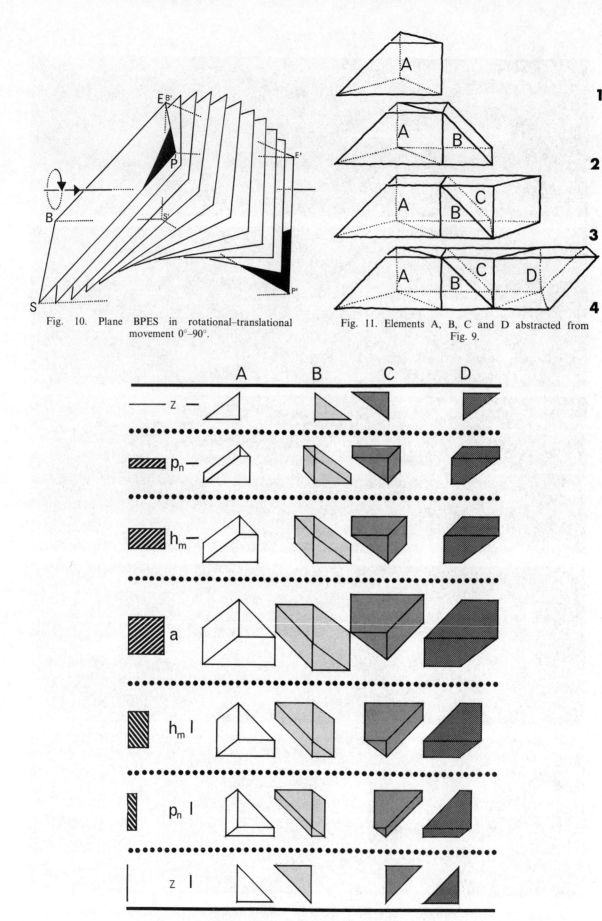

Fig. 10. Plane BPES in rotational–translational movement 0°–90°.

Fig. 11. Elements A, B, C and D abstracted from Fig. 9.

Fig. 12. Seven selected types from the infinite number of elements.

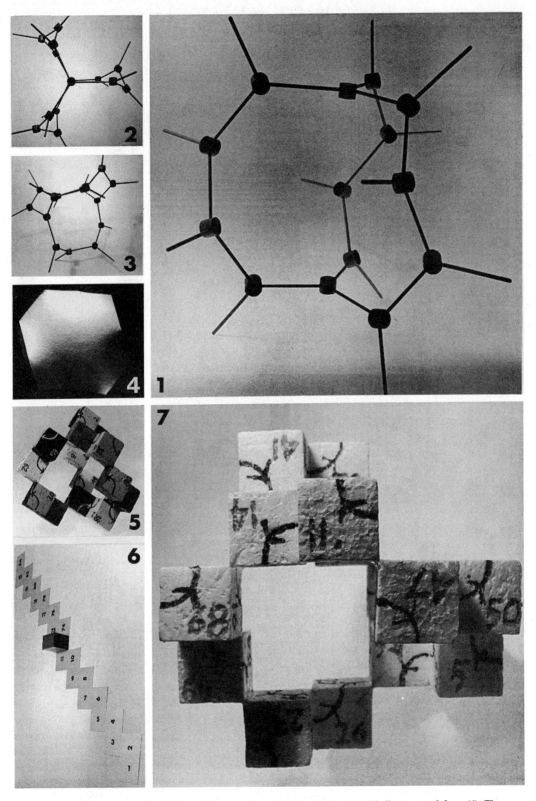

Fig. 13. The structure of a trihedron. Fourteen interrelated points in space (**1–3**) generate 3 faces (**4**). The points could be imagined as the centres of 14 cubes (**5, 7**). A helical path through 84 faces of the cubes unfolds as a continuous chain (**6**).

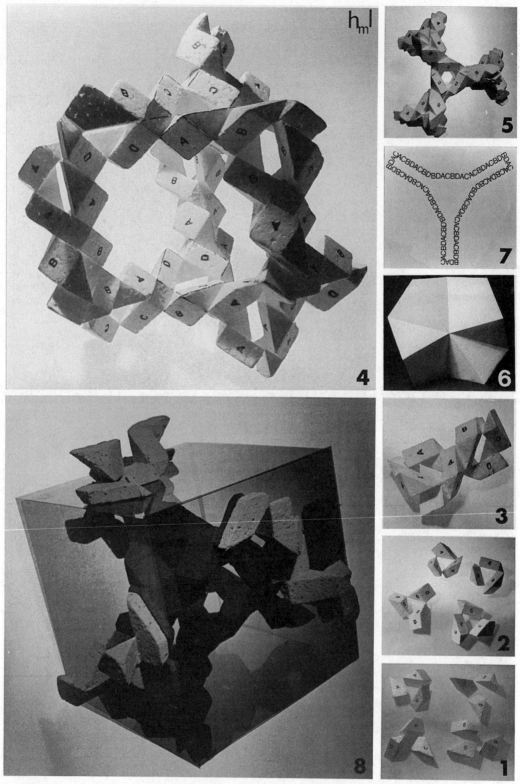

Fig. 14. Genesis of a trihedron. Since the four elements—A, B, C and D, each having a separate colour (different orientations)—are used to form a hierarchical growth of a *tri*hedron, 12 elements have to be used, 3 of each colour (**1**). The 7 configurations (**3**), each having 2 points, form a trihedron. The connected configurations form hcs following code ACBD (different combinations possible). The elements form a chain of 84 letters (**7**). **8**, A trihedron inscribed in a cube.

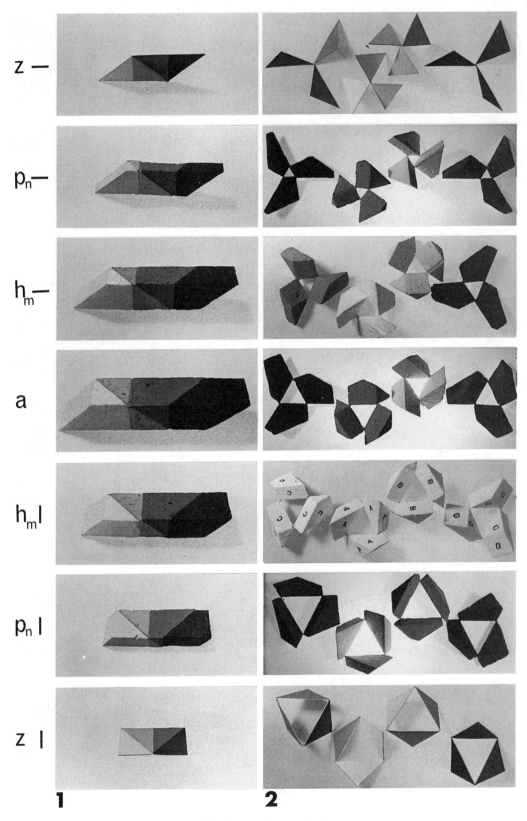

z —

p_n —

h_m —

a

h_m l

p_n l

z l

1

2

Fig. 15—*continued overleaf*

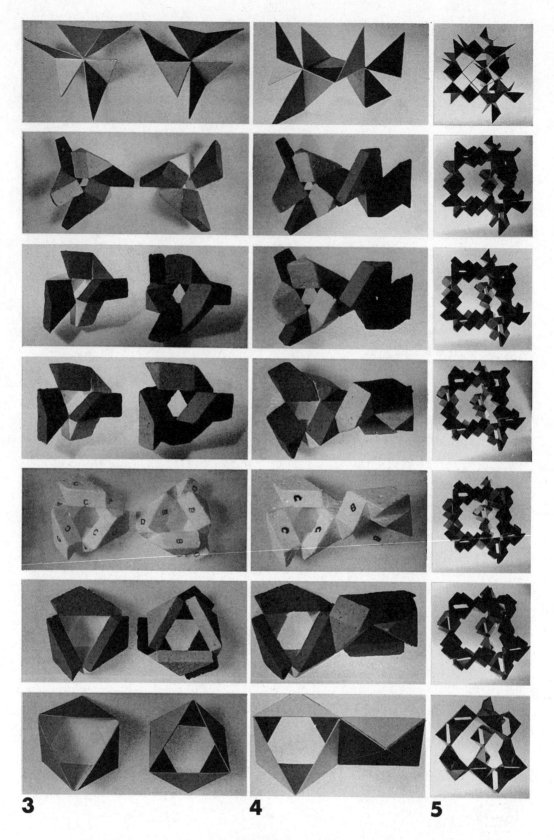

Fig. 15. Hierarchichal growth of trihedrons: from elements (**1**) to sub-components (**2**), components (**3**) and configurations (**4**).

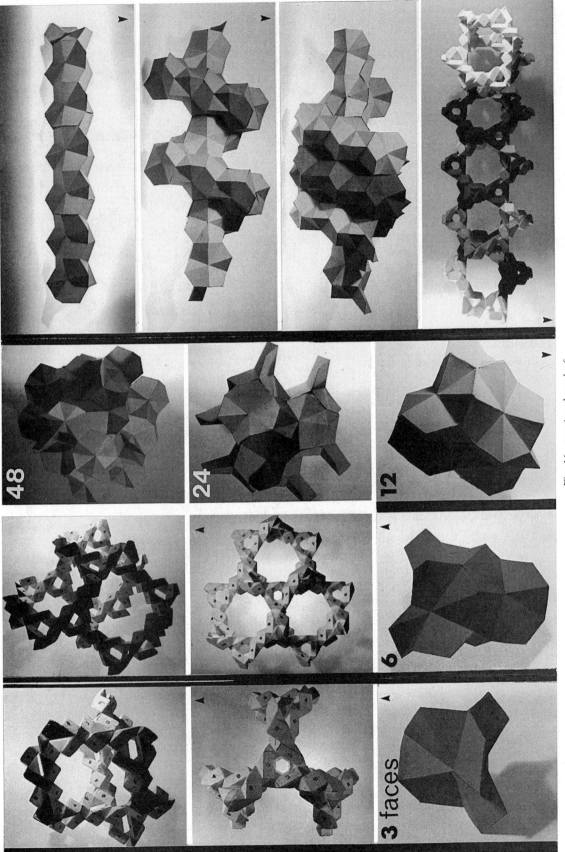

Fig. 16—*continued overleaf*

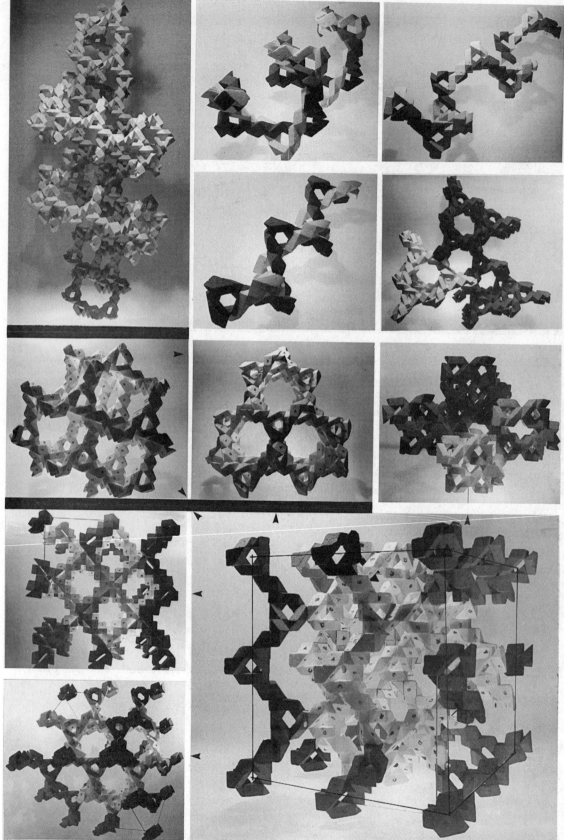

Fig. 16. Close-packing of trihedrons and hcs. A trihedral dodecahedron (12 faces) is inscribed in a cube (Fig. 16.2, lower right-hand corner); some parts, in the darker tone, are added to complete the cube. A few selected double-helical abstractions are shown; also helical growth of trihedrons—as volumes and helical patterns.

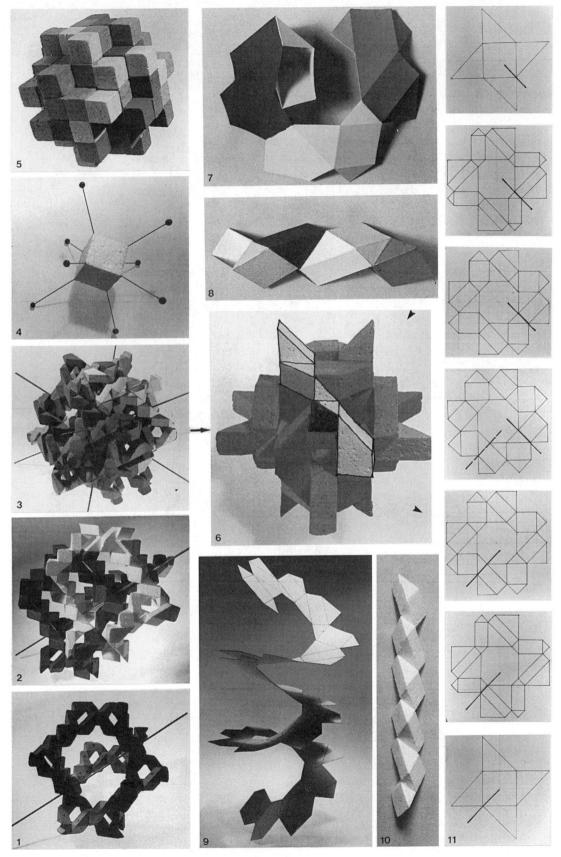

Fig. 17—*continued overleaf*

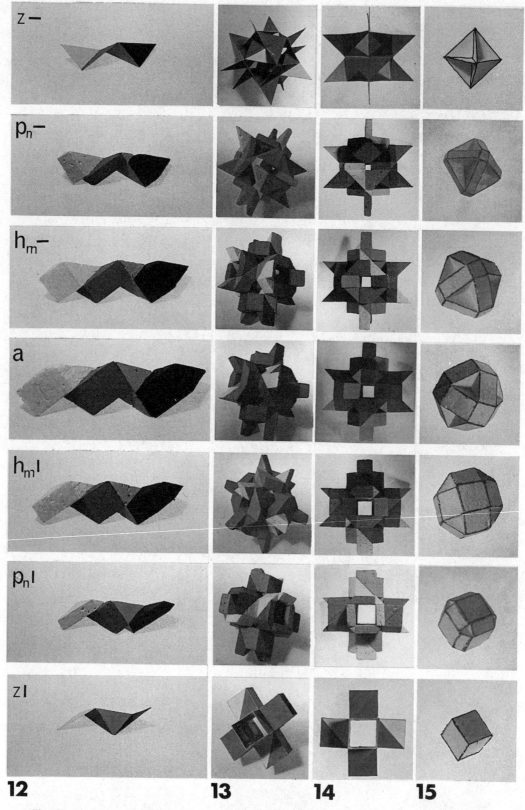

Fig. 17. Interpenetration of trihedrons and emergence of other configurations. Along the axis of a trihedron (**1**), another trihedron (**2**, light tone) penetrates it. Continuing penetrations along other axes (**4**) result in the formation of 8 trihedrons (**3**), shown in **5**. A new configuration is abstracted (**6**), consisting of 12 hc (**12**) each built of 4 elements. These spatial hcs could be constructed from flat patterns (**7**, **8**, **11**). Several connected flat patterns form a continuous hc. The new configuration (**6**), shown in **13** and **14**, encloses empty spaces, materialized in **15**.

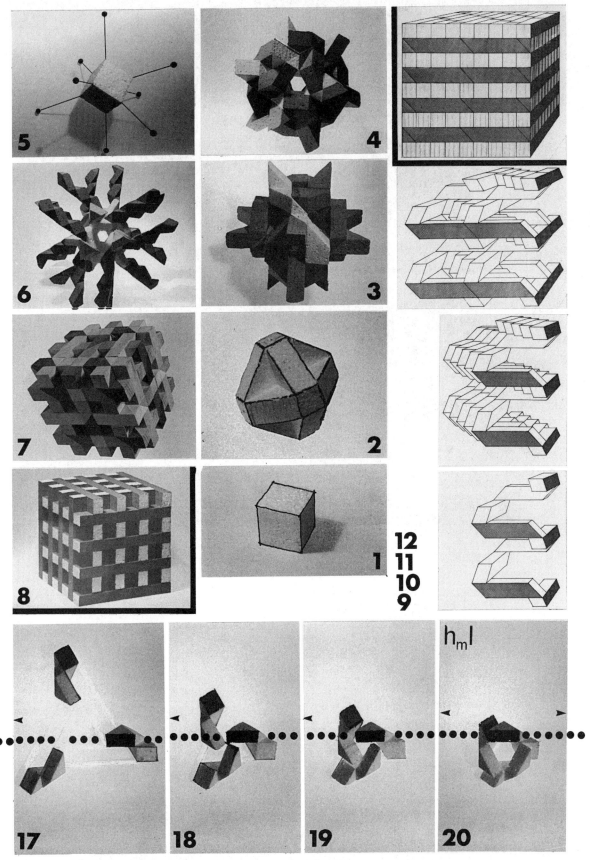

5

4

6

3

7

2

8

1

12
11
10
9

h_ml

17

18

19

20

Fig. 18—*continued overleaf*

Fig. 18. Present and future. **1–12**, answer to the question under Fig. 2: internal hcs, as a result of arranging bars in the two structures; **13**, plan-view of the large helical development in **16**; **14** and **15**, abstractions from larger patterns; **16**, unfolded hc—84 type zI elements. The game is not finished, new transformations are in the process of emerging; **17–24**, the beginning of infinite changes of a sub-component. Structured hcs, themselves illusions, float in space in ordered trihedral relationships!

CONCLUSIONS

1. Both developments—in ft and st—are self-organizing systems. My own role has been that of a midwife: coating the glass sheets (the field) with the carbon black in which the patterns are revealed; cutting the various building blocks; accepting the emergence by chance of the code ACBD, and painting the elements in four different tones. Then, guided by the ultimate restraint—one structure, only one method of connection and one type of elements I let the form (pattern) emerge! What is all this for? To be surprised, to be able to wonder.

2. The genesis of an element by simultaneous translation and 90° rotation applies only to the type "a" elements (Fig. 8). To achieve a prototype—informed emptiness—for all other types a 180° rotation should be exercised (Fig. 19). It seems nature prefers symmetry, but ambidextrous symmetry. We too, being symmetrical, draw with the right or left hand. I knew an architectural student who had pencils in both hands and used them!

3. Indeed, our actions on external things change their intrinsic relationships [38]. The simultaneous translational and rotational movements structuring the forms shown in Fig. 1 have developed internal helical configurations (Figs 18.1–18.12).

4. Why has this infinite variety and diversity under the strictest constraints been achieved, as described in these pages? Guided by ignorance (my strength?), the answer to myself is: the participation of real time at the birth of the element. The EXPERIENCE of time during the helical movement is inscribed in the memory bank of the element as a source of energy. It enables infinite growth.

5. It has been observed that a duckling or other newborn follows the first thing she or he sees when opening the the eyes the first time: the mother (as designed by nature), an overshoe or whatever else is nearby at that moment.

   My case is similar! The obsession with helical configurations originates in my first experience: I remember that helical movement leaving the world of my mother and beginning my own, now filled with hc! Another example of symmetry?

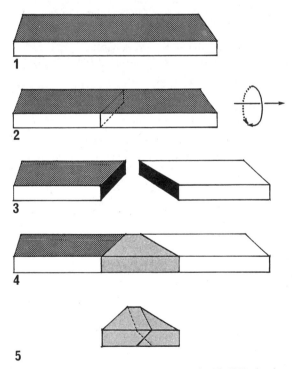

Fig. 19. Genesis of an ambidextrous double element, result of 180° simultaneous rotational and translational movement.

Acknowledgements—As always I am most thankful to all my students, teachers, friends, colleagues and authors who, unknowingly, helped me to find my own way, the only way. It is difficult not to mention Professor Gyorgy Kepes, who not only introduced me to the new landscape in art and science but also initiated the exhibition of my work at MIT (1965), invited me to participate in the exhibition "Light as Creative Medium" at Harvard University (1965) and gave permission to use a photograph of that exhibition; to Ivars Davis for constructive suggestions and technology for manufacturing "the building blocks" used in this research; and to Alison Brooks for the editing.

REFERENCES

1. J. Zvilna, Rotation and form. In *Proc. Int. Ass. Shell and Spatial Structures*, Concordia Univ., Montreal, Quebec (1976).
2. J. Sallis, *Phenomenology and the Return to Beginnings*. Duquesne Univ. Press, Pittsburgh, Pa (1973).
3. G. Kepes, *Language of Vision*. Theobald, Chicago, Ill. (1951).
4. C. S. Smith, Structural hierarchy in science, art, and history. In *On Aesthetics in Science* (Ed. J. Wechsler). MIT Press, Cambridge, Mass. (1978).
5. O. L. Reiser, *The Integration of Human Knowledge*. Extending Horizons Books, Porte Sargent, Boston, Mass. (1958).
6. L. L. Whyte (Ed.), *Aspects of Form*, 2nd edn. Lund Humphries, London (1968).
7. D'Arcy Thompson, *On Growth and Form*. Cambridge Univ. Press, Cambs. (1963; 1st edn, 1917).
8. F. Capra, *The Tao of Physics*. Shambhala, Boulder, Colo. (1975).
9. A. I. T. Chang, *The Tao Architecture*. Princeton Univ. Press, N.J. (1981; 1st edn, 1956).
10. R. G. H. Siu. *The Tao of Science*. MIT Press, Cambridge, Mass. (1957).
11. G. S. Bolen, *The Tao of Psychology*. Harper & Row, San Francisco, Calif. (1979).
12. C.-Y. Chang, *Creativity and Taoism*. Julian Press, New York (1963).
13. J. T. Bonner, *Morphogenesis*. Atheneum, New York (1963).
14. H. Read, *The Origins of Form in Art*. Thames & Hudson, London (1965).
15. T. Munro, *Form and Style in the Arts*. Case Reserve Univ. Press, Cleveland, Ohio (1970).
16. H. Lalvani, *Transpolyhedra*. The author, New York (1977).
17. E. Laszlo, *Introduction to Systems Philosophy*. Gordon & Breach, New York (1972).
18. E. Jantsch, *The Self-organizing Universe*. Pergamon Press, Oxford (1980).
19. L. Moholy-Nagy, *Vision in Motion*. Theobald, Chicago, Ill. (1956).
20. G. Kepes (Ed.), *The Nature and Art of Motion*. Braziller, New York (1965).
21. R. L. Hall with V. E. Cobey, *The World as Crystallized Movement*. *Main Currents in Modern Thought*, Vol. 31, No. 1. New Rochelle, New York (1974).
22. A. Mickunas, *The Primacy of Movement*. *Main Currents in Modern Thought*, Vol 31, No. 1. New Rochelle, New York (1974).
23. H. Hertel, *Structure, Form and Movement*. Reinhold, New York (1966).
24. W. Lewis, *Time and Western Man*. Beacon Press, Boston, Mass. (1957).
25. J. W. Dunne, *An Experiment with Time*. Faber & Faber, London (1927).
26. J. J. C. Smart (Ed.), *Problems of Space and Time*. Macmillan, New York (1964).
27. T. Gold (Ed.), *The Nature of Time*, 2nd Edn. Cornell Univ. Press, Ithaca, N.Y. (1974; 1st edn, 1967).
28. G. Kubler, *The Shape of Time*. Yale Univ. Press, New Haven, Conn. (1962).
29. M. N. McMorris, *Time and Reality in Eliot and Einstein*. *Main Currents in Modern Thought*, Vol. 9, No. 5. New Rochelle, New York (1973).
30. N. Calder, *Einstein's Universe*. Greenwich House, New York, (1979).
31. N. A. Kozyrev, *Possibility of Experimental Study of the Properties of Time* U.S.. Dept of Commerce, Washington, D.C. (1968).
32. M. D. Akhundov, *Conceptions of Space and Time*. MIT Press, Cambridge, Mass. (1986).
33. J. Zvilna, Colored symmetries in space–time. *Comput. Math. Applic.* **12B**, 895–911 (1986). Reprinted in *Symmetry: Unifying Human Understanding* (Ed. I. Hargittai). Pergamon Press, Oxford (1986).
34. T. Izutsu, The nexus of ontological events: a Buddhist view of reality. In *Eranos 1980 Yearbook*. Eranos Foundation, Switzerland (1981).
35. L. Tzu, *Wisdom of Ancient China*. The New American Library, (1955).
36. C. A. Van Peursen, *Phenomenology and Reality*. Duquesne University Press, Pittsburgh, Pa (1972).
37. P. J. Davis and D. Park (Eds), *No Way, The Nature of the Impossible*. Freeman, San Francisco, Calif. (1987).
38. J. Piaget, *Structuralism*. Basic Books, New York (1970).

Computers Math. Applic. Vol. 17, No. 4–6, pp. 1067–1072, 1989
Printed in Great Britain. All rights reserved

THE JOY OF SYMMETRY†

I. Hargittai

Hungarian Academy of Sciences and Eötvös University, Pf. 117, Budapest, H-1431, Hungary

It was a pleasant surprise for me one evening near the end of January in Budapest to receive a phone call from Jim Boggs. He told me that Herman Mark had withdrawn from being the banquet speaker, and instead, Jim was asking me to come over and give this speech. I was, of course, overwhelmed with joy. The question of topic came up, and Jim, knowing about my infatuation with symmetry, suggested symmetry as a possible general theme. By thought association I added my joy to it, hence the title of this presentation.

The importance of thought association will return later in this discourse. For the moment, however, let us remain with the question of this title. The joy part of it, for which I feel responsibility, is true. I was feeling happiness thinking about this evening while preparing for it in Budapest and I am feeling it now as well. I am, however, less than certain about the symmetry part of it. True, I like symmetry, but only to a certain extent. It may be nice, it may be pleasant, but having too much of it may also be boring and irritating. There are various statements pertaining to this, from relating too much symmetry to death, to simply saying that perfect symmetry is not suitable for human habitat. The opposite, i.e. asymmetry, may look more interesting, has more information content, but, of course this talk is not about asymmetry. Generally, however, I think that it is when symmetry is less than perfect, is somewhat distorted, that it really starts appealing to us.

I don't suppose that anybody is wondering how does the topic of symmetry come to a molecular structure meeting. It is one of the cementing concepts in science, and in particular in molecular sciences. My own interest in symmetry has certainly originated from my interest in the structure of molecules. I even remember the moment when I started wondering about symmetry and this wondering has not yet stopped. The biggest surprise I have experienced is the capability of this concept to relate such a diversity of fields and topics, and I am most grateful to it for opening up directions to me which I had not had interest in before, or to which I had even been hostile.

Some people may certainly be wondering, however, about the relationship between "The Joy of Symmetry" and this beautiful dinner party. The title of my presentation may remind us of the title of that bestseller *The Joy of Cooking*. When this comparison came to my mind I was a little discouraged, for cooking implies active participation whereas symmetry may imply passive observation rather than real involvement.

However, when we consider less than perfect symmetries, our participation is also needed. We are certainly not going to discuss geometrical symmetry tonight. Instead, we shall be dealing with symmetries which can only with a certain amount of magnanimity be called symmetries. This is a serious matter. The symmetries in chemical systems, for example, do not belong to geometrical symmetries but to what we may term material symmetries. It is our ability to geometrize, or to symmetrize, if you will, that makes the model of symmetry applicable to real systems. The rigor of geometrical symmetry is then gone and is replaced by some vagueness and fuzziness.

The Joy of Symmetry. This title may invite other thought associations as well. But then some people cannot be helped, as they thought-associate everything with what their mind is occupied by anyway. A few years ago I gave a talk about the importance of small structural differences in Edinburgh. Before the lecture, a distinguished chemist, with a name appearing in every monograph in his field, asked me, "This lecture of yours, will it be about sex?"

†Banquet speech at the *12th Austin Symposium on Molecular Structure*, 29 February–2 March 1988; organized by Professor J. E. Boggs, Department of Chemistry, The University of Texas, Austin, TX 78712, U.S.A.

One more thought about the history of this banquet speech; as you see, I find it difficult to get over this experience. A certain symmetry, or rather the asymmetry of flowing time, I feel in my role today. It was not very long ago, in 1971, that I was giving an invited talk to a crystallographic meeting in Manchester, England. Not only was it my first ever invited talk at a meeting, I also happened to be the youngest invited speaker at that meeting. Such a state of affairs, however, did not last long. In subsequent meetings, where I have happened to be an invited speaker, I no longer had the same distinction. Instead, I have always belonged to the medium age range of the speakers. Until this meeting, that is. Tonight I am once again in a unique state being the substitute speaker for 92-year-old Herman Mark.

For the benefit of those outside the electron diffraction field, Herman Mark was the initiator of gas-phase electron diffraction in 1930, and his associate Raymund Wierl carried out the first such experiment under his direction. Wierl, junior to Mark, died soon afterwards, but Mark survived and is very active even today. He has made a significant name in polymer sciences. He lives in Brooklyn, N.Y., and goes to work at Brooklyn Polytech every day when he is in town. Mark was born in 1895 in Vienna, lived in Germany at the time of starting gas electron diffraction, then returned to Vienna, and came to this country in the year of the Anschluss. His mother was born in Budapest, and Mark also visited Budapest as a small child at the end of the last century.

It happens that later in this speech I shall be making reference to another world known figure who himself was born in Hungary, then moved to Vienna, became a writer in the German language, then moved on westward and became a famous writer in the English language. The name is Arthur Koestler. With this I am not trying to say that all important people are Hungarian or of Hungarian origin, or at least related to Hungarians. Not at all. I have been told that other nations have also contributed to the progress of mankind. And, conversely, I would like to quote the statement which was displayed on the desk of Alexander Korda, the giant of the film industry. "It is not sufficient just to be Hungarian".

But let us return to our proper topic, symmetry. First of all, I would like to show you a beautiful Portugese stamp (Fig. 1) which displays a symmetrical pattern of a sixteenth century decoration. It was found on a tile. Such tilings are quite common on the facades of houses in Portugal. Besides, there is a whole museum of tiles in Lisbon. Let us, for a moment, disregard the central motif, and recognize the four mirror planes of the rest of the pattern. Including then the central motif into our consideration, the whole pattern is seen to have four-fold rotational symmetry only, and no mirror planes.

The medieval pattern on the Portugese stamp is surprisingly complete. Only some of the corners, and in particular the lower left edge show some damage. As we look at the pattern, and examine its symmetry, our eyes and mind skip involuntarily over the damages and imperfections, and see the pattern in its entirety, as a whole, and as if it were complete. This is exactly what I meant when I mentioned our ability to geometrize.

I find the patterns especially intriguing that have rotational symmetry only and no symmetry planes. I would like to show you yet another such pattern, again from Portugese tiles (Fig. 2). Whereas this pattern can be described by point group symmetry, its infinite repetition is described by space group symmetry (Fig. 3). As such patterns are often used for decorating walls, their space groups are even called wallpaper groups. There have been studies of how various space groups influence the mood of people, what is our perception of the various symmetries used for decorations. Some have argued that decorations possessing rotational symmetry only tend to

Fig. 1

Fig. 2

Fig. 3

Fig. 4

Fig. 5

Fig. 6

induce the feeling of motion in us, and facilitate moving even crowds of people. On the other hand, decorations with symmetry planes, especially symmetry planes perpendicular to each other, freeze us, prevent people from moving around. It has been suggested that patterns with rotational symmetry only should be used for decorating dancing halls, and patterns with perpendicular symmetry planes should be used for decorating halls of important meetings.

A few years ago a fellow electron diffractionist, Shuzo Shibata gave me a beautiful Japanese slide on which there is a nice goat in front of the strikingly beautiful Mount Fuji (Fig. 4). I presume that anybody would find this picture of interest, but for a symmetrologist, this picture is a gold mine.

The goat has bilateral symmetry, and the same bilateral symmetry is characteristic for the whole animal kingdom as well as for humans. This bilateral symmetry originates from the fact that for animals, left and right are equivalent, whereas neither up and down, nor front and back are equivalent. This is the result of the translational motion along the Earth's surface so typical for animals. There are, of course, exceptions. I saw a double-headed dog a few years ago in Brussels (Fig. 5). It was not like the double-headed eagle of ancient coats of arms. Rather, this dog had a second head in place of its tail.

Mount Fuji, on the other hand, has cylindrical symmetry. All directions are equivalent around its vertical axis. There is no left and right, nor front and behind, only up and down are different. Many plants have the same cylindrical symmetry originating from their mode of life which fixes them to a certain place and lets them grow only upward.

Left and right symmetry is the most common of all symmetries, and in our everyday language it is often considered a synonym for symmetry. The conservation of left and right has for a long time been considered to be of fundamental importance in physics. We all have also learned about Buridan's ass who was prepared to exchange its life for the conservation of parity. However, more recently this parity has been shown to distort, and I have seen a German student poster advocating survival above parity (Fig. 6).

We find a lot of symmetry in architecture and municipal planning. The Tower of the University of Texas and the whole view of the central mall of the campus radiate symmetry (Fig. 7). What a relief then, especially on a hot summer day, when the operating fountain somewhat decreases this very high symmetry (Fig. 8). The beautiful new architecture in Dallas has certainly reduced symmetries without diminishing aesthetic pleasure (Figs 9 and 10).

At this point, I would like to introduce the most intriguing kind of symmetry, that is antisymmetry. For symmetry itself I did not even try to give a definition. Everybody knows what symmetry is. I myself used to think that I knew, until I started my symmetry studies. However, for antisymmetry, let us quote a definition to simplify our job further. Accordingly, "Operations of antisymmetry transform objects possessing two possible values of a given property from one value to the other".

Antisymmetry is the fundamental property of wave functions in the description of the electronic structure of atoms and molecules, but it is also much more common in our everyday world than we would first think. If we walk over to the campus of the University of Texas, at the northern foot of the Tower there are some Coke machines. There is a machine with the regular drink, another with the regular drink and yet another with the diet version. The operation transforming one

Fig. 7

Fig. 8

Fig. 9

Fig. 10

regular Coke machine into the other regular Coke machine is an identity operation (Fig. 11). However, the operation transforming one regular Coke machine into a diet Coke machine is an antiidentity operation (Fig. 12). In this case the property which has two possible values is the sugar content of the drink.

Antimirror reflections are common. Thus, for example, antisymmetry relates the two polyhedra colored in the opposite way (Fig. 13). Antisymmetry relates the girl and the elephant on the poster advertising a dancing class in Vienna (Fig. 14). An antisymmetry plane was thought to separate two world leaders in 1984 (Fig. 15), whereas a similar double portrait in 1988 might have used a simpler symmetry operation. Coloring changes express antisymmetry of a prize-winning Moscow poster from 1987 referring to Perestroika (Fig. 16).

Fig. 11

Fig. 12

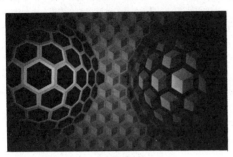

Fig. 13   Fig. 14

Human relations are full of symmetry and even fuller in antisymmetry. Hungarian author Karinthy has interpreted the eternal puzzle of the man/woman relationship by antisymmetry:

> "Man and woman.
> No wonder, they can't understand each other.
> Each wants something different.
> The man wants the woman,
> The woman wants the man."

And now about Koestler. One of his major works is entitled *The Act of Creation*. Koestler made an attempt to inquire into the biological origins of mental creativity and he started by studying the phenomenon of laughter. He said,

> "Humor is an elusive thing, so is the rainbow;
> Yet the study of coloured spectra provided clues to the elementary structure of matter."

This elusiveness fits nicely the vagueness and fuzziness of our approach to symmetry in the nongeometrical world. Koestler has introduced the concept of so-called *bisociation*. Bisociation may be considered the antisymmetric partner of association. According to Koestler, the connection in association is made between thoughts on the same plane, whereas bisociation refers to such connection of thoughts from different planes.

A story, as an example, will clarify the matter perhaps better than any definition:

> "A Marquis at the court of Louis XIV enters his wife's boudoir and finds her in the arms of a Bishop. The Marquis then walks calmly to the window and goes through the motions of blessing the people in the street.
> 'What are you doing?'—cries his anguished wife.
> 'Monseigneur is performing my functions',—replies the Marquis,—'so I am performing his.'".

Fig. 15   Fig. 16

The common association of thoughts would call for the Marquis's killing the Bishop, or his own wife, or both. The Marquis's reaction is quite unexpected, though it follows a certain logic of its own and qualifies his action for bisociation.

Let me quote at this point from a lecture a couple of days ago by Professor Nat Bauld who told the story of bringing Michael Dewar, the world famous theoretical organic chemist, to the University of Texas 25 years ago. He remembered that the UT football team was No. 1 nationally that year and the Chemistry Department was very proud of it. Bringing Michael Dewar to UT, Bauld said, was the first step in building up a Chemistry Department of which the UT football team could be proud. The first reference to the football team here is a simple thought association, while the last one appears in a bisociation.

Speaking about creativity, bisociation of different techniques and different fields has often led to strikingly new discoveries in the history of the sciences. Thus, there may indeed be something to bisociation.

Let us, however, return to another of Koestler's antisymmetric associations:

> "The Prince travelling through his domains, noticed a man in the cheering crowd who bore a striking resemblance to himself. He beckoned him over and asked:
>   'Was your mother ever employed in my palace?'
>   'No Sire',—the man replied.—'But my father was.'"

Here simple association would have meant a simple yes since it was common practice for feudal lords to have bastards, whereas feudal ladies were not supposed to have bastards.

And a very simple antisymmetric association, i.e. bisociation after Koestler,

> "What is a sadist?
> A sadist is a person who is kind to a masochist."

From the classical definition of symmetry, i.e. correspondence in size, shape and position of parts that are on opposite sides of a dividing line or center, we have come a long, long way. It is up to our tolerance how far we wish to go. Again, it is our ability to geometrize, to smooth over rough edges, to make complete a picture that has missing parts, or in other words, to discover the inherent, underlying symmetries even where there is none according to more rigorous criteria. This ability is an important tool in scientific research, facilitating the observation of trends, general phenomena, common characteristics, relationships, and creating classifications. However, it is then not less important to see the differences, to feel the rough edges, to be able to distinguish.

Let me quote at this point Michael Faraday about his childhood, after the theoretical chemist Coulson:

> "Do not suppose that I was a very deep thinker, and was marked a precocious person. I was a lively imaginative person, and could believe in the Arabian Nights as easily as in the Encyclopedia. But facts were important to me, and saved me".

To this Coulson adds,

> "It is when symmetry interprets facts that it serves its purpose: and then it delights us because it links our study of chemistry with another world of the human spirit—the world of order, pattern, beauty, satisfaction. But facts come first. Symmetry encompasses much—but not quite all!".